APPLIED STRENGTH OF MATERIALS

APPLIED STRENGTH OF MATERIALS

Fifth Edition

Robert L. Mott, P.E.
University of Dayton

Upper Saddle River, New Jersey
Columbus, Ohio

Library of Congress Cataloging-in-Publication Data
Mott, Robert L.
Applied strength of materials/Robert L. Mott.—5th ed.
p. cm.
Includes bibliographical references and index.
ISBN-13: 978-0-13-236849-0
ISBN-10: 0-13-236849-8
1. Strength of materials. I. Title
TA405.M883 2008
620.1′12—dc22

2007027383

Editor in Chief: Vernon R. Anthony
Editor: Eric Krassow
Editorial Assistant: Nancy Kesterson
Project Manager: Kevin Happell
Design Coordinator: Diane Ernsberger
Cover Designer: Jeff Vanik
Cover art: Getty Images
Production Manager: Deidra Schwartz
Director of Marketing: David Gesell
Executive Marketing Manager: Derril Trakalo
Marketing Assistant: Les Roberts

This book was set in Times Roman by Aptara, Inc. It was printed and bound by Courier-Kendallville, Inc. The cover was printed by Phoenix Color Corp.

Pearson Education Ltd.
Pearson Education Singapore Pte. Ltd.
Pearson Education Canada, Ltd.
Pearson Education—Japan
Pearson Education Australia Pty. Limited
Pearson Education North Asia Ltd.
Pearson Educación de Mexico, S.A. de C.V.
Pearson Education Malaysia Pte. Ltd.

10 9 8 7 6 5 4 3
ISBN-13: 978-0-13-236849-0
ISBN-10: 0-13-236849-8

Contents

NOTE: Every effort has been made to provide accurate and current Internet information in this book. However, the Internet and information posted on it are constantly changing, and it is inevitable that some of the Internet addresses listed in this textbook will change.

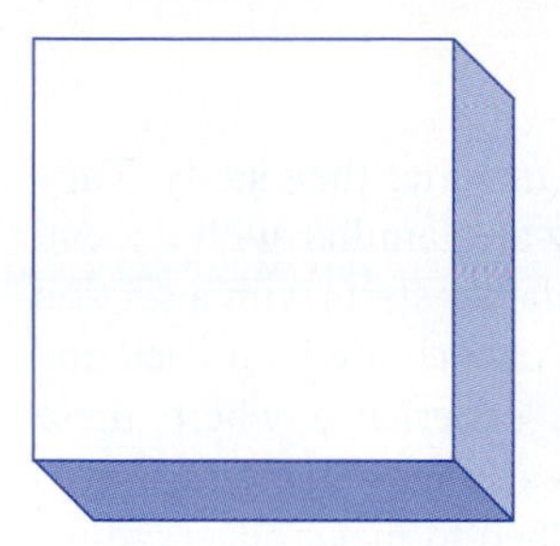

Preface

Objectives of the Book

Applied Strength of Materials, Fifth Edition, provides comprehensive coverage of the important topics in strength of materials with an emphasis on applications, problem solving, and design of structural members, mechanical devices, and systems. The book is written for the student in a course called Strength of Materials, Mechanics of Materials, or Solid Mechanics in an engineering technology program at the baccalaureate or associate degree level or in an applied engineering program.

This book provides good readability for the student, appropriate coverage of the principles of strength of materials for the faculty member teaching the subject, and a problem-solving and design approach that is useful for the practicing designer or engineer. Educational programs in the mechanical, civil, construction, architectural, industrial, and manufacturing fields will find the book suitable for an introductory course in strength of materials.

Style

This text emphasizes the *applications* of the principles of strength of materials to mechanical, manufacturing, structural, and construction problems while providing a firm understanding of those principles. At the same time, the *limitations* on the use of analysis techniques are emphasized to ensure that they are applied properly. Both analysis and design approaches are used in the book.

Units are a mixture of SI metric and U.S. Customary units, in keeping with the dual usage evident in U.S. industry and construction.

Prerequisites

Students should be able to apply the principles of statics prior to using this book. For review, there is a summary of the main techniques of the analysis of forces and momentum in the Appendix. Several example problems are included that are similar to the statics needed to solve practice problems in this book.

While not essential, it is recommended that students have completed an introductory course in calculus. As called for by accrediting agencies, calculus is used to develop the key principles and formulas used in this book. The application of the formulas and most problem-solving and design techniques can be accomplished without the use of calculus.

Features of the Book

The Big Picture. Students should see the relevance of the material they study. They should be able to visualize where devices and systems that they are familiar with depend on the principles of strength of materials. For this reason each chapter starts with a section called *The Big Picture*. Here the basic concepts developed in the chapter are identified and students are asked to think about examples from their own experience where these concepts are used. Sometimes they are asked to explore new things on their own to discover how a product works or how it can fail. They are coached to make observations about the behavior of common mechanical devices, vehicles, industrial machinery, consumer products, and structures. Educational philosophy indicates that students learn better and retain more when such methods are employed.

Activity-Based Learning. Activity-based learning activities are integrated into the popular Big Picture section, a successful feature in all previous editions. The activity can be used independently by the students, by the instructor as a classroom demonstration, or a combination of these approaches. These activities allow the instructor and the students to extend the Big Picture dialog into hands-on experiences that give an enhanced appreciation and greater physical feel for the phenomena involved. Activities can help students from different disciplines work together and learn from each other. The activities are generally simple and can be completed in a short amount of time with inexpensive materials and quick setups. The emphasis is on qualitative appreciation of the physical phenomena with a modest amount of measurement involved. Educational research has shown that students learn better when they are personally involved in activities as opposed to listening to lectures. Furthermore, retention of abilities learned is improved along with greater ability to transfer learning to new and different applications.

Problem-Solving Techniques. Students must be able to solve real problems, complete the necessary calculations, manipulate units in equations, seek appropriate data, and make good design decisions. The example problems in this book are designed to help students master these processes. In addition, students must learn to communicate the results of their work to others in the field. One important means of communication is the presentation of the problem solutions in an orderly, well-documented manner using established methods. Example problems are set off with a distinctive graphic design and type font. Readers are guided in the process of formulating an approach to problem solving that includes:

a. Statement of the objective of the problem
b. Summary of the given information
c. Definition of the analysis technique to be used
d. Detailed development of the results with all of the equations used and unit manipulations
e. At times, comments on the solution to remind the reader of the important concepts involved and to judge the appropriateness of the solution
f. At times, comments present alternate approaches or improvements to the machine element or structural member being analyzed or designed

The reader's thought process is carried beyond the requested answer into a critical review of the result. With this process, designers gain good habits of organization when solving their own problems.

Design Approaches. This text provides extensive information about guidelines for design of mechanical devices and structural members than in most books on this subject. The design approaches are based on another book of mine, *Machine Elements in Mechanical Design*, Fourth Edition, 2004, from Prentice Hall. Learning about design in addition to analysis increases the usefulness of the book to students and professional users. There are some students who will not go on to a following course that emphasizes design. They should get some introduction to the principles of design in the introductory course in strength of materials. For those who do proceed to a design course, they should enter that course with a higher level of capability.

Design Properties of Materials. Chapter 2 includes extensive information and discussion on the proper application of engineering materials of many types, both metallic and nonmetallic. There is an extensive introduction to the nature of composite materials given along with commentary throughout the book on the application of composites to various kinds of load-carrying members. Information about the advantages of composites relative to traditional structural materials such as metals, wood, concrete, and plastics are given. The reader is encouraged to seek more education and experience to learn the unique analysis and design techniques required for the proper application of composite materials. Such materials are, in fact, tailored to a specific application, and general tables of material properties are not readily available.

Chapter 2 also includes a new section on materials selection based on the landmark publication *Materials Selection in Mechanical Design,* 3rd ed., by Michael F. Ashby, published by Elsevier-Butterworth-Heinemann (2005).

End-of-Chapter Problems. There is an extensive set of problems for student practice at the end of each chapter. The problems are typically organized around the main topics in the chapter. In general, they are presented in a graded manner with simpler problems followed by more comprehensive problems. There are many additional problems at the end of most chapters for practice, review, and design.

Extensive Appendix. To complement the use of design approaches, the Appendix provides additional information on material properties, geometry of common areas and commercially available structural shapes, stress concentration factors, formulas for beam deflection, conversion factors, and many others. This allows for a wider variety of problems in the book and for creating tests and projects. It adds to the realism of the book and gives the student practice in looking for the necessary information to solve a problem or to complete a design.

This edition includes a significant amount of additional Appendix data in SI metric units. All commercially available section property data for structural shapes include separate tables of SI data in addition to the formerly included U.S. Customary unit data. The SI data are taken from the latest versions of publications by the American Institute of Steel Construction (AISC). The SI data tables and the U.S. data tables are coordinated so students and instructors can quickly compare the designations and specific data from the two systems. Problems stated in SI metric data should be solved using the SI property data; instructors can develop their own quiz and exam problems completely in the SI system.

An entirely new table has been added on property data for mechanical tubing to supplement the standard pipe data from AISC and to offer designers of mechanical devices or manufacturing applications a wider variety of sizes of hollow circular sections, particularly on the smaller part of the size spectrum.

Appendix data for plastics, structural steel, titanium, and nickel-based alloys have been greatly expanded in this edition with corresponding expansion of the discussions about these materials in Chapter 2.

Electronic Aids to Problem Solving and Design

Most chapters include computer assignments along with suggestions for the use of spreadsheets, computer programs, computer algebra software, and graphing calculators pertinent to strength of materials. Such electronic aids, when used to supplement the basic understanding of the principles presented in the book, lead to a deeper appreciation of those principles and their application to simple and complex problems. Examples of spreadsheets are given in the chapters on column analysis and pressure vessels.

Lists of Internet sites that are pertinent to the topics covered in the chapter are included at the end of most chapters. The sites direct users to a wealth of additional resources to supplement the text. Some sites are for vendors of commercially available products or materials while others provide additional examples of analyses done in the book or more in-depth discussion of certain topics. Users are encouraged to perform their own searches to explore other aids to understanding of strength of materials.

The accompanying CD-ROM contains helpful software to help students understand several important principles of strength of materials. A set of twelve learning aids reinforces the book's presentation of key topics, with dynamic, colorful illustrations and examples worked out in a step-by-step manner. A notation is given in the margin of the sections of the book where these learning aids are pertinent.

A powerful beam analysis software package is included that is useful for five chapters of the book dealing with shearing force and bending moments in beams, centroids and moments of inertia, stresses due to bending, shearing stresses in beams, statically indeterminate beams, and beam deflections (Chapters 5–9).

Instructors and students are cautioned about proper use of the beam analysis software:

> ***Users of computer software must have a solid understanding of the principles of analysis and design to ensure that analysis results and design decisions are based on reliable foundations. We recommend that the beam analysis software be used only after mastering pertinent principles on which the software is based using careful study and manual techniques.***

The advantage of using the software after appropriate prior study is that many additional and more complex practice problems can be solved in the allocated time to give students more in-depth understanding of the behavior of beams for a variety of beam geometries, beam materials, loading patterns, and support conditions. They can pursue several alternative designs and work toward more optimum results.

An online Instructor's Manual is also available to instructors. To access the online Instructor's Manual, go to **www.prenhall.com.** Instructors can search for a text by author, title, ISBN, or by selecting the appropriate discipline from the pull down menu at the top of the catalog home page. To access supplementary materials online, instructors need to request an instructor access code. Go to **www.prenhall.com,** click the **Instructor Resource Center** link, and then click **Register Today** for an instructor access code. Within 48 hours of registering you will receive a confirming e-mail including an instructor access code. Once you have received your code, go to the site and log on for full instructions on downloading the materials that you wish to use.

Adjustments to Format from Previous Edition

Users of previous editions of this book will find a significant amount of reordering of the coverage of some topics. Guided by intensive feedback from users, the revised arrangement is more streamlined. Some highlights of these changes are:

- Chapter 1, Basic Concepts in Strength of Materials, has been reduced in size to focus on the most cogent introductory material. Several sections on material properties, stress, and strain were relocated into Chapters 2 and 3.
- The coverage of deformation due to axial stresses has been integrated into Chapter 3 on Direct Stresses instead of being in a separate chapter.
- All topics on combined stresses have been consolidated into a single chapter (Chapter 10).
- The discussions of continuous beams and the theorem of three moments have been included with the chapter on Shearing Forces and Bending Moments in Beams (Chapter 5). Other topics related to statically indeterminate beams have been integrated into Chapter 9, Deflection of Beams.
- The introduction of the area property of section modulus has been included in the chapter on Centroids and Moments of Inertia of Areas (Chapter 6). This topic is expanded upon in Chapter 7, Stress Due to Bending.

Enhanced Visual Attractiveness. The addition of a second color makes the book more visually appealing; illustrations, graphs, and tables are easier to use and interpret. Many illustrations have been improved by the addition of three-dimensional graphics, greater realism, and more effective use of shading as well as the introduction of color.

Suggested Course Outlines. The online Instructor's Manual that accompanies this text contains suggested outlines for courses as support for instructors in the variety of program types that have used the previous editions of this book or who are initially adopting it.

Acknowledgments

I appreciate the feedback provided by both students and instructors who have used the earlier editions of this book. I am also grateful to my colleagues at the University of Dayton. I would like to thank the participants of a focus group that provided input for the revision of this book: Janice Chambers, Portland Community College; Janak Dave, University of Cincinnati; David Dvorak, University of Maine; Frank Gourley, West Virginia University Institute of Technology; and Jack Zecher, Indiana University–Purdue University at Indianapolis (IUPUI). I would also like to thank the reviewers of this edition: Joana Finegan, Central Michigan University; Robert Michael, Pennsylvania State University, Erie; and Thomas Roberts, Milwaukee Area Technical College for their helpful suggestions for improvement. I hope this edition has implemented those suggestions in a manner consistent with the overall approach of the book.

Special recognition is given to Professor Jack Zecher, Indiana University–Purdue University at Indianapolis (IUPUI), for his expert development of the software provided on the CD that accompanies this book. The twelve interactive learning modules and the beam analysis software blend well with the text and offer students and instructors valuable supplemental resources.

Robert L. Mott
University of Dayton

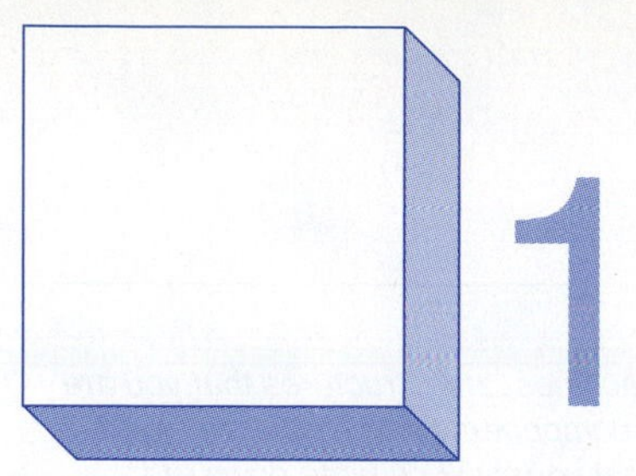

1

Basic Concepts in Strength of Materials

The Big Picture

Basic Concepts in Strength of Materials

Discussion Map

- Products, machines, and structures must be designed to be safe and to provide satisfactory performance during the intended use.
- Safety is paramount. The load-carrying components must not fracture during use.
- Excessive deformation is another form of failure.
- Buckling, occurring when the shape of a load-carrying member becomes unstable, must be avoided.
- You will learn about the basic nature of stresses and strains in this course.
- You will be able to recognize several types of stresses created by different loading and support situations.
- You will analyze situations where more than one kind of stress is experienced by a load-carrying member at the same time.
- Design requires that you determine the shape and size of a load-carrying member and specify the material from which it is to be made.
- You will learn how to design safe load-carrying components of machines and structures.

Discover

Think about products, machines, and structures that you are familiar with that contain components that must carry loads safely. For each device that you think of, write down the following information:

The basic function or purpose of the device.

The description and sketches of its primary components, particularly those that are subjected to significant forces.

The material from which each component is made. Is it a metal or plastic? What kind of metal? What kind of plastic? Is it some other material?

How is each component supported within the product, machine, or structure?

How are forces applied to the component?

What would be the consequence if the component broke?

How much deformation would cause the component to be incapable of performing its desired function?

Consider products around your home; parts of your bicycle, car, or motorcycle; buildings; toys; amusement park rides; aircraft and space vehicles; ships; manufacturing machinery; construction equipment; agricultural machinery; and others.

Discuss these products and systems with your colleagues and with your course instructor or facilitator.

Here are some examples of mechanical and structural systems and how they relate to the material you will study in this book.

1. In your home, the floors must be strong and stiff to support the loads due to people, furniture, and appliances [Figure 1–1(a)]. A typical floor is comprised of a series of joists that are supported on walls or beams, a subfloor on top of the joists, and the finished floor. These elements act together to provide a rigid support system. Pitched roofs employ trusses to span long distances between support walls and to provide the support for the roof sheathing and shingles while remaining fairly lightweight and using materials efficiently. Chairs and tables must be designed to support people and other materials safely and stably. Even in the refrigerator, the shelves must be designed to support heavy milk and juice jugs while being lightweight and allowing the free movement of cooled air over the food. In the garage, you might have a stepladder, a garage door opener, a lawn mower, and shovels, all of which carry forces when they are used. What other examples can you find around the home?

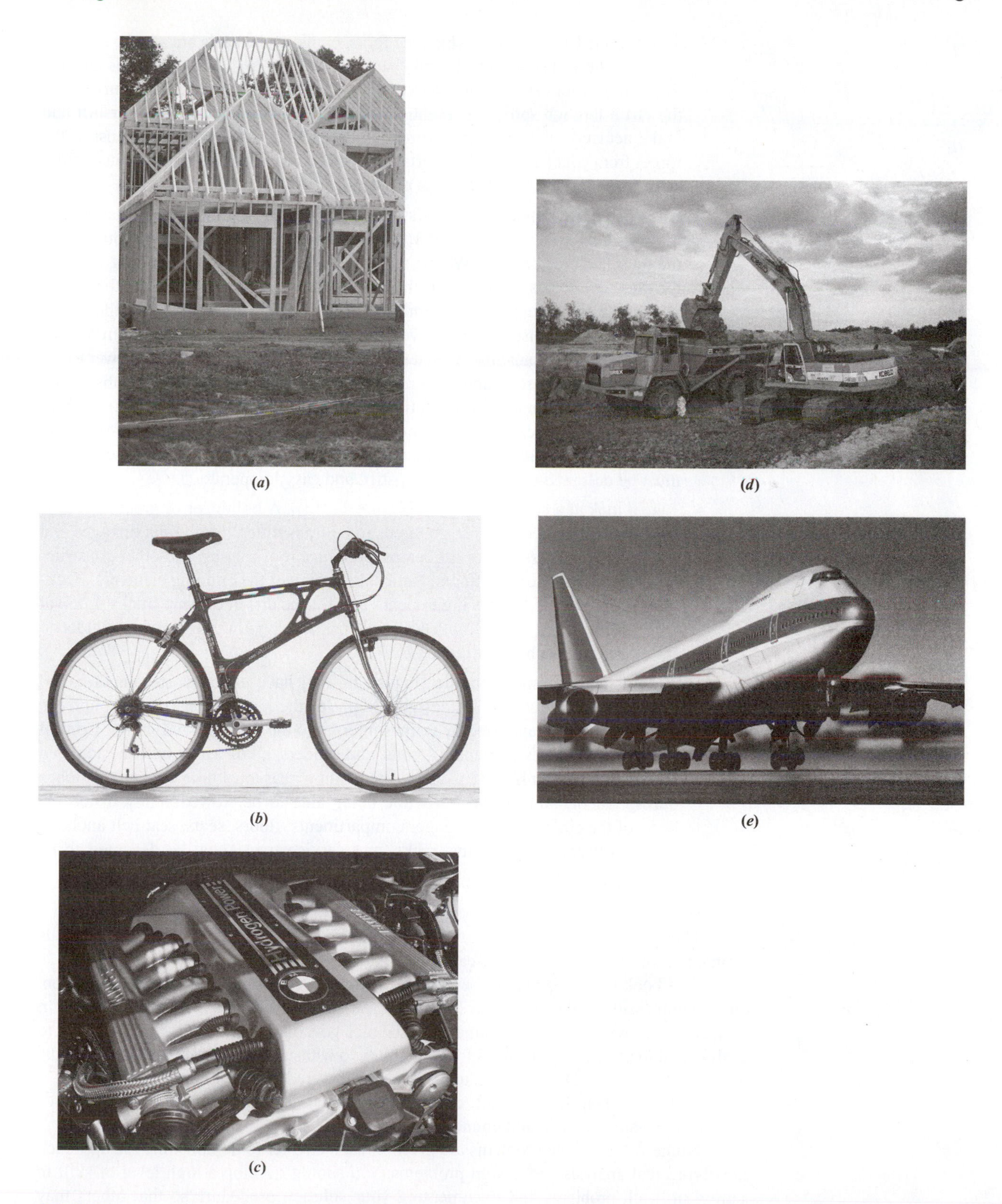

FIGURE 1–1 Examples of mechanical and structural systems. (Source: ***(a)*** The Stock Connection; ***(b)*** Doring Kindersley Media Library; ***(c)*** Photo Researchers, Inc.; ***(d)*** The Stock Connection; ***(e)*** Doring Kindersley Media Library.)

2. Your bicycle has a frame to which the forks, crank, and wheels are attached [Figure 1–1(b)]. The seat support and handlebars also connect to the frame. Pedals take the forces from your feet and cause a torque to be applied to the crank that in turn drives the chain through sprockets, eventually driving the rear wheel. The gearshift and brake actuation linkages have numerous levers, links, and pins that transmit the forces from your hands to do important tasks. The wheels and their axles must be designed to be strong and stiff and to resist repeated force applications.
3. An automobile contains hundreds of mechanical parts that support, latch, and move critical systems [Figure 1–1(c)]. Starting by opening the door, you engage the door latch mechanism. When you close the door, the latch reengages, keeping the door safely closed. You insert the key and turn the ignition, causing the starter to turn the engine until it starts and runs on its own. Inside the engine, the crank, connecting rods, pistons, and valve mechanisms operate in synchronism to provide power to the transmission and then to the drive shaft that delivers power to the wheels. As the car rolls down the road, its suspension elements, shock absorbers, struts, and brakes manage the motion of the chassis and isolate disturbing vibrations from the passenger compartment. Inside the car you may switch on the windshield wipers, adjust your seat position, or open a window. All of these devices must be designed to be safe, strong, stiff, and easy to operate.
4. Take a look at a construction site [Figure 1–1(d)]. A bulldozer or grader is leveling out the land, requiring large forces to be transmitted through its linkages and hydraulic actuators. Cranes lift beams, columns, roof trusses, and other construction materials to elevated floors. Notice the design of the building elements. Backhoes and front-end loaders dig trenches and lift the dirt into dump trucks. Look at their mechanisms. The dump truck itself has a heavy-duty hydraulic cylinder to lift its bed for discharging the load.
5. Aircraft and the space shuttle [Figure 1–1(e)] have efficient structures called monocoque or stressed-skin designs. Many of the loads are carried through the skin of the craft, supported by frames internal to the fuselage and wings. Control of the craft depends on flaps, ailerons, and rudders, all of which must be actuated by hydraulic systems and linkages. The landing gear must take tremendous loads during takeoff and touchdown while being able to be stowed neatly inside the body of the craft. In the passenger compartments, floors, seats, seat belt anchors, door hardware, storage bins, racks, and service trays must be designed to be strong, safe, and lightweight.

These are but a few of the many examples of situations where you may use the skills from strength of materials in your career.

This book is organized to present basic stress analysis first, considering direct tension, direct compression, and direct shear. Torsional shear stresses and stresses in beams follow. In each case, you will learn the fundamental principles governing these kinds of stress, how to analyze real load-carrying members for their ability to withstand such stress, and to design the members themselves. Deformations under loads are covered along with the stress analysis.

You will then learn about combined stresses, statically indeterminate beams, columns, pressure vessels, and connections.

Numerous example problems are given in each chapter to demonstrate the approach to solving real analysis and design problems. You should develop a high level of skill in setting up such problems and documenting your solution procedure so that others may understand and evaluate your work.

1–1 OBJECTIVE OF THIS BOOK—TO ENSURE SAFETY

It is essential that any product, machine, or structure be safe and stable under the loads exerted on it during any foreseeable use. The analysis and design of such devices or structures to ensure safety is the primary objective of this book.

Failure of a component of a structure can occur in several ways:

1. The material of the component could fracture completely.
2. The material may deform excessively under load so that the component is no longer suitable for its purpose.
3. The structure could become unstable and buckle, and thus be unable to carry the intended loads.

Examples of these failure modes should help you to understand the importance of your learning the principles of applied strength of materials as presented in this book.

Preventing Failure by Fracture. Figure 1–2 shows two rods carrying a heavy casting. Imagine that you are the person responsible for designing the rods. Certainly you would want to ensure that the rods were sufficiently strong so they will not break and allow the casting to fall, possibly causing great damage or injury to people. As the designer of the rods, what information would you need? What design decisions must you make? Following is a partial list of questions you should ask.

FIGURE 1–2 Two rods supporting a heavy casting.

1. What is the weight and physical size of the casting?
2. Where is its center of gravity? This is important so you can decide where to place the points of attachment of the rods to the casting.
3. How will the rods be attached to the casting and to the support system at the top?
4. Of what material will the rods be made? What is its strength?
5. What will be the shape and size of the cross section of the rods?
6. How will the load of the casting be initially applied to the rods: slowly, with shock or impact, or with a jerking motion?
7. Will the rods be used for many cycles of loading during their expected life?

Knowing these factors will permit you to design the rods to be safe; that is, so they will not break under the anticipated service conditions. This will be discussed in more detail in Chapters 1 and 3.

Preventing Excessive Deformation. Gears are used in mechanical drives to transmit power in such familiar systems as the transmission of a truck, the drive for a conveyor belt, or the spindle of a machine tool. For proper operation of the gears, it is essential that they are properly aligned so the teeth of the driving gear will mesh smoothly with those of the mating gear. Figure 1–3 shows two shafts carrying gears in mesh. The shafts are supported in bearings that are in turn mounted rigidly in the housing of the transmission. When the gears are transmitting power, forces are developed that tend to push them apart. These forces are resisted by the shafts so that they are loaded as shown in Figure 1–4. The action of the forces perpendicular to the shafts tends to cause them to bend, which would cause the gear teeth to become misaligned. Therefore, the shafts must be designed to keep deflections at the gears to a small, acceptable level. Of course, the shafts must also be designed to be safe under the applied loads. In this type of loading, the shafts are considered to be *beams*. Chapters 7–9 discuss the principles of the design of beams for strength and deflection.

Stability and Buckling. A structure may collapse because one of its critical support members is unable to hold its shape under applied loads, even though the material does not fail by fracturing. An example is a long, slender post or column subjected to a downward, compressive load. At a certain critical load, the column will *buckle*. That is, it will suddenly bend or bow out, losing its preferred straight shape. When this happens, if the load remains applied, the column will totally collapse. Figure 1–5 shows a sketch of such a column that is relatively long with a thin rectangular cross section. You can demonstrate the buckling of this type of column using a simple ruler or meter stick. To prevent buckling, you must

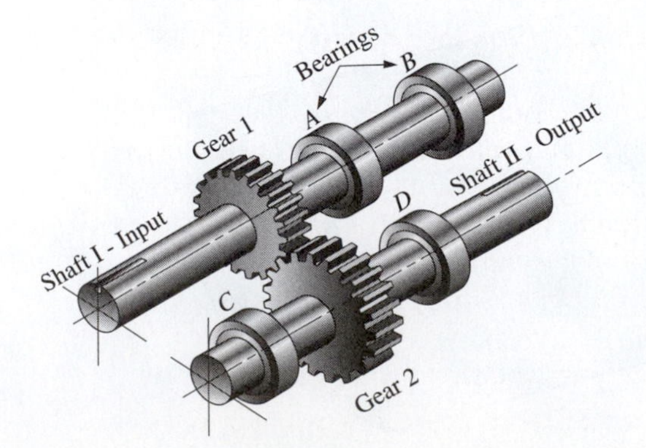

FIGURE 1–3 Two shafts carrying gears in mesh.

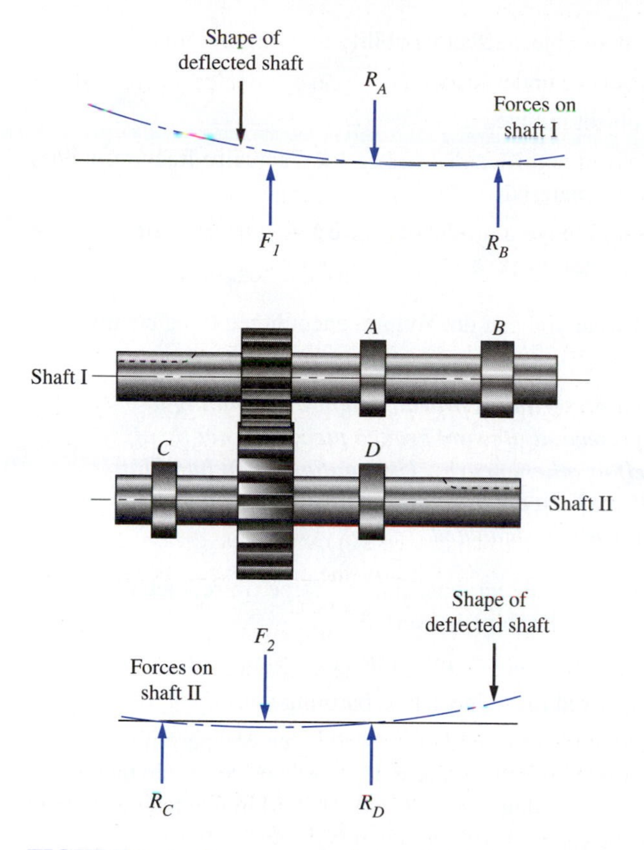

FIGURE 1–4 Forces on shafts I and II of Figure 1–3 with resulting deflection of shafts.

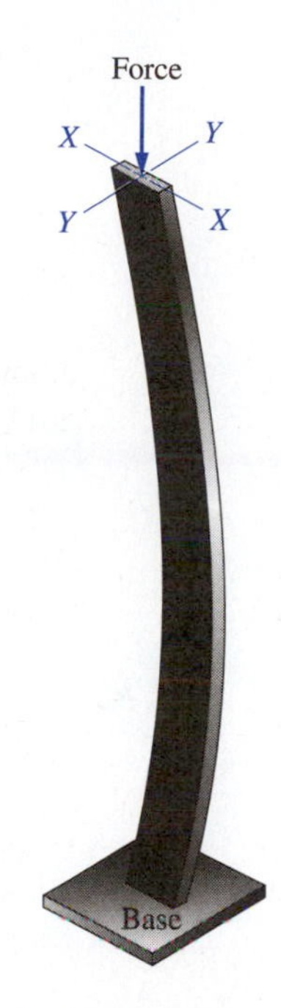

FIGURE 1–5 A slender column in compression, illustrating elastic instability or buckling.

be able to specify an appropriate material, shape, and size for the cross section of a compression member of a given length so it will remain straight under the expected loads. Chapter 11 presents the analysis and design of columns.

In summary, design and analysis, using the principles of strength of materials, are required to ensure that a component is safe with regard to *strength, rigidity,* and *stability*. It is the objective of this book to help you gain the ability to design and analyze load-carrying components of structures and machines that will be safe and suitable for their intended functions.

Activity Chapter 1—Basic Concepts of Strength of Materials

Let's break something!

In the Big Picture you explored many situations in which the principles of strength of materials are applied to create safe structures and products that perform well in their intended applications. When applied properly, well-designed products will not break.

However, useful learning can happen when you purposely cause something to break. This destruction may give you many insights, such as:

1. How loads are applied
2. How load-carrying members are supported

3. How the shape and size of an object affect its ability to withstand forces
4. How different materials behave under load
5. Where the failure of an object initiates
6. How things typically deform significantly before they actually fracture, unless they are made from a brittle material
7. How, sometimes, you would judge a product to have failed because of excessive deformation even if it does not break

Here are some examples of what you can do. You are encouraged to be creative and find others. But be careful—

Ensure that proper personal protection is provided before attempting to break anything. Wear eye protection. Be sure broken pieces cannot fly off and cause injury to yourself or others nearby. Use caution when handling knives, screwdrivers, or other sharp objects. Protect furniture, windows, walls, or other objects that could be damaged.

a. **Direct tension failure**—Occurs when an axial pulling force overcomes the tensile strength of the material on which the force is exerted.
   - Gather a piece of string such as what you might use for flying a kite.
   - Attach one end to a strong fixed location such as by tying it securely to a post.
   - Arrange a way to take hold of the other end of the string that will permit you to pull it slowly with steadily increasing force until it breaks. Perhaps you could tie the free end around a fairly large wooden dowel, say 20 mm (1.0 in) in diameter and about 250 mm (10 in) long, so you can pull with both hands [Figure 1–6(a)].
   - Ask a colleague to observe the string as you increase the force. (Again, be careful!)
   - Have the colleague describe any event that indicates when the string is about to fail; for example, when one small strand breaks.
   - Try to observe the initiation of the failure yourself by relaxing the force.
   - If you have a camera available, take a photograph of the failure area.
   - Then resume applying the force, slowly increasing it until the string breaks completely. Observe the progression of the failure.
   - After the string breaks, lay it on a table and carefully observe the nature of the broken ends.
   - Document your process and the appearance of the failure.

b. **Direct compression failure**—Occurs when a direct axial pushing force causes the material to crush.
   - The piece selected for compression failure should be quite short with a broad cross-sectional area in order for direct compression failure to occur with little or no bending or buckling prior to failure [Figure 1–6(b)]. Note that longer, more slender shapes in compression are called *columns* and fail in a radically different manner.
   - Try a short length, say 25 mm (1.0 in) or less, of a cylindrical wooden dowel.
     - A good way to apply the compression load is with a bench vise like those found in a home shop, an automotive garage, or an industrial workplace.
     - Align the dowel carefully so the force is applied along the axis of the cylinder.
     - Observe the failure, typically a significant amount of crushing, probably near one of the ends. The wood fibers tend to keep the dowel from splitting completely.

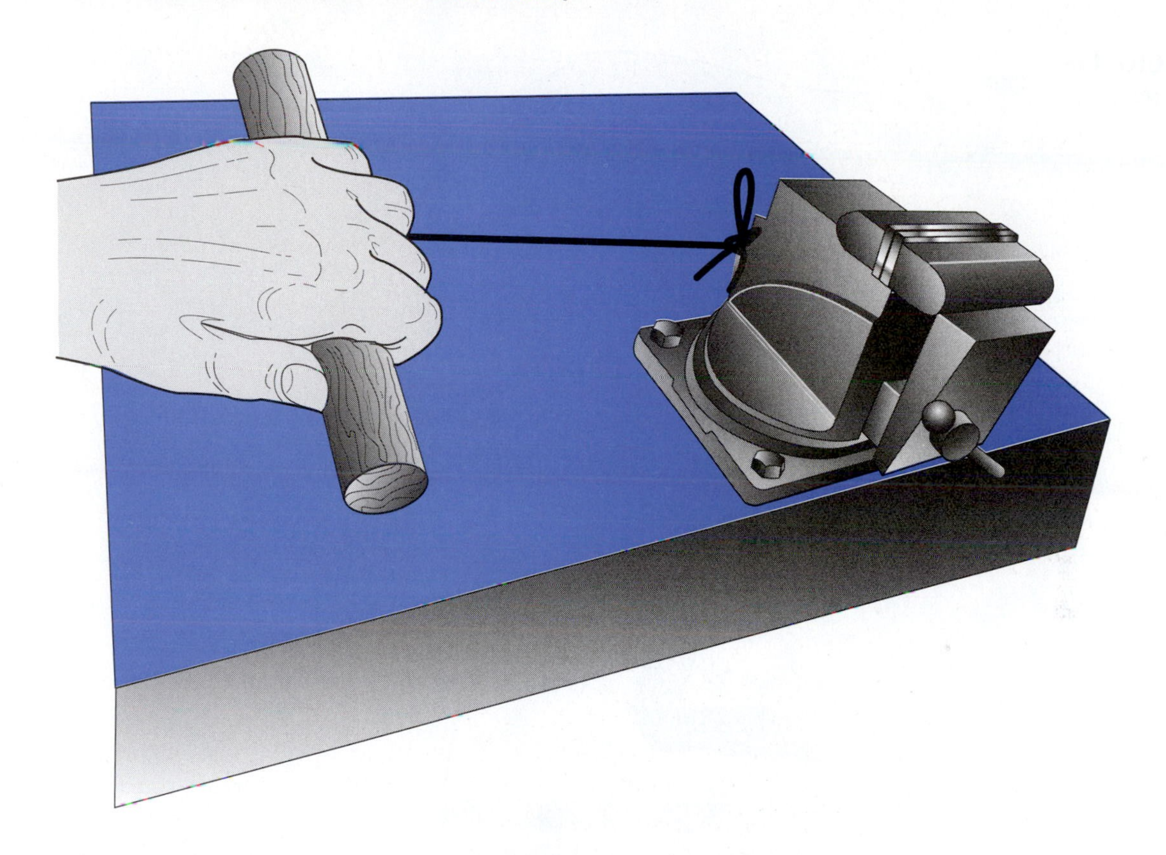

(*a*)

(*b*)

FIGURE 1–6 Examples of simple setups to demonstrate a variety of load-carrying members and kinds of stresses produced by different load and support conditions.

FIGURE 1–6
Continued

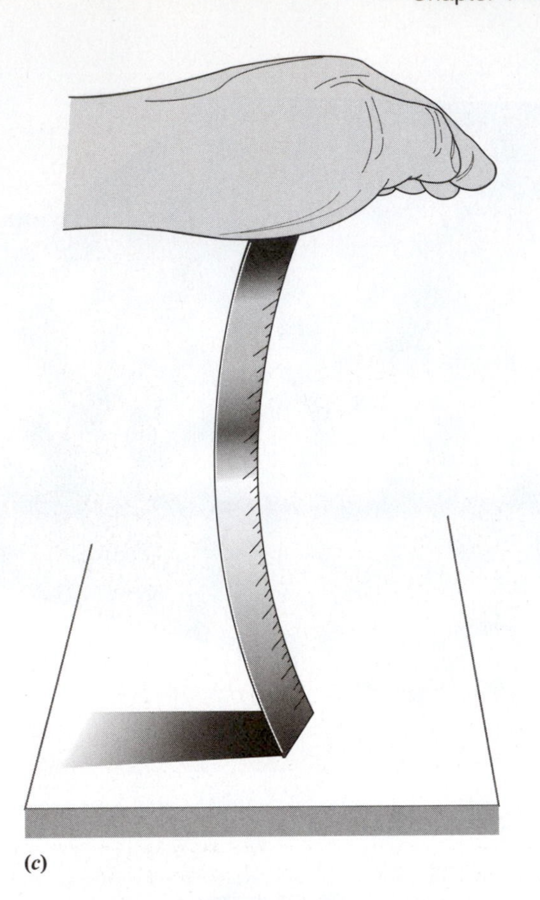

(c)

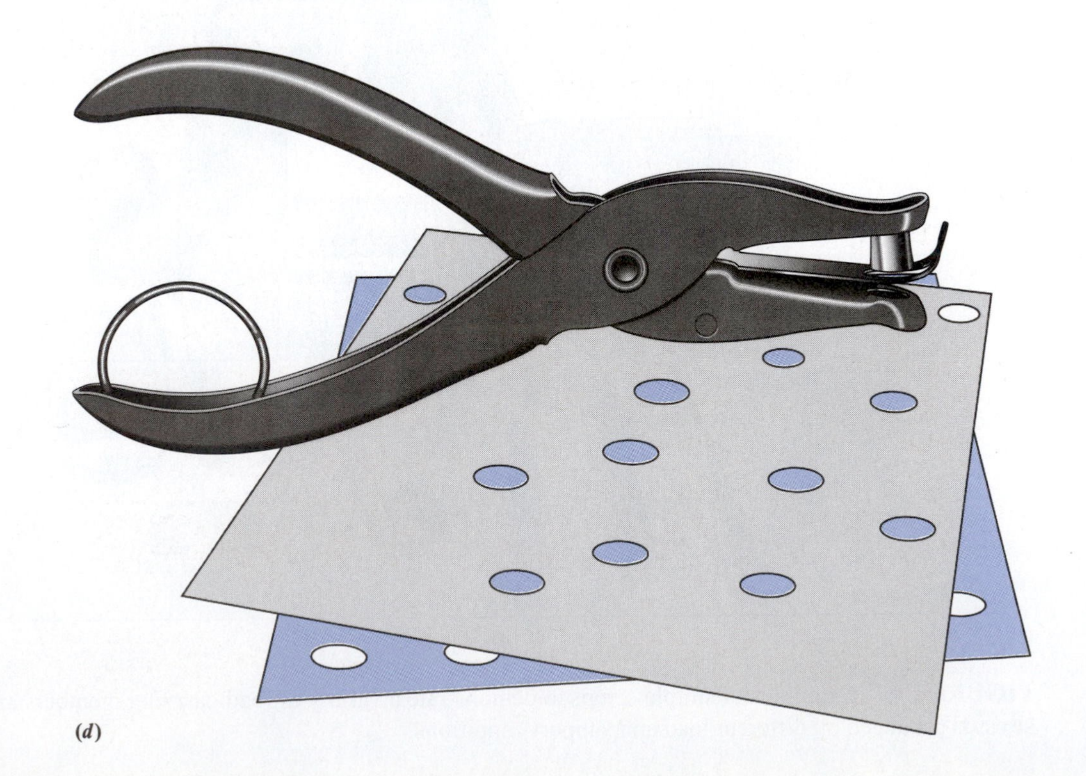

(d)

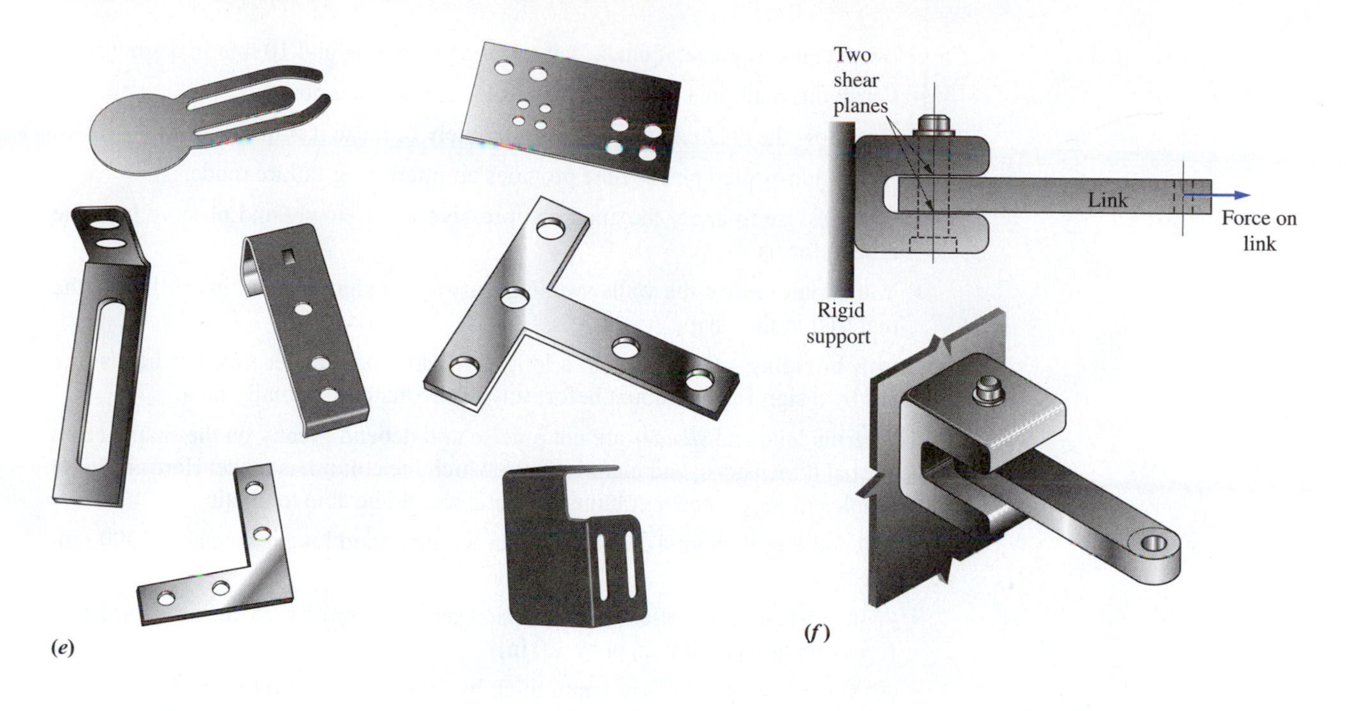

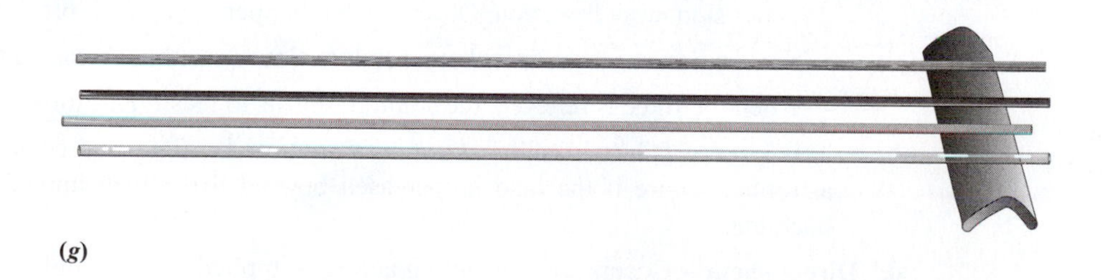

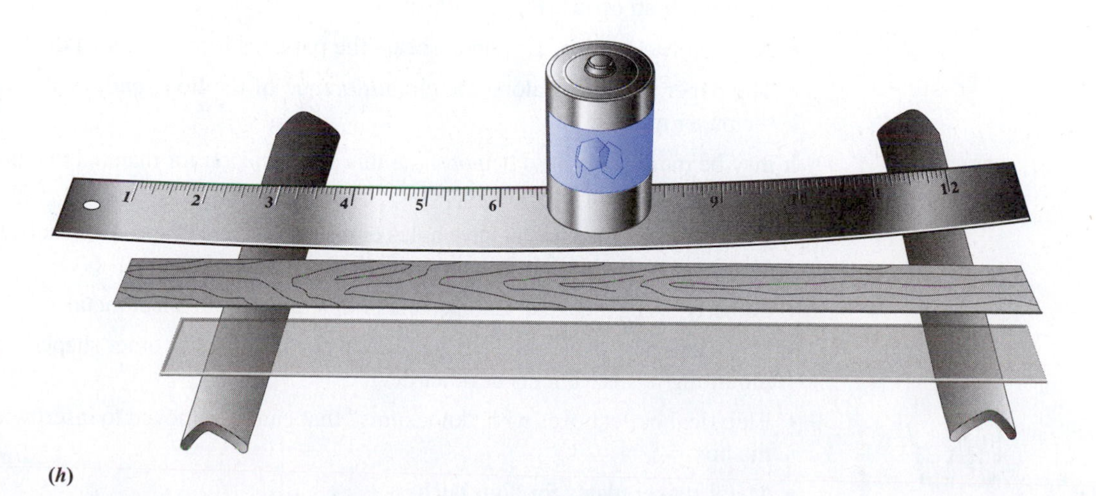

FIGURE 1–6 *Continued*

- Now get a short piece of chalk, say about 25 mm long and 10 mm in diameter.
  - Crush the chalk in a vise as described for the wooden dowel.
  - Note how the chalk does fracture completely because it is quite brittle.
- A short thin-walled plastic tube provides an interesting failure mode.
  - Use the vise to apply the axial compressive force slowly and observe how the tube deforms.
  - You should notice the walls swelling outward, a sign of local instability of the material in this shape.

c. **Column buckling**—Occurs when a long, straight, and slender member bends and buckles by a significant amount before any of the material actually fails.

- The terms *long* and *slender* are not precise and depend greatly on the shape, cross sectional dimensions, and material from which the column is made. Here are some examples of long slender columns that you should be able to obtain:
  - A wood dowel about 4.5 mm ($\frac{3}{16}$ in) in diameter and longer than about 300 mm (12.0 in)
  - A thin wood meter stick (cross section about 6 mm by 25 mm) or yardstick (cross section about 0.25 in by 1.0 in)
  - A thin strip of plastic, say 1 mm thick by 20 mm wide and longer than 50 mm; even a plastic knife like those for picnics could work.
- Try to apply slowly an axial compressive force to the column, being careful that it does not skid away from you. Observe what happens [Figure 1–6(c)].
  - You should notice that the column bends noticeably without breaking the material.
  - When the force is relieved, the column returns to its straight form.
- This kind of failure is called *elastic instability* or *buckling,* and could lead to catastrophic failure if the load is increased beyond that which initially causes the buckling.

d. **Direct shear**—Occurs when a cutting action is applied as you would see with scissors, tin snips, or punches.

- A simple example is a hole punch for writing paper like those you may see in a classroom or an office [Figure 1–6(d)].
  - As you press a lever, the punch shears the paper as it passes into a die below.
  - The paper is sheared along the circumference of the hole, and the thickness of the paper must be cut.
- It may be more difficult to demonstrate this phenomenon for material that is thicker or stronger than simple writing paper.
- Try to find components that have holes or other punched shapes to visualize where direct shear is used in familiar products. Examples are [Figure 1–6(e)]:
  - Some types of brackets, clamps, and straps made from sheet metal
  - Metal cabinets with round holes punched for fasteners or other shapes to permit mounting of instruments or other devices
  - Electrical outlet boxes with "knockouts" that can be removed to insert wires into the box
  - Metal striker plates for door latches
  - Perforated sheet metal often used for decorative purposes

- Another example of direct shear occurs in a type of hinge called a *clevis* [Figure 1–6(f)]. A cylindrical pin passes through end pieces connected to one component and through a hole in another component.
  - As a force is applied to one of the moving components, the pin is subjected to a direct shearing stress that tends to cut it across its cross section.

e. **Torsional shear stress**—Occurs when a load-carrying member is twisted about its long axis due to an applied torque at one or more points along the member, resisted by a reaction torque at one or more other points.

- One of the most frequent applications in which torsional shear stress occurs is in a power transmission shaft. For example, the drive shaft for a vehicle or an industrial machine receives power from a prime mover such as an engine or electric motor. The power is transmitted along the shaft, typically having the shape of a relatively long solid or hollow cylinder, and delivers it to an axle to propel the vehicle, or to a piece of rotating machinery such as a conveyor, machine tool, pump, compressor, food mixer, coffee grinder, or any of numerous other types of mechanical devices. See Figure 1–3.
- The shafts of the mechanical devices listed above are designed to transmit safely the power required to perform the desired function without breaking or twisting a noticeable amount. Therefore, you would not normally be able to apply sufficient torque to break the shaft or to observe significant torsional deformation (twisting). So let's try to find some smaller cylindrical rods or tubes that can be twisted and/or broken when you apply a torque with your hands.
- When discussing columns earlier, we suggested using a small-diameter wood dowel, about 4.5 mm ($\frac{3}{16}$ in). With a length of dowel approximately 500 mm (20 in) or longer, you should be able to hold one end steady, apply a torque to the other end, and observe a significant angle of twist between the fixed and free ends. This is an example of producing torsional shear stress. Try to find other similar examples made from plastic rod or tubing, or small-diameter steel wire such as from a coat hanger [Figure 1–6(g)]. Long foam cylinders sometimes seen at swimming pools are also good.
- You may use the vise again to hold one end of a longer, larger diameter steel rod, say about 6 mm (0.25 in) diameter by about 550 mm (22 in) long. Then use a strong tool such as locking pliers to grab onto the opposite end and twist the rod. Notice how the applied torque must increase with the larger diameter and stiffer material.
- Note that shapes other than cylindrical rods or tubes could also be placed in torsion. However, their behavior and the resulting stresses are radically different, as will be discussed in Chapter 4. For example, find any flat, thin, rectangular piece such as a ruler, paint stirring stick, or meter stick and twist it with your hands. If the piece is fairly long and thin, you should be able to produce a large deformation without breaking it.

f. **Bending stress**—Occurs when a member carries a load perpendicular to its long axis while being supported in a stable manner, thereby "bending" the member, now called a "beam."

- A simple example is a flat wood stick from an ice cream bar. Set the stick on a pair of simple supports near the ends and press down near the middle. You should be able to deflect the stick (bend the beam) easily. Note the curved, bent shape taken by the beam as it is loaded. With moderate loading the beam should return to its original shape after you remove the load, indicating that the deflection is elastic.

- Other shapes you could use that are easily obtained are a meter stick or yardstick, paint stirring stick, heavy card stock, thin plastic strips, or thin sheet metal such as might be used in a metal ruler. Apply the loads slowly to ensure that the beam does not break or deform permanently if you want to keep it [Figure 1–6(h)].
- However, if you have a beam example that is expendable, go ahead now and increase the load and observe what happens. That depends mainly on the kind of material.
- A wood stick will bend significantly and then begin to shatter or split. Make note of where the failure starts and the appearance of the wood after it breaks.
- A thin metal beam will also bend a lot while still returning to its initial shape. Then you should notice that it takes a permanent deformed shape without breaking into two separate pieces. It is said to have "yielded" where some of the material has been stressed beyond its yield strength, as described in Chapters 2 and 3.

1–2 OBJECTIVES OF THIS CHAPTER

In this chapter we present the basic concepts in strength of materials that will be expanded on in later chapters. After completing this chapter, you should be able to:

1. Use correct units for quantities encountered in the study of strength of materials in both the SI metric unit system and the U.S. Customary unit system.
2. Use the terms *mass* and *weight* correctly and be able to compute the value of one when the value of the other is given.
3. Define *stress.*
4. Define *direct normal stress,* both tensile and compressive.
5. Represent normal stresses on stress elements.
6. Define *normal strain.*
7. Define *direct shear stress* and the terms *single shear* and *double shear.*
8. Represent shear stresses on stress elements.
9. Recognize standard structural shapes and standard screw threads and use data for their geometric properties.
10. Describe several approaches to experimental and computational stress analysis.

1–3 PROBLEM-SOLVING PROCEDURE

The study of strength of materials and the practice of stress analysis and design inherently require problem solving. It is important for you to establish good habits of organization in the method used to solve problems and for reporting them in a clear attractive format. This will help you communicate your solution to others and when you need to refer back to a previously solved problem.

The example problems in this book require the following procedure:

a. The original statement of the problem is given.
b. Restate the primary objective of the problem.
c. Give a summary of the pertinent information and data. This is useful to help you decide what is known and what is to be found. It also serves as a convenient place to find data when needed later in the problem solution.

d. Write a general statement of the analysis technique to be used to solve the problem. State any assumptions here also.

e. Complete a detailed development of the results with all of the equations used, the insertion of the values for pertinent data, and the manipulation of units for the results. Conversion factors may be required to produce the final result in appropriate units.

f. Calculate the value of all expected results. In this book, we typically report results with three computed digits of precision and appropriate units. A greater precision is carried through the problem and rounding is done at the end. We consider all given data to be exact.

g. Comment on the solution to clarify details and to critically review the problem. Is the result reasonable? Are there alternative analysis techniques that could have been used?

h. Are there additional analyses that would be desirable to ensure a more robust solution? If it is a design problem, specify a convenient size for key dimensions, a standard shape for the load-carrying member, or a suitable material from which to make the member.

Importance of Knowledge of Statics. An accurate application of the principles of strength of materials requires that the loads to be carried by the member being analyzed or designed be known with a high degree of confidence. The loads may be any combination of gravity forces, direct axial forces, shearing forces, directly applied moments that tend to cause rotation of a member or structure, or torsional moments (torques) that are applied to a shaft to transmit power. It is expected that you have a good background in the study of physics mechanics and its application through a course in statics. You should be able to prepare complete free-body diagrams of structures or individual elements, apply equations of equilibrium, and solve for reactions at supports. If needed, refer to Appendix A–27 for a review of some of the fundamental principles of statics and an example problem showing their application.

1–4 BASIC UNIT SYSTEMS

Computations required in the application of strength of materials involve the manipulation of several sets of units in equations. For numerical accuracy it is of great importance to ensure that consistent units are used in the equations. Throughout this book units will be carried with the applicable numbers.

Because of the present transition in the United States from the traditional U.S. Customary units to metric units, both are used in this book. It is expected that persons entering or continuing an industrial career within the next several years will be faced with having to be familiar with both systems. On the one hand, many new products such as automobiles and business machines are being built using metric dimensions. Thus components and manufacturing equipment will be specified in those units. However, the transition is not being made uniformly in all areas. Designers will continue to deal with such items as structural steel, aluminum, and wood whose properties and dimensions are given in English units in standard references. Also, designers, sales and service people, and those in manufacturing must work with equipment that has already been installed and that was made to U.S. Customary system dimensions. Therefore, it seems logical that persons now working in industry should be capable of working and thinking in both systems.

The formal name for the U.S. Customary unit system is the English Gravitational Unit System (EGU). The metric system, which has been adopted internationally, is called by the French name Systéme International d'Unités, the International System of Units, abbreviated SI in this book. See Reference 5.

TABLE 1–1 Basic quantities in the SI metric unit system.

Quantity	SI unit	Other metric units
Length	meter (m)	millimeter (mm)
Time	second (s)	minute (min), hour (h)
Force	newton (N)	$kg \cdot m/s^2$
Mass	kilogram (kg)	$N \cdot s^2/m$
Temperature	kelvin (K)	degrees Celsius (°C)
Angle	radian (rad)	degree (deg)

TABLE 1–2 Basic quantities in the U.S. Customary unit system.

Quantity	U.S. Customary unit	Other U.S. units
Length	foot (ft)	inch (in)
Time	second (s)	minute (min), hour (h)
Force	pound (lb)	kip*
Mass	slug	$lb \cdot s^2/ft$
Temperature	degrees Fahrenheit (°F)	
Angle	degree (deg)	radian (rad)

*1.0 kip = 1000 pounds. The name is derived from the term *kilopound.*

In most cases, the problems in this book are worked either in the U.S. Customary unit system or the SI system rather than mixing units. In problems for which data are given in both unit systems, it is most desirable to change all data to the same system before completing the problem solution. Appendix A–26 gives conversion factors for use in performing conversions.

The basic quantities for any unit system are length, time, force, mass, temperature, and angle. Table 1–1 lists the units for these quantities in the SI unit system, and Table 1–2 lists the quantities in the U.S. Customary unit system.

Prefixes for SI Units. In the SI system, prefixes should be used to indicate orders of magnitude, thus eliminating digits and providing a convenient substitute for writing powers of 10, as generally preferred for computation. Prefixes representing steps of 1000 are recommended. Those usually encountered in strength of materials are listed in Table 1–3. Table 1–4 shows how computed results should be converted to the use of the standard prefixes for units.

1–5 RELATIONSHIP AMONG MASS, FORCE, AND WEIGHT

Force and mass are separate and distinct quantities. Weight is a special kind of force.

Mass refers to the amount of the substance in a body.

Force is a push or pull effort exerted on a body either by an external source or by gravity.

Weight is the force of gravitational pull on a body.

Mass, force, and weight are related by Newton's law:

$$\text{force} = \text{mass} \times \text{acceleration}$$

We often use the symbols F for force, m for mass, and a for acceleration. Then,

$$F = m \times a \quad \text{or} \quad m = F/a$$

TABLE 1–3 Prefixes for SI units.

Prefix	SI symbol	Factor
giga	G	$10^9 = 1\ 000\ 000\ 000$
mega	M	$10^6 = 1\ 000\ 000$
kilo	k	$10^3 = 1\ 000$
milli	m	$10^{-3} = 0.001$
micro	μ	$10^{-6} = 0.000\ 001$

TABLE 1–4 Proper method of reporting computed quantities.

Computed result	Reported result
0.005 48 m	5.48×10^{-3} m, or 5.48 mm
12 750 N	12.75×10^3 N, or 12.75 kN
34 500 kg	34.5×10^3 kg, or 34.5 Mg (megagrams)

When the pull of gravity is involved in the calculation of the weight of a mass, a takes the value of g, the acceleration due to gravity. Then, using W for weight,

Weight–Mass Relationship

$$W = m \times g \quad \text{or} \quad m = W/g \tag{1–1}$$

We will use the following values for g:

$$\text{SI units: } g = 9.81 \text{ m/s}^2 \qquad \text{U.S. units: } g = 32.2 \text{ ft/s}^2$$

Units for Mass, Force, and Weight. Tables 1–1 and 1–2 show the preferred units and some other convenient units for mass and force in both the SI and U.S. unit systems. The units for force are also used as the units for weight.

The newton (N) in the SI unit system is named in honor of Sir Isaac Newton, and it represents the amount of force required to give a mass of 1.0 kg an acceleration of 1.0 m/s^2. Equivalent units for the newton can be derived as follows using Newton's law with units only:

$$F = m \times a = \text{kg}\cdot\text{m/s}^2 = \text{newton}$$

In the U.S. Customary unit system, the unit for force is defined to be the pound, while the unit of mass (slug) is derived from Newton's law as follows:

$$m = \frac{F}{a} = \frac{\text{lb}}{\text{ft/s}^2} = \frac{\text{lb}\cdot\text{s}^2}{\text{ft}} = \text{slug}$$

The conversion of weight and mass is illustrated in the following example problems.

Example Problem 1–1 (SI system)

A hoist lifts 425 kg of concrete. Compute the weight of the concrete that is the force exerted on the hoist by the concrete.

Solution

Objective: Compute the weight of a mass of concrete.

Given: $m = 425$ kg

Analysis: $W = m \times g$; $g = 9.81$ m/s^2

Results: $W = 425 \text{ kg} \times 9.81 \text{ m/s}^2 = 4170 \text{ kg}\cdot\text{m/s}^2 = 4170 \text{ N}$

Comment: Thus 425 kg of concrete weighs 4170 N.

Example Problem 1–2 (U.S. system)

A hopper of coal weighs 8500 lb. Determine its mass.

Solution

Objective: Compute the mass of a hopper of coal.

Given: $W = 8500$ lb

Analysis: $m = W/g$; $g = 32.2$ ft/s^2

Results: $m = 8500 \text{ lb}/32.2 \text{ ft/s}^2 = 264 \text{ lb}\cdot\text{s}^2/\text{ft} = 264 \text{ slugs}$

Comment: Thus 8500 lb of coal has a mass of 264 slugs.

TABLE 1–5 Density and specific weight units.

	U.S. Customary	SI Metric
Density	slugs/ft^3	kg/m^3
Specific weight	lb/ft^3	N/m^3

Density and Specific Weight. To characterize the mass or weight of a material relative to its volume, we use the terms *density* and *specific weight,* defined as follows:

Density is the amount of mass per unit volume of a material.

Specific weight is the amount of weight per unit volume of a material.

We will use the Greek letter ρ (rho) as the symbol for density. For specific weight we will use γ (gamma).

The units for density and specific weight are summarized in Table 1–5. Other conventions are sometimes used, often leading to confusion. For example, in the United States, density is occasionally expressed in lb/ft^3 or lb/in^3. Two interpretations are used for this. One is that the term implies *weight density* with the same meaning as specific weight. Another is that the quantity *lb* is meant to be *pound mass* rather than *pound weight,* and the two have equal numerical values when g is the standard value.

1–6 THE CONCEPT OF STRESS

The study of strength of materials depends on an understanding of the principles of *stress* and *strain* produced by applied loads on a structure or a machine and the members that make up such systems. These principles are introduced here and applied to relatively simple types of loading with an emphasis on *analysis.* That is, the loads and the geometry of the members are given in the problems, and your task is to analyze the resulting stress within the members and the deformation that is caused by the stress.

We use *direct loading* here to develop the concept of stress and, later, the concept of strain. By the term *direct stress* we refer to cases where the total applied force is shared equally by all parts of the cross section of the load-carrying member. The types of loading considered in this chapter are

- Direct axial loads
- Direct shearing forces
- Bearing loads

Later chapters introduce other types of loading such as bending loads on beams and torsional loads on shafts where the stresses are not uniform across the cross section.

Applications are shown for such real members as

- A rod supporting a heavy load that tends to pull the rod with a tensile force [Figure 1–2]
- Short blocks supporting heavy loads that tend to crush them with compressive forces [Figure 1–7(b)]
- A pin that carries a load acting perpendicular to its axis that tends to cut the pin across one or more of its cross sections (called a *shearing force*) [Figure 1–6(f)]
- A floor on which a leg of a heavy machine is set, tending to cause an indentation in the floor (called a *bearing load*) [Figure 1–7(c)]

FIGURE 1–7 Illustration of compressive stress and bearing stress.

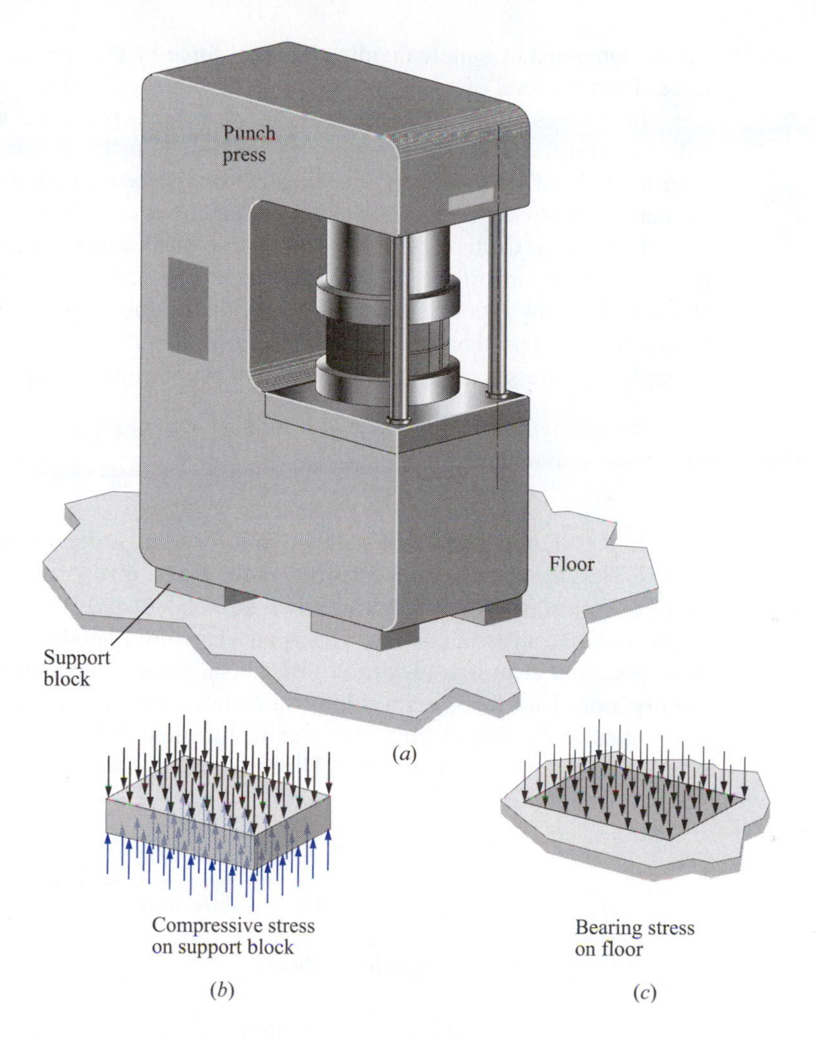

In later chapters, we introduce the concept of *design* in which the objective is to specify the material from which a member is to be made or the detailed dimensions of the member to ensure that it is safe and will perform its intended function. This requires an understanding of the ability of the material to withstand the applied forces without failure (breaking) or excessive deformation. This is really where the term *strength of materials* comes into play. Chapter 2 discusses the design properties of materials and material selection.

Understanding the meaning of *stress* in a load-carrying member, as given below, is of utmost importance in studying strength of materials.

Stress is the internal resistance offered by a unit area of the material from which a member is made to an externally applied load.

We are concerned with what happens *inside* a load-carrying member. We must determine the magnitude of *force* exerted on each *unit area* of the material. The concept of stress can be expressed mathematically as

Definition of Stress

$$\text{stress} = \frac{\text{force}}{\text{area}} = \frac{F}{A} \tag{1–2}$$

In some cases, as described in the following section on *direct normal stress,* the applied force is shared uniformly by the entire cross section of the member. In such cases stress can be computed by simply dividing the total force by the area of the part that resists the force. Then the level of stress will be the same at any point in any cross section.

In other cases, such as for *stress due to bending* presented in Chapter 7, the stress will vary at different positions within the cross section. Then, it is essential that you consider the level of stress *at a point*. Typically, your objective is to determine at which point the maximum stress occurs and what is its magnitude.

In the U.S. Customary unit system, the typical unit for force is the pound and the most convenient unit of area is the square inch. Thus stress is indicated in lb/in^2, abbreviated psi. Typically encountered stress levels in machine design and structural analysis are several thousand psi. For this reason, the unit of kip/in^2 is often used, abbreviated ksi. For example, if a stress is computed to be 26 500 psi, it might be reported as

$$\text{stress} = \frac{26\,500\ \text{lb}}{\text{in}^2} \times \frac{1\ \text{kip}}{1000\ \text{lb}} = \frac{26.5\ \text{kip}}{\text{in}^2} = 26.5\ \text{ksi}$$

In the SI unit system, the standard unit for force is the newton and area is in square meters. Thus the standard unit for stress is the N/m^2, given the name *pascal* and abbreviated Pa. Typical levels of stress are several million pascals, so the most convenient unit for stress is the megapascal or MPa. This is convenient for another reason. In calculating the cross-sectional area of load-carrying members, measurements of dimensions in mm are usually used. Then the stress would be in N/mm^2 and it can be shown that this is numerically equal to the unit of MPa. For example, assume that a force of 15 000 N is exerted over a square area 50 mm on a side. The resisting area would be $2500\ \text{mm}^2$ and the resulting stress would be

$$\text{stress} = \frac{\text{force}}{\text{area}} = \frac{15\,000\ \text{N}}{2500\ \text{mm}^2} = \frac{6.0\ \text{N}}{\text{mm}^2}$$

Converting this to pascals would produce

$$\text{stress} = \frac{6.0\ \text{N}}{\text{mm}^2} \times \frac{(1000)^2\ \text{mm}^2}{\text{m}^2} = 6.0 \times 10^6\ \text{N/m}^2 = 6.0\ \text{MPa}$$

In summary, ***the unit of*** N/mm^2 ***is identical to the MPa.***

1–7 DIRECT NORMAL STRESS

One of the most fundamental types of stress that exists is the *normal stress,* indicated by the lowercase Greek letter σ (sigma), in which the stress acts perpendicular, or normal, to the cross section of the load-carrying member. If the stress is also uniform across the resisting area, the stress is called a *direct normal stress*.

Normal stresses can be either compressive or tensile.

> ***A compressive stress is one that tends to crush the material of the load-carrying member and to shorten the member itself.***
>
> ***A tensile stress is one that tends to stretch the member and pull the material apart.***

The equation for direct normal stress follows from the basic definition of stress because the applied force is shared equally across the entire cross section of the member carrying the force. That is,

FIGURE 1–8
Example of direct compressive stress.

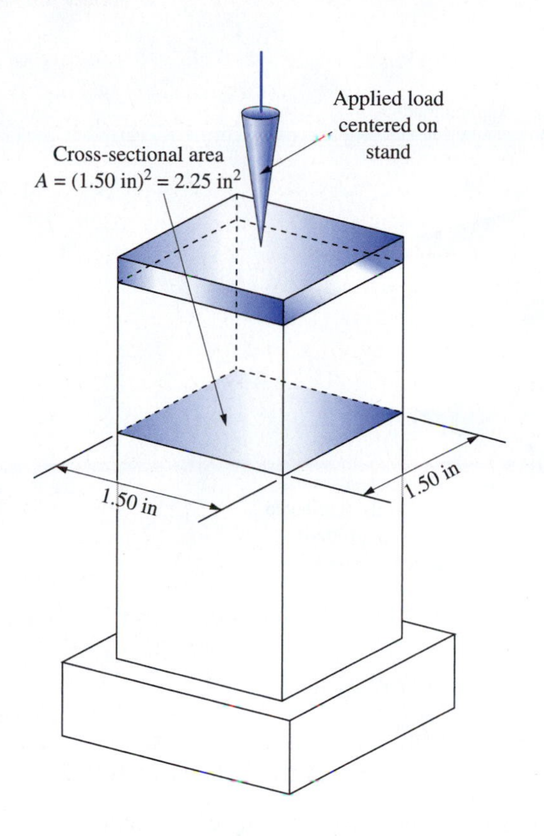

Direct Normal Stress

$$\text{Direct normal stress} = \sigma = \frac{\text{Applied force}}{\text{Area of cross section}} = \frac{F}{A} \tag{1–3}$$

The area of the cross section of the load-carrying member is taken perpendicular to the line of action of the force.

An example of a member subjected to compressive stress is shown in Figure 1–8. The support stand is designed to be placed under heavy equipment during assembly, and the weight of the equipment tends to crush the square shaft of the support, placing it in compression.

Example Problem 1–3 Figure 1–8 shows a support stand designed to carry downward loads. Compute the stress in the square shaft at the upper part of the stand for a load of 27 500 lb. The line of action of the applied load is centered on the axis on the shaft, and the load is applied through a thick plate that distributes the force to the entire cross section of the stand.

Solution

Objective Compute the stress in the upper part of the stand.

Given Load $= F = 27\,500$ lb; load is centered on the stand.
Cross section is square; dimension of each side is 1.50 in.

Analysis On any cross section of the stand, there must be an internal resisting force that acts upward to balance the downward applied load. The internal force is distributed over the cross-sectional area, as shown in Figure 1–9. Each small unit area of the cross section would support the same part of the total load. The stress produced in the square shaft tends to crush the material and it is therefore a compressive stress. Equation (1–3) can be used to compute the magnitude of the stress.

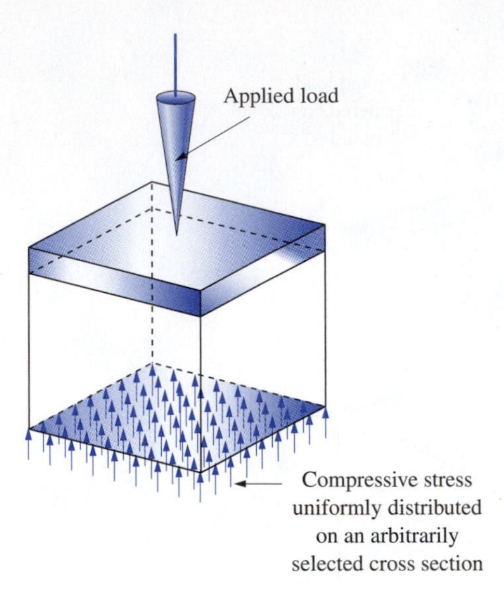

FIGURE 1–9 Compressive stress on an arbitrary cross section of support stand shaft.

Results stress $= \sigma =$ force/area $= F/A$ (compressive)

$$A = (1.50 \text{ in})^2 = 2.25 \text{ in}^2$$
$$\sigma = F/A = 27\,500 \text{ lb}/2.25 \text{ in}^2 = 12\,222 \text{ lb/in}^2 = 12\,222 \text{ psi}$$

Comment This level of stress would be present on any cross section of the square shaft between its ends.

An example of a member subjected to a tensile load is shown in Figure 1–2. The two rods suspend a heavy casting from a crane and the weight of the casting tends to stretch the rods and pull them apart.

Example Problem 1–4 Figure 1–2 shows two circular rods carrying a casting weighing 11.2 kN. If each rod is 12.0 mm in diameter and the two rods share the load equally, compute the stress in the rods.

Solution Objective Compute the stress in the support rods.

Given Casting weighs 11.2 kN. Each rod carries half the load.
Rod diameter $= D =$ 12.0 mm.

Analysis Direct tensile stress is produced in each rod. Use Equation (1–3).

Results $F = 11.2 \text{ kN}/2 = 5.60 \text{ kN}$ or 5600 N on each rod
Area $= A = \pi D^2/4 = \pi(12.0 \text{ mm})^2/4 = 113 \text{ mm}^2$
$\sigma = F/A = 5600 \text{ N}/113 \text{ mm}^2 = 49.5 \text{ N/mm}^2 = 49.5 \text{ MPa}$

LESSON 1

Comment Figure 1–10 shows an arbitrarily selected part of the rod with the applied load on the bottom and the internal tensile stress distributed uniformly over the cut section.

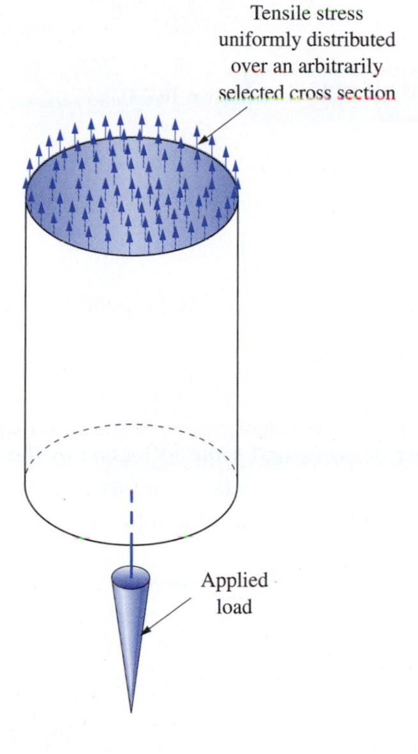

FIGURE 1–10 Tensile stress on an arbitrary cross section of a circular rod.

1–8 STRESS ELEMENTS FOR DIRECT NORMAL STRESSES

The illustrations of stresses in Figures 1–9 and 1–10 are useful for visualizing the nature of the internal resistance to the externally applied force, particularly for these cases in which the stresses are uniform across the entire cross section. In other cases it is more convenient to visualize the stress condition on a small (infinitesimal) element. Consider a small cube of material anywhere inside the square shaft of the support stand shown in Figure 1–8. There must be a net compressive force acting on the top and bottom faces of the cube, as shown in Figure 1–11(a). If the faces are considered to be unit areas, these forces can be considered to be the stresses acting on the faces of the cube. Such a cube is called a *stress element*.

Because the element is taken from a body in equilibrium, the element itself is also in equilibrium and the stresses on the top and bottom faces are the same. A simple stress element like this one is often shown as a two-dimensional square element rather than the three-dimensional cube, as shown in Figure 1–11(b).

Similarly, the tensile stress on any element of the rod in Figure 1–2 can be shown as in Figure 1–12 with the stress vector acting outward from the element. Note that either

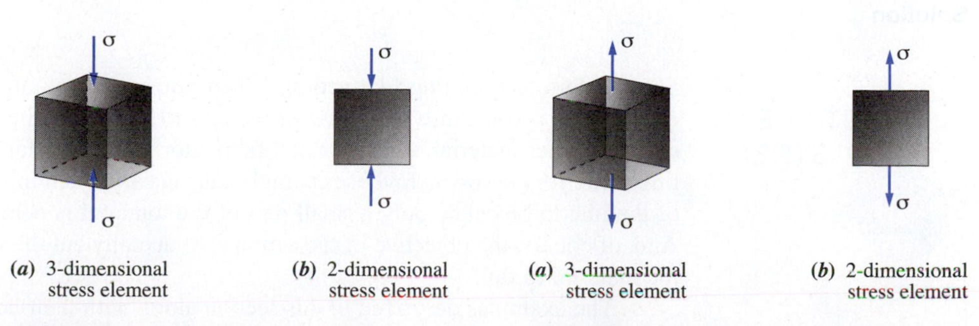

FIGURE 1–11 Stress element for compressive stresses.

FIGURE 1–12 Stress element for tensile stresses.

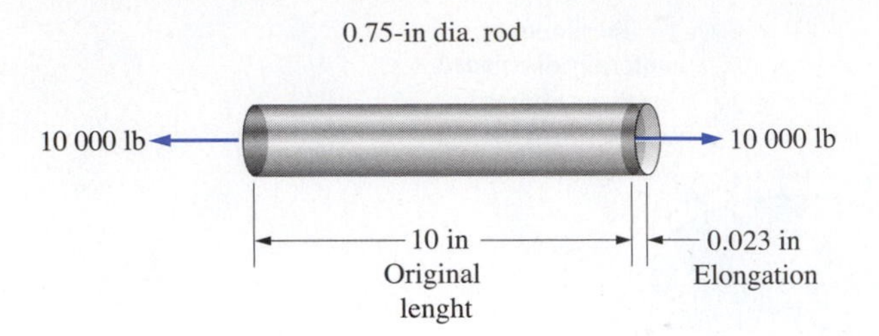

FIGURE 1–13
Elongation of a bar in tension.

compressive or tensile stresses are shown acting perpendicular (normal) to the surface of the element.

1–9 THE CONCEPT OF STRAIN

Any load-carrying member deforms under the influence of the load applied. The square shaft of the support stand in Figure 1–8 gets shorter as the heavy equipment is placed on the stand. The rods supporting the casting in Figure 1–2 get longer as the casting is hung onto them.

The total deformation of a load-carrying member can, of course, be measured. It will also be shown later how the deformation can be calculated.

Figure 1–13 shows an axial tensile force of 10 000 lb applied to an aluminum bar that has a diameter of 0.75 in. Before the load was applied, the length of the bar was 10.000 in. After the load is applied, the length is 10.023 in. Thus the total deformation is 0.023 in.

Strain, also called *unit deformation,* is found by dividing the total deformation by the original length of the bar. The lowercase Greek letter epsilon (ϵ) is used to denote strain:

Definition of Strain

$$\text{strain} = \epsilon = \frac{\text{total deformation}}{\text{original length}} \tag{1–4}$$

For the case shown in Figure 1–13,

$$\epsilon = \frac{0.023 \text{ in}}{10.000 \text{ in}} = 0.0023 \text{ in/in}$$

Strain could be said to be dimensionless because the units in the numerator and denominator could be canceled. However, it is better to report the units as in/in or mm/mm to maintain the definition of deformation per unit length of the member. Later chapters contain more about strain and deformation.

1–10 DIRECT SHEAR STRESS

Shear refers to a cutting-like action. When you use common household scissors, often called *shears,* you cause one blade of the pair to slide over the other to cut (shear) paper, cloth, or other material. A sheet metal fabricator uses a similar shearing action when cutting metal for ductwork. In these examples, the shearing action progresses along the length of the line to be cut so only a small part of the total cut is being made at any given time. And, of course, the objective of the action is to actually cut the material. That is, you *want* the material to fail.

The examples described in this section along with their accompanying figures illustrate several cases where *direct shear* is produced. That is, the applied shearing force is resisted uniformly by the area of the part in shear, producing a unifc m level of shearing

FIGURE 1–14
Illustration of direct shear stress in a punching operation.

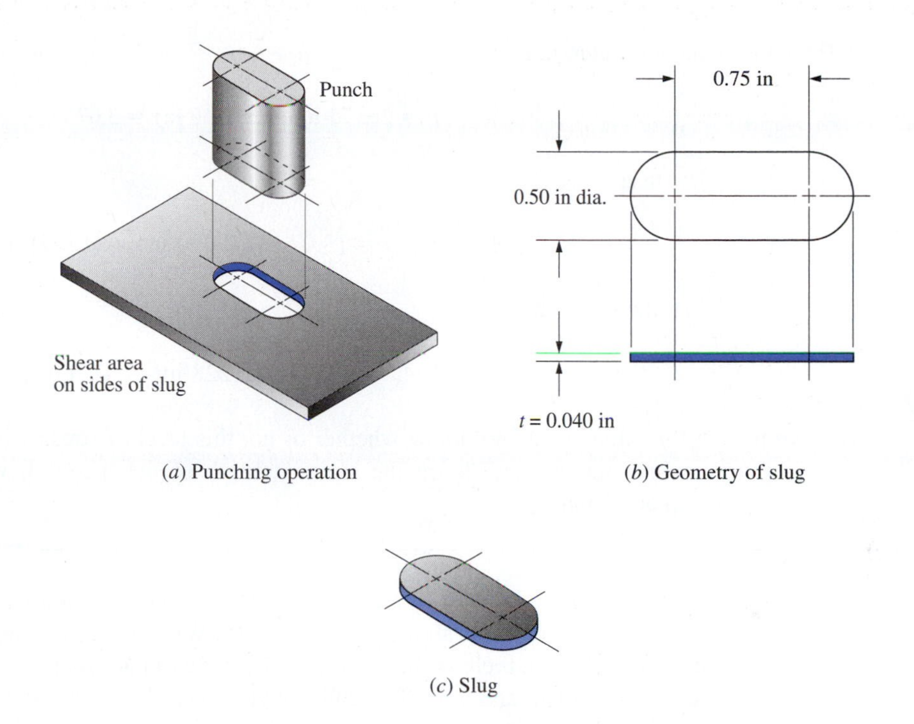

(*a*) Punching operation

(*b*) Geometry of slug

(*c*) Slug

force across the area. The symbol used for shear stress is τ, the lowercase Greek letter *tau*. Then the direct shear stress can be computed from

Direct Shear Stress

$$\text{direct shear stress} = \tau = \frac{\text{applied force}}{\text{shear area}} = \frac{F}{A_s} \qquad \textbf{(1–5)}$$

Figure 1–14 shows a punching operation where the objective is to actually cut one part of the material from the other. The punching action produces a slot in the flat sheet metal. The part removed in the operation is sometimes called a slug. Many different shapes can be produced by punching with either the slug or the sheet having the hole being the desired part. Normally, the punching operation is designed so the entire shape is punched out at the same time. Therefore, the shearing action occurs along the *sides* of the slug. The area in shear for this case is computed by multiplying the length of the perimeter of the cut shape by the thickness of the sheet. That is, for a punching operation,

Shear Area for Punching

$$A_s = \text{perimeter} \times \text{thickness} = p \times t \qquad \textbf{(1.6)}$$

Example Problem 1–5 For the punching operation shown in Figure 1–14, compute the shear stress in the material if a force of 1250 lb is applied through the punch. The thickness of the material is 0.040 in.

Solution

Objective: Compute the shear stress in the material.

Given: $F = 1250$ lb; shape to be punched shown in Fig. 1–14; $t = 0.040$ in.

Analysis: The sides of slug are placed in direct shear resisting the applied force. Use Equations (1–4) and (1–5).

Results The perimeter, p, is

$$p = 2(0.75 \text{ in}) + \pi(0.50 \text{ in}) = 3.07 \text{ in}$$

The shear area,

$$A_s = p \times t = (3.07 \text{ in})(0.040 \text{ in}) = 0.1228 \text{ in}^2$$

Then the shear stress is

LESSON 3

$$\tau = F/A_s = 1250 \text{ lb}/0.1228 \text{ in}^2 = 10\ 176 \text{ psi}$$

Comment At this time, we do not know whether or not this level of stress will cause the slug to be punched out; it depends on the shear strength of the material, which is discussed in Chapters 2 and 3.

Single Shear. A pin or a rivet is often inserted into a cylindrical hole through mating parts to connect them, as shown in Figure 1–15. When forces are applied perpendicular to the axis of the pin, there is the tendency to cut the pin across its cross section, producing a shear stress. This action is often called *single shear* because a single cross section of the pin resists the applied shearing force. In this case, the pin is usually designed so the shear stress is below the level that would cause the pin to fail. Chapter 3 contains more about allowable stress levels.

Double Shear. When a pin connection is designed as shown in Figure 1–16, there are two cross sections to resist the applied force. In this arrangement the pin is in *double shear*.

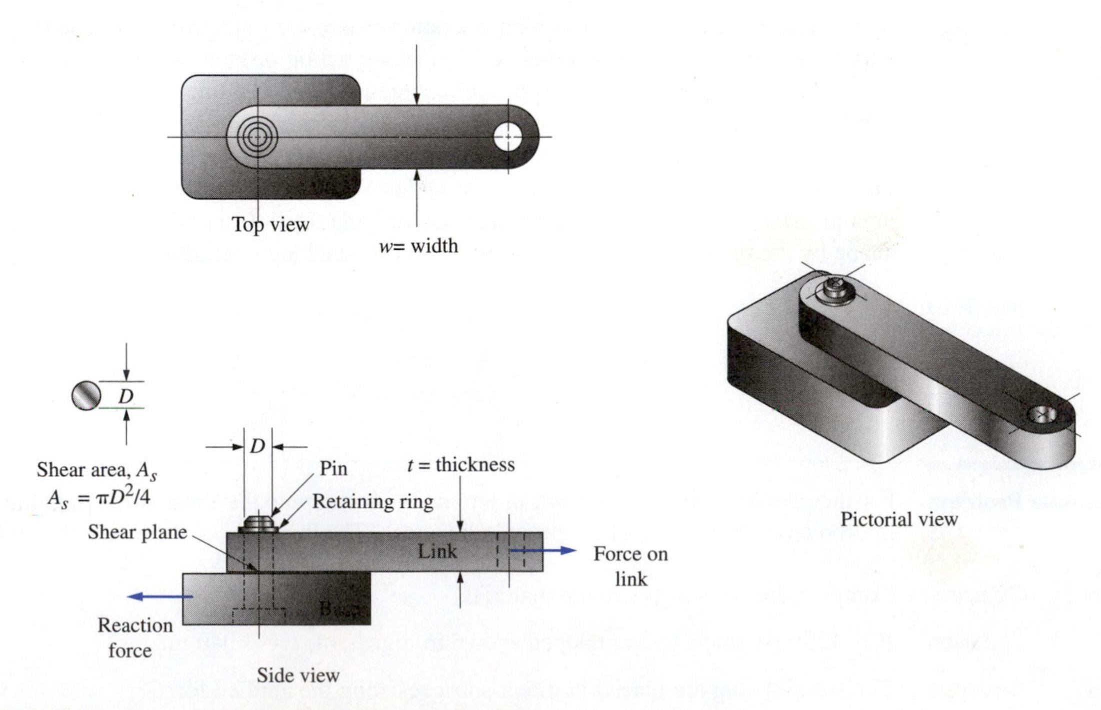

FIGURE 1–15 Pin connection illustrating single shear.

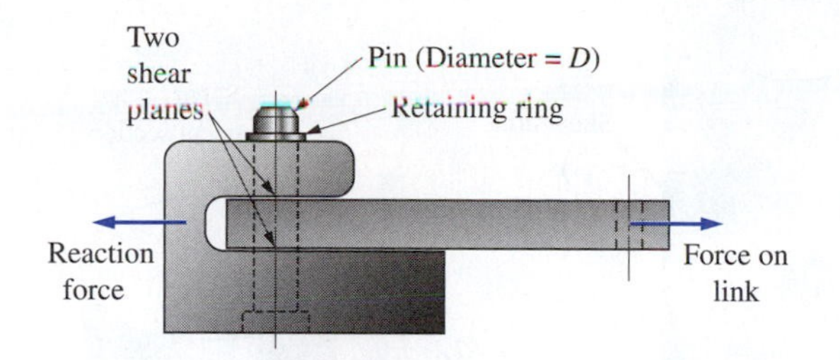

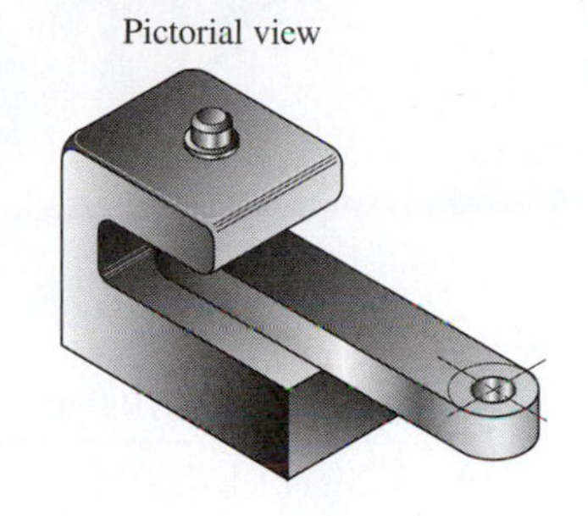

FIGURE 1–16 Pin connection illustrating double shear.

Example Problem 1–6 The force on the link in the simple pin joint shown in Figure 1–15 is 3550 N. If the pin has a diameter of 10.0 mm, compute the shear stress in the pin.

Solution Objective: Compute the shear stress in the pin.

Given: $F = 3550$ N; $D = 10.0$ mm

Analysis: Pin is in direct shear with one cross section of the pin resisting all of the applied force (single shear). Use Equation (1–4).

Results: The shear area, A_s, is

$$A_s = \frac{\pi D^2}{4} = \frac{\pi(10.0\text{ mm})^2}{4} = 78.5\text{ mm}^2$$

Then the shear stress is

$$\tau = \frac{F}{A_s} = \frac{3550\text{ N}}{78.5\text{ mm}^2} = 45.2\text{ N/mm}^2 = 45.2\text{ MPa}$$

Comment: This stress is shown in Figure 1–17 on a cut section of the pin.

Example Problem 1–7 If the pin joint just analyzed was designed as shown in Figure 1–16, compute the shear stress in the pin.

Solution Objective: Compute the shear stress in the pin.

Given: $F = 3550$ N; $D = 10.0$ mm (same as in Ex. Prob. 1–6).

Analysis: Pin is in direct shear with two cross sections of the pin resisting the applied force (double shear). Use Equation (1–4).

FIGURE 1–17 Direct shear stress in a pin.

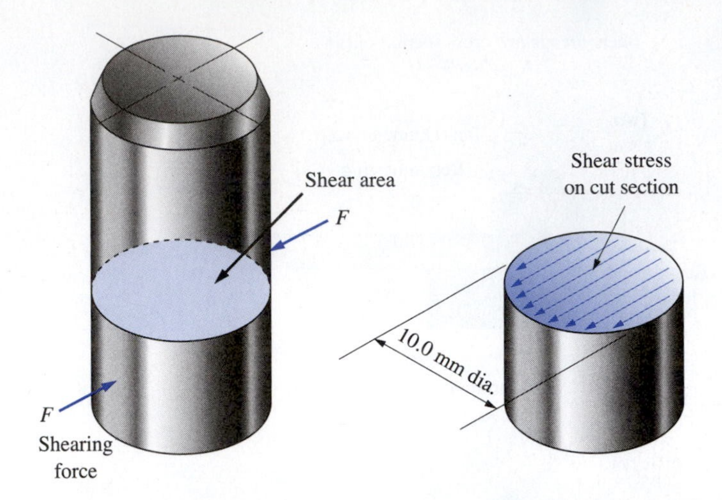

Results The shear area, A_s, is

$$A_s = 2\left(\frac{\pi D^2}{4}\right) = 2\left[\frac{\pi(10.0\text{ mm})^2}{4}\right] = 157\text{ mm}^2$$

The shear stress in the pin is

$$\tau = \frac{F}{A_s} = \frac{3550\text{ N}}{157\text{ mm}^2} = 22.6\text{ N/mm}^2 = 22.6\text{ MPa}$$

LESSON 2

Comment The resulting shear stress is $\frac{1}{2}$ of the value found for single shear.

Keys. Figure 1–18 shows an important application of shear in mechanical drives. When a power transmitting element, such as a gear, chain sprocket, or belt pulley, is placed on a shaft, a key is often used to connect the two and permit the transmission of torque from one to the other.

The torque produces a tangential force at the interface between the shaft and the inside of the hub of the mating element. The torque is reacted by the moment of the force on the key times the radius of the shaft. That is, $T = F(D/2)$. Then the force is $F = 2T/D$.

In Figure 1–18, we have shown the force F_1 exerted by the shaft on the left side of the key. On the right side, an equal force F_2 is the reaction exerted by the hub on the key. This pair of forces tends to cut the key, producing a shear stress. Note that the shear area, A_s, is a rectangle with dimensions $b \times L$. The following example problem illustrates the computation of direct shear stress in a key.

Example Problem 1–8 Figure 1–18 shows a key inserted between a shaft and the mating hub of a gear. If a torque of 1500 lb·in is transmitted from the shaft to the hub, compute the shear stress in the key. For the dimensions of the key, use $L = 0.75$ in; $h = b = 0.25$ in. The diameter of the shaft is 1.25 in.

Solution

Objective Compute the shear stress in the key.

Given $T = 1500$ lb·in; $D = 1.25$ in; $L = 0.75$ in; $h = b = 0.25$ in

Analysis The key is in direct shear. Use Equation (1–4).

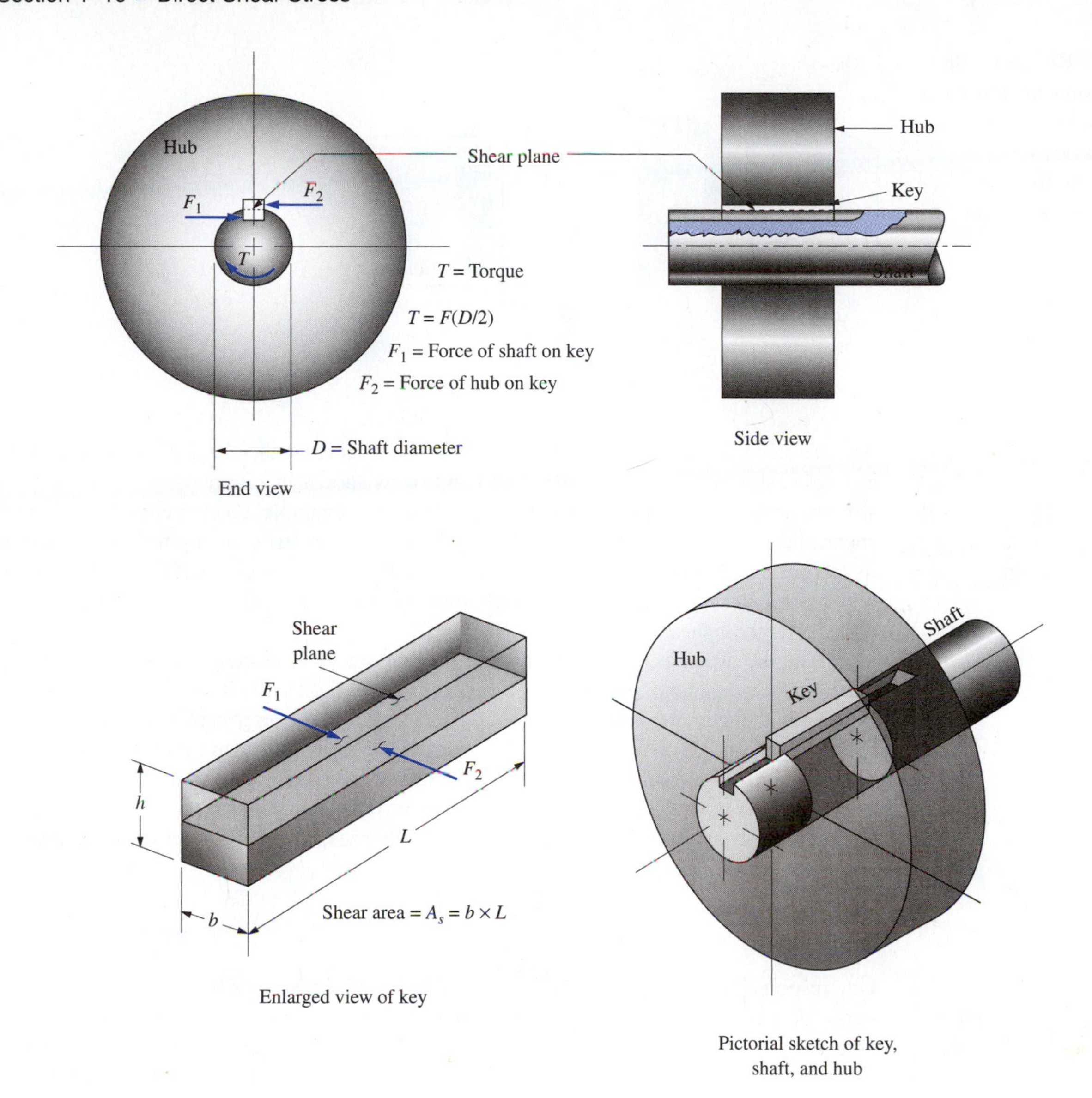

FIGURE 1–18 Direct shearing action on a key between a shaft and the hub of a gear, pulley, or sprocket in a mechanical drive system.

Results Shear area: $A_s = b \times L = (0.25 \text{ in})(0.75 \text{ in}) = 0.1875 \text{ in}^2$. The force on the key is produced by the action of the applied torque. The torque is reacted by the moment of the force on the key times the radius of the shaft. That is, $T = F(D/2)$. Then the force is

$$F = 2T/D = (2)(1500 \text{ lb}\cdot\text{n})/(1.25 \text{ in}) = 2400 \text{ lb}$$

Then the shear stress is

$$\tau = F/A_s = 2400 \text{ lb}/0.1875 \text{ in}^2 = 12\,800 \text{ psi}$$

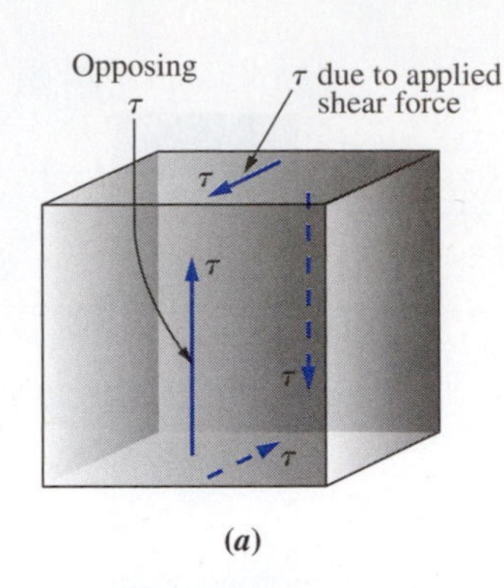

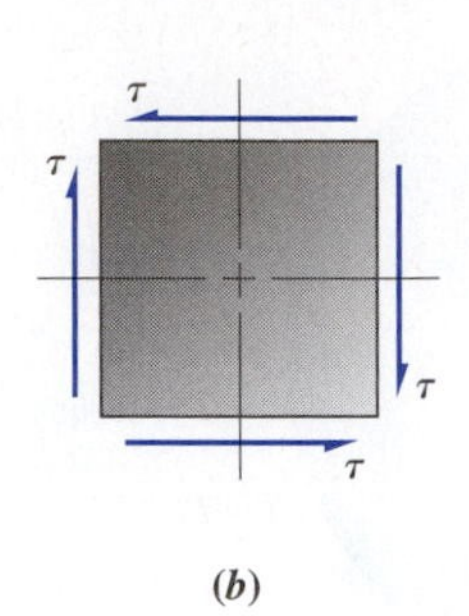

FIGURE 1–19 Stress element showing shear stress. (a) Three-dimensional stress element. (b) Two-dimensional stress element.

1–11 STRESS ELEMENTS FOR SHEAR STRESSES

An infinitesimally small cubic element of the material from the shear plane of any of the examples shown in Section 1–10 would appear as shown in Figure 1–19 with the shear stresses acting parallel to the surfaces of the cube. For example, an element taken from the shear plane of the key in Figure 1–18 would have a shear stress acting toward the left on its top surface. For equilibrium of the element with regard to horizontal forces, there must be an equal stress acting toward the right on the bottom surface. This is the cutting action characteristic of shear.

But the two stress vectors on the top and bottom surfaces cannot exist alone because the element would tend to rotate under the influence of the couple formed by the two shearing forces acting in opposite directions. To balance that couple, a pair of equal shear stresses is developed on the vertical sides of the stress element, as shown in Figure 1–19(a).

The stress element is often drawn in the two-dimensional form shown in Figure 1–19(b). Note how the stress vectors on adjacent faces tend to meet at the corners. Such stress elements are useful in the visualization of stresses acting at a point within a material subjected to shear.

1–12 PREFERRED SIZES AND STANDARD SHAPES

One responsibility of a designer is to specify the final dimensions for load-carrying members. After completing the analyses for stress and deformation (strain), minimum acceptable values for dimensions are known that will ensure that the member will meet performance requirements. The designer then typically specifies the final dimensions to be standard or convenient values that will facilitate the purchase of materials and the manufacture of the parts. This section presents some guides to aid in these decisions.

Preferred Basic Sizes. When the component being designed will be made to the designer's specifications, it is recommended that final dimensions be specified from a set of *preferred basic sizes*. Appendix A–2 lists such data for fractional inch dimensions, decimal inch dimensions, and metric dimensions.

American Standard Screw Threads. Threaded fasteners and machine elements having threaded connections are manufactured according to standard dimensions to ensure interchangeability of parts and to permit convenient manufacture with standard machines and tooling. Appendix A–3 gives the dimensions of American Standard Unified threads. Sizes smaller than $\frac{1}{4}$ in are given numbers from 0 to 12, while fractional-inch sizes are specified for $\frac{1}{4}$ in and larger sizes. Two series are listed: UNC is the designation for coarse threads and UNF designates fine threads. Standard designations are illustrated as follows with a few examples.

6–32 UNC (number size 6, 32 threads per inch, coarse thread)
12–28 UNF (number size 12, 28 threads per inch, fine thread)
$\frac{1}{2}$–13 UNC (fractional size $\frac{1}{2}$ in, 13 threads per inch, coarse thread)
$1\frac{1}{2}$–12 UNF (fractional size $1\frac{1}{2}$ in, 12 threads per inch, fine thread)

Given in the tables are the basic major diameter (D), number of threads per inch (n), and the tensile stress area found from

$$A_t = 0.7854\left(D - \frac{0.9743}{n}\right)^2 \quad \textbf{(1.7)}$$

When a threaded member is subjected to direct tension, the tensile stress area is used to compute the average tensile stress. It corresponds to the smallest area that would be produced by a transverse cut across the threaded rod. For convenience, some standards use the root area or gross area and adjust the value of the allowable stress.

Metric Screw Threads. Appendix A–3 gives similar dimensions for metric threads. Standard metric thread designations are of the form

M10×1.5

where M stands for metric
10 is the basic major diameter in mm
1.5 is the pitch between adjacent threads in mm

Thus the designation shown would denote a metric thread with a basic major diameter of 10.0 mm and a pitch of 1.5 mm. Note that pitch = $1/n$.

Standard Wood Beams. Appendix A–4 gives the dimensions and section properties for many standard sizes of wood beams. Note that the nominal size is simply the "name" of the beam, and it relates to the approximate rough size prior to finishing. Actual finished dimensions are significantly smaller than the nominal sizes. For example, a common "2 × 4" board is actually 1.5 in wide and 3.5 in high. Also note the sketch of the orientation of the beams for the standard designation of the X- and Y-axes. When used as a beam in bending, the long dimension should be vertical for maximum strength and stiffness.

Steel Structural Shapes. Steel manufacturers provide a large array of standard structural shapes that are efficient in the use of material and that are convenient for specification and installation into building structures or machine frames. Included, as shown in Table 1–6, are standard angles (L-shapes), channels (C-shapes), wide-flange beams (W-shapes), American Standard beams (S-shapes), structural tubing, and pipe called *hollow structural sections* (HSS). Note that the W-shapes and S-shapes are often referred to in general conversation as "I-beams" because the shape of the cross section looks like the capital letter I.

Appendix tables A–5 through A–9 give geometric properties of selected structural shapes that cover a fairly wide range of sizes. Note that many more sizes are available as presented in Reference 2. The appendix tables give data for the area of the cross section (A), the weight per foot of length, the location of the centroid of the cross section, the moment of inertia (I), the section modulus (S), and the radius of gyration (r). Some of these properties may be new to you at this time and they will be defined as needed later in the book. The values of I and S are important in the analysis and design of beams. For column analysis, I and r are needed.

TABLE 1–6 Designations for steel and aluminum shapes.

Name of shape	Shape	Symbol	Example designation
Angle		L	L4×3×1/2 Appendix A–5 Metric: L102×76×12.7
Channel		C	C15×50 Appendix A–6 Metric: C380×74
Wide-flange beam		W	W14×43 Appendix A–7 Metric: W360×64
American Standard beam		S	S10×35 Appendix A–8 Metric: S250×52
Structural tubing—square			HSS4×4×1/4 Appendix A–9 Metric: HSS 102×102×6.4
Structural tubing—rectangular			HSS6×4×1/4 Appendix A–9 Metric: HSS 152×102×6.4
Pipe			PIPE 4 STD 4-inch Schedule 40 Appendix A–12 Metric: Pipe 102 STD
Aluminum Association channel		C	C4×1.738 Appendix A–10 Metric: C102×2.586
Aluminum Association I-beam		I	I8×6.181 Appendix A–11 Metric: I203×9.197

Metrication Applied to Structural Shapes. Metrication in the construction industry is progressing even though it is expected that, in the United States, use of standard structural shapes that have been produced for many years to basic dimensions expressed in the U.S. Customary unit system (pound-foot-second system) will remain dominant in the near future. It is important for you to become aware of the trend toward more use of SI metric units in design and fabrication of structures while maintaining the ability to use U.S. units.

Appendix tables A–5 through A–9 include separate pages for U.S. units and SI units for L-shapes, W-shapes, S-shapes, C-shapes, and HSS shapes. In each table, the first column indicates a reference letter for each size of the given shape. This facilitates the correlation of the data for a specific shape between the U.S. and SI units. All of the tables listing SI data have been *soft-converted* from the U.S. units. This means that the long-used standard shapes made to U.S. unit conventions have not been changed. For the SI tables, the designations and the properties of the cross sections have been converted to their metric equivalent using the definitions given in Section 1–4 of this chapter and the conversion factors listed in Appendix A–26. A complete table of equivalencies between the U.S. and SI designations is included in Reference 2. The metric nomenclature for structural shapes is included in Reference 3. Some observations are discussed here.

1. Linear dimensions given in inches are typically converted to the nearest millimeter. However, small dimensions, such as the thickness of the legs of an angle (L-shape), are rounded to the nearest tenth of a mm. The factor used is: 1.0 in = 25.4 mm.
2. Converted nominal sizes are typically rounded to the nearest 10 mm. It is important to use actual dimensions of critical features of shapes in calculations. Nominal sizes are merely the *names* of the shapes suggesting the approximate size.
3. The U.S. designation system lists *weight per foot* (lb/ft) for each shape. More precisely, these data are *mass per foot,* where the units are lb_m/ft, read as *pounds mass per foot.* Common usage often does not include the subscript *m.* For uses on or near the surface of the earth, the numerical value of mass in lb_m is the same as the value for weight (the force due to gravity), often expressed in lb_f or *pounds force.*
4. Conversion of the mass per unit length from lb_m/ft to SI units of kg/m applies the factor,

$$\frac{1.488 \text{ kg/m}}{\text{lb}_\text{m}/\text{ft}}$$

For example, a beam shape listed as weighing 60 lb/ft is converted to SI units as:

$$60 \text{ lb/ft } [(1.488 \text{ kg/m})/(1.0 \text{ lb/ft})] = 89.28 \text{ kg/m}$$

The result is typically rounded to the nearest whole kg/m value, except for very small numbers. Thus this value is reported as 89 kg/m in the table.

5. The weight (force) per unit length is often desired when considering the force applied by one member to another. Then the mass per unit length must be converted to force per unit length by multiplying by $g = 9.81 \text{ m/s}^2$, the acceleration due to gravity. This follows from the relationship $w = mg$ discussed in Section 1–4. For example, for the beam described in Item 4 above, the weight per unit length is:

$$89.28 \text{ kg/m}[9.81 \text{ m/s}^2] = (876 \text{ kg-m/s}^2)/\text{m} = 876 \text{ N/m}$$

Larger values may be listed in kN/m. See Table A–7(SI) in the Appendix.

Steel Angles (L-Shapes). Appendix A–5 shows sketches of the typical shapes of steel angles having equal or unequal leg lengths. Called L-shapes because of the appearance of the cross section, angles are often used as tension members of trusses and towers, framing members for machine structures, lintels over windows and doors in construction, stiffeners for large plates used in housings and beams, brackets, and ledge-type supports for equipment. Some refer to these shapes as "angle iron." The standard designation takes the form shown below, using one example size:

$$L4 \times 3 \times 1/2$$

where L refers to the L-shape, 4 is the length of the longer leg, 3 is the length of the shorter leg, and 1/2 is the thickness of the legs. Dimensions are in inches.

Section property data (I and S) are given for the X-axis and the Y-axis as shown in the drawing. These data are used in later chapters (5–11), and it is essential that values for the proper axis are used in analyzing beams and columns incorporating L-shapes. The intersection of the X- and Y-axes locate the centroid of the shape, and the listed dimensions y and x are used to define that position.

Additional data are given for the radius of gyration, r, of the L-shape relative to the Z-axis, defined as the axis having the smallest value of r. This is important to the analysis of column loading of the shape as described in Chapter 11. The column tends to buckle about the Z-axis that passes through the centroid and has an angle of inclination of α with respect to the Y-axis.

The corresponding SI designation for the L4×3×1/2 shape is,

$$L102 \times 76 \times 12.7$$

This indicates the longer leg length of 102 mm, the shorter leg length of 76 mm, and a thickness of 12.7 mm. See Appendix A–5(SI).

American Standard Channels (C-Shapes). See Appendix A–6 for the appearance of channels and their geometric properties. Channels are used in applications similar to those described for angles. The flat web and the two flanges provide a generally stiffer shape than angles that are more resistant to bending and twisting under load.

The sketch at the top of the table shows that channels have tapered flanges and webs with constant thickness. The slope of the flange taper is approximately 2 in in 12 in, and this makes it difficult to attach other members to the flanges. Special tapered washers are available to facilitate fastening. Note the designation of the X- and Y-axes in the sketch, defined with the web of the channel vertical, which gives it the characteristic "C" shape. This is most important when using channels as beams or columns. The X-axis is located on the horizontal axis of symmetry while the dimension x, given in the table, locates the Y-axis relative to the back of the web. The centroid is at the intersection of the X- and Y-axes.

The form of the standard designation for channels is

$$C15 \times 50$$

where C indicates that it is a standard C-shape
15 is the nominal (and actual) depth in inches with the web vertical
50 is the weight per unit length in lb/ft

The corresponding SI designation for this shape is,

$$C380 \times 74$$

This indicates the nominal depth of 380 mm for the channel and a mass per unit length of 74 kg/m. From Appendix A–6(SI) we find the weight per unit length to be 0.730 kN/m and the actual depth to be 381 mm.

Wide-Flange Shapes (W-Shapes). Refer to Appendix A–7. This is the most common shape used for beams, as will be discussed in Chapters 7–9. W-shapes have relatively thin webs and somewhat thicker, flat flanges with constant thickness. Most of the area of the cross section is in the flanges, farthest away from the horizontal centroidal axis (X-axis), thus making the moment of inertia very high for a given amount of material. (See Chapter 6.)

Note that the properties of moment of inertia and section modulus are very much higher with respect to the X-axis than they are for the Y-axis. Therefore, W-shapes are typically used in the orientation shown in the sketch in Appendix A–7. Also, these shapes are best when used in pure bending without twisting because they are quite flexible in torsion.

The standard designation for W-shapes carries much information. Consider the example,

W14×43

where W indicates that it is a W-shape
14 is the nominal depth in inches
43 is the weight per unit length in lb/ft

The term *depth* is the standard designation for the vertical height of the cross section when placed in the orientation shown in Appendix A–7. Note from the data in the table that the actual depth is often different from the nominal depth. For the W14×43, the actual depth is 13.7 in.

The corresponding SI designation for this shape is,

W360×64

This indicates the nominal depth of 360 mm for the W-beam and a mass per unit length of 64 kg/m. From Appendix A–7(SI) we find the weight per unit length to be 0.628 kN/m and the actual depth to be 348 mm.

American Standard Beams (S-Shapes). Appendix A–8 shows the properties for S-shapes. Much of the discussion given for W-shapes applies to S-shapes as well. Note that, again, the weight per foot of length is included in the designation such as the S10×35 that weighs 35 lb/ft. For most, but not all, of the S-shapes, the actual depth is the same as the nominal depth. The flanges of the S-shapes are tapered at a slope of approximately 2 in in 12 in, similar to the flanges of the C-shapes. The X- and Y-axes are defined as shown with the web vertical.

For the S10×35 shape discussed here, the corresponding SI designation is:

S250×52

This indicates the nominal depth of 250 mm for the S-beam and a mass per unit length of 52 kg/m. From Appendix A–8(SI) we find the weight per unit length to be 0.511 kN/m and the actual depth to be 254 mm.

Often wide-flange shapes (W-shapes) are preferred over S-shapes because of their relatively wide flanges, the constant thickness of the flanges, and generally higher section properties for a given weight and depth.

Structural Tubing (Square and Rectangular). See Appendix A–9 for the appearance and properties for steel structural tubing, also called *hollow structural section* (HSS). These shapes are usually formed from flat sheet and welded along the length. The section properties account for the corner radii. Note the sketches showing the X- and Y-axes. The standard designation takes the form,

$$\text{HSS}6 \times 4 \times 1/4$$

where HSS indicates that it is a standard hollow structural section
6 is the depth of the longer side in inches
4 is the width of the shorter side in inches
1/4 is the nominal wall thickness in inches

Square and rectangular tubing are very useful in machine structures because they provide good section properties for members loaded as beams in bending and for torsional loading (twisting) because of the closed cross section. The flat sides often facilitate fastening of members together or the attachment of equipment to the structural members. Some frames are welded into an integral unit that functions as a stiff space-frame. Square tubing makes an efficient section for columns.

Note that the design wall thickness is somewhat smaller than the nominal. The nominal $\frac{1}{4}$-in wall has a design thickness of 0.233 in. A nominal $\frac{1}{2}$-in wall has a design thickness of 0.465 in.

Appendix A–9 lists two torsional constants, J and C, for the square and rectangular HSS shapes. These are discussed in Chapter 4 in the section called *Torsion of noncircular sections.* The term J is the polar moment of inertia related to the stiffness of the section in resisting twisting. The term C is used for computing the shearing stress in the shape when subjected to torsion. Because these shapes perform well as columns, the radius of gyration is given for each member size because it relates to the tendency for a column to buckle as described in Chapter 11.

For the HSS shape discussed above (HSS6×4×1/4), the corresponding SI designation is:

$$\text{HSS}152 \times 102 \times 6.4$$

This indicates the depth of 152 mm, the width of 102 mm, and nominal wall thickness of 6.4 mm. From Appendix A–9(SI) we find the mass per unit length to be 23.2 kg/m and the weight per unit length to be 228 N/m. The design wall thickness is 5.92 mm.

Pipe and Round Tubing. Hollow circular sections, commonly called *pipe,* are very efficient for use as beams, torsion members, and columns. The placement of the material uniformly away from the center of the pipe enhances the moment of inertia for a given amount of material and gives the pipe uniform properties with respect to all axes through the center of the cross section. The closed cross-sectional shape gives it high strength and stiffness in torsion as well as in bending.

Appendix A–12 gives the properties for a representative few sizes of pipe that may be used for structural applications in construction and mechanical design. Included are the 1designation nomenclature, actual dimensions, area, moment of inertia, section modulus, radius of gyration, and the torsional constants J and Z_p.

There are several types of round hollow members in hundreds of sizes available from which to choose as detailed in References 2, 6, 7, and 8. See also Internet sites 1–11. The following is a summary of common types.

- Construction applications are typically specified from round ***hollow structural sections*** (HSS) ranging in outside diameter from 1.660 in (42.2 mm) to 20.0 in (508 mm) as detailed in Reference 2. A total of 59 different outside diameters are produced, each with multiple wall thicknesses, providing a total of 139 specific sizes of round HSS sections.
  - An example designation for this type of member is: HSS4.000×0.250, indicating an actual outside diameter of 4.000 in and a wall thickness of 0.250 in.
  - The SI equivalent for this member is designated as: HSS101.6×6.4, indicating an actual outside diameter of 101.6 mm and a wall thickness of 6.4 mm.
- ***Steel pipe used for structural application*** in construction is available in three classes, *Standard Weight (Std), Extra Strong (X-Strong* or *XS),* and *Double-Extra Strong (XX-Strong* or *XXS).* Each series has the same set of outside diameters with different increasing wall thicknesses and decreasing inside diameters for the XS and XXS sizes. Designations are related to the *nominal diameters* that range from $\frac{1}{2}$ in to 12 in (13 mm to 310 mm). Actual outside and inside diameters are significantly different from the nominal sizes as indicated in Appendix A–12, and they must be used for design and analysis.
  - An example designation for construction pipe is PIPE4STD, indicating a nominal 4-in size and standard weight. This pipe has an actual outside diameter of 4.500 in, a wall thickness of 0.237 in, and an inside diameter of 4.026 in.
  - The SI equivalent pipe is called PIPE102STD, indicating a nominal size of 102 mm, an actual outside diameter of 114.3 mm, a wall thickness of 6.02 mm, and an inside diameter of 102.3 mm.
  - Note that the complete set of construction pipe is included within the round HSS grouping described above. The example pipe, PIPE4STD, can also be designated as HSS4.500×0.237 for U.S. units or HSS114.3×6 for SI units.
- ***Steel pipe is produced primarily for carrying fluids*** and commonly distributed by plumbing supply centers are designated by different *schedules,* ranging from lightweight Schedule 5 to heavyweight Schedule 160. The most common schedules are 40, 80, and 160, and the majority of these sizes are identical to the STD, XS, and XXS classes described for construction uses. See References 6 and 7 for specific designations and sizes. Steel pipe is available in nominal sizes from $\frac{1}{8}$ in to 42 in. In Appendix A–12, all of the listed sizes are Schedule 40. Those from $\frac{1}{2}$-in to 10-in can also be designated with the PIPE notation described above for construction applications.
  - The pipe called PIPE4STD can also be named 4-in Schedule 40 pipe, and this form will typically be used in this book.
  - The SI equivalent designations for fluid-handling pipe have not yet been defined.
- ***Mechanical tubing*** is used for numerous applications in mechanical design, and its size ranges and designations are significantly different from round HSS or pipe. They are also available in many materials such as carbon steel, alloy steel, stainless steel, brass, copper, titanium, tantalum, and aluminum. Appendix A–13 and A–13(SI) give a few examples of mechanical tubing designated by outside diameter and wall thickness with the wall thickness corresponding to a standard called *Birmingham Wall Gauge* or BWG. Many more size and wall gauges are available as seen in Internet site 9. See Reference 7 and Internet sites 1–3 and 9–11 for other examples of commercially available steel and copper tubing.

- ***Plastic pipe and tubing*** are also commercially available in many materials and sizes and are useful for mechanical design. Some plastic pipe is made to the same dimensions as Schedule 40 steel pipe. Tubing is available in a wide range of diameters and wall thicknesses. See Internet sites 4–6 and 11.

Aluminum Association Standard Channels and I-Beams. Appendixes A–10 and A–11 give the dimensions and section properties of channels and I-beams developed by the Aluminum Association (Reference 1). These are extruded shapes having uniform thicknesses of the webs and flanges with generous radii where they meet. The proportions of these sections are somewhat different from those of the rolled steel sections described earlier. The extruded form offers advantages in the efficient use of material and in the joining of members. This book will use the following forms for the designation of aluminum sections:

$$C4\times1.738 \quad \text{or} \quad I8\times6.181$$

where C or I indicates the basic section shape
4 or 8 indicates the depth of the shape when in the orientation shown
1.738 or 6.181 indicates the weight per unit length in lb/ft

Tables A–10(SI) and A–11(SI) give the soft-converted sizes for Aluminum Association channels and I-beam shapes. The format for the designation is the similar to that for U.S. units. The example for a channel, C4×1.738, becomes

$$C102\times2.586$$

where 102 is the depth in mm
2.586 is the mass per unit length in kg/m
From Table A–10(SI) we find the weight per unit length, 25.37 N/m

The example for an I-beam shape, I8×6.181, becomes

$$I203\times9.197$$

where 203 is the depth in mm
9.197 is the mass per unit length in kg/m
From Table A–10(SI) we find the weight per unit length, 90.22 N/m

1–13 EXPERIMENTAL AND COMPUTATIONAL STRESS ANALYSIS

The primary goal of this book is to help you develop skills to determine the magnitude of stresses analytically. But there are cases in which that is difficult or impossible. Examples include:

- Components that have a complex shape for which the calculation of areas and other geometrical factors needed for stress analysis is difficult.
- The loading patterns and the manner of support may be complex, leading to an inability to compute the appropriate loads at a point of interest.
- Combinations of complex shape and loading may make it difficult to determine where the maximum stress occurs.

In such situations, experimental or computational stress analysis techniques can be used to determine the magnitude and location of critical stresses. A variety of techniques are available as discussed next.

FIGURE 1–20 Photoelastic model of a flat bar with an axial load and a central hole. (Source: Measurements Group, Inc., Raleigh, NC)

Photoelastic Stress Analysis. When a component is basically a two-dimensional shape with a constant thickness, a model of the shape can be produced using a special material that permits the visualization of the stress distribution within the component. The material is typically a transparent plastic that is illuminated while being loaded.

Figure 1–20 shows an example in which a flat bar with a centrally located hole is subjected to an axial load. Variations in the stresses within the bar show as variations in black and white lines and areas in the loaded component. The stress in the main part of the bar away from the hole is nearly uniform and that part shows as a dark field. But in the vicinity of the hole, bands of dark and light areas, called *fringes,* appear, indicating that a gradient in stress levels occurs around the hole.

Indeed, the presence of the hole causes higher stress to occur not only because material has been removed from the bar, but also because the change in cross-sectional shape causes *stress concentrations* to occur. Any case in which a change in the cross-sectional shape or dimensions occur will produce stress concentrations. Knowing the optical characteristics of the photoelastic material allows the determination of the stress at any point.

Data and analytical techniques for determining the maximum stress in the vicinity of stress concentrations are presented in future chapters. Additional photoelastic models will be shown there.

Photoelastic Coatings. For more complex three-dimensional shapes it is impossible to model the geometry in the rigid photoelastic material. Also, it is often desired to measure the stress distribution on real solid components such as an engine block, a valve body, or the structure of a large machine. For such cases, a special photoelastic coating material can be applied to the surface of the part. It is molded to follow contours and adhered to the surface. See Figure 1–21(a).

When the component is loaded, the strains at the surface are transferred to the coating material. A polarized light is then directed at the material, and fringes are evident that indicate the gradients in stresses in the actual part. See Figure 1–21(b). From these observations, you can determine the areas where the maximum stresses occur and compute their approximate values. Then, knowing the general stress distribution and the locations of the highest stress levels, more precise testing can be done if greater accuracy is desired. Strain gaging, described next, is typically used at this time.

Strain Gaging. One of the most frequently used experimental stress analysis devices is the electrical resistance strain gage. It is a very thin metal foil grid made from a strain-sensitive material, such as constantan, with an insulating backing. One style is shown in Figure 1–22(a). The gage is carefully applied with a special adhesive to the surface of the component where critical stresses are likely to occur. Figure 1–22(b) shows a typical installation.

FIGURE 1–21
(a) Photoelastic coating being applied to a complex casting.
(b) Evaluating a stress pattern in a photoelastic material using polarized light. (Source: Measurements Group, Inc., Raleigh, NC)

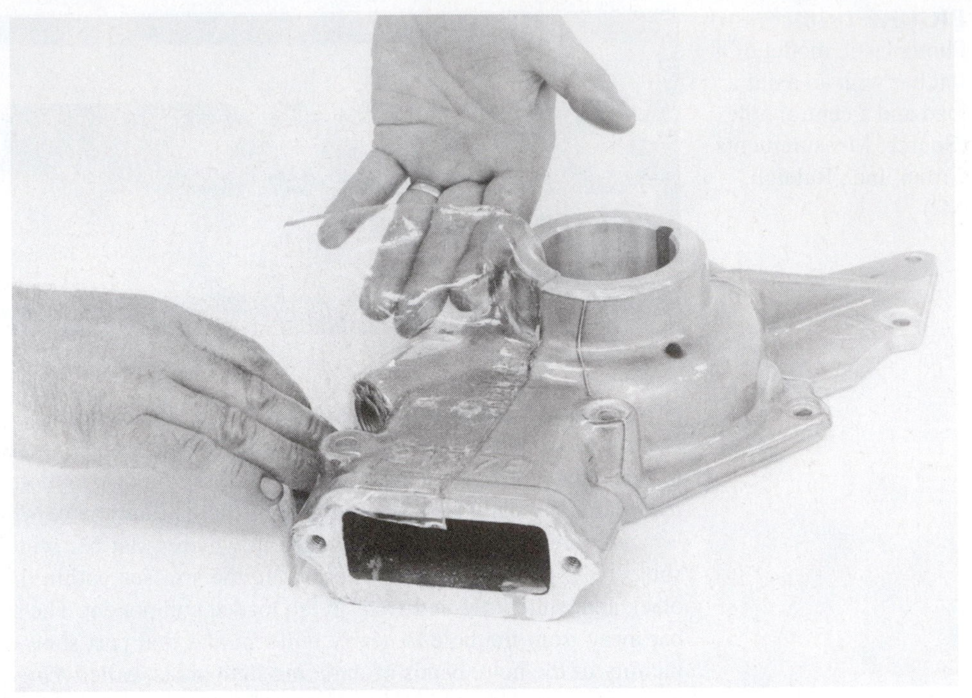

(*a*)

(*b*)

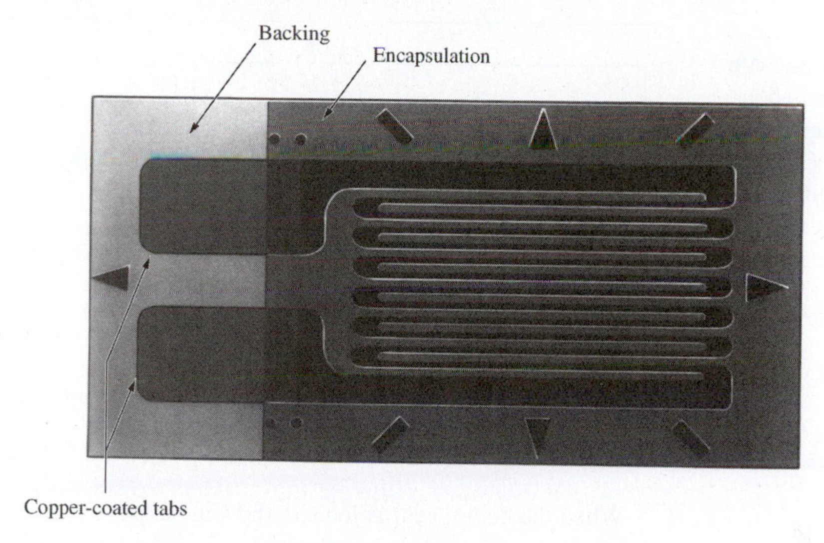

(*a*) Typical geometry

(*b*) Strain gage mounted on a flat bar

FIGURE 1–22 Strain gage—typical geometry and installation. (Source: Measurements Group, Inc., Raleigh, NC)

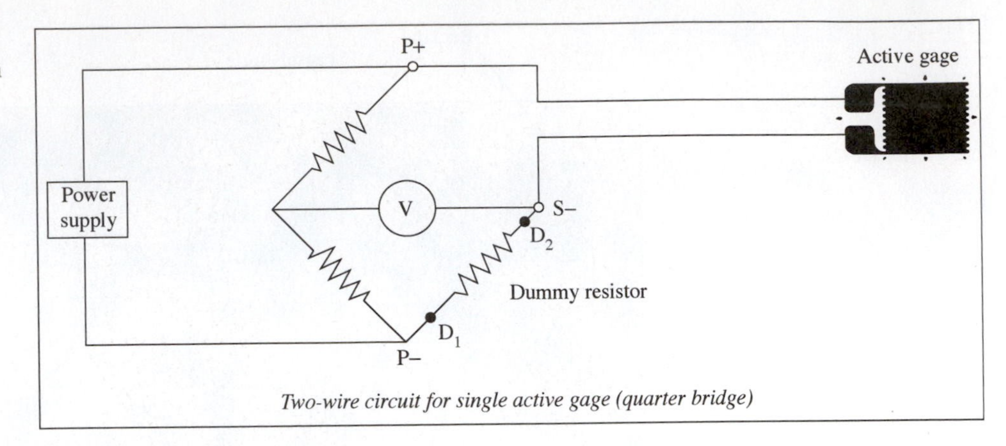

FIGURE 1–23 Wheatstone bridge with strain gage installed in one arm. (Source: Measurements Group, Inc., Raleigh, NC)

When the component is loaded, the gage experiences the same strains as the surface. The resistance of the gage changes in proportion to the applied strain. The gage is then connected into an electrical measurement circuit called a *Wheatstone bridge* as illustrated in Figure 1–23. The bridge has four arms, and the strain gage is connected as one arm. The other arms contain dummy resistors of the same nominal resistance. When the strain gage is subjected to stain, its resistance changes causing the voltage measured across the diagonal of the bridge to change. The voltage reading can be converted to strain when a calibration factor called the *gage factor* is applied. For one-directional strain, as due to direct axial tension, you can use Hooke's law to determine stress from,

$$\sigma = E\epsilon$$

where ϵ is the strain
E is the tensile modulus of elasticity of the material for the part being tested
σ is the normal stress, either tension or compression

Other styles of strain gages and other patterns of installation are used for cases that involve more complex stress conditions. Some of these will be discussed in later chapters.

Finite Element Analysis. Finite element analysis (FEA) is a computational technique that allows the determination of stresses and deflections in complex computer models of load-carrying components. The model is created on a computer-aided design system using a three-dimensional solid modeling approach. Then the model is subdivided into many small, but finite, elements. Simulated loads can be applied, and support conditions can be established on the computer model. Also, material property data such as the modulus of elasticity and Poisson's ratio are entered. Figure 1–24 is an example of a finite element model.

The FEA software uses a matrix analysis approach that computes the effect of forces acting at the boundaries of each element. The result is the integrated effect of the externally applied loads and the supports, yielding computed values for stresses and deflections throughout the component. The compiled results are typically displayed in color graphics that clearly show points of maximum stress and deflection.

The designer can then evaluate the relative safety of the stress levels and proceed to optimize the design to take full advantage of the strength and stiffness of the material from which the component is made. Much more efficient, lighter, and cost-effective designs can

FIGURE 1–24 Finite element analysis model. (Source: Measurements Group, Inc., Raleigh, NC)

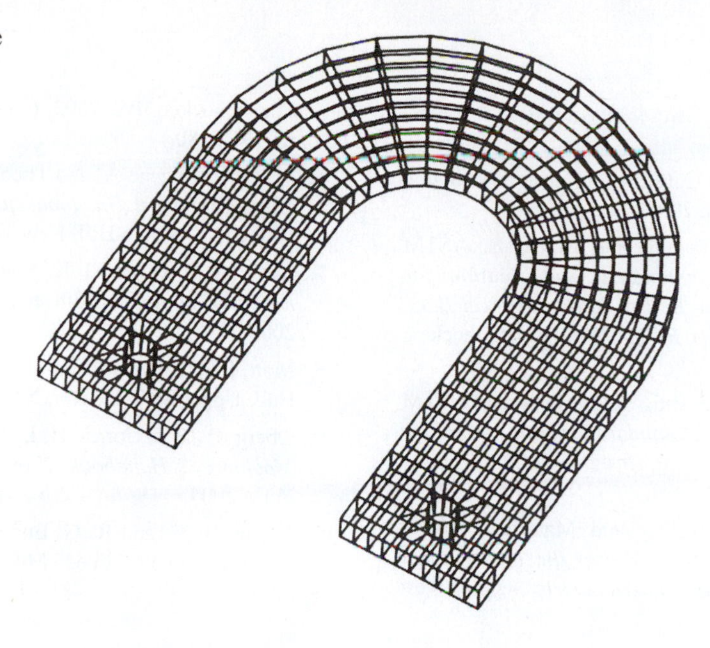

be made using this technique. And the process can be done fairly quickly without making physical models. Often several design iterations are completed in a single day, leading to rapid product development.

INTERNET SITES

1. TubeNet www.tubenet.org/tubes.shtml A listing of dimensions, properties, and suppliers of pipe and tubing in steel, copper, titanium, tantalum, and other materials.
2. Stainless Tubular Products www.stainlesstubular.com A supplier of stainless steel pipe and tubing.
3. Copper Development Association www.copper.org A professional association of the copper industry; the site offers a large amount of data on sizes and physical characteristics of copper tubing. A *Copper Tube Handbook* or parts of it can be downloaded from the site.
4. Plastics Pipe Institute www.plasticpipe.org An association representing all segments of the plastics piping industry; the site includes a list of manufacturers of plastic pipe from which much data for sizes, materials, and application information can be found.
5. Charter Plastics www.charterplastics.com A supplier of polyethylene plastic pipe and tubing.
6. Expert Piping Supply www.expertpiping.com A supplier of copper, steel, PVC, CPVC, polyethylene, and polypropylene pipe in a wide range of diameters and wall thicknesses.
7. The Piping Tool Box www.piping-toolbox.com A site containing data for pipe dimensions, pipe standards, and related topics.
8. American Institute of Steel Construction (AISC) www.aisc.org A trade association that serves the structural steel design community and the construction industry. Publisher of the *AISC Steel Construction Manual.*
9. Webco Mechanical Tubing Products www.webcoindustries.com/tubing/mechanical/ Manufacturer of mechanical tubing in round, square, and rectangular forms made from a variety of carbon and alloy steels.
10. Central Steel & Wire Co. www.centralsteel.com Supplier of a variety of metals (carbon and alloy steel, stainless, aluminum, copper, brass, bronze) in many shapes (structural shapes, strip, sheets, plates, bars, tubing [round, square, rectangular], pipe, wire).
11. Ryerson, Inc. www.ryerson.com Supplier of a wide variety of metals (carbon and alloy steel, stainless, nickel alloys, aluminum, copper, brass, bronze) in many shapes (structural shapes, strip, sheets, plates, bars, tubing [round, square, rectangular], pipe, and wire. Also supplies plastics in sheet, rod, and tubing.
12. Vishay Micro-Measurements www.vishay.com Producer of precision strain gages and associated instruments and accessories.

REFERENCES

1. Aluminum Association, *Aluminum Design Manual,* Washington, DC, 2005.
2. American Institute of Steel Construction, *Steel Construction Manual,* 13th ed., Chicago, 2005.
3. American Society for Testing and Materials, ASTM Standard A6/A6M-05a, *Standard Specification for General Requirements for Rolled Structural Steel Bars, Plates, Shapes, and Sheet Piling,* West Conshohocken, PA, 2005.
4. American Society for Testing and Materials, ASTM Standard E621-94(1999)el, *Standard Practice for the Use of Metric (SI) Units in Building Design and Construction,* West Conshohocken, PA, 1999.
5. American Society for Testing and Materials, ASTM Standard SI-10, *Standard for Use of the International System of Units (SI): The Modern Metric System,* West Conshohocken, PA, 2002. (Note: This standard replaces ASTM E380.)
6. Avallone, Eugene A. and Theodore Baumeister III, eds., *Marks' Standard Handbook for Mechanical Engineers,* 10th ed., McGraw-Hill, New York, 1996.
7. Budynas, R. G. and J. K. Nisbett, *Shigley's Mechanical Engineering Design,* 8th ed., McGraw-Hill, New York, 2007.
8. Mott, R. L., *Applied Fluid Mechanics,* 6th ed., Prentice-Hall, Upper Saddle River, NJ, 2006.
9. Oberg, E., F. D. Jones, H. L. Horton, and H. H. Ryffell, *Machinery's Handbook,* 27th ed., Industrial Press, New York, 2004.
10. Young, W. C. and R. G. Budynas, *Roark's Formulas for Stress and Strain,* 7th ed., McGraw-Hill, New York, 2002.

PROBLEMS

Definitions

1–1. Define *mass* and state the units for mass in both the U.S. Customary unit system and the SI metric unit system.

1–2. Define *weight* and state its units in both systems.

1–3. Define *stress* and state its units in both systems.

1–4. Define *direct normal stress*.

1–5. Explain the difference between compressive stress and tensile stress.

1–6. Define *direct shear stress*.

1–7. Explain the difference between single shear and double shear.

1–8. Draw a stress element subjected to direct tensile stress.

1–9. Draw a stress element subjected to direct compressive stress.

1–10. Draw a stress element subjected to direct shear stress.

1–11. Define *normal strain* and state its units in both systems.

1–12. Define *shearing strain* and state its units in both systems.

1–13. Define *density* and *specific weight* and give the appropriate units for each in both the U.S. Customary system and the SI system.

1–14. A stress computation result is 48 625 951 N/m^2. Express the result in preferred SI notation to four significant numbers.

1–15. A stress computation result is 23.7389 N/mm^2. Express the result in preferred SI notation to three significant figures.

Mass–Weight Conversions

1–16. A truck carries 1800 kg of gravel. What is the weight of the gravel in newtons?

1–17. A four-wheeled truck having a total mass of 4000 kg is sitting on a bridge. If 60% of the weight is on the rear wheels and 40% is on the front wheels, compute the force exerted on the bridge at each wheel.

1–18. A total of 6800 kg of a bulk fertilizer is stored in a flat-bottomed bin having side dimensions 5.0 m $\times$ 3.5 m. Compute the loading on the floor in newtons per square meter, or pascals.

1–19. A mass of 25 kg is suspended by a spring that has a spring scale of 4500 N/m. How much will the spring be stretched?

1–20. Measure the length, width, and thickness of this book in millimeters.

1–21. Determine your own weight in newtons and your mass in kilograms.

1–22. Express the weight found in Problem 1–16 in pounds.

1–23. Express the forces found in Problem 1–17 in pounds.

1–24. Express the loading in Problem 1–18 in pounds per square foot.

1–25. For the data in Problem 1–19, compute the weight of the mass in pounds, the spring scale in pounds per inch, and the stretch of the spring in inches.

1–26. A cast iron base for a machine weighs 2750 lb. Compute its mass in slugs.

1–27. A roll of steel hanging on a scale causes a reading of 12 800 lb. Compute its mass in slugs.

1–28. Determine your own weight in pounds and your mass in slugs.

Unit Conversions

1–29. A pressure vessel contains a gas at 1200 psi. Express the pressure in pascals.

1–30. A structural steel has an allowable stress of 21 600 psi. Express this in pascals.

1–31. The stress at which a material will break under a direct tensile load is called the *ultimate strength*. The range of ultimate strengths for aluminum alloys ranges from about 14 000 to 76 000 psi. Express this range in pascals.

1–32. An electric motor shaft rotates at 1750 rpm. Express the rotational speed in radians per second.

1–33. Express an area of 14.1 in^2 in the units of square millimeters.

1–34. An allowable deformation of a certain beam is 0.080 in. Express the deformation in millimeters.

1–35. A base for a building column measures 18.0 in by 18.0 in on a side and 12.0 in high. Compute the cross-sectional area in both square inches and square millimeters. Compute the volume in cubic inches, cubic feet, cubic millimeters, and cubic meters.

1–36. Compute the area in square inches of a rod having a diameter of 0.505 in. Then convert the result to square millimeters.

Direct Tensile and Compressive Stresses

1–37.M Compute the stress in a round bar subjected to a direct tensile force of 3200 N if the diameter of the bar is 10 mm.

1–38.M Compute the stress in a rectangular bar having cross-sectional dimensions of 10 mm by 30 mm if a direct tensile force of 20 kN is applied.

1–39.E A link in a mechanism for an automated packaging machine is subjected to a tensile force of 860 lb. If the link is square, 0.40 in on a side, compute the stress in the link.

1–40.E A circular rod, with a diameter of $\frac{3}{8}$ in, supports a heater assembly weighing 1850 lb. Compute the stress in the rod.

1–41.M A shelf is being designed to hold crates having a total mass of 1840 kg. Two support rods like those shown in Figure P1–41 hold the shelf. Each rod has a diameter of 12.0 mm. Assume that the center of gravity of the crates is at the middle of the shelf. Compute the stress in the middle portion of the rods.

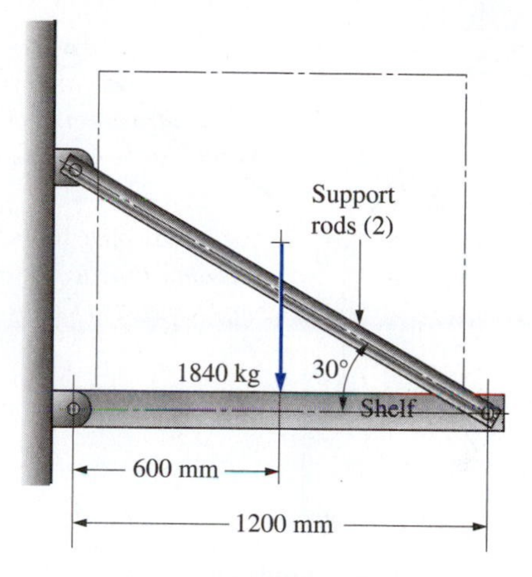

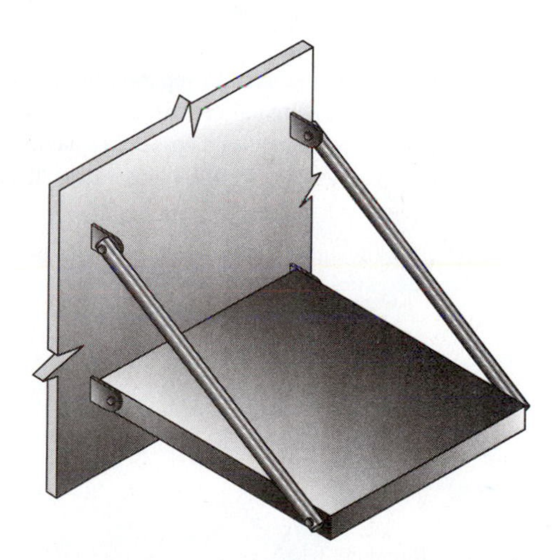

FIGURE P1–41 Shelf support for Problem 1–41.

1–42.E A concrete column base is circular, with a diameter of 8 in, and carries a direct compressive load of 70 000 lb. Compute the compressive stress in the concrete.

1–43.E Three short, square, wood blocks, $3\frac{1}{2}$ in on a side, support a machine weighing 29 500 lb. Compute the compressive stress in the blocks.

1–44.M A short link in a mechanism carries an axial compressive load of 3500 N. If it has a square cross section, 8.0 mm on a side, compute the stress in the link.

1–45.M A machine having a mass of 4200 kg is supported by three solid steel rods arranged as shown in Figure P1–45. Each rod has a diameter of 20 mm. Compute the stress in each rod.

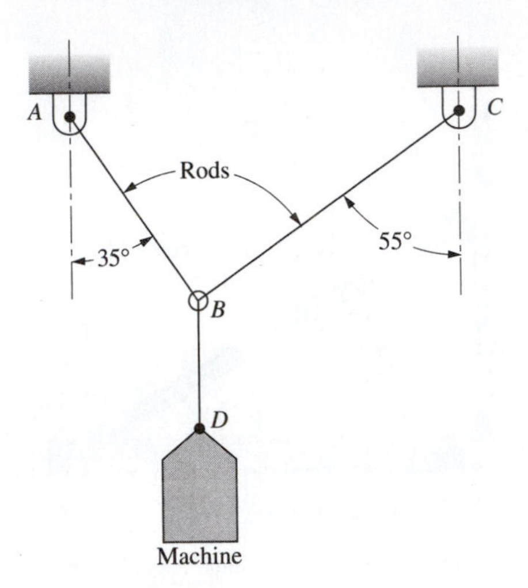

FIGURE P1–45 Support rods for Problem 1–45.

1–46.M A centrifuge is used to separate liquids according to their densities using centrifugal force. Figure P1–46 illustrates one arm of a centrifuge having a bucket at its end to hold the liquid. In operation, the bucket and the liquid have a mass of 0.40 kg. The centrifugal force has the magnitude in newtons of

$$F = 0.010\ 97 \cdot m \cdot R \cdot n^2$$

where m = rotating mass of bucket and liquid (kilograms)
R = radius to center of mass (meters)
n = rotational speed (revolutions per minute) = 3000 rpm

Compute the stress in the round bar. Consider only the force due to the container.

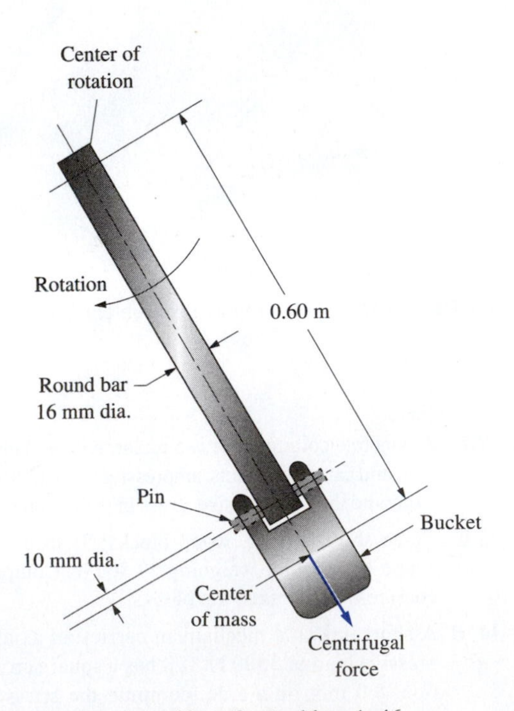

FIGURE P1–46 Centrifuge for Problem 1–46.

1–47.M A square bar carries a series of loads as shown in Figure P1–47. Compute the stress in each segment of the bar. All loads act along the central axis of the bar.

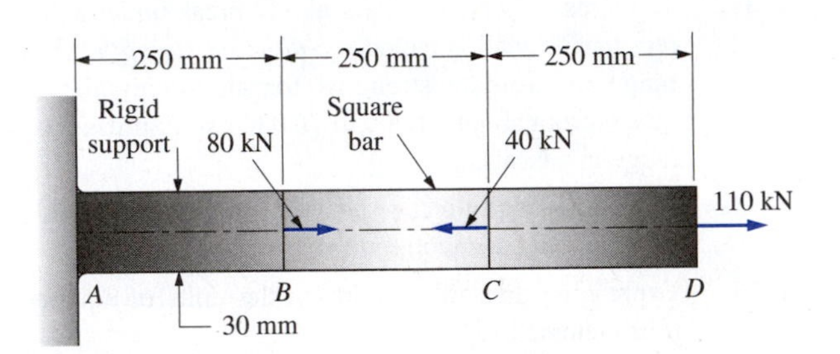

FIGURE P1–47 Square carrying axial loads for Problem 1–47.

1–48.M Repeat Problem 1–47 for the circular bar in Figure P1–48.

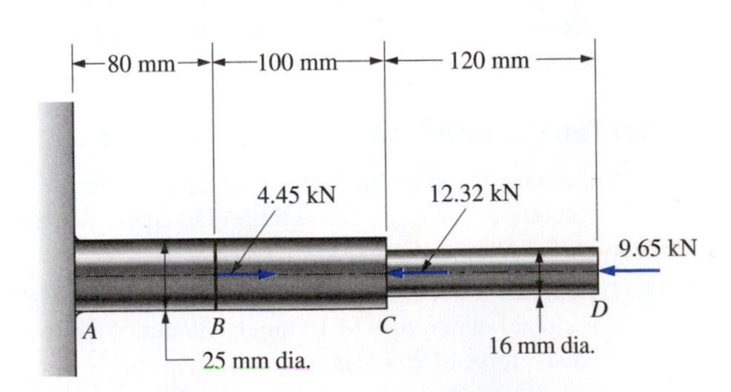

FIGURE P1–48 Circular carrying axial loads for Problem 1–48.

1–49.E Repeat Problem 1–47 for the pipe in Figure P1–49. The pipe is $1\frac{1}{2}$ in schedule 40 steel pipe.

1–50.E Compute the stress in member *BD* shown in Figure P1–50 if the applied force *F* is 2800 lb.

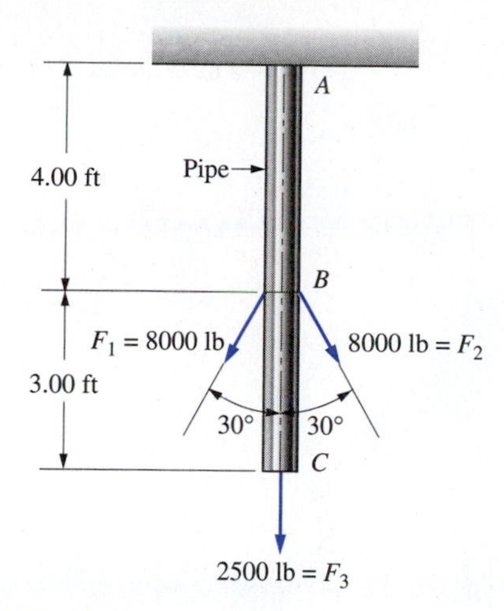

FIGURE P1–49 Pipe for Problem 1–49.

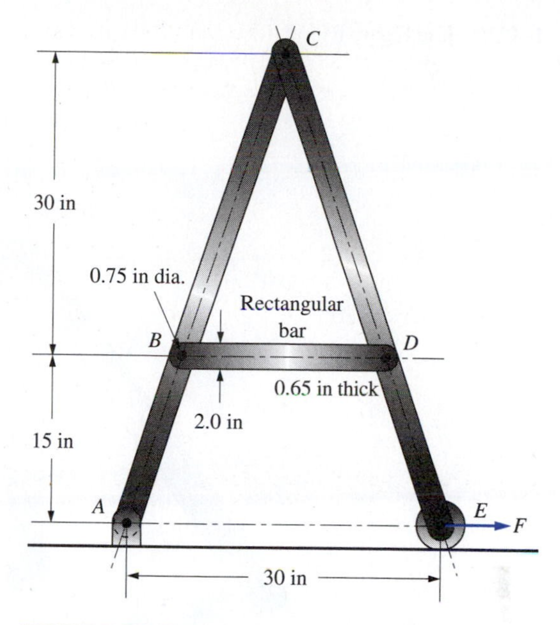

FIGURE P1–50 Frame for Problem 1–50.

For Problems 51 and 52 using the trusses shown in Figures P1–51 and P1–52, compute the forces in all members and the stresses in the midsection, away from any joint. Refer to the Appendix for the cross-sectional area of the members indicated in the figures. Consider all joints to be pinned.

1–51.M Use Figure P1–51.

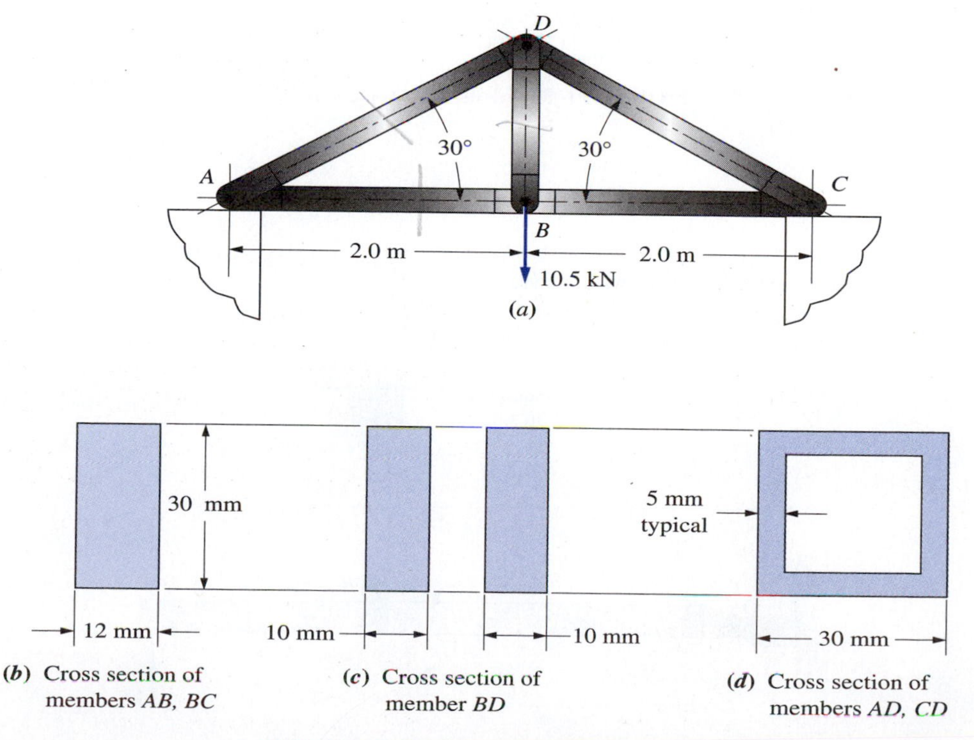

FIGURE P1–51 Truss for Problem 1–51.

1–52.E Use Figure P1–52.

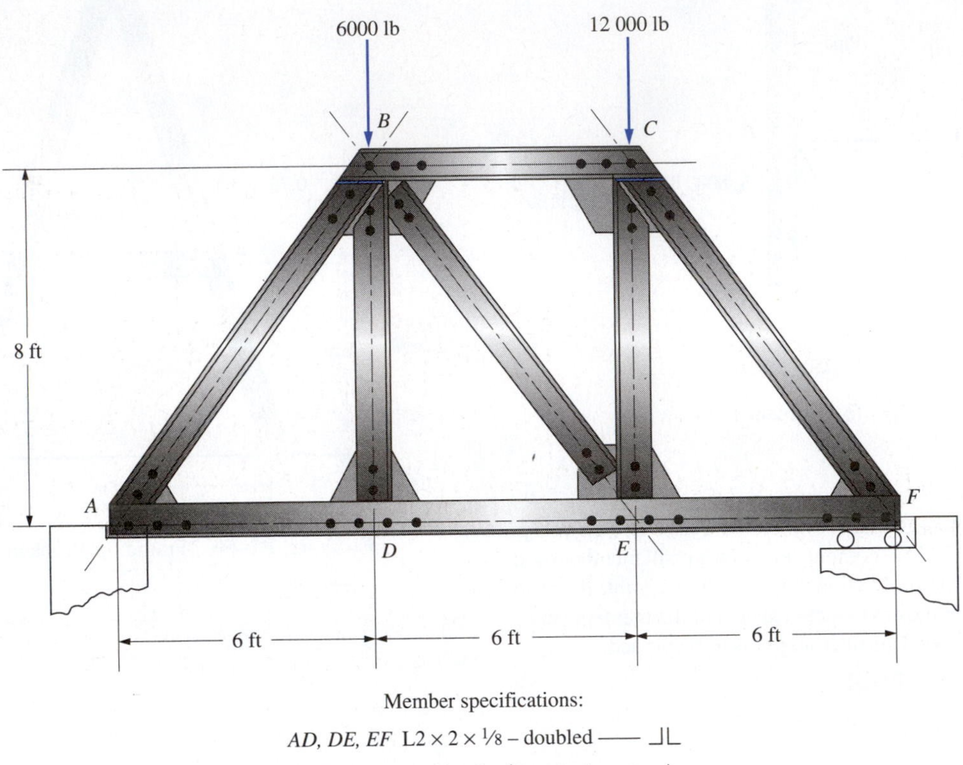

FIGURE P1–52 Truss for Problem 1–52.

1–53.M Find the tensile stress in member *AB* shown in Figure P1–53.

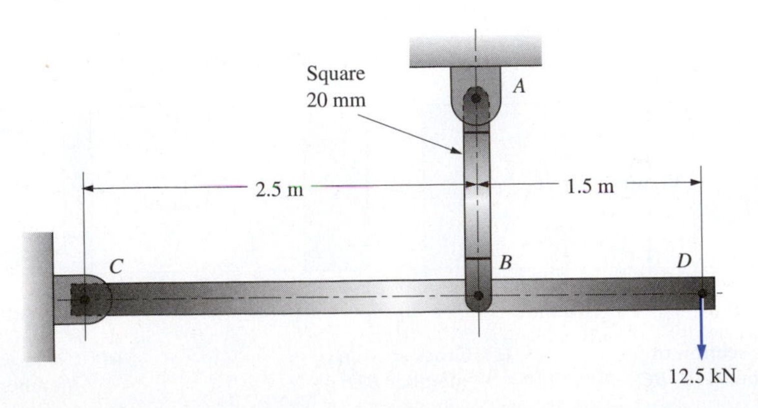

FIGURE 1–53 Support for Problem 1–53.

1–54.E Figure P1–54 shows the shape of a test specimen used to measure the tensile properties of metals (as described in Chapter 2). An axial tensile force is applied through the threaded ends, and the test section is the reduced-diameter part near the middle. Compute the stress in the middle portion when the load is 12 600 lb.

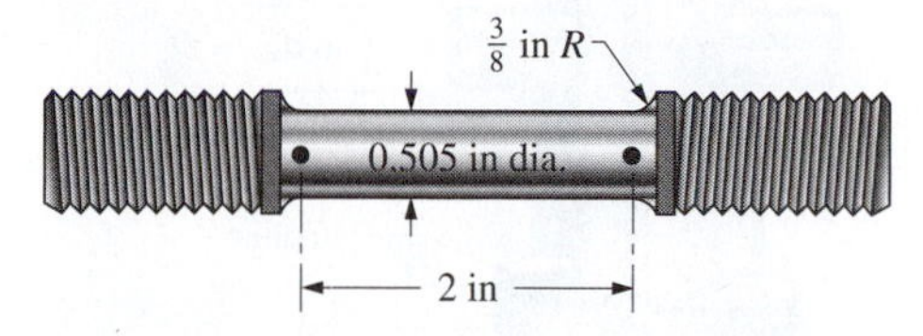

FIGURE P1–54 Tensile test specimen for Problem 1–54.

1–55.E A short compression member has the cross section shown in Figure P1–55. Compute the stress in the member if a compressive force of 52 000 lb is applied in line with its centroidal axis.

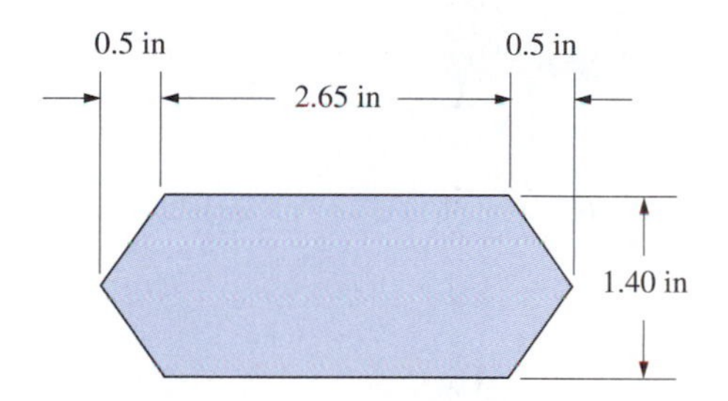

FIGURE P1–55 Short compression member for Problem 1–55.

1–56.M A short compression member has the cross section shown in Figure P1–56. Compute the stress in the member if a compressive force of 640 kN is applied in line with its centroidal axis.

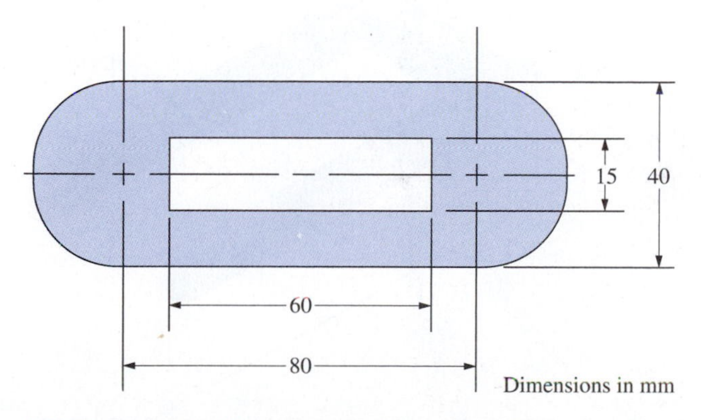

FIGURE P1–56 Short compression member for Problem 1–56.

Direct Shearing Stresses

1–57.M A pin connection like that shown in Figure 1–15 is subjected to a force of 16.5 kN. Determine the shear stress in the 12.0-mm-diameter pin.

1–58.M In a pair of pliers, the hinge pin is subjected to direct shear, as indicated in Figure P1–58. If the pin has a diameter of 3.0 mm and the force exerted at the handle, F_h, is 55 N, compute the stress in the pin.

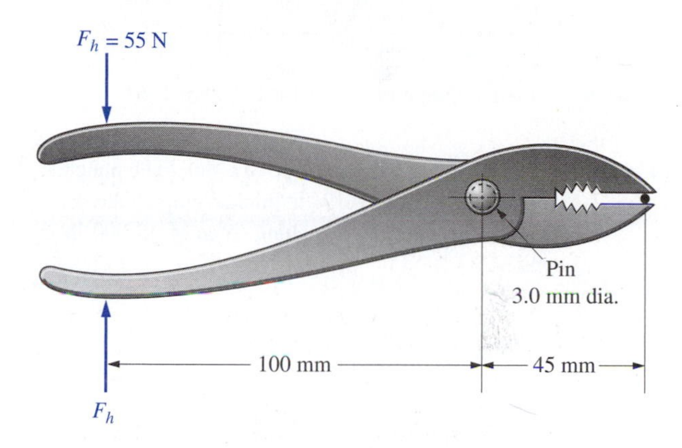

FIGURE P1–58 Pliers for Problem 1–58.

1–59.M For the centrifuge shown in Figure P1–46 and the data from Problem 1–46, compute the shear stress in the pin between the bar and the bucket.

1–60.E A notch is made in a piece of wood, as shown in Figure P1–60, in order to support an external load F of 1800 lb. Compute the shear stress in the wood.

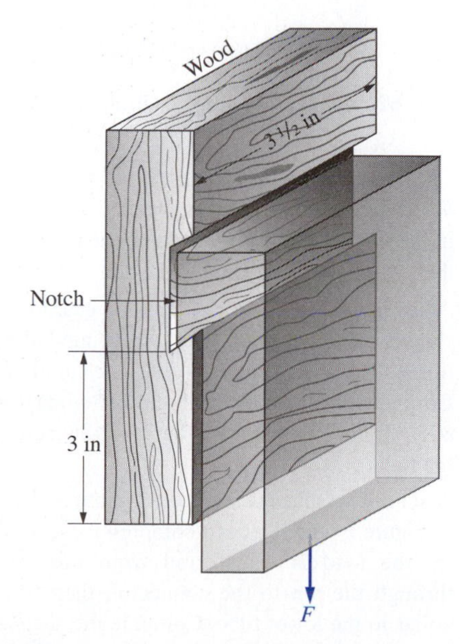

FIGURE P1–60 Notched wood block loaded in shear for Problem 1–60.

1–61.M Figure P1–61 shows the shape of a slug to be punched from a sheet of aluminum 5.0 mm thick. Compute the shear stress in the aluminum if a punching force of 38.6 kN is applied.

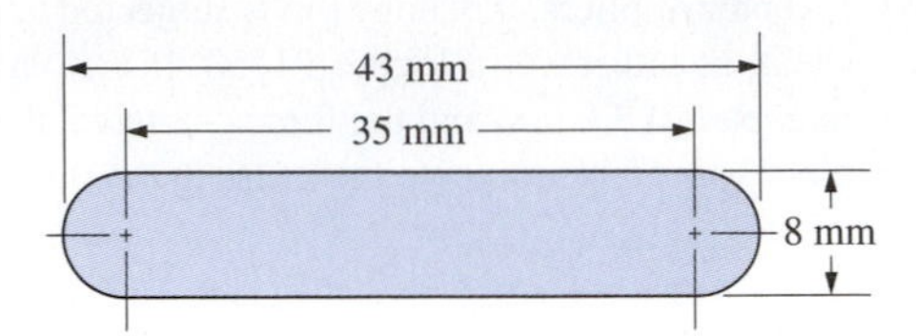

FIGURE P1–61 Shape of a slug for Problem 1–61.

1–62.E Figure P1–62 shows the shape of a slug to be punched from a sheet of steel 0.194 in thick. Compute the shear stress in the steel if a punching force of 45 000 lb is applied.

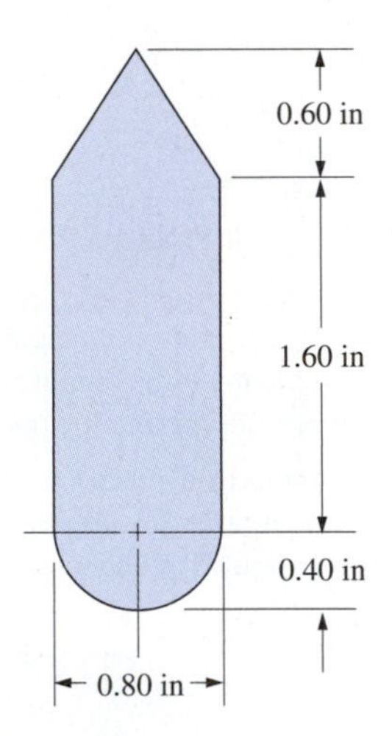

FIGURE P1–62 Shape of a slug for Problem 1–62.

1–63.M The key in Figure 1–18 has the dimensions $b = 10$ mm, $h = 8$ mm, and $L = 22$ mm. Determine the shear stress in the key when 95 N·m of torque is transferred from the 35-mm-diameter shaft to the hub.

1–64.E A key is used to connect a hub of a gear to a shaft, as shown in Figure 1–18. It has a rectangular cross section with $b = \frac{1}{2}$ in and $h = \frac{3}{8}$ in. The length is 2.25 in. Compute the shear stress in the key when it transmits 8000 lb·in of torque from the 2.0-in-diameter shaft to the hub.

1–65.E A set of two tubes is connected in the manner shown in Figure P1–65. Under a compressive load of 20 000 lb, the load is transferred from the upper tube through the pin to the connector, then through the collar to the lower tube. Compute the shear stress in the pin and in the collar.

1–66.E A small hydraulic crane, like that shown in Figure P1–66, carries an 800-lb load. Determine the shear

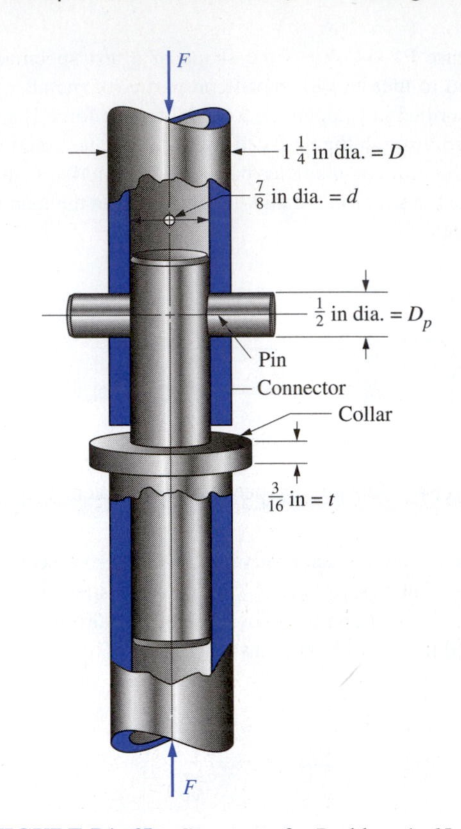

FIGURE P1–65 Connector for Problem 1–65.

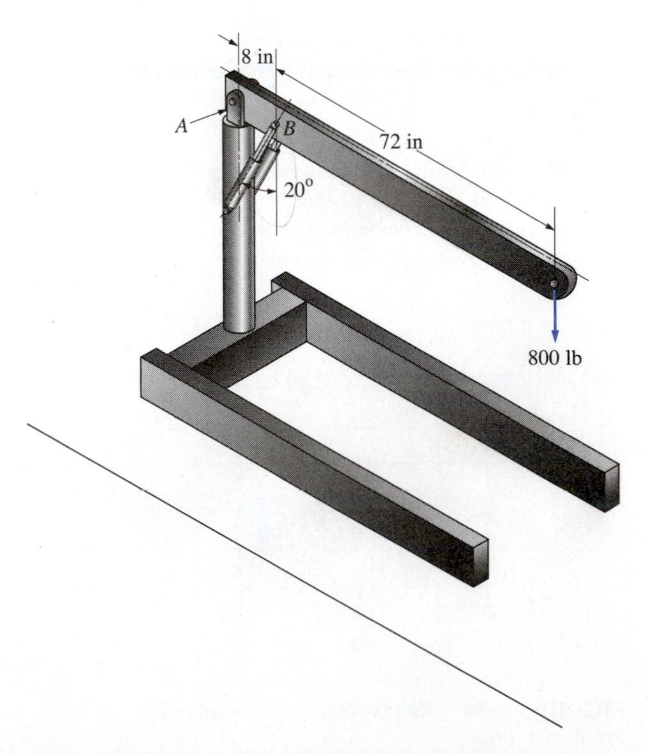

FIGURE P1–66 Hydraulic crane for Problem 1–66.

stress that occurs in the pin at B which is in double shear. The pin diameter is $\frac{3}{8}$ in.

1–67.M A ratchet device on a jack stand for a truck has a tooth configuration as shown in Figure P1–67. For a load of 88 kN, compute the shear stress at the base of the tooth.

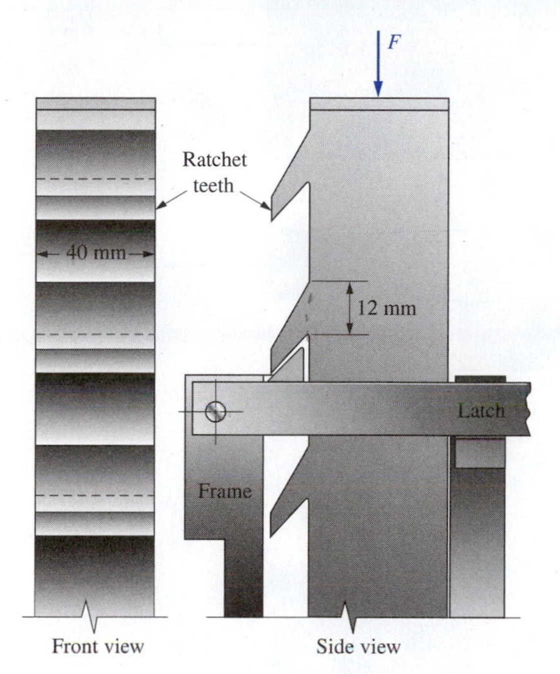

FIGURE P1–67 Ratchet for a jack stand for Problem 1–67.

1–68.M Figure P1–68 shows an assembly in which the upper block is brazed to the lower block. Compute the shear stress in the brazing material if the force is 88.2 kN.

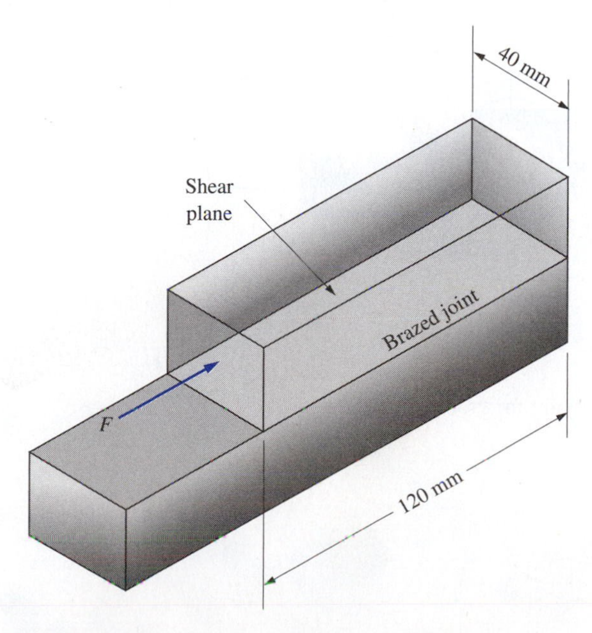

FIGURE P1–68 Brazed components for Problem 1–68.

1–69.M Figure P1–69 shows a bolt subjected to a tensile load. One failure mode would be if the circular shank of the bolt pulled out from the head, a shearing action. Compute the shear stress in the head for this mode of failure if a force of 22.3 kN is applied.

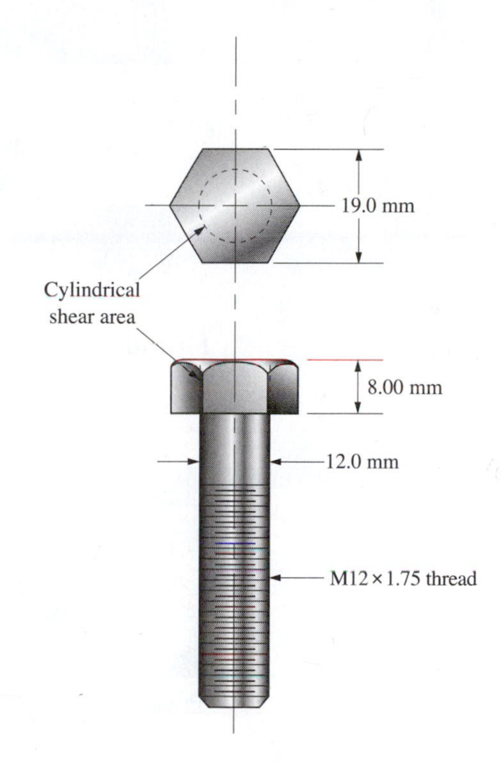

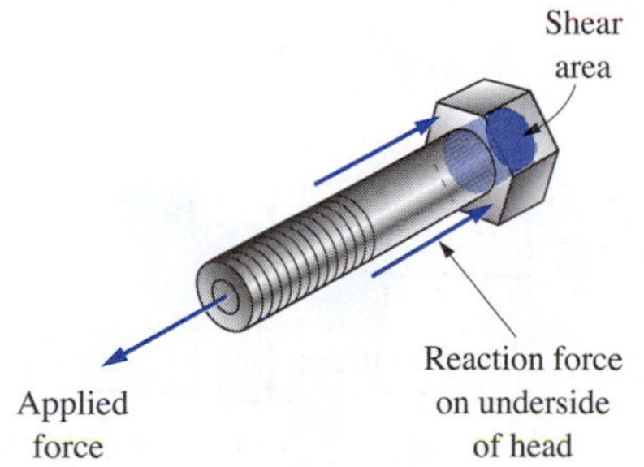

FIGURE P1–69 Bolt for Problem 1–69.

1–70.M Figure P1–70 shows a riveted lap joint connecting two steel plates. Compute the shear stress in the rivets due to a force of 10.2 kN applied to the plates.

1–71.M Figure P1–71 shows a riveted butt joint with cover plates connecting two steel plates. Compute the shear stress in the rivets due to a force of 10.2 kN applied to the plates.

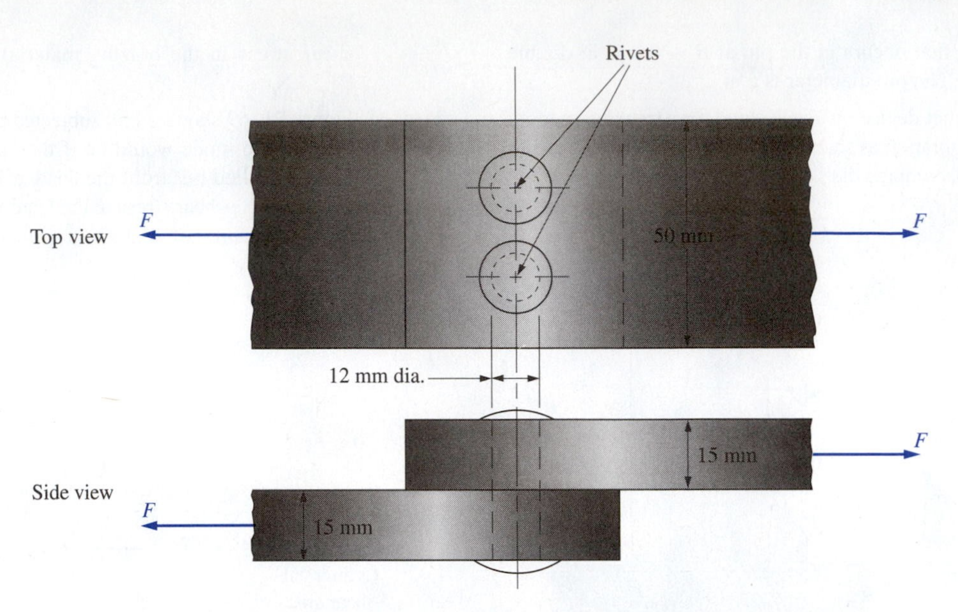

FIGURE P1–70 Riveted lap joint for Problem 1–70.

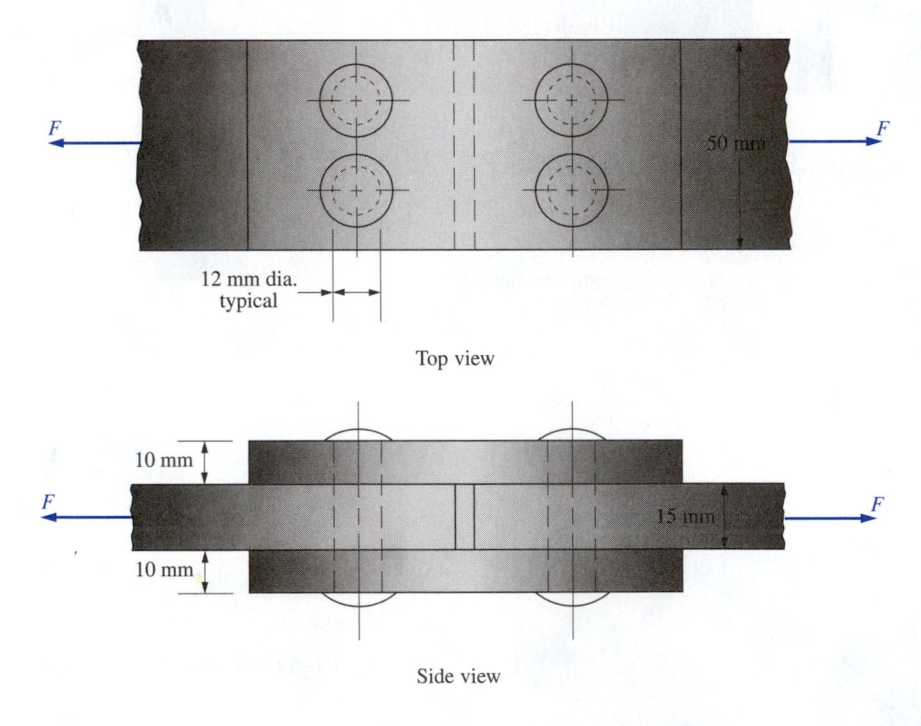

FIGURE P1–71 Riveted butt joint for Problem 1–71.

2

Design Properties of Materials

The Big Picture

Design Properties of Materials

Discussion Map

- As you design machines, structures, or consumer products, you must understand how materials react to externally applied loads.
- You will see how the relationships between stress and strain for a given material are critical to this understanding.
- The strength of a material is taken as the value for which it fails by fracture or by excessive deformation called *yielding*.
- You will become familiar with the ways material properties are listed in the Appendix of this book and in numerous other industrial sources of data.
- You will get an overview of the principal design properties of many metals and nonmetals.

Activity

Have you ever observed carefully a product that failed because a critical component had broken? Describe the product, how it failed, and the appearance of the material in the vicinity of the break. What was the material? Was it a metal or nonmetal?

Look for examples made from metals, plastics, concrete, or other materials. Compare the nature of the failures for each material in terms of the appearance of the fractured material.

Does it look as if it broke suddenly, somewhat how chalk breaks when it is dropped onto a floor? Or did it deform noticeably near the break before actually coming apart?

Why did it break? How was it being used when it broke? Was it a momentary overload, impact, vibration, or repeated flexure?

Study the material properties listed in Appendices A–14 to A–20 of this book. Note the different kinds of steels, cast iron, aluminum, wood, and plastics. Look for the strength ratings called ultimate strength, yield strength, and shear strength. Other properties include modulus of elasticity, percent elongation (a measure of ductility), and density.

Safety, the ultimate goal of design and stress analysis, requires that materials do not fail, thus rendering a product unfit for its intended purpose. Failure can take several forms. The primary types of failure considered in this book are the following:

1. The material of some component could fracture completely.
2. The material may deform excessively under load so the component no longer performs satisfactorily.
3. A structure or one of its components could become unstable and buckle, and thus be unable to carry the intended loads.

The study of strength of materials requires a knowledge of how external forces and moments affect the stresses and deformations developed in the material of a load-carrying member. In order to put this knowledge to practical use, however, a designer needs to know how such stresses and deformations can be withstood safely by the material. Thus, material properties as they relate to design must be understood along with the analysis required to determine the magnitude of stresses and deformations.

In this chapter we present information concerning the materials most frequently used to make components for structures and mechanical devices, emphasizing the design properties of the materials rather than their metallurgical structure or chemical composition. Although it is true that a thorough knowledge of the structure of materials is an aid to a designer, it is most important to know how the materials behave when carrying loads. This is the behavior on which we concentrate in this chapter. See References 4, 9, 12, and 29.

First, we discuss the design properties of materials that affect the performance of a product. Then a wide variety of metals are described including carbon and alloy steel, stainless steel, structural steel, cast iron, aluminum, copper, brass, bronze, zinc, magnesium, titanium, and some of the nickel alloys.

Nonmetals presented include wood, concrete, plastics, and composites. The manner in which the behavior of these materials differs from that of metals is discussed, along with some of their special properties.

2–1 OBJECTIVES OF THIS CHAPTER

After completing this chapter, you should be able to:

1. List typical uses for engineering materials.
2. Define the strength properties of metals, *ultimate tensile strength, yield point, yield strength, elastic limit,* and *proportional limit.*
3. Define *Hooke's law* and the stiffness properties of metals, *modulus of elasticity in tension and modulus of elasticity in shear*.
4. Define *Poisson's ratio* and give its value for typical materials.
5. Describe ductile and brittle behavior of materials.
6. Define *percent elongation* and describe its relationship to the ductility of materials.
7. Describe the Unified Numbering System (UNS) for metals and alloys.
8. Describe the four-digit designation system for steels.
9. Describe the important properties of carbon steels, alloy steels, stainless steels, and structural steels.
10. Describe the four-digit designation system for wrought and cast aluminum alloys.
11. Describe the aluminum temper designations.
12. Describe the design properties of copper, brass, bronze, zinc, magnesium, and titanium.
13. Describe the design properties of cast irons, including gray iron, ductile iron, austempered ductile iron, white iron, and malleable iron.
14. Describe the design properties of wood, concrete, plastics, and composites.

2–2 DESIGN PROPERTIES OF MATERIALS

Material selection requires consideration of many factors, such as,

strength	stiffness	ductility	weight
fracture toughness	machinability	workability	weldability
appearance	stability	cost	availability

compatibility with the environment in which the product must operate

More is said about the general nature of material selection process later in this chapter. Relative to the study of strength of materials, the primary emphasis is on *strength, stiffness,* and *ductility.* The types of strength considered most frequently are *tensile strength, compressive strength,* and *yield strength.* Tensile and yield strengths are considered first.

Tensile and Yield Strengths. Reference data listing the mechanical properties of metals will almost always include the *tensile strength* and *yield strength* of the metal. Comparison of the actual stresses in a part with the tensile or yield strength of the material from which the part is made is the usual method of evaluating the suitability of the

FIGURE 2–1 Universal testing machine for obtaining stress-strain data for materials. (Source: Instron Corporation, Norwood, MA)

material to carry the applied loads safely. Chapter 3 and subsequent chapters discuss more about the details of stress analysis.

The tensile strength and yield strength are determined by testing a sample of the material in a tensile-testing machine such as the one shown in Figure 2–1. (See Reference 8 for the methods used.) A round bar or flat strip is placed in the upper and lower jaws. Figure 2–2 shows a photograph of a typical tensile test specimen. A pulling force is applied slowly and steadily to the sample, stretching it until it breaks. During the test, a graph is made that shows the relationship between the stress in the sample and the strain or unit deformation.

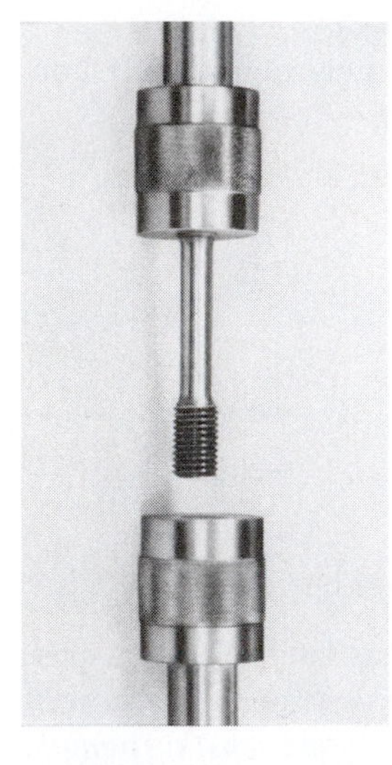

FIGURE 2–2 Tensile test specimen mounted in a holder. (Source: Tinius Olsen Testing Machine Co., Inc., Willow Grove, PA)

A typical stress–strain diagram for a low-carbon steel is shown in Figure 2–3. It can be seen that during the first phase of loading, the plot of stress versus strain is a straight line, indicating that stress is directly proportional to strain. After point *A* on the diagram, the curve is no longer a straight line. This point is called the *proportional limit*. As the load on the sample is continually increased, a point called the *elastic limit* is reached, marked *B* in Figure 2–3. At stresses below this point, the material will return to its original size and shape if the load is removed. At higher stresses, the material is permanently deformed. The *yield point* is the stress at which a noticeable elongation of the sample occurs with no apparent increase in load. The yield point is at *C* in Figure 2–3, about 36 000 psi (248 MPa). Applying still higher loads after the yield point has been reached causes the curve to rise again. After reaching a peak, the curve drops somewhat until finally the sample breaks, terminating the plot. The highest apparent stress taken from the stress–strain diagram is called the *tensile strength*. In Figure 2–3 the tensile strength would be about 53 000 psi (365 MPa).

The fact that the stress–strain curve in Figure 2–3 drops off after reaching a peak tends to indicate that the stress level decreases. Actually, it does not; the *true stress* continues to rise until ultimate failure of the material. The reason for the apparent decrease in stress is that the plot taken from a typical tensile test machine is actually *load versus elongation* rather than *stress versus strain*. The vertical axis is converted to stress by dividing the load (force) on the specimen by the *original* cross-sectional area of the specimen.

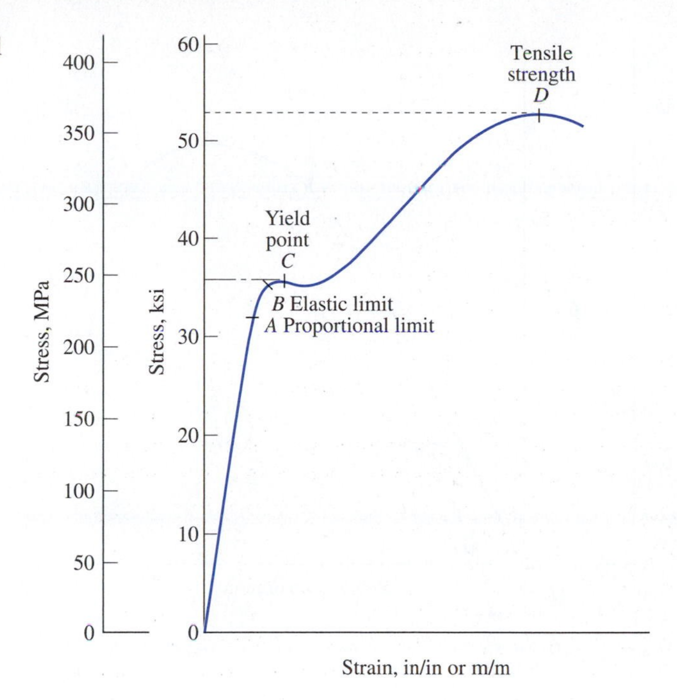

FIGURE 2–3 Typical stress–strain curve for steel.

When the specimen nears its breaking load, there is a reduction in diameter and consequently a reduction in the cross-sectional area. The reduced area requires a lower force to continue stretching the specimen, even though the actual stress in the material is increasing. This results in the dropping curve shown in Figure 2–3. Because it is very difficult to monitor the decreasing diameter, and because experiments have shown that there is little difference between the true maximum stress and that found from the peak of the *apparent stress* versus strain curve, the peak is accepted as the tensile strength of the material.

A summary of the definitions of key strength properties of steels follows:

The proportional limit is the value of stress on the stress–strain curve at which the curve first deviates from a straight line.

The elastic limit is the value of stress on the stress–strain curve at which the material has deformed plastically; that is, it will no longer return to its original size and shape after removing the load.

The yield point is the value of stress on the stress–strain curve at which there is a significant increase in strain with little or no increase in stress.

The tensile strength is the highest value of apparent stress on the stress–strain curve.

Many metals do not exhibit a well-defined yield point like that in Figure 2–3. Some examples are high-strength alloy steels, aluminum, and titanium. However, these materials do in fact yield, in the sense of deforming a sizable amount before fracture actually occurs. For these materials, a typical stress–strain diagram would look like the one shown in Figure 2–4. The curve is smooth with no pronounced yield point. For such materials, the yield strength is defined by a line like *M-N* drawn parallel to the straight-line portion of the test curve.

Point *M* is usually determined by finding that point on the strain axis representing a strain of 0.002 in/in. This point is also called the point of 0.2% offset. The point *N,* where the offset line intersects the curve, defines the yield strength of the material, about 55 000 psi

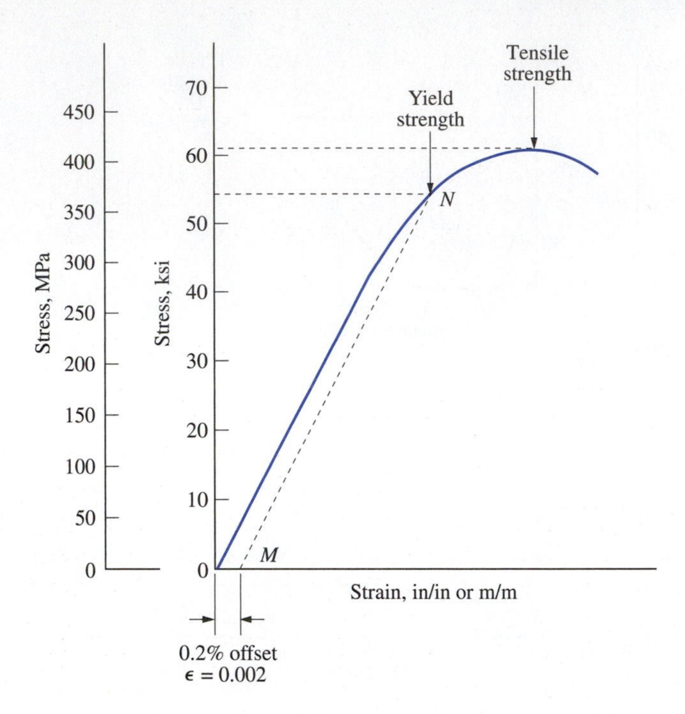

FIGURE 2–4 Typical stress–strain curve for aluminum.

in Figure 2–4. The ultimate strength is at the peak of the curve, as was described before. *Yield strength* is used in place of yield point for these materials.

The units for the offset on the strain axis can be in any system. Recall that strain itself is sometimes called *unit deformation,* a ratio of the elongation of the material at a specific load to its original length in the unloaded condition. Thus strain is actually dimensionless. If measurements were being taken directly in the SI metric system, the strain would be m/m, meters of elongation per meter of original length. Or, mm/mm could be used.

The value of the offset, 0.2%, is typical for most commonly used metals. Other values may be used if 0.2% does not give reliable or convenient results. However, the value of 0.2% is assumed unless otherwise stated.

In summary, for many materials that do not exhibit a pronounced yield point, the definition of *yield strength* is as follows.

The yield strength is the value of stress on the stress–strain curve at which a straight line drawn from a strain value of 0.002 in/in (or m/m) and parallel to the straight portion of the stress–strain curve intersects the curve.

Compressive Strength. The stress–strain behavior of most wrought metals is nearly the same in compression as it is in tension. This is because the material has a nearly uniform, homogeneous structure throughout. When a material behaves in a similar manner regardless of the direction of the loads, it is referred to as an *isotropic* material. For isotropic materials, then, separate compression tests are not usually performed and separate data are not reported for compressive strength.

But many materials exhibit different behavior and strength in compression than in tension. This is called *anisotropic* behavior. Examples are many cast metals, some plastics, concrete, wood, and composites. You should seek data for both compressive strength and tensile strength for such materials.

Stiffness. It is frequently necessary to determine how much a part will deform under load to ensure that excessive deformation does not destroy the usefulness of the part. This can occur at stresses well below the yield strength of the material, especially in very long members or in high-precision devices. Stiffness of a material is a function of its *modulus of elasticity,* sometimes called *Young's modulus*.

The modulus of elasticity, E, is a measure of the stiffness of a material determined by the slope of the straight-line portion of the stress–strain curve. It is the ratio of the change of stress to the corresponding change in strain.

This can be stated mathematically as

$$E = \frac{\text{stress}}{\text{strain}} = \frac{\sigma}{\epsilon} \tag{2–1}$$

Therefore, a material having a steeper slope on its stress–strain curve will be stiffer and will deform less under load than a material having a less steep slope. Figure 2–5 illustrates this concept by showing the straight-line portions of the stress–strain curves for steel, titanium, aluminum, and magnesium. It can be seen that if two otherwise identical parts were made of steel and aluminum, respectively, the aluminum part would deform about three times as much when subjected to the same load.

The design of typical load-carrying members in machines and structures is such that the stress is below the proportional limit; that is, in the straight-line portion of the stress–strain curve. Here we define *Hooke's law:*

When the level of stress in a material under load is below the proportional limit and there is a straight-line relationship between stress and strain, it is said that Hooke's law applies.

Many of the formulas used for stress analysis are based on the assumption that Hooke's law applies. This concept is also useful for experimental stress analysis techniques

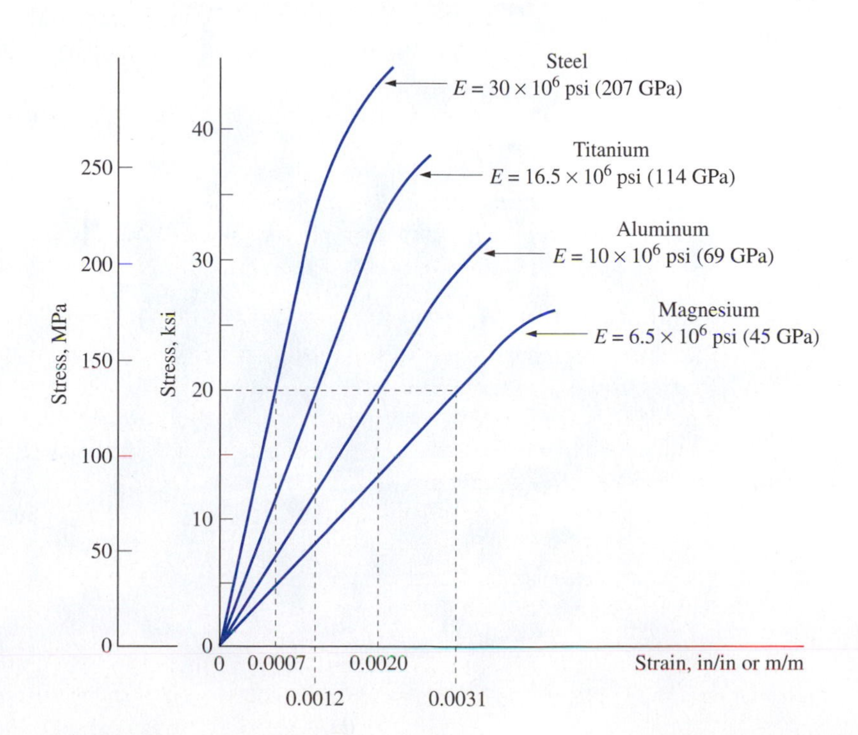

FIGURE 2–5
Modulus of elasticity for different metals.

in which strain is measured at a point. The corresponding stress at the point can be computed from a variation of Equation (2–1),

$$\sigma = E\epsilon \tag{2–2}$$

Equation (2–2) is valid only where strain occurs in only one direction. This is called *uniaxial strain,* and it applies to members subjected to axial tension or compression and to beams in pure bending. When stresses occur in two directions (*biaxial stress*), an additional effect of the second stress must be considered. Biaxial stress is discussed in Chapter 10.

Ductility. When metals break, their fracture can be classified as either ductile or brittle. A ductile material will stretch and yield prior to actual fracture, causing a noticeable decrease in the cross-sectional area at the fractured section. Conversely, a brittle material will fracture suddenly with little or no change in the area at the fractured section. Ductile materials are preferred for parts that carry repeated loads or are subjected to impact loading because they are usually more resistant to fatigue failure and because they are better at absorbing the impact energy.

Ductility in metals is usually measured during the tensile test by noting how much the material has elongated permanently after fracture. At the start of the test, a set of gage marks is scribed on the test sample as shown in Figure 2–6. Most tests use 2.000 in or 50.0 mm

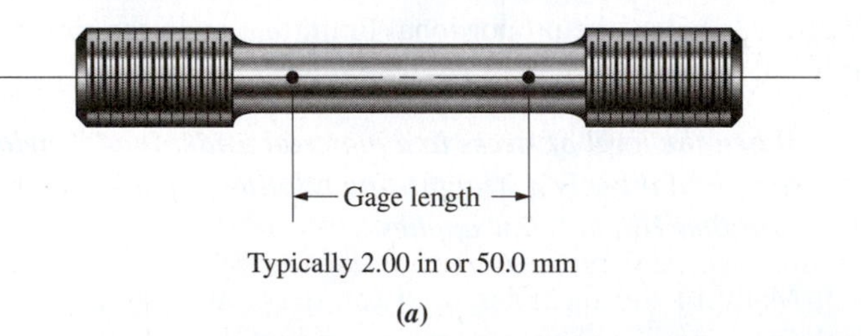

(*a*)

(*b*)

FIGURE 2–6 Gage length for tensile test specimen. (a) Gage length marked on a test specimen. (b) Test specimen in a fixture used to mark gage length. (Source: Tinius Olsen Testing Machine Co., Inc., Willow Grove, PA)

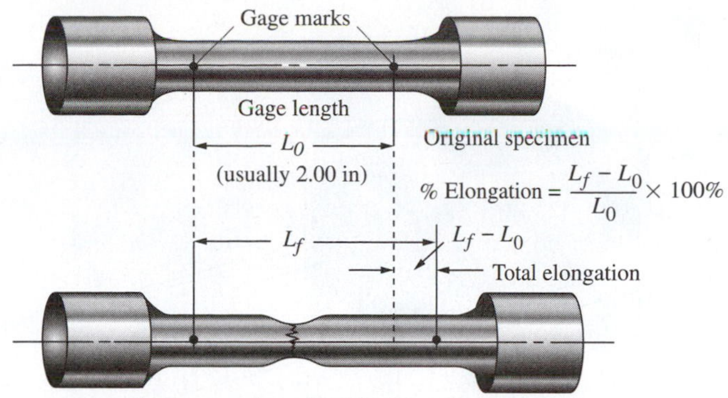

FIGURE 2–7 Measurement of percent elongation.

for the gage length as shown in the figure. Very ductile structural steels sometimes use 8.000 in or 200.0 mm for the gage length. After the sample has been pulled to failure, the broken parts are fitted back together and the distance between the marks is again measured. See Figure 2–7. From these data, the *percent elongation* is computed from

Percent Elongation

$$\text{percent elongation} = \frac{\text{final length} - \text{gage length}}{\text{gage length}} \times 100\% \qquad \textbf{(2–3)}$$

A metal is considered to be *ductile* if its percent elongation is greater than about 5.0%. A material with a percent elongation under 5.0% is considered to be *brittle* and does not exhibit the phenomenon of yielding. Failure of such materials is sudden, without noticeable deformation prior to ultimate fracture. In most structural and mechanical design applications, ductile behavior is desirable, and the percent elongation of the material should be significantly greater than 5.0%. A high-percentage elongation indicates a highly ductile material.

In summary, the following definitions are used to describe ductility for metals:

The percent elongation is the ratio of the plastic elongation of a tensile specimen after ultimate failure within a set of gage marks to the original length between the gage marks. It is one measure of ductility.

A ductile material is one that can be stretched, formed, or drawn to a significant degree before fracture. A metal that exhibits a percent elongation greater than 5.0% is considered to be ductile.

A brittle material is one that fails suddenly under load with little or no plastic deformation. A metal that exhibits a percent elongation less than 5.0% is considered to be brittle.

Virtually all wrought forms of steel and aluminum alloys are ductile. But the higher-strength forms tend to have lower ductility and the designer is often forced to compromise strength and ductility in the specification of a material. Gray cast iron, many forms of cast aluminum, and some high-strength forms of wrought or cast steel are brittle.

Poisson's Ratio. Figure 2–8 shows a more complete understanding of the deformation of a member subjected to normal stress. The element shown is taken from the bar in tension. The tensile force on the bar causes an elongation in the direction of the applied force, as would be expected. But at the same time, the width of the bar is being shortened. Thus a simultaneous elongation and contraction occurs in the stress element. From the elongation, the axial strain can be determined, and from the contraction, the lateral strain can be determined.

FIGURE 2–8
Illustration of Poisson's ratio for an element in tension.

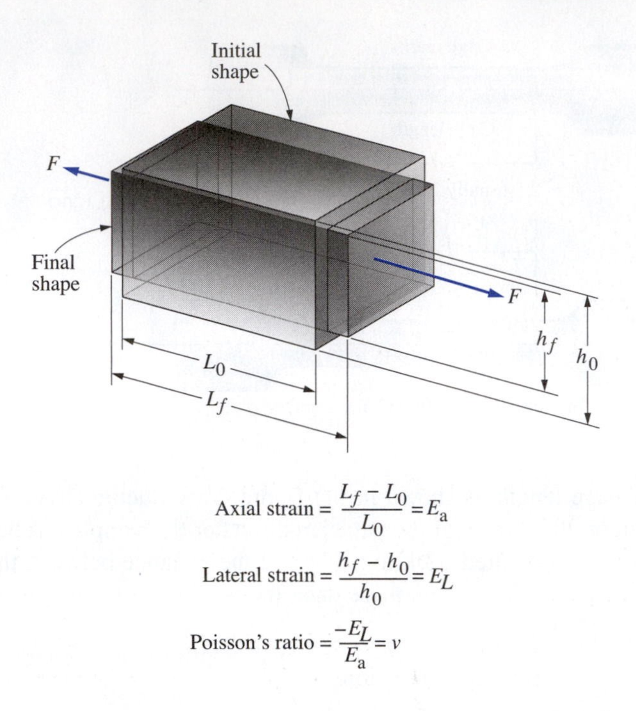

We define the term *Poisson's ratio* as the ratio of the amount of lateral strain to axial strain. That is,

$$\text{Poisson's ratio} = \nu = \frac{\text{Lateral strain}}{\text{Axial strain}} = \frac{-\epsilon_L}{\epsilon_a} \qquad \textbf{(2–4)}$$

The negative sign on the lateral strain is introduced to ensure that Poisson's ratio is a positive number when strains are computed as indicated in Figure 2–8. Poisson's ratio is also said to be the absolute value of the strain ratio.

Most commonly used metallic materials have a Poisson's ratio value between 0.25 and 0.35. For concrete, ν varies widely depending on the grade and on the applied stress, but it usually falls between 0.1 and 0.25. Elastomers and rubber may have Poisson's ratios approaching 0.50. Approximate values for Poisson's ratio are listed in Table 2–1.

Poisson's ratio is used in the calculation of contact stress, sometimes called *Hertz stress,* when forces are transferred across very small areas. Examples shown later in Figure 3–14 include cylinders or spheres acting on flat plates, spherical ball bearings acting on their races, and the contact area between two gear teeth. The complete analysis of such stresses requires

TABLE 2–1 Approximate values of Poisson's ratio, ν.

Concrete	0.10–0.25	Aluminum (most alloys)	0.33
Glass	0.24	Brass	0.33
Ductile iron	0.27	Copper	0.33
Gray cast iron	0.21	Zinc	0.33
Plastics	0.20–0.40	Phosphor bronze	0.35
Carbon and alloy steel	0.29	Magnesium	0.35
Stainless steel (18-8)	0.30	Lead	0.43
Titanium	0.30	Rubber, elastomers	~0.50

Notes: Values are approximate and vary somewhat with specific composition.
Rubber and elastomers approach a limiting value of 0.50.

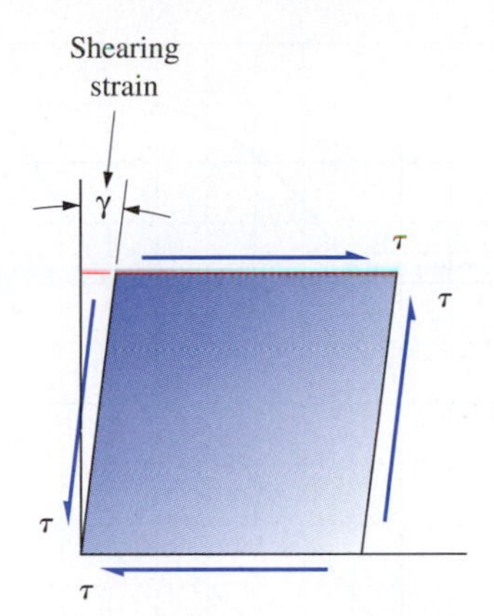

FIGURE 2–9
Shearing strain shown on a stress element.

a detailed description of the deformation that occurs around the small area of contact. Principles of elasticity must be employed, and the deformations are significantly influenced by the Poisson's ratio of the materials of the two members in contact. See References 7 and 10 in Chapter 1.

LESSON 4

Some types of experimental stress analysis also use Poisson's ratio. For example, electrical resistance strain gages sense the local strains on the surface of load-carrying parts. Precise analysis requires that the principal strains in two perpendicular directions be measured with the results used to compute the stress at the location of the gages.

Shearing Strain. Earlier discussions of strain in Chapter 1 described normal strain because it is caused by the normal tensile or compressive stress developed in a load-carrying member. Under the influence of a shear stress, shearing strain would be produced.

Figure 2–9 shows a stress element subjected to shear. The shearing action on parallel faces of the element tends to deform it angularly, as shown to an exaggerated degree. The angle γ (gamma), measured in radians, is the *shearing strain*. Only very small values of shearing strain are encountered in practical problems, and thus the dimensions of the element are only slightly changed.

Modulus of Elasticity in Shear.

> ***The ratio of the shearing stress to the shearing strain is called the* modulus of elasticity in shear *or the* modulus of rigidity, *and is denoted by* G.**

That is,

$$G = \frac{\text{shearing stress}}{\text{shearing strain}} = \frac{\tau}{\gamma} \tag{2–5}$$

G is a property of the material and is related to the tensile modulus and Poisson's ratio by

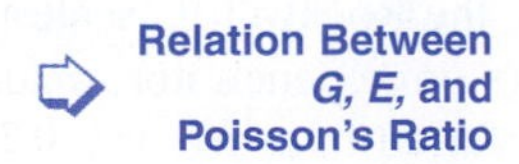

$$G = \frac{E}{2(1 + \nu)} \tag{2–6}$$

The calculation of torsional deflection of shafts in Chapter 4 requires the use of G.

Flexural Strength and Flexural Modulus. Other stiffness and strength measures often reported, particularly for plastics, are called the *flexural strength and flexural modulus.* As the name implies, a specimen of the material is loaded as a beam in flexure (bending) with data taken and plotted for load versus deflection. From these data and from knowledge

FIGURE 2–10 Hardness tester and conversions. (a) Rockwell hardness tester. (b) Hardness conversions and relations to tensile strength for steels. (Source: Instron Corporation, Norwood, MA)

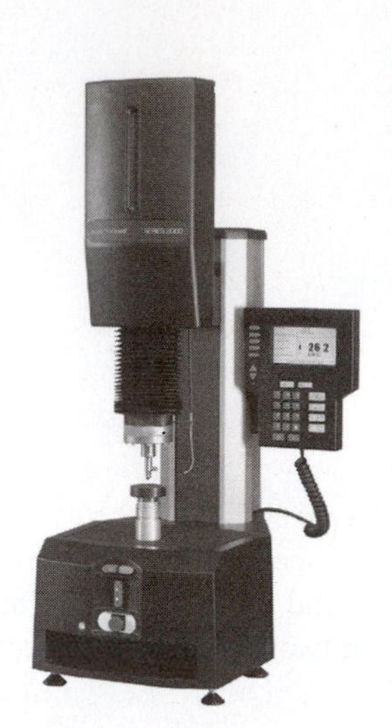

(*a*) Rockwell hardness tester.

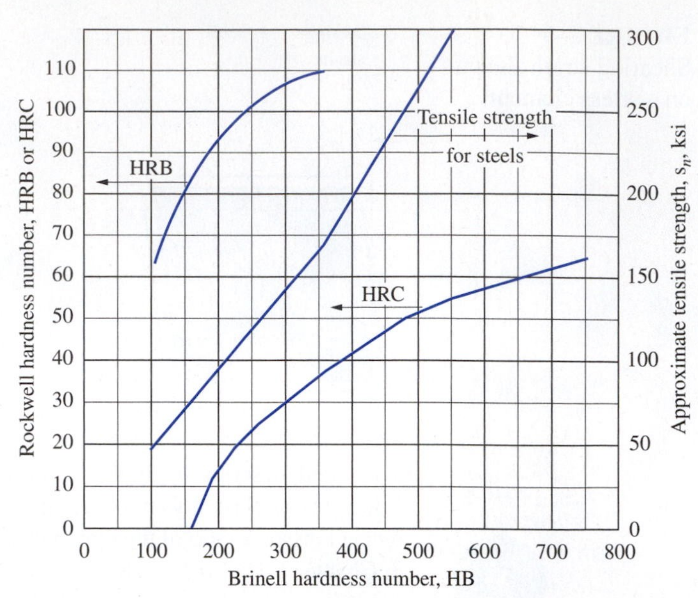

(*b*) Hardness conversions and relations to tensile strength for steels.

of the geometry of the specimen, stress and strain can be computed. The ratio of stress to strain is a measure of the flexural modulus. ASTM standard D 790[1] defines the complete method. Note that the values are significantly different from the tensile modulus because the stress pattern in the specimen is a combination of tension and compression. The data are useful for comparing the strength and stiffness of different materials when a load-carrying part is subjected to bending in service. ISO standard 178 describes a similar method for determining flexural properties.

Hardness. The resistance of a material to indentation by a penetrator is an indication of its *hardness*. Several types of devices, procedures, and penetrators measure hardness; the Brinell hardness tester and the Rockwell hardness tester are most frequently used for machine elements. For steels, the Brinell hardness tester employs a hardened steel ball 10 mm in diameter as the penetrator under a load of 3000-kg force. The load causes a permanent indentation in the test material, and the diameter of the indentation is related to the Brinell hardness number, which is abbreviated HB. The actual quantity being measured is the load divided by the contact area of the indentation. For steels, the value of HB ranges from approximately 100 for an annealed, low-carbon steel to more than 700 for high-strength, high-alloy steels in the as-quenched condition. In the high ranges, above HB 500, the penetrator is sometimes made of tungsten carbide rather than steel. For softer metals, a 500-kg load is used.

The Rockwell hardness tester uses a hardened steel ball with a $\frac{1}{16}$-in diameter under a load of 100-kg force for softer metals, and the resulting hardness is listed as Rockwell B, R_B, or HRB. For harder metals, such as heat-treated alloy steels, the Rockwell C scale is used. A load of 150-kg force is placed on a diamond penetrator (a *brale* penetrator) made in a sphero-conical shape. Rockwell C hardness is sometimes referred to as R_C or HRC. Many other Rockwell scales are used.

The Brinell and Rockwell methods are based on different parameters and lead to quite different numbers. However, since they both measure hardness, there is a correlation between them, as noted in Figure 2–10. It is also important to note that, especially

[1]ASTM International. *Standard Test Method for Flexural Properties of Unreinforced and Reinforced Plastics and Electrical Insulating Materials, Standard D790*. West Conshohocken. PA: ASTM International, 2003.

TABLE 2–2 Comparison of hardness scales with tensile strength for steels.

Material and condition	Hardness			Tensile strength	
	HB	HRB	HRC	ksi	MPa
1020 annealed	121	70		60	414
1040 hot-rolled	144	79		72	496
4140 annealed	197	93	13	95	655
4140 OQT 1000	341	109	37	168	1160
4140 OQT 700	461		49	231	1590

for highly hardenable alloy steels, there is a nearly linear relationship between the Brinell hardness number and the tensile strength of the steel, according to the equation

Approximate Relationship between Harndess and Strength for Steel

$$0.50(\text{HB}) = \text{approximate tensile strength (ksi)} \quad \textbf{(2–7)}$$

This relationship is shown in Figure 2–10. Hardness in a steel indicates wear resistance as well as strength.

To compare the hardness scales with the tensile strength, consider Table 2–2. Note that there is some overlap between the HRB and HRC scales. Normally, HRB is used for the softer metals and ranges from approximately 60 to 100, whereas HRC is used for harder metals and ranges from 20 to 65. Using HRB numbers above 100 or HRC numbers below 20 is not recommended. Those shown in Table 2–2 are for comparison purposes only.

Whereas Rockwell B and C scales are used for substantial components made from steels, other Rockwell hardness scales can be used for different materials as listed in Table 2–3. See Internet site 27 for a downloadable report on Rockwell hardness measurement of metallic materials.

For plastics, rubbers, and elastomers, it is typical to use the Shore or IRHD (International rubber hardness degree) methods. Table 2–4 lists some of the more popular

TABLE 2–3 Rockwell hardness scales and their uses.

Scale	Symbol	Typical uses
A	HRA	Thin steel, hardened steel with shallow case
B	HRB	Softer steels such as low carbon, annealed or hot rolled; softer aluminum, copper, cast irons
C	HRC	Harder steels such as heat-treated alloy steels, tool steels, titanium
D	HRD	Medium-depth case-hardened steels, harder cast irons
E	HRE	Harder bearing metals, aluminum, magnesium, cast irons
F	HRF	Soft thin sheet metals, annealed copper and zinc alloys
G	HRG	Beryllium copper, phosphor bronze, softer cast irons
H	HRH	Aluminum, zinc, lead
K	HRK	Softer bearing metals, plastics, rubbers, and other soft materials. Scales L, M, P, R, S, V, and α also available
15 N	HR15N	Similar to A scale but for thinner materials or thin case hardening
30 N	HR30N	Similar to C scale but for thinner materials or thin case hardening
45 N	HR45N	Similar to D scale but for thinner materials or thin case hardening
15 T	HR15T	Similar to B scale but for thinner materials
30 T	HR30T	Similar to F scale but for thinner materials
45 T	HR45T	Similar to G scale but for thinner materials

Note: Measurement devices use different indenter sizes and shapes and applied forces. For details, consult standards ASTM E18, ISO 6508.

TABLE 2–4 Hardness measurement for plastics, rubbers, and elastomers.

Shore method		
Scale of durometer	Symbol	Typical uses
A	Shore A	Soft vulcanized natural rubbers, elastomers (e.g., neoprene), leather, wax, felt
B	Shore B	Moderately hard rubber as used for printer rolls and platens
C	Shore C	Medium hard rubbers and plastics
D	Shore D	Hard rubbers and plastics such as vinyl sheets, plexiglas, laminate countertops
DO	Shore DO	Very dense textile windings
O	Shore O	Soft rubber such as artgum; textile windings
OO	Shore OO	Low density textile windings; sponge rubber
OOO	Shore OOO	Soft plastic foams
T	Shore T	Medium density textiles on spools
M	Shore M	Rubber O-rings and thin sheet rubber

IRHD method (International rubber hardness degree)	
Name	Typical uses
IRHD Micro	Small: O-rings, small components, thin materials
IRHD Macro L	Soft: Readings up to 35 IRHD L
IRHD Macro N	Normal: Readings from 30 IRHD N to 100 IRHD N

Note: Measurement devices use different indenter sizes and shapes and applied forces. For details, consult standards ASTM D2240, ISO 868, or DIN 53505.

scales. Also used for some plastics are the Rockwell R, L, M, E, K, and α scales. These vary by the size and geometry of the indenter and the applied force.

Other hardness measurement methods include Vickers, Knoop, Universal, and rebound hardness. See Internet sites 24 and 26 for discussions of these methods and their uses.

Toughness, Impact Energy. *Toughness* is the ability of a material to absorb applied energy without failure. Parts subjected to suddenly applied loads, shock, or impact need a high level of toughness. Several methods are used to measure the amount of energy required to break a particular specimen made from a material of interest. The energy absorption value from such tests is often called *impact energy* or *impact resistance.* However, it is important to note that the actual value is highly dependent on the nature of the test sample, particularly its geometry. It is not possible to use the test results in a quantitative way when making design calculations. Rather, the impact energy for several candidate materials for a particular application can be compared with each other as a qualitative indication of their toughness. The final design should be tested under real service conditions to verify its ability to survive safely during expected use.

For metals and plastics, two methods of determining impact energy, *Izod* and *Charpy,* are popular, with data often reported in the literature from vendors of the material. Figure 2–11(a) and (b) show sketches of the dimensions of standard specimens and the manner of loading. In each method, a pendulum with a heavy mass carrying a specially designed striker is allowed to fall from a known height. The striker contacts the specimen with a high velocity at the bottom of the pendulum's arc; therefore, the pendulum possesses a known amount of kinetic energy. Part (c) shows a tester. The specimen is typically broken during the test, taking some of the energy from the pendulum but allowing it to pass through the test area. The testing machine is configured to measure the final height to which the pendulum swings and to indicate the amount of energy removed. That value

FIGURE 2–11 Impact testing devices and equipment. (Source: Instron Corporation, Norwood, MA)

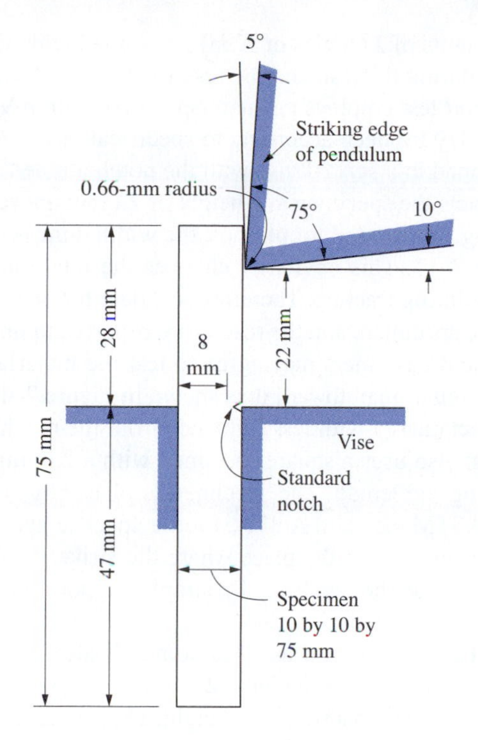

(*a*) Izod test specimen and striker (side view)

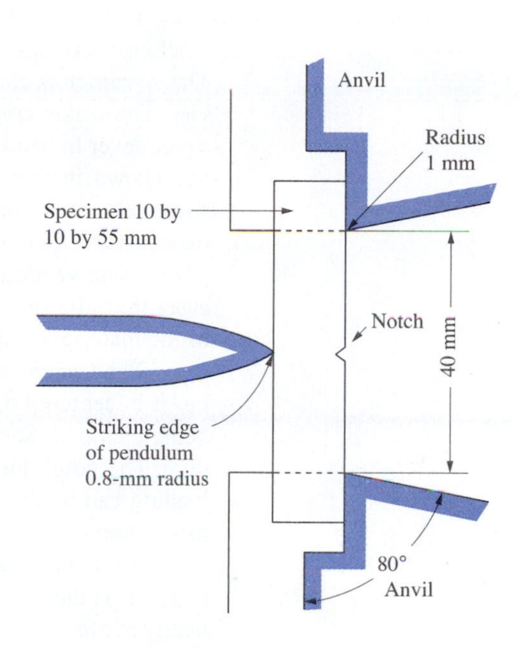

(*b*) Charpy test specimen and striker (top view)

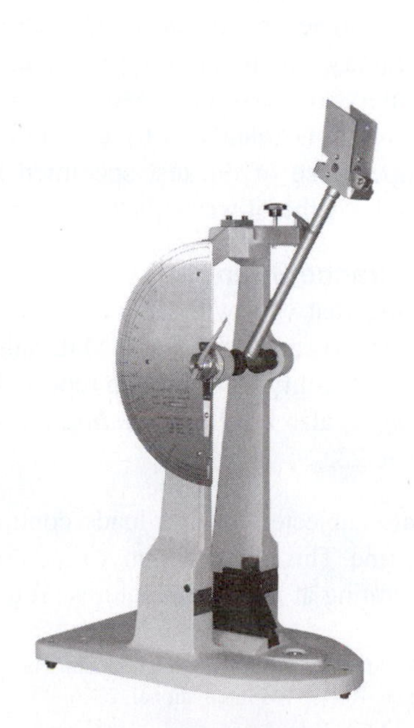

(*c*) Pendulum-type impact tester

(*d*) Drop-type impact tester

is reported in energy units of J (Joules or N·m) or Some highly ductile metals and many plastics do not break during the test, and the result is then reported as *No Break.*

The standard *Izod* test employs a square specimen with a V-shaped notch carefully machined 2.0 mm (0.079 in) deep according to specifications in ASTM standard D 256.[2] The specimen is clamped in a special vise with the notch aligned with the top edge of the vise. The striker contacts the specimen at a height of 22 mm above the notch, loading it as a cantilever in bending. When used for plastics, the width dimension can be different from that shown in Figure 2–11. This obviously changes the total amount of energy that the specimen will absorb during fracture. Therefore, the data for impact energy are divided by the actual width of the specimen, and the results are reported in units of N·m/m or ft·lb/in. Also, some vendors and customers may agree to test the material with the notch facing away from the striker rather than toward it as shown in Figure 2–11. This gives a measure of the material's impact energy with less influence from the notch.

The *Charpy* test also uses a square specimen with a 2.0 mm (0.079 in) deep notch, but it is centered along the length. The specimen is placed against a rigid anvil without being clamped. See ASTM standard A 370[3] for the specific geometry and testing procedure. The notch faces away from the place where the striker contacts the specimen. The loading can be described as the bending of a simply supported beam. The Charpy test is most often used for testing metals.

Another impact testing method used for some plastics, composites, and completed products is the *drop-weight* tester. See Figure 2–11(d). Here a known mass is elevated vertically above the test specimen to a specified height. Thus, it has a known amount of potential energy. Allowing the mass to fall freely imparts a predictable amount of kinetic energy to the specimen clamped to a rigid base. The initial energy, the manner of support, the specimen geometry, and the shape of the striker (called a *tup*) are critical to the results found. One standard method, described in ASTM D 3763[4], employs a spherical tup with a diameter of 12.7 mm (0.50 in). The tup usually pierces the specimen. The apparatus is typically equipped with sensors that measure and plot the load versus deflection characteristics dynamically, giving the designer much information about how the material behaves during an impact event. Summary data reported typically include maximum load, deflection of the specimen at the point of maximum load, and the energy dissipated up to the maximum load point. The energy is calculated by determining the area under the load-deflection diagram. The appearance of the test specimen is also described, indicating whether fracture occurred and whether it was a ductile or brittle fracture.

Fatigue Strength or Endurance Strength. Parts subjected to repeated applications of loads or to stress conditions that vary with time over several thousands or millions of cycles fail because of the phenomenon of *fatigue.* Materials are tested under controlled cyclic loading to determine their ability to resist such repeated loads. The resulting data are reported as the *fatigue strength,* also called the *endurance strength* of the material. See Reference 25 and Internet site 23.

Creep. When materials are subjected to high loads continuously, they may experience progressive elongation over time. This phenomenon, called *creep,* should be considered for most plastics and metals operating at high temperatures. You should check for creep when

[2]ASTM International. *Standard Test Methods for Determining the Izod Pendulum Impact Resistance of Plastics, Standard D256.* West Conshohocken. PA: ASTM International. 2006.

[3]ASTM International. *Standard Test Methods and Definitions for Mechanical Testing of Steel Products, Standard A370.* West Conshohocken. PA: ASTM International, 2005.

[4]ASTM International. *Standard Test Methods for High Speed Puncture of Plastics Using Load and Displacement Sensors, Standard D3763.* West Conshohocken, PA: ASTM International, 2002.

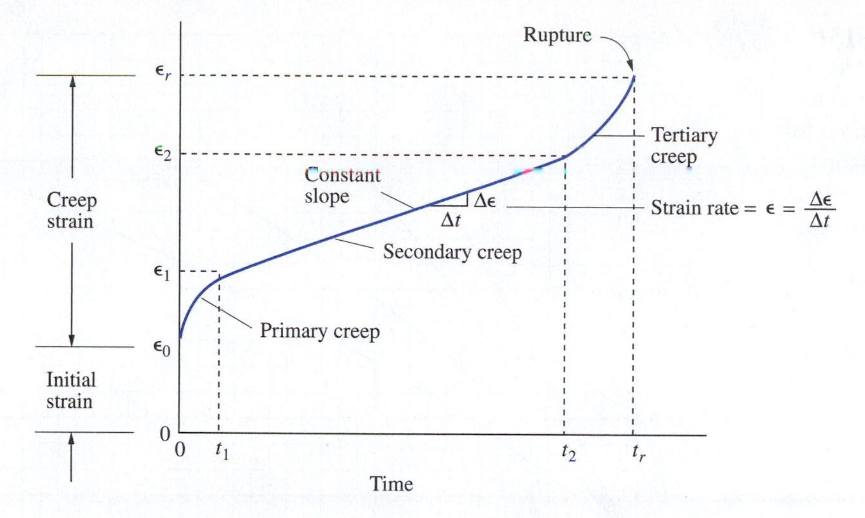

FIGURE 2–12
Typical creep behavior.

the operating temperature of a loaded metal member exceeds approximately 0.3 (T_m), where T_m is the melting temperature expressed as an absolute temperature. Creep can be important for critical members in internal combustion engines, furnaces, steam turbines, gas turbines, nuclear reactors, or rocket engines. The stress can be tension, compression, flexure, or shear.

Figure 2–12 shows the typical behavior of metals that creep. The vertical axis is the creep strain, in units such as in/in or mm/mm, over that which occurs initially as the load is applied. The horizontal axis is time, typically measured in hours because creep develops slowly over a long term. During the primary portion of the creep strain versus time curve, the rate of increase in strain initially rises with a rather steep slope that then decreases. The slope is constant (straight line) during the secondary portion of the curve. Then the slope increases in the tertiary portion that precedes the ultimate fracture of the material.

Creep is measured by subjecting a specimen to a known steady load, possibly through application of a dead weight, while the specimen is heated and maintained at a uniform temperature. Data for strain versus time are taken at least into the secondary creep stage and possibly all the way to fracture to determine the creep rupture strain. Testing over a range of temperatures gives a family of curves that are useful for design.

Creep can occur for many plastics even at or near room temperature. Figure 2–13 shows one way that creep data are displayed for plastic materials. It is a graph of applied stress versus strain in the member with data shown for a specific temperature of the specimen. The curves show the amount of strain that would be developed within the specified times at increasing stress levels. For example, if this material were subjected to a constant stress of 5.0 MPa for 5000 hours, the total strain would be 1.0%. That is, the specimen would elongate by an amount 0.01 times the original length. If the stress were 10.0 MPa for 5000 hours, the total strain would be approximately 2.25%. The designer must take this creep strain into account to ensure that the product performs satisfactorily over time.

Density. *Density* is defined as the mass per unit volume of a material. Its usual units are kg/m^3 in the SI and lb/in^3 in the U.S. Customary unit system, where the pound unit is taken to be pounds-mass. The Greek letter rho (ρ) is the symbol for density.

In some applications, the term *specific weight* or *weight density* is used to indicate the weight per unit volume of a material. Typical units are N/m^3 in the SI and lb/in^3 in the U.S. Customary unit system, where the pound is taken to be pounds-force. The Greek letter gamma (γ) is the symbol for specific weight.

Coefficient of Thermal Expansion. The *coefficient of thermal expansion* is a measure of the change in length of a material subjected to a change in temperature. It is defined

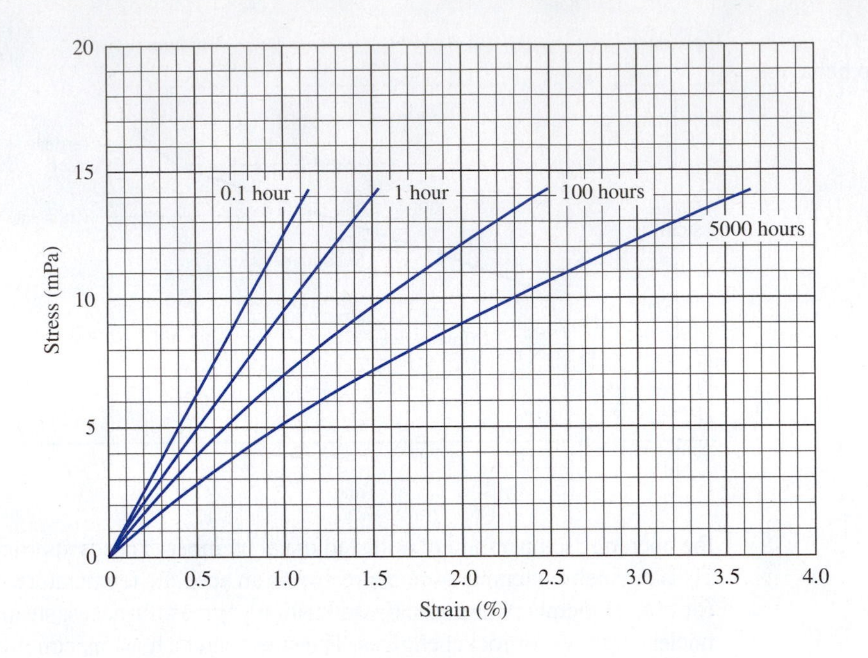

FIGURE 2–13 Example of stress versus strain as a function of time for nylon 66 plastic at 23°C (73°F). (Source: DuPont Polymers, Wilmington, DE)

by the relation

$$\alpha = \frac{\text{change in length}}{L_o(\Delta T)} = \frac{\text{strain}}{(\Delta T)} = \frac{\epsilon}{(\Delta T)} \tag{2–8}$$

where L_o = original length
ΔT = change in temperature

Virtually all metals and plastics expand with increasing temperature, but different materials expand at different rates. For machines and structures containing parts of more than one material, the different rates can have a significant effect on the performance of the assembly and on the stresses produced.

Classification of Metals and Alloys. Various industry associations take responsibility for setting standards for the classification of metals and alloys. Each has its own numbering system, convenient to the particular metal covered by the standard. But this leads to confusion at times when there is overlap between two or more standards and when widely different schemes are used to denote the metals. Order has been brought to the classification of metals by the use of the Unified Numbering System (UNS) as defined in the Standard E 527-83 (2003), **Standard Practice for Numbering Metals and Alloys (UNS),** by the American Society for Testing and Materials (ASTM). (See Reference 3.) Besides listing materials under the control of ASTM itself, the UNS coordinates designations of:

The Aluminum Association (AA)
American Iron and Steel Institute (AISI)
Copper Development Association (CDA)
Society of Automotive Engineers (SAE)

The primary series of numbers within UNS are listed in Table 2–5 along with the organization having responsibility for assigning numbers within each series.

Many alloys within UNS retain the familiar numbers from the systems used for many years by the individual association. For example, the following section describes the

TABLE 2–5 Unified numbering system (UNS).

Number series	Types of metals and alloys	Responsible organization
Nonferrous metals and alloys		
A00001–A99999	Aluminum and aluminum alloys	AA
C00001–C99999	Copper and copper alloys	CDA
E00001–E99999	Rare earth metals and alloys	ASTM
L00001–L99999	Low melting metals and alloys	ASTM
M00001–M99999	Miscellaneous nonferrous metals and alloys	ASTM
N00001–N99999	Nickel and nickel alloys	SAE
P00001–P99999	Precious metals and alloys	ASTM
R00001–R99999	Reactive and refractory metals and alloys	SAE
Z00001–Z99999	Zinc and zinc alloys	ASTM
Ferrous metals and alloys		
D00001–D99999	Steels, mechanical properties specified	SAE
F00001–F99999	Cast irons and cast steels	ASTM
G00001–G99999	Carbon and alloy steels (includes former SAE carbon and alloy steels)	AISI
H00001–H99999	H-steels; specified hardenability	AISI
J00001–J99999	Cast steels (except tool steels)	ASTM
K00001–K99999	Miscellaneous steels and ferrous alloys	ASTM
S00001–S99999	Heat and corrosion resistant (stainless) steels	ASTM
T00001–T99999	Tool steels	AISI

four-digit designation system of the AISI for carbon and alloy steels. Figure 2–14 shows two examples: AISI 1020, a plain carbon steel, and AISI 4140, an alloy steel. These steels would carry the UNS designations G10200 and G41400, respectively.

2–3 STEEL

The term *steel* refers to alloys of iron and carbon and, in many cases, other elements. Because of the large number of steels available, they will be classified in this section as carbon steels, alloy steels, stainless steels, and structural steels. See Reference 5.

For *carbon steels* and *alloy steels,* a four-digit designation code is used to define each alloy. Figure 2–14 shows the significance of each digit. The four digits would be the

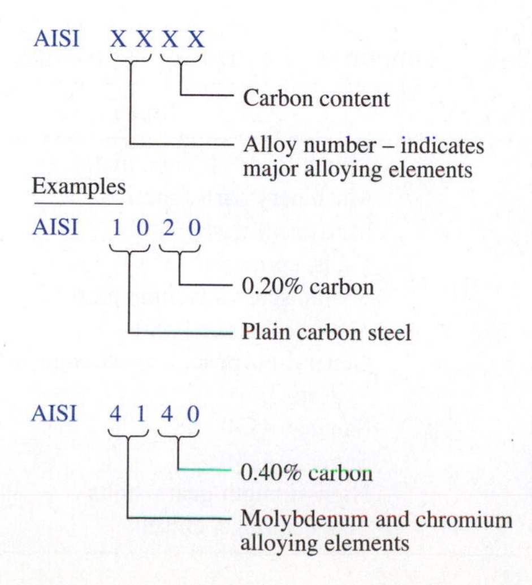

FIGURE 2–14 Steel designation system.

TABLE 2–6 Major alloying elements in steel alloys.

Steel AISI No.	Alloying elements	Steel AISI No.	Alloying elements
10xx	Plain carbon	46xx	Molybdenum-nickel
11xx	Sulfur (free-cutting)	47xx	Molybdenum-nickel-chromium
13xx	Manganese	48xx	Molybdenum-nickel
14xx	Boron	5xxx	Chromium
2xxx	Nickel	6xxx	Chromium-vanadium
3xxx	Nickel-chromium	8xxx	Nickel-chromium-molybdenum
4xxx	Molybdenum	9xxx	Nickel-chromium-molybdenum (except 92xx)
41xx	Molybdenum-chromium	92xx	Silicon-manganese
43xx	Molybdenum-chromium-nickel		

same for steels classified by the American Iron and Steel Institute (AISI) and the Society of Automotive Engineers (SAE). Classification by the American Society for Testing and Materials (ASTM) will be discussed later.

Usually, the first two digits in a four-digit designation for steel will denote the major alloying elements, other than carbon, in the steel. The last two digits denote the average percent (or points) of carbon in the steel. For example, if the last two digits are 40, the steel would have about 0.4% carbon content. Carbon is given such a prominent place in the alloy designation because, in general, as carbon content increases, the strength and hardness of the steel also increases. Carbon content usually ranges from a low of 0.1% to about 1.0%. It should be noted that while strength increases with increasing carbon content, the steel also becomes less ductile.

Table 2–6 shows the major alloying elements, which correspond to the first two digits of the steel designation. Table 2–7 lists some common alloys along with the principal uses for each.

Conditions for Steels. The mechanical properties of carbon and alloy steels are very sensitive to the manner in which they are formed and to heat-treating processes. Appendix A–14 lists the ultimate strength, yield strength, and percent elongation for a variety of steels in a variety of conditions. Note that these are typical or example properties and may not be relied on for design. Material properties are dependent on many factors, including section size, temperature, actual composition, variables in processing, and fabrication techniques. It is the responsibility of the designer to investigate the possible range of properties for a material and to design load-carrying members to be safe regardless of the combination of factors present in a given situation.

TABLE 2–7 Common steel alloys and typical uses.

Steel AISI No.	Typical uses
1020	Structural steel, bars, plate
1040	Machinery parts, shafts
1050	Machinery parts
1095	Tools, springs
1137	Shafts, screw machine parts (free-cutting alloy)
1141	Shafts, machined parts
4130	General-purpose, high-strength steel; shafts, gears, pins
4140	Same as 4130
4150	Same as 4130
5160	High-strength gears, bolts
8760	Tools, springs, chisels

Generally, the more severely a steel is worked, the stronger it will be. Some forms of steel, such as sheet, bar, and structural shapes, are produced by *hot rolling* while still at an elevated temperature. This produces a relatively soft, low-strength steel, which has a very high ductility and is easy to form. Rolling the steel to final form while at or near room temperature is called *cold rolling* and produces a higher strength and somewhat lower ductility. Still higher strength can be achieved by *cold drawing,* drawing the material through dies while it is at or near room temperature. Thus, for these three popular methods of producing steel shapes, the cold-drawn (CD) form results in the highest strength, followed by the cold-rolled (CR) and hot-rolled (HR) forms. This can be seen in Appendix A–14 by comparing the strength of the same steel, say AISI 1040, in the hot-rolled and cold-drawn conditions.

Alloy steels are usually heat-treated to develop specified properties. (See Reference 7.) Heat treatment involves raising the temperature of the steel to above about 1450 to 1650°F (depending on the alloy) and then cooling it rapidly by quenching in either water or oil. After quenching, the steel has a high strength and hardness, but it may also be brittle. For this reason a subsequent treatment called *tempering* (or *drawing*) is usually performed. The steel is reheated to a temperature in the range 400 to 1300°F and then cooled. The effect of tempering an alloy steel can be seen by referring to Figure 2–15. Thus the properties of a heat-treated steel can be controlled by specifying a tempering temperature. In Appendix A–14, the condition of heat-treated alloys is described in a manner like OQT 400. This means that the steel was heat-treated by quenching in oil and then tempered at 400°F. Similarly, WQT 1300 means water quenched and tempered at 1300°F.

The properties of heat-treated steels at tempering temperatures of 400 and 1300°F show the total range of properties that the heat-treated steel can have. However, typical practice

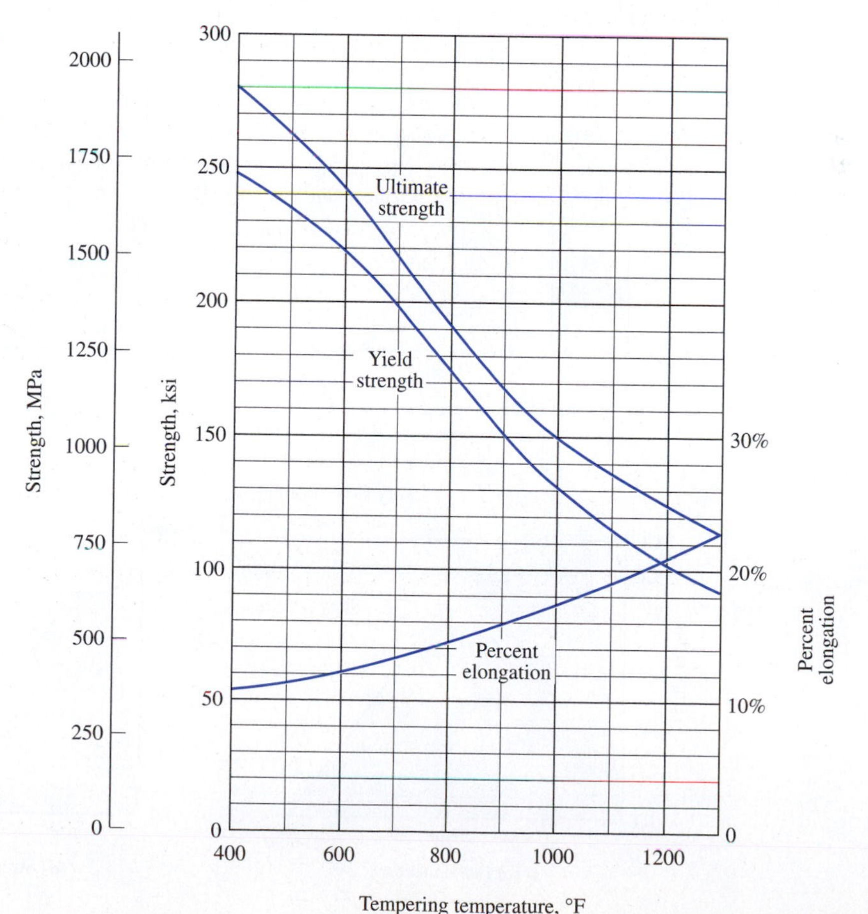

FIGURE 2–15 Effect of tempering temperature on the strength and ductility of an alloy steel.

would specify tempering temperatures no lower than 700°F because steels tend to be too brittle at the lower tempering temperatures. The properties of several alloys are listed in Appendix A–14 at tempering temperatures of 700°F, 900°F, 1100°F, and 1300°F to give you a feeling for the range of strengths available. These alloys are good choices for selecting materials in problems in later chapters. Strengths at intermediate temperatures can be found by interpolation.

Annealing and *normalizing* are thermal treatments designed to soften steel, give it more uniform properties, make it easier to form, or relieve stresses developed in the steel during such processes as welding, machining, or forming. Two of the types of annealing processes used are full anneal and stress-relief anneal. Figure 2–16 illustrates these thermal treatment processes, along with quenching and tempering.

Normalizing of steel starts by heating it to approximately the same temperature (called the *upper critical temperature*) as would be required for through-hardening by quenching, as described before. But rather than quenching, the steel is cooled in still air to room temperature. This results in a uniform, fine-grained structure, improved ductility, better impact resistance, and improved machinability.

Full annealing involves heating to above the upper critical temperature followed by very slow cooling to the lower critical temperature and then cooling in still air to room temperature. This is one of the softest forms of the steel and makes it more workable for shearing, forming, and machining.

FIGURE 2–16 Heat treatments for steel.

Note:
RT = room temperature
LC = lower critical temperature
UC = upper critical temperature

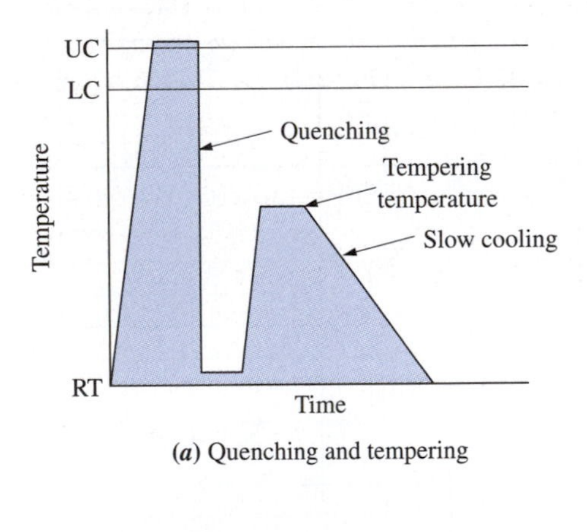

(*a*) Quenching and tempering

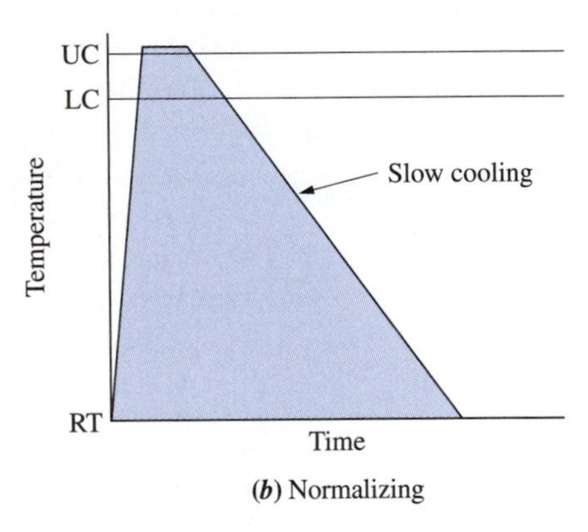

(*b*) Normalizing

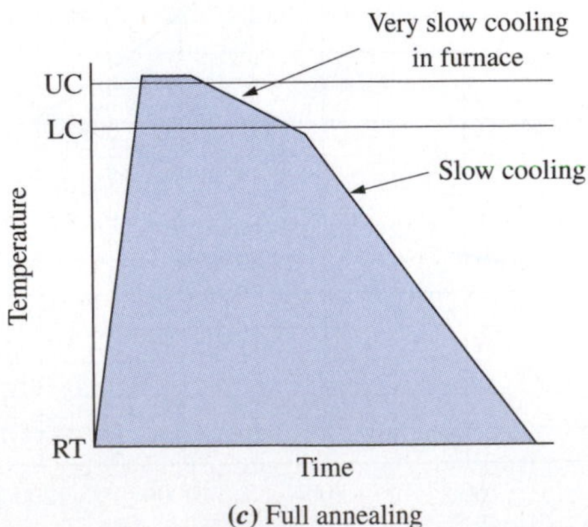

(*c*) Full annealing

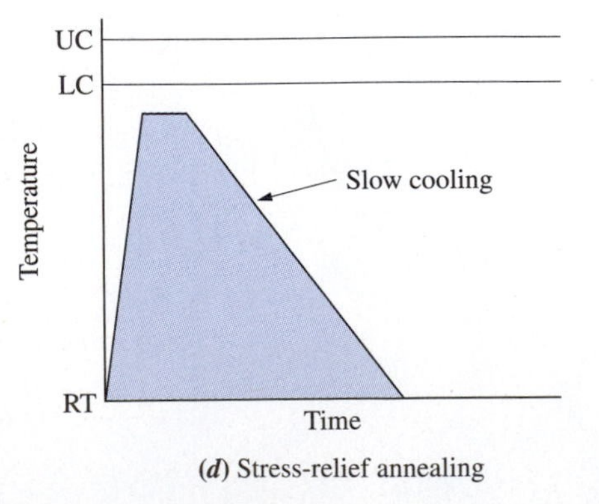

(*d*) Stress-relief annealing

Stress-relief annealing consists of heating to below the lower critical temperature, holding to achieve uniform temperature throughout the part, and then cooling to room temperature. This relieves residual stresses and prevents subsequent distortion.

Stainless Steels. *Stainless steels* get their name because of their corrosion resistance. The primary alloying element in stainless steels is chromium, being present at about 17% in most alloys. A minimum of 10.5% chromium is used, and it may range as high as 27%.

Although over 40 grades of stainless steel are available from steel producers, they are usually categorized into three series containing alloys with similar properties. The properties of some stainless steels are listed in Appendix A–15.

The 200 and 300 series steels have high strength and good corrosion resistance. They can be used at temperatures up to about 1200°F with good retention of properties. Because of their structure, these steels are essentially nonmagnetic. Good ductility and toughness and good weldability make them useful in chemical processing equipment, architectural products, and food-related products. They are not hardenable by heat treatment, but they can be strengthened by cold working. The range of cold working is typically given as *quarter hard, half hard, three-quarters hard,* and *full hard,* with strength increasing with higher hardness. But ductility decreases with increasing hardness. Appendix A–15 shows the properties of some stainless steel alloys at two conditions, annealed and full hard, indicating the extremes of strengths available. The annealed condition is sometimes called *soft*.

The AISI 400 series steels are used for automotive trim and for chemical processing equipment such as acid tanks. Certain alloys can be heat-treated so they can be used as knife blades, springs, ball bearings, and surgical instruments. These steels are magnetic.

Precipitation hardening steels, such as 17-4PH and PH13-8Mo, are hardened by holding at an elevated temperature, about 900 to 1100°F (480 to 600°C). Such steels are generally classed as *high-strength stainless steels* having yield strengths about 180 000 psi (1240 MPa) or higher and are used in aerospace, engines, and other application where the high strength and corrosion resistance are required.

Structural Steels. *Structural steels* are produced in the forms of sheet, plate, bars, tubing, and structural shapes such as I-beams, wide-flange beams, channels, and angles. The American Society for Testing and Materials (ASTM) assigns a number designation to these steels, which is the number of the standard that defines the required minimum properties. Appendix A–16 lists frequently used grades of structural steels and their properties. See also Reference 3 and Internet site 6.

A very popular steel for structural applications is ASTM A36, a carbon steel used for many commercially available shapes, plates, and bars. It has a minimum yield strength of 36 ksi (248 MPa), is weldable, and is used in bridges, buildings, and for general structural purposes.

The W-shapes used widely in building construction and other industrial structures are now commonly made from ASTM A992, one of several grades of high-strength, low-alloy (HSLA) steels. With a minimum yield strength of 50 ksi (345 MPa), it allows the use of lighter beams, as compared with the formerly used ASTM A36 steel, in many applications with consequent significant cost savings. It should be noted that virtually all steels have the same modulus of elasticity, which is an indication of the stiffness of the material. Therefore, it is critical to assess the deflection of a beam in addition to its strength.

Another HSLA grade of structural steel that is growing in use is ASTM A913, Grade 65, having a minimum yield strength of 65 ksi (448 MPa). Its use in heavy column sections and some critical beam or truss applications has produced weight and cost savings for major structures such as the professional football stadium in Houston, TX and a high-rise office tower in New York City. This steel is also available in Grades 50, 60, and 70 with corresponding minimum yield strengths in ksi.

ASTM A242, in Grades 42, 46, and 50, is another HSLA steel that is produced as shapes, plates, and bars for general structural uses. An additional advantage for this alloy is its corrosion resistance, about four times that of plain carbon steel, leading many to refer to it as a *weathering steel.* All three grades are available in W-shapes; Grade 50 is most commonly available for other rolled shapes. As indicated in Appendix A–16, the available grades for plates and bars depend on thickness.

ASTM A514 is a high-strength alloy steel, heat treated by quenching and tempering, and produced as plates and bars. Thicknesses up to 2.5 in have a yield strength of 100 ksi (690 MPa). Larger thicknesses are rated at a minimum yield strength of 90 ksi (620 MPa).

Another general-purpose HSLA structural steel is ASTM A572, available as all types of shapes, plates, and bars. Grades 42, 50, 55, 60, and 65 are all used for shapes. All plates and bars up to 8 in thick are available in Grade 42; up to 4 in for Grade 50; up to 2 in for Grade 55; and up to 1.25 in for Grades 60 and 65.

Hollow structural sections (HSS), sometimes called *structural tubing,* are either round, square, or rectangular and are typically made from ASTM A501 (hot formed) or ASTM A500 (cold formed) in a variety of strength grades. When produced as *pipe,* ASTM A53 Grade B with a yield strength of 35 ksi (240 MPa) is specified. See Appendices A–9 and A–12 for a sampling of sizes of HSS shapes.

In summary, structural steel products come in many forms and a wide range of properties. Careful selection of a suitable type of steel considers the required strength, cost, and availability. Table 2–8 lists the preferred material specifications along with others for

TABLE 2–8 Structural steel grades available for typical shapes, plates, and bars.

		Yield strength		Available steels for listed applications					
ASTM designation	Grade	ksi	MPa	W shapes	S, C, L shapes	Pipe	Rectangular or square HSS	Round HSS	Plates and bars
A36		36	248	A	P	—	—	—	P: up to 8 in
A53		35	241	—	—	P	—	—	—
A242		42	290	A	—	—	—	—	A: 1.5 in to 4 in
		46	317	A	—	—	A	—	A: 0.75 to 1.5 in
		50	345	A	A	—	A	—	A: up to 0.75 in
A500	B	42	290	—	—	—	—	P	—
		46	317	—	—	—	P	—	—
	C	46	317	—	—	—	—	A	—
		50	345	—	—	—	A	—	—
A501		36	248	—	—	—	A	A	—
A514		90	621	—	—	—	—	—	A: 2.5 in to 6 in
		100	690	—	—	—	—	—	A: up to 2.5 in
A572	42	42	290	A	A	—	—	—	A: up to 6 in
	50	50	345	A	A	—	—	—	A: up to 4 in
	55	55	379	A	A	—	—	—	A: up to 2 in
	60	60	414	A	A	—	—	—	A: up to 1.25 in
	65	65	448	A	A	—	—	—	A: up to 1.25 in
A913	65	65	448	A	A	—	—	—	—
A992		50	345	P	—	—	—	—	—

Notes: Adapted from Reference 1, Chapter 1. Other materials, grades, and shapes available.
P = preferred material specification
A = Available—Check with supplier
— = Not available

which the availability should be confirmed prior to specification. See Reference 2 in Chapter 1 for a more complete list of preferred and available structural steels for a wider variety of types and sizes of members.

2–4 CAST IRON

The attractive properties of cast iron include low cost, good wear resistance, good machinability, and the ability to be cast into complex shapes. Five varieties will be discussed here: gray iron, ductile iron, austempered ductile iron, white iron, and malleable iron.

Gray Iron. *Gray iron* is used in automotive engine blocks, machinery bases, brake drums, and large gears. It is usually specified by giving a grade number corresponding to the minimum ultimate tensile strength. For example, grade 20 gray cast iron has a minimum ultimate strength of 20 000 psi (138 MPa); grade 60 has $s_u = 60\,000$ psi (414 MPa), and so on. The usual grades available are from 20 to 60. Gray iron is somewhat brittle, so that yield strength is not usually reported as a property. An outstanding feature of gray iron is that its compressive strength is very high, about three to five times as high as the tensile strength. This should be taken into account in design, especially when a part is subjected to bending stresses, as discussed in Chapter 7.

Because of variations in the rate of cooling after the molten cast iron is poured into a mold, the actual strength of a particular section of a casting is dependent on the thickness of the section. Figure 2–17 illustrates this for grade 40 gray iron. The range of in-place strength may range from as high as 52 000 psi (359 MPa) to as low as 27 000 psi (186 MPa).

Ductile Iron. *Ductile iron* differs from gray iron in that it does exhibit yielding, has a greater percent elongation, and has generally higher tensile strength. Grades of ductile iron are designated by a three-number system such as 80-55-6. The first number indicates the minimum ultimate tensile strength in ksi; the second is the yield strength in ksi; and the third is the percent elongation. Thus grade 80-55-6 has an ultimate strength of 80 000 psi, a yield strength of 55 000 psi, and a percent elongation of 6%. Uses for ductile iron include crankshafts and heavily loaded gears.

Austempered Ductile Iron. *Austempered ductile iron (ADI)* has higher strength and better ductility than standard ductile irons, as can be seen in Appendix A–17. This allows parts to be smaller and lighter and makes ADI desirable for uses such as automotive gears, crankshafts, and structural members for construction and transportation equipment, replacing wrought or cast steels.

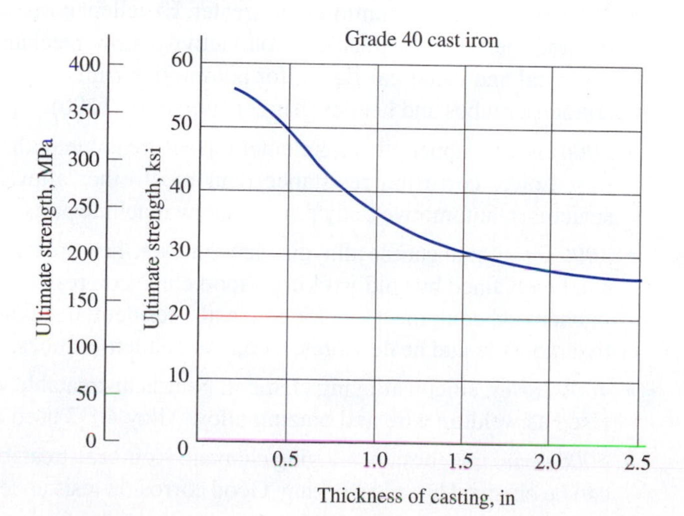

FIGURE 2–17 Strength versus thickness for grade 40 gray cast iron.

Austempering can increase the strength of ductile iron by nearly a factor of 2. First the castings are heated to between 1500°F and 1700°F (816°C to 927°C) and held to achieve a uniform structure. Then they are quenched rapidly to a lower temperature, 450°F to 750°F (232°C to 400°C), and held again. After several hours of isothermal soaking, the castings are allowed to cool to room temperature.

White Iron. *White iron* is produced by rapidly chilling a casting of either gray iron or ductile iron during the solidification process. The chilling is typically applied to selected areas, which become very hard and have a high wear resistance. The chilling does not allow the carbon in the iron to precipitate out during solidification, resulting in the white appearance. Areas away from the chilling medium solidify more slowly and acquire the normal properties of the base iron. One disadvantage of the chilling process is that the white iron is very brittle.

Malleable Iron. *Malleable iron* is used in automotive and truck parts, construction machinery, and electrical equipment. It does exhibit yielding, has tensile strengths comparable to ductile iron, and has ultimate compressive strengths that are somewhat higher than ductile iron. Generally, a five-digit number is used to designate malleable iron grades. For example, grade 40010 has a yield strength of 40 000 psi (276 MPa) and a percent elongation of 10%.

Appendix A–17 lists the mechanical properties of several grades of gray iron, ductile iron, ADI, and malleable iron.

2–5 ALUMINUM

Alloys of aluminum are designed to achieve optimum properties for specific uses. Some are produced primarily as sheet, plate, bars, or wire. Standard structural shapes and special sections are often extruded. Several alloys are used for forging, while others are special casting alloys. Appendix A–18 lists the properties of selected aluminum alloys. See also References 1 and 6.

Aluminum in wrought form uses a four-digit designation to define the several alloys available. The first digit indicates the alloy group according to the principal alloying element. The second digit denotes a modification of the basic alloy. The last two digits identify a specific alloy within the group. A brief description of the seven major series of aluminum alloys follows.

- *1000 Series,* 99.0% aluminum or greater. Excellent corrosion resistance, workability, and thermal and electrical conductivity. Low mechanical properties. Used in chemical and electrical fields, for automotive trim (alloy 1100), and for extruded condenser tubes and heat exchanger fins (alloy 1200).
- *2000 series,* copper alloying element. Heat-treatable with high mechanical properties. Lower corrosion resistance than most other alloys. Used in aircraft skins structures, automotive body panels, screw machine parts, fasteners, and seat shells.
- *3000 series,* manganese alloying element. Non-heat-treatable, but moderate strength can be obtained by cold working. Good corrosion resistance and workability. Used in chemical equipment, cooking utensils, residential siding, storage tanks, automotive radiators and heater cores, trim, and condenser tubes.
- *4000 series,* silicon alloying element. Non-heat-treatable with a low melting point. Used as welding wire and brazing alloy. Alloy 4032 used as pistons.
- *5000 series,* magnesium alloying element. Non-heat-treatable, but moderate strength can be obtained by cold working. Good corrosion resistance and weldability. Used in

marine service, pressure vessels, auto trim, builder's hardware, welded structures, TV towers, drilling rigs, truck bumpers, heat shields, wheels, and various brackets on engines.

- *6000 series,* silicon and magnesium alloying elements. Heat-treatable to moderate strength. Good corrosion resistance, formability, and weld-ability. Used as heavy-duty structures, truck and railroad equipment, pipe, furniture, architectural extrusions, machined parts, and forgings. Automotive applications include suspension parts, bumper assembly parts, driveshafts, yokes, brake cylinders and pistons, screw machine parts, fasteners, and luggage racks. Alloy 6061 is one of the most versatile available.
- *7000 series,* zinc alloying element. Heat-treatable to very high strength. Relatively poor corrosion resistance and weldability. Used mainly for aircraft structural members. Alloy 7075 is among the highest strength alloys available. It is produced in most rolled, drawn, and extruded forms and is also used in forgings. Automotive applications include seat tracks, bumper reinforcements, headrest support bars, and condenser fins.

Aluminum Temper Designations. Since the mechanical properties of virtually all aluminum alloys are very sensitive to cold working or heat treatment, suffixes are applied to the four-digit alloy designations to describe the temper. The most frequently used temper designations are described as follows:

- *O temper.* Fully annealed to obtain high ductility. Annealing makes most alloys easier to form by bending or drawing. Parts formed in the annealed condition are frequently heat-treated later to improve properties.
- *H temper,* strain-hardened. Used to improve the properties of non-heat-treatable alloys such as those in the 1000, 3000, and 5000 series. The H is always followed by a two- or three-digit number to designate a specific degree of strain hardening or special processing. The second digit following the H ranges from 0 to 8 and indicates a successively greater degree of strain hardening, resulting in higher strength. Appendix A–18 lists the properties of several aluminum alloys. Referring to alloy 3003 in that table shows that the yield strength is increased from 18 000 psi (124 MPa) to 27 000 psi (186 MPa) as the temper is changed from H12 to H18.
- *T temper,* heat-treated. Used to improve strength and achieve a stable condition. The T is always followed by one or more digits indicating a particular heat treatment. For wrought products such as sheet, plate, extrusions, bars, and drawn tubes, the most frequently used designations are T4 and T6. The T6 treatment produces higher strength but generally reduces workability. In Appendix A–18 several heat-treatable alloys are listed in the O, T4, and T6 tempers to illustrate the change in properties.

Cast aluminum alloys are designated by a modified four-digit system of the form XXX.X, in which the first digit indicates the main alloy group according to the major alloying elements. Table 2–9 shows the groups. The second two digits indicate the specific alloy within the group or indicate the aluminum purity. The last digit, after the decimal point, indicates the product form: 0 for castings, and 1 or 2 for ingots.

Aluminum is also sensitive to the manner in which it is produced, the size of the section, and temperature. Appendix A–18 lists typical properties and cannot be relied on for design. References 1, 3, and 6 give extensive data on minimum strengths.

Cast aluminum is used for large structural members and housings, engine blocks, cylinder heads, manifolds, internal engine parts, steering gears, transmission housings and internal parts, wheels, and fuel metering devices.

TABLE 2–9 Cast aluminum alloy groups.

Group	Major alloying elements
1XX.X	99% or greater aluminum
2XX.X	Copper
3XX.X	Silicon, copper, magnesium
4XX.X	Silicon
5XX.X	Magnesium
6XX.X	(Unused series)
7XX.X	Zinc
8XX.X	Tin
9XX.X	Other elements

2–6 COPPER, BRASS, AND BRONZE

Copper and copper alloys have good corrosion resistance, are readily fabricated, and have an attractive appearance. They are typically classified as *copper, brass,* or *bronze,* depending on the amounts and types of alloying elements employed. Cold working is used to modify the strength and hardness properties of these metals, ranging from *soft, quarter hard, half hard, three-quarters hard, hard, extra hard, spring,* and *extra spring* tempers, depending on the percent of cold working during processing. The properties of four types of copper-based materials are listed in Appendix A–15 and discussed next. Internet site 11 allows searching through a large database of copper materials.

Copper. The name *copper* is most often used to denote virtually the pure metal having 99% or more copper. The C14500 material has nominally 99.5% copper and 0.55% tellurium, and its typical uses include electrical connectors, motor parts that carry electric current, screw machine products, plumbing fittings, and sprinkler heads.

Beryllium copper (C17200) has nominally 98.1% copper and 1.9% beryllium, thus departing from the strict definition of pure copper. The beryllium adds significant strength, leading to uses such as fuse clips, relay parts, fasteners, Bourdon tubing for pressure gages, pump parts, valves, a variety of machined components, springs, and forgings.

Brass. *Brasses* are alloys of copper and zinc with lead sometimes added to improve machinability. The C36000 material is a free-cutting brass having 61.5% copper, 35.4% zinc, and 3.1% lead. Automotive uses include thermostat parts, fluid connectors, sensor bodies, and threaded inserts for plastic parts. Other industrial uses include builders' hardware, fasteners, faucet components valve seats, nozzles, gears, and screw machine parts. Other brasses are used in marine applications, ammunition cases, home furnishings, and heat exchanger tubing.

Bronze. Copper, alloyed with tin and a wide variety of elements such as lead, zinc, and phosphorous, fall into the *bronze* family of materials. Phosphor bronze (C54400) has 88% copper, 4% tin, 4% lead, 4% zinc, and 0.25% phosphorous. Its good strength, ductility, and wear properties lead to uses such as gears, shafts, bearings, thrust washers, screw machine parts, electrical connectors, and parts for valves.

2–7 ZINC, MAGNESIUM, TITANIUM, AND NICKEL-BASED ALLOYS

Zinc. *Zinc* has moderate strength and toughness and excellent corrosion resistance. It is used in wrought forms such as rolled sheet and foil and drawn rod or wire. Dry-cell battery cans, builder's hardware, and plates for photoengraving are some of the major applications.

Many zinc parts are made by die casting because the melting point is less than 800°F (427°C), much lower than other die-casting metals. The as-cast finish is suitable for many applications, such as business machine parts, pump bodies, motor housings, and frames for light-duty machines. Where a decorative appearance is required, electroplating with nickel and chromium is easily done. Such familiar parts as lamp housings and auto body moldings are made in this manner. See Internet site 13.

Zinc alloys typically contain aluminum and a small amount of magnesium. Some alloys include copper or nickel. The performance of the final products can be very sensitive to small amounts of other elements, and maximum limits are placed on the content of iron, lead, cadmium, and tin in some alloys.

The most widely used zinc casting alloy is called *alloy No. 3,* sometimes referred to as *Zamak 3*. It has 4% aluminum and 0.035% magnesium. Another is called *Zamak 5* and it also contains 4% aluminum with 0.055% magnesium and 1.0% copper. A group of alloys having higher aluminum content are the ZA-alloys, with ZA-8, ZA-12, and ZA-27 being very popular. The strength properties of cast ZA-12 zinc alloy are listed in Appendix A–15.

Magnesium. *Magnesium* is the lightest metal commonly used in load-carrying parts. Its density of only 0.066 lb/in^3 (1830 kg/m^3) is only about one-fourth that of steel and zinc, one-fifth that of copper, and two-thirds that of aluminum. It has moderate strength and lends itself well to applications where the final fabricated weight of the part or structure should be light. Ladders, hand trucks, conveyor parts, portable power tools, and lawn-mower housings use magnesium. In the automotive industry, body parts, blower wheels, pump bodies, and brackets are often made of magnesium. In aircraft, its lightness makes this metal attractive for floors, frames, fuselage skins, and wheels. The stiffness (modulus of elasticity) of magnesium is low, which is an advantage in parts where impact energy must be absorbed. Also, its lightness results in low-weight designs when compared with other metals on an equivalent rigidity basis. See Appendix A–15 for properties of one cast magnesium alloy.

Titanium. *Titanium* has very high strength, and its density is only about half that of steel. Although aluminum has a lower density, titanium is superior to both aluminum and most steels on a strength-to-weight basis. It retains a high percentage of its strength at elevated temperatures and can be used up to about 1000°F (538°C). Most applications of titanium are in the aerospace industry in engine parts, fuselage parts and skins, ducts, spacecraft structures, and pressure vessels. Because of its corrosion resistance and high-temperature strength, the chemical industries use titanium in heat exchangers and as a lining for processing equipment. High cost is a major factor to be considered.

Appendix A–15 shows the properties of one example of each of the four major types of titanium; pure alpha titanium, alpha alloy, beta alloy, and alpha-beta alloy. In general, the alpha-beta and beta alloys are the stronger forms, and they are heat-treatable to permit close tailoring of properties to specific uses, considering formability, machinability, forgeability, weldability, corrosion resistance, high-temperature strength, and creep resistance in addition to the typical room temperature strength, stiffness, and ductility data. The term "aged" indicates a final heat treatment, and more complete references for aging should be consulted to assess the range of properties that can be achieved.

Nickel-Based Alloys. *Nickel-based alloys* are often specified for applications where high corrosion resistance and/or high strength at elevated temperatures are required. Examples are turbine engine components, rocket motors, furnace parts, chemical processing systems, marine-based systems, aerospace systems, valves, pumps, heat exchangers, food processing equipment, pharmaceutical processing systems, nuclear power systems, and air pollution equipment.

Appendix A–15 gives typical properties of three nickel-based alloys listed according to their UNS designations in the NXXXXX series. Most nickel-based alloys are not heat treated; however, cold working is employed to increase strength as is evident for the N04400 alloy in Appendix A–15. For alloys N06600 and N06110, strength data are provided for room temperature and several elevated temperatures up to 1200°F (649°C). Some alloys are usable up to 2000°F (1093°C). Alloys are often grouped under the categories of *corrosion resistant* or *heat resistant.* While nickel makes up 50% or more of these alloys, the major alloying elements include chromium, molybdenum, cobalt, iron, copper, and niobium. See Internet sites 28–30 for some producers of these alloys. Note the brand names offered by these companies and the extensive set of property data available on their websites.

Creep performance is an important property for the nickel-based alloys because they often operate under significant loads and stresses at high temperatures where creep rupture is a possible failure mode. Also, the modulus of elasticity decreases with increasing temperature, which can result in unacceptable deformations that must be considered in the design phase. Producers publish creep and elastic modulus versus temperature data for their materials in addition to the basic strength data.

2–8 NONMETALS IN ENGINEERING DESIGN

Wood and concrete are widely used in construction. Plastics and composites are found in virtually all fields of design, including consumer products, industrial equipment, automobiles, aircraft, and architectural products. To the designer, the properties of strength and stiffness are of primary importance with nonmetals, as they are with metals. Because of the structural differences in the nonmetals, their behavior is quite different from the metals.

Wood, concrete, composites, and many plastics have structures that are *anisotropic.* This means that the mechanical properties of the material are different, depending on the direction of the loading. Also, because of natural chemical changes, the properties vary with time and often with climatic conditions. The designer must be aware of these factors.

2–9 WOOD

Since wood is a natural material, its structure is dependent on the way it grows and not on manipulation by human beings, as is the case in metals. The long, slender, cylindrical shape of trees results in an internal structure composed of longitudinal cells. As the tree grows, successive rings are added outside the older wood. Thus the inner core, called heartwood, has different properties than the sapwood, near the outer surface.

The species of the wood also affects its properties, as different kinds of trees produce harder or softer, stronger or weaker wood. Even in the same species variability occurs because of different growing conditions, such as differences in soil and amount of sun and rain.

The cellular structure of the wood gives it the grain that is so evident when sawn into boards and timber. The strength of the wood is dependent on whether it is loaded perpendicular to or parallel to the grain. Also, going across the grain, the strength is different in a radial direction than in a tangential direction with respect to the original cylindrical tree stem from which it was cut.

Another important variable affecting the strength of wood is moisture content. Changes in relative humidity can vary the amount of water absorbed by the cells of the wood.

Appendix A–19 gives data for typical properties of three species of wood often used in construction: Douglas fir, hemlock, and southern pine. Notice the three grades listed for each. Grade No. 1 is the highest grade, representing clear wood with few if any defects such as knots. Grade No. 2 is common construction lumber that may have occasional knots or other defects. Grade 3 is a poorer grade with generally noticeable knots and large variations in grain. The allowable strength values decrease dramatically from Grade 1 to Grade 3. The grades, and thus the allowable strengths, are determined by using standard rules adopted by the U.S. Forest Products Laboratory. See References 19 and 31.

The tension and compression properties are used in Chapter 3. Bending and horizontal shear strengths are considered in Chapter 7 on stresses in beams. Notice the very

TABLE 2–10 Machine-graded lumber.

	Allowable stresses (parallel to grain)						Modulus of elasticity	
	Bending		Tension		Compressive			
Grade	psi	MPa	psi	MPa	psi	MPa	psi	GPa
Machine-stress rated lumber (MSR)								
1350f-1.3E	1350	9.3	750	5.2	1600	11.0	1.3×10^6	9.0
1800f-1.6E	1800	12.4	1175	8.1	1750	12.1	1.6×10^6	11.0
2400f-2.0E	2400	16.5	1925	13.3	1975	13.6	2.0×10^6	13.8
2850f-2.3E	2850	19.7	2300	15.9	2150	14.8	2.3×10^6	15.9
Machine-evaluated lumber (MEL)								
M-10	1400	9.7	800	5.5	1600	11.0	1.2×10^6	8.3
M-14	1800	12.4	1000	6.9	1750	12.1	1.7×10^6	11.7
M-21	2300	15.9	1400	9.7	1950	13.4	1.9×10^6	13.1
M-24	2700	18.6	1800	12.4	2100	14.5	1.9×10^6	13.1

Source: Forest Products Society, 1997.

low strength in shear, only 70 to 95 psi (0.48 to 0.66 MPa). This represents the prevalent mode of failure for wood beams as discussed further in Chapter 8.

Machine-Graded Structural Lumber. Alternate grading systems for lumber use machine-grading processes to sort lumber into grades related to strength and stiffness properties of the wood. Two such systems are *machine-stress rated (MSR)* and *machine-evaluated lumber (MEL)*. Table 2–10 shows representative grades from each system. More grades are available and not all grades are used for all products. Users should discuss details of rating systems with vendors before specifying particular products.

2–10 CONCRETE

The components of concrete are cement and an aggregate. The addition of water and the thorough mixing of the components tend to produce a uniform structure with cement coating all the aggregate particles. After curing, the mass is securely bonded together. Some of the variables involved in determining the final strength of the concrete are the type of cement used, the type and size of aggregate, and the amount of water added.

A higher quantity of cement in concrete yields a higher strength. Decreasing the quantity of water relative to the amount of cement increases the strength of the concrete. Of course, sufficient water must be added to cause the cement to coat the aggregate and to allow the concrete to be poured and worked before excessive curing takes place. The density of the concrete, affected by the aggregate, is also a factor. A mixture of sand, gravel, and broken stone is usually used for construction grade concrete.

Concrete is graded according to its compressive strength, which varies from about 2000 psi (14 MPa) to 7000 psi (48 MPa). The tensile strength of concrete is extremely low, and it is common practice to assume that it is zero. Of course, reinforcing concrete with steel bars allows its use in beams and wide slabs since the steel resists the tensile loads.

Concrete must be cured to develop its rated strength. It should be kept moist for at least 7 days, at which time it has about 60% to 70% of its rated compressive strength. Although its strength continues to increase for years, the strength at 28 days is often used to determine its rated strength.

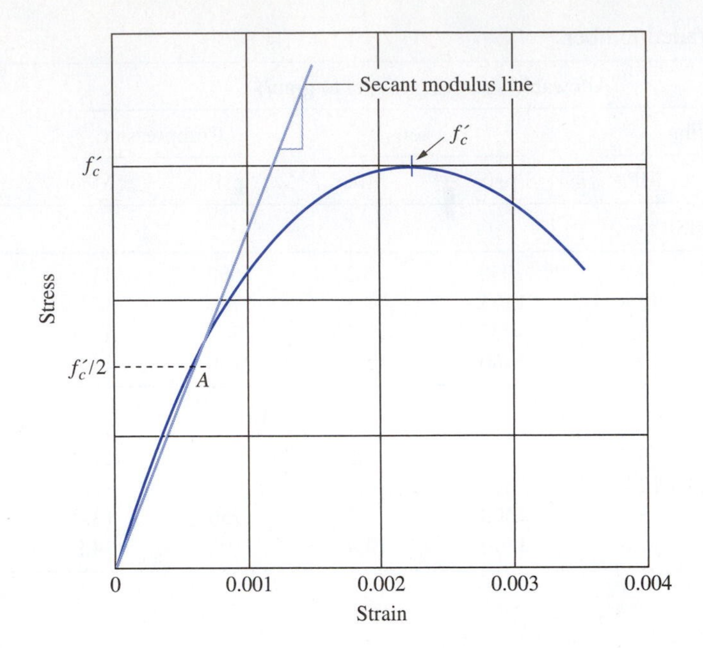

FIGURE 2–18 Typical stress–strain curve for concrete.

Testing of concrete can be done using the universal testing machine similar to that shown in Figure 2–1, arranged to apply a compressive load. The test specimen is typically a 6-in diameter cylinder having a length of 12 in (150 mm × 300 mm). The load is slowly increased until compressive fracture occurs, tracking the stress and strain continuously during the test. Figure 2–18 shows a typical stress–strain curve for concrete. Note the following:

1. The stress–strain curve has a continuous curvature, reaching a peak at a strain of approximately 0.002.
2. The peak of the curve is taken as the ultimate compressive strength, or rated strength, of the concrete, called f'_c in the construction industry.
3. The lack of a straight-line portion of the stress–strain curve requires a special definition of the modulus of elasticity for concrete. The American Concrete Institute (ACI) defines the *secant modulus,* as shown in Figure 2–18, to be the slope of the line from the origin through the intersection on the curve with the stress level of $f'_c/2$ (point A).
4. Failure occurs at a strain of approximately 0.003 to 0.004, depending on the ultimate strength of the concrete. Failure may occur as shearing along angular lines within the specimen or as splitting along nearly vertical planes.

Figure 2–19 shows typical stress–strain curves for several grades of concrete. An important observation from these curves is that the secant modulus varies significantly, being higher for the higher strength grades.

The specific weight of concrete, sometimes called the *unit weight,* ranges from about 90 lb/ft^3 to 160 lb/ft^3 (14.1 kN/m^3 to 25.1 kN/m^3) with the kind of aggregate having a major influence. When using SI units, mass density is often called for, which ranges from 1.44 kg/m^3 to 2.56 kg/m^3. For common construction-grade concrete the nominal values are:

Specific weight:	150 lb/ft^3	or	23.6 kN/m^3
Density:	150 lb$_m$/ft^3	or	2.40 kg/m^3

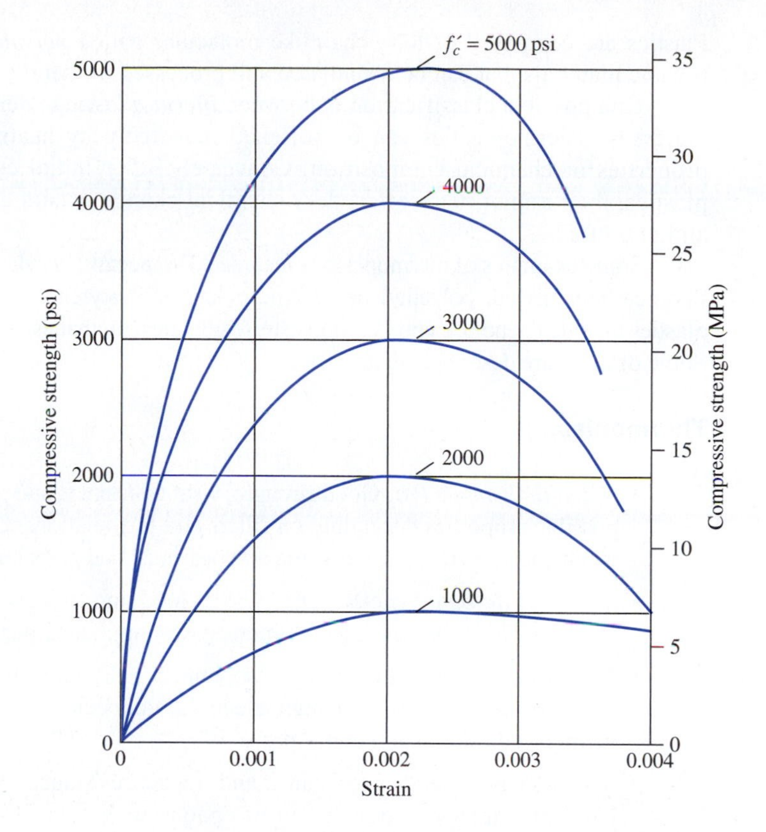

FIGURE 2–19 Typical stress–strain curves for a variety of concrete grades.

The modulus of elasticity is somewhat dependent on the specific weight and the rated strength. According to the American Concrete Institute, an estimate of the modulus can be computed from

$$E_c = 33\gamma^{3/2}\sqrt{f'_c} \tag{2–9}$$

where E_c = Modulus of elasticity in compression, psi
γ = Specific weight, lb/ft^3
f'_c = Rated compressive strength of the concrete, psi

Using $\gamma = 150$ lb/ft^3, the range of expected values for modulus of elasticity, computed from Equation (2–9), is shown in Table 2–11.

The allowable working stresses in concrete are typically 25% of the rated 28-day strength. For example, concrete rated at 2000 psi (14 MPa) would have an allowable stress of 500 psi (3.4 MPa).

TABLE 2–11 Modulus of elasticity for concrete.

Rated strength, f'_c		Modulus of elasticity, E_c	
psi	MPa	psi	GPa
2000	13.8	2.7×10^6	18.6
3000	20.7	3.3×10^6	22.7
4000	27.6	3.8×10^6	26.2
5000	34.5	4.3×10^6	29.6
6000	41.4	4.7×10^6	32.4
7000	48.3	5.1×10^6	35.2

2–11 PLASTICS

Plastics are composed of long chainlike molecules called *polymers*. They are synthetic organic materials that can be formulated and processed in literally thousands of ways.

One possible classification is between *thermoplastic* materials and *thermosetting* materials. Thermoplastics can be softened repeatedly by heating with no change in properties or chemical composition. Conversely, after initial curing of thermosetting plastics, they cannot be resoftened. A chemical change occurs during curing with heat and pressure.

Some examples of thermoplastics include ABS, acetals, acrylics, cellulose acetate, TFE fluorocarbons, nylon, polyethylene, polypropylene, polystyrene, and vinyls. Thermosetting plastics include phenolics, epoxies, polyesters, silicones, urethanes, alkyds, allyls, and aminos. Some of these are described next.

Thermoplastics

- *Nylon (Polyamide PA):* Good strength, wear resistance, and toughness; wide range of possible properties depending on fillers and formulations. Used for structural parts, mechanical devices such as gears and bearings, and parts needing wear resistance.
- *Acrylonitrile-butadiene-Styrene (ABS):* Good impact resistance, rigidity, moderate strength. Used for housings, helmets, cases, appliance parts, pipe, fittings.
- *Polycarbonate:* Excellent toughness, impact resistance, and dimensional stability. Used for cams, gears, housings, electrical connectors, food processing products, helmets, and pump and meter parts.
- *Acrylic:* Good weather resistance and impact resistance; can be made with excellent transparency or translucent or opaque with color. Used for glazing, lenses, signs, and housings.
- *Polyvinyl chloride (PVC):* Good strength, weather resistance, and rigidity. Used for pipe, electrical conduit, small housings, ductwork, and moldings.
- *Polyimide:* Good strength and wear resistance; very good retention of properties at elevated temperatures up to 500°F. Used for bearings, seals, rotating vanes, and electrical parts.
- *Acetal:* High strength, stiffness, hardness, and wear resistance; low friction; good weather resistance and chemical resistance. Used for gears, bushings, sprockets, conveyor parts, and plumbing products. [Generic name: poly-oxy-methylene (POM).]
- *Polyurethane elastomer:* A rubberlike material with exceptional toughness and abrasion resistance; good heat resistance and resistance to oils. Used for wheels, rollers, gears, sprockets, conveyor parts, and tubing.
- *Thermoplastic polyester resin (PET):* Polyethylene terephthalate (PET) resin with fibers of glass and/or mineral. Very high strength and stiffness, excellent resistance to chemicals and heat, excellent dimensional stability, and good electrical properties. Used for pump parts, housings, electrical parts, motor parts, auto parts, oven handles, gears, sprockets, and sporting goods.
- *Polyether-ester elastomer:* Flexible plastic with excellent toughness and resilience, high resistance to creep, impact, and fatigue under flexure, good chemical resistance. Remains flexible at low temperatures and retains good properties at moderately elevated temperatures. Used for seals, belts, pump diaphragms, protective boots, tubing, springs, and impact absorbing devices. High modulus grades can be used for gears and sprockets.

Thermosets

- *Phenolic:* High rigidity, good moldability and dimensional stability, very good electrical properties. Used for load-carrying parts in electrical equipment, switchgear, terminal strips, small housings, handles for appliances and cooking utensils, gears, and structural and mechanical parts. Alkyd, allyl, and amino thermosets have properties and uses similar to those of the phenolics.
- *Polyester:* Known as *fiberglass* when reinforced with glass fibers; high strength and stiffness, good weather resistance. Used for housings, structural shapes, and panels.

Special Considerations for Selecting Plastics. A particular plastic is often selected for a combination of properties, such as light weight, flexibility, color, strength, stiffness, chemical resistance, low friction characteristics, or transparency. Table 2–12 lists the primary plastic materials used for six different types of applications. References 4, 11, 14, 16, 29, and 30 provide an extensive comparative study of the design properties of plastics.

While most of the same definitions of design properties described in Section 2–2 of this chapter can be used for plastics as well as metals, a significant amount of additional information is typically needed to specify a suitable plastic material. Some of the special characteristics of plastics follow. Use data from Appendix A–20 for problem solutions in this book requiring the properties of selected plastics. There is a wide range of properties among the many formulations of plastics even within a given class. Consult the extensive amount of design guidance available from vendors of the plastic materials. See Internet sites 14–19.

1. Most properties of plastics are highly sensitive to temperature. In general, tensile strength, compressive strength, elastic modulus, and impact failure energy decrease significantly as the temperature increases.

TABLE 2–12 Applications of plastic materials.

Applications	Desired properties	Suitable plastics
Housings, containers, ducts	High impact strength, stiffness, low cost, formability, environmental resistance, dimensional stability	ABS, polystyrene, polypropylene, polyethylene, cellulose acetate, acrylics
Low friction—bearings, slides	Low coefficient of friction; resistance to abrasion, heat, corrosion	TFE fluorocarbons, nylon, acetals
High-strength components, gears, cams, rollers	High tensile and impact strength, stability at high temperatures, machinable	Nylon, phenolics, TFE-filled acetals
Chemical and thermal equipment	Chemical and thermal resistance, good strength, low moisture absorption	Fluorocarbons, polypropylene, polyethylene, epoxies, polyesters, phenolics
Electrostructural parts	Electrical resistance, heat resistance, high impact strength, dimensional stability, stiffness	Allyls, alkyds, aminos, epoxies, phenolics, polyesters, silicones
Light-transmission components	Good light transmission in transparent and translucent colors, formability, shatter resistance	Acrylics, polystyrene, cellulose acetate, vinyls

2. Many plastics absorb a considerable amount of moisture from the environment and exhibit dimensional changes and degradation of strength and stiffness properties as a result.
3. Components that carry loads continuously must be designed to accommodate creep or relaxation as described in Section 2–2.
4. The behavior of plastics under repeated loading (fatigue), shock, and impact is highly variable, and many grades are specially formulated for good performance in such environments. Data for toughness as indicated by Izod, Charpy, or drop testing should be acquired from the vendor along with specific fatigue strengths of the materials.
5. Processing methods can have large effects on the final dimensions and properties of parts made from plastics. Molded plastics shrink significantly during solidification and curing. Parting lines produced where mold halves meet may affect strength. The rate of solidification may be widely different in a given part depending on the section thicknesses, the complexity of the shape and the location of sprues that deliver molten plastic into the mold. The same material can produce different properties depending on whether it is processed by injection molding, extrusion, blow molding, or machining from a solid block or bar. See Reference 18.
6. Resistance to chemicals, weather, direct sunlight, and other environmental conditions must be checked.
7. Plastics may exhibit a change in properties as they age, particularly when subjected to elevated temperatures.
8. Flammability and electrical characteristics must be considered. Some plastics are specially formulated for high flammability ratings as called for by Underwriters Laboratory and other agencies.
9. Plastics used for food storage or processing must meet U.S. Food and Drug Administration standards.

2–12 COMPOSITES

Composites are materials having two or more constituents blended in a way that results in mechanical or adhesive bonding between the materials. To form a composite, a filler material is distributed in a matrix so that the filler reinforces the matrix. Typically, the filler is a strong, stiff material while the matrix has a relatively low density. When the two materials bond together, much of the load-carrying ability of the composite is produced by the filler material. The matrix serves to hold the filler in a favorable orientation relative to the manner of loading and to distribute the loads to the filler. The result is a somewhat optimized composite that has high strength and high stiffness with low weight. See References 10, 15, 17, 20, 21, 23, 26, and 27.

A virtually unlimited variety of composite materials can be produced by combining different matrix materials with different fillers in different forms and in different orientations. Even wood and concrete, discussed earlier in this chapter, are technically examples of composites. However, the common use of the term refers to those materials using polymers, metals, or ceramics as matrix materials with a wide variety of filler materials to form composites.

Examples of Finished Products Made from Composite Materials. The number and variety of applications for composite materials is large and growing. The following items are but a sampling of these applications.

Consumer products and recreation: Sporting goods such as tennis rackets, snow skis, snowboards, water skis, surfboards, baseball bats, hockey sticks, vaulting

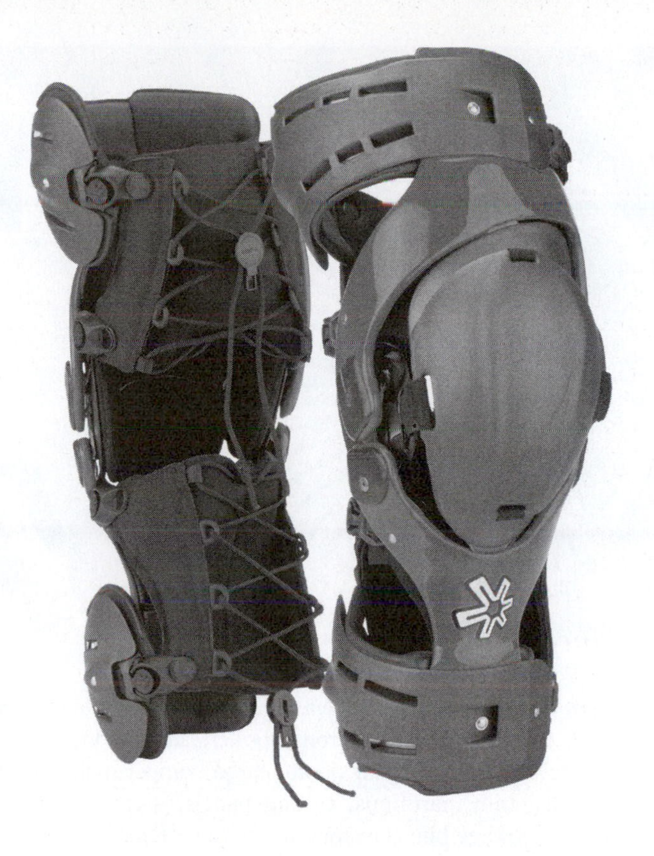

FIGURE 2–20 Knee brace with several components made from composites. (Source: A&P Technology, Cincinnati, OH)

poles, and golf clubs; numerous products having the familiar fiberglass housings and panels; boat hulls and other onboard equipment; medical systems and prosthetic devices. Figure 2–20 shows a knee brace with several components made from composites.

Ground transportation equipment: Bicycle frames, wheels, and seats; automotive and truck body panels and support structures, air ducts, air bags, driveshafts, springs for high-performance sports cars and trucks, floor pans, pickup truck beds, and bumpers. Figure 2–21 shows a high-performance bicycle employing a composite frame and other structural parts.

FIGURE 2–21 High-performance bicycle employing a composite frame and other structural parts. (Source: A&P Technology, Cincinnati, OH)

FIGURE 2–22 Stator vanes for a turbine engine made from composite materials. (Source: A&P Technology, Cincinnati, OH)

Aircraft and aerospace systems: Fuselage panels and internal structural elements, wings, control surfaces (ailerons, spoilers, tails, rudders), floor systems, engine cowls and nacelles, landing gear doors, cargo compartment structure and fittings, interior sidewalls, trim, partitions, ceiling panels, dividers, environmental control system ducting, stowage bins, lavatory structure systems and fixtures, airfoils (blades) in the compressor section of turbine engines, rocket nozzles, helicopter rotors, propellers, and onboard tanks for storing water and wastewater. Figures 2–22 and 2–23 show turbine engine compressor blades and some of the tooling used to manufacture them.

Industrial facilities: Storage tanks and pressure vessels for chemical, agricultural, and petroleum processing, piping for chemicals and through corrosive environments,

FIGURE 2–23 Forming press for manufacturing stator vanes for a turbine engine, shown in Figure 2–22. (Source: A&P Technology, Cincinnati, OH)

septic systems, wastewater treatment facilities, chemical cleaning and plating systems, pulp and paper making equipment, portable tanks for trucks and railroad applications, environmental treatment equipment, protective clothing and helmets, food processing and storage systems, mining, and material handling systems.

Electrical and electronic systems: Printed circuit boards, printed wiring boards, surface mount cards, packaging of electronic components, and switching system components.

Building construction: Structural shapes, exterior panels, roofing and decking systems, doors, window frames, equipment housings, gutters and downspouts, cooling towers, bridges and walkways, piping systems, and ductwork.

Classifications of Composite Materials by Matrix. One method of classifying composite materials is by the type of matrix material. Three general classifications are used as described next, along with typical matrix materials, uses, and matrix-filler combinations.

Polymer matrix composites (PMC)

Thermoplastics: polyethylenes, polyamides (nylons), polystyrenes, polypropylenes, polycarbonates, polyetheretherketones (PEEK), polyphenylene sulfides (PPS), polyvinyl chloride (PVC)

Thermosets: polyesters, epoxies, phenolics, polyimides (PI), vinyl esters, silicones

PMCs are used for their high strength and stiffness, low density, and relatively low cost in aerospace, automotive, marine, chemical, electrical, and sporting applications. Common PMC composites include polyester-glass (conventional fiberglass), epoxy-glass, polyimide-glass, epoxy-aramid, epoxy-carbon, PEEK-carbon, and PPS-carbon.

Metal matrix composites (MMC): aluminum (Al), titanium (Ti), magnesium (Mg), iron (Fe), copper (Cu), nickel (Ni), and alloys of these metals with themselves and with molybdenum (Mo), cesium (Ce), boron (B).

MMCs are preferred for high strength, high stiffness, abrasion resistance, dimensional stability, electrical and thermal conductivity, ability to operate in high temperatures, and toughness and are applied typically in aerospace and engine applications. Examples of MMC composites include Al-SiC (silicon carbide), Ti-SiC, Al-B, Al-C (carbon), Al-graphite, Mg-SiC, and Al-Al_2O_3 (aluminum oxide).

Ceramic matrix composites (CMC): silicon carbide, silicon nitride, alumina, zirconia, glass-ceramic, glass, carbon, graphite.

CMCs are preferred for high strength, high stiffness, high fracture toughness relative to ceramics alone, ability to operate at high temperatures, and low thermal expansion and are attractive for furnaces, engines, and aerospace applications. Common CMC composites include carbon-carbon (C-C), silicon carbide-carbon (SiC-C), silicon carbide-silicon carbide (SiC-SiC), glass ceramic-silicon carbide, silicon carbide-lithium aluminosilicate (SiC-LAS), and silicon carbide-calcium aluminosilicate (SiC-CAS). Where the same basic material is listed as both the matrix and the filler, the filler is of a different form such as whiskers, chopped fibers, or strands to achieve the preferred properties.

Forms of Filler Materials. Many forms of filler materials are used as listed here.

- Continuous fiber strand consisting of many individual filaments bound together
- Chopped strands in short lengths (0.75 to 50 mm or 0.03 to 2.00 in)

FIGURE 2–24 Fabrics and braided sleeves made from glass and carbon fibers used for composite materials. The products shown in Figures 2–20 to 2–22 are made with these filler materials. (Source: A&P Technology, Cincinnati, OH)

- Chopped longer strands randomly spread in the form of a mat
- Chopped longer strands aligned with the principal directions of the load path
- Roving: A group of parallel strands
- Woven fabric made from roving or strands
- Metal filaments or wires
- Solid or hollow microspheres
- Metal, glass, or mica flakes
- Single crystal whiskers of materials such as graphite, silicon carbide, and copper

Figure 2–24 shows several styles of woven and braided fabric formed into tubular shapes and sleeving adaptable to making structural tubing, driveshafts, turbine blades, bats, housings, components for prosthetic devices, surgical devices, and many other composite-based products. Both glass and carbon fibers can be employed. The products shown in Figures 2–20 to 2–22 are made with these braided sleeves.

Types of Filler Materials. Fillers, also called fibers, come in many types based on both organic and inorganic materials. Some of the more popular fillers are listed below.

- Glass fibers in five different types:
  A-glass: good chemical resistance because it contains alkalis such as sodium oxide
  C-glass: special formulations for even higher chemical resistance than A-glass
  E-glass: widely used glass with good electrical insulating ability and good strength
  S-glass: high strength, high temperature glass
  D-glass: better electrical properties than E-glass
- Quartz fibers and high-silica glass: good properties at high temperatures up to 2000°F (1095°C)

- Carbon fibers made from PAN-base carbon (PAN is polyacrylonitrile): approximately 95% carbon with very high modulus of elasticity
- Graphite fibers: greater than 99% carbon and even higher modulus of elasticity than carbon; the stiffest fibers typically used in composites
- Boron coated onto tungsten fibers: good strength and higher modulus of elasticity than glass
- Silicon carbide coated onto tungsten fibers: strength and stiffness similar to boron/tungsten but with higher temperature capability
- Aramid fibers: a member of the polyamide family of polymers; higher strength and stiffness with lower density as compared with glass; very flexible (aramid fibers produced by the DuPont company carry the name *Kevlar*)

Advantages of Composites. Designers typically seek to produce products that are safe, strong, stiff, lightweight, and highly tolerant of the environment in which the product will operate. Composites often excel in meeting these objectives when compared to alternative materials such as metals, wood, and unfilled plastics. Two parameters that are used to compare materials are *specific strength* and *specific modulus,* defined as,

Specific strength is the ratio of the tensile strength of a material to its specific weight.

Specific modulus is the ratio of the modulus of elasticity of a material to its specific weight.

Because the modulus of elasticity is a measure of the stiffness of a material, the specific modulus is sometimes called *specific stiffness*.

Although obviously not a length, both of these quantities have the *unit* of length, derived from the ratio of the units for strength or modulus of elasticity and the units for specific weight. In the U.S. Customary system, the units for tensile strength and modulus of elasticity are lb/in^2 while specific weight (weight per unit volume) is in lb/in^3. Thus, the unit for specific strength or specific modulus is inches. In the SI metric system, strength and modulus are expressed in N/m^2 (Pascals) while specific weight is in N/m^3. Then the unit for specific strength or specific modulus is meters.

Table 2–13 gives comparisons of the specific strength and specific stiffness of selected composite materials with certain steel, aluminum, and titanium alloys. Figure 2–25 shows a comparison of these materials using bar charts. Figure 2–26 is a plot of these data with specific strength on the vertical axis and specific modulus on the horizontal axis. When weight is critical, the ideal material would lie in the upper right part of this chart. Note that data in these charts and figures are for composites having the filler materials aligned in the most favorable direction to withstand the applied loads.

Advantages of composites can be summarized as follows:

1. Specific strengths for composite materials can range as high as five times those of high strength steel alloys.
2. Specific modulus values for composite materials can be as high as eight times those for either steel, aluminum, or titanium alloys.
3. Composite materials typically perform better than steel or aluminum in applications where cyclic loads are encountered leading to the potential for fatigue failure.
4. Where impact loads and vibrations are expected, composites can be specially formulated with materials that provide high toughness and a high level of damping.
5. Some composites have much higher wear resistance than metals.

TABLE 2–13 Comparison of specific strength and specific modulus for selected material.

Material	Tensile strength, s_u (ksi)	Specific weight, γ (lb/in^3)	Specific strength (in)	Specific modulus (in)
Steel ($E = 30 \times 10^6$ psi)				
AISI 1020 HR	55	0.283	0.194×10^6	1.06×10^8
AISI 5160 OQT 700	263	0.283	0.929×10^6	1.06×10^8
Aluminum ($E = 10.0 \times 10^6$ psi)				
6061-T6	45	0.098	0.459×10^6	1.02×10^8
7075-T6	83	0.101	0.822×10^6	0.99×10^8
Titanium ($E = 16.5 \times 10^6$ psi)				
Ti-6Al-4V Quenched and aged at 1000°F	160	0.160	1.00×10^6	1.03×10^8
Glass/epoxy composite ($E = 4.0 \times 10^6$ psi)				
34% fiber content	114	0.061	1.87×10^6	0.66×10^8
Aramid/epoxy composite ($E = 11.0 \times 10^6$ psi)				
60% fiber content	200	0.050	4.0×10^6	2.20×10^8
Boron/epoxy composite ($E = 30.0 \times 10^6$ psi)				
60% fiber content	270	0.075	3.60×10^6	4.00×10^8
Graphite/epoxy composite ($E = 19.7 \times 10^6$ psi)				
62% fiber content	278	0.057	4.86×10^6	3.45×10^8
Graphite/epoxy composite ($E = 48 \times 10^6$ psi)				
Ultrahigh modulus	160	0.058	2.76×10^6	8.28×10^8

6. Careful selection of the matrix and filler materials can provide superior corrosion resistance.
7. Dimensional changes due to changes in temperature are typically much less for composites than for metals. More on this topic is include in Chapter 3 in which the property *coefficient of thermal expansion* is defined.
8. Because composite materials have properties that are highly directional, designers can tailor the placement of reinforcing fibers in directions that provide the required strength and stiffness under the specific loading conditions to be encountered.
9. Composite structures can often be made in complex shapes in one piece, thus reducing the number of parts in a product and the number of fastening operations required. The elimination of joints typically improves the reliability of such structures as well.
10. Composite structures are typically made in their final form directly or in a near-net shape, thus reducing the number of secondary operations required.

Limitations of Composites. Designers must balance many properties of materials in their designs while simultaneously considering manufacturing operations, costs, safety, life, and service of the product. The following list gives some of the major concerns when using composites.

1. Material costs for composites are typically higher than for many alternative materials.
2. Fabrication techniques are quite different from those used to shape metals. New manufacturing equipment may be required along with additional training for production operators.
3. The performance of products made from some composite production techniques is subject to a wider range of variability than for most metal fabrication techniques.

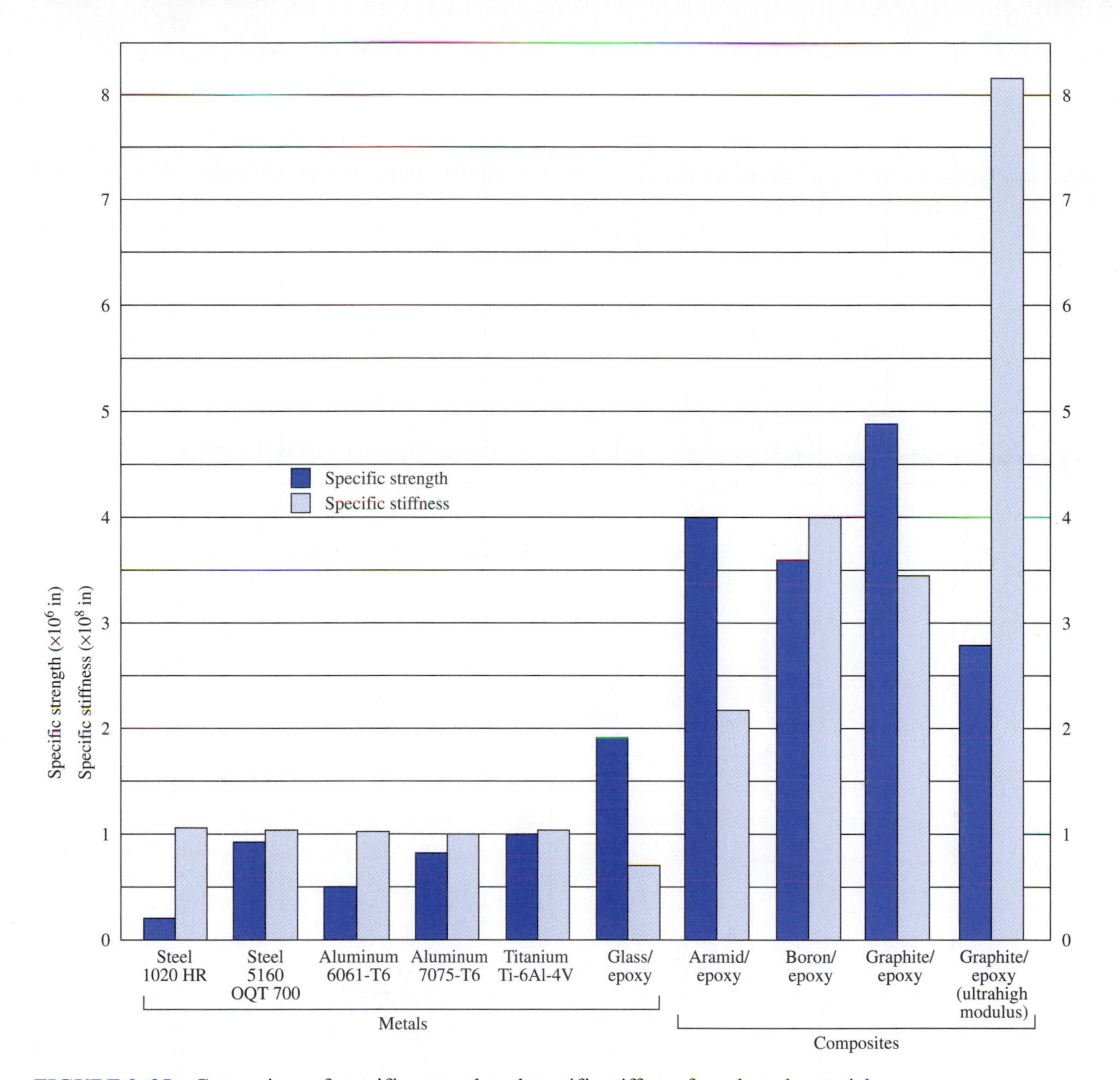

FIGURE 2–25 Comparison of specific strength and specific stiffness for selected materials.

4. The operating temperature limit for composites having a polymeric matrix is typically 500°F (260°C). [But ceramic or metal matrix composites can be used at higher temperatures, up to 1500°C (2700°F), as found in engines.]
5. The properties of composite materials are not isotropic. This means that properties vary dramatically with the direction of the applied loads. Designers must account for these variations to ensure safety and satisfactory operation under all expected types of loading.
6. At this time, many designers lack an understanding of the behavior of composite materials and the details of predicting failure modes. While major advancements have been made in certain industries such as the aerospace and recreational

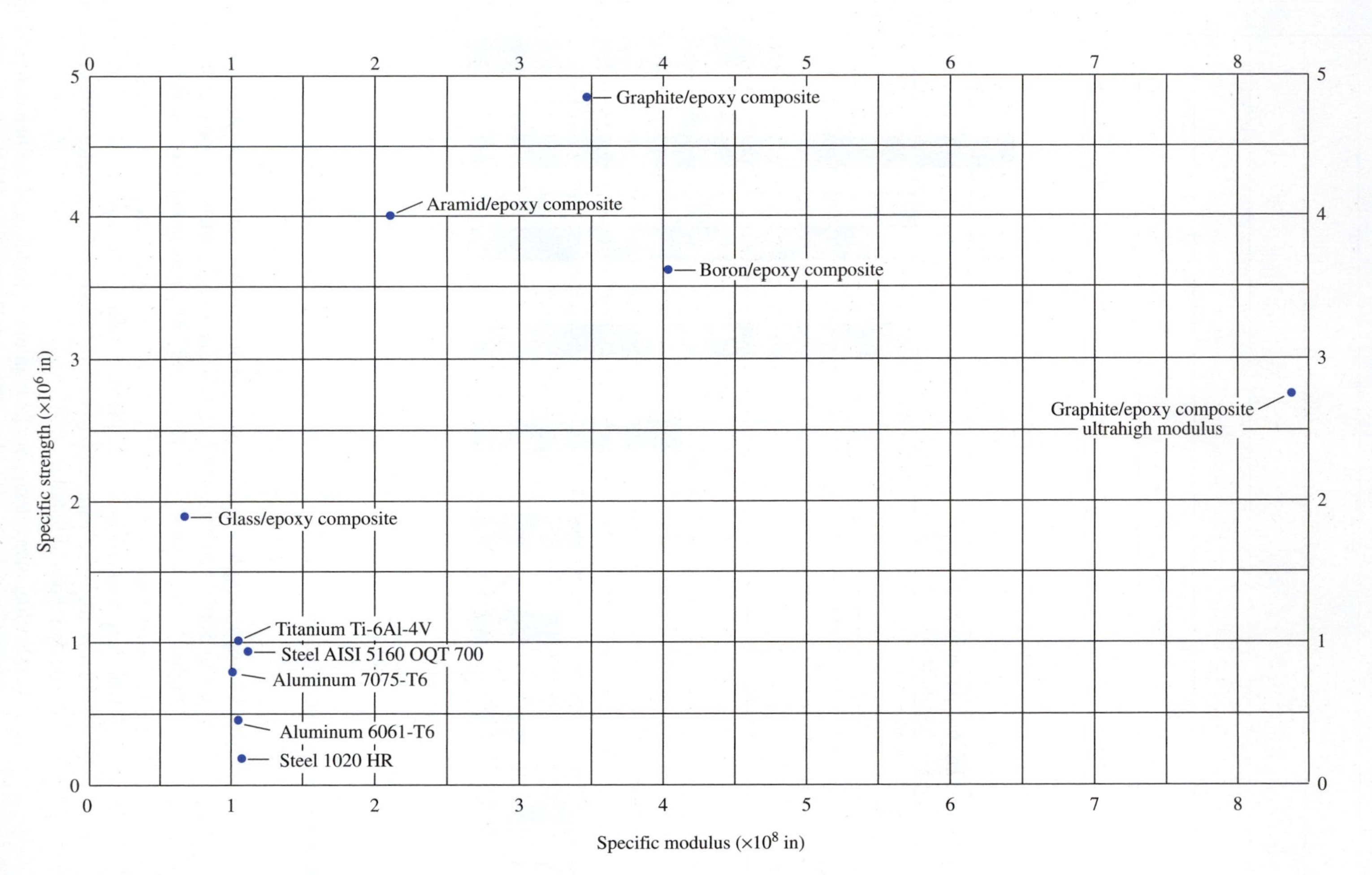

FIGURE 2–26 Specific strength versus specific modulus for selected metals and composites.

equipment fields, there is a need for more general understanding about designing with composite materials.

7. The analysis of composite structures requires the detailed knowledge of more properties of the materials than would be required for metals.
8. Inspection and testing of composite structures is typically more complicated and less precise than for metal structures. Special nondestructive techniques may be required to ensure that there are no major voids in the final product that could seriously weaken the structure. Testing of the complete structure may be required rather than testing a sample of the material because of the interaction of different parts on each other and because of the directionality of the material properties.
9. Repair and maintenance of composite structures is a serious concern. Some of the initial production techniques require special environments of temperature and pressure that may be difficult to reproduce in the field when damage repair is required. Bonding of a repaired area to the parent structure may also be difficult.

Directional Nature of the Properties of Composites. In general, a given material can have three kinds of behaviors with regard to the relationships among its strength and stiffness properties and the directions of load applications. These behaviors are called *isotropic, anisotropic,* and *orthotropic*.

Isotropic behavior means that the elastic response of the material is the same regardless of the direction of the applied load. Homogeneous materials such as most wrought metals (steel, aluminum, copper, titanium, etc.) are typically analyzed as isotropic materials. Many of the stress and deformation analysis techniques used in this book are based on the assumption of isotropic behavior. Material properties data for isotropic materials typically include a single value for tensile strength, modulus of elasticity, Poisson's ratio, and other properties.

Anisotropic behavior means that the elastic response of the material is different in all directions. A material that is a random, highly nonuniform aggregate of constituents would be expected to exhibit anisotropic behavior. Thus it would not be correct to use the basic analysis techniques developed in this book because they are based on the assumption of isotropic behavior. For components and structures made from such materials, testing under actual loading conditions is advised to verify suitability for a particular application.

Orthotropic behavior means that the material properties are different in three mutually perpendicular directions. This is the behavior that is most often assumed for composite materials, particularly those made from laminated construction as described next. To fully describe the behavior of an orthotropic material requires the determination of strength, modulus of elasticity (stiffness), and Poisson's ratio in each of the mutually perpendicular directions. Typically, these directions are called 1, 2, and 3. See Reference 27.

Figure 2–27 shows a segment of a composite material made with the reinforcing fibers aligned in a particular direction. In such cases, the 1-direction is aligned in that direction. Then the 2-direction is typically taken perpendicular to the direction of the fibers

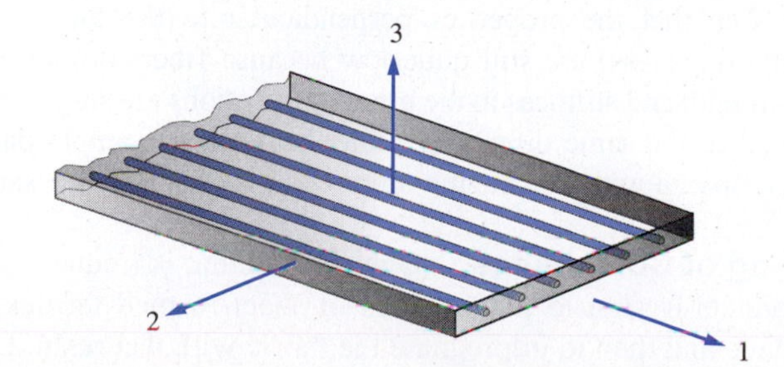

FIGURE 2–27 Unidirectional composite showing direction of principal axes.

TABLE 2–14 Examples of the effect of laminate construction on strength and stiffness.

	Tensile strength				Modulus of elasticity			
	Longitudinal		Transverse		Longitudinal		Transverse	
Laminate type	ksi	MPa	ksi	MPa	10^6 psi	GPa	10^6 psi	GPa
Unidirectional	200	1380	5	34	21	145	1.6	11
Quasi-isotropic	80	552	80	552	8	55	8	55

in the plane of the surface of the component. The 3-direction is then perpendicular to that plane. You should be able to see that the properties of the material in the three directions would be quite different from one another, based on the way in which the fibers contribute to the resistance of the applied loads.

Laminated Composite Construction. Many structures made from composite materials are made from several layers of the basic material containing both the matrix and the reinforcing fibers. The manner in which the layers are oriented relative to one another affects the final properties of the completed structure.

As an illustration, consider that each layer is made from a set of parallel strands of the reinforcing filler material, such as E-glass fibers, embedded in the resin matrix, such as polyester. In this form, the material is sometimes called a *prepreg,* indicating that the filler has been preimpregnated with the matrix prior to forming the structure and curing the assembly.

To produce the maximum strength and stiffness in a particular direction, several layers or plies of the prepreg could be laid on top of one another with all of the fibers aligned in the direction of the expected tensile load. This is called a *unidirectional laminate*. After curing, the laminate would have a very high strength and stiffness when loaded in the direction of the strands, called the *longitudinal* direction. However, the resulting product would have a very low strength and stiffness in the direction perpendicular to the fiber direction, called the *transverse* direction. If any off-axis loads are encountered, the part may fail or deform significantly. Table 2–14 gives sample data for a unidirectional laminated carbon/epoxy composite.

To overcome the lack of off-axis strength and stiffness, laminated structures should be made with a variety of orientations of the layers. One popular arrangement is shown in Figure 2–28. Naming the longitudinal direction of the surface layer the *0° ply,* this structure is referred to as

$$0°, 90°, +45°, -45°, -45°, +45°, 90°, 0°$$

The symmetry and balance of this type of layering technique results in more nearly uniform properties in two directions. The term *quasi-isotropic* is sometimes used to describe such a structure. Note that the properties perpendicular to the faces of the layered structure (through the thickness) are still quite low because fibers do not extend in that direction. Also, the strength and stiffness in the primary directions are somewhat lower than if the plies were aligned in the same direction. Table 2–14 shows sample data for a quasi-isotropic laminate compared with one having unidirectional fibers in the same matrix.

Processing of Composites. One method that is frequently used to produce composite products is first to place layers of sheet-formed fabrics on a form having the desired shape and then to impregnate the fabric with wet resin. Each layer of fabric can

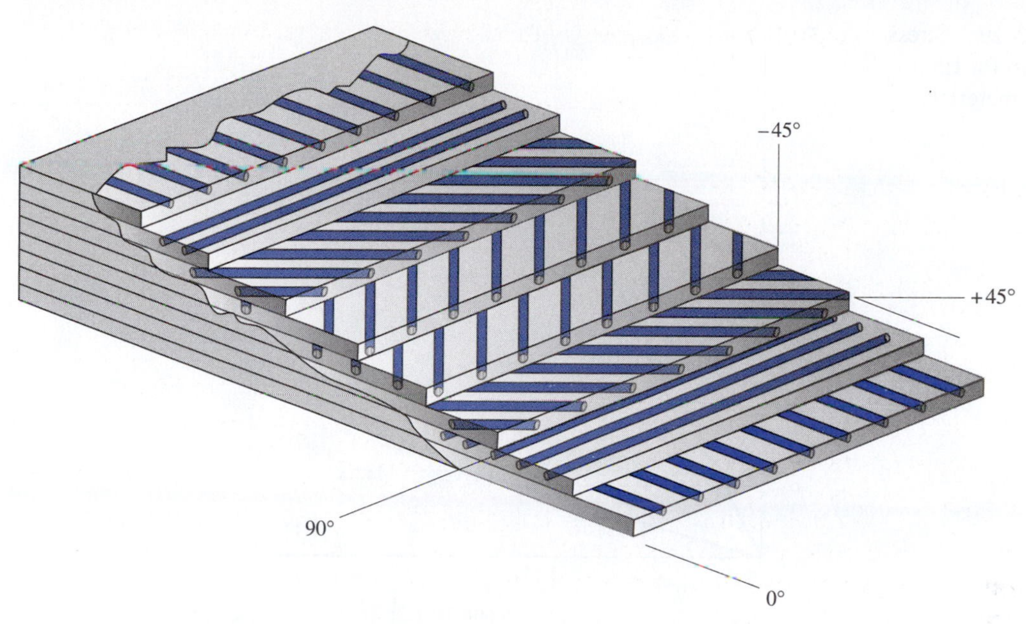

FIGURE 2–28 Multilayer laminated composite construction designed to produce quasi-isotropic properties.

be adjusted in its orientation to produce special properties of the finished article. After the lay-up and resin impregnation are completed, the entire system is subjected to heat and pressure while a curing agent reacts with the base resin to produce cross-linking that binds all of the elements into a three-dimensional, unified structure. The polymer binds to the fibers and holds them in their preferred position and orientation during use.

An alternative method of fabricating composite products starts with a process of preimpregnating the fibers with the resin material to produce strands, tape, braids, or sheets. The resulting form, called a *prepreg,* can then be stacked into layers or wound onto a form to produce the desired shape and thickness. The final step is the curing cycle as described for the wet process.

Polyester-based composites are often produced as *sheet-molding compounds (SMC)* in which preimpregnated fabric sheets are placed into a mold and shaped and cured simultaneously under heat and pressure. Large body panels for automotive applications can be produced in this manner.

Pultrusion is a process in which the fiber reinforcement is coated with resin as it is pulled through a heated die to produce a continuous form in the desired shape. This process is used to produce rod, tubing, structural shapes (I-beams, channels, angles, and so on), tees, and hat sections used as stiffeners in aircraft structures.

Filament winding is used to make pipe, pressure vessels, rocket motor cases, instrument enclosures, and odd-shaped containers. The continuous filament can be placed in a variety of patterns, including helical, axial, and circumferential, to produce desired strength and stiffness characteristics.

Predicting Composite Properties. The following discussion summarizes some of the important variables needed to define the properties of a composite. The subscript c refers to the composite, m refers to the matrix, and f refers to the fibers. The strength and stiffness of a composite material depend on the elastic properties of the fiber and matrix

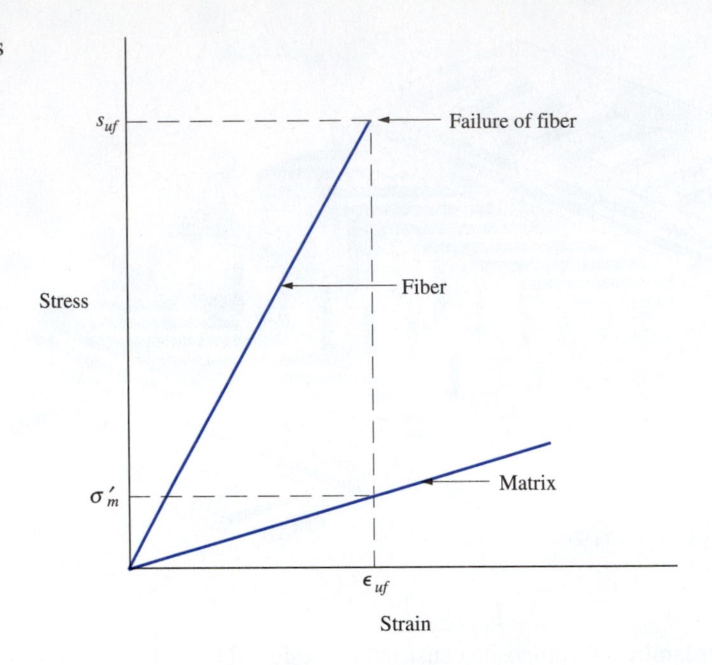

FIGURE 2–29 Stress versus strain for fiber and matrix materials.

components. But another parameter is the relative volume of the composite composed of fibers, V_f, and that composed of the matrix material, V_m. That is,

$$V_f = \text{Volume fraction of fiber in the composite}$$
$$V_m = \text{Volume fraction of matrix in the composite}$$

Note that for a unit volume, $V_f + V_m = 1$. Then, $V_m = 1 - V_f$.

We will use an ideal case to illustrate the way in which the strength and stiffness of a composite can be predicted. Consider a composite with unidirectional continuous fibers aligned in the direction of the applied load. The fibers are typically much stronger and stiffer than the matrix material. Furthermore, the matrix will be able to undergo a larger strain before fracture than the fibers can. Figure 2–29 shows these phenomena on a plot of stress versus strain for the fibers and the matrix. We will use the following notation for key parameters from Figure 2–29.

$$s_{uf} = \text{Ultimate strength of fiber}$$
$$\epsilon_{uf} = \text{Strain in the fiber corresponding to its ultimate strength}$$
$$\sigma'_m = \text{Stress in the matrix at the same strain as } \epsilon_{uf}$$

The ultimate strength of the composite, s_{uc}, is at some intermediate value between s_{uf} and σ'_m, depending on the volume fraction of fiber and matrix in the composite, That is,

$$s_{uc} = s_{uf}V_f + \sigma'_m V_m \quad \textbf{(2–10)}$$

At any lower level of stress, the relationship among the overall stress in the composite, the stress in the fibers, and the stress in the matrix follows a similar pattern.

$$\sigma_c = \sigma_f V_f + \sigma_m V_m \quad \textbf{(2–11)}$$

Figure 2–30 illustrates this relationship on a stress–strain diagram.

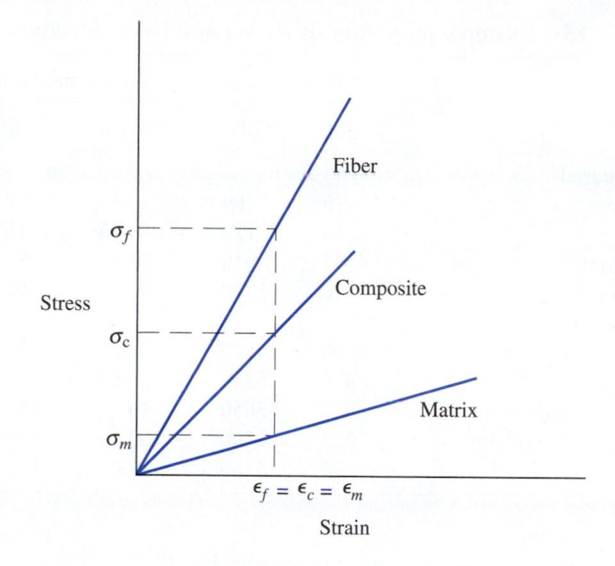

FIGURE 2–30 Relationship among stresses and strains for a composite and its fiber and matrix materials.

Both sides of Equation (2–6) can be divided by the strain at which these stresses occur. And, since for each material $\sigma/\epsilon = E$, the modulus of elasticity for the composite can be shown as,

$$E_c = E_f V_f + E_m V_m \tag{2–12}$$

The density of a composite can be computed in a similar fashion.

$$\rho_c = \rho_f V_f + \rho_m V_m \tag{2–13}$$

Density is defined as *mass per unit volume*. A related property, specific weight, is defined as *weight per unit volume* and is denoted by the symbol γ (Greek letter gamma). The relationship between density and specific weight is simply $\gamma = \rho g$, where g is the acceleration due to gravity. Multiplying each term in Equation (2–8) by g gives the formula for the specific weight of a composite,

$$\gamma_c = \gamma_f V_f + \gamma_m V_m \tag{2–14}$$

The form of Equations (2–12) to (2–14) is often called *the rule of mixtures*.

Table 2–15 lists example values for the properties of some matrix and filler materials. Remember that wide variations can occur in such properties, depending on the exact formulation and the condition of the materials.

Example Problem 2–1 Compute the expected properties of ultimate tensile strength, modulus of elasticity, and specific weight of a composite made from unidirectional strands of carbon-PAN fibers in an epoxy matrix. The volume fraction of fibers is 30%. Use data from Table 2–15.

Solution Objective: Compute the expected values of s_{uc}, E_c, and γ_c for the composite.

Given: Matrix–epoxy: $s_{um} = 18$ ksi; $E_m = 0.56 \times 10^6$ psi; $\gamma_m = 0.047$ lb/in^3.
Fiber–carbon-PAN: $s_{uf} = 470$ ksi; $E_f = 33.5 \times 10^6$ psi; $\gamma_f = 0.064$ lb/in^3.
Volume fraction of fiber, $V_f = 0.30$. And, $V_m = 1.0 - 0.30 = 0.70$.

TABLE 2–15 Example properties of matrix and filler materials.

	Tensile strength		Tensile modulus		Specific weight	
	ksi	MPa	10^6 psi	GPa	lb/in^3	kN/m^3
Matrix materials:						
Polyester	10	69	0.40	2.76	0.047	12.7
Epoxy	18	124	0.56	3.86	0.047	12.7
Aluminum	45	310	10.0	69	0.100	27.1
Titanium	170	1170	16.5	114	0.160	43.4
Filler materials:						
S-glass	600	4140	12.5	86.2	0.09	24.4
Carbon–PAN	470	3240	33.5	231	0.064	17.4
Carbon–PAN (high-strength)	820	5650	40	276	0.065	17.7
Carbon (high-modulus)	325	2200	100	690	0.078	21.2
Aramid	500	3450	19.0	131	0.052	14.1

Analysis and Results Ultimate tensile strength, s_{uc}, computed from Eq. (2–5)

$$s_{uc} = s_{uf}V_f + \sigma'_m V_m$$

To find σ'_m we first find the strain at which the fibers would fail at s_{uf}. Assume that the fibers are linearly elastic to failure. Then,

$$\epsilon_f = s_{uf}/E_f = (470 \times 10^3 \text{ psi})/(33.5 \times 10^6 \text{ psi}) = 0.014$$

At this same strain, the stress in the matrix is

$$\sigma'_m = E_m\epsilon = (0.56 \times 10^6 \text{ psi})(0.014) = 7840 \text{ psi}$$

Then, in Eq. (2–5),

$$s_{uc} = (470\ 000 \text{ psi})(0.30) + (7840 \text{ psi})(0.70) = 146\ 500 \text{ psi}$$

Modulus of elasticity computed from Eq. (2–7),

$$E_c = E_fV_f + E_mV_m = (33.5 \times 10^6)(0.30) + (0.56 \times 10^6)(0.70)$$
$$E_c = 10.4 \times 10^6 \text{ psi}$$

Specific weight computed from Eq. (2–9),

$$\gamma_c = \gamma_fV_f + \gamma_mV_m = (0.064)(0.30) + (0.047)(0.70) = 0.052 \text{ lb/in}^3$$

Summary of Results

$$s_{uc} = 146\ 500 \text{ psi}$$
$$E_c = 10.4 \times 10^6 \text{ psi}$$
$$\gamma_c = 0.052 \text{ lb/in}^3$$

Comment Note that the resulting properties for the composite are intermediate between those for the fibers and the matrix.

2–13 MATERIALS SELECTION

One of the most important tasks for a designer is the specification of the material from which any individual component of a product is to be made. The decision must consider a huge number of factors, many of which have been discussed in this chapter.

The process of material selection must commence with a clear understanding of the functions and design requirements for the product and the individual component. Then, the designer should consider the interrelationships among the following:

- The functions of the component
- The component's shape
- The material from which the component is to be made
- The manufacturing process used to produce the component

Overall requirements for the performance of the component must be detailed. This includes, for example:

- The nature of the forces applied to the component
- The types and magnitudes of stresses created by the applied forces
- The allowable deformation of the component at critical points
- Interfaces with other components of the product
- The environment in which the component is to operate
- Physical size and weight of the component
- Aesthetics expected for the component and the overall product
- Cost targets for the product as a whole and this component in particular
- Anticipated manufacturing processes available

A much more detailed list may be made with more knowledge of specific conditions.

From the results of the exercises described previously, you should develop a list of key material properties that are important. Examples often include:

1. Strength as indicated by ultimate tensile strength, yield strength, compressive strength, fatigue strength, shear strength, and others
2. Stiffness as indicated by the tensile modulus of elasticity, shear modulus of elasticity, or flexural modulus
3. Weight and mass as indicated by specific weight or density
4. Ductility as indicated by the percent elongation
5. Toughness as indicated by the impact energy (Izod, Charpy, etc.)
6. Creep performance data
7. Corrosion resistance and compatibility with the environment
8. Cost for the material
9. Cost to process the material

A list of candidate materials should then be created using your knowledge of the behavior of several material types, successful similar applications, and emerging materials technologies. A rational decision analysis should be applied to determine the most suitable types of materials from the list of candidates. This could take the form of a matrix in which data for the properties just listed for each candidate material are entered and ranked. An analysis of the complete set of data will aid in making the final decision.

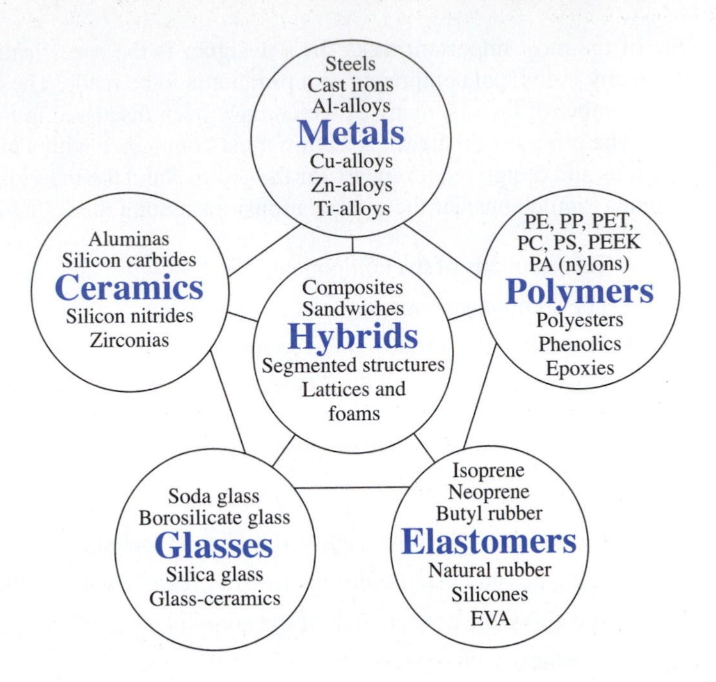

FIGURE 2–31 Classifications of materials. (Source: Granta Design, Cambridge, United Kingdom)

More comprehensive materials selection processes are described in References 4–6, 9, 16, and 29.

Figure 2–31 is a broad overview of the types of materials from which a designer may choose for a particular application. Basically it shows five primary classes: metals, polymers, ceramics, glasses, and elastomers, along with hybrids that combine two or more materials to achieve specialized properties. Composites, discussed in Section 2–12, are an obvious and increasingly popular example of hybrid materials. Also included as hybrids are fabricated structures such as sandwich panels employing very light but relatively stiff inner cores between strong outer skins that carry much of the load. See Figure 2–32. The cores may be foams, honeycombs, corrugated sheets, or other such materials. The resulting structure is an example of selecting materials to optimize the performance of a given component.

However, the sheer number of different materials from which to choose makes material selection a daunting task. Specialized approaches, as described in Reference 4, offer significant guidance in the selection process. Furthermore, computer software is available to permit rapid searching on many parameters to produce lists of candidate materials with quantitative data about their performance, cost, producibility, or other important criteria. Two relatively simple examples of selection processes help to illustrate the method. See Internet site 31.

Some components of a structure or a machine may be limited by strength where avoiding failure is primary. In addition, it may be desirable to minimize the weight of the component as in aerospace applications or where a building designer wants a low total weight of a structure for cost savings and to decrease loads on a foundation. In this case, arranging material property data in the form shown in Figure 2–33 would help to visualize possible choices. Here the vertical axis is the yield strength of the materials and the horizontal axis is the density. Desiring a high strength and low weight, the designer would look toward the upper left of the chart. The shaded areas show the range of strength and density properties for the different classes of materials.

An alternate approach may be applied where stiffness of a component is primary, as in designing a floor for a passenger aircraft or an apartment building. Occupants want a stiff, rigid floor that does not flex noticeably. Again, the designer may want to minimize weight. Figure 2–34 shows a chart of modulus of elasticity, E, versus density for a variety

FIGURE 2–32
Laminated panels with lightweight cores.

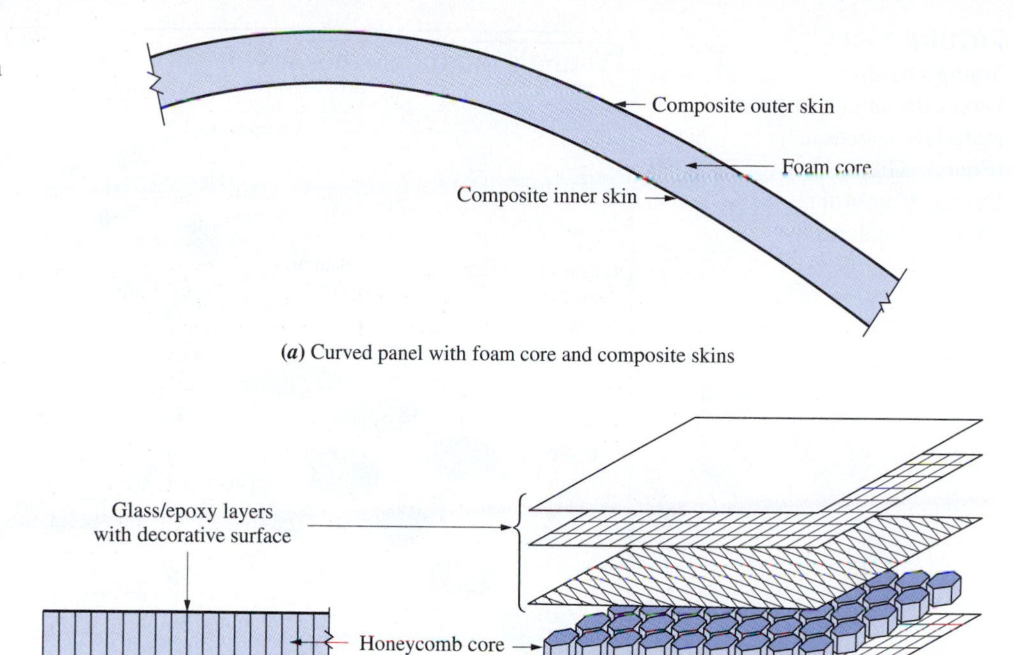

FIGURE 2–33
Strength versus density chart for materials selection. (Source: Granta Design, Cambridge, United Kingdom)

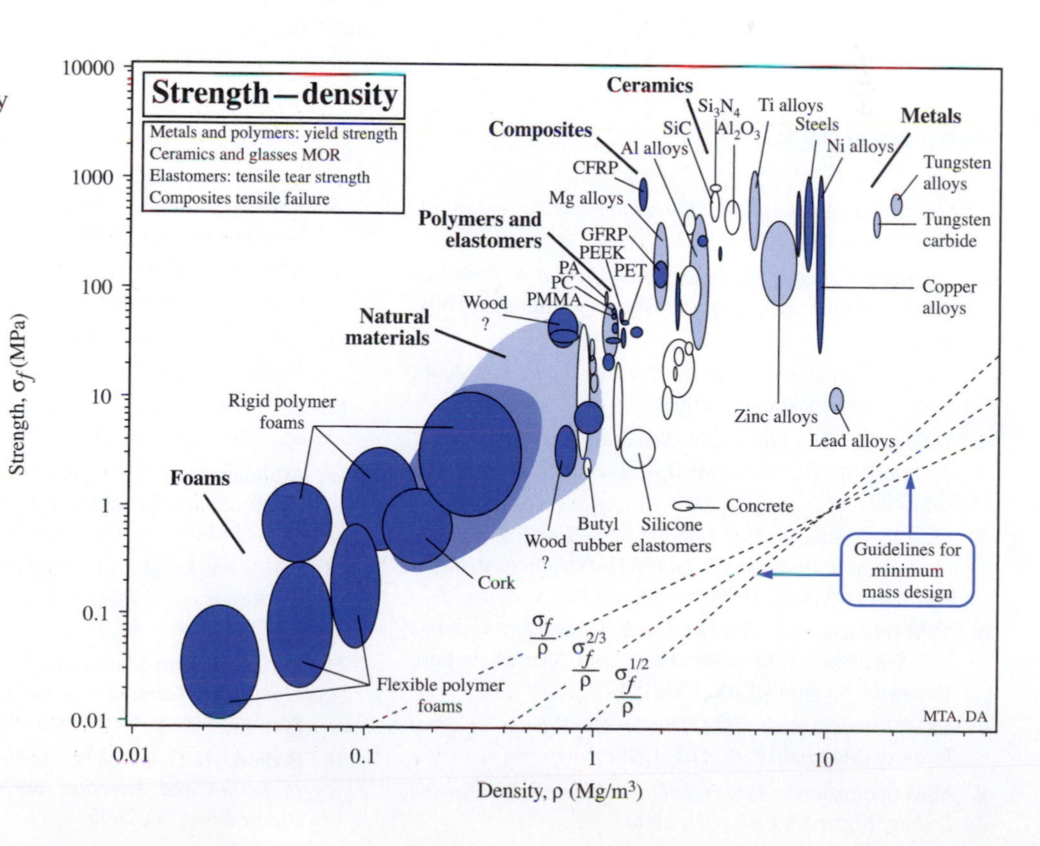

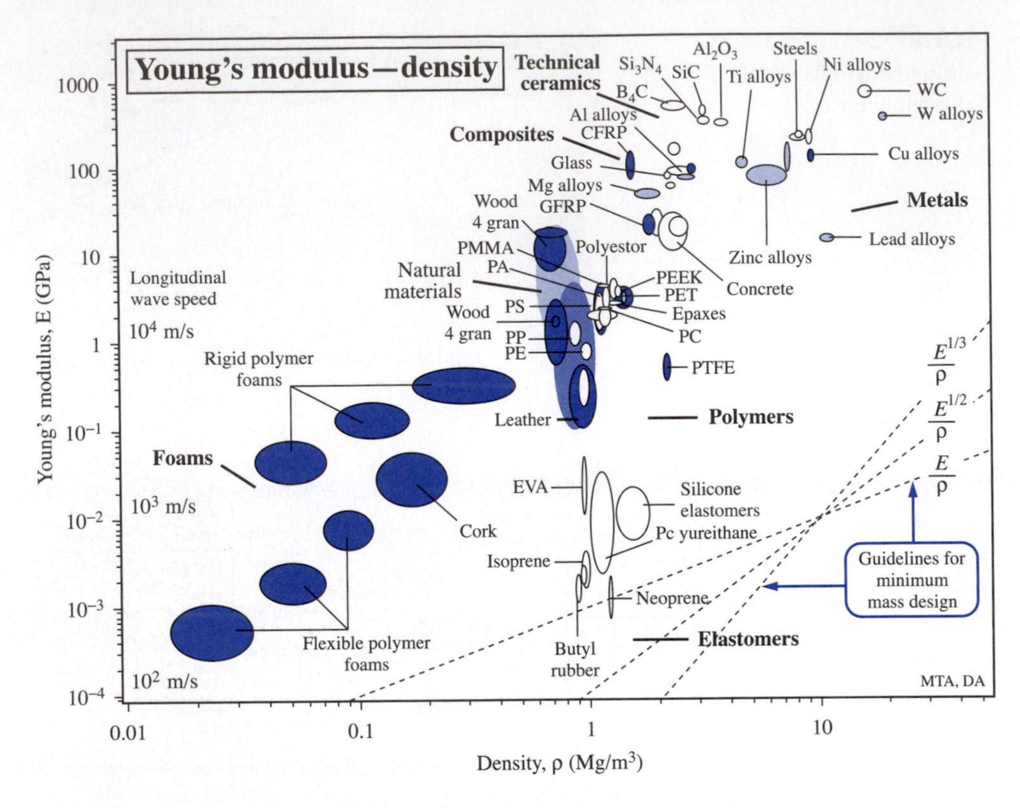

FIGURE 2–34 Young's modulus versus density chart for materials selection. (Source: Granta Design, Cambridge, United Kingdom)

of materials. Here, again, the designer would look to the upper left for desirable materials. More detailed charts show more specific examples within these areas, such as specific steel alloys within the larger "Metals" region.

Numerous other parametric searches can be constructed using the selection software.

REFERENCES

1. Aluminum Association, *Aluminum Design Manual,* Washington, DC, 2005.
2. American Concrete Institute, *2006 Manual of Concrete Practice,* American Concrete Institute, Farmington Hills, MI, 2006.
3. ASTM International, *Annual Book of Standards,* 2006, West Conshohocken, PA, 2006.
4. Ashby, M. F., *Materials Selection in Mechanical Design,* 3rd ed., Butterworth-Heinemann, Burlington, MA, 2005.
5. ASM International, *ASM Handbook, Volume 1: Properties and Selection: Irons, Steels, and High-Performance Alloys,* Materials Park, OH, 1990.
6. ASM International, *ASM Handbook, Volume 2: Properties and Selection: Nonferrous Alloys and Special-Purpose Materials,* Materials Park, OH, 1990.
7. ASM International, *ASM Handbook, Volume 4: Heat Treating,* Materials Park, OH, 1991.
8. ASM International, *ASM Handbook, Volume 8: Mechanical Testing,* Materials Park, OH, 2000.
9. ASM International, *ASM Handbook, Volume 20: Materials Selection and Design,* Materials Park, OH, 1997.
10. ASM International, *ASM Handbook, Volume 21. Composites,* Materials Park, OH, 2001.
11. ASM International, *Engineering Plastics, Engineered Materials Handbook, Volume 2,* Materials Park, OH, 1988.
12. ASM International, *Metals Handbook Desk Edition,* 2nd ed., Materials Park, OH, 1998.
13. Avallone, Eugene A. and Theodore Baumeister III, eds., *Marks' Standard Handbook for Mechanical Engineers,* 10th ed., McGraw-Hill, New York, 1996.
14. Berins, Michael L., ed., *Plastics Engineering Handbook of the Society of the Plastics Industry, Inc.,* 5th ed., Chapman and Hall, New York, 1995.
15. Bertholet, Jean-Marie and Michael J. Cole, *Composite Materials: Mechanical Behavior and Structural Analysis,* Springer Verlag, New York, 1999.
16. Budinski, K. G. and M. K. Budinski, *Engineering Materials: Properties and Selection,* 8th ed., Prentice Hall, Upper Saddle River, NJ, 2005.

17. Bunsell, A. R. and J. Renard, *Fundamentals of Fibre Reinforced Composite Materials,* CRC Press, Boca Raton, FL, 2005.
18. DuPont Engineering Polymers, *Design Handbook for DuPont Engineering Polymers: General Design Principles,* The DuPont Company, Wilmington, DE, 2002.
19. Faherty, Keith F. and Thomas G. Williamson, *Wood Engineering and Construction Handbook,* 3rd ed., McGraw-Hill, New York, 1999.
20. Gay, D., D. Hoa, and S. W. Tsai, *Composite Materials: Design and Applications,* CRC Press, Boca Raton, FL, 2002.
21. Mallick, P. K., *Composites Engineering Handbook,* Marcel Dekker, New York, 1997.
22. The Masonry Society, *Masonry Designer's Guide,* 4th ed., The Masonry, Society, Boulder, CO, 2003.
23. Mazumdar, S., *Composites Manufacturing: Materials, Product, and Process Engineering,* CRC Press, Boca Raton, FL, 2001.
24. Mehta, P. Kumar and P. J. M. Monteiro, *Concrete,* 3rd ed., McGraw-Hill, New York, 2006.
25. Mott, Robert L., *Machine Elements in Mechanical Design,* 4th ed., Prentice-Hall, Upper Saddle River, NJ, 2004.
26. Richardson, T., *Composites: A Design Guide,* Industrial Press, New York, 1987.
27. Staab, George H., *Laminar Composites,* Butterworth-Heinemann, Boston, 1999.
28. Shackelford, J. F. and W. Alexander, *The CRC Materials Science and Engineering Handbook,* 3rd ed., CRC Press, Boca Raton, FL, 2000.
29. Shackelford, J. F., W. Alexander, and J. S. Park, *The CRC Practical Handbook of Materials Selection,* CRC Press, Boca Raton, FL, 1995.
30. Strong, A. Brent., *Plastics: Materials and Processing,* 3rd ed., Prentice-Hall, Upper Saddle River, NJ, 2006.
31. U.S. Department of Agriculture Forest Products Laboratory, *Wood Handbook—Woods as an Engineering Material,* Forest Products Laboratory, Madison, WI, 1999.

INTERNET SITES

1. AZoM.com (The A to Z of Materials) www.azom.com Materials information resource for the design community. No cost, searchable databases for metals, ceramics, polymers, and composites. Can also search by keyword, application, or industry type.
2. Matweb www.matweb.com Database of material properties for many metals, plastics, ceramics, and other engineering materials.
3. ASM International www.asm-intl.org The society for materials engineers and scientists; a worldwide network dedicated to advancing industry, technology, and applications of metals and other materials.
4. TECHstreet www.techstreet.com A store for purchasing standards for the metals industry.
5. SAE International www.sae.org The Society of Automotive Engineers, the engineering society for advancing mobility on land or sea, in air or space. A resource for technical information used in designing self-propelled vehicles. Offers standards on metals, plastic, and other materials along with components and subsystems of vehicles.
6. ASTM International www.astm.org Formerly known as the American Society for Testing and Materials. Develops and sells standards for material properties, testing procedures, and numerous other technical standards.
7. American Iron and Steel Institute www.steel.org AISI develops industry standards for steel materials and products made from steel. Steel product manuals and industry standards are made available through the Iron & Steel Society (ISS), listed separately.
8. Association for Iron & Steel Technology www.aist.org Advances the technical development, production, processing and application of iron and steel. Provides publications, conferences, and technical meetings for the iron and steel community.
9. Aluminum Association www.aluminum.org The association of the aluminum industry. Provides numerous publications that can be purchased.
10. Alcoa, Inc. www.alcoa.com A producer of aluminum and fabricated products. Website can be searched for properties of specific alloys.
11. Copper Development Association www.copper.org Provides a large searchable database of properties of wrought and cast copper, copper alloys, brasses, and bronzes. Allows searching for appropriate alloys for typical industrial uses based on several performance characteristics.
12. Metal Powder Industries Federation www.mpif.org The international trade association representing the powder metal producers. Standards and publications related to the design and production of products using powder metals.
13. INTERZINC www.interzinc.com A market development and technology transfer group dedicated to increasing awareness of zinc casting alloys. Provides design assistance,

alloy selection guide, alloy properties, and descriptions of casting alloys.

14. RAPRA Technology Limited www.rapra.net Comprehensive information source for the plastics and rubber industries. Formerly Rubber and Plastics Research Association. This site also hosts the Cambridge Engineering Selector, a computerized resource using the materials selection methodology of M. F. Ashby. See Reference 4.
15. DuPont Plastics www.plastics.dupont.com Information and data on DuPont plastics and their properties. Searchable database by type of plastic or application.
16. PolymerPlace.com www.polymerplace.com Information resource for the polymer industry.
17. PlasticsUSA.com www.plasticsusa.com An Internet portal providing technical information for the polymer industry and end-user industries including automotive, construction, electrical, medical, packaging, and polyurethane sectors. Includes a plastics materials database, information on trade names, and links to suppliers of plastics materials and processing equipment.
18. Plastics Technology Online www.ptonline.com An online source of information about the plastics industry and processes such as injection molding, extrusion, blow molding, thermoforming, foam, materials, tooling, and auxiliary equipment.
19. Society of Plastics Engineers www.4spe.org SPE promotes scientific and engineering knowledge and education about plastics and polymers worldwide.
20. U.S. Department of Agriculture, Forest Products Laboratory www.fpl.fs.fed.us Research organization devoted to producing information useful to the forest products industry and users of wood products.
21. American Concrete Institute www.aci-int.org Publisher of technical documents related to advancing concrete knowledge, including the 2006 *Manual of Concrete Practice.*
22. The Masonry Society www.masonrysociety.org A professional association dedicated to the advancement of knowledge on masonry and publisher of *Masonry Designer's Guide.*
23. MTS Systems Corporation www.mts.com A supplier of materials testing products including universal testing machines, dynamic and fatigue testing machines for metals, plastics, biomaterials, wood, fibers, and others.
24. Instron Corporation www.instron.com A supplier of materials testing products including universal testing machines, dynamic and fatigue testing machines, thermal shock testers, hardness (Wilson and Shore brands), impact testers, crack propagation and fracture toughness testers.
25. Tinius Olsen, Inc. www.tiniusolsen.com A supplier of materials testing products including static universal testing machines, impact testers, torsion testers, and special equipment for the plastics and food industries. The site has a useful Resource Center tab providing quick reference information.
26. Zwick/Roell Group www.zwickroell.com A supplier of materials testing products including universal testing machines, dynamic and fatigue testing machines, impact testers, and specialized automated systems. The Indentec Hardness Testing Machines, Ltd., unit provides a wide range of hardness testers.
27. National Institute of Standards and Technology (NIST)—Special Publication 960–5, *Rockwell Hardness Measurement of Metallic Materials,* by Samuel R. Low. May be downloaded from: http://www.metallurgy.nist.gov/reports/recommendedpractice/SP960 5.pdf
28. Special Metals Corporation www.specialmetals.com Producer of nickel-based alloys with brand names INCONEL, INCOLOY, NIMONIC, UDIMET, MONEL, and NILO.
29. Allegheny Ludlum Corporation www.alleghenyludlum.com Producer of nickel-based alloys with brand names AL, ALLCORR, and ALTEMP.
30. Haynes International, Inc. www.haynesintl.com Producer of nickel- and cobalt-based alloys with brand names HASTELLOY, HAYNES, and ULTIMET.
31. Granta Material Intelligence www.grantadesign.com Producer and marketer of the EduPack materials selection software using concepts developed by Professor Michael F. Ashby.

PROBLEMS

2–1. Name four kinds of metals commonly used for load-carrying members.

2–2. Name 11 factors that should be considered when selecting a material for a product.

2–3. Define *ultimate tensile strength.*

2–4. Define *yield point.*

2–5. Define *yield strength.*

2–6. When is yield strength used in place of yield point?

2–7. Define *stiffness.*

2–8. What material property is a measure of its stiffness?

2–9. State Hooke's law.

2–10. What material property is a measure of its ductility?

2–11. How is a material classified as to whether it is ductile or brittle?

2–12. Name four types of steels.

2–13. What does the designation AISI 4130 for a steel mean?

2–14. What are the ultimate strength, yield strength, and percent elongation of AISI 1040 hot-rolled steel? Is it a ductile or a brittle material?

2–15. Which has a greater ductility: AISI 1040 hot-rolled steel or AISI 1020 hot-rolled steel?

2–16. What does the designation AISI 1141 OQT 700 mean?

2–17.E If the required yield strength of a steel is 150 ksi, could AISI 1141 be used? Why?

2–18.M What is the modulus of elasticity for AISI 1141 steel? For AISI 5160 steel?

2–19.E A rectangular bar of steel is 1.0 in by 4.0 in by 14.5 in. How much does it weigh in pounds?

2–20.M A circular bar is 50 mm in diameter and 250 mm long. How much does it weigh in newtons?

2–21.M If a force of 400 N is applied to a bar of titanium and an identical bar of magnesium, which would stretch more?

2–22. Name four types of structural steels and list the yield point for each.

2–23. What does the aluminum alloy designation 6061-T6 mean?

2–24.E List the ultimate strength, yield strength, modulus of elasticity, and density for 6061-O, 6061-T4, and 6061-T6 aluminum.

2–25. List five uses for bronze.

2–26. List three desirable characteristics of titanium as compared with aluminum or steel.

2–27. Name five varieties of cast iron.

2–28. Which type of cast iron is usually considered to be brittle?

2–29.E What are the ultimate strengths in tension and in compression for ASTM A48 grade 40 cast iron?

2–30. How does a ductile iron differ from gray iron?

2–31.E List the allowable stresses in bending, tension, compression, and shear for No. 2 grade Douglas fir.

2–32.E What is the normal range of compressive strengths for concrete?

2–33. Describe the difference between thermoplastic and thermosetting materials.

2–34. Name three suitable plastics for use as gears or cams in mechanical devices.

2–35. Describe the term *composite*.

2–36. Name five basic types of materials that are used as a matrix for composites.

2–37. Name five different thermoplastics used as a matrix for composites.

2–38. Name three different thermosetting plastics used as a matrix for composites.

2–39. Name three metals used as a matrix for composites.

2–40. Describe nine forms that filler materials take when used in composites.

2–41. Discuss the differences among *strands, roving,* and *fabric* as different forms of fillers for composites.

2–42. Name seven types of filler materials used for composites.

2–43. Name five different types of glass fillers used for composites and describe the primary features of each.

2–44. Which of the commonly used filler materials has the highest stiffness?

2–45. Which filler materials should be considered for high temperature applications?

2–46. What is a common brand name for aramid fibers?

2–47. Define *specific strength* of a composite.

2–48. Define *specific modulus* of a composite.

2–49. List ten advantages of composites when compared to metals.

2–50. List nine limitations of composites.

2–51. From the data for selected materials in Table 2–13, list the ten materials in order of specific strength from highest to lowest. For each, compute the ratio of its specific strength to that for AISI 1020 HR steel.

2–52. From the data for selected materials in Table 2–13, list the ten materials in order of specific modulus from highest to lowest. For each, compute the ratio of its specific modulus to that for AISI 1020 HR steel.

2–53. Describe a unidirectional laminate and its general strength and stiffness characteristics.

2–54. Describe a quasi-isotropic laminate and its general strength and stiffness characteristics.

2–55. Compare the generally expected specific strength and stiffness characteristics of a quasi-isotropic laminate with a unidirectional laminate.

2–56. Describe a laminated composite that carries the designation 0°, +45°, −45°, −45°, +45°, 0°.

2–57. Describe a laminated composite that carries the designation 0°, +30°, +45°, +45°, +30°, 0°.

2–58. Define the term *volume fraction of fiber* for a composite.

2–59. Define the term *volume fraction of matrix* for a composite.

2–60. If a composite has a volume fraction of fiber of 0.60, what is the volume fraction of matrix?

2–61. Write the equation for the expected ultimate strength of a composite in terms of the properties of its matrix and filler materials.

2–62. Write the equations for the *rule of mixtures* as applied to a unidirectional composite for the stress in the composite, its modulus of elasticity, its density, and its specific weight.

2–63.M Compute the expected properties of ultimate strength, modulus of elasticity, and specific weight of a composite made from unidirectional strands of high-strength carbon-PAN fibers in an epoxy matrix. The volume fraction of fibers is 50%. Compute the specific strength and specific stiffness. Use data from Table 2–15.

2–64.M Repeat Problem 2–63 with high modulus carbon fibers.

2–65.M Repeat Problem 2–63 with aramid fibers.

Problems 2–66 to 2–77

Use Figure P2–66 for all problems. For the given data in the problem statement and data you read from the indicated stress–strain curve, determine the following properties of the material:

a) Yield strength. State whether the yield point is used to determine this value or if the 0.2% offset method is used.

b) Ultimate tensile strength.

c) Proportional limit.

d) Elastic limit.

e) Modulus of elasticity for the range of stress in which Hooke's law applies.

f) Percent elongation. [The gage length for each test is 2.00 in.]

g) State whether the material is ductile or brittle.

h) Examine the results and judge the kind of metal used to determine the test data.

i) Compare your results and find a particular alloy from Appendix tables A–14 to A–18 that has similar properties.

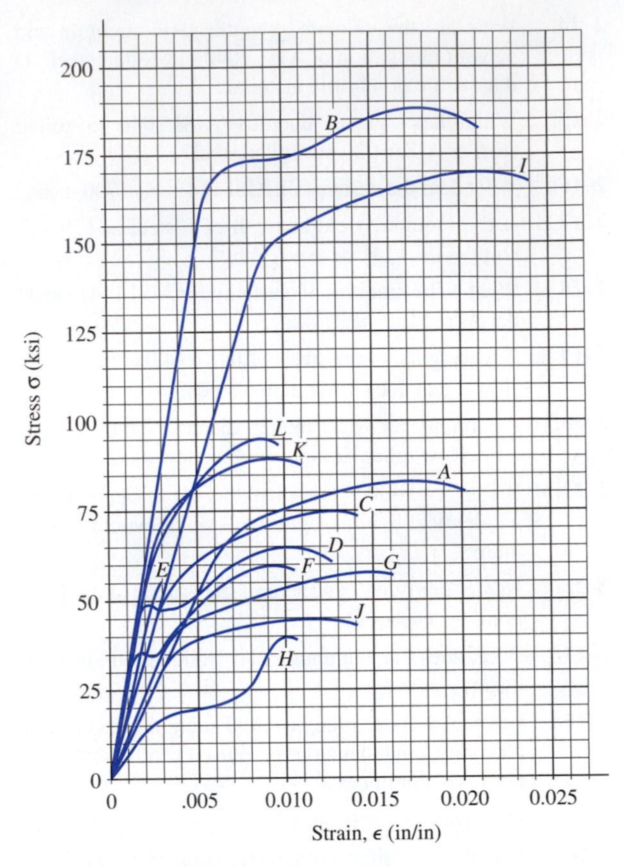

FIGURE P2–66 Stress–strain curves.

2–66. Use Curve A from Figure P2–66. Final length between gage marks = 2.22 in.

2–67. Use Curve B from Figure P2–66. Final length between gage marks = 2.30 in.

2–68. Use Curve C from Figure P2–66. Final length between gage marks = 2.30 in.

2–69 Use Curve D from Figure P2–66. Final length between gage marks = 2.72 in.

2–70. Use Curve E from Figure P2–66. Final length between gage marks = 2.01 in.

2–71. Use Curve F from Figure P2–66. Final length between gage marks = 2.42 in.

2–72. Use Curve G from Figure P2–66. Final length between gage marks = 2.10 in.

2–73. Use Curve H from Figure P2–66. Final length between gage marks = 2.10 in.

2–74. Use Curve I from Figure P2–66. Final length between gage marks = 2.16 in.

2–75. Use Curve J from Figure P2–66. Final length between gage marks = 2.34 in.

2–76. Use Curve K from Figure P2–66. Final length between gage marks = 2.30 in.

2–77. Use Curve L from Figure P2–66. Final length between gage marks = 2.04 in.

3

Direct Stress, Deformation, and Design

The Big Picture

Direct Stresses, Deformation, and Design

Discussion Map

- You will now build on your ability to compute stresses that you learned in Chapter 1 and develop skills in the *design of load-carrying members*.
- In design, you either specify or determine by calculation a suitable material from which to make the member, its shape, and the dimensions required to carry a given load safely.
- You will consider the design of members under direct stresses: axial tensile stress, axial compressive stress, bearing stress, and direct shear stress.
- You will learn how to select a reasonable *design factor, N,* and to apply it to the appropriate material properties to ensure that the member experiences a safe level of stress during its use.
- Different kinds of loading will be discussed: static loads, repeated loads, impact, and shock.
- You will learn to consider stress concentrations for axially loaded members in which abrupt changes in cross section occur.
- In addition, you will learn how to compute the deformation of axially loaded members due to both stress and thermal expansion.

Discover

Reflect back on the Big Picture discussion you had for Chapter 1. There you identified components of consumer products, structures, and machines that you are familiar with.

You thought about the structure of a home, furniture, appliances, bicycles, automobiles, construction equipment, commercial buildings, aircraft, and space vehicles. What else did you think of?

Now, focus on the kinds of loads some of those examples are subjected to.

Which are subjected to loads that do not vary significantly with time, called static loads? An example might be a beam in the basement of your home that holds the house structure above. The dead weight of the structure does not vary over time. What other examples can you find that are subjected to static loads?

Which are subjected to repeated loads? This is the case when a load is applied and removed many times during the expected life of the component. Some components may also experience reversing loads that are alternately stressed in tension and then in compression. Consider parts of your car. For example, the functional parts of the door latch experience high loads and stresses each time the latch is engaged. The loads are removed when the latch is disengaged. That cycle of loading and unloading is repeated many thousands of times during the life of the car. What other examples can you identify?

Which items on your list are subjected to shock and impact? Here the load is applied suddenly and sharply. An ideal example is a nail being struck by a hammer or a portable music player being dropped on the floor. Consider how a baseball bat or tennis racket responds when you hit a long line drive or a strong baseline shot. Can you think of more examples?

Can you see how the three types of loads just described require different criteria for designing safe structures and components? In this chapter you will learn how to specify suitable materials to carry such loads. You will also learn how to compute a safe design stress and how to determine the required shape and dimensions of load-carrying parts so they will not experience stresses above those levels.

In Chapter 1 the concept of direct stress was presented together with examples of the calculation of direct tensile stress, direct compressive stress, direct shear stress, and bearing stress. The emphasis was on the understanding of the basic phenomena, units, terminology, and the magnitude of stresses encountered in typical structural and mechanical applications. Nothing was said about the acceptability of the stress levels that were computed or about the design of members to carry a given load.

In this chapter the primary emphasis is on *design* in which you, as the designer, must make decisions about whether or not a proposed design is satisfactory; what the shape and size of the cross section of a load-carrying member should be; and what material the member should be made from.

Activity Chapter 3: Direct Stress, Deformation, and Design

Set up a system for holding a fine metal wire firmly by one end while providing a means of applying a direct axial tension force to the other. Figure 3–1 shows a commercial test device that accomplishes these functions. A universal tensile testing machine as shown in Figure 2–1 could be used if available. Perform the following:

1. Measure the diameter of the wire and the initial, unloaded length between the point of support and the place where the load is applied.
2. Apply loads in small increments, measuring the amount that the wire elongates for each load.
3. Compute the stress in the wire at each total load applied.
4. Divide each elongation measurement by the initial length of the wire to determine the strain in the wire.
5. Plot a graph of the stress in the wire on the vertical axis versus the elongation, showing also the strain on the horizontal axis.
6. After several points have been plotted (and before the wire breaks!), determine the slope of the line on the graph by dividing the change in stress by the change in strain over some convenient part of the graph where the plot is best approximated as a straight line.
7. The slope computed is a measure of the tensile modulus of elasticity, E, for the metal.

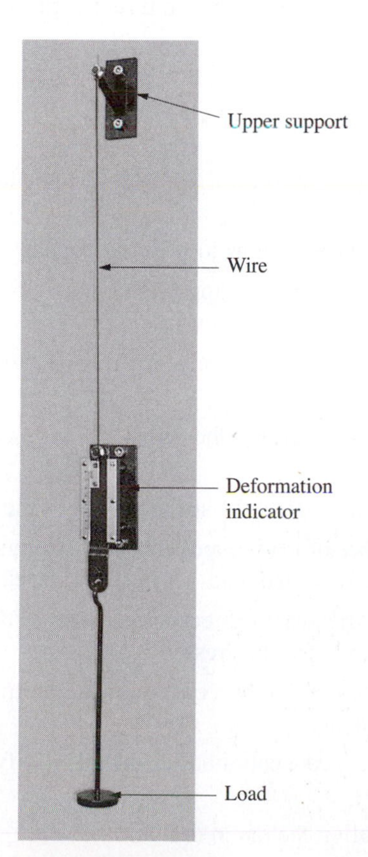

FIGURE 3–1 Testing device for tensile loading of wire. (Source: P. A. Hilton Ltd/Hi-Tech, Hampshire, England.)

8. Compare the computed value of E with that reported in tables if you know what the wire material is. If not, try to identify the type of metal by comparing it with values listed for some of the metals in Appendixes 14–18.
9. Observe that the straight-line relationship on the graph can be stated mathematically as:
   a. $E = \Delta\sigma/\Delta\varepsilon =$ Change in stress/change in strain
   b. $\sigma = E\varepsilon$ or, Stress = Modulus of elasticity × strain
10. Now, continue to add more load until the wire breaks.
11. Using the maximum load that the wire held before breaking, compute the maximum stress in the wire.
12. Of course, we do not normally want to break the wire. Therefore, assuming that we want to limit the load to a safe value no more than one-half of the stress at which it broke, compute that stress.
13. Summarize the results by reporting:
    a. Modulus of elasticity of the metal from which the wire is made.
    b. Maximum stress at which the wire broke.
    c. Stress at one-half of the maximum, calling this the *allowable stress* or the *design stress*.
    d. Total elongation at the load for which the design stress is produced in the wire

This simple test illustrates many of the main concepts you use in this chapter. In addition to stress and deformation caused by direct axial loads, you will study related topics dealing with bearing stress, contact stress, and direct shearing stresses.

3–1 OBJECTIVES OF THIS CHAPTER

After completing this chapter, you should be able to:

1. Describe the conditions that must be met for satisfactory application of the direct stress formulas.
2. Define *design stress* and tell how to determine an acceptable value for it.
3. Define *design factor* and select appropriate values for it depending on the conditions present in a particular design.
4. Discuss the relationship among the terms *design stress, allowable stress,* and *working stress.*
5. Discuss the relationship among the terms *design factor, factor of safety,* and *margin of safety.*
6. Describe 11 factors that affect the specification of the design factor.
7. Describe various types of loads experienced by structures or machine members, including static load, repeated load, impact, and shock.
8. Design members subjected to direct tensile stress, direct compressive stress, direct shear stress, and bearing stress.
9. Evaluate deformations caused by axial stress and thermal expansion and include them in design and analysis.
10. Determine when stress concentrations exist and specify suitable values for stress concentration factors.
11. Use stress concentration factors in design.

3–2 DESIGN OF MEMBERS UNDER DIRECT TENSION OR COMPRESSION

In Chapter 1 the direct stress formula was developed and stated as follows:

$$\sigma = \frac{F}{A} \tag{3–1}$$

where σ = direct normal stress: tension or compression
F = direct axial load
A = cross-sectional area of member subjected to F

For Equation (3–1) to be valid, the following conditions must be met:

1. The loaded member must be straight.
2. The loaded member must have a uniform cross section over the length under consideration.
3. The material from which the member is made must be homogeneous.
4. The load must be applied along the centroidal axis of the member so there is no tendency to bend it.
5. Compression members must be short so that there is no tendency to buckle. (See Chapter 11 for the special analysis required for long, slender members under compressive stress and for the method to decide when a member is to be considered long or short.)

It is important to recognize that the concept of stress refers to the internal resistance provided by a *unit area,* that is, an infinitely small area. Stress is considered to act at a point and may, in general, vary from point to point in a particular body. Equation (3–1) indicates that for a member subjected to direct axial tension or compression, the stress is uniform across the entire area if the five conditions are met. In many practical applications the minor variations that could occur in the local stress levels are accounted for by carefully selecting the allowable stress, as discussed later.

3–3 DESIGN NORMAL STRESSES

Failure occurs in a load-carrying member when it breaks or deforms excessively, rendering it unacceptable for the intended purpose. Therefore, it is essential that the level of applied stress never exceed the ultimate tensile strength or the yield strength of the material. Consideration of excessive deformation without yielding is discussed later in this chapter.

Design stress is that level of stress which may be developed in a material while ensuring that the loaded member is safe.

To compute design stress, two factors must be specified: the *design factor N* and the *property of the material on which the design will be based.* Usually, for metals, the design stress is based on either the yield strength s_y or the ultimate strength s_u of the material.

***The* design factor N *is a number by which the reported strength of a material is divided to obtain the* design stress σ_d.**

A variety of symbols are used in different fields for the strength properties of materials. In this book we use the following:

s_y = Yield strength of a material
s_u = Ultimate tensile strength of a material, or simply tensile strength

The term *yield strength* will be used regardless of whether the value was obtained from observing the yield point or using the offset technique as described in Chapter 2.

Other references may use the symbols σ_y and σ_u for these values. In the building construction field, particularly for steel building frames and members governed by the American Institute of Steel Construction (AISC), the symbols F_y and F_u are used.

The following equations can be used to compute the design stress for a certain value of N:

Design Stress

$$\sigma_d = \frac{s_y}{N} \quad \text{based on yield strength} \tag{3–2}$$

or

$$\sigma_d = \frac{s_u}{N} \quad \text{based on ultimate strength} \tag{3–3}$$

The value of the design factor is normally determined by the designer, using judgment and experience. In some cases, codes, standards, or company policy may specify design factors or design stresses to be used. When the designer must determine the design factor, his or her judgment must be based on an understanding of how parts may fail and the factors that affect the design factor. Sections 3–4, 3–5, and 3–6 give additional information about the design factor and about the choice of methods for computing design stresses.

Other references may use the term *factor of safety* in place of *design factor.* Also, *allowable stress* or *working stress* may be used in place of *design stress.* The choice of terms used in this book emphasizes the role of the designer in specifying the design stress.

Theoretically, a material could be subjected to a stress up to s_y before yield would occur. This condition corresponds to a value of the design factor of $N = 1$ in Equation (3–2). Similarly, with a design factor of $N = 1$ in Equation (3–3), the material would be on the brink of ultimate fracture. Thus $N = 1$ is the lowest value we can consider.

In this book we use the concept of design stresses and design factors as opposed to the margin of safety.

3–4 DESIGN FACTOR

Many different aspects of the design problem are involved in the specification of the design factor. In some cases the precise conditions of service are not known. The designer must then make conservative estimates of the conditions, that is, estimates that would cause the resulting design to be on the safe side when all possible variations are considered. The final choice of a design factor depends on the following 12 conditions.

Codes and Standards. If the member being designed falls under the jurisdiction of an existing code or standard, obviously the design factor or design stress must be chosen to satisfy the code or standard. Examples of standard-setting bodies are:

American Institute of Steel Construction (AISC): buildings, bridges, and similar structures using steel

Aluminum Association (AA): buildings, bridges, and similar structures using aluminum

American Society of Mechanical Engineers (ASME): boilers, pressure vessels, shafting, and numerous other mechanical components.

State building codes: buildings, bridges, and similar structures affecting the public safety

Department of Defense—Military Standards: aerospace vehicle structures and other military products

American National Standards Institute (ANSI): a wide variety of products

American Gear Manufacturers Association (AGMA): gears and gear systems

It is the designer's responsibility to determine which, if any, standards or codes apply to the member being designed and to ensure that the design meets those standards.

Material Strength Basis. Most designs using metals are based on either yield strength or ultimate strength or both, as stated previously. This is because most theories of metal failure show a strong relationship between the stress at failure and these material properties. Also, these properties will almost always be reported for materials used in engineering design. The value of the design factor will be different, depending on which material strength is used as the basis for design, as will be shown later.

Type of Material. A primary consideration with regard to the type of material is its ductility. The failure modes for brittle materials are quite different from those for ductile materials. Since brittle materials such as gray cast iron do not exhibit yielding, designs are always based on ultimate strength. Generally a metal is considered to be brittle if its percent elongation in a 2-in gage length is less than 5%. Except for highly hardened alloys, virtually all steels are ductile. Except for castings, aluminum is ductile. Other material factors that can affect the strength of a part are its uniformity and the confidence in the stated properties.

Manner of Loading. Three main types of loading can be identified. A *static load* is one that is applied to a part slowly and gradually and that remains applied, or at least is applied and removed only infrequently during the design life of the part. *Repeated loads* are those that are applied and removed several thousand times during the design life of the part. Significant fluctuations in load without completely unloading the part are also considered to be repeated loads. Under repeated loading a part fails by the mechanism of fatigue at a stress level much lower than that which would cause failure under a static load. This calls for the use of a higher design factor for repeated loads than for static loads. Parts subject to *impact* or *shock* require the use of a large design factor for two reasons. First, a suddenly applied load causes stresses in the part that are several times higher than those that would be computed by standard formulas. Second, under impact loading the material in the part is usually required to absorb energy from the impacting body. The certainty with which the designer knows the magnitude of the expected loads also must be considered when specifying the design factor.

Possible Misuse of the Part. In most cases the designer has no control over actual conditions of use of the product he or she designs. Legally, it is the responsibility of the designer to consider any reasonably foreseeable use or *misuse* of the product and to ensure the safety of the product. The possibility of an accidental overload on any part of a product must be considered.

Complexity of Stress Analysis. As the manner of loading or the geometry of a structure or a part becomes more complex, the designer is less able to perform a precise analysis of the stress condition. Thus the confidence in the results of stress analysis computations has an effect on the choice of a design factor.

Environment. Materials behave differently in different environmental conditions. Consideration should be given to the effects of temperature, humidity, radiation, weather, sunlight, and corrosive atmospheres on the material during the design life of the part.

TABLE 3–1 Size effect for AISI 4140 OQT 1100 Steel.

Specimen size		Tensile strength		Yield strength		Percent elongation
in	mm	ksi	MPa	ksi	MPa	% in 2 in
0.50	12.7	158	1089	149	1027	18
1.00	25.4	140	965	135	931	20
2.00	50.8	128	883	103	710	22
4.00	101.6	117	807	87	600	22

Size Effect, Sometimes Called *Mass Effect*. Metals exhibit different strengths as the cross-sectional area of a part varies. Most material property data were obtained using standard specimens about 0.50 in (12.7 mm) in diameter. Parts with larger sections usually have lower strengths. Parts of smaller size, for example drawn wire, have significantly higher strengths. An example of the size effect is shown in Table 3–1.

Quality Control. The more careful and comprehensive a quality control program is, the better a designer knows how the product will actually appear in service. With poor quality control, a larger design factor should be used.

Hazard Presented by a Failure. The designer must consider the consequences of a failure to a particular part. Would a catastrophic collapse occur? Would people be placed in danger? Would other equipment be damaged? Such considerations may justify the use of a higher than normal design factor.

Cost. Compromises must usually be made in design in the interest of limiting cost to a reasonable value under market conditions. Of course, where danger to life or property exists, compromises should not be made that would seriously affect the ultimate safety of the product or structure.

Market Segment in Which the Part Is to Be Used. Normally you will be aware of the use for the part you are designing, and this can affect your decision on the appropriate design factor. To use a low design factor requires that the loads, material properties, and manufacturing considerations are well known. Lack of confidence in any of these parameters should lead you to specify a larger design factor. Gaining that confidence may require a significant amount of additional research, stress analysis, quality control, and testing, all of which are expensive. The aerospace industry typically invests in the necessary research and analysis to justify a low design factor so the resulting part is as small and light as practical. Conversely, designers of special manufacturing equipment and some heavy-duty construction or agricultural equipment sometimes use larger design factors because of the inability of finding accurate data on the conditions of use.

3–5 DESIGN APPROACHES AND GUIDELINES FOR DESIGN FACTORS

Experience in design and knowledge about the conditions discussed in the preceding section must be applied to determine a design factor for a particular situation. Ultimately, it is the designer's responsibility to set the design factor to ensure safety of the component being designed while also achieving a cost-effective design.

In this chapter you will find several guidelines for specifying a design factor for direct stresses that will be used in this book. The guidelines are not precise and are based on average conditions. Additional guidelines are presented in future chapters for other kinds of stresses such as torsional shear stresses and stresses due to bending.

TABLE 3–2 Design stress guidelines—direct normal stresses.

Manner of loading	Ductile material	Brittle material
Static	$\sigma_d = s_y/2$	$\sigma_d = s_u/6$
Repeated	$\sigma_d = s_u/8$	$\sigma_d = s_u/10$
Impact or shock	$\sigma_d = s_u/12$	$\sigma_d = s_u/15$

It is wasteful to purposely overdesign a component. However, there are times when uncertainty about the actual conditions of service warrant using a more conservative choice for a design factor than those given in the guidelines.

Also, there are numerous codes and standards that should be consulted in certain industries. Among these codes are those pertaining to building and construction, piping and pressure vessels, military, and aerospace. It is your responsibility to investigate whether the product or system you are designing is controlled by such codes and standards. A small sample of the codes for the application of steel or aluminum in buildings is presented in the next section.

The guidelines presented here are somewhat simplified in order to concentrate on the basic kinds of stresses dealt with in this book. They are generally applicable to homogeneous, isotropic metals. Additional study is advised to extend your understanding of more complex components and structures, nonisotropic materials, and more complex kinds of loading. Particular additional study should be given to repeated loads (called fatigue loading), shock, and impact. All of the references listed at the end of this chapter provide information for such additional study.

Design Factor Guidelines for Direct Normal Stresses. Table 3–2 includes guidelines for selecting design factors for problems in this book for which the component being designed or analyzed is subjected to direct normal stresses, tension, or compression.

The use of design factors and outlines of typical design approaches are summarized here. The specific approach used depends on the goal of the problem. Is the goal to evaluate the relative safety of a given design? Is it to specify a suitable material from which to make a component? Is it to determine the required shape and dimensions of the component when the loading is known and the material has been specified?

Case A To Evaluate the Safety of a Given Design.

Given

a) The magnitude and type of loading on the component of interest.

b) The material, including its condition, from which the component is made.

c) The shape and dimensions of critical geometry of the component.

Find Whether or not the component is reasonably safe.

Method

1. Identify the kind of stress produced by the given loading.
2. Determine the applicable stress analysis technique.
3. Complete the stress analysis to determine the maximum expected stress, σ_{max}, in the component.

4. Determine the yield strength, ultimate tensile strength, and percent elongation for the material. Decide if the material is ductile (percent elongation >5%) or brittle (percent elongation <5%).
5. Determine the appropriate design stress relationship. For direct normal stresses, use σ_d from either Equation (3–2) or (3–3).
6. Set $\sigma_{max} = \sigma_d$ and solve for the resulting design factor, N.

7a. When the design is based on yield strength:

$$\sigma_{max} = \sigma_d = s_y/N$$
$$N = s_y/\sigma_{max}$$

7b. When the design is based on ultimate tensile strength:

$$\sigma_{max} = \sigma_d = s_u/N$$
$$N = s_u/\sigma_{max}$$

8. Compare the resulting value of the design factor with recommended guidelines, considering Table 3–2 and all factors discussed in the preceding section.
9. If the actual design factor is lower than the recommended value, redesign should be done to increase the resulting design factor.
10. If the actual design factor is significantly higher than the recommended value, you should redesign the component to achieve a more cost-effective design that uses less material.

Case B To Specify a Suitable Material from Which to Make a Component.

Given

a) The magnitude and type of loading on the component of interest.

b) The shape and dimensions of critical geometry of the component.

Find The material, including its condition, from which to make the component.

Method

1. Identify the kind of stress produced by the given loading.
2. Determine the applicable stress analysis technique.
3. Complete the stress analysis to determine the maximum expected stress, σ_{max}, in the component.
4. Specify a reasonable design factor from recommended guidelines, considering all factors discussed in the preceding section.
5. Determine the appropriate design stress relationship. For direct normal stresses, use σ_d from either Equation (3–2) or (3–3) using the guidelines in Table 3–2.
6. Set $\sigma_{max} = \sigma_d$ and solve for the required strength of the material.

7a. When the design is based on yield strength:

$$\sigma_{max} = \sigma_d = s_y/N$$
$$\text{Required } s_y = N(\sigma_{max})$$

7b. When the design is based on ultimate tensile strength:

$$\sigma_{max} = \sigma_d = s_u/N$$
$$\text{Required } s_u = N(\sigma_{max})$$

8. Specify a suitable material that has the required strength. Consider also the ductility of the material. If the loading is repeated, shock, or impact, a highly ductile material is recommended.

Case C To Determine the Shape and Dimensions of the Component.

Given

a) The magnitude and type of loading on the component of interest.

b) The material, including its condition, from which the component is to be made.

Find The shape and dimensions of critical geometry of the component.

Method

1. Determine the yield strength, ultimate strength, and percent elongation for the selected material. Decide if the material is ductile (percent elongation $>5\%$) or brittle (percent elongation $<5\%$).
2. Specify an appropriate design factor considering the type of loading, the type of material, the conditions listed in the preceding section, and recommended guidelines. For direct normal stresses, use Table 3–2.
3. Compute the design stress from Equation (3–2) or (3–3).

$$\sigma_d = s_y/N \text{ based on yield strength}$$
$$\sigma_d = s_u/N \text{ based on ultimate tensile strength}$$

4. Write the equation for the expected maximum stress in the component. For direct normal stresses,

$$\sigma_{max} = F/A$$

5. Set $\sigma_{max} = \sigma_d$ and solve for the required cross-sectional area.

$$\sigma_{max} = \sigma_d = F/A$$
$$\text{Required } A = F/\sigma_d$$

6. Determine the minimum required dimensions of the cross-sectional area to achieve the necessary total area. This is dependent on the shape you choose to make the component. It may be solid circular, square, or rectangular, a hollow tube, a standard structural shape such as an angle, or some special shape of your own design.
7. Specify convenient dimensions from the list of preferred basic sizes listed in Appendix A–2.

Case D To Determine the Allowable Load on a Component.

Given

a) The type of loading on the component of interest.

b) The material, including its condition, from which the component is to be made.

c) The shape and dimensions of the component.

Find The allowable load on the component.

Method

1. Determine the yield strength, ultimate strength, and percent elongation for the selected material. Decide if the material is ductile (percent elongation $>5\%$) or brittle (percent elongation $<5\%$).
2. Specify an appropriate design factor considering the type of loading, the type of material, the conditions listed in the preceding section, and recommended guidelines. For direct normal stresses, use Table 3–2.
3. Compute the design stress from Equation (3–2) or (3–3).

$$\sigma_d = s_y/N \text{ based on yield strength}$$
$$\sigma_d = s_u/N \text{ based on ultimate tensile strength}$$

4. Write the equation for the expected maximum stress in the component. For direct normal stresses,

$$\sigma_{max} = F/A$$

5. Set $\sigma_{max} = \sigma_d$ and solve for the maximum allowable load.

$$\sigma_{max} = \sigma_d = F/A$$
$$\text{Maximum allowable } F = \sigma_d(A)$$

The following section gives additional discussion for specific combinations of material type and kinds of loading. Five example problems illustrate the application of the design approaches described in this section.

3–6 METHODS OF COMPUTING DESIGN STRESS

An important factor to be considered when computing the design stress is the manner in which a part may fail when subjected to loads. In this section we discuss failure modes relevant to parts subjected to tensile and compressive loads. Other kinds of loading are discussed later.

The failure modes and the consequent methods of computing design stresses can be classified according to the type of material and the manner of loading. Ductile materials, having more than 5% elongation, exhibit somewhat different modes of failure than do brittle materials. Static loads, repeated loads, and shock loads produce different modes of failure.

Ductile Materials under Static Loads. Ductile materials will undergo large plastic deformations when the stress reaches the yield strength of the material. Under most conditions of use, this would render the part unfit for its intended use. Therefore, for ductile materials subjected to static loads, the design stress is usually based on yield strength. That is,

$$\sigma_d = \frac{s_y}{N}$$

As indicated in Table 3–2, a design factor of $N = 2$ would be a reasonable choice under average conditions.

Ductile Materials under Repeated Loads. Under repeated loads, ductile materials fail by a mechanism called *fatigue.* The level of stress at which fatigue occurs is lower than the yield strength. By testing materials under repeated loads, the stress at which failure will occur can be measured. The terms *fatigue strength* or *endurance strength* are used to denote this stress level. However, fatigue-strength values are often not available. Also, factors such as surface finish, the exact pattern of loading, and the size of a part have a marked effect on the actual fatigue strength. To overcome these difficulties, it is often convenient to use a high value for the design factor when computing the design stress for a part subjected to repeated loads. It is also recommended that the ultimate strength be used as the basis for the design stress because tests show that there is a good correlation between fatigue strength and the ultimate strength. Therefore, for ductile materials subjected to repeated loads, the design stress can be computed from

$$\sigma_d = \frac{s_u}{N}$$

A design factor of $N = 8$ would be reasonable under average conditions. Also, stress concentrations, which are discussed in Section 3–11, must be accounted for since fatigue failures often originate at points of stress concentrations.

Where data are available for the endurance strength of the material, the design stress can be computed from

$$\sigma_d = \frac{s_n}{N}$$

where s_n is the symbol for endurance strength. A design factor N of 3 to 4 is recommended. See Reference 6.

Ductile Materials under Impact or Shock Loading. The failure modes for parts subjected to impact or shock loading are quite complex. They depend on the ability of the material to absorb energy and the flexibility of the part. Large design factors are recommended, because of the general uncertainty of stresses under shock loading. In this book we will use

$$\sigma_d = \frac{s_u}{N}$$

with $N = 12$ for ductile materials subjected to impact or shock loads.

Brittle Materials. Since brittle materials do not exhibit yielding, the design stress must be based on ultimate strength. That is,

$$\sigma_d = \frac{s_u}{N}$$

with $N = 6$ for static loads, $N = 10$ for repeated loads, and $N = 15$ for impact or shock loads. These design factors are higher than for ductile materials because of the sudden manner of failure exhibited by brittle materials.

Design Stresses from Selected Codes. Table 3–3 gives a summary of specifications for design stresses for structural steel as defined by the American Institute of Steel Construction (AISC) and by the Aluminum Association for aluminum alloys. These data pertain to members loaded in tension under static loads such as those found in building-type structures. See References 1 and 2 for a more detailed discussion of these specifications.

TABLE 3–3 Design stress from selected codes—direct normal stresses—static loads on building-like structures.

Structural Steel (AISC): Allowable Stress Design (ASD)
$\sigma_d = s_y/1.67 = 0.60\,s_y$ or $\sigma_d = s_u/2.00 = 0.50\,s_u$ whichever is lower
Aluminum (Aluminum Association):
$\sigma_d = s_y/1.65 = 0.61\,s_y$ or $\sigma_d = s_u/1.95 = 0.51\,s_u$ whichever is lower

The design stresses for steel listed in Table 3–3 relate to the *allowable stress design* method (ASD) that had been the standard for many years. In recent years, the AISC has completed the implementation of a revised approach for the design of structural members for bridges, buildings, and other structures called *load and resistance factor design* (LRFD). Complete information about this approach is included in Reference 2, and the details are included in courses that follow strength of materials in programs in civil engineering, civil engineering technology, construction technology, construction management, architecture, and related programs. This book continues the use of ASD because it introduces the fundamental concepts of design without the very large amount of detail required to fully apply the LRFD method. The basic philosophy of LRFD includes the application of factors, γ, to the various types of loads that can occur singly or in combination in buildings, bridges, and other such structures. The factors were developed after extensive research on how dead loads (D), live loads (L), wind loads (W), earthquake loads (E), and roof, snow, or rainwater loads (L_r, S, R) interact on a statistical basis over a 50-year lifetime. Also included are a set of *resistance factors*, ϕ, for different types of loading such as tensile yielding, tensile fracture, compression, flexure, and shear yielding. These are accumulated and considered in the mathematical inequality,

$$\sum \gamma_i Q_i \leq \phi R_n$$

where R_n is the nominal strength or resistance provided by the load carrying member. The left side of this equation represents the summation of all of the load effects (forces, moments, etc.) Q_i experienced by the member multiplied by the corresponding factor that depends on the specific combination of loads. Therefore, the sum of the factored loads must be less than or equal to the resistance modified by specified factors that are defined in the LRFD manual.

Similarly, for programs such as mechanical engineering and engineering technology, manufacturing engineering and engineering technology, and industrial technology, additional detail on design methodology is typically included in courses that follow strength of materials. For example, see Reference 7.

Example Problem 3–1

A structural support for a machine will be subjected to a static tensile load of 16.0 kN. It is planned to fabricate the support from a square rod made from AISI 1020 hot-rolled steel. Specify suitable dimensions for the cross section of the rod.

Solution

Objective: Specify the dimensions of the cross section of the rod.

Given: $F = 16.0\text{ kN} = 16\,000\text{ N}$ static load.
Material: AISI 1020 HR; $s_y = 331$ MPa; 36% elongation (ductile).
(Data from Appendix A–14)

Analysis: Use Case C from Section 3–5.
Let $\sigma = \sigma_d = s_y/2$ (Table 3–2; ductile material, static load).
Stress analysis: $\sigma = F/A$; then required area $= A = F/\sigma_d$.
But $A = a^2$ (a = dimension of side of square).
Minimum allowable dimension $a = \sqrt{A}$.

Results: $\sigma_d = s_y/2 = 331\text{ MPa}/2 = 165.5\text{ MPa} = 165.5\text{ N/mm}^2$
Required area: $A = F/\sigma_d = (16\,000\text{ N})/(165.5\text{ N/mm}^2) = 96.7\text{ mm}^2$.
Minimum dimension a: $a = \sqrt{A} = \sqrt{96.7\text{ mm}^2} = 9.83$ mm.
Specify: $a = 10$ mm (Appendix A–2; preferred size).

Example Problem 3–2

A tensile member for a roof truss for a building is to carry a static axial tensile load of 19800 lb. It has been proposed that a standard, equal-leg structural steel angle be used for this application using ASTM A36 structural steel. Use the AISC code. Specify a suitable angle from Appendix A–5.

Solution

Objective: Specify a standard equal-leg steel angle.

Given: $F = 19\,800$ lb static load.
Material: ASTM A36; $s_y = 36\,000$ psi; $s_u = 58\,000$ psi.
(Data from Appendix A–16)

Analysis: Use Case C from Section 3–5.
Let $\sigma = \sigma_d = 0.60 s_y$ or $\sigma_d = 0.50 s_u$ (Table 3–3).
Stress analysis: $\sigma = F/A$; then required area $= A = F/\sigma_d$.

Results: $\sigma_d = 0.60\, s_y = 0.60\,(36\,000 \text{ psi}) = 21\,600$ psi
or $\sigma_d = 0.50\, s_u = 0.50\,(58\,000 \text{ psi}) = 29\,000$ psi
Use lower value; $\sigma_d = 21\,600$ psi.
Required area: $A = F/\sigma_d = (19\,800 \text{ lb})/(21\,600 \text{ lb/in}^2) = 0.917 \text{ in}^2$.
This is the minimum allowable area.
Specify: L2 × 2 × 1/4 steel angle (Appendix A–5; lightest section).
$A = 0.944 \text{ in}^2$; weight = 3.21 lb/ft.

Example Problem 3–3

A machine element in a packaging machine is subjected to a tensile load of 36.6 kN which will be repeated several thousand times over the life of the machine. The cross section of the element is 12 mm thick and 20 mm wide. Specify a suitable material from which to make the element.

Solution

Objective: Specify a material for a machine element.

Given: $F = 36.6$ kN = 36600 N repeated load.
Cross section of machine element: rectangle; 12 mm × 20 mm.

Analysis: Use Case B from Section 3–5.
Ductile material desirable for repeated loading.
Let $\sigma = \sigma_d = s_u/8$ (Table 3–2). Then required $s_u = 8\sigma$.
Stress analysis: $\sigma = F/A$.

Results: Area $= A = (12 \text{ mm})(20 \text{ mm}) = 240 \text{ mm}^2$
$\sigma = F/A = (36\,600 \text{ N})/(240 \text{ mm}^2) = 152.5 \text{ N/mm}^2 = 152.5$ MPa
Required ultimate strength: $s_u = 8\sigma = 8\,(152.5 \text{ MPa}) = 1220$ MPa.
Specify: AISI 4140 OQT 900 steel (Appendix A–14).
$s_u = 1289$ MPa; 15% elongation; adequate strength, good ductility.

Comment: Other materials could be selected. Required strength indicates that a heat-treated alloy steel is required. The one selected has the highest percent elongation of any listed in Appendix A–14. If the size of the element could be made somewhat larger, the required strength would be lower and a less costly steel may be found.

Example Problem 3–4

Figure 3–2 shows a design for the support for a heavy machine that will be loaded in axial compression. Gray cast iron, grade 20, has been selected for the support. Specify the allowable static load on the support.

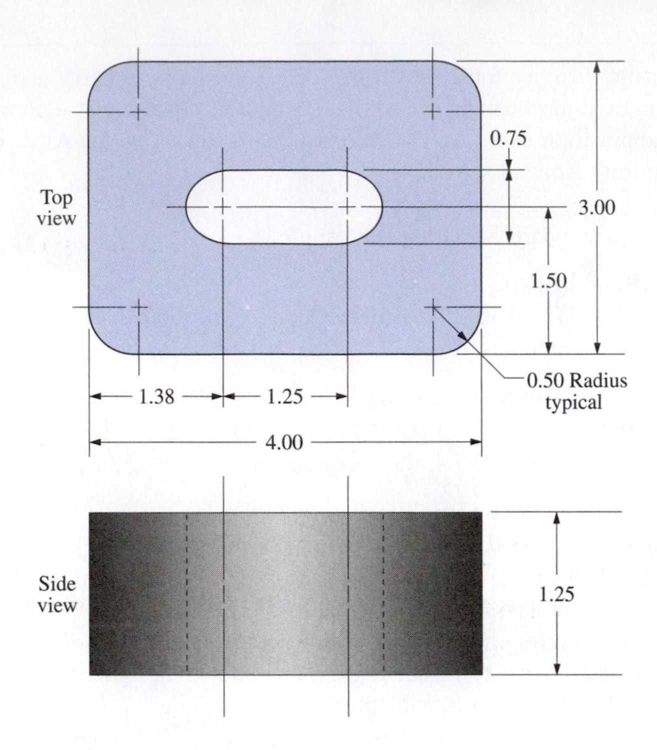

FIGURE 3–2 Machine support for Example Problem 3–4.

Solution Objective Specify the allowable static axial compression load on the support.

Given Material: Gray cast iron, grade 20; $s_u = 80$ ksi in compression (Table A–17); material is brittle. Assume load will be static. Shape of support in Figure 3–2. Compression member is *short* so no buckling occurs.

Analysis Use Case D from Section 3–5.
Stress analysis: $\sigma = F/A$; area computed from Figure 3–2.
Let $\sigma = \sigma_d = s_u/N$; use $N = 6$ (Table 3–2).
Then, allowable, $F = \sigma_d$ (A)

Results $\sigma_d = s_u/6 = 80\,000 \text{ psi}/6 = 13\,300 \text{ psi}$
The cross section of the support is the same as the top view. The net area can be calculated by taking the area of a 3.00 in by 4.00 in rectangle and subtracting the area of the slot and the four corner fillets.

$$\text{Rectangle: } A_R = (3.00 \text{ in})(4.00 \text{ in}) = 12.00 \text{ in}^2$$

$$\text{Slot: } A_S = (0.75)(1.25) + \frac{\pi(0.75)^2}{4} = 1.38 \text{ in}^2$$

The area of each fillet can be computed by the difference between the area of a square with sides equal to the radius of the corner (0.50 in) and a quarter circle of the same radius. Then

$$\text{Fillet: } A_F = r^2 - \frac{1}{4}(\pi r^2)$$

$$A_F = (0.50)^2 - \frac{1}{4}[\pi(0.50)^2] = 0.0537 \text{ in}^2$$

Then the total area is

$$A = A_R - A_S - 4A_F = 12.00 - 1.38 - 4(0.0537) = 10.41\ \text{in}^2$$

We now have the data needed to compute the allowable load.

$$F = A\sigma_d = (10.41\ \text{in}^2)(13\,300\ \text{lb/in}^2) = 138\,500\ \text{lb}$$

This completes the example problem.

3–7 ELASTIC DEFORMATION IN TENSION AND COMPRESSION MEMBERS

Deformation refers to some change in the dimensions of a load-carrying member. Being able to compute the magnitude of deformation is important in the design of precision mechanisms, machine tools, building structures, and machine structures.

An example of where deformation is important is shown in Figure 3–3, in which circular steel tie rods are attached to a C-frame punch press. The tie rods are subjected to tension when in operation. Since they contribute to the rigidity of the press, the amount that they deform under load is something the designer needs to be able to determine.

To develop the relationship from which deformation can be computed for members subjected to axial tension or compression, some concepts from Chapter 1 must be reviewed. *Strain* is defined as the ratio of the total deformation to the original length of a member. (See Figure 3–4.) Using the symbols ϵ for strain, δ for total deformation, and L for length, the formula for strain becomes

$$\epsilon = \frac{\delta}{L} \tag{3–4}$$

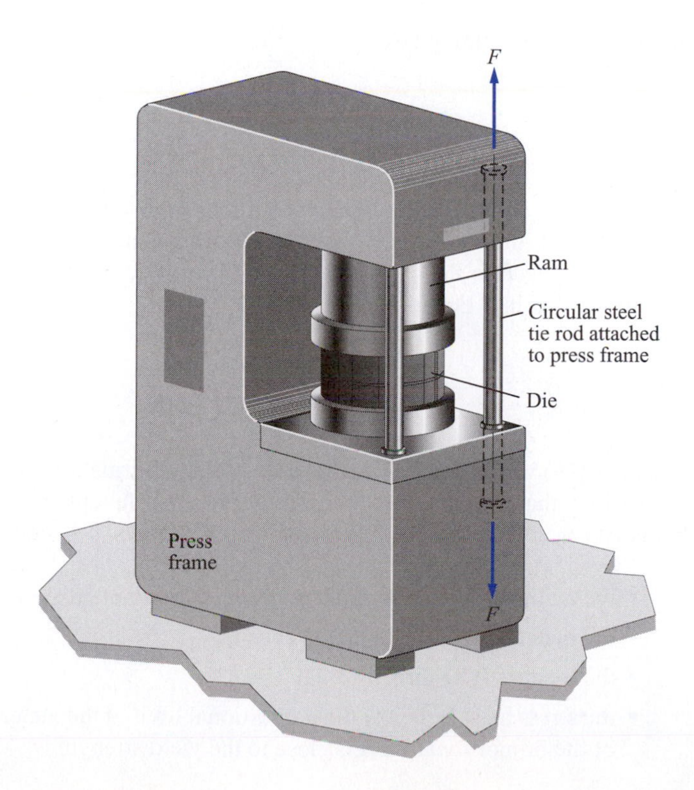

FIGURE 3–3 C-frame press for Example Problem 3–5.

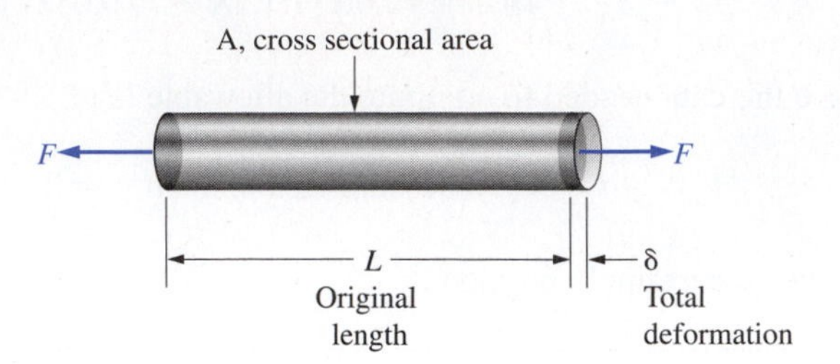

FIGURE 3–4 Illustration of strain.

The stiffness of a material is a function of its modulus of elasticity, E, defined as

$$E = \frac{\text{stress}}{\text{strain}} = \frac{\sigma}{\epsilon} \quad \textbf{(3–5)}$$

Solving for strain gives

$$\epsilon = \frac{\sigma}{E} \quad \textbf{(3–6)}$$

Now Equations (3–4) and (3–6) can be equated:

$$\frac{\delta}{L} = \frac{\sigma}{E} \quad \textbf{(3–7)}$$

Solving for deformation gives

$$\delta = \frac{\sigma L}{E} \quad \textbf{(3–8)}$$

Since this formula applies to members that are subjected to either direct tensile or compressive forces, the direct-stress formula can be used to compute the stress σ. That is, $\sigma = F/A$, where F is the applied load and A is the cross-sectional area of the member. Substituting this into Equation (3–8) gives

$$\delta = \frac{\sigma L}{E} = \frac{FL}{AE} \quad \textbf{(3–9)}$$

Equation (3–9) can be used to compute the total deformation of any load-carrying member, provided that it meets the conditions defined for direct tensile and compressive stress. That is

- the member must be straight and have a constant cross section
- the material must be homogeneous
- the load must be directly axial
- the stress must be below the proportional limit of the material. Recall that the value of the proportional limit is close to the yield strength, s_y.

Example Problem 3–5 The tie rods in the press in Figure 3–3 are made of the steel alloy AISI 5160 OQT 900. Each rod has a diameter of 2.00 in and an initial length of 68.5 in. An axial tensile load of 40 000 lb is exerted on each rod during operation of the press. Compute the deformation of the rods. Check also if the strength of the material is adequate.

Solution Objective Compute the deformation of the tie rods.

Given Rods are steel, AISI 5160 OQT 900; $s_y = 179$ ksi, $s_u = 196$ ksi, 12% elongation
Diameter $= D = 2.00$ in. Length $= L = 68.5$ in. Axial force $= F = 40\,000$ lb.

Analysis Equation (3–9) will be used to compute deformation. The stress in the rods must be checked to ensure that it is below the proportional limit and safe under repeated shock loading.

Results ***Axial tensile stress:*** $\sigma = F/A$.

$$\text{Area} = A = \frac{\pi D^2}{4} = \frac{\pi(2.0\text{ in})^2}{4} = 3.14\text{ in}^2.$$

$$\text{Then, } \sigma = \frac{40\,000\text{ lb}}{3.14\text{ in}^2} = 12\,700\text{ psi.}$$

Therefore, the stress is well below the proportional limit.

For shock loading in a ductile material, Table 3–2 recommends the following design stress. The percent elongation of 12% is in the ductile range.

$$\sigma_d = s_u/12 = 196\text{ ksi}/12 = 16.33\text{ ksi} = 16\,330\text{ psi}$$

Because the actual expected stress is below the design stress, the bar should be safe.

Axial deformation: Use Equation (3–9). All data are known except the modulus of elasticity, E. From the footnotes of Appendix A–14 we find $E = 30 \times 10^6$ psi. Then,

$$\delta = \frac{FL}{AE} = \frac{(40\,000\text{ lb})(68.5\text{ in})}{(3.14\text{ in}^2)(30 \times 10^6\text{ lb/in}^2)} = 0.029\text{ in}$$

Comment For a precision mechanical press, the deformation of 0.029 in (0.74 mm) may be high. An analysis of the complete press system would be advisable. If found to be excessive, the diameter of the bar could be increased, noting that the deformation is inversely proportional to the area of the bar.

Example Problem 3–6 A large pendulum is composed of a 10.0-kg ball suspended by an aluminum wire having a diameter of 1.00 mm and a length of 6.30 m. The aluminum is the alloy 7075-T6. Compute the elongation of the wire due to the weight of the 10-kg ball.

Solution Objective Compute the elongation of the wire.

Given Wire is aluminum alloy 7075-T6; diameter $= D = 1.00$ mm.
Length $= L = 6.30$ m; ball has a mass of 10.0 kg.

Analysis The force on the wire is equal to the weight of the ball, which must be computed from $w = mg$. Then the stress in the wire must be checked to ensure that it is below the proportional limit. Finally, because the stress will then be known, Equation (3–8) will be used to compute the elongation of the wire.

Results ***Force on the wire:*** $F = w = m \cdot g = (10.0\ \text{kg})(9.81\ \text{m/s}^2) = 98.1\ \text{N}.$

Axial tensile stress: $\sigma = F/A$

$$A = \frac{\pi D^2}{4} = \frac{\pi(1.00\ \text{mm})^2}{4} = 0.785\ \text{mm}^2$$

$$\sigma = \frac{F}{A} = \frac{98.1\ \text{N}}{0.785\ \text{mm}^2} = 125\ \text{N/mm}^2 = 125\ \text{MPa}$$

Appendix A–18 lists the yield strength of 7075-T6 aluminum alloy to be 503 MPa. The stress is well below the proportional limit.

The stress can be considered to be steady for a slow-moving pendulum, and the aluminum is ductile, having 11% elongation. From Table 3–2, the design stress can be computed from

$$\sigma_d = s_y/2 = 503\ \text{MPa}/2 = 251\ \text{MPa}$$

Therefore, the wire is safe.

Elongation: For use in Equation (3–8), all data are known except the modulus of elasticity, E. The footnote of Appendix A–18 lists the value of $E = 72\ \text{GPa} = 72 \times 10^9\ \text{Pa}$. Then,

$$\delta = \frac{\sigma L}{E} = \frac{(125\ \text{MPa})(6.30\ \text{m})}{72\ \text{GPa}} = \frac{(125 \times 10^6\ \text{Pa})(6.30\ \text{m})}{72 \times 10^9\ \text{Pa}}$$

$$\delta = 10.9 \times 10^{-3}\ \text{m} = 10.9\ \text{mm}$$

Comment What do you see around you right now that has a dimension similar to 10.9 mm (0.429 in)? Measure the thickness of one of your fingers. Certainly the design of the system containing the pendulum in this example problem would have to take this deflection into account.

Example Problem 3–7 A tension link in a machine must have a length of 610 mm and will be subjected to a repeated axial load of 3000 N. It has been proposed that the link be made of steel and that it have a square cross section. Determine the required dimensions of the link if the elongation under load must not exceed 0.05 mm.

Solution Objective Determine the required dimensions of the square cross section of the link to limit the elongation, δ, to 0.05 mm or less.

Given Axial loading on the link $= F = 3000$ N; length $= L = 610$ mm.
Link will be steel; then $E = 207\ \text{GPa} = 207 \times 10^9\ \text{N/m}^2$. (Appendix A–14)

Analysis In Equation (3–9) for the axial deformation, let $\delta = 0.05$ mm. Then all other data are known except for the cross-sectional area, A. We can solve for A, which is the minimum acceptable cross-sectional area of the link. Let each side of the square cross section be d. Then $A = d^2$ and we can compute the minimum acceptable value for d from $d = \sqrt{A}$. After specifying a convenient size for d, we must check to ensure that the stress is safe and below the proportional limit.

Results ***Required area:*** Solving for A from Equation (3–9) and substituting values gives

$$A = \frac{FL}{E\delta} = \frac{(3000\ \text{N})(610\ \text{mm})}{(207 \times 10^9\ \text{N/m}^2)(0.05\ \text{mm})} = 176.8 \times 10^{-6}\ \text{m}^2$$

Converting to mm^2

$$A = 176.8 \times 10^{-6}\,m^2 \times \frac{(10^3 mm)^2}{m^2} = 176.8\,mm^2$$

and

$$d = \sqrt{A} = \sqrt{176.8\,mm^2} = 13.3\,mm$$

Appendix A–2 lists the next larger preferred size to be 14.0 mm.
The actual cross-sectional area is $A = d^2 = (14.0\,mm)^2 = 196\,mm^2$.

Stress: $\sigma = F/A = 3000\,N/196\,mm^2 = 15.3\,N/mm^2 = 15.3\,MPa$.
For a repeated load, Table 3–2 recommends that the design stress be $\sigma_d = s_u/8$. Letting $\sigma_d = \sigma$, the required value for the ultimate strength is σ

$$s_u = 8(\sigma) = 8(15.3\,MPa) = 123\,MPa$$

Comment Referring to Appendix A–14, we can see that virtually any steel has an ultimate strength far greater than 123 MPa. Unless there were additional design requirements, we should specify the least expensive steel. Then we will specify:

$d = 14.0$ mm
AISI 1020 hot-rolled steel should be low cost. $s_u = 448$ MPa

Note that this design was limited by the allowable elongation and that the resulting stress is relatively low

---

Deformation for Members Carrying Multiple Loads or Having Differing Properties. In Example Problems 3–5 through 3–7, the entire member of interest was uniform in material and cross section, and it was subjected to the same magnitude of axial load throughout. Under such conditions, Equation (3–9) can be used directly to compute the total deformation. In fact, we can say that this equation can *only* be used when all factors, F, L, E, and A, are constant over the section of interest.

LESSON 6

When any factor in Equation (3–9) is different over the length of a given member, the member must be divided into segments for which all factors are the same. Then you can use *superposition* to determine the total deformation of the member. The principle of superposition states that the overall effect of multiple actions on a member is the algebraic sum of the effects of individual components of those actions.

Example Problem 3–8 that follows illustrates the approach to such problems.

Example Problem 3–8 Figure 3–5 shows a steel pipe being used as a bracket to support equipment through cables attached as shown. Select the smallest standard schedule 40 steel pipe that will limit the stress to no more than 18 000 psi. Then for the pipe selected, determine the total downward deflection of point C at the bottom of the pipe as the loads are applied.

Solution Objective Specify a suitable standard schedule 40 steel pipe size and determine the elongation of the pipe.

Given Loading in Figure 3–5; $F_1 = F_2 = 8000$ lb (two forces); $F_3 = 2500$ lb.
Pipe length from A to B: $L_{A-B} = 4.00$ ft(12 in/ft) $= 48.0$ in.
Pipe length from B to C: $L_{B-C} = 3.00$ ft(12 in/ft) $= 36.0$ in.
Maximum allowable stress $= 18\,000$ psi; $E = 30 \times 10^6$ psi (steel).

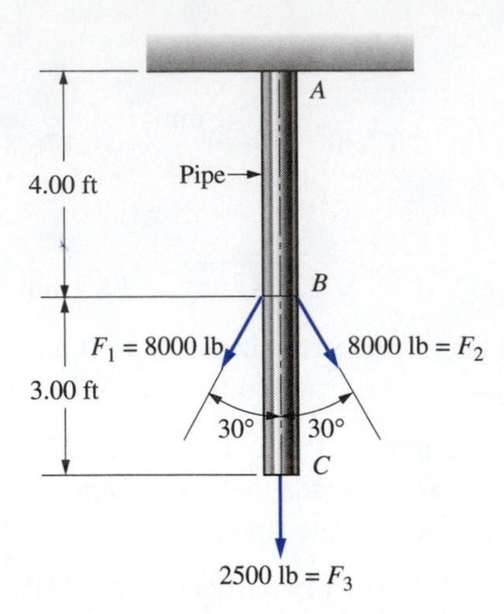

FIGURE 3–5 Pipe for Example Problem 3–8.

Analysis The maximum axial tensile load on the pipe is the sum of F_3 plus the vertical components of each of the 8000 lb forces. This occurs throughout the part of the pipe from A to B. The size of pipe, and the resulting cross-sectional area, must result in a stress in that section of 18 000 psi or less. From B to C, the axial tensile load is $F_{B-C} = 2500$ lb. Because the load is different in the two sections, the computation of the elongation of the pipe must be done in two separate calculations. That is,

$$\delta_C = \delta_{total} = \delta_{A-B} + \delta_{B-C}$$

Results ***Axial loads:***

$$F_{A-B} = F_3 + F_1 \cos 30° + F_2 \cos 30°$$

But

$$F_1 = F_2 = 8000 \text{ lb}$$

Then

$$F_{A-B} = 2500 \text{ lb} + 2(8000 \text{ lb}) \cos 30°$$
$$F_{A-B} = 16400 \text{ lb}$$
$$F_{B-C} = F_3 = 2500 \text{ lb}$$

Stress analysis and calculation of required cross-sectional area: Letting $\sigma = 18000$ psi, the required cross-sectional area of the metal in the pipe is

$$A = \frac{F_{A-B}}{\sigma} = \frac{16400 \text{ lb}}{18000 \text{ lb/in}^2} = 0.911 \text{ in}^2$$

From Appendix A–12, listing the properties of steel pipe, the standard size with the next larger cross-sectional area is the 2-in schedule 40 pipe with $A = 1.075 \text{ in}^2$. As discussed in Chapter 1, this same pipe would be called PIPE2STD in the construction industry, indicating a 2-in nominal pipe size and standard wall thickness.

Deflection:

$$\delta_C = \delta_{total} = \delta_{A-B} + \delta_{B-C}$$

$$\delta_{A-B} = \left(\frac{F_{A-B}L_{A-B}}{AE}\right) = \frac{(16400 \text{ lb})(48 \text{ in})}{(1.075 \text{ in}^2)(30 \times 10^6 \text{ lb/in}^2)} = 0.024 \text{ in}$$

$$\delta_{B-C} = \left(\frac{F_{B-C}L_{B-C}}{AE}\right) = \frac{(2500 \text{ lb})(36 \text{ in})}{(1.075 \text{ in}^2)(30 \times 10^6 \text{ lb/in}^2)} = 0.003 \text{ in}$$

Then

$$\delta_C = \delta_{A-B} + \delta_{B-C} = 0.027 \text{ in}$$

Comment In summary, when the bracket shown in Figure 3–5 is made from a standard 2-in schedule 40 steel pipe, point C moves downward 0.027 in under the influence of the applied loads.

3–8 DEFORMATION DUE TO TEMPERATURE CHANGES

A machine or a structure could undergo deformation or be subjected to stress by changes in temperature in addition to the application of loads. Bridge members and other structural components see temperatures as low as −30°F(−34°C) to as high as 110°F (43°C) in some areas. Vehicles and machinery operating outside experience similar temperature variations. Frequently, a machine part will start at room temperature and then become quite hot as the machine operates. Examples are parts of engines; furnaces; metal-cutting machines; rolling mills; plastics molding and extrusion equipment; food-processing equipment; air compressors; hydraulic and pneumatic devices; and high-speed automation equipment.

As a metal part is heated, it tends to expand. If the expansion is unrestrained, the dimensions of the part will grow but no stress will be developed in the metal. However, in some cases the part is restrained, preventing the change in dimensions. Under such circumstances, stresses will occur.

Different materials change dimensions at different rates when subjected to temperature changes. Most materials expand with increasing temperature, but a few contract and some virtually stay the same size. The *coefficient of thermal expansion* governs the thermal deformation and thermal stress experienced by a material.

> ***The coefficient of thermal expansion, α, is the property of a material that indicates the amount of unit change in dimension with a unit change in temperature.***

The lowercase Greek letter alpha, α, is used as the symbol for the coefficient of thermal expansion.

The units for α are derived from its definition. Stated slightly differently, α is the measure of the change in length of a material per unit length for a 1.0-degree change in temperature. The units, then, for α in the U.S. Customary system would be

$$\text{in}/(\text{in}\cdot{}^\circ\text{F}) \quad \text{or} \quad 1/{}^\circ\text{F} \quad \text{or} \quad {}^\circ\text{F}^{-1}$$

In SI units, α would be

$$\text{m}/(\text{m}\cdot{}^\circ\text{C}) \quad \text{or} \quad \text{mm}/(\text{mm}\cdot{}^\circ\text{C}) \quad \text{or} \quad 1/{}^\circ\text{C} \quad \text{or} \quad {}^\circ\text{C}^{-1}$$

For use in computations, the last form of each unit type is most convenient. However, the first form will help you remember the physical meaning of the term.

It follows from the definition of the coefficient of thermal expansion that the change in length δ of a member can be computed from the equation

$$\delta = \alpha \cdot L \cdot \Delta t \tag{3–10}$$

where L = original length of the member
Δt = change in temperature

TABLE 3–4 Coefficients of thermal expansion, α, for some metals, plate glass, wood, and concrete.

Material	α	
	°F^{-1}	°C^{-1}
Steel, AISI		
1020	6.5×10^{-6}	11.7×10^{-6}
1040	6.3×10^{-6}	11.3×10^{-6}
4140	6.2×10^{-6}	11.2×10^{-6}
Structural steel	6.5×10^{-6}	11.7×10^{-6}
Gray cast iron	6.0×10^{-6}	10.8×10^{-6}
Stainless steel		
AISI 301	9.4×10^{-6}	16.9×10^{-6}
AISI 430	5.8×10^{-6}	10.4×10^{-6}
AISI 501	6.2×10^{-6}	11.2×10^{-6}
Aluminum alloys		
2014	12.8×10^{-6}	23.0×10^{-6}
6061	13.0×10^{-6}	23.4×10^{-6}
7075	12.9×10^{-6}	23.2×10^{-6}
Brass, C36000	11.4×10^{-6}	20.5×10^{-6}
Bronze, C22000	10.2×10^{-6}	18.4×10^{-6}
Copper, C14500	9.9×10^{-6}	17.8×10^{-6}
Magnesium, ASTM AZ63A-T6	14.0×10^{-6}	25.2×10^{-6}
Titanium, Ti-6 A1-4V	5.3×10^{-6}	9.5×10^{-6}
Plate glass	5.0×10^{-6}	9.0×10^{-6}
Wood (pine)	3.0×10^{-6}	5.4×10^{-6}
Concrete	6.0×10^{-6}	10.8×10^{-6}

Table 3–4 gives representative values for the coefficient of thermal expansion for several metals, plate glass, pine wood, and concrete. The actual value for any material varies somewhat with temperature. Those reported in Table 3–4 are approximately average values over the range of temperatures from 32°F(0°C) to 212°F (100°C).

Table 3–5 gives values for α for selected plastic materials. Note that the actual values depend strongly on temperature and on the inclusion of any filler material in the plastic

TABLE 3–5 Coefficients of thermal expansion, α, for selected plastics.

Material	α	
	°F^{-1}	°C^{-1}
ABS–resin	53×10^{-6}	95.4×10^{-6}
ABS/glass	16×10^{-6}	28.8×10^{-6}
Acetal–resin	45×10^{-6}	81.0×10^{-6}
Acetal/glass	22×10^{-6}	39.6×10^{-6}
Nylon 66–resin	45×10^{-6}	81.0×10^{-6}
Nylon 66/glass	13×10^{-6}	23.4×10^{-6}
Polycarbonate–resin	37×10^{-6}	66.6×10^{-6}
Polycarbonate/glass	13×10^{-6}	23.4×10^{-6}
Polyester–resin	53×10^{-6}	95.4×10^{-6}
Polyester/glass	12×10^{-6}	21.6×10^{-6}
Polystyrene–resin	36×10^{-6}	64.8×10^{-6}
Polystyrene/glass	19×10^{-6}	34.2×10^{-6}

TABLE 3–6 Coefficients of thermal expansion, α, for selected composites.

	α			
	Longitudinal		Transverse	
Material	°F^{-1}	°C^{-1}	°F^{-1}	°C^{-1}
E-glass/epoxy–unidirectional	3.5×10^{-6}	6.30×10^{-6}	11.0×10^{-6}	19.8×10^{-6}
Aramid/epoxy–unidirectional	-1.1×10^{-6}	-1.98×10^{-6}	38.0×10^{-6}	68.4×10^{-6}
Carbon/epoxy–unidirectional	0.05×10^{-6}	0.09×10^{-6}	9.0×10^{-6}	16.2×10^{-6}
Carbon/epoxy–quasi-isotropic	1.6×10^{-6}	2.88×10^{-6}	1.6×10^{-6}	2.88×10^{-6}

resin. For each plastic listed, the approximate values for α are given for the unfilled resin and for one filled with 30% glass.

Composites were described in Chapter 2 as materials that combine a matrix with reinforcing fibers made from a variety of materials such as glass, aramid polymer, carbon, or graphite. The matrix materials can be polymers such as polyester or epoxy, ceramics, or some metals such as aluminum. The value for α for the fibers is typically much smaller than for the matrix. In addition, there are numerous ways in which the fibers can be placed in the matrix. The coefficient of thermal expansion for composites, therefore, is very difficult to generalize.

Table 3–6 gives representative values for a few forms of composites. Refer to Chapter 2 for the description of the terms *unidirectional* and *quasi-isotropic*. Particularly with unidirectional placement of the fibers in the matrix, there is a dramatic difference in the value of the coefficient of thermal expansion as a function of the orientation of the material. In the longitudinal direction, aligned with the fibers, the low value of α for the fibers tends to produce a low overall value. However, in the transverse direction, the fibers are not very effective, and the overall value of α is much higher. Notice, too, that for the particular unidirectional aramid/epoxy composite listed, the value of α is actually negative, meaning that this composite gets *smaller* with increasing temperature.

Example Problem 3–9 A rod made from AISI 1040 steel is used as a link in a steering mechanism of a large truck. If its nominal length is 56 in, compute its change in length as the temperature changes from −30°F to 110°F.

Solution

Objective: Compute the change in length for the link.

Given: Link is made from AISI 1040 steel; length = L = 56 in.
Original temperature = $t_1 = -30$°F.
Final temperature = $t_2 = 110$°F.

Analysis: Use Equation (3–10). From Table 3–4, $\alpha = 6.3 \times 10^{-6}$°F^{-1}.

$$\Delta t = t_2 - t_1 = 110\text{°F} - (-30\text{°F}) = 140\text{°F}$$

Results: $\delta = \alpha \cdot L \cdot \Delta t = (6.3 \times 10^{-6}\text{°F}^{-1})(56\text{ in})(140\text{°F}) = 0.049\text{ in}$

Comment: The significance of this amount of deformation would have to be evaluated within the overall design of the steering mechanism for the truck. Approximately twelve sheets of standard bond paper stacked together have a thickness of 0.049 in.

Example Problem 3–10 A pushrod in the valve mechanism of an automotive engine has a nominal length of 203 mm. If the rod is made of AISI 4140 steel, compute the elongation due to a temperature change from −20°C to 140°C.

Solution

Objective: Compute the change in length for the pushrod.

Given: Link is made from AISI 4140 steel; length = L = 203 mm.
Original temperature = $t_1 = -20°C$.
Final temperature = $t_2 = 140°C$.

Analysis: Use Equation (3–10). From Table 3–4, $\alpha = 11.2 \times 10^{-6} °C^{-1}$.

$$\Delta t = t_2 - t_1 = 140°C - (-20°C) = 160°C$$

Results: $\delta = \alpha \cdot L \cdot \Delta T = (11.2 \times 10^{-6} °C^{-1})(203\ mm)(160°C) = 0.364\ mm$

Comment: It would be important to account for this expansion in the design of the valve mechanism. It could cause noise due to loose-fitting parts or high stresses if the expansion is restrained.

Example Problem 3–11 An aluminum frame of 6061 alloy for a window is 4.350 m long and holds a piece of plate glass 4.347 m long when the temperature is 35°C. At what temperature would the aluminum and glass be the same length?

Solution

Objective: Compute the temperature at which the aluminum frame and the glass would be the same length.

Given: Aluminum is 6061 alloy; from Table 3–4 we find that $\alpha_a = 23.4 \times 10^{-6} °C^{-1}$.
For plate glass, $\alpha_g = 9.0 \times 10^{-6} °C^{-1}$.
At $t_1 = 35°C$; $L_{a1} = 4.350$ m; $L_{g1} = 4.347$ m.

Analysis: The temperature would have to decrease in order for the aluminum and glass to reach the same length, since aluminum contracts at a greater rate than glass. As the temperature decreases, the change in temperature Δt would be the same for both the aluminum and the glass. After the temperature change, the length of the aluminum would be

$$L_{a2} = L_{a1} - \alpha_a \cdot L_{a1} \cdot \Delta t$$

where the subscript a refers to the aluminum, 1 refers to the initial condition, and 2 refers to the final condition. The length of the glass would be

$$L_{g2} = L_{g1} - \alpha_g \cdot L_{g1} \cdot \Delta t$$

But when the glass and the aluminum have the same length,

$$L_{a2} = L_{g2}$$

Then

$$L_{a1} - \alpha_a \cdot L_{a1} \cdot \Delta t = L_{g1} - \alpha_g \cdot L_{g1} \cdot \Delta t$$

Solving for Δt gives

$$\Delta t = \frac{L_{a1} - L_{g1}}{\alpha_a \cdot L_{a1} - \alpha_g \cdot L_{g1}}$$

Results

$$\Delta t = \frac{4.350\text{ m} - 4.347\text{ m}}{(23.4 \times 10^{-6}{}^\circ\text{C}^{-1})(4.350\text{ m}) - (9.0 \times 10^{-6}{}^\circ\text{C}^{-1})(4.347\text{ m})}$$

$$\Delta t = \frac{0.003}{(0.000102) - (0.000039)}{}^\circ\text{C} = 48^\circ\text{C}$$

$$\text{Then; } t_2 = t_1 - \Delta t = 35^\circ\text{C} - 48^\circ\text{C} = -13^\circ\text{C}$$

Comment Since this is well within the possible ambient temperature for a building, a dangerous condition could be created by this window. The window frame and glass would contract without stress until a temperature of −13°C was reached. If the temperature continued to decrease, the frame would contract faster than the glass and would generate stress in the glass. Of course, if the stress was great enough, the glass would fracture, possibly causing injury. The window should be reworked so there is a larger difference in size between the glass and the aluminum frame.

3–9 THERMAL STRESS

In the preceding section, parts that were subjected to changes in temperature were unrestrained, so that they could grow or contract freely. If the parts were held in such a way that deformation was resisted, stresses would be developed.

Consider a steel structural member in a furnace that is heated while the members to which it is attached are kept at a lower temperature. Assuming the ideal case, the supports would be considered rigid and immovable. Thus all expansion of the steel member would be prevented.

If the steel part were allowed to expand, it would elongate by an amount $\delta = \alpha \cdot L \cdot \Delta t$. But since it is restrained, this represents the apparent total deformation in the steel. Then the unit strain would be

$$\epsilon = \frac{\delta}{L} = \frac{\alpha \cdot L \cdot \Delta t}{L} = \alpha(\Delta t) \quad \textbf{(3–11)}$$

The resulting stress in the part can be found from

$$\sigma = E\epsilon$$

or

Thermal Stress

$$\sigma = E\alpha(\Delta t) \quad \textbf{(3–12)}$$

This stress would occur without the addition of external forces. You must check to ensure that the resulting stress level will not cause yielding or fracture of the material. Also, for relatively long, slender members in compression, the restraint of expansion can cause the members to buckle due to column action (see Chapter 11).

Example Problem 3–12 A steel structural member in a furnace is made from AISI 1020 steel and undergoes an increase in temperature of 95°F while being held rigid at its ends. Compute the resulting stress in the steel.

Solution Objective Compute the thermal stress in the steel.

Given Steel is AISI 1020; from Table 3–4, $\alpha = 6.5 \times 10^{-6}{}^\circ\text{F}^{-1}$.
$E = 30 \times 10^6$ psi; $\Delta t = 95^\circ\text{F}$.

Analysis Use Equation (3–12); $\sigma = E\alpha(\Delta t)$.

Results $\sigma = (30 \times 10^6 \text{ psi})(6.5 \times 10^{-6}{}^\circ\text{F}^{-1})(95^\circ\text{F}) = 18\,500 \text{ psi}$

Comment Appendix A–14 shows the yield strength of annealed AISI 1020 steel, its weakest form, to be 43 000 psi. Therefore, the structural member would be safe from yielding. But column buckling should also be checked because the stress is compressive.

Example Problem 3–13 An aluminum rod of alloy 2014-T6 in a machine is held at its ends while being cooled from 95°C. At what temperature would the tensile stress in the rod equal half of the yield strength of the aluminum if it is originally at zero stress?

Solution Objective Compute the temperature when $\sigma = s_y/2$.

Given Aluminum is alloy 2014-T6; from Table 3–4, $\alpha = 23.0 \times 10^{-6}{}^\circ\text{C}^{-1}$.
From Appendix A–18, $s_y = 414$ MPa; $E = 73$ GPa; $t_1 = 95^\circ$C.

Analysis Use Equation (3–12) and solve for Δt.

$$\sigma = E\alpha(\Delta t)$$

$$\Delta t = \frac{\sigma}{E\alpha}$$

Let the stress be

$$\sigma = \frac{s_y}{2} = \frac{414 \text{ MPa}}{2} = 207 \text{ MPa}$$

Results

$$\Delta t = \frac{\sigma}{E\alpha} = \frac{207 \text{ MPa}}{(73 \text{ GPa})(23.0 \times 10^{-6}{}^\circ\text{C}^{-1})}$$

$$\Delta t = \frac{207 \times 10^6 \text{ Pa}}{(73 \times 10^9 \text{ Pa})(23.0 \times 10^{-6}{}^\circ\text{C}^{-1})} = 123^\circ\text{C}$$

Since the rod had zero stress when its temperature was 95°C, the temperature at which the stress would be 207 MPa would be

$$t = 95^\circ\text{C} - 123^\circ\text{C} = -28^\circ\text{C}$$

Comment The stress level of one-half of the yield strength is a reasonable design stress for a statically loaded member. Thus, if the temperature were to be lower than −28°C, the safety of the rod would be jeopardized.

Thermal Deformation and Stress under Partial Restraint. There are many practical cases in which a member is originally free to expand due to temperature changes, but after some increase in temperature, external restraints begin to be applied.

Consider the situation shown in Figure 3–6. A brass bar made from C36000 brass, hard, is initially installed in a rigid frame with a small clearance. When the temperature is increased, a series of two actions takes place. First, the bar will expand as the temperature increases until the bar just touches the inside of the frame. For this part of the process, there is no stress produced in the bar. As the bar becomes restrained by the frame, no further expansion can occur and thermal stresses develop during the final part of the temperature rise.

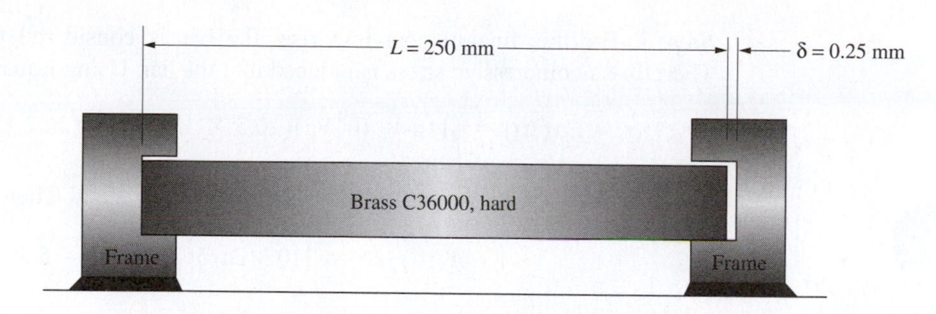

FIGURE 3–6 Brass bar for Example Problem 3–14.

Example Problem 3–14 illustrates how such a problem can be approached.

Example Problem 3–14 The brass bar shown in Figure 3–6 is part of a conveyor that transports components into an oven. Initially, when the temperature is 15°C, there is a total clearance of 0.25 mm between the end of the bar and the inside of the frames on both sides. Describe what happens when the temperature increases from 15°C to 90°C. Consider the frames to be rigid and that they do not change dimension as the temperature rises.

Solution

Objective: Describe the behavior of the bar as the temperature rises.

Given: Initial length of bar: $L = 250$ mm
Initial gap: $\delta = 0.25$ mm
Material of bar: Brass C36000, hard. $\alpha = 20.5 \times 10^{-6}{}^\circ\text{C}^{-1}$ (Table 3–4).

$$s_y = 310 \text{ MPa. } E = 110 \text{ GPa (Appendix A–15)}$$

Analysis: ***Step 1.*** First determine what temperature rise will cause the bar to expand by 0.25 mm, bringing the end of the bar just into contact with the frame. Equation (3–10) can be used.

Step 2. Then determine how much additional temperature rise occurs from that point up to when the temperature is 90°C.

Step 3. Equation (3–12) can then be used to compute the stress developed in the bar during the final temperature rise. The safety of this stress should be evaluated.

Results: ***Step 1.*** The amount of temperature rise to elongate the bar 0.25 mm:

$$\delta = \alpha L(\Delta t_1)$$

Solving for Δt_1 gives

$$\Delta t_1 = \frac{\delta}{\alpha L} = \frac{0.25 \text{ mm}}{(20.5 \times 10^{-6}{}^\circ\text{C}^{-1})(250 \text{ mm})} = 48.8^\circ\text{C}$$

Step 2. The temperature at which this occurs is

$$t_2 = t_1 + \Delta t_1 = 15^\circ\text{C} + 48.8^\circ\text{C} = 63.8^\circ\text{C}$$

Then the additional temperature rise to the maximum temperature is

$$\Delta t_2 = 90^\circ\text{C} - 63.8^\circ\text{C} = 26.2^\circ\text{C}$$

Step 3. For this final temperature rise, the bar is considered to be fully restrained. Therefore a compressive stress is induced into the bar. Using Equation (3–12),

$$\sigma = E\alpha(\Delta t) = (110 \times 10^9 \text{ Pa})(20.5 \times 10^{-6\circ}\text{C}^{-1})(26.2^\circ\text{C}) = 59.08 \text{ MPa}$$

Let's check this stress against yielding. Let $\sigma = \sigma_d = s_y/N$. Then,

$$N = s_y/\sigma = 310 \text{ MPa}/59.08 \text{ MPa} = 5.25$$

LESSON 7

This is a sufficiently high design factor to ensure that yielding will not occur. However, column buckling should be checked (see Chapter 11).

3–10 MEMBERS MADE OF MORE THAN ONE MATERIAL

When two or more materials in a load-carrying member share the load, a special analysis is required to determine what portion of the load each material takes. Consideration of the elastic properties of the materials is required.

Figure 3–7 shows a steel pipe filled with concrete and used to support part of a large structure. The load is distributed evenly across the top of the support. We want to determine the stress in both the steel and the concrete.

Two concepts must be understood in deriving the solution to this problem.

1. The total load F is shared by the steel and the concrete such that $F = F_s + F_c$.
2. Under the compressive load F, the composite support deforms and the two materials deform in equal amounts. That is, $\delta_s = \delta_c$.

Now since the steel and the concrete were originally the same length,

$$\frac{\delta_s}{L} = \frac{\delta_c}{L}$$

But

$$\frac{\delta_s}{L} = \epsilon_s \quad \text{and} \quad \frac{\delta_c}{L} = \epsilon_c$$

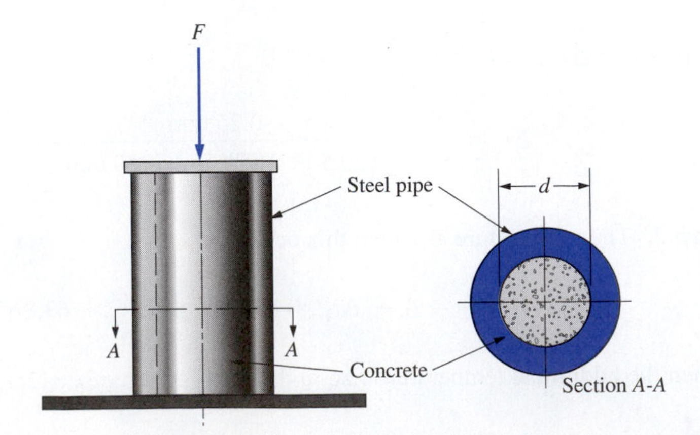

FIGURE 3–7 Steel and concrete post.

Then the strains in the two materials are equal.

$$\epsilon_s = \epsilon_c$$

Also, from the definition of modulus of elasticity,

$$\epsilon_s = \frac{\sigma_s}{E_s} \quad \text{and} \quad \epsilon_c = \frac{\sigma_c}{E_c}$$

Then

$$\frac{\sigma_s}{E_s} = \frac{\sigma_c}{E_c}$$

Solving for σ_s yields

$$\sigma_s = \frac{\sigma_c E_s}{E_c} \quad \textbf{(3–13)}$$

This equation gives the relationship between the two stresses.

Now consider the loads,

$$F_s + F_c = F$$

Since both materials are subject to axial stress,

$$\sigma_s = F_s/A_s \quad \text{and} \quad \sigma_c = F_c/A_c$$

or,

$$F_s = \sigma_s A_s \quad \text{and} \quad F_c = \sigma_c A_c$$

where A_s and A_c are the areas of the steel and concrete, respectively. Then

$$\sigma_s A_s + \sigma_c A_c = F \quad \textbf{(3–14)}$$

Substituting Equation (3–13) into Equation (3–14) gives

$$\frac{A_s \sigma_c E_s}{E_c} + \sigma_c A_c = F$$

Now, solving for σ_c gives

$$\sigma_c = \frac{F E_c}{A_s E_s + A_c E_c} \quad \textbf{(3–15)}$$

Equations (3–13) and (3–15) can now be used to compute the stresses in the steel and the concrete.

Example Problem 3–15 For the support shown in Figure 3–7, the pipe is a standard 6-in schedule 40 steel pipe completely filled with concrete. If the load F is 155 000 lb, compute the stress in the concrete and the steel. For steel use $E = 30 \times 10^6$ psi. For concrete use $E = 3.3 \times 10^6$ psi for a rated strength of $s_c = 3000$ psi (see Section 2–10).

Solution Objective Compute the stress in the concrete and the steel.

Given Load $= F = 155\,000$ lb; $E_s = 30 \times 10^6$ psi; $E_c = 3.3 \times 10^6$ psi.
From Appendix A–12, for a 6-in schedule 40 pipe:
$A_s = 5.581\ \text{in}^2$; inside diameter $= d = 6.065$ in.

Analysis Use Equation (3–15) to compute the stress in the concrete, σ_c. Then use Equation (3–13) to compute σ_s. All data are known except A_c. But,

$$A_c = \frac{\pi d^2}{4} = \frac{\pi(6.065\ \text{in})^2}{4} = 28.89\ \text{in}^2$$

Results Then in Equation (3–15),

$$\sigma_c = \frac{(155\,000\ \text{lb})(3.3 \times 10^6\ \text{psi})}{(5.581\ \text{in}^2)(30 \times 10^6\ \text{psi}) + (28.89\ \text{in}^2)(3.3 \times 10^6\ \text{psi})} = 1946\ \text{psi}$$

Using Equation (3–13) gives

$$\sigma_s = \frac{\sigma_c E_s}{E_c} = \frac{(1946\ \text{psi})(30 \times 10^6\ \text{psi})}{3.3 \times 10^6\ \text{psi}} = 17\,696\ \text{psi}$$

Comment These stresses are fairly high. If it were desired to have at least a design factor of 2.0 based on the yield strength of the steel and 4.0 on the rated strength of the concrete, the required strengths would be,

$$\text{Steel: } s_y = 2(17\,696\ \text{psi}) = 35\,392\ \text{psi}$$

$$\text{Concrete: Rated } \sigma_c = 4(1946\ \text{psi}) = 7784\ \text{psi}$$

The steel could be similar to AISI 1020 annealed or any stronger condition. A rated strength of 3000 psi for the concrete would not be satisfactory.

Summary and Generalization. The analysis presented for Example Problem 3–14 can be generalized for any situation in which loads are shared by two or more members of two different materials provided all members undergo equal strains. Replacing the subscripts s and c in the preceding analysis by the more general subscripts 1 and 2, we can state Equations (3–15) and (3–13) in the forms,

$$\sigma_2 = \frac{FE_2}{A_1E_1 + A_2E_2} \tag{3–16}$$

LESSON 8

$$\sigma_1 = \frac{\sigma_2 E_1}{E_2} \tag{3–17}$$

Composite Compression Members in Construction. The use of concrete-filled HSS members or concrete-encased column sections is common in building construction.

AISC reference manuals list the design strength in axial compression (in kips) of such members as a function of their effective length in extensive sets of tables. Both direct compressive stress and column behavior (discussed later in this book) are considered. Covered are:

- Square and rectangular steel HSS shapes from $2\frac{1}{2}$ in $\times$ $2\frac{1}{2}$ in to 20 in $\times$ 12 in (63.5 mm $\times$ 63.5 mm to 508 mm $\times$ 304.8 mm) and all common wall thicknesses. Included are four combinations of the strength of steel (ASTM A500, Grade B) and the rated strength of concrete:
  - Yield strength of steel = 42 ksi (290 MPa) and $f'_c = 4$ ksi (27.6 MPa)
  - Yield strength of steel = 42 ksi (290 MPa) and $f'_c = 5$ ksi (34.5 MPa)
  - Yield strength of steel = 46 ksi (317 MPa) and $f'_c = 4$ ksi (27.6 MPa)
  - Yield strength of steel = 46 ksi (317 MPa) and $f'_c = 5$ ksi (34.5 MPa)
- Round steel HSS (ASTM A500, Grade B) from 4.00 in to 20.00 in outside diameter (101.6 mm to 508 mm) and all common wall thicknesses. Two combinations of steel and concrete strength are included.
  - Yield strength of steel = 42 ksi (290 MPa) and $f'_c = 4$ ksi (27.6 MPa)
  - Yield strength of steel = 42 ksi (290 MPa) and $f'_c = 5$ ksi (34.5 MPa)
- Steel pipe (ASTM A53 Grade B) from 3 in to 12 in nominal size (75 mm to 310 mm) and all common wall thicknesses; Standard, XS, and XXS. These are combined with two rated strengths of concrete:
  - Yield strength of steel = 35 ksi (241 MPa) and $f'_c = 4$ ksi (27.6 MPa)
  - Yield strength of steel = 35 ksi (241 MPa) and $f'_c = 5$ ksi (34.5 MPa)

3–11 STRESS CONCENTRATION FACTORS FOR DIRECT AXIAL STRESSES

In defining the method for computing stress due to a direct tensile or compressive load on a member, it was emphasized that the member must have a uniform cross section in order for the equation $\sigma = F/A$ to be valid. The reason for this restriction is that where a change in the geometry of a loaded member occurs, the actual stress developed is higher than would be predicted by the standard equation. This phenomenon is called *stress concentration* because detailed studies reveal that localized high stresses appear to concentrate around sections where geometry changes occur.

Figure 3–8 illustrates the case of stress concentration for the example of an axially loaded round bar in tension that has two diameters with a step between them. Note that there is a small fillet at the base of the step. Its importance will be discussed later. Below the sketch of the stepped bar is a graph of stress versus position on the bar. At section 1, where the bar diameter is D and well away from the step, the stress can be computed from

$$\sigma_1 = F/A_1 = F/(\pi D^2/4)$$

At section 2, where the bar diameter has the smaller value of d, the stress is

$$\sigma_2 = F/A_2 = F/(\pi d^2/4)$$

Then the plot of stress versus position might be expected to appear as the straight lines with an abrupt step at the location of the change in diameter. But tests would show that the actual stress distribution would look more like the curved line: blending with the two straight lines well away from the step, but rising much higher near the step itself.

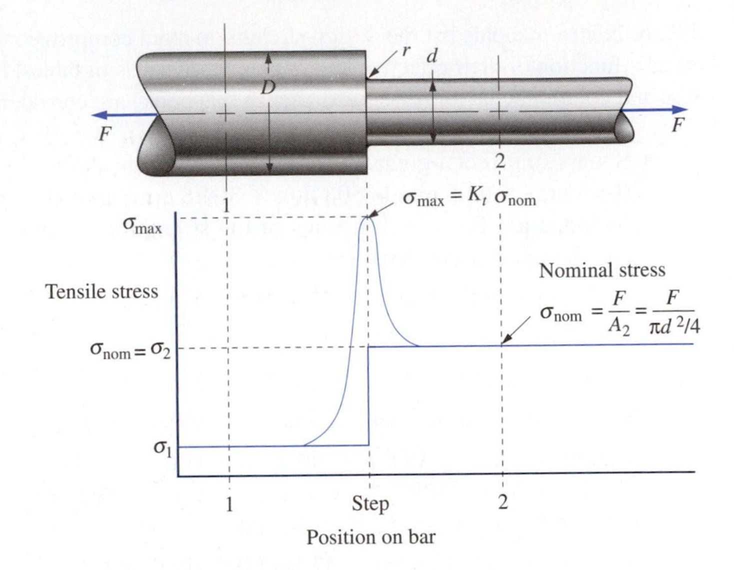

FIGURE 3–8 Stress distribution near a change in geometry.

To account for the higher-than-predicted stress at the step, we modify the direct stress formula to include a *stress concentration factor,* K_t, to produce the form shown in Equation (3–18).

$$\sigma_{max} = K_t \sigma_{nom} \tag{3–18}$$

where, in this case, the nominal stress is based on the smaller section 2. That is,

$$\sigma_{nom} = \sigma_2 = F/A_2 = F/(\pi d^2/4)$$

Then the value of K_t represents the factor by which the actual stress is higher than the nominal stress computed by the standard formula.

Figure 3–9 shows a photoelastic model of another case that demonstrates the phenomenon of stress concentration. The transparent plastic model is a straight rectangular bar of constant thickness carrying an axial load. Near the middle of the bar there is a hole on the axis of the member. The nominal stress in the vicinity of the hole naturally increases because of the removal of material. But the actual maximum stress near the hole is higher than the nominal stress as indicated by the concentration of dark fringes. Appendix A–22–4, curve A, provides data for the magnitude of K_t for this case. The value is dependent on the ratio of the hole diameter to the full width of the bar.

FIGURE 3–9 Photoelastic model showing stress concentration at a hole in an axially loaded flat bar. (Source: Measurements Group, Inc., Raleigh, NC)

Stress concentrations are most damaging for dynamic loading such as repeated loads, impact, or shock. Indeed, fatigue failures most often occur near locations of stress concentrations with small local cracks that grow with time until the remaining section is unable to withstand the load. Under static loading, the high stress near the discontinuity may cause local yielding that would result in a redistribution of the stress to an average value less than the yield strength, and thus the part would still be safe.

Values of Stress Concentration Factors. The magnitude of the stress concentration factor, K_t, depends on the geometry of the member in the vicinity of the discontinuity. Most data have been obtained experimentally by carefully measuring the maximum stress, σ_{max}, using experimental stress analysis techniques such as strain gaging or photoelasticity. (See Section 1–13 in Chapter 1). Computerized approaches using finite element analysis could also be used. Then, the value of K_t is computed from

Stress Concentration Factor

$$K_t = \sigma_{max}/\sigma_{nom} \quad \textbf{(3–19)}$$

where σ_{nom} is the stress that would be computed at the section of interest without considering the stress concentration. In the case now being discussed, direct tensile stress, $\sigma_{nom} = F/A$.

Appendix A–22 contains several charts that can be used to determine the value of K_t for a variety of geometries. See References 8 and 10 for many more cases.

A–22–1: Axially loaded round bar in tension with a circular groove

A–22–2: Axially loaded round bar in tension with a step and a fillet

A–22–3: Axially loaded flat plate in tension with a step and a fillet

A–22–4: Flat plate with a central hole

A–22–5: Round bar with a transverse hole

The chart in Appendix A–22–1 shows the typical pattern for presenting values for stress concentration factors. The vertical axis gives the value of K_t itself. The pertinent geometry factors are the diameter of the full round section, D, the diameter at the base of the groove, d_g, and the radius of the circular groove, r. From these data, *two* parameters are computed. The horizontal axis is the ratio of r/d_g. The family of curves in the chart is for different values of the ratio of D/d_g. The normal use of such a chart when the complete geometry is known is to enter the chart at the value of r/d_g, then project vertically to the curve of D/d_g, then horizontally to the vertical axis to read K_t. Interpolation between curves in the chart is often necessary. Note that the nominal stress for the grooved round bar is based on the stress at the *bottom of the groove,* the smallest area in the vicinity. While this is typical, it is important for you to know the basis of the nominal stress for any given stress concentration chart. The grooves with a circular bottom are often used to distribute oil or other lubricants to a shaft.

The chart in Appendix A–22–2 for the stepped round bar has three geometry factors: the larger diameter, D, the smaller diameter, d, and the radius of the fillet, r, at the step where the change of diameter takes place. Note that the value of K_t rises rapidly for small values of the fillet radius. As a designer, you should provide the largest practical radius for such a fillet to maintain a relatively low maximum stress at the step.

An important use for the chart in Appendix A–22–2 is the analysis of stress concentration factors for round bars having retaining ring grooves, as shown in Figure 3–10. The typical geometry of the groove, specified by the ring manufacturer, is shown in Figure 3–11. The bottom of the groove is flat and the fillet at each end is quite small to provide a sizeable

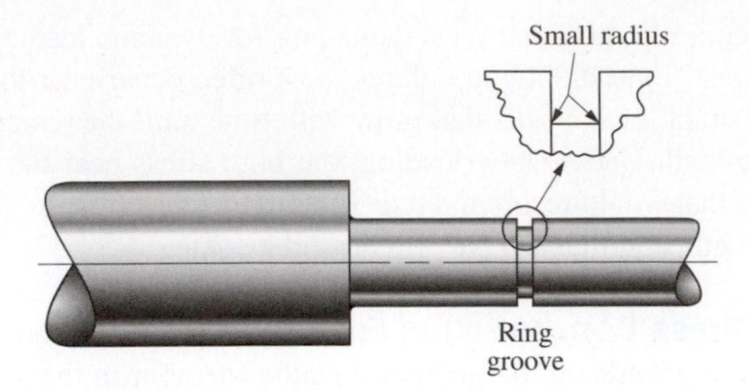

FIGURE 3–10 Stepped shaft with a ring groove.

vertical surface against which to seat the ring. The result is that the ring groove acts like two steps close together. Then Appendix A–22–2 can be used to determine the value of K_t. Sometimes the geometry of the ring groove will result in K_t values that are off the top of the chart. In such cases, an estimate of $K_t = 3.0$ is reasonable but additional data should be sought.

The chart in Appendix A–22–4 contains three curves, all related to a flat plate with a central hole. Curve A is for the case where the plate is subjected to direct tensile stress across its entire cross section in the vicinity of the hole. (See Figure 3–9). Curve B is for the case where a close-fitting pin is inserted in the hole and the tensile load is applied through the pin. The resulting stress concentration factors are somewhat larger because of the more concentrated load. Curve C is for the case of the plate in bending, and this will be discussed in Chapter 7. In each case, however, note that the nominal stress is based on the *net section* through the plate at the location of the hole. For tensile loading, the *net area* is used for σ_{nom}. That is,

$$\sigma_{\text{nom}} = F/A_{\text{net}} = F/(w - d)t$$

where w = width of the plate
t = thickness
d = diameter of the hole

The chart in Appendix A–22–5 for the round bar with a transverse hole contains a variety of data for different loading patterns: tension, bending, and torsion. For now we are concerned with only Curve A for the case of axial tension. Torsion is discussed in Chapter 4 and bending in Chapter 7. Also note that the stress concentration factor is based on the *gross section,* not on the net section at the hole. This means that K_t includes the effects of both the removal of material and the discontinuity, resulting in values that are quite high. However, it makes the use of the chart much easier for you, the designer, because you do not have to compute the area of net section.

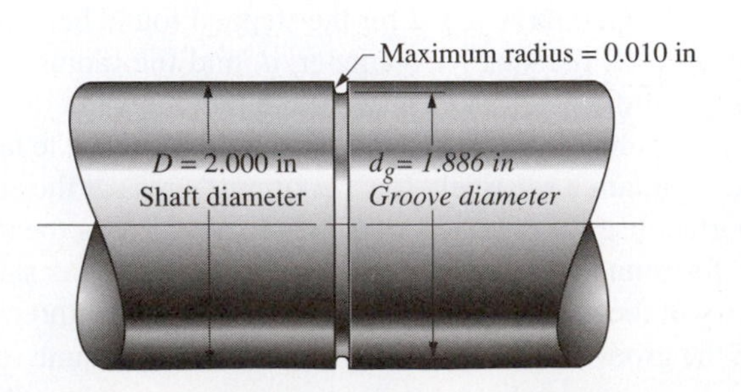

FIGURE 3–11 Sample geometry for a retaining ring groove in a round bar.

Example Problem 3–16 The stepped bar shown in Figure 3–8 is subjected to an axial tensile force of 12 500 lb. Compute the maximum tensile stress in the bar for the following dimensions:

$$D = 1.50 \text{ in}; \quad d = 0.75 \text{ in}; \quad r = 0.060 \text{ in}$$

Solution

Objective: Compute the maximum tensile stress.

Given: $F = 12\,500$ lb; $D = 1.50$ in; $d = 0.75$ in; $r = 0.060$ in

Analysis: Because of the change in diameter, use Equation (3–18).
Use the chart in Appendix A–22–2 to find the value of K_t using r/d and D/d as parameters.

Results:
$\sigma_{max} = K_t \sigma_{nom}$
$\sigma_{nom} = \sigma_2 = F/A_2 = F/(\pi d^2/4) = (12\,500 \text{ lb})/[\pi(0.75 \text{ in})^2/4]$
$\sigma_{nom} = 28\,294 \text{ lb/in}^2$
$r/d = 0.06/0.75 = 0.080$ and $D/d = 1.50/0.75 = 2.00$
Read $K_t = 2.12$ from Appendix A–22–2.
Then $\sigma_{max} = K_t \sigma_{nom} = 2.12\,(28\,294 \text{ psi}) = 59\,983$ psi.

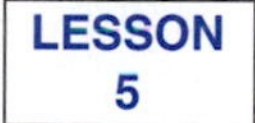

LESSON 5

Comment: The actual maximum stress of approximately 60 000 psi is more than double the value that would be predicted by the standard formula.

3–12 BEARING STRESS

When one solid body rests on another and transfers a load normal to it, the form of stress called *bearing stress* is developed at the surfaces in contact. Similar to direct compressive stress, the bearing stress, called σ_b, is a measure of the tendency for the applied force to crush the supporting member.

Bearing stress is computed in a manner similar to direct normal stresses:

Bearing Stress

$$\sigma_b = \frac{\text{applied load}}{\text{bearing area}} = \frac{F}{A_b} \qquad \textbf{(3–20)}$$

For flat surfaces in contact, the bearing area is simply the area over which the load is transferred from one member to the other. If the two parts have different areas, the smaller area is used. Another requirement is that the materials transmitting the loads must remain nearly rigid and flat in order to maintain their ability to carry the loads. Excessive deflection will reduce the effective bearing area.

Figure 3–12 shows an example from building construction in which bearing stress is important. A hollow steel column 4.00 in square rests on a thick steel plate 6.00 in square. The plate rests on a concrete pier, which in turn rests on a base of gravel. These successively larger areas are necessary to limit bearing stresses to reasonable levels for the materials involved.

Example Problem 3–17 Refer to Figure 3–12. The square steel tube carries 30 000 lb of axial compressive force. Compute the compressive stress in the tube and the bearing stress between each mating surface. Consider the weight of the concrete pier to be 338 lb.

Solution

Objectives: Compute the compressive stress in the tube. Compute the bearing stress at each surface.

Given: Load $F = 30\,000$ lb compression. Weight of pier $= 338$ lb. Geometry of members shown in Figure 3–12.

FIGURE 3–12
Bearing stress example.

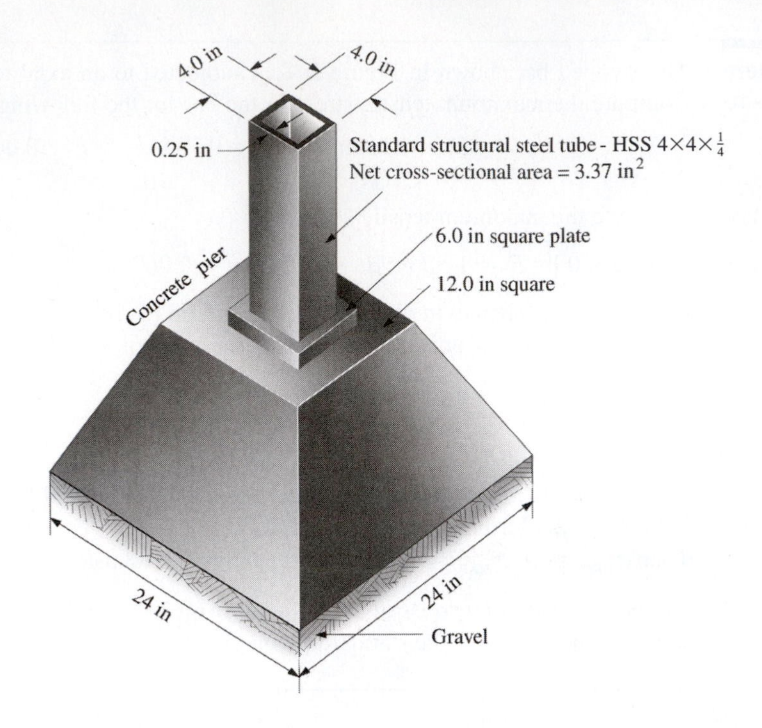

Analysis Tube: Use direct compressive stress formula. Bearing stresses: Use Equation (3–20) for each pair of mating surfaces.

Results *Compressive stress in the tube*: (area $= A = 3.37\ \text{in}^2$)

$$\sigma = F/A = 30\,000\ \text{lb}/3.37\ \text{in}^2 = 8902\ \text{psi}$$

Bearing stress between tube and square plate: This will be equal in magnitude to the compressive stress in the tube because the cross-sectional area of the tube is the smallest area in contact with the plate. Then,

$$\sigma_b = 8902\ \text{psi}$$

Bearing stress between plate and top of concrete pier: The bearing area is that of the square plate because it is the smallest area at the surface.

$$\sigma_b = F/A_b = 30\,000\ \text{lb}/(6.00\ \text{in})^2 = 833\ \text{psi}$$

Bearing stress between the pier and the gravel: The bearing area is that of a square, 24 in on a side. Add 338 lb for the weight of the pier.

$$\sigma_b = F/A_b = 30\,338\ \text{lb}/(24.00\ \text{in})^2 = 52.7\ \text{psi}$$

Comment Allowable bearing stresses are discussed later.

Bearing Stresses in Pin Joints. Frequently in mechanical and structural design, cylindrical pins are used to connect components together. One design for such a connection is shown in Figure 3–13. When transferring a load across the pin, the bearing stress between the pin and each mating member should be computed.

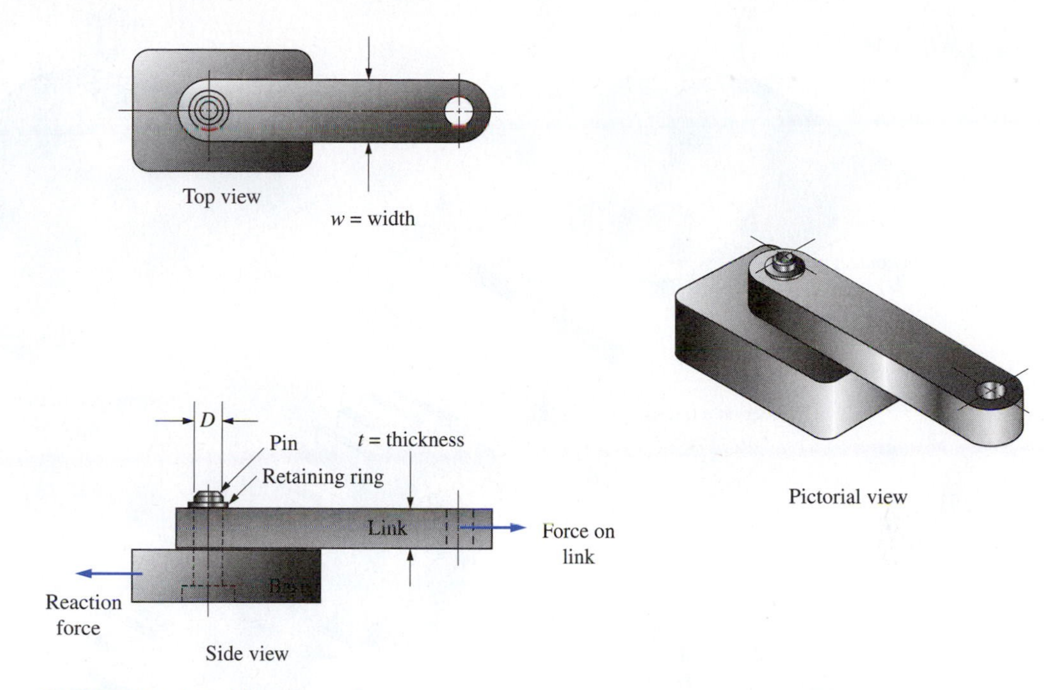

FIGURE 3–13 Pin connection.

The effective bearing area for the cylindrical pin in a close-fitting hole requires that the *projected area* be used, computed as the product of the diameter (D) of the pin and the length (L) of the surface in contact. That is,

$$A_b = D \times L \tag{3–21}$$

Example Problem 3–18 Refer to Figure 3–13. Compute the bearing stress between the 10.0 mm-diameter pin and the mating hole in the link. The force applied to the link is 3550 N. The thickness of the link is 15.0 mm and its width is 25.0 mm.

Solution

Objective: Compute the bearing stress between the mating surfaces of the pin and the inside of the hole in the link.

Given: Load $= F = 3550$ N. $t = 15.0$ mm; $w = 25.0$ mm; $D = 10.0$ mm. Geometry of members shown in Figure 3–13.

Analysis: Bearing stresses: Use Equation (3–20) for each pair of mating surfaces. Use the projected area of the hole for the bearing area.

Results: *Between the pin and the link:* $L = t =$ thickness of the link. From Equation (3–21),

$$A_b = D \times t = (10.0\,\text{mm})(15.0\,\text{mm}) = 150\,\text{mm}^2$$

Then the bearing stress is

$$\sigma_b = \frac{3550\,\text{N}}{150\,\text{mm}^2} = 23.7\,\text{N/mm}^2 = 23.7\,\text{MPa}$$

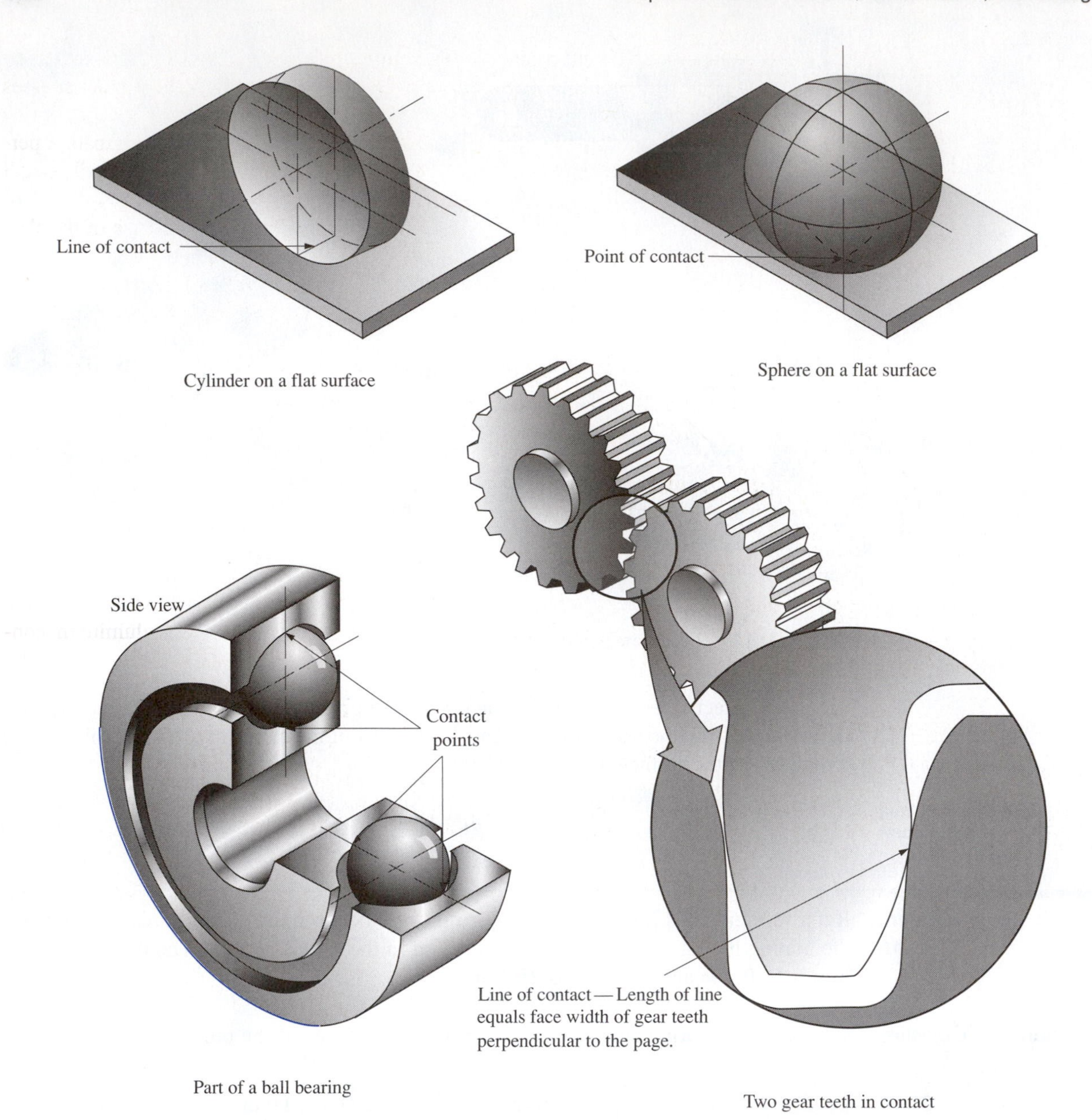

FIGURE 3–14 Examples of load-carrying members subjected to contact stress.

Contact Stress. Cases of bearing stress considered earlier in this section are those in which *surfaces* are in contact and the applied force is distributed over a relatively large area. When the load is applied over a very small area, the concept of *contact stress* must be used.

Examples of contact stress situations are shown in Figure 3–14 and include the following examples:

- cylindrical roller on a flat plate such as a steel railroad wheel on a rail
- spherical ball on a flat plate such as a ball transfer device to move heavy loads
- spherical ball on curved plate such as a ball bearing rolling in its outer race
- two convex curved surfaces such as gear teeth in contact.

The detailed analyses of contact stresses, sometimes called *Hertz stresses*, are not developed in this book. But it is important to recognize that the magnitude of contact stresses can be very high.

Consider the case of a spherical ball on a flat plate carrying a downward load. A perfectly spherical surface will contact a flat plane at only a single, infinitesimally small point. Then, when applying the bearing stress relationship, $\sigma_b = F/A_b$, the magnitude of the area approaches *zero*. Then the stress approaches *infinity*. Actually, because of the elasticity of the materials in contact, there will be some deformation and the contact area becomes a finite, but small, circular area. But the local stress will still be very large. For this reason, load-carrying members subjected to contact stresses are typically made from very hard, high-strength materials.

Similarly, when a cylindrical roller contacts a flat plate, the contact is theoretically a line having a zero width. Therefore, the bearing area is theoretically zero. The elasticity of the materials will produce an actual bearing area that is a narrow rectangle, again resulting in a finite, but large, contact stress. The special case of steel rollers on steel plates is discussed next. See Reference 10 for more detailed analysis.

3–13 DESIGN BEARING STRESS

Bearing stress, defined in Section 3–12, is a localized phenomenon created when two load-carrying parts are placed in contact. This section describes several special cases involving different materials and different geometries of the mating surfaces. Steel, aluminum, concrete, masonry, and soils are discussed.

For some applications a *design bearing stress,* σ_{bd}, is used, defined as

$$\sigma_{bd} = s_y/N \quad \text{or} \quad \sigma_{bd} = Cs_y$$

where N is a design factor and C is a specified coefficient. Note that $C = 1/N$. If the mating parts have different strengths, the lower yield strength should be used. The actual bearing stress must be less than the design bearing stress.

For some applications it is more convenient to specify an *allowable bearing load,* R_a. Then the actual contact load must be lower than the allowable bearing load.

Steel. According to the AISC (see Reference 2), the allowable bearing stress in steel for flat surfaces or on the projected area of pins in reamed, drilled, or bored holes is

Design Bearing Stress for Steel

$$\sigma_{bd} = 0.90s_y \tag{3–22}$$

When rollers or rockers are used to support a beam or other load-carrying member to allow for expansion of the member, the bearing stress is dependent on the diameter of the roller or rocker, d, and its length, L. The stress is inherently very high because the load is carried on only a small rectangular area. Theoretically, the contact between the flat surface and the roller is simply a line; but because of the elasticity of the materials, the actual area is rectangular.

Instead of specifying an allowable bearing stress, the AISC standard (see Reference 2) calls for the computation of the allowable bearing load, R_a, from

Allowable Bearing Load for Steel Roller

$$R_a = (s_y - 13)(0.03)(d)(L) \tag{3–23}$$

where s_y is in ksi
d and L are in inches and $d \leq 25$ in
R_a is in kips

When using SI metric units, Equation (3–23) becomes,

$$R_a = (s_y - 90)(3.0 \times 10^{-5})(d)(L) \quad \textbf{(3–24)}$$

where s_y is in MPa
d and L are in mm and $d \leq 635$ mm
R_a is in kN

Example Problem 3–19 A short beam, shown in Figure 3–15, is made from a rectangular steel bar, 1.25 in thick and 4.50 in high. At each end, the length resting on a steel plate is 2.00 in. If both the bar and the plate are made from ASTM A36 structural steel, compute the maximum allowable load, W, which could be carried by the beam, based on only the bearing stress at the supports. The load is centered between the supports.

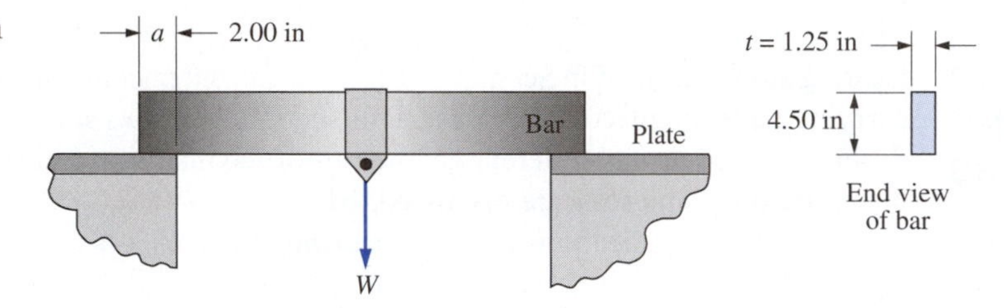

FIGURE 3–15 Beam for Example Problem 3–19.

Solution Objective Compute the allowable load, W, based on only bearing stress.

Given Loading in Figure 3–15.
Bearing area at each end: 2.00 in by 1.25 in (that is, $A_b = ta$).
Material: ASTM A36 structural steel ($s_y = 36\,000$ psi).

Analysis Design bearing stress: $\sigma_{bd} = R_a/A_b$
Where R_a is the allowable reaction at the support and $R_a = W/2$.
Design stress Equation (3–22):

$$\sigma_{bd} = 0.90\, s_y = 0.90\,(36\,000 \text{ psi}) = 32\,400 \text{ psi}$$

Then, $R_a = A_b \sigma_{bd}$ and $W = 2R_a$.

Results Bearing area: $A_b = ta = (1.25 \text{ in})(2.00 \text{ in}) = 2.50 \text{ in}^2$.

$R_a = A_b \sigma_{bd} = (2.50 \text{ in}^2)(32\,400 \text{ lb/in}^2) = 81\,000$ lb.

$W = 2R_a = 2(81\,000 \text{ lb}) = 162\,000 \text{ lb} = 162$ kip

Comment This is a very large load, so it is unlikely that bearing is the mode of failure for this beam.

Example Problem 3–20 An alternative proposal for supporting the bar described in Example Problem 3–19 is shown in Figure 3–16. To allow for expansion of the bar, the left end is supported on a 2.00 in-diameter roller made from AISI 1040 CD steel. Compute the allowable load, W, for this arrangement.

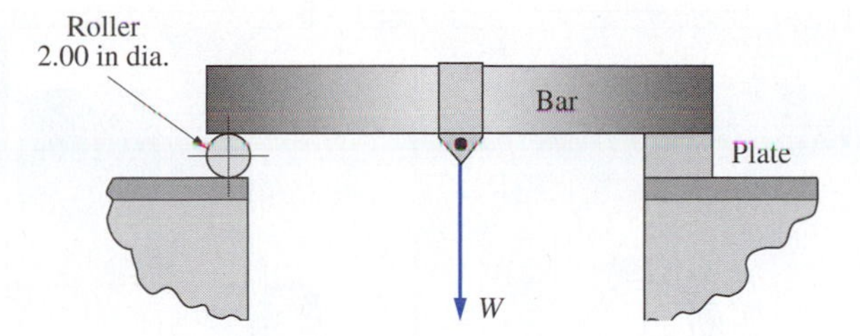

FIGURE 3–16 Beam for Example Problem 3–20.

Solution Objective Compute the allowable load on the beam.

Given Loading in Figure 3–16.
Width of the beam (t = 1.25 in) rests on the roller (d = 2.00 in).
Beam material: ASTM A36 structural steel (s_y = 36 ksi).
Roller material: AISI 1040 CD (s_y = 71 ksi).

Analysis At the roller support, Equation (3–23) applies to determine the allowable load. Note that $W = 2R_a$. Use s_y = 36 ksi, the weaker of the two materials in contact.

Results The allowable bearing load is

$$R_a = (36 - 13)(0.03)(2.00)(1.25) = 1.73 \text{ kip}$$

This would be the allowable reaction at each support. The total load is

$$W = 2R_a = 2(1.73 \text{ kip}) = 3.45 \text{ kip}$$

Comment Note that this is significantly lower than the allowable load for the surfaces found in Example Problem 3–19. Bearing at the roller support may, indeed, limit the load that could be safely supported.

Aluminum. The Aluminum Association (see Reference 1) bases the allowable bearing stresses on aluminum alloys for flat surfaces and pins on the *bearing yield strength.*

$$\sigma_{bd} = \frac{\sigma_{by}}{2.48} \tag{3–25}$$

The minimum values for bearing yield strength are listed in Reference 1. But many references, including the appendix tables in this book, do not include these data. An analysis of the data shows that for most aluminum alloys, the bearing yield strength is approximately 1.60 times larger than the tensile yield strength. Then Equation (3–25) can be restated as

$$\sigma_{bd} = \frac{1.60s_y}{2.48} = 0.65s_y \tag{3–26}$$

We will use this form for design bearing stress for aluminum in this book.

Example Problem 3–21 A rectangular bar is used as a hanger, as shown in Figure 3–17. Compute the allowable load on the basis of bearing stress at the pin connection if the bar and the clevis members are made from 6061-T4 aluminum. The pin is to be made from a stronger material.

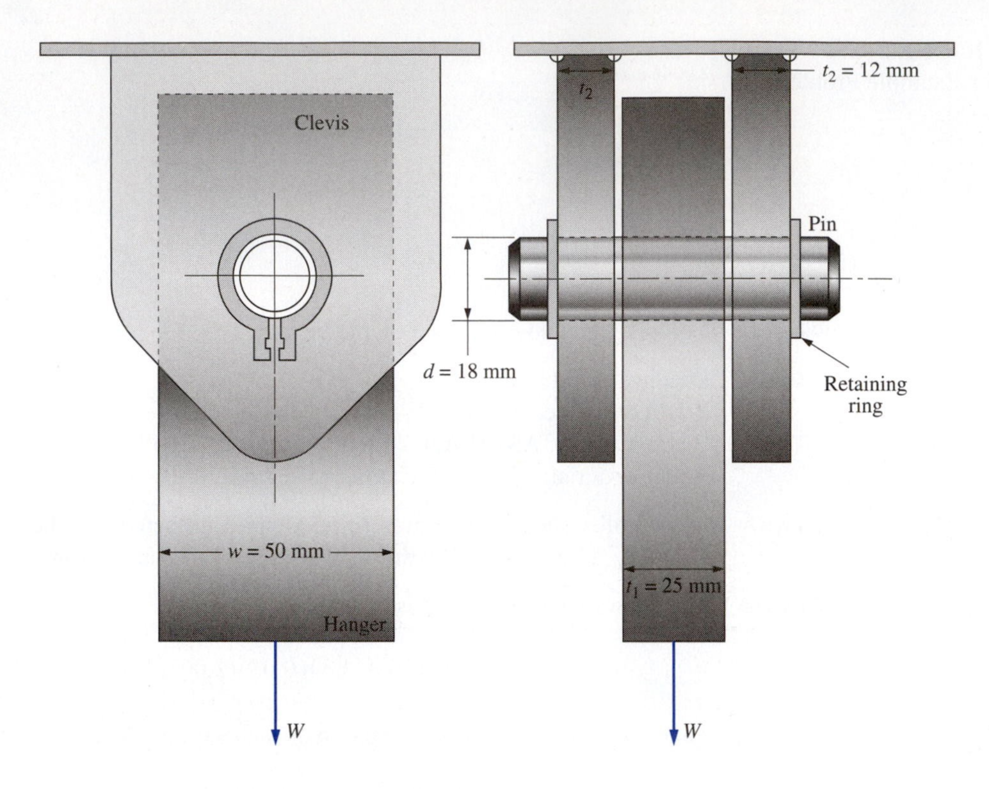

FIGURE 3–17 Hanger for Example Problem 3–21.

Solution

Objective Compute the allowable load on the hanger.

Given Loading in Figure 3–17. Pin diameter $= d = 18$ mm.
Thickness of the hanger $= t_1 = 25$ mm; width $= w = 50$ mm.
Thickness of each part of clevis $= t_2 = 12$ mm.
Hanger and clevis material: aluminum 6061-T4 ($s_y = 145$ MPa).
Pin is stronger than hanger or clevis.

Analysis For cylindrical pins in close-fitting holes, the bearing stress is based on the *projected* area in bearing, found from the diameter of the pin times the length over which the load is distributed.

$$\sigma_b = \frac{F}{A_b} = \frac{F}{dL}$$

Let $\sigma_b = \sigma_{bd} = 0.65\ s_y$ for aluminum 6061-T4.
Bearing area for hanger: $A_{b1} = t_1\, d = (25\text{ mm})(18\text{ mm}) = 450\text{ mm}^2$.
This area carries the full applied load, W.
For each side of clevis: $A_{b2} = t_2 d = (12\text{ mm})(18\text{ mm}) = 216\text{ mm}^2$.
This area carries $1/2$ of the applied load, $W/2$.
Because A_{b2} is less than $1/2$ of A_{b1}, bearing on the clevis governs.

Results $\sigma_{bd} = 0.65\ s_y = 0.65\ (145\text{ MPa}) = 94.3\text{ MPa} = 94.3\text{ N/mm}^2$
$\sigma_b = \sigma_{bd} = (W/2)/A_{b2}$
Then, $W = 2(A_{b2})(\sigma_{bd}) = 2(216\text{ mm}^2)(94.3\text{ N/mm}^2) = 40\,740$ N.

Comment This is a very large force, and other failure modes for the hanger would have to be analyzed. Failure could occur by shear of the pin or tensile failure of the hanger bar or the clevis.

TABLE 3–7 Allowable bearing stresses on masonry and soils for use in this book.

Material	Allowable bearing stress, σ_{bd} psi	MPa
Sandstone and limestone	400	2.76
Brick in cement mortar	250	1.72
Solid hard rock	350	2.41
Shale or medium rock	140	0.96
Soft rock	70	0.48
Hard clay or compact gravel	55	0.38
Soft clay or loose sand	15	0.10

Concrete: $\sigma_{bd} = Kf'_c = (0.34\sqrt{A_2/A_1})f'_c$ (But maximum $\sigma_{bd} = 0.68f'_c$)

Where: f'_c = Rated strength of concrete
A_1 = Bearing area
A_2 = Full area of the support

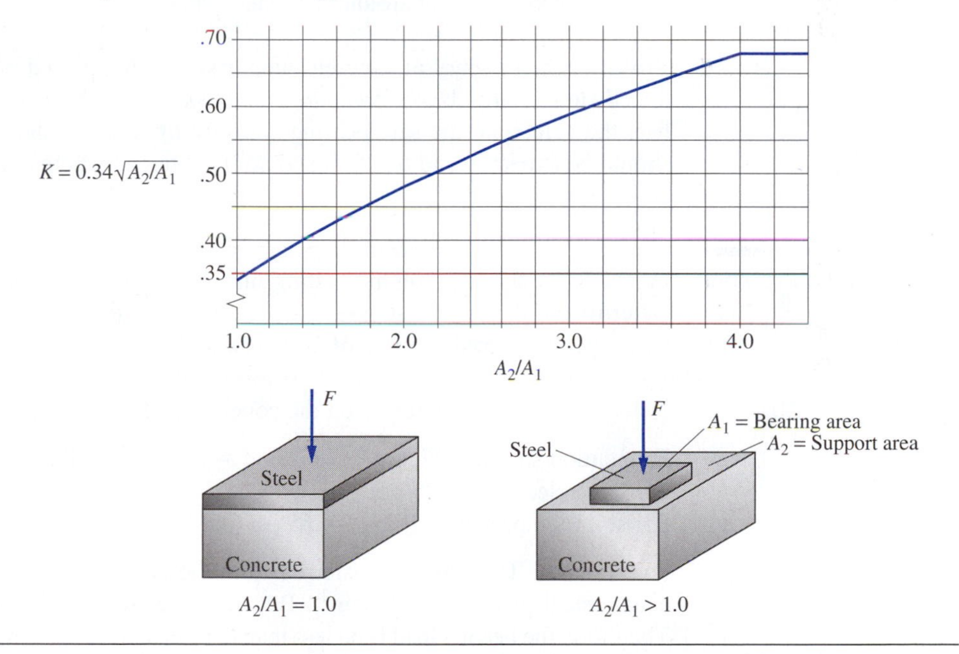

Masonry. The lower part of a support system for a building is often made from concrete, brick, or stone. Loads transferred to such supports usually require consideration of bearing stresses because the strengths of these materials are relatively low compared with metals. It should be noted that the actual strengths of the materials should be used whenever possible because of the wide variation of properties. Also, some building codes list allowable bearing stresses for certain kinds of masonry.

In the absence of specific data, this book uses the allowable bearing stresses shown in Table 3–7.

Note that the allowable bearing stress on concrete is dependent on the rated strength of the concrete and on how the load is applied. The bearing area is taken to be just the area of the steel plate, A_1, that transfers the load to the concrete. The area of the concrete is called A_2 and it can be equal to or greater than A_1. The ratio of A_2/A_1 is a factor in calculating the allowable bearing stress as shown in Table 3–7.

When bearing is on the full area of the concrete (A_2/A_1 is 1.0), the allowable bearing stress is

$$\sigma_{bd} = 0.34 f'_c$$

where f'_c = rated strength of the concrete typically 2000 psi to 7000 psi (14 MPa to 48 MPa).

A higher allowable bearing stress can be used if the area of the concrete is larger than the area of the steel bearing plate. This is because the concrete for a moderate distance beyond the plate is stressed and resists some of the load. When $A_2/A_1 > 1.0$,

$$\sigma_{bd} = 0.34\sqrt{A_2/A_1} f'_c = K f'_c$$

Table 3–7 shows a graph of K versus the ratio A_2/A_1. Note that the maximum allowable bearing stress is $0.68 f'_c$. This corresponds to the area ratio $A_2/A_1 = 4.0$. The support area A_2 must be symmetrical around the plate.

Soils. The masonry or concrete supports are often placed on soils to transfer the loads directly to the earth. *Marks' Standard Handbook for Mechanical Engineers* (see Reference 3) lists the values for the safe bearing capacity of soils, as shown in Table 3–7. Variations should be expected, and test data should be obtained where possible.

Example Problem 3–22 Figure 3–18 shows a column resting on a foundation and carrying a load of 26 000 lb. Determine if the bearing stresses are acceptable for the concrete and the soil. The concrete has a specified rated strength of 2000 psi, and the soil is compact gravel.

Solution

Objective: Determine if bearing stresses on the concrete and the soil are safe.

Given: Foundation shown in Figure 3–18: Load = F = 26 000 lb.
For concrete: f'_c = 2000 psi.
For soil (compact gravel): σ_{bd} = 55 psi (Table 3–7).

Analysis and Results: For concrete: The load is transferred from the column to the concrete through the 12 in-square steel plate (A_1 = 144 in^2). But the concrete pier is 18 in square (A_2 = 324 in^2). Therefore, the bearing load is on less than the full area of the concrete. Then, from Table 3–7,

$$\sigma_{bd} = K f'_c = 0.34\sqrt{\frac{A_2}{A_1}} f'_c = 0.34\sqrt{\frac{324}{144}} f'_c = 0.510 f'_c$$

Note that $K < 0.70$. Then,

$$\sigma_{bd} = 0.510(2000\text{ psi}) = 1020\text{ psi}$$

The bearing stress exerted on the concrete by the steel plate at the base of the column is

$$\sigma_b = \frac{F}{A_b} = \frac{26\,000\text{ lb}}{(12\text{ in})^2} = 180\text{ psi} < \sigma_{bd}$$

Thus, the bearing stress on the concrete is acceptable.

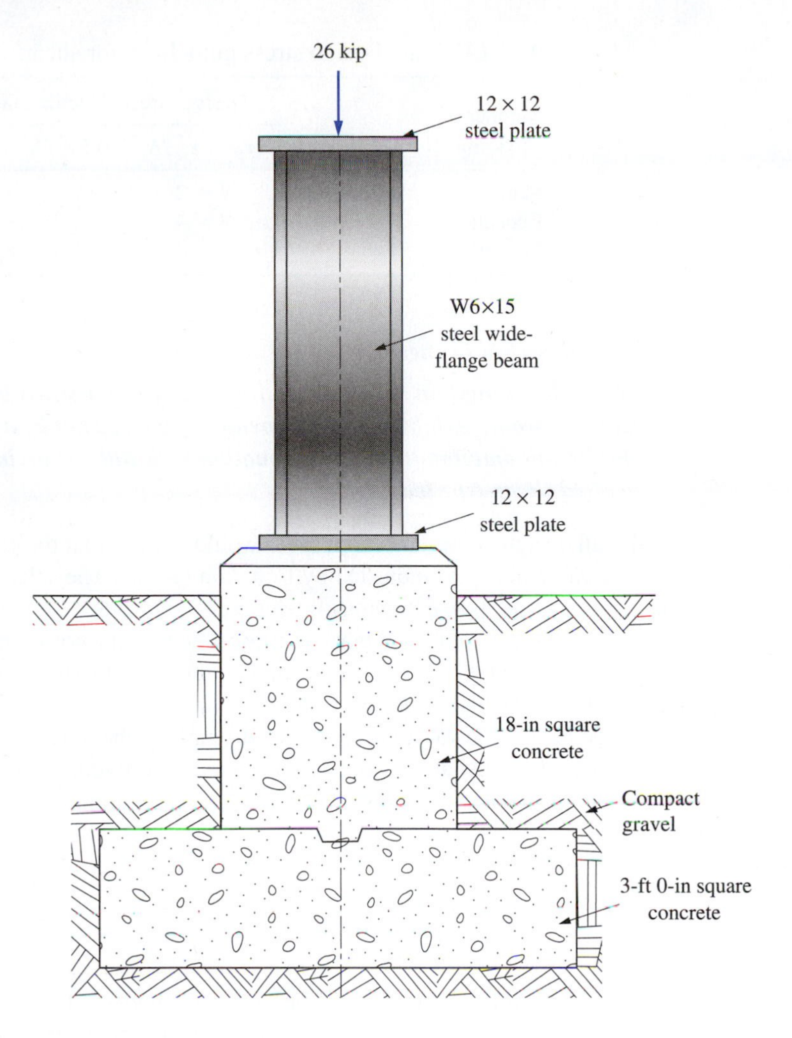

FIGURE 3–18 Column foundation for Example Problem 3–22.

For the soil (gravel) at the base of the foundation:

$$\sigma_b = \frac{F}{A_b} = \frac{26\,000\ \text{lb}}{(36\ \text{in})^2} = 20.1\ \text{psi}$$

This is acceptable because the allowable bearing stress for compact gravel is 55 psi as shown in Table 3–7.

3–14 DESIGN SHEAR STRESS

When members are subjected to shear stresses, as discussed in Chapter 1, Section 1–10, design must be based on the *design shear stress,* τ_d.

Design Shear Stress

$$\tau_d = \frac{S_{ys}}{N} \quad \text{based on the yield strength in shear} \tag{3–27}$$

TABLE 3–8 Design stress guidelines for shear.

Manner of loading	Design stress–Ductile materials $\tau_d = s_{ys}/N = 0.5\,s_y/N = s_y/2N$	
Static	Use $N = 2$	$\tau_d = s_y/4$
Repeated	Use $N = 4$	$\tau_d = s_y/8$
Impact	Use $N = 6$	$\tau_d = s_y/12$

Yield strength in shear.

The yield strength in shear, s_{ys}, is the level of shear stress at which the material would exhibit the phenomenon of yield. That is, it would undergo a significant amount of shear deformation with little or no increase in applied shear-type loading.

Virtually all designs for members in shear would require that the actual shear stress be well below the value of s_{ys}, as indicated by Equation (3–27). The selection of the design factor is taken from Table 3–8. Also consult Section 3–4 for other considerations in the selection of a design factor. Conditions that are more severe than normally encountered or where there is a significant amount of uncertainty about the magnitude of loads or material properties would justify higher design factors.

Of course, if the values of the yield strength in shear are available, they can be used in the design stress equations. But unfortunately, such values are frequently not reported, and it is necessary to rely on estimates. For the yield strength in shear, a frequently used estimate is

Estimate for Yield Strength in Shear

$$s_{ys} = \frac{s_y}{2} = 0.5\,s_y \tag{3–28}$$

This value is taken from the observation of a typical tensile test in which the shear stress is one-half of the direct tensile stress. This phenomenon, related to the *maximum shear stress theory of failure,* is somewhat conservative and will be discussed further in Chapter 10.

Ultimate strength in shear.

The ultimate strength in shear, s_{us}, is the level of shear stress at which the material would actually fracture.

There are some practical applications of shear stress where fracture of the shear-loaded member is *intended* and, therefore, an estimate of s_{us} is needed. Examples include the *shear pin* often used as an element in the drive train of machines having expensive components. Figure 3–19 shows a propeller drive shaft for a boat in which the torque from the drive shaft is transmitted through the pin to the hub of the propeller. The pin must be designed to transmit a level of torque that is typically encountered in moving the boat through the water. However, if the propeller should encounter an obstruction such as a submerged log, it would be desirable to have the inexpensive pin fail rather than the costly propeller. See Example Problem 3–23.

Another example where an estimate of the ultimate strength in shear is needed is the case of the punching operation as described in Chapter 1 and shown in Section 1–9. Here, the punch is expected to completely cut (shear) the desired part from the larger sheet of material. Therefore, the sheared sides of the part must be stressed up to the ultimate strength in shear.

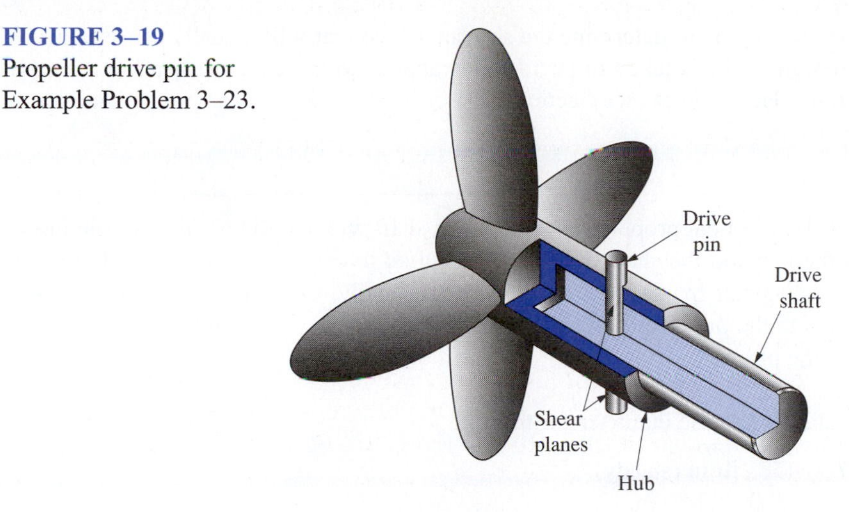

FIGURE 3–19 Propeller drive pin for Example Problem 3–23.

When data for the ultimate strength in shear are known, they should be used. For example, Appendix A–17 gives some values for cast irons and Appendix A–18 gives data for aluminum alloys. But, for occasions when published data are not available, estimates can be computed from the relationships in Table 3–9, taken from Reference 5.

Brittle Materials. Design shear stresses for brittle materials are based on the ultimate strength in shear because they do not exhibit yielding. A higher design factor should be used than for ductile materials because the materials are often less consistent in structure. However, published data on acceptable design factors are lacking. It is recommended that testing be performed on actual prototypes of shear-loaded members made from brittle materials.

Design Approaches. The general design approaches for components subjected to shearing stresses are very similar to those outlined in Cases A–D in Section 3–5. But the design stress in shear, τ_d, should be used in place of the design normal stress, σ_d. You should adapt the design approaches from Section 3–5 to outline the solution procedure depending on whether your goal is to:

a. evaluate the safety of a given design
b. specify a material for a component in shear
c. specify the shape and size of the component to resist applied shearing forces
d. determine the allowable shearing force on a component

See Table 3–8 for guidelines for design factors when ductile materials are used.

TABLE 3–9 Estimates for the ultimate strength in shear.

Formula	Material
$s_{us} = 0.65\ s_u$	Aluminum alloys
$s_{us} = 0.82\ s_u$	Steel
$s_{us} = 0.90\ s_u$	Malleable iron and copper alloys
$s_{us} = 1.30\ s_u$	Gray cast iron

When doing an analysis to determine the shearing force that will actually cause failure in shear, the design factor is taken to be 1.0. This applies to the case of shearing a plug of material from a larger sheet (see Section 1–9) or the case of a shear pin as analyzed in Example Problem 3–24.

Example Problem 3–23 Figure 3–19 shows a boat propeller mounted on a shaft with a cylindrical drive pin inserted through the hub and the shaft. The torque required to drive the propeller is 1575 lb·in and the shaft diameter inside the hub is 3.00 in. Usually, the torque is steady, and it is desired to design the pin to be safe for this condition. Specify a suitable material and the diameter for the pin.

Solution

Objective: Specify a material and the diameter for the pin.

Given: Torque = T = 1575 lb·in (steady).
Shaft diameter = D = 3.00 in.

Analysis:

1. The pin would be subjected to direct shear at the interface between the shaft and the inside of the hub, as shown in Figure 3–20. The applied torque from the shaft results in two equal forces acting perpendicular to the axis of the pin on opposite sides of the shaft, forming a couple. That is,

$$T = FD$$

   Then, $F = T/D$.
2. A material with a moderate-to-high strength is desirable so that the pin will not be overly large. Also, it should have good ductility because of the likelihood of mild shock loading from time to time. Several materials could be chosen.
3. Stress analysis: $\tau = F/A$, where $A = \pi d^2/4$. This is one cross-sectional area for the pin.
4. Design shear stress: $\tau_d = s_y/4$ (Table 3–8).
5. Let $\tau = \tau_d$. Then the required $A = F/\tau_d$ and the required d is

$$d = \sqrt{4\,A/\pi}$$

6. Specify a convenient standard size for the pin.

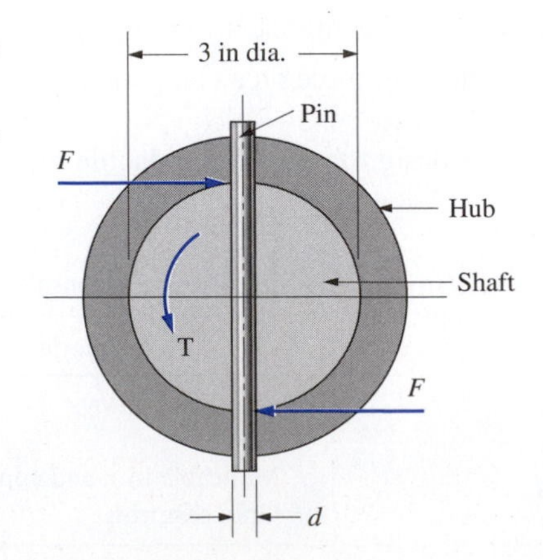

FIGURE 3–20 Cross section through propeller hub and shaft.

Results

1. $F = T/D = (1575\ \text{lb·in})/(3.00\ \text{in}) = 525\ \text{lb}$
2. Specify AISI 1020 cold drawn as a trial ($s_y = 64000$ psi; 15% elongation). Note that the pin would have to be protected from corrosion.

3, 4, 5. $\tau_d = s_y/4 = 64000\ \text{psi}/4 = 16\,000\ \text{psi}$

$A = F/\tau_d = (525\ \text{lb})/(16\,000\ \text{lb/in}^2) = 0.0328\ \text{in}^2$

$d = \sqrt{4A/\pi} = \sqrt{4(0.0328\ \text{in}^2)/\pi} = 0.204\ \text{in}$

6. From Appendix A–2, specify $d = 0.250$ in ($\frac{1}{4}$ in).
Material: AISI 1020 cold drawn steel.

Comment The size seems reasonable compared to the diameter of the shaft.

When the drive pin designed in Example Problem 3–23 is subjected to an overload such as that seen when striking a log, it is desirable for the pin to shear rather than to damage the propeller. The following example problem considers this situation.

Example Problem 3–24 Compute the torque required to shear the pin designed in Example Problem 3–23 and shown in Figures 3–19 and 3–20.

Solution

Objective Compute the torque required to shear the pin.

Given Design as shown in Figures 3–19 and 3–20.
$D = 3.00$ in; $d = 0.250$ in.
Material: AISI 1020 cold drawn steel. $s_u = 75000$ psi.

Analysis The analysis would be similar to the reverse of that for Example Problem 3–23, as summarized below.

1. The pin material would fail when $\tau = s_{us} = 0.82\ s_u$ (Table 3–9).
2. $\tau = F/A$. Then $F = \tau A = (0.82\ s_u)A$.
   Also, $A = \pi d^2/4$.
3. $T = FD$

Results

1. $s_{us} = 0.82\ s_u = 0.82\ (75\,000\ \text{psi}) = 61\,500\ \text{psi}$
2. $A = \pi d^2/4 = A = \pi\ (0.250\ \text{in})^2/4 = 0.0491\ \text{in}^2$
   $F = \tau A = (61\,500\ \text{lb/in}^2)(0.0491\ \text{in}^2) = 3019\ \text{lb}$
3. $T = FD = (3019\ \text{lb})(3.00\ \text{in}) = 9057\ \text{lb·in}$

Comment Compared with the normally applied torque, this value is quite high. The ratio of normal torque to that required to shear the pin is

$$\text{Ratio} = 9057/1575 = 5.75$$

This indicates that the pin would not likely shear under anticipated conditions. However, it may be too high to protect the propeller. Testing of the propeller should be done.

REFERENCES

1. Aluminum Association, *Aluminum Design Manual,* Washington, DC, 2005.
2. American Institute of Steel Construction, *Steel Construction Manual,* 13th ed., Chicago, IL, 2005.
3. Avallone, E. A. and T. Baumeister III, eds., *Marks' Standard Handbook for Mechanical Engineers,* 10th ed., 1996.
4. Budynas, R. G. and J. K. Nisbett, *Shigley's Mechanical Engineering Design,* 8th ed., McGraw-Hill, New York, 2007.
5. Deutschman, A. D., W. J. Michels, and C. E. Wilson, *Machine Design Theory and Practice,* Macmillan, New York, 1975.
6. Juvinall, R. C. and K. M. Marshek, *Fundamentals of Machine Component Design,* 4th ed., Wiley, New York, 2005.
7. Mott, R. L., *Machine Elements in Mechanical Design,* 4th ed., Prentice Hall, Upper Saddle River, NJ, 2004.
8. Pilkey, W., *Peterson's Stress Concentration Factors,* Wiley, New York, 1997.
9. Spotts, M. F., T. E. Shoup and L. E. Hornberger, *Design of Machine Elements,* 8th ed., Prentice Hall, Upper Saddle River, NJ, 2004.
10. Young, W. C. and R. G. Budynas, *Roark's Formulas for Stress and Strain,* 7th ed., McGraw-Hill, New York, 2002.

PROBLEMS

Direct Tension or Compression

3–1.M Specify a suitable aluminum alloy for a round bar having a diameter of 10 mm subjected to a static direct tensile force of 8.50 kN.

3–2.M A rectangular bar having cross-sectional dimensions of 10 mm by 30 mm is subjected to a direct tensile force of 20.0 kN. If the force is to be repeated many times, specify a suitable steel material.

3–3.E A link in a mechanism for an automated packaging machine is subjected to a direct tensile force of 1720 lb, repeated many times. The link is square, 0.40 in on a side. Specify a suitable steel for the link.

3–4.E A circular steel rod having a diameter of $\frac{3}{8}$ in supports a heater assembly and carries a static tensile load of 1850 lb. Specify a suitable structural steel for the rod.

3–5.E A tension member in a wood roof truss is to carry a static tensile force of 5200 lb. It has been proposed to use a standard 2 × 4 made from southern pine, No. 2 grade. Would this be acceptable?

3–6.E For the data in Problem 3–5, suggest an alternative design that would be safe for the given loading. A different size member or a different material may be specified.

3–7.E A guy wire for an antenna tower is to be aluminum, having an allowable stress of 12 000 psi. If the expected maximum load on the wire is 6400 lb, determine the required diameter of the wire.

3–8.M A hopper having a mass of 1150 kg is designed to hold a load of bulk salt having a mass of 6350 kg. The hopper is to be suspended by four rectangular straps, each carrying one-fourth of the load. Steel plate with a thickness of 8.0 mm is to be used to make the straps. What should be the width in order to limit the stress to 70 MPa?

3–9.M A shelf is being designed to hold crates having a total mass of 1840 kg. Two support rods like that shown in Figure P3–9 will hold the shelf. Assume that the center of gravity of the crates is at the middle of the shelf. Specify the required diameter of the circular rods to limit the stress to 110 MPa.

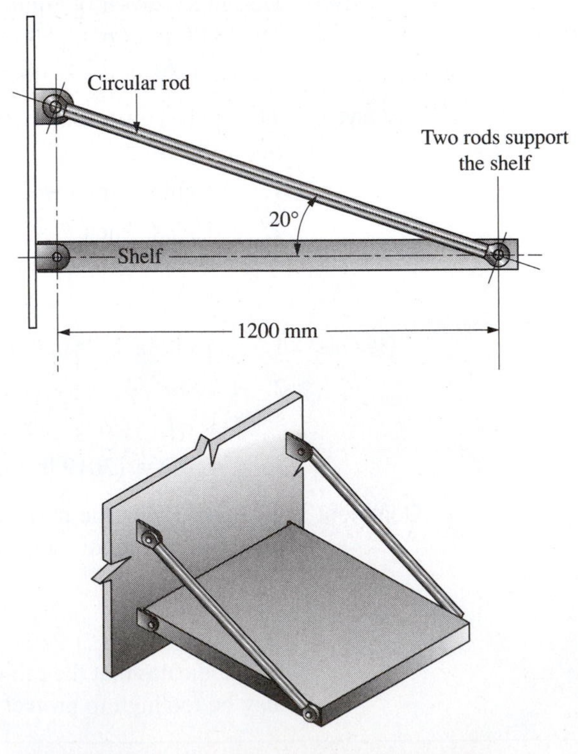

FIGURE P3–9 Shelf support rods for Problem 3–9.

3–10.E A concrete column base is circular, with a diameter of 8.0 in, and carries a static direct compressive load of 70 000 lb. Specify the required rated strength of the concrete according to the recommendations in Section 2–10.

3–11.E Three short wood blocks made from standard 4 × 4 posts support a machine weighing 29 500 lb and share the load equally. Specify a suitable type of wood for the blocks.

3–12.M A circular pier to support a column is to be made of concrete having a rated strength of 3000 psi (20.7 MPa). Specify a suitable diameter for the pier if it is to carry a direct compressive load of 1.50 MN.

3–13.E An aluminum ring has an outside diameter of 12.0 mm and an inside diameter of 10 mm. If the ring is short and is made of 2014-T6, compute the force required to produce ultimate compressive failure in the ring. Assume that s_u is the same in both tension and compression.

3–14.M A wooden cube 40 mm on a side is made from No. 2 hemlock. Compute the allowable compressive force that could be applied to the cube either parallel or perpendicular to the grain.

3–15.E A round bar of ASTM A242 structural steel is to be used as a tension member to stiffen a frame. If a maximum static load of 4000 lb is expected, specify a suitable diameter of the rod.

3–16.E A portion of a casting made of ASTM A48, grade 20 gray cast iron has the shape shown in Figure P3–16 and is subjected to compressive force in line with the centroidal axis of the section. If the member is short and carries a load of 52 000 lb, compute the stress in the section and the design factor.

3–17.M A part for a truck suspension system is to carry a compressive load of 135 kN with the possibility of shock loading. Malleable iron ASTM A220 grade 45008 is to be used. The cross section is to be rectangular with the long dimension twice the short dimension. Specify suitable dimensions for the part.

3–18.E A rectangular plastic link in an office printer is to be made from glass-filled acetal copolymer (see Appendix A–20). It is to carry a tensile force of 110 lb. Space limitations permit a maximum thickness for the link of 0.20 in. Specify a suitable width of the link if a design factor of 8 based on the tensile strength of the plastic is expected.

3–19.M Figure P3–19 shows the cross section of a short compression member that is to carry a static load of 640 kN. Specify a suitable material for the member.

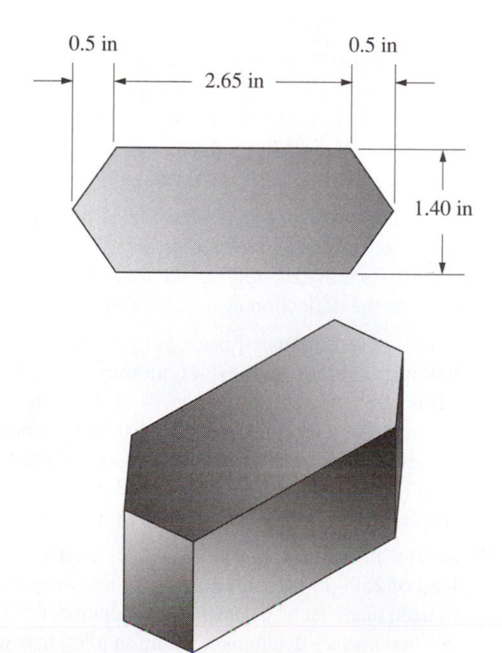

FIGURE P3–16 Short compression member for Problem 3–16.

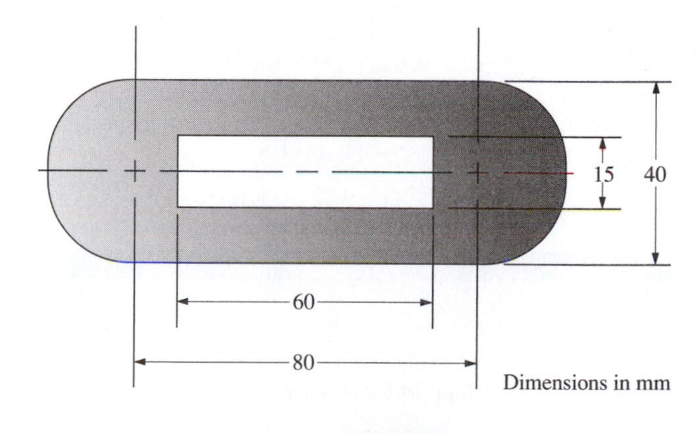

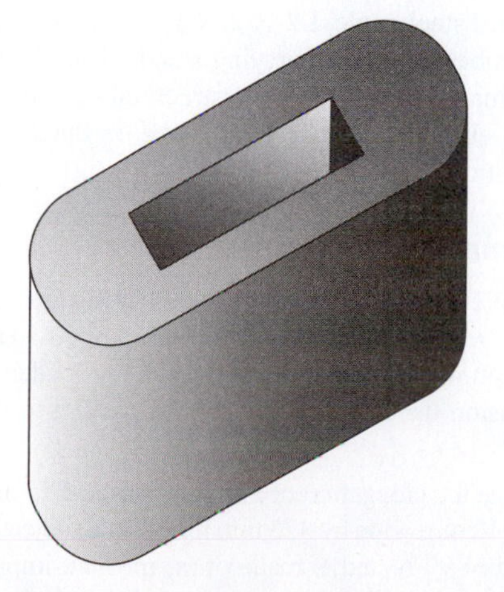

FIGURE P3–19 Short compression member for Problem 3–19.

3–20.M Figure P3–20 shows a bar carrying several static loads. If the bar is made from ASTM A36 structural steel, is it safe?

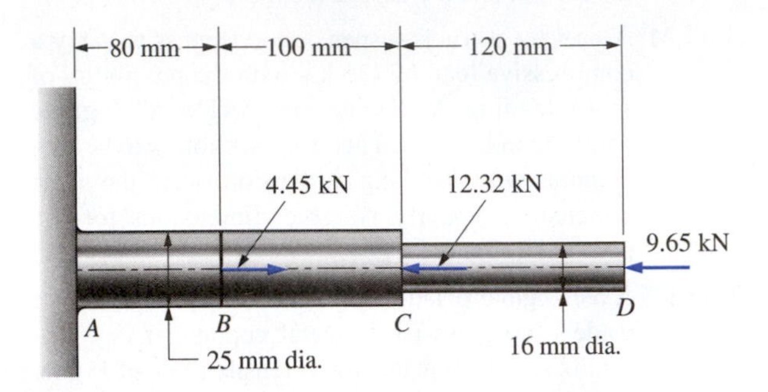

FIGURE P3–20 Bar carrying axial loads for Problem 3–20.

3–21.M In Figure P3–21, specify a suitable aluminum alloy for the member *AB* if the load is to be repeated many times. Consider only the square part near the middle of the member.

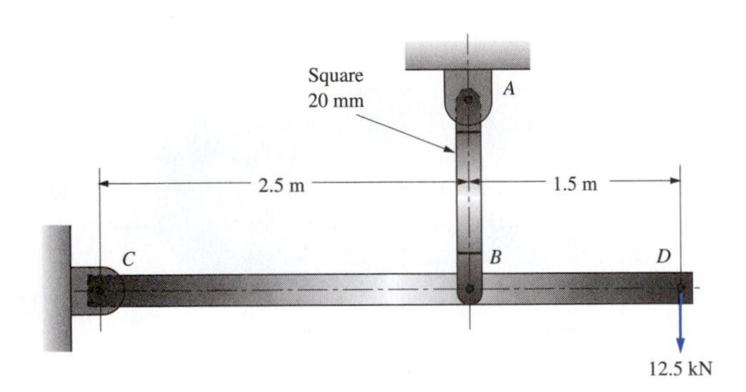

FIGURE P3–21 Support for Problem 3–21.

3–22.E A standard steel angle, L2 × 2 × $\frac{1}{4}$, serves as a tension member in a truss carrying a static load. If the angle is made from ASTM A36 structural steel, compute the allowable tensile load based on the AISC specifications.

Elastic Deformation

3–23.E A post of No. 2 grade hemlock is 6.0 ft long and has a square cross section with side dimensions of 3.50 in. How much would it be shortened when it is loaded in compression up to its allowable load parallel to the grain?

3–24.M Determine the elongation of a strip of plastic 0.75 mm thick by 12 mm wide by 375 mm long if it is subjected to a load of 90 N and is made of (a) medium-impact ABS or (b) phenolic (see Appendix A–20).

3–25.E A hollow aluminum cylinder made of 2014-T4 has an outside diameter of 2.50 in and a wall thickness of 0.085 in. Its length is 14.5 in. What axial compressive force would cause the cylinder to shorten by 0.005 in? What is the resulting stress in the aluminum?

3–26.E A metal bar is found in a stock bin and it appears to be made from either aluminum or magnesium. It has a square cross section with side dimensions of 0.25 in. Discuss two ways that you could determine what the material is.

3–27.M A tensile member is being designed for a car. It must withstand a repeated load of 3500 N and not elongate more than 0.12 mm in its 630-mm length. Use a design factor of 8 based on ultimate strength, and compute the required diameter of a round rod to satisfy these requirements using (a) AISI 1020 hot-rolled steel, (b) AISI 4140 OQT 700 steel, and (c) aluminum alloy 6061-T6. Compare the mass of the three options.

3–28.M A steel bolt has a diameter of 12.0 mm in the unthreaded portion. Determine the elongation in a length of 220 mm if a force of 17.0 kN is applied.

3–29.M In an aircraft structure, a rod is designed to be 1.25 m long and have a square cross section 8.0 mm on a side. Determine the amount of elongation that would occur if it is made of (a) titanium Ti-6A1-4V and (b) AISI 501 OQT 1000 stainless steel. The load is 5000 N.

3–30.E A tension member in a welded steel truss is 13.0 ft long and subjected to a force of 35 000 1b. Choose an equal-leg angle made of ASTM A36 steel that will limit the stress to 21 600 psi. Then compute the elongation in the angle due to the force. Use $E = 29.0 \times 10^6$ psi for structural steel.

3–31.E A link in a mechanism is a rectangular steel bar that is subjected alternately to a tensile load of 450 lb and a compressive load of 50 lb. Its dimensions are: length = 8.40 in, width = 0.25 in, thickness = 0.125 in. Compute the elongation and compression of the link.

3–32.E A cylindrical steel bar is attached at the top and is subjected to three axial forces, as shown in Figure P3–32. Its cross section has an area of 0.50 in^2. Determine the deflection of the free end.

3–33.E A link in an automated packaging machine is a hollow tube made from 6061-T6 aluminum. Its dimensions are: outside diameter = 1.250 in, inside diameter = 1.126 in, length = 36.0 in. Compute the force required to produce a deflection of the bar of 0.050 in. Would the stress produced by the force just found be safe if the load is applied repeatedly?

3–34.E A tension member in a truss is subjected to a static load of 2500 lb. Its dimensions are: length = 8.75 ft, outside diameter = 0.750 in, inside diameter = 0.563 in. First specify a suitable aluminum alloy that would be safe. Then compute the elongation of the member.

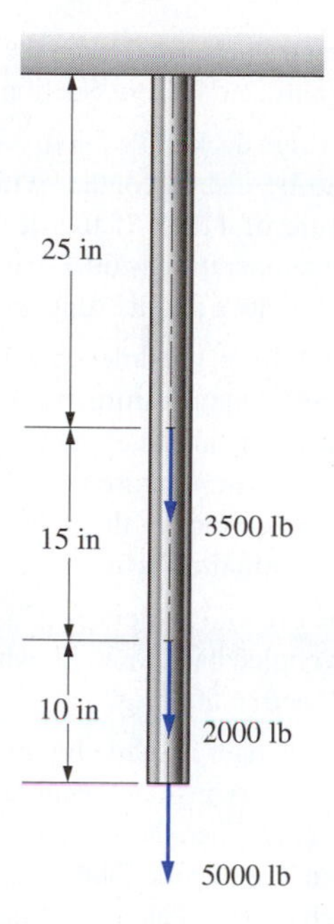

FIGURE P3–32 Bar in axial tension for Problem 3–32.

3–35.M A hollow 6061-T4 aluminum tube, 40 mm long, is used as a spacer in a machine and is subjected to an axial compressive force of 18.2 kN. The tube has an outside diameter of 56.0 mm and an inside diameter of 48.0 mm. Compute the deflection of the tube and the resulting compressive stress.

3–36.E A guy wire is made from AISI 1020 CD steel and has a length of 135 ft. Its diameter is 0.375 in. Compute the stress in the wire and its deflection when subjected to a tensile force of 1600 lb.

3–37.M Compute the total elongation of the bar shown in Figure P3–37 if it is made from titanium Ti-6AI-4V.

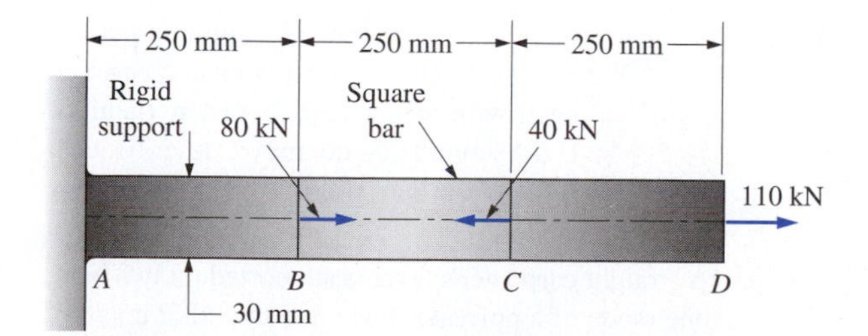

FIGURE P3–37 Bar carrying axial loads for Problem 3–37.

3–38.E During a test of a metal bar it was found that an axial tensile force of 10 000 lb resulted in an elongation of 0.023 in. The bar had these original dimensions: length = 10.000 in, diameter = 0.750 in. Compute the modulus of elasticity of the metal. What kind of metal was it probably made from?

3–39.M The bar shown in Figure P3–39 carries three loads. Compute the deflection of point D relative to point A. The bar is made from standard acrylic plastic.

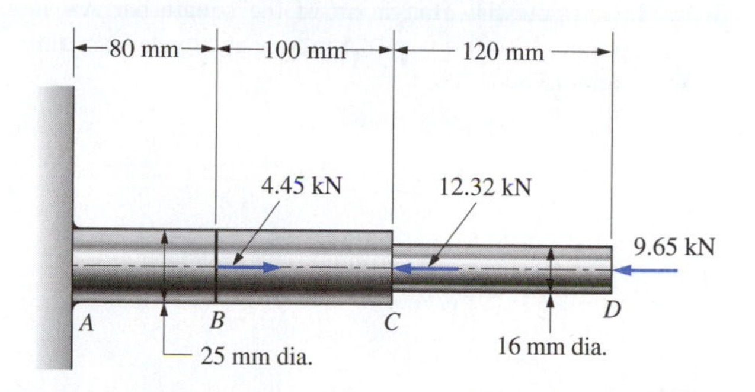

FIGURE P3–39 Bar carrying axial loads for Problem 3–39.

3–40.E A column is made from a cylindrical concrete base supporting a standard HSS $4 \times 4 \times \frac{1}{2}$ steel tube, with a length of 8.60 ft. The base is 3.0 ft long and has a diameter of 8.00 in. First, specify the concrete from Section 2–10 with a rated strength suitable to carry a compressive load of 64 000 lb. Then, assuming that no buckling occurs, compute the total amount that the column would be shortened.

3–41.E A 10.5 ft length of 14-gauge copper electrical wire (C14500, hard) is rigidly attached to a beam at its top. The wire diameter is 0.064 in. How much would it stretch if a person weighing 120 lb hung from the bottom? How much would it stretch if the person weighs 200 lb?

3–42.E A measuring tape used by carpenters is 25.00 ft long and is made from flat strip steel with these dimensions: width = 0.750 in, thickness = 0.006 in. Compute the elongation of the tape and the stress in the steel if a tensile force of 25.0 lb is applied.

3–43.E A wooden post is made from a standard 4×4 (Appendix A–4). If it is made from No. 2 grade southern pine, compute the axial compressive load it could carry before reaching its allowable stress in compression parallel to the grain. Then, if the post is 10.75 ft long, compute the amount that it would be shortened under that load.

3–44.M Ductile iron, ASTM A536, grade 60-40-18, is formed into a hollow square shape, 200 mm outside dimension and 150 mm inside dimension. Compute the load that would produce an axial compressive stress in the iron of 200 MPa. Then, for that load, compute the amount that the member would be shortened from its original length of 1.80 m.

3–45.M A brass wire (C36000, hard) has a diameter of 3.00 mm and is initially 3.600 m long. At this condition, the lower end, with a plate for applying a load, is 6.0 mm from the floor. How many kilograms of lead would have to be added to the plate to cause it to just touch the floor? What would be the stress in the wire at that time?

3–46.M Compute the elongation of the square bar AB in Figure P3–46 if it is 1.25 m long and made from aluminum 6061-T6.

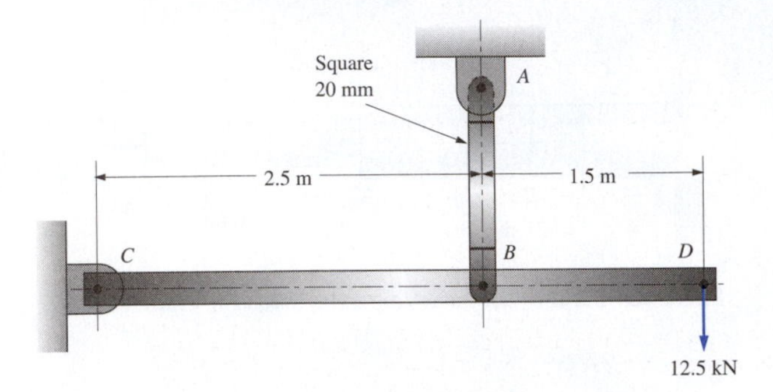

FIGURE P3–46 Support for Problem 3–46.

Thermal Deformation

3–47.E A concrete slab in a highway is 80 ft long. Determine the change in length of the slab if the temperature changes from −30°F to + 110°F.

3–48.M A steel rail for a railroad siding is 12.0 m long and made from AISI 1040 HR steel. Determine the change in length of the rail if the temperature changes from −34°C to +43°C.

3–49.M Determine the stress that would result in the rail described in Problem 3–48 if it were completely restrained from expanding.

3–50.M The pushrods that actuate the valves on a six-cylinder engine are AISI 1040 steel and are 625 mm long and 8.0 mm in diameter. Calculate the change in length of the rods if their temperature varies from −40°C to +116°C and the expansion is unrestrained.

3–51.M If the pushrods described in Problem 3–50 were installed with zero clearance with other parts of the valve mechanism at 25°C, compute the following:

(a) The clearance between parts at −40°C.

(b) The stress in the rod due to a temperature rise to 116°C.

Assume that mating parts are rigid.

3–52.E A bridge deck is made as one continuous concrete slab to 140 ft long at 30°F. Determine the required width of expansion joints at the ends of the bridge if no stress is to be developed when the temperature varies from +30°F to +110°F.

3–53.E When the bridge deck of Problem 3–52 was installed, the width of the expansion joint at each end was only 0.25 in. What stress would be developed if the supports are rigid? For the concrete use f'_c = 4000 psi and find E from Section 2–10.

3–54.E For the bridge deck in Problem 3–52, assume that the deck is to be just in contact with its support at the temperature of 110°F. If the deck is to be installed when the temperature is 60°F, what should the gap be between the deck and its supports?

3–55.M A ring of AISI 301 stainless steel is to be placed on a shaft having a temperature of 20°C and a diameter of 55.200 mm. The inside diameter of the ring is 55.100 mm. To what temperature must the ring be heated to make it 55.300 mm in diameter and thus allow it to be slipped onto the shaft?

3–56.M When the ring of Problem 3–55 is placed on the shaft and then cooled back to 20°C, what tensile stress will be developed in the ring?

3–57.M A heat exchanger is made by arranging several brass (C36000) tubes inside a stainless steel (AISI 430) shell. Initially, when the temperature is 10°C, the tubes are 4.20 m long and the shell is 4.50 m long. Determine how much each will elongate when heated to 85°C.

3–58.E In Alaska, a 40-ft section of AISI 1020 steel pipe may see a variation in temperature from −50°F when it is at ambient temperature to +140°F when it is carrying heated oil. Compute the change in the length of the pipe under these conditions.

3–59.M A square bar of magnesium is 30 mm on a side and 250.0 mm long at 20°C. It is placed between two rigid supports set 250.1 mm apart. The bar is then heated to 70°C while the supports do not move. Compute the resulting stress in the bar.

3–60.M A square rod, 8.0 mm on a side, is made from AISI 1040 cold-drawn steel and has a length of 175 mm. It is placed snugly between two unmoving supports with no stress in the rod. Then the temperature is increased by 90°C. What is the final stress in the rod?

3–61.E A square bar is made from 6061-T4 aluminum alloy. At 75°F its length is 10.500 in. It is placed between rigid supports with a distance between them of 10.505 in. If the supports do not move, describe what would happen to the bar when its temperature is raised to 400°F.

3–62.E A straight carpenter's level is supported on two bars; one made of a polyester resin and the other made of titanium Ti-6A1-4V. The distance between the bars is 24.00 in. At a temperature of 65°F, the level is perfectly level and each bar has a length of 30.00 in. What would be the angle of tilt of the level when the temperature is increased to 212°F?

3–63.E When manufactured, a steel (AISI 1040) measuring tape was exactly 25.000 ft long at a temperature of 68°F. Compute the error that would result if the tape is used at −15°F.

3–64.M Figure P3–64 shows two bars of different materials separated by 0.50 mm when the temperature is 20°C. At what temperature would they touch?

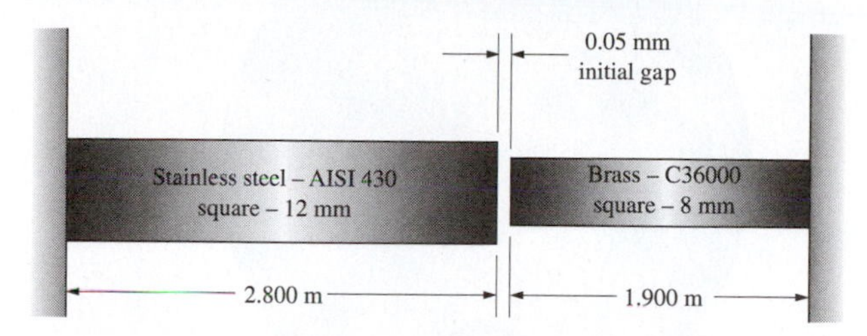

FIGURE P3–64 Problem 3–64.

3–65.M A stainless steel (AISI 301) wire is stretched between rigid supports so that a stress of 40 MPa is induced in the wire at a temperature of 20°C. What would be the stress at a temperature of −15°C?

3–66.M For the conditions described in Problem 3–65, at what temperature would the stress in the wire be zero?

Members Made from Two Materials

3–67.M A short post is made by welding steel plates into a square, as shown in Figure P3–67, and then filling the area inside with concrete. Compute the stress in the steel and in the concrete if $b = 150$ mm, $t = 10$ mm, and the post carries an axial load of 900 kN. See Section 2–10 for concrete properties. Use $f'_c = 6000$ psi.

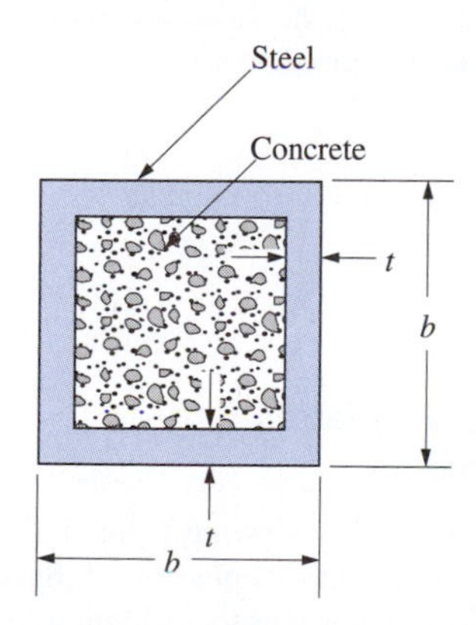

FIGURE P3–67 Post for Problems 3–67, 3–68, and 3–69.

3–68.E A short post is made by filling a standard $6 \times 6 \times \frac{1}{2}$ steel tube with concrete, as shown in Figure P3–67. The steel has an allowable stress of 21 600 psi. The concrete has a rated strength of 6000 psi but, in this application, the stress is to be limited to 1500 psi. See Section 2–10 for the modulus of elasticity for the concrete. Compute the allowable load on the post.

3–69.E A short post is being designed to support an axial compressive load of 500 000 lb. It is to be made by welding $\frac{1}{2}$-in thick plates of A36 steel into a square and filling the area inside with concrete, as shown in Figure P3–67. It is required to determine the dimension of the side of the post b in order to limit the stress in the steel to no more than 21 600 psi and in the concrete to no more than 1500 psi. See Section 2–10 for concrete properties. Use $f'_c = 6000$ psi.

3–70.M Two disks are connected by four rods, as shown in Figure P3–70. All rods are 6.0 mm in diameter and have the same length. Two rods are steel ($E = 207$ GPa), and two are aluminum ($E = 69$ GPa). Compute the stress in each rod when an axial force of 11.3 kN is applied to the disks.

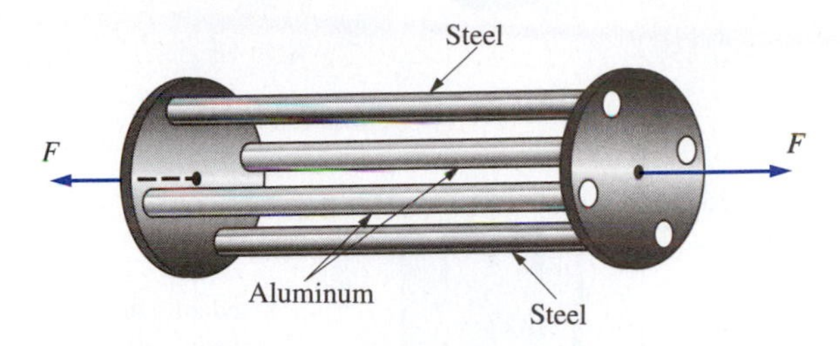

FIGURE P3–70 Problem 3–70.

3–71.M An array of three wires is used to suspend a casting having a mass of 2265 kg in such a way that the wires are symmetrically loaded (see Figure P3–71). The outer two wires are AISI 430 stainless steel, full hard. The middle wire is hard beryllium copper, C17200. All three wires have the same diameter and length. Determine the required diameter of the wires

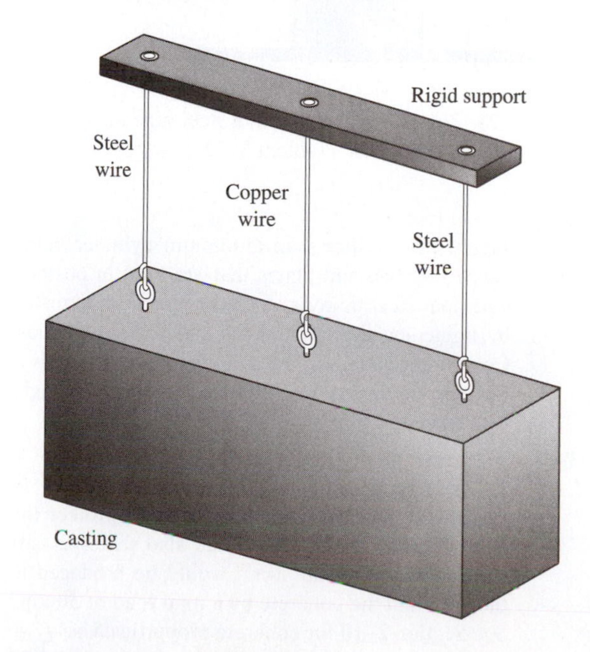

FIGURE P3–71 Problem 3–71.

if none is to be stressed beyond one-half of its yield strength.

3–72.M Figure P3–72 shows a load being applied to an inner cylindrical member that is initially 0.12 mm longer than a second concentric hollow pipe. What would be the stress in both members if a total load of 350 kN is applied?

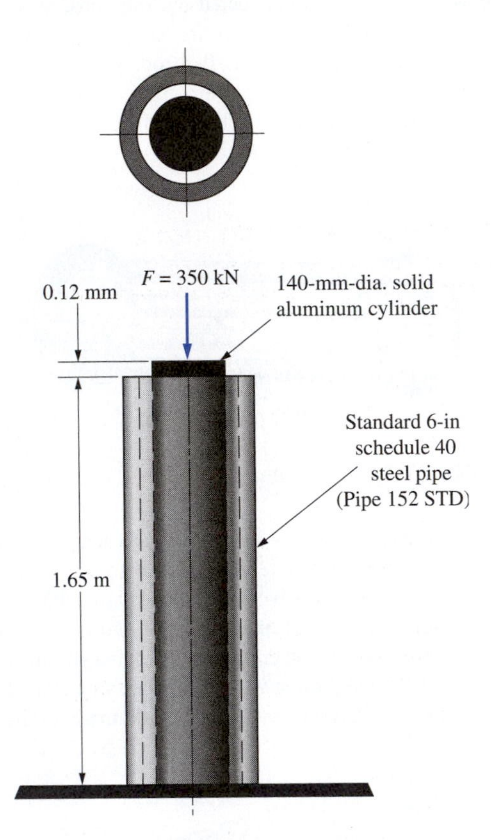

FIGURE P3–72 Aluminum bar in a steel pipe under an axial compression load for Problem 3–72.

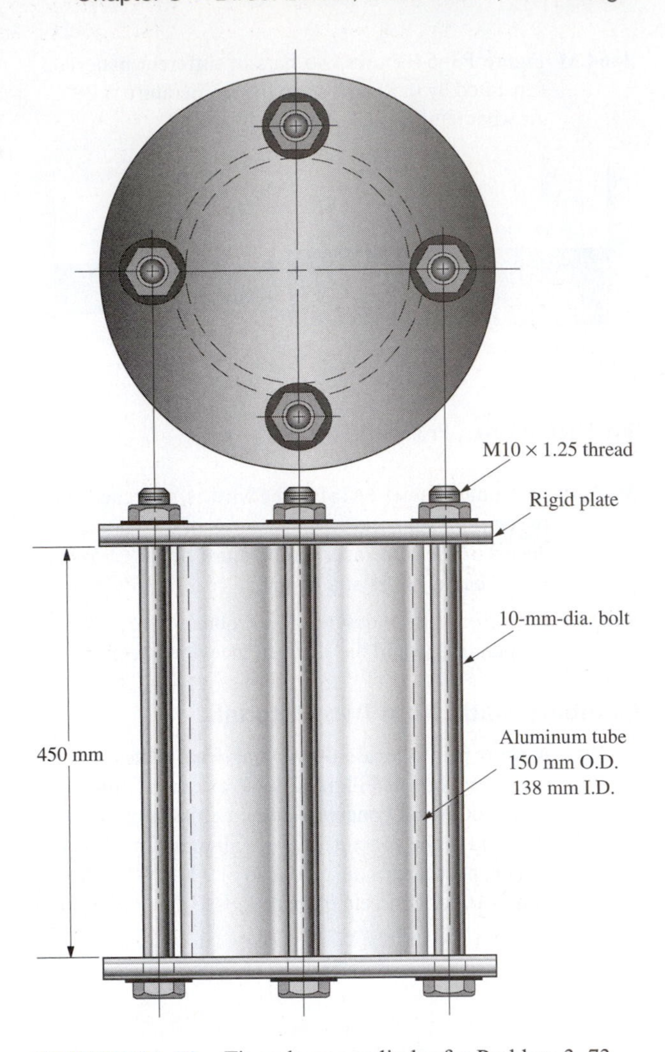

FIGURE P3–73 Tie rods on a cylinder for Problem 3–73.

3–73.M Figure P3–73 shows an aluminum cylinder being capped by two end plates that are held in position with four steel tie rods. A clamping force is created by tightening the nuts on the ends of the tie rods. Compute the stress in the cylinder and the tie rods if the nuts are turned one full turn from the hand-tight condition.

3–74.E A column for a building is made by encasing a W6 × 15 wide-flange shape in concrete, as shown in Figure P3–74. The concrete helps to protect the steel from the heat of a fire and also shares in carrying the load. What stress would be produced in the steel and the concrete by a total load of 50 kip? See Section 2–10 for concrete properties. Use $f_c' =$ 2000 psi.

Additional Deformation and Design Problems

3–75. A 4 × 4 wood post having a length of 4.25 m is made from No. 2 southern pine wood. Specify the maximum load in newtons the column would support and not exceed the allowable stress in compression parallel to the grain. Assume no buckling occurs. Then compute the total deformation of the post when subjected to this magnitude of load.

3–76. An aluminum bar (6061-T4) is initially 225.00 mm long at a temperature of 20°C. It is placed between two immovable end plates that are 225.50 mm apart. The bar is then heated to 205°C. Describe the sequence of events that occurs as the heating takes place in terms of deformation and stress. Evaluate the acceptability of the final condition.

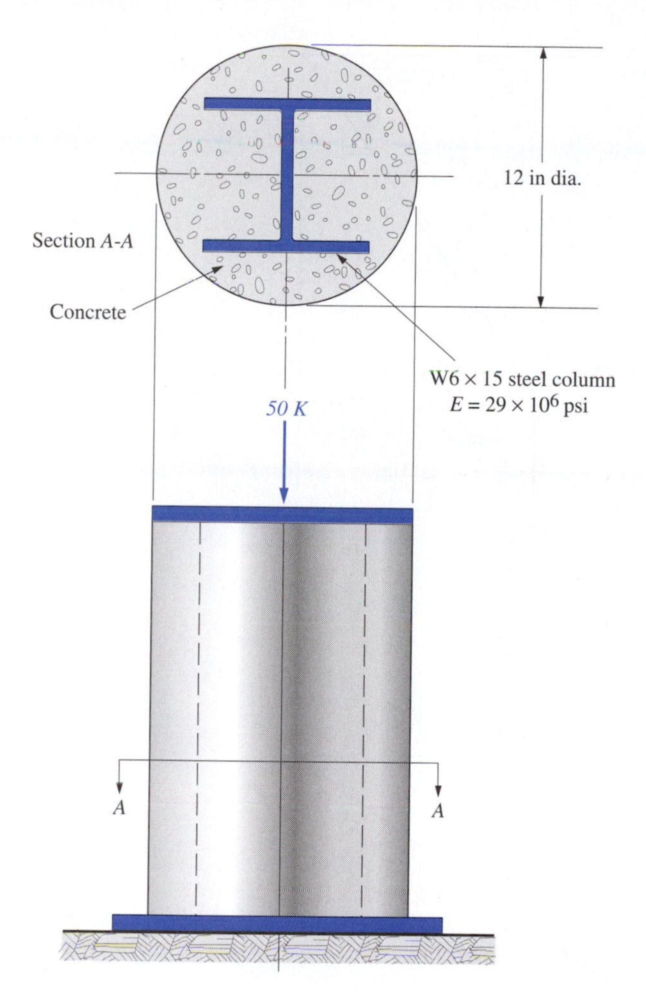

FIGURE P3–74 Steel column encased in concrete for Problem 3–74.

3–77. A rectangular bar of aluminum (2014-T4) has a length of 2.400 m and cross-sectional dimensions of 25 mm by 50 mm when the temperature is 20°C.

(a) At what temperature would the length be 2.405 m if the bar is unrestrained?

(b) If the temperature is increased an additional 30°C while the bar is restrained from growing lengthwise, compute the resulting stress in the bar.

(c) Is the stress level safe?

3–78. A tensile rod in a mechanism is made from AISI 4140 OQT 1300 steel. It has a rectangular cross section, 30 mm × 20 mm and a length of 700 mm. It is to be used in the following manner: (a) it must not elongate more than 0.50 mm under load and (b) it must be safe when subjected to a repeated axial tensile load. Specify the maximum permissible load on the rod.

3–79. Refer to Figure P3–79. Compute the total elongation of the bar if it is made from the aluminum alloy 2014-T6.

3–80. Figure P3–80 shows two rods, AC and BC, supporting a load at C. Each rod is steel and has a diameter of 8.00 mm. Compute the elongation of rods AC and BC when a load having a mass of 680 kg is suspended from C.

3–81. Figure P3–81 shows two rods, AB and BC, supporting a load at B. Each rod is steel and has a diameter of 10.00 mm. Compute the elongation of rods AB and BC when a load having a mass of 4200 kg is suspended from B.

3–82 to 3–90. A standard tensile test specimen has a straight cylindrical center section with a diameter of 0.505 in. Gage marks are inscribed 2.000 in apart when the specimen is in the unloaded condition. Compute the distance between the gage marks when the given material is subjected to an axial stress equal to 90% of the listed yield strength from the Appendix tables. Assume that these stresses are below the proportional limit for the materials. Also compute force required to create this stress level and the strain in the area between the gage marks.

3–82. AISI 1040 cold-drawn steel

3–83. AISI 5160 OQT 700 steel

3–84. AISI 501 OQT 1000 stainless steel

3–85. C17200 hard beryllium copper

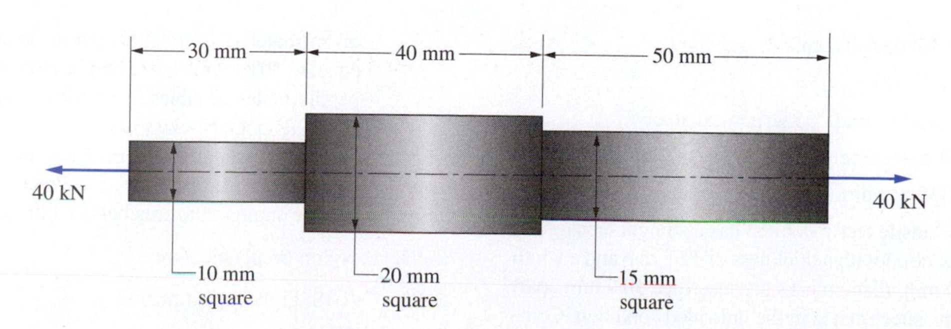

FIGURE P3–79 Bar for Problem 3–79.

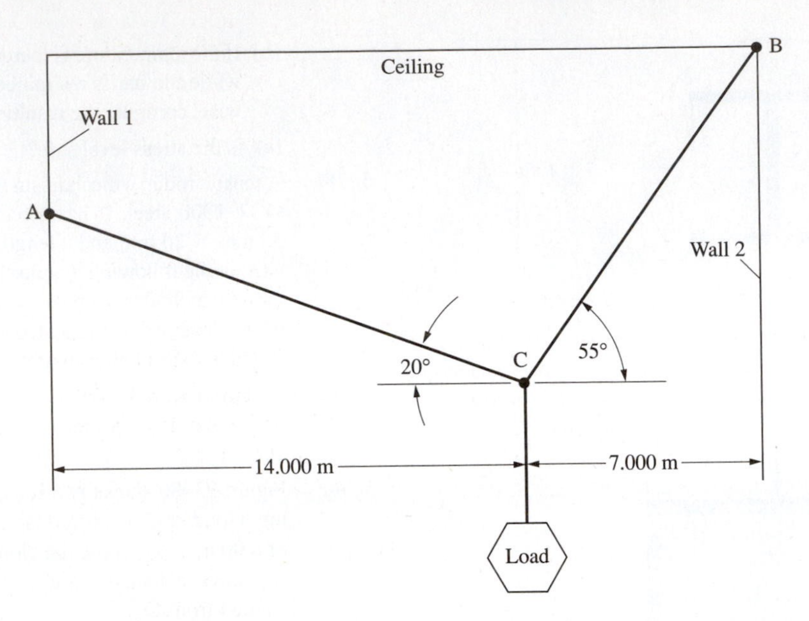

FIGURE P3–80 Support system for Problem 3–80.

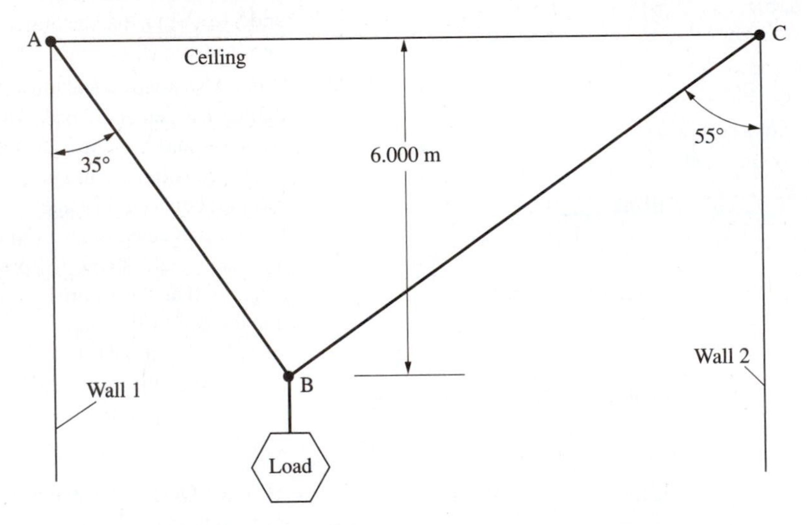

FIGURE P3–81 Support system for Problem 3–81.

3–86. ASTM AZ 63A-T6 cast magnesium

3–87. ZA 12 cast zinc

3–88. ASTM A572 Grade 65 structural steel

3–89. Grade 4 austempered ductile iron

3–90. 5154-H38 aluminum alloy

3–91 to 3–99. A tensile test specimen has a straight rectangular center section with a thickness of 12.5 mm and a width of 16.0 mm. Gage marks are inscribed 50.0 mm apart when the specimen is in the unloaded condition. Compute the distance between the gage marks when the given material is subjected to an axial tensile stress equal to 50% of the listed tensile strength from the Appendix tables or tables in Chapter 2. Assume that the materials obey Hooke's law and that these stresses are below the proportional limit for the materials. Also compute the force required to create this stress level and the strain in the area between the gage marks.

3–91. Nylon 66 plastic, day

3–92. ABS, high-impact plastic

3–93. Acetal copolymer plastic

3–94. Polyurethane elastomer

3–95. Phenolic plastic

3–96. Glass/epoxy composite (Table 2–13)

3–97. Aramid/epoxy composite (Table 2–13)

3–98. Graphite/epoxy, 62% fiber (Table 2–13)

3–99. Graphite/epoxy composite ultra high modulus (Table 2–13)

Stress Concentrations for Direct Axial Stresses

3–100.M A circular rod with a diameter of 40.0 mm has a groove cut to a diameter of 35 mm. A full radius of 3.0 mm is produced at the bottom of the groove. Compute the maximum stress in the rod when an axial tensile force of 46 kN is applied.

3–101.E A circular rod with a diameter of 1.50 in has a groove cut to a diameter of 1.25 in. A full radius of 0.12 in is produced at the bottom of the groove. Compute the maximum stress in the rod when an axial tensile force of 10 300 lb is applied.

3–102.E A circular rod with a diameter of 0.40 in has a groove cut to a diameter of 0.35 in. A full radius of 0.040 in is produced at the bottom of the groove. Compute the maximum stress in the rod when an axial tensile force of 1250 lb is applied.

3–103.M A circular rod with a diameter of 10.0 mm has a groove cut to a diameter of 8.0 mm. A full radius of 1.20 mm is produced at the bottom of the groove. Compute the maximum stress in the rod when an axial tensile force of 5500 N is applied.

3–104.E A flat plate has a width of 2.50 in and a thickness of 0.400 in. Part of the plate is machined to a width of 2.20 in using a milling cutter having a diameter of 0.250 in. Compute the maximum stress in the plate when an axial tensile force of 17 500 lb is applied.

3–105.M A flat plate has a width of 60.0 mm and a thickness of 10.0 mm. Part of the plate is machined to a width of 55 mm using a milling cutter having a diameter of 6.00 mm. Compute the maximum stress in the plate when an axial tensile force of 75 kN is applied.

3–106.M A flat plate has a width of 25.0 mm and a thickness of 3.0 mm. Part of the plate is machined to a width of 22 mm using a milling cutter having a diameter of 5.00 mm. Compute the maximum stress in the plate when an axial tensile force of 6800 N is applied.

3–107.E A flat plate has a width of 0.80 in and a thickness of 0.120 in. Part of the plate is machined to a width of 0.50 in using a milling cutter having a diameter of 0.200 in. Compute the maximum stress in the plate when an axial tensile force of 1800 lb is applied.

3–108.M A circular rod with a diameter of 50.0 mm is cut down to a diameter of 40 mm with a cutter having a nose radius of 6.0 mm. Compute the maximum stress in the rod when an axial tensile force of 230 kN is applied.

3–109.E A circular rod with a diameter of 2.50 in is cut down to a diameter of 1.75 in with a cutter having a nose radius of 0.25 in. Compute the maximum stress in the rod when an axial tensile force of 48.0 kips is applied.

3–110.E A circular rod with a diameter of 0.38 in is cut down to a diameter of 0.32 in with a cutter having a nose radius of 0.02 in. Compute the maximum stress in the rod when an axial tensile force of 375 lb is applied.

3–111.M A circular rod with a diameter of 10.0 mm is cut down to a diameter of 8.00 mm with a cutter having a nose radius of 0.50 mm. Compute the maximum stress in the rod when an axial tensile force of 1600 N is applied.

3–112.E A flat plate has a width of 2.50 in and a thickness of 0.400 in. A hole with a diameter of 1.75 in is bored through the plate. Compute the maximum stress in the plate when an axial tensile force of 14 200 lb is applied.

3–113.M A flat plate has a width of 60 mm and a thickness of 8.00 mm. A hole with a diameter of 40.0 mm is bored through the plate. Compute the maximum stress in the plate when an axial tensile force of 65.0 kN is applied.

3–114.M A flat plate has a width of 18.0 mm and a thickness of 2.50 mm. A hole with a diameter of 8.00 mm is bored through the plate. Compute the maximum stress in the plate when an axial tensile force of 2250 N is applied.

3–115.E A flat plate has a width of 0.60 in and a thickness of 0.088 in. A hole with a diameter of 0.25 in is bored through the plate. Compute the maximum stress in the plate when an axial tensile force of 475 lb is applied.

3–116.M A circular rod with a diameter of 50.0 mm has a 20 mm diameter hole cut transversely through it. Compute the maximum stress in the rod when an axial tensile force of 120 kN is applied.

3–117.E A circular rod with a diameter of 2.0 in has a 0.75 in diameter hole cut transversely through it. Compute the maximum stress in the rod when a 22 500 lb axial tensile force is applied.

3–118.E A circular rod with a diameter of 0.63 in has a 0.35 in diameter hole cut transversely through it. Compute the maximum stress in the rod when a 2800 lb axial tensile force is applied.

3–119.M A circular rod with a diameter of 12.0 mm has a 7.25 mm diameter hole cut transversely through it. Compute the maximum stress in the rod when an axial tensile force of 7500 N is applied.

3–120.M Figure P3–120 shows a circular shaft subjected to a repeated axial tensile load of 25 kN. The shaft is made from AISI 4140 OQT 1100 steel. Determine the design factor at the hole and at the fillet.

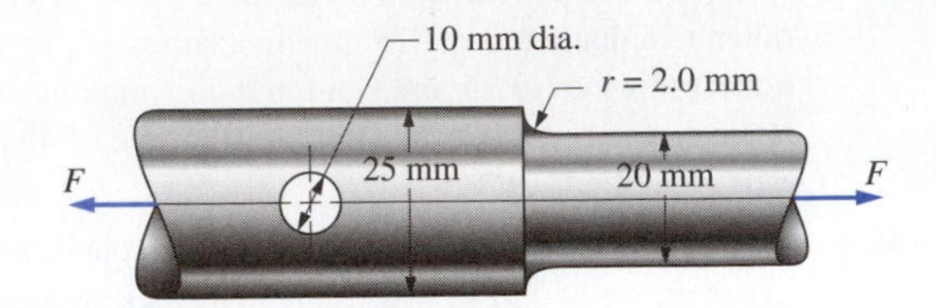

FIGURE P3–120 Shaft for Problem 3–120.

3–121.M A valve stem in an automotive engine is subjected to an axial tensile load of 900 N due to the valve spring, as shown in Figure P3–121. Compute the maximum stress in the stem at the place where the spring force acts against the shoulder.

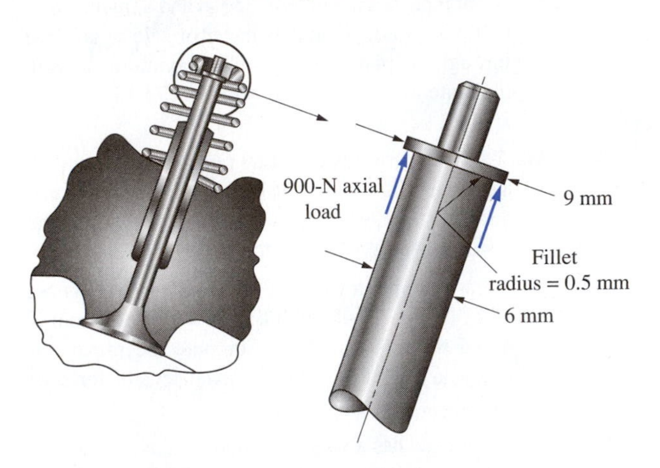

FIGURE P3–121 Valve stem for Problem 3–121.

3–122. A part of a tie rod in a steering linkage has the shape shown in Figure P3–122. As the machine cycles, it exerts a repeated direct tensile load of 8.25 kN on the tie rod. Compute the expected maximum tensile stress on the rod and specify a suitable material from which to make it.

3–123.M A round shaft has two grooves in which rings are placed to retain a gear in position, as shown in Figure P3–123. If the shaft is subjected to an axial tensile force of 36 kN, compute the maximum tensile stress in the shaft.

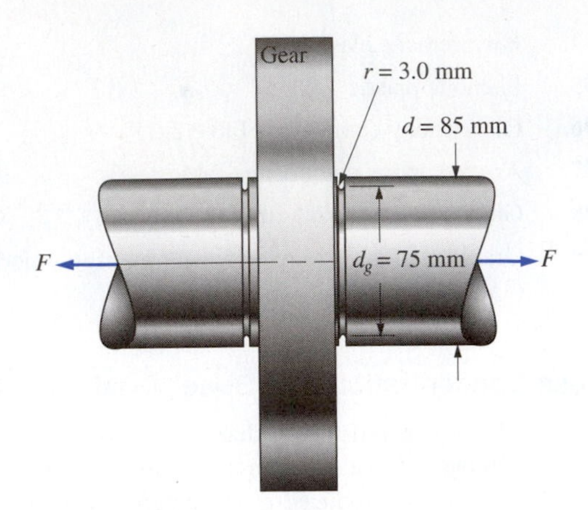

FIGURE P3–123 Shaft for Problem 3–123.

3–124.E Figure P3–124 shows the proposed design for a tensile rod. The larger diameter is known, $D = 1.00$ in, along with the hole diameter, $a = 0.50$ in. It has also been decided that the stress concentration factor at the fillet is to be 1.7. The smaller diameter, d, and the fillet radius, r, are to be specified such that the stress at the fillet is the same as that at the hole.

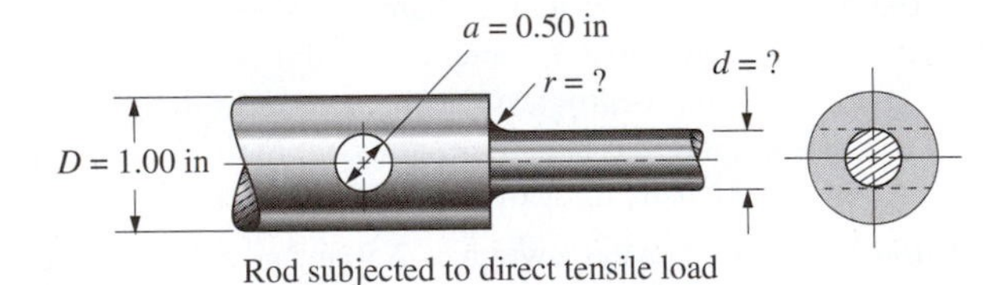

FIGURE P3–124 Tensile rod for Problem 3–124.

3–125. A machine member shown in Figure P3–125 is made from AISI 1141 OQT 1100 steel. Determine the allowable repeated axial tensile force that can be applied. The force is applied through a pin in the end holes.

3–126. Refer to Figure P3–126. Specify a suitable material for the bar shown if the applied force F is 12.6 kN. Mild shock is expected.

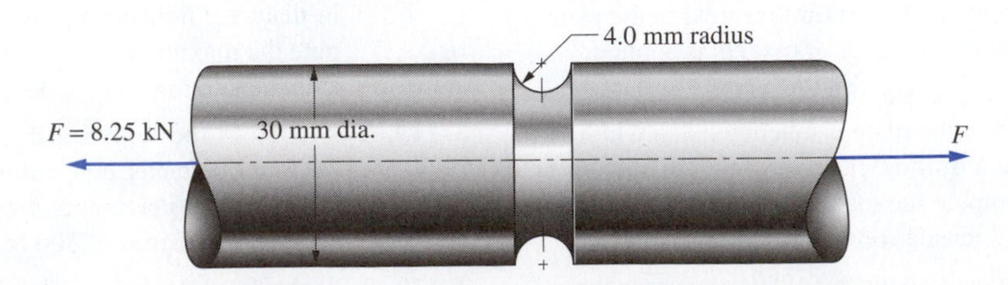

FIGURE P3–122 Part of a tie-road for Problem 3–122.

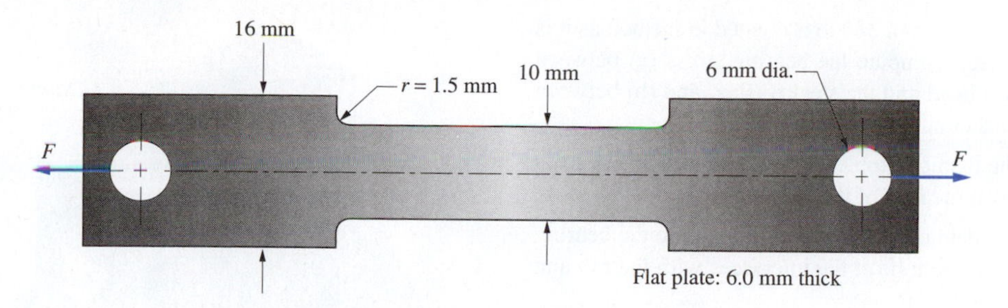

FIGURE P3–125 Machine member for Problem 3–125.

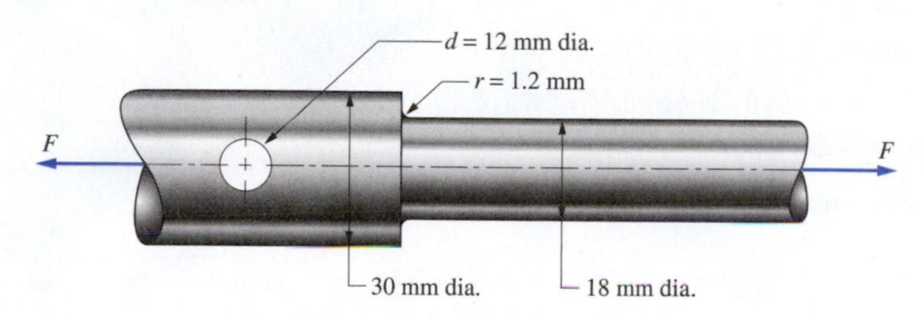

FIGURE 3–126 Bar for Problem 3–126.

Bearing Stress

3–127.E Compute the bearing stresses at the mating surfaces *A*, *B*, *C*, and *D*, in Figure P3–127.

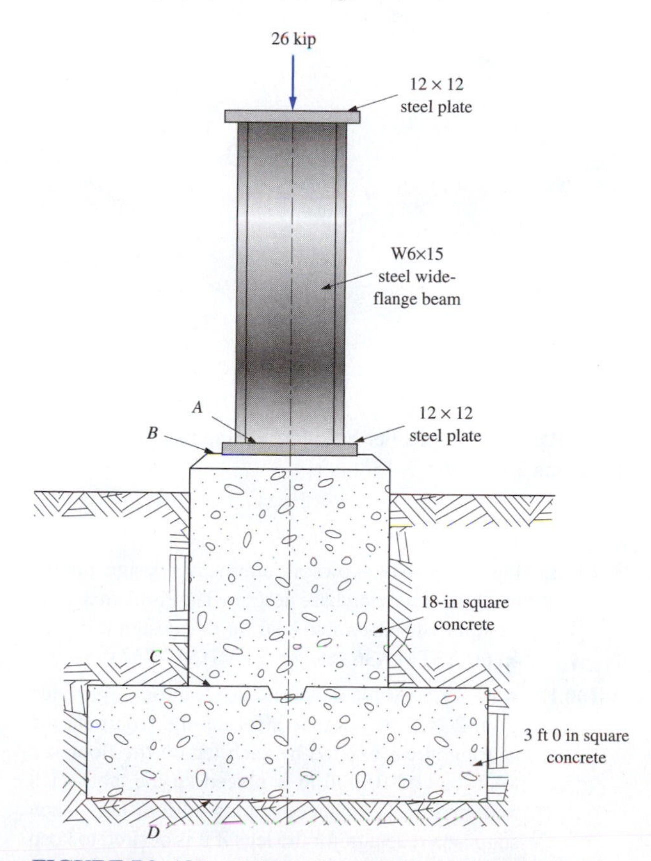

FIGURE P3–127 Column foundation for Problem 3–127.

3–128. A 2-in schedule 40 steel pipe is used as a leg for a machine. The load carried by the leg is 2350 lb.

(a) Compute the bearing stress on the floor if the pipe is left open at its end.

(b) Compute the bearing stress on the floor if a flat plate is welded to the bottom of the pipe having a diameter equal to the outside diameter of the pipe.

3–129.E A bolt and washer are used to fasten a wooden board to a concrete foundation as shown in Figure P3–129.

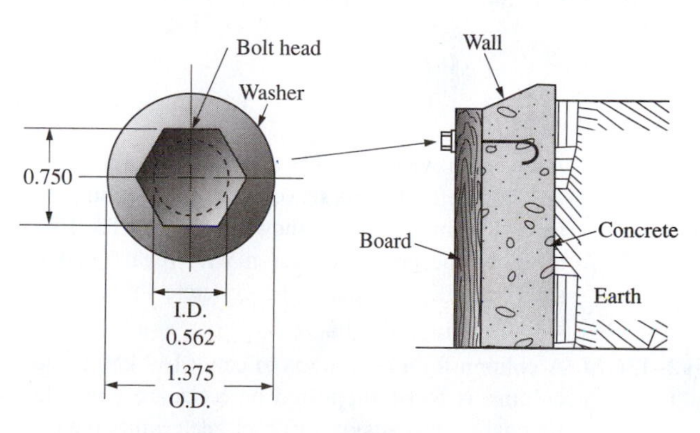

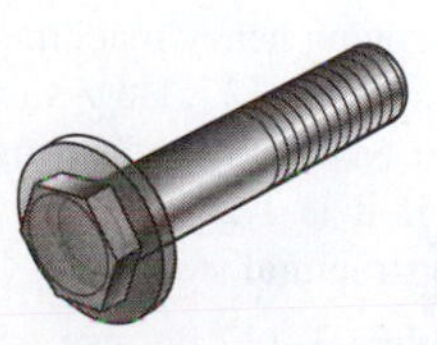

FIGURE P3–129 Bolt and washer for Problem 1–129.

A tensile force of 385 lb is created in the bolt as it is tightened. Compute the bearing stress (a) between the bolt head and the steel washer, and (b) between the washer and the board.

3–130.E For the data of Problem 3–64, compute the bearing stress on the side of the key.

3–131.E For the data of Problem 3–65, compute the bearing stress on the tube at the interfaces with the pin and the collar.

3–132.M For the data of Problem 3–70, compute the bearing stress on the rivets.

3–133.M For the data of Problem 3–71, compute the bearing stress on the rivets.

3–134.M The heel of a woman's shoe has the shape shown in Figure P3–134. If the force on the heel is 535 N, compute the bearing stress on the floor.

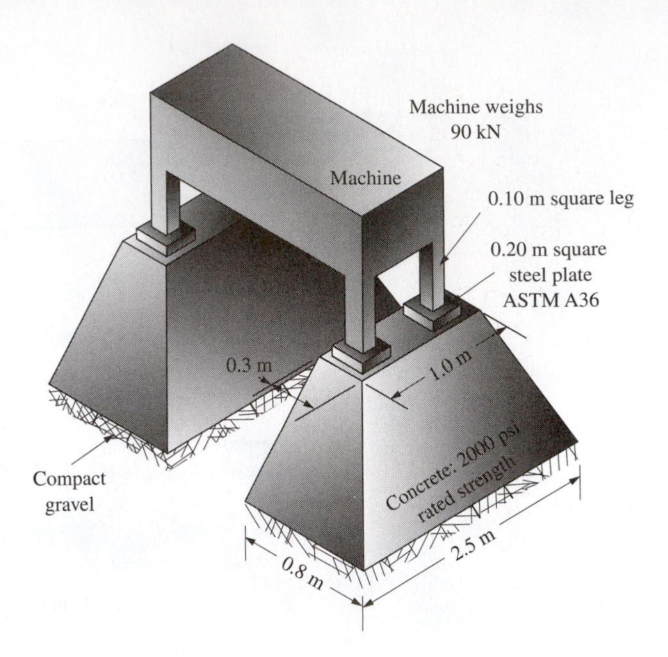

FIGURE P3–135 Machine supports for Problem 3–135.

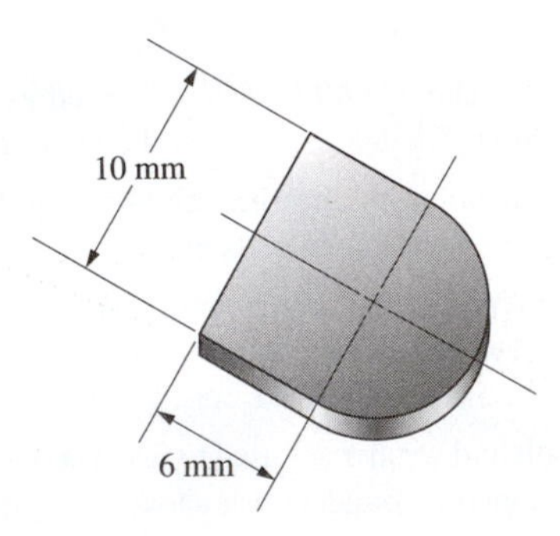

FIGURE P3–134 Shoe heel for Problem 3–134.

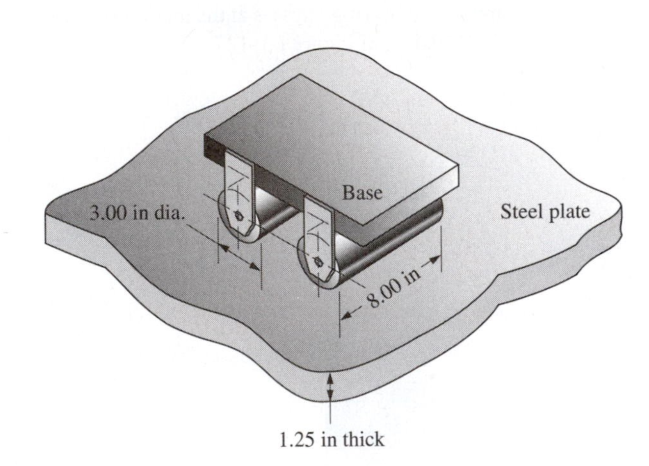

FIGURE P3–137 Roller base for moving machinery for Problems 3–137 and 3–138.

3–135.M A machine weighs 90 kN and rests on four legs. Two concrete supports serve as the foundation. The load is symmetrical, as shown in Figure P3–135. Evaluate the proposed design with regard to the safety of the steel plate, the concrete foundation, and the soil for bearing.

3–136.M A column for a building is to carry 160 kN. If the column is to be supported on a square concrete foundation that sits on soft rock, determine the required dimensions of the foundation block.

3–137.E A base for moving heavy machinery is designed as shown in Figure P3–137. How much load could the base support based on the bearing capacity of the steel plate if it is 1.25 in thick and made from ASTM A36 structural steel?

3–138.E Repeat Problem 3–137 but use ASTM A242 high-strength low-alloy steel plate.

3–139.E Figure P3–139 shows an alternative design for the machinery moving base described in Problem 3–137. Compute the allowable load for this design if it rests on (a) ASTM A36 steel or (b) ASTM A242 steel.

3–140.E A heavy table for industrial use has four legs made from $2 \times 2 \times \frac{1}{4}$ square steel tubing. Compute the bearing stress exerted by each leg on the floor if a total load of 10 000 lb is placed on the table such that the load is balanced among all four legs. Then suggest a redesign for the legs if it is desired to keep the bearing stress below 400 psi.

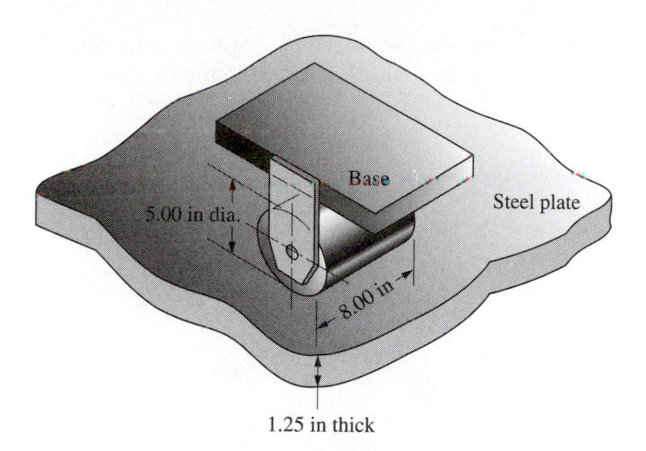

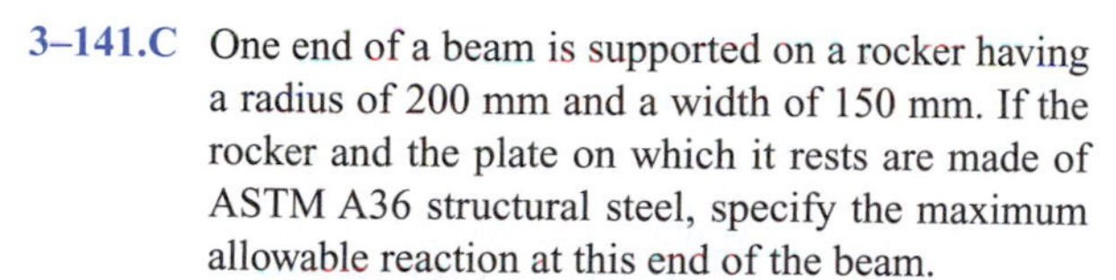

FIGURE P3–139 Roller base for moving machinery for Problem 3–139.

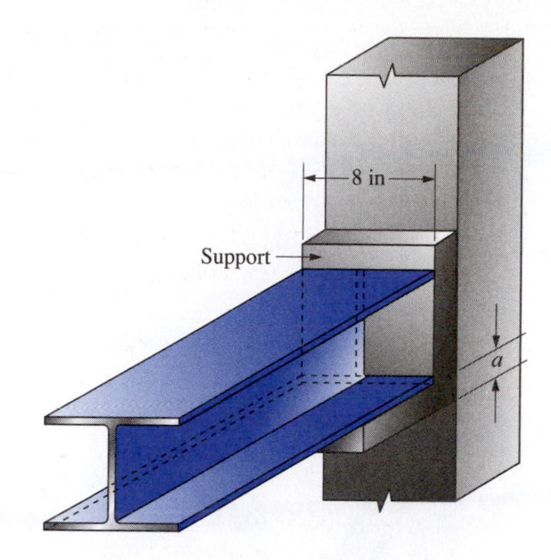

FIGURE P3–147 Beam support for Problem 3–147.

3–141.C One end of a beam is supported on a rocker having a radius of 200 mm and a width of 150 mm. If the rocker and the plate on which it rests are made of ASTM A36 structural steel, specify the maximum allowable reaction at this end of the beam.

3–142. The base of a special machine has four legs made from standard L3 × 3 × $\frac{1}{4}$ steel angles. Compute the bearing stress exerted by the legs on the floor if the total weight of the machine is 28 500 lb and the load is equally divided among the four legs.

3–143. Redesign the bottom of the machine legs of Problem 3-142 to permit them to sit on a concrete floor and to satisfy the bearing stress requirements shown in Table 3–7. The concrete has a rated strength of 3 000 psi.

3–144. A wood block having a cross section of 3.50 × 7.50 in is to be placed on a bed of compact gravel. Compute the maximum allowable compressive load that can be applied to the block so as not to exceed the bearing strength of the gravel.

3–145. A column is made from a standard 4-inch schedule 40 steel pipe. Compute the maximum axial compressive load it can take and not exceed the allowable bearing stress on a concrete floor. The concrete has a rated strength of 4000 psi.

3–146. Add a steel bearing plate to the bottom of the column described in Problem 3–145 to allow an axial compressive load at least 10.0 times as much as that determined for the column without a plate.

Shearing Stress

3–147.E A support for a beam is made as shown in Figure P3–147. Determine the required thickness of the projecting ledge a if the maximum shear stress is to be 6000 psi. The load on the support is 21 000 lb.

3–148.M The lower control arm on an automotive suspension system is connected to the frame by a round steel pin 16 mm in diameter. Two sides of the arm transfer loads from the frame to the arm, as sketched in Figure P3–148. How much shear force could the pin withstand if it is made of AISI 1040 cold-drawn steel and a design factor of 6 based on the yield strength in shear is desired?

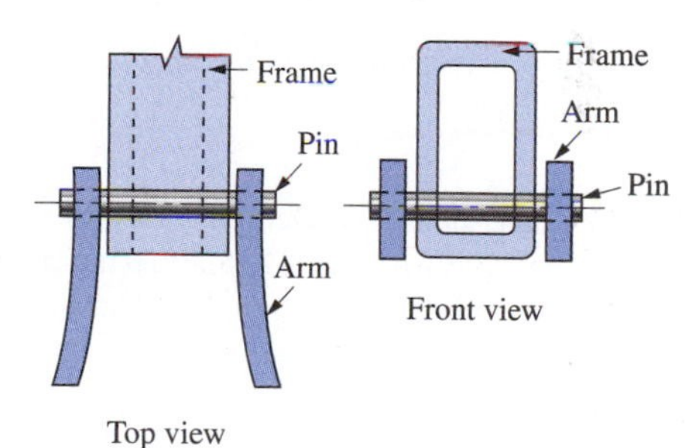

FIGURE P3–148 Pin for automotive suspension system for Problem 3–148.

3–149.M A centrifuge is used to separate liquids according to their densities using centrifugal force. Figure P3–149 illustrates one arm of a centrifuge having a bucket at its end to hold the liquid. In operation, the bucket and the liquid have a mass of 0.40 kg. The centrifugal force has the magnitude in newtons of

$$F = 0.01097 \cdot m \cdot R \cdot n^2$$

where m = rotating mass of bucket and liquid (kg)
R = radius to center of mass (meters)
n = rotational speed (rpm)

The centrifugal force places the pin holding the bucket in direct shear. Compute the stress in the pin due to a rotational speed of 3000 rpm. Then, specify a suitable steel for the pin, considering the load to be repeated.

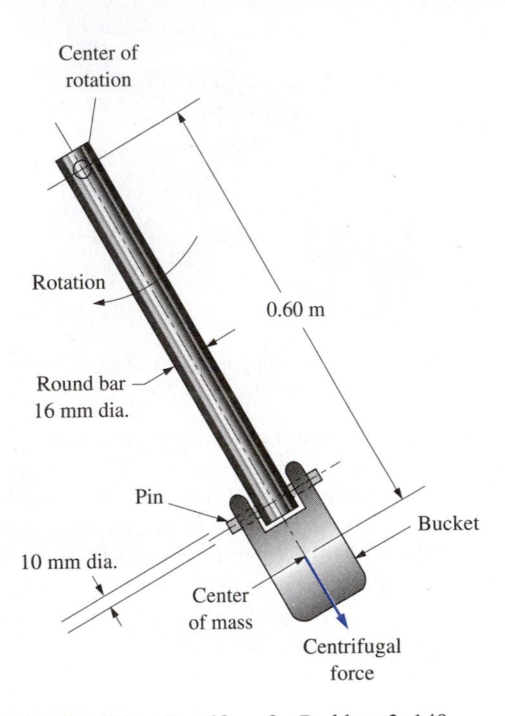

FIGURE P3–149 Centrifuge for Problem 3–149.

3–150.M A circular punch is used to punch a 20.0-mm-diameter hole in a sheet of AISI 1020 hot-rolled steel having a thickness of 8.0 mm. Compute the force required to punch out the slug.

3–151.M Repeat Problem 3–150, using aluminum 6061-T4.

3–152.M Repeat Problem 3–150, using C14500 copper, hard.

3–153.M Repeat Problem 3–150, using AISI 430 full hard stainless steel.

3–154.M Determine the force required to punch a slug the shape shown in Figure P3–154 from a sheet of AISI 1020 hot-rolled steel having a thickness of 5.0 mm.

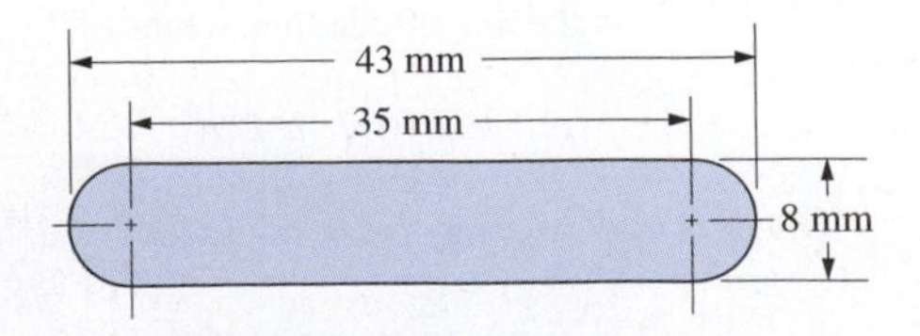

FIGURE P3–154 Shape of a slug for Problem 3–154.

3–155.E Determine the force required to punch a slug the shape shown in Figure P3–155 from a sheet of aluminum 3003-H18 having a thickness of 0.194 in.

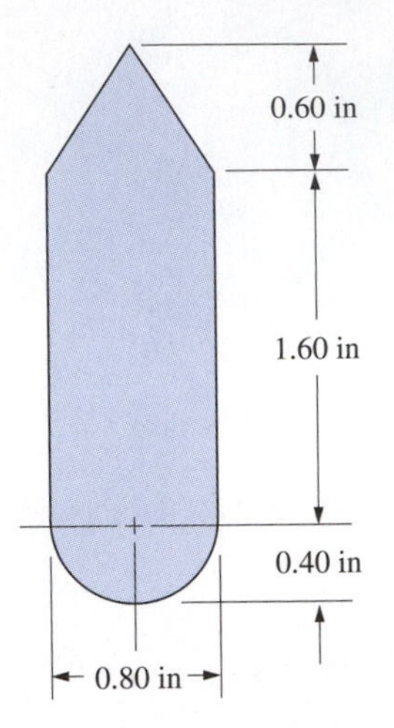

FIGURE P3–155 Shape of a slug for Problem 3–155.

3–156.E A notch is made in a piece of wood, as shown in Figure P3–156, to support an external load of 1800 lb. Compute the shear stress in the wood. Is the notch safe? (See Appendix A–19.)

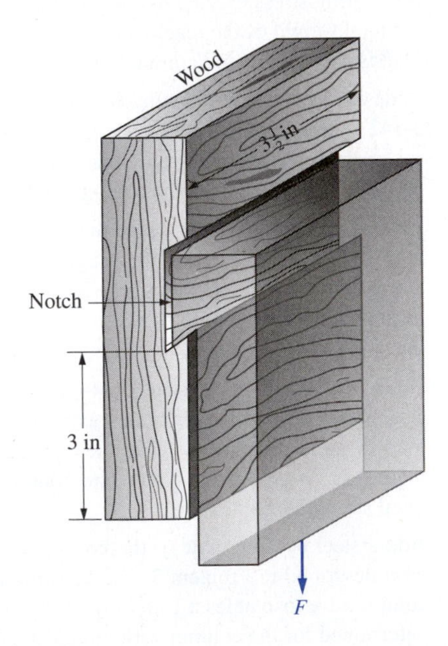

FIGURE P3–156 Notched wood block loaded in shear for Problem 3–156.

3–157.E Compute the force required to shear a straight edge of a sheet of AISI 1040 cold-drawn steel having a thickness of 0.105 in. The length of the edge is 7.50 in.

3–158.E Repeat Problem 3–157 using AISI 5160 OQT 700 steel.

3–159.E Repeat Problem 3–157 using AISI 301 full hard stainless steel.

3–160.E Repeat Problem 3–157 using C36000 brass, hard.

3–161.E Repeat Problem 3–157, using aluminum 5154-H32.

3–162.M For the lever shown in Figure P3–162, called a *bellcrank,* compute the required diameter of the pin A if the load is repeated and the pin is made from C17200 beryllium copper, hard. The loads are repeated many times.

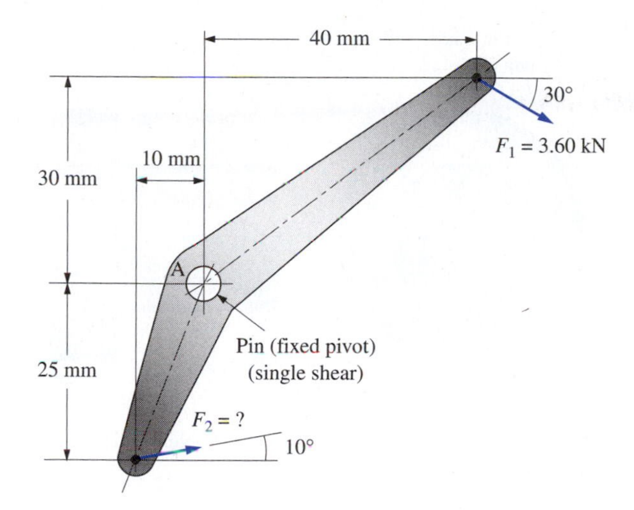

FIGURE P3–162 Bellcrank for Problem 3–162.

3–163.M For the structure shown in Figure P3–163, determine the required diameter of each pin if it is made from AISI 1020 cold-drawn steel. Each pin is in double shear and the load is static.

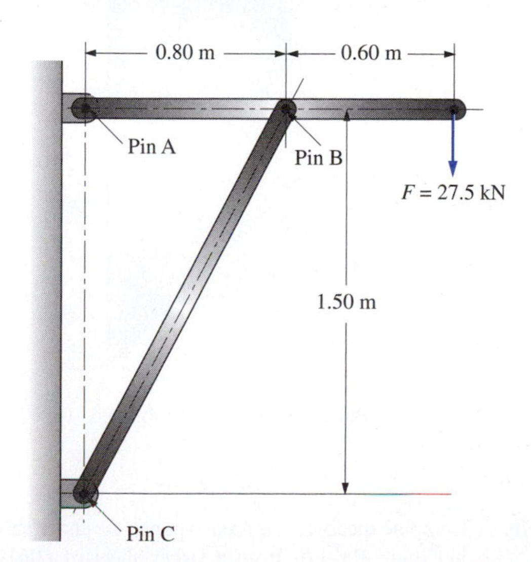

FIGURE P3–163 Pin-connected structure for Problem 3–163.

3–164.M For the structure shown in Figure P3–164, determine the required diameter of each pin if it is made from ASTM A572 high-strength low-alloy columbium-vanadium structural steel, grade 50. Each pin is in double shear and the load is static.

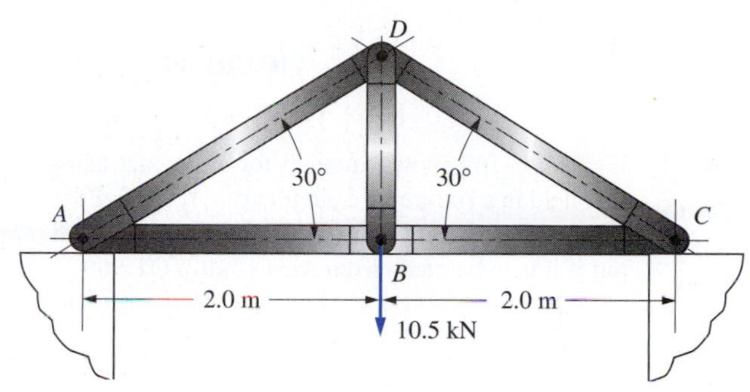

FIGURE P3–164 Truss for Problem 3–164.

3–165.E A pry bar as shown in Figure P3–165 is used to provide a large mechanical advantage for lifting heavy machines. An operator can exert a force of 280 lb on the handle. Compute the lifting force and the shear stress on the wheel axle.

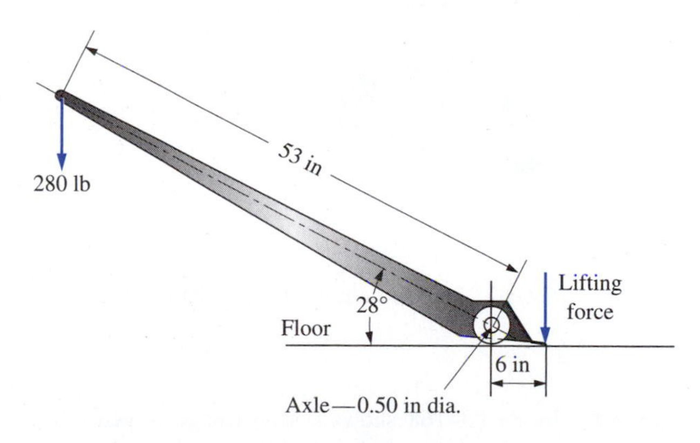

FIGURE P3–165 Pry bar for Problem 3–165.

3–166. Figure P3–166 shows a steel strap with holes and slots punched from it. The entire strap is also punched from a larger sheet of 3003-H12 aluminum alloy, 1.40 mm thick. Compute the total force required to produce the part if all shearing is done in one stroke.

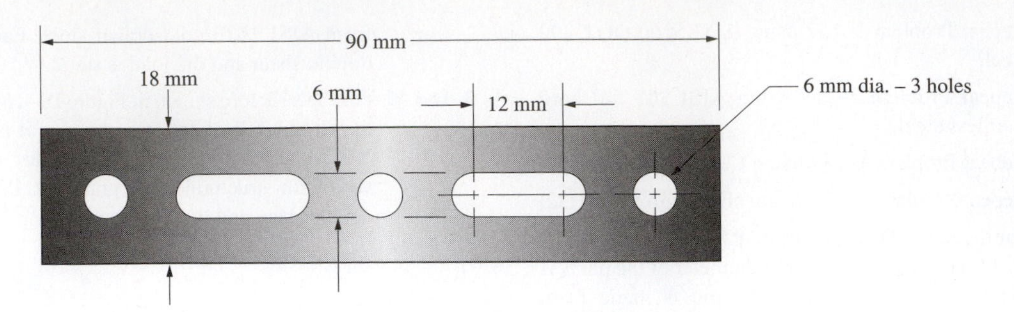

FIGURE P3–166 Steel Strap for Problem 3–166.

3–167.E Figure P3–167 shows an anvil for an impact hammer held in a fixture by a circular pin. If the force is to be 500 lb, specify a suitable diameter for the steel pin if it is to be made from AISI 1040 WQT 900.

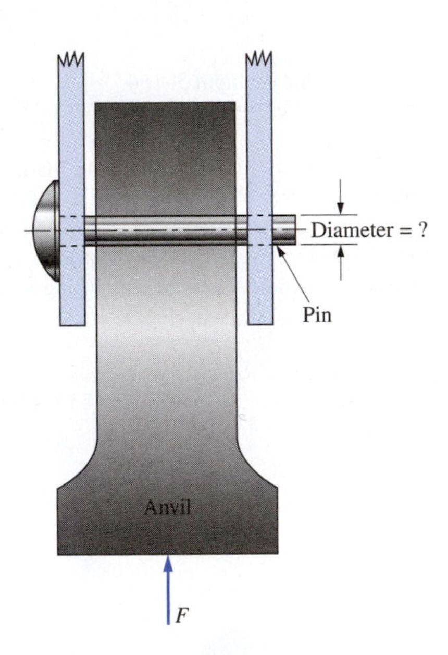

FIGURE P3–167 Impact hammer for Problem 3–167.

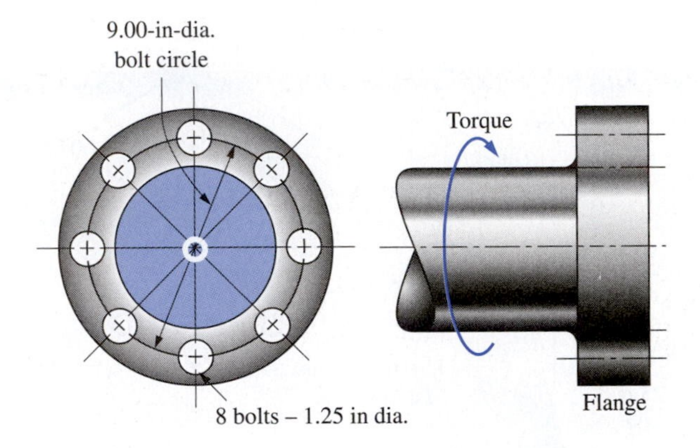

FIGURE P3–168 Flange coupling for Problem 3–168.

3–168.E Figure P3–168 shows a steel flange forged integrally with the shaft, which is to be loaded in torsion. Eight bolts serve to couple the flange to a mating flange. Assume that each bolt carries an equal load. Compute the maximum permissible torque on the coupling if the shear stress in the bolts is not to exceed 6000 psi.

3–169. Compute the force required to completely punch out the shape shown in Figure P3–169 from a large sheet of steel having a thickness of 0.085 in. The material is AISI 1020 cold-drawn steel.

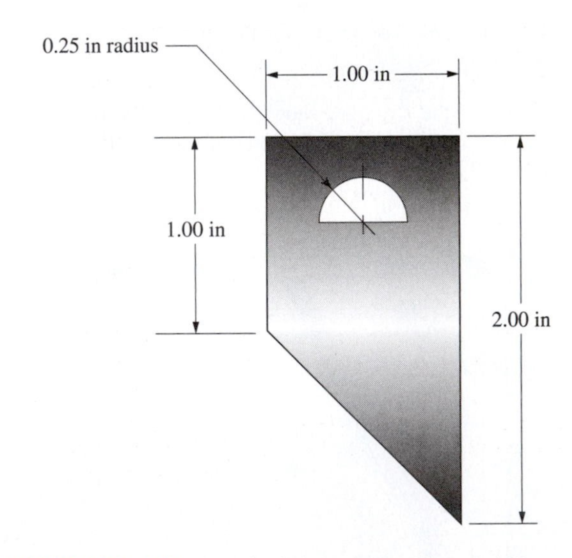

FIGURE P3–169 Punched shape for Problem 3–169.

3–170. Compute the force required to punch the shape shown in Figure P3–170 from a larger sheet of 6061-T4 aluminum sheet having a thickness of 0.10 in.

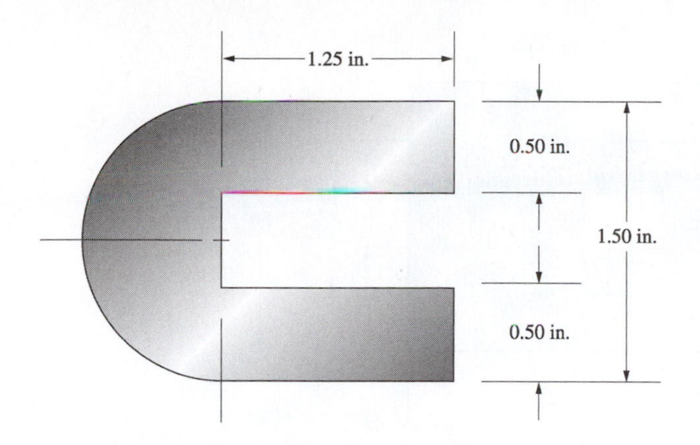

FIGURE P3–170 Punched shape for Problem 3–170.

3–171. Compute the force required to punch out a slug having the shape shown in Figure P3–171 from a sheet of the aluminum alloy 3003-H18. The thickness of the sheet is 3.0 mm.

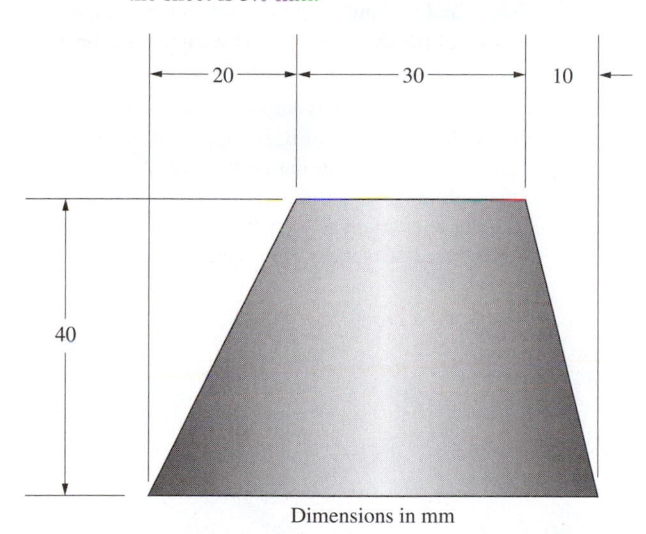

FIGURE P3–171 Punched shape for Problem 3–171.

3–172. Compute the force required to punch out a slug having the shape shown in Figure P3–172 from a sheet of AISI 1040 cold-drawn steel having a thickness of 1.60 mm.

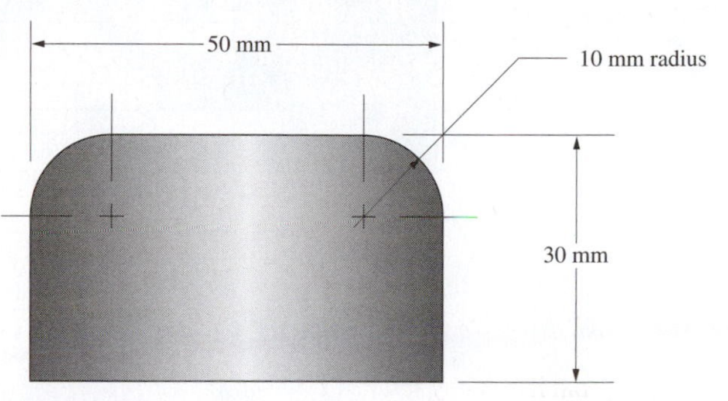

FIGURE 3–172 Punched shape for Problem 3–172.

3–173. Figure P3–173 shows the shape of the blade for a utility knife to be sheared from a sheet of AISI 1080 OQT 900 steel. The blade is 0.80 mm thick. Compute the force required to punch the blade shape from a larger sheet.

3–174. Compute the force required to punch out the shaded shape shown in Figure P3–174 from the larger sheet of the aluminum alloy 5154-H38 having a thickness of 2.00 mm.

3–175. Use the shape shown in Figure P3–174 and other data from Problem 3-174. However, the design of the punch is different. It first punches the central hole and the two rectangular slots. Then it punches the outside edges of the shape. Determine the press force required to produce the part with this punch design.

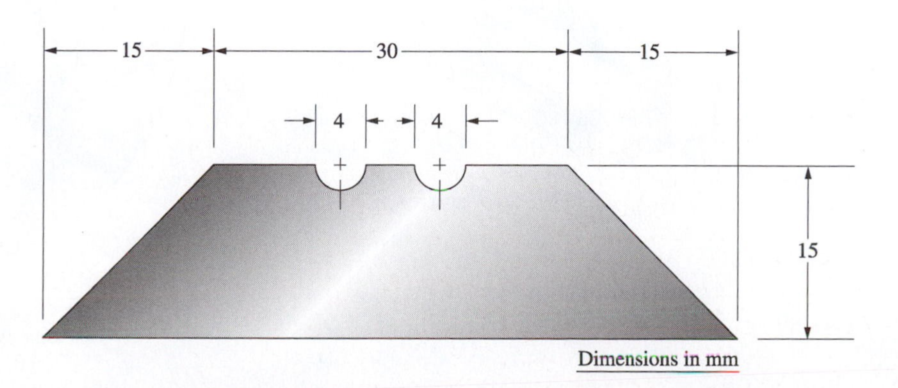

FIGURE P3–173 Utility knife blade for Problem 3–173.

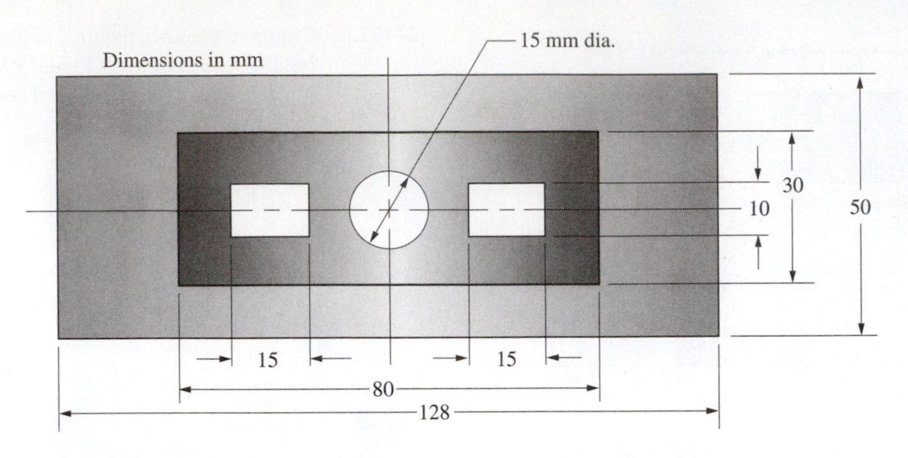

FIGURE P3–174 Punched shape for Problems 3–174 and 3–175.

Problems with More Than One Kind of Direct Stress and Design Problems

3–176.E A gear transmits 5000 lb·in of torque to a circular shaft having a diameter of 2.25 in. A square key 0.50 in on a side connects the shaft to the hub of the gear, as shown in Figure P3–176. The key is made from AISI 1020 cold drawn steel. Determine the required length of the key, L, to be safe for shear and bearing. Use a design factor of 2.0 based on yield for shear and the AISC allowable bearing stress.

3–177.E A section of pipe is supported by a saddle-like structure which, in turn, is supported on two steel pins, as illustrated in Figure P3–177. If the load on the saddle is 42 000 lb, determine the required diameter and length of the pins. Use AISI 1040 cold-drawn steel. Consider both shear and bearing.

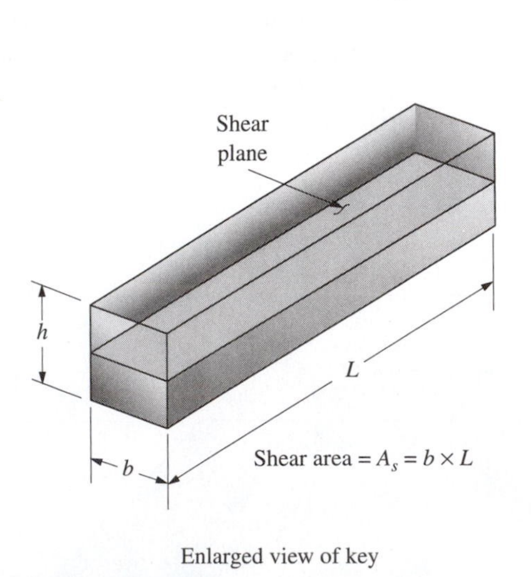

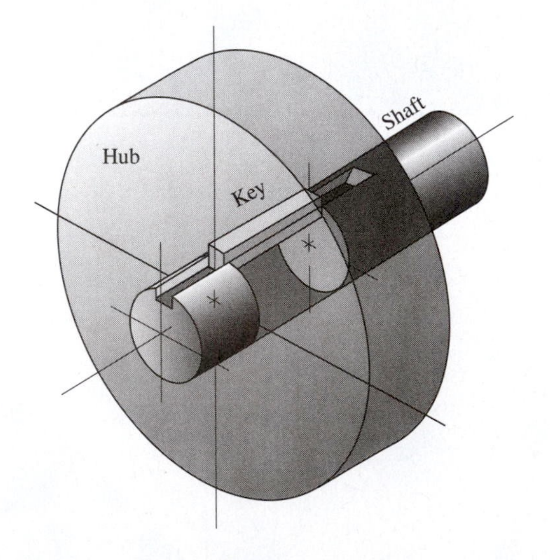

FIGURE P3–176 Key for Problem 3–176.

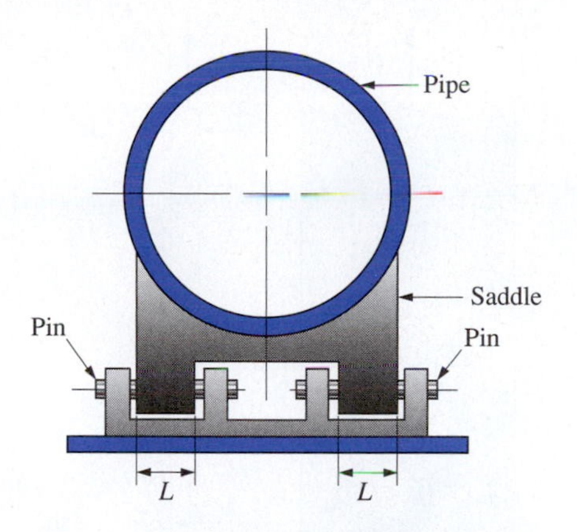

FIGURE P3–177 Pipe saddle for Problem 3–177.

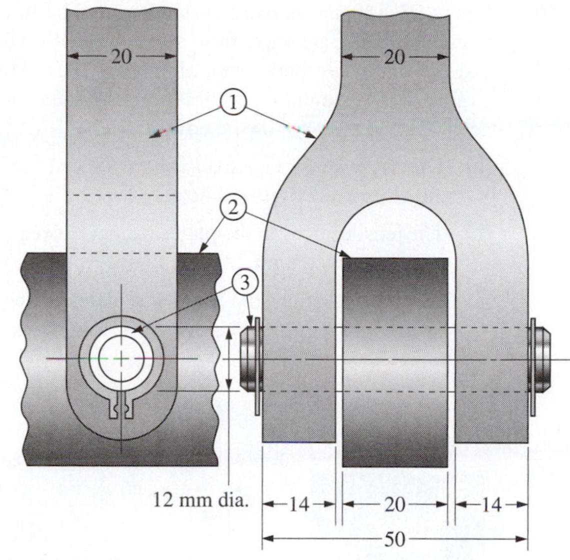

FIGURE P3–178 Clevis for beam support for Problem 3–178.

3–178.M Figure P3–178 shows the design of the lower end of the member *AB* in Figure P3–21. Use the load shown in Figure P3–21 and assume that it is static. Member 1 is made from aluminum alloy 6061-T4; member 2 is made from aluminum alloy 2014-T4; pin 3 is made from aluminum alloy 2014-T6. Perform the following analyses:

(a) Evaluate member 1 for safety in tension in the area of the pin holes.

(b) Evaluate member 1 for safety in bearing at the pin.

(c) Evaluate member 2 for safety in bearing at the pin.

(d) Evaluate the pin 3 for safety in bearing.

(e) Evaluate the pin 3 for safety in shear.

3–179.E Figure P3–179 illustrates a type of engineering chain used for conveying applications. All components are made from AISI 1040 cold-drawn steel. Evaluate the allowable tensile force (repeated) on the chain with respect to:

(a) Shear of the pin

(b) Bearing of the pin on the side plates

(c) Tension in the side plates.

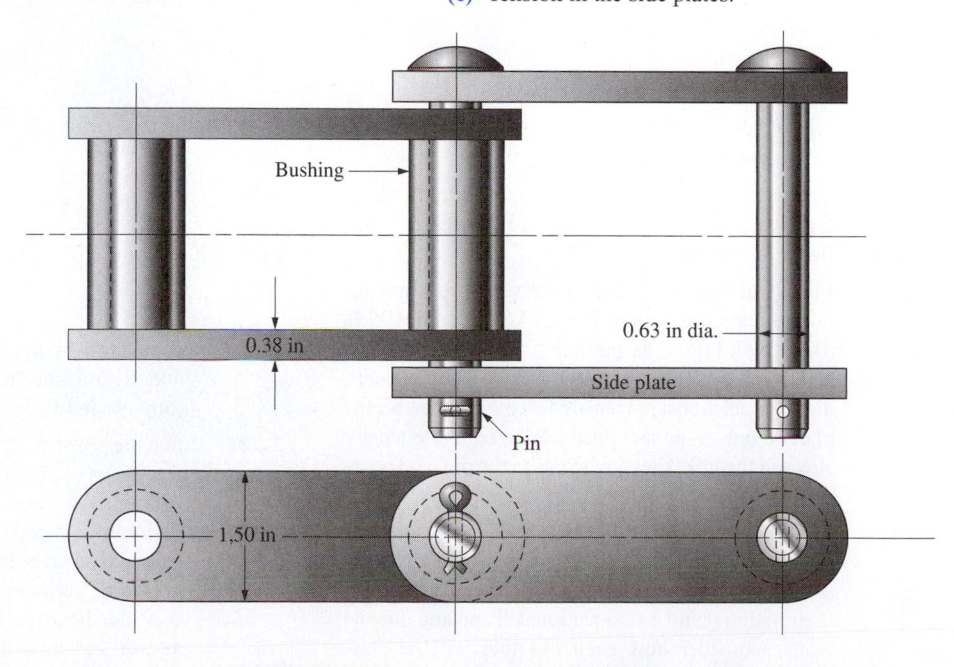

FIGURE P3–179 Conveyor chain for Problem 3–179.

3–180. Figure P3–180 shows two plates connected by two rivets. The plates are made from aluminum 6061-T6 and the rivets are made from aluminum 2014-T4. Evaluate the maximum allowable force on the connection to meet the following design criteria:

(a) The shear stress in a rivet cannot exceed $\frac{1}{4}$ of the ultimate strength in shear.

(b) The tensile stress in the plates cannot exceed $\frac{1}{3}$ of the yield strength.

(c) The bearing stress on either the plates or the rivets cannot exceed the design bearing stress from Equation 3–26.

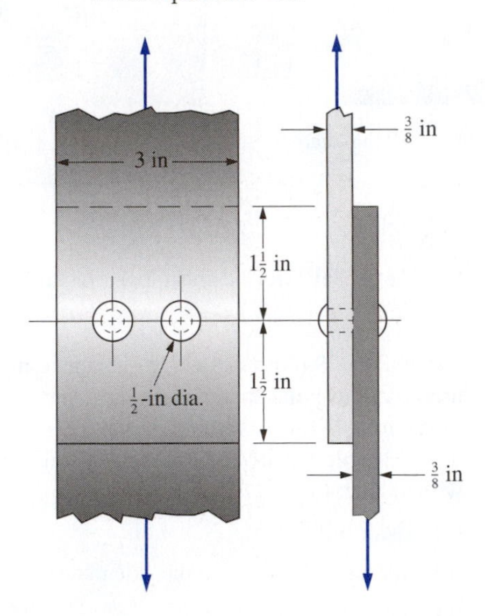

FIGURE P3–180 Riveted plates for Problem 3–180.

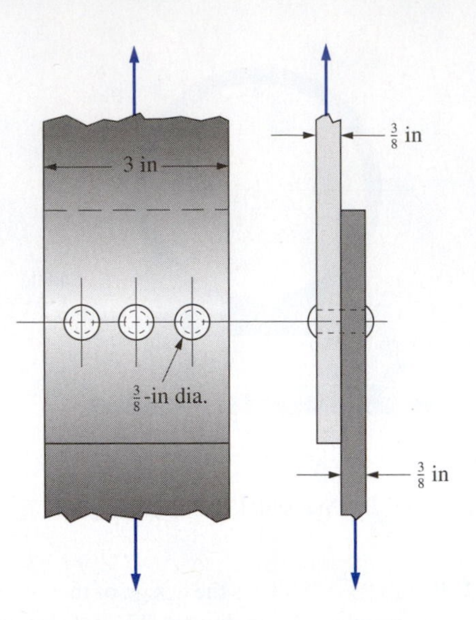

FIGURE P3–181 Riveted plates for Problem 3–181.

3–181. Repeat Problem 3–180 for the connection shown in Figure P3–181. The materials and the design criteria are the same.

3–182. Repeat Problem 3–180 for the connection shown in Figure P3–182. The materials and the design criteria are the same.

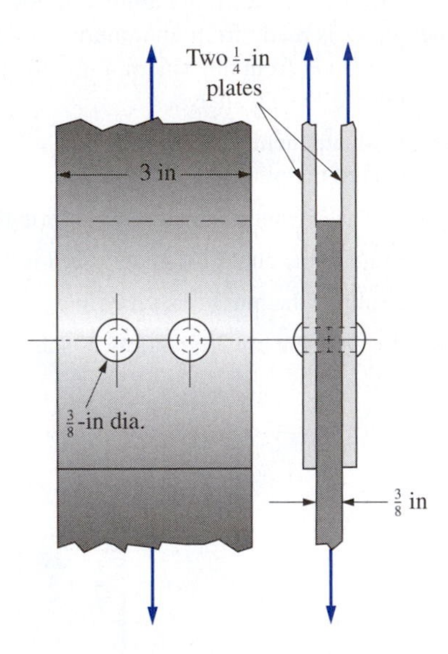

FIGURE P3–182 Riveted plates for Problem 3–182.

3–183. Link A in Figure P3–183 is connected to a mating link through a pin joint. Link A is made from AISI 4140 OQT 1100. The pin has a diameter of 0.75 in and is made from AISI 1141 OQT 1100. Compute the allowable repeated tensile force F that can be applied to link A for safe operation in regard to tensile stress in the link, shearing stress in the pin, and bearing stress between the pin and the hole in link A.

3–184. For the truss sketched in Figure P3–184, compute the forces in all members. Then, for those members subjected to tensile forces, propose a design for their material and cross-sectional shape and dimensions. Consider how each member will be connected to adjacent members at the pin joints. Also consider how the loads can be applied at joints C and E and how the supports can be provided at joints A and B.

3–185. For the truss sketched in Figure P3–185, compute the forces in all members. Then, for those members subjected to tensile forces, propose a design for their material and cross-sectional shape and dimensions. Consider how each member will be connected to adjacent members at the pin joints. Also consider how the loads can be applied at joints F and G and how the supports can be provided at joints A and D.

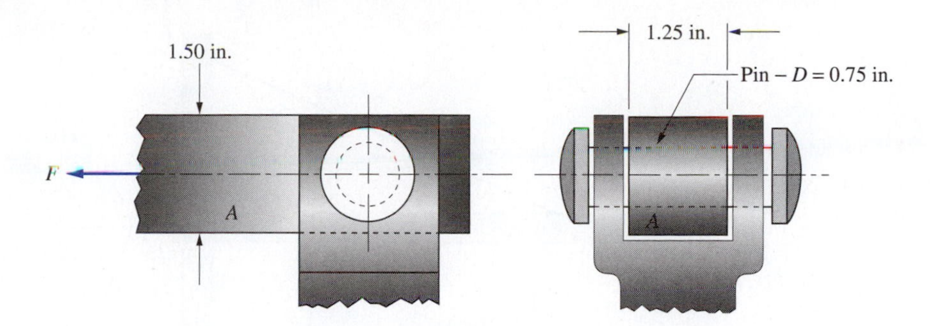

FIGURE P3–183 Connection for Problem 3–183.

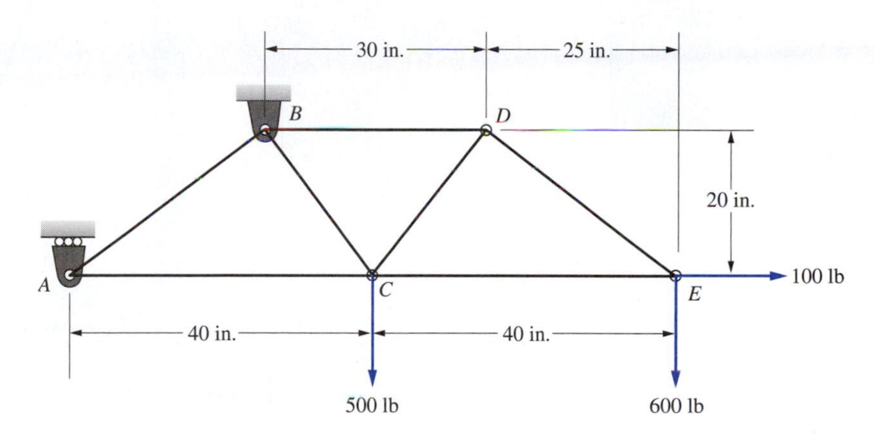

FIGURE P3–184 Truss for Problem 3–184.

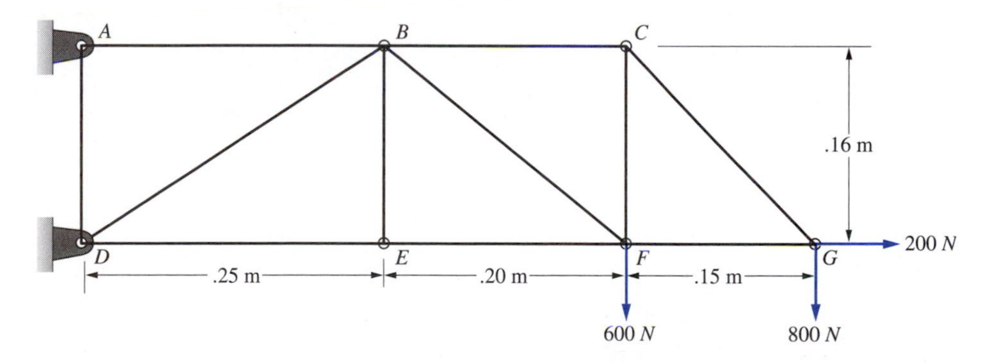

FIGURE P3–185 Truss for Problem 3–185.

3–186. See Figure P3–186. The load is 34.0 kN and it will be applied with moderate shock.

- **(a)** Compute the forces in the support rods AB and AE.
- **(b)** Design the support rods, specifying the material and a suitable diameter.
- **(c)** Compute the shearing forces in all six pins.
- **(d)** Design the pins to be safe in shear, specifying the material and a suitable diameter.
- **(e)** Check the pins you have designed for acceptable bearing stress. Assume that the mating material through which the holes are produced is stronger than the pin material. Specify the minimum required thickness of the mating part to provide adequate bearing area.

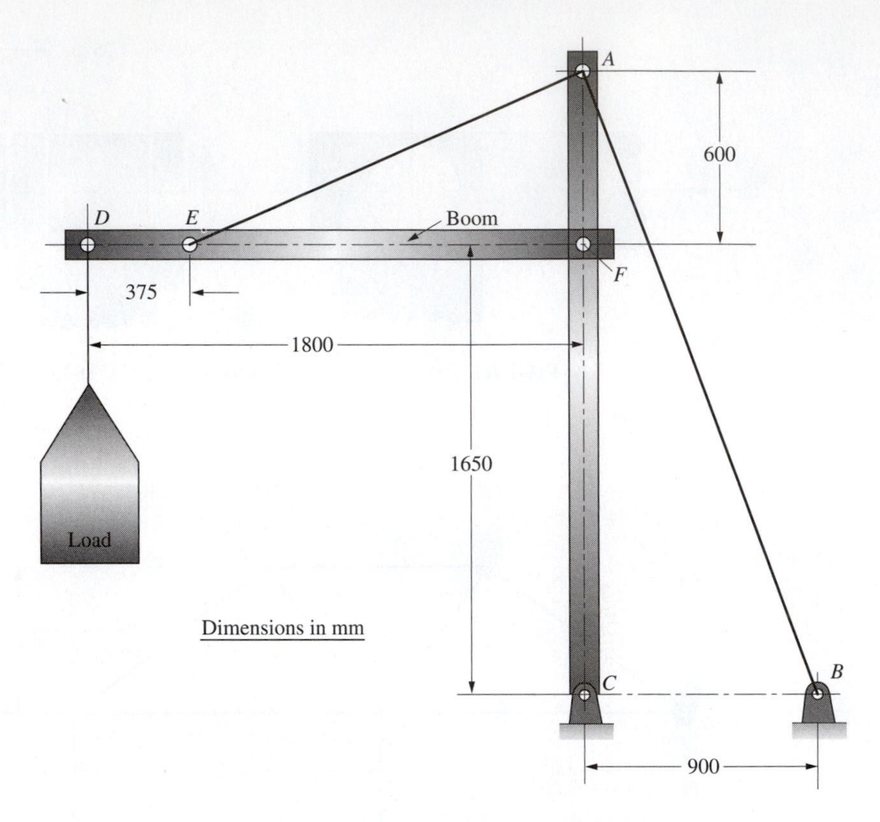

FIGURE P3–186 Structure for Problem 3–186.

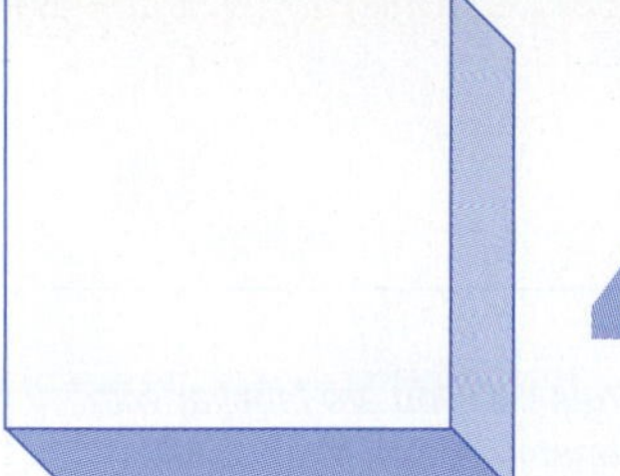

4

Torsional Shear Stress and Torsional Deformation

The Big Picture

Torsional Shear Stress and Torsional Deformation

Discussion Map

- In this chapter, you will learn how to compute the level of shearing stress experienced by members carrying a torque.
- You will learn how to use the torsional shear stress equation for this calculation.
- You will also learn how to compute the amount of torsional deformation in a torsionally loaded member.
- You will use the angle of twist equation.

Discover

Think about products or machines that you are familiar with that have rotating shafts that transmit power. Make a list, describing the source of the power and the path that the power takes as it is delivered to the point of ultimate use. See Figure 4–1, the gear-type speed reducer example, on the next page.

Sometimes torque is applied to a member that does not rotate. Can you think of an example? Consider the socket wrench shown in Figure 4–4 that is being used to tighten a bolt. The bolt rotates at first when it is being driven down to clamp two members together. Then, when the bolt is finally tight, you are applying a torque through the extension shaft to the head of the bolt. In fact, proper assembly of bolted connections requires that a specified torque be applied to the head of each bolt.

When any member is subjected to a torque, it demonstrates some amount of twisting where one end is rotated with respect to the other. The amount of this elastic deformation must often be determined by analytical methods to ensure that the member has sufficient rigidity.

Can you think of other examples where torsional rigidity is an important design and performance parameter?

Torsion refers to the loading of a member that tends to cause it to rotate or twist. Such a load is called a *torque, rotational moment, twisting moment,* or *couple*. When a torque is applied to a member, *shearing stress* is developed within the member and *torsional deformation* is created, resulting in an angle of twist of one end of the member relative to the other.

A familiar example of a torsional member is a circular shaft being driven by an electric motor and delivering power through a gear to a mating gear. Figure 4–1 shows a drawing of a possible layout. Its operation is described next.

The motor uses electrical energy to produce a driving torque in its output shaft. The motor shaft is connected to the input shaft of the gear-type speed reducer through a flexible coupling. The coupling allows the transmission of torque while eliminating any tendency to bend the shaft or to cause axial loads to develop. When the motor rotates while developing torque, it transmits power to the shafts of the system shown.

The input shaft is supported on two antifriction bearings and it carries the small gear A, called a *pinion,* between the two bearings. The torque applied to the shaft from the coupling is transmitted down the shaft to the gear, where a reaction torque exists in the form of a resistance offered by the mating gear B.

The power is then transferred to gear B, causing it to rotate. When the gears are operating at a constant speed and transmitting power, there is a uniform level of torque in the input shaft from the coupling to the gear. That part of the input shaft from gear A to the outboard bearing experiences virtually no torque because the bearing provides very little resistance.

The mating gear B receives the power from the pinion A and delivers it into the output shaft. The power then travels down the output shaft, through another flexible coupling, to the driven machine. This may be a fan, a conveyor, a machine tool, or some other device that uses the power to perform some useful work. Thus the output shaft experiences a

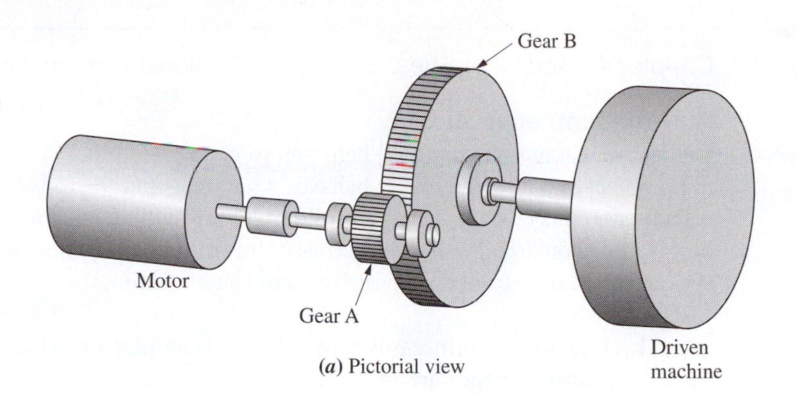

(*a*) Pictorial view

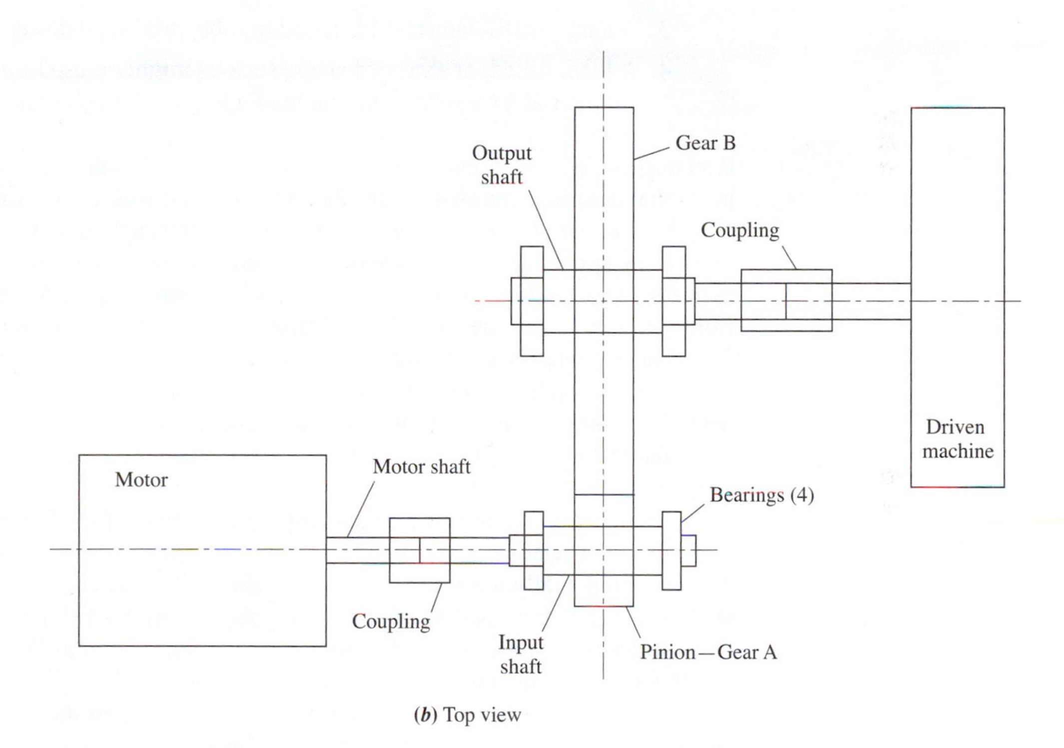

(*b*) Top view

FIGURE 4–1 General layout of saw drive with single-stage gear reducer.

torque from gear B to the coupling. The driveshaft for the driven machine also is subjected to that same level of torque.

You should try to visualize the flow of power and torque described here. Each part of each shaft that experiences an applied torque will develop internal shearing stresses. You must learn how to compute the level of such stresses as a function of the applied torque and the geometry of the shaft using the torsional shear stress equation.

You will also learn how to compute the amount of torsional deformation in a torsionally loaded member. You will use the *angle of twist equation* that relates the applied torque, the length of the member, its geometry, and the modulus of elasticity of its material to the angle of twist.

The goal of these analyses is to ensure that the loaded member is safe for the applied torsional load and sufficiently rigid to perform properly in service.

Activity Chapter 4—Torsional Shear Stress and Torsional Deformation

A. Nonquantitative Activity

Gather some materials that will help you to get a good "feel" for how a torque is applied to an object and how the object behaves when resisting the torque. Most of the examples discussed in this chapter are either solid or hollow cylinders. However, some attention will be given to noncircular shapes such as solid or hollow squares, rectangles, hexagons, or special rolled or extruded shapes. Here are some examples:

1. A relatively thin cardboard tube such as that on which paper towels or gift-wrapping paper are rolled
2. A flexible foam cylinder such as a swim toy or a cushion
3. A long, small-diameter plastic rod or tube, or a wood dowel
4. A long, rolled, or extruded shape such as might be used for a closet door rail, a curtain rod, or a part of the frame of a toy, a cabinet, or an appliance

Recognizing that the terms *long, small-diameter, flexible,* and *thin* are imprecise, look for items that deform somewhat easily when manipulated with your own hands.

Grasp one of your cylindrical items by the ends as illustrated in Figure 4–2. Then twist it in opposing directions with each hand while observing how it deforms. The cylinder in part (a) of Figure 4–2 is not subjected to twisting. A rectangular grid, made from circumferential and longitudinal lines, is illustrated on its surface. Part (b) of the figure shows what happens when twisting is applied. Note how the longitudinal lines warp as the cylinder twists, while the circumferential lines remain mostly unchanged. If feasible, make a similar grid on your item. These observations are useful when considering the formulas we use to calculate stresses and deformations in members subjected to torsion.

Using a thin cardboard tube and applying the twisting action, notice how it is quite stiff in torsion relative to the weight or general strength of the cardboard. If you do not have a commercially made tube, find a piece of stiff paper card stock about 5 in × 11 in (125 mm × 275 mm) and roll it into a tube approximately 1.0 in to 1.5 in (25 mm to 35 mm) in diameter. You might roll it around a cylindrical broom handle to form the tube. Apply tape along the entire length. The tube will be stiff in torsion even though you may be able to crush it easily. Now, carefully cut the tube along its length and notice how it becomes extremely flexible in torsion. This illustrates the importance of a continuous circular cross section for good torsional rigidity.

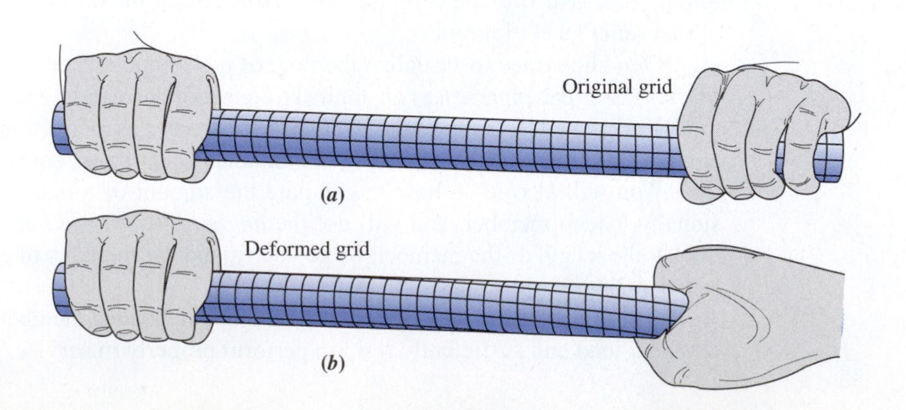

FIGURE 4–2 Twisting of a cylindrical rod.

B. Quantitative Activity

For quantitative measurements, a rigid structure is required to hold fixed one end of a member such as a cylindrical rod or a tube while allowing the other end to be held in place but free to rotate as a torque is applied to that end. A means of measuring the applied torque and the torsional deformation (angle of twist) is required.

Figure 4–3 shows a commercially available apparatus that facilitates such a test and demonstration. Torque is applied by hanging known weights at the end of an arm attached to the end of the rod that is allowed to rotate. The applied torque is calculated simply from the basic definition of torque,

$$T = Fr$$

where T = Applied torque
F = Applied force at the end of the arm
r = Perpendicular distance from the line of action of the force to the center of the rod

The angular deformation is measured on a fixed protractor scale with an indicator affixed to the rod.

Design an experiment from which you can record the following data:

1. General description of the rod or tube being subjected to the twisting action
2. Material from which the rod is made
3. Diameter of the rod along with the inside diameter if a hollow tube is used
4. Length between where the rod is held fixed and where the torque is applied
5. A table of data listing the following for each data point:
   a. Incremental applied force on the arm
   b. Length of the torque arm
   c. Computed torque ($T = Fr$)
   d. Angle of rotation

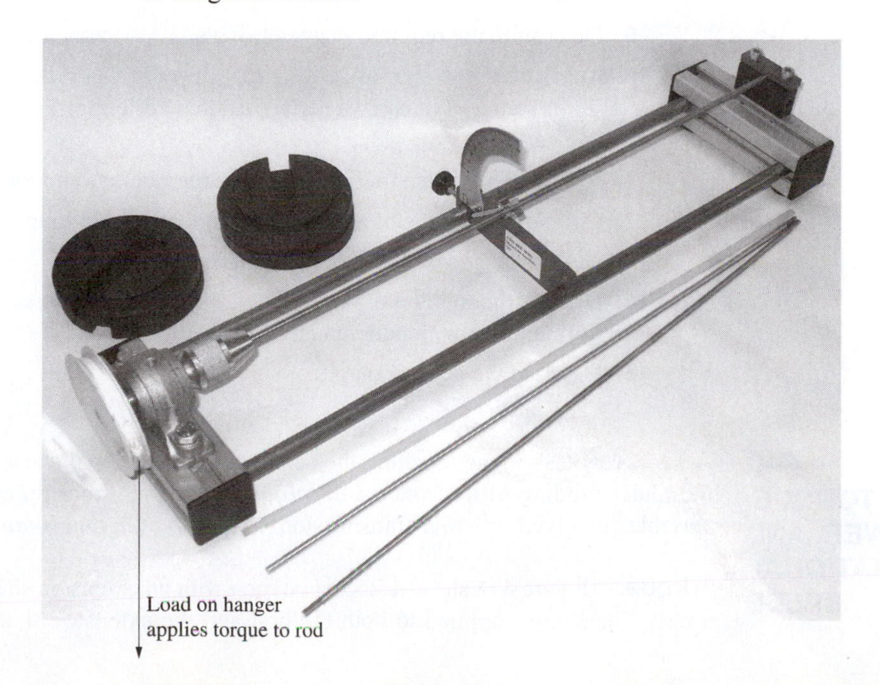

FIGURE 4–3 Torsion testing device. (Source: P. A. Hilton Ltd./Hi-Tech, Hampshire, England)

A variety of deliverables from the testing can be produced:

1. Graph of angle of twist versus the applied torque
2. Record and description of any failure of the rod or tube that may have occurred, such as excessive bending, fracture, tearing, crushing, or wrinkling, and the load at which the failure occurred
3. Maintenance of a record of the data for use after the principles of torsional shear stress and torsional deformation analysis are developed later in this chapter. Then appropriate calculations of geometrical and material properties of the rod or tube can be made, and computed stress and deformation can be compared with the measured data.

4–1 OBJECTIVES OF THIS CHAPTER

After completing this chapter, you should be able to:

1. Define *torque* and compute the magnitude of torque exerted on a member subjected to torsional loading.
2. Define the relationship among the three critical variables involved in power transmission: power, torque, and rotational speed.
3. Manipulate the units for power, torque, and rotational speed in both the SI metric system and the U.S. Customary system.
4. Compute the maximum shear stress in a member subjected to torsional loading.
5. Define the *polar moment of inertia* and compute its value for solid and hollow round shafts.
6. Compute the shear stress at any point within a member loaded in torsion.
7. Specify a suitable design shear stress for a member loaded in torsion.
8. Define the *polar section modulus* and compute its value for solid and hollow round shafts.
9. Determine the required diameter of a shaft to carry a given torque safely.
10. Compare the design of solid and hollow shafts on the basis of the mass of the shafts required to carry a certain torque while limiting the torsional shear stress to a certain design value.
11. Apply stress concentration factors to members in torsion.
12. Compute the angle of a twist of a member loaded in torsion.
13. Define the *shear modulus of elasticity.*
14. Discuss the method of computing torsional shear stress and torsional deflection for members with noncircular cross sections.
15. Describe the general shapes of members having relatively high torsional stiffness.

4–2 TORQUE, POWER, AND ROTATIONAL SPEED

A necessary task in approaching the calculation of torsional shear stress and deflection is the understanding of the concept of *torque* and the relationship among the three critical variables involved in power transmission: *torque, power,* and *rotational speed.*

Torque. Figure 4–4 shows a socket wrench with an extension shaft being used to tighten a bolt. The *torque,* applied to both the bolt and the extension shaft, is the product of the

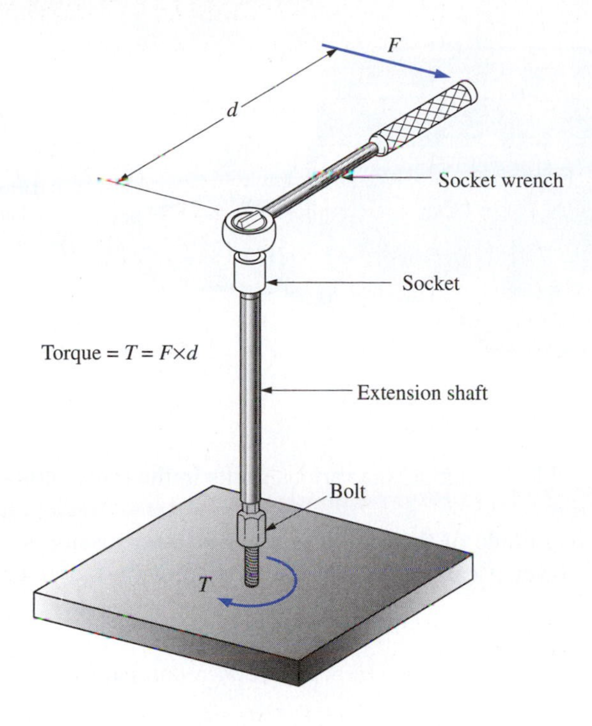

FIGURE 4–4 Wrench applying a torque to a bolt.

applied force and the distance from the line of action of the force to the axis of the bolt. That is,

Torque

$$\text{torque} = T = F \times d \quad \textbf{(4–1)}$$

Thus torque is expressed in the units of *force times distance,* which is N·m in the SI metric system and lb·in or lb·ft in the U.S. Customary system.

Example Problem 4–1 For the wrench in Figure 4–4, compute the magnitude of the torque applied to the bolt if a force of 50 N is exerted at a point 250 mm out from the axis of the socket.

Solution

Objective: Compute the torque on the wrench.

Given: Setup in Figure 4–4. $F = 50$ N. $d = 250$ mm = torque arm length

Analysis: Use Equation (4–1), $T = F \times d$

Results: $\text{Torque} = T = (50\,\text{N})(250\,\text{mm}) \times \dfrac{1\,\text{m}}{1000\,\text{mm}} = 12.5\,\text{N}\cdot\text{m}$

Comment: The wrench is applying a torque of 12.5 N·m to the bolt.

Power.

Power is defined as the rate of transferring energy.

Figure 4–5 shows a drive system for a boat. Power developed by the engine flows through the transmission and the driveshaft to the propeller, where it drives the boat forward. The

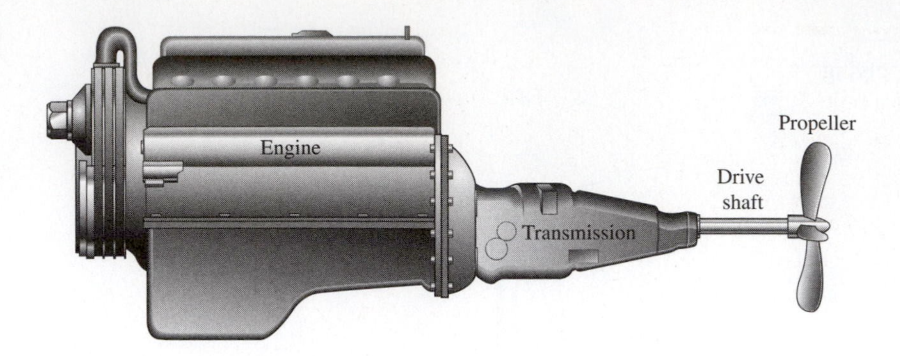

FIGURE 4–5 Drive system for a boat.

crankshaft inside the engine, the various shafts in the transmission, and the driveshaft are all subjected to torsion.

The magnitude of the torque in a power transmission shaft is dependent on the amount of power it carries and on the speed of rotation, according to the following relation:

 **Power**

$$\text{power} = \text{torque} \times \text{rotational speed}$$

$$P = T \times n \tag{4–2}$$

This is a very useful relationship because if any two values, P, n, or T, are known, the third can be computed.

Careful attention must be paid to units when working with torque, power, and rotational speed. Appropriate units in the SI metric system and the U.S. Customary system are reviewed next.

Power in SI Metric System Units. In the SI metric system, the *joule* is the standard unit for energy and it is equivalent to the N·m, the standard unit for torque. That is,

$$1.0\ \text{J} = 1.0\ \text{N}\cdot\text{m}$$

Then power is defined as

SI Units for Power

$$\text{power} = \frac{\text{energy}}{\text{time}} = \frac{\text{joule}}{\text{second}} = \frac{\text{J}}{\text{s}} = \frac{\text{N}\cdot\text{m}}{\text{s}} = \text{watt} = \text{W} \tag{4–3}$$

Note that 1.0 J/s is defined to be 1.0 watt (1.0 W). The watt is a rather small unit of power, so the kilowatt (1.0 kW = 1000 W) is often used.

The standard unit for rotational speed in the SI metric system is *radians per second,* rad/s. Frequently, however, rotational speed is expressed in revolutions per minute, rpm. The conversion required is illustrated below, converting 1750 rpm to rad/s.

$$n = \frac{1750\ \text{rev}}{\text{min}} \times \frac{2\pi\ \text{rad}}{\text{rev}} \times \frac{1\ \text{min}}{60\ \text{s}} = 183\ \text{rad/s}$$

When using n in rad/s in Equation (4–2), the radian is considered to be *no unit at all,* as illustrated in the following example problem.

Example Problem 4–2 The driveshaft for the boat shown in Figure 4–5 transmits 95 kW of power while rotating at 525 rpm. Compute the torque in the shaft.

Solution

Objective: Compute the torque in the shaft.

Given: $P = 95\text{ kW} = 95\,000\text{ W} = 95\,000\text{ N}\cdot\text{m/s}$; $n = 525$ rpm

Analysis: Equation (4–2) will be solved for T and used to compute torque.

$$P = Tn; \quad \text{then, } T = P/n$$

But n must be in rad/s, found as follows:

$$n = \frac{525\text{ rev}}{\text{min}} \times \frac{2\pi\text{ rad}}{\text{rev}} \times \frac{1\text{ min}}{60\text{ s}} = 55.0\text{ rad/s}$$

Results: The torque is

$$T = \frac{P}{n} = \frac{95\,000\text{ N}\cdot\text{m}}{\text{s}} \times \frac{1}{55.0\text{ rad/s}} = 1727\text{ N}\cdot\text{m}$$

Comment: Note that the radian unit is ignored in such calculations.

Power in U.S. Customary Units. Typical units for torque, power, and rotational speed in the U.S. Customary unit system are

T = torque (lb·in)

n = rotational speed (rpm)

P = power (horsepower, hp)

Note that 1.0 hp = 6600 lb·in/s. Then the unit conversions required to ensure consistent units are

$$\text{power} = T(\text{lb}\cdot\text{in}) \times n\left(\frac{\text{rev}}{\text{min}}\right) \times \frac{1\text{ min}}{60\text{ s}} \times \frac{2\pi\text{ rad}}{\text{rev}} \times \frac{1\text{ hp}}{6600\text{ lb}\cdot\text{in/s}}$$

or

$$\text{power} = \frac{Tn}{63\,000} \tag{4–4}$$

Example Problem 4–3 Compute the power, in the unit of horsepower, being transmitted by a shaft if it is developing a torque of 15 000 lb·in and rotating at 525 rpm.

Solution

Objective: Compute the power transmitted by the shaft.

Given: $T = 15\,000$ lb·in; $n = 525$ rpm

Analysis Equation (4–4) will be used directly because T and n are in the proper units of lb·in and rpm. Power will be in horsepower.

Results The power is

$$P = \frac{Tn}{63\,000} = \frac{(15\,000)(525)}{63\,000} = 125 \text{ hp}$$

Torque Sensors. There are times when mechanical designers and engineers need to measure torque in product development laboratories or to build torque sensing into operating equipment. Internet site 4 lists some commercial manufacturers of torque sensors that cover a wide range of torque values from 5 mN·m to over 225 000 N·m (0.044 lb·in to 2×10^6 lb·in). The sensors incorporate transducers to produce an electronic digital readout or analog signal to indicate torque values. Internet site 5 is an article discussing the history, current status, and future projections for torque-sensing technology.

4–3 TORSIONAL SHEAR STRESS IN MEMBERS WITH CIRCULAR CROSS SECTIONS

When a member is subjected to an externally applied torque, an internal resisting torque must be developed in the material from which the member is made. The internal resisting torque is the result of stresses developed in the material.

Figure 4–6 shows a circular bar subjected to a torque, T. Section N would be rotated relative to section M as shown. If an element on the surface of the bar were isolated, it would be subjected to shearing forces on the sides parallel to cross sections M and N, as shown. These shearing forces result in shear stresses on the element. For the stress element to be in equilibrium, equal shearing stresses must exist on the top and bottom faces of the element.

The shear stress element shown in Figure 4–6 is fundamentally the same as that shown in Figure 1–19 in the discussion of direct shear stress. While the manner in which the stresses are created differs, the nature of torsional shear stress is the same as direct shear stress when an infinitesimal element is considered.

When the circular bar is subjected to the externally applied torque, the material in each cross section is deformed in a manner such that the fibers on the outside surface experience the maximum strain. At the central axis of the bar, no strain at all is produced. Between the center and the outside, there is a linear variation of strain with radial position r.

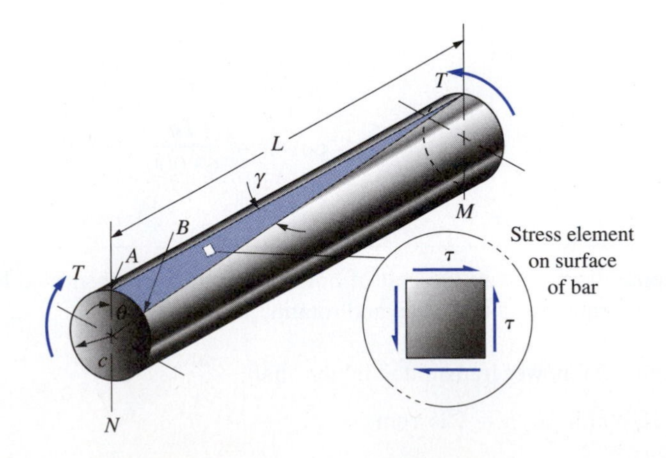

FIGURE 4–6
Torsional shear stress in a circular bar.

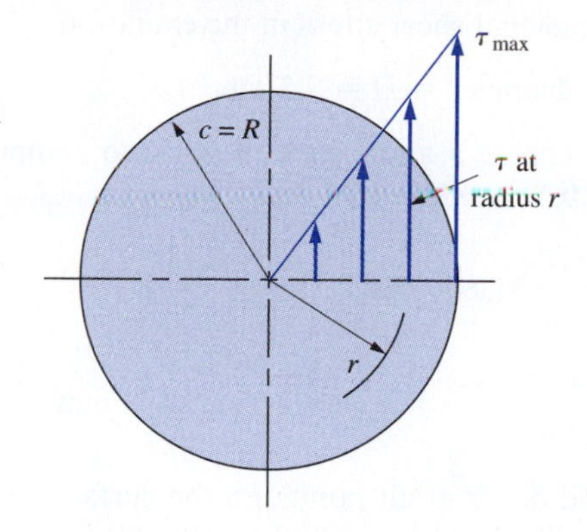

FIGURE 4–7 Distribution of shear stress on a cross section of the bar.

Because stress is directly proportional to strain, we can say that the maximum stress occurs at the outside surface, that there is a linear variation of stress with radial position r, and that a zero stress level occurs at the center. Figure 4–7 illustrates these observations.

The derivation will be shown in the next section for the formula for the maximum shear stress on the outer surface of the bar. For now, we state the *torsional shear stress formula* as

Torsional Shear Stress Formula

$$\tau_{max} = \frac{Tc}{J} \qquad \textbf{(4–5)}$$

where T = applied torque at the section of interest
c = radius of the cross section
J = polar moment of inertia of the circular cross section

The formula for J for a solid circular cross section is developed in Section 4–6. At this time, we show the results of the derivation as,

Polar Moment of Inertia for Circular Bar

$$J = \frac{\pi D^4}{32} \qquad \textbf{(4–6)}$$

where D is the diameter of the shaft; that is, $D = 2R$.

Because of the linear variation of stress and strain with position in the bar as shown in Figure 4–7, the stress, τ, at any radial position, r, can be computed from

Shear Stress at Any Radius

$$\tau = \tau_{max}\frac{r}{c} \qquad \textbf{(4–7)}$$

Equations (4–5), (4–6), and (4–7) can be used to compute the shear stress at any point in a circular bar subjected to an externally applied torque. The following example problems illustrate the use of these equations.

Example Problem 4–4 For the socket wrench extension shown in Figure 4–4, compute the maximum torsional shear stress in the middle portion where the diameter is 9.5 mm. The applied torque is 10.0 N·m.

Solution Objective Compute the maximum torsional shear stress in the extension.

Given Torque = T = 10.0 N·m; diameter = D = 9.5 mm

Analysis Use Equation (4–6) to compute J and Equation (4–5) to compute the maximum shear stress. Also, $c = D/2 = 9.5\text{ mm}/2 = 4.75\text{ mm}$.

Results

$$J = \frac{\pi D^4}{32} = \frac{\pi(9.5\text{ mm})^4}{32} = 800\text{ mm}^4$$

$$\tau_{max} = \frac{Tc}{J} = \frac{(10\text{ N}\cdot\text{m})(4.75\text{ mm})}{800\text{ mm}^4} \times \frac{10^3\text{mm}}{\text{m}} = 59.4\text{ N/mm}^2 = 59.4\text{ MPa}$$

Comment This level of stress would occur at all points on the surface of the circular part of the extension.

Example Problem 4–5 Calculate the maximum torsional shear stress that would develop in a solid circular shaft, having a diameter of 1.25 in, if it is transmitting 125 hp while rotating at 525 rpm.

Solution Objective Compute the maximum torsional shear stress in the shaft.

Given Power = P = 125 hp; rotational speed = n = 525 rpm
Shaft diameter = D = 1.25 in

Analysis Solve Equation (4–4) for the torque, T. Use Equation (4–6) to compute J and Equation (4–5) to compute the maximum shear stress. Also, $c = D/2 = 1.25\text{ in}/2 = 0.625\text{ in}$.

Results Equation (4–4),

$$\text{Power} = P = \frac{Tn}{63\,000}$$

Solving for the torque T gives

$$T = \frac{63\,000P}{n}$$

Recall that this equation will give the value of the torque directly in lb·in when P is in horsepower and n is in rpm. Then

$$T = \frac{63\,000(125)}{525} = 15\,000\text{ lb}\cdot\text{in}$$

$$J = \frac{\pi D^4}{32} = \frac{\pi(1.25\text{ in})^4}{32} = 0.240\text{ in}^4$$

Then

$$\tau_{max} = \frac{Tc}{J} = \frac{(15\,000\text{ lb}\cdot\text{in})(0.625\text{ in})}{0.240\text{ in}^4} = 39\,100\text{ psi}$$

Comment This level of stress would occur at all points on the surface of the shaft.

4–4 DEVELOPMENT OF THE TORSIONAL SHEAR STRESS FORMULA

The standard form of the torsional shear stress formula for a circular bar subjected to an externally applied torque was shown as Equation (4–5), and its use was illustrated in Example Problems 4–4 and 4–5. This section will show the development of that formula. Reference should be made to Figures 4–6 and 4–7 for the general nature of the torsional loading and a visualization of the effect the torque has on the behavior of the circular bar.

In this development, it is assumed that the material for the bar behaves in accordance with Hooke's law; that is, stress is directly proportional to strain. Also, the properties of the bar are homogeneous and isotropic; that is, the material reacts the same regardless of the direction of the applied loads. Also, it is assumed that the bar is of constant cross section in the vicinity of the section of interest.

Considering two cross sections M and N at different places on the bar, section N would be rotated through an angle θ relative to section M. The fibers of the material would undergo a strain that would be maximum at the outside surface of the bar and vary linearly with radial position to zero at the center of the bar. Because, for elastic materials obeying Hooke's law, stress is proportional to strain, the maximum stress would also occur at the outside of the bar, as shown in Figure 4–7. The linear variation of stress, τ, with radial position in the cross section, r, is also shown. Then, by proportion using similar triangles,

$$\frac{\tau}{r} = \frac{\tau_{\max}}{c} \tag{4–8}$$

Then the shear stress at any radius can be expressed as a function of the maximum shear stress at the outside of the shaft,

$$\tau = \tau_{\max} \times \frac{r}{c} \tag{4–9}$$

It should be noted that the shear stress τ acts uniformly on a small ring-shaped area, dA, of the shaft, as illustrated in Figure 4–8. Now since force equals stress times area, the force on the area dA is

$$dF = \tau dA = \underbrace{\tau_{\max}\frac{r}{c}}_{\text{stress}} \times \underbrace{dA}_{\text{area}}$$

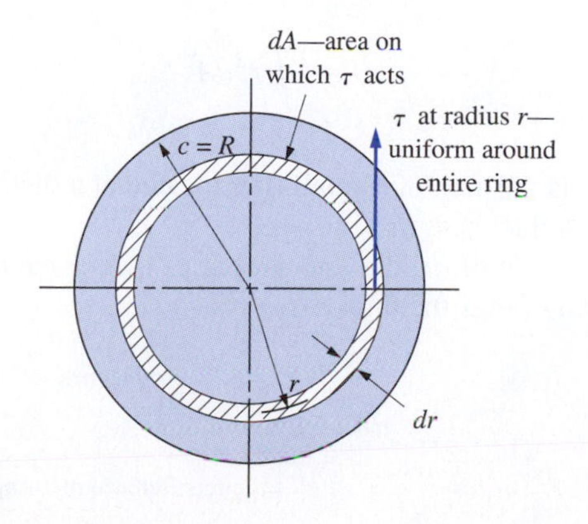

FIGURE 4–8 Shear stress τ at radius r acting on the area dA.

The next step is to consider that the torque dT developed by this force is the product of dF and the radial distance to dA. Then

$$dT = dF \times r = \underbrace{\tau_{max}\frac{r}{c}dA}_{\text{force}} \times \underbrace{r}_{\text{radius}} = \tau_{max}\frac{r^2}{c}dA$$

This equation is the internal resisting torque developed on the small area dA. The total torque on the entire area would be the sum of all the individual torques on all areas of the cross section. The process of summing is accomplished by the mathematical technique of integration, illustrated as follows:

$$T = \int_A dT = \int_A \tau_{max}\frac{r^2}{c}dA$$

In the process of integration, constant terms such as τ_{max} and c can be brought outside the integral sign, leaving

$$T = \frac{\tau_{max}}{c}\int_A r^2 dA \tag{4–10}$$

In mechanics, the term $\int r^2 dA$ is given the name *polar moment of intertia* and is identified by the symbol J. Equation (4–10) can then be written

$$T = \tau_{max}\frac{J}{c}$$

or

$$\tau_{max} = \frac{Tc}{J} \tag{4–11}$$

The method of evaluating J is developed in the next section.

Equation (4–11), which is identical to Equation (4–5), can be used to compute the maximum shear stress on a circular bar subjected to torsion. The maximum shear stress occurs anywhere on the outer surface of the bar.

4–5 POLAR MOMENT OF INERTIA FOR SOLID CIRCULAR BARS

Refer to Figure 4–8 showing a solid circular cross section. To evaluate J from

$$J = \int r^2 dA$$

it must be seen that dA is the area of a small ring located at a distance r from the center of the section and having a thickness dr.

For a small magnitude of dr, the area is that of a strip having a length equal to the circumference of the ring times the thickness.

$$dA = \underbrace{2\pi r}_{\text{circumference of a ring at the radius } r} \times \overbrace{dr}^{\text{thickness of the ring}}$$

Then the polar moment of inertia for the entire cross section can be found by integrating from $r = 0$ at the center of the bar to $r = R$ at the outer surface.

$$J = \int_0^R r^2 dA = \int_0^R r^2(2\pi r)dr = \int_0^R 2\pi r^3 dr = \frac{2\pi R^4}{4} = \frac{\pi R^4}{2}$$

It is usually more convenient to use diameter rather than radius. Then since $R = D/2$,

$$J = \frac{\pi(D/2)^4}{2} = \frac{\pi D^4}{32} \tag{4–12}$$

4–6 TORSIONAL SHEAR STRESS AND POLAR MOMENT OF INERTIA FOR HOLLOW CIRCULAR BARS

It will be shown later that there are many advantages to using a hollow circular bar, as compared with a solid bar, to carry a torque. This section covers the method of computing the maximum shear stress and the polar moment of inertia for a hollow bar.

Figure 4–9 shows the basic geometry for a hollow bar. The variables are:

R_i = inside radius
D_i = inside diameter
R_o = outside radius = c
D_o = outside diameter

The logic and details of the development of the torsional shear stress formula as shown in Section 4–4 apply as well to a hollow bar as to the solid bar. The difference between them is in the evaluation of the polar moment of inertia, as will be shown later. Therefore, Equation (4–5) or (4–11) can be used to compute the maximum torsional shear stress in either the solid or the hollow bar.

Also, as illustrated in Figure 4–9, the maximum shear stress occurs at the outer surface of the bar, and there is a linear variation of stress with radial position inside the bar.

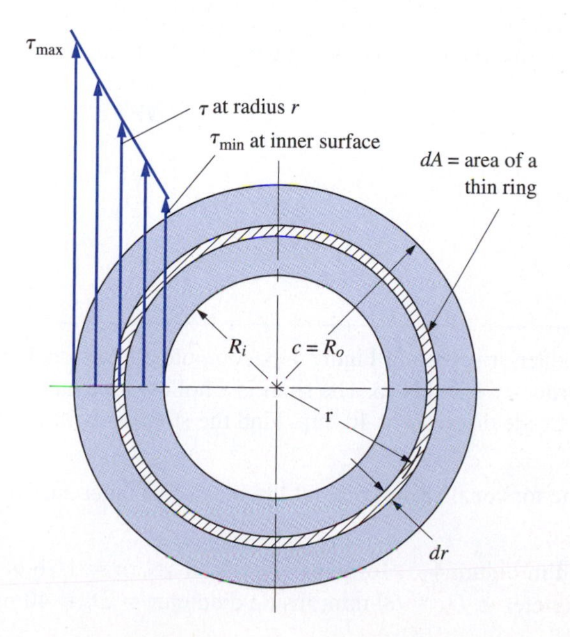

FIGURE 4–9 Notation for variables used to derive J for a hollow round bar.

The minimum shear stress occurs at the inside surface. The shear stress at any radial position can be computed from Equation (4–7) or (4–9).

Polar Moment of Inertia for a Hollow Bar. The process of developing the formula for the polar moment of inertia for a hollow bar is similar to that used for the solid bar. Refer again to Figure 4–9 for the geometry. Starting with the basic definition of the polar moment of inertia,

$$J = \int r^2 dA$$

as before, $dA = 2\pi r\,dr$. But for the hollow bar, r varies only from R_i to R_o. Then

$$J = \int_{R_i}^{R_o} r^2(2\pi r)dr = 2\pi \int_{R_i}^{R_o} r^3 dr = \frac{2\pi(R_o^4 - R_i^4)}{4}$$

$$J = \frac{\pi}{2}(R_o^4 - R_i^4)$$

Substituting $R_o = D_o/2$ and $R_i = D_i/2$ gives

Polar Moment of Inertia for a Hollow Bar

$$J = \frac{\pi}{32}(D_o^4 - D_i^4) \quad (4\text{–}13)$$

This is the equation for the polar moment of inertia for a hollow circular bar.

Summary of Relationships for Torsional Shear Stresses in Hollow Circular Bars.

Maximum Shear Stress

$$\tau_{max} = \frac{Tc}{J} \quad (4\text{–}11)$$

τ_{max} occurs at the outer surface of the bar, where c is the radius of the bar.

Shear Stress at Any Radial Position r

$$\tau = \tau_{max}\frac{r}{c} = \frac{Tr}{J} \quad (4\text{–}9)$$

Polar Moment of Inertia for Hollow Bars

$$J = \frac{\pi}{32}(D_o^4 - D_i^4) \quad (4\text{–}13)$$

Example Problem 4–6 For the propeller driveshaft of Figure 4–5, compute the torsional shear stress when it is transmitting a torque of 1.76 kN·m. The shaft is a hollow tube having an outside diameter of 60 mm and an inside diameter of 40 mm. Find the stress at both the outer and inner surfaces.

Solution Objective: Compute the torsional shear stress at the outer and inner surfaces of the hollow propeller driveshaft.

Given: Shaft shown in Figure 4–5. Torque $= T = 1.76$ kN·m $= 1.76 \times 10^3$ N·m.
Outside diameter $= D_o = 60$ mm; inside diameter $= D_i = 40$ mm.

Analysis The final calculation for the torsional shear stress at the outer surface will be made using Equation (4–11). Equation (4–9) will be used to compute the stress at the inner surface. The polar moment of inertia will be computed using Equation (4–13). And $c = D_o/2 = 30$ mm.

Results At the outer surface,

$$\tau_{max} = \frac{Tc}{J}$$

$$J = \frac{\pi}{32}(D_o^4 - D_i^4) = \frac{\pi}{32}(60^4 - 40^4)\text{ mm}^4 = 1.02 \times 10^6 \text{ mm}^4$$

$$\tau_{max} = \frac{Tc}{J} = \frac{(1.76 \times 10^3 \text{ N}\cdot\text{m})(30 \text{ mm})}{1.02 \times 10^6 \text{ mm}^4} \times \frac{10^3 \text{ mm}}{\text{m}}$$

$$\tau_{max} = 51.8 \text{ N/mm}^2 = 51.8 \text{ MPa}$$

At the inner surface, $r = D_i/2 = 40 \text{ mm}/2 = 20$ mm.

$$\tau = \tau_{max}\frac{r}{c} = 51.8 \text{ MPa} \times \frac{20 \text{ mm}}{30 \text{ mm}} = 34.5 \text{ MPa}$$

Comment You should visualize these stress values plotted on the cross section shown in Figure 4–9.

4–7 DESIGN OF CIRCULAR MEMBERS UNDER TORSION

In a design problem, the loading on a member is known, and what is required is to determine the geometry of the member to ensure that it will carry the loads safely. Material selection and the determination of design stresses are integral parts of the design process. *The techniques developed in this section are for circular members only, subjected only to torsion.* Of course, both solid and hollow circular members are covered. Torsion in noncircular members is covered in a later section of this chapter. The combination of torsion with bending and axial loads is presented in later chapters.

The basic torsional shear stress equation, Equation (4–11), was expressed as

$$\tau_{max} = \frac{Tc}{J} \quad \textbf{(4–11)}$$

In design, we can substitute a certain design stress τ_d for τ_{max}. As in the case of members subjected to direct shear stress and made of ductile materials, the design stress is related to the yield strength of the material in shear. That is,

$$\tau_d = \frac{s_{ys}}{N}$$

where N is the design factor chosen by the designer based on the manner of loading. Table 4–1 can be used as a guide to determine the value of N.

Where the data for s_{ys} are not available, the value can be estimated as $s_y/2$. This will give a reasonable, and usually conservative, estimate for ductile metals, especially steel. Then

Design Shear Stress

$$\tau_d = \frac{s_{ys}}{N} = \frac{s_y}{2N} \quad \textbf{(4–14)}$$

TABLE 4–1 Design factors and design shear stresses for ductile metals.

Manner of loading	Design factor	Design shear stress $\tau_d = s_y/2N$
Static or steady torsion	2	$\tau_d = s_y/4$
Repeated torsion	4	$\tau_d = s_y/8$
Torsional impact or shock	6	$\tau_d = s_y/12$

The torque, T, would be known in a design problem. Then, in Equation (4–11), only c and J are left to be determined. Notice that both c and J are properties of the geometry of the member that is being designed. For solid circular members (shafts), the geometry is completely defined by the diameter. It has been shown that

$$c = \frac{D}{2}$$

and

$$J = \frac{\pi D^4}{32}$$

It is now convenient to note that if the quotient J/c is formed, a simple expression involving D is obtained.

In the study of strength of materials, the term J/c is given the name *polar section modulus,* and the symbol Z_p is used to denote it.

Polar Section Modulus—Solid Shafts

$$Z_p = \frac{J}{c} = \frac{\pi D^4}{32} \times \frac{1}{D/2} = \frac{\pi D^3}{16} \qquad \textbf{(4–15)}$$

Substituting Z_p for J/c in Equation (4–11) gives

Maximum Shear Stress

$$\tau_{\text{max}} = \frac{T}{Z_p} \qquad \textbf{(4–16)}$$

To use this equation in design, we can let $\tau_{\text{max}} = \tau_d$ and then solve for Z_p.

Required Polar Section Modulus

$$Z_p = \frac{T}{\tau_d} \qquad \textbf{(4–17)}$$

Equation (4–17) will give the required value of the polar section modulus of a circular shaft to limit the torsional shear stress to τ_d when subjected to a torque T. Then Equation (4–15) can be used to find the required diameter of a solid circular shaft. Solving for D gives us

$$D = \sqrt[3]{\frac{16 Z_p}{\pi}} \qquad \textbf{(4–18)}$$

If a hollow shaft is to be designed,

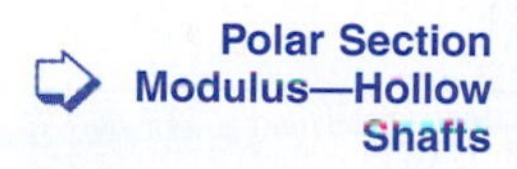
Polar Section Modulus—Hollow Shafts

$$Z_p = \frac{J}{c} = \frac{\pi}{32}(D_o^4 - D_i^4) \times \frac{1}{D_o/2}$$

$$Z_p = \frac{\pi}{16}\frac{D_o^4 - D_i^4}{D_o} \quad \textbf{(4–19)}$$

In this case, one of the diameters *or* the relationship between the two diameters would have to be specified in order to solve for the complete geometry of the hollow shaft.

PROGRAMMED EXAMPLE PROBLEM

The completion of this problem is presented in a programmed format. You should answer each question as it is posed before looking at the next panel beyond the line across the page. This process is intended to involve you in the decision-making activities required of a designer.

Example Problem 4–7

The final drive to a conveyor that feeds coal to a railroad car is a shaft loaded in pure torsion and carrying 800 N·m of torque. A proposed design calls for the shaft to have a solid circular cross section. Complete the design by first specifying a suitable steel for the shaft and then specifying the diameter.

Solution

Objective

1. Specify a suitable steel for the shaft.
2. Specify the shaft diameter.

Given

Applied torque = T = 800 N·m
Shaft drives a coal conveyor.

Analysis

First, as an aid in selecting a suitable material, what manner of loading will the shaft experience in service?

The drive for a coal conveyor is likely to experience very rough service as coal is dumped onto the conveyor. Therefore, the design should be able to accommodate impact and shock loading. What properties should the steel for the shaft possess?

A highly ductile material should be used because such materials withstand shock loading much better than more brittle materials. The steel should have a moderately high strength so that the required diameter of the shaft is reasonable. It may be important to choose a steel with good machinability because the shaft is likely to require machining during its manufacture. What is a typically used measure of ductility for steels?

Chapter 2 mentions that the *percent elongation* for a steel is an indication of its ductility. To withstand impact and shock, a steel having a value somewhat higher than 10% elongation should be specified. Now specify a suitable steel.

There are many steels that can be used satisfactorily. Let's specify AISI 1141 OQT 1300. List pertinent data from Appendix A–14.

You should have found that s_y = 469 MPa and that the 28% elongation indicates a high ductility. Also note, as stated in Chapter 2, the 1100 series steels have good machinability because of a relatively high sulfur content in the alloy.

We will be using Equations (4–16), (4–17), and (4–18) to continue the design process with the ultimate goal of specifying a suitable diameter for the shaft. We know the applied torque is 800 N·m. The next task should be to determine an acceptable design shear stress. How will you do that?

Table 4–1 calls for $\tau_d = s_y/2N$ with $N = 6$; that is, $\tau_d = s_y/12$. Then,

$$\tau_d = s_y/12 = 469 \text{ MPa}/12 = 39.1 \text{ MPa} = 39.1 \text{ N/mm}^2$$

What is the next step?

We can use Equation (4–17) to compute the required value of the polar section modulus for the cross section of the shaft. Do that now.

You should have the required $Z_p = 20.5 \times 10^3 \text{ mm}^3$, found from

$$Z_p = \frac{T}{\tau_d} = \frac{800 \text{ N}\cdot\text{m}}{39.1 \text{ N/mm}^2} \times \frac{10^3 \text{ mm}}{\text{m}} = 20.5 \times 10^3 \text{ mm}^3$$

What is the next step?

We can compute the minimum acceptable diameter for the shaft using Equation (4–18). Do that now.

You should have $D_{\min} = 47.1$ mm, found from

$$D = \sqrt[3]{\frac{16Z_p}{\pi}} = \sqrt[3]{\frac{16(20.5 \times 10^3) \text{ mm}^3}{\pi}} = 47.1 \text{ mm}$$

It would be appropriate to specify a convenient size for the shaft diameter that is slightly larger than this value. Use Appendix A–2 as a guide and specify a diameter.

Specifying $D = 50$ mm is preferred.

Summary of Results The shaft will be made from AISI 1141 OQT 1300 steel with a diameter of 50 mm.

Comment The maximum shear stress at the outer surface of the 50-mm diameter shaft is actually less than the design stress because we specified a preferred diameter slightly greater than the minimum required diameter of 47.1 mm. Let's now compute the actual maximum stress in the shaft. First, we will compute the polar section modulus for the 50-mm diameter shaft.

$$Z_p = \frac{\pi D^3}{16} = \frac{\pi(50)^3 \text{ mm}^3}{16} = 24.5 \times 10^3 \text{ mm}^3$$

Then, the maximum shear stress is

$$\tau_{\max} = \frac{T}{Z_p} = \frac{800 \text{ N}\cdot\text{m}}{24.5 \times 10^3 \text{ mm}^3} \times \frac{10^3 \text{ mm}}{1 \text{ m}} = 32.6 \text{ N/mm}^2 = 32.6 \text{ MPa}$$

4–8 COMPARISON OF SOLID AND HOLLOW CIRCULAR MEMBERS

We will now demonstrate, by example, that hollow shafts are more efficient than solid shafts. Here the term *efficiency* is used as a measure of the mass of material in a shaft required to carry a given torque with a given shear stress level. The following example problem shows the design of a hollow shaft with a slightly larger outside diameter that has the same maximum shear stress as the 50-mm diameter solid shaft just designed. Then, the mass of the hollow shaft is compared with that of the solid shaft.

PROGRAMMED EXAMPLE PROBLEM

The completion of this problem is presented in a programmed format. You should answer each question as it is posed before looking at the next panel beyond the line across the page. This process is intended to involve you in the decision-making activities required of a designer.

Example Problem 4–8

An alternative design for the shaft described in Example Problem 4–7 would be to use a hollow tube for the shaft. Assume that a tube having an outside diameter of 60 mm is available in the same material as specified for the solid shaft (AISI 1141 OQT 1300). Compute what maximum inside diameter the tube can have that would result in the same stress in the steel as the 50-mm solid shaft.

Solution

Objective: Compute the maximum allowable inside diameter for the hollow shaft.

Given: From Example Problem 4–7, maximum shear stress = τ_{max} = 32.6 MPa. D_o = 60 mm. Applied torque = T = 800 N·m.

Analysis: Because torsional shear stress is inversely proportional to the polar section modulus, it is necessary that the hollow tube have the same value for Z_p as does the 50-mm diameter solid shaft. That is, $Z_p = 24.5 \times 10^3\ \text{mm}^3$. What is the formula for Z_p for a hollow shaft?

$$Z_p = \frac{\pi}{16}\frac{D_o^4 - D_i^4}{D_o}$$

The outside diameter, D_o, is known to be 60 mm. We can then solve for the required inside diameter, D_i. Do that now.

You should have

$$D_i = \left(D_o^4 - \frac{16Z_pD_o}{\pi}\right)^{1/4}$$

Now compute the maximum allowable inside diameter.

$$D_i = \left[(60)^4 - \frac{(16)(24.5 \times 10^3)(60)}{\pi}\right]^{1/4} \text{mm} = 48.4\,\text{mm}$$

Summary of Results: Final design of the hollow shaft:
D_o = 60 mm; D_i = 48.4 mm.
Material: AISI 1141 OQT 1300 steel.
Maximum shear stress at outer surface = τ_{max} = 32.6 MPa.

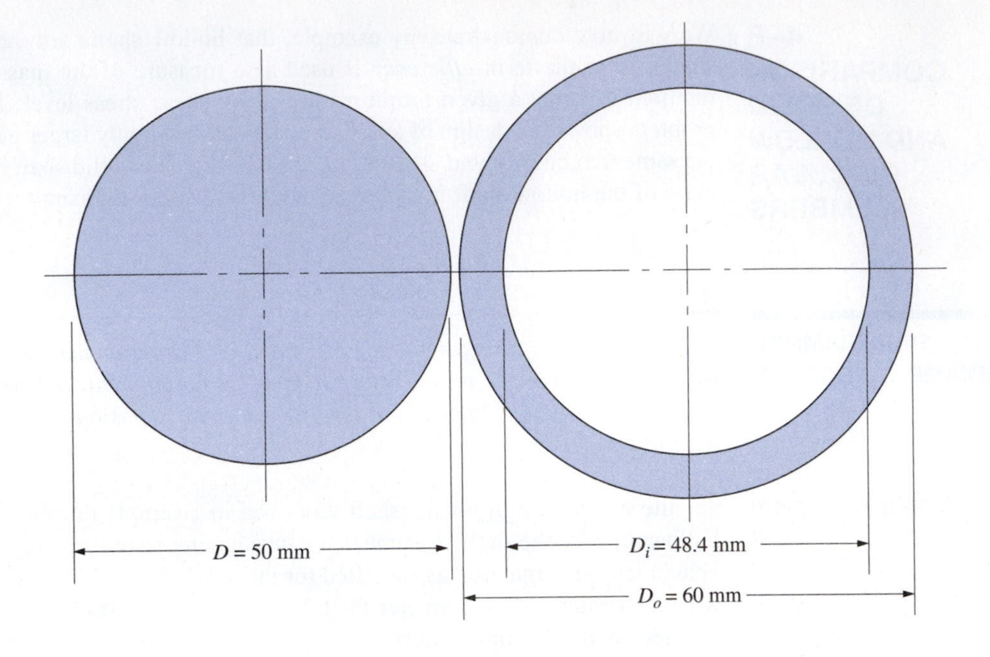

FIGURE 4–10 Comparison of solid and hollow shafts for Example Problem 5–8.

Comment Figure 4–10 shows a comparison of the hollow shaft with the solid shaft having the same maximum shear stress. The designs are drawn full size. It appears to the eye that the hollow shaft uses less material. The following section will demonstrate that this is true.

Masses of Solid and Hollow Shafts. In many design situations, economy of material usage is a major criterion of performance for a product. In aerospace applications, every reduction in the mass of the aircraft or space vehicle allows increased payload. Automobiles achieve higher fuel economy when they are lighter. Also, since raw materials are purchased on a price per unit mass basis, a lighter part generally costs less.

Providing economy of material usage for load-carrying members requires that all the material in the member be stressed to a level approaching the safe design stress. Then every portion is carrying its share of the load.

Example Problems 4–7 and 4–8 can be used to illustrate this point. Recall that both designs shown in Figure 4–10 result in the same maximum torsional shear stress in the steel shaft. The hollow shaft is slightly larger in outside diameter, but it is the *volume* of metal that determines the mass of the shaft. Consider a length of shaft 1.0 m long. For the solid shaft, the volume is the cross-sectional area times the length.

$$V_S = A_S L = \frac{\pi D^2}{4} L$$

$$V_S = \frac{\pi(50\text{ mm})^2}{4} \times 1.0\text{ m} \times \frac{1\text{ m}^2}{(10^3\text{ mm})^2} = 1.96 \times 10^{-3}\text{ m}^3$$

The mass is the volume times the density, ρ. Appendix A–14 gives the density of steel to be 7680 kg/m^3. Then the mass of the solid shaft is $V_S \times \rho$.

$$M_S = 1.96 \times 10^{-3}\text{ m}^3 \times 7680\text{ kg/m}^3 = 15.1\text{ kg}$$

For the hollow shaft the volume is

$$V_H = A_H L = \frac{\pi}{4}(D_o^2 - D_i^2)(L)$$

$$V_H = \frac{\pi}{4}(60^2 - 48.4^2)\ \text{mm}^2 \times 1.0\ \text{m} \times \frac{1\ \text{m}^2}{(10^3\ \text{mm})^2}$$

$$V_H = 0.988 \times 10^{-3}\ \text{m}^3$$

The mass of the hollow shaft is $V_H \times \rho$.

$$M_H = 0.988 \times 10^{-3}\ \text{m}^3 \times 7680\ \text{kg/m}^3 = 7.58\ \text{kg}$$

Thus it can be seen that the hollow shaft has almost exactly *one-half the mass* of the solid shaft, even though both are subjected to the same stress level for a given applied torque.

A generalization of this analysis can be made. For comparison of the mass of a hollow shaft to that of a solid shaft having the same length, L, and material density, ρ, we can compute the ratio of the two masses from,

$$\frac{M_H}{M_S} = \frac{V_H \rho}{V_S \rho} = \frac{A_H L \rho}{A_S L \rho} = \frac{A_H}{A_S}$$

The ratio of the masses is equal to the ratio of the cross-sectional areas of the shafts.

The reason for the hollow shaft being lighter is that a greater portion of its material is being stressed to a higher level than in the solid shaft. Figure 4–7 shows a sketch of the stress distribution in the solid shaft. The maximum stress, 32.6 MPa, occurs at the outer surface. The stress then varies linearly with the radius for other points within the shaft to *zero* at the center. Therefore the material near the middle of the shaft is not being used efficiently.

Contrast this with the sketch of the hollow shaft in Figure 4–9. Again the stress at the outer surface is the maximum, 32.6 MPa. The stress at the inner surface of the hollow shaft can be found from Equation (4–6).

$$\tau = \tau_{\text{max}}\frac{r}{c}$$

At the inner surface, $r = R_i = D_i/2 = 48.4\ \text{mm}/2 = 24.2\ \text{mm}$. Also, $c = R_o = D_o/2 = 60\ \text{mm}/2 = 30\ \text{mm}$. Then

$$\tau = 32.6\ \text{MPa}\frac{24.2}{30} = 26.3\ \text{MPa}$$

The stress at points between the inner and outer surfaces varies linearly with the radius to each point. Thus it can be seen that all of the material in the hollow shaft shown in Figure 4–9 is being stressed to a fairly high but safe level. This illustrates why the hollow section requires less material.

Of course, the specific data used in the previous illustration cannot be generalized to all problems. However, for torsional loading of circular members, a hollow section can be designed that is lighter than a solid section while subjecting the material to the same maximum torsional shear stress.

Caution for Thin Hollow Shafts. You should approach with care the tendency for making the wall thickness of hollow shafts thinner and thinner to gain efficiency of the use of material. There is a limit to which the wall thickness can be thinned before it becomes unstable and begins to buckle locally and wrinkle. As that occurs, the shaft would likely collapse suddenly.

Reference 5 presents the following equations as estimates for the level of torsional shear stress that will cause buckling of the wall of a thin-walled tubular shaft.

Note that the assumption that the wall is thin typically requires that the ratio of the mean radius of the wall to the wall thickness is 10 or greater. You should check this before applying the equations.

The primary variables involved are:

τ' = Approximate torsional shear stress at which buckling will occur
E = Modulus of elasticity of the shaft material
t = Wall thickness of the shaft
l = Length of the shaft
r = Radius of the shaft

The choice of which equation to use is based on the value of the ratio of l/r.

a) If $l/r > 6.6\sqrt[4]{1 - \nu^2}\sqrt{r/t}$:

$$\tau' = 0.272\frac{E}{(1 - \nu^2)^{3/4}}(t/r)^{3/2}$$

See Table 2–1 in Chapter 2 for estimates of the values for v of common metals. If v is approximately 0.3 as it is for steel, stainless steel, titanium, aluminum, and brass, this equation becomes:

$$\tau' = 0292\,E(t/r)^{3/2}$$

b) If $5 < l/r < 6.6\sqrt[4]{1 - \nu^2}\sqrt{r/t}$:

$$\tau' = E(t/l)^2\left[1.8 + \sqrt{1.2 + 0.201(l/\sqrt{tr})^3}\right]$$

c) If $l/r < 5$:

$$\tau' = 0.10\,E\,(t/r) + 5E(t/l)^2$$

4–9 STRESS CONCENTRATIONS IN TORSIONALLY LOADED MEMBERS

Torsionally loaded members, particularly power transmission shafts, are often made with changes in the geometry at various positions. Figure 4–11 shows an example. This is a part of a shaft where a power transmission element, such as a gear, would be mounted. The bore in the hub of the gear would have a diameter that would allow it to slide over the right part of the shaft where the shaft diameter is $d = 25$ mm. A square or rectangular key would be placed in the keyseat and there would be a corresponding keyway in the hub of the gear so it could pass over the key. The gear would then be moved onto the shaft from the right until it stopped against the shoulder at Section 2, created by the increase in the shaft diameter to $D = 40$ mm. To keep the gear in position, a retaining ring is inserted into the ring groove at Section 4.

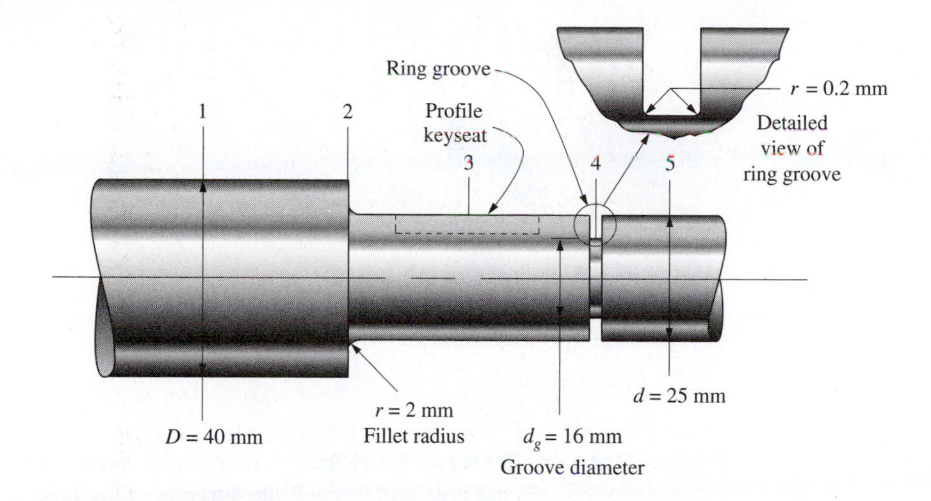

FIGURE 4–11 Shaft with stress concentrations.

Changes in the cross section of a member loaded in torsion cause the local stress near the changes to be higher than would be predicted by using the torsional shear stress formula. The actual level of stress in such cases is determined experimentally. Then a *stress concentration factor* is determined, which allows the maximum stress in similar designs to be computed from the relationship

$$\tau_{\max} = K_t \tau_{\text{nom}} = K_t(T/Z_p) \tag{4–20}$$

The term τ_{nom} is the nominal stress due to torsion, which would be developed in the parts if the stress concentration were not present. Thus the standard torsional shear stress formulas [Equations (4–5) and (4–16)] can be used to compute the nominal stress. The value of K_t is a factor by which the actual maximum stress is greater than the nominal stress.

Referring again to Figure 4–11, note that there would be several levels of stress at different places along the length of the bar, even if the applied torque is the same throughout. The differing diameters and the presence of stress concentrations cause the varying stress levels. The stress at Section 1, where $D = 40$ mm, would be relatively low because there is a large diameter and a correspondingly large polar section modulus. At Section 2, the diameter of the shaft reduces to $d = 25$ mm, and the step produces a stress concentration that tends to raise the local stress level. Then the keyseat at Section 3 sets up a different stress concentration. At Section 4, two major factors occur that both tend to increase the local stress. Cutting the ring groove reduces the diameter to $d_g = 16$ mm and also produces two closely spaced steps with relatively small fillet radii at the bottom of the groove. At Section 5, well away from the ring groove, the stress would be equal to the nominal stress in the 25-mm diameter shaft. Example Problem 4–10 illustrates all of these situations by performing actual calculations of the stresses at all five sections of the shaft.

Before the example problem, let's look at the nature of stress concentration factors. The following list of Appendix charts gives data for several typical cases.

Appendix A–22–5: Round bar with transverse hole in torsion

Appendix A–22–6: Grooved round bar in torsion

Appendix A–22–7: Stepped round bar in torsion

Appendix A–22–11: Shafts with keyseats

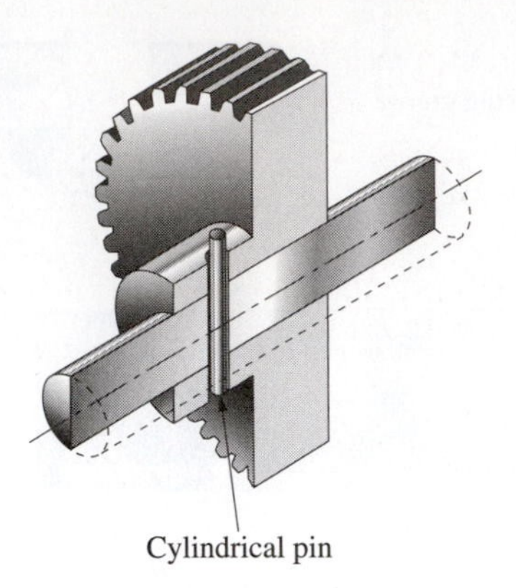

FIGURE 4–12 Round shaft with a transverse hole used to allow connection of a gear to the shaft with a cylindrical pin.

Round Bar with Transverse Hole. One purpose for drilling a hole in a shaft is to insert a pin through the hole and through the corresponding hole in the hub of a machine element such as a gear, pulley, or chain sprocket. See Figure 4–12. The pin serves to locate the machine element axially on the shaft while also transmitting torque from the shaft to the element or from the element to the shaft. The hole in the shaft is an abrupt change in geometry and it causes a stress concentration. Appendix A–22–5 is a chart for this case from which K_t can be determined. Curve C is for the case of torsionally loaded shafts. Note that the formula for the nominal stress in the shaft is based on the full, gross, circular cross section of the shaft.

Grooved Round Bar. Round-bottomed grooves are cut into round bars for the purpose of installing seals or for distributing lubricating oil around a shaft. See Figure 4–13. The stress concentration factor is dependent on the ratio of the shaft diameter to the diameter of the groove, and on the ratio of the groove radius to the base diameter of the groove. The groove is cut with a tool having a rounded nose to produce the round-bottomed groove. The radius should be as large as possible to minimize the stress concentration factor. Note that the nominal stress is based on the diameter *at the base of the groove.* See Appendix A–22–6.

Stepped Round Bar. Shafts are often made with two or more diameters, resulting in a stepped shaft like that shown in Appendix A–22–7 and Figure 4–11. The face of the step provides a convenient means of locating one side of an element mounted on the shaft, such as a bearing, gear, pulley, or chain sprocket. Care should be exercised in defining the radius at the bottom of the step, called the *fillet radius.* Sharp corners are to be avoided, as they cause extremely high stress concentration factors. The radius should be as large as possible while being compatible with the elements mounted on the shaft.

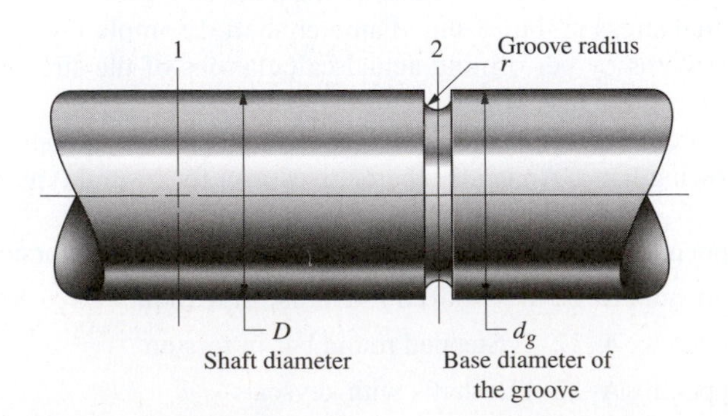

FIGURE 4–13 Shaft with a circular groove.

Retaining rings seated in grooves cut into the shaft are often used to locate machine elements, as shown in Figure 4–11. The grooves are typically flat bottomed with small radii at the sides. Some designers treat such grooves as two steps on the shaft close together and use the stepped shaft chart (Appendix A–22–7) to determine the stress concentration factor. Because of the small radius at the base of the groove, the relative radius is often quite small, resulting in values of K_t so high they are off the chart. In such cases a value of $K_t = 3.0$ is sometimes used.

Shafts with Keyseats. Power transmission elements typically transmit torque to and from shafts through keys fitted into keyseats cut into the shaft, as shown in Figure 4–14. The V-belt pulley mounted on the end of the motor shaft shown is an

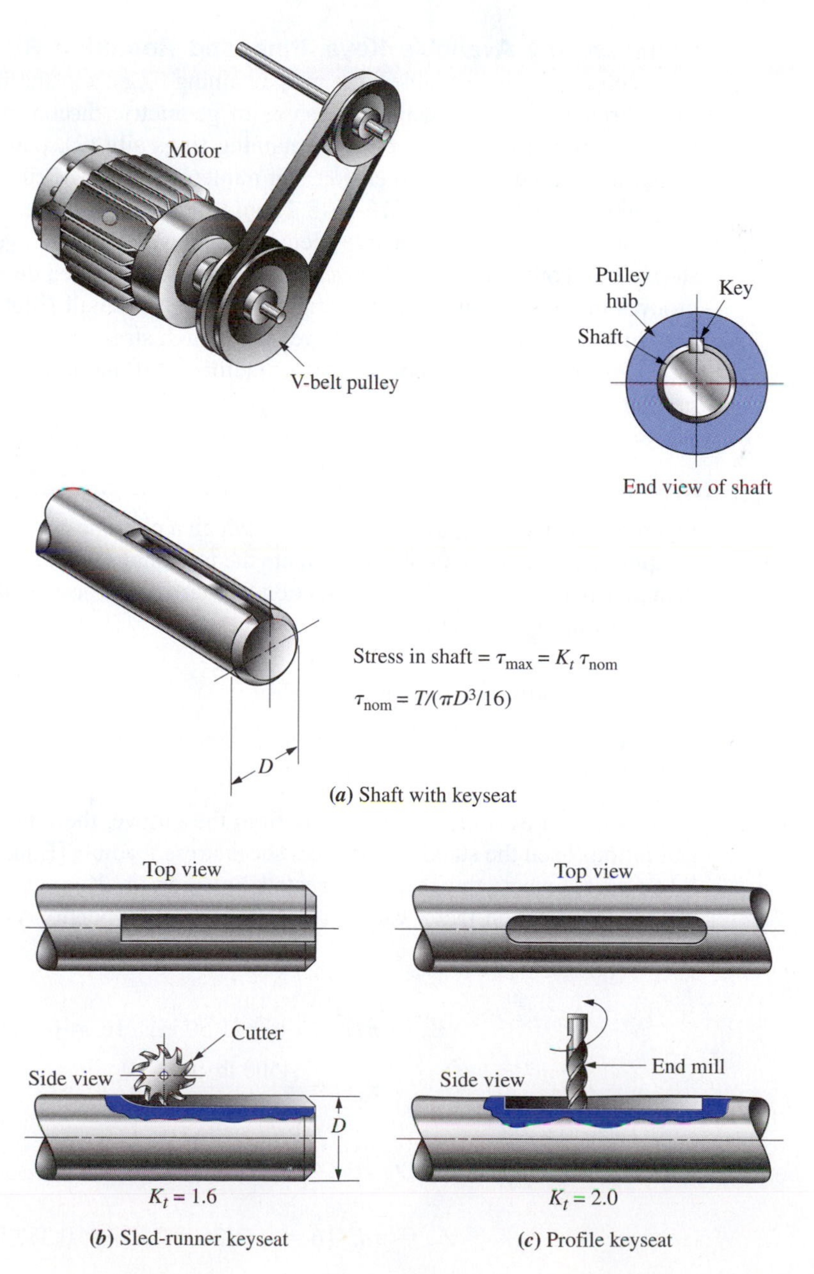

FIGURE 4–14 Stress concentration factors for keyseats. (a) Typical application. (b) Sled-runner type keyseat made with a circular milling cutter. (c) Profile type keyseat made with an end mill.

example. Two types of keyseats are in frequent use: the *sled-runner* and the *profile* keyseats.

A circular milling cutter having a thickness equal to the width of the keyseat is used to cut the sled-runner keyseat, typically on the end of a shaft, as shown in Figure 4–14(b). As the cutter ends its cut, it leaves a gentle radius, as shown in the side view, resulting in $K_t = 1.6$ as a design value.

A profile keyseat is cut with an end mill having a diameter equal to the width of the keyseat. Usually used at a location away from the ends of the shaft, it leaves a square corner at the ends of the keyseat when viewed from the side, as shown in Figure 4–14(c). This is more severe than the sled-runner and a value of $K_t = 2.0$ is used. Note that the stress concentration factors account for both the removal of material from the shaft and the change in geometry.

Commercially Available Keys, Pins, and Retaining Rings. Internet sites 1–3 show commercially available keys, pins, retaining rings, and shafts with keyseats that have been discussed in this section as causes of geometric discontinuities that lead to stress concentrations in shafting. For the retaining rings site (2), pages can be accessed from an online catalog giving dimensions for numerous styles of rings for shaft sizes ranging from approximately $\frac{1}{8}$ in to 10 in (3.2 mm to 254 mm). Style Basic 5100 is commonly used on shafts to secure active elements such as bearings, gears, sprockets, pulleys, sheaves, and other devices. The overall geometry and detailed dimensions are given for the grooves in the shaft in which the rings seat. Note the small fillet radius called out for the base of the groove that gives rise to relatively high stress concentration factors.

The use of stress concentration factors is illustrated in the following example problems.

Example Problem 4–9 Figure 4–13 shows a portion of a shaft in which a circular groove has been machined. For an applied torque of 4500 lb·in, compute the torsional shear stress at Section 1 in the full-diameter part of the shaft and at Section 2 where the groove is located. Use $r = 0.10$ in, $D = 1.50$ in, $d_g = 1.25$ in.

Solution

Objective: Compute the stress at Sections 1 and 2.

Given: Applied torque $= T = 4500$ lb·in. Shaft geometry in Figure 4–13.
$r = 0.10$ in, $D = 1.50$ in, $d_g = 1.25$ in.

Analysis: Assuming that Section 1 is well away from the groove, there is no significant stress concentration. Then the standard torsional shear stress formula [Equation (4–16)] can be used. At Section 2 where the groove is located, Equation (4–20) must be used. The value of the stress concentration factor can be determined from Appendix A–22–6.

Results: At Section 1: $\tau_{max} = T/Z_p$.

$$Z_p = \pi D^3/16 = \pi(1.50\text{ in})^3/16 = 0.663\text{ in}^3$$

$$\tau_{max} = \frac{4500\text{ lb}\cdot\text{in}}{0.663\text{ in}^3} = 6790\text{ psi}$$

At Section 2: $\tau_{max} = K_t T/Z_p$

$$Z_p = \pi d_g^3/16 = \pi(1.25\text{ in})^3/16 = 0.383\text{ in}^3$$

To evaluate K_t, two ratios must be computed, as called for in Appendix A–22–6.

$$D/d_g = (1.50\text{ in})/(1.25\text{ in}) = 1.20$$
$$r/d_g = (0.10\text{ in})/(1.25\text{ in}) = 0.08$$

Then, reading from the chart in Appendix A–22–6, $K_t = 1.55$.

We can now compute the maximum shear stress.

$$\tau_{max} = \frac{K_t T}{Z_p} = \frac{(1.55)(4500\text{ lb}\cdot\text{in})}{0.383\text{ in}^3} = 18\,200\text{ psi}$$

Comment Note that the stress at the groove is 2.6 times higher than that in the full-diameter part of the shaft. One reason is the reduction in the diameter at the groove. Also, the use of the stress concentration factor is essential to predict the actual maximum stress level at the groove.

PROGRAMMED EXAMPLE PROBLEM *The stress in each section will be analyzed separately in a panel set off by a horizontal line across the page. You should perform the indicated calculations before looking at the results shown.*

Example Problem 4–10 Figure 4–11 shows a portion of a shaft where a gear is to be mounted. The gear will be centered over the keyseat at Section 3. It will rest against the shoulder at Section 2 and be held in position with a retaining ring placed in the groove at Section 4. A repeated torque of 20 N·m is applied throughout the shaft. Compute the maximum shear stress in the shaft at Sections 1, 2, 3, 4, and 5. Then specify a suitable steel material for the shaft.

Solution Objective

1. Compute the stresses at Sections 1, 2, 3, 4, and 5.
2. Specify a suitable steel for the shaft.

Given Shaft geometry shown in Figure 4–11. $T = 20$ N·m repeated.

Analysis In each case, the analysis requires the application of Equation (4–20).

$$\tau_{max} = K_t T/Z_p$$

The torque will always be taken to be 20 N·m. You must evaluate the stress concentration factor and the appropriate polar section modulus for each section. Note that $K_t = 1.0$ where there is no change in the geometry. Compute the stress at Section 1.

Section 1. There is no change in geometry, so $K_t = 1.0$. The shaft diameter is $D = 40$ mm. Then,

$$Z_p = \frac{\pi D^3}{16} = \frac{\pi(40\text{ mm})^3}{16} = 12\,570\text{ mm}^3$$

$$\tau_1 = \frac{20\text{ N}\cdot\text{m}}{12\,570\text{ mm}^3} \times \frac{10^3\text{ mm}}{\text{m}} = 1.59\frac{\text{N}}{\text{mm}^2} = 1.59\text{ MPa}$$

Compute the stress at Section 2.

Section 2. The stepped shaft and the shoulder fillet produce a stress concentration here that must be evaluated using Appendix A–22–7. The polar section modulus must be based on the smaller diameter; d = 25 mm. The results are

$$\tau_{\text{nom}} = \frac{T}{\pi d^3/16} = \frac{20\ \text{N}\cdot\text{m}}{[\pi(25)^3/16]\ \text{mm}^3} \times \frac{10^3\ \text{mm}}{\text{m}}$$
$$= 6.52\ \text{N/mm}^2 = 6.52\ \text{MPa}$$

The value of K_t depends on the ratios D/d and r/d.

$$\frac{D}{d} = \frac{40\ \text{mm}}{25\ \text{mm}} = 1.60$$
$$\frac{r}{d} = \frac{2\ \text{mm}}{25\ \text{mm}} = 0.08$$

Then from Appendix A–22–7, K_t = 1.45. Then

$$\tau_2 = (1.45)(6.52\ \text{MPa}) = 9.45\ \text{MPa}$$

Now compute the stress at Section 3.

Section 3. The profile-type keyseat presents a stress concentration factor of 2.0. The nominal stress is the same as that computed at the shoulder fillet. Then

$$\tau_3 = K_t\tau_{\text{nom}} = (2.0)(6.52\ \text{MPa}) = 13.04\ \text{MPa}$$

Compute the stress at Section 4.

Section 4. Section 4 is the location of the ring groove. Here the nominal stress is computed on the basis of the root diameter of the groove.

$$\tau_{\text{nom}} = \frac{T}{\pi d_g^3/16} = \frac{20\ \text{N}\cdot\text{m}}{[\pi(16)^3/16]\ \text{mm}^3} \times \frac{10^3\ \text{mm}}{\text{m}} = 24.9\frac{\text{N}}{\text{mm}^2}$$
$$\tau_{\text{nom}} = 24.9\ \text{MPa}$$

The value of K_t depends on d/d_g and r/d_g.

$$\frac{d}{d_g} = \frac{25\ \text{mm}}{16\ \text{mm}} = 1.56$$
$$\frac{r}{d_g} = \frac{0.2\ \text{mm}}{16\ \text{mm}} = 0.013$$

Referring to Appendix A–22–7, the stress concentration factor is off the chart. This is the type of case for which K_t = 3.0 is reasonable.

$$\tau_4 = K_t\tau_{\text{nom}} = (3.0)(24.9\ \text{MPa}) = 74.7\ \text{MPa}$$

Compute the stress at Section 5.

Section 5. Section 5 is in the smaller portion of the shaft, where no stress concentration occurs. Then $K_t = 1.0$ and

$$\tau = \frac{T}{Z_p} = \frac{T}{\pi d^3/16}$$

Notice that this is identical to the nominal stress computed for Sections 2 and 3. Then at Section 5,

$$\tau_5 = 6.52 \text{ MPa}$$

Summary of Results A wide range of stress levels exists in the vicinity of the place on the shaft where the gear is to be mounted.

$$\tau_1 = 1.59 \text{ MPa} \quad D = 40 \text{ mm.} \quad K_t = 1.0.$$
$$\tau_2 = 9.45 \text{ MPa} \quad d = 25 \text{ mm.} \quad K_t = 1.45. \text{ Step.}$$
$$\tau_3 = 13.04 \text{ MPa} \quad d = 25 \text{ mm.} \quad K_t = 2.0. \text{ Keyseat.}$$
$$\tau_4 = 74.7 \text{ MPa} \quad d_g = 16 \text{ mm.} \quad K_t = 3.0. \text{ Ring groove.}$$
$$\tau_5 = 6.52 \text{ MPa} \quad d = 25 \text{ mm.} \quad K_t = 1.0.$$

The specification of a suitable material must be based on the stress at Section 4 at the ring groove. Let the design stress, τ_d, be equal to that stress level and determine the required yield strength of the material.

You should have a required yield strength of $s_y = 598$ MPa. For the repeated torque, $N = 4$ is recommended in Table 4–1, resulting in

$$\tau_d = s_y/2N = s_y/8$$

Then, solving for s_y gives,

$$s_y = 8(\tau_d) = 8(74.7 \text{ MPa}) = 598 \text{ MPa}$$

Specify a suitable material.

From Appendix A–14, two suitable steels for this requirement are AISI 1040 WQT 900 and AISI 4140 OQT 1300. Both have adequate strength and a high ductility as measured by the percent elongation. Certainly, other alloys and heat treatments could be used.

Comment Review the results of this example problem. It illustrates the importance of considering the details of the design of a shaft at any local area where stress concentrations may occur.

4–10 TWISTING—ELASTIC TORSIONAL DEFORMATION

Stiffness in addition to strength is an important design consideration for torsionally loaded members. The measure of torsional stiffness is the angle of twist of one part of a shaft relative to another part when a certain torque is applied.

In mechanical power transmission applications, excessive twisting of a shaft may cause vibration problems, which would result in noise and improper synchronization of moving parts. One guideline for torsional stiffness is related to the desired degree of precision, as listed in Table 4–2 (see References 1 and 3).

TABLE 4–2 Recommended torsional stiffness: angle of twist per unit length.

Application	Torsional deflection	
	deg/in	rad/m
General machine part	1×10^{-3} to 1×10^{-2}	6.9×10^{-4} to 6.9×10^{-3}
Moderate precision	2×10^{-5} to 4×10^{-4}	1.4×10^{-5} to 2.7×10^{-4}
High precision	1×10^{-6} to 2×10^{-5}	6.9×10^{-7} to 1.4×10^{-5}

In structural design, load-carrying members are sometimes loaded in torsion as well as tension or bending. The rigidity of the structure then depends on the torsional stiffness of the components. Any load applied off from the axis of a member and transverse to the axis will produce torsion. This section will discuss twisting of circular members, both solid and hollow. Noncircular sections will be covered in a later section. It is very important to note that the behavior of an open-section shape such as a channel or angle is much different from that of a closed section such as a pipe or rectangular tube. In general, the open sections have very low torsional stiffness.

Angle of Twist of a Circular Member. Consider the shaft shown in Figure 4–15. One end of the shaft, Section M, is held fixed while a torque T is applied to the other end. Under these conditions the shaft will twist between the two ends through an angle θ.

The derivation of the angle-of-twist formula depends on some basic assumptions about the behavior of a circular member when subjected to torsion. As the torque is applied, an element of length L along the outer surface of the member, which was initially straight, rotates through a small angle γ (gamma). Likewise, a radius of the member in a cross section rotates through a small angle θ. In Figure 4–15, the rotations γ and θ are both related to the arc length AB on the surface of the bar. From geometry, for small angles, the arc length is the product of the angle in radians and the distance from the center of the rotation. Therefore, the arc length AB can be expressed as either

$$AB = \gamma L$$

or

$$AB = \theta c$$

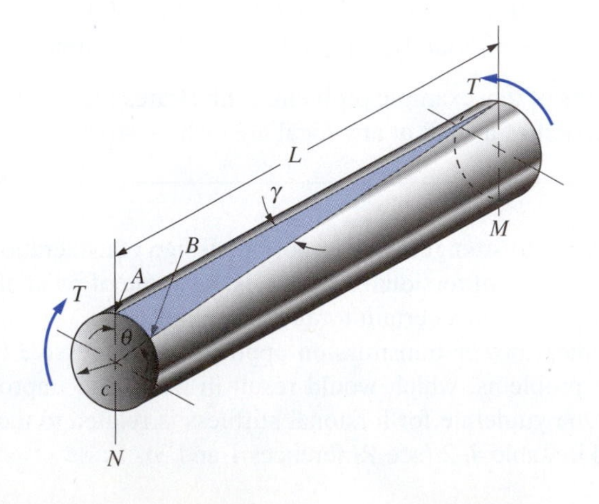

FIGURE 4–15 Torsional deformation in a circular bar.

where c is the outside radius of the bar. These two expressions for the arc length AB can be equated to each other,

$$\gamma L = \theta c$$

Solving for γ gives

$$\gamma = \frac{\theta c}{L} \tag{4–21}$$

The angle γ is a measure of the maximum shearing strain in an element on the outer surface of the bar. It was discussed in Chapter 2 that the shearing strain, γ, is related to the shearing stress, τ, by the modulus of elasticity in shear, G. That was expressed as Equation (2–5),

$$G = \frac{\tau}{\gamma} \tag{2–5}$$

At the outer surface, then,

$$\tau = G\gamma$$

But the torsional shear stress formula [Equation (4–11)] states

$$\tau = \frac{Tc}{J}$$

Equating these two expressions for τ gives

$$G\gamma = \frac{Tc}{J}$$

Now, substituting from Equation (4–21) for γ, we obtain

$$\frac{G\theta c}{L} = \frac{Tc}{J}$$

We can now cancel c and solve for θ:

Angle of Twist

$$\theta = \frac{TL}{JG} \tag{4–22}$$

The resulting angle of twist, θ, is in radians. When consistent units are used for all terms in the calculation, all units will cancel, leaving a dimensionless number. This should be interpreted as the angle, θ, in radians.

Use of Angle of Twist Formula.

- Equation (4–22) can be used to compute the angle of twist of one section of a circular bar, either solid or hollow, with respect to another section where L is the distance

TABLE 4–3 Shear modulus of elasticity, *G*.

Material	Shear modulus, G	
	GPa	psi
Plain carbon and alloy steels	80	11.5×10^6
Stainless steel type 304	69	10.0×10^6
Aluminum 6061-T6	26	3.75×10^6
Beryllium copper	48	7.0×10^6
Magnesium	17	2.4×10^6
Titanium alloy	43	6.2×10^6

between them, provided that the torque, T, the polar moment of inertia, J, and the shear modulus of elasticity, G, are the same over the entire length, L.

- If any of these factors vary in a given problem, the bar can be subdivided into segments over which they are constant to compute angles of rotation for those segments. Then the computed angles can be combined algebraically to get the total angle of twist. This principle, called *superposition,* will be illustrated in Example Problem 4–13.
- The shear modulus of elasticity, G, is a measure of the torsional stiffness of the material of the bar. Table 4–3 gives values for G for selected materials.

Example Problem 4–11 Determine the angle of twist in degrees between two sections 250 mm apart in a steel rod having a diameter of 10 mm when a torque of 15 N·m is applied. Figure 4–15 shows a sketch of the arrangement.

Solution

Objective: Compute the angle of twist in degrees.

Given: Applied torque $= T = 15$ N·m. Circular bar: diameter $= D = 10$ mm. Length $= L = 250$ mm. (Figure 4–15)

Analysis: Equation (4–22) can be used. Compute $J = \pi D^4/32$. $G = 80\text{ GPa} = 80 \times 10^9\text{ N/m}^2$ (Table 4–3).

Result: $\theta = \dfrac{TL}{JG}$

The value of J is

$$J = \frac{\pi D^4}{32} = \frac{\pi (10\text{ mm})^4}{32} = 982\text{ mm}^4$$

Then

$$\theta = \frac{TL}{JG} = \frac{(15\text{ N}\cdot\text{m})(250\text{ mm})}{(982\text{ mm}^4)(80 \times 10^9\text{ N/m}^2)} \times \frac{(10^3\text{ mm})^3}{1\text{ m}^3} = 0.048\text{ rad}$$

Note that all units cancel. Expressing the angle in degrees,

$$\theta = 0.048\text{ rad} \times \frac{180\text{ deg}}{\pi\text{ rad}} = 2.73\text{ deg}$$

Example Problem 4–12 Determine the required diameter of a round shaft made of aluminum alloy 6061-T6 if it is to twist not more than 0.08 deg in 1.0 ft of length when a torque of 75 lb·in is applied.

Solution

Objective: Compute the required diameter, D, of the round shaft.

Given: Applied torque $= T = 75$ lb·in. Length $= L = 1.0$ ft $= 12$ in.
Maximum angle of twist $= 0.08$ deg. Aluminum 6061-T6.

Analysis: Equation (4–22) can be solved for J because J is the only term involving the unknown diameter, D. Then, solve for D from $J = \pi D^4/32$
$G = 3.75 \times 10^6$ psi (Table 4–3).

Results:

$$\theta = \frac{TL}{JG}$$

$$J = \frac{TL}{\theta G}$$

The angle of twist must be expressed in radians.

$$\theta = 0.08 \text{ deg} \times \frac{\pi \text{ rad}}{180 \text{ deg}} = 0.0014 \text{ rad}$$

Then

$$J = \frac{TL}{\theta G} = \frac{(75 \text{ lb·in})(12 \text{ in})}{(0.0014)(3.75 \times 10^6 \text{ lb/in}^2)} = 0.171 \text{ in}^4$$

Now since $J = \pi D^4/32$,

$$D = \left(\frac{32J}{\pi}\right)^{1/4} = \left[\frac{(32)(0.171 \text{ in}^4)}{\pi}\right]^{1/4} = 1.15 \text{ in}$$

Comment: This is the minimum acceptable diameter. You should specify a convenient, preferred size, say 1.25 in. The resulting angle of twist will then be less than 0.08 deg over a 1.0 ft length.

Deformation in Shafts with Multiple Levels of Torque. When a single shaft carries two or more elements that apply torque to it, it is necessary to determine the level of torque in all segments of the shaft. The use of free-body diagrams of parts of the shaft is helpful for this process.

We can use the principle of equilibrium to determine the level of torque that must exist internal to the shaft at any section by imagining that the shaft is cut at a section of interest. Then the externally applied torque values on the remaining part of the shaft to either side of the cut must be balanced by the internal torque in the shaft at the cut. This technique is illustrated in Example Problem 4–13.

Let's define a notation convention for such problems. First, to report the torque applied to a shaft by an active element such as a gear or pulley, refer to that torque with a single subscript that designates the section where the torque is applied to the shaft. For example, the torque applied to a section called A will be called T_A.

To report the internal torque in the shaft between two sections of interest, refer to that torque with a double subscript that identifies the end-points of the segment for which that value applies. For example, to report the torque in the segment between sections B and C on a shaft, call it T_{BC}.

The problem also illustrates the calculation of the angle of twist for each segment of the shaft for which the values of torque and polar moment of inertia are constant. Then the net resulting angle of twist over the total length of the shaft is computed from the algebraic sum of the angles of twist in the various segments.

Stresses in Shafts with Multiple Levels of Torque. The torsional shear stress also varies with position in the shaft when either the torque in the shaft or the size of the shaft changes. Example Problem 4–14 demonstrates the method of analyzing such shafts.

PROGRAMMED EXAMPLE PROBLEM

The completion of this example problem will be shown in programmed format. You should perform each indicated operation before moving to the next panel.

Example Problem 4–13

Figure 4–16 shows a steel shaft to which three disks are attached. The shaft is fixed against rotation at its left end, but free to rotate in a bearing at its right end. Each disk is 300 mm in diameter. Downward forces act at the outer surfaces of the disks so that torques are applied to the rod. Determine the angle of twist of section A relative to the fixed section E.

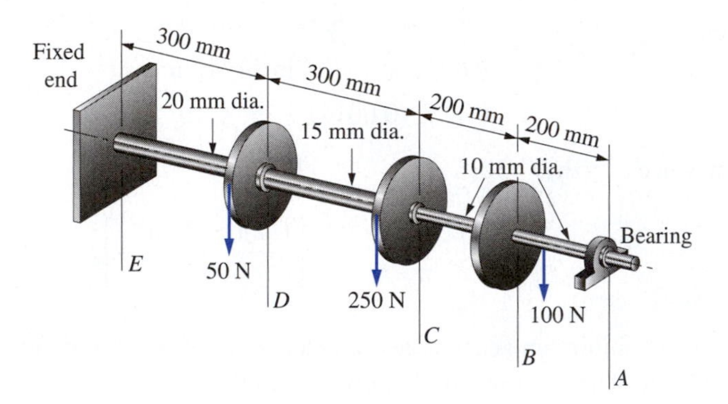

FIGURE 4–16 Shaft for Example Problem 4–13.

Solution

Objective: Compute the angle of twist of the shaft at A relative to E.

Given: Shaft is steel. $G = 80$ GPa. (Table 4–3)
Geometry of shaft and loading is shown in Figure 4–16.
For each disk, diameter $= D = 300$ mm. Radius $= R = 150$ mm.

Analysis: The shaft shown in Figure 4–16 has a total length of 1000 mm or 1.0 m. But there are four segments of the shaft having different lengths, diameters, or applied levels of torque. Therefore, Equation (4–22) must be applied to each segment separately to compute the angle of twist for each. Then, the total angle of twist in section A relative to E will be the algebraic sum of the four angles.

The first operation is to compute the magnitude and direction of the applied torques at each disk, B, C, and D. Do that now. Recall the definition of torque from Equation (4–1).

For directions, assume a view point along the rod from the right end. The magnitude of the torque on each disk is the product of the force acting on the periphery of the disk times the radius of the disk. Then, considering clockwise to be positive, torque on disk B, clockwise:

$$T_B = (100\text{ N})(150\text{ mm}) = 15\,000\text{ N}\cdot\text{mm} = 15\text{ N}\cdot\text{m (CW)}$$

torque on disk C, counterclockwise:

$$T_C = -(250\text{ N})(150\text{ mm}) = -37\,500\text{ N}\cdot\text{mm} = -37.5\text{ N}\cdot\text{m (CCW)}$$

torque on disk D, counterclockwise:

$$T_D = -(50\text{ N})(150\text{ mm}) = -7500\text{ N}\cdot\text{mm} = -7.5\text{ N}\cdot\text{m (CCW)}$$

Now determine the level of torque in each segment of the shaft. You should visualize a free-body diagram of any part of the shaft between the ends of each segment by "cutting" the shaft and computing the magnitude of the torque applied to the shaft to the right of the cut section. The internal torque in the shaft must be equal in magnitude and opposite in direction to the externally applied torque to maintain equilibrium.

It is suggested that you start at the right end at A. The bearing allows free rotation of the shaft at that end. Then move to the left and compute the torque for segments AB, BC, CD, and DE. What is the level of torque in segment AB?

For the segment AB, up to, but not including B, the torque in the shaft is zero because the bearing allows free rotation. This is illustrated in Figure 4–17 that shows the part of the shaft from section B to the end of the shaft at A. Visualize a cut in the shaft anywhere between B and A and isolate the cut section to the right as a free-body. Because there is no externally applied torque, the internal torque in the shaft everywhere between B and A must be zero. That is, $T_{AB} = 0$.

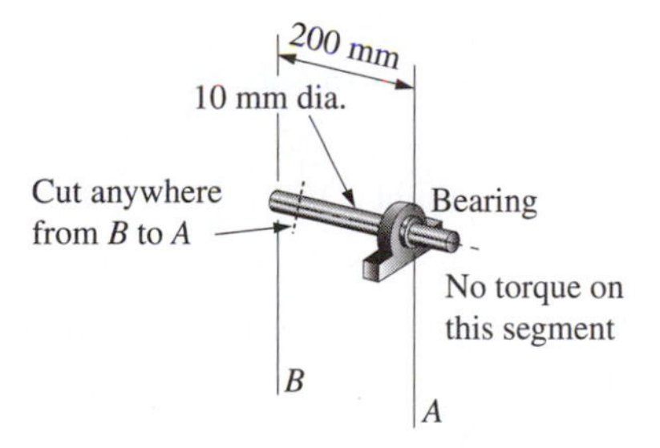

FIGURE 4–17 Free-body diagram of segment AB for the shaft in Example Problem 4–13.

Now, consider the torque applied by the disk at B and determine the torque in the segment BC.

Cutting the shaft anywhere to the right of C in the segment BC would result in an externally applied torque of 15 N·m clockwise, due to the torque on disk B. Figure 4–18 shows the free-body diagram of that part of the shaft to the right of section C after it is assumed to be cut. Only disk B applies an external torque to the shaft. The internal torque in the shaft at the cut must balance the torque on disk B.

Therefore, the torque throughout the segment BC is

$$T_{BC} = 15\text{ N}\cdot\text{m (CW)}$$

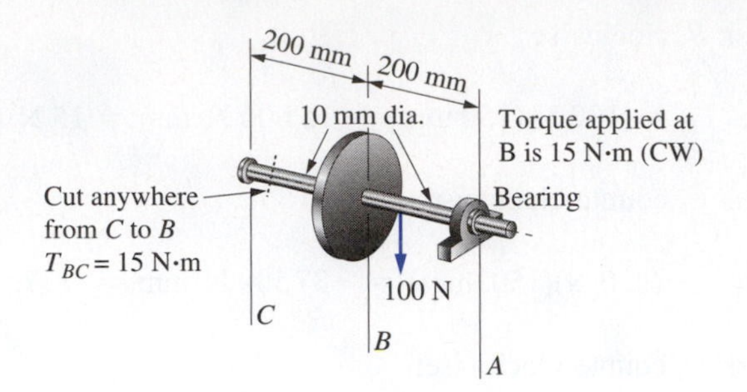

FIGURE 4–18 Free-body diagram of segment to right of section C for shaft in Example Problem 4–13.

We will consider this torque to be clockwise (CW) and positive because it tends to cause a clockwise rotation of the rod.

Now determine the torque in segment CD, called T_{CD}.

Cutting the shaft anywhere between C and D would result in both T_C and T_B acting on the shaft to the right of the cut section. See Figure 4–19. But they act in opposite sense, one clockwise and one counterclockwise. Thus the net torque applied to the shaft is the difference between them. That is,

$$T_{CD} = -T_C + T_B = -37.5 \text{ N·m} + 15 \text{ N·m} = -22.5 \text{ N·m (CCW)}$$

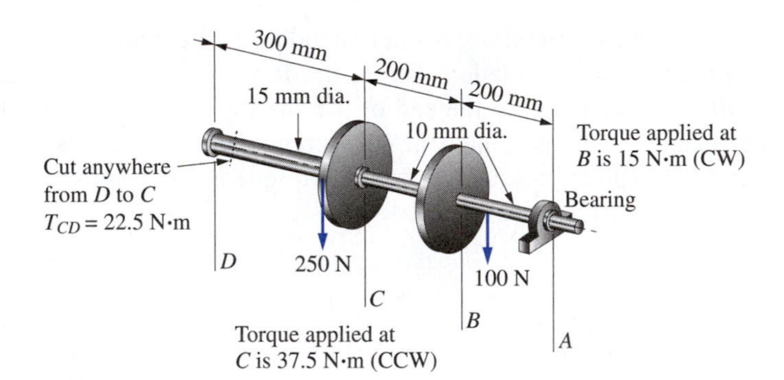

FIGURE 4–19 Free-body diagram of segment to right of section D for shaft in Example Problem 4–13.

Continue this process for the final segment, DE.

Between D and E in the shaft, the torque is the resultant of all the applied torques at D, C, and B. See Figure 4–20. Then,

$$T_{DE} = -T_D - T_C + T_B = -7.5 \text{ N·m} - 37.5 \text{ N·m} + 15 \text{ N·m} = -30 \text{ N·m (CCW)}$$

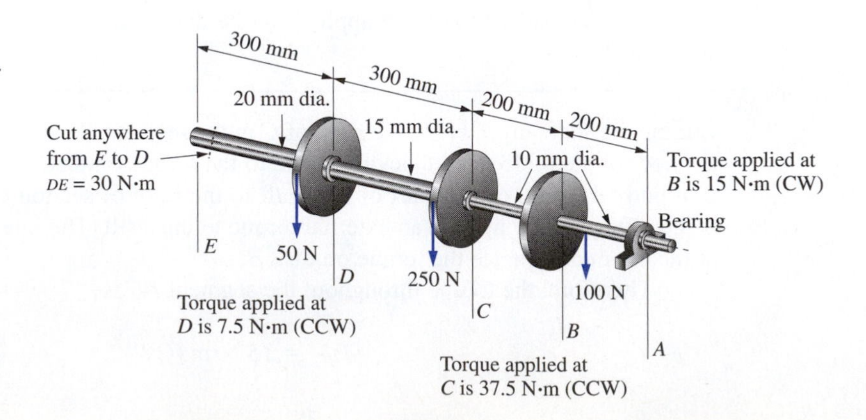

FIGURE 4–20 Free-body diagram for segment to right of section E for shaft in Example Problem 4–13.

The fixed support at E must be capable of providing a reaction torque of 30 N·m to maintain the shaft in equilibrium.

In summary, the distribution of torque in the shaft can be shown in graphical form as in Figure 4–21. Notice that the applied torques T_B, T_C, and T_D are the *changes* in torque that occur at B, C, and D but that they are not necessarily the magnitudes of the torque *in the shaft* at those points.

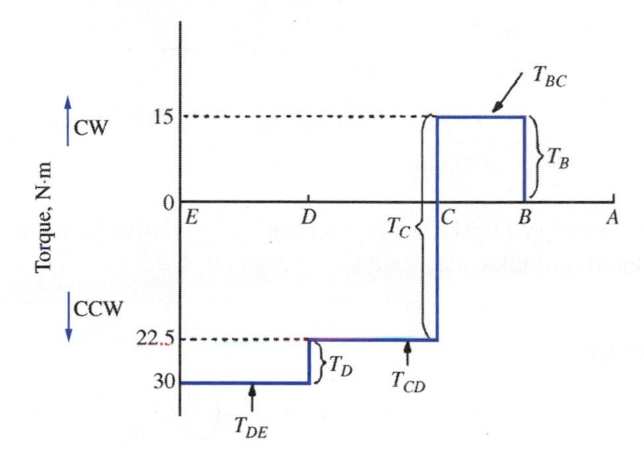

FIGURE 4–21 Torque distribution in shaft for Example Problem 4–13.

Now, compute the angle of twist in each segment by applying Equation (4–22) successively. Start with segment AB.

Segment AB

$$\theta_{AB} = T_{AB}\left(\frac{L}{JG}\right)_{AB}$$

Since $T_{AB} = 0$, $\theta_{AB} = 0$. There is no twisting of the shaft between A and B.

Now continue with segment BC.

Segment BC

$$\theta_{BC} = T_{BC}\left(\frac{L}{JG}\right)_{BC}$$

We know $T_{BC} = 15$ N·m, $L = 200$ mm, and $G = 80$ GPa for steel. For the 10-mm-diameter shaft,

$$J = \frac{\pi D^4}{32} = \frac{\pi(10\text{ mm})^4}{32} = 982\text{ mm}^4$$

Then

$$\theta_{BC} = \frac{(15\text{N}\cdot\text{m})(200\text{ mm})}{(982\text{ mm}^4)(80 \times 10^9\text{N/m}^2)} \times \frac{(10^3)^3\text{ mm}^3}{\text{m}^3} = 0.038\text{ rad}$$

This means that section B is rotated 0.038 rad clockwise relative to section C, since θ_{BC} is the total angle of twist in the segment BC.

Continue with segment CD.

Segment CD

$$\theta_{CD} = T_{CD}\left(\frac{L}{JG}\right)_{CD}$$

Here $T_{CD} = -22.5$ N·m, $L = 300$ mm, and the shaft diameter is 15 mm. Then

$$J = \frac{\pi D^4}{32} = \frac{\pi(15\text{ mm})^4}{32} = 4970\text{ mm}^4$$

$$\theta_{CD} = \frac{-(22.5\text{ N}\cdot\text{m})(300\text{ mm})}{(4970\text{ mm}^4)(80 \times 10^9\text{ N/m}^2)} \times \frac{(10^3)^3\text{ mm}^3}{\text{m}^3} = -0.017\text{ rad}$$

Section C is rotated 0.017 rad counterclockwise relative to section D.
Finally, complete the analysis for segment DE.

Segment DE

$$\theta_{DE} = T_{DE}\left(\frac{L}{JG}\right)_{DE}$$

Here $T_{DE} = -30$ N·m, $L = 300$ mm, and $D = 20$ mm. Then

$$J = \frac{\pi D^4}{32} = \frac{\pi(20\text{ mm})^4}{32} = 15\,700\text{ mm}^4$$

$$\theta_{DE} = \frac{-(30\text{ N}\cdot\text{m})(300\text{ mm})}{(15\,700\text{ mm}^4)(80 \times 10^9\text{ N/mm}^2)} \times \frac{(10^3)^3\text{ mm}^3}{\text{m}^3} = -0.007\text{ rad}$$

Section D is rotated 0.007 rad counterclockwise relative to E. The final operation is to compute the total angle of twist from E to A by summing the angles of twist for all segments algebraically. Do that now.

Total angle of twist from E to A

$$\begin{aligned}\theta_{AE} &= \theta_{AB} + \theta_{BC} - \theta_{CD} - \theta_{DE} \\ &= 0 + 0.038 - 0.017 - 0.007 = 0.014\text{ rad}\end{aligned}$$

Summary and Comment It should help your visualization of what is happening in the shaft throughout its length by plotting a graph of the angle of twist as a function of position. This is done in Figure 4–22 by setting the zero reference point as E. The straight line from E to D shows the linear change in angle with position to the value of -0.007 rad (counterclockwise). From there, the angle grows by an additional -0.017 rad between D and C. In segment BC, the relative rotation is clockwise with a magnitude of 0.038 rad, ending at the final value of 0.014 rad at B. And, because there is no torque in the segment AB, the angle of rotation remains at that value.

Example Problem 4–14 Compute the torsional shear stress that occurs in each segment of the shaft shown in Figure 4–16 and described in Example Problem 4–13.

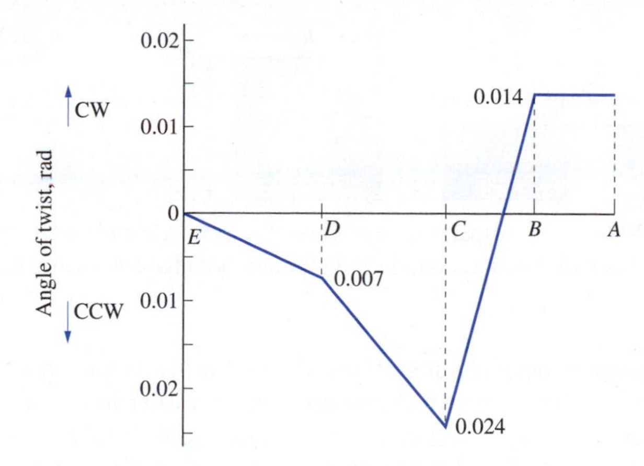

FIGURE 4–22 Angle of twist versus position on the shaft for Example Problem 4–13.

Solution Objective Compute the torsional shear stress.

Given Shaft in Figure 4–16 having multiple levels of torque and different diameters.

Analysis The torsional shear stress equation is used.

$$\tau = Tc/J$$

We will apply this equation over each segment where all factors are the same. The shaft will be broken into segments *AB*, *BC*, *CD*, and *DE*. Some data are taken from the results of Example Problem 4–13.

Results ***Segment AB:*** No torque is transmitted in this segment. Therefore,

$$\tau_{AB} = 0$$

Segment BC: $T_{BC} = 15$ N·m. $D = 10$ mm. $c = 5$ mm. $J_{BC} = 982\ \text{mm}^4$. Then

$$\tau_{BC} = \frac{T_{BC}\,c}{J_{BC}} = \frac{(15\ \text{N}\cdot\text{m})(5\ \text{mm})}{982\ \text{mm}^4}\,\frac{1000\ \text{mm}}{\text{m}} = 76.4\ \text{N/mm}^2 = 76.4\ \text{MPa}$$

Segment CD: $T_{CD} = 22.5$ N·m. $D = 15$ mm. $c = 7.5$ mm. $J_{CD} = 4970\ \text{mm}^4$. Then

$$\tau_{CD} = \frac{T_{CD}c}{J_{CD}} = \frac{(22.5\ \text{N}\cdot\text{m})(7.5\ \text{mm})}{4970\ \text{mm}^4}\,\frac{1000\ \text{mm}}{\text{m}} = 34.0\ \text{N/mm}^2 = 34.0\ \text{MPa}$$

Segment DE: $T_{DE} = 30$ N·m. $D = 20$ mm. $c = 10$ mm. $J_{DE} = 15\,700\ \text{mm}^4$. Then

$$\tau_{DE} = \frac{T_{DE}\,c}{J_{DE}} = \frac{(30\ \text{N}\cdot\text{m})(10\ \text{mm})}{15\,700\ \text{mm}^4}\,\frac{1000\ \text{mm}}{\text{m}} = 19.1\ \text{N/mm}^2 = 19.1\ \text{MPa}$$

LESSON 9

Comment A summary of the stress distribution in the shaft follows.

$$\tau_{AB} = 0 \qquad \tau_{BC} = 76.4\ \text{MPa} \qquad \tau_{CD} = 34.0\ \text{MPa} \qquad \tau_{DE} = 19.1\ \text{MPa}$$

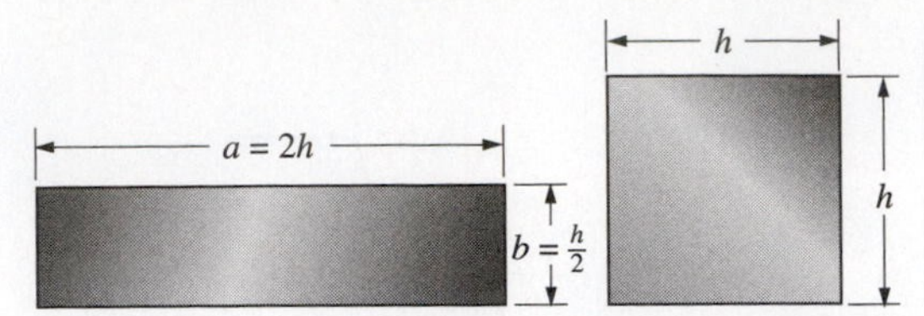

FIGURE 4–23 Comparison of stiffness for rectangle and square sections in torsion. Square is two times stiffer than rectangle even though both have the same area: $ab = h^2$.

4–11 TORSION IN NONCIRCULAR SECTIONS

The behavior of noncircular sections when subjected to torsion is vastly different from that of circular sections, for which the discussions earlier in this chapter applied. There is a large variety of shapes that can be imagined, and the analysis of stiffness and strength is different for each. The development of the relationships involved will not be done here. Compilations of the pertinent formulas occur in References 1 to 5, and a few are given in this section.

Some generalizations can be made. Solid sections having the same cross-sectional area are stiffer when their shape more closely approaches a circle (see Figure 4–23). Conversely, a member made up of long, thin sections that do not form a closed, tube-like shape are very weak and flexible in torsion. Examples of flexible sections are common structural shapes such as wide-flange beams, standard I-beams, channels, angles, and tees, as illustrated in Figure 4–24. Pipes, solid bars, and structural rectangular tubes have high rigidity, or stiffness (see Figure 4–25).

An interesting illustration of the lack of stiffness of open, thin sections is shown in Figure 4–26. The thin plate (a), the angle (b), and the channel (c) have the same thickness and cross-sectional area, and all have nearly the same torsional stiffness. Likewise, if the thin plate were formed into a circular shape (d), but with a slit remaining, its stiffness would remain low. However, closing the tube completely as in Figure 4–25(a) by welding or by drawing a seamless tube would produce a relatively stiff member. Understanding these comparisons is an aid to selecting a reasonable shape for members loaded in torsion.

Figure 4–27 shows torsional properties for seven cases of noncircular cross sections that are commonly encountered in machine design and structural analysis. We will use the same equations for the maximum torsional shear stress, τ_{max} [Equation (4–16)], and torsional deformation, δ [Equation (4–22)].

$$\tau_{max} = \frac{T}{Z_p} \tag{4–16}$$

$$\theta = \frac{TL}{GJ} \tag{4–22}$$

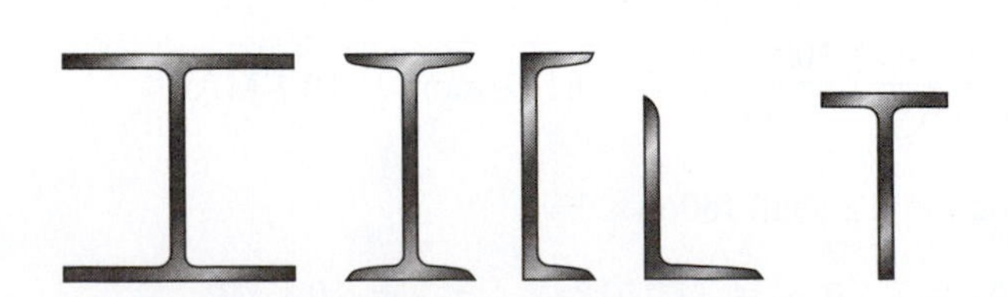

FIGURE 4–24 Torsionally flexible sections.

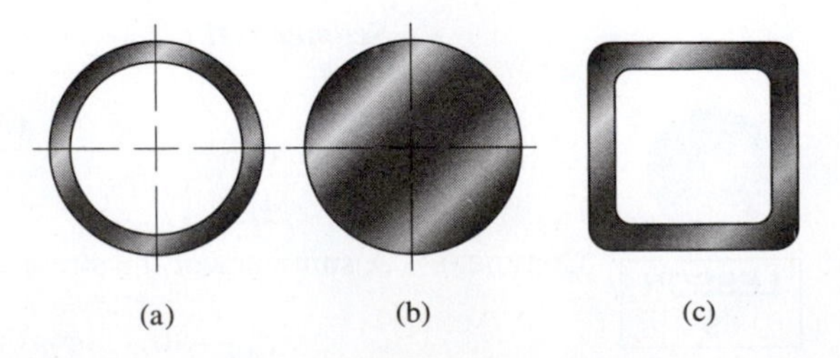

FIGURE 4–25 Torsionally stiff sections.

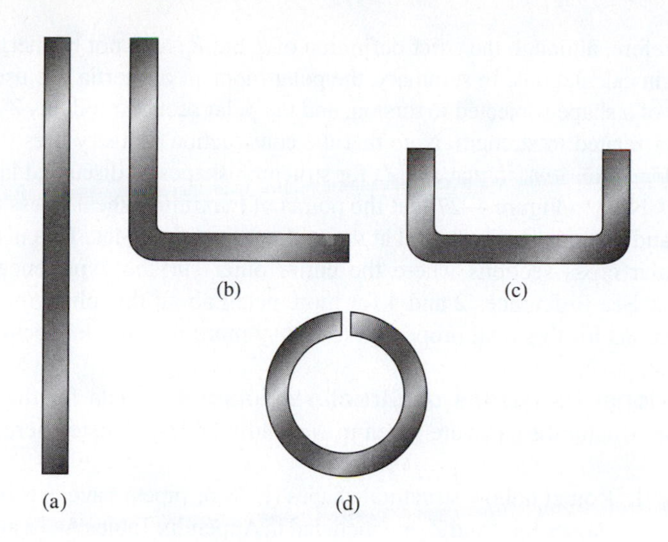

FIGURE 4–26 Sections having nearly equal (and low) torsional stiffness.

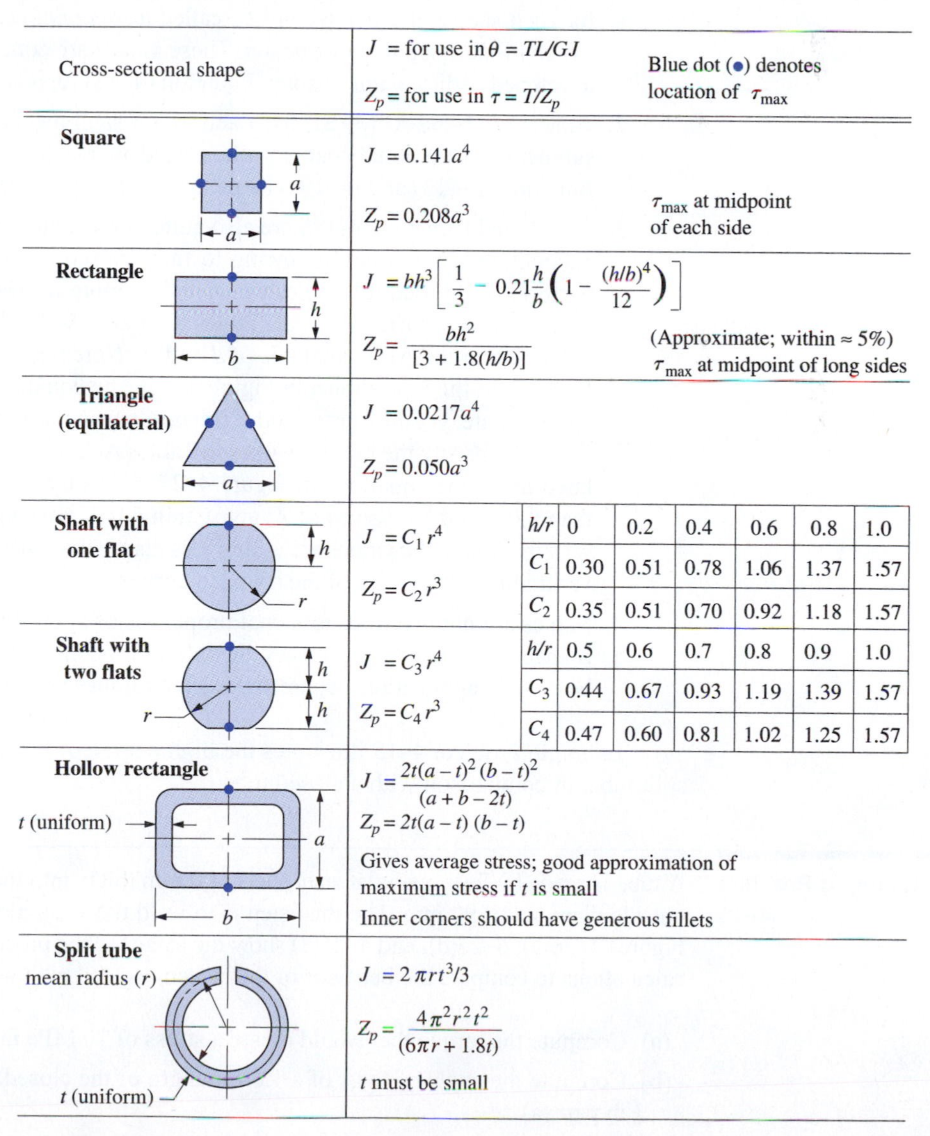

Cross-sectional shape	J = for use in $\theta = TL/GJ$; Z_p = for use in $\tau = T/Z_p$	Blue dot (•) denotes location of τ_{max}
Square	$J = 0.141a^4$; $Z_p = 0.208a^3$	τ_{max} at midpoint of each side
Rectangle	$J = bh^3\left[\frac{1}{3} - 0.21\frac{h}{b}\left(1 - \frac{(h/b)^4}{12}\right)\right]$; $Z_p = \frac{bh^2}{[3 + 1.8(h/b)]}$	(Approximate; within ≈ 5%) τ_{max} at midpoint of long sides
Triangle (equilateral)	$J = 0.0217a^4$; $Z_p = 0.050a^3$	
Shaft with one flat	$J = C_1 r^4$; $Z_p = C_2 r^3$	see table below
Shaft with two flats	$J = C_3 r^4$; $Z_p = C_4 r^3$	see table below
Hollow rectangle	$J = \frac{2t(a-t)^2(b-t)^2}{(a+b-2t)}$; $Z_p = 2t(a-t)(b-t)$	Gives average stress; good approximation of maximum stress if t is small. Inner corners should have generous fillets
Split tube	$J = 2\pi r t^3/3$; $Z_p = \frac{4\pi^2 r^2 t^2}{(6\pi r + 1.8t)}$	t must be small

Shaft with one flat:

h/r	0	0.2	0.4	0.6	0.8	1.0
C_1	0.30	0.51	0.78	1.06	1.37	1.57
C_2	0.35	0.51	0.70	0.92	1.18	1.57

Shaft with two flats:

h/r	0.5	0.6	0.7	0.8	0.9	1.0
C_3	0.44	0.67	0.93	1.19	1.39	1.57
C_4	0.47	0.60	0.81	1.02	1.25	1.57

FIGURE 4–27 Methods for determining values for J and Z_p for several types of cross sections. (Source: *Machine Elements in Mechanical Design,* 4th ed., Robert L. Mott, copyright © Prentice Hall, Upper Saddle River, N.J., 2004. Reprinted by permission of the publisher)

Therefore, although the strict definition of J and Z_p may not be met, we will use those symbols in calculations. In summary, the polar moment of inertia J is used to represent the stiffness of a shape subjected to torsion, and the polar section modulus Z_p is used to represent the factor related to strength. Note that the construction industry uses the symbol C, called the *torsional constant,* in place of Z_p for structural shapes as discussed later in this chapter.

Note in Figure 4–27 that the points of maximum shear stress are indicated by a large dot and that they typically fall at the midpoints of flat sides. This is radically different from circular cross sections where the entire outer surface experiences the maximum shear stress. See References 2 and 4 for more detail about this phenomenon. Reference 5 gives equations for torsional properties for many more noncircular sections.

Torsional Properties of Structural Shapes. Data for the torsional properties of some structural shapes are given in Appendix tables, as listed here:

1. Round hollow structural shapes (HSS or pipes) have excellent torsional resistance. Values for J and Z_p are included in Appendix Tables A–12 and A–12 (SI). AISC data for such shapes use the symbol C, called a *torsional constant,* in place of Z_p for structural design and construction. These values are computed using the equations developed earlier in this chapter, Equation (4–13) for J and (4–19) for Z_p or C.
2. Tables A–13 and A–13 (SI) list J and Z_p for steel mechanical tubing that is often subjected to torsional loading when used in machines and vehicle structures. Equations (4–13) and (4–19) can be used to compute these values as well.
3. Square and rectangular HSS are also quite good for use where torsional loading is encountered in a structure, owing to their closed cross sections. Square shapes perform better than rectangles because they more nearly simulate the round HSS shapes that are virtually ideal. Tables A–9 and A–9 (SI) give torsional property values using the AISC symbols of J and C. Note that the square and rectangular HSS are of the type shown in Figure 4–27. The equations for J and Z_p (or C) are approximate, yet they give good estimates, particularly when the wall thickness is thin. Comparing the tabular values in Tables A–9 and A–9 (SI) with values calculated using the equations in Figure 4–27 shows that values for J are within less than 3.0% and the values of Z_p are within 4.0%. For most of the shapes with $\frac{1}{4}$ in wall thickness, the values are within less than 1.0%. Differences may be caused by variations in the shape of the rounded corners.
4. Values are not given for torsional properties of open shapes such as L, C, W, and S shapes because they are very flexible in torsion, and it is advised that they not be used in applications where significant torques are encountered.

Example Problem 4–15 illustrates the high degree of flexibility of an open shape, a split tube, in comparison with a closed tube.

Example Problem 4–15

A tube is made by forming a flat steel sheet, 4.0 mm thick, into the circular shape having an outside diameter of 90 mm. The final step is to weld the seam along the length of the tube. Figures 4–26(a), 4–26(d), and 4–25(a) show the stages of the process. Perform the following calculations to compare the behavior of the closed, welded tube with that of the open tube.

(a) Compute the torque that would create a stress of 10 MPa in the closed, welded tube.

(b) Compute the angle of twist of a 1.0-m length of the closed tube for the torque found in part (a).

(c) Compute the stress in the open tube for the torque found in part (a).

(d) Compute the angle of twist of a 1.0-m length of the open tube for the torque found in part (a).

(e) Compare the stress and deflection of the open tube with those of the closed tube.

Solution *The solution will be shown in a programmed format with each part of the solution, (a)–(e), given as a separate panel. Each part can be approached as a different problem with* Objective, Given, Analysis, *and* Results *sections.*

Complete part (a) now.

Objective Compute the torque on the closed tube that would create a torsional shear stress of 10 MPa.

Given Tube is steel. $D_o = 90$ mm. Wall thickness $= t = 4.0$ mm.
$D_i = D_o - 2t = 90\text{ mm} - 2(4.0\text{ mm}) = 82$ mm.

Analysis Use the maximum shear stress Equation (4–11) and solve for T.

Results

$$\tau_{max} = \frac{Tc}{J} \tag{4–11}$$

Then

$$T = \frac{\tau_{max}J}{c}$$

We can compute J from Equation (4–13),

$$J = \frac{\pi}{32}(D_o^4 - D_i^4) \tag{4–13}$$

Using $D_o = 90\text{ mm} = 0.09$ m and $D_i = 82\text{ mm} = 0.082$ m,

$$J = \frac{\pi}{32}(0.09^4 - 0.082^4)\text{m}^4 = 2.00 \times 10^{-6}\text{ m}^4$$

Now, letting $\tau_{max} = 10\text{ MPa} = 10 \times 10^6\text{ N/m}^2$, we have

$$T = \frac{\tau_{max}J}{c} = \frac{(10 \times 10^6\text{ N/m}^2)(2.00 \times 10^{-6}\text{ m}^4)}{0.045\text{ m}} = 444\text{ N}\cdot\text{m}$$

That is, a torque of 444 N·m applied to the closed welded tube would produce a maximum torsional shear stress of 10 MPa in the tube. Note that this is a very low stress level for steel.

Complete part (b) of the problem.

Objective For the closed tube used in (a), compute the angle of twist.

Given $J = 2.00 \times 10^{-6}\text{ m}^4$. Length $= L = 1.0$ m. Torque $= T = 444$ N·m.
$G = 80$ GPa.

Analysis Use Equation (4–22) to compute the angle of twist, θ.

Results

$$\theta = \frac{TL}{GJ} = \frac{(444\text{ N}\cdot\text{m})(1.0\text{ m})}{(80 \times 10^9\text{ N/m}^2)(2.00 \times 10^{-6}\text{ m}^4)} = 0.00278\text{ rad}$$

Converting θ to degrees gives

$$\theta = 0.00278\text{ rad}\frac{180\text{ deg}}{\pi\text{ rad}} = 0.159\text{ deg}$$

Again, note that this is a very small angle of twist.

Consider a tube shaped as in parts (a) and (b) except that it is not closed, as shown in Figure 4–26(d). Compute the torsional shear stress in the open tube due to the torque of 444 N·m.

Objective Compute the shear stress in the open tube.

Given Torque = T = 444 N·m. D_o = 90 mm. Wall thickness = t = 4.0 mm.

Analysis Equation (4–16) can be used to compute the maximum shear stress for the open tube before it is welded, treating it as a noncircular cross section. The formula for Z_p is given in Figure 4–27.

$$Z_p = \frac{4\pi^2 r^2 t^2}{6\pi r + 1.8t}$$

Results The mean radius is

$$r = \frac{D_o}{2} - \frac{t}{2} = \frac{90 \text{ mm}}{2} - \frac{4 \text{ mm}}{2} = 43 \text{ mm}$$

Then

$$Z_p = \frac{4\pi^2(43)^2(4)^2}{6\pi(43) + 1.8(4)} \text{mm}^3 = 1428 \text{ mm}^3$$

The stress in the open tube is, then,

$$\tau_{max} = \frac{T}{Z_p} = \frac{444 \text{ N}\cdot\text{m}}{1428 \text{ mm}^3} \frac{1000 \text{ mm}}{\text{m}} = 311 \text{ MPa}$$

Now, complete part (d) by computing the angle of twist of the open tube.

Objective Compute the angle of twist of the open tube for T = 444 N·m.

Given Length = L = 1.0 m. Torque = T = 444 N·m. G = 80 GPa.
D_o = 90 mm. Wall thickness = t = 4.0 mm.

Analysis The angle of twist for the open tube can be computed using Equation (4–22). Figure 4–27 gives us the formula for the torsional rigidity constant, J.

Results Using r = 43 mm = 0.043 m and t = 4 mm = 0.004 m yields

$$J = \frac{2\pi r t^3}{3} = \frac{2\pi(0.043)(0.004)^3}{3} = 5.764 \times 10^{-9} \text{ m}^4$$

Then the angle of twist is

$$\theta = \frac{TL}{GJ} = \frac{(444 \text{ N}\cdot\text{m})(1.0 \text{ m})}{(80 \times 10^9 \text{ N/m}^2)(5.764 \times 10^{-9} m^4)} = 0.963 \text{ rad}$$

Converting θ to degrees, we obtain

$$\theta = 0.963 \text{ rad} \frac{180 \text{ deg}}{\pi \text{ rad}} = 55.2 \text{ deg}$$

Now complete the final part (e), comparing the open and closed tubes.

Objective Compare the stress and deflection of the open tube with those of the closed tube.

Given From parts (a), (b), (c), and (d):

Stress in the closed tube $= \tau_c = 10$ MPa.
Angle of twist of the closed tube $= \theta_c = 0.159$ deg.
Stress in the open tube $= \tau_o = 311$ MPa.
Angle of twist of the open tube $= \theta_o = 55.2$ deg.

Analysis Compute the ratio of stresses τ_o/τ_c and the ratio of the angles of twist θ_o/θ_c.

Results Ratio of stresses:

$$\frac{\tau_o}{\tau_c} = \frac{311 \text{ MPa}}{10 \text{ MPa}} = 31.1 \text{ times greater}$$

Ratio of angle of twist:

$$\frac{\theta_o}{\theta_c} = \frac{55.2 \text{ deg}}{0.159 \text{ deg}} = 347 \text{ times greater}$$

Comment These comparisons dramatically show the advantage of using closed sections to carry torsion.

In addition, the actual stress level computed for the open tube is probably greater than the allowable stress for many steels. Recall that the design shear stress is

$$\tau_d = \frac{0.5 s_y}{N}$$

Then the required yield strength of the material to be safe at a stress level of 311 MPa and a design factor of 2.0 is

$$s_y = \frac{N\tau_d}{0.5} = \frac{2(311 \text{ MPa})}{0.5} = 1244 \text{ MPa}$$

Only a few highly heat-treated steels listed in Appendix A–14 have this level of yield strength.

REFERENCES

1. Blodgett, O. W., *Design of Weldments,* James F. Lincoln Arc Welding Foundation, Cleveland, OH, 1963.
2. Boresi, A. P. and R. J. Schmidt, *Advanced Mechanics of Materials,* 6th ed., John Wiley, New York, 2002.
3. Mott, R. L., *Machine Elements in Mechanical Design,* 4th ed., Prentice Hall, Upper Saddle River, NJ, 2004.
4. Young, W. C. and R. D. Cook, *Advanced Mechanics of Materials,* 2nd ed., Prentice Hall, Upper Saddle River, NJ, 1999.
5. Young, W. C. and R. G. Budynas, *Roark's Formulas for Stress and Strain,* 7th ed., McGraw-Hill, New York, 2002.

INTERNET SITES

1. Driv-Lok, Inc. www.driv-lok.com Manufacturer of a wide variety of parallel keys and press-fit fastening devices such as grooved pins, dowel pins, spring pins, and studs. Includes data for key size and materials available for keys.
2. Truarc Company, LLC. www.truarc.com Manufacturer of a wide variety of retaining rings for industrial, commercial, military, and consumer products. Online catalog data give groove dimensions and details for a large range of sizes. Model 5100 Basic rings are suggested for shaft applications.
3. Keystone Manufacturing www.trukey.com Manufacturer of keys and shafting with premilled keyways. Includes tables of key sizes for different sizes of shafts and material properties for shafting.
4. Sensors Portal www.sensorsportal.com/HTML/SENSORS/TorqueSens_Manuf.htm The Sensors Portal site indentifies numerous manufacturers of sensors for many kinds of measurement, including torque measurement.
5. S&T e-Digest http://www.sensorsportal.com/HTML/DIGEST/march_06/P_59.pdf An electronic version of *Sensors & Transducers Magazine.* The listed page is an article called "Evolution and Future of Torque Measurement Technology," by Dr. W. Krimmel, published March 27, 2006.

PROBLEMS

4–1.M Compute the torsional shear stress that would be produced in a solid circular shaft having a diameter of 20 mm when subjected to a torque of 280 N·m.

4–2.M A hollow shaft has an outside diameter of 35 mm and an inside diameter of 25 mm. Compute the torsional shear stress in the shaft when it is subjected to a torque of 560 N·m.

4–3.E Compute the torsional shear stress in a shaft having a diameter of 1.25 in when carrying a torque of 1550 lb·in.

4–4.E A steel tube is used as a shaft carrying 5500 lb·in of torque. The outside diameter is 1.75 in, and the wall thickness is $\frac{1}{8}$ in. Compute the torsional shear stress at the outside and the inside surfaces of the tube.

4–5.M A movie projector drive mechanism is driven by a 0.08-kW motor whose shaft rotates at 180 rad/s. Compute the torsional shear stress in its 3.0-mm-diameter shaft.

4–6.M The impeller of a fluid agitator rotates at 42 rad/s and requires 35 kW of power. Compute the torsional shear stress in the shaft that drives the impeller if it is hollow and has an outside diameter of 40 mm and an inside diameter of 25 mm.

4–7.E A driveshaft for a milling machine transmits 15.0 hp at a speed of 240 rpm. Compute the torsional shear stress in the shaft if it is solid with a diameter of 1.44 in. Would the shaft be safe if the torque is applied with shock and it is made from AISI 4140 OQT 1300 steel?

4–8.E Repeat Problem 4–7 if the shaft contains a profile keyseat.

4–9.E Figure P4–9 shows the end of the vertical shaft for a rotary lawn mower. Compute the maximum torsional shear stress in the shaft if it is transmitting 7.5 hp to the blade when rotating 2200 rpm. Specify a suitable steel for the shaft.

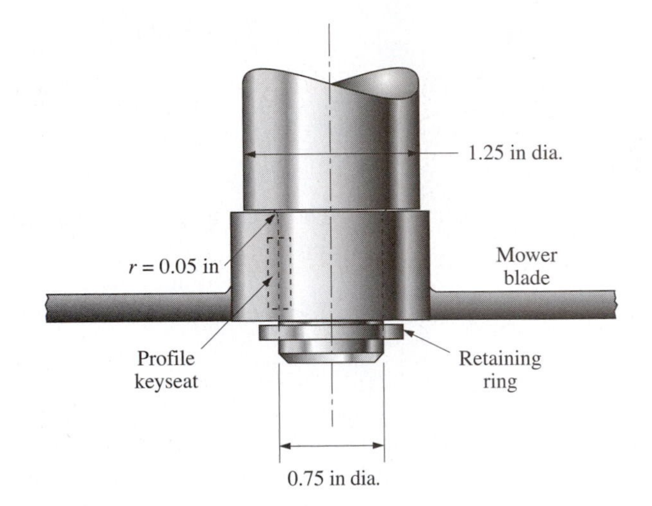

FIGURE P4–9 Shaft for Problem 4–9.

4–10.E Figure P4–10 shows a stepped shaft in torsion. The larger section also has a hole drilled through.

(a) Compute the maximum shear stress at the step for an applied torque of 7500 lb·in.

(b) Determine the largest hole that could be drilled in the shaft and still maintain the stress near the hole at or below that at the step.

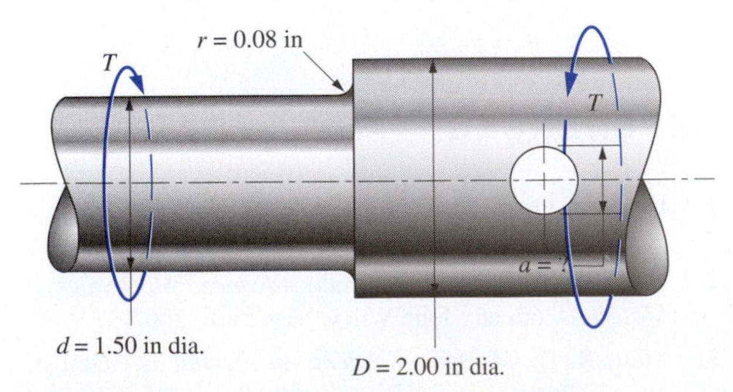

FIGURE P4–10 Shaft for Problem 4–10.

4–11.M Compute the torsional shear stress and the angle of twist in degrees in an aluminum tube, 600 mm long, having an inside diameter of 60 mm and an outside diameter of 80 mm when subjected to a steady torque of 4500 N·m. Then specify a suitable aluminum alloy for the tube.

4–12.M Two designs for a shaft are being considered. Both have an outside diameter of 50 mm and are 600 mm long. One is solid but the other is hollow with an inside diameter of 40 mm. Both are made from steel. Compare the torsional shear stress, the angle of twist, and the mass of the two designs if they are subjected to a torque of 850 N·m.

4–13.M Determine the required inside and outside diameters for a hollow shaft to carry a torque of 1200 N·m with a maximum torsional shear stress of 45 MPa. Make the ratio of the outside diameter to the inside diameter approximately 1.25.

4–14.E A gear driveshaft for a milling machine transmits 7.5 hp at a speed of 240 rpm. Compute the torsional shear stress in the 0.860-in-diameter solid shaft.

4–15.E The input shaft for the gear drive described in Problem 4–14 also transmits 7.5 hp, but rotates at 1140 rpm. Determine the required diameter of the input shaft to give it the same stress as the output shaft.

4–16.E Determine the stress that would result in a $1\frac{1}{2}$-in schedule 40 steel pipe if a plumber applies a force of 80 lb at the end of a wrench handle 18 in long.

4–17.E A rotating sign makes 1 rev every 5 s. In a high wind, a torque of 30 lb·ft is required to maintain rotation. Compute the power required to drive the sign. Also compute the stress in the final driveshaft if it has a diameter of 0.60 in. Specify a suitable steel for the shaft to provide a design factor of 4 based on yield strength in shear.

4–18.M A short, cylindrical bar is welded to a rigid plate at one end, and then a torque is applied at the other. If the bar has a diameter of 15 mm and is made of AISI 1020 cold-drawn steel, compute the torque that must be applied to it to subject it to a stress equal to its yield strength in shear. Use $s_{ys} = s_y/2$.

4–19.E A propeller driveshaft on a ship is to transmit 2500 hp at 75 rpm. It is to be made of AISI 1040 WQT 1300 steel. Use a design factor of 6 based on the yield strength in shear. The shaft is to be hollow, with the inside diameter equal to 0.80 times the outside diameter. Determine the required diameter of the shaft.

4–20.E If the propeller shaft of Problem 4–19 was to be solid instead of hollow, determine the required diameter. Then compute the ratio of the weight of the solid shaft to that of the hollow shaft.

4–21.M A power screwdriver uses a shaft with a diameter of 5.0 mm. What torque can be applied to the screwdriver if the limiting stress due to torsion is 80 MPa?

4–22.M An extension for a socket wrench similar to that shown in Figure 4–4 has a diameter of 6.0 mm and a length of 250 mm. Compute the stress and angle of twist in the extension when a torque of 5.5 N·m is applied. The extension is steel.

4–23.M Compute the angle of twist in a steel shaft 15 mm in diameter and 250 mm long when a torque of 240 N·m is applied.

4–24.M Compute the angle of twist in an aluminum tube that has an outside diameter of 80 mm and an inside diameter of 60 mm when subjected to a torque of 2250 N·m. The tube is 1200 mm long.

4–25.E A steel rod with a length of 8.0 ft and a diameter of 0.625 in is used as a long wrench to unscrew a plug at the bottom of a pool of water. If it requires 40 lb·ft of torque to loosen the plug, compute the angle of twist of the rod.

4–26.E For the rod described in Problem 4–25, what must the diameter be if only 2.0 deg of twist is desired when 40 lb·ft of torque is applied?

4–27.M Compute the angle of twist of the free end relative to the fixed end of the steel bar shown in Figure P4–27.

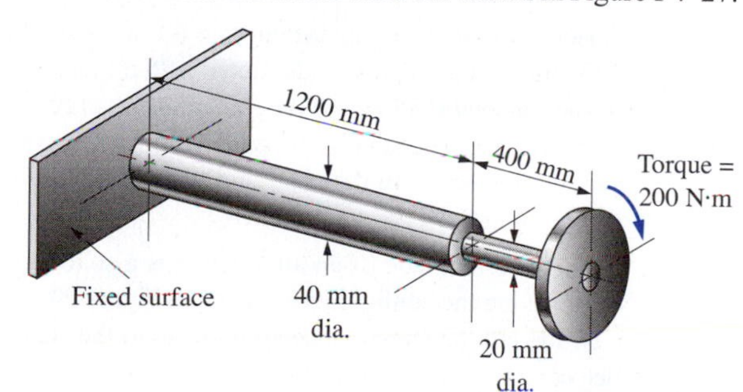

FIGURE P4–27 Bar for Problem 4–27.

4–28.M A meter for measuring torque uses the angle of twist of a shaft to indicate torque. The shaft is to be 150 mm long and made of 6061-T6 aluminum alloy. Determine the required diameter of the shaft if it is desired to have a twist of 10.0 deg when a torque of 5.0 N·m is applied to the meter. For the shaft thus designed, compute the torsional shear stress and then compute the resulting design factor for the shaft. Is it satisfactory? If not, what would you do?

4–29.M A beryllium copper wire having a diameter of 1.50 mm and a length of 40 mm is used as a small torsion bar in an instrument. Determine what angle of twist would result in the wire when it is stressed to 250 MPa.

4–30.M A fuel line in an aircraft is made of a titanium alloy. The tubular line has an outside diameter of 18 mm and an inside diameter of 16 mm. Compute the stress in the tube if a length of 1.65 m must be twisted through an angle of 40 deg during installation. Determine the design factor based on the yield strength in shear if the tube is Ti-6Al-4V, aged.

4–31.M For the shaft shown in Figure P4–31, compute the angle of twist of pulleys *B* and *C* relative to *A*. The steel shaft has a diameter of 35 mm throughout its length. The torques are $T_1 = 1500$ N·m, $T_2 = 1000$ N·m, $T_3 = 500$ N·m. The lengths are $L_1 = 500$ mm, $L_2 = 800$ mm.

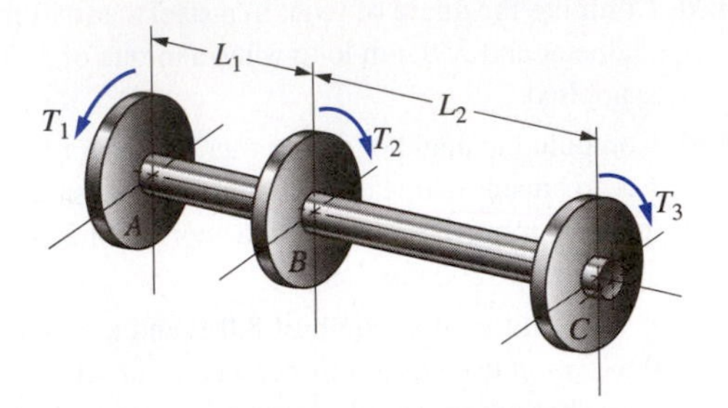

FIGURE P4–31 Shaft for Problem 4–31.

4–32.M A torsion bar in a light truck suspension is to be 820 mm long and made of steel. It is subjected to a torque of 1360 N·m and must be limited to 2.2 deg of twist. Determine the required diameter of the solid round bar. Then compute the stress in the bar.

4–33.M A steel driveshaft for an automobile is a hollow tube 1525 mm long. Its outside diameter is 75 mm, and its inside diameter is 55 mm. If the shaft transmits 120 kW of power at a speed of 225 rad/s, compute the torsional shear stress in the shaft and the angle of twist of one end relative to the other.

4–34.M A part of a rear axle of an automobile is a solid steel shaft having the configuration shown in Figure P4–34. Considering the stress concentration due to the shoulder, compute the torsional shear stress in the axle when it rotates at 70.0 rad/s, transmitting 60 kW of power.

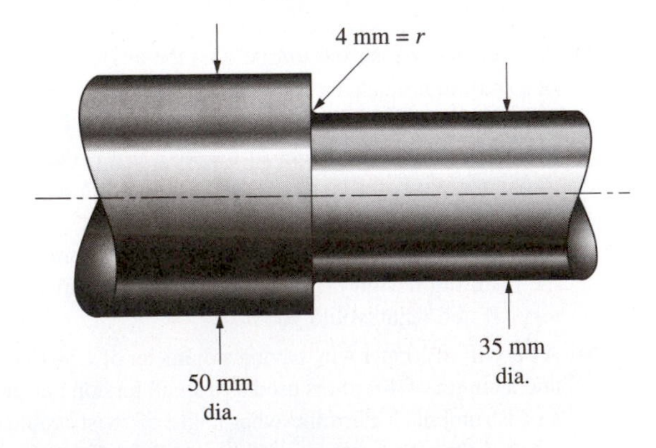

FIGURE P4–34 Axle for Problem 4–34.

4–35.M Part of an output shaft from an automotive transmission has the configuration shown in Figure P4–35. If the shaft is transmitting 105 kW at 220 rad/s, compute the maximum torsional shear stress in the shaft. Account for the stress concentration at the place where the speedometer gear is located.

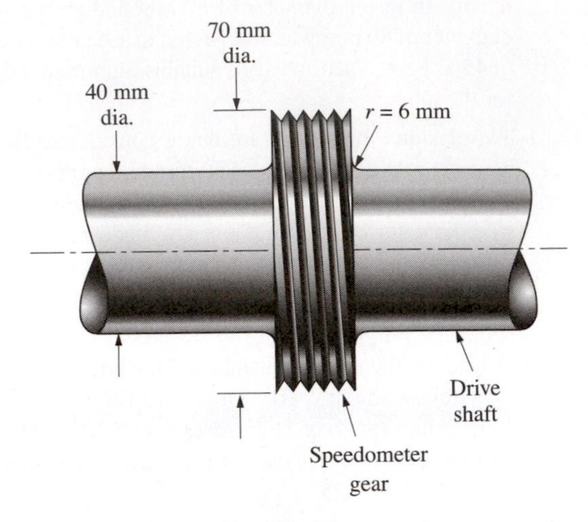

FIGURE P4–35 Shaft for Problem 4–35.

The indicated figures for Problems 4–36 through 4–39 show portions of shafts from power transmission equipment. Compute the maximum repeated torque that could be safely applied to each shaft if it is to be made from AISI 1141 OQT 1100 steel.

4–36.M Use Figure P4–36.

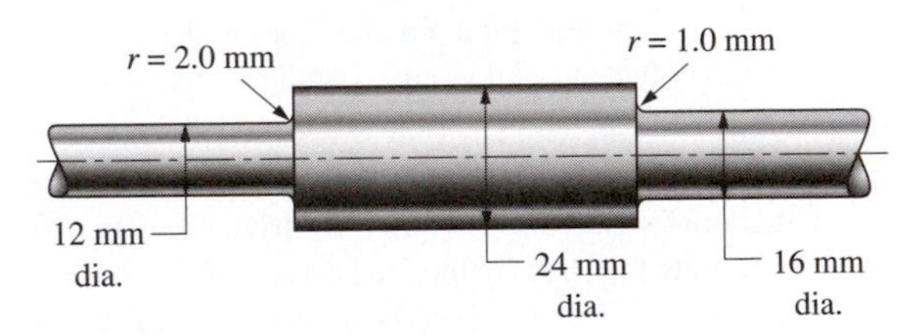

FIGURE P4–36 Shaft for Problem 4–36.

4–37.E Use Figure P4–37.

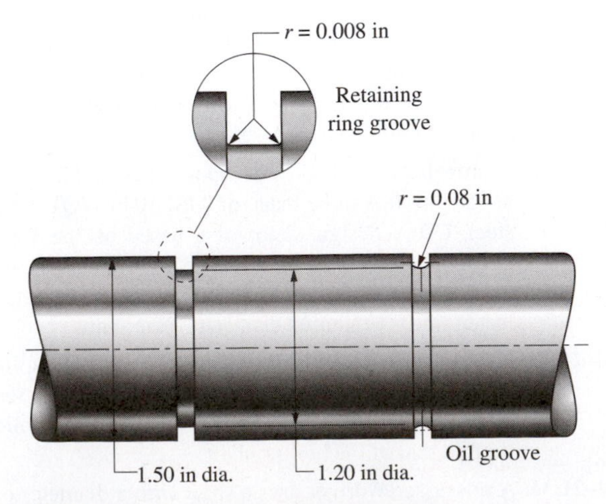

FIGURE P4–37 Shaft for Problem 4–37.

4–38.M Use Figure P4–38.

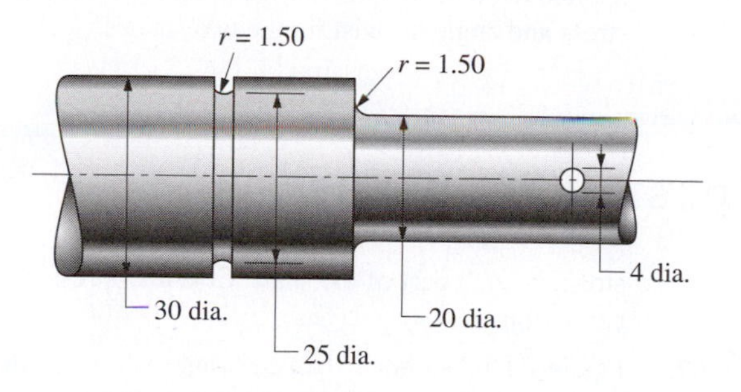

FIGURE P4–38 Shaft for Problem 4–38.

4–39.E Use Figure P4–39.

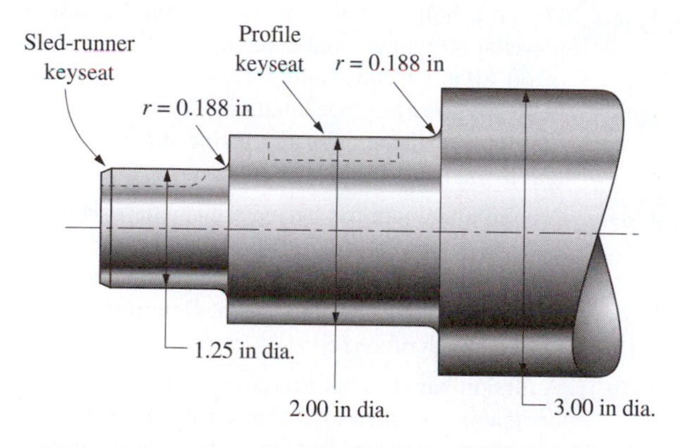

FIGURE P4–39 Shaft for Problem 4–39.

Noncircular Sections

4–40.M Compute the torque that would produce a torsional shear stress of 50 MPa in a square steel rod 20 mm on a side.

4–41.M For the rod described in Problem 4–40, compute the angle of twist that would be produced by the torque found in the problem over a length of 1.80 m.

4–42.E Compute the torque that would produce a torsional shear stress of 7500 psi in a square aluminum rod 1.25 in on a side.

4–43.E For the rod described in Problem 4–42, compute the angle of twist that would be produced by the torque found in the problem over a length of 48 in.

4–44.E Compute the torque that would produce a torsional shear stress of 7500 psi in a rectangular aluminum bar 1.25 in thick by 3.00 in wide.

4–45.E For the bar described in Problem 4–44, compute the angle of twist that would be produced by the torque found in the problem over a length of 48 in.

4–46.M An extruded aluminum bar is in the form of an equilateral triangle 30 mm on a side. What torque is required to cause an angle of twist in the bar of 0.80 deg over a length of 2.60 m?

4–47.M For the triangular bar described in Problem 4–46, what stress would be developed in the bar when carrying the torque found in the problem?

4–48.E As shown in Figure P4–48, a portion of a steel shaft has a flat machined on one side. Compute the torsional shear stress in both the circular section and the one with the flat when a torque of 850 lb·in is applied.

4–49.E For the steel shaft shown in Figure P4–48, compute the angle of twist of one end relative to the other if a torque of 850 lb·in is applied uniformly along the length.

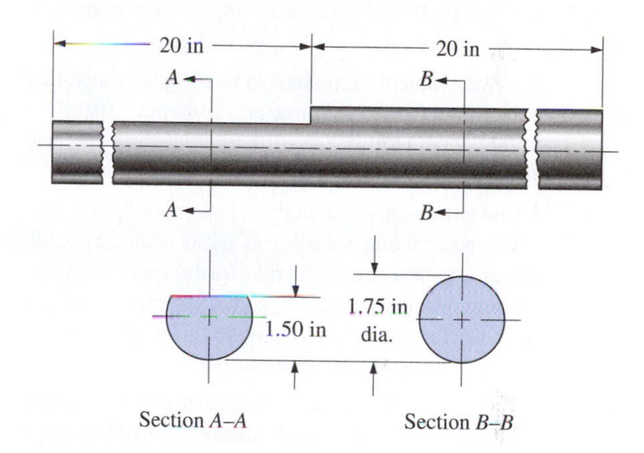

FIGURE P4–48 Problems 4–48 to 4–51.

4–50.E Repeat Problem 4–48 with all the data the same except that two flats are machined on the shaft, resulting in a total measurement across the flats of 1.25 in.

4–51.E Repeat Problem 4–49 for a shaft having two flats, resulting in a total measurement across the flats of 1.25 in.

4–52.M A square stud 200 mm long and 8 mm on a side is made from titanium, Ti-6Al-4V, aged. What angle of twist will result when a wrench is applying a pure torque that causes the stress to equal the yield strength of the material in shear?

4–53.E A standard square structural steel tube has cross-sectional dimensions of $4 \times 4 \times \frac{1}{4}$ in and is 8.00 ft long. Compute the torque required to twist the tube 3.00 deg.

4–54.E For the tube in Problem 4–53, compute the maximum stress in the tube when it is twisted 3.00 deg. Would this be safe if the tube is made from ASTM A501 structural steel and the load was static?

4–55.E Repeat Problem 4–53 for a rectangular tube $6 \times 4 \times \frac{1}{4}$.

4–56.E Repeat Problem 4–54 for a rectangular tube, $6 \times 4 \times \frac{1}{4}$.

4–57.E A standard 6-in schedule 40 steel pipe has approximately the same cross-sectional area as a square tube, $6 \times 6 \times \frac{1}{4}$, and thus both would weigh about the same for a given length. If the same torque were applied to both, compare the resulting torsional shear stress and angle of twist for the two shapes.

ADDITIONAL PRACTICE, REVIEW, AND DESIGN PROBLEMS

4–58. Compute the torsional shear stress in a solid round shaft carrying 125 kW of power while rotating at 1150 rpm. The shaft has a diameter of 35 mm.

4–59. For the shaft in Problem 4–58, specify a suitable steel if the power is smooth and steady.

4–60. For the shaft in Problem 4–58, specify a suitable steel if the power is repeated.

4–61. For the shaft in Problem 4–58, specify a suitable steel if the power is applied with shock.

4–62. A power transmission shaft is to carry a steady 12.0 horsepower while rotating at 1150 rpm. Specify a suitable diameter for the solid circular shaft if it is to be made from AISI 1040 WQT 1100 steel.

4–63. A power transmission shaft is to carry a steady 20.0 horsepower while rotating at 3450 rpm. The shaft drives a special machine that forms sheet steel and some shock loading is expected. Specify a suitable diameter for the solid circular shaft if it is to be made from AISI 4140 OQT 1300 steel.

4–64. An alloy steel shaft has an outside diameter of 100 mm. A central hole of 60 mm diameter is bored in part of its length as shown in Figure P4–64. Compute the shearing stress in the hollow section if the stress in the solid section is 200 MPa.

4–65. For the shaft described in Problem 4–64, compute the angle of twist of the right end with respect to the left end.

4–66. Figure P4–66 shows a shaft carrying three pulleys that rotates at 1750 rpm. Pulley A delivers 15 kW to a mating pulley that drives a fan. Pulley C delivers 20 kW to a different mating pulley that drives a conveyor. All power comes into the shaft through pulley B. Considering only torsion, compute the shearing stress in each part of the shaft. Consider stress concentrations.

4–67. Figure P4–67 shows a shaft carrying three gears that rotates at 1150 rpm. Gear A delivers 20 kW to a mating gear that drives a mixer. Gear C delivers 12 kW to a different mating gear that drives a circular saw. All power comes into the shaft through gear B. Considering only torsion, compute the shearing stress in each part of the shaft. Consider stress concentrations.

4–68. Design a hollow steel shaft to transmit 225 kW of power at 80 rpm without exceeding a shearing stress of 60 MPa. Consider only torsion and assume that there are no stress concentrations. Make the ratio of the outside diameter to the inside diameter to be approximately 1.25.

4–69. A solid aluminum rod having a diameter of 4.0 mm is to be twisted through $\frac{1}{2}$ revolution (180 degrees) while acting as a torsion bar. The torsional shear stress must not exceed 150 MPa. Determine the required length of the rod.

4–70. A torsion bar is to be made from a hollow titanium bar that has a length of 200 mm. The ratio of the outside diameter to the inside diameter is to be approximately 1.50. The desired torsional stiffness is to be 0.015 degree of rotation per 1.0 N·m of applied torque. Specify the outside diameter and the inside diameter of the bar.

4–71. For the bar designed in Problem 4–70, compute the torsional shear stress that would occur in the bar when it is rotated 10.0 degrees.

4–72. For the bar designed in Problem 4–70, compute the angle of twist when the torsional shear stress in the bar has a design factor of 3 based on yield strength in shear. The material for the bar is AISI 4141 OQT 1100 steel.

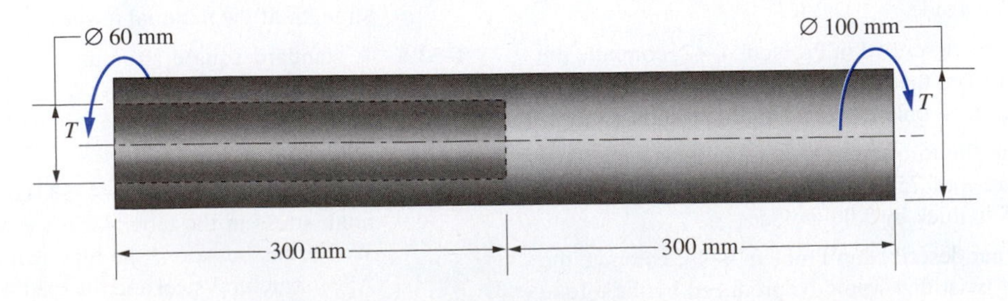

FIGURE P4–64 Shaft for Problems 4–64 and 4–65.

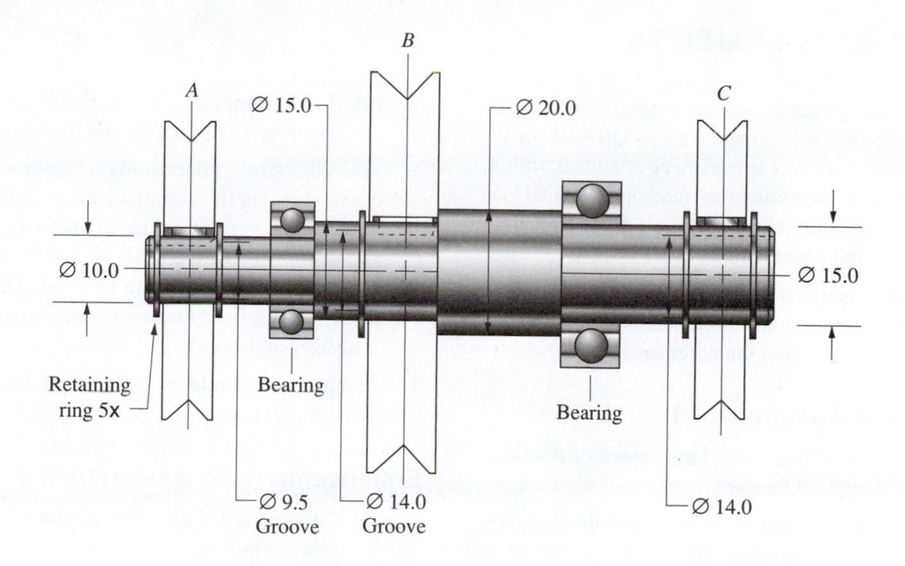

Fillet radius = 0.50 mm at each step
Dimensions in mm
Sled-runner keyseat at *A*, *C*
Profile keyseat at *B*

FIGURE P4–66 Shaft for Problem 4–66.

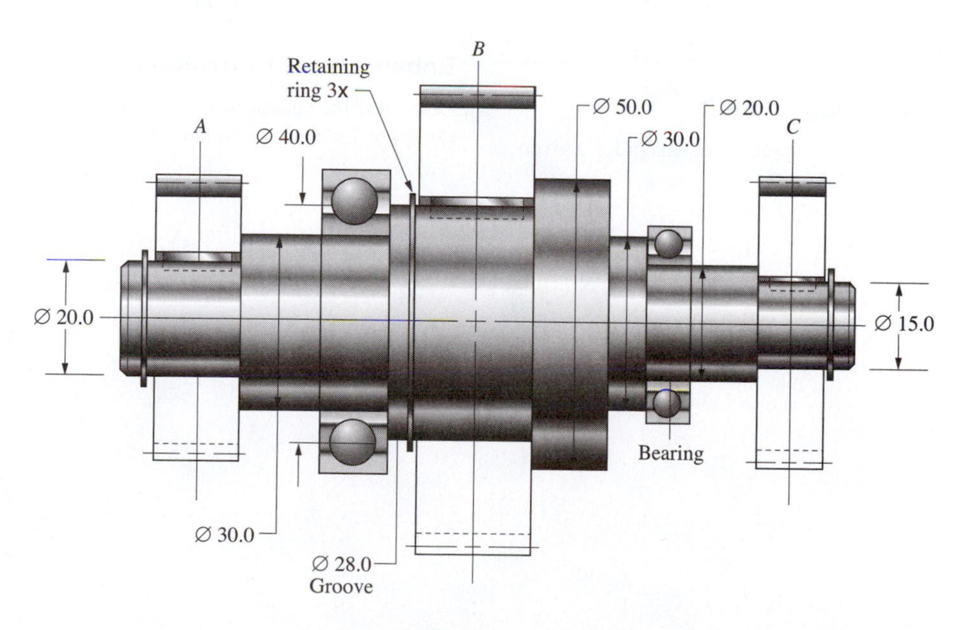

Profile keyseats at *A*, *B*, *C*
Fillet radius = 1.00 mm at each step
Dimensions in mm

FIGURE P4–67 Shaft for Problem 4–67.

COMPUTER ASSIGNMENTS

Note: Any of these assignments can be completed as a computer program in a technical language such as QBASIC or C. Or they could be designed as a spreadsheet or solved with a computer algebra system or some other mathematics analysis software. Be sure to consider the units of all factors entered into the problems and indicate the units of all results.

1. Given the need to design a solid circular shaft for a given torque, a given material yield strength, and a given design factor, compute the required diameter for the shaft.

Enhancements to Assignment 1

(a) For a given power transmitted and speed of rotation, compute the applied torque.

(b) Include a table of materials from which the designer can select. Then automatically look up the yield strength.

(c) Include the design factor table, Table 4–1. Then prompt the designer to specify the manner of loading only and determine the appropriate design factor from the table within the program.

2. Repeat assignment 1, except design a hollow circular shaft. Three possible solution procedures exist:

(a) For a given outside diameter, compute the required inside diameter.

(b) For a given inside diameter, compute the required outside diameter.

(c) For a given ratio of D_i/D_o, find both D_i and D_o.

Enhancements to Assignment 2

(a) Compute the mass of the resulting design for a given length and material density.

(b) If the computer has graphics capability, draw the resulting cross section and dimension it.

3. Enter the stress concentration factor curves into a program, allowing the automatic computation of K_t for given factors such as fillet radius, diameter ratio, hole diameter, and so on. Any of the cases shown in Appendix A–22–5, A–22–6, or A–22–7 could be used. This program could be run by itself or made an enhancement of other stress analysis programs.

4. Compute the angle of twist from Equation (4–22) for given T, L, G, and J.

Enhancements to Assignment 4

(a) Compute J for given dimensions for the shaft, either solid or hollow.

(b) Include a table of values for G, from Table 4–3, in the program.

5. Compute the required diameter of a solid circular shaft to limit the angle of twist to a specified amount.

6. Compute the angle of twist for one end of a multisection shaft relative to the other, similar to Example Problem 4–13. Allow the lengths, diameters, materials, and torques to be different in each section.

7. Write a program to compute the values for the effective section modulus, Z_p, and the torsional stiffness constant, J, from Figure 4–27 for any or all cases.

Enhancement to Assignment 7

Determine the equations for C_1, C_2, C_3, and C_4 in terms of the ratio, h/r, for shafts with flats. Use a curve-fitting routine.

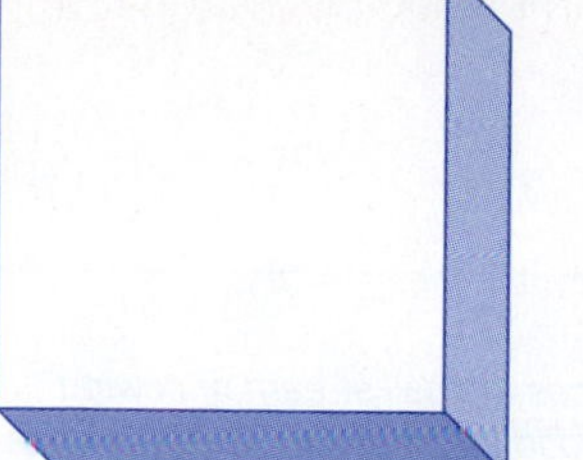

5

Shearing Forces and Bending Moments in Beams

The Big Picture

Shearing Forces and Bending Moments in Beams

Discussion Map

- Much of the discussion in the next six chapters deals with beams. You should be familiar with the following definition of a beam:

  A beam is a member that carries loads transversely, that is, perpendicular to its long axis.

- In this chapter you will learn how to characterize the types of beams and the loadings placed on them.
- In later chapters, you will learn how to compute the magnitude of stresses in beams and to design them to ensure that the beams are safe.
- You will also learn in later chapters how to compute the deflection of a beam under a given loading pattern.

Discover

Imagine that you are trying to cross a small stream of flowing water. You find a wide plank of wood that is long enough to span from your side of the stream to the other side. You lay it across and get ready to step onto the plank. Then you might think to yourself, "Is this plank strong enough to carry my weight as I cross the creek?" You recognize that when you walk onto the plank it is going to bend downward. You likely know that it will bend more as you move toward the middle of the span. Will it bend so much that your feet will dip into the water? Will it actually break under your weight and plunge you in all the way?

Unless you know how to analyze the stresses in beams, and how to compute their deflections, you are left to your own judgment about the safety of the plank. This simple example should give you some understanding of the need for learning how to analyze the behavior of a beam.

What other examples of beams can you think of?

Consider the construction of your home, commercial buildings, or bridges.

Consider recreation equipment and exercise devices.

Look carefully at children's toys and see if some of their parts carry loads perpendicular to their long axes.

Check out a car or a truck and see if you can tell which parts will bend under the load of the vehicle itself or from loads applied to them during operation.

Consider chairs and other seating systems. Look at the flat, plank-like seating in the bleachers at a football, basketball, or baseball game.

Look at equipment used in highway construction or in building any kind of structure. Most equipment must lift and carry heavy loads or push or pull with large forces.

How does your list of beam examples compare with the following items?

- In the basement of your home you might have a long steel beam extending across the room or along the middle of the length of the building to hold the joists of the floor above.
- The joists themselves are beams carrying the load from the floor to the supporting steel beam or to the walls of the foundation.
- Even the flooring itself acts as a beam carrying your weight and the weight of furniture and appliances that apply loads between the joists.
- The tops of doors and windows have strong beams, called lintels, to hold up the structure above these openings in walls.
- The sides of stairs, called stringers, act as beams as they carry the weight of a person climbing the steps.
- A baseball bat or a tennis racket acts like a special kind of beam, called a cantilever, that is held fixed at one end by your hands and that applies a force to the ball out near the end of the racket or bat.
- The balance beam in gymnastics is an obvious example, and you could also include the parallel bars.

- Did you consider the basketball hoop as a beam? When a player stuffs a dunk and hangs on to the rim, it certainly bends down. It has to be strong and stiff enough to withstand that load without breaking or permanently bending.
- Some people exercise by rowing, pulling on oars to move the boat through the water. You apply a pulling force on the handles against the resistance of the oar blade in the water, and the oar bends with respect to the oarlock attached to the boat.
- Did you look for examples of beams at a swimming pool? The diving board is a beam that must flex with just the right feel while remaining strong for thousands of dives. The steps of the ladder used to climb to the high board are beams.
- Certain kinds of car, truck, and trailer axles are beams that transfer the loads from the vehicle to the wheels and then to the ground.
- At construction sites for new buildings, you should be able to see numerous horizontal beams being erected that will carry the loads of each floor of the building to the vertical columns and on to the foundation.

In this chapter you will learn how to compute how the loads on a beam result in internal forces and moments that must be resisted. The stress and deflection analyses follow in later chapters.

Activity Chapter 5—Shearing Forces and Bending Moments in Beams

To comprehend clearly the behavior of beams in bending, you must visualize what is happening *within the fibers of the beam.* This is difficult because you cannot observe directly how each small particle of the material of the beam is affected by the action of loads on the beam. However, a simple experiment will provide some insight and allow you to deduce the nonvisible phenomena of *shearing forces and bending moments* to which this chapter is devoted. It will lead to further understanding of related phenomena in the following several chapters.

Setup. Acquire or fabricate a simple beam and two supports for it. The beam should be sufficiently flexible that you can bend it with small forces using only your hands, but it should be stiff enough to demonstrate the natural curvature that a beam exhibits under load. Possibilities are beams with lengths of approximately 200 mm to 1000 mm long (8 in to 36 in), 25 mm wide (1.0 in), and made from:

1. A thin flat piece of wood about 2 mm to 3 mm thick (0.08 in to 0.12 in)
2. Dense cardboard of similar dimensions as described in item 1. You may glue several thinner strips together to achieve the proper thickness and stiffness. Spread the glue evenly across the entire surfaces of contact to ensure that the composite piece acts as a single unit.
3. A strip of flat sheet aluminum about 1.6 mm thick (0.06 in)
4. A strip of flat sheet steel about 0.5 mm thick (0.02 in)
5. A strip of flat plastic sheet about 2.0 mm thick (0.08 in)

Several of these shapes and materials are approximated by inexpensive rulers, straightedges, and meter sticks available in office supply or home improvement stores.

The two supports should be stable when laid on a flat surface and should support the beam at least 15 mm (0.6 in) off the surface. An ideal example is a small angle with leg lengths from 20 mm to 30 mm (0.75 in to 1.25 in) made from wood, plastic, or metal. Place the legs down to create a "knife-edge" on which to support the beam.

Also acquire a good straightedge about 75 mm to 150 mm long (3.0 in to 6.0 in).

Activity A

Start by simply picking up the beam and applying forces that tend to bend it while observing the shape that the beam assumes and the direction of the forces required to cause the beam to bend. Bring the beam to a steady equilibrium after each action. If possible, sketch the beam in its bent shape and indicate where the supporting forces or grasping actions were applied. How many ways can you do that? Possibilities are:

1. Grasp the beam with the thumb and two or more fingers of each hand somewhat apart from each other and rotate the beam to a bent shape.
2. Place the thumb of each hand on the bottom surface of the beam and one finger of each hand outboard from the thumbs. Then push down with the fingers.
3. Place the thumb of each hand on the bottom surface of the beam and one finger of each hand inboard from the thumbs. Then push down with the fingers.

Steps 4–7 may be easier if you have two rigid blocks about 75 mm to 100 mm long (3.0 in to 4.0 in) on each side of the beam and grasp the beam like a "sandwich."

4. Grasp one end of the beam firmly with one hand. Then push up or down with a finger of the other hand.
5. Grasp both ends of the beam firmly and rotate the beam by rotating your wrists. Try different combinations of the direction of rotation of each hand.
6. Grasp both ends of the beam firmly and rotate the beam by moving one hand up or down relative to the other, keeping your wrists steady.
7. After performing step 6, release your grasp of one hand and observe what happens to the shape of the beam. As you release your grip, try to maintain the position of the beam where it was.

Observations and Conclusions. You should observe the following:

1. Some actions produced a continuous curvature of the beam in one direction.
2. Other actions produced curvatures that are partly concaved upward and partly concaved downward.
3. When parts of the beam were outside the points of support or grasp, that part of the beam was straight.
4. The firm grasping controlled not only the position of the beam but also the curvature so that the slope of the beam was inline with the grasp at the point where it was clamped.
5. When a part of a bent beam was released from a firm grasp, its shape assumed a continuous, one-direction curvature regardless of its shape while being grasped.
6. If a supporting force or a firm grasp was removed completely, the beam would continue to move back to its originally straight form.

Activity B

1. Set two simple supports on a flat surface and apart so that approximately $\frac{1}{4}$ of the length of the beam overhangs each support.
2. Place the beam on the supports.
3. Apply a downward force with one finger to the midpoint of the beam between the supports.
4. Use the straightedge to observe the curvature of the beam between the supports and in each overhang.

Observations and Conclusions. What do you observe from this simple experiment? Here are some things to consider:

1. The downward force applied to the beam at its middle causes the beam to bend in a curved shape between the supports that is concaved upward. This illustrates *positive bending.*
2. The parts of the beam overhanging the supports are straight, not curved.
3. The maximum deflection occurs at the middle of the beam where the load is applied.
4. The deflection is below the level of the supports between the two supports, but upward in the overhangs outside of the supports.

Activity C

1. Set the supports on a flat surface and apart so that approximately $\frac{1}{4}$ of the length of the beam overhangs each support.
2. Apply a downward force at the end of both overhangs and note the shape of the beam.
3. Now apply only one force at one end and observe that the reaction force at the opposite support must be redirected to be downward rather than upward as before.
4. Apply one force between the supports and one at the end of one of the overhangs. Observe the shape of the deflected beam and sketch it.
5. Now apply three forces, one downward on each overhang and one downward between the supports. Observe the shape of the deflected beam and sketch it.

Activity D

This activity requires that you now think about how the material within the beam must act to cause the beam to behave in the observed manner, depending on the way it was supported and loaded.

1. Repeat Activity B and hold the downward force, F, in position. Figure 5–1(a) shows the beam in its original straight, undeflected form with the supports at points A and C and the force, F, applied at B.
2. Sketch the shape of the deflected beam as shown in Figure 5–1(b). With a small straightedge, verify that the deflected beam shape is curved between A and C while the overhang to the left of A remains straight.
3. What direction must the forces act at each support? (***Answer:*** upward)
4. What is the magnitude of the force at each support? (***Answer:*** $F/2$)
5. If the beam were to be cut at any place, D, between the left support and the middle of the beam where the load is applied, what would have to happen at the cut end of

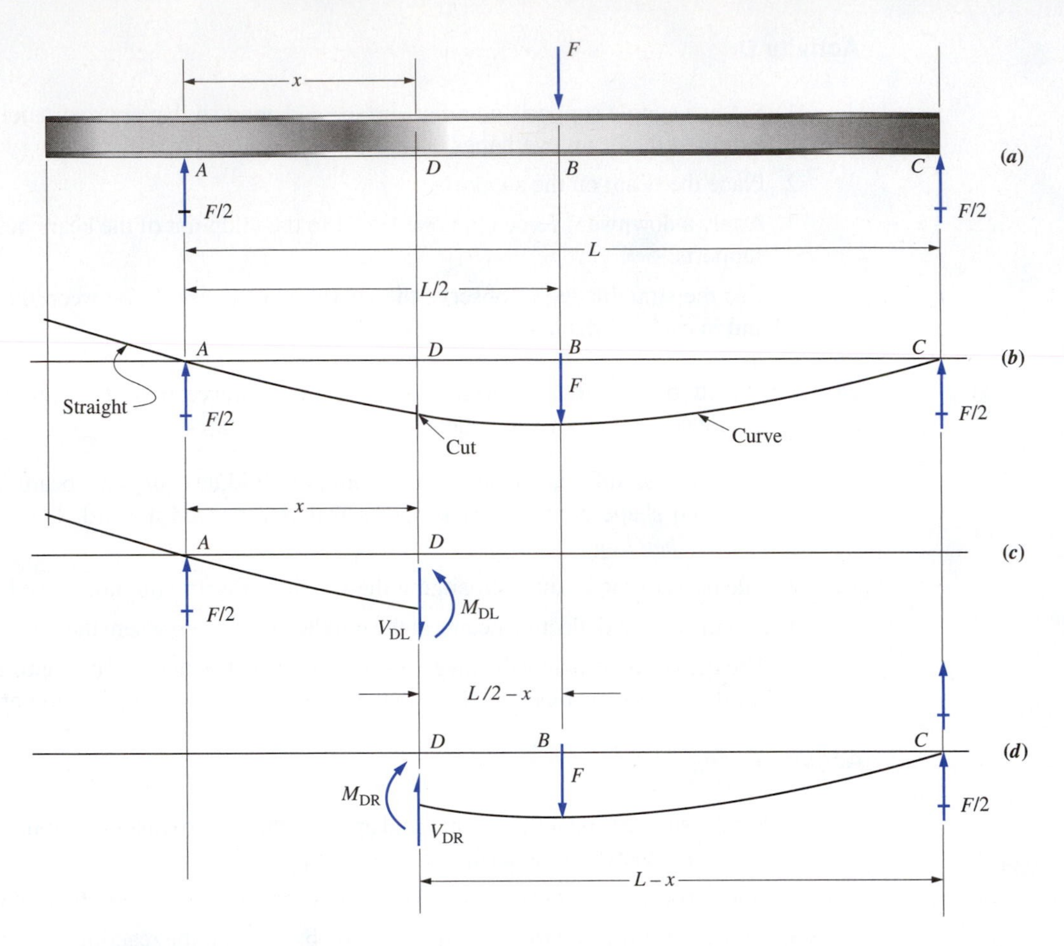

FIGURE 5–1 Simply supported beam with concentrated load at midspan and an overhang—illustration of internal shearing force and bending moment.

each part of the beam if the beam was to be kept in the same position and shape as before it was cut? Refer to Figure 5–1(c) and (d) showing the deflected beam with a cut made at point D, located an arbitrary distance x from the left support.

a. ***Answer*** for what happens to the *left cut section* using that part only as a free-body diagram [Figure 5–1(c)]:

   i. The upward reaction force of $F/2$ remains applied at the support. Then to maintain equilibrium of forces in the vertical direction, a downward force of $F/2$ must be exerted on the cut section. This force is shown and called V_{DL}, the shearing force at D for the left cut section. In the complete beam, this force had been applied by the part of the beam to the right that was cut away.

   ii. To maintain the part of the beam to the left of the cut in its formerly curved shape, a counterclockwise moment must be applied at the cut, called M_{DL}, the moment at D for the left cut section.

   iii. To maintain rotational equilibrium of the part of the beam to the left of the cut, the moment must be equal in magnitude to the couple created by the reaction ($F/2$) and the internal shearing force at the cut ($F/2$) acting at a distance of x apart. That moment is

$$M_{DL} = (F/2)(x) = Fx/2 \text{ counterclockwise}$$

iv. Now, the cut section is in equilibrium for forces in the vertical direction and for moments. Furthermore, the moment at the cut section maintains the curved shape of the deflected beam between the support and the cut section, while allowing the part of the beam beyond the overhang to remain straight.

b. ***Answer*** for what happens to the *right cut section* using that part only as a free-body diagram [Figure 5–1(d)]:

i. The same logic applies as for the left cut section. To maintain equilibrium of vertical forces, an upward shearing force must exist in the beam, called V_{DR}, the shearing force at D for the right cut section.

ii. The magnitude of the shearing force can be found by summing forces in the vertical direction for the right cut section.

$$\Sigma F_v = 0 = F/2 + V_{DR} - F$$
$$V_{DR} = F - F/2 = F/2$$

iii. To maintain the curved shape of the beam, a clockwise moment must be applied at the cut, called M_{DR}, the moment at D for the right section.

iv. To maintain rotational equilibrium of the part of the beam to the right of the cut, the moment must be equal to the sum of the moments due to the other two forces: F acting at $(L/2 - x)$ from the cut; and $F/2$ acting at $(L - x)$ from the cut. Then,

$$M_{DR} = F(L/2 - x) - (F/2)(L - x) = Fx/2 \text{ clockwise}$$

c. In summary, parts (c) and (d) of Figure 5–1 show the free-body diagrams of the left and right cut sections. Note that the shearing forces and bending moments at the cut on the two parts form action-reaction pairs, equal in magnitude but opposite in direction.

In this chapter, you will apply processes such as this to more complex beams with a variety of loading and support conditions. You will learn procedures to determine the magnitude of shearing forces and bending moments in a more direct fashion and to draw complete shearing force and bending moment diagrams that display their variations across the entire length of the beam. The ability to perform these analyses is a critical *first step* to computing the stresses and deformations in a beam, as will be discussed in the following chapters.

5–1 OBJECTIVES OF THIS CHAPTER

After completing this chapter, you should be able to:

1. Define the term *beam* and recognize when a load-carrying member is a beam.
2. Describe several kinds of beam loading patterns: *concentrated loads, uniformly distributed loads, linearly varying distributed loads,* and *concentrated moments*.
3. Describe several kinds of beams according to the manner of support: *simple beam, overhanging beam, cantilever,* and *composite beam* having more than one part.
4. Draw free-body diagrams for beams and parts of beams showing all external forces and reactions.
5. Compute the magnitude of reaction forces and moments and determine their directions.
6. Define *shearing force* and determine the magnitude of shearing force anywhere within a beam.

7. Draw free-body diagrams of *parts* of beams and show the internal shearing forces.
8. Draw complete shearing force diagrams for beams carrying a variety of loading patterns and with a variety of support conditions.
9. Define *bending moment* and determine the magnitude of bending moment anywhere within a beam.
10. Draw free-body diagrams of *parts* of beams and show the internal bending moments.
11. Draw complete bending moment diagrams for beams carrying a variety of loading patterns and with a variety of support conditions.
12. Use the *laws of beam diagrams* to relate the load, shearing force, and bending moment diagrams to each other and to draw the diagrams.
13. Draw free-body diagrams of parts of composite beams and structures and draw the shearing force and bending moment diagrams for each part.
14. Properly consider *concentrated moments* in the analysis of beams.
15. Use the *theorem of three moments* to analyze continuous beams having three or more supports and carrying any combination of concentrated and distributed loads.

5–2 BEAM LOADING, SUPPORTS, AND TYPES OF BEAMS

Recall the definition of a beam.

A beam is a member that carries loads transversely, that is, perpendicular to its long axis.

When analyzing a beam to determine reactions, internal shearing forces, and internal bending moments, it is helpful to classify the manner of loading, the type of supports, and the type of beam.

Beams are subjected to a variety of loading patterns, including,

Normal concentrated loads
Inclined concentrated loads
Uniformly distributed loads
Varying distributed loads
Concentrated moments

Support types include,

Simple, roller-type support
Pinned support
Fixed support

Beam types include,

Simply supported beams; or simple beams
Overhanging beams
Cantilever beams; or cantilevers
Compound beams
Continuous beams

Understanding all of these terms will help you communicate features of beam designs and perform the required analyses. Following are brief descriptions of each with illustrations to help you visualize them.

Loading Patterns In later sections we will show that the nature of the loading pattern determines the variation of the shearing force and bending moment along the length of the beam. Here we define the five most frequently encountered loading patterns and give examples of each. More complex loading patterns can often be analyzed by considering them to be combinations of two or more of these basic types.

Normal concentrated loads

A normal concentrated load is one that acts perpendicular (normal) to the major axis of the beam at only a point or over a very small length of the beam.

Figure 5–2(a) shows the typical manner of representing a beam carrying normal concentrated loads. Each load is shown as a vector acting on the beam, perpendicular to its long axis. Part (b) is a sketch of a situation that would produce concentrated loads. The weight of the pipes and their contents determines the magnitudes of the loads.

While we often visualize loads acting downward due to gravity, actual loads can act in any direction. Particularly in mechanical machinery, forces produced by linkages, actuators, springs, clamps, and other devices can act in any direction. Figure 5–3 shows a simple example.

Normal concentrated loads tend to cause pure bending of the beam. Most of the problems in this chapter will feature this type of loading. The analysis of the bending stresses produced is presented in Chapter 7.

Inclined concentrated loads

An inclined concentrated load is one that acts effectively at a point but whose line of action is at some angle to the main axis of the beam.

FIGURE 5–2 Simple beam with normal concentrated loads.

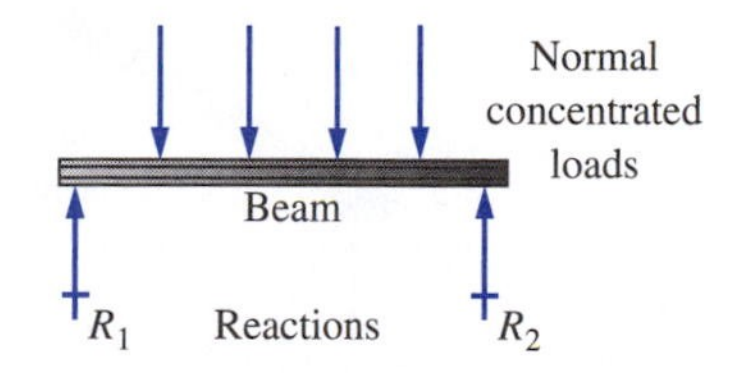

(*a*) Schematic representation of beam, loads, and reactions

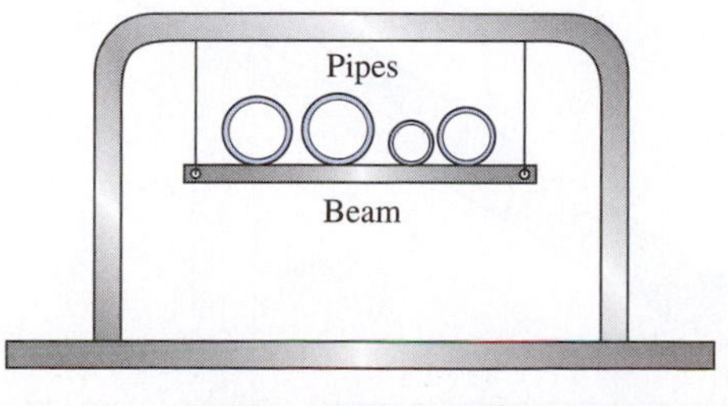

(*b*) Pictorial representation of beam and loads

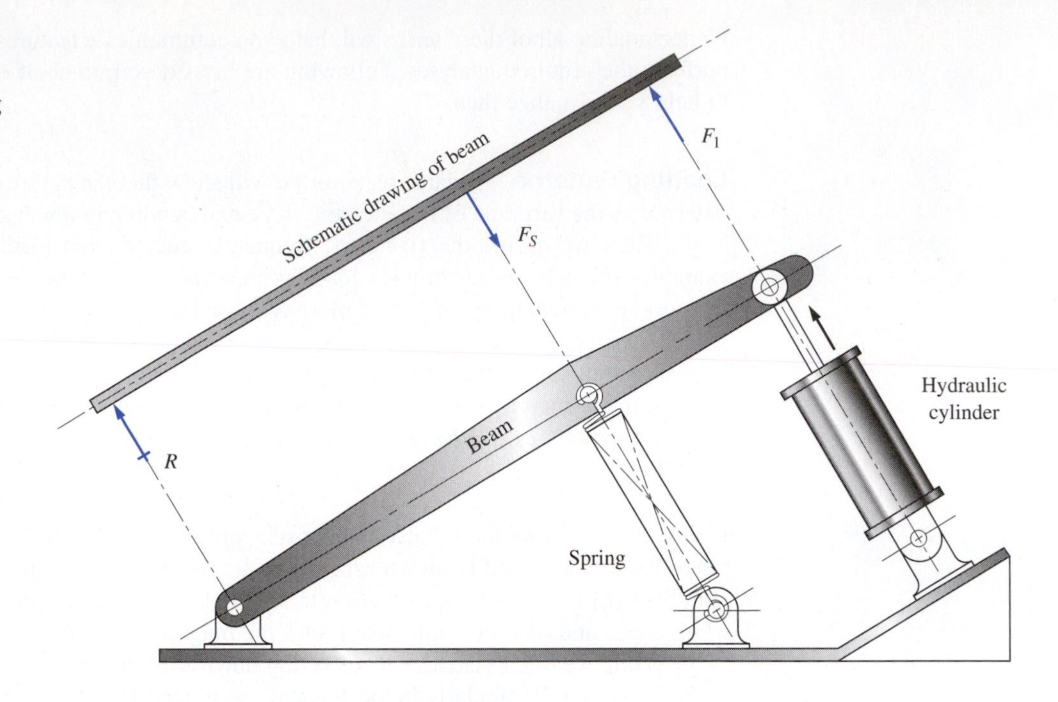

FIGURE 5–3 Machine lever as a simple beam carrying normal concentrated loads.

Figure 5–4 shows an example of an inclined concentrated load. The inclined load exerted by the spring causes a combination of bending and axial stresses to be developed in the beam. Chapter 10 presents the analysis techniques for such a loading.

Uniformly distributed loads
Loads of constant magnitude acting perpendicular to the axis of a beam over a significant part of the length of the beam are called uniformly distributed loads.

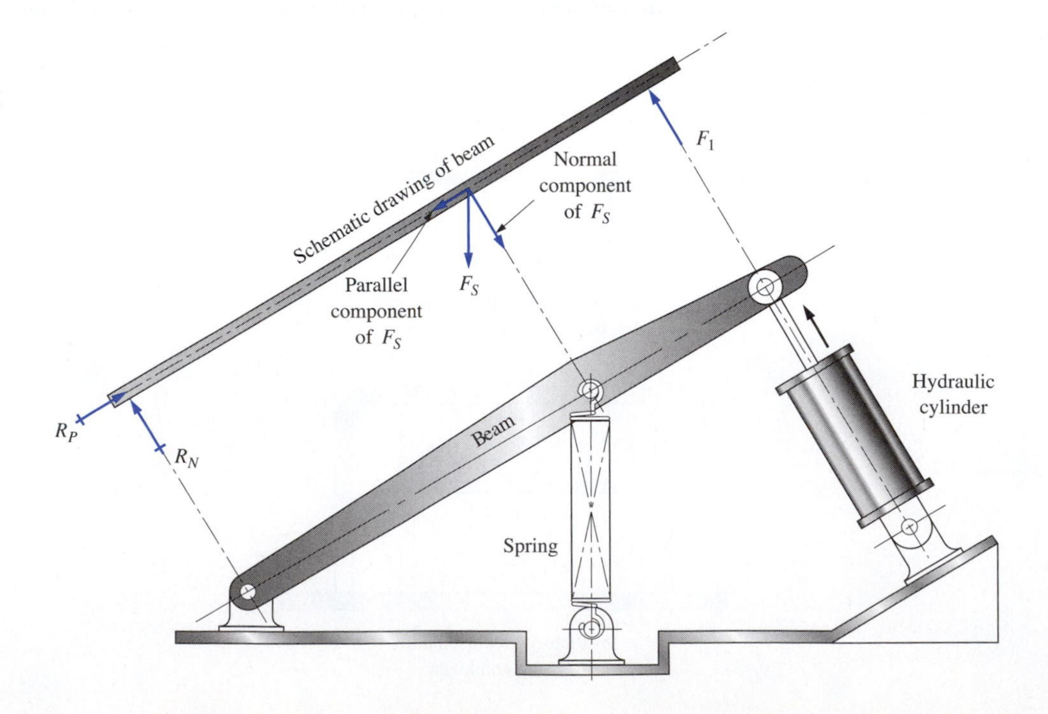

FIGURE 5–4 Machine lever as a simple beam carrying normal concentrated loads and an inclined load.

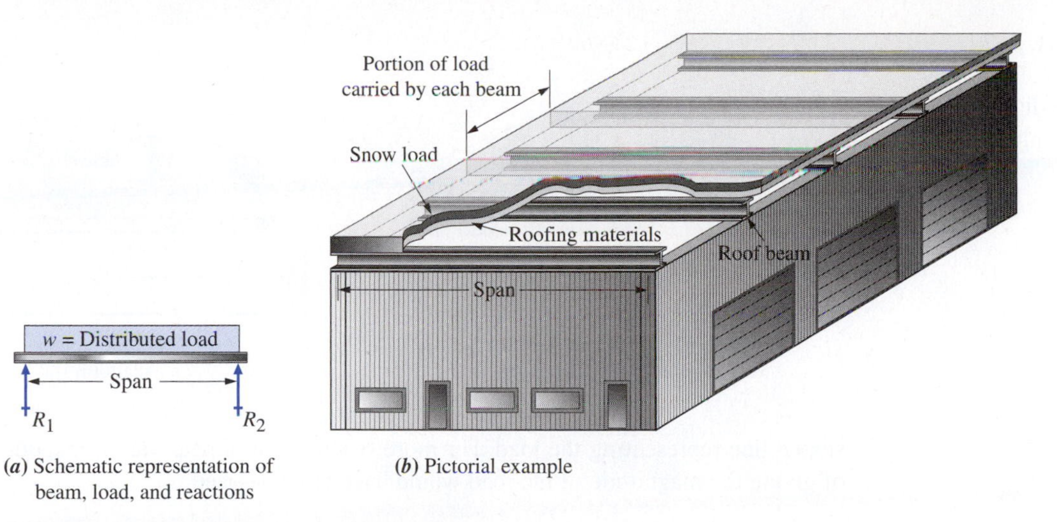

(*a*) Schematic representation of beam, load, and reactions

(*b*) Pictorial example

FIGURE 5–5 Simple beam with uniformly distributed load.

An example of this type of load would be the weight of wet snow of uniform depth on a roof carried by a horizontal roof beam. Also, the materials that make up the roof structure itself are often uniformly distributed. Figure 5–5 illustrates such a loading pattern and shows the manner we use in this book to represent uniformly distributed loads in problems. The blue-shaded rectangular area defines the extent of the load along the length of the beam. The magnitude of the load is called w, in units of force per unit length. Typical units would be lb/in, kN/m, or K/ft. Recall that 1 K = 1 kip = 1000 lb.

For example, if the loading on the beam shown in Figure 5–5 was $w = 150$ lb/ft, you should visualize that each 1.0 ft of length of the beam carries 150 lb of load. If the span is 10.0 ft, the total load is 1500 lb.

Varying distributed loads
Loads of varying magnitude acting perpendicular to the axis of a beam over a significant part of the length of the beam are called varying distributed loads.

Figures 5–6 and 5–7 show examples of structures carrying varying distributed loads. When the load varies linearly, we quantify such loads by giving the value of w at each end of the

FIGURE 5–6 Example of linearly varying distributed load on a cantilever.

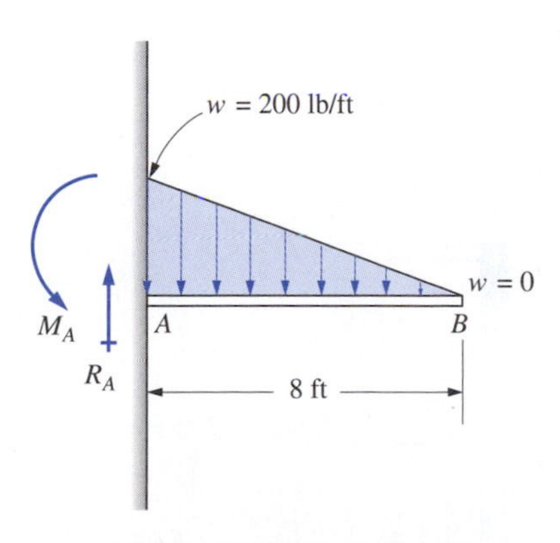

(*a*) Schematic representation of beam, load, reaction force, and moment with sample data

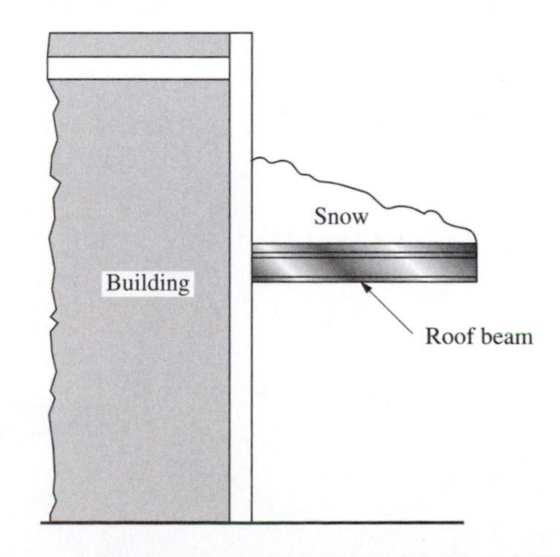

(*b*) Pictorial example – snow load on a projecting roof

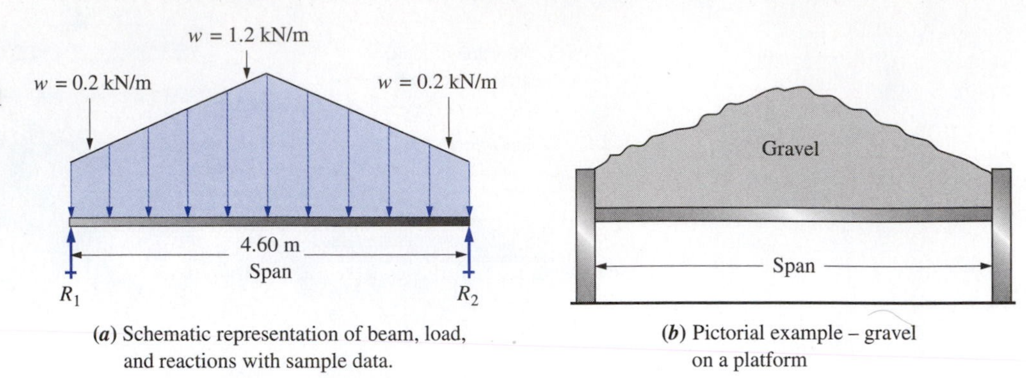

(*a*) Schematic representation of beam, load, and reactions with sample data.

(*b*) Pictorial example – gravel on a platform

FIGURE 5–7 Example of linearly varying distributed load on a simple beam.

sloped line representing the load. For more complex, nonlinear variations, other schemes of giving the magnitude of the load would have to be devised.

Concentrated Moments. A moment is an action that tends to cause rotation of an object. Moments can be produced by a pair of parallel forces acting in opposite directions; this is called a *couple*. The action on a crank or a lever also produces a moment.

When a moment acts on a beam at a point in a manner that tends to cause it to undergo pure rotation, it is called a concentrated moment.

Figure 5–8 shows an example of a concentrated moment. The forces acting at the ends of the vertical arms form a couple and tend to twist the beam into the shape shown. The fact that the two forces making up the couple are equal and opposite results in no net horizontal force being applied to the beam.

Concentrated moments can also be applied to a beam by any force acting parallel to its axis with a line of action some distance from the axis. This is illustrated in Figure 5–9, which represents a shaft carrying a helical gear. The force F acts parallel to the axis of its shaft and tends to bend it in the plane of the page. The difference here is that there is an unbalanced horizontal force also applied to the beam that must be reacted to at one of the supports.

FIGURE 5–8 Concentrated moment on a compound beam.

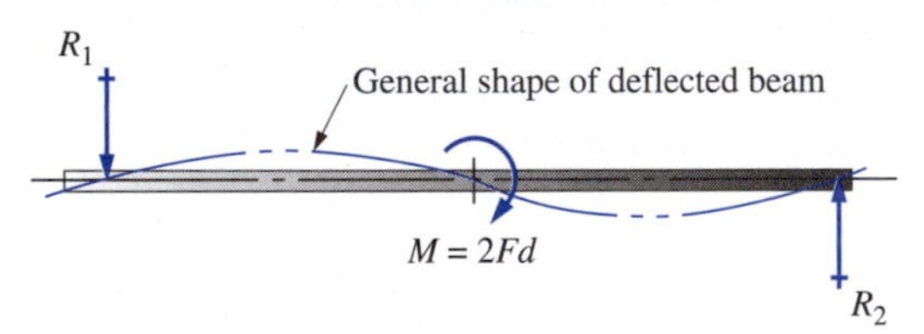

(*a*) Schematic representation of horizontal part of compound beam carrying a concentrated moment

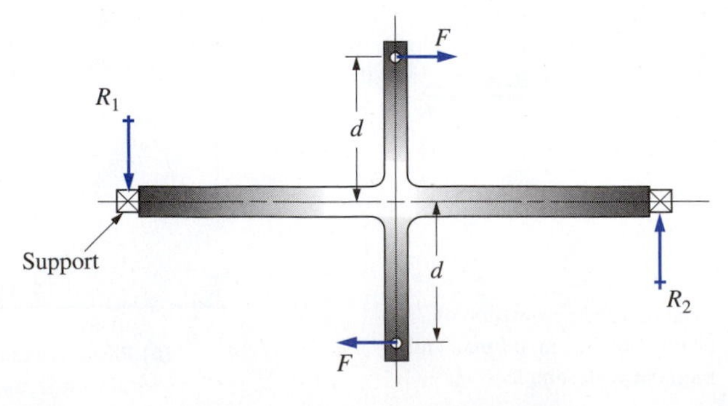

(*b*) Compound beam—pictorial example

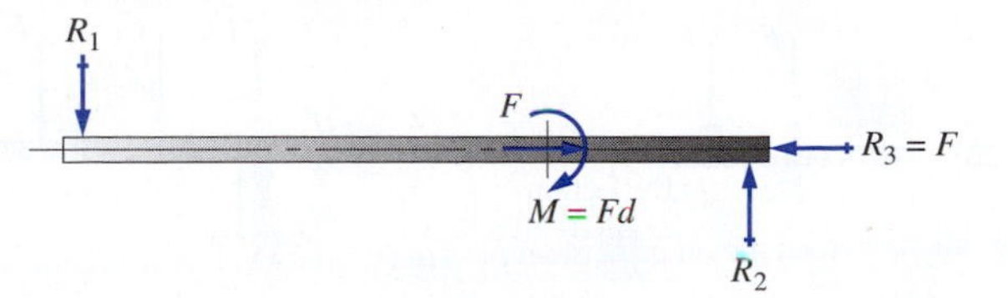

(*a*) Schematic representation of horizontal part of compound beam showing concentrated moment and horizontal reaction

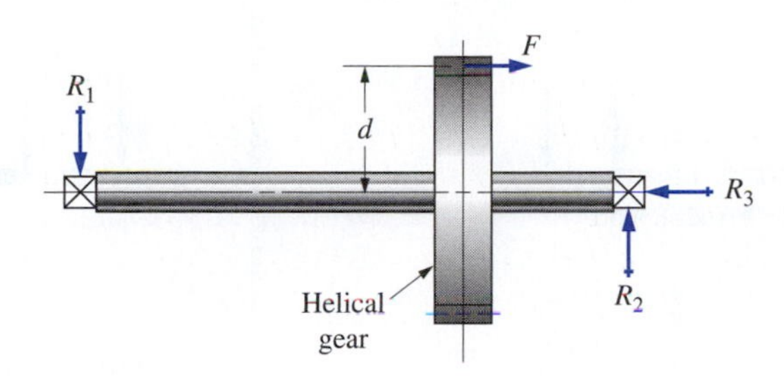

(*b*) Compound beam—pictorial example

FIGURE 5–9 Concentrated moment on a compound beam.

Support Types. All beams must be supported in a stable manner to hold them in equilibrium. All externally applied loads and moments must be reacted to by one or more supports. Different types of supports offer different types of reaction capability.

Simple support or roller support
A simple support is one that can resist only forces acting perpendicular to the beam.

One of the best illustrations of simple supports is the pair of theoretically frictionless rollers shown at the ends of the beam in Figure 5–10(a). They provide upward support against the downward action of the load on the beam. As the beam tends to bend under the influence of the applied loading and the reactions, the bending deformation would not be resisted by the rollers. But if there were any horizontal components of the load, the rollers would roll and the beam would be unrestrained. Therefore, using two rollers alone is not adequate.

Pinned support ***An example of a pinned support is a hinge that can resist forces in two directions but which allows rotation about the axis of the pin in the hinge.***

Figure 5–10(b) shows the same beam as in Figure 5–10(a) with the roller at the left end replaced by a pinned support. This system would provide adequate support while allowing the beam to bend freely. Any horizontal force would be resisted in the pinned joint.

Fixed support
A fixed support is one that is held solidly such that it resists forces in any direction and also prohibits rotation of the beam at the support.

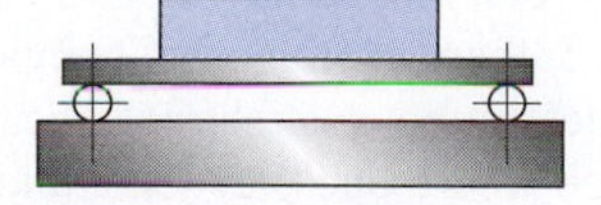

(*a*) Beam on two rollers

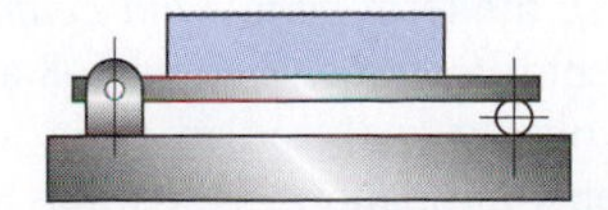

(*b*) Beam with pinned support and one roller

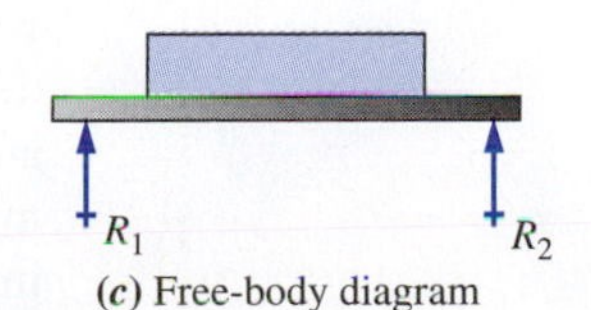

(*c*) Free-body diagram for (*a*) or (*b*)

FIGURE 5–10 Examples of simple supports.

FIGURE 5–11
Beams with fixed supports.

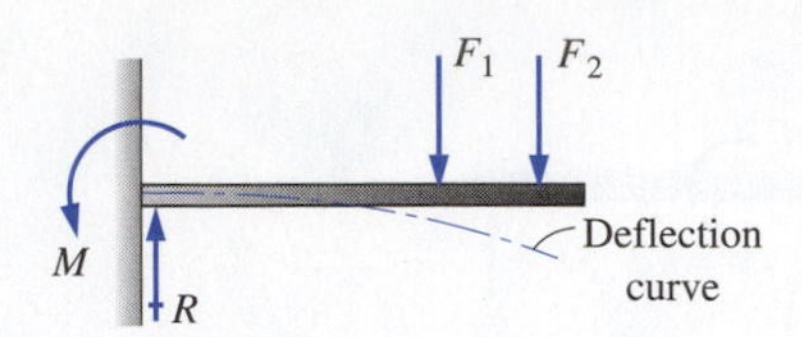

(*a*) Schematic representation of fixed support for a cantilever

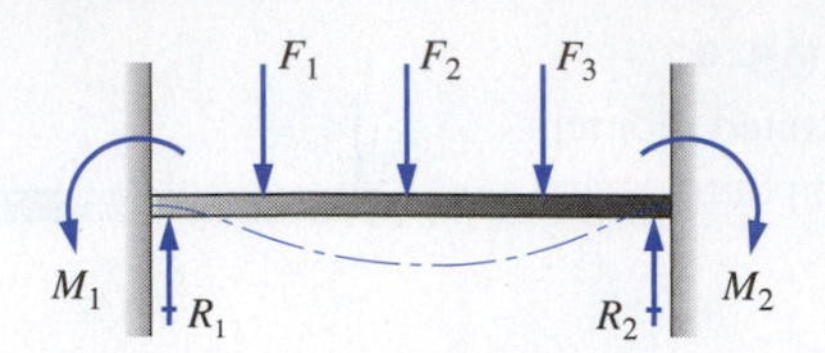

(*c*) Schematic representation of beam with two fixed supports

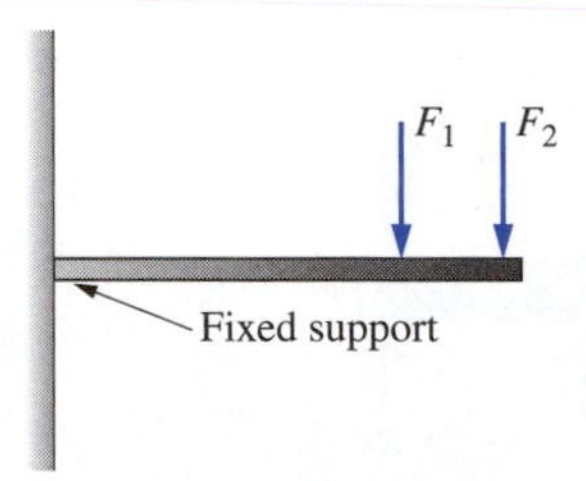

(*b*) Pictorial representation

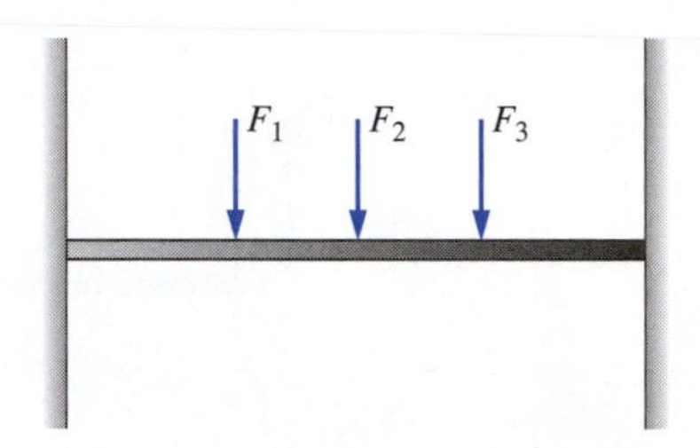

(*d*) Pictorial representation

One manner of creating a fixed support is to produce a close-fitting, socket-like hole in a rigid structure into which the end of the beam is inserted. The fixed support resists moments as well as forces because it prevents rotation. Figure 5–11 shows two examples of the use of fixed supports.

Beam Types. The type of beam is indicated by the types of supports and their placement.

Simple Beam. A simple beam is one that carries only loads acting perpendicular to its axis and that is supported only at its ends by simple supports acting perpendicular to the axis. Figure 5–1 is an example of a simple beam. When all loads act downward, the beam would deflect in the classic bent, concave upward shape. This is referred to as positive bending.

Overhanging Beam. An overhanging beam is one in which the loaded beam extends outside the supports. Figure 5–12 is an example of an overhanging beam. The loads on the overhangs tend to bend them downward, producing negative bending.

Cantilever. A cantilever has only one end supported, as shown in Figure 5–13 depicting a crane boom attached rigidly to a stiff vertical column. It is essential that the support be fixed because it must provide vertical support for the externally applied loads along with a moment reaction to resist the moment produced by the loads. Figures 5–6 and 5–11(a) are other examples of cantilevers.

Compound Beam. While the beams depicted so far were single, straight members, we will use the term *compound beam* to refer to one having two or more parts extending in different directions. Figures 5–8 and 5–9 are examples of compound beams. Such beams are typically analyzed in parts to determine the internal shearing forces and bending moments throughout. Often, the place where one part joins another is a critical point of interest.

FIGURE 5–12
Overhanging beam.

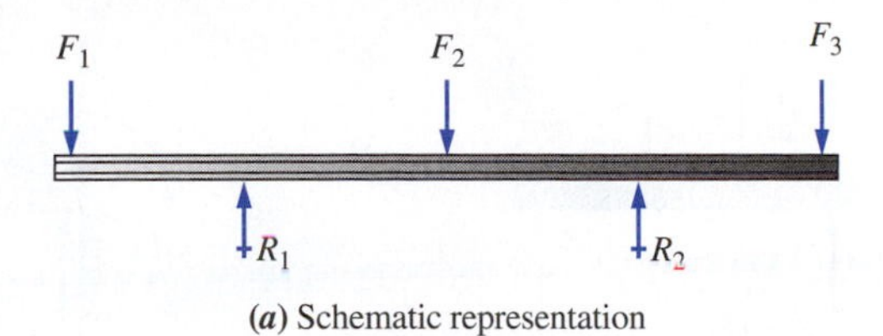

(*a*) Schematic representation

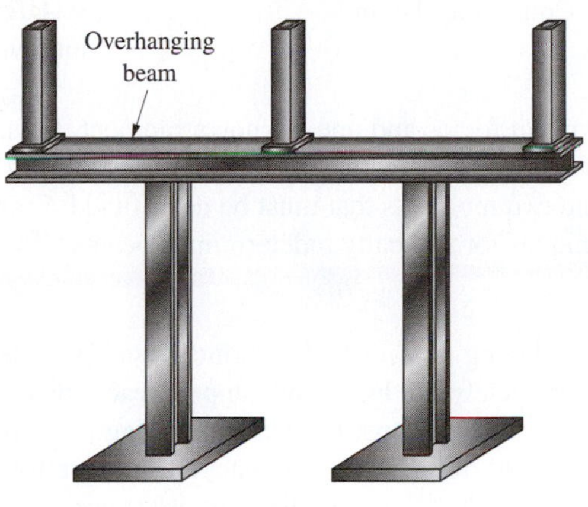

(*b*) Pictorial example

Continuous Beam. Except for the beam shown in Figure 5–11, parts (c) and (d), beams discussed before had one or two supports and only two unknown reactions. The principles of statics allow us to compute all reaction forces and moments from the classical equations of equilibrium because there are two unknowns and two independent equations available from which to solve for those unknowns. Such beams are called *statically determinate*. In contrast, *continuous beams* have additional supports, requiring different approaches to analyze the reaction forces and moments. These are called *statically indeterminate* beams. Figure 5–14 shows an example of a continuous beam on three supports. Figure 5–15 shows a beam with one fixed end and one simply supported end. Notice that

FIGURE 5–13
Cantilever beam.

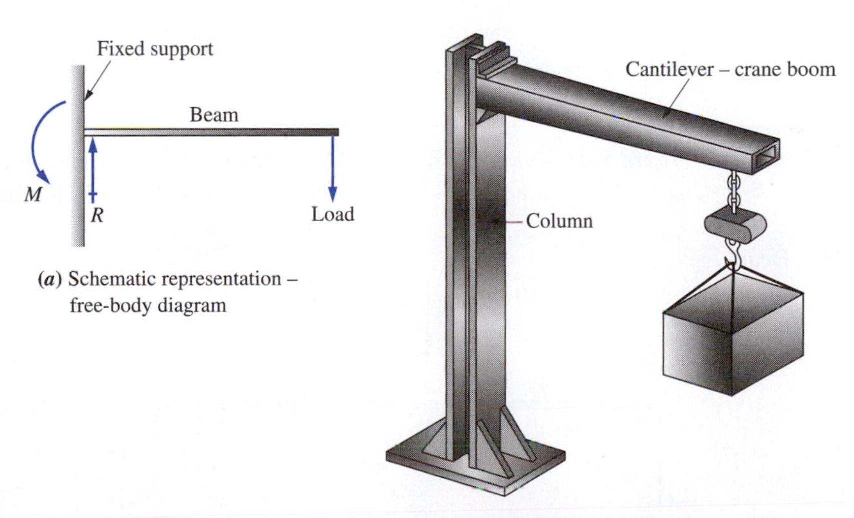

(*a*) Schematic representation – free-body diagram

(*b*) Pictorial example

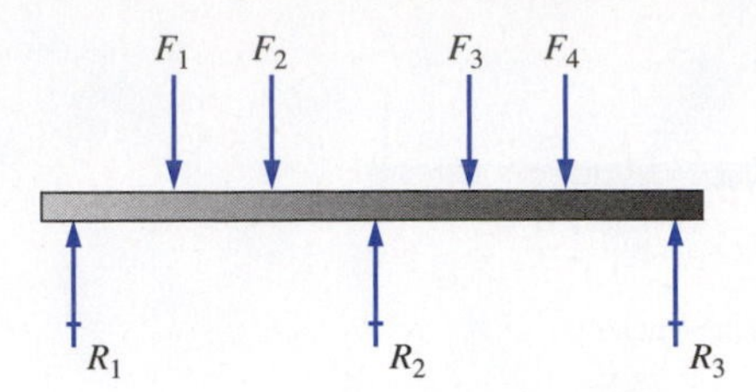

FIGURE 5–14 Continuous beam on three supports.

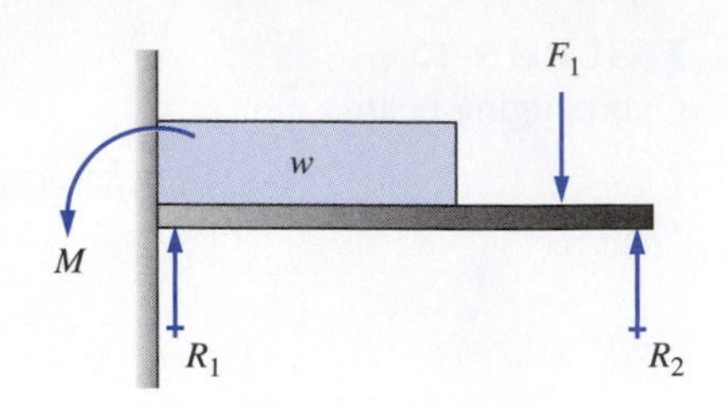

FIGURE 5–15 Supported cantilever.

there are two unknown forces and one unknown moment. Figures 5–11(c) and (d) show a beam with two fixed ends which is also statically indeterminate because there are two reaction forces and two moments that must be determined. Sections 5–12 and 9–6 present the analysis techniques for statically indeterminate beams.

5–3 REACTIONS AT SUPPORTS

The first step in analyzing a beam to determine its safety under a given loading arrangement is to show completely the loads and support reactions on a free-body diagram. It is very important to be able to construct free-body diagrams from the physical picture or description of the loaded beam. This was done in each case for Figures 5–1 through 5–15.

After constructing the free-body diagram, it is necessary to compute the magnitude of all support reactions. It is assumed that the methods used to find reactions were studied previously. Therefore, only a few examples are shown here as a review and as an illustration of the techniques used throughout this book.

The following general procedure is recommended for solving for reactions on simple or overhanging beams.

Guidelines for Solving for Reactions

1. Draw the free-body diagram.
2. Use the equilibrium equation $\Sigma M = 0$ by summing moments about the point of application of one support reaction. The resulting equation can then be solved for the other reaction.
3. Use $\Sigma M = 0$ by summing moments about the point of application of the second reaction to find the first reaction.
4. Use $\Sigma F = 0$ to check the accuracy of your calculations.

Example Problem 5–1

Figure 5–16 shows the free-body diagram for the beam carrying pipes, which was originally shown in Figure 5–2. Compute the reaction forces in the support rods.

FIGURE 5–16 Beam loading.

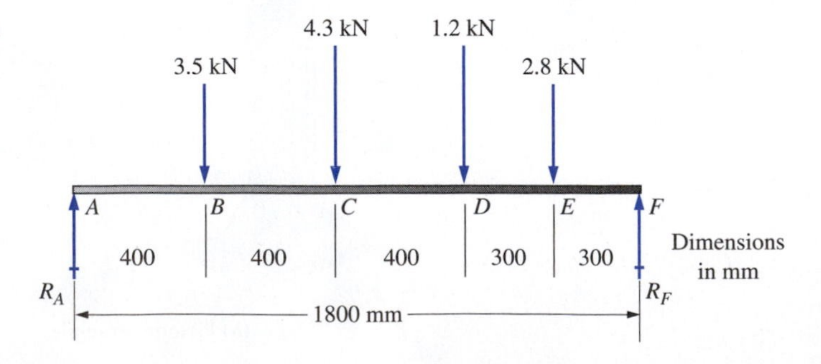

Solution Objective Compute the reaction forces at the ends of the beam.

Given Pictorial of the beam shown in Figure 5–2. Free-body diagram showing the loading is Figure 5–16. Loads are applied at points labeled B, C, D, and E. Reactions act at points A and F and are called R_A and R_F.

Analysis The *Guidelines for solving for reactions* will be used to compute the reactions. Figure 5–16 is the free-body diagram, so we will start with Step 2.

Results To find the reaction R_F, sum moments about points A.

$$\sum M_A = 0 = 3.5(400) + 4.3(800) + 1.2(1200) + 2.8(1500) - R_F(1800)$$

Note that all forces are in kN and distances in mm. Each moment term has the units of kN·mm. Now solve for R_F.

$$R_F = \frac{[3.5(400) + 4.3(800) + 1.2(1200) + 2.8(1500)]\ \text{kN}\cdot\text{mm}}{1800\ \text{mm}} = 5.82\ \text{kN}$$

Now, to find R_A, sum moments about point F.

$$\sum M_F = 0 = 2.8(300) + 1.2(600) + 4.3(1000) + 3.5(1400) - R_A(1800)$$

$$R_A = \frac{[2.8(300) + 1.2(600) + 4.3(1000) + 3.5(1400)]\ \text{kN}\cdot\text{mm}}{1800\ \text{mm}} = 5.98\ \text{kN}$$

Now use $\Sigma F = 0$ for the vertical direction as a check.

Downward forces: $(3.5 + 4.3 + 1.2 + 2.8)\ \text{kN} = 11.8\ \text{kN}$

Upward reactions: $(5.82 + 5.98)\ \text{kN} = 11.8\ \text{kN}$ (check)

Comment Remember to show the reaction forces R_A and R_F at their proper points on the beam and in the proper directions.

Example Problem 5–2 Compute the reactions on the beam shown in Figure 5–17(a).

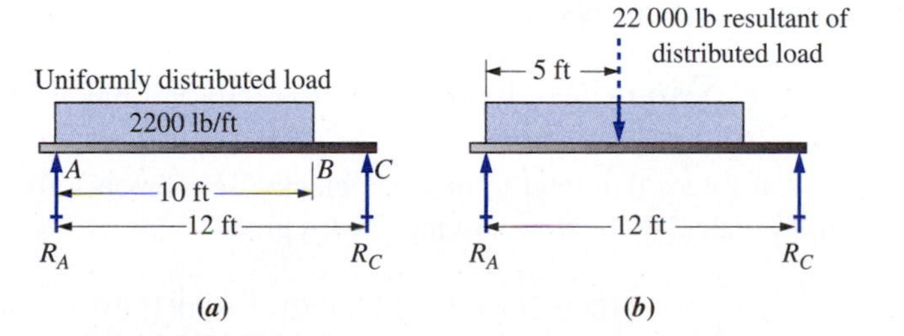

FIGURE 5–17 Beam loading.

Solution Objective Compute the reaction forces at the ends of the beam.

Given Pictorial of the beam shown in Figure 5–17(a). Distributed load of 2200 lb/ft is applied over 10 ft at the left end of the beam. Reactions act at points A and C and are called R_A and R_C.

Analysis The *Guidelines for solving for reactions* will be used to compute the reactions. Figure 5–17(b) is an equivalent free-body diagram with the resultant of the distributed load shown acting at the centroid of the load.

Results

$$\sum M_A = 0 = 22\,000\ \text{lb}(5\ \text{ft}) - R_C(12\ \text{ft})$$

$$R_C = \frac{22\,000\ \text{lb}(5\ \text{ft})}{12\ \text{ft}} = 9167\ \text{lb}$$

$$\sum M_C = 0 = 22\,000\ \text{lb}(7\ \text{ft}) - R_A(12\ \text{ft})$$

$$R_A = \frac{22\,000\ \text{lb}(7\text{ft})}{12\ \text{ft}} = 12\,833\ \text{lb}$$

Finally, as a check, in the vertical direction,

$$\sum F = 0$$

Downward forces: 22 000 lb

Upward forces: $R_A + R_C = 12\,833 + 9167 = 22\,000\ \text{lb}$ (check)

Comment *Note that the resultant is used only to find reactions.* Later, when we find shearing forces and bending moments, the distributed load itself must be used.

Example Problem 5–3 Compute the reactions for the overhanging beam in Figure 5–18.

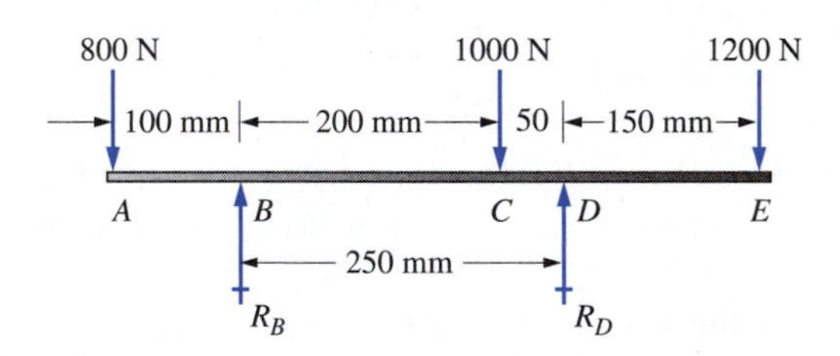

FIGURE 5-18 Beam loading.

Solution

Objective Compute the reaction forces at points B and D.

Given Loading on the beam shown in Figure 5–18. Reactions are R_B and R_D.

Analysis The *Guidelines for solving for reactions* will be used to compute the reactions.

Results First, summing moments about point B,

$$\sum M_B = 0 = 1000(200) - R_D(250) + 1200(400) - 800(100)$$

Notice that forces that tend to produce clockwise moments about B are considered positive in this calculation. Now solving for R_D gives

$$R_D = \frac{[1000(200) + 1200(400) - 800(100)]\ \text{N}\cdot\text{mm}}{250\ \text{mm}} = 2400\ \text{N}$$

Summing moments about D will allow computation of R_B.

$$\sum M_D = 0 = 1000(50) - R_B(250) + 800(350) - 1200(150)$$

$$R_B = \frac{[1000(50) + 800(350) - 1200(150)]\ \text{N}\cdot\text{mm}}{250\ \text{mm}} = 600\ \text{N}$$

Check with $\Sigma F = 0$ in the vertical direction:

$$\text{Downward forces:} \quad (800 + 1000 + 1200)\text{ N} = 3000\text{ N}$$

$$\text{Upward forces:} \quad R_B + R_D = (600 + 2400)\text{ N} = 3000\text{ N} \quad \text{(check)}$$

Comment In summary, the left reaction, R_B, is 600 N and the right reaction, R_D, is 2400 N.

Example Problem 5–4 Compute the reactions for the cantilever beam in Figure 5–19.

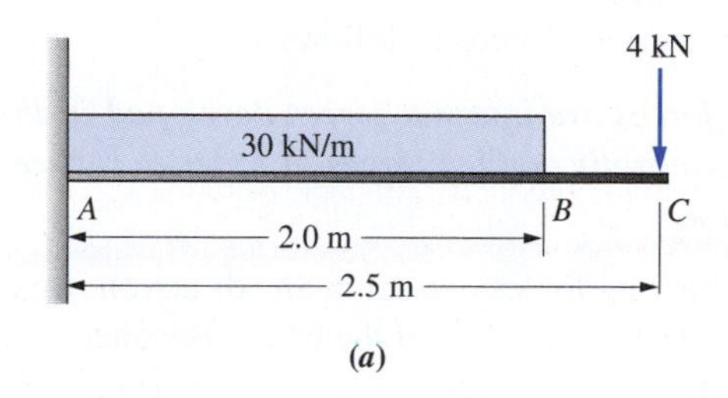

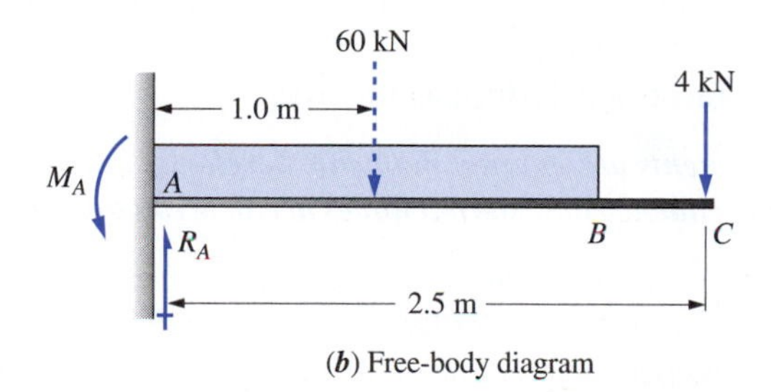

FIGURE 5–19 Beam loading.

Solution

Objective Compute the reactions at point A at the wall.

Given Loading on the beam shown in Figure 5–19(a).

Analysis The *Guidelines for solving for reactions* will be used.

In the case of cantilever beams, the reactions at the wall are composed of an upward force R_A that must balance all downward forces on the beam and a reaction moment M_A that must balance the tendency for the applied loads to rotate the beam. These are shown in Figure 5–19(b). Also shown is the resultant, 60 kN, of the distributed load.

Results Then, by summing forces in the vertical direction, we obtain

$$R_A = 60\text{ kN} + 4\text{ kN} = 64\text{ kN}$$

Summing moments about point A yields

$$M_A = 60\text{ kN}(1.0\text{ m}) + 4\text{ kN}(2.5\text{ m}) = 70\text{ kN}\cdot\text{m}$$

Comment The reaction at the support for the cantilever shown in Figure 5–19 includes a vertical force, $R_A = 64$ kN and a counterclockwise moment, $M_A = 70$ kN·m.

Again, note that the resultant of the distributed load is used only for finding reactions. The original distributed load acting over its entire length must be considered when analyzing shearing forces and bending moments in later sections.

5–4 SHEARING FORCES AND BENDING MOMENTS FOR CONCENTRATED LOADS

Chapters 7 and 8 will show that two kinds of stresses are developed in a beam, *shearing stresses* and *bending stresses*. Shearing stress is developed within the beam because of internal shearing forces that tend to shear, or cut, the beam. Bending stress is developed by internal bending moments that tend to bend the beam into a curved shape. The internal shearing forces and bending moments are produced in reaction to external forces and moments applied to the beam. Recall what you did in the activity.

Prior to studying the actual shearing stresses and bending stresses, this chapter helps you to visualize and to compute the values for the internal shearing forces and bending moments themselves. Each is dependent on the nature of the applied loads on the beam and the way it is supported.

We define shearing forces as follows:

Shearing forces are internal forces developed in the material of a beam to balance externally applied forces in order to secure equilibrium of all parts of the beam.

While internal shearing forces can act in any direction, we need to consider first those that act perpendicular to the long axis of the beam. Because we typically visualize beams to be oriented horizontally and to carry loads acting vertically downward, these internal shearing forces will usually act vertically. Thus, we refer to them as *vertical shearing forces,* indicated by the symbol, V. Of course, beams can actually be oriented in any direction as illustrated in Figures 5–3 and 5–4.

Bending moments are defined as follows:

Bending moments are internal moments developed in the material of a beam to balance the tendency for external forces to cause rotation of any part of the beam.

Bending moments cause the beam to assume its characteristic curved, or "bent," shape. As you did in the activity at the start of this chapter, it is desirable to use a flexible flat beam to visualize the gross behavior of a beam in response to different loading patterns and support conditions.

Figure 5–20 shows a simply supported beam with a single concentrated load acting at the midpoint of its length, similar to that shown in Figure 5–1 and used in the activity. Our goal now is to develop data and observations from which we can prepare complete plots of the variations in shearing forces and bending moments across the entire length of the beam. We use the free-body diagram approach as we did in the activity, considering segments equally spread along the length of the beam at points A, B, C, D, and E. Parts (b), (c), (d), and (e) of Figure 5–20 show the resulting free-body diagrams, using only segments to the left of the cuts at the given positions.

The diagrams in parts (b) and (c) result in vertical shearing forces acting downward at the cut section. We now establish the following sign convention for shearing forces:

> Internal shearing forces acting *downward* are considered to be *positive.*
> Internal shearing forces acting *upward* are considered to be *negative.*

We can also generalize the method used to find the magnitude of the shearing force at any section of a beam as follows:

> The magnitude of the shearing force in any part of a beam is equal to the algebraic sum of all external forces acting to the left of the section of interest.

FIGURE 5–20 Free-body diagrams used to find shearing forces and bending moments.

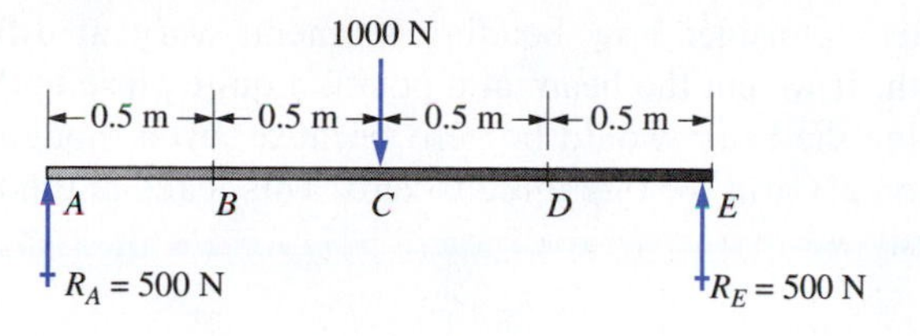

(*a*) Beam loading

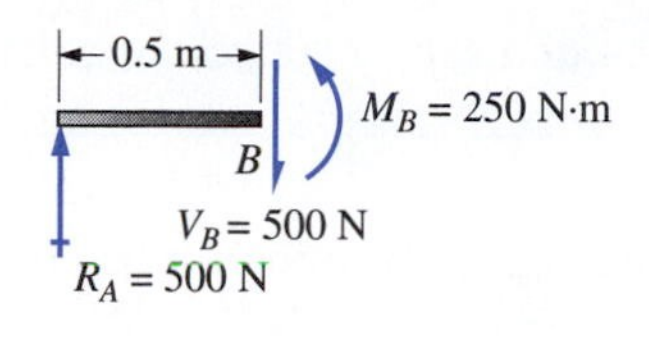

(*b*)

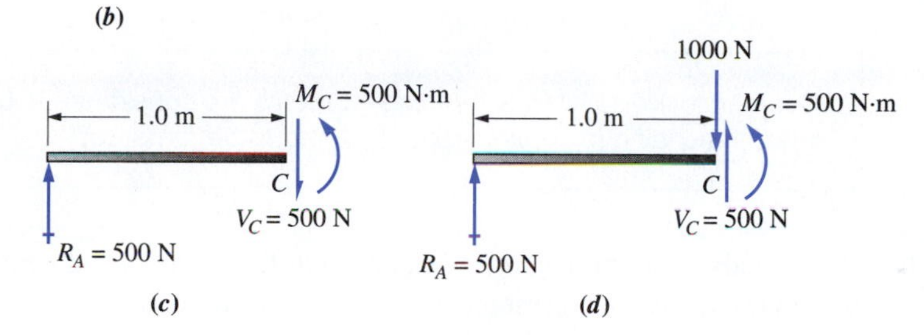

(*c*) (*d*)

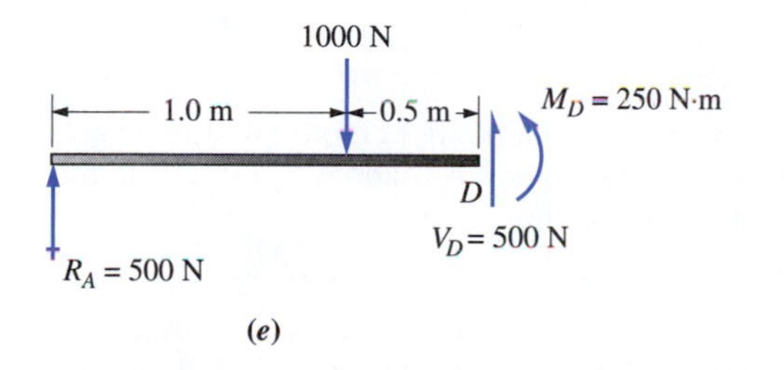

(*e*)

As you study parts (b) and (c) of Figure 5–20, it should be clear that if you consider a cut section anywhere between position A and position C, the internal shearing force would be 500 N. The general rule demonstrated here is:

> On any segment of a beam where no loads are applied, the value of the shearing force remains constant.

Note that diagrams (c) and (d) are both taken at position C. However, one is just before the point of application of the 1000 N force, and the other is just beyond where the force is applied. The result is that the internal vertical shearing force abruptly changes direction (sign) from (downward) positive to (upward) negative at C where the concentrated force is applied. The following rule can now be stated:

> A concentrated load (or reaction) on a beam causes an abrupt change in the shearing force in the beam by an amount equal to the magnitude of the load and in the direction of the load.

Now let's consider how bending moments vary at different points along the beam's length. If we cut the beam at a position quite close to A at the left support, the moment inside the beam would be zero because, even though the reaction force acts there, the moment arm for this force is zero. This leads to the following rule for bending moments:

> The bending moments at the ends of a simply supported beam are zero.

We now establish the following sign convention for bending moments:

> Counterclockwise internal bending moments are considered to be positive. Clockwise internal bending moments are considered to be negative.

Study diagrams (b), (c), (d), and (e) in Figure 5–20 and demonstrate to yourself how the values of the bending moment were determined by summing moments about the position of the cut section as was done in the activity earlier. The general rule is then:

> The magnitude of the internal bending moment at any section of a beam is equal to the algebraic sum of the moments, taken with respect to the cut section, of all forces acting to the left of the section of interest.

We can summarize the results from the analyses at the various cut sections of the beam.

Section A:	$V_A = 500\text{ N}$	$M_A = 0$
Section B:	$V_B = 500\text{ N}$	$M_B = 250\text{ N}\cdot\text{m}$
Section C to left of load:	$V_C = 500\text{ N}$	$M_C = 500\text{ N}\cdot\text{m}$
Section C to right of load:	$V_C = -500\text{ N}$	$M_C = 500\text{ N}\cdot\text{m}$
Section D:	$V_D = -500\text{ N}$	$M_D = 250\text{ N}\cdot\text{m}$
Section E:	$V_E = -500\text{ N}$	$M_E = 0\text{ N}\cdot\text{m}$

It is useful to plot these values on diagrams as shown in Figure 5–21. The top diagram is called the *load diagram*. The middle one is the *shearing force diagram* and the lower one is the *bending moment diagram*. The three diagrams should be aligned as shown to emphasize the relationships of the shearing forces and the bending moments to positions along the length of the beam. *You should also label the vertical axes as shown to communicate to any user of the diagrams exactly what values are being plotted and in what units they are expressed.*

Review, now, the various rules given in shaded boxes during this analysis. Note how the effects of those rules are illustrated in the diagrams of Figure 5–21. Also notice that the bending moment plots as a straight line with a positive slope between *A* and *C*. It then plots as a straight line with a negative slope between *C* and *E*. These observations lead to other general rules:

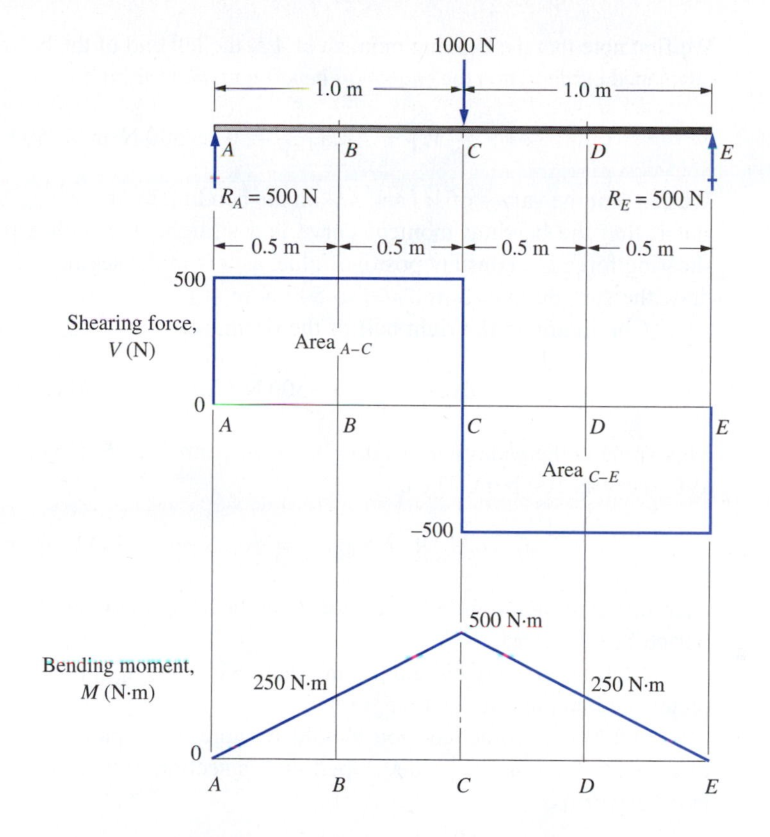

FIGURE 5–21 Shearing force and bending moment diagrams.

> The plot of the bending moment curve will be a straight line over those segments of the beam where the shearing force has a constant value. If the shearing force in the segment is positive, the bending moment curve will have a constant positive slope. Conversely, if the shearing force in the segment is negative, the bending moment curve will have a constant negative slope.

One more rule can be developed that allows the preparation of the bending moment diagram directly from the data in the shearing force diagram. We state the rule first, then we apply it to the beam loading shown in Figure 5–21. Then we show the mathematical basis for the rule.

> The change in bending moment between two points on a beam is equal to the area under the shearing force curve between the same two points.

In most cases, the areas required to implement this rule are simply computed. This rule, called the *area rule,* can be applied over a segment of any length in a beam to determine the change in bending moment.

Consider the segment *A–C* in Figure 5–21. The shearing force is a constant value of 500 N over the 1.0 m length of this segment. Then the area under the shearing force curve is,

$$\text{Area}_{A-C} = (500\text{ N})(1.0\text{ m}) = 500\text{ N}\cdot\text{m}$$

We first note that the bending moment at A at the left end of the beam is zero based on a rule developed earlier. Then the *change in bending moment* from A to C is 500 N·m. Then,

$$M_C = M_A + \text{Area}_{A-C} = 0 + 500\,\text{N·m} = 500\,\text{N·m}$$

We can plot the values of M_A and M_C on the bending moment diagram. Then an earlier rule stated that the bending moment curve is a straight line with a positive slope when the shearing force is a constant positive value, as it is in the segment A–C. You can then simply draw the straight line from 0 at A to 500 N·m at C.

Continuing to the right half of the beam, segment C–E, we compute the area to be

$$\text{Area}_{C-E} = (-500\,\text{N})(1.0\,\text{m}) = -500\,\text{N·m}$$

This value is the *change in bending moment* from C to E. However, the bending moment at C starts as 500 N·m. Then,

$$M_E = M_C + \text{Area}_{C-E} = 500\,\text{N·m} - 500\,\text{N·m} = 0\,\text{N·m}$$

Also demonstrated here is the rule that the bending moment at the ends of a simply supported beam is zero.

Plotting M_C and M_E and joining them with a straight line with a negative slope completes the bending moment diagram.

After some practice, you should be able to prepare the three beam diagrams by directly applying the rules developed in this section without drawing the individual free-body diagrams.

Development of the Area Rule for Bending Moment Diagrams. In general, the area rule can be stated as

$$dM = V\,dx \tag{5–1}$$

where dM is the change in moment due to a shearing force V acting over a small length segment dx. Figure 5–22 illustrates Equation (5–1).

Over a larger length segment, the process of integration can be used to determine the total change in moment over the segment. Between two points, A and C,

$$\int_{M_A}^{M_C} dM = \int_{x_A}^{x_C} V\,dx \tag{5–2}$$

If the shearing force V is constant over the segment, as it is in segment A–C in Figure 5–22, Equation (5–2) becomes

$$\int_{M_A}^{M_C} dM = V\int_{x_A}^{x_C} dx \tag{5–3}$$

Completing the integration gives

$$M_C - M_A = V(x_C - x_A) \tag{5–4}$$

This result is consistent with the rule stated previously. Note that $M_C - M_A$ is the *change* in moment between points A and C. The right side of Equation (5–4) is the area under the shearing force curve between A and C.

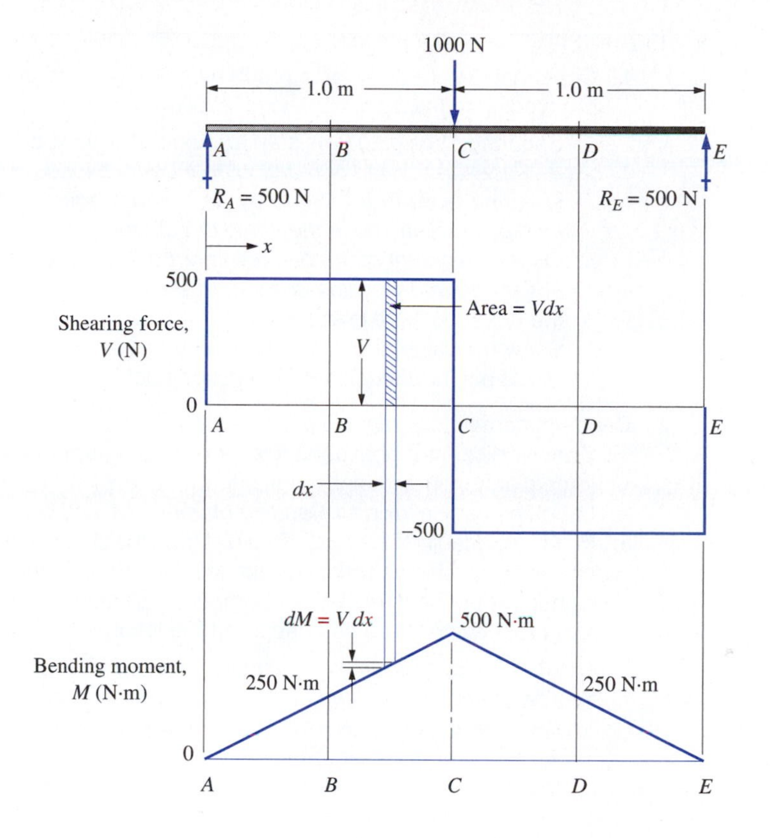

FIGURE 5–22 Shearing force and bending moment diagrams—area rule.

5–5 GUIDELINES FOR DRAWING BEAM DIAGRAMS FOR CONCENTRATED LOADS

The principles developed and illustrated in the preceding section can be generalized into a set of guidelines that can be used to draw the load, shearing force, and bending moment diagrams for virtually any simply supported beam carrying concentrated loads.

The guidelines are listed in the following shaded box. You should review them now and relate them to the rules developed in Section 5–4. Then we will apply the guidelines to new and more complex problems.

Guidelines for drawing the load, shearing force, and bending moment diagrams for simply supported beams carrying only concentrated loads.

A. Load diagram.
   1. Draw the beam and show the loads and reaction forces acting on it.
   2. Label points on the beam where loads or reactions act with letters.
   3. Solve for the magnitudes and directions of the reaction forces and show them on the load diagram.

B. Shearing force diagram.
   1. Draw vertical and horizontal axes for the diagram in relation to the drawing of the beam loading diagram in a manner similar to that shown in Figure 5–21.
   2. Label the vertical axis as *Shearing force, V*, and give the units for force.
   3. Project lines from each applied load or reaction force on the beam loading diagram down to the shearing force diagram. Label points of interest for reference. Label points where loads or reaction forces act as done in the load diagram.

4. Construct the shearing force graph by starting at the left end of the beam and proceeding to the right, applying the following rules.
   a. *Shearing force diagrams start and end at zero at the ends of the beam.*
   b. *A concentrated load or reaction on a beam causes an abrupt change in the shearing force in the beam by an amount equal to the magnitude of the load or reaction and in the same direction.*
   c. *On any segment of the beam where no loads are applied, the value of the shearing force remains constant, resulting in a straight horizontal line on the shearing force diagram.*
   d. Show the value of the shearing force at key points on the diagram, generally at points where loads or reactions act.

C. Bending moment diagram.
1. Draw vertical and horizontal axes for the diagram in relation to the shearing force diagram in a manner similar to that shown in Figure 5–21.
2. Label the vertical axis as *Bending Moment, M,* and give the units for moment.
3. Project vertical lines at points of interest from the shearing force diagram, including all points used for the shearing force diagram, down to the bending moment axis.
4. Construct the bending moment graph by starting at the left end of the beam and proceeding to the right, applying the following rules.
   a. *At the ends of a simply supported beam, the bending moment is zero.*
   b. *The change in bending moment between two points on a beam is equal to the area under the shearing force curve between those two points. Thus, when the area under the shearing force curve is positive (above the axis), the bending moment is increasing, and vice versa.*
   c. *The maximum bending moment occurs at a point where the shearing force curve crosses its zero axis.*
   d. *On a section of the beam where no loads are applied, the bending moment diagram will be a straight line.*
   e. *The slope of the bending moment curve at any point is equal to the magnitude of the shearing force at that point.*
   f. Label the values for the bending moment on the graph at each point of interest.

Example Problem 5–5 Draw the complete shearing force and bending moment diagrams for the beam shown in Figure 5–23.

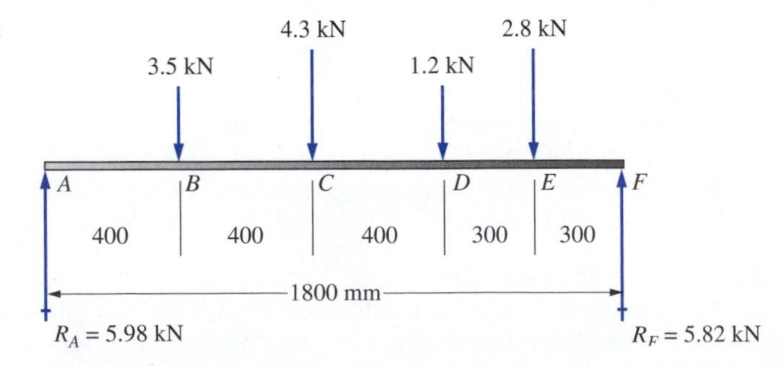

FIGURE 5–23 Beam loading.

Solution Objective Draw the complete shearing force and bending moment diagrams.

Given The beam loading shown in Figure 5–23. It is a simply supported beam carrying all concentrated loads. This is the same beam for which the reaction forces were computed in Example Problem 5–1. The results of Part A of the guidelines are shown in Figure 5–23.

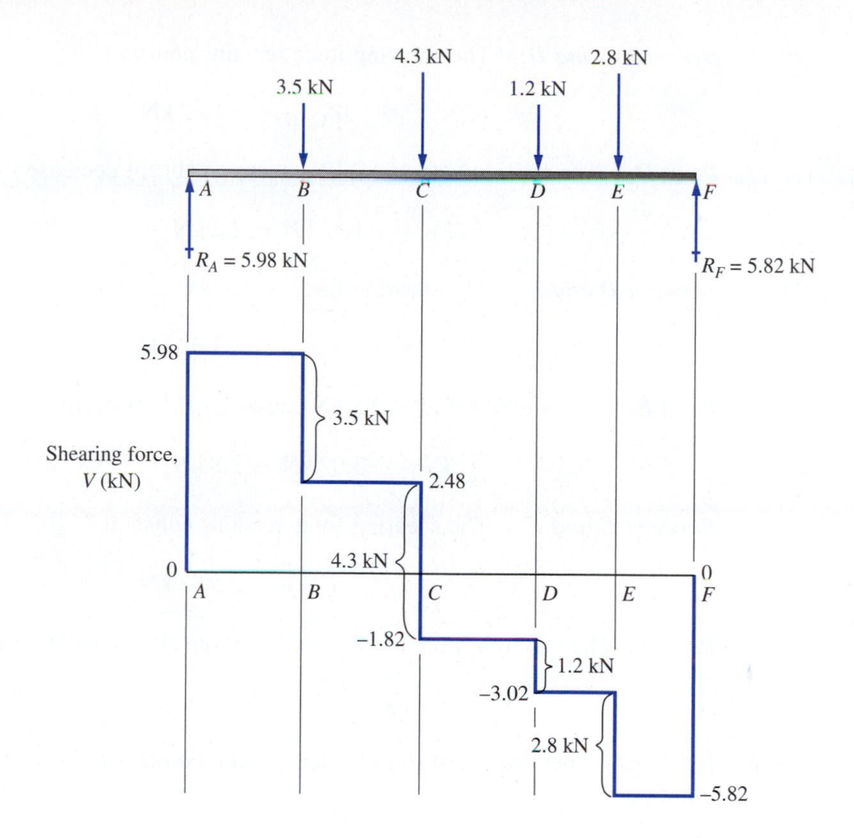

FIGURE 5–24 Shearing force diagram.

Analysis The *guidelines for drawing shearing force and bending moment diagrams for simply supported beams carrying only concentrated loads* will be used.

Results Part B The completed shearing force diagram is shown in Figure 5–24. The process used to determine each part of the diagram is described.

Point A: The reaction R_A is encountered immediately. Then,

$$V_A = R_A = 5.98 \text{ kN}$$

Between A and B: Because there are no loads applied, the shearing force remains constant. That is,

$$V_{A-B} = 5.98 \text{ kN}$$

Point B: The applied load of 3.5 kN causes an abrupt decrease in V.

$$V_B = 5.98 \text{ kN} - 3.5 \text{ kN} = 2.48 \text{ kN}$$

Between B and C: The shearing force remains constant.

$$V_{B-C} = 2.48 \text{ kN}$$

Point C: The applied load of 4.3 kN causes an abrupt decrease in V.

$$V_C = 2.48 \text{ kN} - 4.3 \text{ kN} = -1.82 \text{ kN}$$

Between C and D: The shearing force remains constant.

$$V_{C-D} = -1.82 \text{ kN}$$

Point D: The applied load of 1.2 kN causes an abrupt decrease in V.

$$V_D = -1.82 \text{ kN} - 1.2 \text{ kN} = -3.02 \text{ kN}$$

Between D and E: The shearing force remains constant.

$$V_{D-E} = -3.02 \text{ kN}$$

Point E: The applied load of 2.8 kN causes an abrupt decrease in V.

$$V_E = -3.02 \text{ kN} - 2.8 \text{ kN} = -5.82 \text{ kN}$$

Between E and F: The shearing force remains constant.

$$V_{E-F} = -5.82 \text{ kN}$$

Point F: The reaction force of 5.82 kN causes an abrupt increase in V.

$$V_F = -5.82 \text{ kN} + 5.82 \text{ kN} = 0$$

Comment Note that values for the shearing forces at key points are shown right on the diagram at those points.

Results Part C The bending moment diagram is drawn directly below the shearing force diagram so the relationship between points on the beam loading diagram can be related to both. See Figure 5–25. It is helpful to extend vertical lines from the points of interest on the beam (points A to F in this example) down to the horizontal axis of the bending moment diagram.

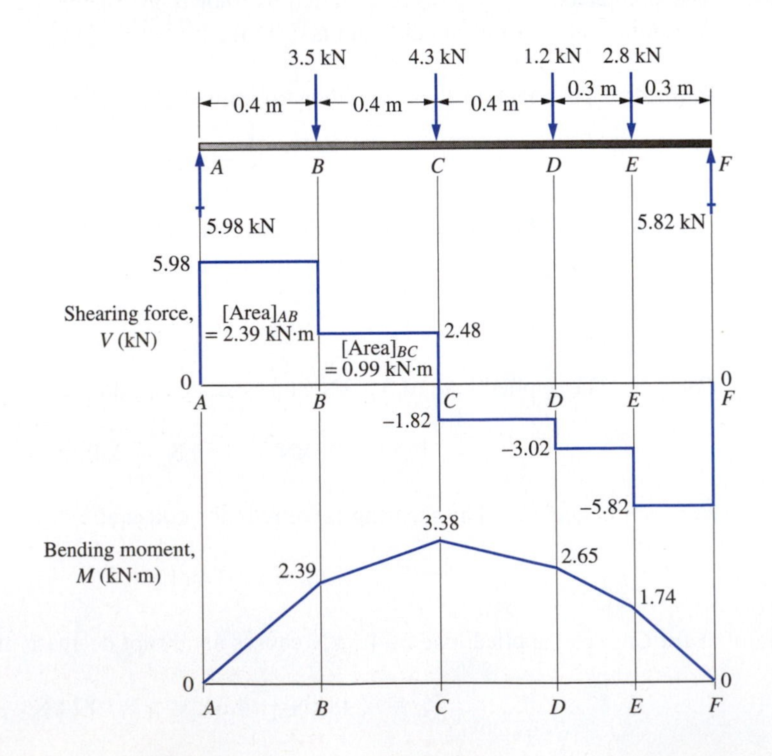

FIGURE 5–25 Shearing force and bending moment diagrams.

It is most convenient to start drawing the bending moment diagram at the left end of the beam and work toward the right, considering each segment separately. For beams like this one having all concentrated loads, the segments should be chosen as those for which the shearing force is constant. In this case, the segments are AB, BC, CD, DE, and EF.

At point A: We use the rule that the bending moment is zero at the ends of a simply supported beam. That is, $M_A = 0$.

Point B: For this and each subsequent point, we apply the area rule. The general pattern is

$$M_B = M_A + [\text{Area}]_{AB}$$

where $[\text{Area}]_{AB}$ = area under the shearing force curve between A and B. Using data from the shearing force curve,

$$[\text{Area}]_{AB} = V_{AB} \times \text{width of the segment } AB$$

But, $V_{AB} = 5.98$ kN over the segment AB that is 0.40 m long. Then, as shown in Figure 5–25,

$$[\text{Area}]_{AB} = 5.98\ \text{kN}(0.40\ \text{m}) = 2.39\ \text{kN}\cdot\text{m}$$

Finally, then,

$$M_B = M_A + [\text{Area}]_{AB} = 0 + 2.39\ \text{kN}\cdot\text{m} = 2.39\ \text{kN}\cdot\text{m}$$

This value is plotted at point B on the bending moment diagram. Then a straight line is drawn from M_A to M_B because the shearing force is constant over that segment. Bending moment values at C, D, E, and F are found in a similar manner. Refer to Figure 5–25.

Point C:

$$M_C = M_B + [\text{Area}]_{BC}$$
$$[\text{Area}]_{BC} = 2.48\ \text{kN}(0.40\ \text{m}) = 0.99\ \text{kN}\cdot\text{m}$$
$$M_C = 2.39\ \text{kN}\cdot\text{m} + 0.99\ \text{kN}\cdot\text{m} = 3.38\ \text{kN}\cdot\text{m}$$

Point D:

$$M_D = M_C + [\text{Area}]_{CD}$$
$$[\text{Area}]_{CD} = -1.82\ \text{kN}(0.40\ \text{m}) = -0.73\ \text{kN}\cdot\text{m}$$
$$M_D = 3.38\ \text{kN}\cdot\text{m} - 0.73\ \text{kN}\cdot\text{m} = 2.65\ \text{kN}\cdot\text{m}$$

Note that $[\text{Area}]_{CD}$ is negative because it is below the axis.

Point E:

$$M_E = M_D + [\text{Area}]_{DE}$$
$$[\text{Area}]_{DE} = -3.02\ \text{kN}(0.30\ \text{m}) = -0.91\ \text{kN}\cdot\text{m}$$
$$M_E = 2.65\ \text{kN}\cdot\text{m} - 0.91\ \text{kN}\cdot\text{m} = 1.74\ \text{kN}\cdot\text{m}$$

Point F:

$$M_F = M_E + [\text{Area}]_{EF}$$
$$[\text{Area}]_{EF} = -5.82\text{ kN}(0.30\text{ m}) = -1.74\text{ kN}\cdot\text{m}$$
$$M_F = 1.74\text{ kN}\cdot\text{m} - 1.74\text{ kN}\cdot\text{m} = 0\text{ kN}\cdot\text{m}$$

Summary and Comment — The bending moment values are shown on the diagram at their respective points so users of the diagram can see the relative values. The fact that $M_F = 0$ is a check on the calculations because the rule for simply supported beams states that the bending moment at F must be zero. The objective of drawing the bending moment diagram is often to locate the point where the maximum bending moment exists. Here we see that $M_C = 3.38$ kN·m is the maximum value.

Example Problem 5–6 Draw the complete shearing force and bending moment diagrams for the beam shown in Figure 5–26.

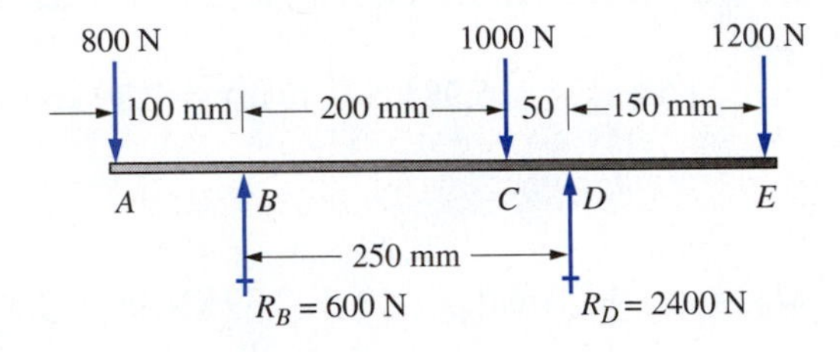

FIGURE 5-26 Beam loading.

Solution

Objective — Draw the complete shearing force and bending moment diagrams.

Given — The beam loading shown in Figure 5–26. It is an overhanging beam carrying all concentrated loads. This is the same beam for which the reaction forces were computed in Example Problem 5–3. The results of Part A of the guidelines are shown in Figure 5–26. $R_B = 600$ N. $R_D = 2400$ N.

Analysis — The *guidelines for drawing shearing force and bending moment diagrams for beams carrying only concentrated loads* will be used.

Results — The steps of Part B of the guidelines were used to prepare the completed shearing force diagram shown in Figure 5–27. The process is described here.

Point A: The applied downward force of 800 N causes the shearing force diagram to drop immediately to −800 N. That is, $V_A = -800$ N.

Between A and B: No loads are applied in this segment. Then $V_{A\text{–}B} = -800$ N.

Point B: The upward reaction force of 600 N acts at B. Then

$$V_B = -800\text{ N} + 600\text{ N} = -200\text{ N}$$

Between B and C: No loads are applied in this segment. Then $V_{B\text{–}C} = -200$ N.

Point C: The applied downward force of 1000 N causes the shearing force diagram to drop immediately by that amount. Then,

$$V_C = -200\text{ N} - 1000\text{ N} = -1200\text{ N}$$

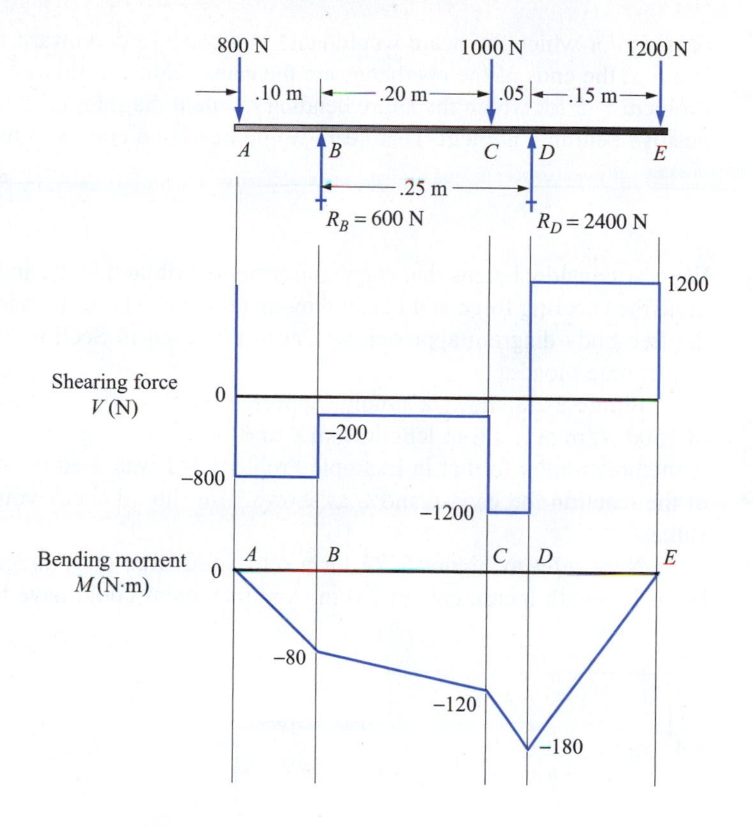

FIGURE 5–27 Shearing force and bending moment diagrams for Example Problem 5–6.

Between C and D: No loads are applied in this segment. Then $V_{C\text{-}D} = -1200$ N.

Point D: The upward reaction force of 2400 N acts at D. Then

$$V_D = -1200\text{ N} + 2400\text{ N} = +1200\text{ N}$$

Between D and E: No loads are applied in this segment. Then $V_{D\text{-}E} = +1200$ N.

Point E: The applied downward force of 1200 N causes the shearing force diagram to drop immediately by that amount. Then,

$$V_E = 1200\text{ N} - 1200\text{ N} = 0\text{ N}$$

The steps of Part C of the guidelines were used to prepare the completed bending moment diagram shown in Figure 5–27. The process is described here.

At Point A: This is a free end of the beam. Therefore, $M_A = 0$ N·m.

We apply the area rule for each subsequent point on the beam.

Point B: $M_B = M_A + [\text{Area}]_{AB} = 0 + (-800\text{ N})(0.10\text{ m}) = -80\text{ N·m}$

Point C: $M_C = M_B + [\text{Area}]_{BC} = -80\text{ N·m} + (-200\text{ N})(0.20\text{ m}) = -120\text{ N·m}$

Point D: $M_D = M_C + [\text{Area}]_{CD} = -120\text{ N·m} + (-1200\text{ N})(0.05\text{ m}) = -180\text{ N·m}$

Point E: $M_E = M_D + [\text{Area}]_{DE} = -180\text{ N·m} + (1200\text{ N})(0.15\text{ m}) = 0\text{ N·m}$

Summary and Comment

When seeking the maximum bending moment, we really look for the maximum absolute value. The maximum bending moment is -180 N·m at point D. Note that the resulting bending moment diagram for this problem lies below the axis. This is indicative of *negative bending* for which the beam would bend in a concave downward fashion. The forces at A and E at the ends of the overhangs are the cause. Contrast this with the result of Example Problem 5–5 for which the entire bending moment diagram is above the axis, indicating a positive bending moment. That beam would bend in a concave upward fashion.

5–6 SHEARING FORCES AND BENDING MOMENTS FOR DISTRIBUTED LOADS

We now consider beams that carry uniformly distributed loads and analyze them to determine the shearing force and bending moment that exists in any part of the beam. Let's use the free-body diagram approach similar to that used in Section 5–5 for beams with only concentrated loads.

Figure 5–28 shows a simply supported beam carrying a uniformly distributed load of 1500 N/m over a 6-m length from A to B. The 3-m length from B to C carries no load. A method similar to that in Example Problem 5–2 was used to determine the magnitude of the reaction forces at A and C as shown. You should check your ability to verify these values.

Now refer to Figure 5–29 for a set of four free-body diagrams of portions of the beam in length increments of 2.0 m. Any increment could have been chosen. The 2.0-m

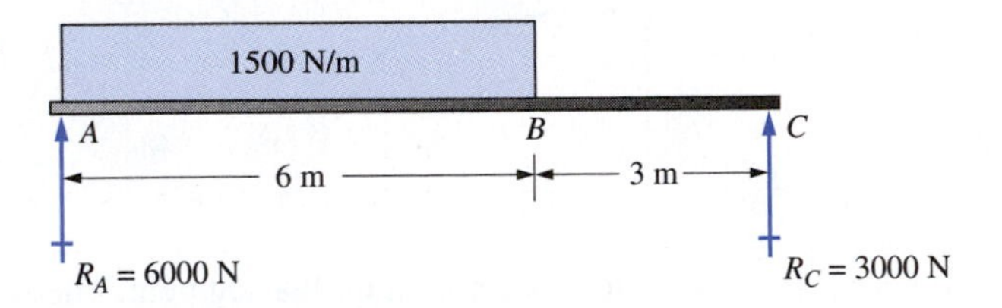

FIGURE 5–28 Beam loading.

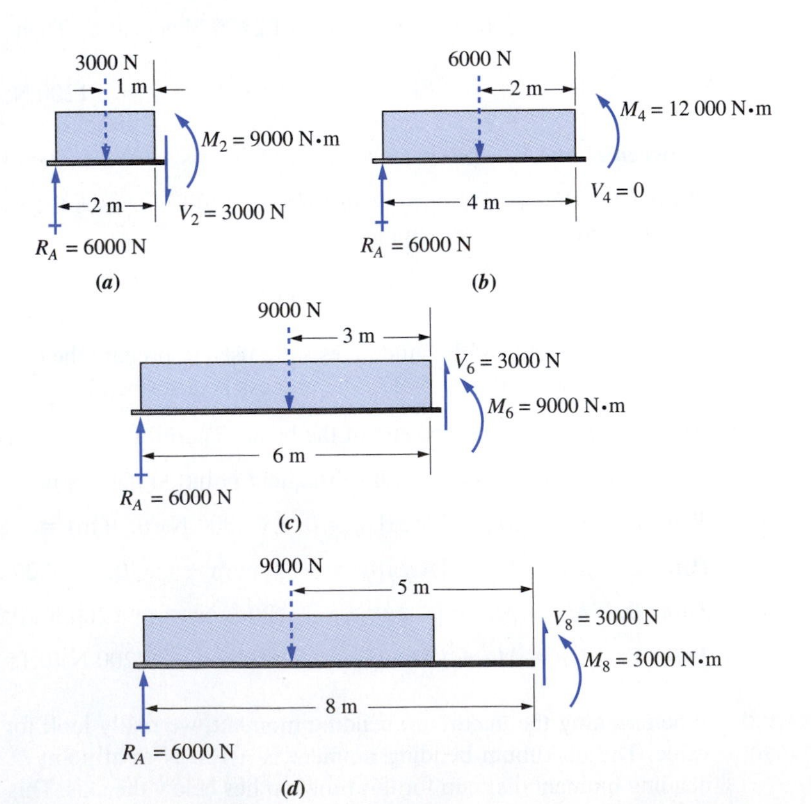

FIGURE 5–29 Free-body diagrams used to find shearing forces and bending moments.

increments provide a sufficient number of points to eventually plot graphs for the shearing force and bending moment diagrams using a method similar to that used in the preceding sections. But the shape of the curves will be different because of the distributed load. Follow along with the development of the values for shearing force and bending moment. Then we generalize the method of producing the diagrams directly without the use of free-body diagrams.

Part (a) of Figure 5–29: This is the free-body diagram of the first 2.0 m segment. The reaction force, $R_A = 6000$ N, acts at the left end of the beam, point A. Then, we determine that the 2.0-m length of the 1500 N/m distributed load exerts a resultant force of 3000 N and it can be visualized to act midway on the segment. Then,

At A:

$$\text{Shearing force} = V_A = 6000\ \text{N}.$$
$$\text{Bending moment} = M_A = 0.$$

At the Cut Section:

$$V_2 = 6000\ \text{N} - 3000\ \text{N} = 3000\ \text{N (downward)}$$
$$M_2 = (6000\ \text{N})(2.0\ \text{m}) - (3000\ \text{N})(1.0\ \text{m}) = 9000\ \text{N}\cdot\text{m (CCW)}$$

The subscript 2 indicates that the cut section is at 2.0 m from point A.

Part (b) of Figure 5–29: This is the segment having a length of 4.0 m. The resultant of the distributed load is 6000 N shown acting midway on the segment. Then,

At the Cut Section:

$$V_4 = 6000\ \text{N} - 6000\ \text{N} = 0$$
$$M_4 = (6000\ \text{N})(4.0\ \text{m}) - (6000\ \text{N})(2.0\ \text{m}) = 12\,000\ \text{N}\cdot\text{m (CCW)}$$

Part (c) of Figure 5–29: This 6.0-m length has a resultant load of 9000 N acting at 3.0 m from A. This is the entire distributed load.

At the Cut Section:

$$V_6 = 6000\ \text{N} - 9000\ \text{N} = -3000\ \text{N (upward)}$$
$$M_6 = (6000\ \text{N})(6.0\ \text{m}) - (9000\ \text{m})(3.0\ \text{m}) = 9000\ \text{N}\cdot\text{m (CCW)}$$

Part (d) of Figure 5–29: This 8.0 m length of the beam carries the entire distributed load having a resultant of 9000 N acting 3.0 m from A as before. There are no loads acting on the final 2.0 m of the segment.

At the Cut Section:

$$V_8 = 6000\ \text{N} - 9000\ \text{N} = -3000\ \text{N (upward)}$$
$$M_8 = (6000\ \text{N})(8.0\ \text{m}) - (9000\ \text{N})(5.0\ \text{m}) = 3000\ \text{N}\cdot\text{m (CCW)}$$

Remember that the bending moment at C at the right end of the beam is zero.

Now we can plot these points as shown in Figure 5–30. We can make some observations about the shape of the shearing force and bending moment curves on the part of the beam that carries the distributed load.

- The shearing force varies linearly with position. The slope of the straight line is equal to the load per unit length of the distributed load. That is, the slope is −1500 N/m. You should observe that each 1.0-m length of the beam carries a load of 1500 N.
- The bending moment increases along a curve over the first 4.0-m length of the beam while the shearing force is positive. Then it decreases along a curve for the next 2.0 m where the shearing force is negative.

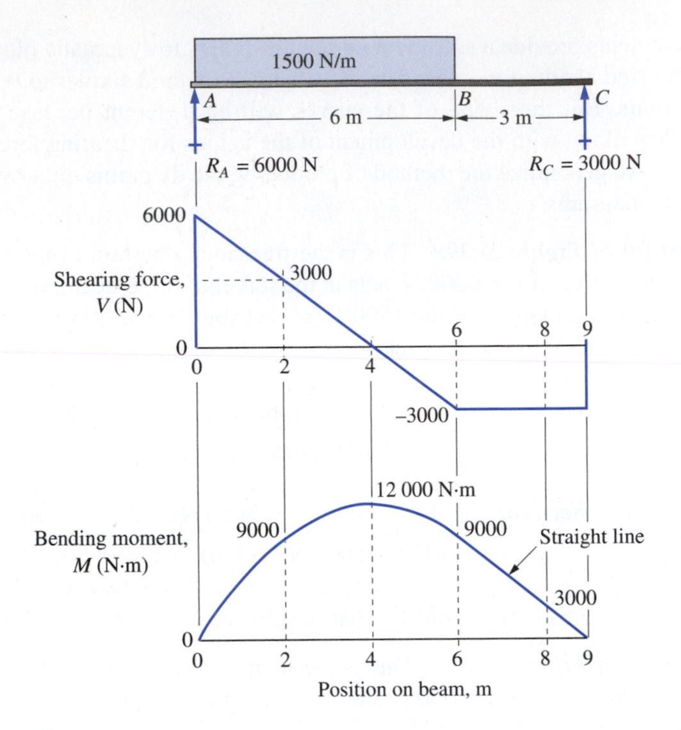

FIGURE 5–30 Complete beam loading, shearing force, and bending moment diagrams.

These observations are consistent with the guidelines for drawing beam diagrams in Section 5–6. We could have used the *area rule* to produce the values for bending moment at any point. See Rule 4b in Section C of the guidelines. For example, consider the first 4.0-m length of the beam. The shearing force starts at 6000 N at A and decreases to zero at the 4.0-m point. The area under that part of the curve is all positive, above the axis of the shearing force diagram. Therefore the bending moment will increase from zero at the free end of the beam at A by an amount equal to the area of the triangle with a height of 6000 N and a length of 4.0 m. That is,

$$M_4 = M_A + [\text{Area}]_{0-4} = 0 + (1/2)(4.0\text{ m})(6000\text{ N}) = 12\,000\text{ N}\cdot\text{m}$$

We could have used the area rule to find the value of the bending moment at B where the distributed load ends. We called this value M_6. Note that the shearing force decreases from zero to -3000 N over this length. Then the bending moment at B is

$$M_B = M_6 = M_4 + [\text{Area}]_{4-6} = 12\,000\text{ N}\cdot\text{m} + (1/2)(2.0\text{ m})(-3000\text{ N}) = 9000\text{ N}\cdot\text{m}$$

The final segment of the beam diagrams can be drawn using the guidelines given before because there are no loads applied over the last 3.0 m of the beam. The shearing force is a constant value of -3000 N. Therefore, the bending moment plots as a straight line with a constant negative slope. The change in bending moment at point 9 is,

$$M_9 = M_6 + [\text{Area}]_{6-9} = 9000\text{ N}\cdot\text{m} + (-3000\text{ N})(3.0\text{ m}) = 0$$

We now state new rules that can be appended to the guidelines for drawing shearing force and bending moment diagrams.

Whenever uniformly distributed loads act on a beam:

1. Over the length of a beam carrying a uniformly distributed load, the shearing force curve is a straight line having a negative slope equal to the amount of the load per unit length.
2. The change in shearing force between any two points on a beam where a uniformly distributed load acts is equal to the resultant of the distributed load over that length. That is, the change in shearing force is:

$$\Delta V = \text{(load per unit length)(length of the segment)}$$

3. Over the length of a beam carrying a uniformly distributed load, the bending moment plots as a curved line where the slope of the line at any point is equal to the magnitude of the shearing force at that point.
4. The change in bending moment between any two points on a beam is equal to the area under the shearing force curve between those two points.

Let's discuss Rule 3, defining the slope of the bending moment curve. Refer to Figure 5–30 again and observe the shape of the bending moment curve.

Starting at the left end of the beam, point A, we observe that the shearing force at that point is 6000 N, the largest value on the beam. Rule 3 states that the bending moment curve should have very high positive slope at that point, and that is how the curve is drawn.

Now consider point 2, 2.0 m from A. The magnitude of the shearing force has decreased to 3000 N. Therefore, the slope of the bending moment curve should be less steep but still positive. In fact, if we progress from A to point 4, the slope of the bending moment curve should start with a high positive slope, and gradually decrease in slope to zero at point 4 where the shearing force crosses the axis. Recall that a line drawn tangent to a point on a curve having a zero slope is horizontal. This identifies the peak of the curve and the value of the bending moment there is a maximum for that particular segment. This logic illustrates Rule 4c in Section C of the guidelines from before. *The maximum bending moment occurs at a point where the shearing force crosses its zero axis.* It is necessary for you to find the point where the shearing force curve crosses the axis. The example problem that follows illustrates the process.

Example Problem 5–7 Draw the complete shearing force and bending moment diagrams for the beam shown in Figure 5–31.

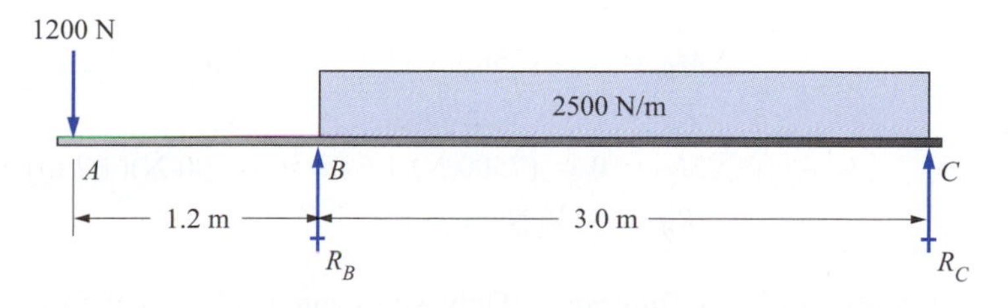

FIGURE 5–31 Beam loading for Example Problem 5–7.

Solution Objective Draw the complete shearing force and bending moment diagrams.

Given Beam loading in Figure 5–31. The beam has a concentrated load at the end of an overhang to the left of the left support and a uniformly distributed load over the span between the two supports.

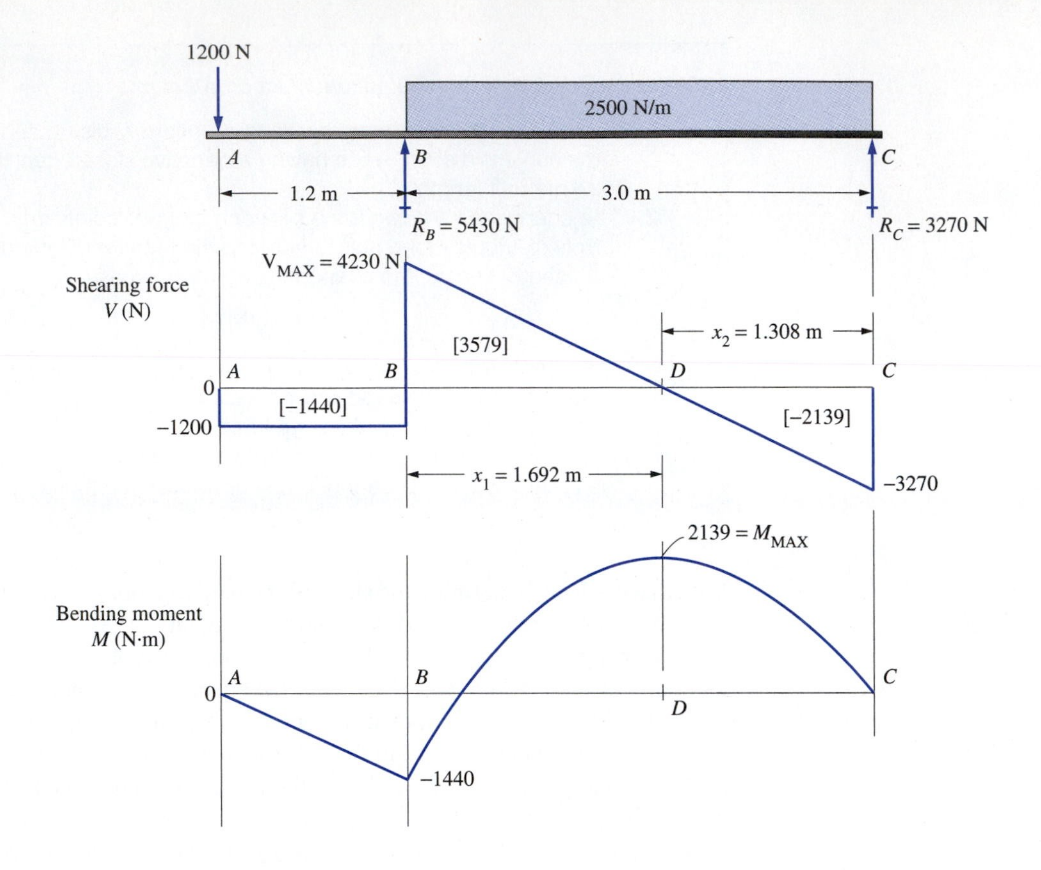

FIGURE 5–32 Results for Example Problem 5–7.

Analysis We will use the guidelines for drawing beam diagrams first presented in Section 5–5 and supplemented with additional guidelines for distributed loads developed in this section.

Results The completed beam diagrams are shown in Figure 5–32. The details are explained next.

Reactions at Supports: The resultant of the distributed load is

$$(2500 \text{ N/m})(3.0 \text{ m}) = 7500 \text{ N}$$

The resultant acts at the midpoint of the distributed load, 1.5 m from B.

$$\begin{aligned} \Sigma M_B &= 0 = (7500 \text{ N})(1.5 \text{ m}) - (1200 \text{ N})(1.2 \text{ m}) - R_C(3.0 \text{ m}) \\ R_C &= 3270 \text{ N upward} \\ \Sigma M_C &= 0 = (7500 \text{ N})(1.5 \text{ m}) + (1200 \text{ N})(4.2 \text{ m}) - R_B(3.0 \text{ m}) \\ R_B &= 5430 \text{ N upward} \end{aligned}$$

Shearing Force Diagram: Only a concentrated load acts from A to B. The shearing force is a constant negative 1200 N to the point just before the application of the reaction force at B.

At B: The reaction at B causes the shearing force to rise to

$$V_B = -1200 \text{ N} + 5430 \text{ N} = 4230 \text{ N}$$

Between B and C: The shearing force curve decreases 2500 N over each 1.0 m, a total of 7500 N for the 3.0 m span. Then,

$$V_C = V_B - 7500\text{ N} = 4230\text{ N} - 7500\text{ N} = -3270\text{ N}$$

The shearing force curve is a straight line with a constant negative slope of −2500 N/m, equal to the amount of load per unit length applied to the beam.

At C: The reaction force at C, 3270 N, brings the shearing force curve back to zero at C.

At D Where the Shearing Force Curve Crosses the Axis: It is necessary to find the location of point D because that is where the maximum bending moment on the beam occurs. The distance from B to D is called x_1 in the figure. This is the distance required for the shearing force to decrease from $V_B = 4230$ N to zero.

We use the fact that the curve is decreasing 2500 N over each 1.0 m of distance from B in a straight-line fashion. Let's define the magnitude of the distributed load as w. Then,

$$w = 2500\text{ N/m}$$

We can solve for x_1 from

$$x_1 = \frac{V_B}{w} = \frac{4230\text{ N}}{2500\text{ N/m}} = 1.692\text{ m}$$

It is appropriate to determine the distance from D to C now also because it will be needed for the bending moment calculations. This is called x_2 in the figure.

$$x_2 = 3.0\text{ m} - x_1 = 3.0\text{ m} - 1.692\text{ m} = 1.308\text{ m}$$

This completes the development of the shearing force curve.

Bending Moment Diagram:

Between A and B: The moment at A is zero because it is at the free end of the beam. The change in moment from A to B is equal to the rectangular area under the shearing force curve between A and B.

$$M_B = M_A + [\text{Area}]_{A-B} = 0 + (-1200\text{ N})(1.2\text{ m}) = -1440\text{ N·m}$$

The curve is a straight line because the shearing force has a constant value in the segment from A to B.

Between B and D: We need to determine the bending moment at D because that will be the maximum value anywhere on the beam. The change in moment from B to D is equal to the triangular area under the shearing force curve between B and D. But the moment has the value of −1440 N·m at B. Therefore, noting that the base of the triangle is $x = 1.692$ m,

$$M_D = M_B + [\text{Area}]_{B-D} = -1440\text{ N·m} + (1/2)(4230\text{ N})(1.692\text{ m}) = 2139\text{ N·m}$$

Study carefully the shape of the curve. At B the shearing force has a high positive value. It then decreases to zero at D. Then the curve starts at B with a high positive slope as indicated by the tangent to the curve. The slope decreases steadily until it becomes zero (horizontal) at D.

Between D and C: The triangular area under the shearing force curve between D and C is the change in bending moment between those two points. Then, noting that the height of the triangle is -3270 N and the base is $x_2 = 1.308$ m,

$$M_C = M_D + [\text{Area}]_{D\text{-}C} = 2139\ \text{N}\cdot\text{m} + (1/2)(-3270\ \text{N})(1.308\ \text{m}) = 0\ \text{N}\cdot\text{m}$$

This is actually a check on the calculations because the bending moment at C, a free end of the simply supported beam, *must* be zero. If any other result is found, there has been some error in the calculations.

The shape of the curve from D to C should, again, be studied carefully. At point D the value of the shearing force is zero, and so the bending moment curve at D has a zero slope. Therefore, it continues from the end of the previous curve drawn from B to D. Then the shearing force becomes negative and gets increasingly negative as it progresses to point C. The curve is concave downward, ending with a rather steep negative slope.

This completes the development of the bending moment diagram.

Comment You should observe some key data from Figure 5–32. You should note the maximum values of the shearing force and the bending moment, and where they occur. In summary, then,

Maximum shearing force: $V_B = 4230$ N
located at point B at the left support.

Maximum bending moment: $M_D = 2139$ N·m
located 1.692 m from the left support.

For some beam types, it is important to know both the maximum positive and the maximum negative values of the bending moment because the material or the shape of the beam may perform differently under each type of moment. In this beam,

Maximum negative bending moment: $M_B = -1440$ N·m

5–7 GENERAL SHAPES FOUND IN BENDING MOMENT DIAGRAMS

You will likely need practice to become proficient at drawing shearing force and bending moment diagrams and applying the guidelines described in Sections 5–4 to 5–6. One technique that is helpful is to learn the general relationships between the shape of the bending moment curve and the shearing force curve in the same segment of the beam.

Figure 5–33 should help you to visualize those relationships. One key skill is to apply correctly guideline Rule 4e in Section C listed in the shaded box in Section 5–5. That guideline states:

> ***The slope of the bending moment curve at any point is equal to the magnitude of the shearing force at that point.***

We have seen examples of many of these shapes for segments of the bending moment diagrams in Figures 5–21, 5–25, 5–27, 5–30, and 5–32.

For example, the shapes from parts 1 and 3 of Figure 5–33 appear in any beam having only concentrated loads such as Figures 5–21, 5–25, and 5–27. These same shapes apply for parts of beams that have only concentrated loads such as the last segment of Figure 5–30 and the first segment of Figure 5–32. In each of these cases, the shearing force has a constant value, resulting in a bending moment curve that has a constant slope. That is, the bending moment curve will be a straight line.

Parts 4, 5, 6, and 7 of Figure 5–33 apply to segments of a beam carrying uniformly distributed loads. For example, the first 6 m of the beam in Figure 5–30 looks like parts 5 and 6. Similarly, in Figure 5–32, the 3.0-m length between the supports looks like parts 5

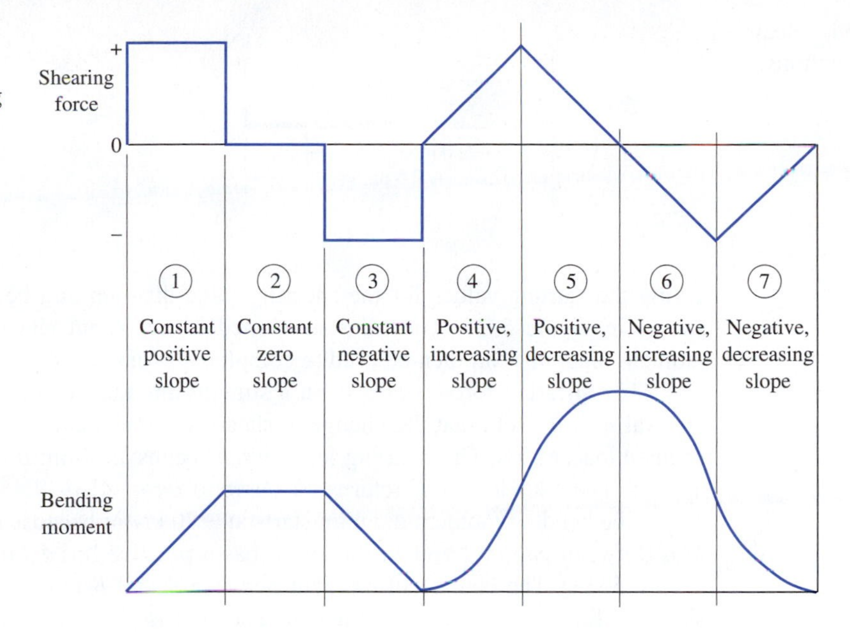

FIGURE 5–33 General shapes for moment curves relating to corresponding shearing force curves.

and 6. In each case, the uniformly distributed load causes the shearing force to decrease by a constant amount per unit length, producing a shearing force curve that starts with a positive value and continues as a straight line with a negative slope. The corresponding bending moment diagram starts with a fairly large positive slope and then curves with a concave downward shape as shown in part 5. If the shearing force curve crosses the zero axis, the bending moment curve reaches a maximum value where the slope is zero. Then, if the shearing force continues into a negative region, as in part 6, the corresponding bending moment will have an increasingly negative slope.

In most applications, and in problems in this book, it is adequate to sketch the general shape of the shearing force and bending moment curves and to indicate the magnitudes at key points of interest. Typically these are the following:

- points where concentrated loads are applied
- points where distributed loads begin and end
- at any point where a maximum or minimum value of shearing force or bending moment may occur.

5–8 SHEARING FORCES AND BENDING MOMENTS FOR CANTILEVER BEAMS

The manner of support of a cantilever beam causes the analysis of its shearing forces and bending moments to be somewhat different from that for simply supported beams. The most notable difference is that, at the place where the beam is supported, it is fixed and can therefore resist moments. Thus, at the fixed end of the beam, the bending moment is not zero, as it was for simply supported beams. In fact, the bending moment at the fixed end of the beam is usually the *maximum*.

Consider the cantilever beam shown in Figure 5–34. Earlier, in Example Problem 5–4, it was shown that the support reactions at point A are a vertical force $R_A = 64$ kN and a counterclockwise moment $M_A = 70$ kN·m. These are equal to the values of the shearing force and bending moment at the left end of the beam. According to convention, the upward reaction force R_A is positive, and the counterclockwise moment M_A is negative,

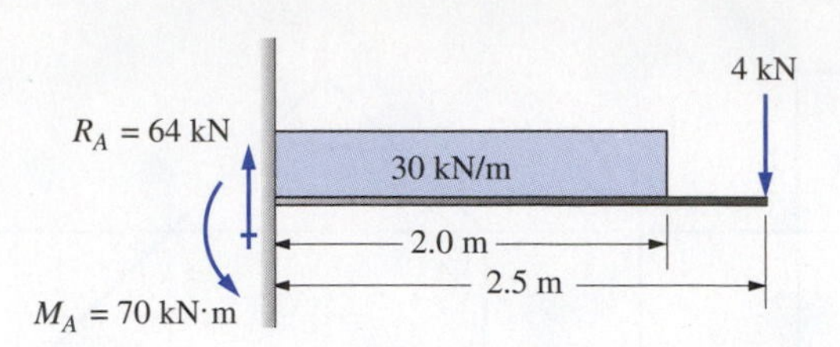

FIGURE 5–34 Beam loading and reactions.

giving the starting values for the shearing force diagram and bending moment diagram shown in Figure 5–35. The rules developed earlier about shearing force and bending moment diagrams can then be used to complete the diagrams.

The shearing force decreases in a straight-line manner from 64 kN to 4 kN in the interval A to B. Note that the change in shearing force is equal to the amount of the distributed load, 60 kN. The shearing force remains constant from B to C, where no loads are applied. The 4-kN load at C returns the curve to zero.

The bending moment diagram starts at -70 kN·m because of the reaction moment M_A. Between points A and B, the curve has a positive but decreasing slope (type 5 in Figure 5–33). The change in moment between A and B is equal to the area under the shearing force curve between A and B. The area is a combination of a rectangle and a triangle.

$$[\text{Area}]_{A-B} = 4\text{ kN }(2\text{ m}) + \frac{1}{2}(60\text{ kN})(2\text{ m}) = 68\text{ kN}\cdot\text{m}$$

Then the moment at B is

$$M_B = M_A + [\text{Area}]_{A-B} = -70\text{ kN}\cdot\text{m} + 68\text{ kN}\cdot\text{m} = -2\text{ kN}\cdot\text{m}$$

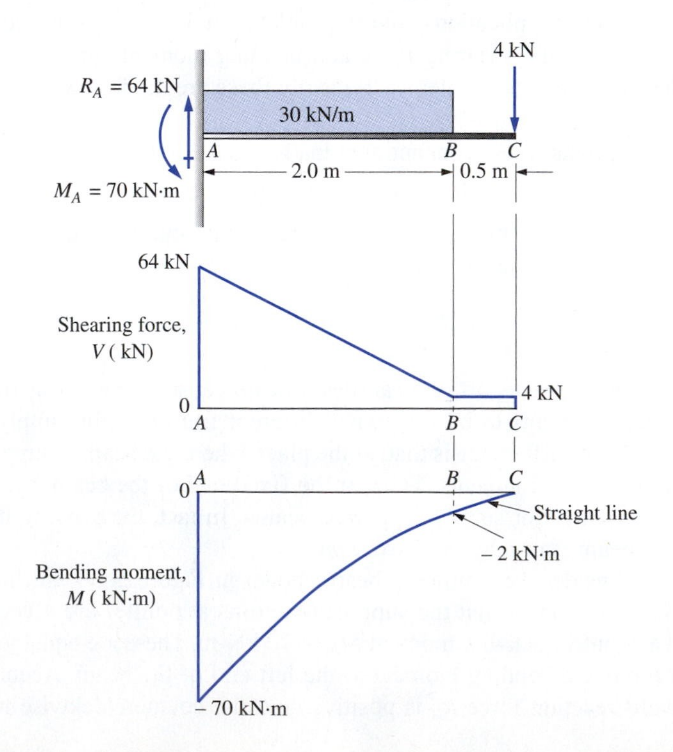

FIGURE 5–35 Complete load, shearing force, and bending moment diagrams.

Finally, from B to C,

$$M_C = M_B + [\text{Area}]_{B-C} = -2\,\text{kN·m} + 4\,\text{kN}(0.5\,\text{m}) = 0$$

Since point C is a *free* end of the beam, the moment there must be zero.

5–9 BEAMS WITH LINEARLY VARYING DISTRIBUTED LOADS

Figures 5–6 and 5–7 in Section 5–2 show two examples of beams carrying linearly varying distributed loads. We demonstrate here the method of drawing the general shape of the shearing force and bending moment diagrams for such beams, and how to determine the magnitude of the maximum shearing force and maximum bending moment. For many practical problems these are the primary objectives. Later, in Section 5–11 a mathematical approach is shown that yields a more complete definition of the shape of the shearing force and bending moment diagrams.

Refer now to Figure 5–36 which shows the load diagram for the cantilever from Figure 5–6. The amount of loading per unit length varies linearly from $w = -200$ lb/ft (downward) at the support point A to w = zero at the right end B. This straight-line curve is called a *first-degree curve* because the loading varies directly with position on the beam, x. For such a loading, the reaction at A, called R_A, is the resultant of the total distributed load, found by computing the area under the triangular-shaped load curve. That is,

$$R_A = \tfrac{1}{2}(-200\,\text{lb/ft})(8\,\text{ft}) = -800\,\text{lb}$$

The bending moment at the support, called M_A, must equal the moment of all of the applied load to the right of A. This can be found by considering the resultant to act at the centroid of the distributed load. For the triangular-shaped load curve, the centroid is $1/3$ of the length of the beam from point A. Calling this distance, x, we can say,

$$x = L/3 = (8\,\text{ft})/3 = 2.667\,\text{ft}$$

Then the moment at A is the product of the resultant times x. That is,

$$M_A = R_A x = (800\,\text{lb})(2.667\,\text{ft}) = 2133\,\text{lb·ft}$$

These values, $R_A = 800$ lb and $M_A = 2133$ lb·ft, are the maximum values for shearing force and bending moment, respectively. In most cases, that is the objective of the analysis. If so, the analysis can be concluded.

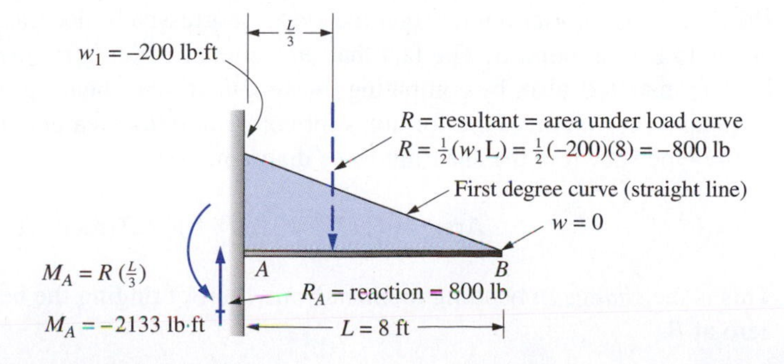

FIGURE 5–36 Load diagram, reaction, and moment for cantilever carrying a linearly varying distributed load.

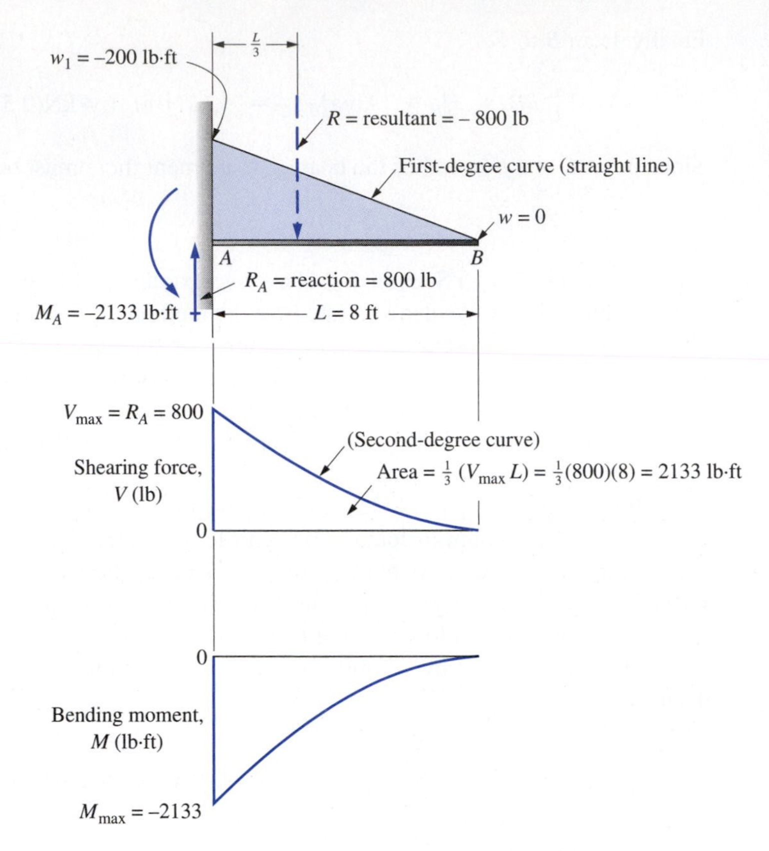

FIGURE 5–37 Load, shearing force, and bending moment diagrams for beam loading in Figure 5–35.

However, if the shapes of the shearing force and bending moment diagrams are desired, they can be sketched using the principles developed earlier in this chapter. Figure 5–37 shows the results. The shearing force diagram starts at A with the value of 800 lb, equal to the reaction, R_A. The value of the shearing force then decreases for points to the right of A as additional loads are applied. Note that the shearing force curve is not a straight line because the amount of loading decreases as we proceed from A toward B. At B the loading is zero, resulting in the zero value of shearing force at B. The slope of the shearing force curve at any point is equal to the amount of loading per unit length at the corresponding point on the load diagram. Thus, the shearing force curve starts with a relatively large negative slope at A and then a progressively smaller negative slope as we approach B. This is generally called a *second-degree curve* because the value varies with the *square* of the distance x.

The bending moment diagram can be sketched by first noting that $M_A = -2133$ lb·ft. The curve has a relatively large positive slope at A because of the large positive value for shearing force at that point. Then the slope progressively decreases, with increasing distance, to zero at point B. The fact that the value of bending moment equals zero at B can be demonstrated, also, by computing the area under the shearing force curve from A to B. Appendix A–1 includes the formulas for computing the area under a second-degree curve of the type shown in the shearing force diagram. That is,

$$\text{Area} = (1/3)(800\text{ lb})(8\text{ ft}) = 2133\text{ lb}\cdot\text{ft}$$

This is the *change* in bending moment from A to B, bringing the bending moment curve to zero at B.

5–10 FREE-BODY DIAGRAMS OF PARTS OF STRUCTURES

Examples considered thus far have been for generally straight beams with all transverse loads, that is, loads acting perpendicular to the main axis of the beam. Many machine elements and structures are more complex, having parts that extend away from the main beam-like part.

For example, consider the simple post with an extended arm, shown in Figure 5–38, composed of vertical and horizontal parts. The vertical post is rigidly secured at its base. At the end of the extended horizontal arm, a downward load is applied. An example of such a loading is a support system for a sign over a highway. Another would be a support post for a basketball hoop in which the downward force could be a player hanging from the rim after a slam dunk. A mechanical design application is a bracket supporting machine parts during processing.

Under such conditions, it is convenient to analyze the structure or machine element by considering each part separately and creating a free-body diagram for each part. At the actual joints between parts, one part exerts forces and moments on the other. Using this approach, you will be able to design each part on the basis of how it is loaded, using the basic principles of beam analysis in this chapter and those that follow.

Post with an Extended Arm. The objective of the analysis is to draw the complete shearing force and bending moment diagrams for the horizontal and vertical parts of the post/arm structure shown in Figure 5–38. The first step is to "break" the arm from the post at the right-angle corner.

Figure 5–39 shows the horizontal arm as a free-body with the applied load, F, acting at its right end. The result appears similar to the cantilever analyzed earlier in this chapter. We know that the arm is in equilibrium as a part of the total structure and, therefore, it must be in equilibrium when considered by itself. Then, at the left end where it joins the vertical post, there must be a force equal to F acting vertically upward to maintain the sum of the vertical forces equal to zero. But the two vertical forces form a couple that tends to rotate the arm in a clockwise direction. To maintain rotational equilibrium, there must be

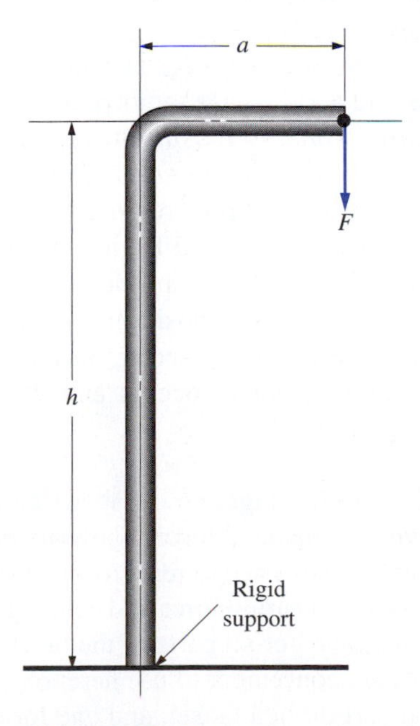

FIGURE 5–38 Post with an extended arm.

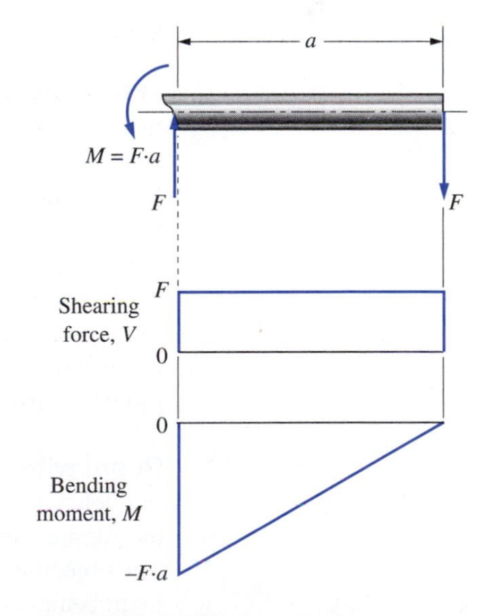

FIGURE 5–39 Free-body, shearing force, and bending moment diagrams for the horizontal arm.

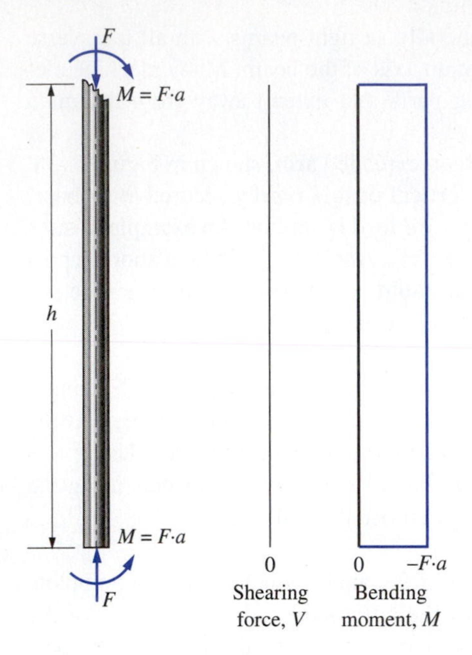

FIGURE 5–40 Free-body shearing force and bending moment diagrams for the vertical post.

FIGURE 5–41 Beam with an L-shaped bracket.

a counterclockwise moment internal to the arm at its left end with a magnitude of $M = F{\cdot}a$, where a is the length of the arm. Having completed the free-body diagram, the shearing force and bending moment diagrams can be drawn as shown in Figure 5–39. The shearing force is equal to F throughout the length of the arm. The maximum bending moment occurs at the left end of the arm where $M = F{\cdot}a$.

The free-body diagram for the vertical post is shown in Figure 5–40. At the top of the post, a downward force and a clockwise moment are shown, exerted on the vertical post by the horizontal arm. *Notice the action-reaction pair that exists at joints between parts. Equal but oppositely directed forces and moments act on the two parts.* Completing the free-body diagram for the post requires an upward force and a counterclockwise moment at its lower end, provided by the attachment means at its base. Finally, Figure 5–40 shows the shearing force and bending moment diagrams for the post, drawn vertically to relate the values to positions on the post. No shearing force exists because there are no transverse forces acting on the post. Where no shearing force exists, no change in bending moment occurs and there is a uniform bending moment throughout the post.

Beam with an L-Shaped Bracket. Figure 5–41 shows an L-shaped bracket extending below the main beam carrying an inclined force. The main beam is supported by simple supports at A and C. Support C is designed to react to any unbalanced horizontal force. The objective is to draw the complete shearing force and bending moment diagrams for the main beam and the free-body diagrams for all parts of the bracket.

Three free-body diagrams are convenient to use here: one for the horizontal part of the bracket, one for the vertical part of the bracket, and one for the main beam itself. But first it is helpful to resolve the applied force into its vertical and horizontal components, as indicated by the dashed vectors at the end of the bracket.

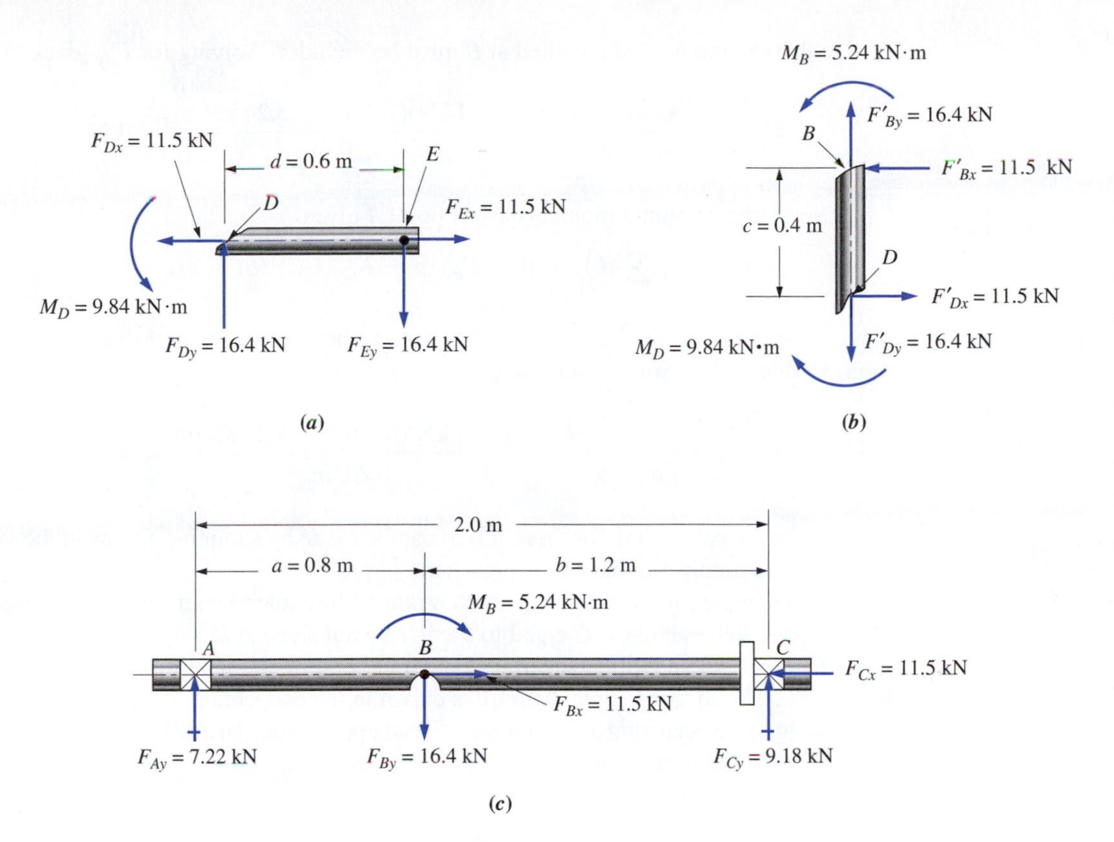

FIGURE 5–42 Free-body diagrams. (a) Free-body diagram for part *DE*. (b) Free-body diagram for part *BD*. (c) Free-body diagram for part *ABC*, the main beam.

Figure 5–42 shows the three free-body diagrams. Starting with the part *DE* shown in (a), the applied forces at *E* must be balanced by the oppositely directed forces at *D* for equilibrium in the vertical and horizontal directions. But rotational equilibrium must be produced by an internal moment at *D*. Summing moments with respect to point *D* shows that

$$M_D = F_{Ey} \cdot d = (16.4\ \text{kN})(0.6\ \text{m}) = 9.84\ \text{kN} \cdot \text{m}$$

In Figure 5–42(b) the forces and moments at *D* have the same values but opposite directions from those at *D* in part (a) of the figure. Vertical and horizontal equilibrium conditions show the forces at *B* to be equal to those at *D*. The moment at *B* can be found by summing moments about *B* as follows:

$$\left(\sum M\right)_B = 0 = M_D - F_{Dx} \cdot c - M_B$$

Then

$$M_B = M_D - F_{Dx} \cdot c = 9.84\ \text{kN} \cdot \text{m} - (11.5\ \text{kN})(0.4\ \text{m}) = 5.24\ \text{kN} \cdot \text{m}$$

Now the main beam *ABC* can be analyzed. Use Figure 5–42(c). The forces and moment are shown applied at *B* with the values taken from point *B* on part *BD*. We must now solve for the reactions at *A* and *C*. First, summing moments about point *C* yields

$$\left(\sum M\right)_C = 0 = F_{By} \cdot b - F_{Ay} \cdot (a + b) - M_B$$

Notice that the moment M_B applied at B must be included. Solving for F_{Ay} gives

$$F_{Ay} = \frac{(F_{By} \cdot b) - M_B}{a + b} = \frac{(16.4 \text{ kN})(1.2 \text{ m}) - 5.24 \text{ kN} \cdot \text{m}}{2.0 \text{ m}} = 7.22 \text{ kN}$$

Similarly, summing moments about point A gives

$$\left(\sum M\right)_A = 0 = F_{By} \cdot a - F_{Cy} \cdot (a + b) + M_B$$

Notice that the moment M_B applied at B is positive because it acts in the same sense as the moment due to F_{By}. Solving for F_{Cy} gives

$$F_{Cy} = \frac{(F_{By} \cdot a) - M_B}{a + b} = \frac{(16.4 \text{ kN})(0.8 \text{ m}) + 5.24 \text{ kN} \cdot \text{m}}{2.0 \text{ m}} = 9.18 \text{ kN}$$

A check on the calculation for these forces can be made by summing forces in the vertical direction and noting that the sum equals zero.

The completion of the free-body diagram for the main beam requires the inclusion of the horizontal reaction at C equal to the horizontal force at B.

Figure 5–43 shows the shearing force and bending moment diagrams for the main beam ABC. The shearing force diagram is drawn in the conventional manner with changes in shearing force occurring at each point of load application. The difference from previous work is in the moment diagram.

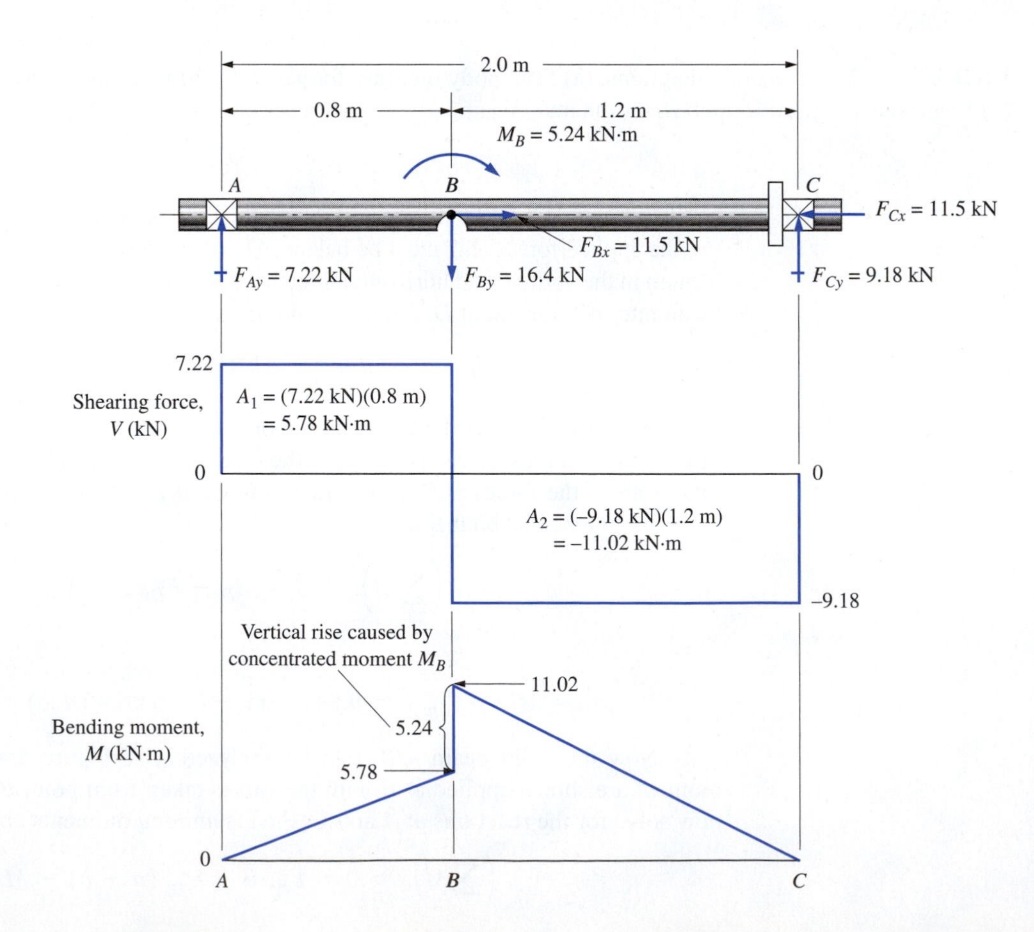

FIGURE 5–43 Shearing force and bending moment diagrams for main beam of Figure 5–41.

The following steps were used:

1. The moment at A equals zero because A is a simple support.
2. The increase in moment from A to B equals the area under the shearing force curve between A and B, 5.78 kN·m.
3. At point B the moment M_B is considered to be a *concentrated moment* resulting in an abrupt change in the value of the bending moment by the amount of the applied moment, 5.24 kN·m, thus resulting in the peak value of 11.02 kN·m. The convention used here is:

> a. When a concentrated moment is *clockwise,* the moment diagram *rises.*
> b. When a concentrated moment is *counterclockwise,* the moment diagram *drops.*

4. Between B and C, the moment decreases to zero because of the negative shearing force and the corresponding negative area under the shearing force curve.

5–11 MATHEMATICAL ANALYSIS OF BEAM DIAGRAMS

For most practical problems, the preparation of the load, shearing force, and bending moment diagrams using the techniques shown earlier in this chapter are adequate and convenient. A wide variety of beam types and loadings can be analyzed with sufficient detail to permit the logical design of the beams to ensure safety and to limit deflections to acceptable values. The methods of accomplishing these objectives are presented in Chapters 7–11.

However, there are some types of loading and some types of design techniques that can benefit from the representation of the load, shearing force, and bending moment diagrams by mathematical equations. This section presents the methods of creating such equations.

The following guidelines are for writing sets of equations that completely define the load, shearing force, and bending moment as a function of the position on the beam.

Guidelines for Writing Beam Diagram Equations

1. Draw the load diagram showing all externally applied loads and reactions.
2. Compute the values for all reactions.
3. Label points along the length of the beam where concentrated loads are applied or where distributed loads begin or end.
4. Draw the shearing force and bending moment diagrams using the techniques shown earlier in this chapter, noting values at the critical points defined in Step 3.
5. Establish conventions for denoting positions on the beam and signs of shearing forces and bending moment. In most cases, we will use the following convention (see Figures 5–20 to 5–32).
   a. Position on the beam will be denoted by the variable x measured from the left end of the beam.
   b. Downward loads will be negative.
   c. A positive shearing force is one that acts downward within the beam at a given section. An alternative way of determining this is to analyze the net external vertical force acting on that part of the beam to the left of the section of interest. If the net external force is upward, the internal shearing force in the beam is positive.

d. A positive bending moment is one that acts counterclockwise within the beam at a given section. A positive bending moment will tend to cause a beam to bend in a concave upward shape, typical of a simply supported beam carrying downward loads between the supports.

6. Consider separately each segment of the beam between the points defined in Step 3. The shearing force curve should be continuous within each segment.
7. If the shearing force diagram consists of all straight lines caused by concentrated or uniformly distributed loads, the fundamental principles of analytic geometry can be used to write equations for the shearing force versus position on the beam for each segment. The resulting equations will be of the form

$$V_{AB} = \text{Constant} \qquad \text{(zero-degree equation)}$$
$$V_{BC} = ax + b \qquad \text{(first-degree equation)}$$

The subscripts define the beginning and end of the segment of interest.

8. If the shearing force diagram contains segments that are curved caused by varying distributed loads, first write equations for the load versus position on the beam. Then, derive the equations for the shearing force versus position on the beam from

$$V_{AB} = \int w_{AB}\,dx + C$$

where w_{AB} is the equation for load in the segment AB as a function of x, and C is a constant of integration. The resulting shearing force equation will be of the second degree or higher, depending on the complexity of the loading pattern. Compute the value of the constants of integration using known values of V at given locations x.

9. Derive equations for the bending moment as a function of position on the beam for each segment, using the method.

$$M_{AB} = \int V_{AB}\,dx + C$$

This is the mathematical equivalent of the *area rule* for beam diagrams used earlier because the process of integration determines the area under the shearing force curve. Compute the value of the constants of integration using known values of M at given locations x.

10. The result at this point is a set of equations for shearing force and bending moment for each segment of the beam. It would be wise to check the equations for accuracy by substituting key values of x for which the shearing force and bending moment are known into the equations to ensure that the correct values for V and M are computed.
11. Determine the maximum values of shearing force and bending moment if they are not already known by substituting values for x in the appropriate equations where the maximum values are expected. Recall the rule that the maximum bending moment will occur at a point where the shearing force curve crosses the x-axis—that is, where $V = 0$.

This procedure is demonstrated by the following four examples.

Simply Supported Beam with a Concentrated Load. The objective is to write the equations for the shearing force and bending moment diagrams for the beam and loading shown in Figure 5–44, using the guidelines given in this section.

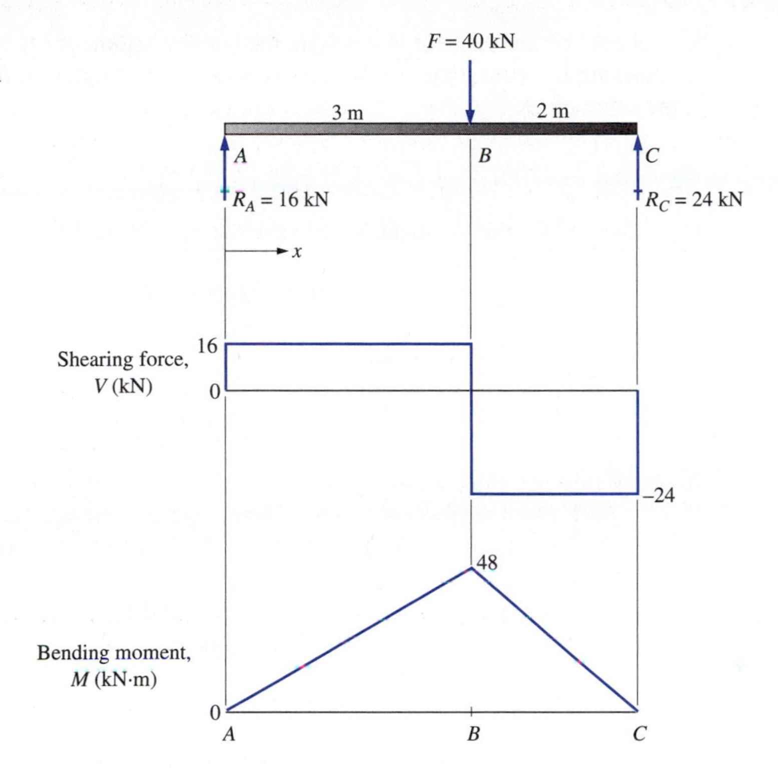

FIGURE 5–44 Simply supported beam with a concentrated load.

Steps 1 to 4 have been completed and shown in Figure 5–44. Points of interest are labeled A at the left support, B at the point of application of the load, and C at the right support. Equations will be developed for the two segments, AB and BC, where AB applies from $x = 0$ to $x = 3$ m and BC applies from $x = 3$ m to $x = 5$ m. Steps 5 and 6 will be as defined in the guidelines.

Step 7 can be applied to write the equations for the shearing force curve as follows:

$$V_{AB} = 16$$
$$V_{BC} = -24$$

The units for shearing force are taken to be kN.

Step 8 does not apply to this example.

Step 9 is now applied to derive the equations for the bending moment in the two segments.

$$M_{AB} = \int V_{AB}\,dx + C = \int 16\,dx + C = 16x + C$$

To evaluate the constant of integration C, we can note that at $x = 0$, $M_{AB} = 0$. Substituting these values into the moment equation gives

$$0 = 16(0) + C$$

Then, $C = 0$. The final equation can now be written as,

$$M_{AB} = 16x$$

As a check, we can see that at $x = 3$ m, the bending moment $M_B = 48$ kN·m, as shown in the bending moment diagram. We can now derive the equation for the bending moment in the segment BC.

$$M_{BC} = \int V_{BC}\,dx + C = \int -24\,dx + C = -24x + C$$

To evaluate C for this segment, we can use the condition at $x = 5$, $M_{BC} = 0$. Then,

$$0 = -24(5) + C$$

Then, $C = 120$. The final equation is

$$M_{BC} = -24x + 120$$

To check this equation, substitute $x = 3$.

$$M_{BC} = -24(3) + 120 = -72 + 120 = 48 \quad \text{(check)}$$

In summary, the equations for the shearing force and bending moment diagrams are:

In the segment AB from $x = 0$ to $x = 3$ m:

$$V_{AB} = 16$$
$$M_{AB} = 16x$$

In the segment BC from $x = 3$ m to $x = 5$ m:

$$V_{BC} = -24$$
$$M_{BC} = -24x + 120$$

The maximum values for the shearing force and bending moment are obvious from the diagrams.

$$V_{max} = -24\text{ kN throughout the segment BC}$$
$$M_{max} = 48\text{ kN}\cdot\text{m at point } B(x = 3\text{ m})$$

Simply Supported Beam with a Partial Uniformly Distributed Load. The objective is to write the equations for the shearing force and bending moment diagrams for the beam and loading shown in Figure 5–45, using the guidelines given in this section. Note that this is the same beam and loading that was shown earlier in Figure 5–30.

Steps 1 to 4 have been completed and shown in Figure 5–45. Points of interest are labeled A at the left support, B at the point where the distributed load ends, and C at the right support. Equations will be developed for the two segments, AB and BC, where AB applies from $x = 0$ to $x = 6$ m and BC applies from $x = 6$ m to $x = 9$ m.

Step 5(b) can be used to write an equation for the loading in the segment AB:

$$w_{AB} = -1500\text{ N/m}$$

Step 7 can be applied to write the equations for the shearing force curve. In the segment AB, the curve is a straight line, so we can write it in the form,

$$V_{AB} = ax + b$$

FIGURE 5–45
Simply supported beam with a partial uniformly distributed load.

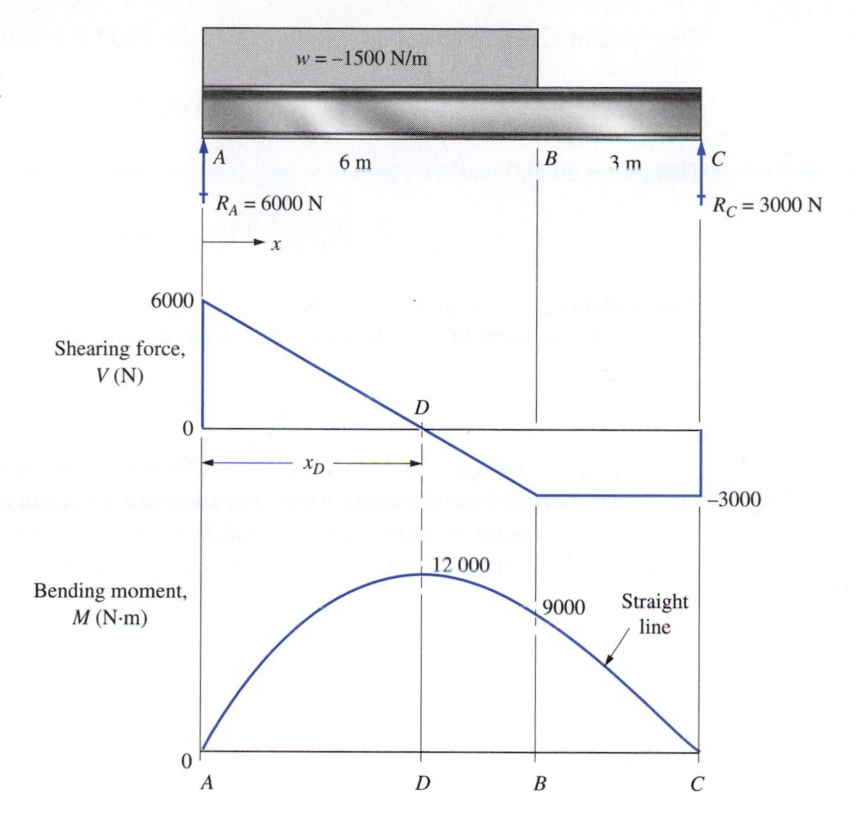

where a is the slope of the line and b is the intercept of the line with the V axis at $x = 0$. A convenient way to determine the slope is to observe that the slope is equal to the amount of loading per unit length for the distributed load. That is, $a = -1500$ N/m. The value of the intercept b can be observed from the shearing force diagram; $b = 6000$ N. Then the final form of the shearing force equation is

$$V_{AB} = -1500x + 6000$$

We can check the equation by substituting $x = 6$ m and computing V_B.

$$V_{AB} = -1500(6) + 6000 = -3000 = V_B$$

This checks with the known value of shearing force at point B.

Note that we could have used Step 8 to determine the equation for V_{AB}. Note that,

$$w_{AB} = -1500 \text{ N/m}$$

Then,

$$V_{AB} = \int w_{AB}\,dx + C = \int -1500\,dx + C = -1500x + C$$

The value of C can be found by substituting $V_{AB} = 6000$ at $x = 0$.

$$6000 = -1500(0) + C$$

Then, $C = 6000$. Finally,

$$V_{AB} = -1500x + 6000$$

This is identical to the previous result.

In the segment BC the shearing force is a constant value,

$$V_{BC} = -3000$$

Before proceeding to determine the equations for the bending moment diagram, recall that a critical point occurs where the shearing force curve crosses the zero axis. That will be a point of maximum bending moment. Let's call this point D and find the value of x_D where $V = 0$ by setting the equation for V_{AB} equal to zero and solving for x_D.

$$V_{AB} = 0 = -1500x_D + 6000$$
$$x_D = 6000/1500 = 4.0 \text{ m}$$

We will use this value later to find the bending moment at D.

Step 9 of the guidelines is now used to determine the equations for the bending moment diagram. First, in the segment AB,

$$M_{AB} = \int V_{AB}\,dx + C = \int (-1500x + 6000)\,dx + C$$
$$M_{AB} = -750x^2 + 6000x + C$$

To evaluate C, note that at $x = 0$, $M_{AB} = 0$. Then, $C = 0$. And

$$M_{AB} = -750x^2 + 6000x$$

We can check the equation by finding M_B at $x = 6$ m.

$$M_B = -750(6)^2 + 6000(6) = 9000 \quad \text{(check)}$$

Also, we need the value of the maximum moment at D where $x = 4.0$ m.

$$M_D = -750(4)^2 + 6000(4) = 12\,000 \quad \text{(check)}$$

For the segment BC:

$$M_{BC} = \int V_{BC}\,dx + C = \int -3000\,dx + C = -3000x + C$$

But, at $x = 9$ m, $M_{BC} = 0$. Then,

$$0 = -3000(9) + C$$

and $C = 27\,000$. Finally,

$$M_{BC} = -3000x + 27\,000$$

Check this equation at point B where $x = 6$ m.

$$M_B = -3000(6) + 27\,000 = -18\,000 + 27\,000 = 9000 \quad \text{(check)}$$

In summary, the equations for the shearing force and bending moment diagrams are: In the segment AB from $x = 0$ to $x = 6$ m:

$$V_{AB} = -1500x + 6000$$
$$M_{AB} = -750x^2 + 6000x$$

In the segment BC from $x = 6$ m to $x = 9$ m:

$$V_{BC} = -3000$$
$$M_{BC} = -3000x + 27\,000$$

The maximum values for the shearing force and bending moment are obvious from the diagrams.

$$V_{max} = 6000 \text{ N at the left end at point } A$$
$$M_{max} = 12\,000 \text{ N·m at point } D \ (x = 4 \text{ m})$$

Cantilever with a Varying Distributed Load. The objective is to write the equations for the shearing force and bending moment diagrams for the beam and loading shown in Figure 5–46, using the guidelines given in this section. Note that this is the same beam and loading shown earlier in Figure 5–37.

There will be only one segment for this example, encompassing the entire length of the beam, because the load, shearing force, and bending moment curves are continuous.

We must first write an equation for the loading that varies linearly from a value of -200 lb/ft at the left end at A to zero at point B where $x = 8$ ft. You must note that the loading is shown on top of the beam acting downward in the manner we are accustomed to viewing loads. But the downwardly directed load is actually negative. To aid in writing the equation, you could redraw the loading diagram in the form of a *graph* of load versus position x, as shown in Figure 5–47. Then we can write the equation for the straight line,

$$w_{AB} = ax + b$$

The slope, a, can be evaluated by the ratio of the change in w over a given distance x. Using the entire length of the beam gives

$$a = \frac{w_1 - w_2}{x_1 - x_2} = \frac{-200 - 0}{0 - 8} = 25$$

The value of $b = -200$ can be observed from the diagram in Figure 5–47. Then the final equation for the loading is

$$w_{AB} = 25x - 200$$

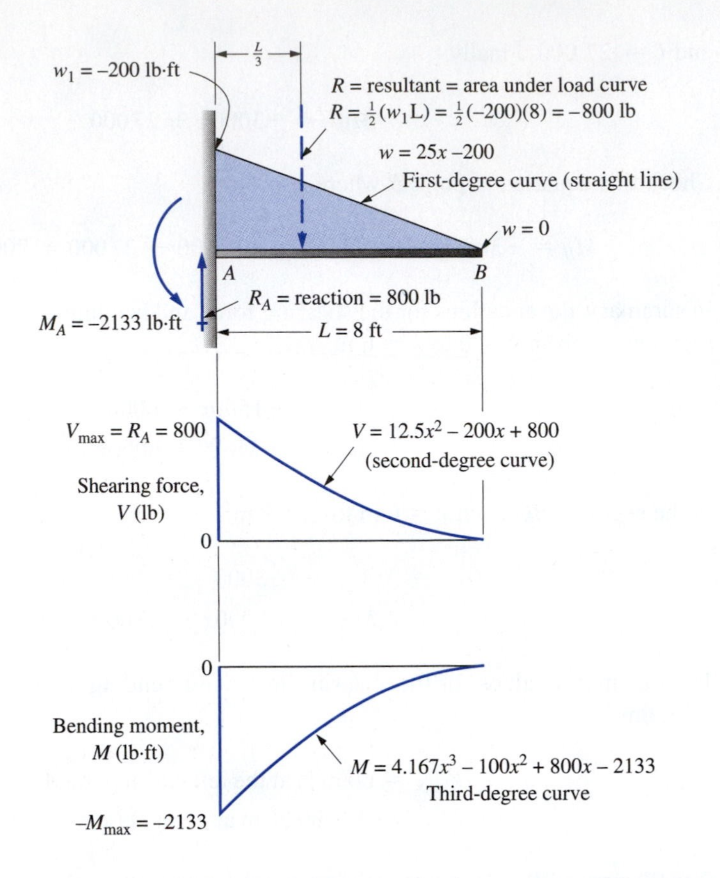

FIGURE 5–46 Cantilever with a varying distributed load.

We can check this equation by evaluating w at $x = 8$ ft at the end of the beam.

$$w_{AB} = 25(8) - 200 = 0 \quad \text{(check)}$$

Now we can derive the equation for the shearing force diagram.

$$V_{AB} = \int w_{AB}\, dx + C = \int (25x - 200)\, dx + C = 12.5x^2 - 200x + C$$

Use the condition that at $x = 0$, $V_{AB} = 800$ to evaluate C.

$$800 = 12.5(0)^2 - 200(0) + C$$

Then, $C = 800$. And the final equation for the shearing force is

$$V_{AB} = 12.5x^2 - 200x + 800$$

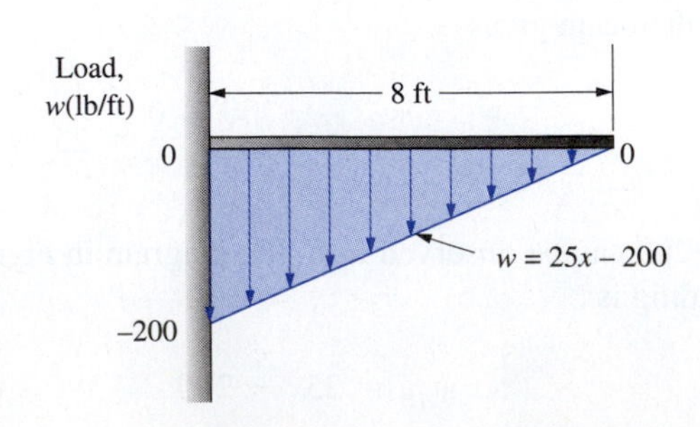

FIGURE 5–47 Alternate display of loading on the beam in Figure 5–46.

We can check this equation by evaluating V at $x = 8$ ft at the end of the beam.

$$V_{AB} = 12.5(8)^2 = 200(8) + 800 = 0 \quad \text{(check)}$$

Now we can derive the equation for the bending moment diagram.

$$M_{AB} = \int V_{AB}\,dx + C = \int (12.5x^2 - 200x + 800)dx + C$$

$$M_{AB} = 4.167x^3 - 100x^2 + 800x + C$$

Using the condition that at $x = 0$, $M_{AB} = -2133$, we can evaluate C.

$$-2133 = 4.167(0)^3 - 100(0)^2 + 800(0) + C$$

Then, $C = -2133$. The final equation for bending moment is

$$M_{AB} = 4.167x^3 - 100x^2 + 800x - 2133$$

We can check this equation by evaluating M at $x = 8$ ft at the end of the beam.

$$M_{AB} = 4.167(8)^3 - 100(8)^2 + 800(8) - 2133 = 0 \quad \text{(check)}$$

In summary, the equations for the load, shearing force, and bending moment diagrams for the beam shown in Figure 5–46 are

$$w_{AB} = 25x - 200 \qquad \text{(a first-degree curve; straight line)}$$
$$V_{AB} = 12.5x^2 - 200x + 800 \qquad \text{(a second-degree curve)}$$
$$M_{AB} = 4.167x^3 - 100x^2 + 800x - 2133 \qquad \text{(a third-degree curve)}$$

Simply Supported Beam with a Varying Distributed Load. The objective is to write the equations for the shearing force and bending moment diagrams for the beam and loading shown in Figure 5–48, using the guidelines given in this section. A pictorial of one manner of creating this loading pattern is shown in Figure 5–7.

Because of the symmetry of the loading, the two reactions will be of equal magnitude. Each will be equal to the area under one-half of the load diagram. Breaking that into a rectangle 0.2 kN/m high by 2.30 m wide and a triangle 1.0 kN/m high and 2.30 m wide, we can compute:

$$R_A = R_C = (0.2)(2.30) + 0.5(1.0)(2.30) = 0.46 + 1.15 = 1.61 \text{ kN}$$

The general shapes of the shearing force and bending moment diagrams are sketched in Figure 5–48. We can reason that the shearing force curve crosses the zero axis at the middle of the beam at $x = 2.30$ m. Therefore, the maximum bending moment will occur at that point also. In principle, the magnitude of the maximum bending moment is equal to the area under the shearing force curve between points A and B. However, the calculation of that area is difficult because the curve is of the second degree and it does not start at its vertex. Then the formulas in Appendix A–1 cannot be used directly. This is one reason for developing the equations for the shearing force and bending moment diagrams.

Let's first write the equation for the load on the left half of the beam from A to B. The amount of loading per unit length starts at -0.20 kN/m (downward) and increases in

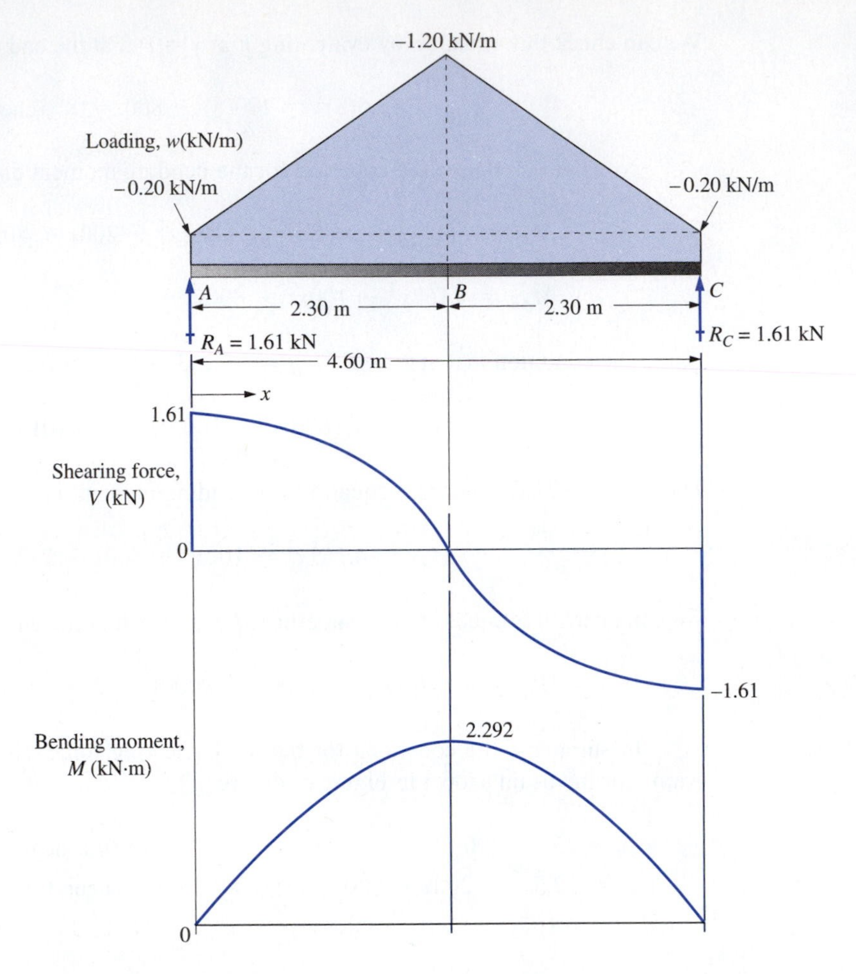

FIGURE 5–48 Simply supported beam with a varying distributed load.

magnitude to −1.20 kN/m. Again, as was done in the preceding example, it may help to draw the load diagram as a graph, as shown in Figure 5–49. Then we can write the equation for the straight line in the form,

$$w_{AB} = ax + b$$

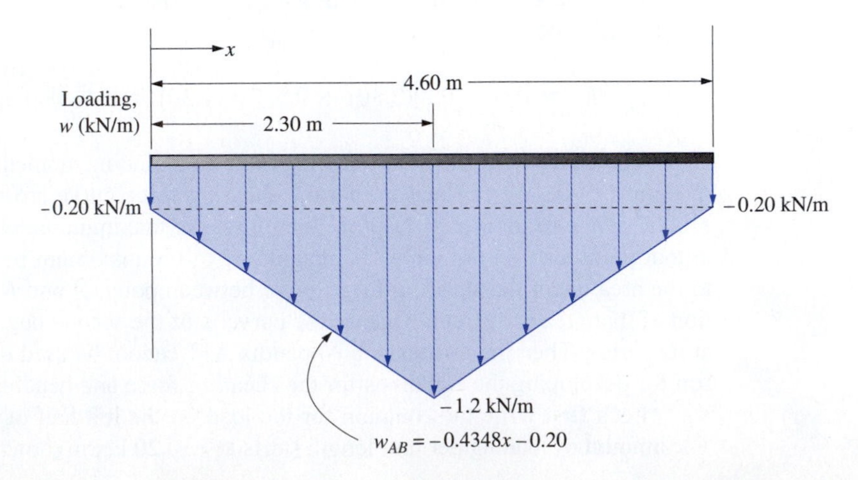

FIGURE 5–49 Alternate display of loading on the beam in Figure 5–48.

The slope, a, can be evaluated by the ratio of the change in w over a given distance x. Using half the length of the beam gives

$$a = \frac{w_1 - w_2}{x_1 - x_2} = \frac{-0.20 - (-1.20)}{0 - 2.30} = -0.4348$$

The value of $b = -0.20$ can be observed from the diagram in Figure 5–48. Then the final equation for the loading is

$$w_{AB} = ax + b = -0.4348x - 0.20$$

We can check this equation at $x = 2.30$ at the middle of the beam.

$$w_{AB} = -0.4348x - 0.20 = 0.4348(2.30) - 0.20 = 1.20\ \text{kN/m} \quad \text{(check)}$$

Now we can derive the equation for the shearing force diagram for the segment AB.

$$V_{AB} = \int w_{AB}\,dx + C = \int (-0.4348x - 0.20)\,dx + C = -0.2174x^2 - 0.20x + C$$

Use the condition that at $x = 0$, $V_{AB} = 1.61$ to evaluate C. Then, $C = 1.61$ and the final form of the equation is

$$V_{AB} = -0.2174x^2 - 0.20x + 1.61$$

We can check this equation at the middle of the beam by substituting $x = 2.30$ m.

$$V_{AB} = -0.2174(2.30)^2 - 0.20(2.30) + 1.61 = 0 \quad \text{(check)}$$

Now, we can derive the equation for the bending moment diagram.

$$M_{AB} = \int V_{AB}\,dx + C = \int (-0.2174x^2 - 0.20x + 1.61)\,dx + C$$

$$M_{AB} = -0.07246x^3 - 0.10x^2 + 1.61x + C$$

Using the condition that at $x = 0$, $M_{AB} = 0$, we can evaluate $C = 0$. And

$$M_{AB} = -0.07246x^3 - 0.10x^2 + 1.61x$$

Checking at $x = 2.30$ m, gives $M_B = 2.292$ kN·m. The equations for the right side of the diagrams could be derived similarly. However, because of the symmetry of the diagrams, the completion of the curves is the mirror image of those already computed. In summary, the equations for the left half of the load, shearing force, and bending moment diagrams are

$$w_{AB} = -0.4348x - 0.20$$
$$V_{AB} = -0.2174x^2 - 0.20x + 1.61$$
$$M_{AB} = -0.07246x^3 - 0.10x^2 + 1.61x$$

The maximum shearing force is 1.61 kN at each support, and the maximum bending moment is 2.292 kN·m at the middle of the beam.

5–12 CONTINUOUS BEAMS—THEOREM OF THREE MOMENTS

A continuous beam on any number of supports can be analyzed by using the *theorem of three moments*. The theorem actually relates the bending moments at three successive supports to each other and to the loads on the beam. For a beam with only three supports, the theorem allows the direct computation of the moment at the middle support. Known end conditions provide data for computing moments at the ends. Then the principles of statics can be used to find reactions.

For beams on more than three supports, the theorem is applied successively to sets of three adjacent supports (two spans), yielding a set of equations that can be solved simultaneously for the unknown moments.

The theorem of three moments can be used for any combination of loads. Special forms of the theorem have been developed for uniformly distributed loads and concentrated loads. These forms will be used in this chapter.

Uniformly Distributed Loads on Adjacent Spans. Figure 5–50 shows the arrangement of loads and the definition of terms applicable to Equation (5–5).

Three-Moment Equation—Distributed Loads

$$M_A L_I + 2M_B(L_1 + L_2) + M_C L_2 = \frac{-w_1 L_1^3}{4} - \frac{w_2 L_2^3}{4} \quad \textbf{(5–5)}$$

The values of w_1 and w_2 are expressed in units of force per unit length such as N/m, lb/ft, etc. The bending moments at the supports A, B, and C are M_A, M_B, and M_C. If M_A and M_C at the ends of the beam are known, M_B can be found from Equation (5–5) directly. Example problems will demonstrate the application of this equation.

The special case in which two equal spans carry equal uniform loads allows the simplification of Equation (5–5). If $L_1 = L_2 = L$ and $w_1 = w_2 = w$, then

$$M_A + 4M_B + M_C = \frac{-wL^2}{2} \quad \textbf{(5–6)}$$

Concentrated Loads on Adjacent Spans. If adjacent spans carry only one concentrated load each, as shown in Figure 5–51, then Equation (5–7) applies.

Three-Moment Equation—Concentrated Loads

$$M_A L_1 + 2M_B(L_1 + L_2) + M_C L_2 = \frac{-P_1 a}{L_1}(L_1^2 - a^2) - \frac{P_2 b}{L_2}(L_2^2 - b^2) \quad \textbf{(5–7)}$$

The special case of two equal spans carrying equal, centrally placed loads allows the simplification of Equation (5–7). If $P_1 = P_2 = P$ and $a = b = L/2$, then,

$$M_A + 4M_B + M_C = 3PL/4 \quad \textbf{(5–8)}$$

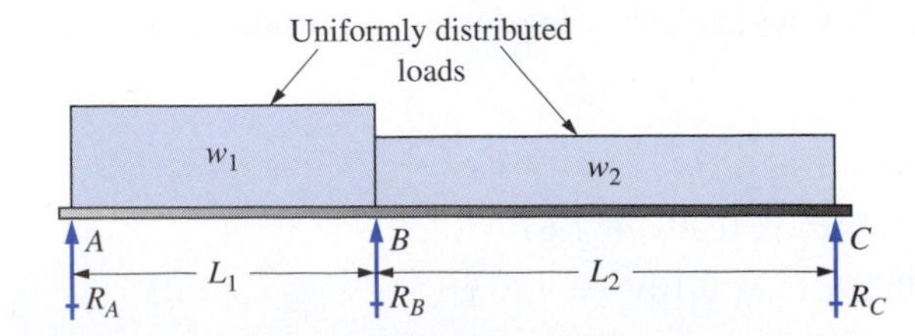

FIGURE 5–50 Uniformly distributed loads on a continuous beam of two spans.

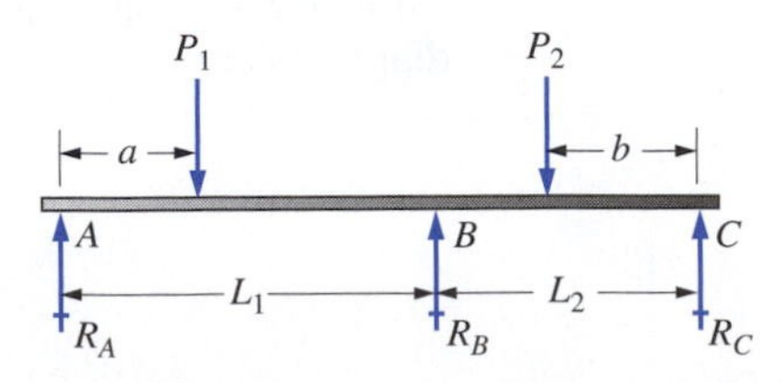

FIGURE 5–51 Continuous beam for two spans with one concentrated load on each span.

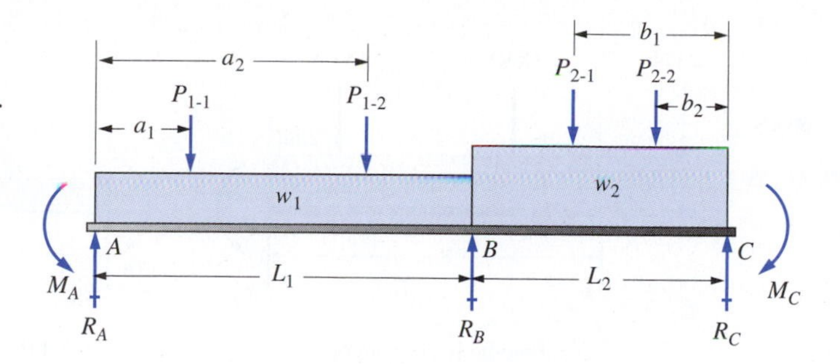

FIGURE 5–52 General notation for terms in Equation (5–9).

Combinations of Uniformly Distributed Loads and Several Concentrated Loads. This is a somewhat general case, allowing each span to carry a uniformly distributed load and any number of concentrated loads, as suggested in Figure 5–52. The general equation for such a loading is a combination of Equations (5–5) and (5–7), given as Equation (5–9).

Three-Moment Equation—General Form

$$M_AL_1 + 2M_B(L_1 + L_2) + M_CL_2 = -\sum\left[\frac{P_ia_i}{L_1}(L_1^2 - a_i^2)\right]_1 - \sum\left[\frac{P_ib_i}{L_2}(L_2^2 - b_i^2)\right]_2 - \frac{w_1L_1^3}{4} - \frac{w_2L_2^3}{4} \quad (5\text{–}9)$$

The term in the bracket with the subscript *1* is to be evaluated for *each* concentrated load on span 1 and then summed together. Similarly, the term having the subscript *2* is repeatedly applied for all loads on span 2. Notice that the distances a_i are measured from the reaction at A for each load on span 1, and the distances b_i are measured from the reaction at C for each load on span 2. The moments at the ends A and C can be due to concentrated moments applied there or to loads on overhangs beyond the supports. Any of the terms in Equation (5–9) may be left out of a problem solution if there is no appropriate load or moment existing at a particular section for which the equation is being written. Furthermore, other concentrated loads could be included in addition to those actually shown in Figure 5–52.

Example Problem 5–8 The loading composed of a combination of distributed loads and concentrated loads, shown in Figure 5–53, is to be analyzed to determine the reactions at the three supports and the complete shearing force and bending moment diagrams. The 17-m beam is to be used as a floor beam in an industrial building.

Solution Objective: Determine the support reactions and draw the shearing force and bending moment diagrams.

Given: Beam loading in Figure 5–53.

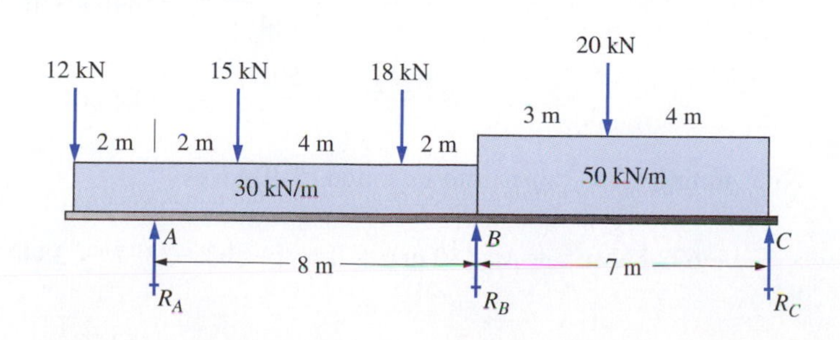

FIGURE 5–53 Beam for Example Problem 5–8.

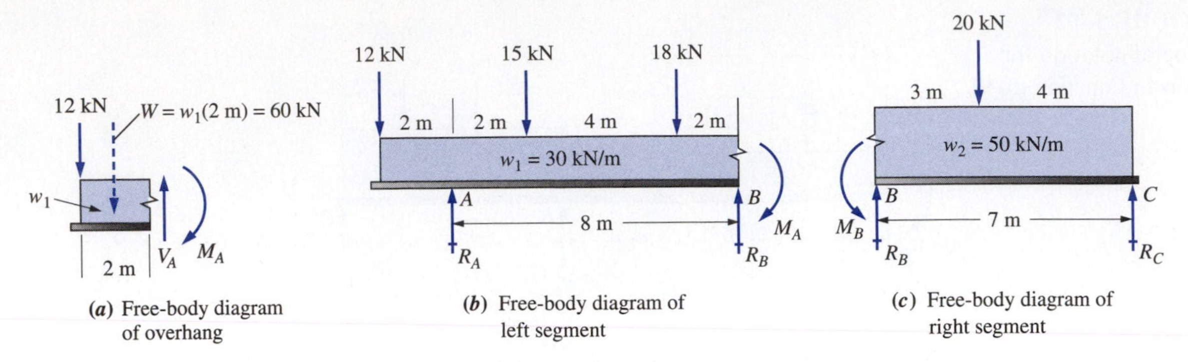

FIGURE 5–54 Free-body diagrams of beam segments used to find reactions R_A, R_B, and R_C.

Analysis Because the loading pattern contains both concentrated loads and uniformly distributed loads, we must use Equation (5–9). The subscript *1* refers to span 1 between supports A and B, and subscript *2* refers to span 2 between supports B and C. We must evaluate the actual magnitude of M_A and M_C to facilitate the solution of Equation (5–9). Because point C is at the end of a simply supported span, $M_C = 0$. At point A we can consider the overhanging part of the beam to the left of A to be a free-body as shown in Figure 5–54(a). Then sum moments about A to find the internal moment in the beam at A. Along with the 12 kN load 2.0 m from A, a resultant of 60 kN of load due to the distributed load acts a distance of 1.0 m from A. Then,

$$M_A = -12\,\text{kN}\,(2\,\text{m}) - 60\,\text{kN}\,(1\,\text{m}) = -84\,\text{kN}\cdot\text{m}$$

Each remaining term in Equation (5–9) will be evaluated.

$$M_A L_1 = -84\,\text{kN}\cdot\text{m}(8\,\text{m}) = -672\,\text{kN}\cdot\text{m}^2$$

$$2M_B(L_1 + L_2) = 2M_B(8 + 7) = M_B(30\,\text{m})$$

$$-\sum\left[\frac{P_i a_i}{L_1}(L_1^2 - a_i^2)\right]_1 = -\frac{15(2)}{8}(8^2 - 2^2) - \frac{18(6)}{8}(8^2 - 6^2)$$

$$= -603\,\text{kN}\cdot\text{m}^2$$

$$-\sum\left[\frac{P_i b_i}{L_2}(L_2^2 - b_i^2)\right]_2 = -\frac{20(4)}{7}(7^2 - 4^2) = -377\,\text{kN}\cdot\text{m}^2$$

$$-\frac{w_1 L_1^3}{4} = -\frac{30(8)^3}{4} = -3840\,\text{kN}\cdot\text{m}^2$$

$$-\frac{w_2 L_2^3}{4} = -\frac{50(7)^3}{4} = -4288\,\text{kN}\cdot\text{m}^2$$

Now putting these values into Equation (5–9) gives

$$-672\,\text{kN}\cdot\text{m}^2 + M_B(30\,\text{m}) + 0 = (-603 - 377 - 3840 - 4288)\,\text{kN}\cdot\text{m}^2$$

Solving for M_B yields

$$M_B = -281 \text{ kN}\cdot\text{m}$$

Support Reactions. Having the three moments at the locations of the support points A, B, and C allows the computation of the reactions at those supports. The solution procedure starts by considering each of the two spans as separate free-bodies, as shown in Figure 5–54(b) and (c). In each case, when the beam is divided at B, the moment M_B is shown acting at the cut section to maintain equilibrium. Then, using the left segment, we can sum moments about point B and solve for the left reaction, R_A.

$$\sum M_B = 0 = 12 \text{ kN}(10 \text{ m}) + 15 \text{ kN}(6 \text{ m}) + 300 \text{ kN}(5 \text{ m}) + 18 \text{ kN}(2 \text{ m})$$

$$-281 \text{ kN}\cdot\text{m} - R_A(8 \text{ m})$$

$$R_A = 183 \text{ kN}$$

Similarly, using the right segment and summing moments about point B allows the determination of the right reaction, R_C.

$$\sum M_B = 0 = 20 \text{ kN}(3 \text{ m}) + 350 \text{ kN}(3.5 \text{ m}) - 281 \text{ kN}\cdot\text{m} - R_C(7 \text{ m})$$

$$R_C = 143 \text{ kN}$$

Now we can use the $\sum F_v = 0$ to find the middle reaction, R_B.

$$\sum F_v = 0 = 12 \text{ kN} + 15 \text{ kN} + 18 \text{ kN} + 20 \text{ kN} + 300 \text{ kN} + 350 \text{ kN}$$

$$-183 \text{ kN} - 143 \text{ kN} - R_B$$

$$R_B = 389 \text{ kN}$$

Shearing Force and Bending Moment Diagrams. We now have the necessary data to draw the complete diagrams, as shown in Figure 5–55.

Comment In summary, the reaction forces are

$$R_A = 183 \text{ kN}$$

$$R_B = 389 \text{ kN}$$

$$R_C = 143 \text{ kN}$$

Figure 5–55 shows that the local maximum positive bending moments occur between the supports, and the local maximum negative bending moments occur at the supports. The overall maximum positive bending moment is 204 kN·m at a point 2.86 m from C where the shearing force curve crosses the zero axis. The actual maximum negative bending moment is -281 kN·m at support B. If a beam of uniform cross section is used, it would have to be designed to withstand a bending moment of 281 kN·m. But notice that this is a rather localized peak on the bending moment diagram. It may prove economical to design the beam to withstand the 204 kN·m bending moment and then add reinforcing plates in the vicinity of support B to increase the section modulus there to a level that would be safe for the 281 kN·m moment. You might notice many highway overpasses designed in this manner.

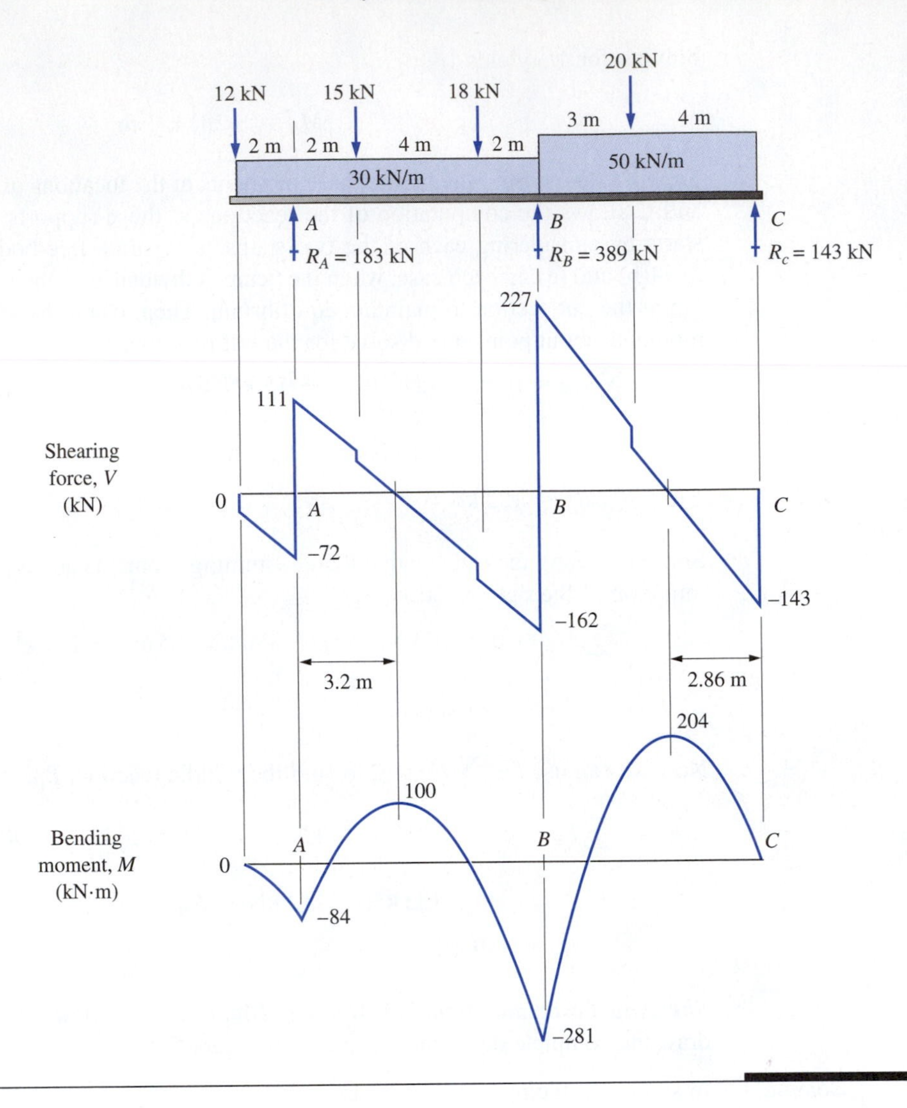

FIGURE 5–55 Shearing force and bending moment diagrams for Example Problem 5–8.

Formulas for Common Loading and Support Patterns. Appendix A–25 gives illustrations and formulas for shearing force and bending moment diagrams for ten special cases for continuous beams. These cases apply to frequently encountered loading and support patterns in building construction, bridge design, structures for machinery or vehicles, and some consumer products. Included are:

1. Supported cantilevers—four loading patterns
2. Fixed end beams—three loading patterns
3. Continuous beams—uniformly distributed loads for two spans, three spans, and four spans.

The tables give:

- Sketches of the resulting shearing force and bending moment diagrams
- The general shape of the beam deflections

- Equations for reactions at supports, and shearing forces and bending moments at critical points
- Equations for beam deflection for most loadings.

See Chapter 9 for more discussion and analysis techniques for beam deflections.

It is important to understand the general shape of the shearing force and bending moment diagrams so you will be able to design safe and efficient beams as discussed in Chapters 7 and 8.

PROBLEMS

Problems for Figures P5–1 to P5–76

Figure P5–1 to P5–76 show a variety of beam types and loading conditions. For the beam in each figure, any or all of the following problem statements can be applied:

1. Compute the reactions at the supports using the techniques shown in Section 5–3.
2. Draw the complete shearing force and bending moment diagrams using the techniques shown in Sections 5–4 to 5–9.
3. Determine the magnitude and location of the maximum absolute value of the shearing force and bending moment.
4. Use the free-body diagram approach shown in Sections 5–4 to 5–6 to determine the internal shearing force and bending moment at any specified point in a beam.
5. Write equations for all segments of the shearing force and bending moment diagrams using the guidelines presented in Section 5–11.

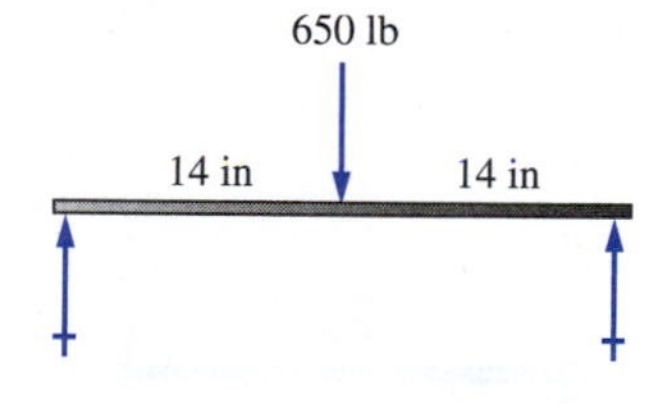

FIGURE P5–1

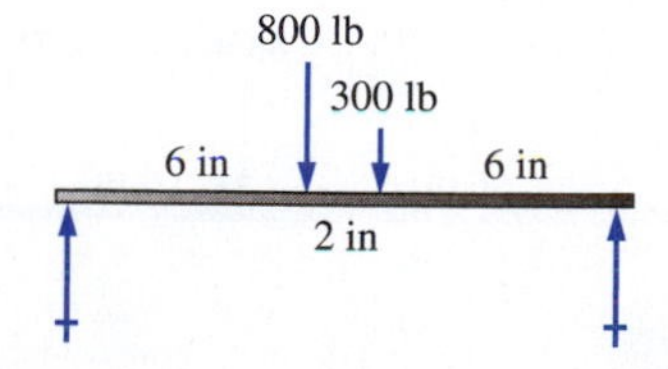

FIGURE P5–2

16 K
4 ft
10 ft

FIGURE P5–3

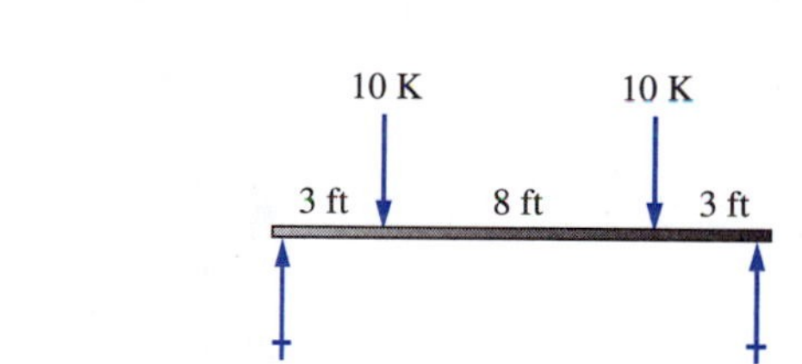

FIGURE P5–4

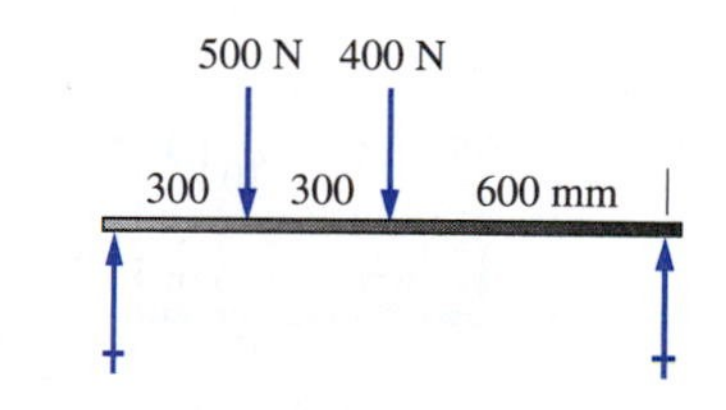

FIGURE P5–5

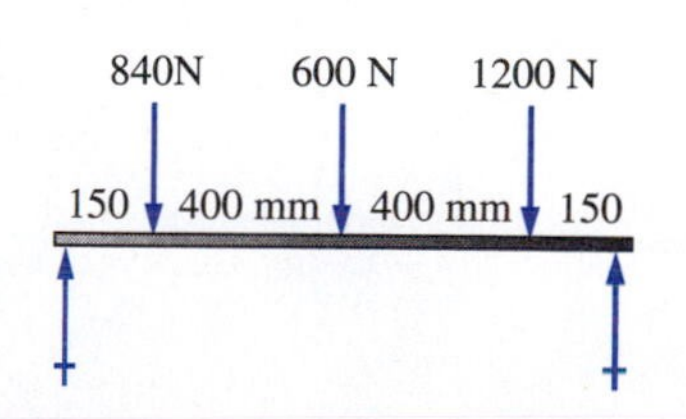

FIGURE P5–6

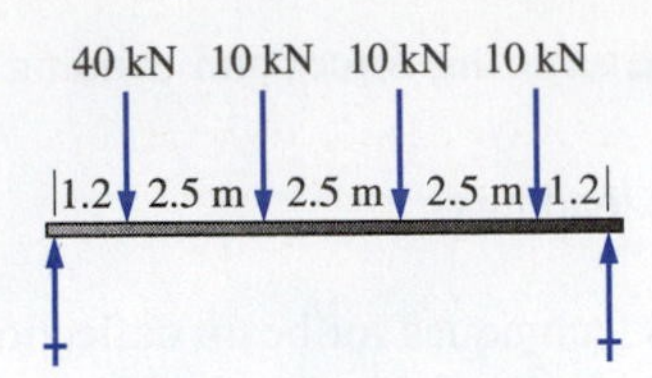

FIGURE P5–7

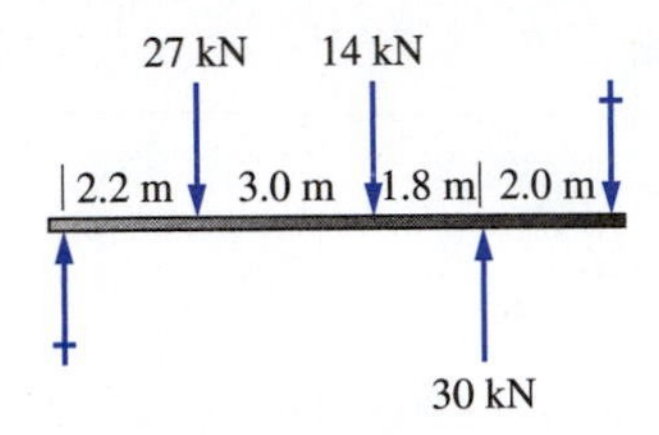

FIGURE P5–8

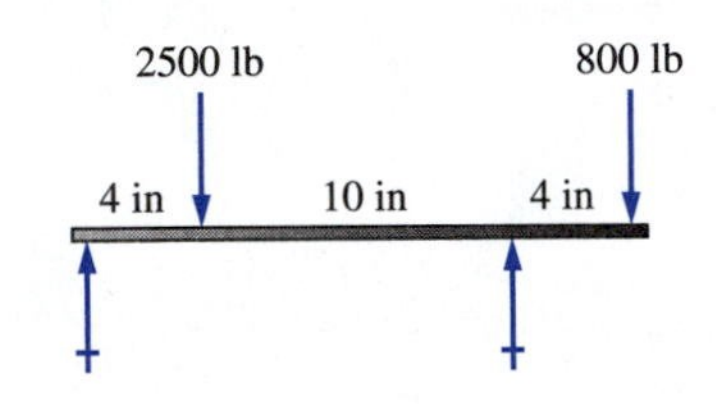

FIGURE P5–9

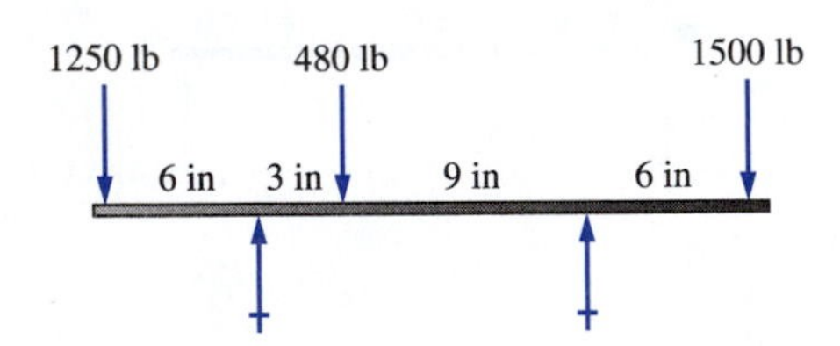

FIGURE P5–10

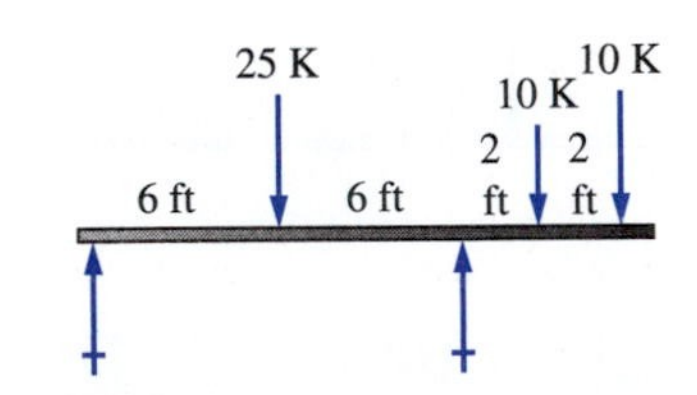

FIGURE P5–11

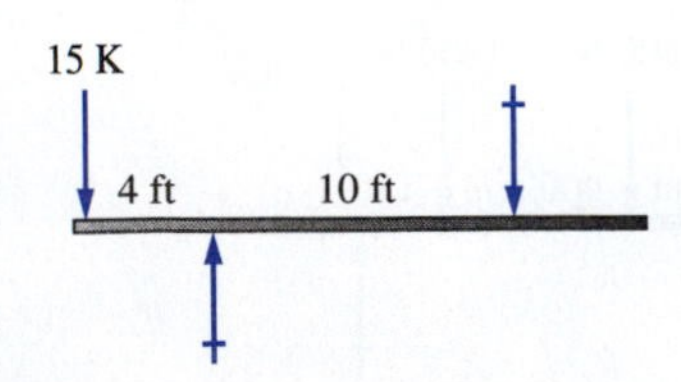

FIGURE P5–12

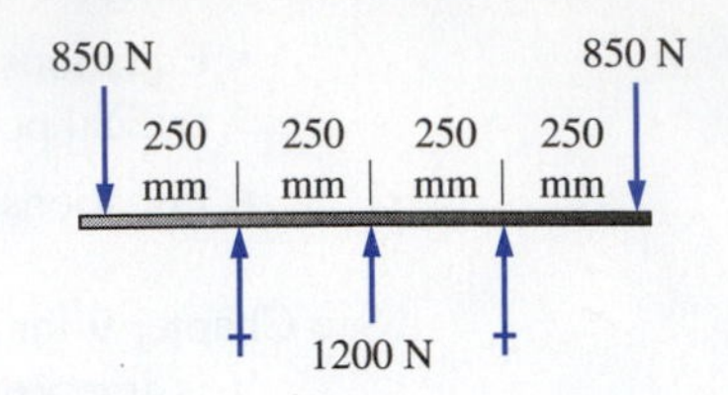

FIGURE P5–13

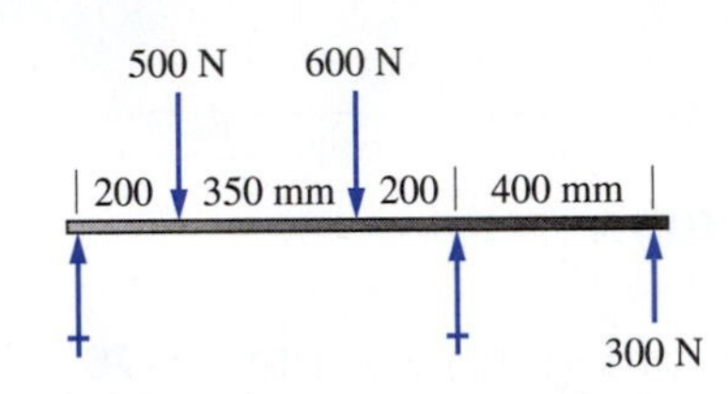

FIGURE P5–14

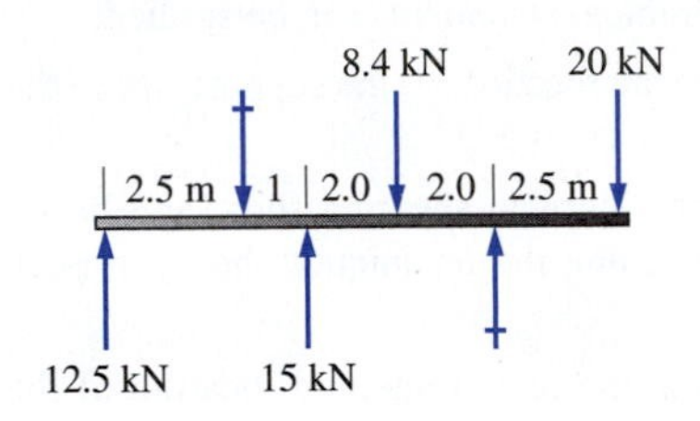

FIGURE P5–15

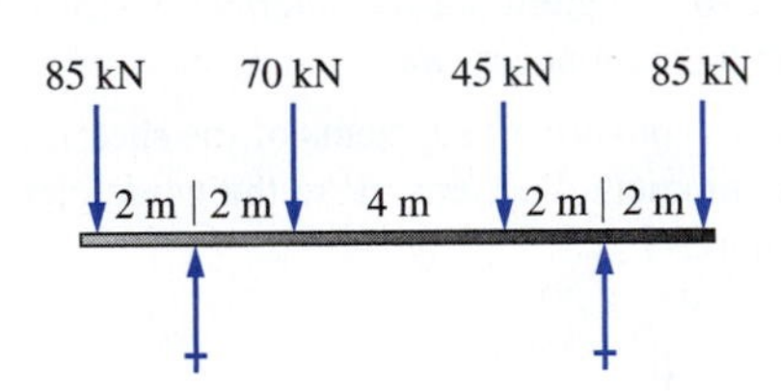

FIGURE P5–16

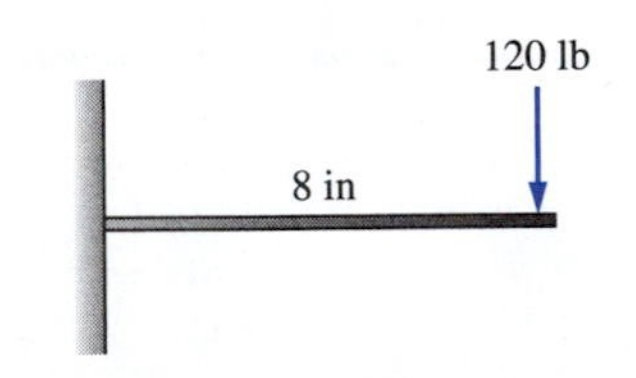

FIGURE P5–17

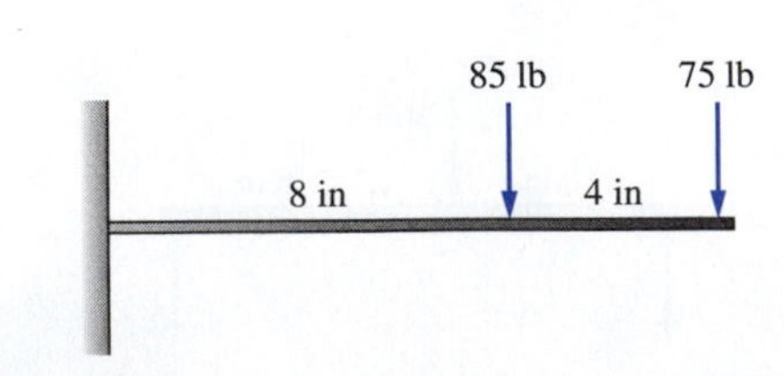

FIGURE P5–18

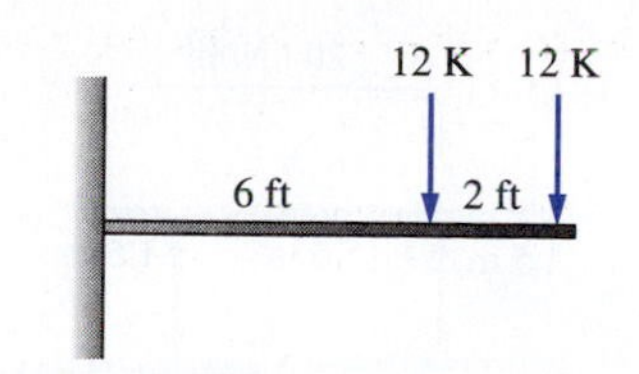

FIGURE P5–19

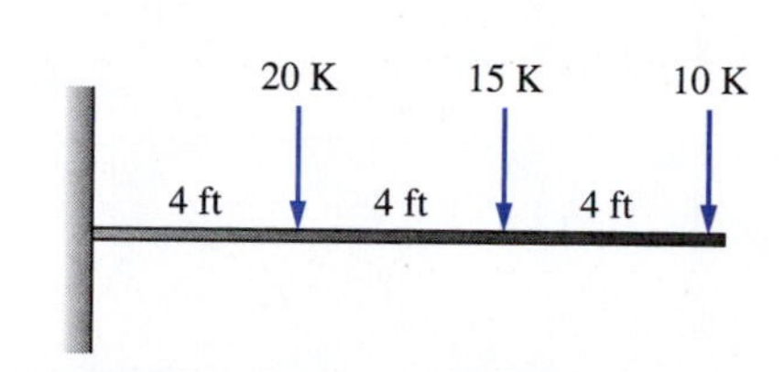

FIGURE P5–20

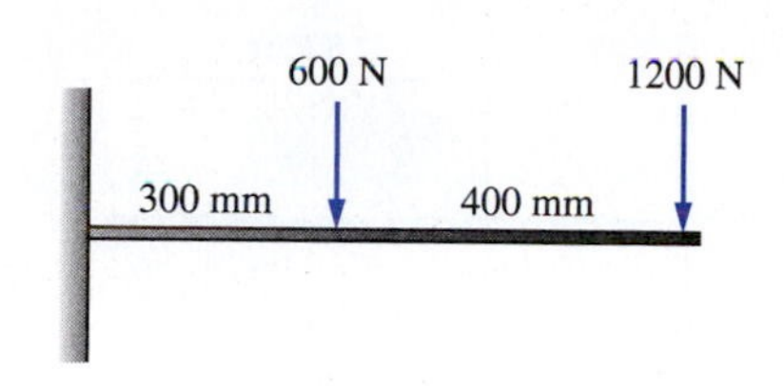

FIGURE P5–21

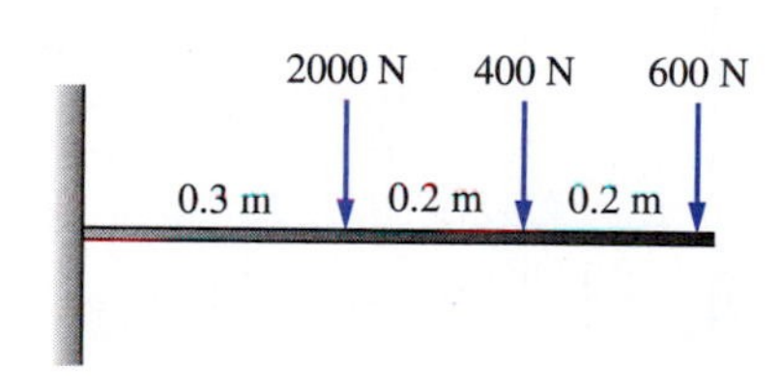

FIGURE P5–22

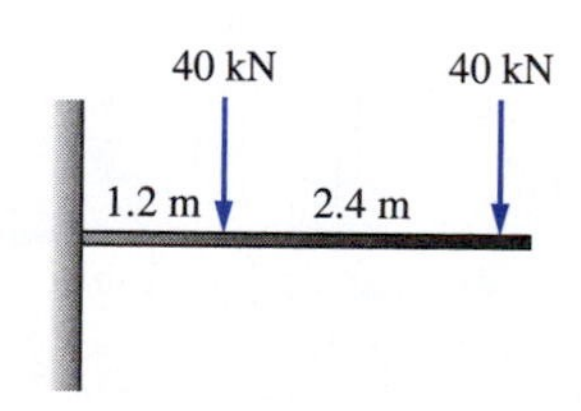

FIGURE P5–23

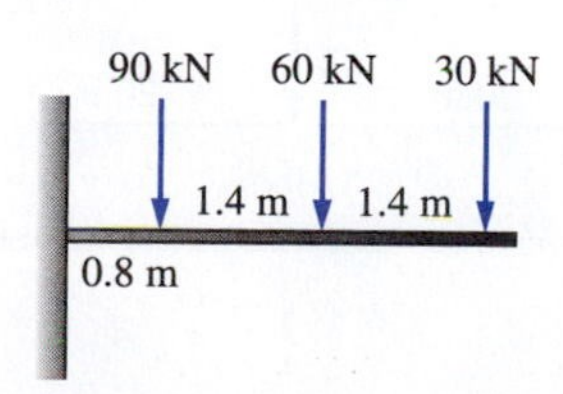

FIGURE P5–24

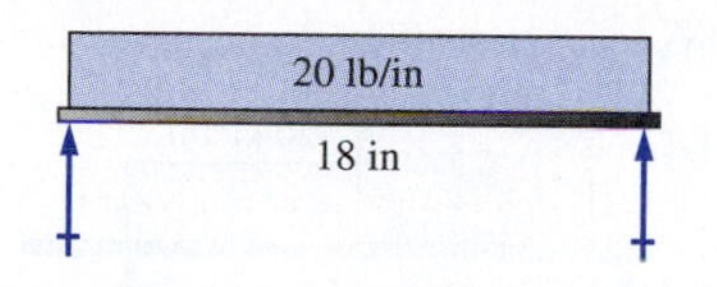

FIGURE P5–25

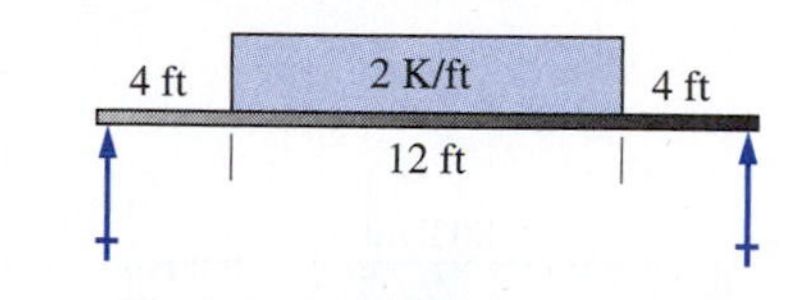

FIGURE P5–26

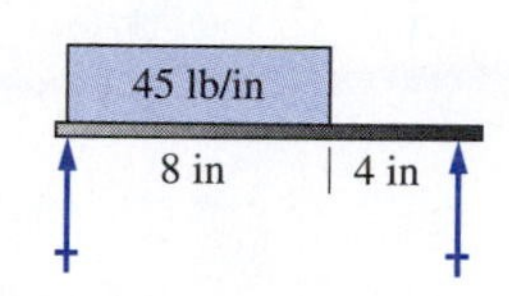

FIGURE P5–27

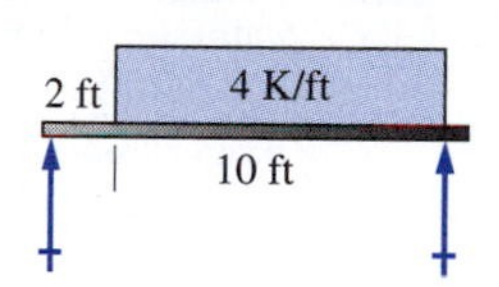

FIGURE P5–28

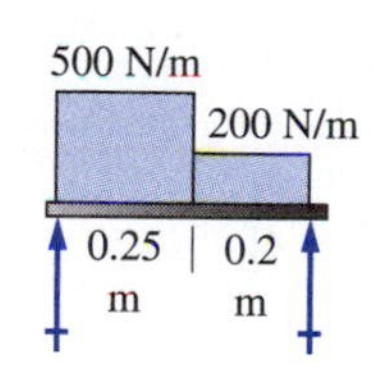

FIGURE P5–29

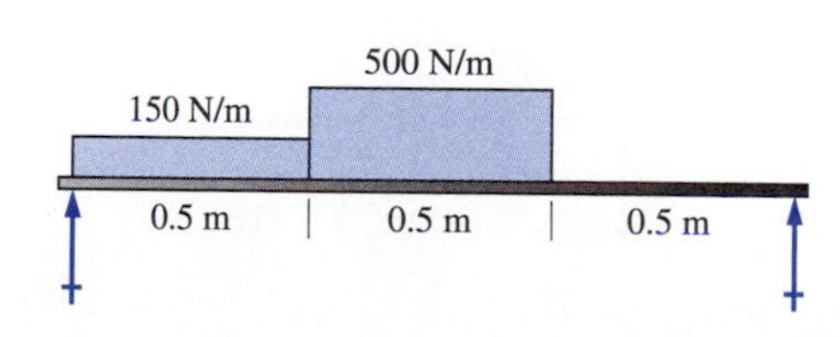

FIGURE P5–30

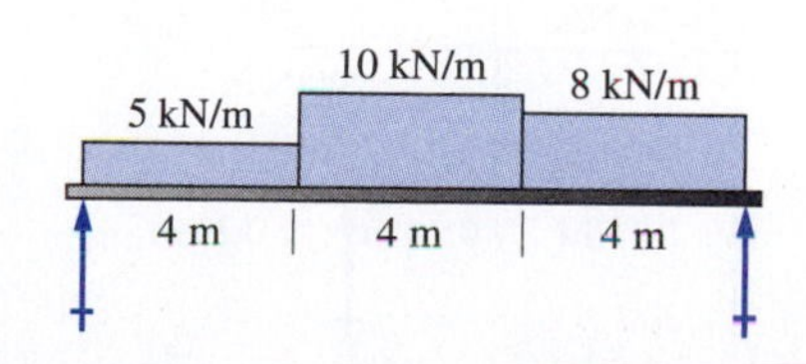

FIGURE P5–31

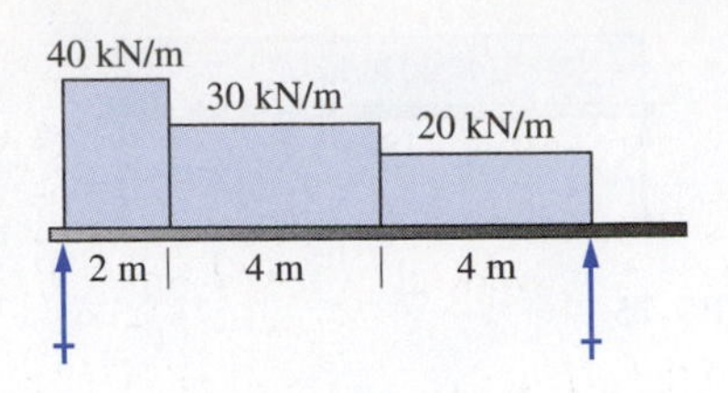

FIGURE P5–32

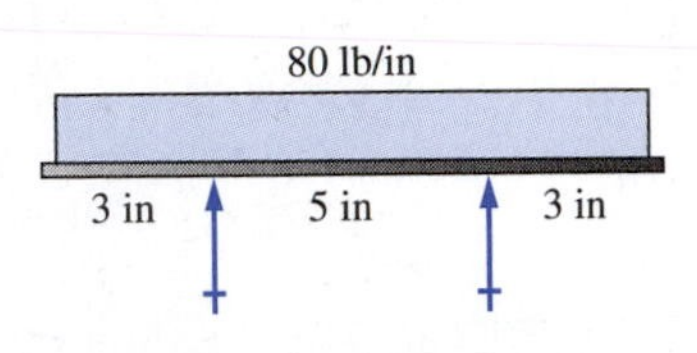

FIGURE P5–33

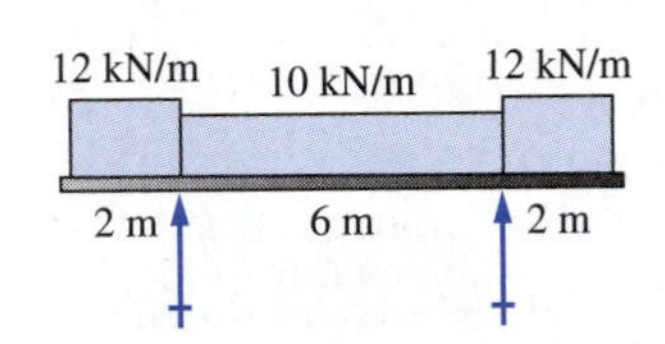

FIGURE P5–34

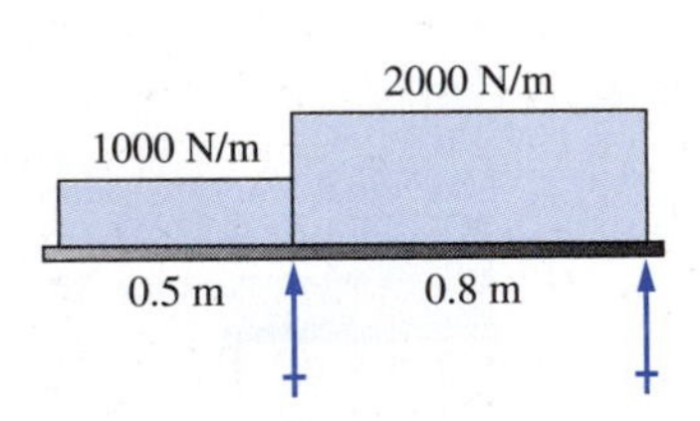

FIGURE P5–35

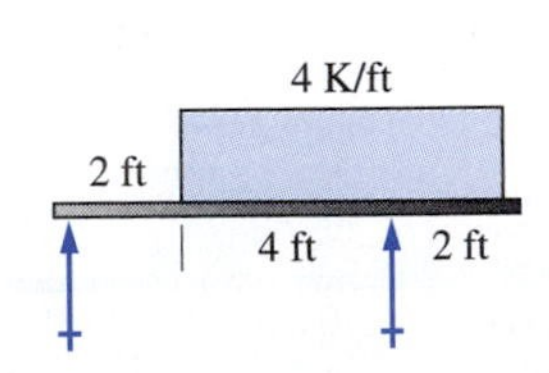

FIGURE P5–36

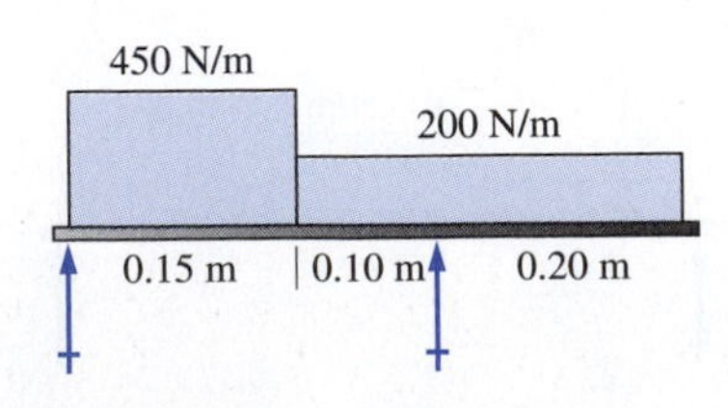

FIGURE P5–37

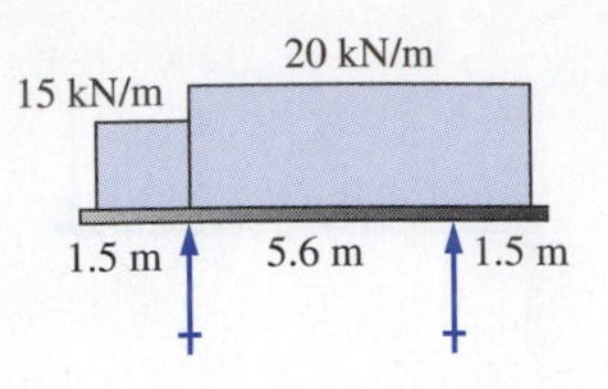

FIGURE P5–38

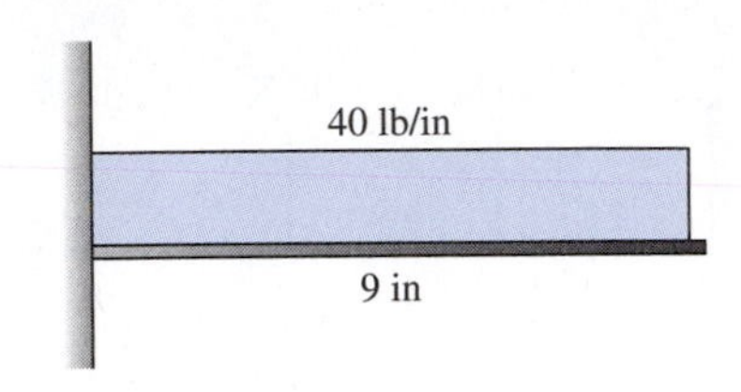

FIGURE P5–39

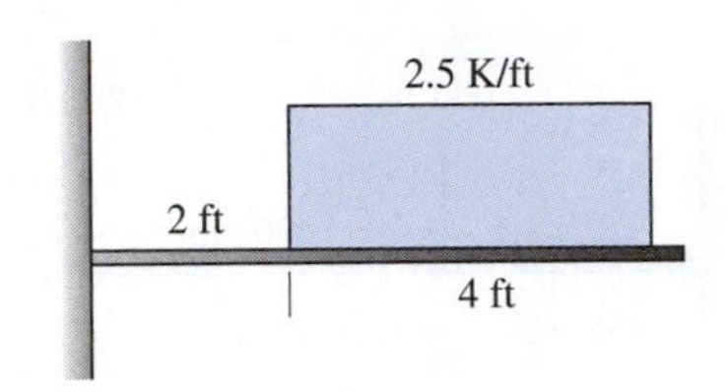

FIGURE P5–40

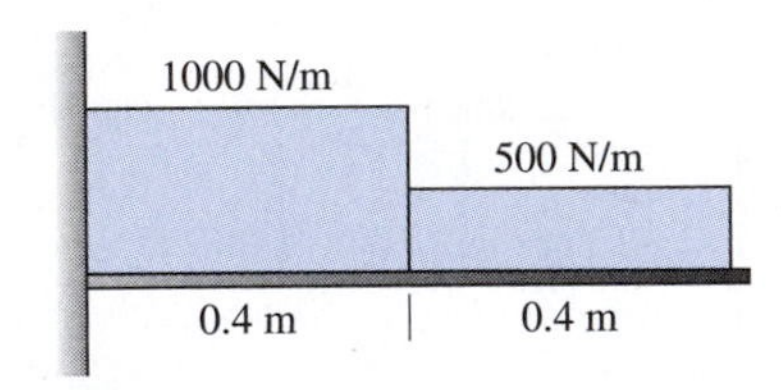

FIGURE P5–41

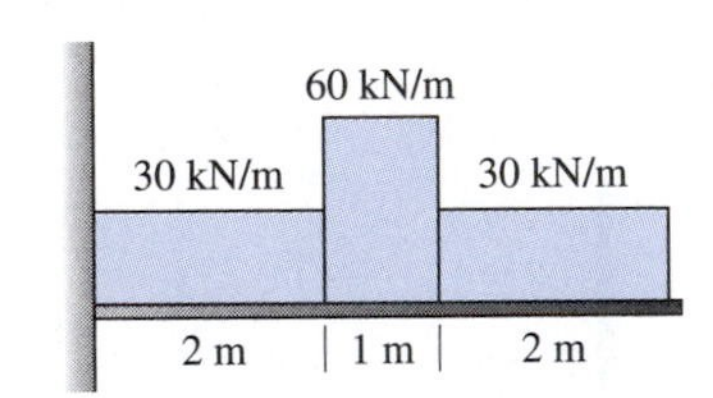

FIGURE P5–42

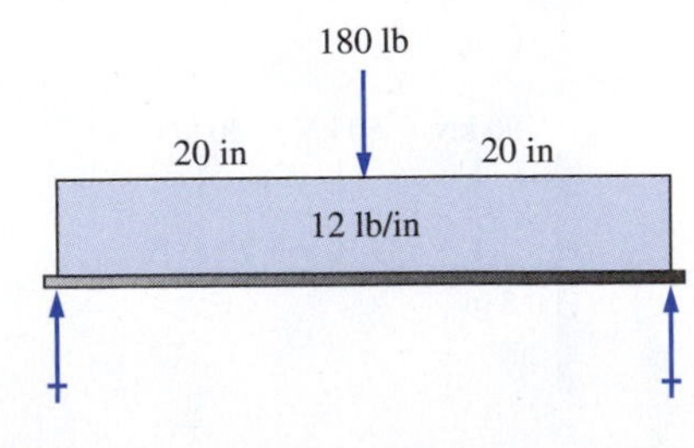

FIGURE P5–43

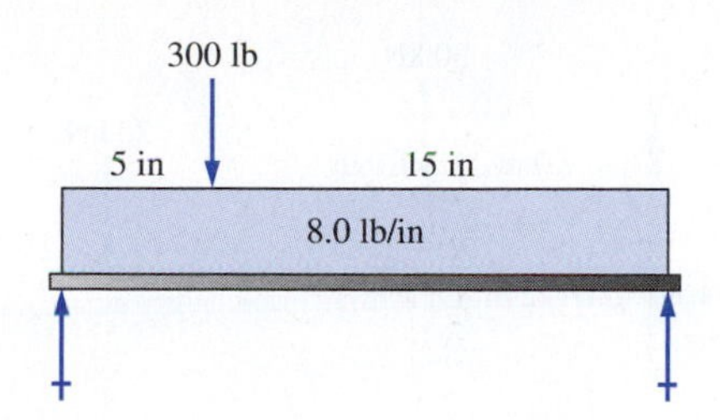

FIGURE P5–44

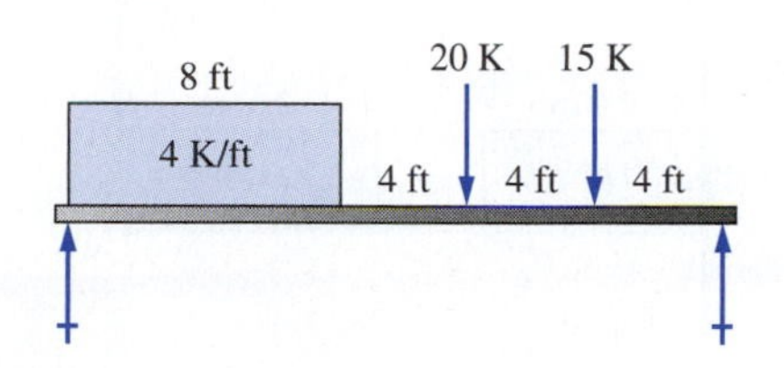

FIGURE P5–45

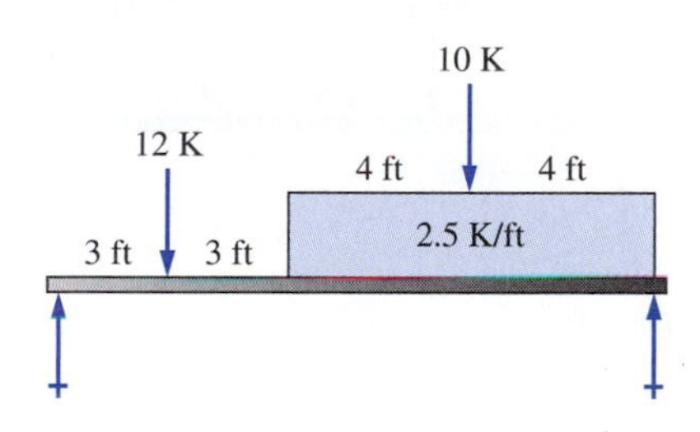

FIGURE P5–46

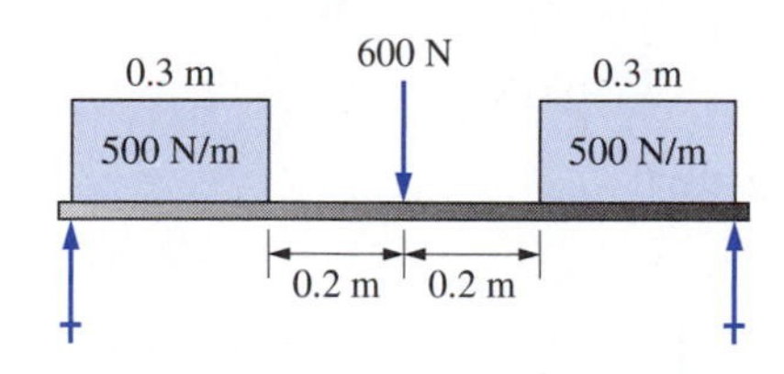

FIGURE P5–47

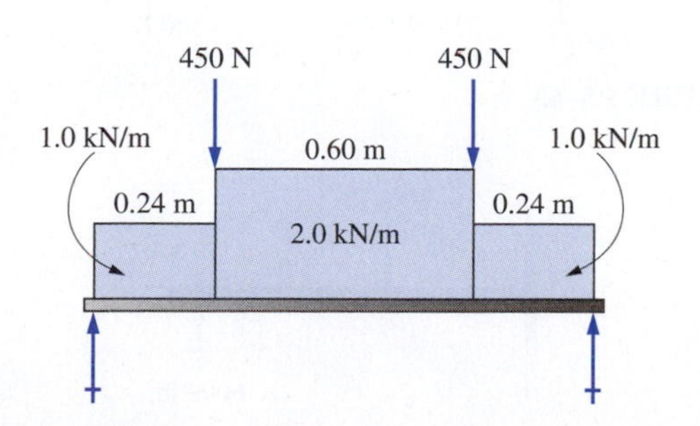

FIGURE P5–48

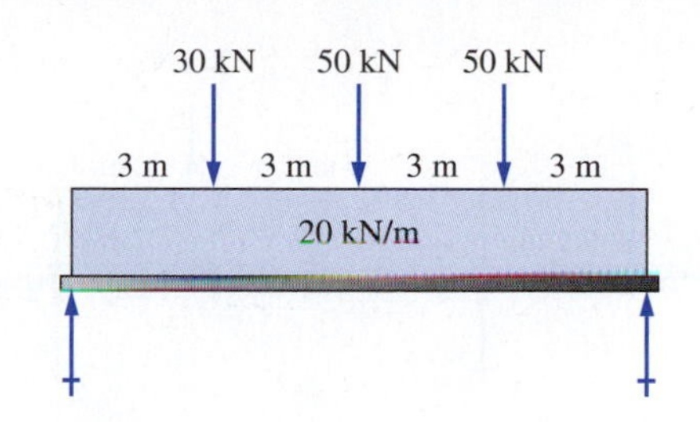

FIGURE P5–49

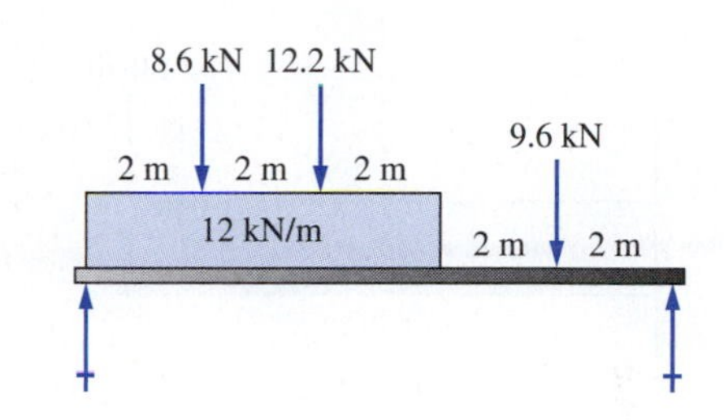

FIGURE P5–50

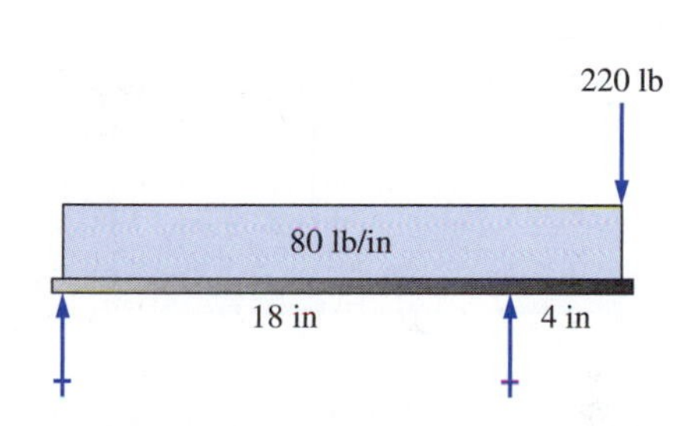

FIGURE P5–51

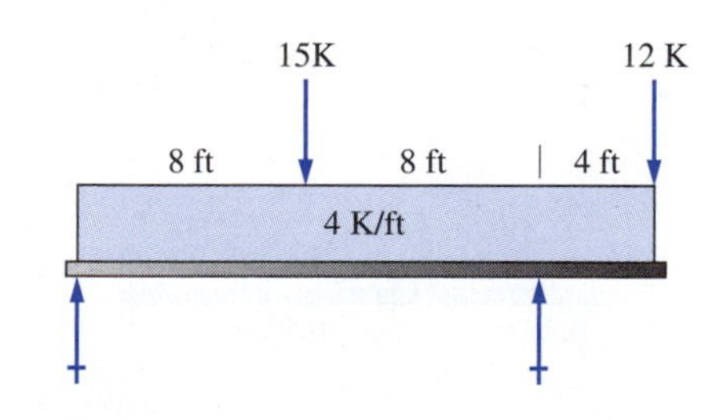

FIGURE P5–52

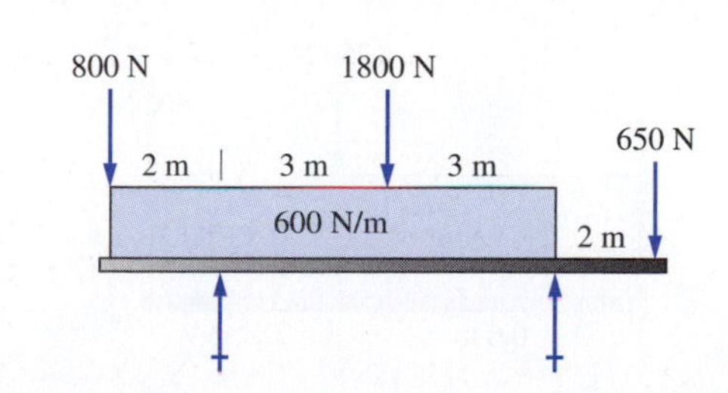

FIGURE P5–53

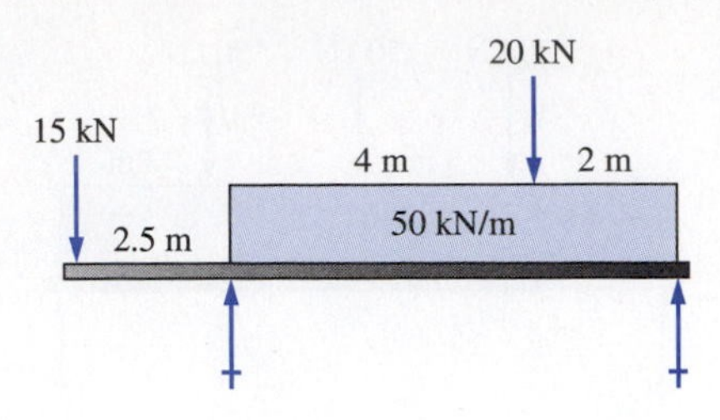

FIGURE P5–54

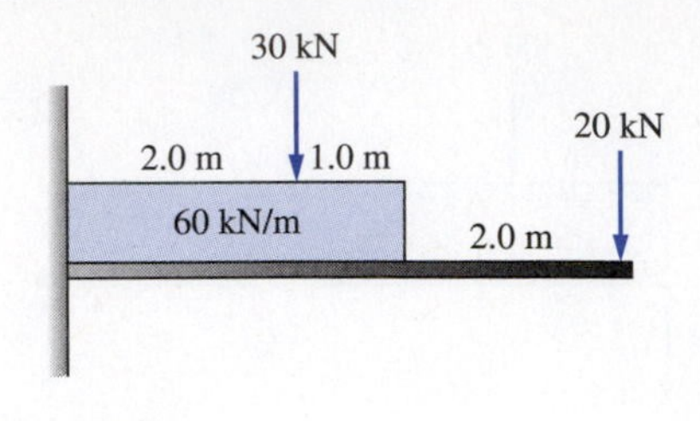

FIGURE P5–59

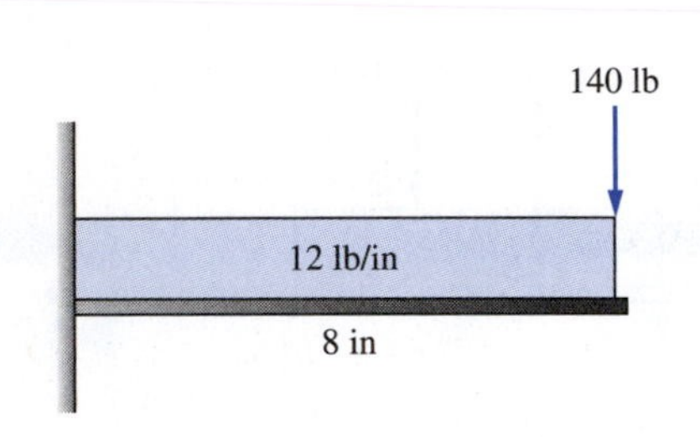

FIGURE P5–55

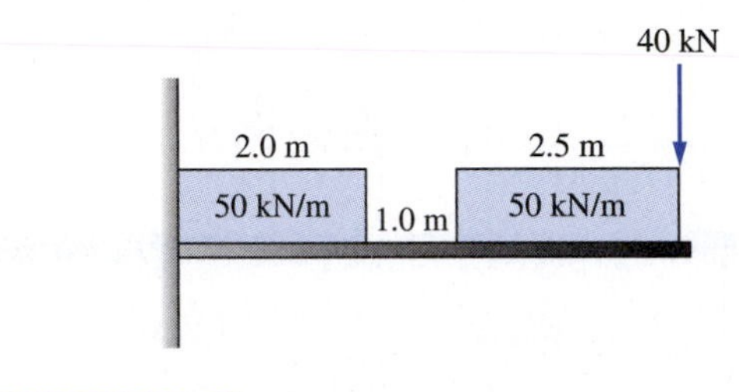

FIGURE P5–60

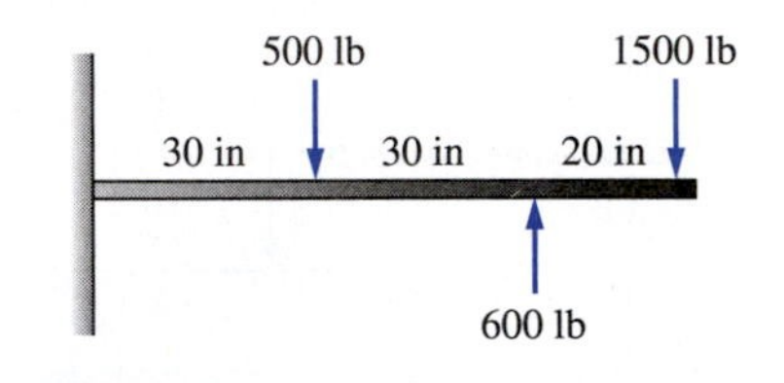

FIGURE P5–61

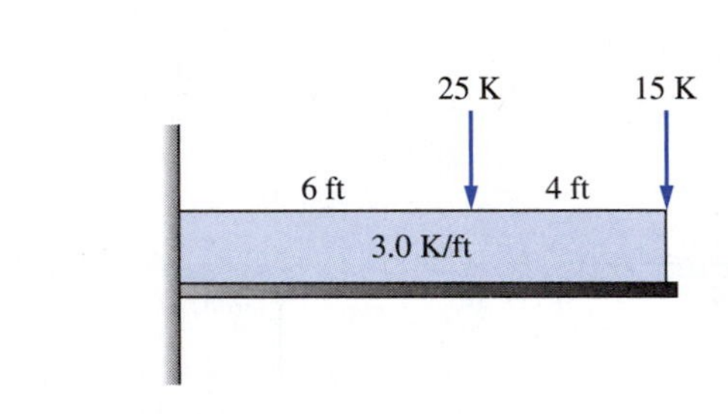

FIGURE P5–56

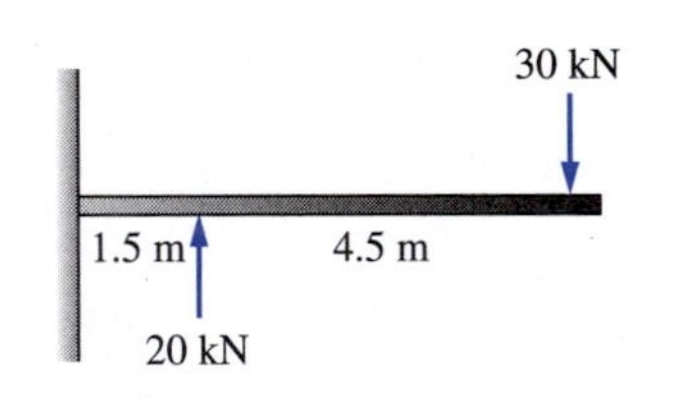

FIGURE P5–62

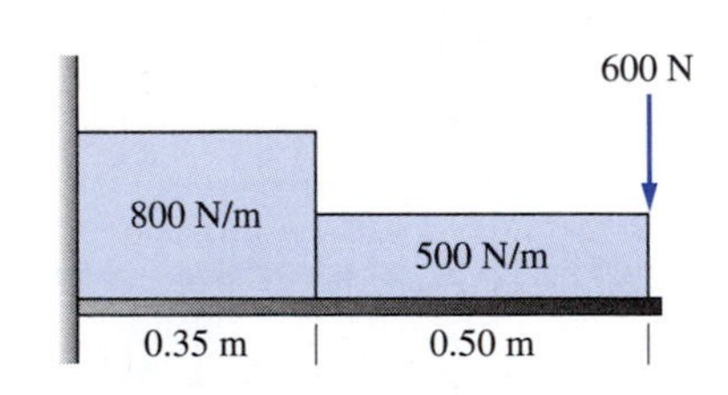

FIGURE P5–57

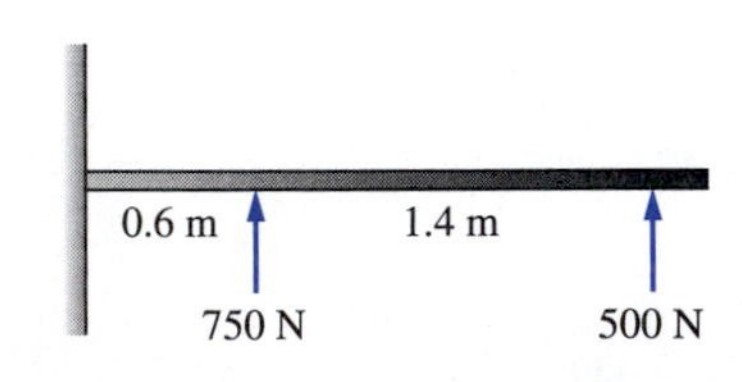

FIGURE P5–63

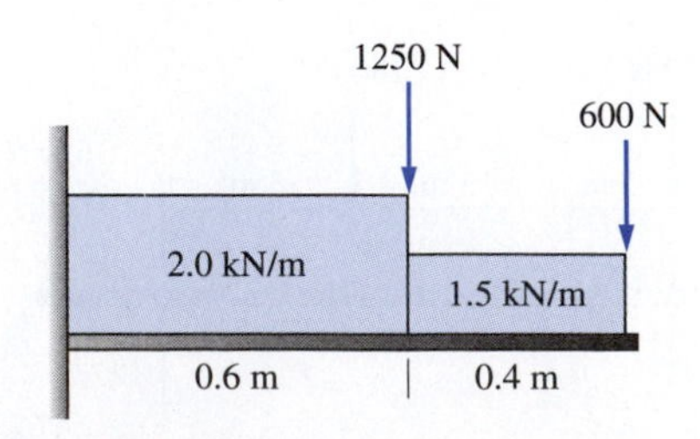

FIGURE P5–58

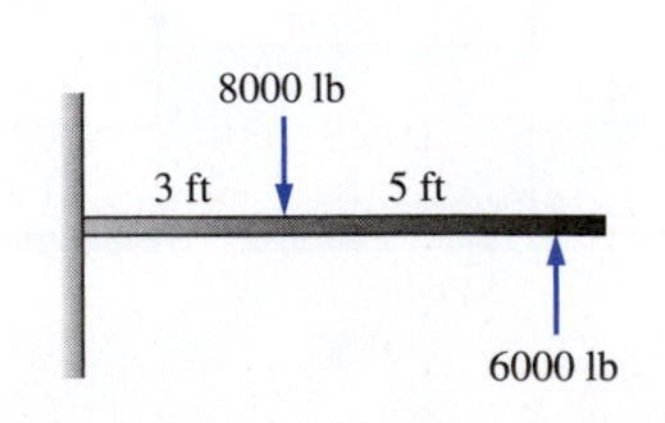

FIGURE P5–64

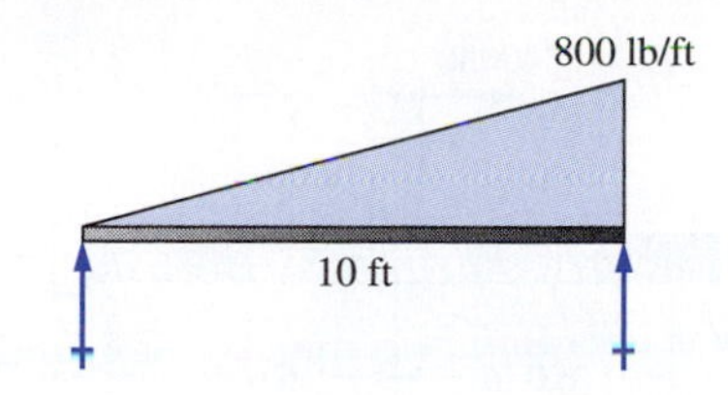

FIGURE P5–65

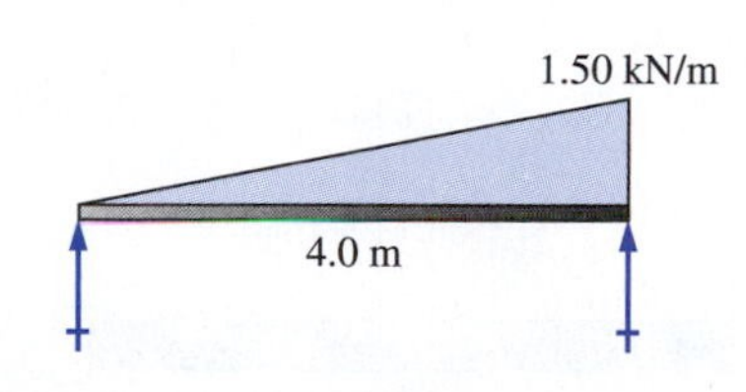

FIGURE P5–66

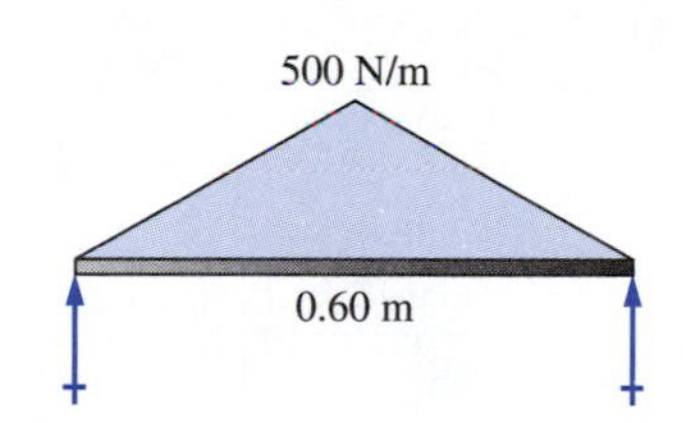

FIGURE P5–67

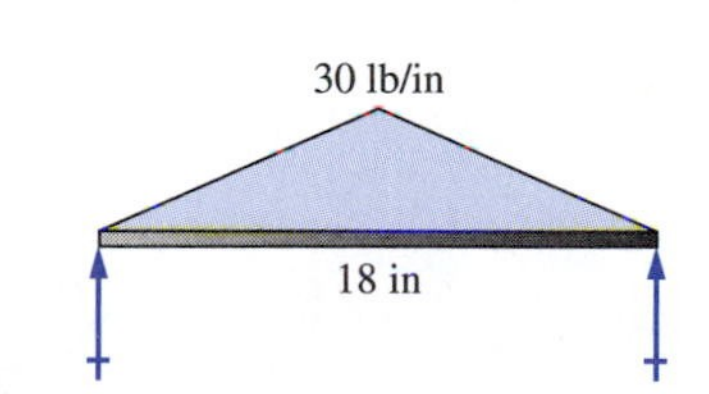

FIGURE P5–68

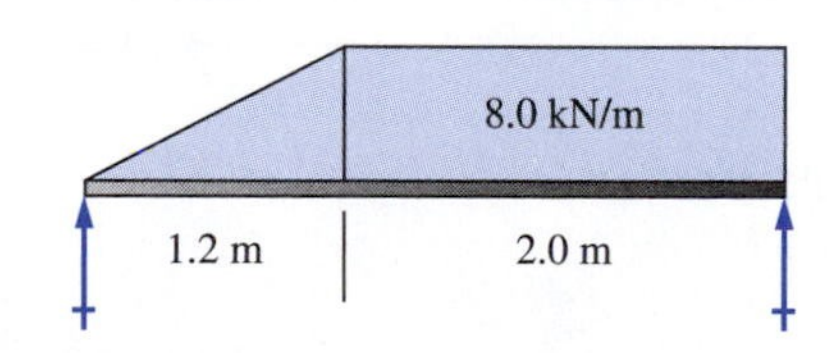

FIGURE P5–69

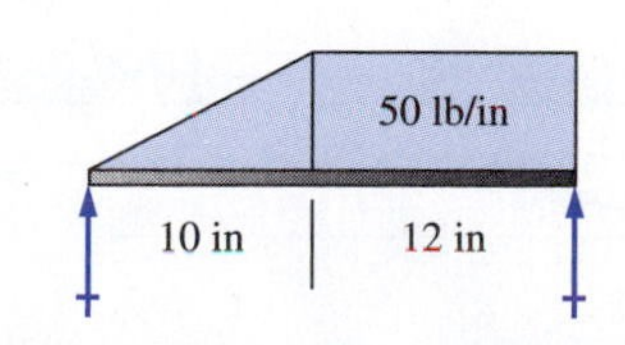

FIGURE P5–70

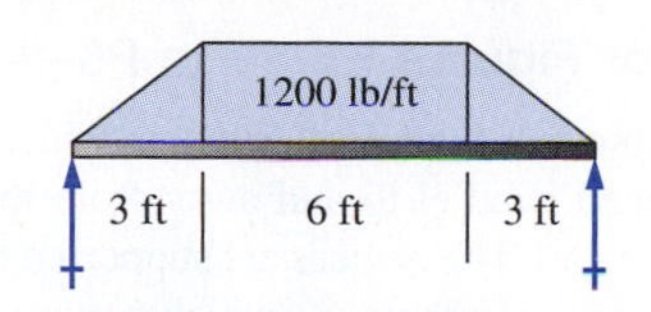

FIGURE P5–71

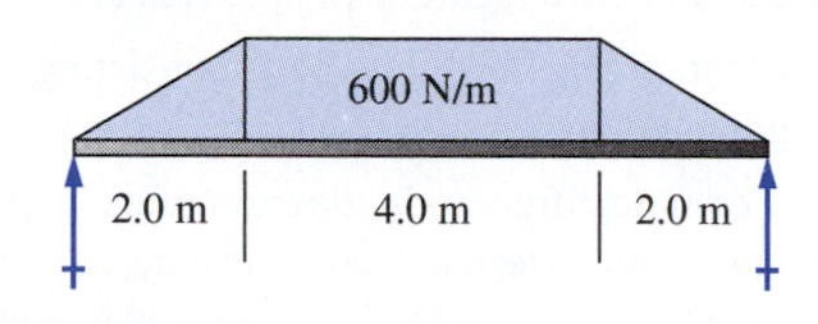

FIGURE P5–72

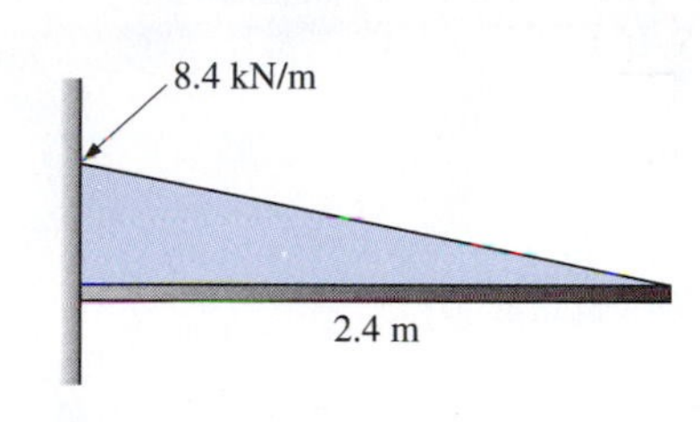

FIGURE P5–73

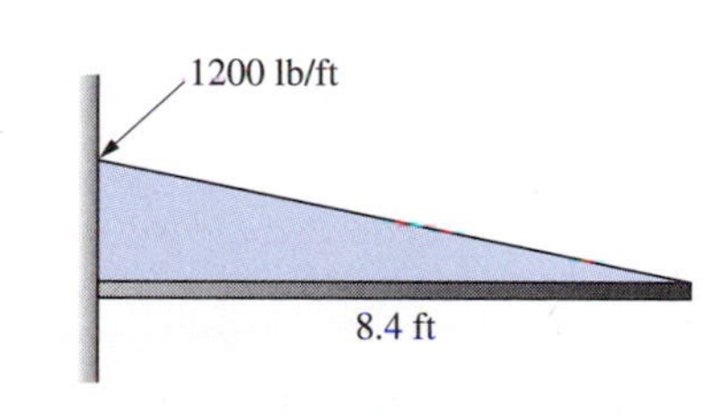

FIGURE P5–74

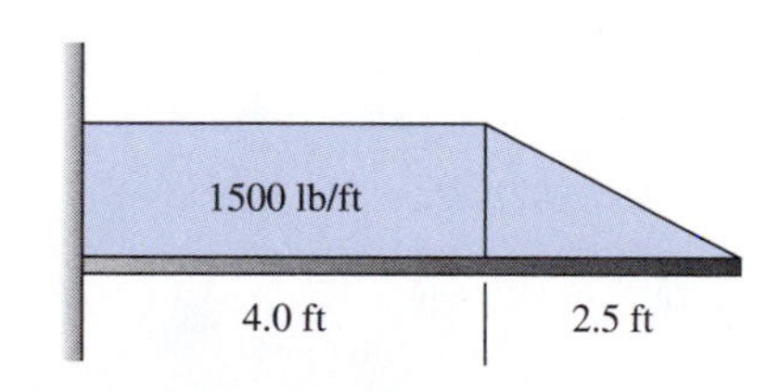

FIGURE P5–75

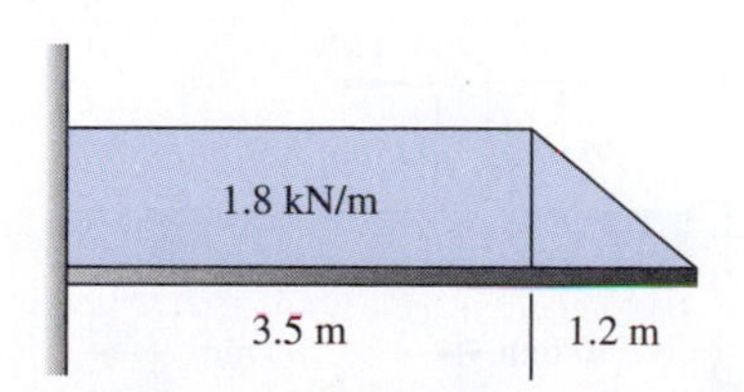

FIGURE P5–76

Problems for Figures P5–77 to P5–84

Each figure shows a mechanical device in which one or more forces are applied parallel to and away from the axis of the main, beam-like part. The devices are supported by bearings at the locations marked with an ×, which can provide reaction forces in any direction perpendicular to the axis of the beam. One of the bearings has the capability of resisting horizontally directed forces. For each figure the objectives are:

1. Break the compound beam into parts consisting of each of the straight components.
2. Show the complete free-body diagram of each component part including all external loads and internal forces and bending moments required to keep the part in equilibrium.
3. For the main, horizontal part only, draw the complete shearing force and bending moment diagrams. Refer to Section 5–10 for examples.

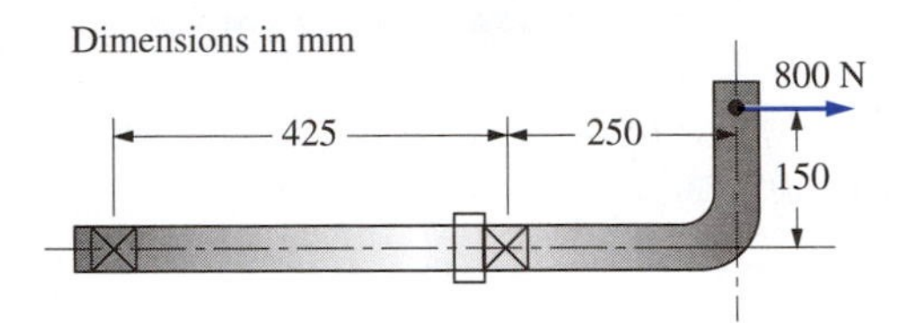

FIGURE P5–77

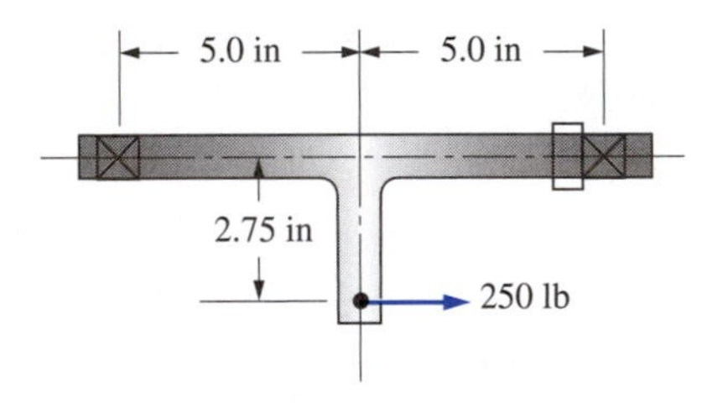

FIGURE P5–78

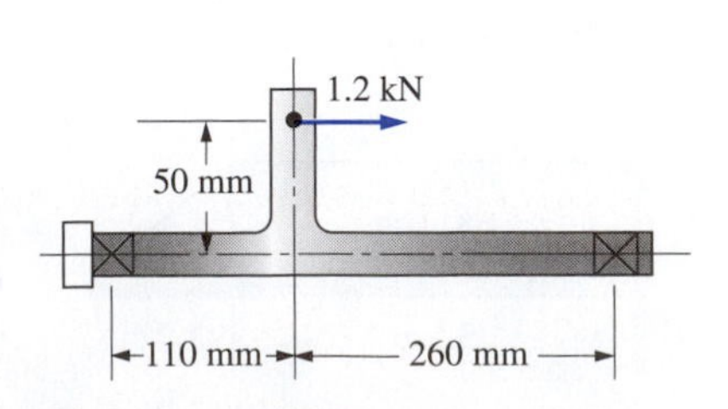

FIGURE P5–79

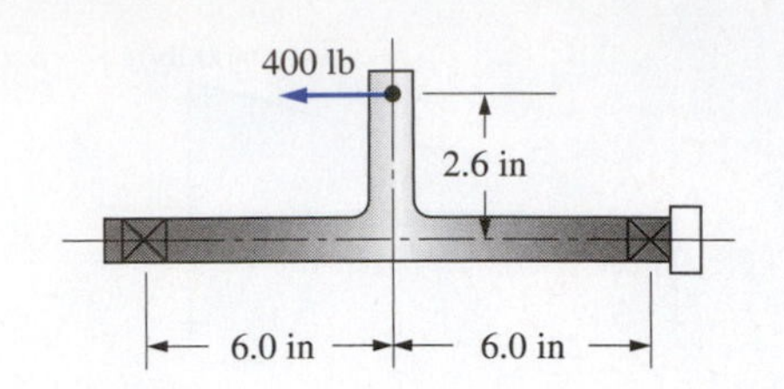

FIGURE P5–80

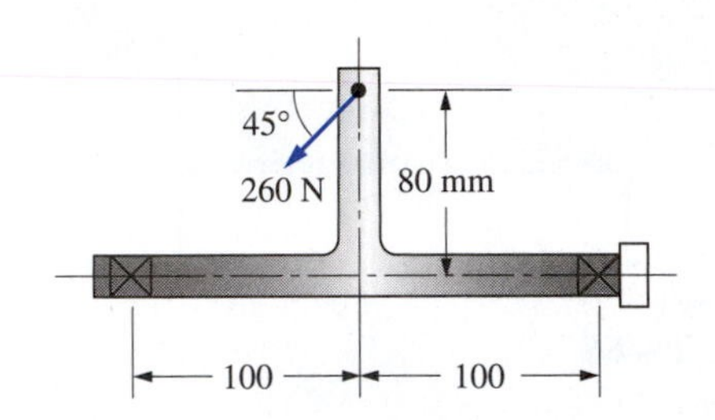

FIGURE P5–81

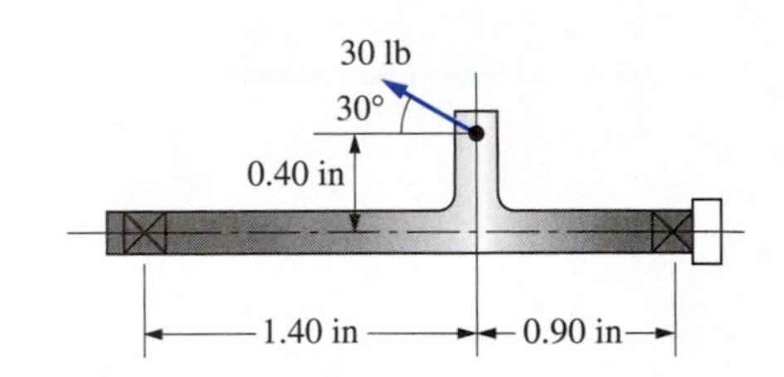

FIGURE P5–82

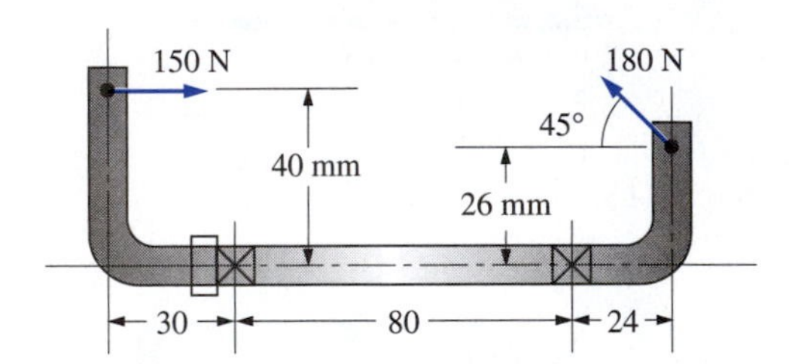

FIGURE P5–83

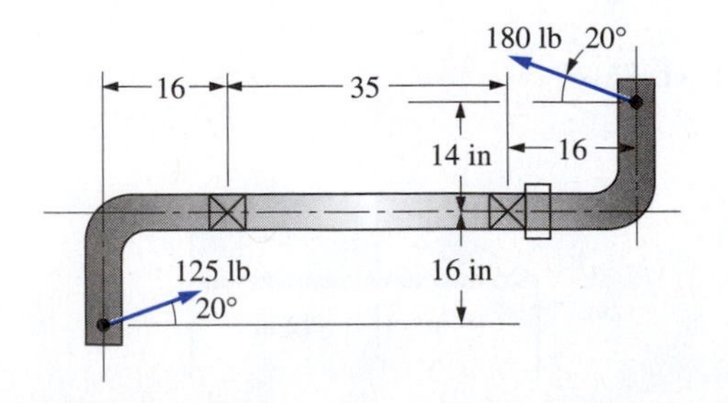

FIGURE P5–84

ADDITIONAL PRACTICE AND REVIEW PROBLEMS

5–85. For the beam shown in Figure P5–8, determine the shearing force and bending moment at a point 4.0 m from the left support using the free-body diagram method.

5–86. For the beam shown in Figure P5–15, determine the shearing force and bending moment at a point 1.0 m to the right of the right support using the free-body diagram method.

5–87. For the beam shown in Figure P5–22, determine the shearing force and bending moment at a point 0.45 m from the wall using the free-body diagram method.

5–88. For the beam shown in Figure P5–35, determine the shearing force and bending moment at a point 0.9 m from the left end of the beam using the free-body diagram method.

5–89. For the beam shown in Figure P5–53, determine the shearing force and bending moment at a point 6.0 m from the left end of the beam using the free-body diagram method.

5–90. For the beam shown in Figure P5–58, determine the shearing force and bending moment at a point 0.8 m from the wall using the free-body diagram method.

5–91. For the beam shown in Figure P5–69, determine the shearing force and bending moment at a point 2.2 m from the left support using the free-body diagram method.

5–92. For the beam shown in Figure P5–76, determine the shearing force and bending moment at a point 4.0 m from the wall using the free-body diagram method.

Problems 5–93 to 5–110

Draw the complete shearing force and bending moment diagrams for the beams shown in the accompanying figures. Report the maximum shearing force, the maximum positive bending moment, and the maximum negative bending moment. Indicate the location of those values on the beam.

Concentrated Loads

P5–93 P5–94 P5–95 P5–96 P5–97 P5–98

Distributed Loads

P5–99 P5–100 P5–101 P5–102 P5–103 P5–104
P5–105 P5–106 P5–107 P5–108 P5–109 P5–110

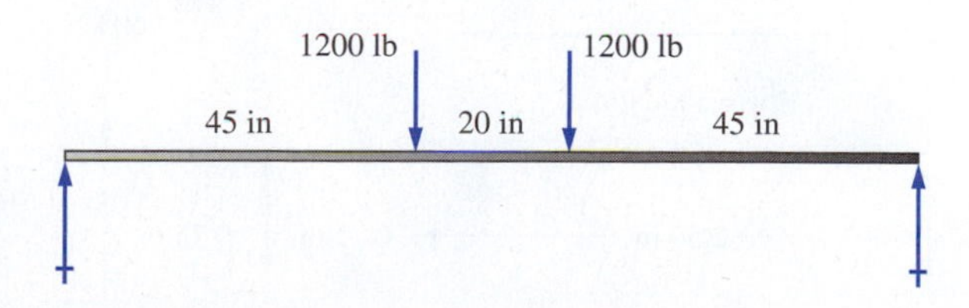

FIGURE P5–93

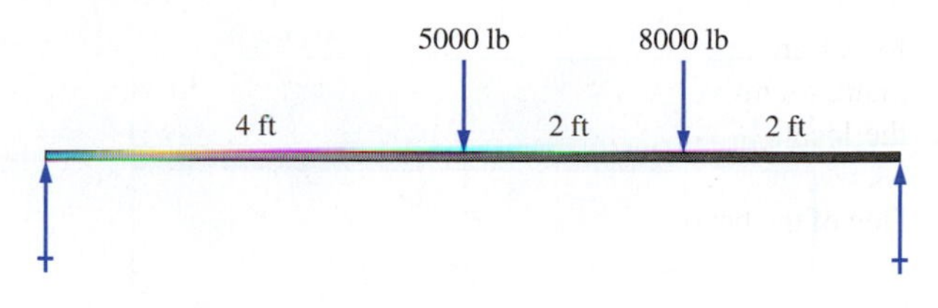

FIGURE P5–94

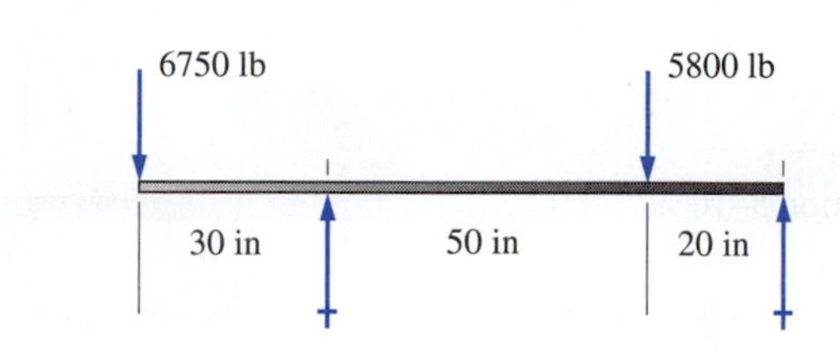

FIGURE P5–95

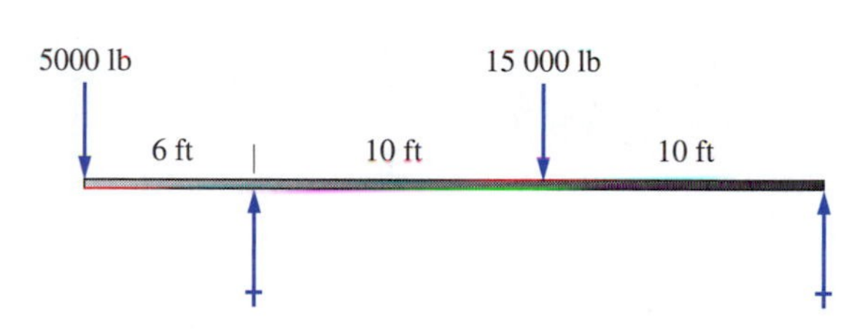

FIGURE P5–96

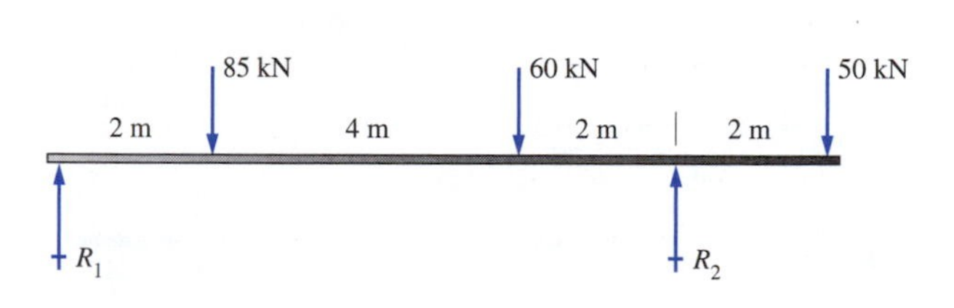

FIGURE P5–97

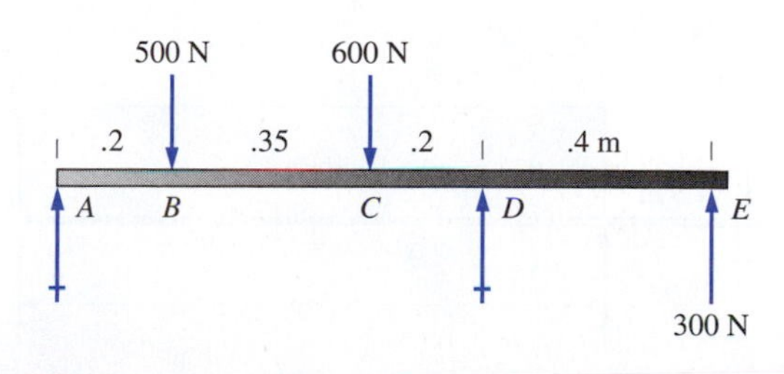

FIGURE P5–98

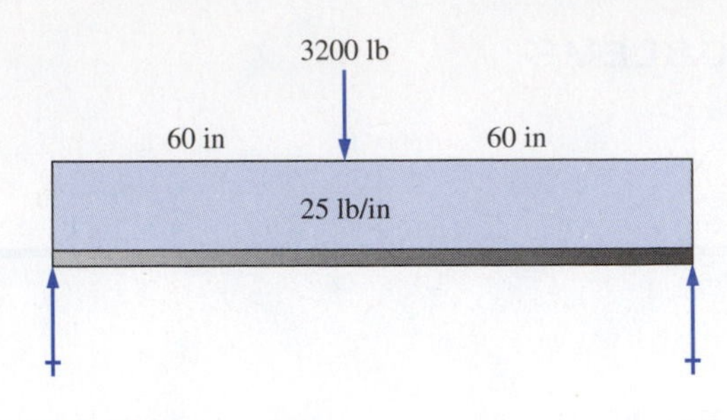

FIGURE P5–99

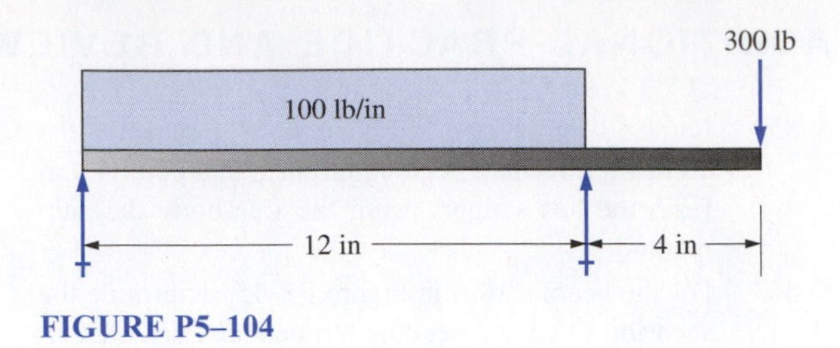

FIGURE P5–104

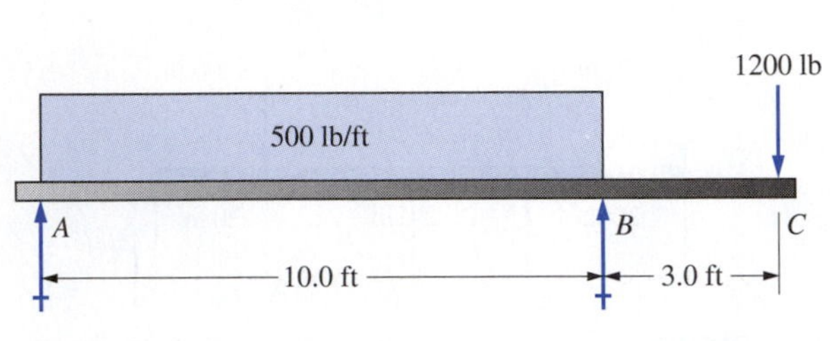

FIGURE P5–100

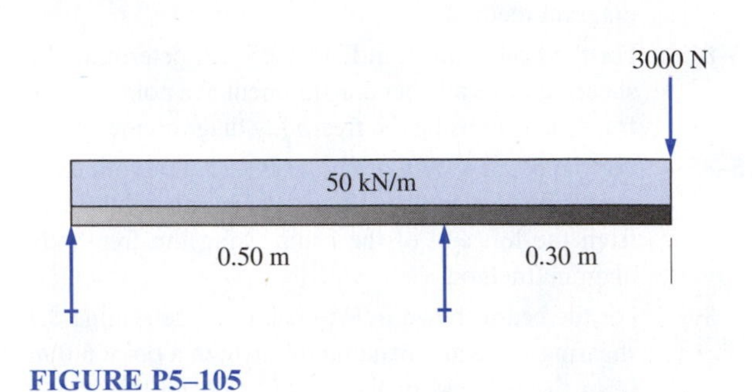

FIGURE P5–105

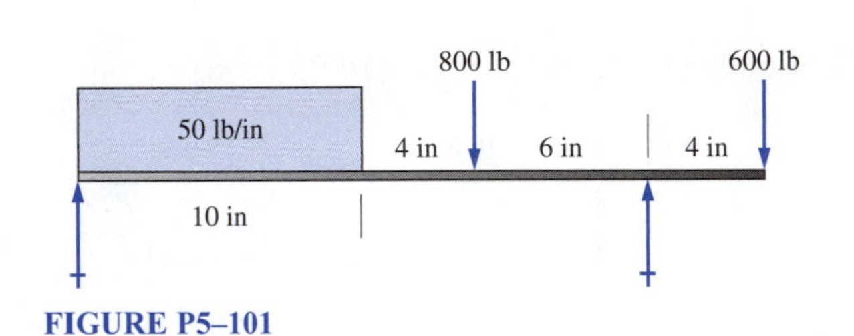

FIGURE P5–101

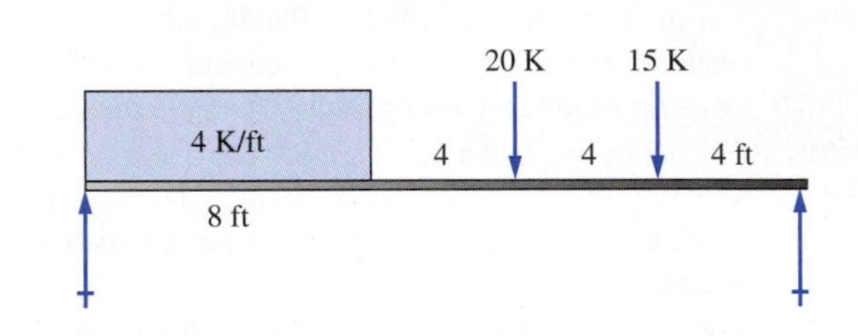

FIGURE P5–106

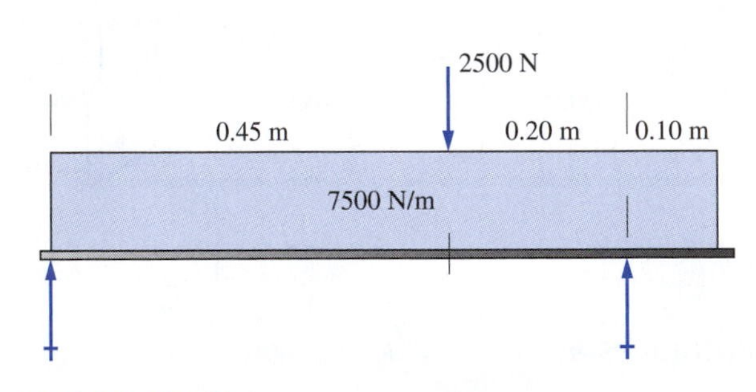

FIGURE P5–102

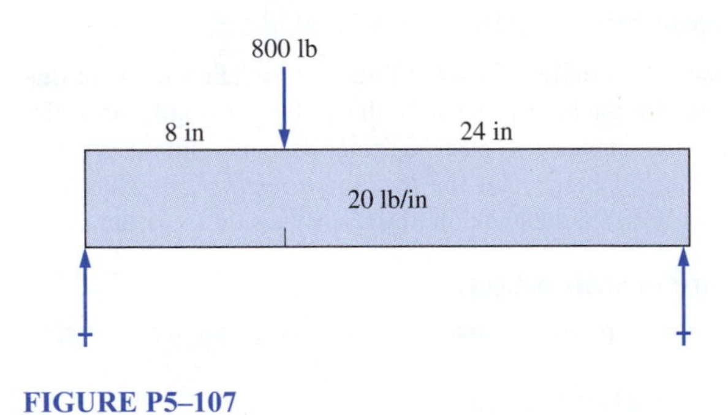

FIGURE P5–107

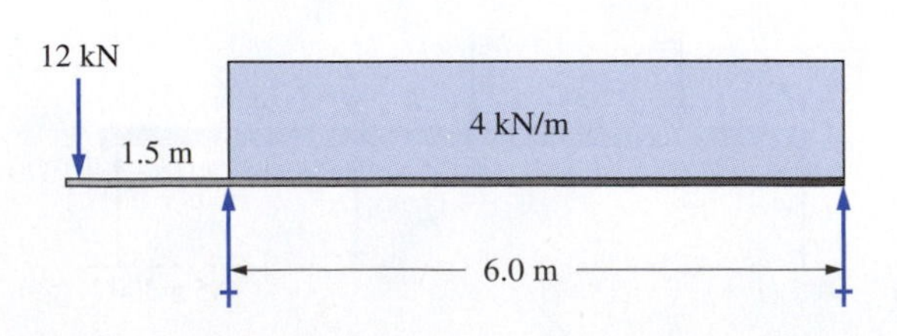

FIGURE P5–103

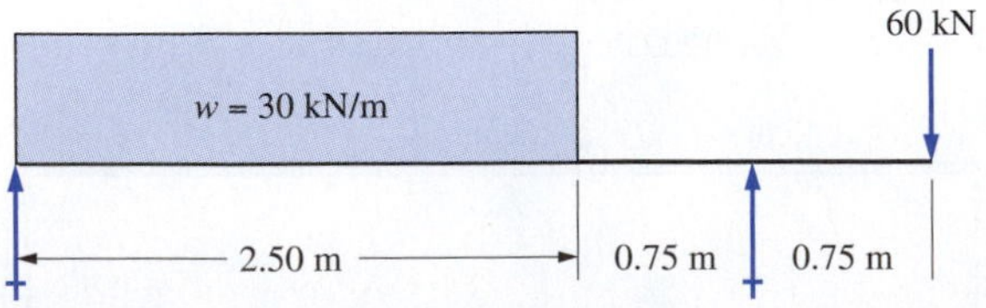

FIGURE P5–108

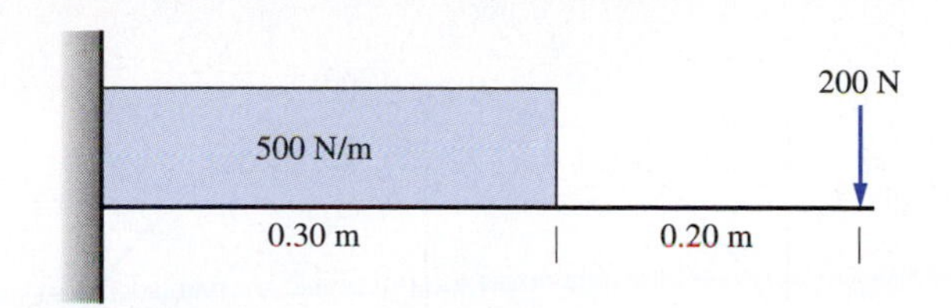

FIGURE P5–109

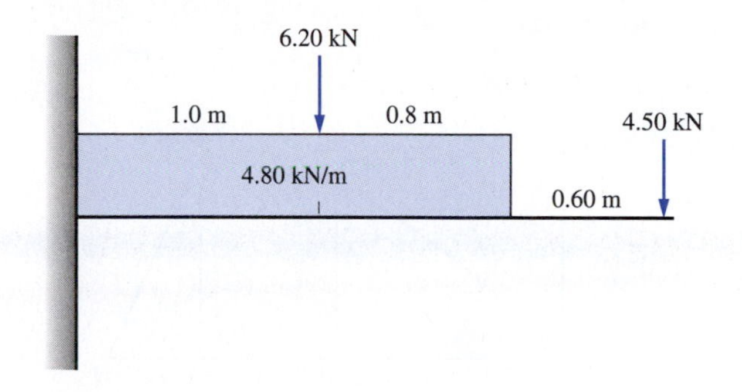

FIGURE P5–110

Problems 5–111 to 5–116

Figures P5–111 to P5–116 show completed shearing force diagrams for beams. Develop the loading diagram and the bending moment diagram for each beam.

Problems 5–117 to 5–122

Figures P5–117 to P5–122 show completed bending moment diagrams for beams. Develop the shearing force diagram and the loading diagram for each beam.

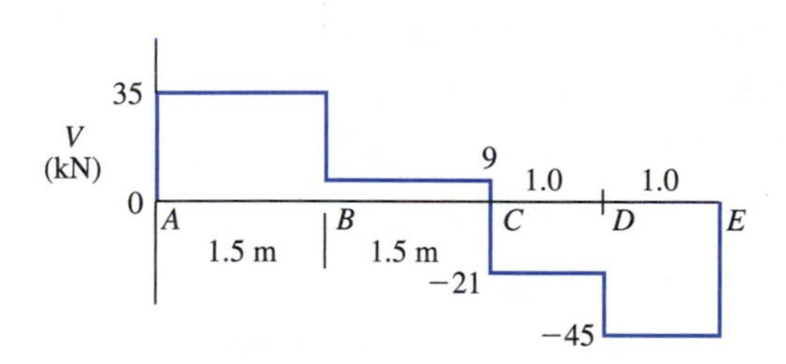

FIGURE P5–111

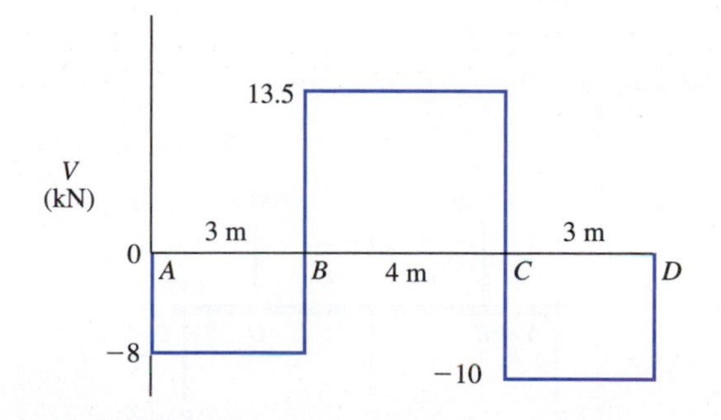

FIGURE P5–112

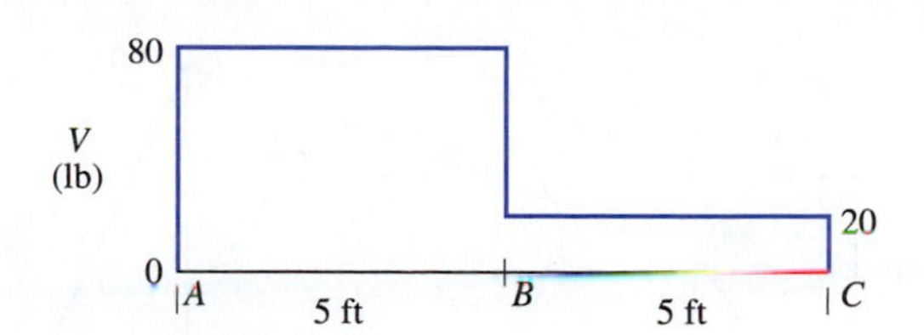

FIGURE P5–113

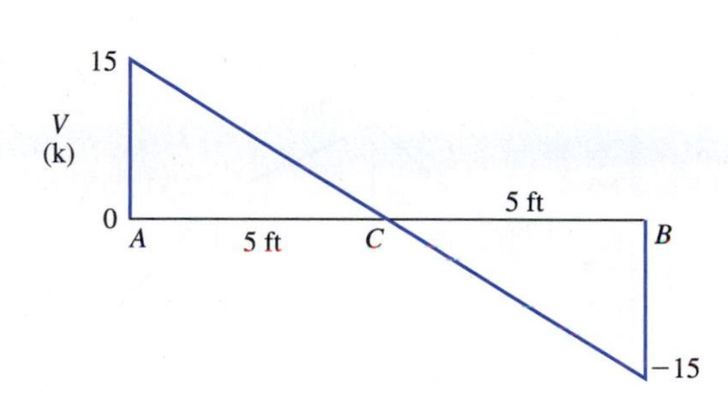

FIGURE P5–114

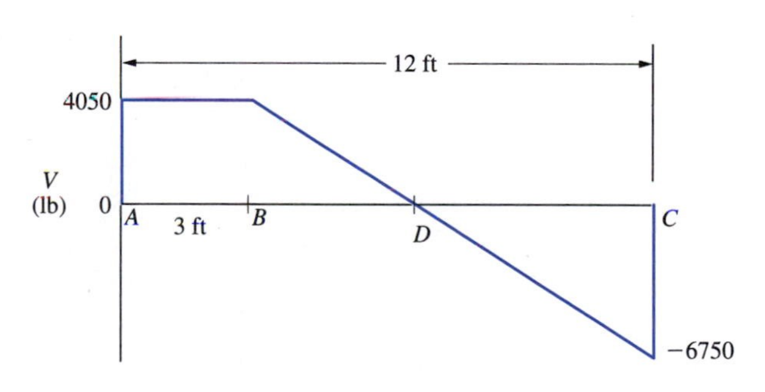

FIGURE P5–115

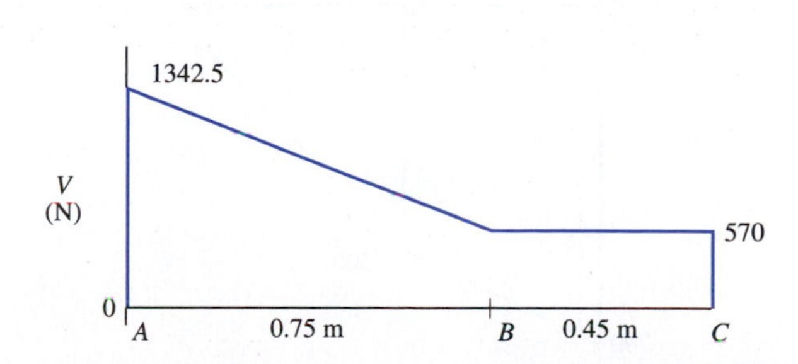

FIGURE P5–116

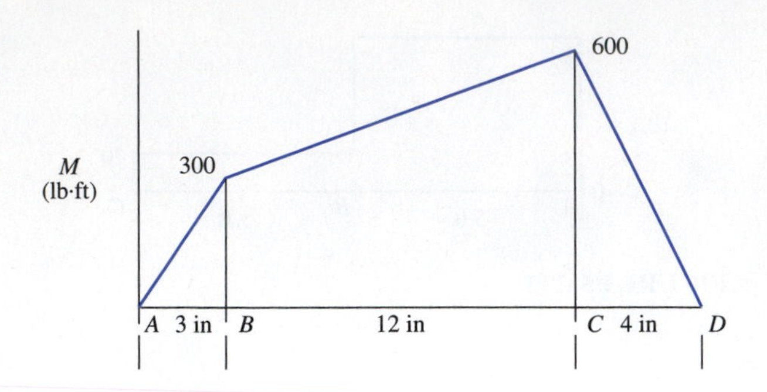

FIGURE P5–117

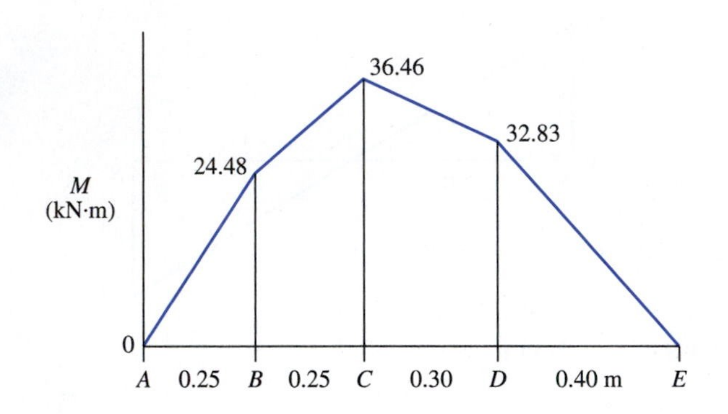

FIGURE P5–118

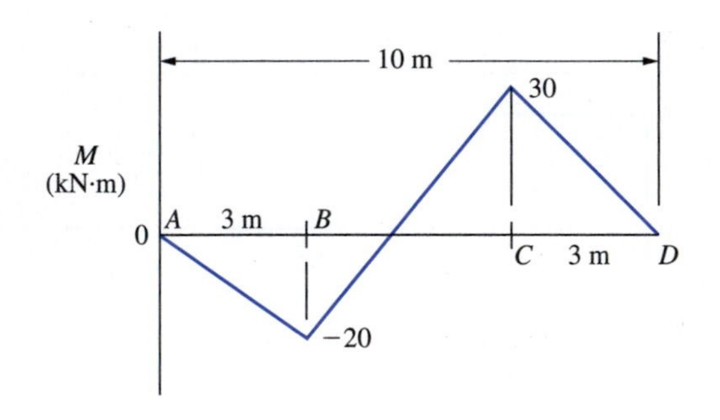

FIGURE P5–119

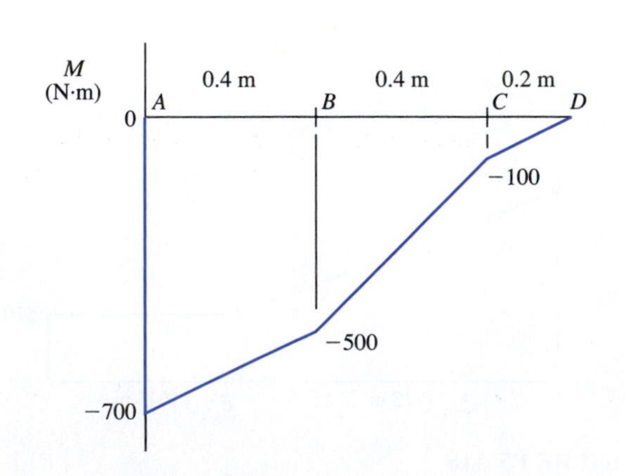

FIGURE P5–120

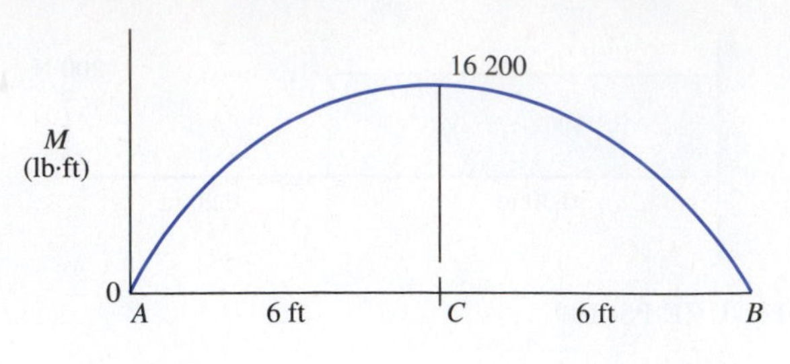

FIGURE P5–121

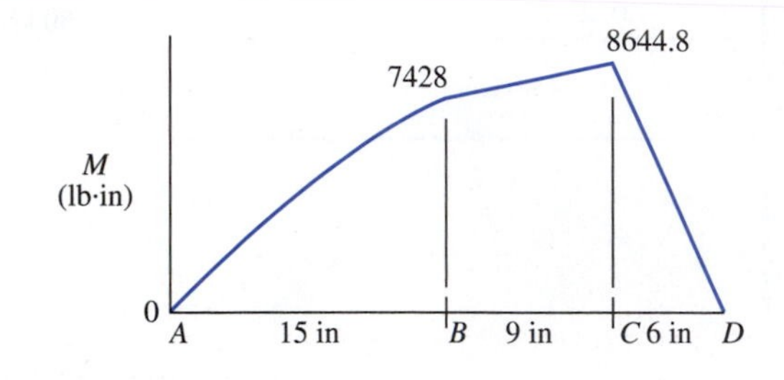

FIGURE P5–122

Continuous Beams—Theorem of Three Moments

For Problems 5–123 through 5–130, use the theorem of three moments to determine the reactions at all supports and draw the complete shearing force and bending moment diagrams. Indicate the maximum shearing force and bending moment for each beam.

5–123.M Use Figure P5–123.

5–124.E Use Figure P5–124.

5–125.E Use Figure P5–125.

5–126.M Use Figure P5–126.

5–127.M Use Figure P5–127.

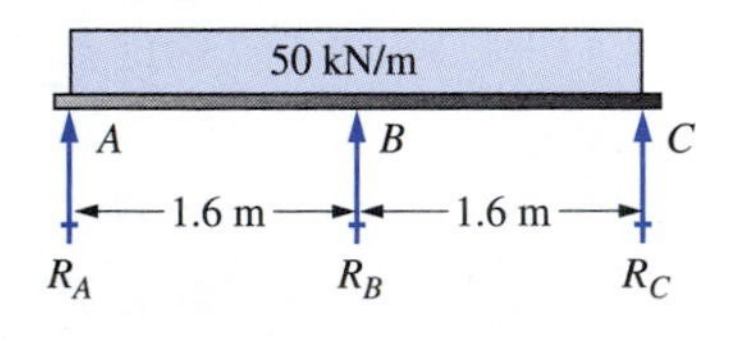

FIGURE P5–123

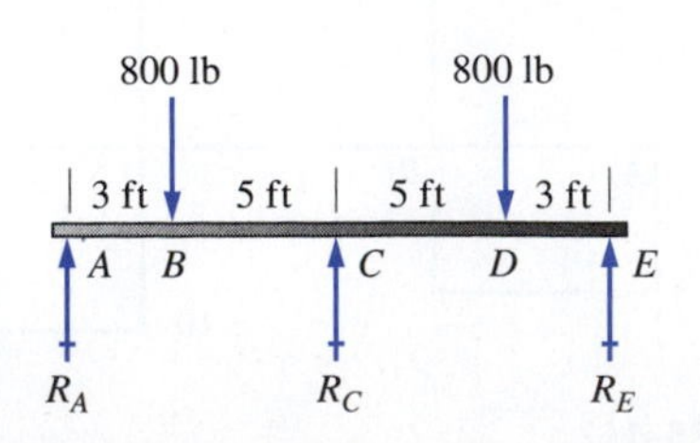

FIGURE P5–124

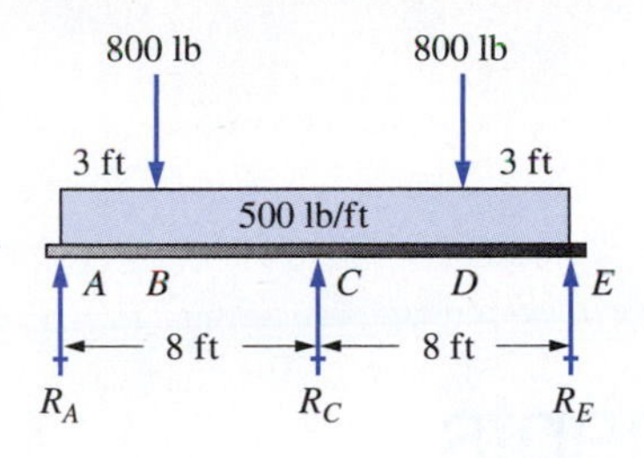

FIGURE P5–125

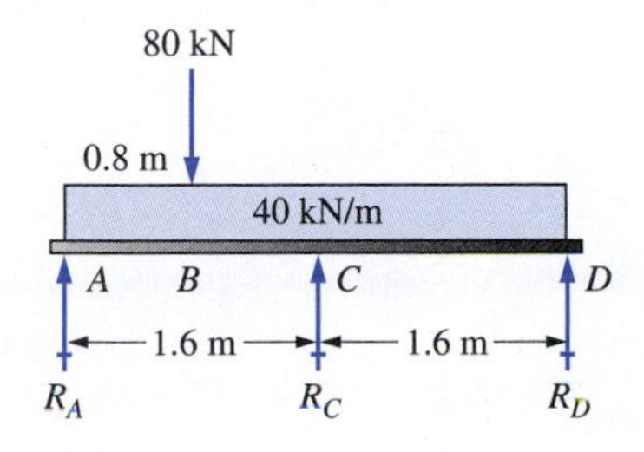

FIGURE P5–126

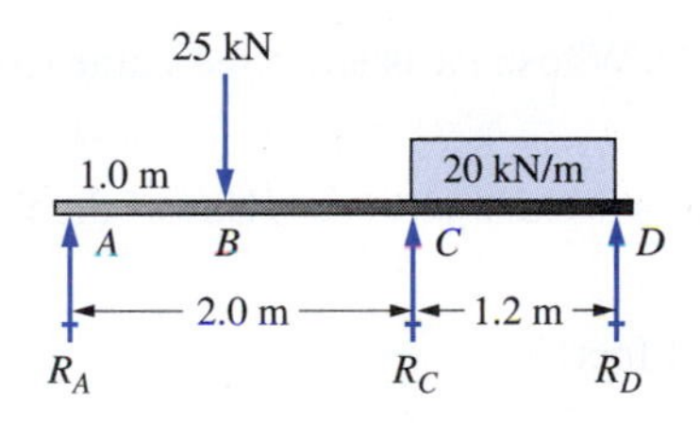

FIGURE P5–127

5–128.M Use Figure P5–128.

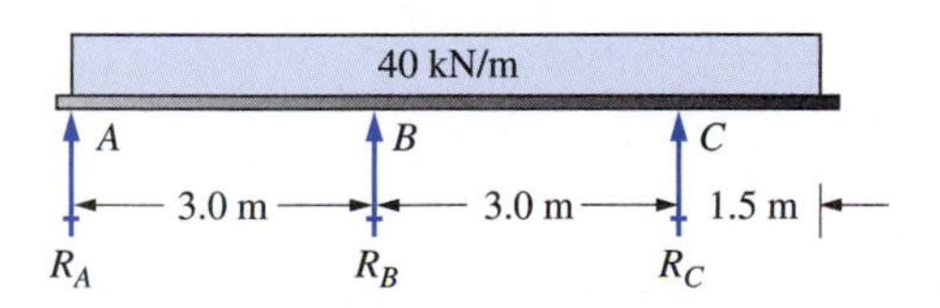

FIGURE P5–128

5–129.M Use Figure P5–129.

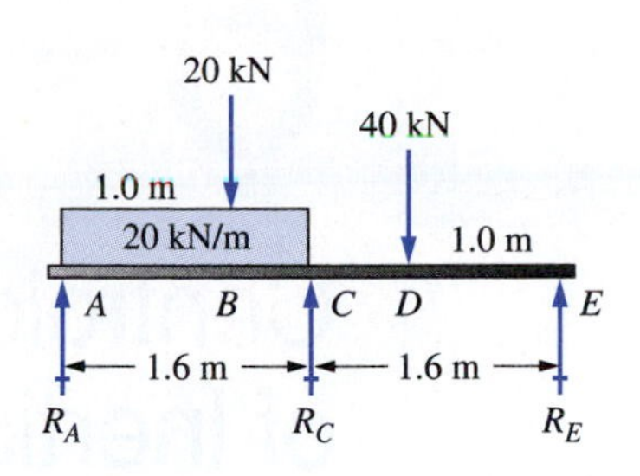

FIGURE P5–129

5–130.M Use Figure P5–130.

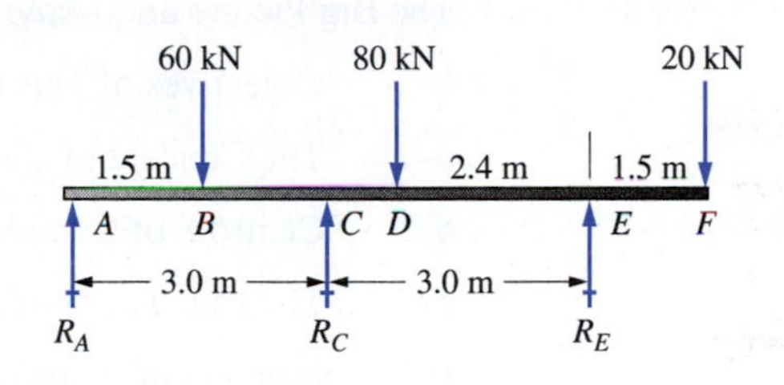

FIGURE P5–130

6

Centroids and Moments of Inertia of Areas

The Big Picture

Centroids and Moments of Inertia of Areas

Discussion Map

- In Chapter 5 you learned to determine the value of shearing forces and bending moments in all parts of beams as part of the requirements for computing shearing stresses and bending stresses in later chapters. This chapter continues this pattern by presenting the properties of the shape of the cross section of the beam, also required for complete analysis of the stresses and deformations of beams.
- The properties of the cross-sectional area of the beam that are of interest here are the *centroid* and the *moment of inertia with respect to the centroidal axis.* If you have already mastered these topics through a study of *statics,* this chapter should present a worthwhile review and a tailoring of the subject to the applications of interest in strength of materials. If you have not studied centroids and moments of inertia, the concepts and techniques presented here will enable you to solve the beam analysis problems throughout this book and in many real design situations.

Discover

Chapter 5 included some examples of the types of beams and the kinds of loads they carry. Review that now, along with the Big Picture, in Section 5–1. As you think of examples of beams that you have seen yourself, can you picture the shapes of their cross sections? What shape would you see if you looked directly into the end of the beam?

Look ahead in this chapter to see the wide variety of shapes that are often specified for the cross section of beams. Figure 6–7 shows a typical "I-beam" as it is called. This is one of the most efficient shapes for a beam (as you will see as you study the next several chapters). In this discussion, efficiency refers to a beam shape that performs well with regard to stresses and deflections while requiring a relatively small amount of material. Such a beam would usually have a lower cost and be lighter than most other shapes.

Look back to Chapter 1 where commercially available structural shapes in steel and aluminum were described. Review Section 1–17, Table 1–7, and Appendices A5 through A13. There you will see sections called wide flange beams (W-shapes); American standard beams (S-shapes); channels (C-shapes); angles (L-shapes); square and rectangular hollow structural shapes (HSS-shapes); and circular hollow pipe and tubing. *All of these shapes are commonly used for beams, either alone or in combination.*

Look ahead to Section 6–8 where we show how to analyze composite sections made from two or more kinds of shapes or add flat plates to critical sections to enhance the ability of the beam to withstand loads and limit deflections.

Have you seen any of these types of beams? Discuss this with colleagues and compare their experiences with yours.

Activity Chapter 6—Centroids and Moments of Inertia of Areas

Let's do an Internet search to find evidence of various shapes used for the cross section of beams and columns. Try the ones described here that refer to the sites listed at the end of this chapter. Each shows examples of structural products, components of machinery, or parts of consumer products that perform as beams or columns and that must be strong, stiff, and attractive. Use your own creativity to find other interesting sites.

1. Site 1 is from a company that produces a wide variety of relatively small precision cold rolled and drawn shapes. Starting with long sections of circular wire or rod, unique, highly accurate shapes are produced by drawing or rolling that perform well as beams while meeting the needs of designers for special functionality such as guiding mating parts along a liner path. Note that some of the shapes are similar to the I-beams included in the Appendix but much smaller. Others have longitudinal ribs or curved shapes that add to the stiffness of the beams. These shapes can be applied in precision manufacturing automation equipment, business machines, consumer products, electrical distribution equipment, and a host of other devices.

2. A product for the building construction industry is illustrated in Site 2. Fabricated from a variety of sawn rectangular wood shapes, the resulting I-shape is strong and lightweight, allowing the building designer to specify floor systems that span long spaces with good strength and stiffness. An additional advantage is that the assembly and arrangement of these joists and the covering plywood sheets provide silent support as people walk across the floor.
3. Site 3 shows some of the special shapes produced by a steel fabricator for use in large structures. Note the combination of angles, channels, bar stock, and other basic forms to produce an efficient truss-like support system for use as floor joists or roof rafters. The joists are sometimes combined with steel decking and a concrete slab covering to produce a strong, stiff composite floor or roof system.
4. Special shapes produced by the extrusion process using aluminum are shown in Site 4. Molten metal is forced through dies with openings having the shapes shown. The resulting products made from these forms can be used in automotive applications, cabinetry, window and door frames, furniture, aircraft, and many other industries. Check out the examples of applications listed on this site. Then you may visit some of the sites of companies that produce the extrusions to learn more about the process and the equipment used to make them.
5. Site 5 shows shapes produced in plastics using extrusion processes similar to those described for Site 4. Nylon, ABS, polycarbonate, PVC, and several other plastics are used for applications in the automotive/transportation, architectural, building construction, aerospace, electronics, and many other industries. Also check out the design, engineering, and manufacturing processes used to produce these shapes.
6. Site 6 shows an example from the ladder industry where rolled or extruded shapes are typically used for the side frames and steps that are subjected to bending in service. Note that the side rails are special channel or I-shaped sections allowing the fastening of the steps. The steps themselves are formed, hollow shapes that provide a flat surface for good footing as the user climbs the ladder, while also offering strong, stiff support. Materials used include aluminum, fiberglass, steel, and wood. Planks used to provide a stable horizontal platform for workers are also provided by this company.
7. Have you noticed the trucks like those shown in Site 7, used by electrical utility installers, tree-trimming teams, painters, and others to raise workers to high points? The articulated booms act as beams while the hydraulic actuators allow smooth, precise positioning close to the worksite. Designers of such trucks must be able to determine the centroids and moments of inertia of these components to ensure the safety of the workers and the durability of the equipment.
8. The product featured in Site 8 provides a versatile set of structural components that can be assembled into frames tailored to a particular application. Made from roll-formed steel, the resulting channel shape provides the strength and stiffness required for the frame while also facilitating the attachment of equipment, side panels, floor plates, and other subsystems. Check out the application examples under the *Tech. Data* tab. The Introduction shows how the system works while illustrating a variety of applications. Then you can explore more specific details under the heading *Mechanical, Architectural, Industrial,* and *Electrical*. Find the detailed data in the online catalog for the variety of sizes

available that list the section properties such as moment of inertia, cross-sectional area, and the weight per unit length. These shapes can be used in some of the problems in this chapter and in following chapters.

9. Site 9 shows an extruded aluminum structural framing system that can be used in similar applications as those described in Site 8. The flat sides and integral channels facilitate assembly and provide a clean, neat appearance of the finished structure. The shapes of the cross sections are optimized for load-carrying ability while being adaptable to build special machines, workstations, automation equipment, mobile units, and storage systems. Check out the online catalogs that describe technical data required for proper application including section properties of area, moment of inertia, and weight per unit length.

Inspection of these sites and those you may have found on your own demonstrate your need to develop the ability to analyze many different shapes to determine the location of their centroids and the magnitudes of their moment of inertia with respect to the centroid. Use the information and exercises in this chapter to acquire and refine your ability to do just that and be prepared to transfer this ability to the following chapters that discuss beams and columns.

6–1 OBJECTIVES OF THIS CHAPTER

After completing this chapter, you should be able to:

1. Define *centroid.*
2. Locate the centroid of simple shapes by inspection.
3. Compute the location of the centroid for complex shapes by treating them as composites of two or more simple shapes.
4. Define *moment of inertia* as it applies to the cross-sectional area of beams.
5. Use formulas to compute the moment of inertia for simple shapes with respect to the centroidal axes of the area.
6. Compute the moment of inertia of complex shapes by treating them as composites of two or more simple shapes.
7. Properly use the *parallel axis theorem* in computing the moment of inertia of complex shapes.
8. Analyze composite beam shapes made from two or more standard structural shapes to determine the resulting centroid location and moment of inertia.
9. Recognize what types of shapes are efficient in terms of providing a large moment of inertia relative to the amount of area of the cross section.

6–2 THE CONCEPT OF CENTROID—SIMPLE SHAPES

The *centroid* of an area is the point about which the area could be balanced if it were supported from that point. The word is derived from the word *center,* and it can be thought of as the geometrical center of an area. For three-dimensional bodies, the term *center of gravity,* or *center of mass,* is used to define a similar point.

For simple areas, such as the circle, the square, the rectangle, and the triangle, the location of the centroid is easy to visualize. Figure 6–1 shows the locations, denoted by *C.* If these shapes were carefully made and the location for the centroid carefully found, the shapes could be balanced on a pencil point at the centroid. Of course, a steady hand is required. How's yours?

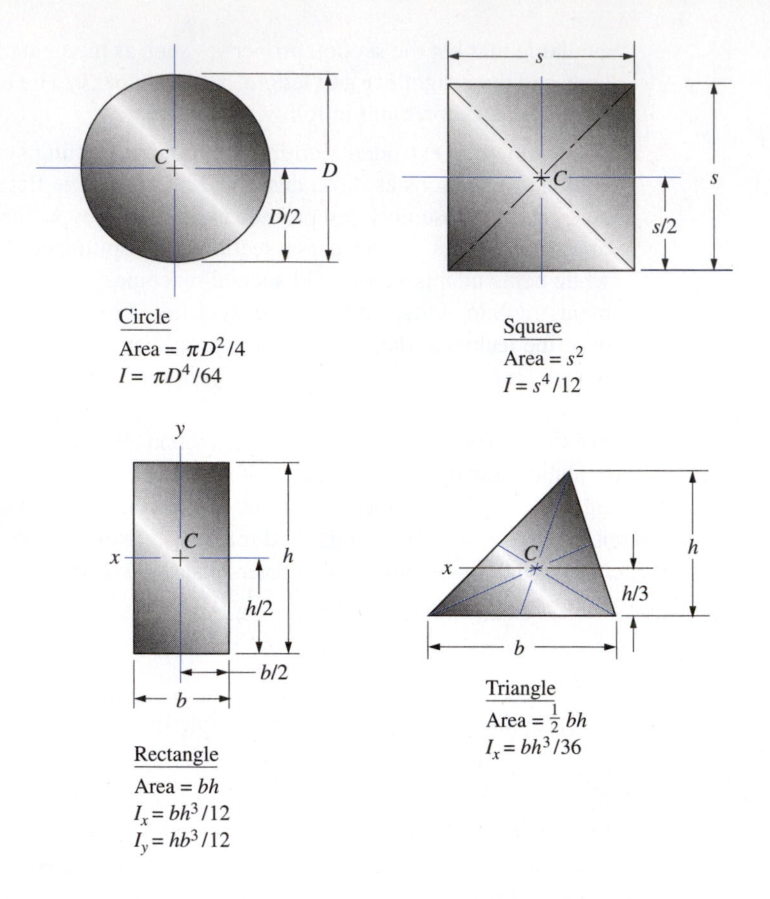

FIGURE 6–1 Properties of simple areas. The centroid is denoted as *C*.

Appendix A–1 gives more data for centroids and other properties of areas for a variety of shapes. See References 1–3 for additional shapes.

6–3 CENTROID OF COMPLEX SHAPES

Most complex shapes can be considered to be made up by combining several simple shapes. This can be used to facilitate the location of the centroid, as will be demonstrated later.

A simple rule for locating centroids can be used for some special combinations of areas:

- If the area has an axis of symmetry, the centroid will lie on that axis.
- If the area has two axes of symmetry, the centroid is at the intersection of these two axes.

Figure 6–2 shows six examples where these rules apply.

Where two axes of symmetry do not occur, the *method of composite areas* can be used to locate the centroid. For example, consider the shape shown in Figure 6–3. It has a vertical axis of symmetry but not a horizontal axis of symmetry. Such areas can be considered to be a composite of two or more simple areas for which the centroid can be found by applying the following principle:

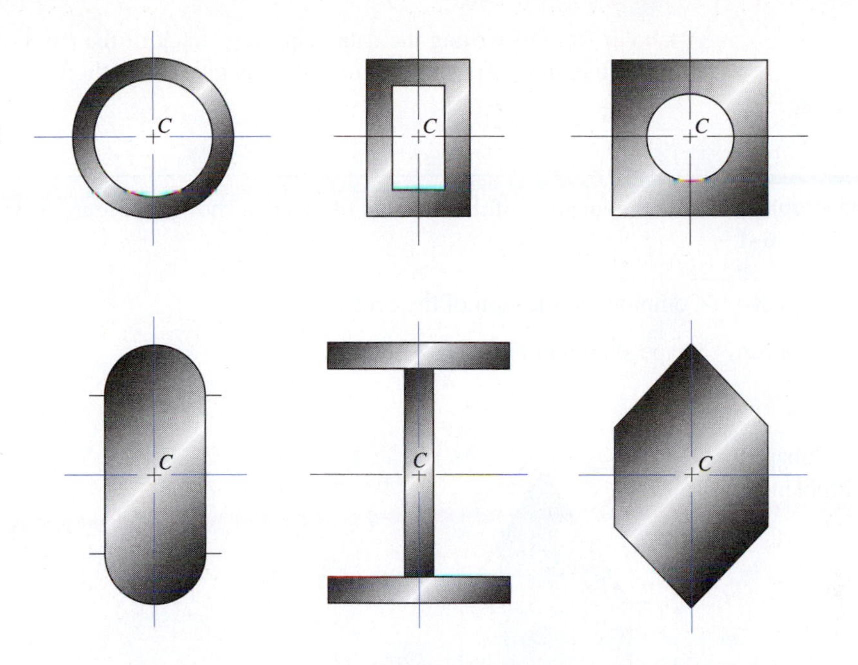

FIGURE 6–2 Composite shapes having two axes of symmetry. The centroid is denoted as C.

> The product of the total area times the distance to the centroid of the total area is equal to the sum of the products of the area of each component part times the distance to its centroid, with the distances measured from the same reference axis.

This principle uses the concept of the *moment of an area,* that is, the product of the area times the distance from a reference axis to the centroid of the area. The principle states:

> The moment of the total area with respect to a particular axis is equal to the sum of the moments of all the component parts with respect to the same axis.

This can be stated mathematically as

$$A_T\overline{Y} = \sum(A_i y_i) \qquad \textbf{(6–1)}$$

where A_T = total area of the composite shape
$\overline{Y}$ = distance to the centroid of the composite shape measured from some reference axis
A_i = area of one component part of the shape
y_i = distance to the centroid of the component part from the reference axis

The subscript i indicates that there may be several component parts, and the product $A_i\,y_i$ for each must be formed and then summed together, as called for in Equation (6–1). Since our objective is to compute $\overline{Y}$, Equation (6–1) can be solved:

$$\overline{Y} = \frac{\sum(A_i y_i)}{A_T} \qquad \textbf{(6–2)}$$

A tabular form of writing the data helps keep track of the parts of the calculations called for in Equation (6–2). An example will illustrate the method.

Example Problem 6–1 Find the location of the centroid of the area shown in Figure 6–3.

Solution Objective Compute the location of the centroid.

Given Shape shown in Figure 6–3.

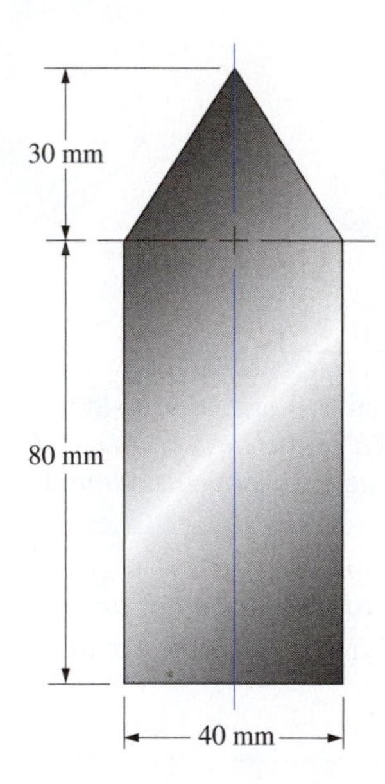

FIGURE 6–3 Shape for Example Problem 6–1.

Analysis Because the shape has a vertical axis of symmetry, the centroid lies on that line. The vertical distance from the bottom of the shape to the centroid will be computed using Equation (6–2). The total area is divided into a rectangle (part 1) and a triangle (part 2), as shown in Figure 6–4. Each part is a simple shape for which the centroid is found using the data from Figure 6–1. The distances to the centroids from the bottom of the shape are shown in Figure 6–4 as y_1 and y_2.

Results The following table facilitates the calculations for data required in Equation (6–2).

Part	A_i	y_i	$A_i y_i$
1	3200 mm^2	40 mm	128 000 mm^3
2	600 mm^2	90 mm	54 000 mm^3
$A_T = 3800\ mm^2$		$\Sigma(A_i y_i) = 182\,000\ mm^3$	

Now $\overline{Y}$ can be computed.

$$\overline{Y} = \frac{\Sigma(A_i y_i)}{A_T} = \frac{182\,000\ \text{mm}^3}{3800\ \text{mm}^2} = 47.9\ \text{mm}$$

FIGURE 6–4 Data used in Example Problem 6–1.

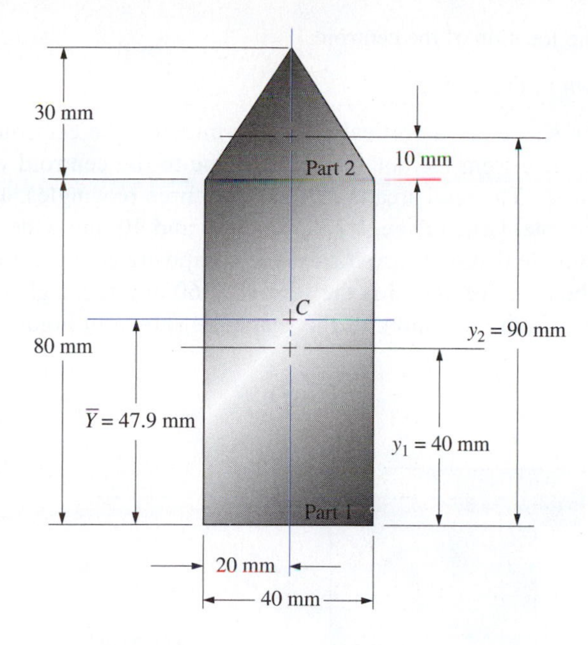

This locates the centroid as shown in Figure 6–4.

Comment — In summary, the centroid is on the vertical axis of symmetry at a distance of 47.9 mm up from the bottom of the shape.

The composite area method works also for sections where parts are removed as well as added. In this case the removed area is considered negative, illustrated as follows.

Example Problem 6–2 Find the location of the centroid of the area shown in Figure 6–5.

FIGURE 6–5 Shape for Example Problem 6–2.

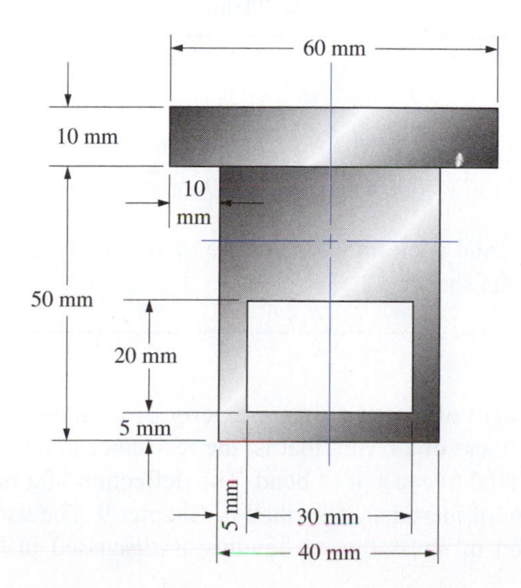

Solution Objective Compute the location of the centroid.

Given Shape shown in Figure 6–5.

Analysis Because the shape has a vertical axis of symmetry, the centroid lies on that line. The vertical distance from the bottom of the shape to the centroid will be computed using Equation (6–2). The total area is divided into three rectangles, as shown in Figure 6–6. Part 1 is the total large rectangle, 50 mm high and 40 mm wide. Part 2 is the 20 mm × 30 mm rectangle that is *removed* from the composite area and the area, A_2, will be considered to be negative. Part 3 is the 10 mm × 60 mm rectangle on top. The distances to the centroids from the bottom of the shape are shown in Figure 6–6 as y_1, y_2, and y_3.

FIGURE 6–6 Data used in Example Problem 6–2.

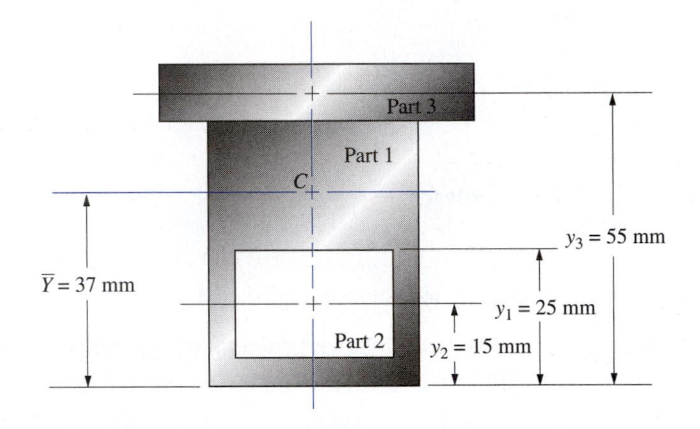

Results The following table facilitates the calculations for data required in Equation (6–2).

Part	A_i	y_i	$A_i y_i$
1	2000 mm^2	25 mm	50 000 mm^3
2	−600 mm^2	15 mm	−9000 mm^3
3	600 mm^2	55 mm	33 000 mm^3
	A_T = 2000 mm^2	$\Sigma(A_i y_i)$	= 74 000 mm^3

Then

$$\bar{Y} = \frac{\sum(A_i y_i)}{A_T} = \frac{74\,000\ \text{mm}^3}{2000\ \text{mm}^2} = 37.0\ \text{mm}$$

Comment In summary, the centroid is on the vertical axis of symmetry at a distance of 37.0 mm up from the bottom of the shape.

6–4 THE CONCEPT OF MOMENT OF INERTIA OF AN AREA

In the study of strength of materials, the property of *moment of inertia of an area* is an indication of the stiffness of a beam, that is, the resistance to deflection of the beam when carrying loads that tend to cause it to bend. The deflection of a beam is inversely proportional to the moment of inertia as described in Chapter 9. The use of the moment of inertia in the calculation of stress due to bending is discussed in Chapter 7 of this book.

Stresses due to vertical shearing forces also depend on moment of inertia and are discussed in Chapter 8.

Some mathematicians and stress analysts use the term *second moment of area* instead of *moment of inertia*. That term is, in fact, more descriptive of the definition of this property in the following discussion. Others call this term *area moment of inertia* or *rectangular moment of inertia* to help distinguish it from *polar moment of inertia* as used in the analysis of torsional shear stress in Chapter 4. There are other uses of the term *moment of inertia* in the study of dynamics where the term refers to a property of a three-dimensional mass.

However, the term *moment of inertia* has become common usage in certain handbooks and industry publications. See References 1–3. Therefore this book will continue to use the term *moment of inertia*. You should understand the context of a problem where this term is needed to ensure that the correct approach is used to determine its value.

Of interest is the moment of inertia of the shape of the *cross section* of the beam. For example, consider the overhanging beam shown in Chapter 5 in Figure 5–12. Its cross section is in the form of an "I," as sketched in Figure 6–7. Because of this shape, such a beam is often referred to as an "I-beam."

Another example is shown in Figure 5–13 where the horizontal boom of the crane assembly is a cantilever whose cross section is a hollow rectangle, as sketched in Figure 6–8. Notice in the original figure, the vertical dimension of the hollow rectangle decreases for sections farther away from the left end where the boom attaches to the support post. Chapter 7 describes why.

Both examples shown in Figures 6–7 and 6–8 represent shapes that are relatively efficient in the use of material to produce large values of their moment of inertia. In most important cases in the study of strength of materials, the moment of inertia of a shape, denoted by the symbol I, is a function of the placement of the area with respect to the *centroidal axis* of the shape, the axis that passes through the centroid of the shape. It is most desirable, from the standpoint of the efficient use of material, to place as much of the material as far away from the centroidal axis as practical. This observation is based on the definition of moment of inertia given here.

The moment of inertia of an area with respect to a particular axis is defined as the sum of the products obtained by multiplying each infinitesimally small element of the area by the square of its distance from the axis.

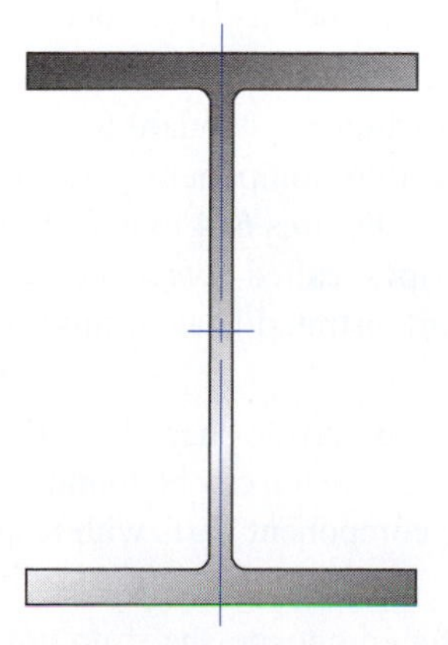

FIGURE 6–7 Typical cross-sectional shape of an I-beam.

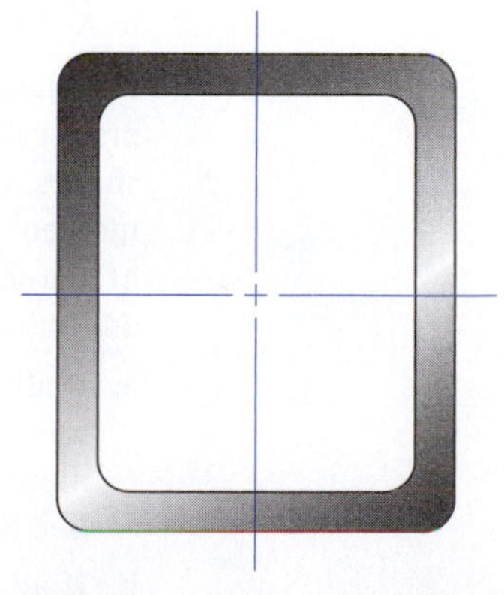

FIGURE 6–8 Typical cross-sectional shape of a hollow rectangular tube.

Thus, you should be able to reason that if much of the area is placed far away from the centroidal axis, the moment of inertia would tend to be large.

The mathematical formula for the moment of inertia, I, follows from the definition. An approximate method involves the process of *summation,* indicated by Σ.

$$I = \sum y^2(\Delta A) \tag{6–3}$$

This would require that the total area be divided into many very small parts represented by ΔA, and that the distance y to the centroid of each part from the axis of interest be determined. Then, the product of y^2 (ΔA) would be computed for each small part, followed by summing all such products. This is a very tedious process, and, fortunately, one that is not used often.

A refinement of the summation method indicated by Equation (6–3) is the process of *integration,* which is the mathematical technique of summing infinitesimal quantities over an entire area. The true mathematical definition of moment of inertia requires the use of integration as follows:

$$I = \int y^2 dA \tag{6–4}$$

Here, the term dA is an area of infinitesimally small size and y, as before, is the distance to the centroid of dA. We will demonstrate the development and use of Equation (6–4) in a later section. However, in many practical problems, it is not necessary to perform the integration process. We show these techniques first.

Looking closely at the definition, you should be able to see that the units for moment of inertia will be *length to the fourth power*. Examples are in^4, m^4, and mm^4.

There are several methods of determining the magnitude of the moment of inertia.

1. For simple shapes it is convenient to use standard formulas that have been derived from the basic definition given earlier. Figure 6–1 shows such formulas for four shapes and Appendix A–1 gives several more. Reference 2 includes a table of formulas for I for 42 different shapes.
2. For standard commercially available shapes such as wide-flange beams (W-shapes), channels (C-shapes), angles (L-shapes), and pipe, values of moment of inertia are tabulated in published references such as Reference 1. See also Appendixes A–4 to A–13.
3. For more complex shapes for which no standard formulas are available, it is often practical to divide the shape into component parts that are themselves simple shapes. Examples are shown in Figures 6–4 to 6–8. The details of calculating the moment of inertia of such shapes, called *composite shapes,* depend on the nature of the shapes and will be demonstrated later in this chapter. Important concepts are stated here.
   a. If all component parts of a composite shape have the same centroidal axis, the total moment of inertia for the shape can be found by adding or subtracting the moments of inertia of the component parts with respect to the centroidal axis. See Section 6–5.
   b. If all component parts of a composite shape do not have the same centroidal axis, the use of a process called the parallel axis theorem is required. See Section 6–6.

4. The fundamental definition of moment of inertia, Equation (6–4), can be used if the geometry of the shape can be represented in mathematical terms that can be integrated. See Section 6–7.
5. Many computer-aided design software systems include automatic calculation of the location of the centroid and the moment of inertia of any closed shape drawn in the system.
6. For the special case of a shape that can be represented as a composite of rectangles having sides perpendicular or parallel to the centroidal axis, a special tabulation technique can be applied that is described in the last section of this chapter. This technique lends itself well to solution by using a programmable calculator, a spreadsheet, or a simple computer program.

6–5 MOMENT OF INERTIA OF COMPOSITE SHAPES WHOSE PARTS HAVE THE SAME CENTROIDAL AXIS

A composite shape is one made up of two or more parts that are themselves simple shapes for which formulas are available to calculate the moment of inertia, I. A special case is when all parts have the same centroidal axis. Then the moment of inertia for the composite shape is found by combining the values of I for all parts according to the following rule:

> If the component parts of a composite area all have the same centroidal axis, the total moment of inertia can be found by adding or subtracting the moments of inertia of the component parts with respect to the centroidal axis. The value of I is added if the part is a positive solid area. If the part is a void, the value of I is subtracted.

Figure 6–9 shows an example of such a shape, composed of a vertical central stem, 30 mm wide and 80 mm high, and two side parts, 30 mm wide and 40 mm high. Notice that all have their own centroidal axis coincident with the centroidal axis x-x for the composite section. The rule just stated can then be used to compute the total value of I for the cross by adding the value of I for each of the three parts. See Example Problem 6–3.

Figure 6–10 shows an example where there is a 35-mm-diameter circular hole removed from a square whose sides measure 50 mm. The circle and the square have the same centroidal axis x-x. The rule can then be used to compute the value of I for the square and then to subtract the value of I for the circle to obtain the total value for I of the composite shape. See Example Problem 6–4.

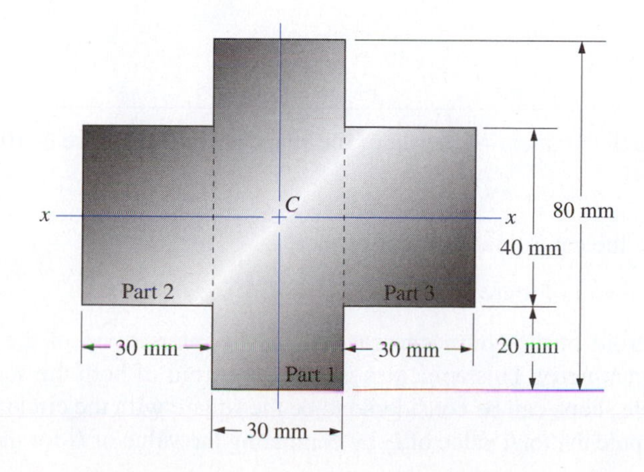

FIGURE 6–9 Shape for Example Problem 6–3.

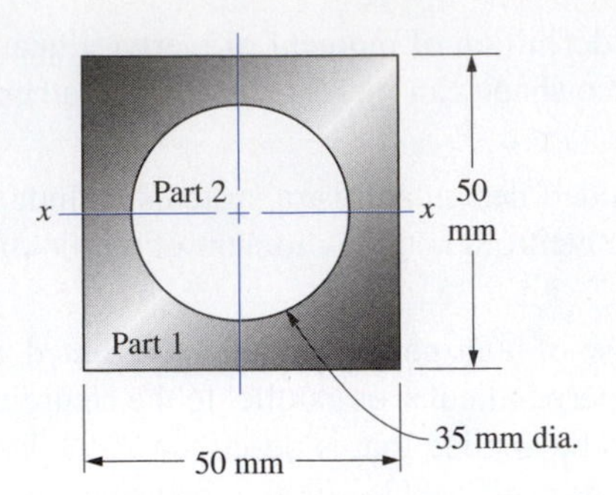

FIGURE 6–10 Shape for Example Problem 6–4.

Example Problem 6–3 Compute the moment of inertia of the cross-shape shown in Figure 6–9 with respect to its centroidal axis.

Solution

Objective: Compute the centroidal moment of inertia.

Given: Shape shown in Figure 6–9.

Analysis: The centroid of the cross-shape is at the intersection of the horizontal and vertical axes of symmetry. Dividing the cross into the three parts shown in the figure results in each part having the same centroidal axis, *x-x*, as the entire composite section. Therefore, we can compute the value of I for each part and sum them to obtain the total value, I_T. That is,

$$I_T = I_1 + I_2 + I_3$$

Results: Referring to Figure 6–1 for the formula for I for a rectangle gives

$$I_1 = \frac{bh^3}{12} = \frac{30(80)^3}{12} = 1.28 \times 10^6\,\text{mm}^4$$

$$I_2 = I_3 = \frac{30(40)^3}{12} = 0.16 \times 10^6\,\text{mm}^4$$

Then

$$I_T = 1.28 \times 10^6\,\text{mm}^4 + 2(0.16 \times 10^6\,\text{mm}^4) = 1.60 \times 10^6\,\text{mm}^4$$

Example Problem 6–4 Compute the moment of inertia of the shape shown in Figure 6–10 with respect to its centroidal axis.

Solution

Objective: Compute the centroidal moment of inertia.

Given: Shape shown in Figure 6–10.

Analysis: The centroid of the composite shape is at the intersection of the horizontal and vertical axes of symmetry. This coincides with the centroid of both the square and the circle. The composite shape can be considered to be the square with the circle removed. Therefore, we can compute the total value of I_T by computing the value of I_1 for the square and subtracting

I_2 for the circle. That is,

$$I_T = I_1 - I_2$$

Results $$I_1 = \frac{s^4}{12} = \frac{(50)^4}{12} = 520.8 \times 10^3\ \text{mm}^4$$

$$I_2 = \frac{\pi D^4}{64} = \frac{\pi(35)^4}{64} = 73.7 \times 10^3\ \text{mm}^4$$

For the composite section,

$$I_T = I_1 - I_2 = 447.1 \times 10^3\ \text{mm}^4$$

6–6 MOMENT OF INERTIA FOR COMPOSITE SHAPES—GENERAL CASE—USE OF THE PARALLEL AXIS THEOREM

When a composite section is composed of parts whose centroidal axes do not lie on the centroidal axis of the entire section, the process of simply summing the values of I for the parts *cannot* be used. It is necessary to employ the *parallel axis theorem.*

The general statement of the parallel axis theorem is

> The moment of inertia of a shape with respect to a certain axis is equal to the sum of the moment of inertia of the shape with respect to its own centroidal axis plus an amount called the **transfer term** computed from Ad^2, where A is the area of the shape and d is the distance from the centroid of the shape to the axis of interest.

This theorem can be applied to compute the total moment of inertia for a general composite shape by using the following procedure. In this case, the axis of interest is the centroidal axis of the composite shape that must be found using the method given in Section 6–3.

General Procedure for Computing the Moment of Inertia for a Composite Shape

1. Divide the composite shape into component parts that are simple shapes for which formulas are available to compute the moment of inertia of the part with respect to its own centroidal axis. Identify the parts as *1, 2, 3,* and so forth.
2. Locate the distance from the centroid of each component part to some convenient reference axis, typically the bottom of the composite section. Call these distances y_1, y_2, y_3, and so forth.
3. Locate the centroid of the composite section using the method given in Section 6–3. Call the distance from the reference axis used in step 2 to the centroid, $\overline{Y}$.
4. Compute the moment of inertia of each part with respect to its own centroidal axis, calling these values, I_1, I_2, I_3, and so forth.
5. Determine the distance from the centroid of the composite shape to the centroid of each part, calling these values d_1, d_2, d_3, and so forth. Note that $d_1 = \overline{Y} - y_1$, $d_2 = \overline{Y} - y_2$, $d_3 = \overline{Y} - y_3$, and so forth. Use the absolute value of each distance.
6. Compute the *transfer term* for each part from $A_i d_i^2$ where A_i is the area of the part and d_i is the distance found in step 5.
7. Compute the total moment of inertia of the composite section with respect to its centroidal axis from

$$I_T = I_1 + A_1 d_1^2 + I_2 + A_2 d_2^2 + I_3 + A_3 d_3^2 + \cdots \quad \textbf{(6–5)}$$

Equation (6–5) is called the *parallel axis theorem* because it defines how to transfer the moment of inertia of an area from one axis to any parallel axis. As applied here, the two axes are the centroidal axis of the component part and the centroidal axis of the composite section. For each part of a composite section, the sum $I + Ad^2$ is a measure of its contribution to the total moment of inertia.

Implementation of the *General procedure for computing the moment of inertia for a composite shape* can be facilitated by preparing a table that is an extension of the one used in Section 6–3 to find the location of the centroid of the shape. The general design of this table follows.

Part	A_i	y_i	$A_i y_i$	I_i	$d_i = \overline{Y} - y_i$	$A_i d_i^2$	$I_i + A_i d_i^2$
1							
2							
3							
$A_T = \Sigma A_i =$		$\Sigma(A_i y_i) =$		$I_T = \Sigma(I_i + A_i d_i^2) =$			

$$\text{Distance to centroid} = \overline{Y} = \frac{\Sigma(A_i y_i)}{A_T} =$$

The use of the table and the general guidelines are demonstrated in Example Problems 6–5, 6–6, and 6–7. The benefit from using this type of table becomes greater as the number of components gets greater. Also, the use of a computer spreadsheet to make the appropriate calculations is very convenient.

Example Problem 6–5 Compute the moment of inertia of the tee shape in Figure 6–11 with respect to its centroidal axis.

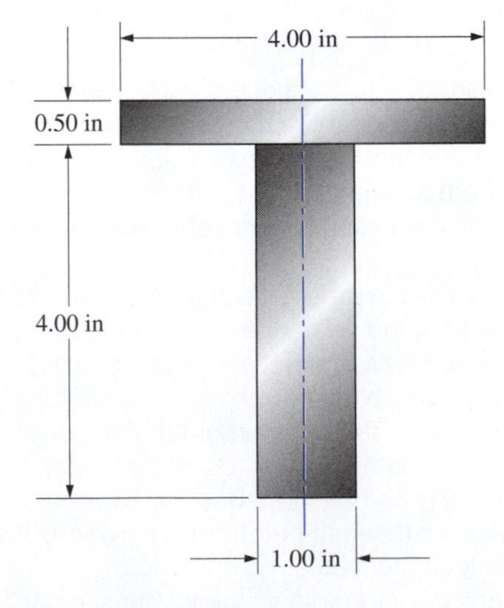

FIGURE 6–11 Shape for Example Problem 6–5.

Solution Objective Compute the moment of inertia.

Given Shape shown in Figure 6–11.

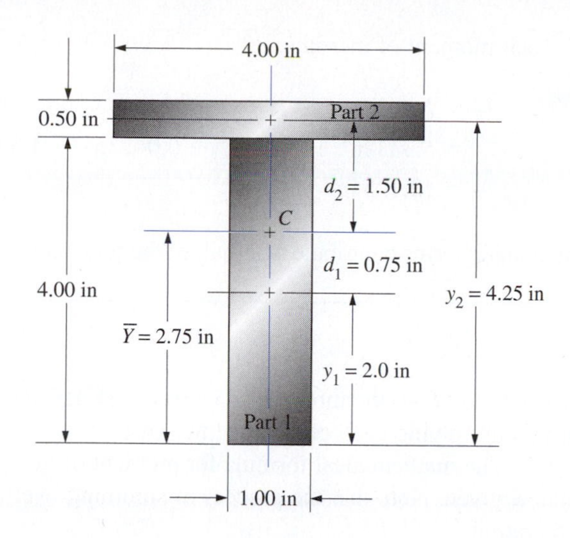

FIGURE 6–12 Data used in Example Problem 6–5.

Analysis Use the **General Procedure** listed in this section. As **step 1,** divide the tee shape into two parts, as shown in Figure 6–12. Part 1 is the vertical stem and part 2 is the horizontal flange.

Results The following table summarizes the complete set of data used to compute the total moment of inertia with respect to the centroid of the tee shape. Some of the data are shown also in Figure 6–12. Comments are given here to show the manner of arriving at certain data.

Part	A_i	y_i	$A_i y_i$	I_i	d_i	$A_i d_i^2$	$I_i + A_i d_i^2$
1	4.0 in^2	2.0 in	8.0 in^3	5.333 in^4	0.75 in	2.25 in^4	7.583 in^4
2	2.0 in^2	4.25 in	8.50 in^3	0.042 in^4	1.50 in	4.50 in^4	4.542 in^4
	$A_T = 6.0$ in^2	$\Sigma(A_i y_i) = 16.5$ in^3					$I_T = 12.125$ in^4

$$\overline{Y} = \frac{\Sigma(A_i y_i)}{A_T} = \frac{16.5\ \text{in}^3}{6.0\ \text{in}^2} = 2.75\ \text{in}$$

Steps 2, 3. The first three columns of the table give data for computing the location of the centroid using the technique shown in Section 6–3. Distances, y_i, are measured upward from the bottom of the tee. The result is $\overline{Y} = 2.75$ in.

Step 4. Both parts are simple rectangles. Then, the values of I are

$$I_1 = bh^3/12 = (1.0)(4.0)^3/12 = 5.333\ \text{in}^4$$
$$I_2 = bh^3/12 = (4.0)(0.5)^3/12 = 0.042\ \text{in}^4$$

Step 5. Distances, d_i, from the overall centroid to the centroid of each part,

$$d_1 = \overline{Y} - y_1 = 2.75\ \text{in} - 2.0\ \text{in} = 0.75\ \text{in}$$
$$d_2 = y_2 - \overline{Y} = 4.25\ \text{in} - 2.75\ \text{in} = 1.50\ \text{in}$$

Step 6. Transfer term for each part,

$$A_1 d_1^2 = (4.0\ \text{in}^2)(0.75\ \text{in})^2 = 2.25\ \text{in}^4$$
$$A_2 d_2^2 = (2.0\ \text{in}^2)(1.50\ \text{in})^2 = 4.50\ \text{in}^4$$

Step 7. Total moment of inertia:

$$I_T = I_1 + A_1 d_1^2 + I_2 + A_2 d_2^2$$
$$= 5.333\ \text{in}^4 + 2.25\ \text{in}^4 + 0.042\ \text{in}^4 + 4.50\ \text{in}^4$$
$$I_T = 12.125\ \text{in}^4$$

Comment Notice that the transfer terms contribute over half of the total value to the moment of inertia.

6–7 MATHEMATICAL DEFINITION OF MOMENT OF INERTIA

As stated in the Section 6–4, the moment of inertia, I, is defined as the sum of the products obtained by multiplying each element of the area by the square of its distance from the reference axis. The mathematical formula for moment of inertia follows from that definition and is now given. Note that the process of summing over an entire area is accomplished by integration.

$$I = \int y^2\,dA \qquad \textbf{(6–4)}$$

Refer to Figure 6–13 for an illustration of the terms in this formula for the special case of a rectangle, for which we want to compute the moment of inertia with respect to its centroidal axis. The small element of area is shown as a thin strip parallel to the centroidal axis where the width of the strip is the total width of the rectangle, b, and the thickness of the strip is a small value, dy. Then the area of the element is

$$dA = b \cdot dy$$

The distance, y, is the distance from the centroidal axis to the centroid of the elemental area as shown. Substituting these values into Equation (6–4) allows the derivation of the formula for the moment of inertia of the rectangle with respect to its centroidal axis. Note that to integrate over the entire area requires the limits for the integral to be from $-h/2$ to $+h/2$.

$$I = \int_{-h/2}^{+h/2} y^2\,dA = \int_{-h/2}^{+h/2} y^2(b \cdot dy)$$

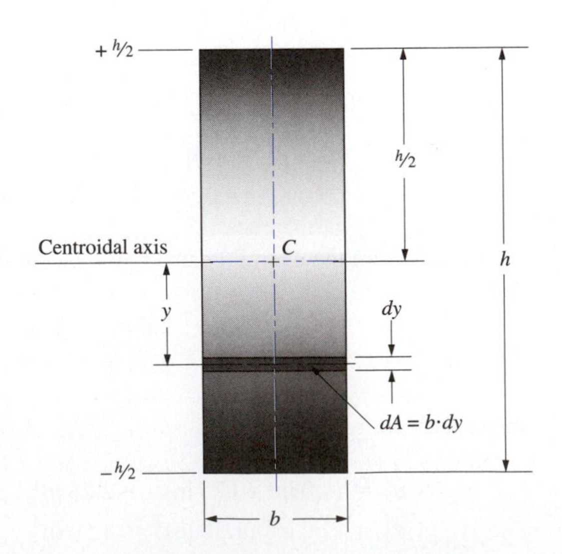

FIGURE 6–13 Data used in derivation of moment of inertia for a rectangle.

Because b is a constant, it can be taken outside the integral, giving

$$I = b\int_{-h/2}^{+h/2} y^2 dy = b\left[\frac{y^3}{3}\right]_{-h/2}^{+h/2}$$

Inserting the limits for the integral gives

$$I = b\left[\frac{h^3}{24} - \frac{(-h)^3}{24}\right] = b\left[\frac{2h^3}{24}\right] = \frac{bh^3}{12}$$

This is the formula reported in the tables. Similar procedures can be used to develop the formulas for other shapes.

6–8 COMPOSITE SECTIONS MADE FROM COMMERCIALLY AVAILABLE SHAPES

In Section 1–12 commercially available structural shapes were described for wood, steel, and aluminum. Properties of representative sizes of these shapes are listed in the following appendix tables:

- Appendix A–4 for wood beams
- Appendix A–5 for structural steel angles
- Appendix A–6 for structural steel channels
- Appendix A–7 for structural steel wide-flange shapes
- Appendix A–8 for structural steel American Standard beams
- Appendix A–9 for structural tubing—square and rectangular
- Appendix A–10 for aluminum standard channels
- Appendix A–11 for aluminum standard I-beams
- Appendix A–12 for standard schedule 40 steel pipe
- Appendix A–13 for steel mechanical tubing

In addition to being very good by themselves for use as beams, these shapes are often combined to produce special composite shapes with enhanced properties.

When used separately, the properties for designing can be read directly from the tables for area, moment of inertia, and pertinent dimensions. When combined into composite shapes, the area and the moment of inertia of the component shapes with respect to their own centroidal axes are needed and can be read from the tables. Also, the tables give the location of the centroid for the shape, which is often needed to determine distances needed to compute the transfer-of-axis term, Ad^2, in the moment of inertia calculation. The following example problems illustrate these processes.

Example Problem 6–6

Compute the moment of inertia of the composite I-beam shape shown in Figure 6–14 with respect to its centroidal axis. The shape is formed by welding a 0.50 in thick by 6.00 in wide plate to both the top and bottom flanges to increase the stiffness of the standard aluminum I-beam.

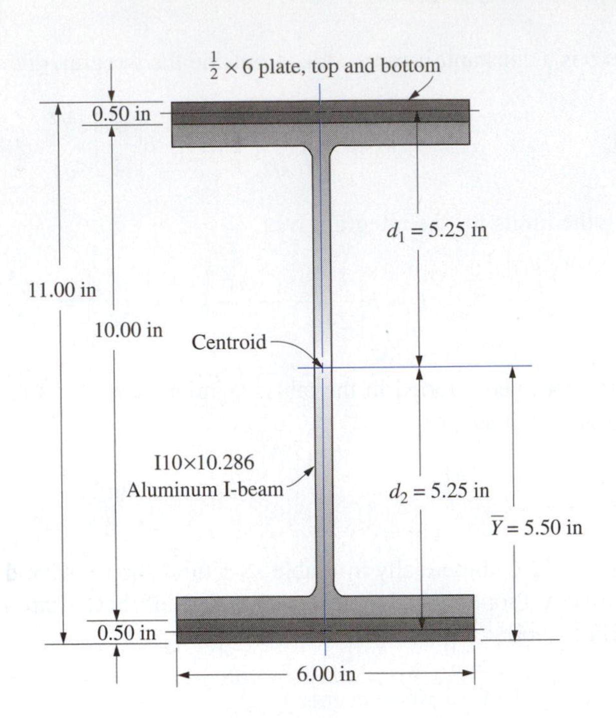

FIGURE 6–14 Data used in Example Problem 6–6.

Solution

Objective: Compute the moment of inertia.

Given: Shape shown in Figure 6–14. For the I10×10.286 beam shape: $I = 155.79 \text{ in}^4$; $A = 8.747 \text{ in}^2$ (from Appendix A–11)

Analysis: Use the **General Procedure** listed earlier in this chapter. As **step 1,** divide the beam shape into three parts. Part 1 is the I-beam; part 2 is the bottom plate; part 3 is the top plate. As **steps 2 and 3,** the centroid is coincident with the centroid of the I-beam because the composite shape is symmetrical. Thus, $\overline{Y} = 5.50$ in, or one-half of the total height of the composite shape. No separate calculation of $\overline{Y}$ is needed.

Results: The following table summarizes the complete set of data used in steps 4–7 to compute the total moment of inertia with respect to the centroid of the beam shape. Some of the data are shown also in Figure 6–14. Comments are given here to show the manner of arriving at certain data.

Part	A_i	y_i	$A_i y_i$	I_i	$d_i = \overline{Y} - y_i$	$A_i d_i^2$	$I_i + A_i d_i^2$
1	8.747	5.50	–	155.79	0	0	155.79
2	3.00	0.25	–	0.063	5.25	82.69	82.75
3	3.00	10.75	–	0.063	5.25	82.69	82.75
$A_T = \Sigma A_i = 14.747 \text{ in}^2$		$\Sigma(A_i y_i) =$	–				$I_T = \Sigma(I_i + A_i d_i^2) = 321.29 \text{ in}^4$

Distance to centroid $= \overline{Y} = \dfrac{\Sigma(A_i y_i)}{A_T} = 5.50$ in (by inspection)

Step 4. For each rectangular plate,

$$I_2 = I_3 = bh^3/12 = (6.0)(0.5)^3/12 = 0.063 \text{ in}^4$$

Step 5. Distance from the overall centroid to the centroid of each part:

$$d_1 = 0.0 \text{ in beacuse the centroids are coincident}$$
$$d_2 = 5.50 - 0.25 = 5.25 \text{ in}$$
$$d_3 = 10.75 - 5.50 = 5.25 \text{ in}$$

Step 6. Transfer term for each part:

$$A_1 d_1^2 = 0.0 \text{ because } d_1 = 0.0$$
$$A_2 d_2^2 = A_3 d_3^2 = (3.00)(5.25)^2 = 82.69 \text{ in}^4$$

Step 7. Total moment of inertia:

$$I_T = I_1 + I_2 + A_2 d_2^2 + I_3 + A_3 d_3^2$$
$$I_T = 155.79 + 0.063 + (3.0)(5.25)^2 + 0.063 + (3.0)(5.25)^2$$
$$I_T = 321.29 \text{ in}^4$$

Comment Notice that the two added plates more than double the total value of the moment of inertia as compared with the original I-beam shape. Also, virtually all of the added value is due to the transfer terms and not to the basic moment of inertia of the plates themselves.

Example Problem 6–7 Compute the moment of inertia of the fabricated I-beam shape shown in Figure 6–15 with respect to its centroidal axis. The shape is formed by welding four standard L4×4×$\frac{1}{2}$ steel angles to a $\frac{1}{2}$ × 16 vertical plate.

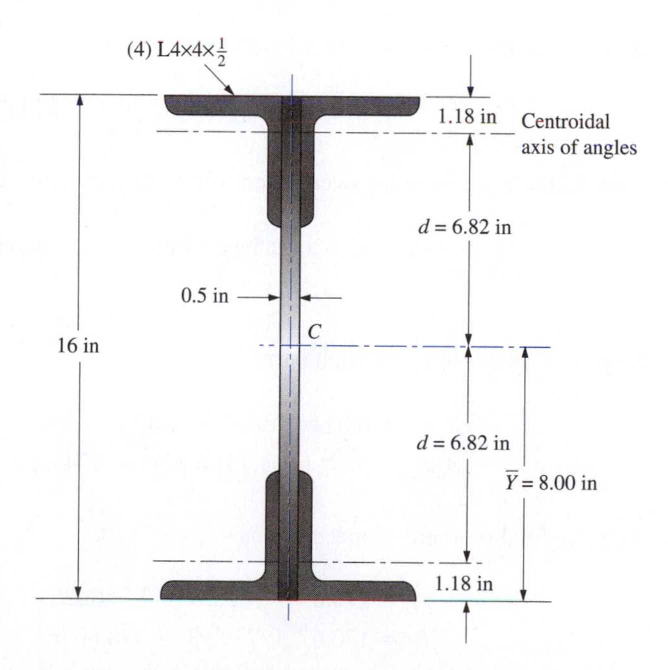

FIGURE 6–15 Data used in Example Problem 6–7.

Solution Objective Compute the moment of inertia.

Given Shape shown in Figure 6–15. For each angle:
$I = 5.52 \text{ in}^4$; $A = 3.75 \text{ in}^2$ (from Appendix A–5)

Analysis Use the **General Procedure** listed earlier in this section. As **step 1,** we can consider the vertical plate to be part 1. Since the angles are all the same and placed an equal distance from the centroid of the composite shape, we can compute key values for one angle and multiply the results by 4. The location of the angles places the flat faces even with the top and bottom of the vertical plate. The location of the centroid of each angle is then 1.18 in from the top or bottom, based on the location of the centroid of the angles themselves, as listed in Appendix A–5. For **steps 2 and 3,** the centroid is coincident with the centroid of the I-beam because the composite shape is symmetrical. Thus, $\bar{Y} = 8.00$ in, or one-half of the total height of the composite shape. No separate calculation of $\bar{Y}$ is needed.

Results The following table summarizes the complete set of data used in steps 4–7 to compute the total moment of inertia with respect to the centroid of the beam shape. Some of the data are shown also in Figure 6–15. Comments are given here to show the manner of arriving at certain data. The second line of the table, shown in italics, gives data for one angle for reference only. Line 3 gives the data for all four angles. Then the final results are found by summing lines 1 and 3.

Part	A_i	y_i	$A_i y_i$	I_i	$d_i = \bar{Y} - y_i$	$A_i d_i^2$	$I_i + A_i d_i^2$
1	8.00	8.00	–	170.67	0	0	170.67
(2) Ref.	*3.75*	*1.18*	–	*5.56*	*6.82*	*174.42*	*179.98* only
4 × (2)	15.00	–	–	22.24	–	697.68	719.93
$A_T = \Sigma A_i = 23.00 \text{ in}^2$		$\Sigma(A_i y_i) =$	–			$I_T = \Sigma(I_i + A_i d_i^2) = 890.60 \text{ in}^4$	

$$\text{Distance to centroid} = \bar{Y} = \frac{\Sigma(A_i y_i)}{A_T} = 8.00 \text{ in (by inspection)}$$

Step 4. For the vertical rectangular plate,

$$I_1 = bh^3/12 = (0.5)(16)^3/12 = 170.67 \text{ in}^4$$

Step 5. Distance from the overall centroid to the centroid of each part:

$$d_1 = 0.0 \text{ in because the centroids are coincident}$$
$$d_2 = 8.00 - 1.18 = 6.82 \text{ in}$$

Step 6. Transfer term for each part:

$$A_1 d_1^2 = 0.0 \text{ because } d_1 = 0.0$$
$$A_2 d_2^2 = A_3 d_3^2 = (3.75)(6.82)^2 = 174.42 \text{ in}^4$$

Step 7. Total moment of inertia:

$$I_T = 170.67 + 4[5.56 + 3.75(6.82)^2]$$
$$I_T = 170.67 + 719.93 = 890.60 \text{ in}^4$$

Comment Approximately 80% of the total value of moment of inertia is contributed by the four angles.

6–9 MOMENT OF INERTIA FOR SHAPES WITH ALL RECTANGULAR PARTS

A method is shown here for computing the moment of inertia of special shapes that can be divided into parts, all of which are rectangles with their sides perpendicular and parallel to the axis of interest. An example would be the tee section analyzed in Example Problem 6–5 and shown in Figure 6–11. The method is somewhat simpler than the method described in Section 6–6 that used the parallel axis theorem, although both methods are based on the same fundamental principles.

The method involves the following steps:

1. Divide the composite section into a convenient number of parts so that each part is a rectangle with its sides perpendicular and parallel to the horizontal axis.
2. For each part, identify the following dimensions:

   b = width

   y_1 = distance from the base of the composite section to the bottom of that part

   y_2 = distance from the base of the composite section to the top of that part

3. Compute the area of each part from the equation,

$$A = b(y_2 - y_1)$$

4. Compute A_T, the sum of the areas for all parts.
5. Compute the moment of the area of each part from the equation,

$$M = b(y_2^2 - y_1^2)/2$$

6. Compute M_T, the sum of the moments for all parts.
7. Compute the location of the centroid relative to the base of the composite section from

$$\overline{Y} = M_T/A_T$$

8. Compute the moment of inertia with respect to the base of the composite section of each part from

$$I_b = b(y_2^3 - y_1^3)/3$$

9. Compute I_{bT}, the sum of the moments of inertia for all parts.
10. Compute the total moment of inertia with respect to the centroid of the composite section from

$$I_c = I_{bT} - A_T\overline{Y}^2$$

This process lends itself very well to automated computation using a programmable calculator, a computer program, or a spreadsheet. As an illustration, Figure 6–16 shows the spreadsheet calculation of the centroidal moment of inertia for the tee section shown in Figure 6–11 and for which the calculation of the moment of inertia was done in Example Problem 6–5 using the parallel axis theorem. The results are, of course, identical. See also Figure 6–17 for data.

Note that there are some blank lines in the spreadsheet because allowance was made for up to six parts for the composite section whereas this one has only two. The spreadsheet could be expanded to include any number of parts.

FIGURE 6–16 Spreadsheet for computing moment of inertia for a shape with all rectangular parts.

MOMENT OF INERTIA FOR A SHAPE WITH ALL RECTANGULAR PARTS

Problem ID: Example Problem 6–5
Enter data for all parts of the composite area in the shaded cells
For each part: b = width; y_1 = distance to bottom of part; y_2 = distance to top of part

	Dimensions			Area	Moment	*I* with respect to base
	b	y_1	y_2	A	M	I_b
	(in)	(in)	(in)	(in^2)	(in^3)	(in^4)
Part:						
1	1.000	0.000	4.000	4.000	8.000	21.333
2	4.000	4.000	4.500	2.000	8.500	36.167
3				0.000	0.000	0.000
4				0.000	0.000	0.000
5				0.000	0.000	0.000
6				0.000	0.000	0.000

Totals: A_T = 6.000 in^2 M_T = 16.500 in^3 I_{bT} = 57.500 in^4

Note: $A = b(y_2 - y_1)$ $M = b(y_2^2 - y_1^2)/2$ $I_b = b(y_2^3 - y_1^3)/3$

Results:

Distance from base to centroid = $\boldsymbol{M_T/A_T = \bar{Y}}$ = **2.75 in**

Moment of inertia with respect to the centroid = $\boldsymbol{I_{bT} - A_T\bar{Y}^2 = I_c}$ = **12.125 in^4**

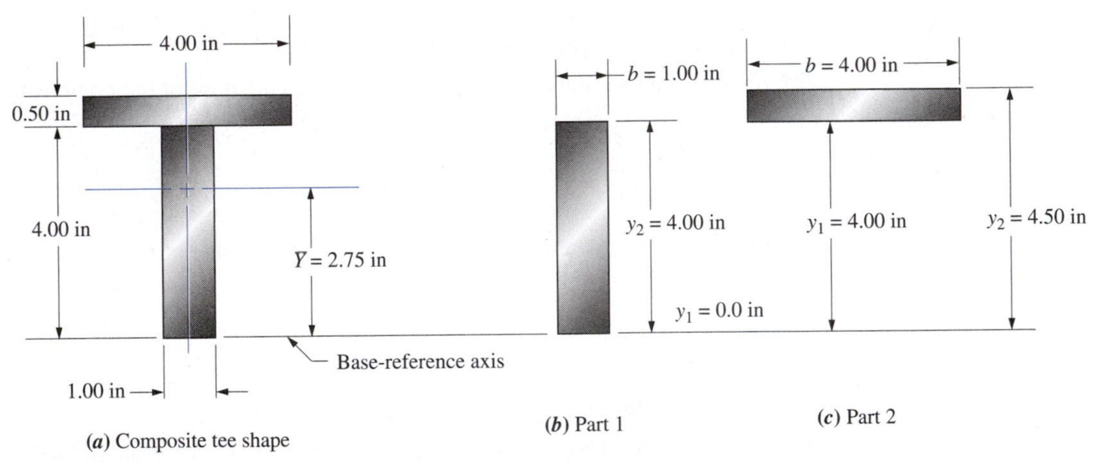

FIGURE 6–17 Tee section for illustration of the method of computing moment of inertia described in Section 6–9.

6–10 RADIUS OF GYRATION

Another property of the area of a cross section is called the *radius of gyration.* It is used most frequently in the study of strength of materials when analyzing or designing columns, relatively long, slender members carrying axial compressive loads.

Columns were mentioned briefly in the activity in Chapter 1 and when direct compressive stresses were defined. There it was emphasized that the simple direct stress formula, $\sigma = F/A$, applies to compression members only if they are "short." Short compression members fail when the direct compressive stress exceeds the yield strength of the material in compression. Long compression members, called columns, fail by elastic instability. That is, they are unable to maintain their initially straight form when a critical compressive load is reached. Instead they begin to bend elastically and, if the load is increased moderately above the critical load, the column will buckle suddenly and collapse, usually with catastrophic consequences.

You may have observed column buckling behavior. Have you ever taken a long, thin rod or a thin strip of wood, plastic, or metal and subjected it to a compressive load? For example, the typical meter stick you probably used in an early science class is a long, slender column. If you supported it on a table or the floor and pushed down on it with a moderate force, it would bend as shown in Figure 6–18. This is an example of elastic instability or buckling. If you release the load just after the onset of buckling, you will notice that the column did not deform permanently. This is evidence that the failure was not in the material, but rather it was because the shape of the column was not able to maintain the straightness of the column. The property of the shape that determines when a column will buckle is its radius of gyration.

However, the method of determining when the compression member is long or short was not given in Chapter 1 because it is dependent on the moment of inertia that is the

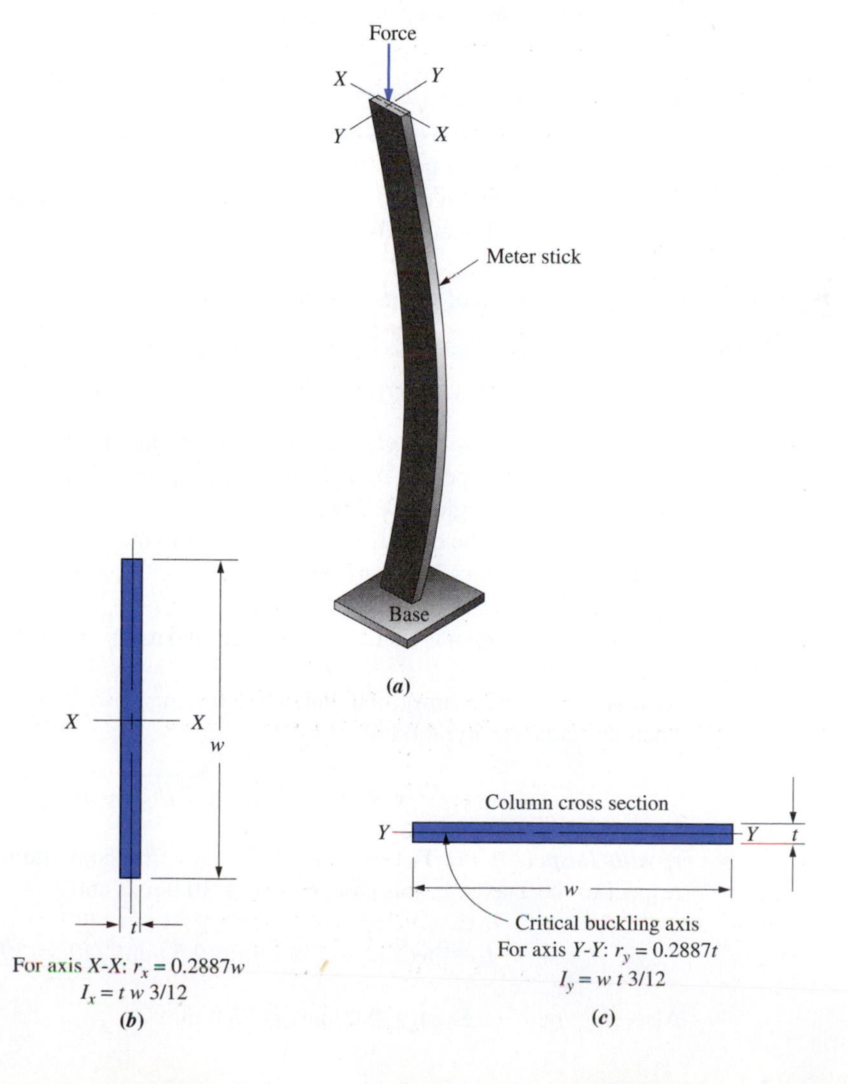

FIGURE 6–18 Illustration of a metal stick exhibiting buckling as a column.

focus of much of this chapter. In fact, it is the *ratio of the length of the member to its radius of gyration* that determines when a member is long or short. That ratio is called the *slenderness ratio.* For now, we only define the radius of gyration. Later, in Chapter 11, it will be used in the analysis of columns.

Definition of Radius of Gyration, *r*. We define the radius of gyration, r, as

$$r = \sqrt{\frac{I}{A}} \qquad \textbf{(6–6)}$$

where I = moment of inertia of the cross section of the column with respect to one of the principal axes
A = area of the cross section

Because both I and A are geometrical properties of the cross section, the radius of gyration, r, is also. Formulas for computing r for several common shapes are given in Appendix A–1. Also, r is listed with other properties for some of the standard structural shapes in the Appendix. For those for which r is not listed, the values of I and A are available and Equation (6–6) can be used to compute r very simply.

Example Problem 6–8 Compute the radius of gyration for the cross section of a meter stick that has a thickness of 2.5 mm and a width of 30 mm. Consider both principal axes, X and Y, as illustrated in Figure 6–18. Use Equation (6–6).

Solution

Objective: Compute the radius of gyration with respect to axes X and Y.

Given: A rectangular cross section. Thickness = t = 2.5 mm. Width = w = 30 mm.

Analysis: Use Equation (6–6). $r = \sqrt{I/A}$

Results: ***r_x with Respect to the X-Axis.*** First, let's compute the moment of inertia with respect to the X-axis. From Appendix A–1, we find the general form of the equation for the moment of inertia of a rectangle to be $I = bh^3/12$. Note that b is the dimension parallel to the axis of interest and h is the dimension perpendicular to the axis of interest. In this case, $b = t =$ 2.5 mm and $h = w =$ 30.0 mm. Then,

$$I_X = tw^3/12 = (2.5\text{ mm})(30.0\text{ mm})^3/12 = 5625\text{ mm}^4$$

Also, $A = tw = (2.5\text{ mm})(30.0\text{ mm}) = 75.0\text{ mm}^2$
Then the radius of gyration is

$$r_X = \sqrt{I_X/A} = \sqrt{(5625\text{ mm}^4)/(75.0\text{ mm}^2)} = 8.66\text{ mm}$$

r_y with Respect to the Y-Axis. Similarly, let's first compute the moment of inertia with respect to the Y-axis. In this case, $b = w =$ 30.0 mm and $h = t =$ 2.5 mm. Then,

$$I_Y = wt^3/12 = (30.0\text{ mm})(2.5\text{ mm})^3/12 = 39.06\text{ mm}^4$$

Also, $A = tw = (2.5\text{ mm})(30.0\text{ mm}) = 75.0\text{ mm}^2$

Then the radius of gyration is

$$r_Y = \sqrt{I_Y/A} = \sqrt{(39.06\ \text{mm}^4)/(75.0\ \text{mm}^2)} = 0.722\ \text{mm}$$

Comment It is obvious that the choice of the axis about which to compute the radius of gyration is critical to the result. Here I_X is much greater than I_Y and therefore, r_X is much greater than r_Y. You should gain an appreciation of the importance of the selection of the axis from this example. You will see the effect of this observation in Chapter 11 on column analysis and design. The result is that the meter stick would buckle about its Y-axis, not the X-axis. It will always buckle about the axis with the smallest radius of gyration.

Alternate Solution Where a formula for computing the radius of gyration is available, it makes, the computation much simpler. For the rectangular section of this problem, Appendix A–1 gives

$$r_X = h/\sqrt{12} = 0.2887\,h$$

and

$$r_Y = b/\sqrt{12} = 0.2887\,b$$

These formulas for r_X and r_Y are easily derived from the basic definition:

$$r_X = \sqrt{I/A} = \sqrt{(bh^3/12)/(bh)} = \sqrt{h^2/12} = h/\sqrt{12}$$

$$r_Y = \sqrt{I/A} = \sqrt{(hb^3/12)/(bh)} = \sqrt{b^2/12} = b/\sqrt{12}$$

In this case, $h = w = 30.0$ mm and $b = t = 2.5$ mm. Then

$$r_X = 0.2887\,w = (0.2887)(30.0\ \text{mm}) = 8.66\ \text{mm}$$

and

$$r_Y = 0.2887\,t = (0.2887)(2.5\ \text{mm}) = 0.722\ \text{mm}$$

These values match those found using Equation (6–6).

Radius of Gyration for Structural Shapes. Review the data in Appendices A–4 through A–13 that give the section properties for a large variety of structural shapes often used in the structures of machines, platforms, or buildings. Some of these tables list the radius of gyration directly. Others list both the moment of inertia, I, and the cross-sectional area, A. Then you can use Equation 6–6 to compute the radius of gyration, r.

It is also necessary to determine the axis for which the radius of gyration is the smallest. For the wide-flange beams (Appendix A–7) and American Standard beams (Appendix A–8), the minimum value of r is that computed with respect to the Y-Y axis; that is,

$$r_{\text{min}} = \sqrt{\frac{I_Y}{A}}$$

Similarly, for rectangular structural tubing (Appendix A–9), the minimum radius of gyration is that with respect to the Y-Y axis. Values for r are listed in the table.

For structural steel angles, called L-shapes, neither the X-X nor the Y-Y axis provides the minimum radius of gyration. As illustrated in Appendix A–5, r_{min} is computed with respect to the Z-Z axis, with the values listed in the table.

For symmetrical sections, the value of r is the same with respect to any principal axis. Such shapes are the solid or hollow circular section and the solid or hollow square section.

6–11 SECTION MODULUS

In the analysis of bending of beams in Chapter 7, it is shown that the bending stress is directly proportional to the bending moment acting on the beam at a given location and inversely proportional to a property of the cross-sectional area called the *section modulus*. Expressed mathematically,

$$\text{Stress due to bending} = \sigma = M/S$$

where M is the bending moment found using the principles discussed in Chapter 5 and S is the section modulus, defined in Chapter 7 to be

$$S = I/c$$

The symbol, c, is taken to be the distance from the centroidal axis of the cross section to the outermost fiber of the beam. Because both I and c are properties of the cross-sectional area of the beam, the section modulus is also a property of the area, and it would typically be used for analysis and design.

Figure 6–19 shows an example illustrating these concepts. The standard steel W-shape, W8×21, has an area moment of inertia of 75.3 in^4 and an overall depth of 8.28 in as given in Appendix A–7. Because of symmetry, we see that the horizontal centroidal axis about which bending occurs is at the middle of the depth. Therefore, the distance from the centroid to the outermost fiber of the beam, c, is

$$c = \text{Depth}/2 = (8.28\text{ in})/2 = 4.14\text{ in}$$

Finally, we can calculate the value of the section modulus from

$$S = I/c = (75.3\text{ in}^4)/4.14\text{ in} = 18.2\text{ in}^3$$

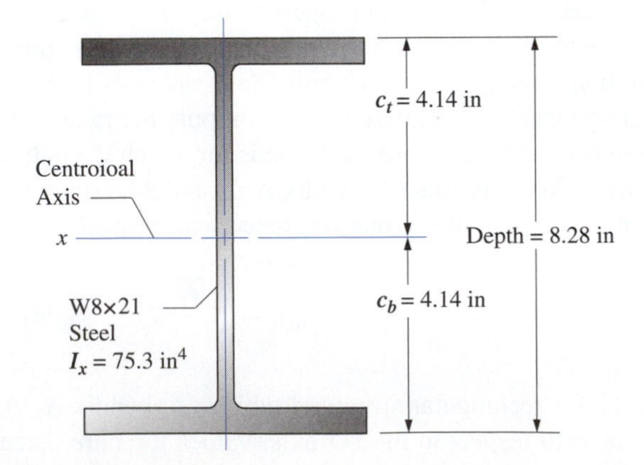

FIGURE 6–19 Illustration of c distance on W-shape beam cross section.

Now we can use this value to compute the stress due to bending. Assume that the beam is subjected to a bending moment of 18 750 lb·ft. Then

$$\sigma = M/S = \frac{18\,750\ \text{lb}\cdot\text{ft}}{18.2\ \text{in}^3}\ \frac{12\ \text{in}}{\text{ft}} = 12\,363\ \text{lb/in}^2 = 12.4\ \text{ksi}$$

Bending produces tensile stress on one outside surface and compressive stress on the other, as will be demonstrated in Chapter 7. For this example, the tensile stress and the compressive stress would have equal magnitude because the c distance is equal to either the top or bottom surface.

Not every section is symmetrical, as demonstrated in the following example problem, resulting in unequal stresses at the top and bottom surfaces of the beam.

Example Problem 6–9 Determine the section modulus for the tee shape used in Example Problem 6–5. Then compute the maximum stress due to bending at the top and bottom surfaces if the bending moment is 8650 lb·in.

Solution The results of Example Problem 6–5, shown in Figure 6–12, include the following:

- Overall height of the section: 4.50 in
- Location of the horizontal centroidal axis: 2.75 in above the base
- Area moment of inertia: 12.125 in^4

We can now observe that the distances from the centroidal axis to the top and bottom surfaces are unequal.

$$c_b = 2.75\ \text{in from the centroid to the bottom surface}$$
$$c_t = 4.50\ \text{in} - 2.75\ \text{in} = 1.75\ \text{in from the centroid to the top surface}$$

Two values of the section modulus can be computed.

Top surface: $S_t = I/c_t = (12.125\ \text{in}^4)/1.75\ \text{in} = 6.93\ \text{in}^3$

Bottom Surface: $S_b = I/c_b = (12.125\ \text{in}^4)/2.75\ \text{in} = 4.41\ \text{in}^3$

We can now compute the stresses at the top and bottom surfaces.

Top surface: $\sigma = M/S = (8650\ \text{lb}\cdot\text{in})/6.93\ \text{in}^3 = 1248\ \text{psi}$

Bottom surface: $\sigma = M/S = (8650\ \text{lb}\cdot\text{in})/4.41\ \text{in}^3 = 1961\ \text{psi}$

Note that the lower value for section modulus gives the higher value of stress and that would typically be the objective of the stress calculation. However, this will be explored in greater detail in Chapter 7 for special cases where the material of a beam has different strengths in tension and in compression.

REFERENCES

1. American Institute of Steel Construction, *Steel Construction Manual,* 13th ed., Chicago, IL, 2005.
2. Oberg, Erik, et al., *Machinery's Handbook,* 27th ed., Industrial Press, New York, 2004.
3. Young, Warren C. and Richard G. Budynas, *Roark's Formulas for Stress and Strain,* 7th ed., McGraw-Hill, New York, 2002.

INTERNET SITES

1. Profiles, Inc. www.profiles-inc.com Manufacturer of custom rolled or drawn shapes for precision components using carbon and alloy steel, copper, brass, and other non-ferrous materials.
2. iLevel by Weyerhaeuser www.ilevel.com/floors Manufacturer of floor joists for building construction made as efficient I-shaped fabrications from simple wood components. Select the TJI® Joists product. The Joist Specifier Guide gives detailed dimensions and application data for the Silent Floor—Trus Joist products.
3. Nucor Corporation—Vulcraft Group www.vulcraft.com Manufacturer of open-web steel joists, joist girders, floor and roof deck shapes, and composite floor joist systems for use in commercial shopping centers, schools, and office buildings.
4. Aluminum Extruders Council www.aec.org/exapps/alumshow.html An industry organization promoting the use of aluminum extrusions in many types of product, building construction, and transportation system. The *Showcase* illustrates in-depth descriptions of successful applications. The *Extrusion Basics* and *Technical Information* tabs describe the extrusion process, alloys used, environmental implications, and a variety of useful resources for instructors and students.
5. American Extruded Plastics, Inc. www.aeplastics.com Manufacturer of extruded plastics for mechanical, architectural, building construction, aerospace, electronics, transportation, and many other industries.
6. Green Bull, Inc. www.greenbullladder.com Manufacturer of ladders made from aluminum, steel, fiberglass, and wood. Also produces extendable planks, platforms, mobile scaffolds, and carts.
7. Altec, Inc. www.altec.com Designer and manufacturer of truck-mounted articulated and extendable aerial platforms for positioning personnel and equipment at elevated sites. Applications to the electrical power distribution, building construction, tree trimming, and other industries.
8. Unistrut Corporation www.unistrut.com Manufacturer of systems of components made from steel or fiberglass that can be assembled in a variety of configurations to produce frames, structures, brackets, storage units, support systems, mobile units, and other devices. Online catalogs provide section property data, application information, assembly instructions, and illustrations of uses.
9. 80/20, Inc. www.8020.net Manufacturer of systems of components made from extruded aluminum that can be assembled in a variety of configurations for use as production equipment, carts, automation devices, workstations, storage units, and a variety of other systems.
10. Paramount Extrusions Company http://paramountextrusions.com/shapes Manufacturer of a variety of standard extrusions such as channels, H-sections, T-sections, hollow tubes (round, square, rectangular), square and rectangular bars, handle sections, and many others for use in office furniture, consumer products, cabinets, and so forth. The catalog gives data that can be useful in problems in this book.
11. Jackson Tube Service, Inc. http://www.jackson-tube.comshapes.htm Manufacturer of a wide variety of tubing shapes and sizes that can be useful in problems in this book.

PROBLEMS

Centroid and Moment of Inertia

For each of the shapes in Figures P6–1 through P6–48, determine the location of the horizontal centroidal axis and the magnitude of the moment of inertia of the shape with respect to that axis using the parallel axis theorem described in Section 6–6.

The method described in Section 6–9 can be applied to compute the moment of inertia for those shapes consisting of two or more rectangular parts whose sides are perpendicular and parallel to the horizontal axis. Included are the shapes in Figures P6–1 through P6–15 and the composite shapes made from standard wood beam shapes in Figures P6–21 through P6–24.

Figures P6–1 through P6–20 are special shapes that might be produced by plastic or aluminum extrusion, by machining from solid bar stock, or by welding separate components.

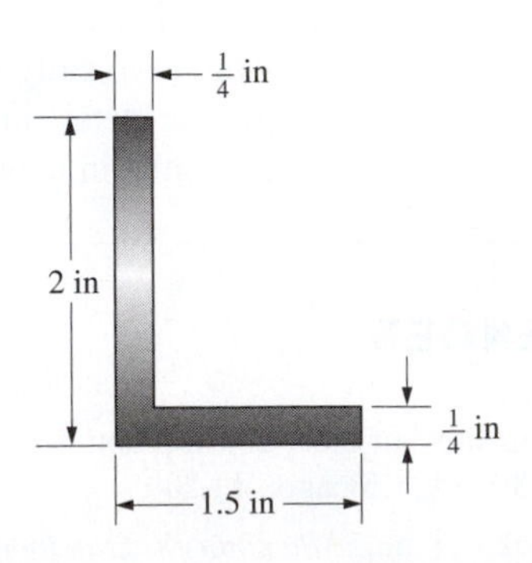

FIGURE P6–1

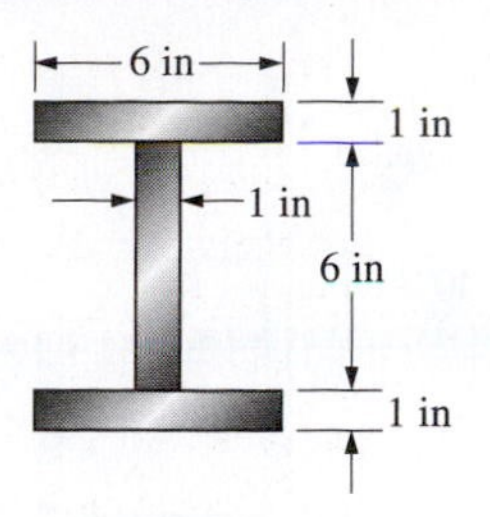

FIGURE P6–2

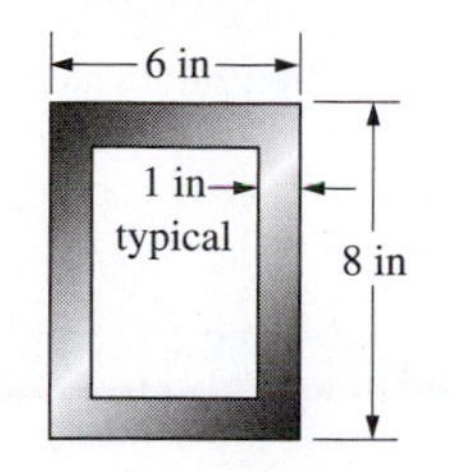

FIGURE P6–3

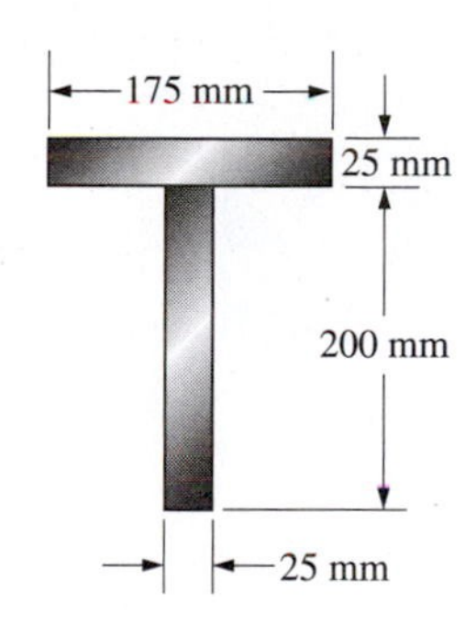

FIGURE P6–4

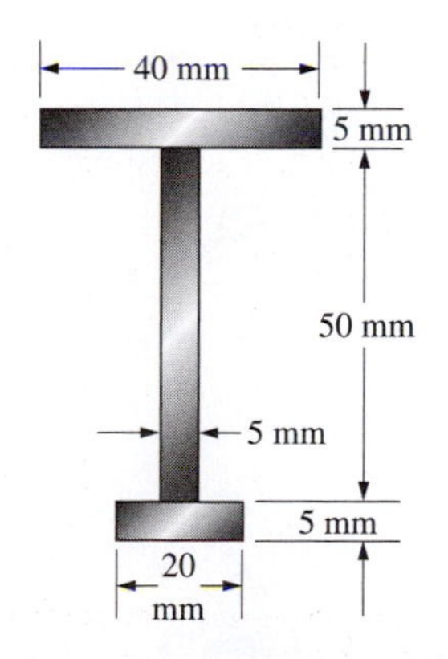

FIGURE P6–5

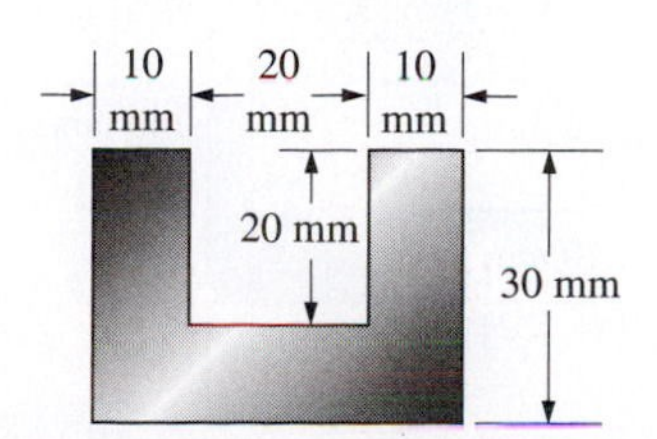

FIGURE P6–6

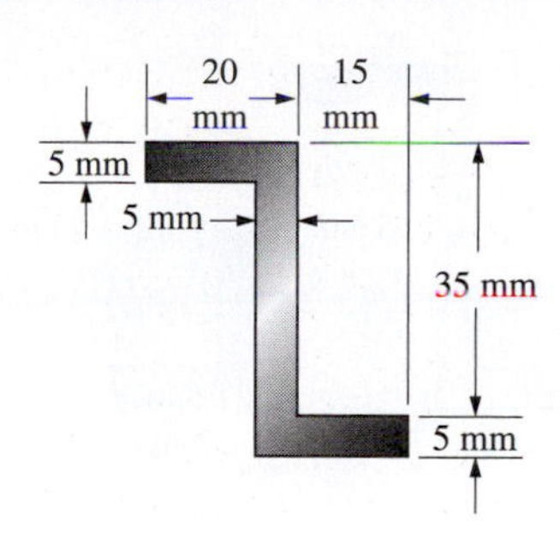

FIGURE P6–7

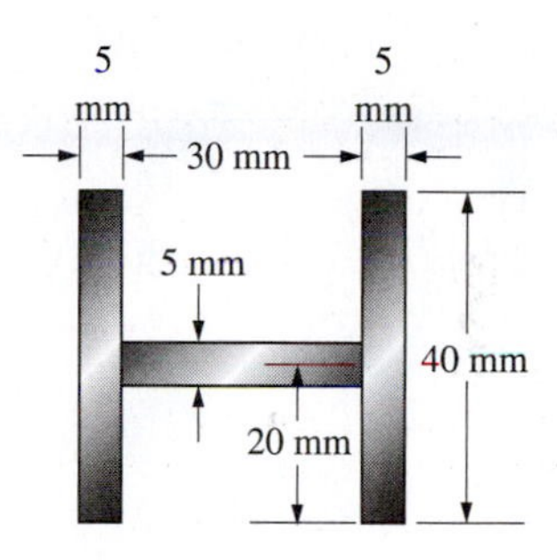

FIGURE P6–8

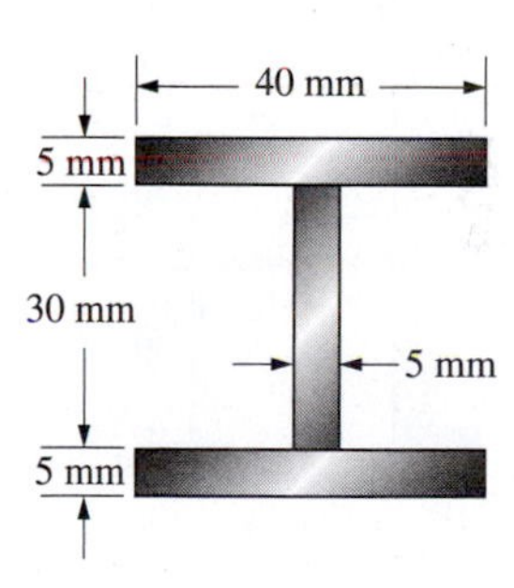

FIGURE P6–9

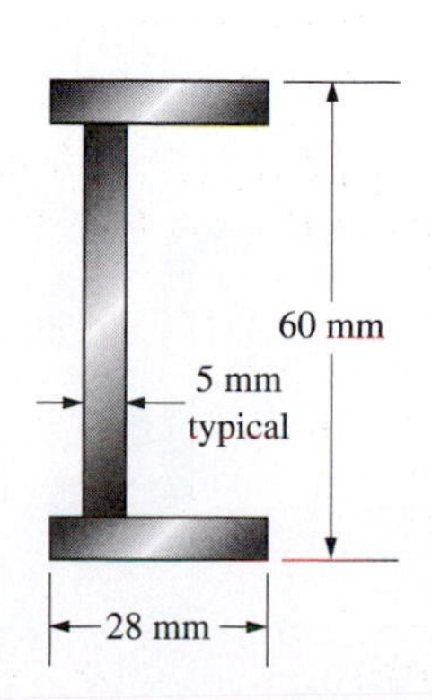

FIGURE P6–10

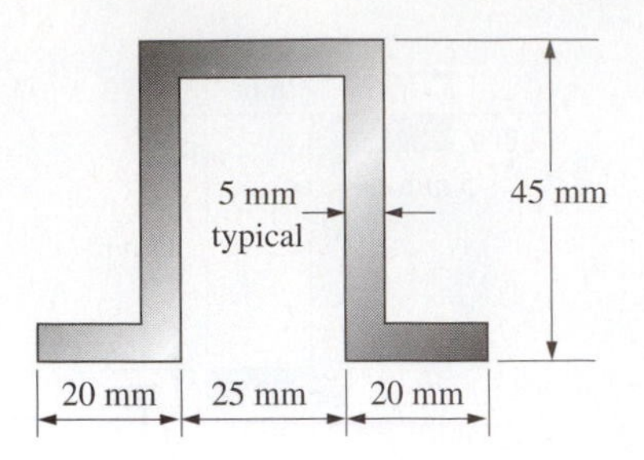

FIGURE P6–11

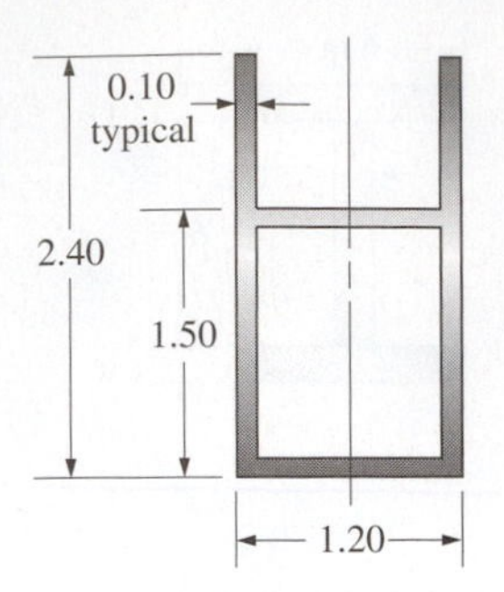

FIGURE P6–15

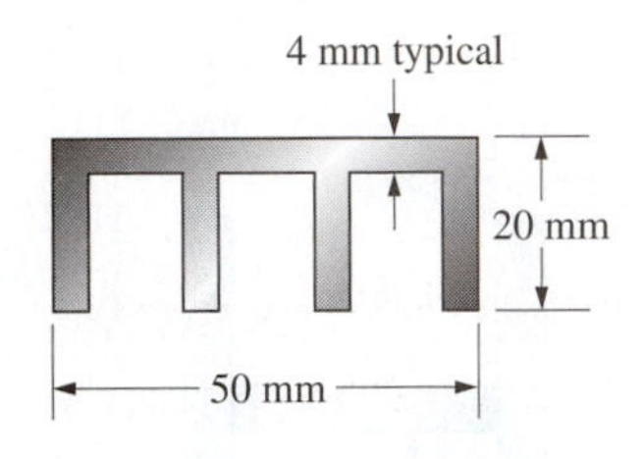

FIGURE P6–12

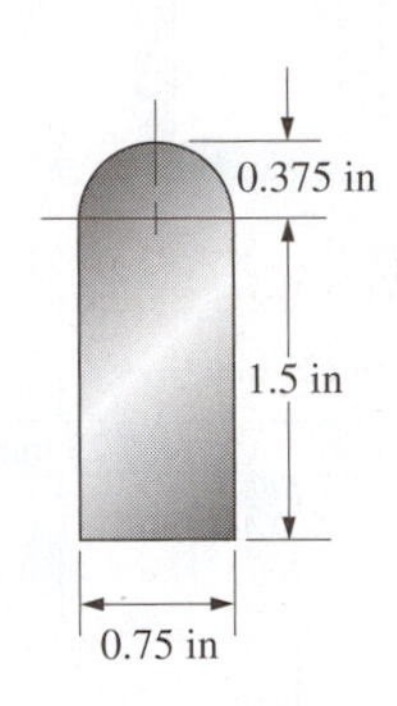

FIGURE P6–16

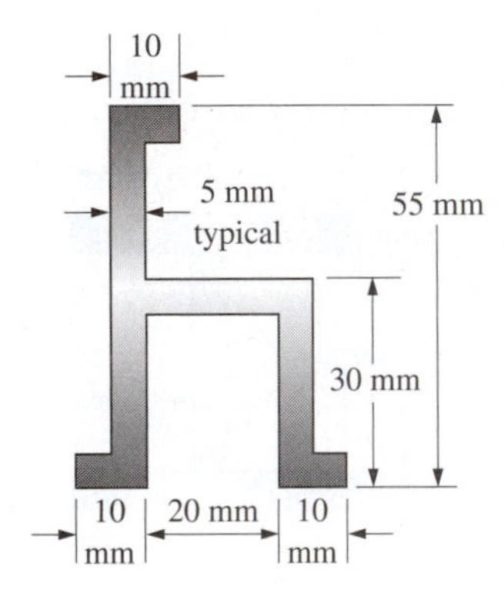

FIGURE P6–13

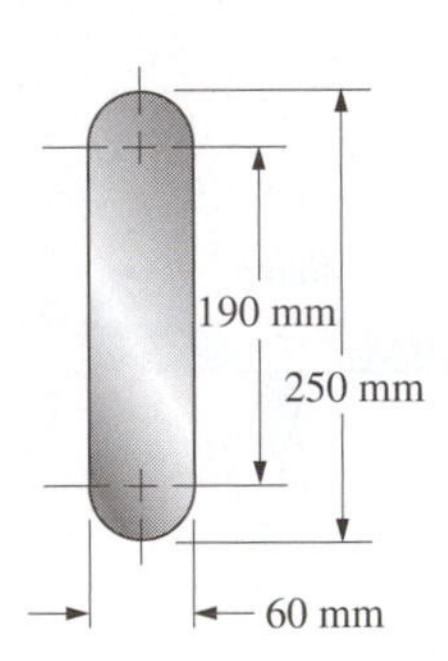

FIGURE P6–17

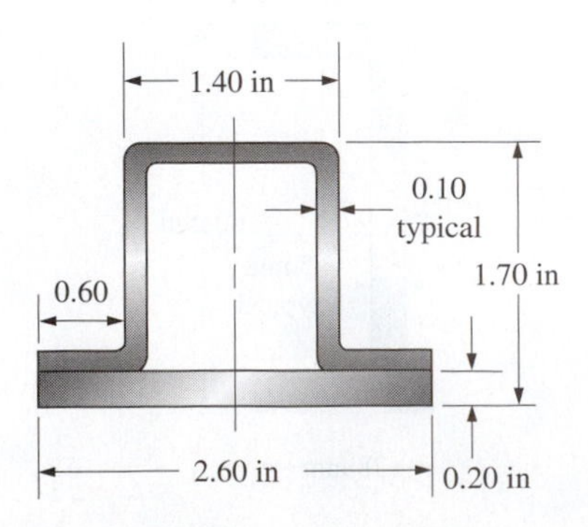

FIGURE P6–14

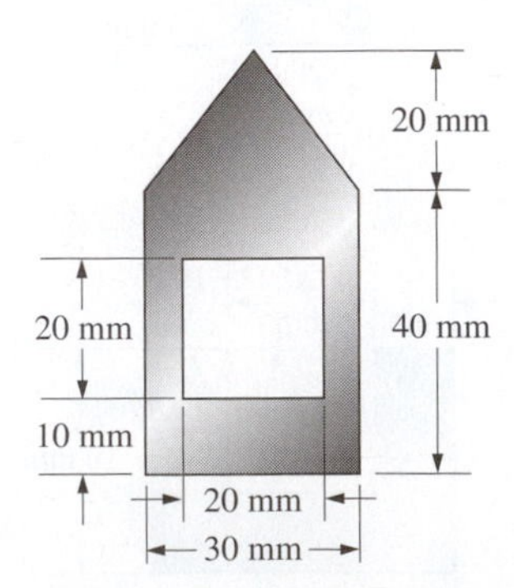

FIGURE P6–18

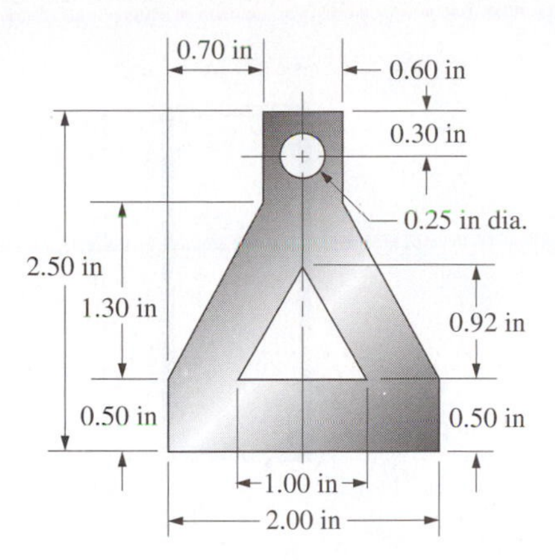

FIGURE P6–19

Dimensions in inches

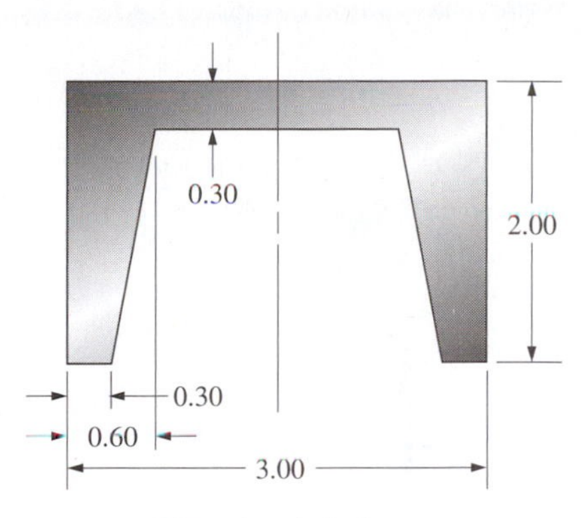

FIGURE P6–20

Figures P6–21 through P6–24 are composite beam sections that can be made by fastening standard wood shapes together by nails, screws, or adhesive.

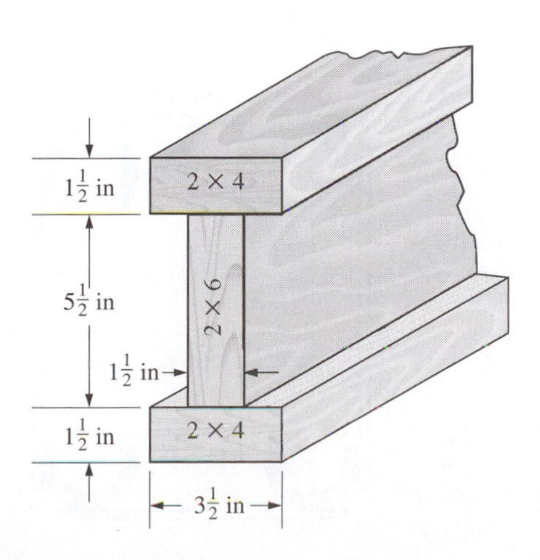

FIGURE P6–21

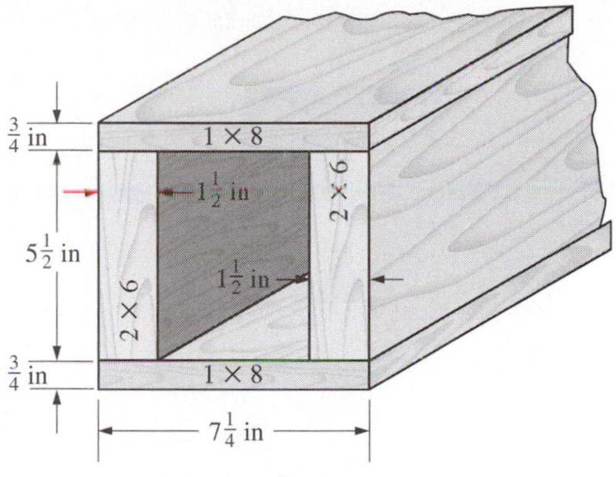

FIGURE P6–22

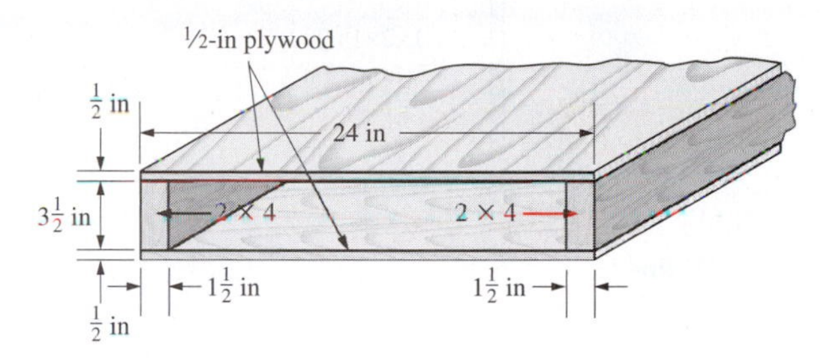

FIGURE P6–23

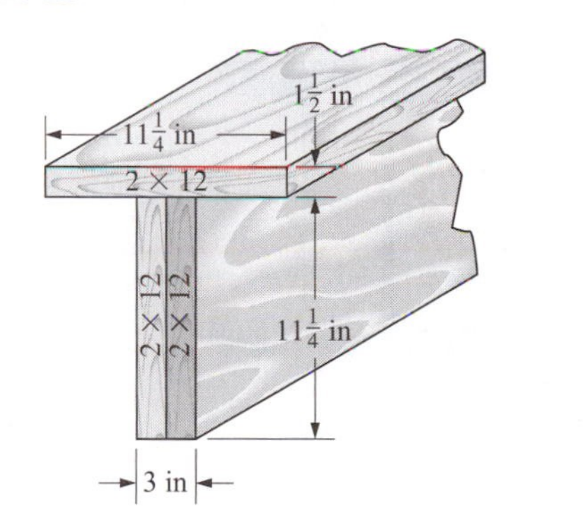

FIGURE P6–24

Figures P6–25 through P6–46 are composite beam sections fabricated from standard steel or aluminum structural shapes.

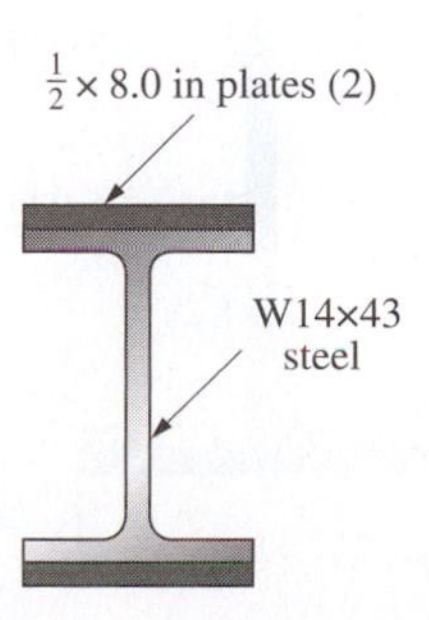

FIGURE P6–25

FIGURE P6–26

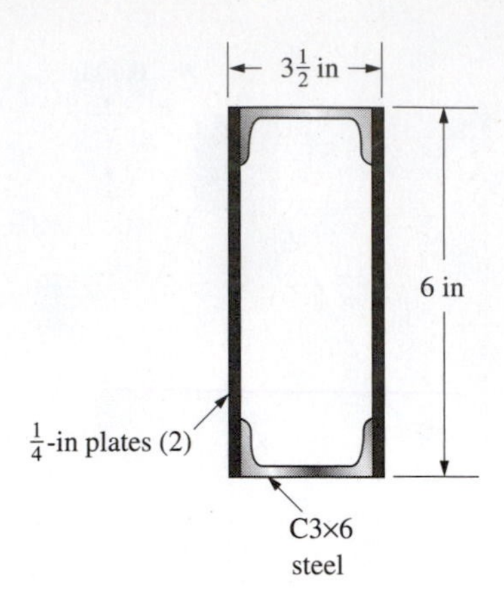

FIGURE P6–30

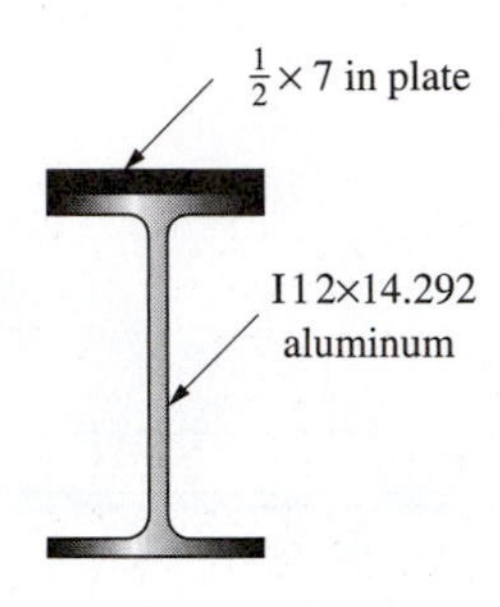

FIGURE P6–27

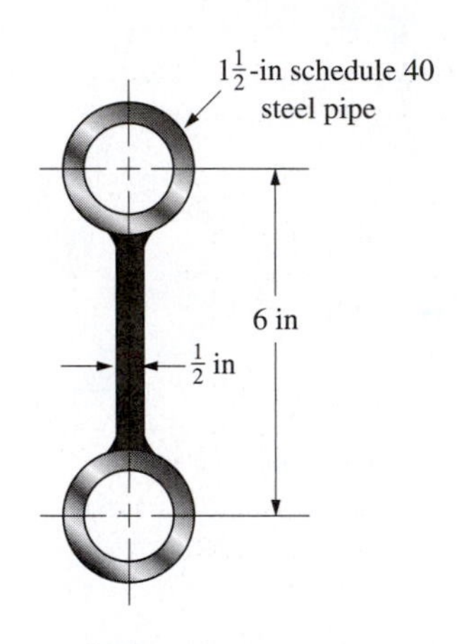

FIGURE P6–31

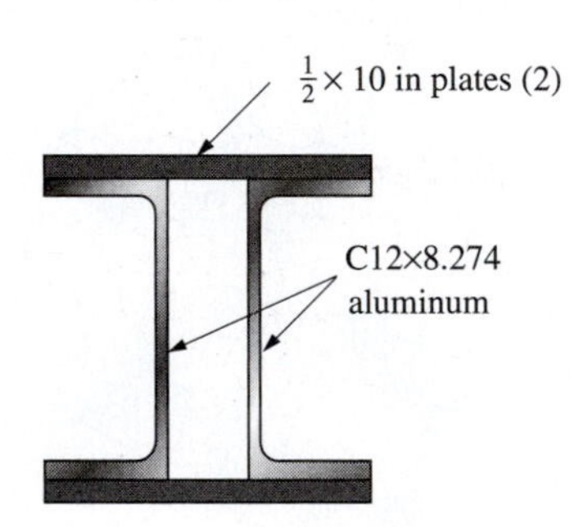

FIGURE P6–28

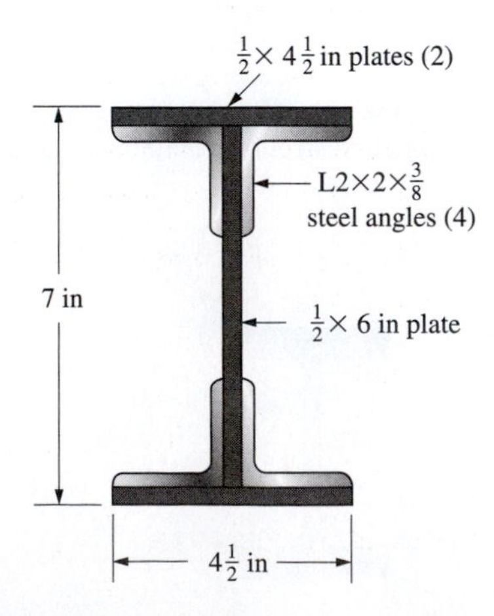

FIGURE P6–29

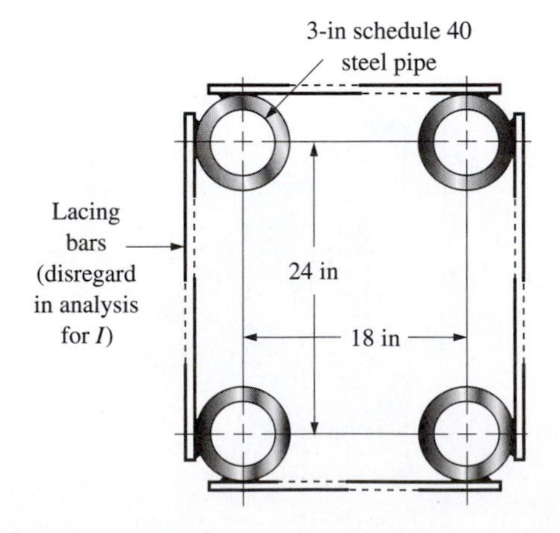

FIGURE P6–32

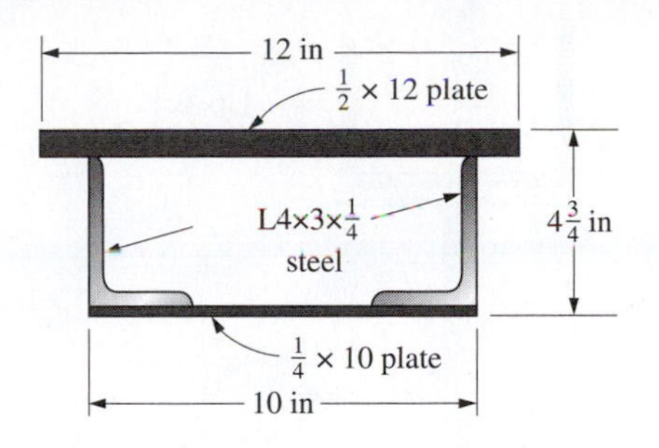

FIGURE P6–33

FIGURE P6–34

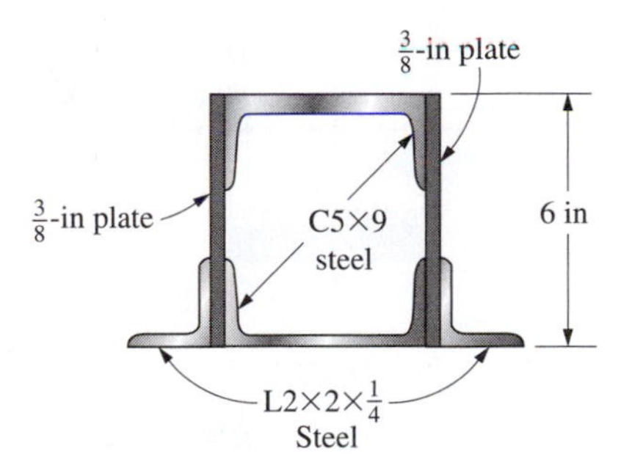

FIGURE P6–35

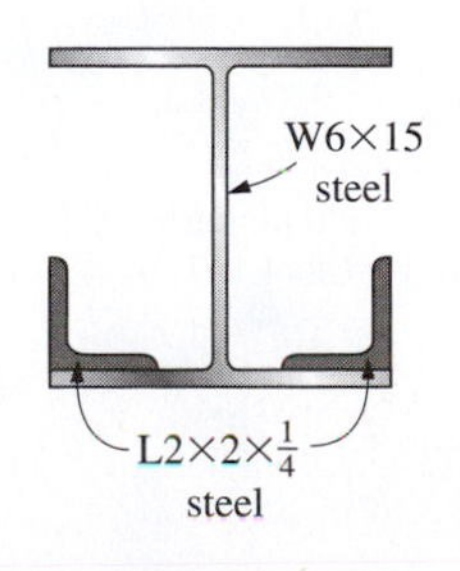

FIGURE P6–36

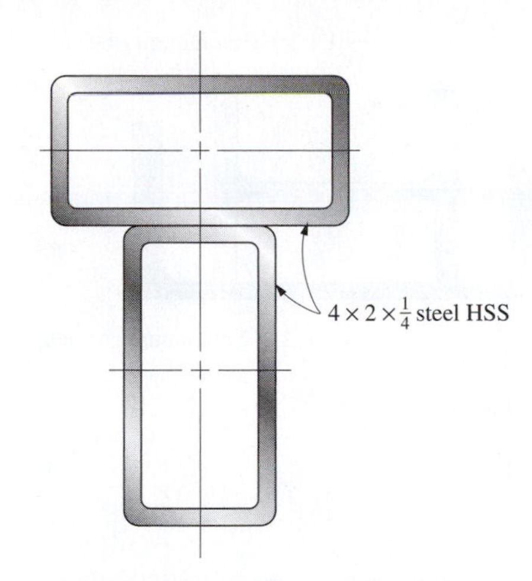

FIGURE P6–37

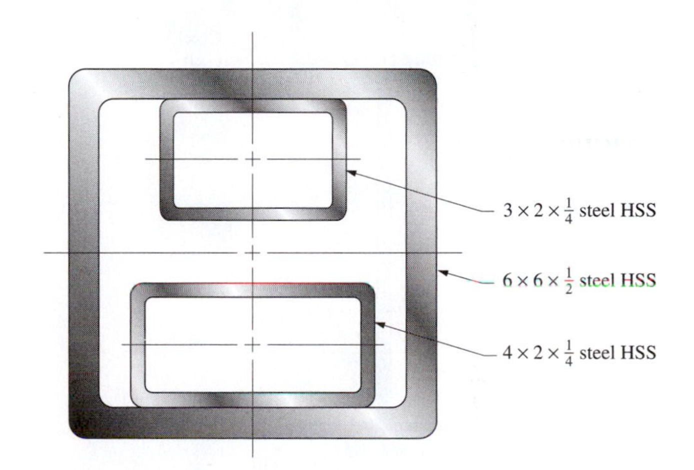

FIGURE P6–38

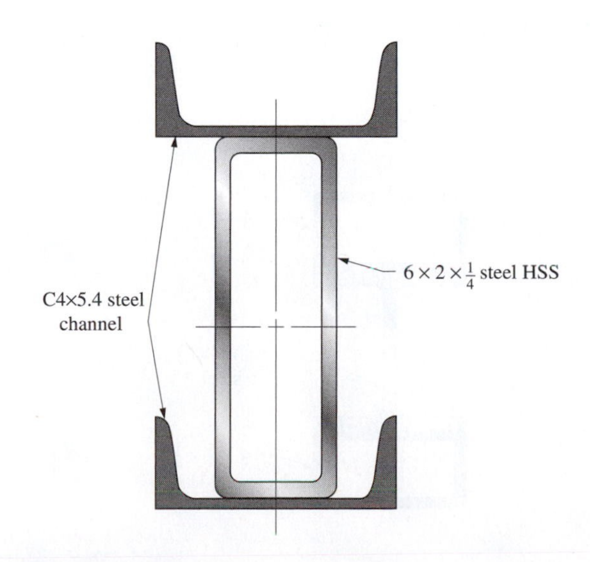

FIGURE P6–39

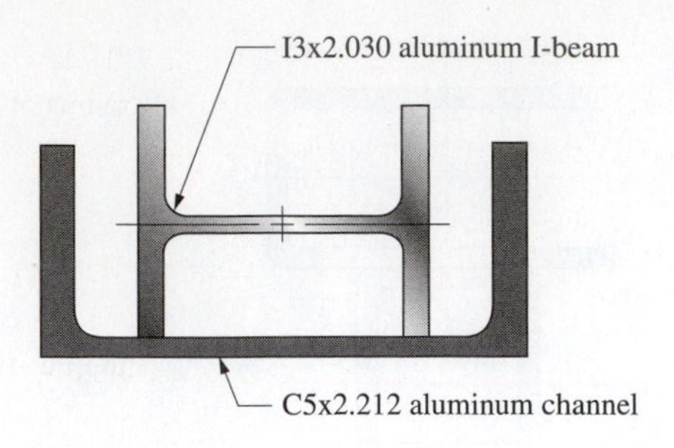

FIGURE P6–40

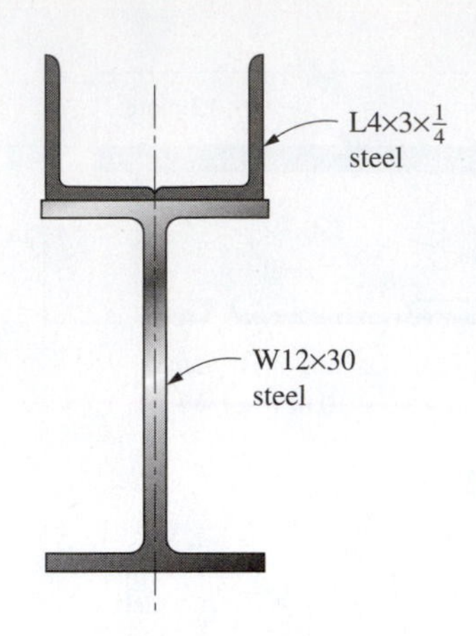

FIGURE P6–43

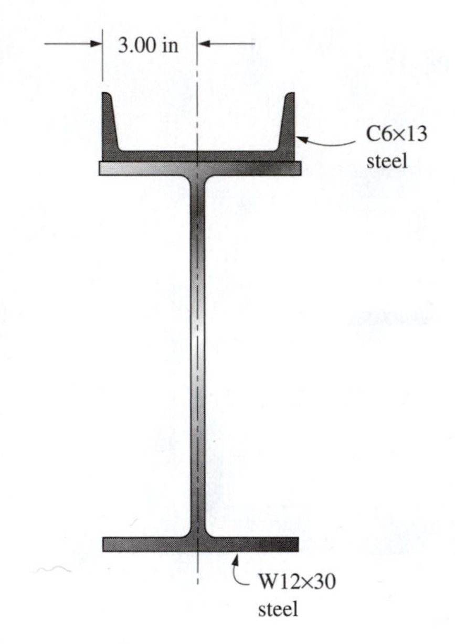

FIGURE P6–41

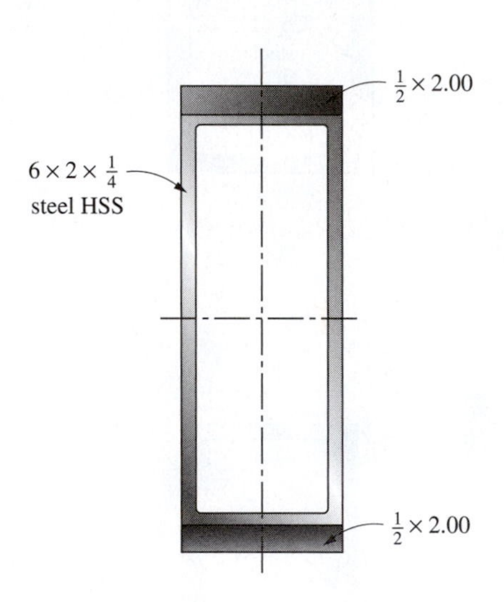

FIGURE P6–44

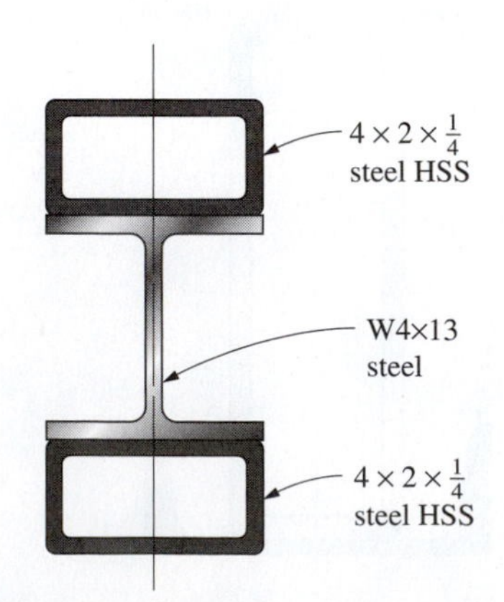

FIGURE P6–42

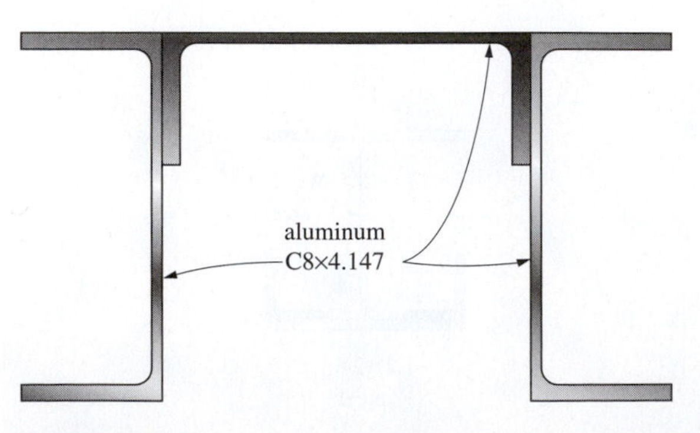

FIGURE P6–45

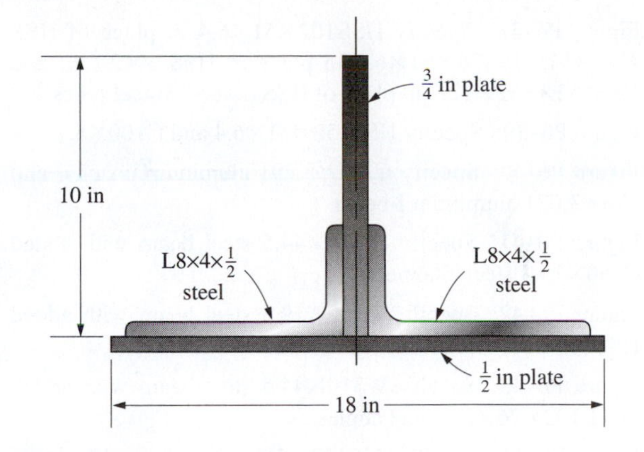

FIGURE P6–46

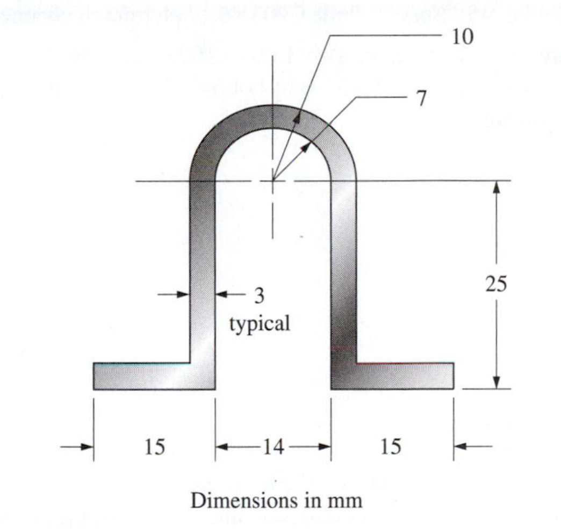

FIGURE P6–47

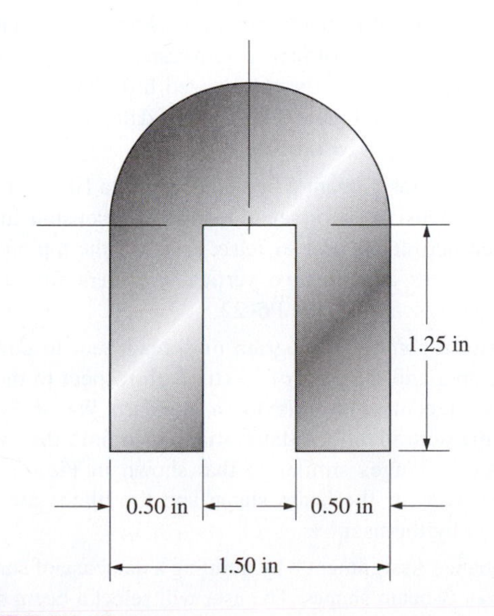

FIGURE P6–48

Radius of Gyration

For Problems 6–49 through 6–66, compute the radius of gyration for the shape in the indicated figure. Use the horizontal centroidal axis for the computation.

6–49. Use Figure P6–2.
6–50. Use Figure P6–3.
6–51. Use Figure P6–4.
6–52. Use Figure P6–5.
6–53. Use Figure P6–6.
6–54. Use Figure P6–8.
6–55. Use Figure P6–9.
6–56. Use Figure P6–11.
6–57. Use Figure P6–12.
6–58. Use Figure P6–14.
6–59. Use Figure P6–15.
6–60. Use Figure P6–16.
6–61. Use Figure P6–17.
6–62. Use Figure P6–21.
6–63. Use Figure P6–22.
6–64. Use Figure P6–23.
6–65. Use Figure P6–24.
6–66. Use Figure P6–25.

For Problems 6–67 through 6–81, compute the radius of gyration for the shape in the indicated figure. Use the vertical centroidal axis for the computation.

6–67. Use Figure P6–2.
6–68. Use Figure P6–3.
6–69. Use Figure P6–4.
6–70. Use Figure P6–5.
6–71. Use Figure P6–16.
6–72. Use Figure P6–17.
6–73. Use Figure P6–21.
6–74. Use Figure P6–22.
6–75. Use Figure P6–23.
6–76. Use Figure P6–24.
6–77. Use Figure P6–25.
6–78. Use Figure P6–26.
6–79. Use Figure P6–27.
6–80. Use Figure P6–42.
6–81. Use Figure P6–44.

Problems for Centroids and Moments of Inertia Using SI Metric Data

The following problems are given with metric data:

P6–4 through P6–13, P6–17 through P6–18, P6–47, P6–51 through P6–57, P6–61, P6–69 through P6–70, P6–72

Additional problems can be expressed in SI metric data by making the following substitutions to earlier given figures. SI metric data for these sections are listed in the SI Units versions of the Appendix tables for properties of the given shapes. Note that the dimensions for rectangular plates have been modified to use preferred metric dimensions for thickness and width according to the data in Appendix A–2.

Figures P6–21 through P6–24: Use metric data from Appendix A–4 for standard dimension lumber.

Figure P6–25: Specify W360×64 steel beam with added 12 mm × 200 mm plates.

Figure P6–26: Specify S300×74 steel beam with added C300×37 steel channel.

Figure P6–27: Specify I305×23.80 aluminum I-beam with added 12 mm × 180 mm plates.

Figure P6–28: Specify C305×12.31 aluminum channel with added 12 mm × 250 mm plates.

Figure P6–29: Specify L51×51×9.5 steel angles with added 12 mm × 110 mm plates top and bottom and 12 mm × 150 mm web.

Figure P6–30: Specify C80×8.9 steel channels with added 6 mm × 150 mm side plates.

Figure P6–31: Specify PIPE 38 STD steel pipes with 12 mm thick web, 150 mm on center.

Figure P6–32: Specify PIPE 75 STD steel pipes spaced apart 600 mm vertically and 450 mm horizontally.

Figure P6–33: Specify L102×76×6.4 steel angles with added 12 mm × 300 mm top plate and 6.0 mm × 250 mm bottom plate. Overall height = 120 mm.

Figure P6–34: Specify C150×19.3 steel channels.

Figure P6–35: Specify C130×13 steel channels with 10 mm × 150 mm side plates and l51×51×6.4 angles.

Figure P6–36: Specify W150×22.5 steel beam with added l51×51×6.4 steel angles.

Figure P6–37: Specify HSS102×51×6.4 steel rectangular tube.

Figure P6–38: Specify HSS102×51×6.4 in place of HSS 4×2×¼; HSS76×51×6.4 in place of HSS 3×2×¼; and HSS152×152×12.7 in place of HSS 6×6×½ steel tubes.

Figure P6–39: Specify HSS152×51×6.4 and C100×8.

Figure P6–40: Specify C127×3.291 aluminum channel and 176×3.021 aluminum I-beam

Figure P6–41: Specify W310×44.5 steel beam with added C150×19.3 steel channel.

Figure P6–42: Specify W100×19.3 steel beam with added HSS102×51×6.4 steel tube.

Figure P6–43: Specify W310×44.5 steel beam with added two L102×76×6.4 steel angles.

Figure P6–44: Specify HSS152×51×6.4 and two 12 mm × 50 mm plates.

Figure P6–45: Specify three C203×6.17 aluminum channels.

Figure P6–46: Specify two L203×102×12.7 steel angles with added 12 mm × 450 mm bottom plate and 20 mm × 250 mm web.

COMPUTER ASSIGNMENTS

1. For a generalized I-shape having equal top and bottom flanges similar to that shown in Figure P6–2, write a computer program or spreadsheet to compute the location of the horizontal centroidal axis, the total area, and the moment of inertia with respect to the horizontal centroidal axis for any set of actual dimensions to be input by the user.
2. For the generalized T-shape similar to that shown in Figure P6–4, write a computer program or spreadsheet to compute the location of the horizontal centroidal axis, the total area, and the moment of inertia with respect to the horizontal centroidal axis for any set of actual dimensions to be input by the user.
3. For the generalized I-shape similar to that shown in Figure P6–5, write a computer program or spreadsheet to compute the location of the horizontal centroidal axis, the total area, and the moment of inertia with respect to the horizontal centroidal axis for any set of actual dimensions to be input by the user.
4. For any generalized shape that can be subdivided into some number of rectangular components with horizontal axes, write a computer program or spreadsheet to compute the location of the horizontal centroidal axis, the total area, and the moment of inertia with respect to the horizontal centroidal axis for any set of actual dimensions to be input by the user. Use the parallel axis theorem.
5. For the generalized hat-section shape similar to that shown in Figure P6–11, write a computer program or spreadsheet to compute the location of the horizontal centroidal axis, the total area, and the moment of inertia with respect to the horizontal centroidal axis for any set of actual dimensions to be input by the user.
6. Given a set of standard dimension lumber, compute the area and moment of inertia with respect to the horizontal centroidal axis for the generalized box shape similar to that shown in Figure P6–22. The data for the lumber should be input by the user.
7. Enhance Assignment 6 by building a data file containing the dimensions of a set of standard dimension lumber. Then permit the user to select sizes for the top and bottom plates and the two vertical members for the box shape shown in Figure P6–22.
8. Write a computer program or spreadsheet to compute the area and moment of inertia with respect to the horizontal centroidal axis for a standard W- or S-beam shape with identical plates attached to both the top and bottom flanges similar to that shown in Figure 6–14. The data for the beam shape and the plates are to be input by the user.
9. Enhance Assignment 8 by building a database of standard W- or S-beam shapes. The user will select a beam shape. The data for the plates are to be input by the user.

10. Given a standard W- or S-beam shape and its properties, write a computer program or spreadsheet to compute the required thickness for plates to be attached to the top and bottom flanges to produce a specified moment of inertia of the composite section, as shown in Figure 6–14. Make the width of the plates equal to the width of the flange. For the resulting section, compute the total area.
11. Using the computer program or spreadsheet written for Assignment 1 for the generalized I-shape, perform an analysis of the area (A), the moment of inertia (I), and the ratio of I to A, as the thickness of the web is varied over a specified range. Keep all other dimensions for the shape the same. Note that the ratio of I to A is basically the same as the ratio of the stiffness of a beam having this shape to its weight, because the deflection of a beam is inversely proportional to the moment of inertia and the weight of the beam is proportional to its cross-sectional area.
12. Repeat Assignment 11 but vary the height of the section while keeping all other dimensions the same.
13. Repeat Assignment 11 but vary the thickness of the flange while keeping all other dimensions the same.
14. Repeat Assignment 11 but vary the width of the flange while keeping all other dimensions the same.
15. Write a computer program or spreadsheet to compute the moment of inertia with respect to the horizontal centroidal axis for any composite shape that can be divided into parts, all of which are rectangular having sides perpendicular and parallel to the horizontal axis, using the method described in Section 6–9. Produce output from the program for any of the shapes in Figures P6–1 through P6–15 and Figures P6–21 through P6–24.

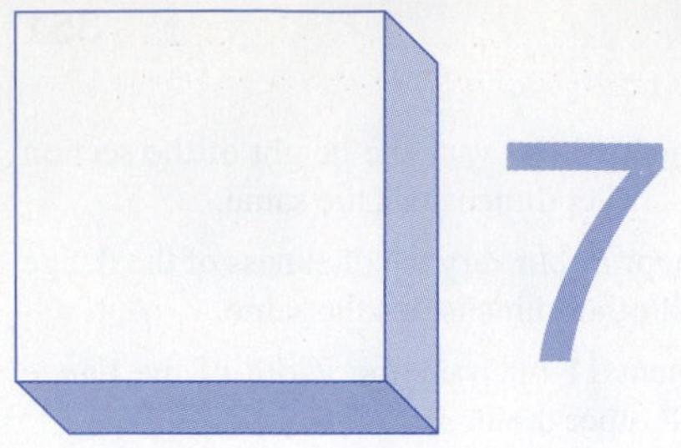

7 Stress Due to Bending

The Big Picture

Stress Due to Bending

Discussion Map

- Beams were defined in Chapter 5 to be load-carrying members on which the loads act perpendicular to the long axis of the beam.
- You have already learned how to determine the bending moment diagram for beams carrying a variety of load types.
- The stress due to bending in the beam is directly proportional to the bending moment.
- The properties of the cross-sectional area of the beam (overall dimensions, centroid location, area moment of inertia, and section modulus) play important roles in the magnitude of the bending stress. In general, an efficient beam cross section has a high area moment of inertia relative to its area and, therefore, it has a low weight.
- In this chapter you develop your ability to determine the stress due to bending and to design efficient beams.

Activity Chapter 7

Gather some materials that you can use to demonstrate the variety of ways that a beam can fail. Activities from earlier chapters, particularly Chapters 1, 5, and 6, called for collecting a similar set of sample beams, and some of them may be useful here. You will be installing them in a simple fixture that supports the beams at two points and that allows a load to be applied.

Alternatively, you could clamp the beam at one end to solid surface and load it as a cantilever. Figure 7–1 shows a commercially available loading frame that would serve well.

The beams that you gather should demonstrate some of the six failure modes illustrated in Figure 7–2, briefly described here. The beams will be destroyed during the tests, so use small, simple, inexpensive samples.

FIGURE 7–1 Beam loading frame. (Source: P. A. Hilton Ltd./ Hi-Tech, Hampshire, England)

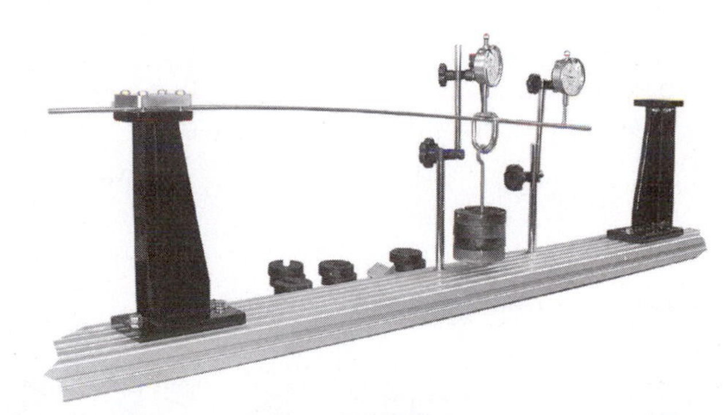

(*a*) Cantilever

(*b*) Simply supported beam

FIGURE 7–2 Failure modes for beams.

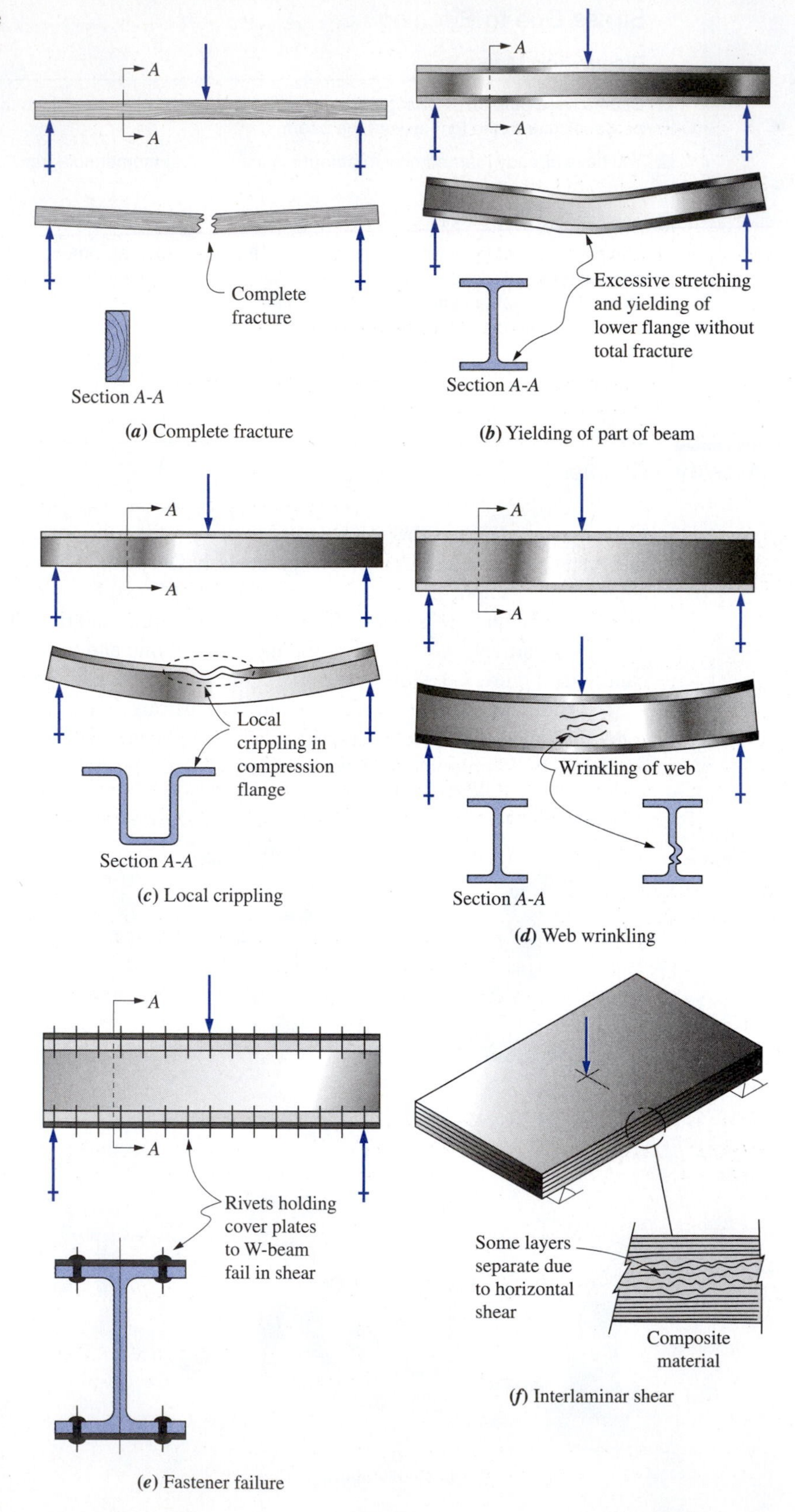

1. ***Complete fracture***—Figure 7–2(a): The material for the beam should be somewhat brittle and the cross section should be small enough to break under low to moderate loads. Chalk is a classic example of such a beam. You could also use a wood pencil, a small wood dowel, or small parts made from some alloys of aluminum, zinc, or magnesium, particularly those that are cast. The beam will break completely and usually suddenly.
2. ***Yielding of part of the beam***—Figure 7–2(b): The material should be ductile so that it will deform prior to any actual tearing of the material. Small metal rods with a variety of cross-sectional shapes, thin flat metal beams, and small plastic beams are examples. You should stop the loading after significant yielding is evident.
3. ***Local crippling***—Figure 7–2(c): This typically occurs when thin extended flanges of a beam are subjected to compressive stress as would be produced by a downward load on the top of a simply supported beam or on the bottom of a cantilever. You should look for a rolled thin sheet metal beam such as a curtain rod as an example.
4. ***Web wrinkling***—Figure 7–2(d): A relatively tall I-shape or box beam with thin vertical webs or sides may fail by wrinkling. This may be difficult to find, but rolled thin sheet metal shapes are good examples.
5. ***Fastener failure***—Figure 7–2(e): Fabricated beams built up from two or more parts may fail by shearing fasteners such as rivets, screws, bolts, pegs, or nails or by exceeding the shear strength of adhesives, brazed joints, or welds.
6. ***Interlaminar shear***—Figure 7–2(f): Components made from composite materials such as fiberglass, carbon/epoxy, or metal-matrix composites are often built up in layers impregnated with the matrix material that is then cured into a cohesive structure. An important failure mode for these materials is the separation of the layers under load such as by bending a panel or a beam. This is called *interlaminar shear.* Try to find such a beam, load it, and watch for the separation of the layers.

You will see during these demonstrations that some of the beam failures are, in fact, *bending failures,* which are discussed in this chapter, whereas some others are *shearing failures,* discussed in the following chapter. Bending and shearing stresses are both inherently present in any beam. Which type of failure actually occurs first depends on the loading pattern, the materials from which the beam is made, the shape of the cross section, and the actual dimensions of that shape.

7–1 OBJECTIVES OF THIS CHAPTER

After completing this chapter, you should be able to:

1. Learn the statement of the *flexure formula* and apply it properly to compute the maximum stress due to bending at the outer fibers of the beam.
2. Compute the stress at any point within the cross section of the beam and describe the variation of stress with position in the beam.
3. Understand the conditions on the use of the flexure formula.
4. Recognize that it is necessary to ensure that the beam does not twist under the influence of the bending loads.

5. Define the *neutral axis* and understand that it is coincident with the centroidal axis of the cross section of the beam.
6. Understand the derivation of the flexure formula and the effect of the moment of inertia on bending stress.
7. Determine the appropriate design stress for use in designing beams.
8. Design beams to carry a given loading safely.
9. Define the *section modulus* of the cross section of the beam.
10. Select standard structural shapes for use as beams.
11. Recognize when it is necessary to use stress concentration factors in the analysis of stress due to bending and apply appropriate factors properly.
12. Define the *flexural center* and describe its proper use in the analysis of stress due to bending.

7–2 THE FLEXURE FORMULA

Beams must be designed to be safe. When loads are applied perpendicular to the long axis of a beam, bending moments are developed inside the beam, causing it to bend. Observe a thin beam. The characteristically curved shape shown in Figure 7–3 is evident. The fibers of the beam near its top surface are shortened and placed in compression. Conversely, the fibers near the bottom surface are stretched and placed in tension.

Taking a short segment of the beam from Figure 7–3, we show in Figure 7–4 how the shape would change under the influence of the bending moments inside the beam. In part (a) the segment is in its initially straight form when it is not carrying a load. Part (b) shows the same segment as it is deformed by the application of the bending moments. Lines that were initially horizontal become curved. The ends of the segment, which were initially straight and vertical, remain straight. However, now they are inclined, having rotated about the centroidal axis of the cross section of the beam. The result is that the material along the top surface has been placed under compression and consequently shortened. Also, the material along the bottom surface has been placed under tension and has elongated.

In fact, all of the material above the centroidal axis is in compression. The maximum shortening (compressive strain) occurs at the top. Because stress is proportional to strain, it can be reasoned that the maximum compressive stress occurs at the top surface. Similarly, all of the material below the centroidal axis is in tension. The maximum elongation (tensile strain) occurs at the bottom, producing the maximum tensile stress.

We can also reason that, if the upper part of the beam is in compression and the lower part is in tension, then there must be some place in the beam where there is no strain at all. That place is called the *neutral axis* and it will be shown later that it is coincident with the *centroidal axis* of the beam. In summary, we can conclude that,

In a beam subjected to a bending moment of the type shown in Figure 7–4, material above the centroidal axis will be in compression with the maximum compressive stress occurring at the top surface.

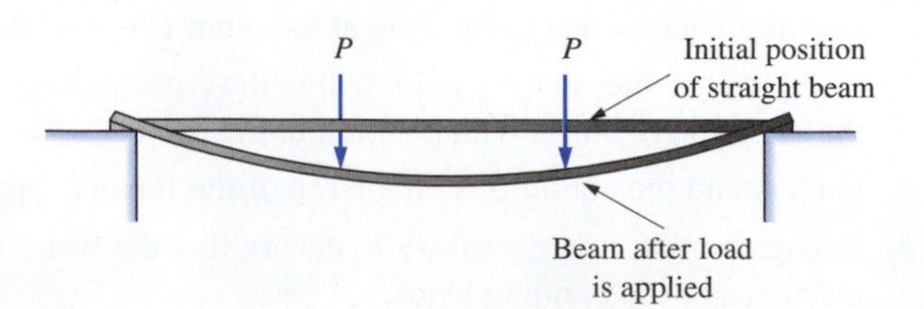

FIGURE 7–3
Example of a beam.

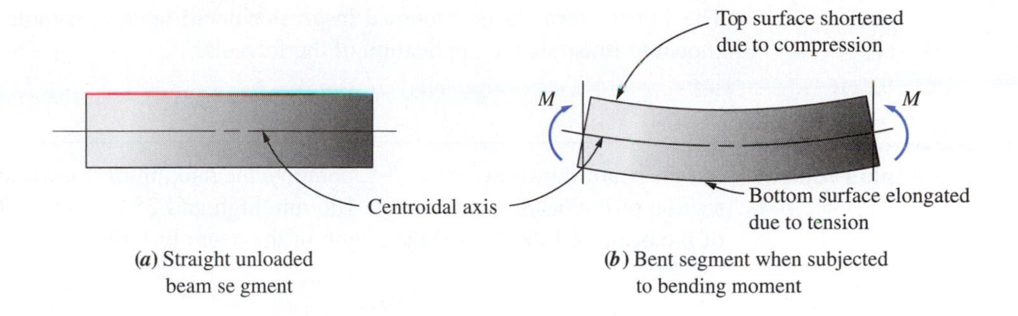

FIGURE 7–4 Influence of bending moment on beam segment.

Material below the centroidal axis will be in tension with the maximum tensile stress occurring at the bottom surface.

Along the centroidal axis itself, there is zero strain and zero stress due to bending. This is called the neutral axis.

In designing or analyzing beams, it is usually the objective to determine the maximum tensile and compressive stress. It can be concluded from this discussion that these maximums are dependent on the distance from the neutral axis (centroidal axis) to the top and bottom surfaces. We will call that distance, c.

The stress due to bending is also proportional to the magnitude of the bending moment applied to the section of interest. The shape and dimensions of the cross section of the beam determine its ability to withstand the applied bending moment. It will be shown later that the bending stress is inversely proportional to the area moment of inertia of the cross section with respect to its horizontal centroidal axis.

We now state the *flexure formula* which can be used to compute the maximum stress due to bending.

Flexure Formula

$$\sigma_{max} = \frac{Mc}{I} \tag{7–1}$$

where σ_{max} = maximum stress at the outermost fiber of the beam
M = bending moment at the section of interest
c = distance from the centroidal axis of the beam to the outermost fiber
I = area moment of inertia of the cross section with respect to its centroidal axis

The following steps are typically taken to apply Equation (7–1).

Guidelines for Applying the Flexure Formula

1. Determine the maximum bending moment on the beam by drawing the shearing force and bending moment diagrams.
2. Locate the centroid of the cross section of the beam.
3. Compute the moment of inertia of the area of the cross section with respect to its centroidal axis.
4. Compute the distance c from the centroidal axis to the top or bottom of the beam, whichever is greater.
5. Compute the stress from the flexure formula,

$$\sigma_{max} = \frac{Mc}{I}$$

The flexure formula is discussed in greater detail later. A sample problem will now be shown to illustrate the application of the formula.

Example Problem 7–1 For the beam shown in Figure 7–5, compute the maximum stress due to bending. The cross section of the beam is a rectangle 100 mm high and 25 mm wide. The load at the middle of the beam is 1500 N, and the length of the beam is 3.40 m.

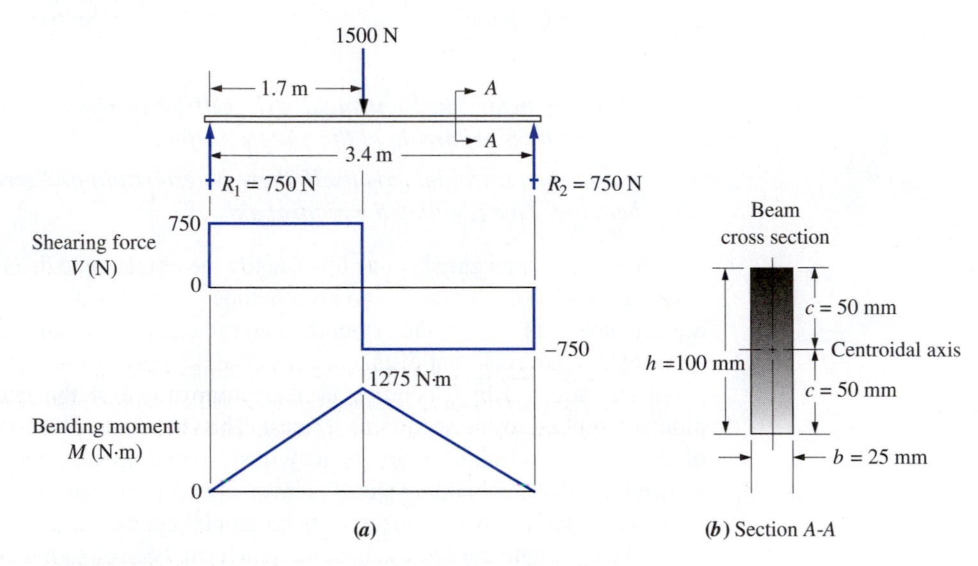

FIGURE 7–5 Beam data for Example Problem 7–1.

Solution

Objective: Compute the maximum stress due to bending.

Given: Beam and loading shown in Figure 7–5.

Analysis: The guidelines defined in this section will be used.

Results:

Step 1. The shearing force and bending moment diagrams have been drawn and included in Figure 7–5. The maximum bending moment is 1275 N·m at the middle of the beam.

Step 2. The centroid of the rectangular cross section is at the intersection of the two axes of symmetry, 50 mm from either the top or the bottom surface of the beam.

Step 3. The moment of inertia of the area for the rectangular shape with respect to the centroidal axis is

$$I = \frac{bh^3}{12} = \frac{25(100)^3}{12} = 2.08 \times 10^6\ \text{mm}^4$$

Step 4. The distance $c = 50$ mm from the centroidal axis to either the top or the bottom surface.

Step 5. The maximum stress due to bending occurs at the top or the bottom of the beam at the point of maximum bending moment. Applying Equation (7–1) gives

$$\sigma_{max} = \frac{Mc}{I} = \frac{(1275\ \text{N}\cdot\text{m})(50\ \text{mm})}{2.08 \times 10^6\ \text{mm}^4} \times \frac{10^3\ \text{mm}}{\text{m}}$$

$$\sigma_{max} = 30.6\ \text{N/mm}^2 = 30.6\ \text{MPa}$$

7–3 CONDITIONS ON THE USE OF THE FLEXURE FORMULA

The proper application of the flexure formula requires the understanding of the conditions under which it is valid, listed as follows:

1. The beam must be straight or very nearly so.
2. The cross section of the beam must be uniform.
3. All loads and support reactions must act perpendicular to the axis of the beam.
4. The beam must not twist while the loads are being applied.
5. The beam must be relatively long and narrow in proportion to its depth.
6. The material from which the beam is made must be homogeneous, and it must have an equal modulus of elasticity in tension and compression.
7. The stress resulting from the loading must not exceed the proportional limit of the material.
8. No part of the beam may fail from instability, that is, from the buckling or crippling of thin sections.
9. The section where the stress is to be computed must not be close to the point of application of concentrated loads.

Although the list of conditions appears to be long, the flexure formula still applies to a wide variety of real cases. Beams violating some of the conditions can be analyzed by using a modified formula or by using a combined stress approach. For example, for condition 2, a change in cross section will cause stress concentrations, which can be handled as described in Section 7–9.

The combined bending and axial stress or bending and torsional stress produced by violating condition 3 are discussed in Chapter 10. If the other conditions are violated, special analyses are required, which are not covered in this book.

Condition 4 is important, and attention must be paid to the shape of the cross section to ensure that twisting does not occur. In general, if the beam has a vertical axis of symmetry and if the loads are applied through that axis, no twisting will result. Figure 7–6 shows some typical shapes used for beams that satisfy condition 4. Conversely, Figure 7–7 shows several that do not satisfy condition 4. In each of these cases, the beam would tend to twist as well as bend as the load is applied in the manner shown. Of course, these sections can

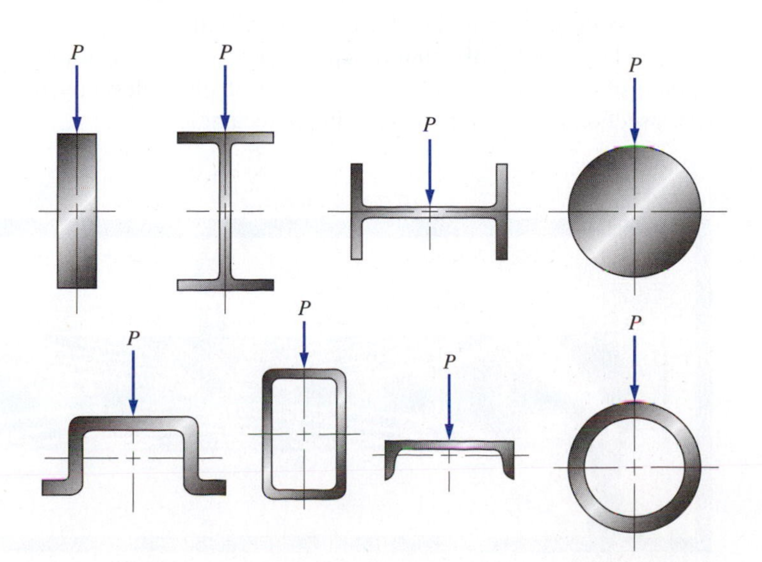

FIGURE 7–6 Example beam shapes with loads acting through an axis of symmetry.

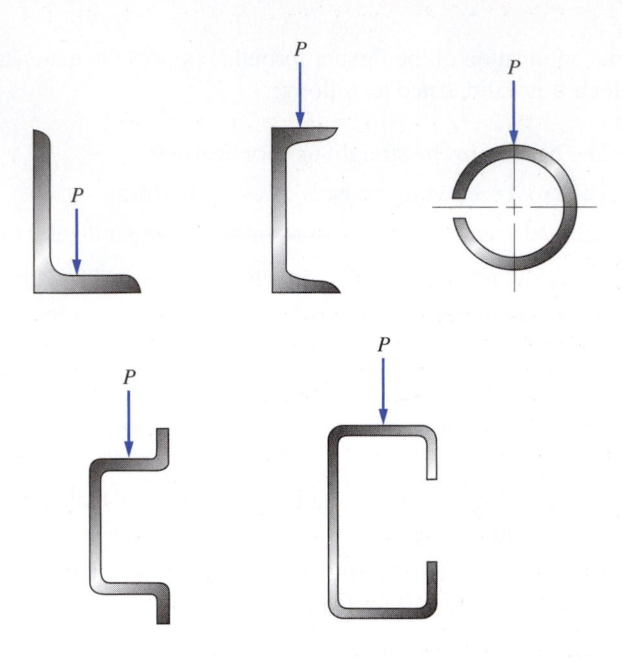

FIGURE 7–7 Example beam shapes with loads not acting through an axis of symmetry, resulting in twisting of the beam.

support some load, but the actual stress condition in them is different from that which would be predicted from the flexure formula. More about these kinds of beams is presented in Section 7–10.

Condition 5 is difficult to quantify in general because of the variety of cross-sectional shapes and support conditions used for beams. Some designers define a relatively long beam as one for which the ratio of length to depth is greater than about 10.

Condition 8 is important because long, thin members and, sometimes, thin sections of members tend to buckle at stress levels well below the yield strength of the material. Such failures are called *instability* and are to be avoided. Frequently, cross braces or local stiffeners are added to beams to relieve the problem of instability. An example can be seen in the wood joist floor construction of many homes and commercial buildings. The relatively slender wood joists are braced near the midpoints to avoid buckling. See References 1–4 and 16 for additional analysis of long slender beams or thin sections of beams.

The need for condition 9 can be visualized by referring to Figure 7–8, which shows a photoelastic model of a simply supported beam carrying a single concentrated load at the middle of the span. Recall the discussion of photoelasticity from Chapter 1. The beam cross section in this example is a simple rectangle.

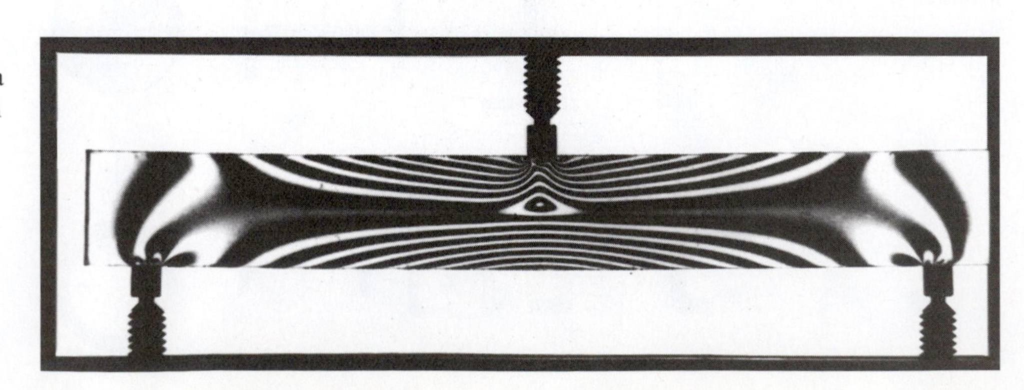

FIGURE 7–8 Photoelastic model of a simply supported beam with a concentrated load. (Source: Measurements Group, Inc., Raleigh, NC)

First, look at the beam between either support and the point of application of the load. Notice how the black and white fringe lines are smooth and approximately evenly spaced from top to bottom on the beam. Also, note that there are more fringe lines near the middle of the span and fewer as you approach the loads. The greater number of fringe lines indicates a higher stress gradient. This demonstrates that the bending stress is proportional to the bending moment that would be maximum under the load and zero at the supports for this loading pattern.

Now look at any cross section of the beam. The next section shows that the stress distribution on a cross section of such a beam has the following characteristics:

- It is zero at the centroidal axis of the cross section of the beam, half way from the top or bottom of the beam.
- It is a maximum tensile stress at the bottom surface.
- It is a maximum compressive stress at the top surface.
- It varies linearly from the top to the bottom.

Starting at the midpoint of the cross section and counting the number of fringes from there to the top or bottom surface would allow you to determine the stress level at that section if you knew the characteristics of the photoelastic material.

Now look at the parts of the beam that are close to the point of application of the load or near the supports. You can see a more complex fringe pattern that indicates high local stresses that result from the concentrated force. You must be sure that the material of the beam can withstand these local stresses by using relatively large bearing areas at these points. Notice that these local effects near the supports dissipate quickly as you move away from them. This is referred to as the *principle of St. Venant,* in recognition of the work this renowned French scientist reported in 1855.

7–4 STRESS DISTRIBUTION ON A CROSS SECTION OF A BEAM

Refer again to Figure 7–4 showing the manner in which a segment of a beam deforms under the influence of a bending moment. The segment assumes the characteristic "bent" shape as the upper fibers are shortened and the lower fibers are elongated. The neutral axis, coincident with the centroidal axis of the cross section of the beam, bends but it is not strained. Therefore, at the neutral axis the stress due to bending is zero.

Figure 7–4 also shows that the ends of the beam segment, which were initially straight and vertical, remain straight. However, they rotate as the bending moment is applied. The linear distance from a point on the initial vertical end line to the correspon-ding point on the rotated end line is an indication of the amount of strain produced at that point in the cross section. It can be reasoned, therefore, that there is a linear variation of strain with position in the cross section as a function of the distance away from the neutral axis. Moving from the neutral axis toward the top of the section results in greater com-pressive strain while moving downward toward the bottom results in greater tensile strain. For materials following Hooke's law, stress is proportional to strain. The resulting stress distribution, then, is as shown in Figure 7–9.

To represent the stress at some point within the cross section, we can express it in terms of the maximum stress by noting the linear variation of stress with distance away from the neutral axis. Calling that distance y, we can write an equation for the stress, σ, at any point as,

$$\sigma = \sigma_{\text{max}} \frac{y}{c} \quad \textbf{(7–2)}$$

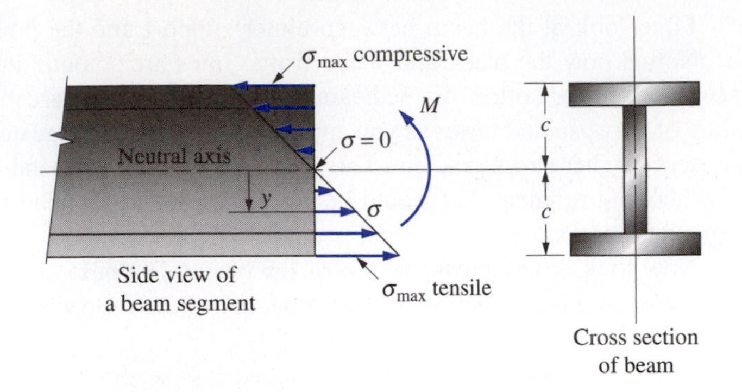

FIGURE 7–9 Bending stress distribution on a symmetrical section.

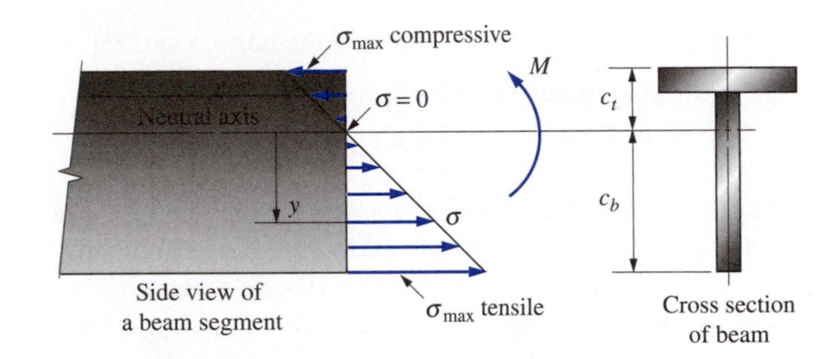

FIGURE 7–10 Bending stress distribution on a nonsymmetrical section.

The general form of the stress distribution shown in Figure 7–9 would occur in any beam section having a centroidal axis equidistant from the top and bottom surfaces. For such cases, the magnitude of the maximum compressive stress would equal the maximum tensile stress.

If the centroidal axis of the section is not the same distance from both the top and bottom surfaces, the stress distribution shown in Figure 7–10 would occur. Still the stress at the neutral axis would be zero. Still the stress would vary linearly with distance from the neutral axis. Now the maximum stress at the bottom of the section is greater than that at the top because it is farther from the neutral axis. Using the distances c_b and c_t as indicated in Figure 7–10, the stresses would be

$$\sigma_{max} = \frac{Mc_b}{I} \quad \text{(tension at the bottom)}$$

$$\sigma_{max} = \frac{Mc_t}{I} \quad \text{(compression at the top)}$$

7–5 DERIVATION OF THE FLEXURE FORMULA

We can better understand the basis for the flexure formula by following the analysis used to derive it. The principles of static equilibrium are used here to show two concepts that were introduced earlier in this chapter but that were stated without proof. One is that the *neutral axis* is coincident with the *centroidal axis* of the cross section. The second is the flexure formula itself and the significance of the moment of inertia of the area of the cross section.

Refer to Figure 7–9, which shows the distribution of stress over the cross section of a beam. The shape of the cross section is not relevant to the analysis and the I-shape is shown merely for example. The figure shows a portion of a beam, cut at some arbitrary section, with an internal bending moment acting on the section. The stresses, some tensile and some compressive, would tend to produce forces on the cut section in the axial direction. Equilibrium requires that the net sum of these forces must be zero. In general, force equals stress times area. Because the stress varies with position on the cross section, it is necessary to look at the force on any small elemental area, dA, and then sum these forces over the entire area using the process of integration. These concepts can be shown analytically as:

Equilibrium condition: $\sum F = 0$

Force on any element of area: $dA = \sigma\, dA$

Total force on the cross-sectional area:

$$\sum F = \int_A \sigma\, dA = 0 \tag{7–3}$$

Now we can express the stress σ at any point in terms of the maximum stress by using Equation (7–2):

$$\sigma = \sigma_{max}\frac{y}{c}$$

where y is the distance from the neutral axis to the point where the stress is equal to σ. Substituting this into Equation (7–3) gives

$$\sum F = \int_A \sigma\, dA = \int_A \sigma_{max}\frac{y}{c}\, dA = 0$$

Because σ_{max} and c are constants, they can be taken outside the integral sign.

$$\sum F = \frac{\sigma_{max}}{c}\int_A y\, dA = 0$$

Neither σ_{max} nor c is zero, so the other factor, $\int_A y\, dA$, must be zero. By definition and as illustrated in Chapter 6,

$$\int_A y\, dA = \overline{Y}(A)$$

where $\overline{Y}$ is the distance to the centroid of the area from the reference axis and A is the total area. Again, A cannot be zero, so, finally, it must be true that $\overline{Y} = 0$. Because the reference axis is the neutral axis, this shows that the neutral axis is coincident with the centroidal axis of the cross section.

The derivation of the flexure formula is based on the principle of equilibrium, which requires that the sum of the moments about any point must be zero. Figure 7–9 shows that a bending moment M acts at the cut section. This must be balanced by the net moment created by the stress on the cross section. Moment is the product of force times the distance from the reference axis to the line of action of the force. As used above,

$$\sum F = \int_A \sigma\, dA = \int_A \sigma_{max}\frac{y}{c}\, dA$$

Multiplying this by distance y gives the resultant moment of the force that must be equal to the internal bending moment M. That is,

$$M = \sum F(y) = \int_A \underbrace{\underbrace{\sigma_{max} \frac{y}{c}}_{\text{stress}} \underbrace{dA}_{\text{area}}}_{\text{force}} \underbrace{(y)}_{\text{moment arm}}$$

Simplifying, we obtain

$$M = \frac{\sigma_{max}}{c} \int_A y^2 \, dA$$

By definition, and as illustrated in Chapter 6, the last term in this equation is the moment of inertia I of the area of the cross section with respect to its centroidal axis.

$$I = \int_A y^2 \, dA$$

Then

$$M = \frac{\sigma_{max}}{c} I$$

Solving for σ_{max} yields

$$\sigma_{max} = \frac{Mc}{I}$$

This is the form of the flexure formula shown earlier as Equation (7–1).

Two example problems are shown next illustrating the use of the flexure formula.

7–6 APPLICATIONS—BEAM ANALYSIS

Example Problem 7–2

The tee section shown in Figure 7–11 is from a simply supported beam that carries a bending moment of 100 000 lb·in due to a load on the top surface. It has been determined that $I = 18.16 \text{ in}^4$. The centroid of the section is 3.25 in up from the bottom of the beam. Compute the stress due to bending in the beam at the six axes a to f indicated in the figure. Then plot a graph of stress versus position in the cross section.

Solution

Objective Compute the bending stress at six axes a–f. Plot a graph of stress versus position in the cross section.

Given $M = 100\ 000$ lb·in. Tee shape of cross section shown in Figure 7–11.
$I = 18.16 \text{ in}^4$. $\bar{Y} = 3.25$ in from bottom of beam.

Analysis Equation (7–1) will be used to compute σ_{max}, which occurs at the bottom of the beam (axis a), because that is the location of the outermost fiber of the beam, farthest from the cen-

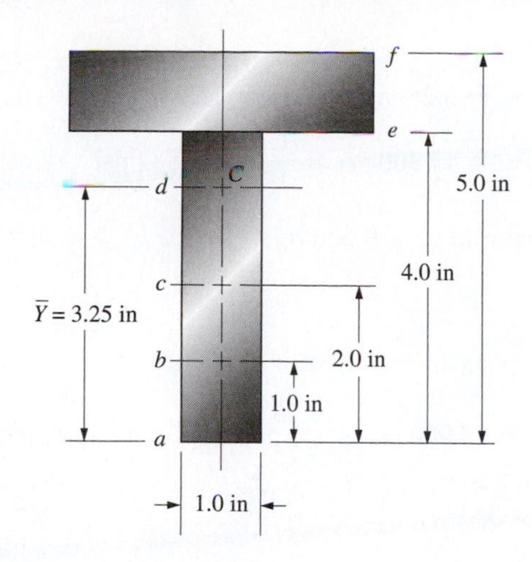

FIGURE 7–11 Tee section for beam in Example Problem 7–2.

troidal axis. Then the stress at other axes will be computed using Equation (7–2), giving results to four significant figures to demonstrate the principle. See Figure 7–12 for values of *y*.

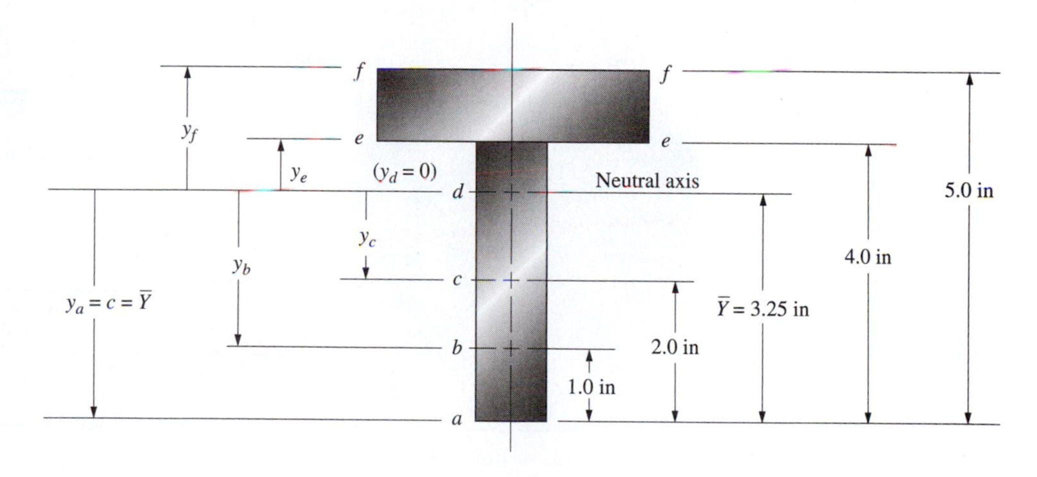

FIGURE 7–12 Data for Example Problem 7–2.

Results **_At Axis a._** In Equation (7–1), use $c = \bar{Y} = 3.25$ in.

$$\sigma_{\max} = \sigma_a = \frac{Mc}{I} = \frac{(100\,000\ \text{lb}\cdot\text{in})(3.25\ \text{in})}{18.16\ \text{in}^4}$$

$$\sigma_a = 17\,900\ \text{psi} \quad \text{(tension)}$$

At Axis b.

$$\sigma_b = \sigma_{\max}\frac{y_b}{c} = \sigma_a\frac{y_b}{c}$$

$$y_b = 3.25\ \text{in} - 1.0\ \text{in} = 2.25\ \text{in}$$

$$\sigma_b = 17\,900\ \text{psi} \times \frac{2.25}{3.25} = 12\,390\ \text{psi} \quad \text{(tension)}$$

At Axis c.

$$y_c = 3.25 \text{ in} - 2.0 \text{ in} = 1.25 \text{ in}$$

$$\sigma_c = 17\,900 \text{ psi} \times \frac{1.25}{3.25} = 6883 \text{ psi} \quad \text{(tension)}$$

At Axis d. At the centroid $y_d = 0$ and $\sigma_d = 0$.

At Axis e.

$$y_e = 4.0 \text{ in} - 3.25 \text{ in} = 0.75 \text{ in}$$

$$\sigma_e = 17\,900 \text{ psi} \times \frac{0.75}{3.25} = 4130 \text{ psi} \quad \text{(compression)}$$

At Axis f.

$$y_f = 5.0 \text{ in} - 3.25 \text{ in} = 1.75 \text{ in}$$

$$\sigma_f = 17\,900 \text{ psi} \times \frac{1.75}{3.25} = 9637 \text{ psi} \quad \text{(compression)}$$

The graph of these data is shown in Figure 7–13.

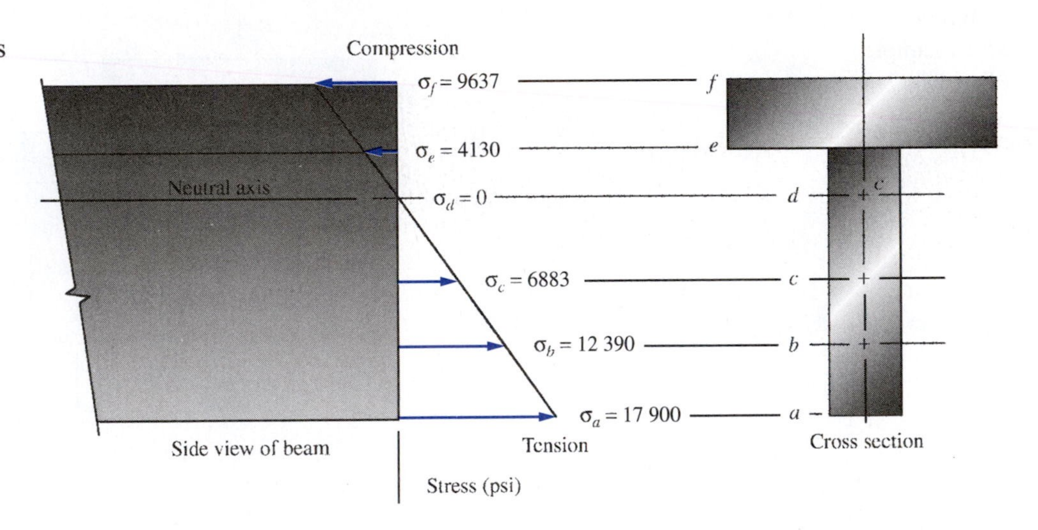

FIGURE 7–13 Stress distribution on the tee section for Example Problem 7–2.

LESSON 10

Comment Notice the linear variation of stress with distance from the neutral axis and that the stresses above the neutral axis are compressive while those below are tensile.

Example Problem 7–3 Figure 7–14 shows the bending moment diagram for a 25-ft long beam in a large machine structure. It has been proposed that the beam be made from a standard W14×43 steel shape. Compute the maximum stress due to bending in the beam.

Solution Objective Compute the maximum stress due to bending.

Given Bending moment diagram shown in Figure 7–14. W14×43 beam shape.

FIGURE 7–14 Bending moment diagram for beam in Example Problem 7–3.

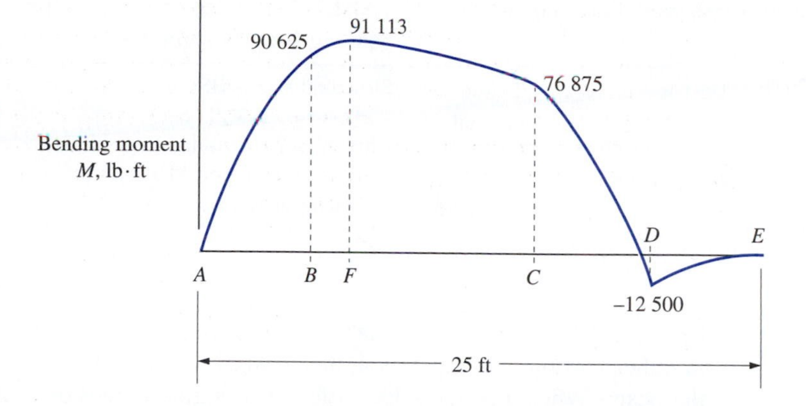

Analysis Use Equation (7–1). In Figure 7–14, identify the maximum bending moment on the beam to be 91 113 lb·ft at point F. Find the values of I and c from the table of properties for W-beam shapes in Appendix A–7.

$$I = 428\ \text{in}^4$$

$$c = \text{depth}/2 = 13.70\ \text{in}/2 = 6.85\ \text{in.}$$

Results

$$\sigma_{max} = \frac{Mc}{I} = 91\,113\ \text{lb}\cdot\text{ft} \times \frac{12\ \text{in}}{\text{ft}} \times \frac{6.85\ \text{in}}{428\ \text{in}^4} = 17\,500\ \text{psi}$$

Comment This maximum stress would occur as a tensile stress on the bottom surface of the beam and as a compressive stress on the top surface at the position F. Because this is a relatively long beam, it must be laterally braced as described in Reference 2.

7–7 APPLICATIONS—BEAM DESIGN AND DESIGN STRESSES

To design a beam, its material, length, placement of loads, placement of supports, and the size and shape of its cross section must be specified. Normally, the length, placement of loads, and placement of supports are given by the requirements of the application. The material specification and the size and shape of the cross section are determined by the designer.

The primary duty of the designer is to ensure the safety of the design. This requires a stress analysis of the beam and a decision concerning the allowable or design stress to which the chosen material may be subjected. The examples presented here will concentrate on these items. Also of interest to the designer are cost, appearance, physical size, weight, compatibility of the design with other components of the machine or structure, and the availability of the material or beam shape.

Two basic approaches will be shown for beam design. One involves the specification of the *material* from which the beam will be made and the general *shape* of the beam (circular, rectangular, W-beam, etc.), with the subsequent determination of the required dimensions of the cross section of the beam. The second requires specifying the *dimensions* and *shape* of the beam and then computing the required strength of a material from which to make the beam. Then, the actual material is specified.

Design Stress for Metals—General Guidelines. When specifying design stresses, it is important to keep in mind that both tensile and compressive stresses are produced in

TABLE 7–1 Design stress guidelines—bending stresses.

Manner of loading	Ductile material	Brittle material
Static	$\sigma_d = s_y/2$	$\sigma_d = s_u/6$
Repeated	$\sigma_d = s_u/8$	$\sigma_d = s_u/10$
Impact or shock	$\sigma_d = s_u/12$	$\sigma_d = s_u/15$

TABLE 7–2 Design stresses from selected codes—bending stresses-static loads on building-like structures.

Structural steel (AISC):
$\sigma_d = s_y/1.5 = 0.66\ s_y$
Aluminum (Aluminum Association):
$\sigma_d = s_y/1.65 = 0.61\ s_y$ or $\sigma_d = s_u/1.95 = 0.51\ s_u$
whichever is lower

beams. If the material is reasonably homogeneous and isotropic having the same strength in either tension or compression, then design is based on the highest stress developed in the beam. When a material has different strengths in tension and compression, as is the case for cast iron or wood, then both the maximum tensile and the maximum compressive stresses must be checked.

The approach used most often in this book to determine design stresses is similar to that first described in Chapter 3, and it would be wise to review that discussion at this time. Table 7–1 lists the design stress guidelines we will use for beams in machines and special structures under conditions where loads and material properties are well known. Larger design factors may be used where greater uncertainty exists. We will use Table 7–1 for problems in this book involving metals, unless stated otherwise.

Design Stresses from Selected Codes. Table 7–2 gives a summary of specifications for design bending stresses as defined by the American Institute of Steel Construction (AISC) for structural steel and by the Aluminum Association (AA) for aluminum alloys. These data pertain to beams under static loads such as those found in building-type structures, and use the allowable stress design method.

Additional analysis is required for the parts of beams under compressive stresses because of the possibility of local buckling, especially in shapes having thin sections or extended flanges. Long beams must also be checked for the possibility of twisting. Lateral supports for the compression flanges of long beams are often required to resist the tendency for the beam to twist. See References 1 and 2 for more detailed discussions of these specifications. Wide-flange beam shapes (W-shapes) should be checked to ensure that they are compact as defined in Reference 2. Noncompact sections require the use of a lower design stress.

Load and Resistance Factor Design (LRFD). LRFD was first discussed in Chapter 3 in the context of design of axially loaded members. A major objective of LRFD is to account for the probability that the maximum dead load and the maximum live load would occur at the same time over the life of the member. As a result, loading on beams must be separated carefully between that portion that is dead load (primarily the weight of the structure itself) and that which is live load (occupancy, snow, rain, wind, and seismic). Each type is treated with different load factors. Factors are also applied to the basic capacity of the beam, typically bending moment. Then the factored loads are compared with the factored resistance factor to ensure the safety of the member. The LRFD beam design process also includes organized steps to check for the possibility of local buckling of the flanges, crippling of the web, the need for lateral bracing, and details of the beam near support points and where concentrated loads are applied. Details of this process are described in References 2 and 3 and are typically taught in following courses such as structural steel design.

Beam Deflection. Failure of a beam to perform to its expected level may occur because of excessive deflection even though the beam has not failed any strength criterion. This topic will be discussed in Chapter 9. At this point, we will only mention that beam deflection limits are often stated in terms of a proportion of the span (L) of the beam between supports. For example, some applications limit deflection to $L/180$, $L/240$, or $L/360$, depending on the desired rigidity of the structure. You will learn from Chapter 9 that beam deflections are, in general, inversely proportional to the moment of inertia (I) of the area of the cross section of the beam. Equation 7–1 shows that the stress is also dependent on I as well as the depth of the beam as it relates to the distance c from the centroidal axis to the outside of the beam. Therefore, you should consider beam design based on strength to be tentative until the deflection can be evaluated.

Design Stresses for Nonmetals. When problems involve nonmetals such as wood, plastics, and composites, the concept of yield strength is not typically used. Furthermore, the strengths listed in tables are often based on statistical averages of many tests. Variations in material composition and structure can lead to variations in strength properties. Whenever possible, the actual material to be used in a structure should be tested to determine its strength.

Appendix A–19 lists *allowable* stress values for three species of wood according to the listed grades for applications in building structures and similar static load uses. If load conditions are very well known, a beam can be loaded up to the listed bending stress values. To the extent that there are uncertainties in the conditions of loading, design factors can be applied to the listed values resulting in lower design stresses. No firm guidelines are given here and testing is advised. We will use the listed allowable stresses unless otherwise stated. See also Reference 7.

The properties of plastics listed in Appendix A–20 can be considered typical for the listed types. It should be noted that there are many variables involved in producing plastics, and it is important to obtain more complete data from manufacturers or by testing the actual material to be used. Also, plastics differ dramatically from one to another in their ability to withstand repeated loading, shock, and impact. In this chapter, we will take the *flexural strength* from Appendix A–20 to be the representative strength of the listed plastics when used in beams. We will assume that failure is imminent at these stress levels. For general, static load cases, we will apply a design factor of $N = 2$ to those values to determine the design stress.

Composites have many advantages when applied to the design of beams because the placement of material can be optimized to provide efficient, lightweight beams. The resulting structure is typically not homogeneous, so the properties are highly anisotropic. Therefore, the flexure formula as stated in Equations (7–1) and (7–2) cannot be relied upon to give accurate values of stress. General approaches to using composites in beams will be discussed later in this chapter.

7–8 SECTION MODULUS AND DESIGN PROCEDURES

The stress analysis will require the use of the flexure formula

$$\sigma_{max} = \frac{Mc}{I}$$

A modified form is desirable for cases in which the determination of the dimensions of a section is to be done. Notice that both the moment of inertia I and the distance c are geometrical properties of the cross-sectional area of the beam. Therefore, the quotient I/c is also a geometrical property. For convenience, we can define a new term, *section modulus*, denoted by the letter S.

$$S = \frac{I}{c} \tag{7–4}$$

The flexure formula then becomes

$$\sigma_{max} = \frac{M}{S} \tag{7–5}$$

This is the most convenient form for use in design. Example problems will demonstrate the use of the section modulus. Appendix A–1 gives formulas for S for some shapes. Appendices A–4 through A–13 include the value of S for structural shapes.

Design Procedures. Here are two approaches to design problems. The first applies when the loading pattern and material are known and the shape and dimensions of the cross section of the beam are to be determined. The second procedure applies when the loading pattern, the shape of the beam cross section, and its dimensions have been specified and the objective is to specify a suitable material for the beam to ensure safety.

A. **Design procedure to determine the required dimensions for a beam.**
Given: Loading pattern and material from which the beam is to be made.
1. Determine the maximum bending moment in the beam, typically by drawing the complete shearing force and bending moment diagrams.
2. Determine the applicable approach for specifying the design stress from Section 7–7.
3. Compute the value of the design stress.
4. Using the flexure formula expressed in terms of the section modulus, Equation (7–5), solve for the section modulus, S. Then let the maximum stress equal the design stress and compute the required minimum value of the section modulus to limit the actual stress to no more than the design stress.
5. For a specially designed beam shape, determine the required minimum dimensions of the shape to achieve the required section modulus. Then specify the next larger convenient dimensions using the tables of preferred sizes in Appendix A–2.
6. To select a standard structural shape such as those listed in Appendices A–4 through A–13, consult the appropriate table of data and specify one having at least the value of the section modulus, S, computed in step 4. Typically, it is recommended that the lightest suitable beam shape be specified because the cost of the beam made from a given material is generally directly related to its weight. Reference 3 includes extensive tables for beam shapes with their section modulus values ordered by the weight of the section to facilitate the selection of the lightest beam. Where space limitations exist, the actual dimensions of the shape must be considered.

B. **Design procedure for specifying a material for a given beam.**
Given: Loading pattern, shape, and dimensions for the beam.
1. Determine the maximum bending moment in the beam, typically by drawing the complete shearing force and bending moment diagrams.
2. Compute the section modulus for the beam cross section.

3. Compute the maximum bending stress from the flexure formula, Equation (7–5).
4. Determine the applicable approach for specifying the design stress from Section 7–7 and specify a suitable design factor.
5. Set the computed maximum stress from step 3 equal to the formula for the design stress.
6. Solve for the required minimum value of the strength of the material, either s_y or s_u.
7. Select the type of material from which the beam is to be made such as steel, aluminum, cast iron, titanium, or copper.
8. Consult the tables of data for material properties such as those in Appendixes A–14 through A–20, and identify a set of candidate materials having at least the required strength.
9. Specify the material to be used, considering any factor appropriate to the application, such as ductility, cost, corrosion potential, ease of fabrication, or weight. For metals, it is essential to specify the condition of the material in addition to the alloy.

Example Problems 7–4 through 7–6 illustrate these approaches.

Example Problem 7–4

A beam is to be designed to carry the static loads shown in Figure 7–15. The cross section of the beam will be rectangular and made from ASTM A36 structural steel plate having a thickness of 1.25 in. Specify a suitable height for the cross section.

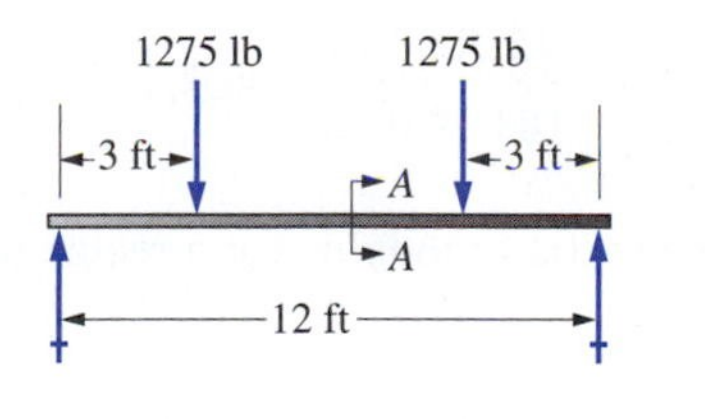

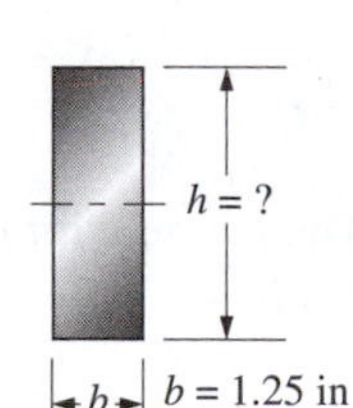

Cross section of beam—Section *A-A*

FIGURE 7–15 Loading and cross section for beam in Example Problem 7–4.

Solution

Objective: Specify the height of the rectangular cross section.

Given: Loading pattern shown in Figure 7–15. ASTM A36 structural steel. Width of the beam to be 1.25 in. Static loads.

Analysis: We will use the design procedure *A* from this section.

Results: ***Step 1.*** Figure 7–16 shows the completed shearing force and bending moment diagrams. The maximum bending moment is 45 900 lb·in between the loads, in the middle of the beam span from $x = 3.0$ ft to 9 ft.

Step 2. From Table 7–1, for static load on a ductile material,

$$\sigma_d = S_y/2$$

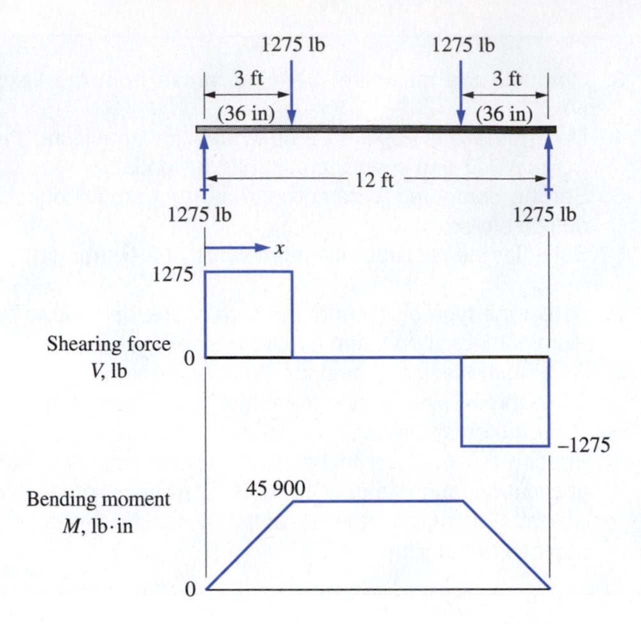

FIGURE 7–16 Load, shearing force, and bending moment diagrams for Example Problem 7–4.

Step 3. From Appendix A–16, $s_y = 36\,000$ psi for ASTM A36 steel. For a static load, a design factor of $N = 2$ based on yield strength is reasonable. Then

$$\sigma_d = \frac{s_y}{N} = \frac{36\,000 \text{ psi}}{2} = 18\,000 \text{ psi}$$

Step 4. The required S is

$$S = \frac{M}{\sigma_d} = \frac{45\,900 \text{ lb}\cdot\text{in}}{18\,000 \text{ lb/in}^2} = 2.55 \text{ in}^3$$

Step 5. The formula for the section modulus for a rectangular section with a height h and a thickness b is

$$S = \frac{I}{c} = \frac{bh^3}{12(h/2)} = \frac{bh^2}{6}$$

For the beam in this design problem, b will be 1.25 in. Then solving for h gives

$$S = \frac{bh^2}{6}$$

$$h = \sqrt{\frac{6S}{b}} = \sqrt{\frac{6(2.55 \text{ in}^3)}{1.25 \text{ in}}}$$

$$h = 3.50 \text{ in}$$

Comment The computed minimum value for h is a convenient size. Specify $h = 3.50$ in. The beam will then be rectangular in shape, with dimensions of 1.25 in × 3.50 in. Note that because the beam is rather long, 12 ft, there may be a tendency for it to deform laterally because of elastic instability. Lateral bracing may be required. Also, deflection should be checked using the methods discussed in Chapter 9.

Example Problem 7–5 The roof of an industrial building is to be supported by wide-flange beams spaced 4 ft on centers across a 20-ft span, as sketched in Figure 7–17 .The roof will be a poured concrete slab, 4 in thick. The design live load on the roof is 200 lb/ft^2. Specify a suitable wide-flange beam that will limit the stress in the beam to the design stress for ASTM A36 structural steel using the AISC specification.

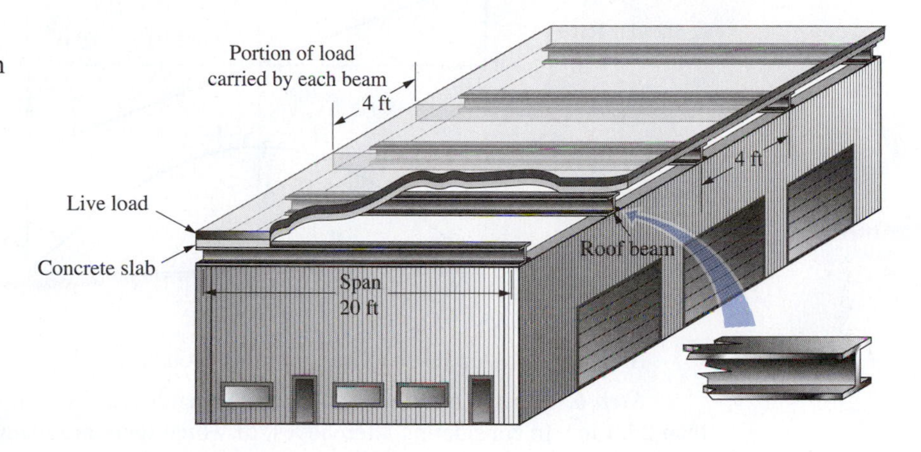

FIGURE 7–17 Roof structure for building in Example Problem 7–5.

Solution Objective Specify a suitable wide-flange beam.

Given Loading pattern in Figure 7–17. Design stress from AISC specifications for ASTM A36 structural steel.

Analysis Design procedure A from this section will be used.

Results ***Step 1.*** We must first determine the load on each beam of the roof structure. Dividing the load evenly among adjacent beams would result in each beam carrying a 4-ft-wide portion of the roof load. In addition to the 200-lb/ft^2 live load, the weight of the concrete slab offers a sizeable load. In Section 2–10 we find that the concrete weighs 150 lb/ft^3. Then each square foot of the roof, 4.0 in thick, would weigh 50 lb. This is called the *dead load*. Then the total loading due to the roof is 250 lb/ft^2. Now, notice that each foot of length of the beam carries 4 ft^2 of the roof. Therefore, the load on the beam is a uniformly distributed load of 1000 lb/ft. Figure 7–18 shows the loaded beam and the shearing force and bending moment diagrams. The maximum bending moment is 50000 lb·ft.

Steps 2 and 3. Table 7–2 calls for the design stress to be

$$\sigma_d = 0.66\, s_y$$

From Appendix A–16, the yield strength of ASTM A36 structural steel is 36 000 psi. Then

$$\sigma_d = 0.66\, s_y = 0.66(36\,000\text{ psi}) = 23\,760\text{ psi}$$

Step 4. In order to select a wide-flange beam, the required section modulus must be calculated.

$$\sigma = \frac{M}{S}$$

$$S = \frac{M}{\sigma_d} = \frac{50\,000\text{ lb}\cdot\text{ft}}{23\,760\text{ lb/in}^2} \times \frac{12\text{ in}}{\text{ft}} = 25.3\text{ in}^3$$

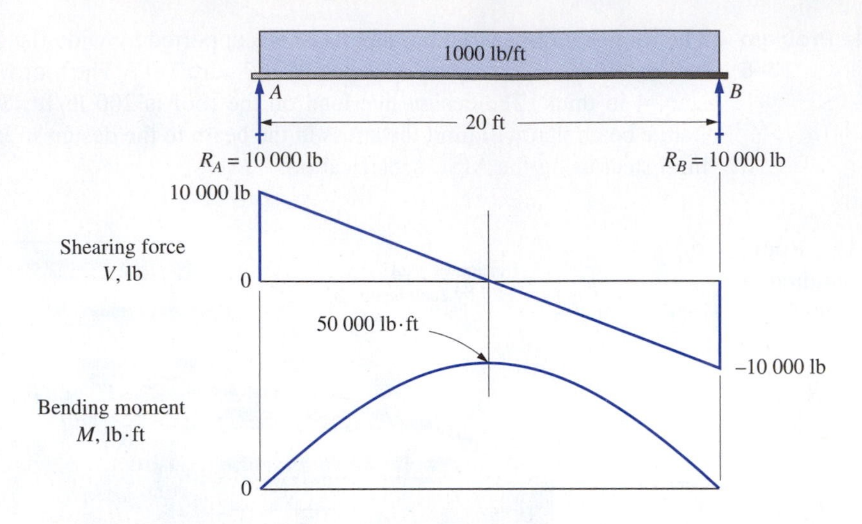

FIGURE 7–18 Load, shearing force, and bending moment diagrams for Example Problem 7–5.

Step 5. This step does not apply in this problem.

Step 6. A beam must be found from Appendix A–7 which has a value of S greater than 25.3 in^3. In considering alternatives, of which there are many, you should search for the lightest beam that will be safe, since the cost of the beam is based on its weight. Some possible beams are

$$W14\times26: \quad S = 35.3\ \text{in}^3, 26\ \text{lb/ft}$$
$$W12\times30: \quad S = 38.6\ \text{in}^3, 30\ \text{lb/ft}$$

Of these, the W14×26 would be preferred because it is the lightest.

Comment In the final design of the structure, lateral bracing would be required as defined in Reference 2. Also, deflection of the beam should be checked using the methods described in Chapter 9.

Example Problem 7–6 A support beam for a conveyor system carries the loads shown in Figure 7–19. Support points are at points A and C. The 20 kN load at B and the 10 kN load at D are to be applied repeatedly many thousands of times. It has been proposed to use a 50 mm diameter circular steel bar for the beam. Specify a suitable steel for the beam.

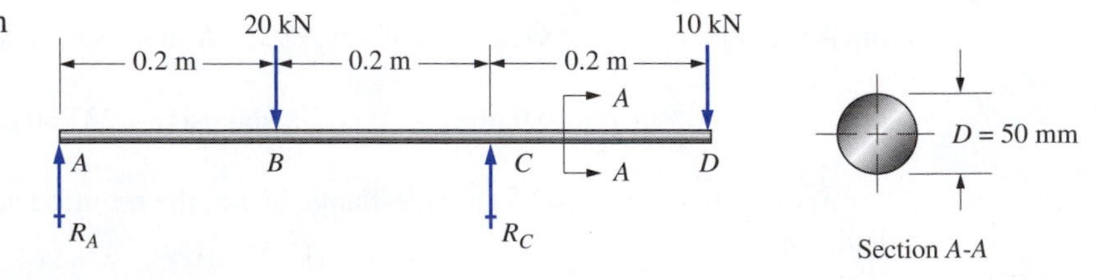

FIGURE 7–19 Beam loading and cross section for beam in Example Problem 7–6.

Solution

Objective Specify a suitable steel.

Given Loading pattern in Figure 7–19. Loads are repeated.
Beam is to be circular, $D = 50$ mm.

Analysis Use the type B procedure given in this section.

Results **Step 1.** Figure 7–20 shows the completed shearing force and bending moment diagrams. The maximum bending moment is 2.00 kN·m at the support at C.

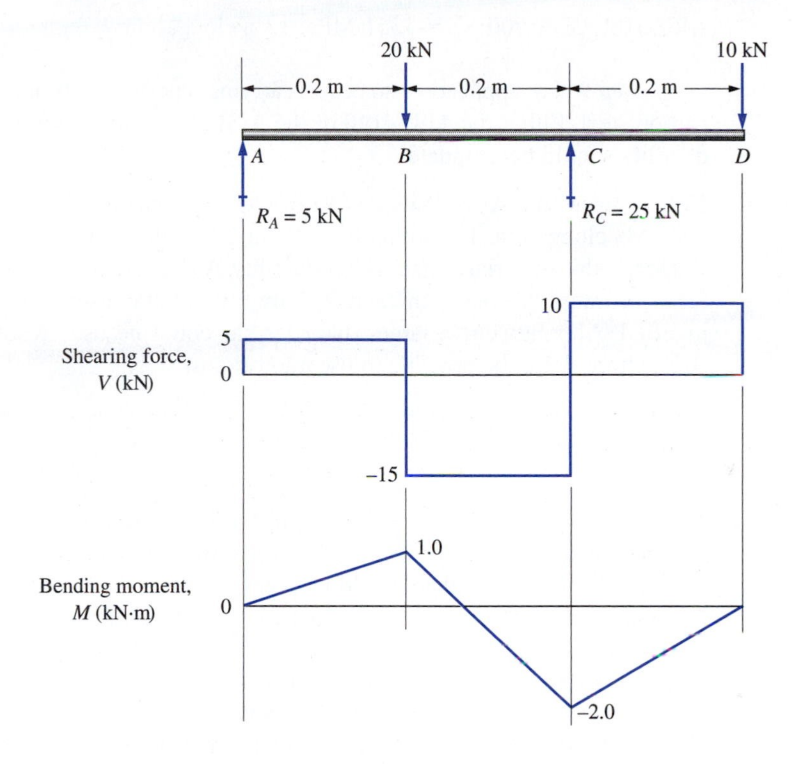

FIGURE 7–20 Shearing force and bending moment diagrams for Example Problem 7–6.

Step 2. Appendix A–1 gives the formula for S for a round bar.

$$S = \pi D^3/32 = \pi(50\text{ mm})^3/32 = 12\,272\text{ mm}^3$$

Step 3. Using Equation (7–5),

$$\sigma_{\max} = \frac{M}{S} = \frac{2.0\text{ kN}\cdot\text{m}}{12\,272\text{ mm}^3} \times \frac{10^3\text{ mm}}{\text{m}} \times \frac{10^3\text{ N}}{\text{kN}}$$

$$\sigma_{\max} = 163\text{ N/mm}^2 = 163\text{ MPa}$$

Step 4. We can use Table 7–1 to determine an appropriate formula for design stress. The steel selected should be highly ductile because of the repeated load. Then we will use $\sigma_d = s_u/8$.

Step 5. Let $\sigma_{\max} = 163\text{ MPa} = \sigma_d = s_u/8$

Step 6. Solving for s_u gives

$$s_u = 8(\sigma_{\max}) = 8(163\text{ MPa}) = 1304\text{ MPa}$$

Step 7. It was decided to use steel.

Step 8. Appendix A–14 lists several common steel alloys. From that table, we can select candidate materials that have good ductility and an ultimate strength of at least 1304 MPa. Four are listed as follows.

AISI 1080 OQT 700; s_u = 1303 MPa; 12% elongation
AISI 1141 OQT 700; s_u = 1331 MPa; 9% elongation
AISI 4140 OQT 700; s_u = 1593 MPa; 12% elongation
AISI 5160 OQT 900; s_u = 1351 MPa; 12% elongation

Step 9. For applications to beams carrying repeated loads, it is typical to use a medium carbon steel. Either the AISI 4140 or the AISI 5160 could be used. With 12% elongation, ductility should be adequate.

Comment Note in Appendix A–14 that AISI 4140 OQT 900 has an ultimate strength of 1289 MPa and 15% elongation. The strength is within 2% of the computed value. It may be suitable to specify this material to gain better ductility. A slight reduction in the design factor would result. Because the values in Table 7–1 are somewhat conservative this would normally be justified. Alternatively, a larger diameter bar could be used resulting in a lower design stress. It may than be possible to use a lower cost steel.

7–9 STRESS CONCENTRATIONS

The conditions specified for valid use of the flexure formula in Section 7–3 included the statement that the beam must have a uniform cross section. Changes in cross section result in higher local stresses than would be predicted from the direct application of the flexure formula. Similar observations were made in earlier chapters concerning direct axial stresses and torsional shear stresses. The use of *stress concentration factors* will allow the analysis of beams that do include changes in cross section.

To visualize one example of stress concentration, refer to Figure 7–21 which shows a photoelastic model of a simply supported beam carrying two identical loads. The loading pattern is similar to that used in Example Problem 7–4 and shown in Figures 7–15 and 7–16. The bending moment is the same on any section in the middle of the beam between the two loads. At the very middle of the beam are two semicircular notches that present a local change in geometry. Two effects occur. The removal of material by the notches reduces the section modulus of the beam at that location and causes the increase in stress. In addition, the sudden change in geometry produces higher stresses in the near vicinity of each notch. This is evidenced by the tightly spaced fringe lines around the notches. Note that the effect of the notches is very local, dissipating a short distance to either side of the notches. Published stress concentration factors allow the calculation of such locally increased stresses.

FIGURE 7–21 Photoelastic model of a beam with stress concentrations due to notches. (Source: Measurements Group, Inc., Raleigh, NC)

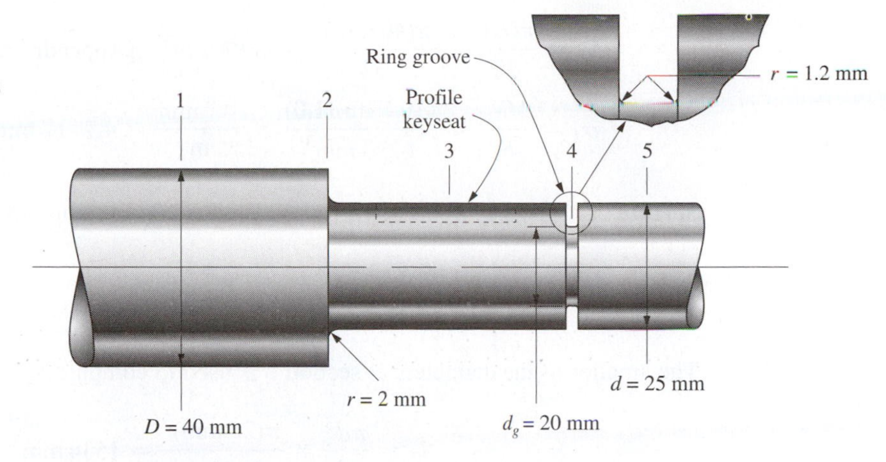

FIGURE 7–22 Portion of a shaft with several changes in cross section producing stress concentrations.

You can also see evidence of the local contact stresses at points of support and at the locations of the applied loads. This was discussed in Section 7–4 as the St. Venant principle in conjunction with the photoelastic model shown in Figure 7–7.

Another example of stress concentrations for a beam is the portion of a circular shaft shown in Figure 7–22. In the design of shafts for carrying power-transmitting elements, the use of steps in the diameter is encountered frequently. Examples were shown in Chapter 4, where torsional shear stresses were discussed. Considering the shaft as a beam subjected to bending moments, there would be stress concentrations at the shoulder (2), the keyseat (3), and the groove (4) as shown in Figure 7–22.

At sections where stress concentrations occur, the stress due to bending would be calculated from a modified form of the flexure formula.

Flexure Formula with Stress Concentration

$$\sigma_{\max} = \frac{McK_t}{I} = \frac{MK_t}{S} \qquad \textbf{(7–6)}$$

The stress concentration factor K_t is found experimentally, with the values reported in graphs such as those in Appendix A–22, cases 4, 5, 8, 9, 10, and 11. See also Reference 10 in Chapter 3 for additional cases.

Example Problem 7–7

Figure 7–22 shows a portion of a round shaft where a gear is mounted. A bending moment of 30 N·m is applied at this location. Compute the stress due to bending at sections 1, 2, 3, 4, and 5.

Solution

Objective: Compute the stress due to bending at sections 1, 2, 3, 4, and 5.

Given: Beam geometry in Figure 7–22. $M = 30$ N·m.

Analysis: Stress concentrations must be considered because of the several changes in geometry in the area of interest. Equation (7–6) will be used to compute the maximum stress at each section. Appendix A–22 is the source of data for stress concentration factors, K_t.

Results: ***Section 1.*** This is the part of the shaft where the diameter is 40 mm and no changes in geometry occur. Therefore the stress concentration factor is 1.0 and the stress is the same as that which would be computed from the flexure formula alone. That is, $\sigma = M/S$.

$$S_1 = \frac{\pi(D)^3}{32} = \frac{\pi(40\text{ mm})^3}{32} = 6283\text{ mm}^3 \quad \text{[Appendix A–1]}$$

$$\sigma_1 = \frac{MK_{t1}}{S_1} = \frac{(30\text{ N}\cdot\text{m})(1.0)}{6283\text{ mm}^3} \times \frac{10^3\text{ mm}}{\text{m}} = 4.77\text{ N/mm}^2 = 4.77\text{ MPa}$$

Section 2. The step in the shaft causes a stress concentration to occur. Then the stress is

$$\sigma_2 = \frac{MK_{t2}}{S_2}$$

The smaller of the diameters at section 2 is used to compute S_2. From Appendix A–1,

$$S_2 = \frac{\pi d^3}{32} = \frac{\pi(25\text{ mm})^3}{32} = 1534\text{ mm}^3$$

The value of K_{t2} depends on the ratios r/d and D/d. (See Appendix A–22–9)

$$\frac{r}{d} = \frac{2\text{ mm}}{25\text{ mm}} = 0.08$$

$$\frac{D}{d} = \frac{40\text{ mm}}{25\text{ mm}} = 1.60$$

From Appendix A–22–9, $K_{t2} = 1.87$. Then

$$\sigma_2 = \frac{MK_t}{S_2} = \frac{(30\text{ N}\cdot\text{m})(1.87)}{1534\text{ mm}^3} \times \frac{10^3\text{ mm}}{\text{m}} = 36.6\text{ N/mm}^2$$

$$\sigma_2 = 36.6\text{ MPa}$$

Section 3. The keyseat causes a stress concentration factor of 2.0 as listed in Appendix A–22–11. S_3 is based on the full 25 mm diameter of the shaft. Therefore $S_3 = S_2 = 1534\text{ mm}^3$. Then

$$\sigma_3 = \frac{MK_{t3}}{S_3} = \frac{(30\text{ N}\cdot\text{m})(2.0)}{1534\text{ mm}^3} \times \frac{10^3\text{ mm}}{\text{m}} = 39.1\text{ N/mm}^2$$

$$\sigma_3 = 39.1\text{ MPa}$$

Section 4. The groove requires the use of Appendix A–22–9 again to find K_{t4}. Note that the nominal stress is based on the root diameter of the groove, d_g. For the groove,

$$\frac{r}{d_g} = \frac{1.2\text{ mm}}{20\text{ mm}} = 0.06$$

$$\frac{d}{d_g} = \frac{25\text{ mm}}{20\text{ mm}} = 1.25$$

Then, $K_{t4} = 1.93$. The section modulus at the root of the groove is

$$S_4 = \frac{\pi d_g^3}{32} = \frac{\pi(20\text{ mm})^3}{32} = 785\text{ mm}^3$$

Now the stress at section 4 is

$$\sigma_4 = \frac{MK_{t4}}{S_4} = \frac{(30\ \text{N}\cdot\text{m})(1.93)}{785\ \text{mm}^3} \times \frac{10^3\text{mm}}{\text{m}} = 73.8\ \text{N/mm}^2$$

$$\sigma_4 = 73.8\ \text{MPa}$$

Section 5. This is the part of the shaft where the diameter is 25 mm and no changes in geometry occur. Therefore the stress concentration factor is 1.0 and $S_5 = 1534\ \text{mm}^3$ as computed for Section 2.

$$\sigma_5 = \frac{MK_{t5}}{S_5} = \frac{(30\ \text{N}\cdot\text{m})(1.0)}{(1534\ \text{mm}^3)} \times \frac{10^3\ \text{mm}}{\text{m}} = 19.6\ \text{N/mm}^2 = 19.6\ \text{MPa}$$

Comment Observe the large variation in stress levels that exist over this relatively small portion of the shaft. Figure 7–23 is a graph showing that variation. The stress at Section 4 is by far the largest because of the small diameter at the base of the groove and the rather high stress concentration factor, K_{t4}. Design of the shaft to determine a suitable material must use this level of stress as the actual maximum stress.

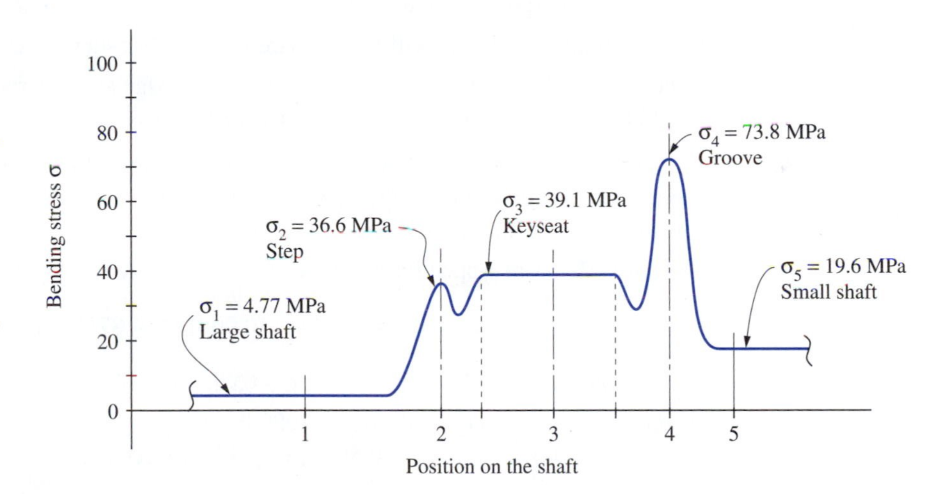

FIGURE 7–23 Variation in bending stress in shaft from Example Problem 7–7.

Example Problem 7–8 Figure 7–24 shows a cantilever bracket carrying a partial uniformly distributed load over the left 10 in and a concentrated load at its right end. The geometry varies at sections *A*, *B*, and *C* as shown. The bracket is made from aluminum 2014-T6, and it is desired to have a minimum design factor of 8 based on the ultimate strength. Evaluate the acceptability of the given design. If any section is unsafe, propose a redesign that will result in a satisfactory stress level. Consider stress concentrations at sections *B* and *C*. The attachment at *A* is blended smoothly such that it can be assumed that $K_t = 1.0$.

Solution Objective Evaluate the beam shown in Figure 7–24 to ensure that the minimum design factor is 8 based on ultimate strength. If not, redesign the beam.

Given Loading and beam geometry in Figure 7–24; Aluminum 2014-T6.

Analysis **1.** The shearing force and bending moment diagrams will be drawn.

2. The design stress will be computed from $\sigma_d = s_u/8$.

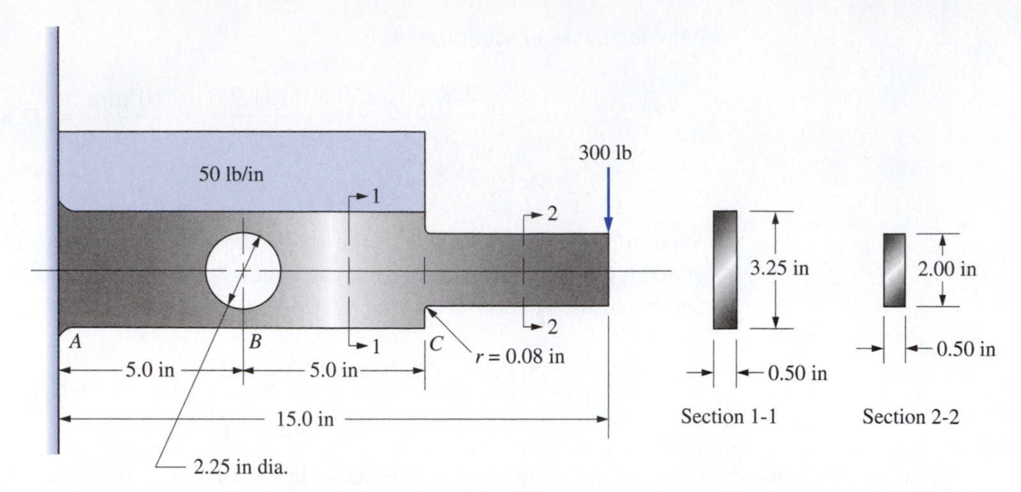

FIGURE 7–24 Beam and loading for Example Problem 7–8.

3. The stress will be computed at sections A, B, and C, considering stress concentrations at B and C. These are the three likely points of failure because of either bending moment or stress concentration. Everywhere else will have a lower bending stress. At each section, the stress will be computed from $\sigma = MK_t/S$, and the values of M, K_t, and S must be determined at each section.
4. The computed stresses will be compared with the design stress.
5. For any section having a stress higher than the design stress, a redesign will be proposed and the stress will be recomputed to verify that it is safe as redesigned.

Results

Step 1. Figure 7–25 shows the completed shearing force and bending moment diagrams. Note that the values of the bending moment at sections B and C have been computed. You should verify the given values.

Step 2. From Appendix A–18, we find $s_u = 70 \text{ ksi} = 70\ 000 \text{ psi}$. Then,

$$\sigma_d = S_u/8 = (70\,000)/8 = 8750 \text{ psi}$$

Steps 3 and 4. At each section, $\sigma = MK_t/S = K_t\,\sigma_{\text{nom}}$.

Section A: $K_t = 1.0$ (given). $M_A = 7000$ lb·in.
Dimensions: $b = 0.50$ in; $h = 3.25$ in; rectangle.

$$S = bh^2/6 = (0.50 \text{ in})(3.25 \text{ in})^2/6 = 0.880 \text{ in}^3$$

$$\sigma = \frac{(7000 \text{ lb}\cdot\text{in})(1.0)}{0.880 \text{ in}^3} = 7953 \text{ psi} < 8750 \text{ psi} \quad \textbf{OK}$$

Section B: $M_B = 3625$ lb·in. Find K_t from Appendix A–22–4.
Dimensions: $t = b = 0.50$ in; $w = h = 3.25$ in; $d = 2.25$ in
$d/w = 2.25/3.25 = 0.692$; then $K_t = 1.40$ (curve C).

$$\sigma = K_t\sigma_{\text{nom}} = \frac{K_t(6Mw)}{(w^3 - d^3)t} = \frac{(1.40)(6)(3625)(3.25)}{[(3.25)^3 - (2.25)^3](0.50)}$$

$$\sigma = 8629 \text{ psi} < 8750 \text{ psi} \quad \textbf{OK}$$

Section C: $M_C = 1500$ lb·in. Find K_t from Appendix A–22–10.
Dimensions: $t = 0.50$ in; $H = 3.25$ in; $h = 2.00$ in; $r = 0.08$ in
$H/h = 3.25/2.0 = 1.625$; $r/h = 0.08/2.00 = 0.04$.
Then, $K_t = 2.40$.

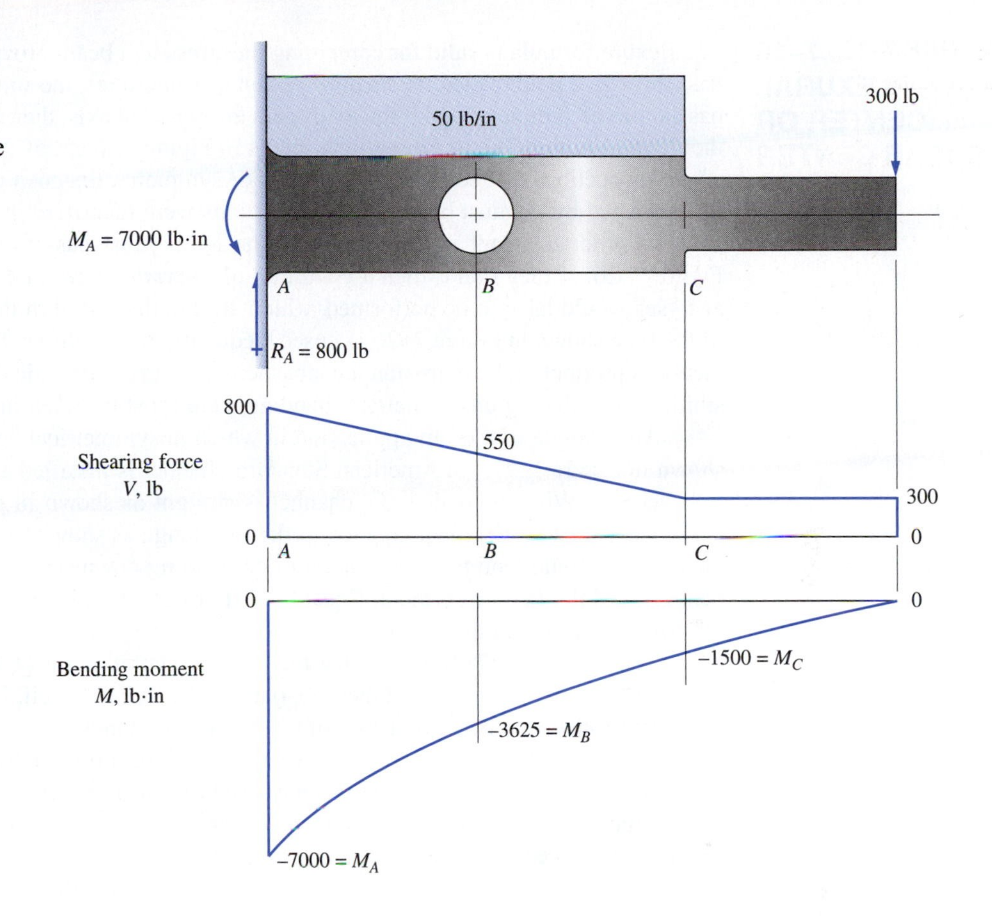

FIGURE 7–25 Shearing force and bending moment diagrams for Example Problem 7–8.

$$\text{Section modulus} = S = th^2/6 = (0.50)(2.00)^2/6 = 0.333 \text{ in}^3$$

$$\sigma = \frac{MK_t}{S} = \frac{(1500 \text{ lb}\cdot\text{in})(2.40)}{0.333 \text{ in}^3} = 10\,800 \text{ psi} \quad \textbf{too high}$$

Step 5. Proposed Redesign at C: Because the stress concentration factor at section C is quite high, increase the fillet radius. The maximum allowable stress concentration factor is found by solving the stress equation for K_t and letting $\sigma = \sigma_d = 8750$ psi. Then,

$$K_t = \frac{S\sigma_d}{M} = \frac{(0.333 \text{ in}^3)(8750 \text{ lb/in}^2)}{1500 \text{ lb}\cdot\text{in}} = 1.94$$

From Appendix A–22–10, the minimum value of $r/h = 0.08$ to limit K_t to 1.94. Then, $r_{\text{min}} = 0.08(h) = 0.08(2.00) = 0.16$.

Let $r = 0.20$ in; $r/h = 0.20/2.00 = 0.10$; $K_t = 1.80$. Then,

$$\sigma = \frac{MK_t}{S} = \frac{(1500 \text{ lb}\cdot\text{in})(1.80)}{0.333 \text{ in}^3} = 8100 \text{ psi} \quad \textbf{OK}$$

Comment This problem is a good illustration of the necessity of analyzing any point within a beam where high stress may occur because of high bending moment, high stress concentration, small section modulus, or some combination of these. It also demonstrates one method of redesigning a beam to ensure safety.

7–10 FLEXURAL CENTER OR SHEAR CENTER

The flexure formula is valid for computing the stress in a beam provided the applied loads pass through a point called the *flexural center,* or sometimes, the *shear center*. If a section has an axis of symmetry and if the loads pass through that axis, then they also pass through the flexural center. The beam sections shown in Figure 7–6 are of this type.

For sections loaded away from an axis of symmetry, the position of the flexural center, indicated by Q, must be found. Such sections were identified in Figure 7–7.

In order to result in pure bending, the loads must pass through Q, as shown in Figure 7–26. If they don't, then a condition of *unsymmetrical bending* occurs and other analyses would have to be performed, which are not discussed in this book. The sections of the type shown in Figure 7–26 are used frequently in structures. Some lend themselves nicely to production by extrusion and are therefore very economical. Because of the possibility of producing unsymmetrical bending, care must be taken in their application.

An example of a beam application in which unsymmetrical bending would occur is shown in Figure 7–27. An American Standard channel is installed as a cantilever with its web vertical. When unloaded, the channel is straight as shown in part (a) of the figure. However, with a vertical load applied to the top flange as shown in part (b) of the figure, the channel would tend to twist in addition to bending downward. Clearly this is not pure bending and the flexure formula, Equation 7–1, would not give an accurate prediction of the stress condition in the channel.

Refer to Figure 7–26 to see that the flexural center, point Q, for a channel with its web vertical, is well to the left of the web, outside the channel itself. In order for pure bending to be produced, the line of action of the applied load must pass through this point. You might design a bracket like that shown in Figure 7–27(c) through which to apply the load.

Notice that you may apply the channel as a beam with the legs down or up and pass the load through the axis of symmetry as shown in part (d) of the figure. Then the bending action would not produce twisting and the flexure formula can be used to compute the bending stress.

The following two example problems demonstrate the method of locating the flexural center, Q.

Example Problem 7–9 Determine the location of the flexural center for the two sections shown in Figure 7–28.

Solution

Objective: Locate the shear center, Q, for the two shapes.

Given: Shapes in Figure 7–28: channel in 7–28(a); hat section in 7–28(b).

Analysis: The general location of the shear center for each shape is shown in Figure 7–26 along with the means of computing the value of e that locates Q relative to specified features of the shapes.

Results: ***Channel Section (a).*** From Figure 7–26, the distance e to the flexural center is

$$e = \frac{b^2h^2t}{4I_x}$$

Note that the dimensions b and h are measured to the middle of the flange or web. Then $b = 40$ mm and $h = 50$ mm. Because of symmetry about the X axis, I_x can be found by the difference between the value of I for the large outside rectangle (54 mm by 42 mm) and the smaller rectangle removed (46 mm by 38 mm).

$$I_x = \frac{(42)(54)^3}{12} - \frac{(38)(46)^3}{12} = 0.243 \times 10^6\ \text{mm}^4$$

FIGURE 7–26
Location of flexural center Q.

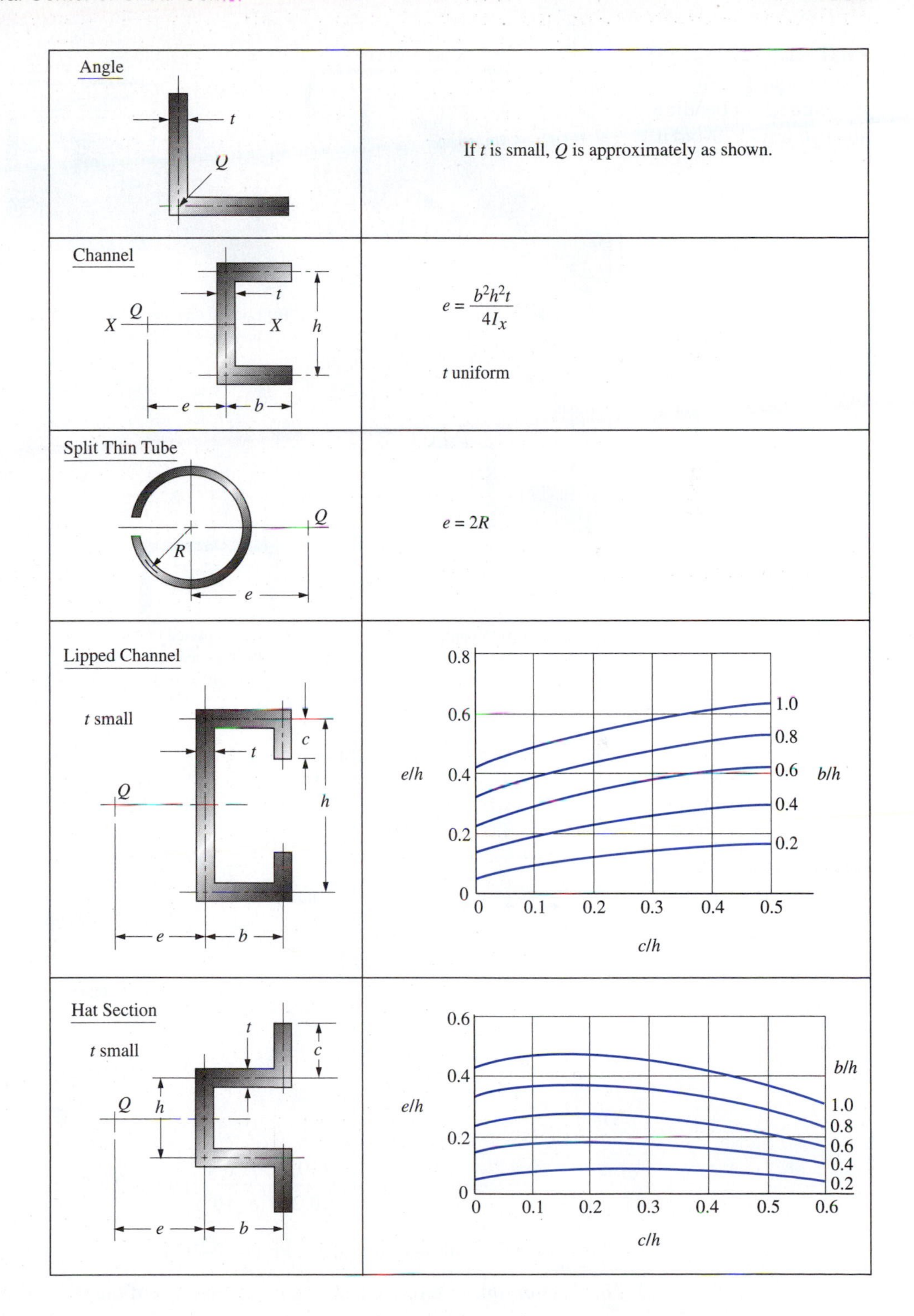

FIGURE 7–27 Illustration of unsymmetrical bending and two ways to avoid it.

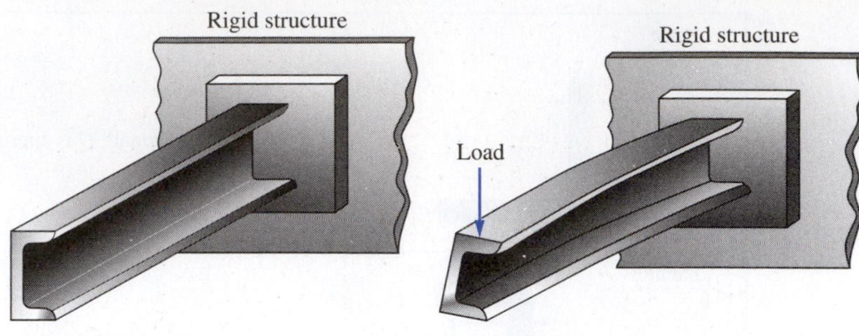

(*a*) Unloaded channel

(*b*) Loaded channel showing twisting due to unsymmetrical bending

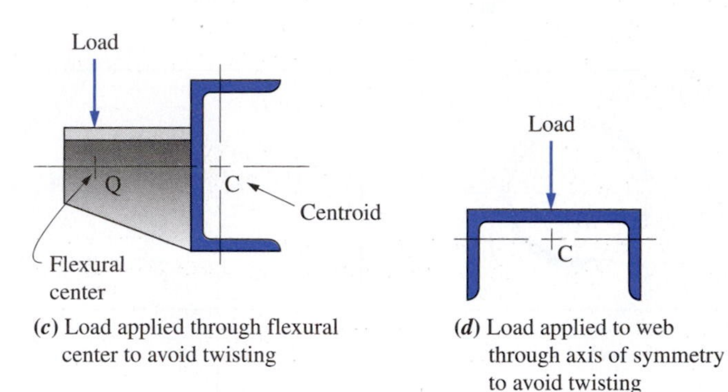

(*c*) Load applied through flexural center to avoid twisting

(*d*) Load applied to web through axis of symmetry to avoid twisting

FIGURE 7–28 Beam sections for which the flexural centers are computed in Example Problem 7–9.

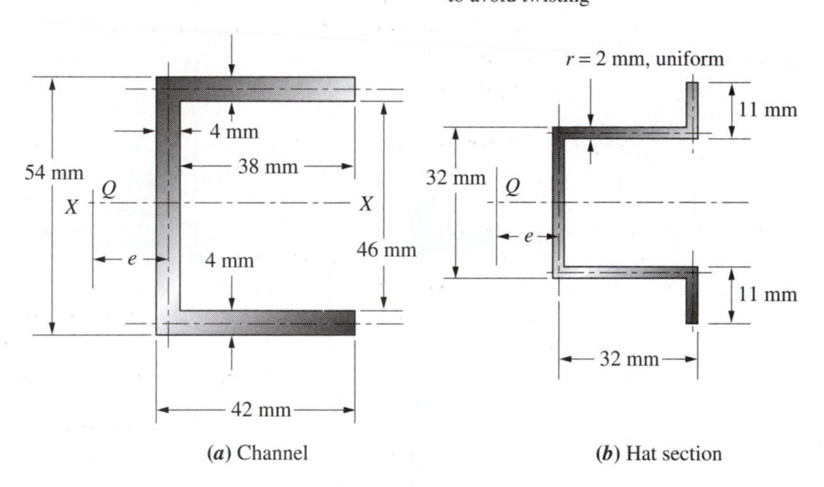

(*a*) Channel

(*b*) Hat section

Then

$$e = \frac{(40)^2(50)^2(4)}{4(0.243 \times 10^6)} \text{mm} = 16.5 \text{ mm}$$

This dimension is drawn to scale in Figure 7–28(a).

Hat Section (b). Here the distance e is a function of the ratios c/h and b/h.

$$\frac{c}{h} = \frac{10}{30} = 0.3$$

$$\frac{b}{h} = \frac{30}{30} = 1.0$$

Then from Figure 7–26, $e/h = 0.45$. Solving for e yields

$$e = 0.45h = 0.45(30 \text{ mm}) = 13.5\text{mm}$$

This dimension is drawn to scale in Figure 7–28(b).

Comment Now, can you devise a design for using either section as a beam and provide for the application of the load through the flexural center Q to produce pure bending?

7–11 PREFERRED SHAPES FOR BEAM CROSS SECTIONS

Recall the discussion earlier in this chapter of the stress distribution in the cross section of a beam characterized by the equations

$$\sigma_{\max} = Mc/I = M/S \quad \text{at the outermost fiber of the beam}$$

$$\sigma = \sigma_{\max}(y/c) \quad \text{at any point at a distance } y \text{ from neutral axis}$$

Figures 7–9 and 7–10 illustrate the stress distribution. It must be understood that these equations apply strictly only to beams made from homogeneous, isotropic materials; that is, those having equal properties in all directions.

Because the larger stresses occur near the top and bottom of the cross section, the material there provides more of the resistance to the externally applied bending moment than material nearer to the neutral axis. It follows that it is desirable to place the larger part of the material away from the neutral axis to obtain efficient use of the material. In this discussion, efficiency refers to maxin izing the moment of inertia and section modulus of the shape for a given amount of material, as indicated by the area of the cross section.

Figure 7–29 shows several examples of efficient shapes for beam cross sections. These illustrations are based on the assumption that the most significant stress is a bending stress caused by loads acting on top of the beam perpendicular to the neutral axis. Examples are shown in Figures 7--5, 7–15, and 7–18. In such cases, it is said that the bending is positive about the horizontal neutral axis. It is also assumed that the material is equally strong in compression and tension.

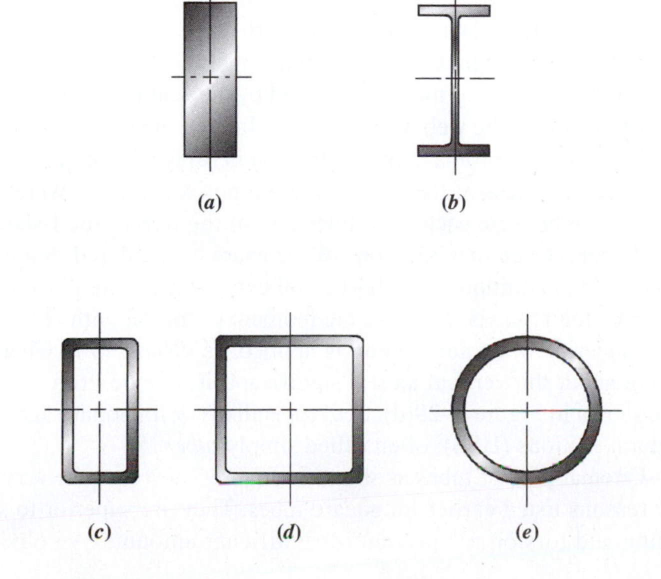

FIGURE 7–29 Efficient shapes for beams.

Starting with the simple rectangular shape in Figure 7–29(a), it is preferred to orient the long dimension vertically as shown because the moment of inertia is proportional to the *cube* of the height of the rectangle, where the height is the dimension perpendicular to the neutral axis. For example, consider the case of a rectangle having dimensions of 40 mm by 125 mm and compare the resulting values of I and S.

	125 mm dimension vertical	40 mm dimension vertical
$I = bh^3/12$	$I_1 = (40)(125)^3/12 = 6.51 \times 10^6\ \text{mm}^4$	$I_2 = (125)(40)^3/12 = 0.667 \times 10^6\ \text{mm}^4$
$S = bh^2/6$	$S_1 = (40)(125)^2/6 = 1.04 \times 10^5\ \text{mm}^3$	$S_2 = (125)(40)^2/6 = 0.333 \times 10^5\ \text{mm}^3$

Comparing these results gives

$$\frac{I_1}{I_2} = \frac{6.51 \times 10^6\ \text{mm}^4}{0.667 \times 10^6\ \text{mm}^4} = 9.76 \quad \frac{S_1}{S_2} = \frac{1.04 \times 10^5\ \text{mm}^3}{0.333 \times 10^5\ \text{mm}^3} = 3.12$$

The comparison of the values of the section modulus, S, is the most pertinent for comparing stresses in beams because it contains both the moment of inertia, I, and the distance, c, to the outermost fiber of the beam cross section. While the section with the long dimension vertical has a moment of inertia almost ten times that with the long dimension horizontal, it is over three times taller, resulting in the improvement in the section modulus of approximately three times. Still, that is a significant improvement.

A related factor in the comparison of beam shapes is that the *deflection* of a beam is inversely proportional to the moment of inertia, I, as will be shown in Chapter 9. Therefore, the taller rectangular beam in the previous example would be expected to deflect only 1/9.76 times as much as the shorter one, only about 10% as much.

The shape shown in Figure 7–29(b) is the familiar "I-beam." Placing most of the material in the horizontal flanges at the top and bottom of the section puts them in the regions of the highest stresses for maximum resistance to the bending moment. The relatively thin vertical web serves to hold the flanges in position and provides resistance to shearing forces, as described in Chapter 8. It would be well to study the proportions of standard steel and aluminum I-shaped sections listed in Appendixes A–7, A–8, and A–11 to get a feel for reasonable flange and web thicknesses. The thickness of that flange which is in compression is critical with regard to buckling when the beam is relatively long. References 1–3 give data on proper proportions.

The tall, rectangular tube shown in Figure 7–29(c) is very similar to the I-shape in its resistance to bending moments caused by vertical loads. The two vertical sides serve a similar function as the web of the I-shape. In fact, the area moment of inertia with respect to the horizontal centroidal axis for the tube in (c) would be identical to that for the I-shape in (b) if the thickness of the top and bottom horizontal parts were equal and if the vertical sides of the tube were each $\frac{1}{2}$ the thickness of the web of the I-shape. The tube is superior to the I-shape when combinations of loads are encountered that cause bending about the vertical axis in addition to the horizontal axis, because the placement of the vertical sides away from the Y-Y axis increases the moment of inertia with respect to that axis. The tube is also superior when any torsion is applied, as discussed in Chapter 4. When torsion or bending about the vertical axis is significant, it may be preferred to use the square tube shape shown in Figure 7–29(d). See Appendix A–9 for square and rectangular steel hollow structural sections (HSS), often called simply *tubes*.

Circular pipe or tubes as shown in Figure 7–29(e) make very efficient beams for the same reasons listed earlier for square tubes. They are superior to square tubes when both bending and torsion are present in significant amounts. An obvious example of where a circular tube is preferred is the case of a rotating shaft carrying both bending and torsional loads such as the drive shaft and the axles of a car or truck. See Appendix A–12 for steel pipe and A–13 for circular mechanical tubing.

Shapes Made from Thin Materials. Economical production of beams having moderate dimensions can be done using rollforming or pressforming of relatively thin flat sheet materials. Aluminum and many plastics are extruded to produce shapes having a uniform cross section, often with thin walls and extended flanges. Examples are shown in Chapter 6, Figures P6–10 through P6–20. Such shapes can be specially adapted to the use of the beam. See if you can identify beam-like members with special shapes around you. In your home you might find such beams used as closet door rails, curtain rods, structures for metal furniture, patio covers or awnings, ladders, parts for plastic toys, tools in the workshop, or parts for appliances or lawn maintenance tools. In your car, look at the windshield wiper arms, suspension members, gearshift levers, linkages or brackets in the engine compartment, and the bumpers. Aircraft structures contain numerous examples of thin-walled shapes designed to take advantage of their very light weight. See also Internet sites for this chapter.

Figure 7–30 shows three examples of extruded or roll-formed shapes found around the home. Part (a) shows a closet door rail where the track for the roller that supports the door is produced as an integral part of the aluminum extrusion. The extruded side rail of an aluminum extension ladder is sketched in part (b). Part (c) shows a portion of the

FIGURE 7–30 Examples of thin-walled beam sections.

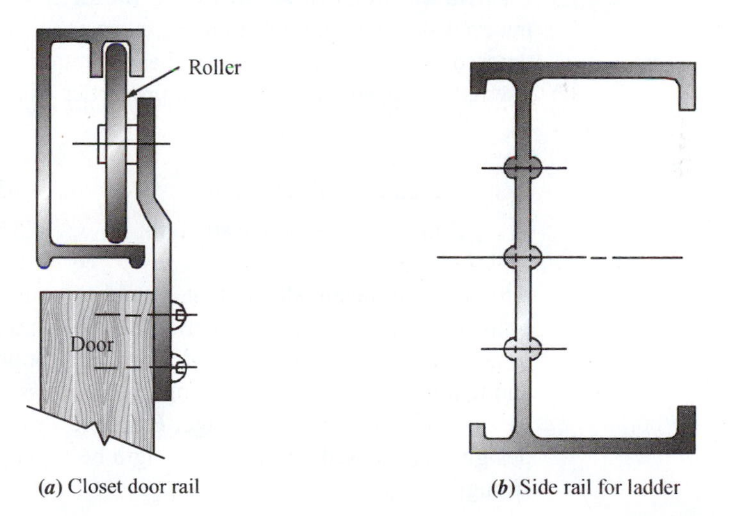

(*a*) Closet door rail

(*b*) Side rail for ladder

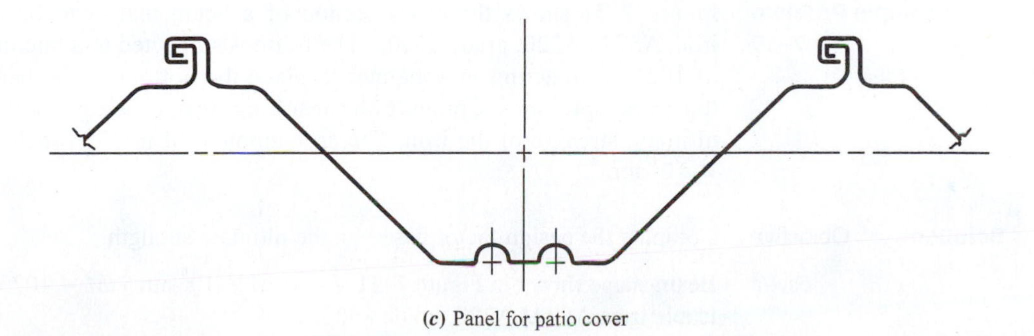

(*c*) Panel for patio cover

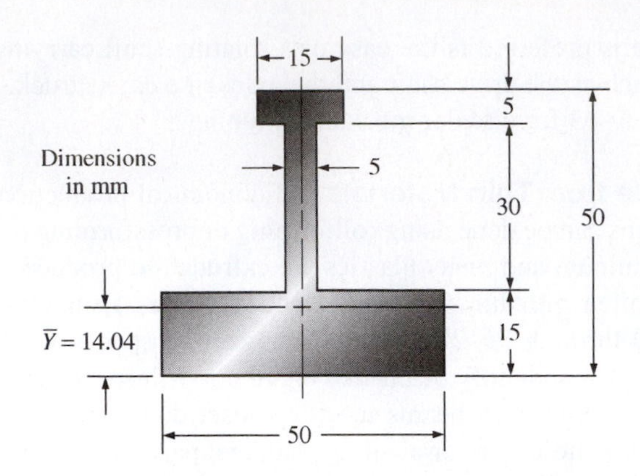

FIGURE 7–31 Nonsymmetrical shape for cross section of beam in Example Problem 7–10.

roll-formed decking for a patio cover made from aluminum sheet only 0.025 in (0.64 mm) thick. The shape is specially designed to link together to form a continuous panel to cover a wide area. Some design features of these sections should be noted. Extended flanges are reinforced with bulb-like projections to provide local stiffness that resists wrinkling or buckling of the flanges. Broad flat areas are stiffened by ribs or roll-formed corrugations, also to inhibit local buckling. References 1–3 provide guidelines for the design of such features.

Beams Made from Anisotropic Materials. The design of beams to be made from materials having different strengths in tension and compression requires special care. Most types of cast iron, for example, have a much higher compressive strength than tensile strength. Appendix A–17 lists the properties of malleable iron ASTM A220, grade 80002 as,

Ultimate tensile strength: $s_u = 655$ MPa (95 ksi)

Ultimate compressive strength: $s_{uc} = 1650$ MPa (240 ksi)

An efficient beam shape that would account for this difference is the modified I-shape shown in Figure 7–31. Because the typical, positive bending moment places the bottom flange in tension, providing a larger bottom flange moves the neutral axis down and tends to decrease the resulting tensile stress in the bottom flange relative to the compressive stress in the top flange. Example Problem 7–10 illustrates the result with the design factor based on tensile strength being nearly equal to that based on compressive strength.

Example Problem 7–10 Figure 7–31 shows the cross section of a beam that is to be made from malleable iron, ASTM A220, grade 80002. The beam is subjected to a maximum bending moment of 1025 N·m, acting in a manner to place the bottom of the beam in tension and the top in compression. Compute the resulting design factor for the beam based on the ultimate strength of the iron. The area moment of inertia for the cross section is 1.80×10^5 mm^4.

Solution Objective Compute the design factor based on the ultimate strength.

Given Beam shape shown in Figure 7–31. $I = 1.80 \times 10^5$ mm^4. $M = 1025$ N·m. Material is malleable iron, ASTM A220, grade 80002.

Analysis Because the beam cross section is not symmetrical, the value of the maximum tensile stress at the bottom of the beam, σ_{tb}, will be lower than the maximum compressive stress at the top, σ_{ct}. We will compute:

$$\sigma_{tb} = Mc_b/I \quad \text{and} \quad \sigma_{ct} = Mc_t/I$$

where $c_b = \overline{Y} = 14.04$ mm and $c_t = 50 - 14.04 = 35.96$ mm. The tensile stress at the bottom will be compared with the ultimate tensile strength to determine the design factor based on tension, N_t, from

$$\sigma_{tb} = s_u/N_t \quad \text{or} \quad N_t = s_u/\sigma_{tb}$$

where $s_u = 655$ MPa from Appendix A–17. Then the compressive stress at the top will be compared with the ultimate compressive strength to determine the design factor based on compression, N_c, from

$$\sigma_{ct} = s_{uc}/N_c \quad \text{or} \quad N_c = s_{uc}/\sigma_{ct}$$

where $s_{uc} = 1650$ MPa from Appendix A–17. The lower of the two values of N will be the final design factor for the beam.

Results At the bottom of the beam,

$$\sigma_{tb} = \frac{Mc_b}{I} = \frac{(1025\ \text{N}\cdot\text{m})(14.04\ \text{mm})}{1.80 \times 10^5\ \text{mm}^4}\cdot\frac{(1000\ \text{mm})}{\text{m}} = 79.95\ \text{MPa}$$

$$N_t = S_u/\sigma_{tb} = 655\ \text{MPa}/79.95\ \text{MPa} = 8.19$$

At the top of the beam,

$$\sigma_{ct} = \frac{Mc_t}{I} = \frac{(1025\ \text{N}\cdot\text{m})(35.96\ \text{mm})}{1.80 \times 10^5\ \text{mm}^4}\cdot\frac{(1000\ \text{mm})}{\text{m}} = 204.8\ \text{MPa}$$

$$N_c = s_{uc}/\sigma_{ct} = 1650\ \text{MPa}/204.8\ \text{MPa} = 8.06$$

Comment The compressive stress at the top of the beam is the limiting value in this problem because the smaller design factor exists there. Note that the two values of the design factor were quite close to being equal, indicating that the shape of the cross section has been reasonably well optimized for the different strengths in tension and compression.

7–12 DESIGN OF BEAMS TO BE MADE FROM COMPOSITE MATERIALS

Composite materials, discussed in Chapter 2, have superior properties when applied to beam design because of the ability to tailor the constituents for the composite and their placement in the beam. Composite processing often allows for unique shapes to be designed that optimize the geometry of the structure with regard to the magnitude and direction of loads to be carried. Combining these features with the inherent advantages of composites in terms of high strength-to-weight and stiffness-to-weight ratios makes them very desirable for use in beams. See References 5, 6, 8, 11, 12, 15, and 17.

The discussion of Section 7–10 applies equally well to the design of composite beams. The designer should select a shape for the beam cross section that is, itself, efficient in resisting bending moments. In addition, the designer can call for the placement of a higher concentration of the stronger, stiffer fibers in the regions where the higher stresses would be expected: namely, at the outermost fibers of the beam farthest from the neutral axis. More plies of a fabric-type filler can be placed in the high-stress regions.

An effective technique for composite beam design is to employ a very light core material for a structure made from a rigid foam or a honeycomb material, covered by relatively thin layers of the strong, stiff fibers in a polymer matrix. If it is known that the bending moments will always act in the same direction, the fibers of the composite can be aligned with the direction of the tensile and compressive stresses in the beam. If the bending moments are expected to act in a variety of directions, a more disperse placement of fibers can be specified or fabric plies can be placed at a variety of angles, as suggested in Figure 2–28.

Care must be exercised in designing and testing composite beam structures because of the multiple possible modes of failure. The structure may fail in the high tensile stress region by failure of the fibers or the matrix or by the disengagement of the fibers from the matrix. Perhaps a more likely failure mode for a laminated composite is interlaminar shear failure in regions of high shear stress near the neutral axis, as sketched in Figure 7–2(f). Failure could also occur in the compressive stress region by local buckling of the shape or by delamination.

When the beam has been designed with the assumption of bending in a certain plane, it is essential that loads are properly applied and that the shape itself promotes pure bending rather than a combination of bending and torsion. The discussion of *flexural center,* Section 7–10, should be reviewed.

The shape and dimensions of the beam cross section can be varied to correspond to the magnitude of the bending moment at various positions in a beam. For example, a cantilever carrying a concentrated load at its end experiences the highest bending moment at the support point, and the magnitude of the bending moment decreases linearly out to the end of the beam. Then the cross section can be deeper at the support and progressively smaller toward the end. A simply supported beam with a load at the center has its highest bending moment at the center, decreasing toward each support. Then the beam can be thicker at the center and thinner toward the ends.

Beams with broad, flat or curved surfaces, such as the wings of an aircraft, must be designed for stiffness of the broad panels as well as for adequate strength. The skin of the panel may have to be supported by internal ribs to break it into smaller areas.

Penetrations in a composite beam must be carefully designed to ensure smooth transfer of loads from one part of the beam to another. If feasible, the placement of penetrations should be in regions of low stress. Similarly, fasteners must be carefully designed to ensure proper engagement in the fibrous composite material. Thickened bosses may be provided where fasteners are to be located. It may be possible to minimize the number of fasteners by clever shaping of the structure such as by molding in brackets integral with the main structure.

In summary, the designer of composite beams must carefully analyze the stress distribution in the beam and attempt to optimize the placement of material to optimize the shape and dimensions of the beam. The designer must visualize the path of load transfer from its point of application to the ultimate point of support.

REFERENCES

1. Aluminum Association, *Aluminum Design Manual,* Washington, D.C., 2005.
2. American Institute of Steel Construction, *Specification for Structural Steel Buildings,* AISC, Chicago, IL, 2005.
3. American Institute of Steel Construction, *Manual of Steel Construction: Load and Resistance Factor Design,* 3rd ed., AISC, Chicago, IL, 2001.
4. American Institute of Steel Construction, *Manual of Steel Construction: Allowable Stress Design,* 9th ed., AISC, Chicago, IL, 1989.
5. ASM International, *ASM Handbook, Volume 21: Composites,* Materials Park, OH, 2001.
6. ASTM International, *The Composite Materials Handbook-MIL 17,* ASTM, West Conshohocken, PA, 2002.

7. Halperin, D. A. and T. G. Bible, *Principles of Timber Design for Architects and Builders,* John Wiley & Sons, New York, 1994.
8. Jang, Bor Z., *Advanced Polymer Composites: Principles and Applications,* ASM International, Materials Park, OH, 1994.
9. Mallick, P. K., *Composites Engineering Handbook,* Marcel Dekker, New York, 1997.
10. McCormac, J. C. and J. Nelson, *Structural Steel Design—LRFD Method,* 4th ed., Prentice Hall, Upper Saddle River, NJ, 2007.
11. Miravete, A., *Optimisation of Composite Structures Design,* Woodhead Publishing, Cambridge, England, 1996.
12. Quinn, J. A., *Composites Design Manual,* Woodhead Publishing, Cambridge, England, 1999.
13. Segui, W. T., T. Ziolkowski, and B. Stenquist, *LRFD Steel Design,* 3rd ed., Thomson-Engineering, Toronto, Ontario, Canada, 2002.
14. Speigel, L. and G. F. Limbrunner, *Applied Structural Steel Design,* 4th ed., Prentice Hall, Englewood Cliffs, NJ, 2001.
15. Staab, George H., *Laminar Composites,* Butterworth-Heinemann, Boston, 1999.
16. Young, W. C. and R. G. Budynas, *Roark's Formulas for Stress and Strain,* 7th ed., McGraw-Hill, New York, 2002.
17. Zweben, Carl, Composite Materials, Part 1, Section 10, *Mechanical Engineers Handbook, Materials and Mechanical Design,* 3rd ed., Myer Kutz, Ed., John Wiley & Sons, New York, 2005.

INTERNET SITES

Any open-ended design problem, where the objective is to specify a suitable beam shape and size of a beam's cross section, could make use of the wide variety of shapes and sizes available from the following sites, repeated from Chapter 6, to supplement data given in the Appendix.

1. Profiles, Inc. www.profiles-inc.com Manufacturer of custom rolled or drawn shapes for precision components using carbon and alloy steel, copper, brass, and other nonferrous materials.
2. iLevel by Weyerhaeuser www.ilevel.com/floors Manufacturer of floor joists for building construction made as efficient I-shaped fabrications from simple wood components. Select the TJI® Joists product. The Joist Specifier Guide gives detailed dimensions and application data for the Silent Floor—Trus Joist products.
3. Nucor Corporation—Vulcraft Group www.vulcraft.com Manufacturer of open-web steel joists, joist girders, floor and roof deck shapes, and composite floor joist systems for use in commercial shopping centers, schools, and office buildings.
4. Unistrut Corporation www.unistrut.com Manufacturer of systems of components made from steel or fiberglass that can be assembled in a variety of configurations to produce frames, structures, brackets, storage units, support systems, mobile units, and other devices. Online catalogs provide section property data, application information, assembly instructions, and illustrations of uses.
5. 80/20, Inc, www.8020.net Manufacturer of systems of components made from extruded aluminum that can be assembled in a variety of configurations for use as production equipment, carts, automation devices, workstations, storage units, and a variety of other systems.
6. Paramount Extrusions Company http://paramountextrusions.com/shapes Manufacturer of a variety of standard extrusions such as channels, H-sections, T-sections, hollow tubes (round, square, rectangular), square and rectangular bars, handle sections, and many others for use in office furniture, consumer products, cabinets, and so forth. The catalog gives data that can be useful in problems in this book.
7. Jackson Tube Service, Inc. www.jackson-tube.com shapes.htm. Manufacturer of a wide variety of tubing shapes and sizes that can be useful for problems in this book.

PROBLEMS

Analysis of Bending Stresses

7–1.M A square bar 30 mm on a side is used as a simply supported beam subjected to a bending moment of 425 N·m. Compute the maximum stress due to bending in the bar.

7–2.M Compute the maximum stress due to bending in a round rod 20 mm in diameter if it is subjected to a bending moment of 120 N·m.

7–3.E A bending moment of 5800 lb·in is applied to a beam having a rectangular cross section with dimensions of 0.75 in × 1.50 in. Compute the maximum bending stress in the beam (a) if the vertical side is 1.50 in, and (b) if the vertical side is 0.75 in.

7–4.E A wood beam carries a bending moment of 15 500 lb·in. It has a rectangular cross section 1.50 in wide by 7.25 in high. Compute the maximum stress due to bending in the beam.

7–5.E The loading shown in Figure P7–5 is to be carried by a W12×16 steel beam. Compute the stress due to bending.

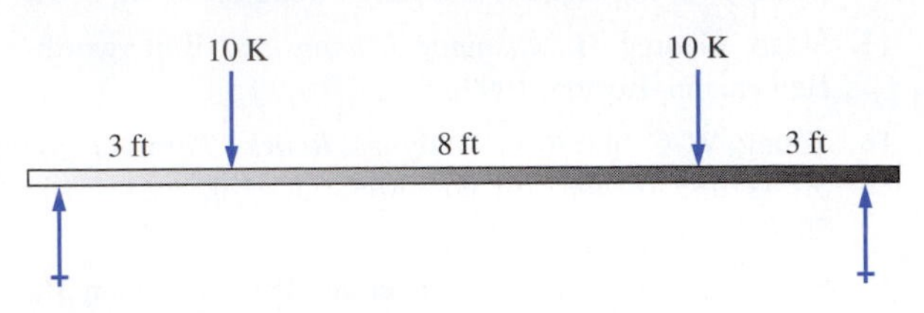

FIGURE P7–5

7–6.E An American Standard beam, S12×35, carries the load shown in Figure P7–6. Compute the stress due to bending.

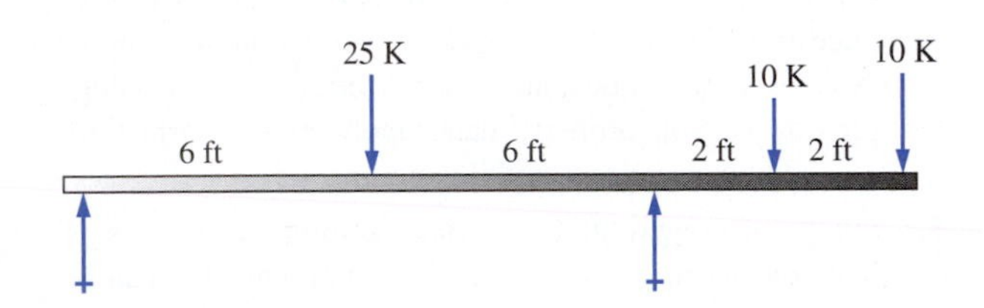

FIGURE P7–6

7–7.E The 24-in long beam shown in Figure P7–7 is an aluminum channel, C4×2.331, positioned with the legs down so that the flat 4-in surface can carry the applied loads. Compute the maximum tensile and maximum compressive stresses in the channel.

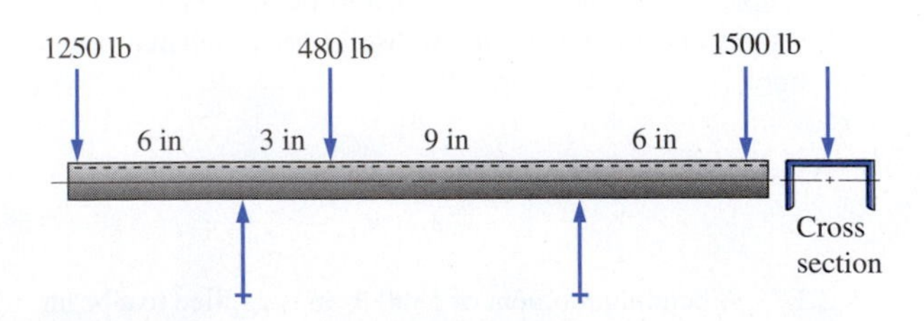

FIGURE P7–7

7–8.E The 650-lb load at the center of the 28-in-long bar shown in Figure P7–8 is carried by a standard steel pipe, $1\frac{1}{2}$-in schedule 40. Compute the stress in the pipe due to bending.

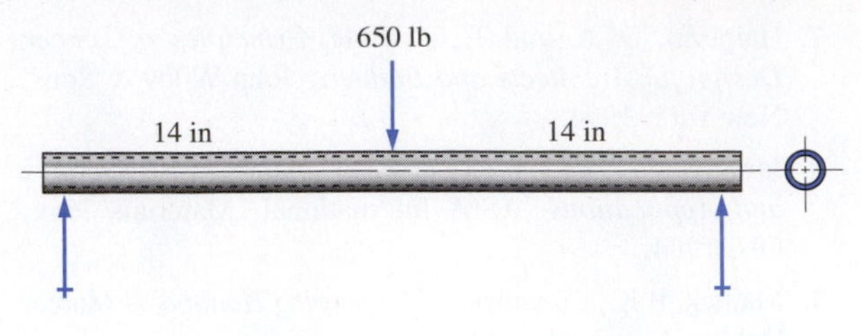

FIGURE P7–8

7–9.M The loading shown in Figure P7–9(a) is to be carried by the fabricated beam shown in Figure P7–9(b). Compute the stress due to bending in the beam.

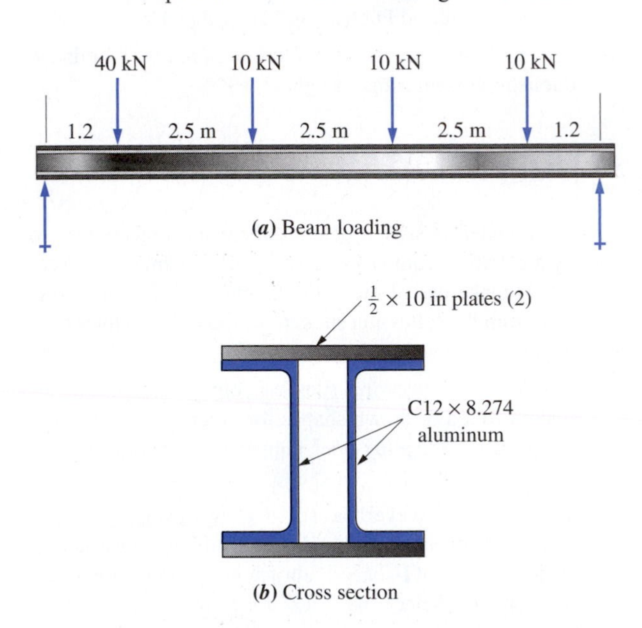

FIGURE P7–9

7–10.C An aluminum I-beam, I9×8.361, carries the load shown in Figure P7–10. Compute the stress due to bending in the beam.

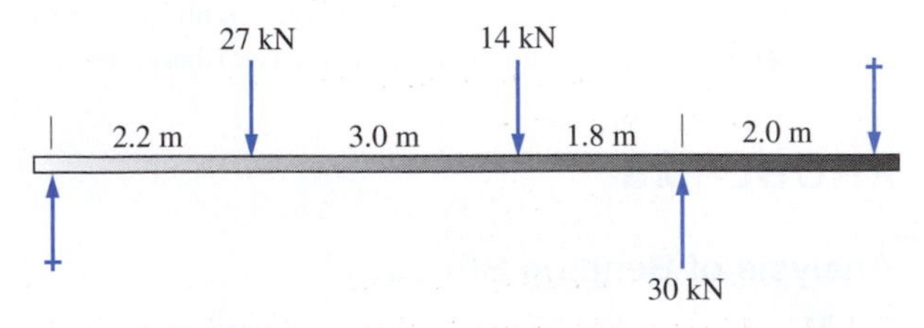

FIGURE P7–10

7–11.E A part of a truck frame is composed of two channel-shaped members, as shown in Figure P7–11. If the moment at the section is 60 000 lb·ft, compute the bending stress in the frame. Assume that the two channels act as a single beam.

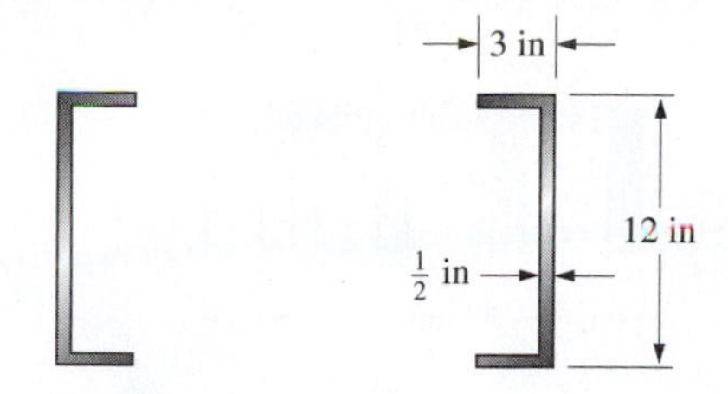

FIGURE P7–11 Truck frame members for Problem 7–11.

Design of Beams

7–12.M Compute the minimum required diameter of a round bar used as a beam to carry a bending moment of 240 N·m with a stress no greater than 125 MPa.

7–13.M A rectangular bar is to be used as a beam subjected to a bending moment of 145 N·m. If its height is to be three times its width, compute the minimum required dimensions of the bar to limit the stress to 55 MPa.

7–14.M The tee section shown in Figure P7–14 is to carry a bending moment of 28.0 kN·m. It is to be made of steel plates welded together. If the load on the beam is a dead load, would AISI 1020 hot-rolled steel be satisfactory for the plates?

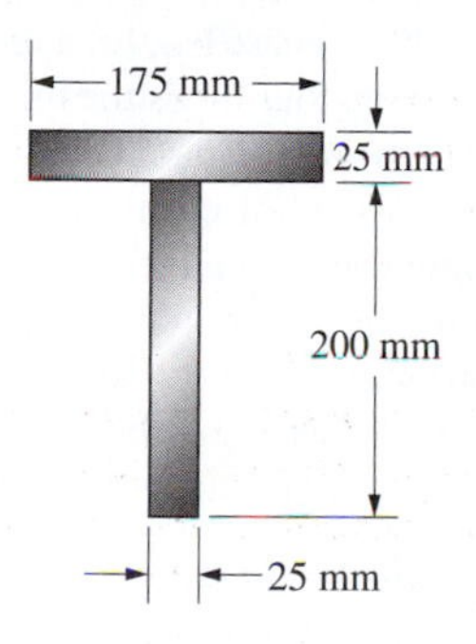

FIGURE P7–14

7–15.M The modified I-section shown in Figure P7–15 to be extruded aluminum. Specify a suitable aluminum alloy if the beam is to carry a repeated load resulting in a bending moment of 275 N·m.

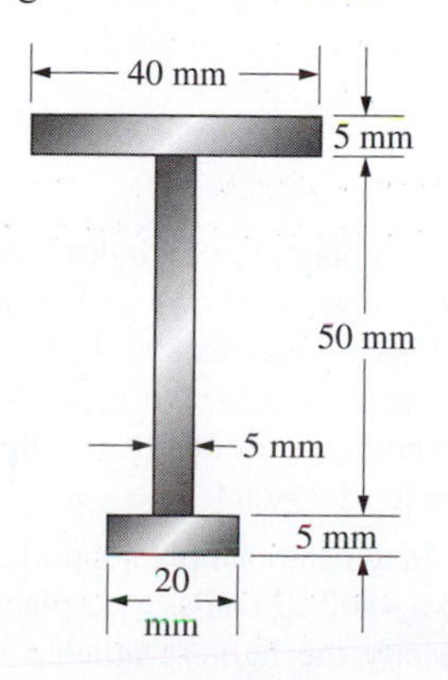

FIGURE P7–15

7–16.E A standard steel pipe is to be used as a chinning bar for personal exercise. The bar is to be 42 in long and simply supported at its ends. Specify a suitable size pipe if the bending stress is to be limited to 10 000 psi and a 280-lb man hangs by one hand in the middle.

7–17.E A pipeline is to be supported above ground on horizontal beams, 14 ft long. Consider each beam to be simply supported at its ends. Each beam carries the combined weight of 50 ft of 48-in-diameter pipe and the oil flowing through it, about 42 000 lb. Assuming the load acts at the center of the beam, specify the required section modulus of the beam to limit the bending stress to 20 000 psi. Then specify a suitable wide flange or American Standard beam.

7–18.E A wood platform is to be made of standard plywood and finished lumber using the cross section shown in Figure P7–18(a). Would the platform be safe if four men, weighing 250 lb each, were to stand 2 ft apart, as shown in Figure P7–18(b)? Consider only bending stresses (see Chapter 8 for shear stresses).

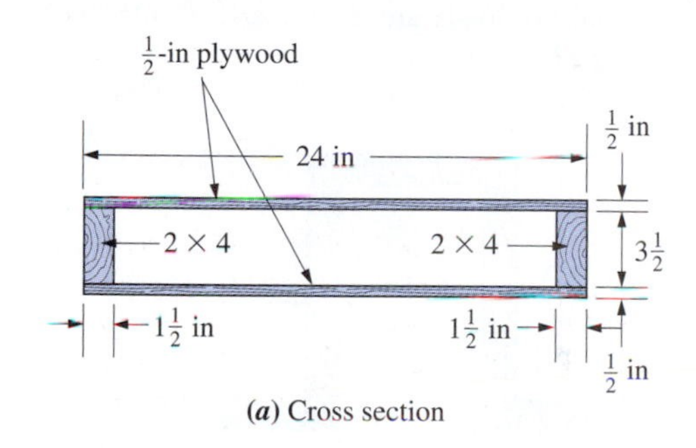

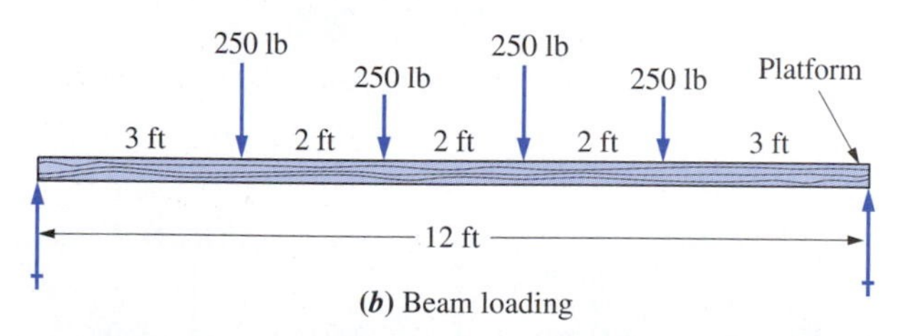

FIGURE P7–18

7–19.E A diving board has a hollow rectangular cross section 30 in wide and 3.0 in thick and is supported as shown in Figure P7–19. Compute the maximum stress due to bending in the board if a 300-lb person stands at the end. Would the board be safe if it were made of extruded 6061-T4 aluminum and the person landed at the end of the board with an impact?

7–20.M The loading shown in Figure P7–20(a) is to be carried by an extruded aluminum hat-section beam having the cross section shown in Figure P7–20(b). Compute the maximum stress due to bending in the

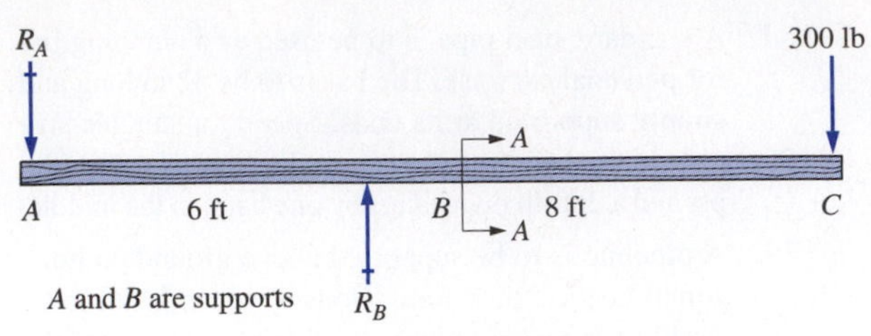

(*a*) Loads on diving board

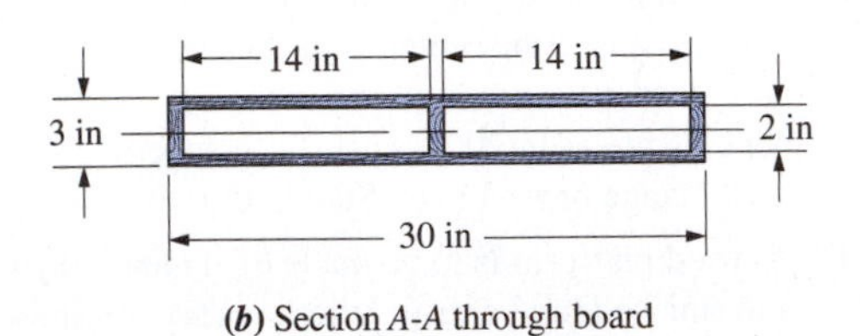

(*b*) Section *A-A* through board

FIGURE P7–19 Diving board for Problem 7–19.

beam. If it is made of extruded 6061-T4 aluminum and the loads are dead loads, would the beam be safe?

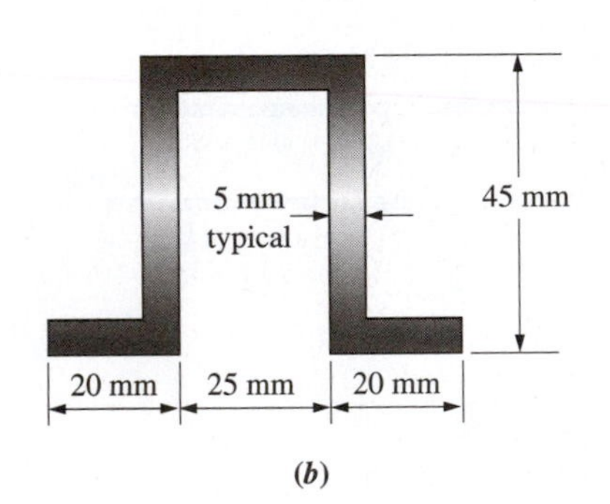

(*b*)

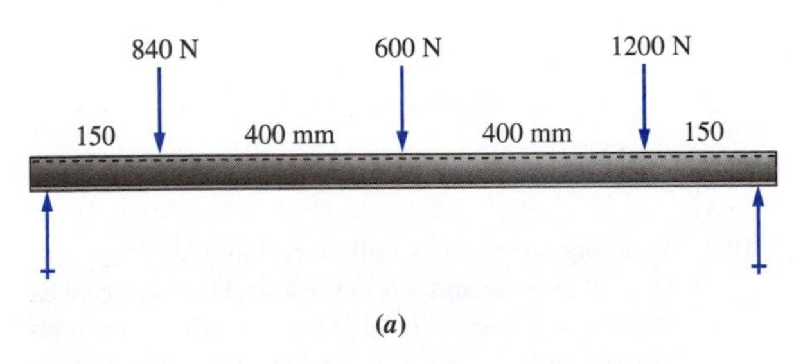

(*a*)

FIGURE P7–20

7–21.M The extruded shape shown in Figure P7–21(a) is to be used to carry the loads shown in Figure P7–21(b), which is a part of a business machine frame. The loads are due to a motor mounted on the frame and can be considered dead loads. Specify a suitable aluminum alloy for the beam.

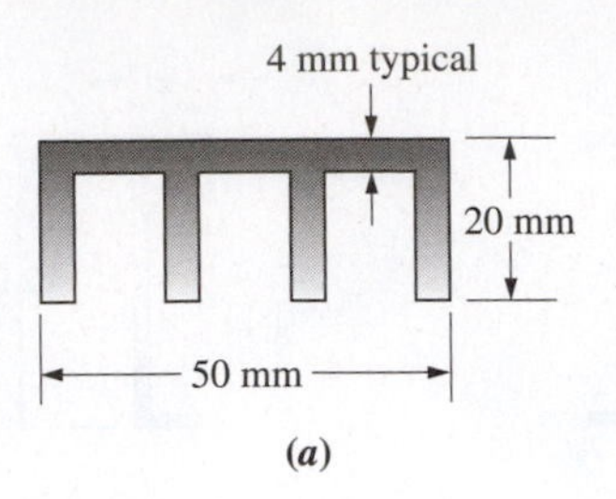

(*a*)

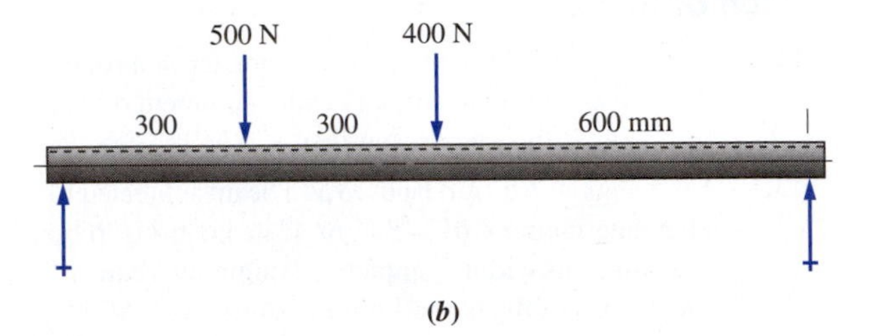

(*b*)

FIGURE P7–21

7–22.M A beam is being designed to support the loads shown in Figure P7–22. The four shapes proposed are (a) a round bar, (b) a square bar, (c) a rectangular bar with the height made four times the thickness, and (d) the lightest American Standard beam. Determine the required dimensions of each proposed shape to limit the maximum stress due to bending to 80 MPa. Then compare the magnitude of the cross-sectional areas of the four shapes. Since the weight of the beam is proportional to its area, the one with the smallest area will be the lightest.

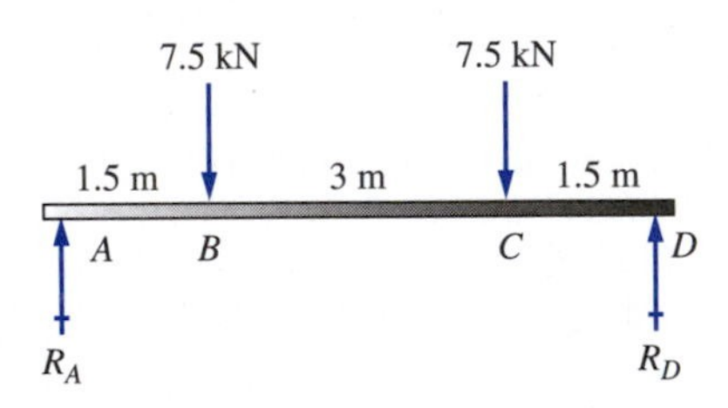

FIGURE P7–22

7–23.E A children's play gym includes a cross beam carrying four swings, as shown in Figure P7–23 Assume that each swing carries 300 lb. It is desired to use a standard steel pipe for the beam, keeping the stress due to bending below 10 000 psi. Specify the suitable size pipe for the beam.

7–24.E A 60 in long beam simply supported at its ends is to carry two 4800 lb loads, each placed 14 in from an end. Specify the lightest suitable steel tube for the beam, either square or rectangular, to produce a

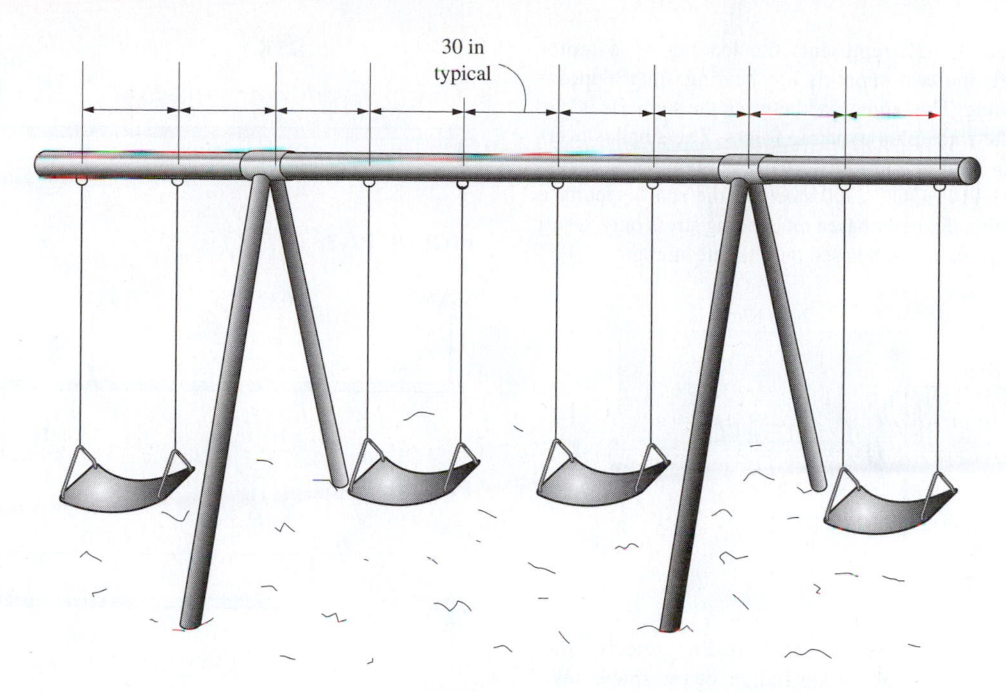

FIGURE P7–23

design factor of 4 based on yield strength. The tube is to be cold formed from ASTM A500, grade B steel.

7–25.E Repeat Problem 7–24, but specify the lightest standard aluminum I-beam from Appendix A–11. The beam will be extruded using alloy 6061-T6.

7–26.E Repeat Problem 7–24, but specify the lightest wide-flange steel shape from Appendix A–7. The beam will be made from ASTM A992 structural steel.

7–27.E Repeat Problem 7–24, but specify the lightest structural steel channel from Appendix A–6. The channel is to be installed with the legs down so that the loads can be applied to the flat back of the web of the channel. The channel will be made from ASTM A36 structural steel.

7–28.E Repeat Problem 7–24, but specify the lightest standard schedule 40 steel pipe from Appendix A–12. The pipe is to be made from ASTM A501 hot-formed steel.

7–29.E Repeat Problem 7–24, but design the beam using any material and shape of your choosing to achieve a safe beam that is lighter than any of the results from Problems 7–24 through 7–28.

7–30.E The shape shown in Figure P7–30 is to be made from extruded plastic and used as a simply supported beam, 12 ft long, to carry two electric cables weighing a total of 6.5 lb/ft of length. Specify a suitable plastic for the extrusion to provide a design factor of 4.0 based on flexural strength.

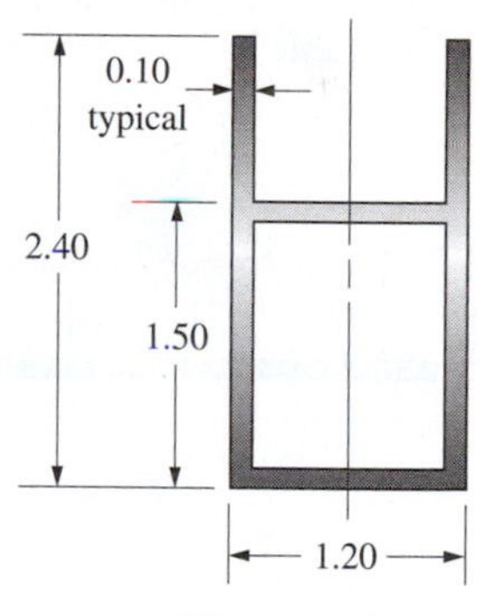

FIGURE P7–30

7–31.C The loading shown in Figure P7–31 represents the load on a floor beam of a commercial building. Determine the maximum bending moment on the beam, and then specify a wide-flange shape that will limit the stress to 150 MPa.

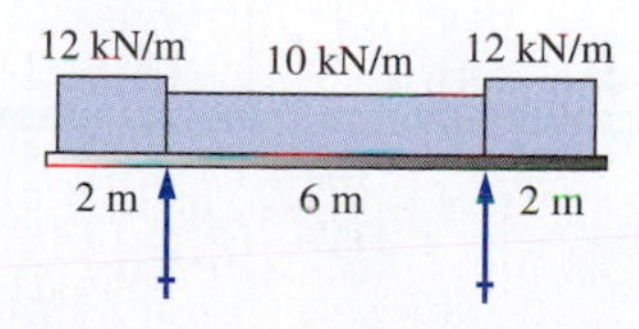

FIGURE P7–31

7–32.M Figure P7–32 represents the loading on a motor shaft; the two supports are bearings in the motor housing. The larger load between the supports is due to the rotor plus dynamic forces. The smaller overhung load is due to externally applied loads. Using AISI 1141 OQT 1300 steel for the shaft, specify a suitable diameter based on bending stress only. Use a design factor of 8 based on ultimate strength.

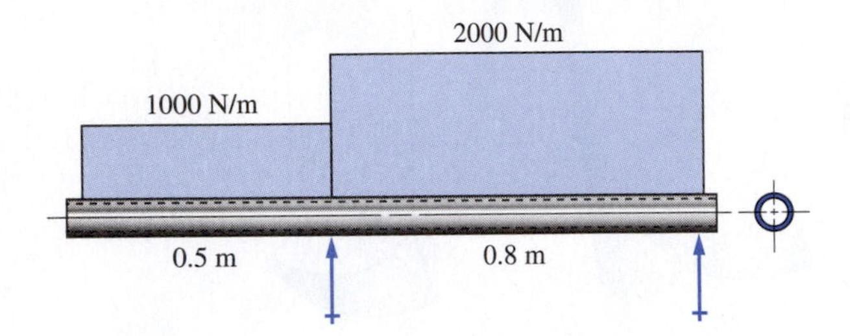

FIGURE P7–32

7–33 to 7–42. Using the indicated loading, specify the lightest standard wide-flange beam shape (W-shape) that will limit the stress due to bending to the allowable design stress from the AISC specification. All loads are static and the beams are made from ASTM A992 structural steel. Each problem number is the same as the corresponding figure number.

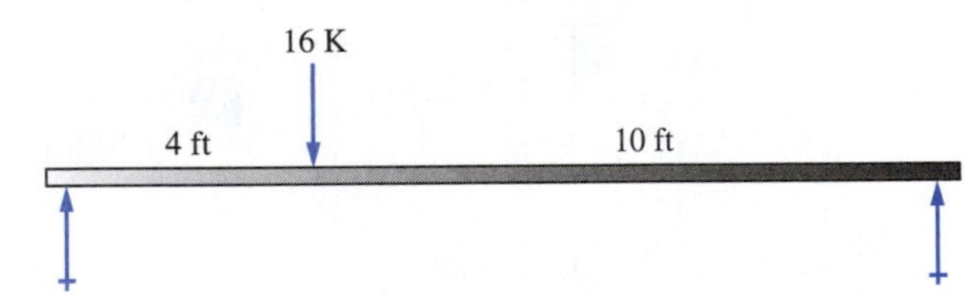

FIGURE P7–33

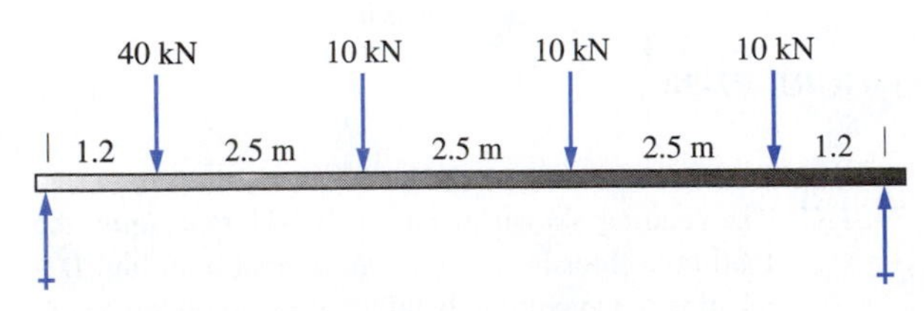

FIGURE P7–34

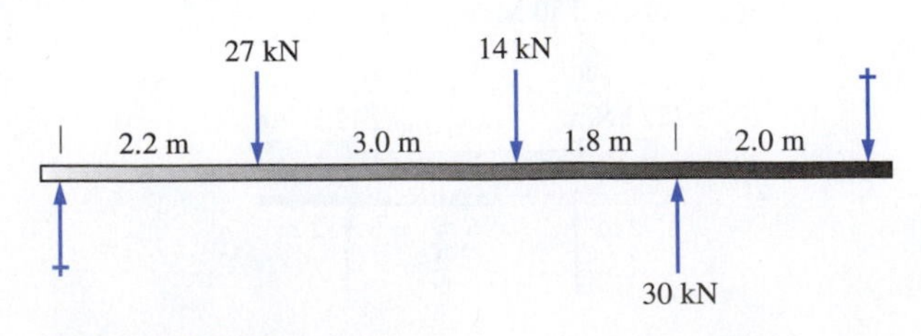

FIGURE P7–35

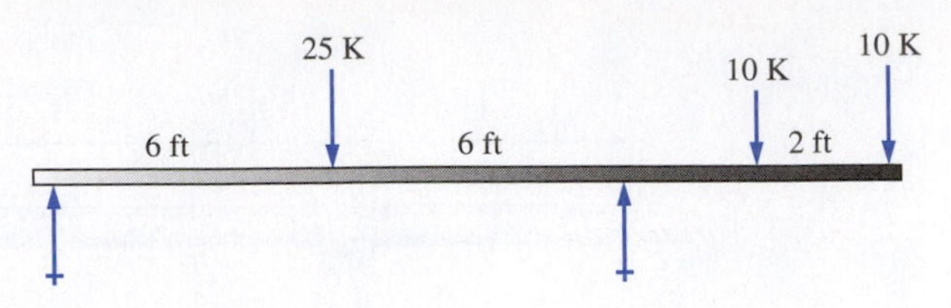

FIGURE P7–36

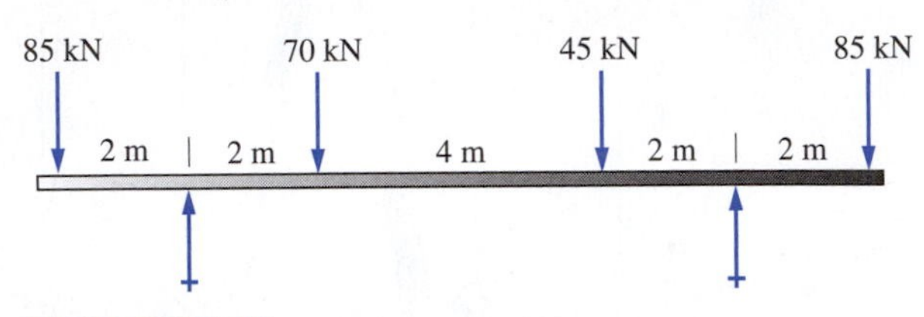

FIGURE P7–37

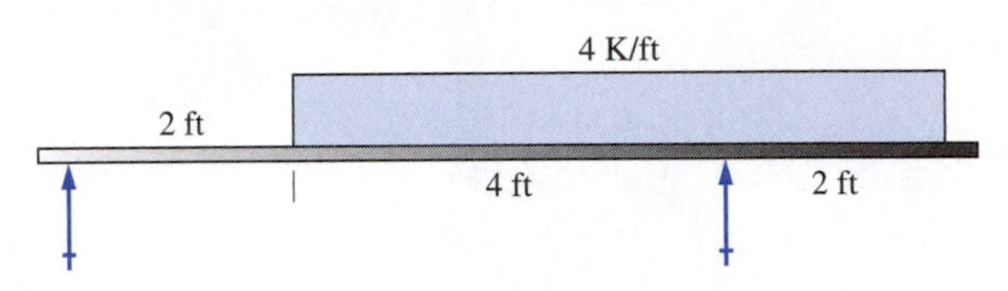

FIGURE P7–38

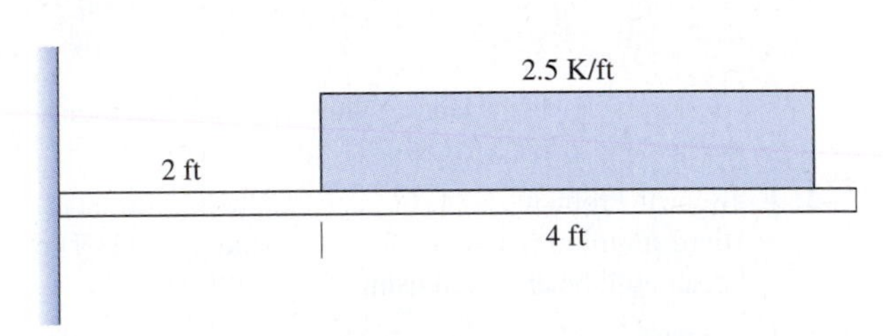

FIGURE P7–39

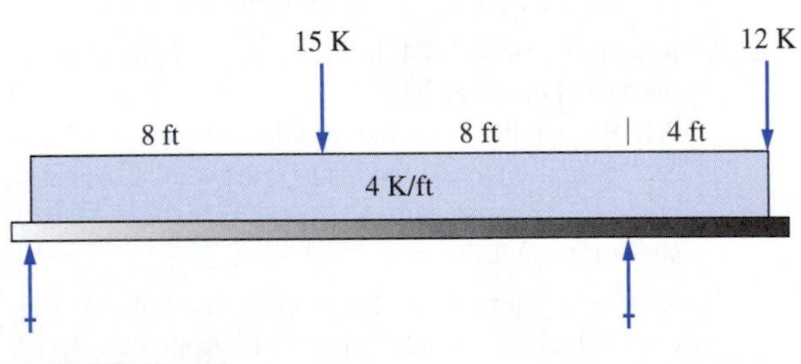

FIGURE P7–40

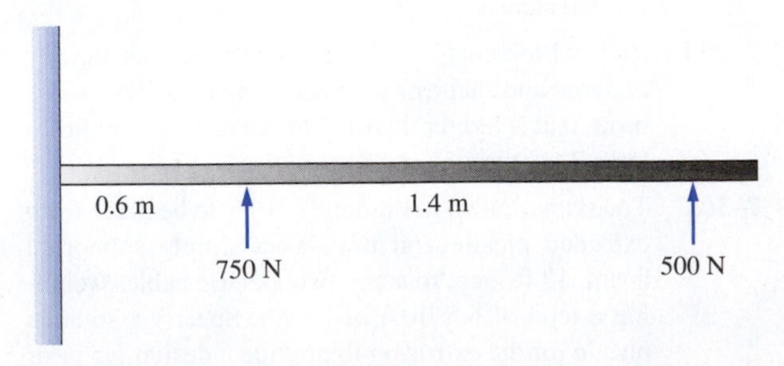

FIGURE P7–41

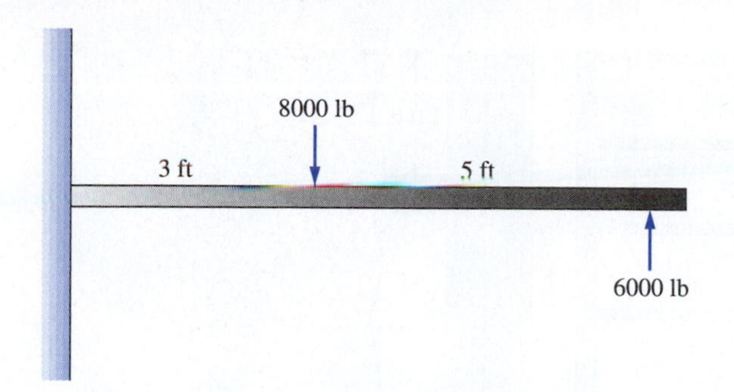

FIGURE 7–42

7–43 to 7–52. Repeat Problems 7–33 through 7–42 but specify the lightest American Standard Beam (S-shape).

7–53 to 7–62. Repeat Problems 7–33 through 7–42 but use ASTM A572 Grade 60 high-strength low-alloy structural steel.

7–63.E A floor joist for a building is to be made from a standard wooden beam selected from Appendix A–4. If the beam is to be simply supported at its ends and carry a uniformly distributed load of 125 lb/ft over the entire 10-ft length, specify a suitable beam size. The beam will be made from No. 2 grade southern pine. Consider only bending stress.

7–64.E A bench for football players is to carry the load shown in Figure P7–64 approximating the case when 10 players, each weighing 300 lb, sit close together, each taking 18 in of the length of the bench. If the cross section of the bench is made as shown in Figure P7–64, would it be safe for bending stress? The wood is No. 2 grade hemlock.

7–65.E A bench is to be designed for football players. It is to carry the load shown in Figure P7–64 approximating the case when 10 players, each weighing 300 lb, sit close together, each taking 18 in of the length of the bench. The bench is to be T-shaped, made from No. 2 grade hemlock, as shown with a 2×12 top board. Specify the required vertical member of the tee if the bench is to be safe for bending stress.

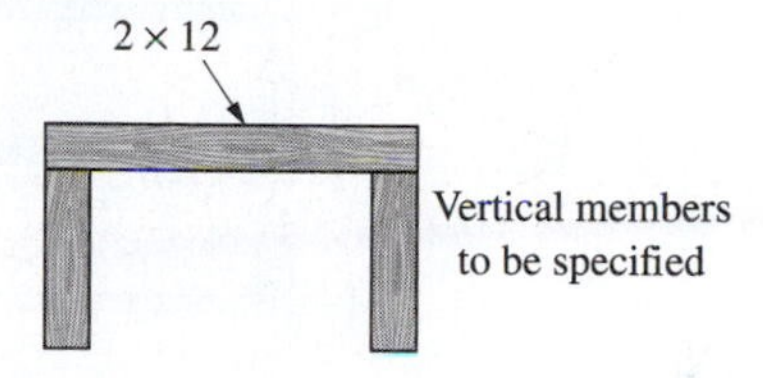

FIGURE P7–66 Shape for bench cross section for Problem 7–66.

7–66.E Repeat Problem 7–65, but use the cross-section shape shown in Figure P7–66.

7–67.E Repeat Problem 7–65, but use any cross-section shape of your choosing made from standard wooden beams from Appendix A–4. Try to achieve a lighter design than in Problem 7–65 or 7–66. Note that a lighter design would have a smaller cross-section area.

7–68.E A wood deck is being designed to carry a uniformly distributed load over its entire area of 100 lb/ft^2. Joists are to be used as shown in Figure P7–68, set 16 in on center. If the deck is to be 8 ft by 12 ft in size,

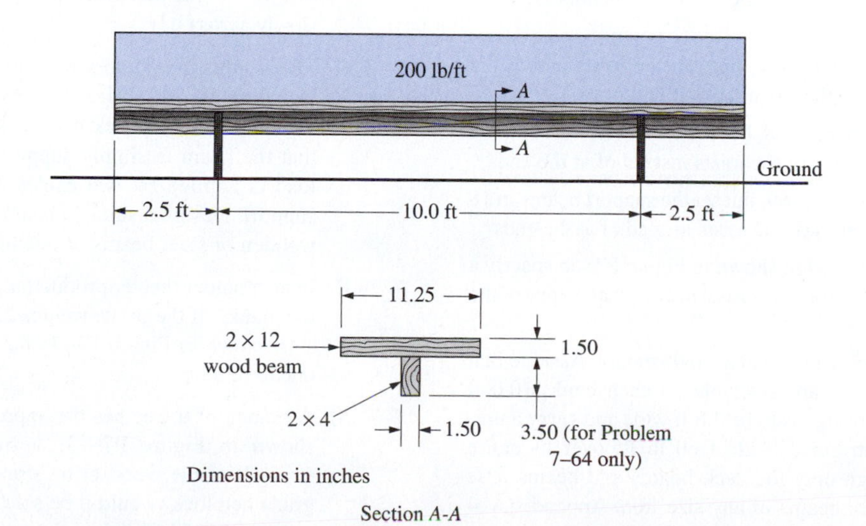

FIGURE P7–64 Bench and load for Problems 7–64, 7–65, 7–66, and 7–67.

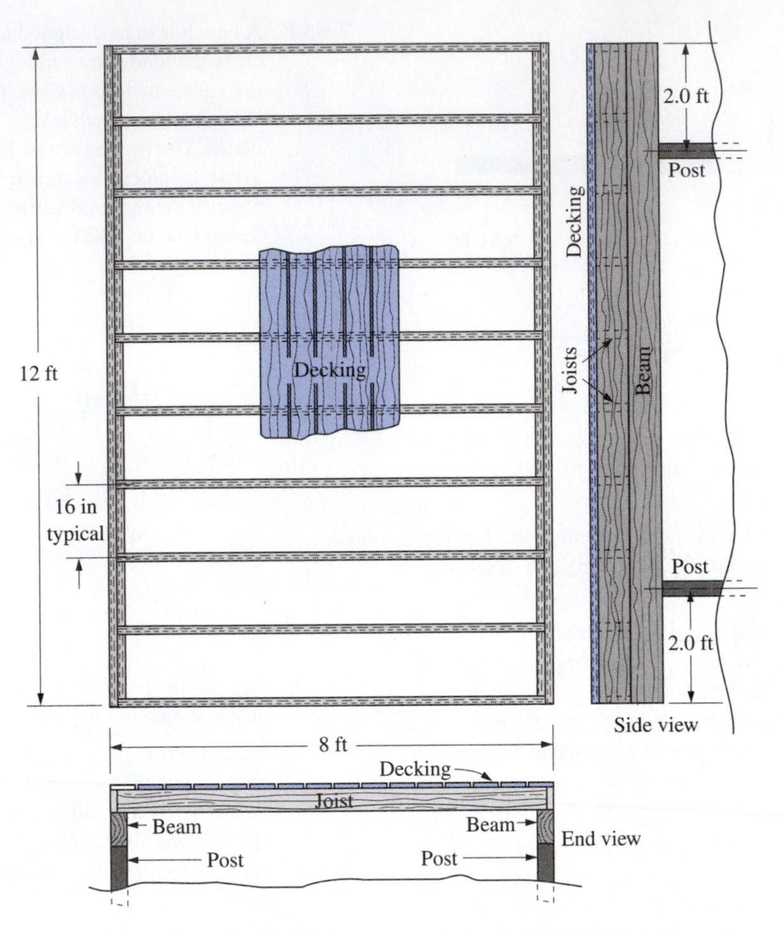

FIGURE P7–68 Deck design for Problem 7–68.

determine the required size for the joists. Use standard wooden beam sections from Appendix A–4 and No. 2 hemlock.

7–69.E Repeat Problem 7–68, but run the joists across the 12-ft length rather than the 8-ft width.

7–70.E Repeat Problem 7–68, but set the support beam in 18 in from the ends of the joists instead of at the ends.

7–71.E Repeat Problem 7–69, but set the support beams in 18 in from the ends of the joists instead of at the ends.

7–72.E For the deck design shown in Figure P7–68 specify a suitable size for the cross beams that support the joists.

7–73.E Design a bridge to span a small stream. Assume that rigid supports are available on each bank, 10.0 ft apart. The bridge is to be 3.0 ft wide and carry a uniformly distributed load of 60 lb/ft^2 over its entire area. Design only the deck boards and beams. Use two or more beams of any size from Appendix A–4 or others of your own design.

7–74.E Would the bridge you designed in Problem 7–73 be safe if a horse and rider weighing 2200 lb walked slowly across it?

7–75.E Millwrights in a factory need to suspend a machine weighing 10 500 lb from a beam having a span of 12.0 ft so that a truck can back under it. Assume that the beam is simply supported at its ends. The load is applied by two cables, each 3.0 ft from a support. Design a suitable beam. Consider standard wooden or steel beams or one of your own design.

7–76.E In an amateur theater production, a pirate is to "walk the plank." If the pirate weighs 220 lb, would the design shown in Figure P7–76 be safe? If not, design one to be safe.

7–77.M A branch of a tree has the approximate dimensions shown in Figure P7–77. Assuming the bending strength of the wood to be similar to that of No. 3 grade hemlock, would it be safe for a person having a mass of 135 kg to sit in the swing?

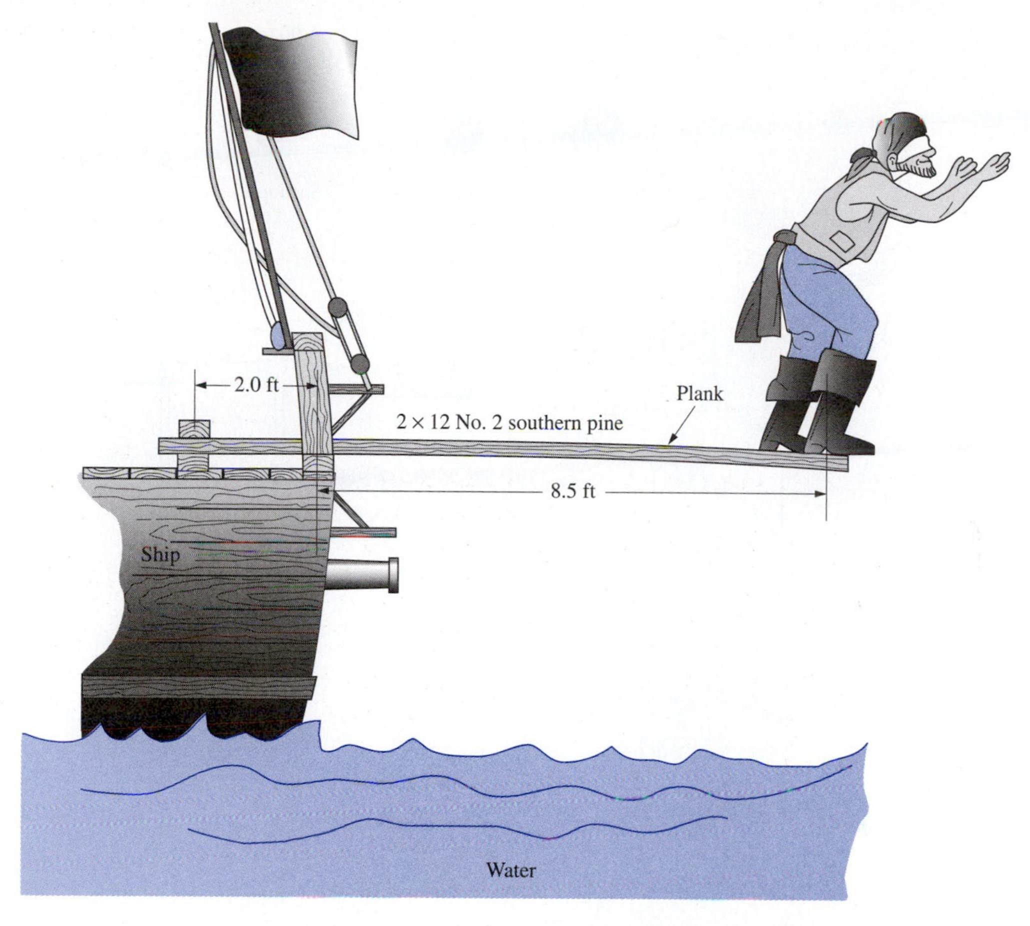

FIGURE P7–76 Pirate walking the plank in Problem 7–76.

7–78.E Would it be safe to use a standard 2×4 made from No. 2 grade southern pine as a lever as shown in Figure P7–78 to lift one side of a machine? If not, what would you suggest be used?

7–79.M Figure P7–79 shows the cross section of an extruded plastic beam made from nylon 66 dry. Specify the largest uniformly distributed load in N/mm the beam could carry if it is simply supported with a span of 0.80 m. The maximum stress due to bending is not to exceed one-half of the flexural strength of the nylon.

7–80.M The I-beam shape in Figure P7–80 is to carry two identical concentrated loads of 2.25 kN each, symmetrically placed on a simply supported beam with a span of 0.60 m. Each load is 0.2 m from an end. Which of the plastics from Appendix A–20 would carry these loads with a stress due to bending no more than one-third of their flexural strength?

7–81.M A bridge for a toy construction set is to be made from acetal copolymer with the flexural strength listed in Appendix A–20. The cross section of one beam is shown in Figure P7–81. What maximum concentrated load could be applied to the middle of the beam if the span was 1.25 m and the ends are simply supported? Do not exceed one-half of the flexural strength of the plastic.

7–82.M A structural element in a computer printer is to carry the load shown in Figure P7–82(a) with the uniformly distributed load representing electronic components mounted on a printed circuit board and the concentrated loads applied from a power supply. It is proposed to make the beam from polycarbonate with the cross section shape shown in Figure P7–82(b). Compute the stress in the beam and compare it with the flexural strength for polycarbonate from Appendix A–20.

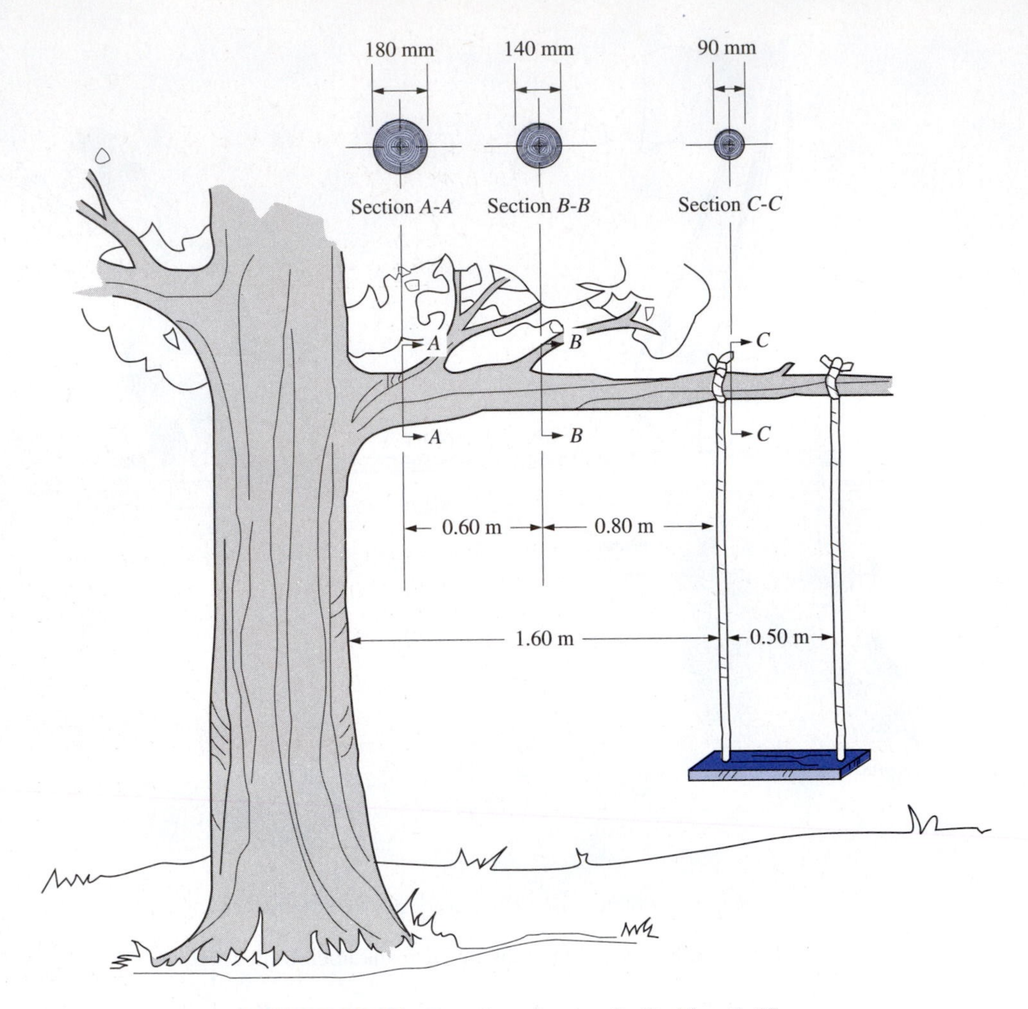

FIGURE P7–77 Branch and swing for Problem 7–77.

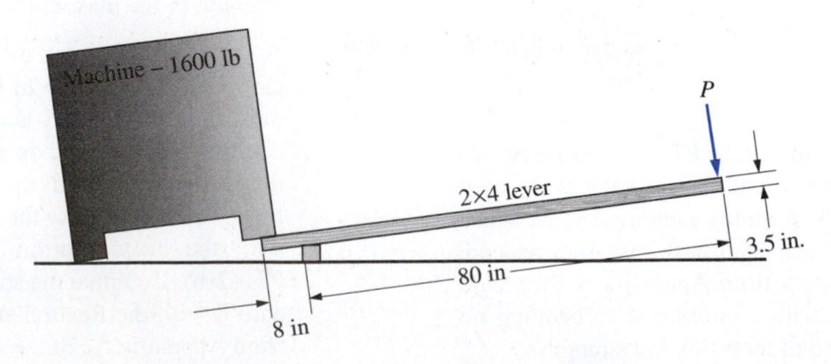

FIGURE P7–78 2×4 used as a lever in Problem 7–78.

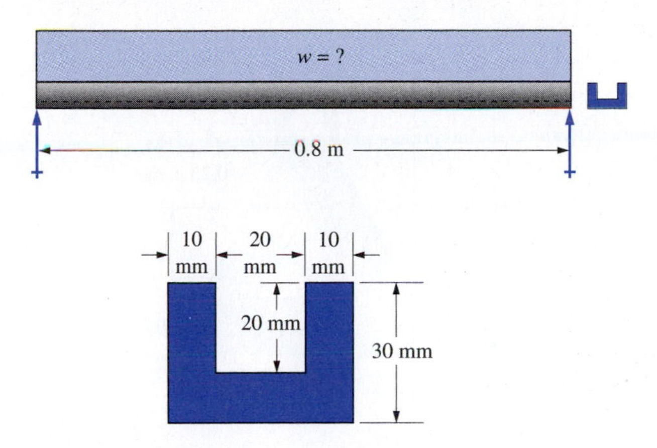

FIGURE P7–79

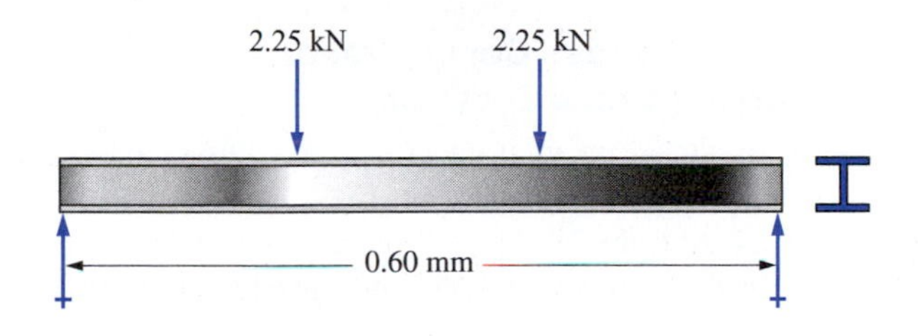

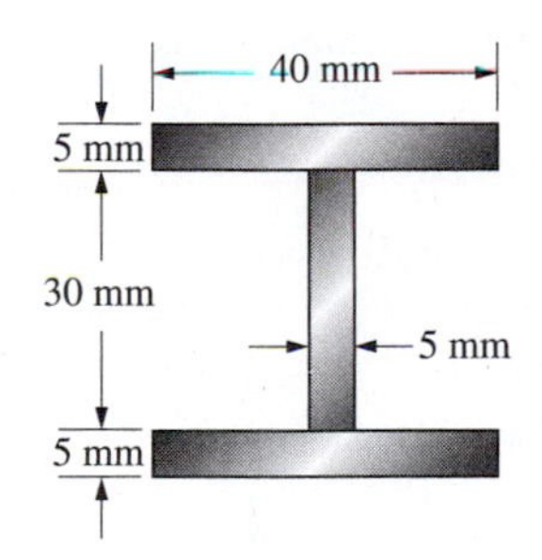

FIGURE P7–80

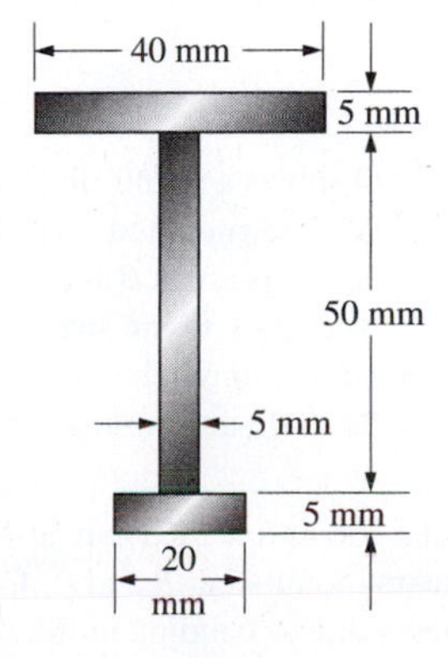

FIGURE P7–81

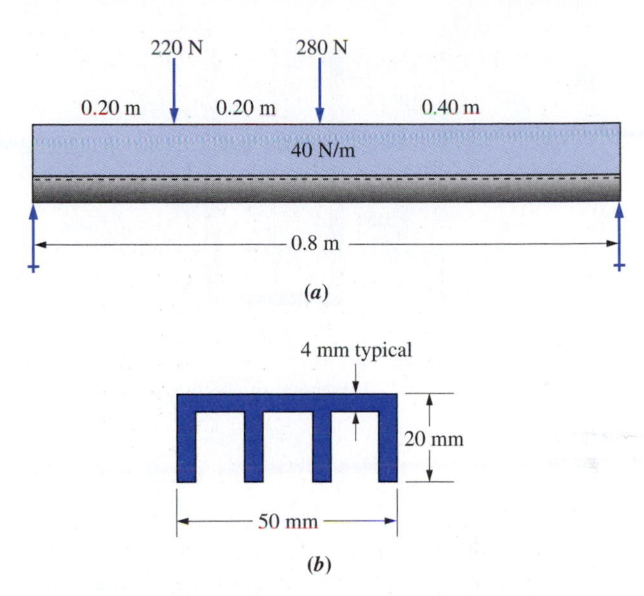

FIGURE P7–82 Beam in a computer printer for Problem 7–82.

7–83.E It is proposed to make the steps for a child's slide from molded polycarbonate with the cross section shown in Figure P7–83. The steps are to be 14.0 in wide and will be simply supported at the ends. What maximum weight can be applied at the center of the step if the stress due to bending must not exceed one-third of the flexural strength of the polycarbonate listed in Appendix A–20?

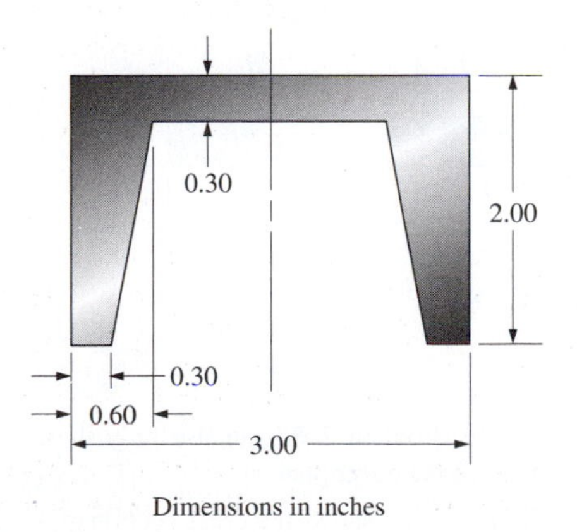

FIGURE P7–83

7–84.E The shape shown in Figure P7–84 is used as a beam for a carport. The span of the beam will be 8.0 ft. Compute the maximum uniformly distributed load that can be carried by the beam if it is extruded from 6061-T4 aluminum. Use a design factor of 2 based on yield strength.

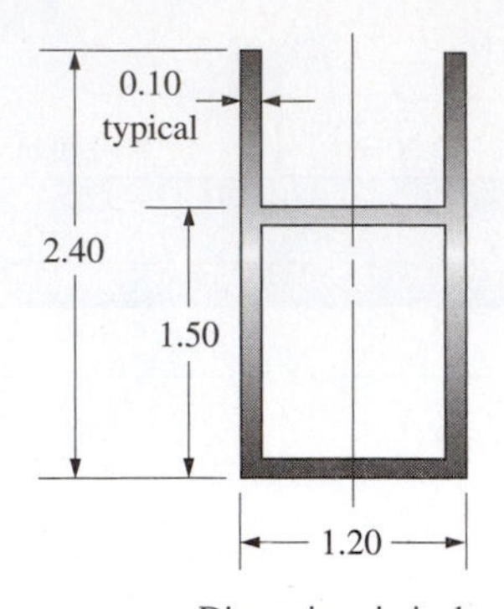

FIGURE P7–84

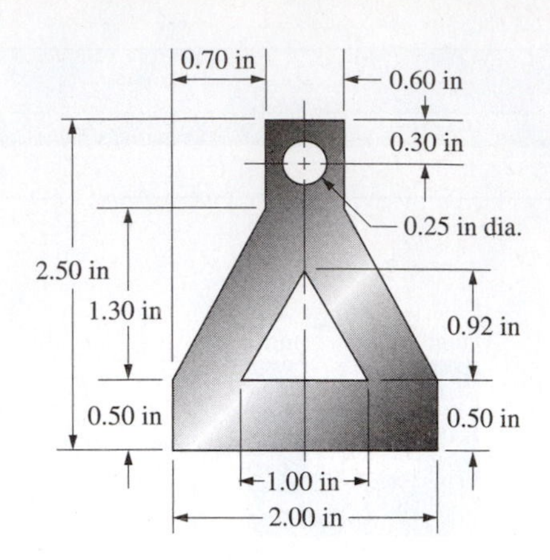

FIGURE P7–87

7–85.E Figure P7–85 shows the cross section of an aluminum beam made by fastening a flat plate to the bottom of a roll-formed hat section. If the beam is used as a cantilever, 24 in long, compute the maximum allowable concentrated load that can be applied at the end if the maximum stress is to be no more than one-eighth of the ultimate strength of 2014-T4 aluminum.

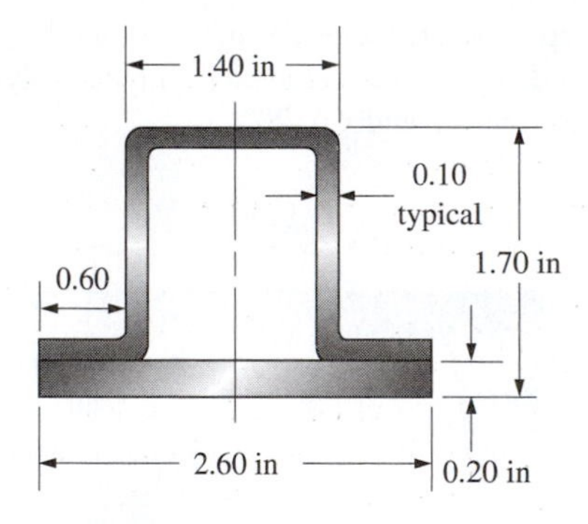

FIGURE P7–85

7–86.E Repeat Problem 7–85 but use only the hat section without the cover plate.

7–87.E Figure P7–87 shows the cross section of a beam that is to be extruded from aluminum 6061-T6 alloy. If the beam is to be used as a cantilever, 42 in long, compute the maximum allowable uniformly distributed load it could carry while limiting the stress due to bending to one-sixth of the ultimate strength of the aluminum.

7–88.E Repeat Problem 7–87, but use Figure P7–83 for the cross section and casting alloy 356.0-T6.

Beams with Stress Concentrations and Varying Cross Sections

7–89.E In Figure P7–89 the 4-in pipe mates smoothly with its support so that no stress concentration exists at D. At C the $3\frac{1}{2}$-in pipe is placed inside the 4-in pipe with a spacer ring to provide a good fit. Then a $\frac{1}{4}$-in smoothly radiused fillet weld is used to secure the section together. Accounting for the stress concentration at the joint, determine how far out point C must be to limit the stress to 20 000 psi. Use Appendix A–22–9 for the stress concentration factor. Is the 4-in pipe safe at D?

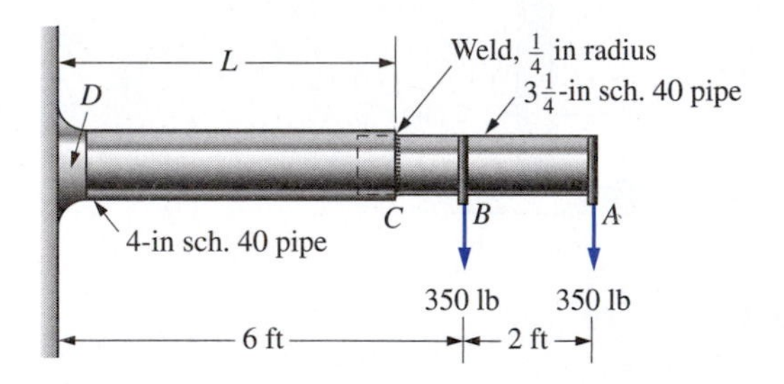

FIGURE P7–89

7–90.M Figure P7–90 shows a round shaft from a gear transmission. Gears are mounted at points A, C, and E. Supporting bearings are at B and D. The forces transmitted from the gears to the shaft are shown, all acting downward. Compute the maximum stress due to bending in the shaft, accounting for stress concentrations.

7–91.M The forces shown on the shaft in Figure P7–91 are due to gears mounted at B and C. Compute the maximum stress due to bending in the shaft.

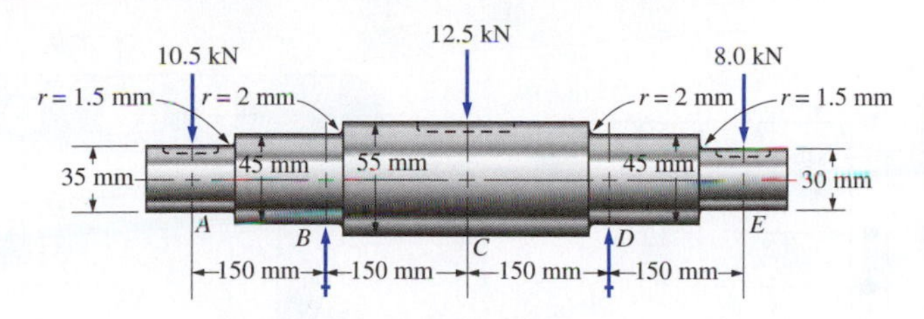

FIGURE P7–90

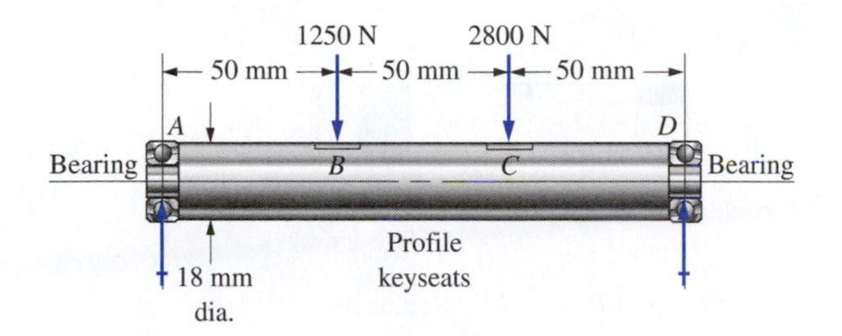

FIGURE P7–91

7–92.E Figure P7–92 shows a machine shaft supported by two bearing at its ends. The two forces are exerted on the shaft by gears. Considering only bending stresses, compute the maximum stress in the shaft and tell where it occurs.

7–93.E Figure P7–93 shows a lever made from a rectangular bar of steel. Compute the stress due to bending at the fulcrum, 20 in from the pivot and at each of the holes in the bar. The diameter of each hole is 0.75 in.

7–94.E Repeat Problem 7–93 but use the diameter of the holes as 1.38 in.

7–95.E In Figure P7–93, the holes in the bar are provided to permit the length of the lever to be changed relative to the pivot. Compute the maximum bending stress in the lever as the pivot is moved to each hole. Use the diameter of the holes as 1.25 in.

7–96.M The bracket shown in Figure P7–96 carries the opposing forces created by a spring. If the force, F, is 2500 N, compute the bending stress at a section such as A-A, away from the holes.

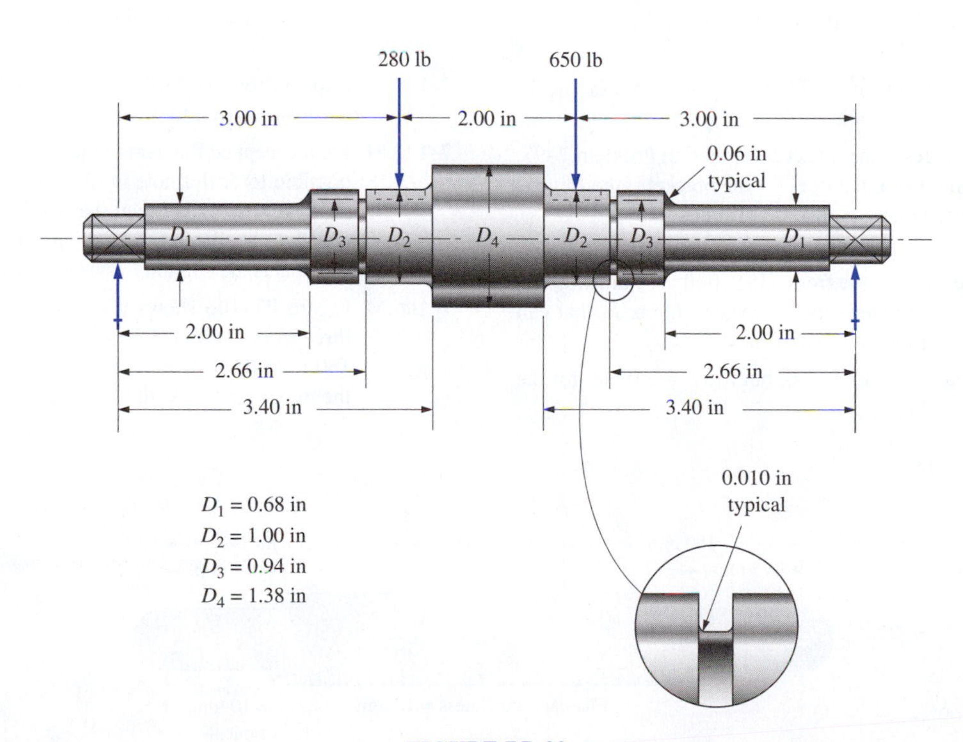

FIGURE P7–92

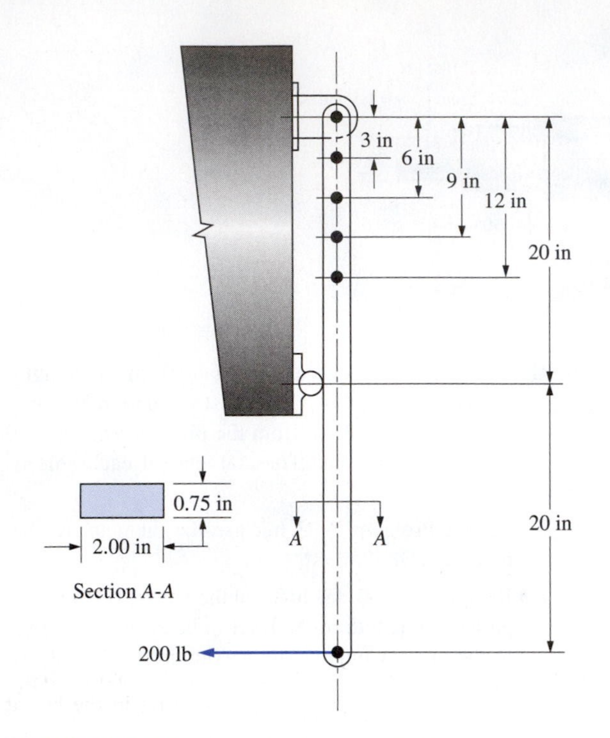

FIGURE P7–93 Lever for Problems 7–93 through 7–95.

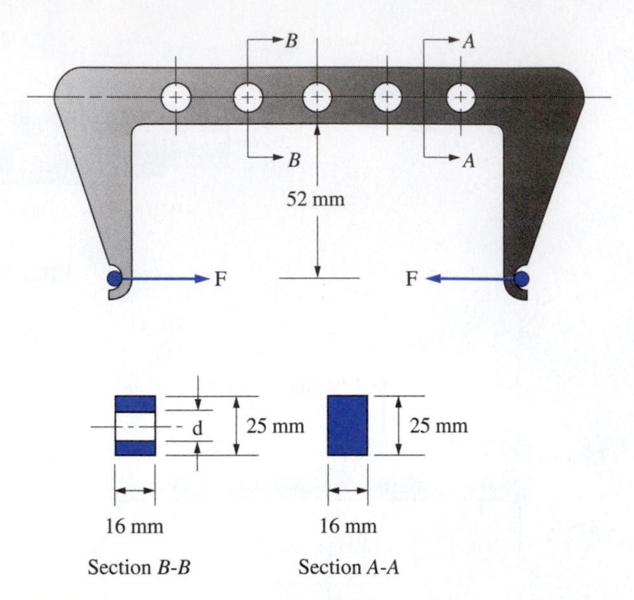

FIGURE P7–96 Bracket for Problems 7–96 through 7–99.

7–97.M If the force, F, in Figure P7–96 is 2500 N, compute the bending stress at a section through the holes, such as B-B. Use $d = 12$ mm for the diameter of the holes.

7–98.M Repeat Problem 7–97 but use $d = 15$ mm for the diameter of the holes.

7–99.M For the resulting stress computed in Problem 7–98, specify a suitable steel for the bracket if the force is repeated many thousands of times.

7–100.M Figure P7–100 shows a stepped flat bar in bending. If the bar is made from AISI 1040 cold-drawn steel, compute the maximum repeated force, F, that can safely be applied to the bar.

7–101.M Repeat Problem 7–100, but use $r = 2.0$ mm for the fillet radius.

7–102.M For the stepped flat bar shown in Figure P7–100, change the 75-mm dimension that locates the step to a value that makes the bending stress at the step equal to that at the point of application of the load.

7–103.M For the stepped flat bar shown in Figure P7–100, change the size of the fillet radius to make the bending stress at the fillet equal to that at the point of application of the load.

7–104.M Repeat Problem 7–100, but change the depth of the bar from 60 mm to 75 mm.

7–105.M For the stepped flat bar in Figure P7–100, would it be possible to drill a hole in the middle of the 60-mm depth of the bar between the two forces without increasing the maximum bending stress in the bar? If so, what is the maximum size hole that can be put in?

7–106.M Figure P7–106 shows a stepped flat bar carrying three concentrated loads. Let $P = 200$ N, $L_1 = 180$ mm, $L_2 = 80$ mm, and $L_3 = 40$ mm. Compute the maximum stress due to bending and state where

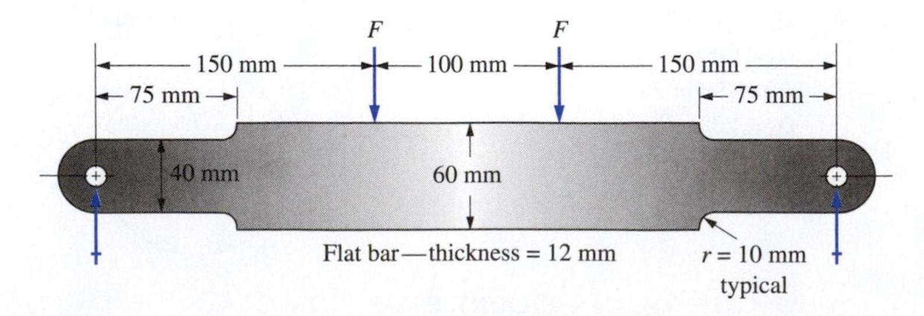

FIGURE P7–100 Stepped flat bar for Problems 7–100 through 7–105.

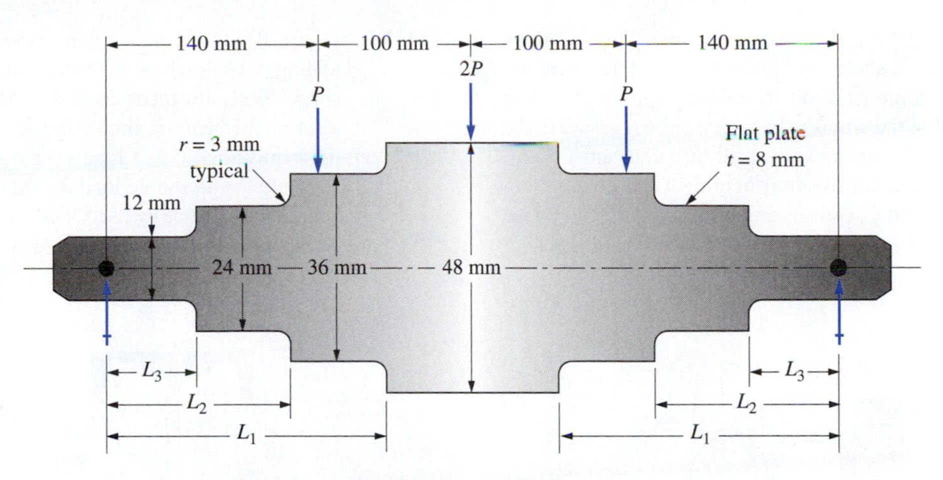

FIGURE P7–106 Stepped flat bar for Problems 7–106 through 7–110.

it occurs. The bar is braced against lateral bending and twisting. Note that the length dimensions in the figure are not drawn to scale.

7–107.M For the data of Problem 7–106, specify a suitable material for the bar to produce a design factor of 8 based on ultimate strength.

7–108.M Repeat Problem 7–107 except use $r = 1.50$ mm for the fillet radius.

7–109.M For the stepped flat bar shown in Figure P7–106, let $P = 400$ N. The bar is to be made from titanium, Ti-6Al-4V, and a design factor of 8 based on ultimate strength is desired. Specify the maximum permissible lengths, L_1, L_2, and L_3, that would be safe.

7–110.M The stepped flat bar in Figure P7–106 is to be made from AISI 4140 OQT 1100 steel. Use $L_1 = 180$ mm, $L_2 = 80$ mm, and $L_3 = 40$ mm. Compute the maximum allowable force P that could be applied to the bar if a design factor of 8 based on ultimate strength is desired.

7–111.M Figure P7–111 shows a flat bar that has a uniform thickness of 20 mm. The depth tapers from $h_1 = 40$ mm to $h_2 = 20$ mm in order to save weight. Compute the stress due to bending in the bar at points spaced 40 mm apart from the support to the load. Then create a graph of stress versus distance from the support. The bar is symmetrical with respect to its middle. Let $P = 5.0$ kN.

7–112.M For the bar shown in Figure P7–111, let $h_1 = 60$ mm and $h_2 = 20$ mm. The bar is to be made from polycarbonate plastic. Compute the maximum permissible load P that will produce a design factor of 4 based on the flexural strength of the plastic. The bar is symmetrical with respect to its middle.

7–113.M In Figure P7–111, the load $P = 1.20$ kN and the bar is to be made from AISI 5160 OQT 1300 steel. Compute the required dimensions h_1 and h_2 that will produce a design factor of 8 based on ultimate strength. The bar is symmetrical with respect to its middle.

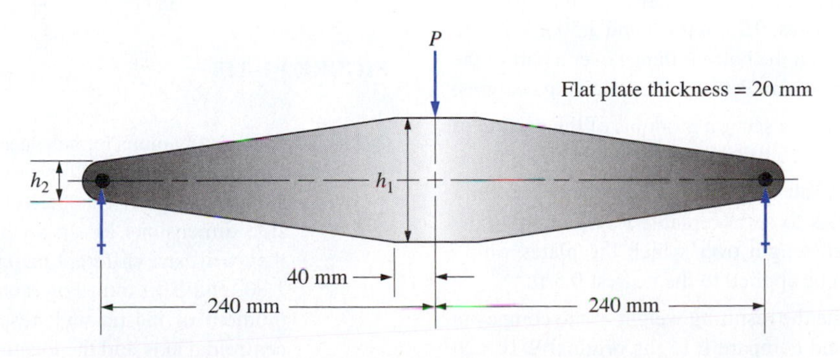

FIGURE P7–111 Tapered flat bar for Problems 7–111 through 7–113.

7–114.E A rack is being designed to support large sections of pipe, as shown in Figure P7–114. Each pipe exerts a force of 2500 lb on the support arm. The height of the arm is to be tapered as suggested in the figure, but the thickness will be a constant 1.50 in. Determine the required height of the arm at sections *B* and *C*, considering only bending stress. Use AISI 1040 hot-rolled steel for the arms and a design factor of 4 based on yield strength.

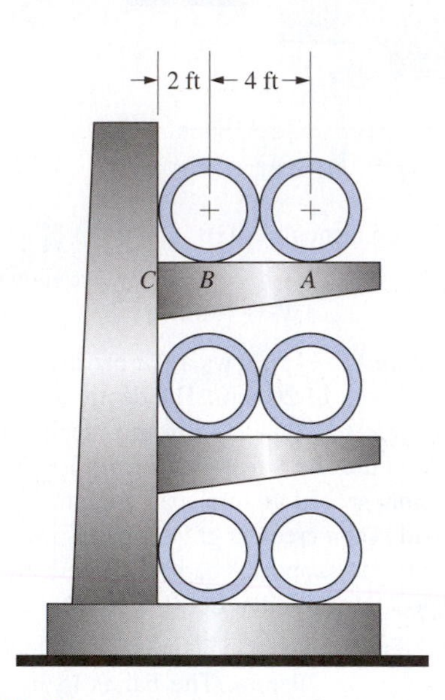

FIGURE P7–114 Pipe storage rack for Problem 7–114.

7–115.E Specify the lightest wide-flange beam shape (W-shape) that can carry a uniformly distributed static load of 2.5 kip/ft over the entire length of a simply supported span, 12.0 ft long. Use the AISC specification and ASTM A992 structural steel.

7–116.E A proposal is being evaluated to save weight for the beam application of Problem 7–115. The result for that problem required a W14×26 beam that would weigh 312 lb for the 12-ft length. A W12×16 beam would only weigh 192 lb but does not have sufficient section modulus *S*. To increase *S*, it is proposed to add steel plates, 0.25 in thick and 3.50 in wide, to both the top and the bottom flange over a part of the middle of the beam. Perform the following analyses:

(a) Compute the section modulus of the portion of the W12×16 beam with the cover plates.

(b) If the result of part (a) is satisfactory to limit the stress to an acceptable level, compute the required length over which the plates would have to be applied to the nearest 0.5 ft.

(c) Compute the resulting weight of the composite beam and compare it to the original W14×26 beam.

7–117.E Figure P7–117 shows a composite beam made by adding a channel to an American Standard beam shape. Both are made from ASTM A36 structural steel. If the beam is simply supported and carries a uniformly distributed load over a span of 15.0 ft, compute the allowable load for the composite beam and for the S-shape by itself. The load is static and the AISC specification for design stress is to be used.

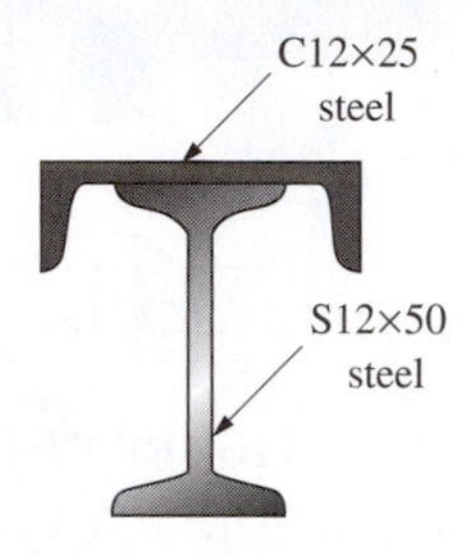

FIGURE P7–117

Flexural Center

7–118.M Compute the location of the flexural center of a channel-shaped member shown in Figure P7–118 measured from the left of the vertical web.

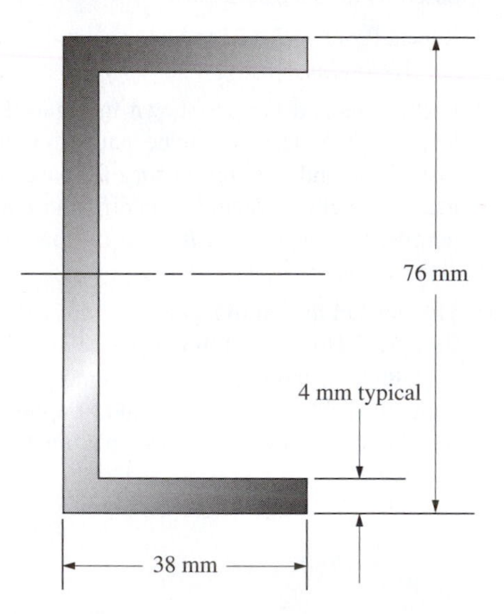

FIGURE P7–118

7–119.M A company plans to make a series of three channel-shaped beams by rollforming them from flat-sheet aluminum. Each channel is to have the same outside dimensions as shown in Figure P7–118, but they will have different material thicknesses, 0.50, 1.60, and 3.00 mm. For each design, compute the moment of inertia with respect to the horizontal centroidal axis and the location of the flexural center, measured from the left face of the vertical web.

7–120.E Compute the location of the flexural center for the hat section shown in Figure P7–120 measured from the left face of the vertical web.

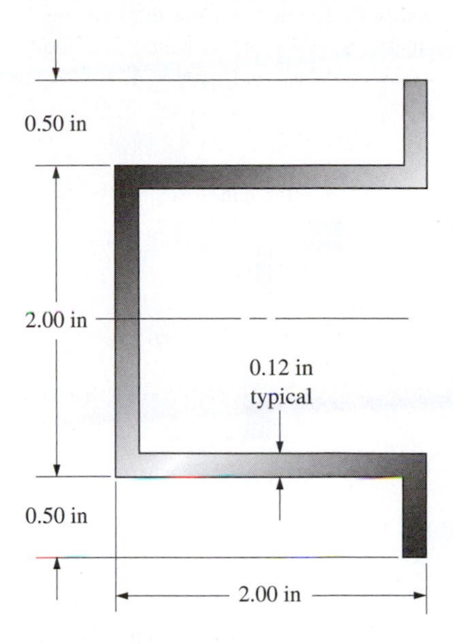

FIGURE P7–120

7–121.E A company plans to make a series of three hat sections by rollforming them from flat-sheet aluminum. Each hat section is to have the same outside dimensions as shown in Figure P7–120, but they will have different material thicknesses, 0.020, 0.063, and 0.125 in. For each design, compute the location of the flexural center, measured from the left face of the vertical web.

7–122.M Compute the location of the flexural center for the lipped channel shown in Figure P7–122, measured from the left face of the vertical web.

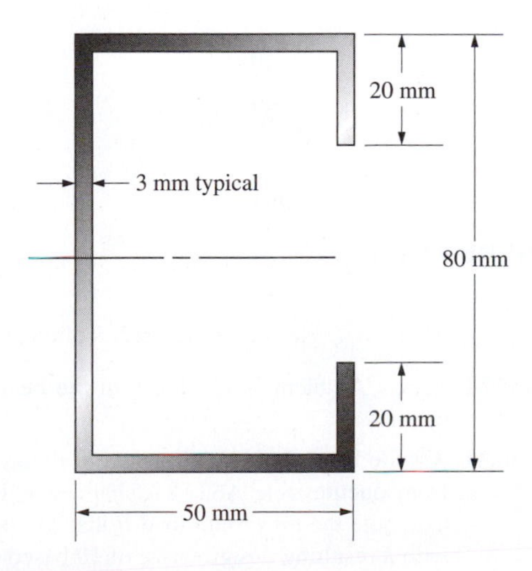

FIGURE P7–122

7–123.M A company plans to make a series of three lipped channels by rollforming them from flat-sheet aluminum. Each channel section is to have the same outside dimensions as shown in Figure P7–122, but they will have different material thicknesses, 0.50, 1.60, and 3.00 mm. For each design, compute the location of the flexural center, measured from the left face of the vertical web.

7–124.M Compute the location of the flexural center of a split, thin tube if it has an outside diameter of 50 mm and a wall thickness of 4 mm.

7–125.E For an aluminum channel C2×0.577 with its web oriented vertically, compute the location of its flexural center. Neglect the effect of the fillets between the flanges and the web.

7–126.M If the hat section shown in Figure P7–126 were turned 90 deg from the position shown, compute the location of its flexural center.

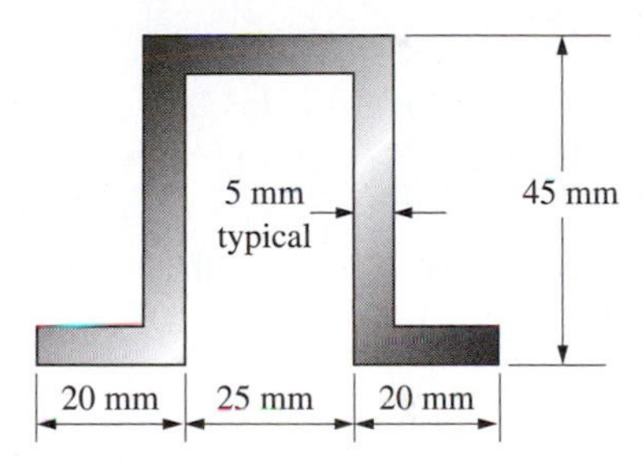

FIGURE P7–126

Beams Made from Anisotropic Materials

7–127.E The beam section shown in Figure P7–127 is to be extruded from 6061-T6 aluminum. The allowable tensile strength is 19 ksi. Because of the relatively thin extended legs on the top, the allowable compressive strength is only 14 ksi. The beam is to span 6.5 ft and will be simply supported at its ends. Compute the maximum allowable uniformly distributed load on the beam.

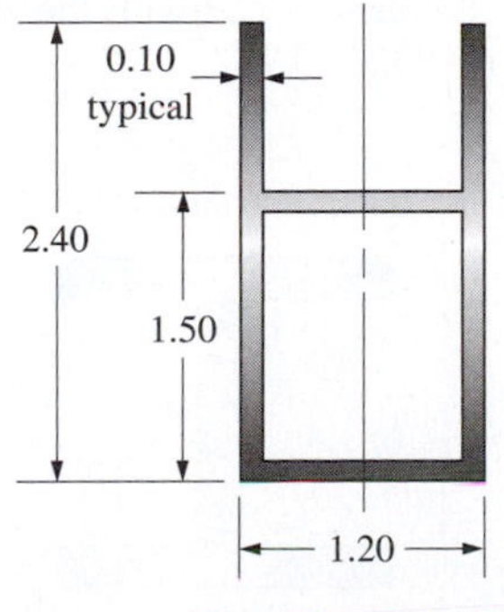

FIGURE P7–127

7–128.E Repeat Problem 7–127 but turn the section upside down. With the legs pointed downward, they are in tension and can withstand 19 ksi. The part of the section in compression at the top is now well supported and can withstand 21 ksi.

7–129.M The shape in Figure P7–129 is to carry a single concentrated load at the center of a 1200-mm span. The allowable strength in tension is 100 MPa, while the allowable strength in compression anywhere is 70 MPa. Compute the allowable load.

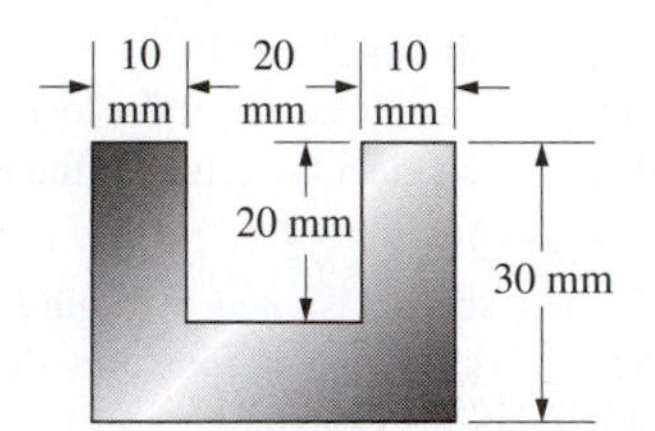

FIGURE P7–129

7–130.M Repeat Problem 7–129 with the section turned upside down.

7–131.M Repeat Problem 7–129 with the shape shown in Figure P7–131.

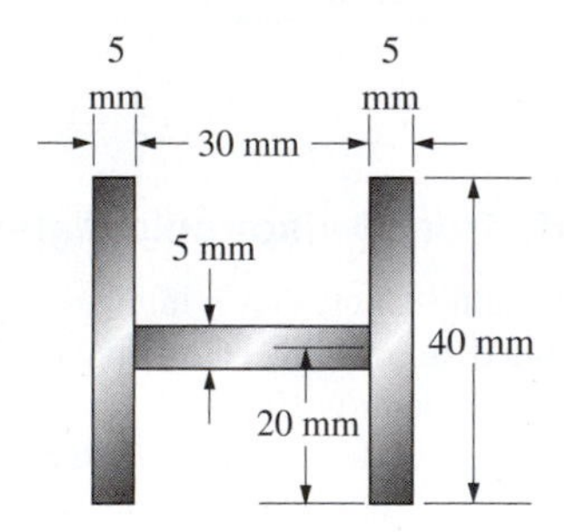

FIGURE P7–131

7–132.M Repeat Problem 7–129 with the shape shown in Figure P7–132.

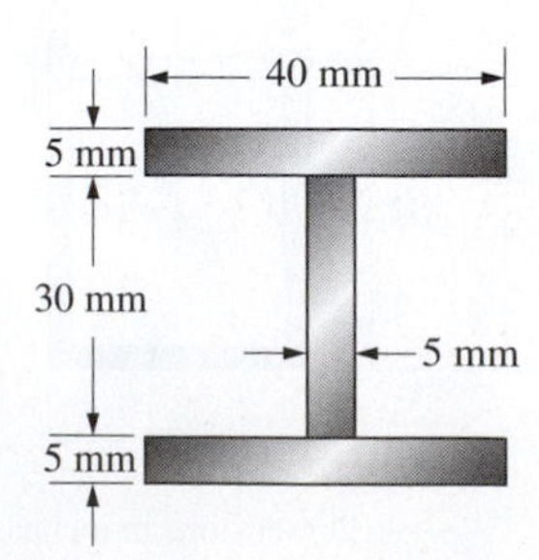

FIGURE P7–132

7–133.M The T-shaped beam cross section shown in Figure P7–133 is to be made from gray cast iron, ASTM A48 Grade 40. It is to be loaded with two equal loads P, 1.0 m from the ends of the 2.80 m long beam. Specify the largest static load P that the beam could carry. Use $N = 4$.

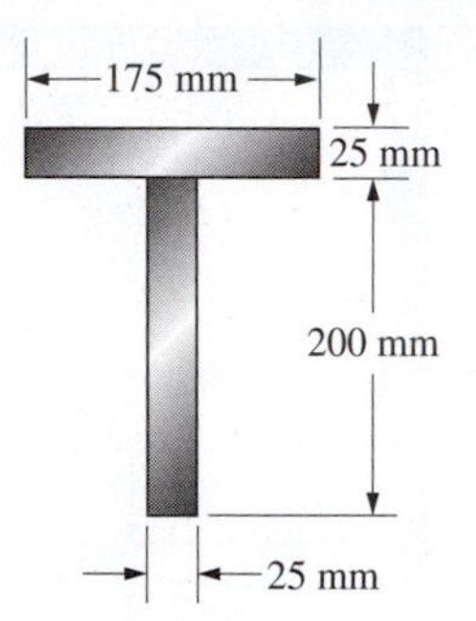

FIGURE P7–133

7–134.M The modified I-beam shape shown in Figure P7–134 is to carry a uniformly distributed static load over its entire 1.20 m length. Specify the maximum allowable load if the beam is made from malleable iron, ASTM A220, class 80002. Use $N = 4$.

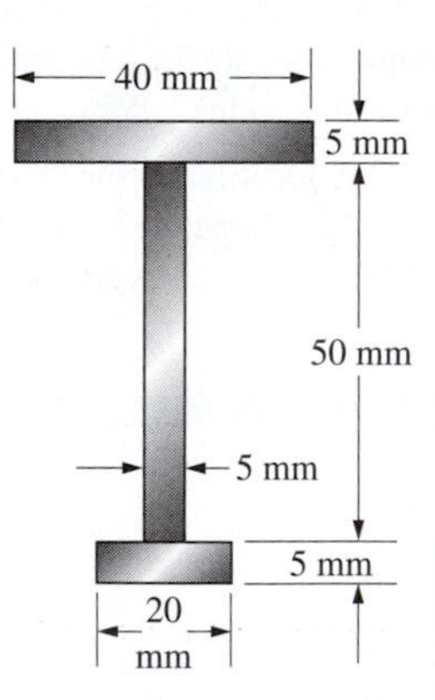

FIGURE P7–134

7–135.M Repeat Problem 7–134 but turn the beam upside down.

7–136.M A wide beam is made as shown in Figure P7–136 from ductile iron, ASTM A536, Grade 120-90-2. Compute the maximum load P that can be carried with a resulting design factor of 10 based on either tensile or compressive ultimate strength.

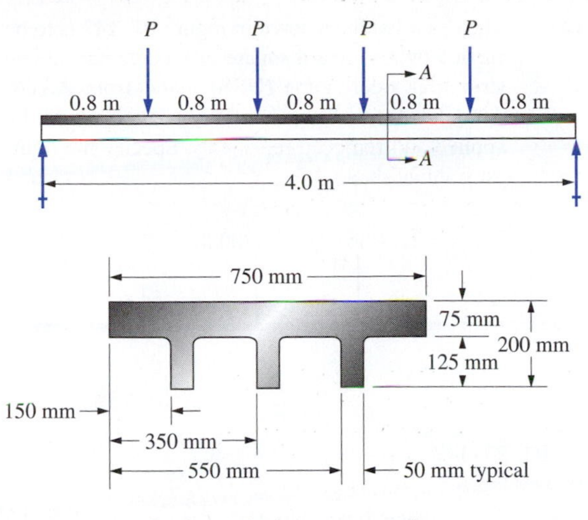

FIGURE P7–136

7–137.M Repeat Problem 7–136 but increase the depth of the vertical ribs by a factor of 2.0.

7–138.M Problems 7–133 through 7–137 illustrate that a beam shape made as a modified I-shape more nearly optimizes the use of the available strength of a material having different strengths in tension and compression. Design an I-shape that has a nearly uniform design factor of 6 based on ultimate strength in either tension or compression when made from gray iron, Grade 20, and which carries a uniformly distributed load of 20 kN/m over its 1.20 m length. (*Note*: You may want to use the computer program written for Assignment 3 at the end of Chapter 6 to facilitate the computations. A trial and error solution may be used.)

ADDITIONAL PRACTICE AND REVIEW PROBLEMS

7–139. A solid circular bar has a diameter of 50 mm and a length of 350 mm. It is installed in the frame of a machine as a cantilever and carries a single, static concentrated load of 2.40 kN at its end. Would the bar be safe if it is made from ASTM A48 gray cast iron, Grade 40? Consider only stress due to bending.

7–140. A beam is to be designed to carry the load shown in Figure 7–15 in Section 7–9. The load is to be applied and removed repeatedly. Specify the lightest standard aluminum I-beam shape for the beam if it is to be made from 6061 T6 aluminum.

7–141. Compute the maximum stress due to bending for the beam shown in Figure P7–141.

7–142. The load shown in Figure P7–142 is carried by a rotating solid circular bar with a diameter of 30.0 mm. The material is AISI 1040 WQT 1300. Is the bar safe in bending?

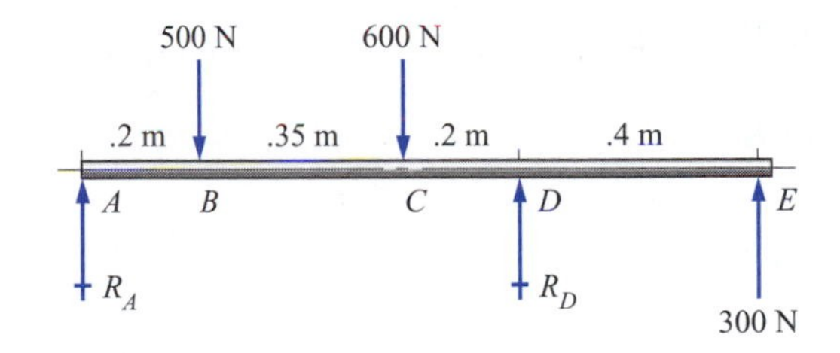

FIGURE P7–142

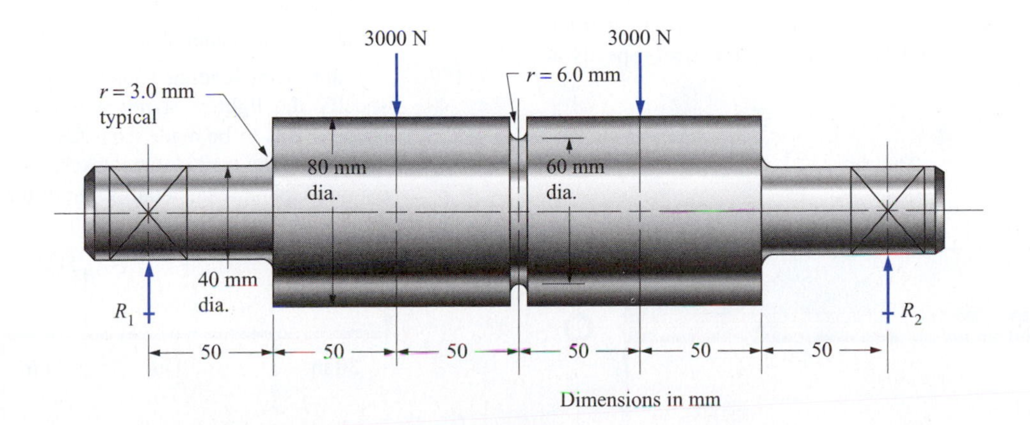

FIGURE P7–141

7–143. Specify the optimum wide flange beam shape to carry the load shown in Figure P7–143 if the beam is made from ASTM A992 structural steel. Use the AISC specifications.

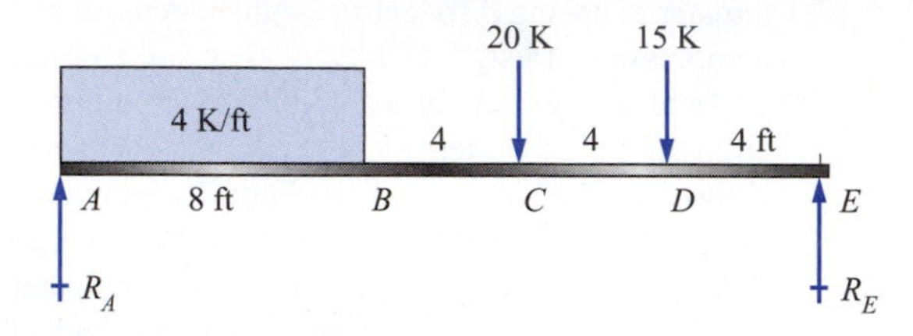

FIGURE P7–143

7–144. If the beam described in Problem 7–143 is to be made from ASTM A572 HSLA, grade 65, specify the optimum beam. Is there an advantage to using the higher strength steel?

7–145. For the beam loading shown in Figure P7–145, specify a suitable square or rectangular hollow structural shape (HSS) if it is made from ASTM A500 grade C steel. Use the AISC specifications.

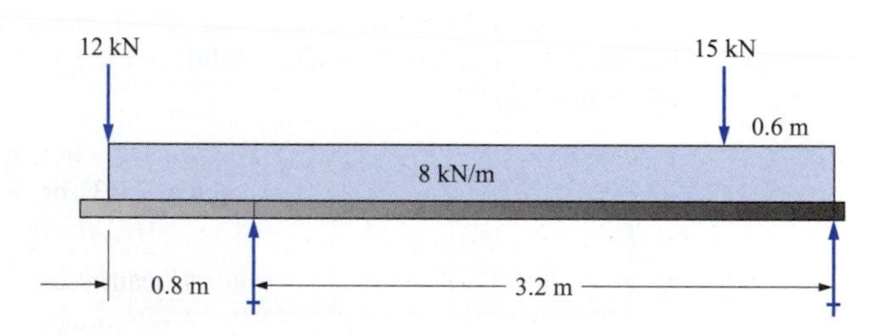

FIGURE P7–145

7–146. The loading shown in Figure P7–146 is to be carried by a beam made from a standard schedule 40 steel pipe. It is desired to have a design factor of 4.0 based on yield strength with the pipe made from cold-formed ASTM A500 grade C steel. Specify a pipe size.

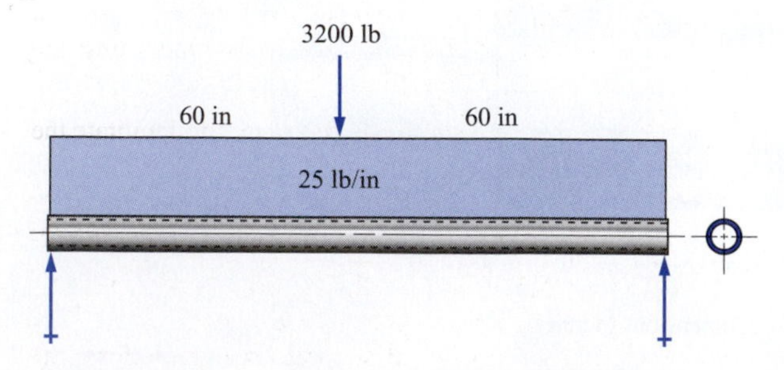

FIGURE P7–146

7–147. The beam loading shown in Figure P7–147 is to be carried by a standard square or rectangular hollow structural steel shape (HSS) made from ASTM A501 hot-formed structural steel. The load will be applied and removed repeatedly. Specify the lightest suitable shape.

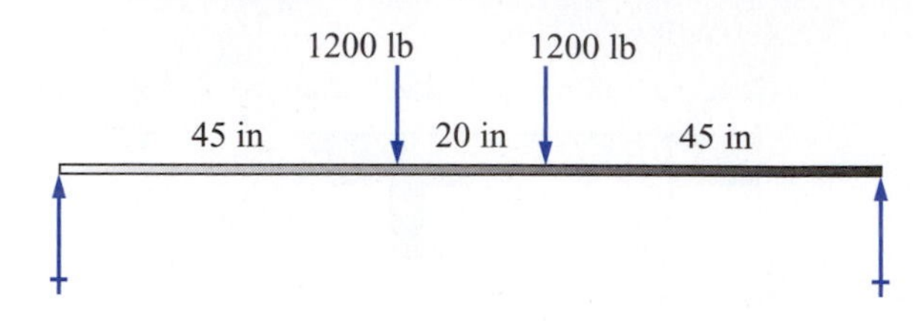

FIGURE P7–147

7–148. A cantilever beam has the cross-sectional shape shown in Figure P7–148. It is 36.0 inches long and it will carry a single concentrated load at its end. A design factor of 4.0 is desired when using ASTM A48 Grade 60 gray cast iron. Compute the allowable load on the beam.

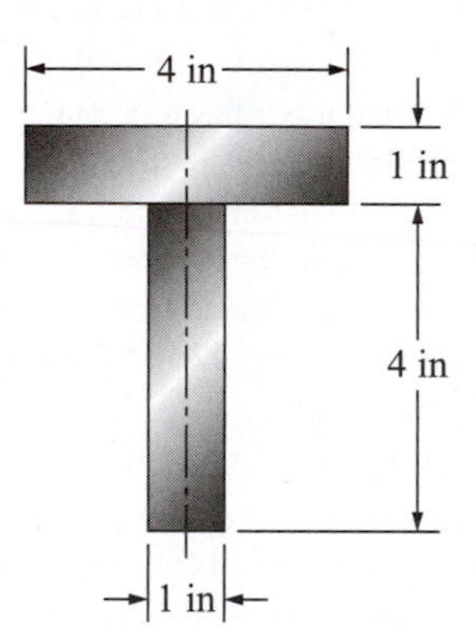

FIGURE P7–148

7–149. Redesign the shape of the beam from Problem 7–148 so that it will carry a minimum of 6000 lb using the same material and design factor.

7–150. For the beam loading shown in Figure P7–150, specify the lightest standard wide flange beam shape if it is to be made from ASTM A992 structural steel. The load is static. Use the AISC specifications. Compute the total weight of the beam.

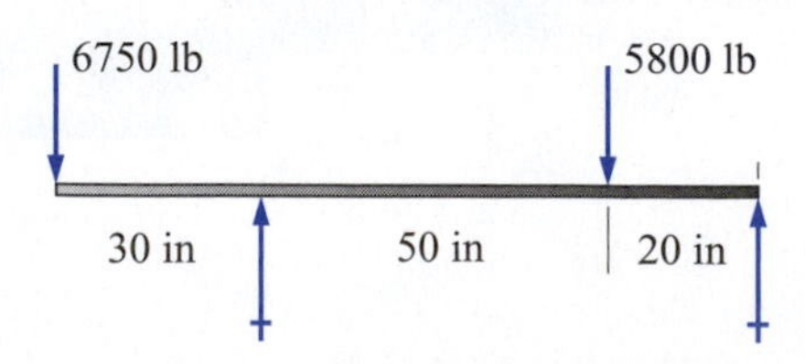

FIGURE P7–150

7–151. An alternate design for the beam to carry the load shown in Figure P7–150 is shown in Figure P7–151. The material for the hollow tube is ASTM A501 hot-formed structural steel and the plates are made from ASTM A36 structural steel. Evaluate whether or not this design is safe. Also, compute its total weight and compare it to the W-shape you specified in Problem 7–150.

7–152. Propose a different design for the beam to carry the loading shown in Figure P7–150 that will be safe but lighter than either design described in Problems 7–150 or 7–151.

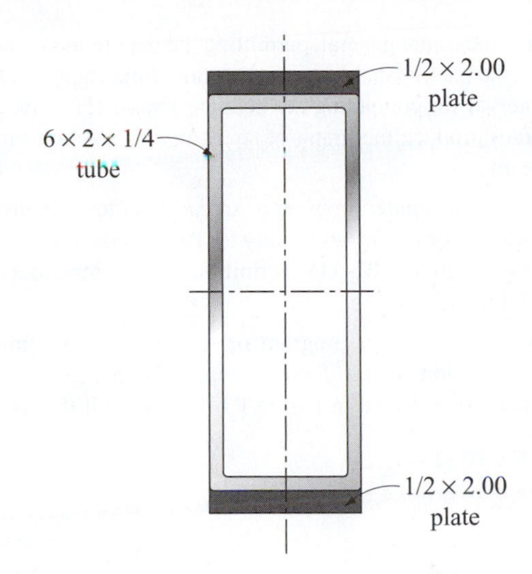

FIGURE P7–151

COMPUTER ASSIGNMENTS

1. Write a program or spreadsheet to compute the maximum bending stress for a simply supported beam carrying a single concentrated load at its center. Allow the operator to input the load, span, and beam section properties. The output should include the maximum bending moment and the maximum bending stress and indicate where the maximum stress occurs.

Enhancements for Assignment 1

(a) For the computed stress, compute the required strength of the material for the beam to produce a given design factor.

(b) In addition to (a), include a table of properties for a selected material such as the data for steel in Appendix A–14. Then search the table for a suitable steel from which the beam can be made.

2. Repeat Assignment 1 except use a uniformly distributed load.
3. Repeat Assignment 1 except the beam is a cantilever with a single concentrated load at its end.
4. Write a program or spreadsheet to compute the maximum bending moment for a simply supported beam carrying a single concentrated load at its center. Allow the operator to input the load and span. Then compute the required section modulus for the cross section of the beam to limit the maximum bending stress to a given level or to achieve a given design factor for a given material. The output should include the maximum bending moment and the required section modulus.

Enhancements for Assignment 4

(a) After computing the required section modulus, have the program complete the design of the beam cross section for a given general shape, such as rectangular with a given ratio of thickness to depth (see Problem 7–13), or circular.

(b) Incude a table of properties for standard beam sections such as any of those in Appendixes A–4 through A–13 and have the program search for a suitable beam section to provide the required section modulus.

5. Repeat Assignment 4 but use a uniformly distributed load.
6. Repeat Assignment 4 but use the load described in Problem 7–22.
7. Repeat Assignment 4 but use any loading pattern assigned by the instructor.
8. Write a computer program or spreadsheet to facilitate the solution of Problem 7–138, including the computation of section properties for the modified I-shape using the techniques of Chapter 6.
9. Write a computer program or spreadsheet to facilitate the solution of problems of the type given in Problem 7–116. Make the program general, permitting the user to input the loading on the beam, the desired beam section properties, and the dimensions of the plates to be added to the basic beam section.
10. Write a computer program or spreadsheet to perform the computations called for in Problem 7–111, but make the

program more general, permitting the user to input values for the load, span, beam cross-section dimensions, and the interval for computing the bending stress. Have the program produce the graph of stress versus position on the beam.

11. Write a computer program or spreadsheet to compute the location for the flexural center for the generalized channel shape in Figure P7–118. Permit the user to input data for all dimensions.

12. Write a computer program or spreadsheet to compute the location for the flexural center for the generalized hat section shown in Figure P7–120. Permit the user to input data for all dimensions. Curve-fitting techniques and interpolation may be used to interpret the graph in Figure 7–27.

13. Write a computer program or spreadsheet to compute the location for the flexural center for the generalized lipped channel shown in Figure P7–122. Permit the user to input data for all dimensions. Curve-fitting techniques and interpolation may be used to interpret the graph in Figure 7–27.

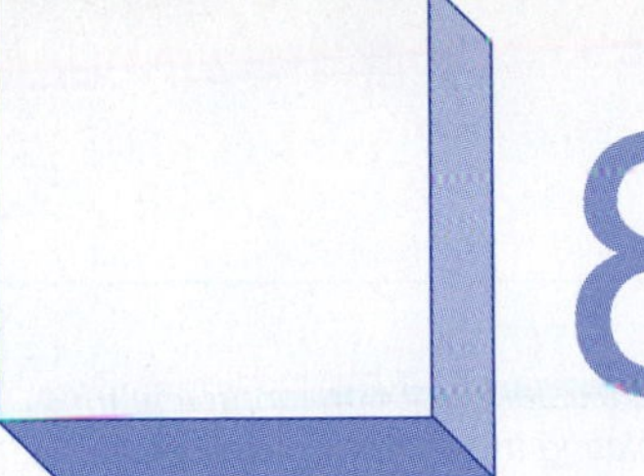

8

Shearing Stresses in Beams

The Big Picture

Shearing Stresses in Beams

Discussion Map

- In Chapter 8 your goal is to continue to develop your ability to analyze and design beams. You will build on your knowledge of shearing forces and bending moments from Chapter 5, moment of inertia from Chapter 6, and bending stresses from Chapter 7. Review all of the examples of beams and loading patterns you discussed in the Big Picture from Chapter 5 and the types of beam cross-sectional shapes from Chapter 6.
- For a beam to be safe to carry a given load, it must satisfy design stress limits for *both bending stresses and shearing stresses.* In most cases the bending stress is the most critical. But shearing stresses must always be checked and, as discussed in Section 8–2, there are some important situations where shearing stresses are the predominant mode of failure. Of course stiffness, or resistance to deflection, is also an important goal when designing a beam.
- Continuing the analysis of beams, this chapter is concerned with the stresses created within a beam due to the presence of shearing forces. As shown in Figure 8–1, shearing forces are visualized to act within the beam on its cross section and to be directed transverse, that is perpendicular, to the axis of the beam. Thus they would tend to create *transverse shearing stresses,* sometimes called *vertical shearing stresses.*
- But if a small stress element subjected to such shearing stresses is isolated, as shown in Figure 8–2, it can be seen that horizontal shearing stresses must also exist in order to cause the element to be in equilibrium. Thus, both vertical and horizontal shearing stresses, having the same magnitude at a given point, are created by shearing stresses in beams.

Activity

Demonstrate for yourself the existence of horizontal shearing stresses in a beam by completing the following exercise.

1. *Make a beam from several thin, flat strips as illustrated in Figure 8–3(a). The width should be narrow relative to the length of the strip. For example, make the length about 125 mm (5.0 inches) and the width about 25 mm (1.0 in).*
2. *An ideal set of material would be five strips cut from standard index cards that measure about 75 mm by 125 mm (3.0 in × 5.0 in). Make at least three sets of five strips each. Other suitable materials are thin plastic, playing cards, or thin sheet metal. You will also need a stapler and some white glue.*
3. *Arrange two simple supports that are about 25 mm (1.0 in) high. Place them approximately 100 mm (4.0 in) apart.*
4. *Obtain a simple weight that is small and that will cause one of the strips of material to deflect all the way to the table, at least 25 mm (1.0 in). A stack of six quarters works well.*
5. *It should not be surprising that a stack of six quarters will deflect one thin strip of card stock at least an inch. But what if you make your beam with two, three, four, or five strips, each placed loosely on top of the others? Try that and document what happens.*
6. *The test in Step 5 likely showed that four or five strips would hold the quarters above the table, but there would still be a significant amount of deflection.*
7. *Now take another stack of five strips and staple them together. Place one staple about 12 mm (0.25 in) from each end and one about 45 mm (1.75 in) from each end, for a total of four staples. This leaves an area in the middle to load this beam design with the stack of quarters. What happens?*
8. *Hey! That works pretty good. The beam deflects just a small amount. How is this beam different from the stack of five loose strips? Figure 8–3(b) demonstrates that the loose strips slide on one another when carrying the load. Each added strip carries a bit more load but the strips are not acting together. The staples in the second stack of five strips hold each strip tightly to the next one and prohibit sliding. Thus the strips act as a unit.*
9. *From this exercise so far, you should be able to visualize that there are horizontal forces that act between the strips of card stock. The loose stack can only resist those forces by friction and that is generally insufficient to keep the strips from sliding. But the staples have adequate strength to resist the horizontal forces.*

10. *In the process, the material of the staples is placed in shear. The stapled beam still deflects a bit because you only used four staples. Some parts of the beam can still slide a small amount. Adding more staples should produce an even stiffer beam.*
11. *Now consider what would happen if the tendency for the strips to slide was restrained over the entire surface of each strip in contact with its adjacent strip. Try that using glue. Now test this beam design and compare its performance with the stapled beam.*
12. *For the glued beam you should notice virtually no deflection when loaded with six quarters. The continuous film of glue prohibits the sliding of the individual strips so the beam acts as a solid, thick bar that is significantly more resistant to deformation than the stacked or stapled designs. In the glued design, the glue must have sufficient shear strength to resist the horizontal shearing stresses created.*
13. *Now, what if you were able to make the beam from a single strip of card stock that has a thickness equal to the glued stack of five thinner strips? You could demonstrate the equivalence of the two designs. See if you can find a thick piece of card stock and try this.*
14. *In conclusion, you should have discovered from this exercise that there are shearing stresses developed within the material of a beam in bending.*

Your primary goal as you study this chapter is to develop the competence to analyze a beam to determine the shearing stresses created in it. You will also apply this skill to evaluate the safety of the beam with respect to shear and to design a safe beam. When using an adhesive, welding, brazing, or mechanical fasteners, such as nails, rivets, bolts, pins, or even staples to hold components of a beam together, you should be able to analyze them to ensure that they have adequate strength.

8–1 OBJECTIVES OF THIS CHAPTER

After completing this chapter, you should be able to:

1. Describe the conditions under which shearing stresses are created in beams.
2. Compute the magnitude of shearing stresses in beams by using the general shear formula.
3. Define and evaluate the *first moment of the area* required in the analysis of shearing stresses.
4. Specify where the maximum shearing stress occurs on the cross section of a beam.
5. Compute the shearing stress at any point within the cross section of a beam.
6. Describe the general distribution of shearing stress as a function of position within the cross section of a beam.
7. Understand the basis for the development of the general shearing stress formula.
8. Describe four design applications where shearing stresses are likely to be critical in beams.

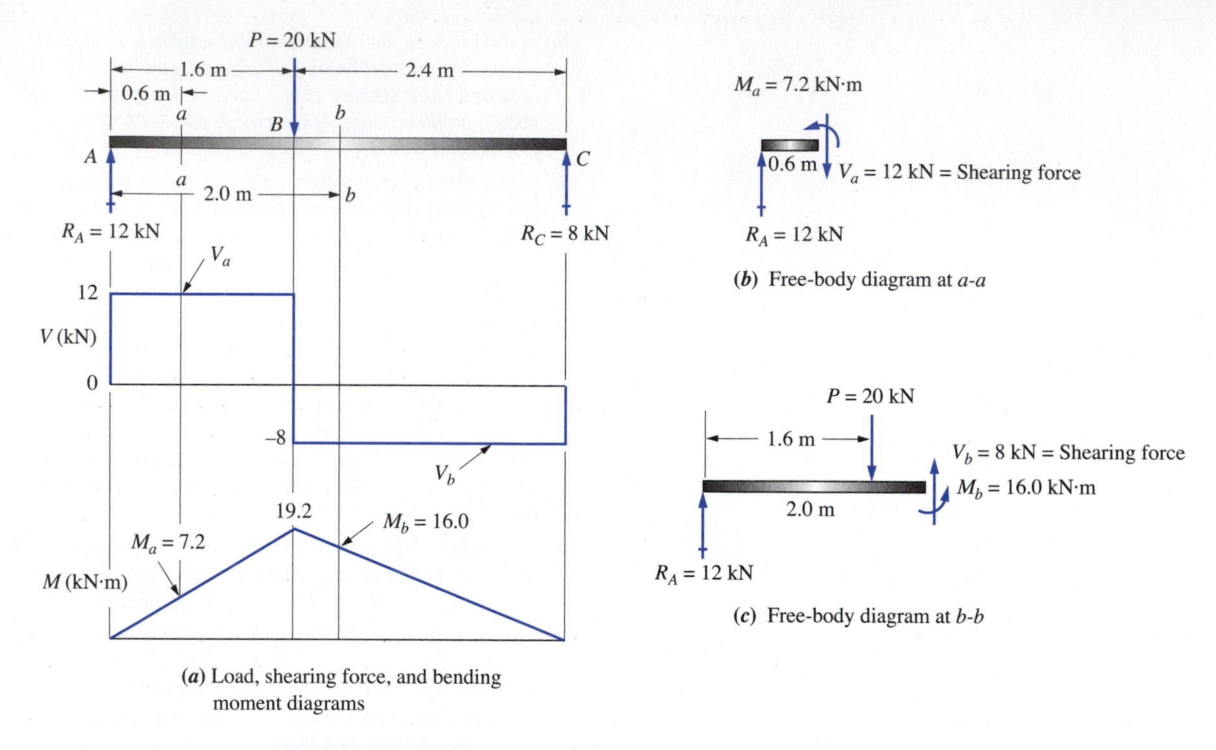

FIGURE 8–1 Shearing forces in beams.

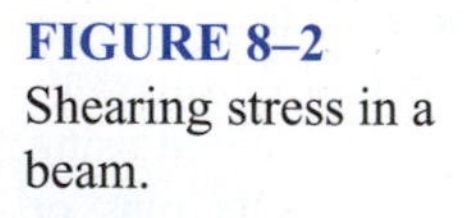

FIGURE 8–2 Shearing stress in a beam.

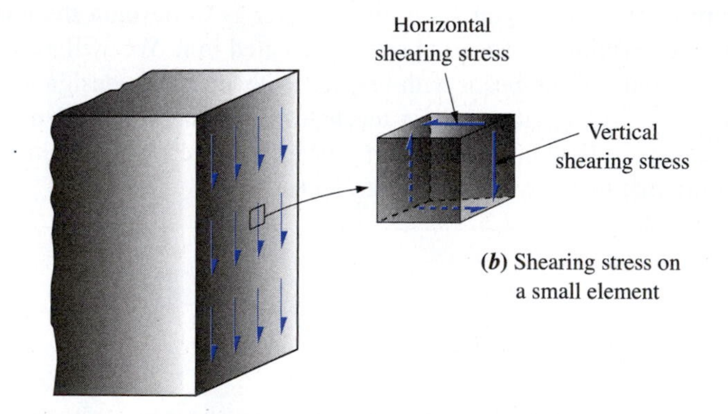

9. Develop and use special shear formulas for computing the maximum shearing stress in beams having rectangular or solid circular cross sections.
10. Understand the development of approximate relationships for estimating the maximum shearing stress in beams having cross sections with tall thin webs or those with thin-walled hollow tubular shapes.
11. Specify a suitable design shearing stress and apply it to evaluate the acceptability of a given beam design.
12. Define *shear flow* and compute its value.
13. Use the shear flow to evaluate the design of fabricated beam sections held together by nails, bolts, rivets, adhesives, welding, or other means of fastening.

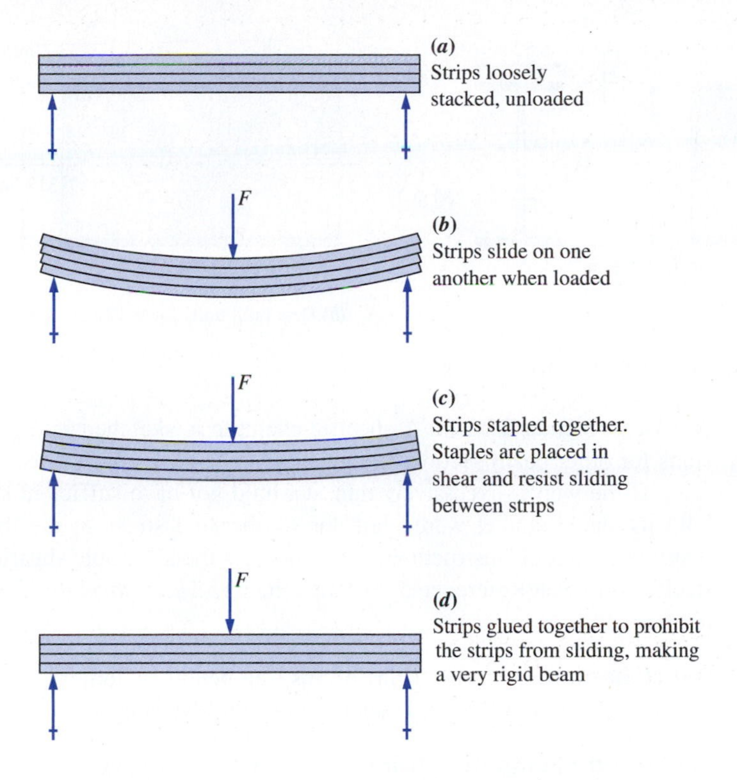

FIGURE 8–3 Illustration of the presence of shearing stress in a beam.

8–2 IMPORTANCE OF SHEARING STRESSES IN BEAMS

Several situations exist in practical design in which the mode of failure is likely to be shearing of a part of a beam or of a means of fastening a composite beam together. Five such situations are described here.

Wooden Beams. Wood is inherently weak in shear along the planes parallel to the grain of the wood. Consider the beam shown in Figure 8–4, which is similar to the joists used in floor and roof structures for wood-frame construction. The grain runs generally parallel to the long axis in commercially available lumber. When subjected to transverse loads, the initial failure in a wooden beam is likely to be by separation along the grain of the wood, due to excessive horizontal shearing stress. Note in Appendix A–19 that the allowable shearing stress in common species of wood ranges from only 70 to 95 psi (0.48 to 0.66 MPa), very low values.

Thin-Webbed Beams. An efficient beam cross section would be one with relatively thick horizontal flanges on the top and bottom with a thin vertical web connecting them together. This generally describes the familiar "I-beam," the wide-flange beam, or the

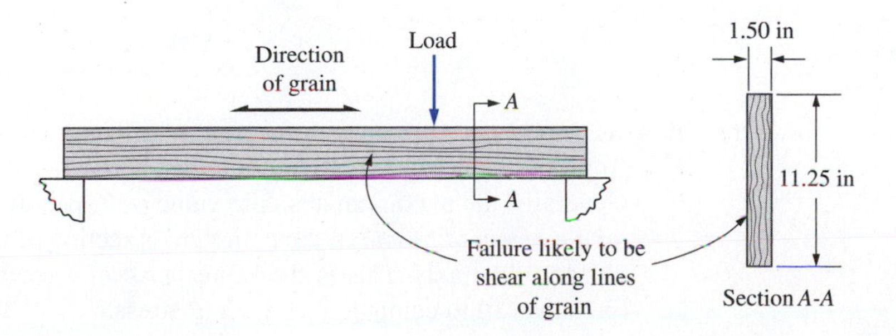

FIGURE 8–4 Shear failure in a wood beam.

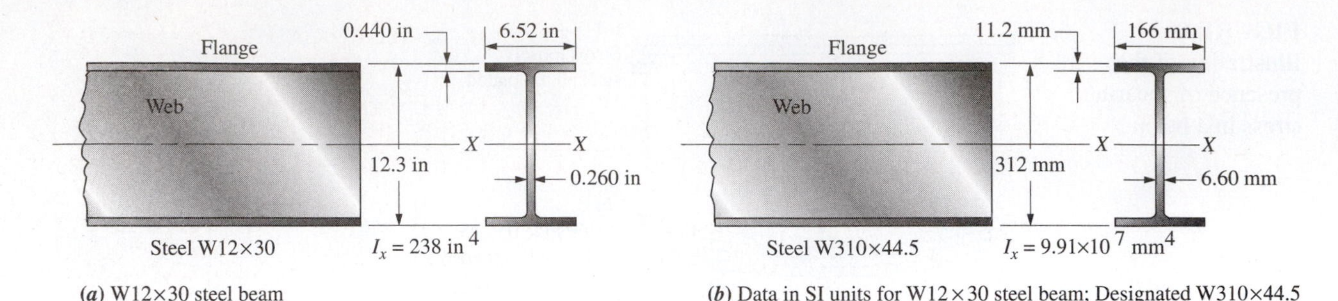

(*a*) W12×30 steel beam

(*b*) Data in SI units for W12 × 30 steel beam; Designated W310×44.5

FIGURE 8–5 Example of thin-webbed beam shape.

American Standard beam. A specific example is sketched in Figure 8–5. Actual dimensions for other beam sections are given in Appendixes A–7, A–8, and A–11.

If the web is excessively thin, it would not have sufficient stiffness and stability to hold its shape, and it would fail due to shearing stress in the thin web. The American Institute of Steel Construction (AISC) defines the allowable shearing stress in the webs of steel beams. See Reference 5. See also the *web shear formula,* defined later in this chapter, Section 8–6.

Short Beams. In very short beams, the bending moment, and therefore the bending stress, is likely to be small. In such beams, the shearing stress may be the limiting stress.

Fastening Means in Fabricated Beams. As shown in Figure 8–3, the fasteners in a composite beam section are subjected to shearing stresses. The concept of *shear flow,* developed later, can be used to evaluate the safety of such beams or to specify the required type, number, and spacing of fasteners to use. Also, beams made of composite materials are examples of fabricated beams. Separation of the layers of the composite, called *interlaminar shear,* is a potential mode of failure.

Stressed Skin Structures. Aircraft and aerospace structures and some ground-based vehicles and industrial equipment are made using a *stressed skin* design. Sometimes called *monocoque* structures, they are designed to carry much of the load in the thin skins of the structure. The method of shear flow is typically used to evaluate such structures, but this application is not developed in this book.

8–3 THE GENERAL SHEAR FORMULA

Presented here is the general shear formula from which you can compute the magnitude of the shearing stress at any point within the cross section of a beam carrying a vertical shearing force. In Section 8–5, the formula itself is developed. You may wish to study the development of the formula along with this section.

The general shear formula is stated as follows:

General Shear Formula

$$\tau = \frac{VQ}{It} \tag{8–1}$$

where V = *vertical shearing force* at the section of interest. The value of V can be found from the shearing force diagram developed as described in Chapter 5. Generally, the maximum absolute value of V, positive or negative, is used.

I = *area moment of inertia* of the entire cross section of the beam with respect to its centroidal axis. This is the same value of I used in the flexure formula ($\sigma = Mc/I$) to compute the bending stress.

t = *thickness* of the cross section taken at the axis where the shearing stress is to be computed.

Q = *first moment,* with respect to the overall centroidal axis, *of the area* of that part of the cross section that lies away from the axis where the shearing stress is to be computed.

In some books Q is called the *statical moment*. We will use the term *first moment of the area* to refer to Q in this book. But you should remember all of the parts of the definition given. To calculate the value of Q, we define it mathematically as

First Moment of the Area

$$Q = A_p\bar{y} \tag{8–2}$$

where A_p = area of that *part* of the cross section that lies away from the axis where the shearing stress is to be computed.

$\bar{y}$ = distance to the centroid of A_p from the centroidal axis of the entire cross section.

Note that Q is the *moment of an area;* that is, area times distance. Therefore, it will have the units of length cubed, such as in^3, m^3, or mm^3.

Careful evaluation of Q is critical to proper use of the general shear formula. It is helpful to draw a sketch of the beam cross section and to highlight the partial area, A_p. Then show the location of the centroid of the partial area on the sketch. Figure 8–6 shows an example for which this has been done. In this example, the objective is to calculate the shearing stress at the axis labeled *a-a*. The shaded area is A_p, shown as that part away from the axis *a-a*.

The following three example problems illustrate the method of computing Q. In each, this is the procedure used.

Method of Computing Q, the First Moment of the Area

1. Locate the centroidal axis for the entire cross section.
2. Draw in the axis where the shearing stress is to be calculated.
3. Identify the partial area A_p away from the axis of interest and shade it for emphasis.

If the partial area A_p is a simple area for which the centroid is readily found by simple calculations, use steps 4–7 to compute Q. Otherwise, use steps 8–11.

4. Compute the magnitude of A_p.
5. Locate the centroid of the partial area.
6. Compute the distance $\bar{y}$ from the centroidal axis of the full section to the centroid of the partial area.
7. Compute $Q = A_p\bar{y}$.

For cases in which the partial area is itself a composite area made up of several component parts, steps 8–11 are used.

8. Divide A_p into component parts that are simple areas and label them A_1, A_2, A_3, and so on. Compute their values.
9. Locate the centroid of each component area.
10. Determine the distances from the centroidal axis of the full section to the centroid of each component area, calling them y_1, y_2, y_3, and so on.
11. Compute $Q = A_p\bar{y}$ from

$$Q = A_p\bar{y} = A_1y_1 + A_2y_2 + A_3y_3 + \cdots \tag{8–3}$$

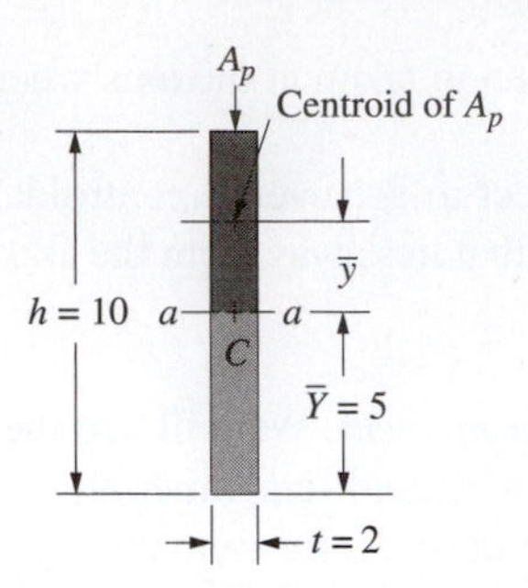

FIGURE 8–6 Example of A_p and $\bar{y}$ for use in computing Q.

Example Problem 8–1 For the rectangular section in Figure 8–6, compute the first moment of the area Q as it would be used in the general shear formula to compute the vertical shearing stress at the section marked *a-a*.

Solution

Objective: Compute the value of Q.

Given: Shape and dimensions of cross section in Figure 8–6.

Analysis: Use the method defined in this section.

Results: ***Step 1.*** The centroidal axis for this section is at its midheight, $h/2$, from the bottom. For this problem, $h/2 = 5.00$ in.

Step 2. The axis of interest is *a-a*, coincident with the centroidal axis for this example.

Step 3. The partial area, A_p, is shown shaded in the figure to be the upper half of the rectangle.

Because the partial area is itself a simple rectangle, steps 4–7 are used to compute Q.

Step 4. The partial area is

$$A_p = t(h/2) = (2.0\text{ in})(5.0\text{ in}) = 10\text{ in}^2$$

Step 5. The centroid of the partial area is at its midheight, 2.5 in above *a-a*.

Step 6. Because the centroidal axis is coincident with the axis *a-a*, $\bar{y} = 2.5$ in.

Step 7. Now Q can be computed.

$$Q = A_p\bar{y} = (10.0\text{ in}^2)(2.5\text{ in}) = 25.0\text{ in}^3$$

Example Problem 8–2 For the I-shaped section in Figure 8–7, compute the first moment of the area Q as it would be used in the general shear formula to compute the vertical shearing stress at the section marked *a-a*.

Solution

Objective: Compute the value of Q.

Given: Shape and dimensions of cross section in Figure 8–7.

Analysis: Use the method defined in this section.

Results: ***Step 1.*** The I-shape is symmetrical and, therefore, the centroidal axis lies at half the height from its base, 5.0 in.

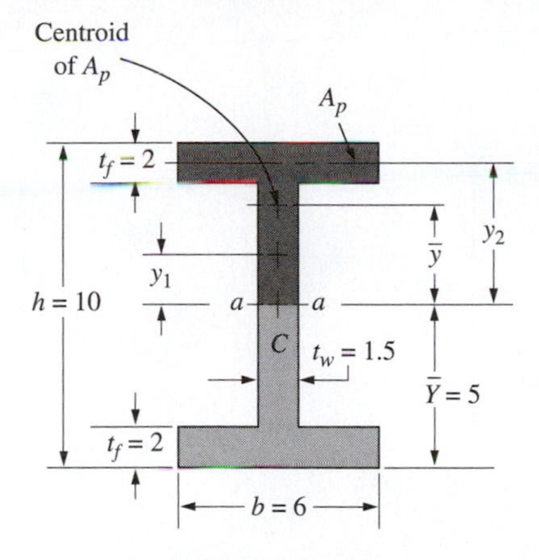

FIGURE 8–7 I-shape for Example Problem 8–2.

Step 2. The axis of interest is *a-a*, coincident with the centroidal axis for this example.

Step 3. The partial area A_p is shown shaded in the figure to be the upper half of the I-shape.

Because the partial area is in the form of a "T", steps 8–11 are used to compute Q.

Step 8. The T-shape is divided into two parts: the upper half of the vertical web is part 1 and the entire top flange is part 2. The magnitudes of these areas are

$$A_1 = \left(\frac{h}{2} - t_f\right)(t_w) = (5.0\text{ in} - 2.0\text{ in})(1.5\text{ in}) = 4.5\text{ in}^2$$

$$A_2 = bt_f = (6.0\text{ in})(2.0\text{ in}) = 12.0\text{ in}^2$$

Step 9. Each part is a rectangle for which the centroid is at its midheight, as shown in the figure.

Step 10. The required distances are

$$y_1 = \frac{1}{2}\left(\frac{h}{2} - t_f\right) = \frac{1}{2}(5.0\text{ in} - 2.0\text{ in}) = 1.5\text{ in}$$

$$y_2 = \left(\frac{h}{2} - \frac{t_f}{2}\right) = (5.0\text{ in} - 1.0\text{ in}) = 4.0\text{ in}$$

Step 11. Using Equation (8–3) gives us

$$Q = A_1y_1 + A_2y_2$$

$$Q = (4.5\text{ in}^2)(1.5\text{ in}) + (12.0\text{ in}^2)(4.0\text{ in}) = 54.75\text{ in}^3$$

Comment Note that it is not necessary to compute the total area A_p or the location of its centroid of the partial area in the case of composite areas. Only the sum of the products of Ay for all parts of A_p is needed.

Example Problem 8–3 For the T-shaped section in Figure 8–8, compute the first moment of the area Q as it would be used in the general shear formula to compute the vertical shearing stress at the section marked *a-a* at the very top of the web, just below where it joins the flange.

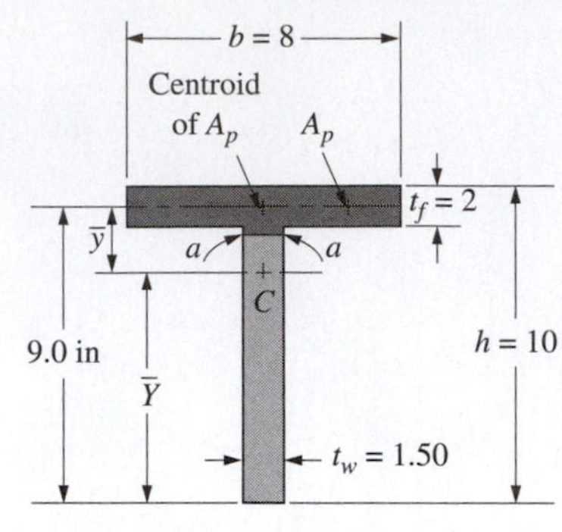

FIGURE 8–8 T-shape for Example Problem 8–3.

Solution

Objective: Compute the value of Q.

Given: Shape and dimensions of cross section in Figure 8–8.

Analysis: Use the method defined in this section.

Results: ***Step 1.*** Locate the centroid of the entire cross section.

$$\bar{Y} = \frac{A_w y_w + A_f y_f}{A_w + A_f}$$

where the subscript w refers to the vertical web and the subscript f refers to the top flange. Then

$$\bar{Y} = \frac{(12)(4) + (16)(9)}{12 + 16} = 6.86 \text{ in}$$

Step 2. The axis of interest, a-a, is at the very top of the web, just below the flange.

Step 3. The partial area above a-a is the entire flange.

Step 4. $A_p = (8 \text{ in})\,(2 \text{ in}) = 16 \text{ in}^2$

Step 5. The centroid of A_p is 1.0 in down from the top of the flange, which is 9.0 in above the base of the tee.

Step 6. $\bar{y} = 9.0 \text{ in} - \bar{Y} = 9.0 \text{ in} - 6.86 \text{ in} = 2.14 \text{ in}$

Step 7. $Q = A_p\bar{y} = (16 \text{ in}^2)(2.14 \text{ in}) = 34.2 \text{ in}^3$

Comment: It should be noted that the value of Q would be the same if the axis of interest a-a were to be taken at the very bottom of the flange just above the web. But the resulting shearing stresses would be markedly different. The thickness of the section, t, would be equal to the entire width of the flange, whereas for the axis a-a used in this problem, the thickness of the web is used. This will be shown later.

Use of the General Shear Formula. Example problems are presented here to illustrate the use of the general shear formula [Equation (8–1)] to compute the vertical shearing stress in a beam. The following procedure is typical of that used in solving such problems.

Guidelines for Computing Shearing Stresses in Beams

The overall objective is to compute the shearing stress at any specified position on the beam at any specified axis within the cross section using the general shear formula,

$$\tau = \frac{VQ}{It} \tag{8–1}$$

1. Determine the vertical shearing force V at the section of interest. This may require preparation of the complete shearing force diagram using the procedures of Chapter 5.
2. Locate the centroid of the entire cross section and draw the neutral axis through the centroid.
3. Compute the moment of inertia of the section with respect to the neutral axis.
4. Identify the axis for which the shearing stress is to be computed and determine the thickness t at that axis. Include all parts of the section that are cut by the axis of interest when computing t.
5. Compute Q, the first moment, with respect to the neutral axis, of the partial area away from the axis of interest. Use the procedure developed in this section.
6. Compute the shearing stress using Equation (8–1).

Example Problem 8–4 Compute the shearing stress at the axis *a-a* for a beam with the rectangular cross section shown in Figure 8–6. The shearing force, V, on the section of interest is 1200 lb.

Solution

Objective: Compute the shearing stress at the axis *a-a*.

Given: Cross section shape and dimensions in Figure 8–6. $V = 1200$ lb.

Analysis: Use the *Guidelines for computing shearing stresses in beams*.

Results:

Step 1. $V = 1200$ lb (given)

Step 2. For the rectangular shape, the centroid is at the midheight, as shown in Figure 8–6, coincident with axis *a-a*. $\overline{Y} = 5.00$ in.

Step 3. $I = bh^3/12 = (2.0)\,(10.0)^3/12 = 166.7\ \text{in}^4$

Step 4. Thickness $= t = 2.0$ in at axis *a-a*.

Step 5. Normally we would compute $Q = A_p\bar{y}$ using the method shown earlier in this chapter. But the value of Q for the section in Figure 8–6 was computed in Example Problem 8–1. Use $Q = 25.0\ \text{in}^3$.

Step 6. Using Equation (8–1),

$$\tau = \frac{VQ}{It} = \frac{(1200\ \text{lb})\,(25.0\ \text{in}^3)}{(166.7\ \text{in}^4)(2.0\ \text{in})} = 90.0\ \text{psi}$$

Example Problem 8–5 Compute the shearing stress at the axes *a-a* and *b-b* for a beam with the T-shaped cross section shown in Figure 8–8. Axis *a-a* is at the very top of the vertical web, just below the flange. Axis *b-b* is at the very bottom of the flange. The shearing force, V, on the section of interest is 1200 lb.

Solution

Objective: Compute the shearing stress at the axes *a-a* and *b-b*.

Given: Cross section shape and dimensions in Figure 8–8. $V = 1200$ lb.

Analysis Use the *Guidelines for computing shearing stresses in beams.*

Results For the axis *a-a*:

Step 1. $V = 1200$ lb (given)

Step 2. This particular T-shape was analyzed in Example Problem 8–3. Use $\overline{Y} = 6.86$ in.

Step 3. We will use the methods of Chapter 6 to compute I. Let the web be part 1 and the flange be part 2. For each part, $I = bh^3/12$ and $d = \overline{Y} - \overline{y}$.

Part	I	A	d	Ad^2	$I + Ad^2$
1	64.00	12.0	2.86	98.15	162.15
2	5.33	16.0	2.14	73.27	78.60
				Total $I =$	240.75 in^4

Step 4. Thickness $= t = 1.5$ in at axis *a-a* in the web.

Step 5. Normally we would compute $Q = A_p\overline{y}$ using the method shown earlier in this chapter. But the value of Q for the section in Figure 8–8 was computed in Example Problem 8–3. Use $Q = 34.2\ \text{in}^3$.

Step 6. Using Equation (8–1),

$$\tau = \frac{VQ}{It} = \frac{(1200\ \text{lb})(34.2\ \text{in}^3)}{(240.75\ \text{in}^4)(1.5\ \text{in})} = 114\ \text{psi}$$

For the axis *b-b*: Some of the data will be the same as at *a-a*.

Step 1. $V = 1200$ lb (given)

Step 2. Again, use $\overline{Y} = 6.86$ in.

Step 3. $I = 240.75\ \text{in}^4$

Step 4. Thickness $= t = 8.0$ in at axis *b-b* in the flange.

Step 5. Again, use $Q = 34.2\ \text{in}^3$. The value is the same as at axis *a-a* because both A_p and $\overline{y}$ are the same.

Step 6. Using Equation (8–1),

$$\tau = \frac{VQ}{It} = \frac{(1200\ \text{lb})(34.2\ \text{in}^3)}{(240.75\ \text{in}^4)(8.0\ \text{in})} = 21.3\ \text{psi}$$

Comment Note the dramatic reduction in the value of the shearing stress when moving from the web to the flange.

8–4 DISTRIBUTION OF SHEARING STRESS IN BEAMS

Most applications require that the maximum shearing stress be determined to evaluate the acceptability of the stress relative to some criterion of design. For most sections used for beams, the maximum shearing stress occurs at the neutral axis, coincident with the centroidal axis, about which bending occurs. The following rule can be used to decide when to apply this observation.

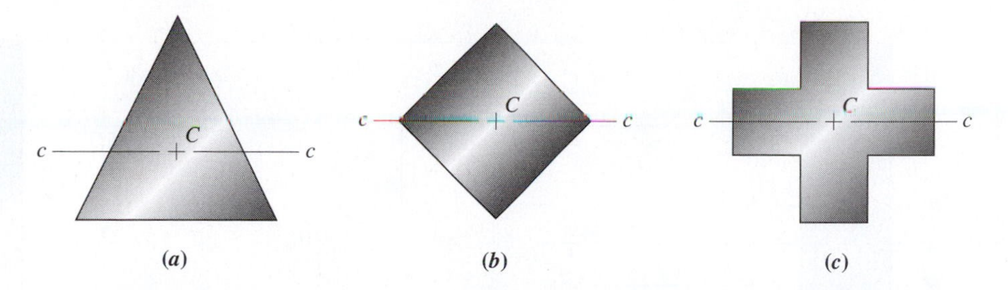

FIGURE 8–9 Beam cross sections for which the maximum shearing stress may not occur at the centroidal axis, *c-c*.

> Provided that the thickness at the centroidal axis is not greater than at some other axis, the maximum shearing stress in the cross section of a beam occurs at the centroidal axis.

Thus the computation of the shearing stress only at the centroidal axis would give the maximum shearing stress in the section, making the computations at other axes unnecessary.

The logic behind this rule can be seen by examining Equation (8–1), the general shear formula. To compute the shearing stress at any axis, the values of the shearing force V and the moment of inertia I are the same. Because the thickness, t, is in the denominator, the smallest thickness would tend to produce the largest shearing stress, as implied in the statement of the rule. But the value of the first moment of the area Q also varies at different axes, decreasing as the axis of interest moves toward the outside of the section. Recall that Q is the product of the partial area A_p and the distance $\bar{y}$ to the centroid of A_p. For axes away from the centroidal axis, the area decreases at a faster rate than $\bar{y}$ increases, resulting in the value of Q decreasing. Thus the maximum value of Q will be that for stress computed at the centroidal axis. It follows that the maximum shearing stress will always occur at the centroidal axis *unless the thickness at some other axis is smaller than that at the centroidal axis*.

The shapes shown in Figures 8–6, 8–7, and 8–8 are all examples that conform to the rule that the maximum shearing stress occurs at the neutral axis because each has its smallest thickness at the neutral axis. Figure 8–9 shows three examples where the rule *does not apply*. In each example, at some axes away from the neutral axis, the thickness is smaller than that at the neutral axis. In such cases, the maximum shearing stress *may* occur at some other axis. Example Problem 8–7 illustrates this observation by analyzing the triangular section.

The solid and hollow circular sections are important examples of where the maximum shearing stress does occur at the neutral axis even though the thickness decreases at other axes. It can be shown that the ratio Q/t continuously decreases for axes away from the neutral axis at the diameter.

The following example problems illustrate the shearing stress distribution in beams of different shapes. Note the comments at the end of each problem for some general conclusions.

Example Problem 8–6 Compute the distribution of shearing stress with position in the cross section for a beam with the rectangular shape shown in Figure 8–6. The actual dimensions are 2.0 in by 10.0 in. Plot the results. The shearing force, V, on the section of interest is 1200 lb.

Solution Objective Compute the shearing stress at several axes and plot τ versus position.

Given Cross section shape and dimensions in Figure 8–6. $V = 1200$ lb.

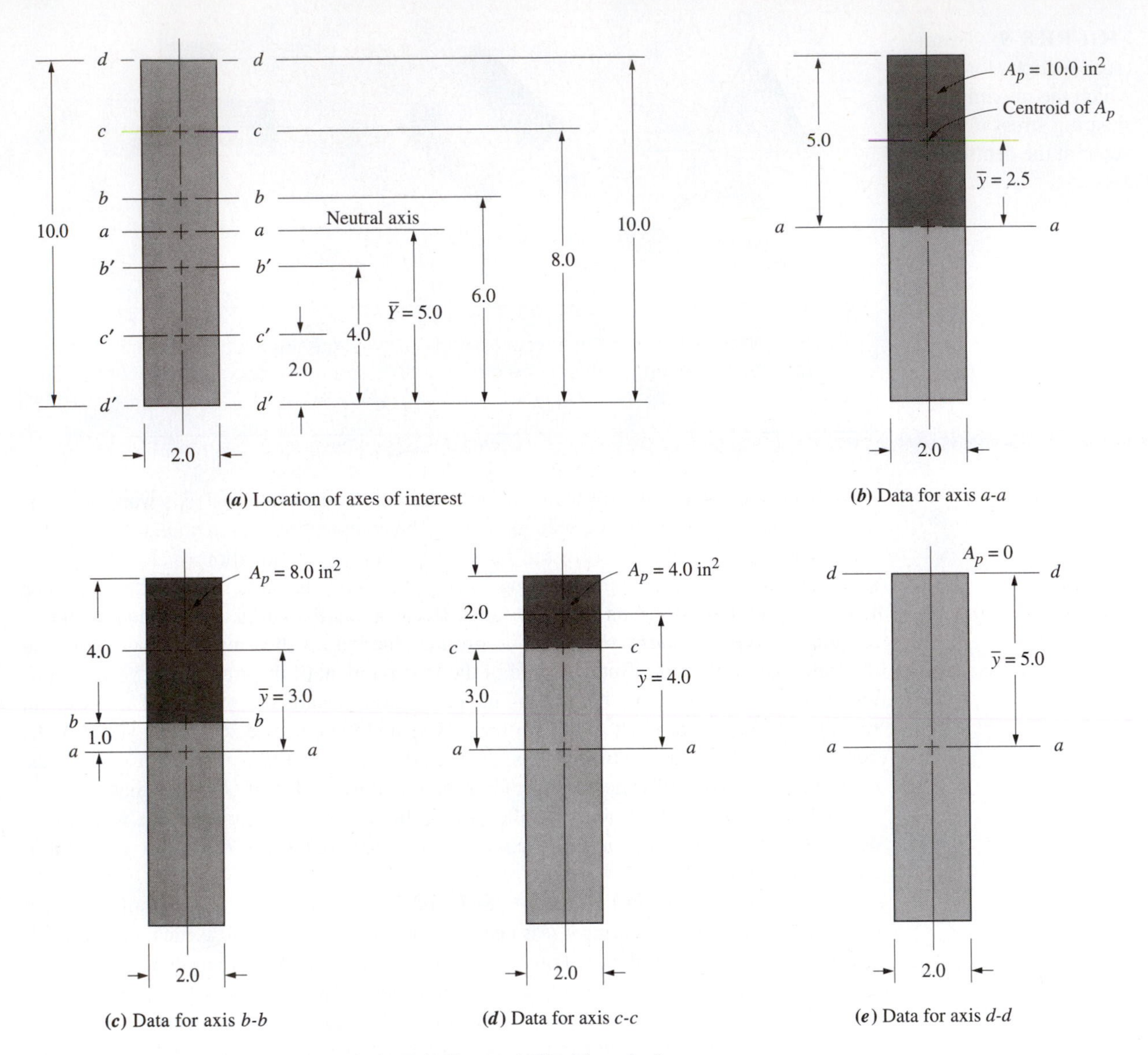

FIGURE 8–10 Data for Q calculations in Example Problem 8–6.

Analysis Use the *Guidelines for computing shearing stresses in beams*. Because the shape is symmetrical with respect to the centroidal axis, we choose to compute the shearing stresses in the upper part at the axes *a-a*, *b-b*, *c-c*, and *d-d*, as shown in Figure 8–10. Then, the values of stresses in the lower part at sections *b′-b′*, *c′-c′*, and *d′-d′* will be the same as the corresponding points above.

Results ***Step 1.*** $V = 1200$ lb (given)

Step 2. For the rectangular shape, the centroid is at the midheight, as shown in Figure 8–6, coincident with axis *a-a*. $\overline{Y} = 5.00$ in.

Step 3. $I = bh^3/12 = (2.0)(10.0)^3/12 = 166.7\ \text{in}^4$

Step 4. Thickness $= t = 2.0$ in at all axes.

Step 5. We will compute $Q = A_p y$ for each axis using the method shown earlier in this chapter. Recall that the value of Q for this section at the centroidal axis was computed in Example Problem 8–1 where we found $Q = 25.0\ \text{in}^3$. A similar calculation is summarized in the table following Step 6, using data from Figure 8–10.

Step 6. Using Equation (8–1), the calculation for shearing stress at the neutral axis *a-a* is shown here.

$$\tau = \frac{VQ}{It} = \frac{(1200\ \text{lb})(25.0\ \text{in}^3)}{(166.7\ \text{in}^4)(2.0\ \text{in})} = 90.0\ \text{psi}$$

The calculation would be the same at the other axes with only the value of Q changing. See the following table.

Axis	V	I	t	A_p	y	$Q = A_p y$	$\tau = VQ/It$
a-a	1200	166.7	2.0	10.0	2.5	25.0	90.0 psi
b-b	1200	166.7	2.0	8.0	3.0	24.0	86.4 psi
c-c	1200	166.7	2.0	4.0	4.0	16.0	57.6 psi
d-d	1200	166.7	2.0	0.0	5.0	0.0	0.0 psi

The results of shearing stress versus position are shown in Figure 8–11 alongside the rectangular section itself.

FIGURE 8–11
Distribution of shearing stress on rectangular section for Example Problem 8–6.

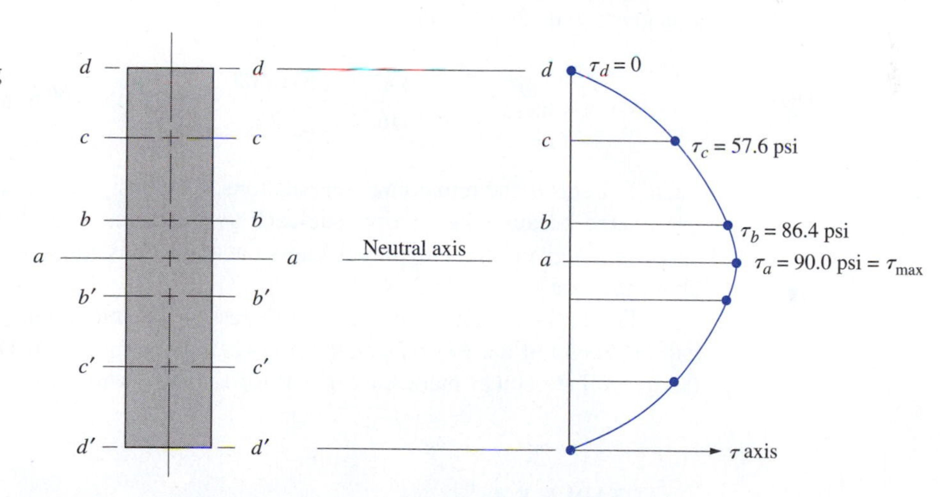

LESSON 11

Comment Note that the maximum shearing stress does occur at the neutral axis as predicted. The variation of shearing stress with position is parabolic, ending with zero stress at the top and bottom surfaces.

Example Problem 8–7 For the triangular beam cross section shown in Figure 8–12, compute the shearing stress that occurs at the axes *a* through *g*, each 50 mm apart. Plot the variation of stress with position on the section. The shearing force is 50 kN.

FIGURE 8–12
Triangular cross section for a beam for which the maximum shearing stress does not occur at the centroidal axis.

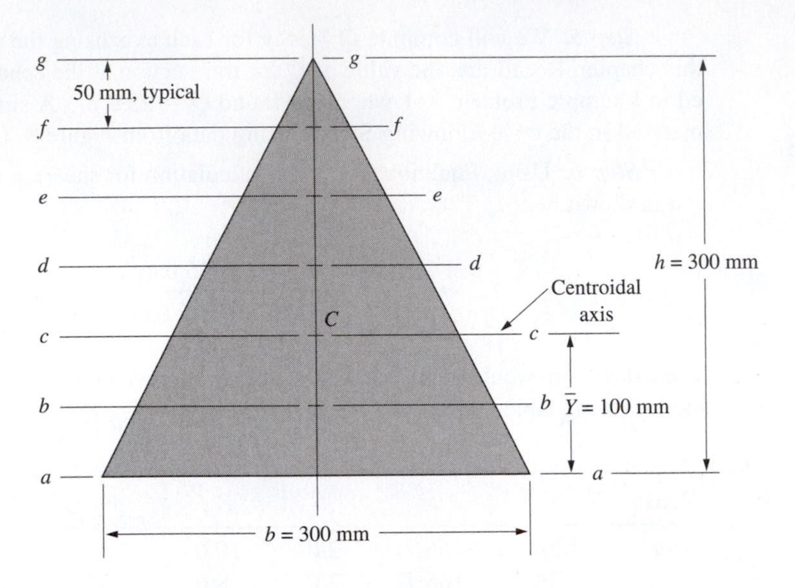

Solution Objective Compute the shearing stress at seven axes and plot τ versus position.

Given Cross-section shape and dimensions in Figure 8–12. $V = 50$ kN.

Analysis Use the *Guidelines for computing shearing stresses in beams.*

Results In the general shear formula, the values of V and I will be the same for all computations. V is given to be 50 kN and

$$I = \frac{bh^3}{36} = \frac{(300)(300)^3}{36} = 225 \times 10^6\,\text{mm}^4$$

Table 8–1 shows the remaining computations. Obviously, the value for Q for axes *a-a* and *g-g* is zero because the area outside each axis is zero. Note that because of the unique shape of the given triangle, the thickness t at any axis is equal to the height of the triangle above the axis.

Figure 8–13 shows a plot of these stresses. The maximum shearing stress occurs at half the height of the section, and the stress at the centroid (at $h/3$) is lower. This illustrates the general statement made earlier that for sections whose minimum thickness does not

TABLE 8–1

Axis	A_p (mm^2)	$\bar{y}$ (mm)	$Q = A_p\bar{y}$ (mm^3)	t (mm)	τ (MPa)
a-a	0	100	0	300	0
b-b	13 750	75.8	1.042×10^6	250	0.92
c-c	20 000	66.7	1.333×10^6	200	1.48
d-d	11 250	100.0	1.125×10^6	150	1.67
e-e	5 000	133.3	0.667×10^6	100	1.48
f-f	1 250	166.7	0.208×10^6	50	0.92
g-g	0	200	0	0	0

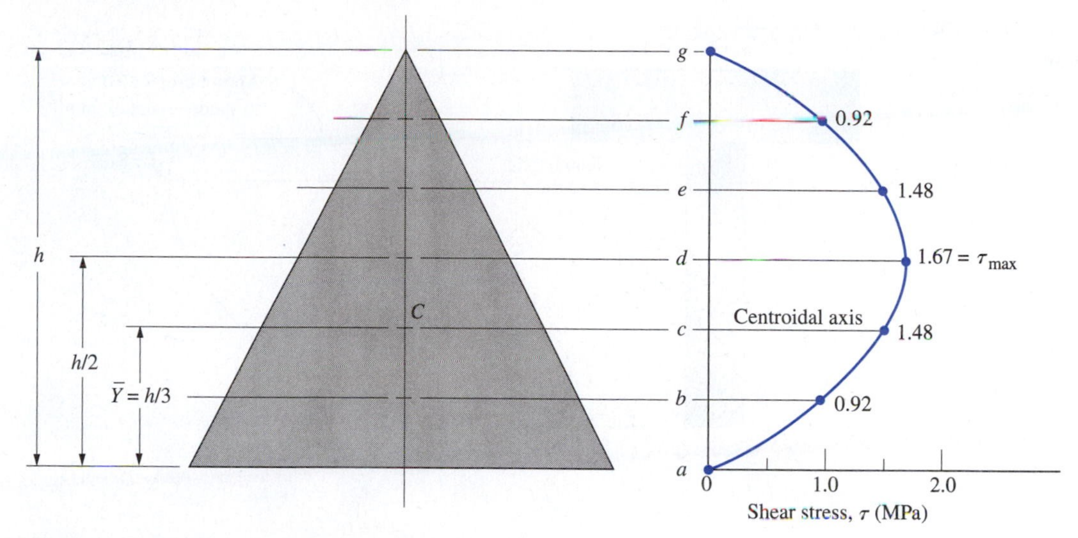

FIGURE 8–13 Shearing stress distribution in the triangular cross section for Example Problem 8–7.

occur at the centroidal axis, the maximum shearing stress may occur at some axis other than the centroidal axis.

Comment One further note can be made about the computations shown for the triangular section. For the axis *b-b*, the partial area A_p was taken as that area *below* *b-b*. The resulting section is the trapezoid between *b-b* and the bottom of the beam. For all other axes, the partial area A_p was taken as the triangular area *above* the axis. The area below the axis could have been used, but the computations would have been more difficult. When computing Q, it does not matter whether the area above or below the axis of interest is used for computing A_p and $\bar{y}$.

By reviewing the results of the example problems worked thus far in this chapter, the following conclusions can be drawn.

Summary of Observations about the Distribution of Shearing Stress in the Cross Section of a Beam

1. The shearing stress at the outside of the section away from the centroidal axis is zero.
2. The maximum shearing stress in the cross section occurs at the centroidal axis provided that the thickness there is no greater than at some other axis.
3. Within a part of the cross section where the thickness is constant, the shearing stress varies in a curved fashion, decreasing as the distance from the centroidal axis increases. The curve is actually a part of a parabola.
4. At an axis where the thickness changes abruptly, as where the web of a tee or an I-shape joins the flange, the shearing stress also changes abruptly, being much smaller in the flange than in the thinner web. See Figure 8–14.

FIGURE 8–14 Stress distribution for shapes with abrupt changes in thickness.

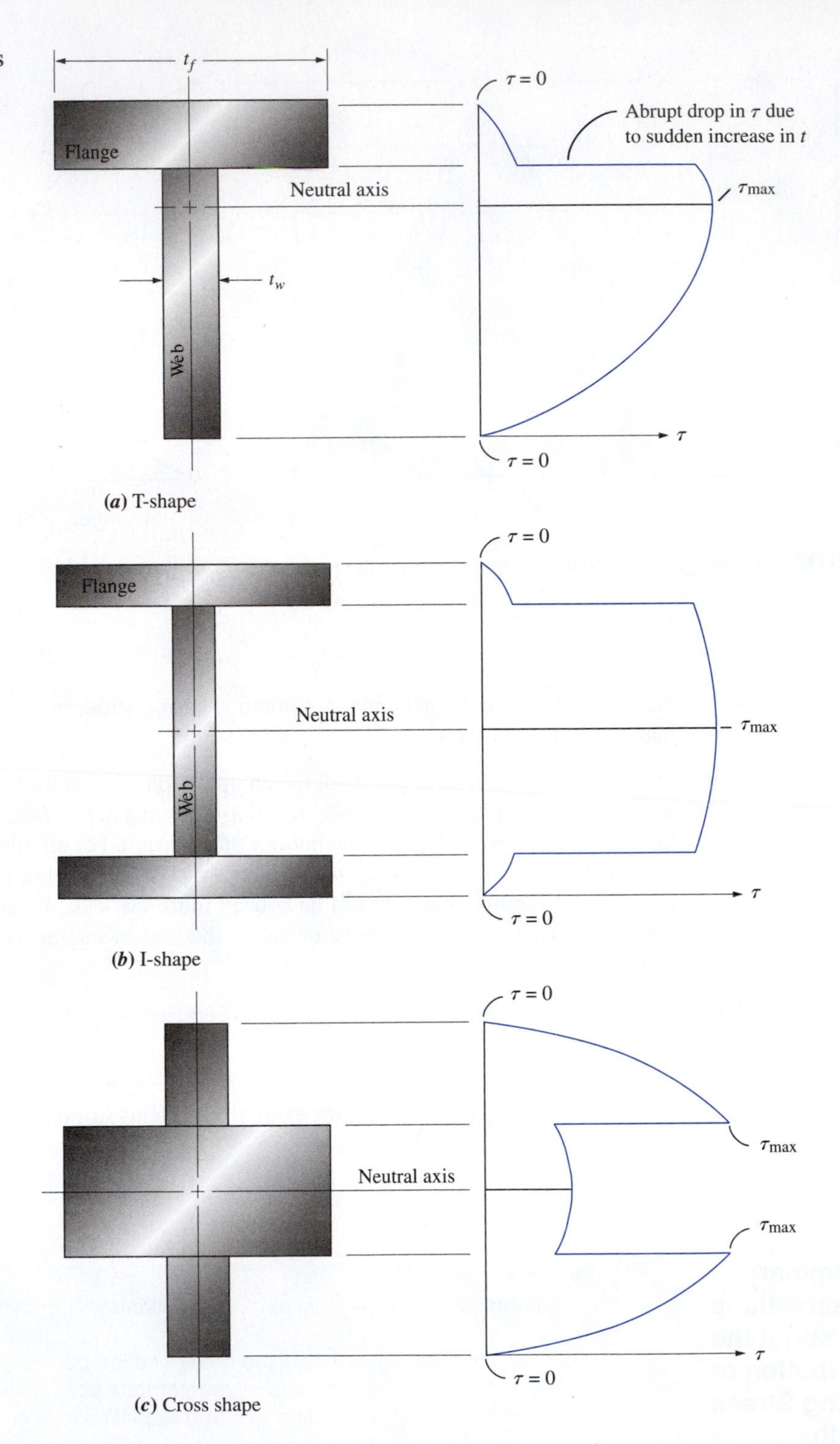

8–5 DEVELOPMENT OF THE GENERAL SHEAR FORMULA

This section presents the background information on the general shear formula. Figure 8–15 shows a beam carrying two transverse loads and the corresponding shearing force and bending moment diagrams to help you visualize certain relationships.

The *moment-area* principle of beam diagrams states that *the change in bending moment between two points on a beam is equal to the area under the shearing force curve between those two points*. For example, consider two points in segment *A-B* of the beam in Figure 8–15, marked x_1 and x_2, a small distance dx apart. The moment at x_1 is M_1 and the moment at x_2 is M_2. Then the moment-area rule states that

$$M_2 - M_1 = V(dx) = dM$$

This can also be stated,

$$V = \frac{dM}{dx} \qquad \textbf{(8–4)}$$

That is, the differential change in bending moment for a differential change in position on the beam is equal to the shearing force occurring at that position.

Equation (8–4) can also be developed by looking at a free-body diagram of the small segment of the beam between x_1 and x_2, as shown in Figure 8–16(a). Since this is a cut section from the beam, the internal shearing forces and bending moments are shown acting on the cut faces. Because the beam itself is in equilibrium, this segment is also. Then the sum of moments about a point in the left face at O must be zero. This gives

$$\sum M_O = 0 = M_1 - M_2 + V(dx) = -dM + V(dx)$$

Or, as shown before,

$$V = \frac{dM}{dx}$$

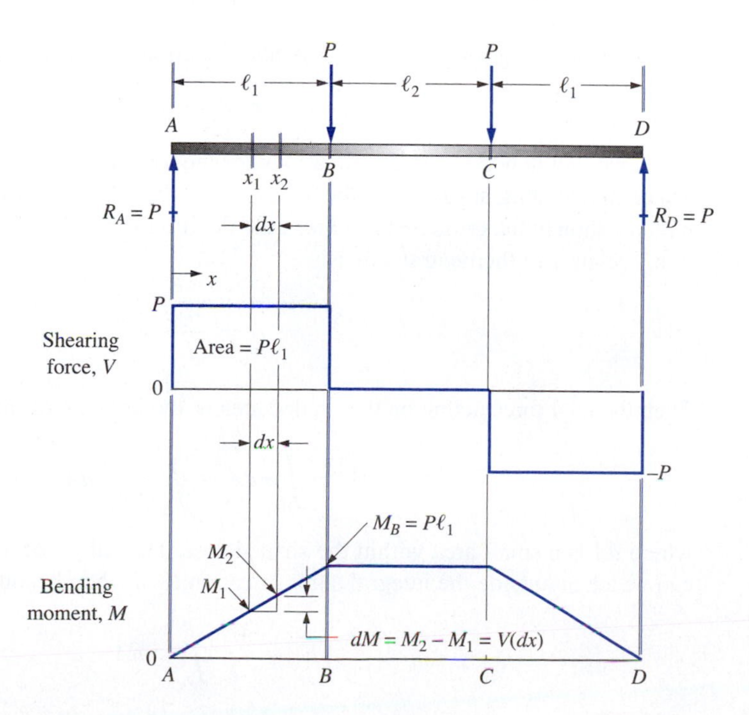

FIGURE 8–15 Beam diagrams used to develop general shear formula.

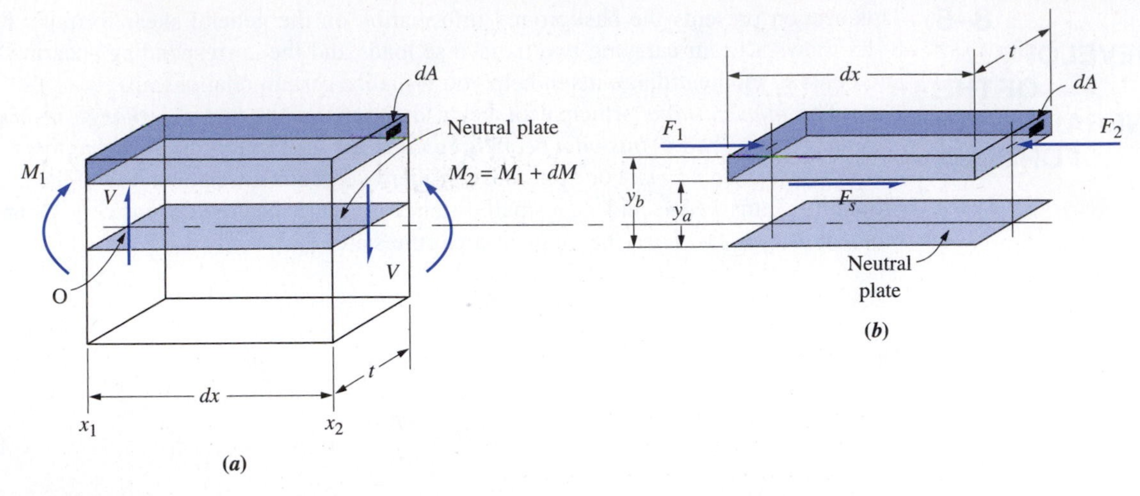

FIGURE 8–16 Forces on a portion of a cut segment of a beam. (a) Free-body diagram of beam segment. (b) Isolated portion of segment.

Any *part* of the beam segment in Figure 8–16(a) must also be in equilibrium. The shaded portion isolated in Figure 8–16(b) is acted on by forces parallel to the axis of the beam. On the left side, F_1 is due to the bending stress acting at that section on the area. On the right side, F_2 is due to the bending stress acting at that section on the area. In general, the values of F_1 and F_2 will be different and there must be a third force acting on the bottom face of the shaded portion of the segment to maintain equilibrium. This is the shearing force, F_s, which causes the shearing stress in the beam. Figure 8–16(b) shows F_s acting on the area $t(dx)$. Then the shearing stress is

$$\tau = \frac{F_s}{t(dx)} \tag{8–5}$$

By summing forces in the horizontal direction, we find

$$F_s = F_2 - F_1 \tag{8–6}$$

We will now develop the equations for the forces F_1 and F_2. Each force is the product of the bending stress times the area over which it acts. But the bending stress varies with position in the cross section. From the flexure formula, the bending stress at any position y relative to the neutral axis is

$$\sigma = \frac{My}{I}$$

Then the total force acting on the shaded area of the left face of the beam segment is

$$F_1 = \int_A \sigma\, dA = \int_{y_a}^{y_b} \frac{M_1 y}{I} dA \tag{8–7}$$

where dA is a small area within the shaded area. The values of M_1 and I are constant and can be taken outside the integral sign. Equation (8–7) then becomes

$$F_1 = \frac{M_1}{I} \int_{y_a}^{y_b} y\, dA \tag{8–8}$$

Now the last part of Equation (8–8) corresponds to the definition of the centroid of the shaded area. That is,

$$\int_{y_a}^{y_b} y\,dA = \bar{y}A_p \qquad \textbf{(8–9)}$$

where A_p is the area of the shaded portion of the left face of the segment and $\bar{y}$ is the distance from the neutral axis to the centroid of A_p. This product of $\bar{y}A_p$ is called the first moment of the area Q in the general shear formula. Making this substitution in Equation (8–8) gives

$$F_1 = \frac{M_1}{I}\int_{y_a}^{y_b} y\,dA = \frac{M_1}{I}\bar{y}A_p = \frac{M_1Q}{I} \qquad \textbf{(8–10)}$$

Similar reasoning can be used to develop the relationship for the force F_2 on the right face of the segment.

$$F_2 = \frac{M_2Q}{I} \qquad \textbf{(8–11)}$$

Substitutions can now be made for F_1 and F_2 in Equation (8–6) to complete the development of the shearing force.

$$F_s = F_2 - F_1 = \frac{M_2Q}{I} - \frac{M_1Q}{I} = \frac{Q}{I}(M_2 - M_1) \qquad \textbf{(8–12)}$$

Earlier we defined $(M_2 - M_1) = dM$. Then

$$F_s = \frac{Q(dM)}{I} \qquad \textbf{(8–13)}$$

Then, in Equation (8–5),

$$\tau = \frac{F_s}{t(dx)} = \frac{Q(dM)}{It(dx)}$$

But, from Equation (8–4), $V = dM/dx$. Then

$$\tau = \frac{VQ}{It}$$

This is the form of the general shear formula [Equation (8–1)] used in this chapter.

8–6 SPECIAL SHEAR FORMULAS

As demonstrated in several example problems, the general shear formula can be used to compute the shearing stress at any axis on any cross section of the beam. However, frequently it is desired to know only the *maximum shearing stress*. For many common shapes used for beams, it is possible to develop special simplified formulas that will give the maximum shearing stress quickly. The rectangle, circle, thin-walled hollow tube, and thin-webbed shapes can be analyzed this way. The formulas are developed in this section.

For all of these section shapes, the maximum shearing stress occurs at the neutral axis. The rectangle and thin-webbed shapes conform to the rule stated in Section 8–5 because

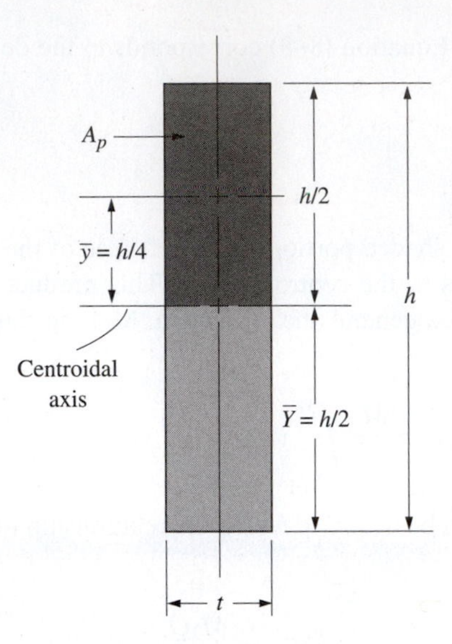

FIGURE 8–17
Rectangular shape.

the thickness at the neutral axis is no greater than at other axes in the section. The circle and the thin-walled tube do not conform to the rule. However, it can be shown that the ratio Q/t in the general shear formula decreases continuously as the axis of interest moves away from the neutral axis, resulting in the decrease in the shearing stress.

Rectangular Shape. Figure 8–17 shows a typical rectangular cross section having a thickness t and a height h. The three geometrical terms in the general shear formula can be expressed in terms of t and h.

$$I = \frac{th^3}{12}$$

$$t = t$$

$$Q = A_p\bar{y} \quad \text{(for area above centroidal axis)}$$

$$Q = \frac{th}{2} \cdot \frac{h}{4} = \frac{th^2}{8}$$

Putting these terms in the general shear formula gives

$$\tau_{\max} = \frac{VQ}{It} = V \cdot \frac{th^2}{8} \cdot \frac{12}{th^3} \cdot \frac{1}{t} = \frac{3}{2}\frac{V}{th}$$

But since th is the total area of the section,

Special Shear Formula for Rectangle

$$\tau_{\max} = \frac{3V}{2A} \quad \textbf{(8–14)}$$

Equation (8–14) can be used to compute exactly the maximum shearing stress in a rectangular beam at its centroidal axis.

Note that $\tau = V/A$ represents the *average* shearing stress on the section. Thus the maximum shearing stress on a rectangular cross section is 1.5 times higher than the average.

Example Problem 8–8 Compute the maximum shearing stress that would occur in the rectangular cross section of a beam like that shown in Figure 8–17. The shearing force is 1000 lb, $t = 2.0$ in, and $h = 8.0$ in.

Solution Using Equation (8–14) yields

$$\tau_{max} = \frac{3V}{2A} = \frac{3(1000 \text{ lb})}{2(2 \text{ in})(8 \text{ in})} = 93.8 \text{ psi}$$

Circular Shape. The special shear formula for the circular shape is developed in a similar manner to that used for the rectangular shape. Equations for Q, I, and t are written in terms of the primary size variable for the circular shape, its diameter. Then the general shear formula is simplified (refer to Figure 8–18).

$$t = D$$

$$I = \frac{\pi D^4}{64}$$

$$Q = A_p\bar{y} \quad \text{(for the semicircle above the centroid)}$$

$$Q = \frac{\pi D^2}{8} \cdot \frac{2D}{3\pi} = \frac{D^3}{12}$$

Then the maximum shearing stress is

$$\tau_{max} = \frac{VQ}{It} = V \times \frac{D^3}{12} \times \frac{64}{\pi D^4} \times \frac{1}{D} = \frac{64V}{12\pi D^2}$$

To refine the equation, factor out a 4 from the numerator and then note that the total area of the circular section is $A = \pi D^2/4$.

$$\tau_{max} = \frac{16(4)V}{12\pi D^2} = \frac{16V}{12A}$$

$$\tau_{max} = \frac{4V}{3A} \tag{8–15}$$

This shows that the maximum shearing stress is 1.33 times higher than the average on the circular section.

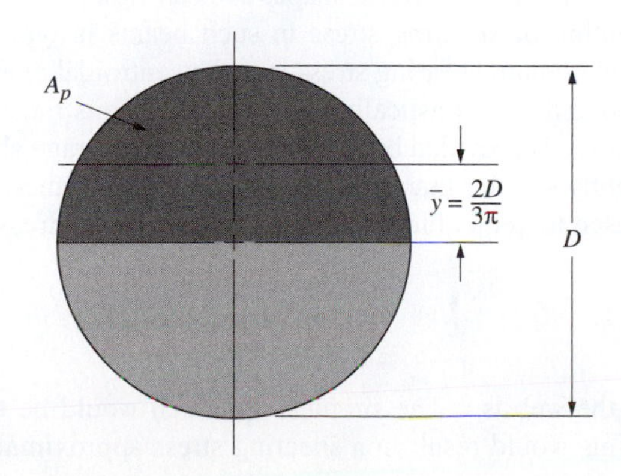

FIGURE 8–18 Circular shape.

Example Problem 8–9 Compute the maximum shearing stress that would occur in a circular shaft, 50 mm in diameter, if it is subjected to a vertical shearing force of 110 kN.

Solution Equation (8–15) will give the maximum shearing stress at the horizontal diameter of the shaft.

$$\tau_{max} = \frac{4V}{3A}$$

But
$$A = \frac{\pi D^2}{4} = \frac{\pi(50\text{ mm})^2}{4} = 1963\text{ mm}^2$$

Then
$$\tau_{max} = \frac{4(110 \times 10^3\text{ N})}{3(1963\text{ mm}^2)} = 74.7\text{ MPa}$$

Hollow Thin-Walled Tubular Shape. Removing material from the center of a circular cross section tends to increase the local value of the shearing stress, especially near the diameter where the maximum shearing stress occurs. Although not giving a complete development here, it is observed that the maximum shearing stress in a thin-walled tube is approximately twice the average. That is,

Special Shear Formula for Thin-Walled Tube

$$\tau_{max} \approx 2\frac{V}{A} \tag{8–16}$$

where A is the total cross-sectional area of the tube.

Example Problem 8–10 Compute the approximate maximum shearing stress that would occur in a 3-in schedule 40 steel pipe if it is used as a beam and subjected to a shearing force of 6200 lb.

Solution Equation (8–16) should be used. From Appendix A–12 we find that the cross-sectional area of the 3-in schedule 40 steel pipe is 2.228 in^2. Then an estimate of the maximum shearing stress in the pipe, occurring near the horizontal diameter, is

$$\tau_{max} \approx 2\frac{V}{A} = \frac{2(6200\text{ lb})}{2.228\text{ in}^2} = 5566\text{ psi}$$

Thin-Webbed Shapes. Structural shapes such as W- and S-beams have relatively thin webs. The distribution of shearing stress in such beams is typically like that shown in Figure 8–19. The maximum shearing stress is at the centroidal axis. It decreases slightly in the rest of the web and then drastically in the flanges. Thus most of the resistance to the vertical shearing force is provided by the web. Also, the average shearing stress in the web would be just slightly smaller than the maximum stress. For these reasons, the *web shear formula* is often used to get a quick estimate of the shearing stress in thin-webbed shapes.

Web Shear Formula for Thin-Webbed Shapes

$$\tau_{max} \approx \frac{V}{A_{web}} = \frac{V}{th} \tag{8–17}$$

The thickness of the web is t. The simplest approach would be to use the full height of the beam for h. This would result in a shearing stress approximately 15% lower than the

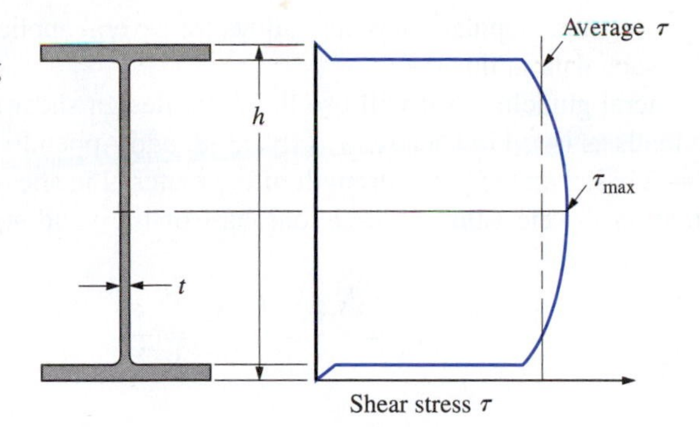

FIGURE 8–19 Distribution of shearing stress in a thin-webbed shape.

actual maximum shearing stress at the centroidal axis for typical beam shapes. Using just the web height between the flanges would produce a closer approximation of the maximum shearing stress, probably less than 10% lower than the actual value. In problems using the web shear formula, we use the full height of the cross section unless otherwise stated.

In summary, for thin-webbed shapes, compute the shearing stress from the web shear formula using the full height of the beam for h and the actual thickness of the web for t. Then, to obtain a more accurate estimate of the maximum shearing stress, increase this value by about 15%.

Example Problem 8–11 Using the web shear formula, compute the shearing stress in a W12×16 beam if it is subjected to a shearing force of 25 000 lb.

Solution In Appendix A–7 for W-beams, it is found that the web thickness is 0.220 in, and the overall depth (height) of the beam is 12.00 in. Then, using Equation (8–17), we have

$$\tau_{max} \approx \frac{V}{th} = \frac{25\,000\ \text{lb}}{(0.220\ \text{in})(12.00\ \text{in})} = 9470\ \text{psi}$$

8–7 DESIGN FOR SHEAR

The design shear stress depends greatly upon the material from which the beam is to be made and on the form of the member subjected to the shearing stress. A limited amount of data is presented in this book, and the reader is advised to check more complete references, such as those listed at the end of the chapter.

Steel. For shear stress in the webs of rolled steel beam shapes, the AISC generally recommends

$$\tau_d = 0.40\, s_y \tag{8–18}$$

However, there are extensive discussions in References 3–5 for special cases of short beams; beams with unusually tall, thin webs; and beams with stiffeners applied either in the vertical or horizontal directions. Careful consideration of these factors is advised. See also References 11 and 13.

Aluminum. The Aluminum Association also provides extensive data for various conditions of loading and beam geometry. For example, Reference 1 gives actual data for allowable shear stress of the more popular aluminum alloys for several applications. It is not practical to summarize such data in this book.

As a general guideline, we will use the same design shear stress for ductile metals under static loads as listed in Chapter 3, Table 3–8, and Appendix A–21. That is, a design factor of $N = 2$ based on the yield strength of the material in shear, s_{ys}, is suggested. Also, an approximation for the value of s_{ys} is one-half of the yield strength in tension, s_y. In summary,

$$\tau_d = \frac{s_{ys}}{N} = \frac{0.5s_y}{N} = \frac{s_y}{2N} \tag{8–19}$$

For $N = 2$,

$$\tau_d = \frac{s_y}{4} = 0.25s_y$$

Wood. For wood beams, data are given in Appendix A–19 for allowable horizontal shear stress. Note that the values are quite low, typically less than 100 psi (0.69 MPa). Shear failure is frequently the limiting factor for wood beams. See also References 9 and 14.

Concrete. Shear strength of concrete is quite low compared with that of most metals. The American Concrete Institute (ACI) specifies that the ultimate shear strength is $2\sqrt{f'_c}$ where f'_c is the rated strength of the concrete, typically ranging from 2000 psi to 7000 psi. Then the shear strength ranges from 89 psi to 167 psi.

Furthermore, whenever a shear stress is created in concrete, a corresponding tensile stress results. As explained in Chapter 10, for example, when pure shear exists on an element, there is a tensile stress of equal magnitude produced on a plane 45 degrees from the original orientation of the element. With the extremely low tensile strength of concrete, tensile failures often occur in zones of high shear stress and they propagate at a 45 degree angle. To counteract this failure mode, the design of concrete beams always includes steel shear reinforcement, typically in the form of stirrups placed perpendicular to the longitudinal reinforcing bars required for bending strength. See Figure 8–20. Welded wire fabric or steel bars bent up at an angle of 30° to 60° are also sometimes used. See References 2 and 12 for design procedures to determine the size, layout, and spacing of the bars.

Masonry. The design of beams using masonry made from brick, stone, and concrete unit blocks must also consider shearing stresses. Similarly, load-bearing walls and shear walls subjected to bending loads from wind forces and forces transferred from floors, roofs, or supported beams must be designed to resist shear. Failure often occurs in the mortar joints along a diagonal in the wall, creating a noticeable zigzag pattern. See Reference 10.

Composite Materials. The design, testing, and analysis of structures made from composite materials are discussed in References 6–8. Composites made using laminated

FIGURE 8–20

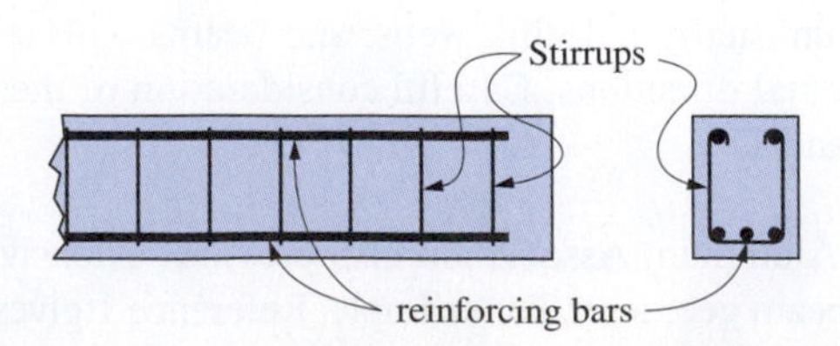

structures described in Chapter 2 often fail by the mechanism of *interlaminar shear,* where the internal shearing stresses cause the separation of layers within the structure. Reference 7 describes a testing method to evaluate a beam's tendency to fail in this manner. With other forms of composite materials, such as pultrusions, sheet molding compounds (SMC), those reinforced with long or short chopped fibers, or those employing woven fabrics or roving, failure can occur when the individual reinforcing fibers separate from the matrix. References 6 and 8 discuss methods of designing structural components to withstand shearing stresses.

8–8 SHEAR FLOW

Built-up sections used for beams, such as those shown in Figures 8–21 and 8–22, must be analyzed to determine the proper size and spacing of fasteners. The discussion in preceding sections showed that horizontal shearing forces exist at the planes joined by the nails, bolts, and rivets. Thus the fasteners are subjected to shear. Usually, the size and material of the fastener will permit the specification of an allowable shearing force on each. Then the beam must be analyzed to determine a suitable spacing for the fasteners that will ensure that all parts of the beam act together.

The term *shear flow* is useful for analyzing built-up sections. Called q, the shear flow is found by multiplying the shearing stress at a section by the thickness at that section. That is,

$$q = \tau t \tag{8–20}$$

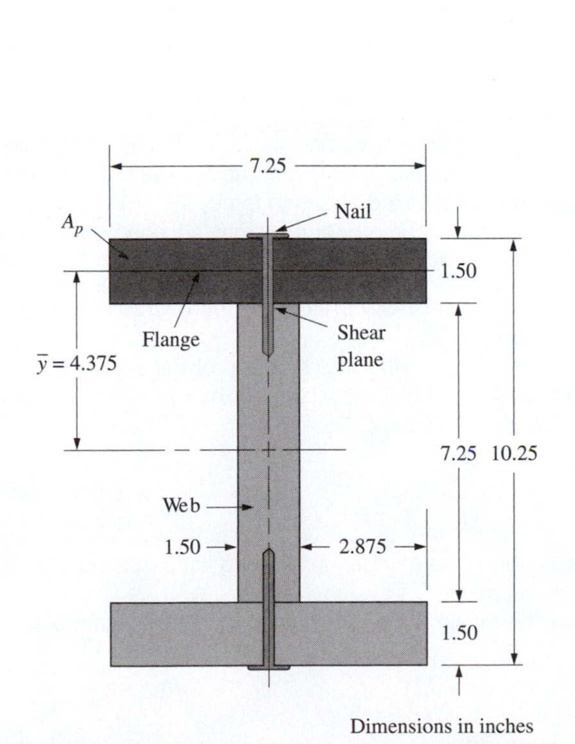

FIGURE 8–21 Beam shape for Example Problem 8–12.

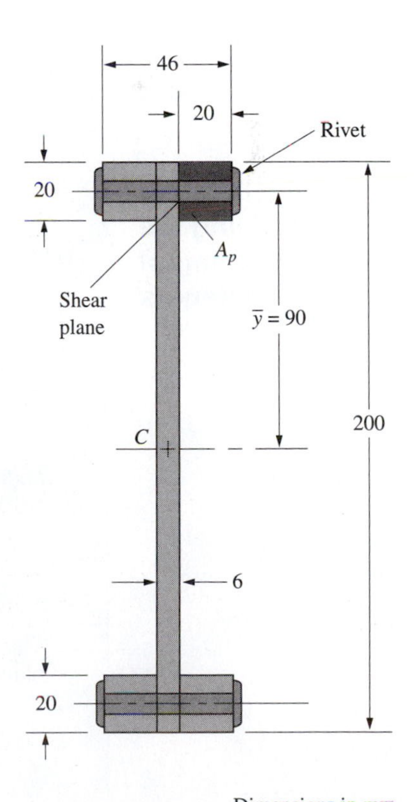

FIGURE 8–22 Built-up beam shape for Example Problem 8–13.

But from the general shear formula,

$$\tau = \frac{VQ}{It}$$

Then

$$q = \tau t = \frac{VQ}{I} \qquad \textbf{(8–21)}$$

The units for q are *force per unit length,* such as N/m, N/mm, or lb/in. The shear flow is a measure of how much shearing force must be resisted at a particular section per unit length. Knowing the shearing force capacity of a fastener then allows the determination of a safe spacing for the fasteners.

For example, if a particular style of nail can safely withstand 150 lb of shearing force, we will define

$$F_{sd} = 150 \text{ lb}$$

Then, if in a particular place on a beam fabricated by nailing boards together with the shear flow computed to be $q = 28.5$ lb/in, the maximum spacing, s_{max} of the nail is

$$s_{max} = \frac{F_{sd}}{q} = \frac{150 \text{ lb}}{28.5 \text{ lb/in}} = 5.26 \text{ in} \qquad \textbf{(8–22)}$$

Guidelines for Specifying the Spacing of Fasteners

The objective is to specify suitable spacing for fasteners to hold parts of a composite beam shape together while resisting an applied vertical shearing force.

Required data: Applied shearing force, V.
Geometry of the cross section of the beam.
Allowable shearing force on each fastener, F_{sd}.

1. Compute the moment of inertia, I, for the entire cross section with respect to its centroidal axis.
2. Compute the value of the first moment of the area, Q, for that part of the cross section outside the fasteners. Use $Q = A_p\bar{y}$ as defined in Section 8–4.
3. Compute the shear flow, q, from

$$q = \frac{VQ}{I}$$

   The result will be the amount of shearing force that must be resisted per unit length along the beam.
4. Compute the maximum allowable spacing of the fasteners, s_{max}, from

$$s_{max} = F_{sd}/q$$

5. Specify a convenient spacing between fasteners less than the maximum allowable spacing.

Example Problem 8–12 Determine the proper spacing of the nails used to secure the flange boards to the web of the built-up I-beam shown in Figure 8–21. All boards are standard 2×8 wood shapes. The nails to be used can safely resist 250 lb of shearing force each. The load on the beam is shown in Figure 8–15 with $P = 500$ lb.

Solution

Objective: Specify a suitable spacing for the nails.

Given: Loading in Figure 8–15. $P = 500$ lb. $F_{sd} = 250$ lb/nail.
Beam shape and dimensions in Figure 8–21.

Analysis: Use the *Guidelines for specifying the spacing of fasteners.*

Results: The maximum shearing force on the beam is 500 lb, occurring between each support and the applied loads.

Step 1. The moment of inertia can be computed by subtracting the two open-space rectangles at the sides of the web from the full rectangle surrounding the I-shape.

$$I = \frac{7.25(10.25)^3}{12} - \frac{2(2.875)(7.25)^3}{12} = 468.0 \text{ in}^4$$

Step 2. At the place where the nails join the boards, Q is evaluated for the area of the top (or bottom) flange board.

$$Q = A_p\bar{y} = (1.5 \text{ in})(7.25 \text{ in})(4.375 \text{ in}) = 47.6 \text{ in}^3$$

Step 3. Then the shear flow is

$$q = \frac{VQ}{I} = \frac{(500 \text{ lb})(47.6 \text{ in}^3)}{468 \text{ in}^4} = 50.9 \text{ lb/in}$$

This means that 50.9 lb of force must be resisted along each inch of length of the beam at the point between the flange and the web boards.

Step 4. Since each nail can withstand 250 lb, the maximum spacing is

$$S_{max} = \frac{F_{sd}}{q} = \frac{250 \text{ lb}}{50.9 \text{ lb/in}} = 4.92 \text{ in}$$

Step 5. A spacing of $s = 4.5$ in would be reasonable.

The principle of shear flow also applies for sections like that shown in Figure 8–22, in which a beam section is fabricated by riveting square bars to a vertical web plate to form an I-shape. The shear flow occurs from the web plate to the flange bars. Thus, when evaluating the statical moment Q, the partial area, A_p, is taken to be the area of one of the flange bars.

Example Problem 8–13 A fabricated beam is made by riveting square aluminum bars to a vertical plate, as shown in Figure 8–22. The bars are 20 mm square. The plate is 6 mm thick and 200 mm high. The rivets can withstand 800 N of shearing force across one cross section. Determine the required spacing of the rivets if a shearing force of 5 kN is applied.

Solution Objective Specify a suitable spacing for the rivets.

Given Shearing force = 5 kN. F_{sd} = 800 N/rivet.
Beam shape and dimensions in Figure 8–22.

Analysis Use the *Guidelines for specifying the spacing of fasteners.*

Results ***Step 1.*** I is the moment of inertia of the entire cross section,

$$I = \frac{6(200)^3}{12} + 4\left[\frac{20^4}{12} + (20)(20)(90)^2\right]$$
$$I = 17.0 \times 10^6\ \text{mm}^4$$

Step 2. Q is the product of $A_p\bar{y}$ for the area *outside* the section where the shear is to be calculated. In this case, the partial area A_p is the 20-mm square area *to the side* of the web. For the beam in Figure 8–22,

$$Q = A_p\bar{y} \quad \text{(for one square bar)}$$
$$Q = (20)(20)(90)\ \text{mm}^3 = 36\,000\ \text{mm}^3$$

Step 3. Then for V = 5 kN,

$$q = \frac{VQ}{I} = \frac{(5 \times 10^3\ \text{N})(36 \times 10^3\ \text{mm}^3)}{17.0 \times 10^6\ \text{mm}^4} = 10.6\ \text{N/mm}$$

Thus a shearing force of 10.6 N is to be resisted for each millimeter of length of the beam.

Step 4. Since each rivet can withstand 800 N of shearing force, the maximum spacing is

$$S_{\text{max}} = \frac{F_{sd}}{q} = \frac{800\ \text{N}}{10.6\ \text{N/mm}} = 75.5\ \text{mm}$$

Step 5. Specify a spacing of s = 75 mm.

REFERENCES

1. Aluminum Association, *Aluminum Design Manual,* Washington, D.C., 2005.
2. American Concrete Institute, *2006 Manual of Concrete Practice,* American Concrete Institute, Farmington Hills, MI, 2006.
3. American Institute of Steel Construction, *Specification for Structural Steel Buildings,* AISC, Chicago, IL, 2005.
4. American Institute of Steel Construction, *Manual of Steel Construction: Load and Resistance Factor Design,* 3rd ed., AISC, Chicago, IL, 2001.
5. American Institute of Steel Construction, *Manual of Steel Construction: Allowable Stress Design,* 9th ed., AISC, Chicago, IL, 1989.
6. ASM International, *ASM Handbook, Volume 21: Composites,* Materials Park, OH, 2001.
7. ASTM International, *Standard D2344—Standard Test Method for Short-Beam Strength of Polymer Matrix Composite Materials and Their Laminates,* West Conshohocken, PA, ASTM International, 2006.
8. ASTM International, *The Composite Materials Handbook—MIL 17,* West Conshohocken, PA, ASTM International, 2002.
9. Halperin, D. A. and T. G. Bible, *Principles of Timber Design for Architects and Builders,* John Wiley & Sons, New York, 1994.

10. The Masonry Society, *Masonry Designer's Guide,* 4th ed., The Masonry Society, Boulder, CO, 2003.

11. McCormac, J. C. and J. Nelson, *Structural Steel Design—LRFD Method,* 4th ed., Prentice Hall, Upper Saddle River, NJ, 2007.

12. Mehta, P. K. and P. J. M. Monteiro, *Concrete,* 3rd ed., McGraw-Hill, New York, 2006.

13. Speigel, L. and G. F. Limbrunner, *Applied Structural Steel Design,* 4th ed., Prentice Hall, Englewood Cliffs, NJ, 2001.

14. U.S. Department of Agriculture Forest Products Laboratory, *Wood Handbook—Wood as an Engineering Material,* Forest Products Laboratory, Madison, WI, 1999.

PROBLEMS

General Shear Formula

For Problems 8–1 through 8–20, compute the shearing stress at the horizontal neutral axis for a beam having the cross-sectional shape shown in the given figure for the given shearing force. Use the general shear formula.

8–1.M Use a rectangular shape having a width of 50 mm and a height of 200 mm. $V = 7500$ N.

8–2.M Use a rectangular shape having a width of 38 mm and a height of 180 mm. $V = 5000$ N.

8–3.E Use a rectangular shape having a width of 1.5 in and a height of 7.25 in. $V = 12\ 500$ lb.

8–4.E Use a rectangular shape having a width of 3.5 in and a height of 11.25 in. $V = 20\ 000$ lb.

8–5.M Use a circular shape having a diameter of 50 mm. $V = 4500$ N.

8–6.M Use a circular shape having a diameter of 38 mm. $V = 2500$ N.

8–7.E Use a circular shape having a diameter of 2.00 in. $V = 7500$ lb.

8–8.E Use a circular shape having a diameter of 0.63 in. $V = 850$ lb.

8–9.E Use the shape shown in Figure P8–9. $V = 1500$ lb.

0.375 in
1.5 in
0.75 in

FIGURE P8–9

8–10.E Use the shape shown in Figure P8–10. $V = 850$ lb.

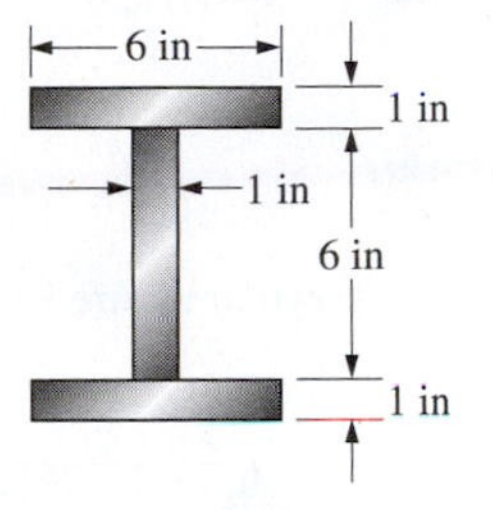

FIGURE P8–10

8–11.E Use the shape shown in Figure P8–11. $V = 850$ lb.

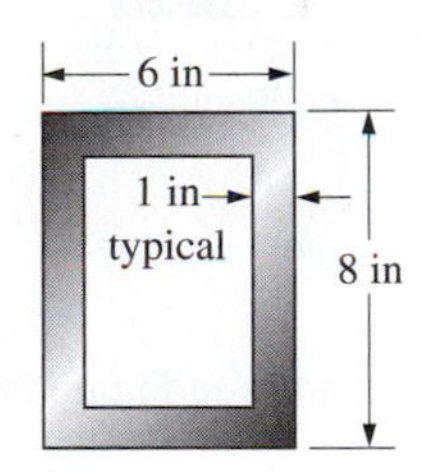

FIGURE P8–11

8–12.M Use the shape shown in Figure P8–12. $V = 112$ kN.

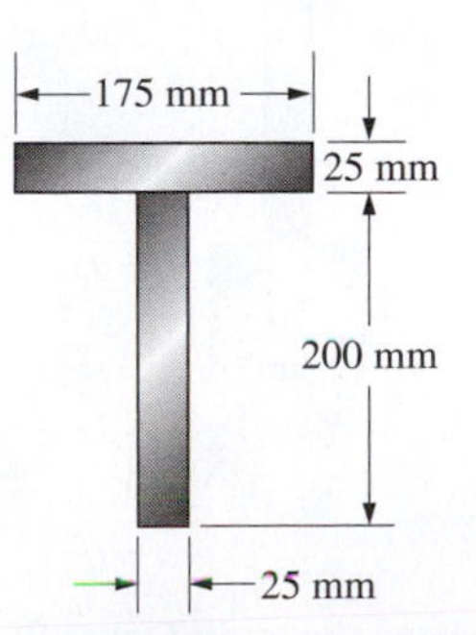

FIGURE P8–12

8–13.M Use the shape shown in Figure P8–13. $V = 71.2$ kN.

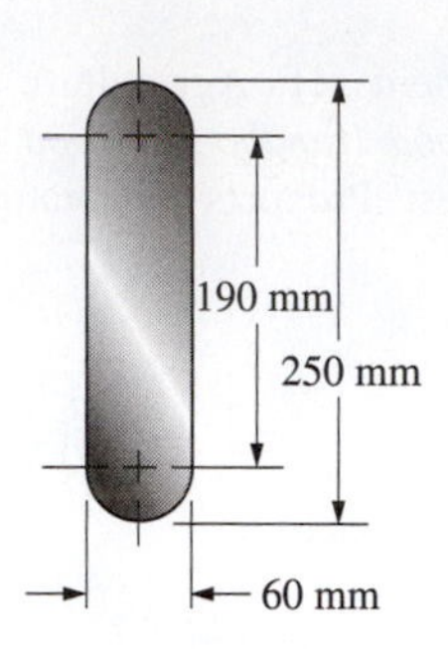

FIGURE P8–13

8–14.M Use the shape shown in Figure P8–14. $V = 1780$ N.

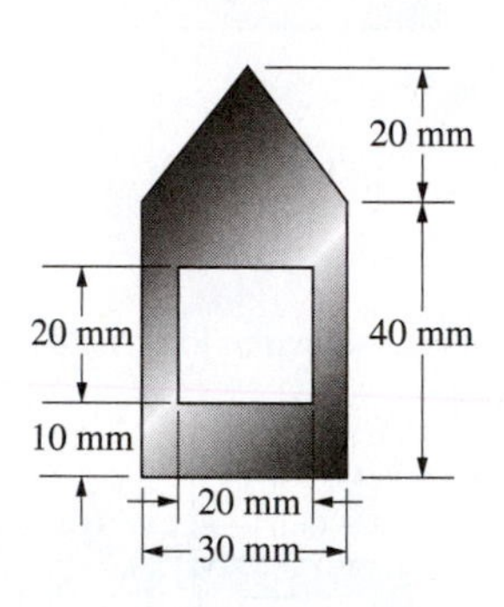

FIGURE P8–14

8–15.M Use the shape shown in Figure P8–15. $V = 675$ N.

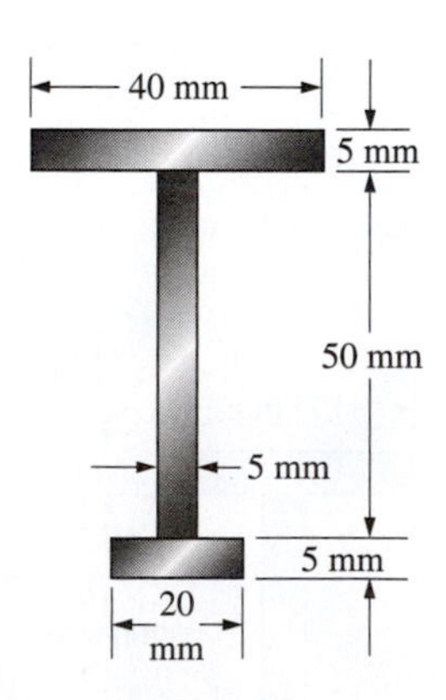

FIGURE P8–15

8–16.M Use the shape shown in Figure P8–16. $V = 2.5$ kN.

FIGURE P8–16

8–17.M Use the shape shown in Figure P8–17. $V = 10.5$ kN.

FIGURE P8–17

8–18.E Use the shape shown in Figure P8–18. $V = 1200$ lb.

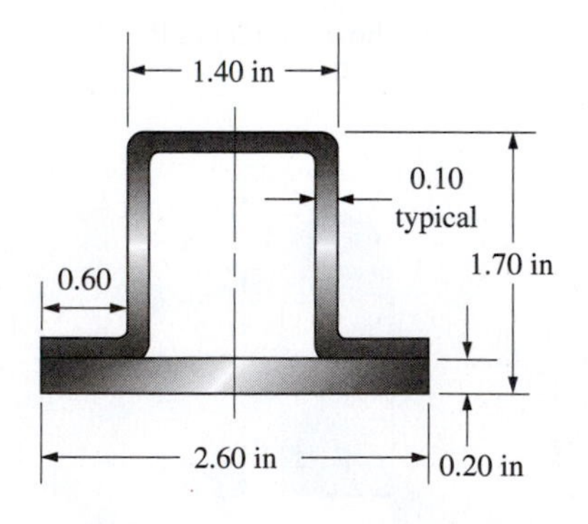

FIGURE P8–18

8–19.E Use the shape shown in Figure P8–19. $V = 775$ lb.

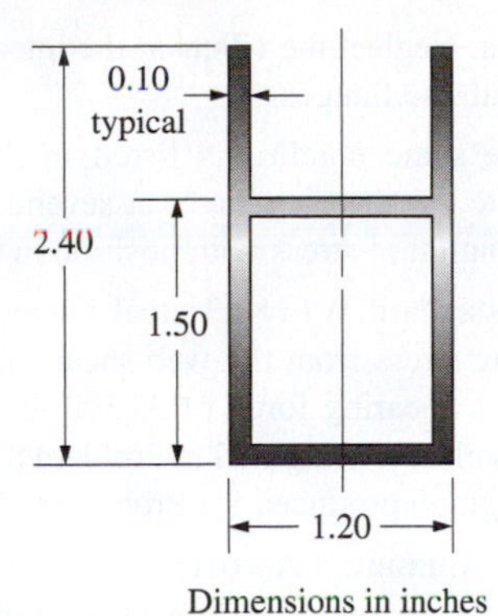

FIGURE P8–19

8–20.E Use the shape shown in Figure P8–20. $V = 2500$ lb.

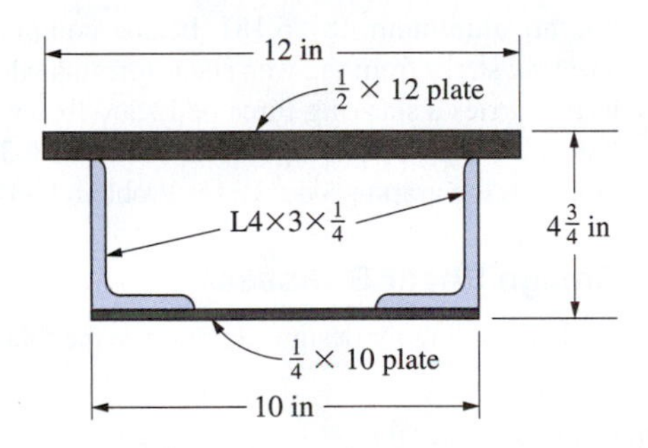

FIGURE P8–20

For Problems 8–21 through 8–30, assume that the indicated shape is the cross section of a beam made from wood having an allowable shearing stress of 70 psi, which is that of No. 2 grade southern pine listed in Appendix A–19. Compute the maximum allowable shearing force for each shape. Use the general shear formula.

8–21.E Use a standard 2 × 4 wooden beam with the long dimension vertical.

8–22.E Use a standard 2 × 4 wooden beam with the long dimension horizontal.

8–23.E Use a standard 2 × 12 wooden beam with the long dimension vertical.

8–24.E Use a standard 2 × 12 wooden beam with the long dimension horizontal.

8–25.E Use a standard 10 × 12 wooden beam with the long dimension vertical.

8–26.E Use a standard 10 × 12 wooden beam with the long dimension horizontal.

8–27.E Use the shape shown in Figure P8–27.

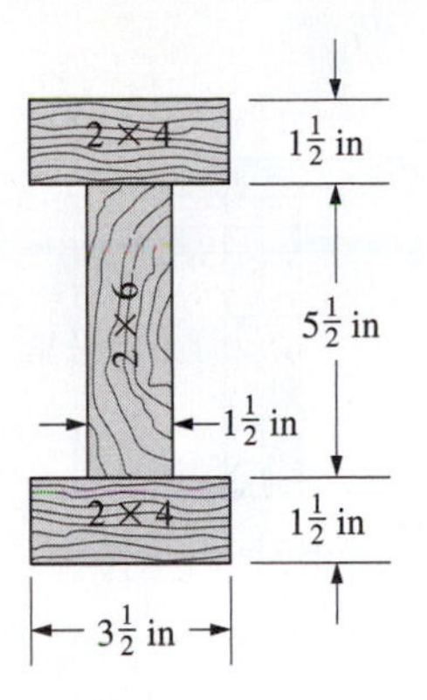

FIGURE P8–27

8–28.E Use the shape shown in Figure P8–28.

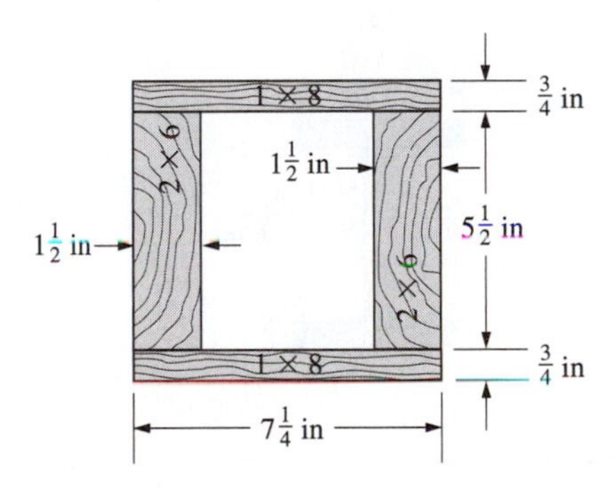

FIGURE P8–28

8–29.E Use the shape shown in Figure P8–29.

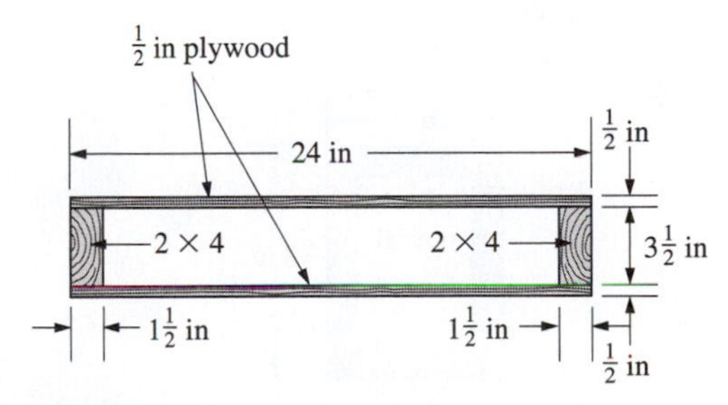

FIGURE P8–29

8–30.E Use the shape shown in Figure P8–30.

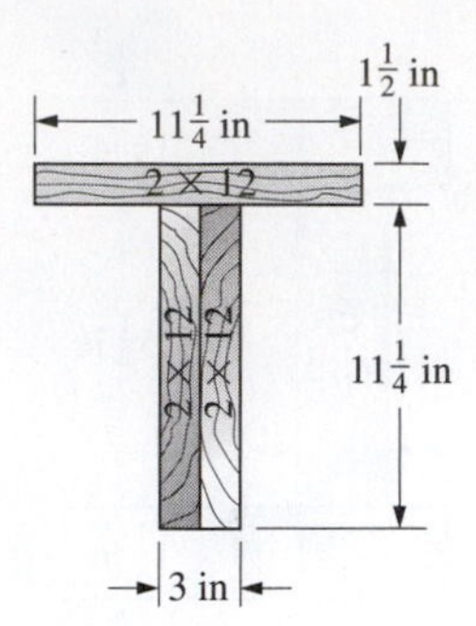

FIGURE P8–30

8–31.E For a beam having the I-shape cross section shown in Figure P8–31, compute the shearing stress on horizontal axes 0.50 in apart from the bottom to the top. At the ends of the web where it joins the flanges, compute the stress in both the web and the flange. Use a shearing force of 500 lb. Then plot the results.

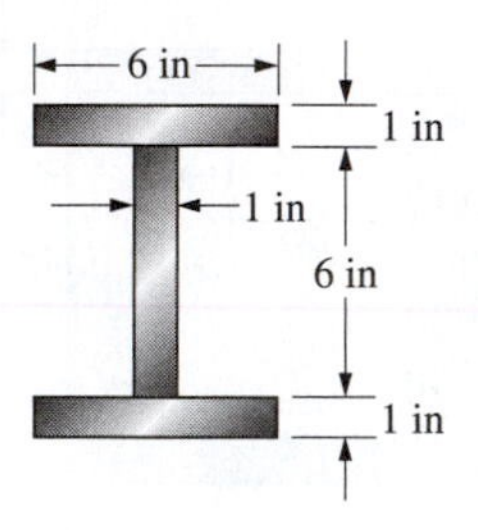

FIGURE P8–31

8–32.E For a beam having the box-shape cross section shown in Figure P8–32, compute the shearing stress on horizontal axes 0.50 in apart from the bottom to the top. At the ends of the vertical sides where they join the flanges, compute the stress in both the web and the flange. Use a shearing force of 500 lb. Then plot the results.

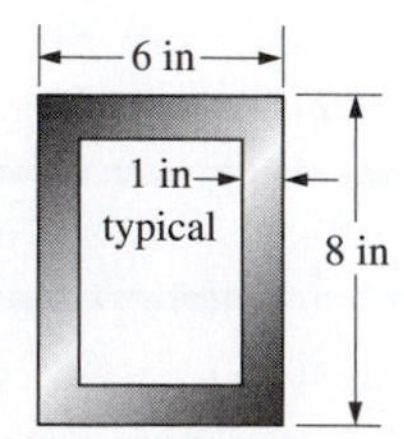

FIGURE P8–32

8–33.E For a standard W14×43 steel beam, compute the shearing stress at the neutral axis when subjected to a shearing force of 33 500 lb. Use the general shear formula. Neglect the fillets at the intersection of the web with the flanges.

8–34.E For the same conditions listed in Problem 8–33, compute the shearing stress at several axes and plot the variation of stress with position in the beam.

8–35.E For a standard W14×43 steel beam, compute the shearing stress from the web shear formula when it carries a shearing force of 33 500 lb. Compare this value with that computed in Problem 8–33 and plot it on the graph produced for Problem 8–34.

8–36.E For an Aluminum Association Standard I8×6.181 beam, compute the shearing stress at the neutral axis when subjected to a shearing force of 13 500 lb. Use the general shear formula. Neglect the fillets at the intersection of the web with the flanges.

8–37.E For the same conditions listed in Problem 8–36, compute the shearing stress at several axes and plot the variation of stress with position in the beam.

8–38.E For an aluminum I8×6.181 beam, compute the shearing stress from the web shear formula when the beam carries a shearing force of 13 500 lb. Compare this value with that computed in Problem 8–36 and plot it on the graph produced for Problem 8–37.

Use of Design Shear Stresses

Note: In problems calling for design stresses, use the following.

For structural steel:

In bending: $\sigma_d = 0.66s_y$

In shear: $\tau_d = 0.4s_y$

For any other metal:

In bending: $\sigma_d = \dfrac{s_y}{N}$

In shear: $\tau_d = 0.5\dfrac{s_y}{N}$

For wood:

Use allowable stresses in Appendix A–19.

8–39.E The loading shown in Figure P8–39 is to be carried by a W12×16 steel beam. Compute the shearing stress using the web shear formula. Also compute the maximum bending stress. Then compare the stresses to the design stresses for ASTM A992 structural steel.

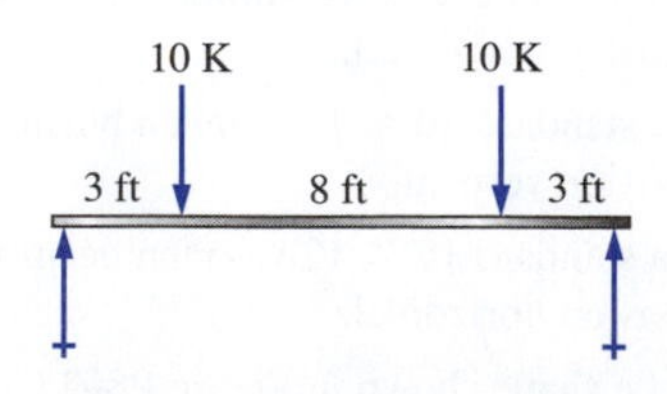

FIGURE P8–39

8–40.E Specify a suitable wide-flange beam to be made from ASTM A992 structural steel to carry the load shown in Figure P8–39 based on the design stress in bending. Then, for the beam selected, compute the shearing stress from the web shear formula and compare it with the design shear stress.

8–41.E Specify a suitable wide-flange beam to be made from ASTM A992 structural steel to carry the load shown in Figure P8–41 based on the design stress in bending. Then, for the beam selected, compute the shearing stress from the web shear formula and compare it with the design shear stress.

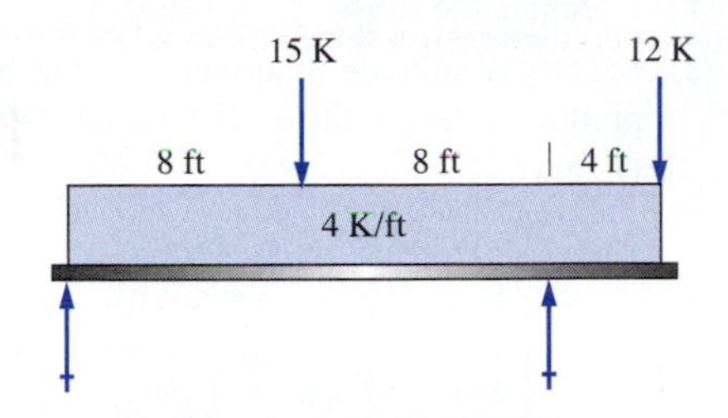

FIGURE P8–41

8–42.C Specify a suitable wide-flange beam to be made from ASTM A992 structural steel to carry the load shown in Figure P8–42 based on the design stress in bending. Then, for the beam selected, compute the shearing stress from the web shear formula and compare it with the design shear stress.

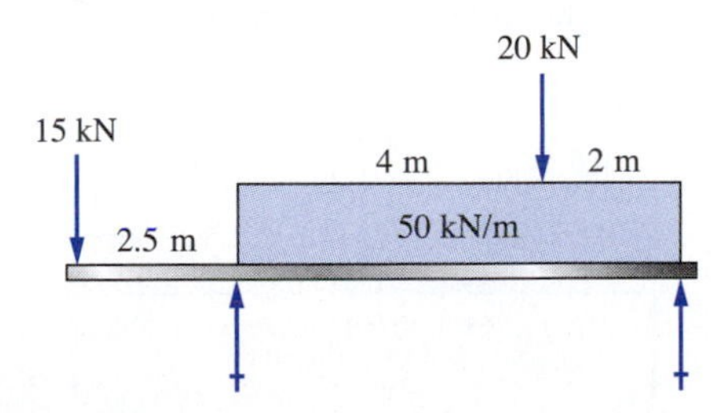

FIGURE P8–42

8–43.E Specify a suitable standard steel pipe from Appendix A–12 to be made from ASTM A53, Grade B steel to carry the load shown in Figure P8–43 based on the design stress in bending with a design factor of 3. Then, for the pipe selected, compute the shearing stress from the special shear formula for hollow tubes and compute the resulting design factor from the design shear stress formula.

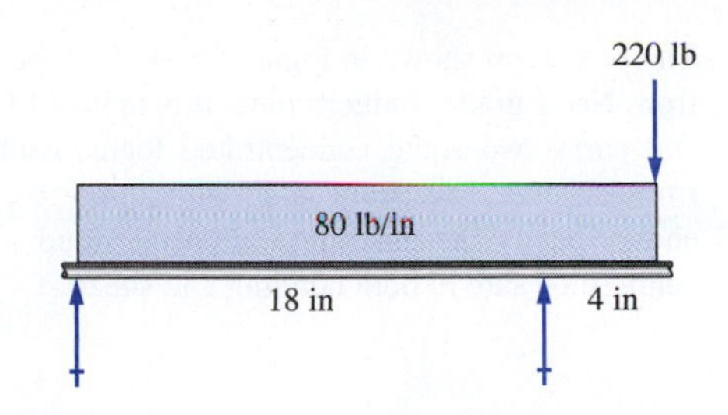

FIGURE P8–43

8–44.E An Aluminum Association standard channel (Appendix A–10) is to be specified to carry the load shown in Figure P8–44 to produce a design factor of 4 in bending. The legs of the channel are to point down. The channel is made from 6061-T6 aluminum. For the channel selected, compute the maximum shearing stress.

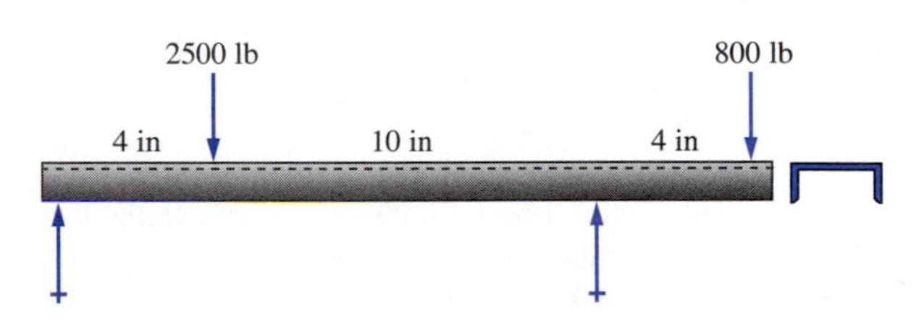

FIGURE P8–44

8–45.E A wooden joist in the floor of a building is to carry a uniformly distributed load of 200 lb/ft over a length of 12.0 ft. Specify a suitable standard wooden beam shape for the joist, made from No. 2 grade hemlock, to be safe in both bending and shear (see Appendices A–4 and A–19).

8–46.C A wooden beam in an outdoor structure is to carry the load shown in Figure P8–46. If it is to be made from No. 3 grade Douglas fir, specify a suitable standard wooden beam to be safe in both bending and shear (see Appendices A–4 and A–19).

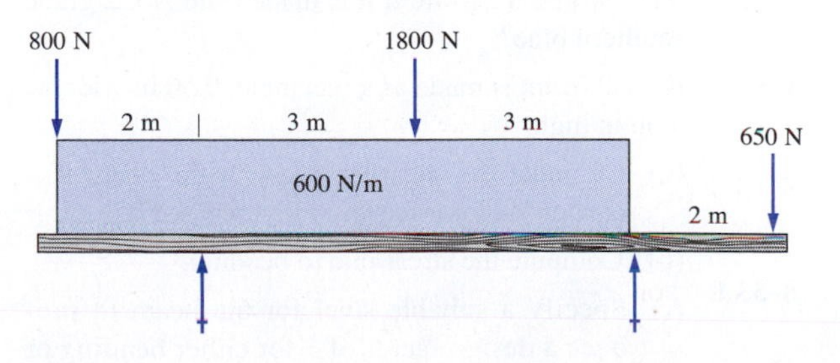

FIGURE P8–46

8–47.E The box beam shown in Figure P8–47 is to be made from No. 1 grade southern pine. It is to be 14 ft long and carry two equal concentrated loads, each 3 ft from an end. The beam is simply supported at its ends. Specify the maximum allowable load for the beam to be safe in both bending and shear.

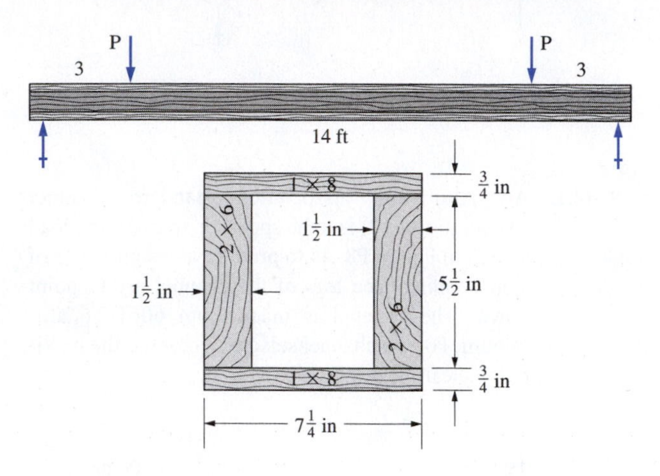

FIGURE P8–47

8–48.C An aluminum I-beam, I229×12.44, carries the load shown in Figure P8–48. Compute the shearing stress in the beam using the web shear formula.

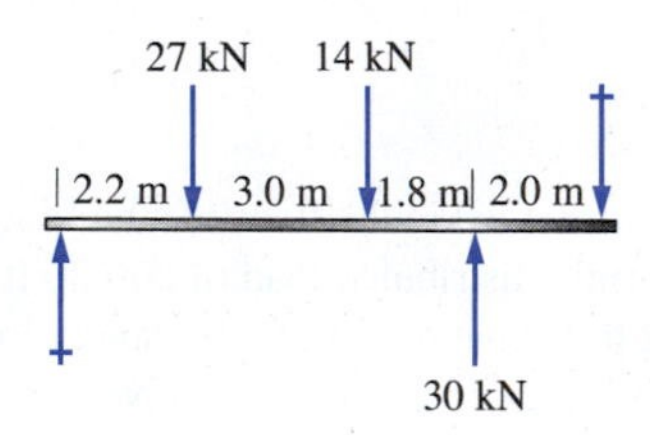

FIGURE P8–48

8–49.C Compute the bending stress for the beam in Problem 8–48.

8–50.E A 2 × 8 wooden floor joist in a home is simply supported, 12 ft long, and carries a uniformly distributed load of 80 lb/ft. Compute the shearing stress in the joist. Would it be safe if it is made from No. 2 grade southern pine?

8–51.E A steel beam is made as a rectangle, 0.50 in wide by 4.00 in high.

(a) Compute the shearing stress in the beam if it carries the load shown in Figure P8–51.

(b) Compute the stress due to bending.

(c) Specify a suitable steel for the beam to produce a design factor of 3 for either bending or shear.

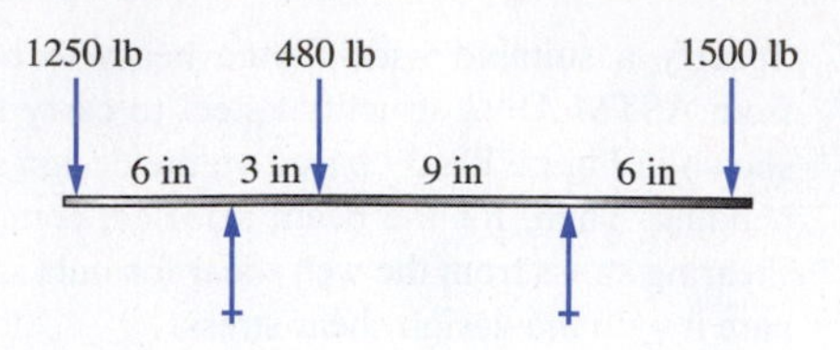

FIGURE P8–51

8–52.M An aluminum beam is made as a rectangle, 16 mm wide by 60 mm high.

(a) Compute the shearing stress in the beam if it carries the load shown in Figure P8–52.

(b) Compute the stress due to bending.

(c) Specify a suitable aluminum for the beam to produce a design factor of 3 for either bending or shear.

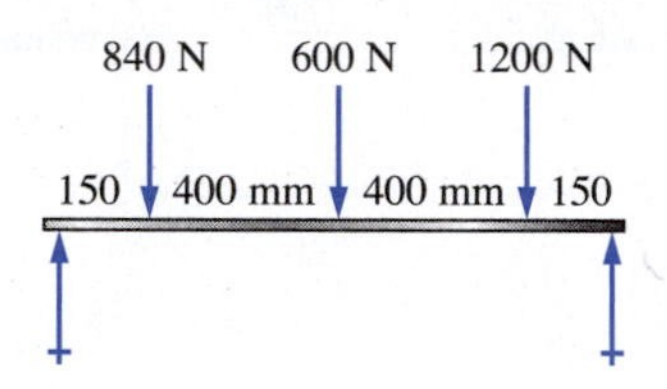

FIGURE P8–52

8–53.M It is planned to use a rectangular bar to carry the load shown in Figure P8–53. Its thickness is to be 12 mm and it is to be made from aluminum 6061-T6. Determine the required height of the rectangle to produce a design factor of 4 in bending based on yield strength. Then compute the shearing stress in the bar and the resulting design factor for shear.

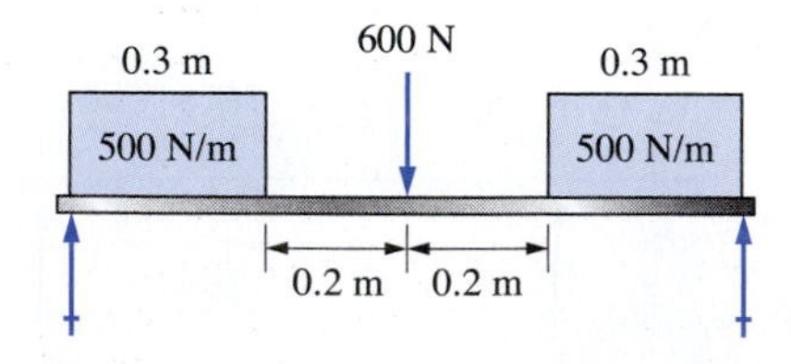

FIGURE P8–53

8–54.M A round shaft, 40 mm in diameter, carries the load shown in Figure P8–54.

(a) Compute the maximum shearing stress in the shaft.

(b) Compute the maximum stress due to bending.

(c) Specify a suitable steel for the shaft to produce a design factor of 4 based on yield strength in either shear or bending.

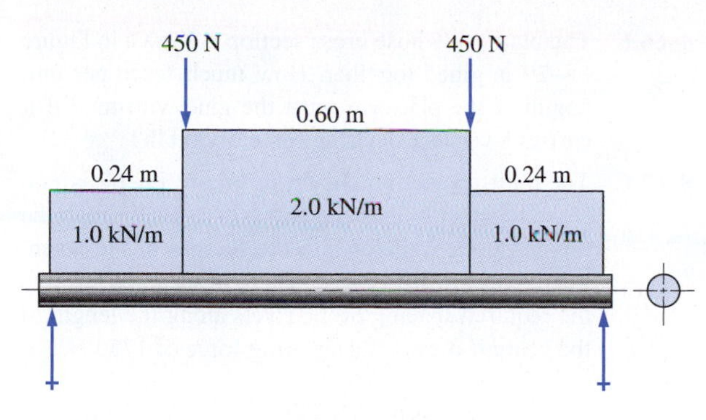

FIGURE P8–54

8–55.M Compute the required diameter of a round bar to carry the load shown in Figure P8–55 while limiting the stress due to bending to 120 MPa. Then compute the resulting shearing stress in the bar and compare it to the bending stress.

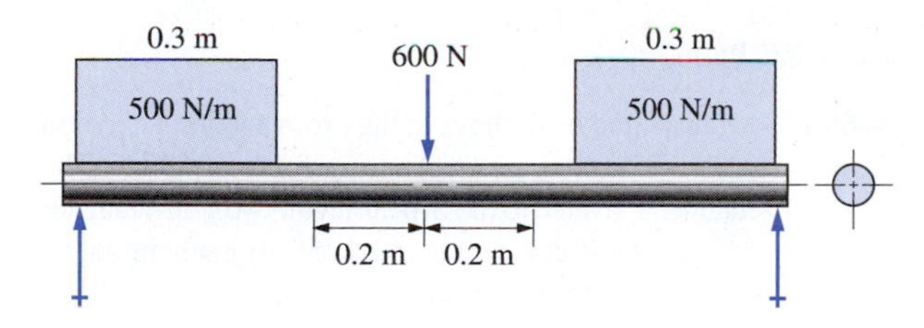

FIGURE P8–55

8–56.E Compute the maximum allowable vertical shearing force on a wood dowel having a diameter of 1.50 in, if the maximum allowable shearing stress is 70 psi.

8–57.E A standard steel pipe is to be selected from Appendix A–12 to be used as a chinning bar in a gym. It is to be simply supported at the ends of its 36-in length. Men weighing up to 400 lb are expected to hang on the bar with either one or two hands at any place along its length. The pipe is to be made from AISI 1020 hot-rolled steel. Specify a suitable pipe to provide a design factor of 6 based on yield strength in either bending or shear.

8–58.E A standard steel pipe is to be simply supported at its ends and carry a single concentrated load of 2800 lb at its center. The pipe is to be made from AISI 1020 hot-rolled steel. The minimum design factor is to be 4 based on yield strength for either bending or shear. Specify a suitable pipe size from Appendix A–12 if the length of the pipe is

(a) 1.5 in
(b) 3.0 in
(c) 4.5 in
(d) 6.0 in

Shear Flow Problems

8–59.E The shape in Figure P8–59 is to be made by gluing the flat plate to the hat section. If the beam made from this section is subjected to a shearing force of 1200 lb, compute the shear flow at the joint. What must be the shearing strength of the glue in psi?

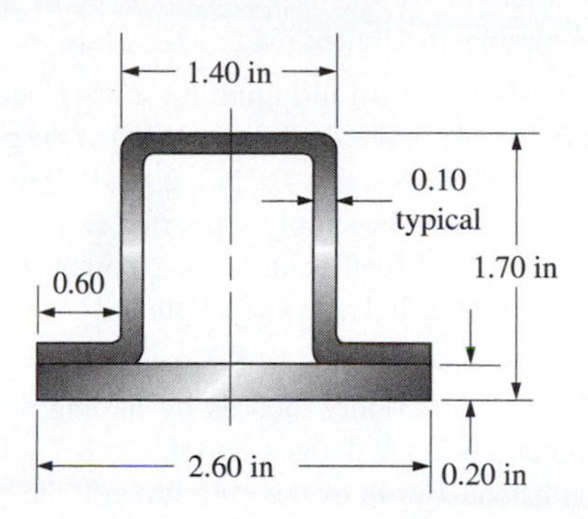

FIGURE P8–59

8–60.E The shape in Figure P8–60 is to be made by using a metal-bonding glue between the S-beam and the web of the channel. Compute the shear flow at the joint and the required shearing strength of the glue for a shearing force of 2500 lb.

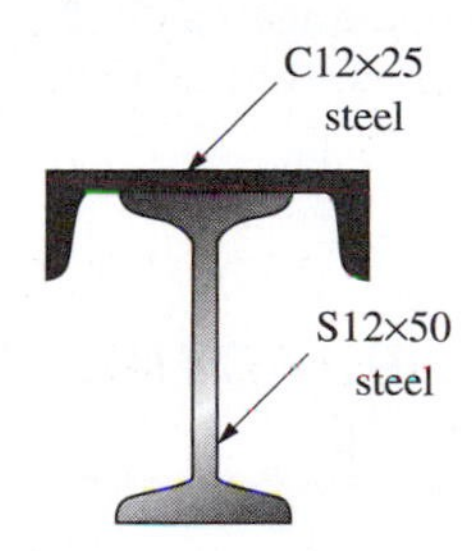

FIGURE P8–60

8–61.E The shape in Figure P8–20 is to be fabricated by riveting the bottom plate to the angles and then welding the top plate to the angles. When used as a beam, there are four potential failure modes: bending stress, shearing stress in the angles, shear in the welds, and shear of the rivets. The shape is to be used as the seat of a bench carrying a uniformly distributed load over a span of 10.0 ft. Compute the maximum allowable distributed load for the following design limits.

(a) The material of all components is aluminum 6061-T4, and a design factor of 4 is required for either bending or shear.

(b) The allowable shear flow on each weld is 1800 lb/in.

(c) The rivets are spaced 4.0 in apart along the entire length of the beam. Each rivet can withstand 600 lb of shear.

8–62.E An alternative design for the bench described in Problem 8–61 is to use the built-up wood T-shape shown in Figure P8–30. The wood is to be No. 3 grade southern pine. One nail is to be driven into each vertical 2 × 12. Each nail can withstand 160 lb in shear, and the nails are spaced 6.0 in apart along the length of the beam. Compute the maximum allowable distributed load on the beam.

8–63.E The shape shown in Figure P8–27 is glued together and the allowable shearing strength of the glue is 800 psi. The components are No. 2 grade Douglas fir. If the beam is to be simply supported and carry a single concentrated load at its center, compute the maximum allowable load. The length is 10 ft.

8–64.E The I-section shown in Figure P8–27 is fabricated from three wooden boards by nailing through the top and bottom flanges into the web. Each nail can withstand 180 lb of shearing force. If the beam having this section carries a vertical shearing force of 300 lb, what spacing would be required between nails?

8–65.E The built-up section shown in Figure P8–28 is nailed together by driving one nail through each side of the top and bottom boards into the $1\frac{1}{2}$-in-thick sides. If each nail can withstand 150 lb of shearing force, determine the required nail spacing when the beam carries a vertical shearing force of 600 lb.

8–66.E The platform whose cross section is shown in Figure P8–29 is glued together. How much force per unit length of the platform must the glue withstand if it carries a vertical shearing force of 500 lb?

8–67.C The built-up section shown in Figure P8–67 is fastened together by passing two $\frac{3}{8}$-in rivets through the top and bottom plates into the flanges of the beam. Each rivet will withstand 2650 lb in shear. Determine the required spacing of the rivets along the length of the beam if it carries a shearing force of 175 kN.

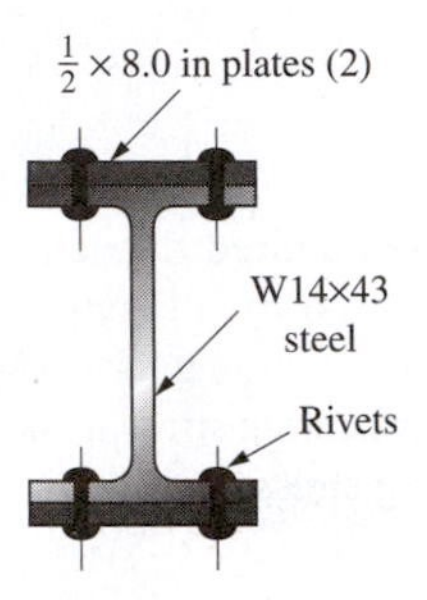

FIGURE P8–67

8–68.E A fabricated beam having the cross section shown in Figure P8–60 carries a shearing force of 50 kN. The channel is riveted to the S-beam with two $\frac{1}{4}$-in-diameter rivets, which can withstand 1750 lb each in shear. Determine the required rivet spacing.

ADDITIONAL PRACTICE AND REVIEW PROBLEMS

Web Shear

For Problems 8–69 through 8–73, compute the shearing stress in the web of the given beam using the web shear formula. Then compare the result with the design shear stress defined in Section 8–7 to determine if the shearing stress is safe.

8–69. A W18×55 shape carries a vertical shearing force of 36.6 K. The material is ASTM A992 structural steel.

8–70. A W18×40 shape carries a vertical shearing force of 36.6 K. The material is ASTM A992 structural steel.

8–71. A W14×26 shape carries a vertical shearing force of 10 000 lb. The material is ASTM A992 structural steel.

8–72. An aluminum 6I×4.692 shape carries a vertical shearing force of 10 000 lb. The material is 6061-T6 aluminum.

8–73. A W10×12 shape carries a vertical shearing force of 6750 lb. The material is ASTM A992 structural steel.

General Shear Formula and Special Shear Formulas

For Problems 8–74 through 8–79, compute the maximum shearing stress using the general shear formula or one of the special shear formulas. Then compare the result with the design shear stress defined in Section 8–7 to determine if the shearing stress is safe.

8–74. A standard rectangular steel tube, HSS6×2×$\frac{1}{4}$, carries a static vertical shearing force of 6750 lb. The material is ASTM A500 grade B structural steel.

8–75. A standard 2 × 8 wood beam made from No. 2 stress-rated southern pine is installed as a cantilever and carries a vertical shearing force of 480 lb.

8–76. The shape shown in Figure P8–29 is made from No. 2 Douglas fir and carries a shearing force of 750 lb.

8–77. A standard rectangular steel tube, HSS8×2×$\frac{1}{4}$, carries a vertical shearing force of 12 000 lb. The material is ASTM A500 grade B structural steel.

8–78. The shape shown in Figure P8–78 carries a vertical shearing force of 1800 lb. The materials are ASTM

A992 structural steel for the W-beam and A500 grade B for the HSS sections.

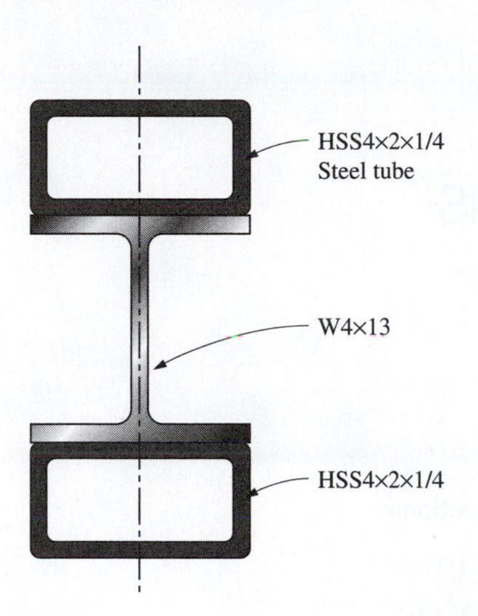

FIGURE P8–78

8–79. A standard C10×6.136 aluminum channel is installed with the legs down and carries a vertical shearing force of 430 lb. The material is 6061-T6 aluminum.

Shear Flow

8–80. For the data of Problem 8–78, compute the shear flow at the interface between the W-shape and the tube.

8–81. The shape shown in Figure P8–29 is to carry a vertical shearing force of 500 lb. The plywood is to be nailed to the 2 × 4 members using nails that can carry a shearing force of 135 lb each. Specify a suitable spacing for the nails.

9 Deflection of Beams

The Big Picture

Deflection of Beams

Discussion Map

- The proper performance of machine parts; the structural rigidity of buildings, vehicles, toys, and consumer products; and the tendency for a component to vibrate are all dependent on limiting deformations or deflections of beams to acceptable values. You will explore methods of analyzing beam deflections in this chapter.
- This topic builds on the competencies you developed in previous chapters.
- You discovered in Chapters 7 and 8 how to determine the normal stresses in beams due to bending moments and to the shearing stresses due to shearing forces that are created in beams.
- You also learned how to design beams to be sufficiently strong to ensure that they would not fail by fracture or yielding of the material of the beam.
- Another potential failure mode is the excessive deformation or deflection of the beam. Now you will explore ways that beams deflect under load and learn several ways for computing the predicted deflections.
- Two classes of beams will be considered:
  - *Statically determinate beams* (Figure 9–1)—those for which all unknown reaction forces and moments can be found using the classic equations of static equilibrium:
    - $\Sigma F = 0$ in any direction
    - $\Sigma M = 0$ about any point
  - *Statically indeterminate beams* (Figure 9–2)—those for which there are too many unknowns to solve by conventional methods of statics. Other analysis techniques are introduced here to enable these types of beams to be analyzed.
- You will also explore the effect of different types of support and loading conditions on the general flexibility or stiffness of beams so you can make sound judgments about which types to employ in your designs.

Activity Statically Determinate Beams

Acquire again the same materials you used for the activities in Chapters 5, 6, 7, and 8 when you explored the general behavior of beams, the shearing forces and bending moments developed in them, their cross-sectional shapes, bending stresses, and shearing stresses. There you used simple, flexible beam examples, a variety of supports, and some means of loading the beams.

Using Figure 9–1 as a guide, set up the various types of beam loading and support conditions described here.

1. Support the beam at its ends, place two concentrated loads near the middle, and observe the resulting shape of the beam. It deflects downward, forming a concave upward curved shape. This is the classical demonstration of positive bending [Figure 9–1(a)].
2. Remove one load and move the other along the beam. Observe how the shape of the deflected beam changes [Figure 9–1(b)].
3. Move one of the supports in from the end of the beam to create an overhang. Then place one load near the end of the overhang. One thing you will need to do is reverse the direction of the support at the opposite end of the beam or else it will rotate up off the support. Then you should notice that the beam deflects

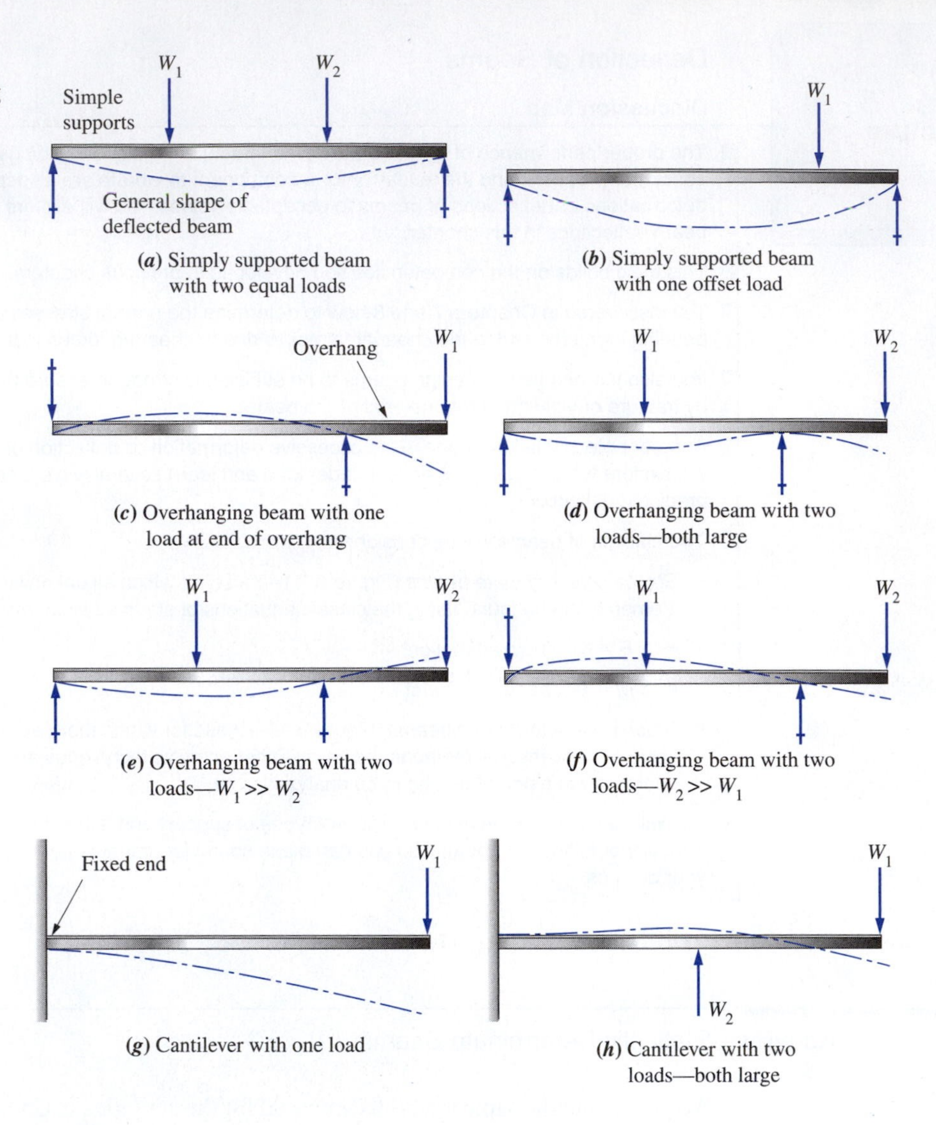

FIGURE 9–1 A variety of beam loading and support conditions.

(*a*) Simply supported beam with two equal loads

(*b*) Simply supported beam with one offset load

(*c*) Overhanging beam with one load at end of overhang

(*d*) Overhanging beam with two loads—both large

(*e*) Overhanging beam with two loads—$W_1 >> W_2$

(*f*) Overhanging beam with two loads—$W_2 >> W_1$

(*g*) Cantilever with one load

(*h*) Cantilever with two loads—both large

into a convex or concave downward shape. This is an example of negative bending [Figure 9–1(c)].

4. Next, use the overhanging beam and place loads both at the end of the overhang and between the supports. Vary the magnitudes and placements of the two loads and see what kinds of shapes for the deflected beam you can produce [Figure 9–1(d), 9–1(e), 9–1(f)].
5. Now configure the beam as a cantilever by clamping it firmly to a table. Perhaps you can use your book and a firm hand to ensure that one end is held fixed. Place loads at various positions on the cantilever and in different directions and observe its shape. Also observe the relative flexibility of the cantilever depending on the shape of the beam, its length, and the placement of the load. You should see that a cantilever is somewhat more flexible than a similar beam of the same length supported at two points [Figure 9–1(g) and 9–1(h)].

FIGURE 9–2
Examples of statically indeterminate beams.

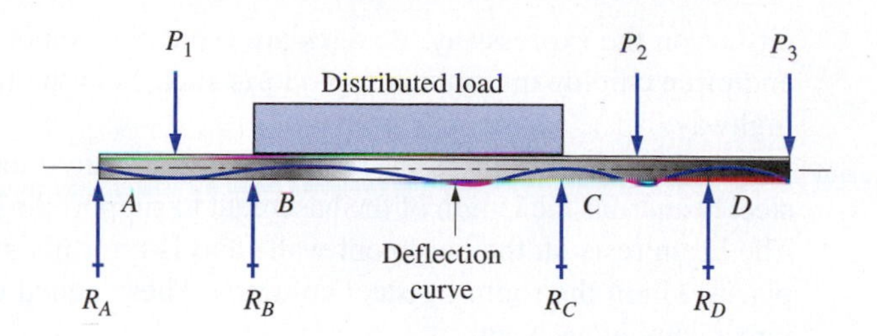

(*a*) Continuous beam—a beam on three or more simple supports

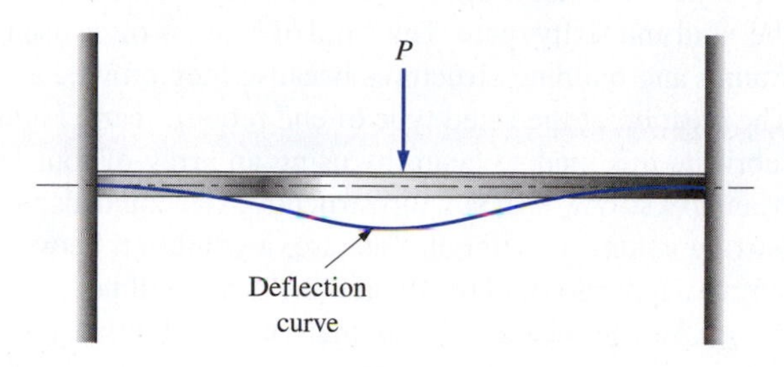

(*b*) Beam with two fixed ends

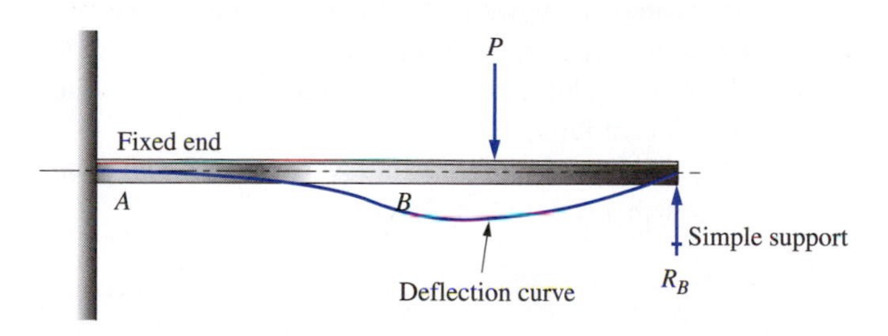

(*c*) Supported cantilever

Activity Statically Indeterminate Beams

Set up demonstrations of the three types of statically indeterminate beams described here using the same materials from the activity for statically determinate beams, but adding additional supports. Use Figure 9–2 as a guide.

1. A *continuous beam* is any beam with more than two supports. The name comes from the observation that the beam is continuous over several supports. Figure 9–2(a) shows one example. Use your beam-building materials to first set up a simply supported beam with supports on each end fairly widely spaced. Apply one or several loads with your hand. Note the flexibility of the beam. Now place a third support under the middle of the beam and apply the loads again. Do you sense the increased stiffness of the beam? Adding a fourth support will increase the stiffness even more.

   Where have you seen continuous beams? There are numerous examples in building structures and highway bridges. Look for them as you drive around the city or on the expressway. Bridges are typically supported on piers at each end and often employ intermediate supports such as in the median strip of a divided highway.

Another example is in the basement of many ranch-style homes. Often a steel beam runs the length of the basement to support the joists for the floor above. The beam rests on the basement walls and is typically supported at one or more places within the room by steel columns. These added columns make the beam into a continuous beam.

2. A *fixed-end beam* is one that is rigidly constrained at both ends to prevent rotation and to support the loads on the beam. See Figure 9–2(b) for a sketch. Notice that the deflected beam shape starts with a horizontal slope at each support because of the rigid end fixity there. This kind of beam is often used in the design of machine frames and building structures because they provide a very stiff, rigid assembly. The creation of the fixed type of end requires care. Figure 9–3 shows one way to fabricate this kind of beam by using an array of four bolts tightly fastening the beam to a strong, stiff support structure. You could also weld the beams to the supporting columns. Without due care, a condition between that of fixed ends and simple supports could result, and the beam will not perform as planned.

   Use your beam-building materials to simulate a fixed-end beam. Then load it and observe its relative stiffness compared with the other designs.

3. A *supported cantilever* is a beam with one fixed end and a second, simple support at the other as shown in Figure 9–2(c). Earlier we observed that a cantilever is one of the most flexible types of beams. The second support contributes a significant amount of additional rigidity. Figure 9–4 shows one way of constructing a supported cantilever.

   Make a supported cantilever yourself with your materials. Compare its stiffness with a simple cantilever.

FIGURE 9–3
Fixed-end beam.

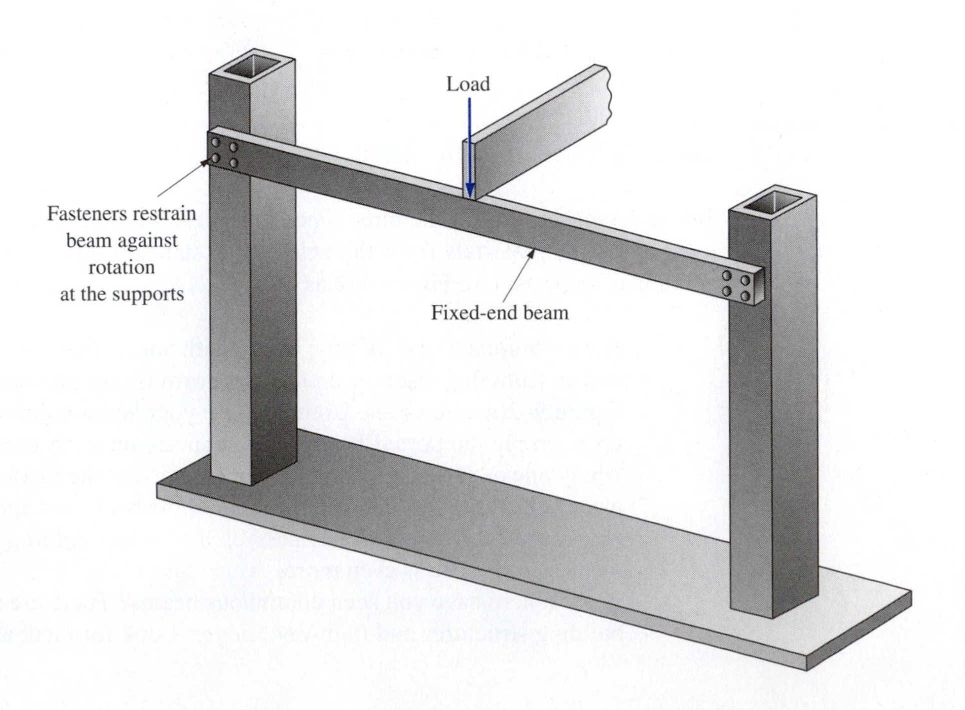

FIGURE 9–4
Supported cantilever.

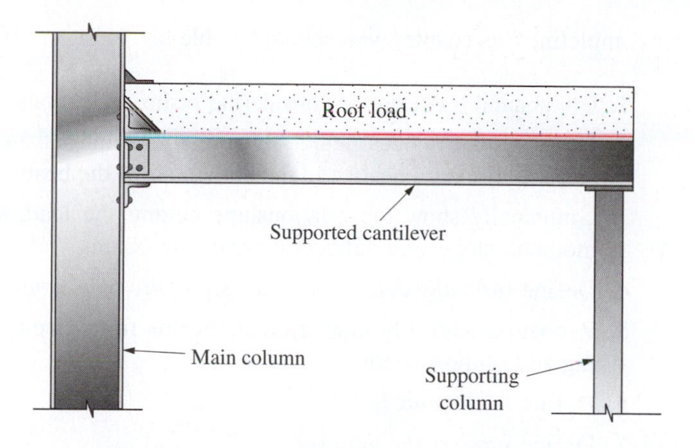

The analyses required for the statically indeterminate beams are somewhat more difficult than for statically determinate beams. This chapter provides several tools you can use to design and analyze such beams. Also included are some comparisons among the various types to help you make judgments about appropriate designs to use.

Methods of Analyzing Beam Deflections. In this chapter we present the principles on which the computation of the deflection of beams is based, along with four popular methods of deflection analysis: the *formula method,* the *superposition method,* the *successive integration method,* and the *moment-area method.*

Each method has its advantages and disadvantages, and your choice of which method to use depends on the nature of the problem.

- The formula method is the simplest, but it depends on the availability of a suitable formula to match the application.
- The superposition method, a modest extension of the formula method, dramatically expands the number of practical problems that can be handled without a significant increase in the complexity of your work.
- The moment-area method is fairly quick and simple, but it is typically used for computing the deflections of only one or a few points on the beam. Its use requires a high level of understanding of the principle of moments and the techniques of preparing bending moment diagrams.
- The successive integration method is perhaps the most general, and it can be used to solve virtually any combination of loading and support conditions for statically determinate beams. However, its use requires the ability to write the equations for the shearing force and bending moment diagrams and to derive equations for the slope and deflection of the beam using integral calculus. The successive integration method results in equations for the slope and deflection for the entire beam and enables the direct determination of the point of maximum deflection. Published formulas were developed using the successive integration or the moment-area method.

Several computer-assisted beam analysis programs are available to reduce the time and computation required to determine the deflection of beams. Although they can relieve the designer of much work, it is recommended that you understand the principles on which they are based before using them. See the list of Internet sites.

9–1 OBJECTIVES OF THIS CHAPTER

After completing this chapter, you should be able to:

1. Understand the need for considering beam deflections.
2. Understand the development of the relationships between the manner of loading and support for a beam and the deflection of the beam.
3. Graphically show the relationships among the load, shearing force, bending moment, slope, and deflection curves for beams.
4. Define *statically determinate* and *statically indeterminate* as applied to beams.
5. Recognize statically indeterminate beams from given descriptions of the loading and support conditions.
6. Define *continuous beam.*
7. Define *supported cantilever.*
8. Define *fixed-end beam.*
9. Use standard formulas to compute the deflection of beams at selected points.
10. Use the principle of superposition along with standard formulas to solve problems of greater complexity.
11. Compare the relative strength and stiffness of beams having different support systems for given loadings.
12. Develop formulas for the deflection of beams for certain cases using the successive integration method.
13. Apply the method of successive integration to beams having a variety of loading and support conditions.
14. Use the moment-area method to solve for the slope and deflection for beams.
15. Write computer programs to assist in using the several methods of beam analysis described in this chapter.

The organization of the chapter allows selective coverage. In general, all the information necessary to use each method is included within that part of the chapter. An exception is that the understanding of the formula method is necessary before using the superposition method.

9–2 THE NEED FOR CONSIDERING BEAM DEFLECTIONS

The spindle of a lathe or drill press and the arbor of a milling machine carry cutting tools for machining metals. Deflection of the spindle or arbor would have an adverse effect on the accuracy that the machine could produce. The manner of loading and support of these machine elements indicate that they are beams, and the approach to computing their deflection will be discussed later in this chapter.

Precision measuring equipment must also be designed to be very rigid. Deflection caused by the application of measuring forces reduces the precision of the desired measurement.

Power-transmission shafts carrying gears must have sufficient rigidity to ensure that the gear teeth mesh properly. Excessive deflection of the shafts would tend to separate the mating gears, resulting in a movement away from the most desirable point of contact between the gear teeth. Noise generation, decreased power-transmitting capability, and increased wear would result. For straight spur gears, it is recommended that the movement between two gears not exceed 0.005 in (0.13 mm). This limit is the *sum* of the movement of the two shafts carrying the mating gears at the location of the gears.

The floors of buildings must have sufficient rigidity to carry expected loads. Occupants of the building should not notice floor deflections. Machines and other equipment require a stable floor support for proper operation. Beams carrying plastered ceilings must not deflect excessively so as not to crack the plaster. A limit of $\frac{1}{360}$ times the span of the beam carrying a ceiling is often used for deflection.

Frames of vehicles, metal-forming machines, automation devices, and process equipment must also possess sufficient rigidity to ensure satisfactory operation of the equipment carried by the frame. The bead of a lathe, the crown of a punch press, the structure of an automatic assembly device, and the frame of a truck are examples.

Vibration is caused by the forced oscillations of parts of a structure or machine. The tendency to vibrate at a certain frequency and the severity of the vibrations are functions of the flexibility of the parts. Of course, *flexibility* is a term used to describe how much a part deflects under load. Vibration problems can be solved by either *increasing* or *decreasing* the stiffness of the part, depending on the circumstances. In either case, an understanding of how to compute deflections of beams is important.

Recommended Deflection Limits. It is the designer's responsibility to specify the maximum allowable deflection for a beam in a machine, frame, or structure. Knowledge of the application should give guidance. In the absence of such guidance, the following limits are suggested in References 2 and 3:

General machine part: $y_{max} = 0.0005$ to 0.003 in/in or mm/mm of beam length.
Moderate precision: $y_{max} = 0.000\,01$ to 0.0005 in/in or mm/mm of beam length.
High precision: $y_{max} = 0.000\,001$ to $0.000\,01$ in/in or mm/mm of beam length.

The numerical values given here could also be expressed as the ratio of the maximum deflection to the length of the beam. For example,

$$y_{max}/L = 0.0005 \text{ to } 0.003 \text{ for a general machine part.}$$
$$y_{max}/L = 0.000\,01 \text{ to } 0.0005 \text{ for moderate precision.}$$
$$y_{max}/L = 0.000\,001 \text{ to } 0.000\,01 \text{ for high precision.}$$

Then, multiplying the number by the length of the beam would produce the recommended maximum deflection.

For example, consider a beam, 18 in (450 mm) long, for a general purpose conveyor. The lower end of the range of recommended deflection is

$$y_{max} = \frac{0.0005 \text{ in}}{\text{in of length}} \times 18 \text{ in} = 0.009 \text{ in}$$

The same calculation applies to measurements in mm.

$$y_{max} = \frac{0.0005 \text{ mm}}{\text{mm of length}} \times 450 \text{ mm} = 0.225 \text{ mm}$$

If it does not adversely affect the operation of the system containing the conveyor, the deflection could be as high as

$$y_{max} = \frac{0.003 \text{ in}}{\text{in of length}} \times 18 \text{ in} = 0.054 \text{ in [slightly less than } \tfrac{1}{16} \text{ in]}$$

Using the metric data gives

$$y_{max} = \frac{0.003 \text{ mm}}{\text{mm of length}} \times 450 \text{ mm} = 1.35 \text{ mm}$$

Now, if the conveyor is part of a high-speed transfer station for an automated assembly machine, these deflections are likely to be excessive because the parts on the conveyor must be accurately positioned at each station. Moving to the "moderate precision" data, the limits would be in the range of

$$y_{max} = \frac{0.000\,01 \text{ in}}{\text{in of length}} \times 18 \text{ in} = 0.000\,18 \text{ in to } y_{max} = \frac{0.0005}{\text{in of length}} \times 18 \text{ in} = 0.009 \text{ in}$$

The corresponding limits using metric measurements are 0.0045 mm to 0.225 mm. These limits are readily achievable and should be applied to any piece of accurate machinery.

What if the beam were part of a precision inertial guidance system for a spacecraft? Here the "high precision" data should be applied. The range of deflection should be

$$y_{max} = \frac{0.000\,001 \text{ in}}{\text{in of length}} \times 18 \text{ in} = 0.000\,018 \text{ in to } y_{max} = \frac{0.000\,01}{\text{in of length}} \times 18 \text{ in} = 0.000\,18 \text{ in}$$

The corresponding limits using metric measurements are 0.000 45 mm to 0.0045 mm. These are, indeed, precision magnitudes for beam deflections.

9–3 GENERAL PRINCIPLES AND DEFINITIONS OF TERMS

To describe graphically the condition of a beam carrying a pattern of loading, five diagrams are used, as shown in Figure 9–5. The first three diagrams were used in earlier work. The *load diagram* is the free-body diagram on which all external loads and support reactions are shown. From that, the *shearing force diagram* was developed, which enables the calculation of shearing stresses in the beam at any section. The *bending moment diagram* is a plot of the variation of bending moment with position on the beam, with the results used for computing the stress due to bending. The horizontal axis of these plots is the position on the beam, called x. It is typical to measure x relative to the left end of the beam, but any reference point can be used.

Deflection Diagram. The last two diagrams are related to the deformation of the beam under the influence of the loads. It is convenient to begin discussion with the last diagram, the *deflection diagram,* because this shows the shape of the deflected beam. Actually, it is the plot of the position of the neutral axis of the beam relative to its initial position. The initial position is taken to be the straight line between the two support points on the unloaded beam. The amount of deflection will be called y, with positive values measured upward. Typical beams carrying downward loads, such as the one shown in Figure 9–5, will result in downward (negative) deflections of the beam.

Slope Diagram. A line drawn tangent to the deflection curve at a point of interest would define the slope of the deflection curve at that point. The slope is indicated as the angle, θ, measured in radians, relative to the horizontal, as shown in Figure 9–5. The plot of the slope as a function of position on the beam is the *slope curve,* drawn below the bending moment curve and above the deflection curve. Note on the given beam that the slope of the left portion

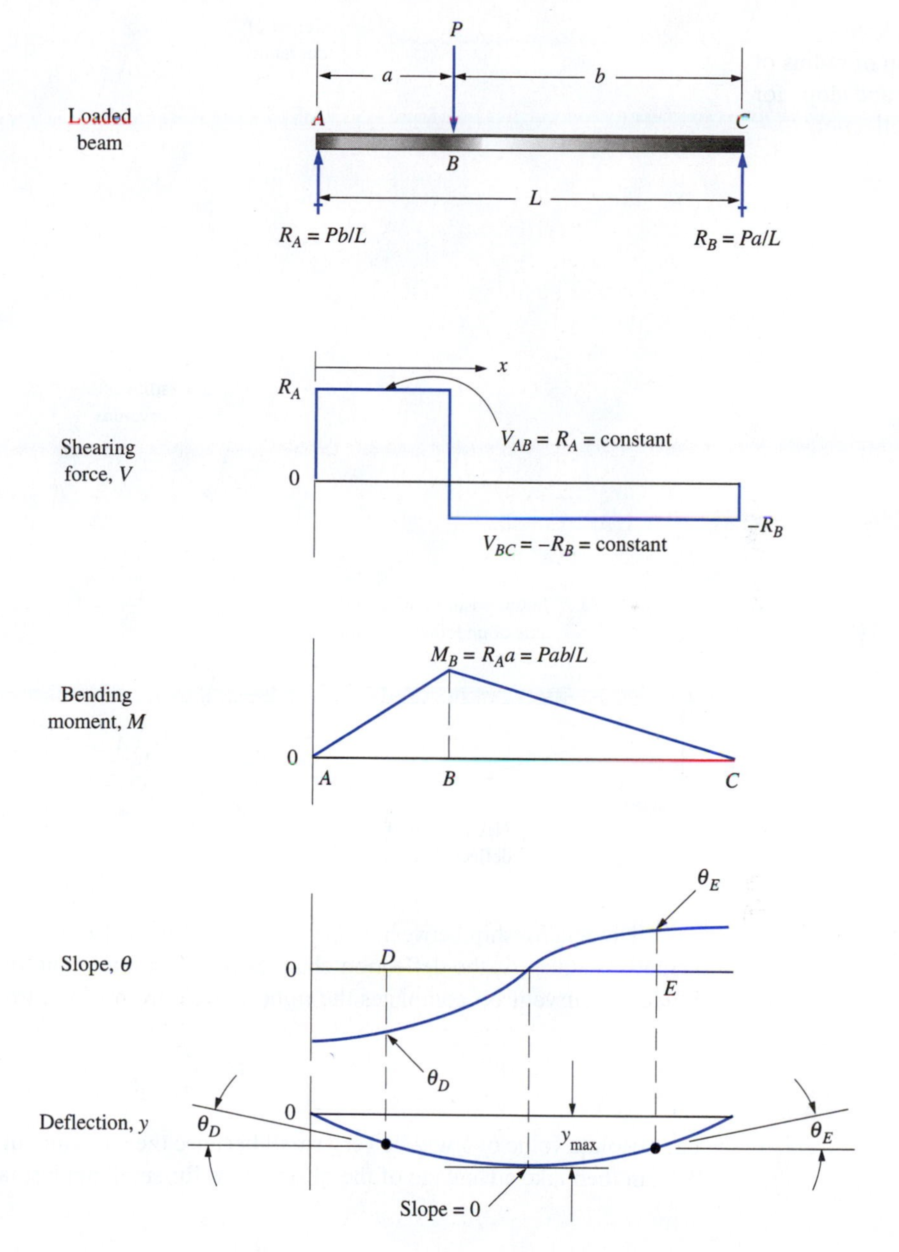

FIGURE 9–5 Five beam diagrams.

of the deflection curve is negative and the right portion has a positive slope. The point where the tangent line is itself horizontal is the point of zero slope and defines the location of the maximum deflection. This observation will be used in the discussion of the moment-area method and the successive integration method later in this chapter.

Radius of Curvature. Figure 9–6 shows the radius of curvature, R, at a particular point. For practical beams, the curvature is very slight, resulting in a very large value for R. For convenience, the shape of the deflection curve is exaggerated to aid in visualizing the principles and the variables involved in the analysis. Remember from analytic geometry that the radius of curvature at a point is perpendicular to the line drawn tangent to the curve at that point.

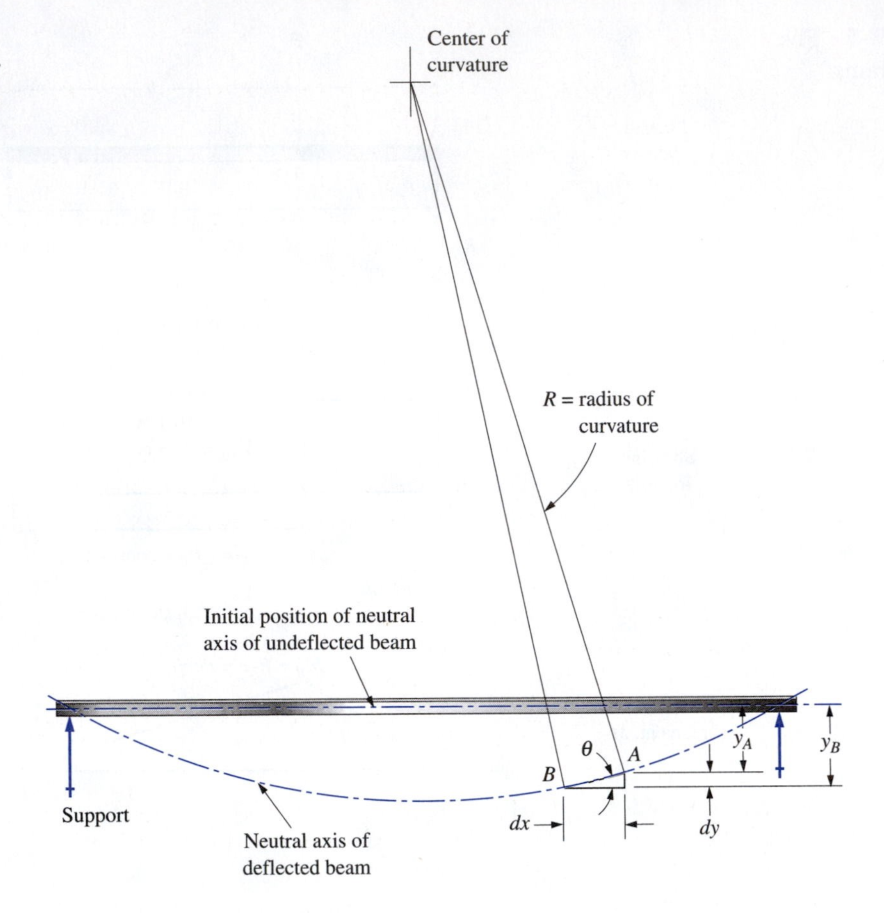

FIGURE 9–6 Illustration of radius of curvature and slope for a beam deflection curve.

The relationship between slope and deflection is also illustrated in Figure 9–6. Over a small distance dx, the deflection changes by a small amount dy. A small portion of the deflection curve itself completes the right triangle from which we can define

$$\tan\theta = \frac{dy}{dx} \tag{9–1}$$

The absolute value of θ will be very small because the curvature of the beam is very slight. We can then take advantage of the observation for small angles, $\tan\theta = \theta$. Then

$$\theta = \frac{dy}{dx} \tag{9–2}$$

It can then be concluded that:

The slope of the deflection curve at a point is equal to the ratio of the change in the deflection to the change in position on the beam.

Beam Stiffness. It will be shown later that the amount of deflection for a beam is inversely proportional to the *beam stiffness,* indicated by the product EI, where

E = modulus of elasticity of the material of the beam

I = moment of inertia of the cross section of the beam with respect to the neutral axis

It may help you to think of these terms as follows. The modulus of elasticity, E, is the *material stiffness*. The moment of inertia, I, is the *shape stiffness* for the cross section of the beam. As noted in Chapter 6, I is defined as the second moment of the area of the cross section. To minimize the deflection of a beam you should choose a material with the highest practical modulus of elasticity. And you should design the shape to have the largest practical moment of inertia. In general, the shape that is efficient for minimizing bending stresses, as discussed in Chapter 7, is also good for minimizing deflections of the beam.

9–4 BEAM DEFLECTIONS USING THE FORMULA METHOD

For many practical configurations of beam loading and support, formulas have been derived that allow the computation of the deflection at any point on the beam. The method of successive integration or the moment-area method, described later, may be used to develop the equations. Appendices A–23, A–24, and A–25 include many examples of formulas for beam deflections.

The deflection formulas are valid only for the cases where the cross section of the beam is uniform for its entire length. Example problems will demonstrate the application of the formulas.

Simply Supported Beams. Appendix A–23 includes ten different conditions of loading on beams that are simply supported, that is, beams having two and only two simple supports. Some are overhanging beams. We have shown earlier that such beams can be analyzed for the values of the reactions using the standard equations of equilibrium. Then the shearing force and bending moment diagrams can be developed using the methods from Chapter 5 from which the stress analysis of the beam can be completed, as discussed in Chapters 7 and 8. It must be understood that both the stress analysis and the deflection analysis of beams should be completed to assess the acceptability of a beam design.

The loading conditions in Appendix A–23 include a single concentrated load, two concentrated loads, a variety of distributed loads, and one case with a concentrated moment. The concentrated moment could be developed in the manner of the examples shown in Section 5–10. The thin phantom line in the diagrams is a sketch of the shape of the deflected beam, somewhat exaggerated. This can help you to visualize where critical points of deflection can be expected.

Note carefully the labeling for loads and dimensions in the diagrams of the beam deflection cases. It is essential that the real beam being analyzed match the general form of a given case and that you identify accurately the variables used in the formulas to the right of the diagrams. For most cases, formulas are given for the maximum deflection to be expected, deflections at the ends of overhangs, and deflections at points of applications of concentrated loads. Some cases include formulas for the deflection at any chosen point.

Take special notice of the general form of the deflection formulas. While some are more complex than others, the following general characteristics can be observed. Understanding these observations can help you make good decisions when designing beams.

1. Deflections are denoted by the variable y and are the change in the position of the neutral axis of the beam from its unloaded condition to the final, loaded condition, measured perpendicular to the original neutral axis.
2. Upward deflections are positive; downward deflections are negative.
3. The variable x, when used, denotes the horizontal position on the beam, measured from one of the supports. In some cases, a second position variable v is indicated, measured from the other support.
4. Deflections are proportional to the load applied to the beam.

5. Deflections are inversely proportional to the *stiffness* of the beam, defined as the product of E, the stiffness of the material from which the beam is made, and I, the moment of inertia of the cross section of the beam, also called *shape stiffness*.
6. Deflections are proportional to the *cube* of some critical length dimension, typically the span between the supports or the length of an overhang.

Cantilevers. Appendix A–24 includes four cases in which cantilever beams carry concentrated loads, distributed loads, or a concentrated moment. The maximum deflection obviously occurs at the free end of the beam. The fixed end constrains the beam against rotation at the support so that the deflection curve has a zero slope there.

Example Problem 9–1

Determine the maximum deflection of a simply supported beam carrying a hydraulic cylinder in a machine used to press bushings into a casting, as shown in Figure 9–7. The force exerted during the pressing operation is 15 kN. The beam is rectangular, 25 mm thick and 100 mm high, and made of steel.

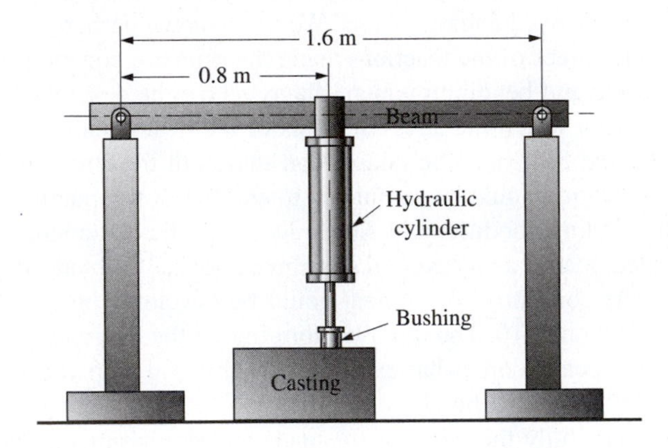

FIGURE 9–7 Beam for Example Problem 9–1.

Solution

Objective: Compute the maximum deflection of the given beam.

Given: System in Figure 9–7. Load = P = 15 kN. Span = L = 1.60 m.
Beam cross section: 25 mm wide by 100 mm high. Steel beam.

Analysis: The given beam can be considered to be a simply supported beam with a concentrated force applied in an upward direction at its center. Case a in Appendix A–23 applies.

Results: Using the formula from Appendix A–23–a, we find the maximum deflection to be

$$y = \frac{-PL^3}{48EI}$$

But P itself is negative, acting upward. From Appendix A–14, for steel, E = 207 GPa = 207×10^9 N/m^2. For the rectangular beam,

$$I = \frac{(25)(100)^3}{12} = 2.083 \times 10^6 \text{ mm}^4$$

Then

$$y = \frac{-PL^3}{48EI} = \frac{-(-15 \times 10^3\,\text{N})(1.6\,\text{m})^3}{48(207 \times 10^9\,\text{N/m}^2)(2.083 \times 10^6\,\text{mm}^4)} \times \frac{(10^3\,\text{mm})^5}{\text{m}^5}$$

$$y = 2.97\,\text{mm}$$

Comment This is a relatively high deflection that could adversely affect the accuracy of the bushing placement operation. It would be useful to evaluate the ratio of the deflection to the span for this beam and compare it to the recommendations in Section 9–2.

$$y_{max}/L = 2.97\,\text{mm}/1600\,\text{mm} = 0.001\,86$$

This is within the range of acceptable deflections for a general machine part. However, to place the bushing accurately in the casting, greater precision is desired. A stiffer beam shape (one with a higher moment of inertia, I) should be considered. Alternatively, the support system could be modified to decrease the span between the supports, a desirable approach because the deflection is proportional to the cube of the length. Assuming that the overall operation of the system permits the redesign of the span to be one-half of the given span (0.80 m), the deflection would be only 0.37 mm, $\frac{1}{8}$ as much as the given design. Checking the ratio of the deflection to the span again gives

$$y_{max}/L = 0.37\,\text{mm}/800\,\text{mm} = 0.000\,46$$

This places the design within the more desirable moderate precision range.

The stress in the beam should also be computed to assess the safety of the design. Figure 9–8 shows the load, shearing force, and bending moment diagrams for the original beam design from which we find the maximum bending moment in the beam to be $M = 6.00$ kN·m. The flexure formula can be used to compute the stress.

$$\sigma = \frac{Mc}{I} = \frac{(6.00\,\text{kN}\cdot\text{m})(50\,\text{mm})}{2.083 \times 10^6\,\text{mm}^4} \cdot \frac{10^3\,\text{N}}{\text{kN}} \cdot \frac{10^3\,\text{mm}}{\text{m}} = 144\,\text{MPa}$$

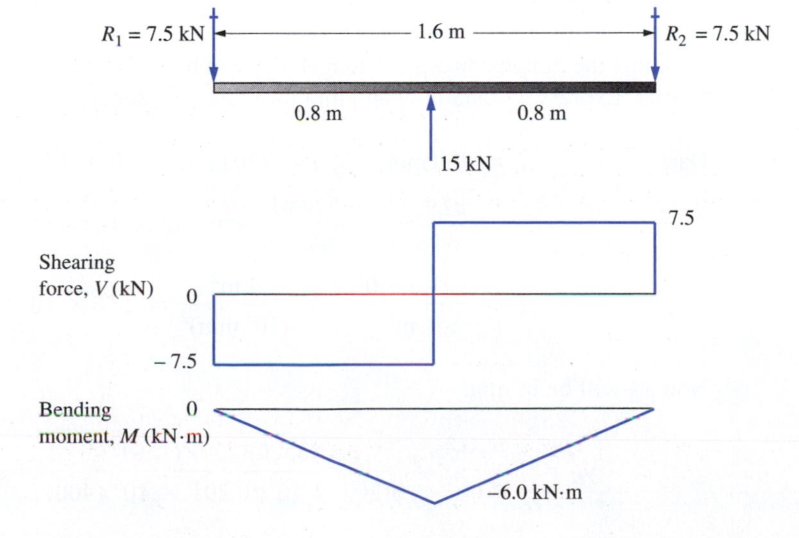

FIGURE 9–8 Beam diagrams for Example Problem 9–1.

This is a relatively high stress level. To continue the analysis, note that the beam would be subjected to repeated bending stress. Therefore, the recommended design stress is

$$\sigma_d = s_u/8$$

Letting $\sigma_d = \sigma$, we can solve for the required ultimate strength.

$$s_u = 8\,\sigma = (8)(144\text{ MPa}) = 1152\text{ MPa}$$

Consulting Appendix A–14 for the properties of selected steels, we could specify AISI 4140 OQT 900 steel that has an ultimate strength of 1281 MPa. This is a fairly expensive, heat-treated steel. A redesign of the type discussed for limiting deflection would reduce the stress and permit the use of a lower-priced steel.

Example Problem 9–2 A round shaft, 45 mm in diameter, carries a 3500-N load, as shown in Figure 9–9. If the shaft is steel, compute the deflection at the load and at the point C, 100 mm from the right end of the shaft. Also compute the maximum deflection.

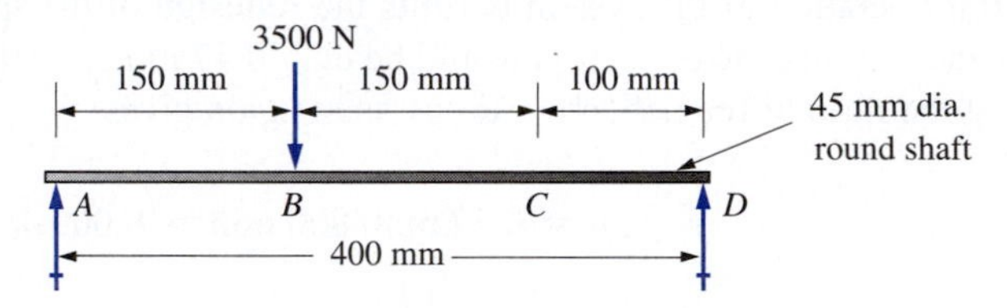

FIGURE 9–9 Shaft for Example Problem 9–2.

Solution

Objective: Compute the deflection at points B and C and at the point of maximum deflection.

Given: Beam in Figure 9–9. Load $= P = 3500$ N
Beam is round shaft; $D = 45$ mm. Steel beam.

Analysis: Appendix A–23–b applies because the beam is simply supported and has a single concentrated load placed away from the middle of the beam.

Results: **At Point *B*.** The following formula applies.

$$y_B = \frac{-Pa^2b^2}{3EIL}$$

Note that the dimension a is the longer segment between the load and one support; b is the shorter. Express all data in N and mm for unit consistency.

Data: $L = 400\text{ mm}; \quad a = 250\text{ mm}; \quad b = 150\text{ mm}$

$$I = \frac{\pi D^4}{64} = \frac{\pi(45\text{ mm})^4}{64} = 0.201 \times 10^6\text{ mm}^4$$

$$E = \frac{207 \times 10^9\text{ N}}{\text{m}^2} \times \frac{1\text{ m}^2}{(10^3\text{ mm})^2} = 207 \times 10^3\text{ N/mm}^2$$

Now y_B will be in mm.

$$y_B = \frac{-Pa^2b^2}{3EIL} = \frac{-(3500)(250)^2(150)^2}{3(207 \times 10^3)(0.201 \times 10^6)(400)} = -0.0985\text{ mm}$$

At Point *C*: Note that point *C* is in the longer segment. Previous data apply. Also, $x = 100$ mm from the right support to *C*.

$$y_c = \frac{Pbx}{6EIL}(L^2 - b^2 - x^2)$$

$$y_c = \frac{-(3500)(150)(100)}{6(207 \times 10^3)(0.201 \times 10^6)(400)}(400^2 - 150^2 - 100^2)$$

$$y_c = -0.0670 \text{ mm}$$

Maximum Deflection: Appendix A–23–b shows that the maximum deflection occurs in the longer segment of the beam at a distance x_1 from the support, where,

$$x_1 = \sqrt{a(L + b)/3} = \sqrt{(250)(400 + 150)/3} = 214 \text{ mm}$$

Then the deflection at that point is

$$y_{max} = \frac{-Pab(L + b)\sqrt{3a(L + b)}}{27(EIL)}$$

$$y_{max} = \frac{-(3500)(250)(150)(400 + 150)\sqrt{3(250)(400 + 150)}}{(27)(207 \times 10^3)(0.201 \times 10^6)(400)}$$

$$y_{max} = -0.103 \text{ mm}$$

Comment The results are shown in Figure 9–10.

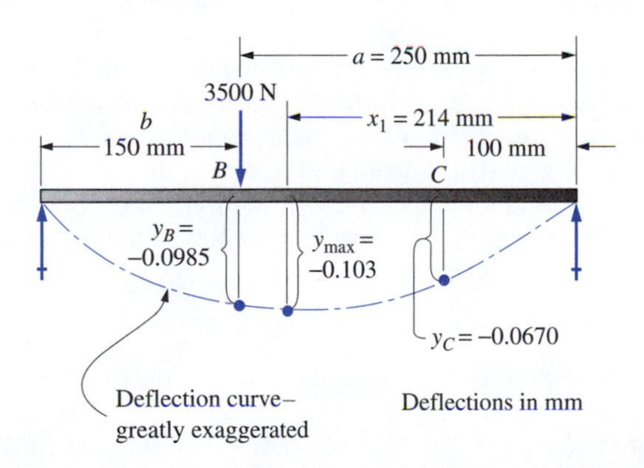

FIGURE 9–10 Deflection curve for beam in Example Problem 9–2.

Formulas for Statically Indeterminate Beams. Appendix A–25 shows several examples of statically indeterminate beams along with formulas for computing the reactions at supports and the deflection at any point on the beams. These formulas can be applied directly in the manner demonstrated for statically determinate beams. Also shown are plots of the shearing force and bending moment diagrams along with the necessary formulas for computing the values at critical points.

The general characteristics of statically indeterminate beams are quite different from the statically determinate designs studied in earlier chapters. This is particularly true for the manner of computing the reaction moments and forces at supports, the distribution of the bending moment with respect to position on the beam, the magnitude of the deflection at various points on the beam, and the general shape of the deflection curve.

The formulas for reactions, shearing force, and bending moments given in Appendix A–25 were derived using the principles discussed in Chapter 5 with due consideration for the statically indeterminate nature of the loading and support conditions. In particular, the *theorem of three moments* from Section 5–12 was used for the continuous beams. The techniques of superposition shown later in this chapter were used to develop the equations for deflections.

As you review the formulas in Appendix A–25, note the following general characteristics.

General Characteristics of Statically Indeterminate Beams

Supported Cantilevers (Cases *a* to *d* in Appendix A–25)

1. The fixed end provides a rigid support that resists any tendency for the beam to rotate. Thus, there is typically a significant bending moment there.
2. The second support is taken to be a simple support. If the simple support is at the free end of the beam, as in Cases *a*, *b*, and *c*, the bending moment there is zero.
3. If the supported cantilever has an overhang, as in Case *d*, the largest bending moment often occurs at the simple support. The shape of the bending moment curve is generally opposite that for the cases without an overhang.
4. There is a point of zero bending moment on a supported cantilever, generally near the fixed end.

Fixed-End Beams (Cases *e*, *f*, and *g* in Appendix A–25)

1. The bending moments at the fixed ends are *not* zero and may be the maximum for the beam.
2. When loads are downward on a fixed-end beam, the bending moments at the ends are *negative,* indicating that the deflection curve near there is *concave downward.*
3. When loads are downward, the bending moments near the middle of fixed-end beams are *positive,* indicating that the deflection curve there is *concave upward.*
4. There are typically two points of zero bending moment in fixed-end beams.
5. The deflection curve for a fixed-end beam has a zero slope at the fixed ends because of the restraint provided there against rotation.
6. The means of fastening a fixed-end beam to its supports must be capable of resisting the bending moments and the shearing forces at these points. Chapter 13 on Connections should be consulted for techniques of designing and analyzing connections that must resist moments.

Continuous Beams (Cases *h*, *i*, and *j* in Appendix A–25)

1. Points of maximum positive bending moment generally occur near the middle of spans between supports.
2. Points of maximum negative bending moment generally occur at the interior supports, and these are often the maximum bending moments.
3. Particularly for long continuous beams, it is often economically desirable to tailor the cross-sectional shape and dimensions to provide reinforcement over the sections that have the larger bending moments to accommodate the locally high values. An example would be to design the main beam shape to withstand the maximum positive bending moment between the supports and then to add reinforcing plates to the top or bottom of the beam near the supports to increase the moment of inertia and section modulus in the regions of high bending moment. Another approach would be to increase the depth of the beam near the supports. Highway overpasses and bridges over rivers often have these design features.
4. Where long continuous beams must be made in sections that are fastened together on site, it may be desirable to place the connection near a point of zero bending moment to simplify the design of the connection.

9–5 COMPARISON OF THE MANNER OF SUPPORT FOR BEAMS

In the following example problems we demonstrate the use of various formulas from Appendices A–23, A–24, and A–25 and also generate data with which to compare the performance of four different types of beam support to accomplish the same objective; that is, to support a given load at a given distance from one or two supports. The comparison is based on the magnitude of the bending stress and deflection in the four beams where each has the same material, shape, and size. The best performing beam, then, is the one with the lower stress and the smaller deflection.

The parameters of the comparisons are the following:

1. The four types of beam support to be compared are:
   a. Cantilever
   b. Simply supported beam
   c. Supported cantilever
   d. Fixed-end beam
2. Each beam is to support a single, static, concentrated load of 1200 lb.
3. The load is to be placed at a distance of 30 in from any support.
4. Each beam will be made from ASTM A36 structural steel for which we will use the following properties: $s_y = 36\,000$ psi; $E = 30 \times 10^6$ psi.
5. The maximum allowable bending stress will be based on the AISC standard,

$$\sigma_d = 0.66 s_y = 0.66(36\,000\text{ psi}) = 23\,760\text{ psi}$$

6. The maximum allowable deflection will be based on a limit of $L/360$, where L is the length of the beam.

It can be shown that, of the four types of beams and support systems to be considered, the cantilever has the poorest performance with regard to stress and deflection. Therefore, we will start the process of comparison by specifying the beam cross section shape and size that will ensure that the cantilever meets the requirements stated in items 5 and 6. Then, the same shape and size will be used for the other three designs. Five example problems follow, generating the results for the desired comparison.

Example Problem 9–3: Design and analysis of the cantilever.

Example Problem 9–4: Analysis of the simply supported beam.

Example Problem 9–5: Analysis of the supported cantilever.

Example Problem 9–6: Analysis of the fixed-end beam.

Example Problem 9–7: Comparison of the four beam designs.

Example Problem 9–3

For the cantilever shown in Figure 9–11, specify the lightest standard steel channel shape that will limit the bending stress to 23 760 psi and limit the maximum deflection to $L/360$. The channel is to be positioned with the legs pointing downward and the flat side of the web on top to provide a surface for mounting the load. For the beam shape selected, compute the actual expected maximum stress and deflection.

Solution Objective: Design the cantilever and compute the resulting stress and deflection.

Given: Beam dimensions and loading as shown in Figure 9–11. Beam shape to be a standard steel channel with the legs pointing downward.

$$\sigma_d = 23\,760\text{ psi}; \quad y_{\max} = L/360 \quad \text{with } L = L_c = 30\text{ in}$$

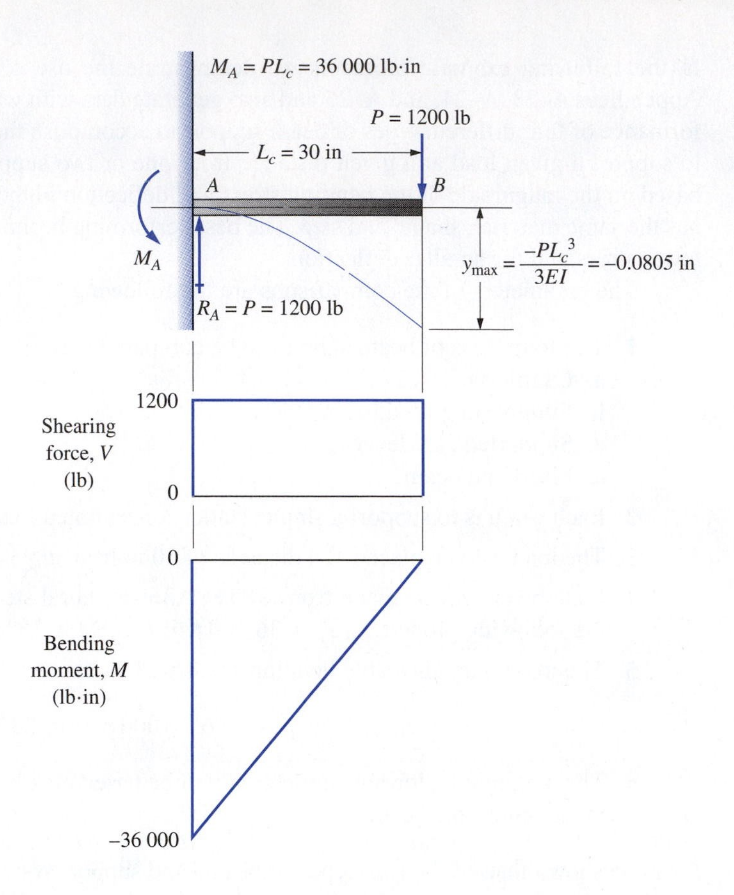

FIGURE 9–11 Cantilever.

Analysis Figure 9–11 contains the loading, shearing force, and bending moment diagrams from which we find that $M_{max} = 36\,000$ lb·in. Then,

$$\sigma = \sigma_d = M/S$$

Stress analysis: For a safe stress, the required section modulus is

$$S = M/\sigma_d = (36\,000 \text{ lb·in})/(23\,760 \text{ lb/in}^2) = 1.515 \text{ in}^3$$

From Appendix A–6, considering the properties with respect to the *Y-Y* axis, we select C12×25 as the lightest suitable section. Its properties are

$$S = 1.88 \text{ in}^3; \qquad I = 4.47 \text{ in}^4; \qquad w = 25 \text{ lb/ft}$$

Deflection: The maximum allowable deflection is

$$y_{max} = -L/360 = -(30 \text{ in})/360 = -0.0833 \text{ in}$$

From Appendix A–24, the formula for the maximum deflection is

$$y_{max} = -PL^3/3EI \qquad \text{at the end of the cantilever}$$

Then the required moment of inertia is

$$I = -PL^3/3Ey_{max}$$

$$I = \frac{(1200\text{ lb})(30\text{ in})^3}{3(30 \times 10^6)(-0.0833\text{ in})} = 4.32\text{ in}^4$$

The previously selected beam shape is satisfactory for deflection.

Actual bending stress:

$$\sigma = M/S = (36\,000\text{ lb}\cdot\text{in})/(1.88\text{ in}^3) = 19\,150\text{ psi}$$

Actual deflection:

$$y_{max} = \frac{-PL^3}{3EI} = \frac{-(1200\text{ lb})(30\text{ in})^3}{3(30 \times 10^6\text{ lb/in}^2)(4.47\text{ in}^4)} = -0.0805\text{ in}$$

Results *Summary of results:*

Beam shape: C12×25 standard steel channel

$$\sigma = 19\,150\text{ psi}$$
$$y_{max} = -0.0805\text{ in}$$

Comment These results will be compared with the other designs in the following example problems.

Example Problem 9–4 The simply supported beam shown in Figure 9–12 is to be made from a standard steel channel, C12×25, with the legs pointing downward. Compute the actual expected maximum stress and deflection and compare them with the results of Example Problem 9–3 for the cantilever.

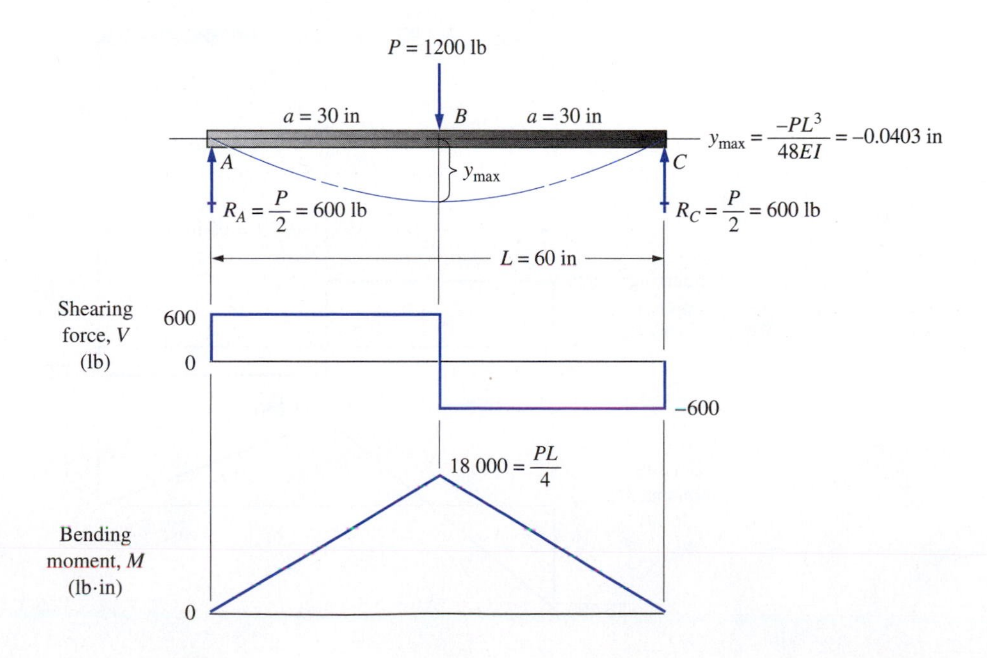

FIGURE 9–12 Simply supported beam.

Solution Objective Compute the maximum stress and deflection. Compare with the cantilever results.

Given Beam dimensions and loading as shown in Figure 9–12. Beam shape is a standard steel channel, C12×25, with the legs pointing downward.
Beam properties: $S = 1.88\ \text{in}^3$; $I = 4.47\ \text{in}^4$

Analysis From Figure 9–12 we find $M_{max} = 18\,000$ lb·in. Then,
Actual bending stress:

$$\sigma = M/S = (18\,000\ \text{lb}\cdot\text{in})/(1.88\ \text{in}^3) = 9575\ \text{psi}$$

Actual deflection:

$$y_{max} = \frac{-PL^3}{48EI} = \frac{-(1200\ \text{lb})(60\ \text{in})^3}{48(30 \times 10^6\ \text{lb/in}^2)(4.47\ \text{in}^4)} = -0.0403\ \text{in}$$

Results *Comparison of results with cantilever:* Using the subscript *1* for the cantilever and *2* for the simply supported beam,

$$\sigma_2/\sigma_1 = (9575\ \text{psi})/(19\,150\ \text{psi}) = 0.50$$
$$y_2/y_1 = (-0.0403\ \text{in})/(-0.0805\ \text{in}) = 0.50$$

Comment These results will be compared with the other designs in Example Problem 9–7.

Example Problem 9–5 The supported cantilever shown in Figure 9–13 is to be made from a standard steel channel, C12×25, with the legs pointing downward. Compute the actual expected maximum stress and deflection and compare them with the results of Example Problem 9–3 for the cantilever.

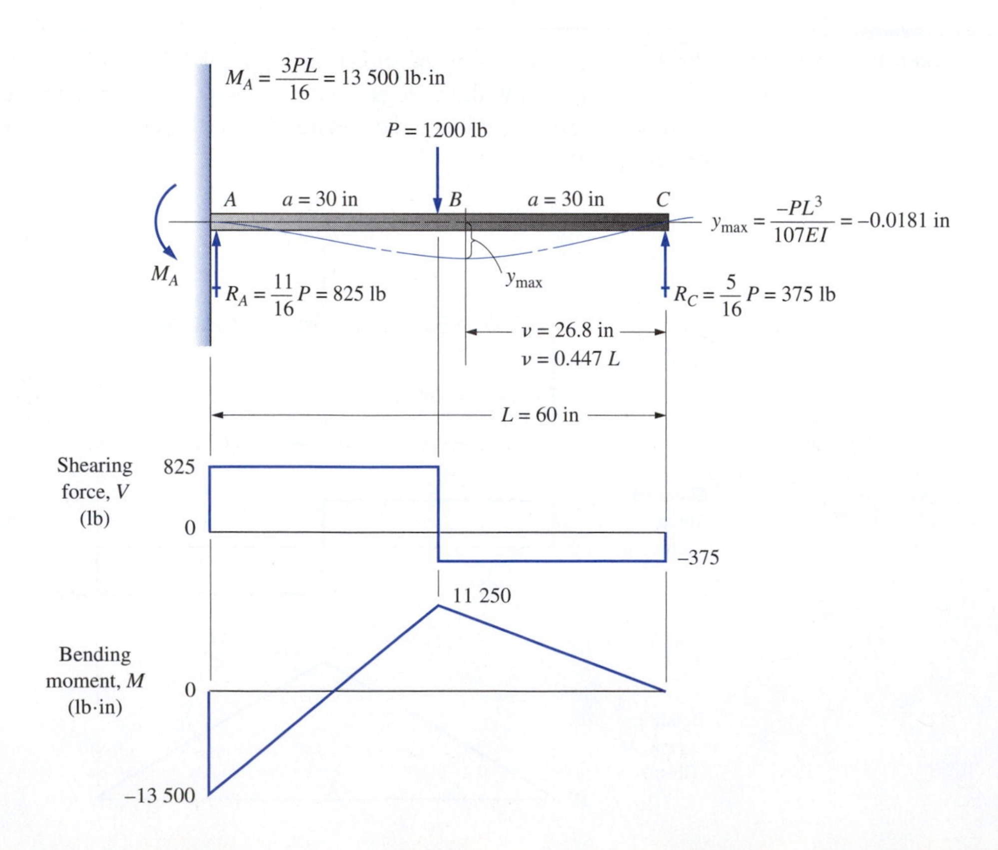

FIGURE 9–13 Supported cantilever.

Solution Objective Compute the maximum stress and deflection. Compare with the cantilever results.

Given Beam dimensions and loading as shown in Figure 9–12. Beam shape is a standard steel channel, C12×25, with the legs pointing downward.
Beam properties: $S = 1.88\ \text{in}^3$; $I = 4.47\ \text{in}^4$

Analysis From Figure 9–12 we find $M_{max} = 13\,500$ lb·in. Then,
Actual bending stress:

$$\sigma = M/S = (13\,500\ \text{lb}\cdot\text{in})/(1.88\ \text{in}^3) = 7180\ \text{psi}$$

Actual deflection: From Appendix A–25(a),

$$y_{max} = \frac{-PL^3}{107EI} = \frac{-(1200\ \text{lb})(60\ \text{in})^3}{107(30\times10^6\ \text{lb/in}^2)(4.47\ \text{in}^4)} = -0.0181\ \text{in}$$

Results *Comparison of results with cantilever.* Using the subscript *1* for the cantilever and *3* for the supported cantilever,

$$\sigma_3/\sigma_1 = (7180\ \text{psi})/(19\,150\ \text{psi}) = 0.375$$
$$y_3/y_1 = (-0.0181\ \text{in})/(-0.0805\ \text{in}) = 0.225$$

Comment These results will be compared with the other designs in Example Problem 9–7.

Example Problem 9–6 The fixed-end beam shown in Figure 9–14 is to be made from a standard steel channel, C12×25, with the legs pointing downward. Compute the actual expected maximum stress and deflection and compare them with the results of Example Problem 9–3 for the cantilever.

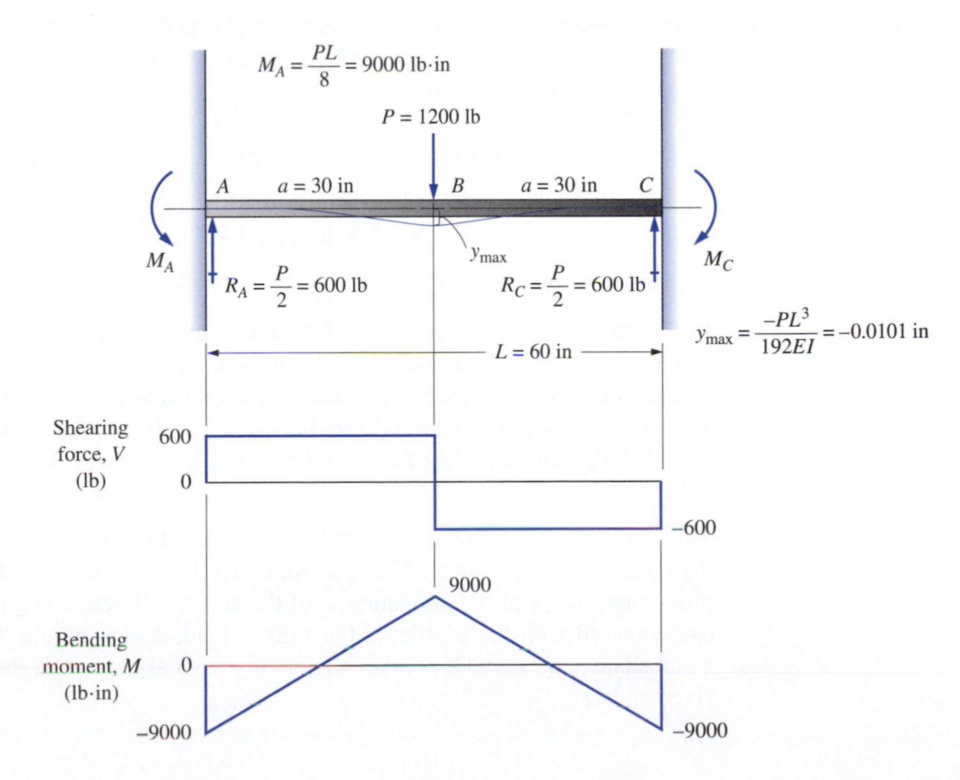

FIGURE 9–14 Fixed-end beam.

Solution Objective Compute the maximum stress and deflection. Compare with the cantilever results.

Given Beam dimensions and loading as shown in Figure 9–14. Beam shape is a standard steel channel, C12×25, with the legs pointing downward.
Beam properties: $S = 1.88\ \text{in}^3$; $I = 4.47\ \text{in}^4$

Analysis From Figure 9–14 we find $M_{max} = 9000$ lb·in. Then,
Actual bending stress:

$$\sigma = M/S = (9000\ \text{lb·in})/(1.88\ \text{in}^3) = 4787\ \text{psi}$$

Actual deflection: From Appendix A–25(e),

$$y_{max} = \frac{-PL^3}{192EI} = \frac{-(1200\ \text{lb})(60\ \text{in})^3}{192(30 \times 10^6\ \text{lb/in}^2)(4.47\ \text{in}^4)} = -0.0101\ \text{in}$$

Results *Comparison of results with cantilever:* Using the subscript *1* for the cantilever and *4* for the fixed-end beam,

$$\sigma_4/\sigma_1 = (4787\ \text{psi})/(19\,150\ \text{psi}) = 0.250$$
$$y_4/y_1 = (-0.0101\ \text{in})/(-0.0805\ \text{in}) = 0.125$$

Comment These results will be compared with the other designs in Example Problem 9–7.

Example Problem 9–7 Compare the behavior of the four beams shown in Figures 9–11 through 9–14 with regard to shearing forces, bending moments, and maximum deflection. Use the results of Example Problems 9–3 through 9–6.

Solution Objective Compare the performance of the four beams.

Given Beam designs shown in Figures 9–11 through 9–14.
Results of Example Problems 9–3 through 9–6.

Analysis Figure 9–15 shows the comparison.

Results In Figure 9–15, the graphs on the left show plots of the deflection, shearing force, and bending moment as a function of position on the beam for the four designs superimposed on each other. In each case, it is obvious that the cantilever produces the highest value by a significant margin. On the right are bar graphs of the relative performance where the values of deflection, shearing force, and bending moment for the cantilever are assigned a value of 1.0.

Comment The results shown in Figure 9–15 indicate that it is desirable to provide two supports rather than one if that is practical. Also, providing two fixed ends for the beam is most desirable, producing only $\frac{1}{8}$ of the deflection, $\frac{1}{2}$ of the shearing force, and $\frac{1}{4}$ of the bending moment as compared with the cantilever for a given load, distance from the load to the support, beam shape, and size. Of course, it must be practical to provide the second support or the fixed ends.

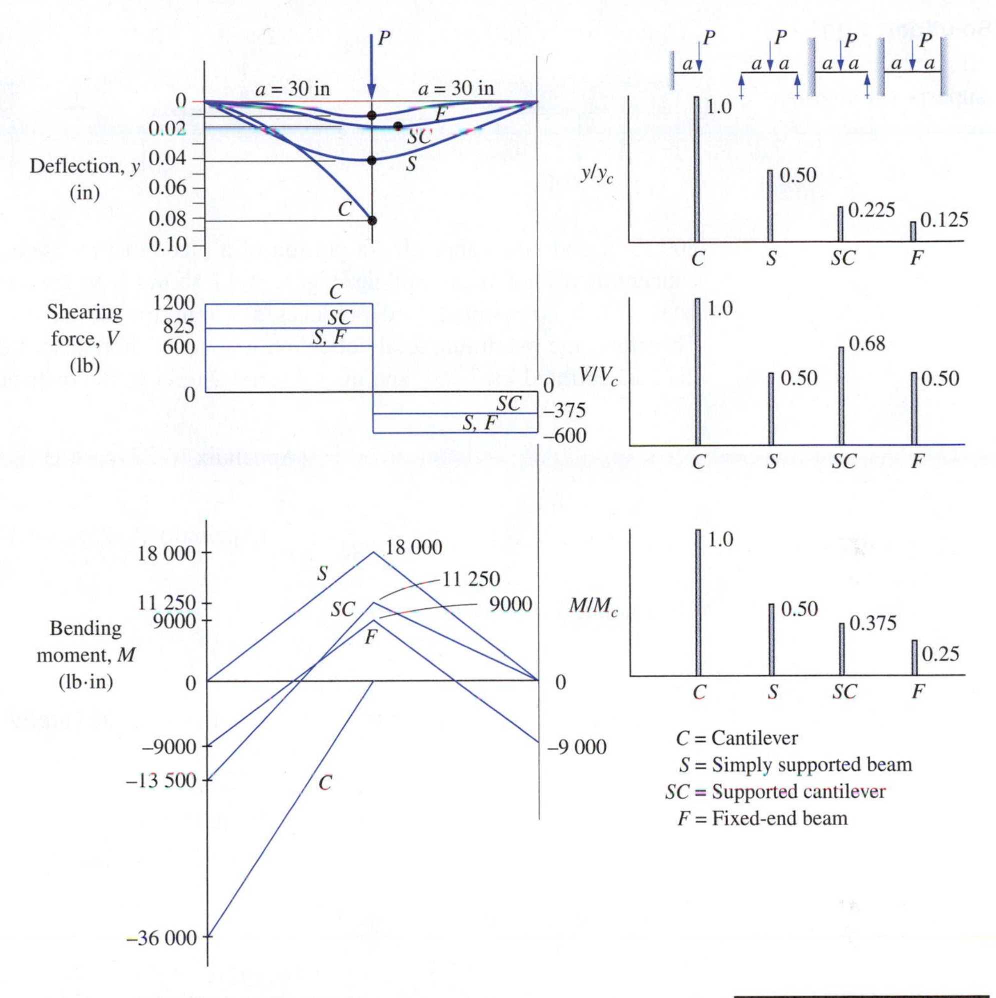

FIGURE 9–15 Comparison of performance of the four beam designs in Figures 9–11 through 9–14.

9–6 SUPERPOSITION USING DEFLECTION FORMULAS

Formulas such as those used in the preceding section are available for a large number of cases of loading and support conditions. Obviously, these cases would allow the solution of many practical beam deflection problems. An even larger number of situations can be handled by the use of the *principle of superposition.*

If a particular loading and support pattern can be broken into components such that each component is like one of the cases for which a formula is available, then the total deflection at a point on the beam is equal to the sum of the deflections caused by each component. The deflection due to one component load is *superposed* on deflections due to the other loads, thus giving the name *superposition*.

An example of where superposition can be applied is shown in Figure 9–16. Represented in the figure is a roof beam carrying a uniformly distributed roof load of

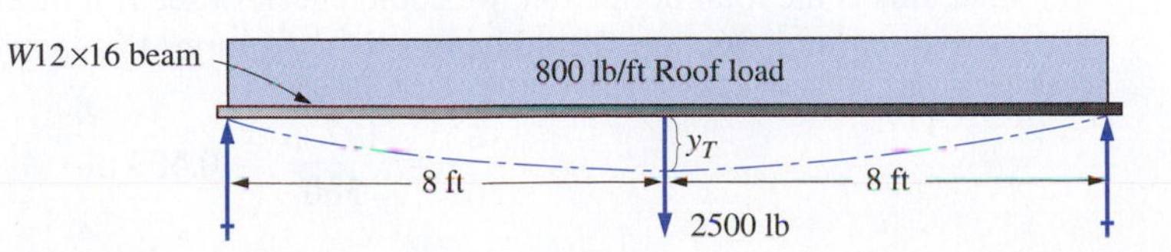

FIGURE 9–16 Roof beam.

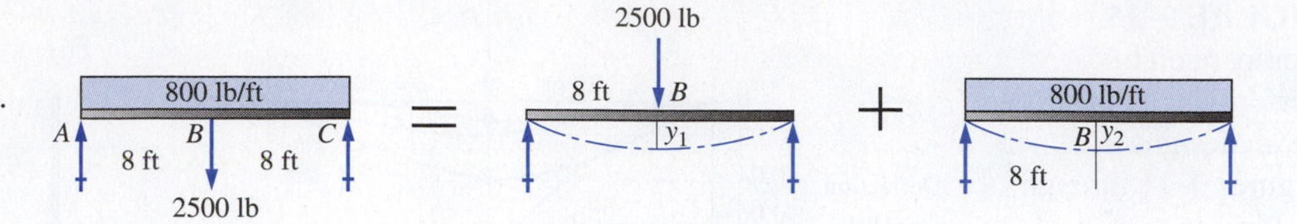

FIGURE 9–17 Illustration of the superposition principle.

800 lb/ft and also supporting a portion of a piece of process equipment that provides a concentrated load at the middle. Figure 9–17 shows how the loads are considered separately. Each component load produces a maximum deflection at the middle, point B. Therefore, the maximum total deflection will occur there also. Let the subscript *1* refer to the concentrated load case and the subscript *2* refer to the distributed load case. Then

$$y_1 = \frac{-PL^3}{48EI} \text{ (Appendix A–23, case a)}$$

$$y_2 = \frac{-5}{384}\frac{WL^3}{EI} \text{ (Appendix A–23, case d)}$$

The total deflection will be

$$y_T = y_1 + y_2$$

The terms L, E, and I will be the same for both cases. Be careful to keep units consistent.

$$L = 16 \text{ ft} \times 12 \text{ in/ft} = 192 \text{ in}$$
$$E = 30 \times 10^6 \text{ psi for steel}$$
$$I = 103 \text{ in}^4 \text{ for W12×16 beam}$$

To compute y_1, let $P = 2500$ lb.

$$y_1 = \frac{-2500(192)^3}{48(30 \times 10^6)(103)} \text{ in} = -0.119 \text{ in}$$

To compute y_2, W is the total resultant of the distributed load.

$$W = (800 \text{ lb/ft})(16 \text{ ft}) = 12\,800 \text{ lb}$$

Then

$$y_2 = \frac{-5(12\,800)(192)^3}{384(30 \times 10^6)(103)} \text{ in} = -0.382 \text{ in}$$

and

$$y_T = y_1 + y_2 = -0.119 \text{ in} - 0.382 \text{ in} = -0.501 \text{ in}$$

Since this is the total deflection, we could check to see if it meets the recommendation that the maximum deflection should be less than $\frac{1}{360}$ times the span of the beam. The span is L.

$$\frac{L}{360} = \frac{192 \text{ in}}{360} = 0.533 \text{ in}$$

The computed deflection was 0.501 in, which is satisfactory.

The superposition principle is valid for any place on the beam, not just at the loads. The following example problem illustrates this.

Example Problem 9–8

Figure 9–18 shows a shaft carrying two gears and simply supported at its ends by bearings. The mating gears above exert downward forces, which tend to separate the gears. Other forces acting horizontally are not considered in this analysis. Determine the total deflection at the gears *B* and *C* if the shaft is steel and has a diameter of 45 mm.

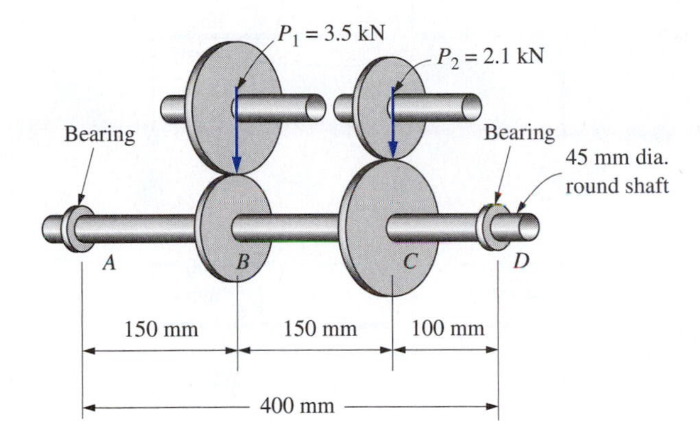

FIGURE 9–18 Shaft for Example Problem 9–8.

Solution

Objective Compute the deflection at points *B* and *C*.

Given Shaft *ABCD* in Figure 9–18. Shaft is steel. $D = 45$ mm.
Load at $B = P_1 = 3.5$ kN $= 3500$ N. Load at $C = P_2 = 2.1$ kN $= 2100$ N.

Analysis The two unsymmetrically placed, concentrated loads constitute a situation for which none of the given cases for beam deflection formulas is valid. However, it can be solved by using the formulas of Appendix A–23–b twice. Considering each load separately, the deflections at *B* and *C* can be found for each. The total, then, would be the sum of the component deflections. Figure 9–19 shows the logic from which we can say

$$y_B = y_{B1} + y_{B2}$$
$$y_C = y_{C1} + y_{C2}$$

where y_B = total deflection at *B*
y_C = total deflection at *C*
y_{B1} = deflection at *B* due to 3.5-kN load alone
y_{C1} = deflection at *C* due to 3.5-kN load alone
y_{B2} = deflection at *B* due to 2.1-kN load alone
y_{C2} = deflection at *C* due to 2.1-kN load alone

For all the calculations, the values of *E*, *I*, and *L* will be needed. These are the same as those in Example Problem 9–2.

$$E = 207 \times 10^3 \text{ N/mm}^2$$
$$I = 0.201 \times 10^6 \text{ mm}^4$$
$$L = 400 \text{ mm}$$

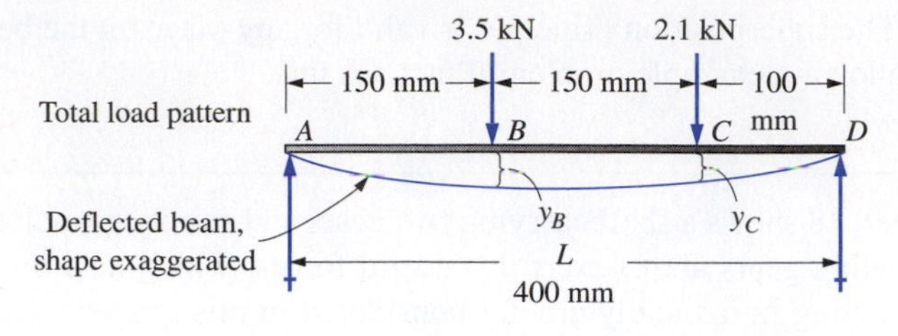

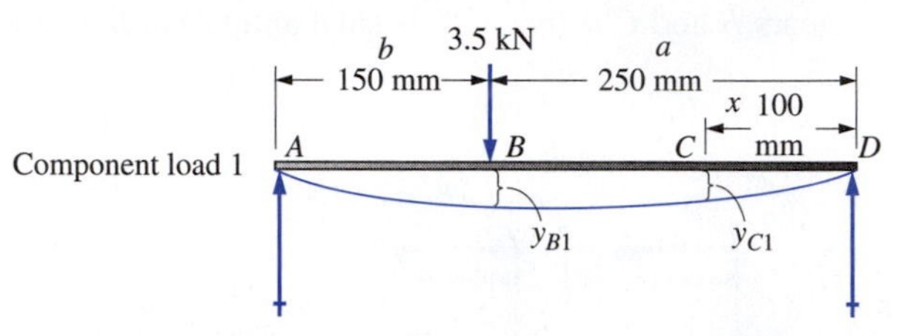

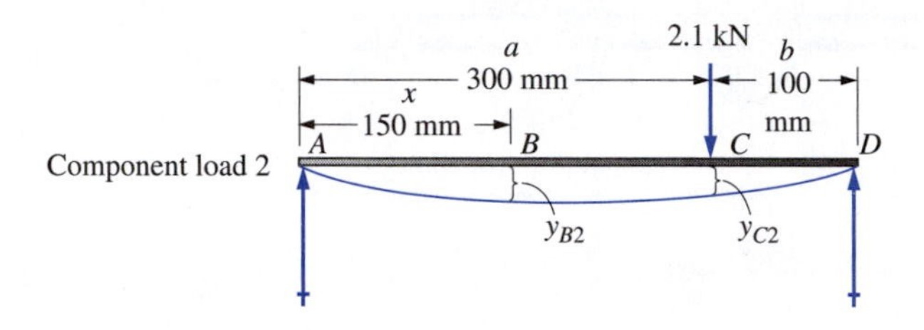

FIGURE 9–19 Superposition logic for deflection of shaft in Figure 9–18.

The product of *EIL* is present in all the formulas.

$$EIL = (207 \times 10^3)(0.201 \times 10^6)(400) = 16.64 \times 10^{12}$$

Now the individual component deflections will be computed. Note the values for the variables a, b, and x are different for each component load. See the data labeled in Figure 9–19.

Results

For component 1: y_{B1} is at the load. y_{C1} is in the longer segment.

$$y_{B1} = \frac{-P_1 a^2 b^2}{3EIL} = \frac{-(3.5 \times 10^3)(250)^2(150)^2}{3(16.64 \times 10^{12})} = -0.0985 \text{ mm}$$

$$y_{C1} = \frac{-P_1 bx}{6EIL}(L^2 - b^2 - x^2)$$

$$y_{C1} = \frac{-(3.5 \times 10^3)(150)(100)}{6(16.64 \times 10^{12})}(400^2 - 150^2 - 100^2) = -0.0670 \text{ mm}$$

For component 2, the load will be 2.1 kN at point C. Then y_{B2} is in the longer segment. y_{C2} is at the load.

$$y_{B2} = \frac{-P_2 bx}{6EIL}(L^2 - b^2 - x^2)$$

$$y_{B2} = \frac{-(2.1 \times 10^3)(100)(150)}{6(16.64 \times 10^{12})}(400^2 - 100^2 - 150^2) = -0.0402 \text{ mm}$$

$$y_{C2} = \frac{-P_2 a^2 b^2}{3EIL} = \frac{-(2.1 \times 10^3)(300)^2(100)^2}{3(16.64 \times 10^{12})} = -0.0378 \text{ mm}$$

Now, by superposition,

$$y_B = y_{B1} + y_{B2} = -0.0985 \text{ mm} - 0.0402 \text{ mm} = -0.1387 \text{ mm}$$
$$y_C = y_{C1} + y_{C2} = -0.0670 \text{ mm} - 0.0378 \text{ mm} = -0.1048 \text{ mm}$$

Comment In Section 9–2 it was observed that a recommended limit for the movement of one gear relative to its mating gear is 0.13 mm. Thus, this shaft is too flexible since the deflection at B exceeds 0.13 mm without even considering the deflection of the mating shaft.

Superposition Applied to Statically Indeterminate Beams. Superposition can be applied to enable redundant reactions and moments in statically indeterminate beams to be determined as demonstrated here.

Consider first the supported cantilever shown in Figure 9–20. Because of the restraint at A and the simple support at B, the unknown reactions include:

1. The vertical force R_B
2. The vertical force R_A
3. The restraining moment M_A

The conditions assumed for this beam are that the supports at A and B are absolutely rigid and at the same level, and that the connection at A does not allow any rotation of the beam at that point. Conversely, the support at B will allow rotation and cannot resist moments.

If the support at B is removed, the beam would deflect downward, as shown in Figure 9–21(a), by an amount y_{B1} due to the load P. Now if the load is removed and the reaction force R_B is applied upward at B, the beam would deflect upward by an amount

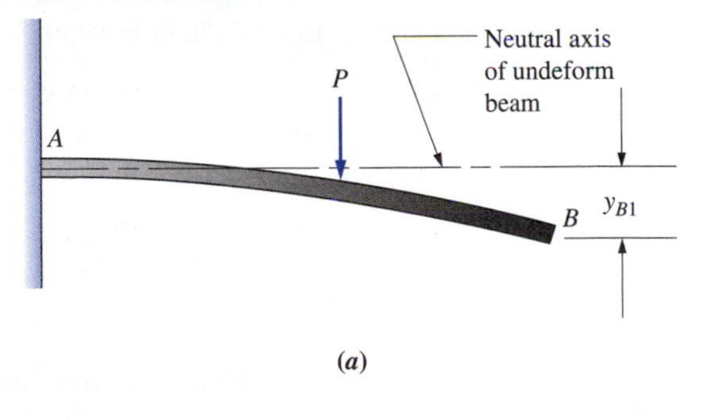

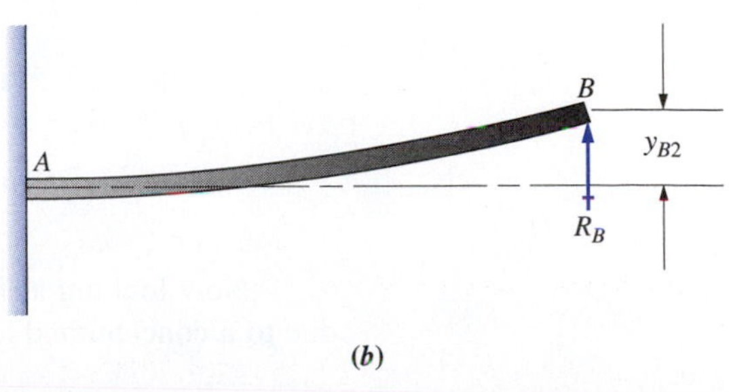

FIGURE 9–21 Superposition applied to the supported cantilever.

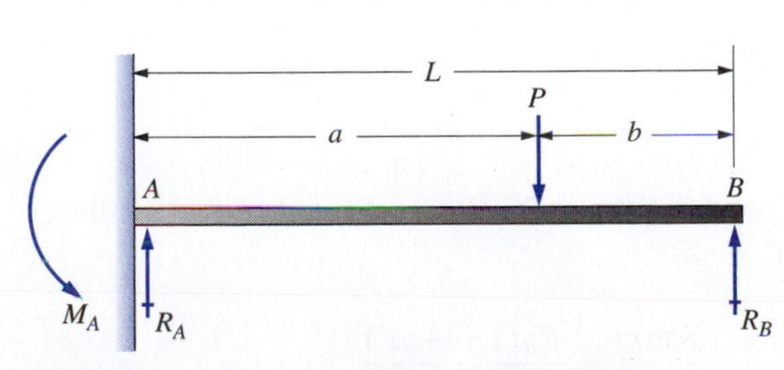

FIGURE 9–20 Supported cantilever.

y_{B2}, as shown in Figure 9–21(b). In reality, of course, both forces are applied, and the deflection at B is *zero*. The principle of superposition would then provide the conclusion that

$$y_{B1} + y_{B2} = 0$$

This equation, along with the normal equations of static equilibrium, will permit the evaluation of all three unknowns, as demonstrated in the example problem that follows. It must be recognized that the principles of static equilibrium are still valid for statically indeterminate beams. However, they are not sufficient to allow a direct solution.

Example Problem 9–9 Determine the support reactions at A and B for the supported cantilever shown in Figure 9–20 if the load P is 2600 N and placed 1.20 m out from A. The total length of the beam is 1.80 m. Then draw the complete shearing force and bending moment diagrams, and design the beam by specifying a configuration, a material, and the required dimensions of the beam. Use a design factor of 8 based on ultimate strength since the load will be repeated often.

Solution

Objective: Determine the support reactions, draw the shearing force and bending moment diagrams, and design the beam.

Given: Beam loading in Figure 9–20. $P = 2600$ N (repeated). $L = 1.80$ m. $a = 1.20$ m. Use a design factor of $N = 8$ based on S_u.

Analysis: Use the superposition method.

Results: *Reaction at B, R_B*

We must first determine the reaction at B using superposition. Previously we observed,

$$y_{B1} + y_{B2} = 0$$

The equation for y_{B1} can be found from the beam deflection formulas in Appendix A–24. As suggested in Figure 9–21(a), the deflection at the end of a cantilever carrying an intermediate load is required. Then

$$y_{B1} = \frac{-Pa^2}{6EI}(3L - a)$$

where $P = 2600$ N
$a = 1.20$ m
$L = 1.80$ m

The values of E and I are still unknown, but we can express the deflection in terms of EI.

$$y_{B1} = \frac{(-2600\text{ N})(1.20\text{ m})^2}{6EI}[3(1.80\text{ m}) - 1.20\text{ m}]$$

$$y_{B1} = \frac{-2621\text{ N}\cdot\text{m}^3}{EI}$$

Now looking at Figure 9–21(b), we need the deflection at the end of the cantilever due to a concentrated load there. Then

$$y_{B2} = \frac{PL^3}{3EI} = \frac{R_B(1.8\text{ m})^3}{3EI} = \frac{R_B(1.944\text{ m}^3)}{EI}$$

The sum of y_{B1} and y_{B2} is zero.

$$\frac{-2621\ \text{N}\cdot\text{m}^3}{EI} + \frac{R_B(1.944\ \text{m}^3)}{EI} = 0$$

The term EI can be canceled out, allowing the solution for R_B.

$$R_B = \frac{2621\ \text{N}\cdot\text{m}^3}{1.944\ \text{m}^3} = 1348\ \text{N}$$

The values of R_A and M_A can now be found using the equations of static equilibrium.

Reaction at A, R_A

$$\sum F = 0 \quad \text{(in the vertical direction)}$$
$$R_A + R_B - P = 0$$
$$R_A = P - R_B = 2600\ \text{N} - 1348\ \text{N} = 1252\ \text{N}$$

Bending moment at A, M_A

Summing moments about point A gives

$$0 = M_A - 2600\ \text{N}\,(1.2\ \text{m}) + 1348\ \text{N}\,(1.8\ \text{m})$$
$$M_A = 693\ \text{N}\cdot\text{m}$$

The positive sign for the result indicates that the assumed sense of the reaction moment in Figure 9–20 is correct. However, this is a negative moment because it causes the beam to bend concave downward near the support A.

Shearing force and bending moment diagrams

The shearing force and bending moment diagrams can now be drawn, as shown in Figure 9–22, using conventional techniques. The maximum bending moment occurs at the load where $M = 809\ \text{N}\cdot\text{m}$.

Beam design

The beam can now be designed. Let's assume that the actual installation is similar to that sketched in Figure 9–23, with the beam welded at its left end and resting on another beam at the right end. A rectangular bar would work well in this arrangement, and a ratio of $h = 3t$ will be assumed. A carbon steel such as AISI 1040, hot rolled, provides an ultimate strength of 621 MPa. Its percent elongation, 25%, suggests good ductility, which will help it resist the repeated loads. The design should be based on bending stress.

$$\sigma = \frac{M}{S}$$

But let

$$\sigma = \sigma_d = \frac{S_u}{N} = \frac{621\ \text{MPa}}{8} = 77.6\ \text{MPa}$$

Then

$$S = \frac{M}{\sigma_d} = \frac{809\ \text{N}\cdot\text{m}}{77.6\ \text{N/mm}^2} \times \frac{10^3\ \text{mm}}{\text{m}} = 10\,425\ \text{mm}^3$$

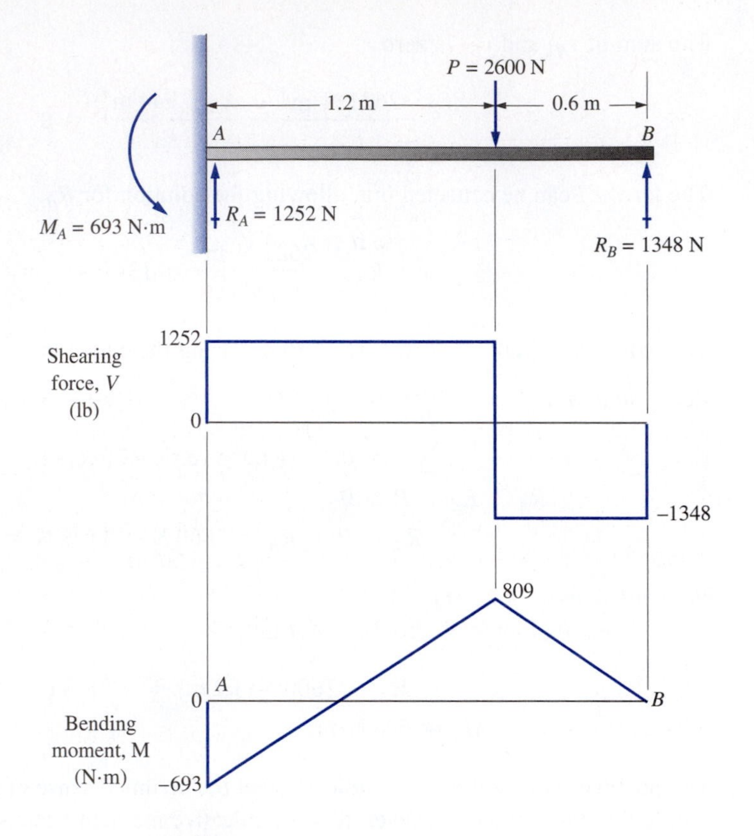

FIGURE 9–22 Shearing force and bending moment diagrams for the supported cantilever in Example Problem 9–9.

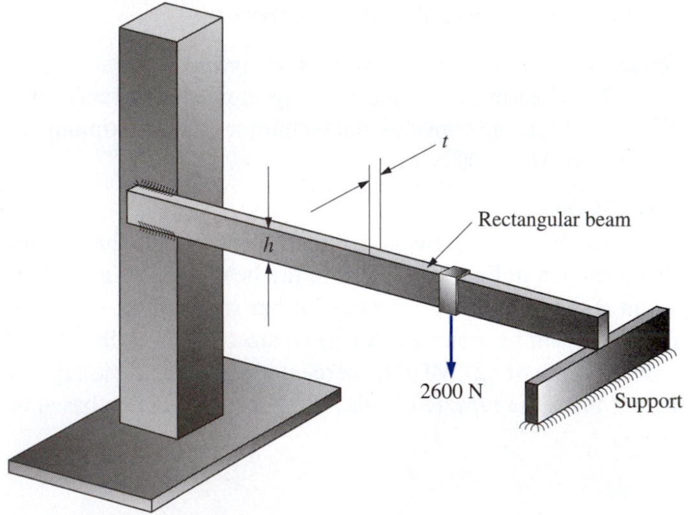

FIGURE 9–23 Physical implementation of a supported cantilever.

For a rectangular bar,

$$S = \frac{th^2}{6} = \frac{t(3t)^2}{6} = \frac{9t^3}{6} = 1.5t^3$$

Then

$$1.5t^3 = 10\,425 \text{ mm}^3$$
$$t = 19.1 \text{ mm}$$

Let's use the preferred size of 20 mm for t. Then

$$h = 3t = 3(20\text{ mm}) = 60\text{ mm}$$

Comment The final design can be summarized as a rectangular steel bar of AISI 1040, hot rolled, 20 mm thick and 60 mm high, welded to a rigid support at its left end and resting on a simple support at its right end. The maximum stress in the bar would be less than 77.6 MPa, providing a design factor of at least 8 based on ultimate strength.

The superposition method can be applied to any supported cantilever beam analysis for which the equation for the deflection due to the applied load can be found. Either the beam deflection formulas like those in the Appendix, the moment-area method, or the successive integration method developed later can be used.

Continuous beams can also be analyzed using superposition. Consider the beam on three supports shown in Figure 9–24. The three unknown support reactions make the beam statically indeterminate. The "extra" reaction R_C can be found using the technique suggested in Figure 9–25. Removing the support at C would cause the deflection y_{C1} downward due to the two 800-lb loads. Case c of Appendix A–23 can be used to find y_{C1}. Then if the loads are imagined to be removed and the reaction R_C is replaced, the upward deflection y_{C2} would result. The formulas of Case a in Appendix A–23 can be used.

Here again, of course, the actual deflection at C is zero because of the unyielding support. Therefore,

$$y_{C1} + y_{C2} = 0$$

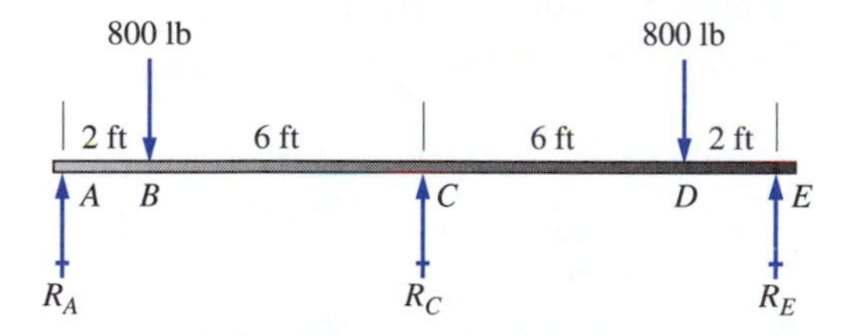

FIGURE 9–24 Continuous beam.

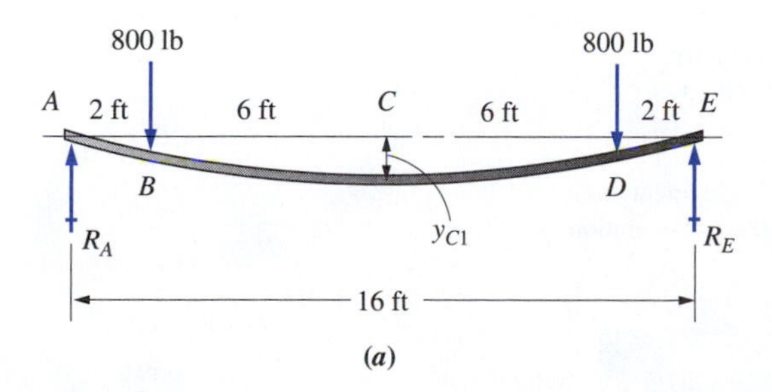

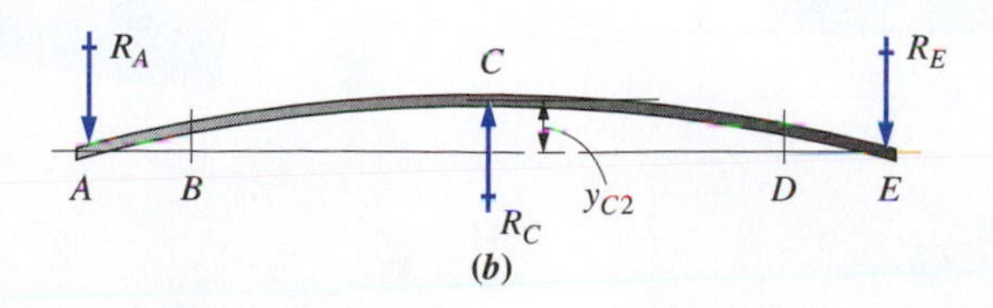

FIGURE 9–25 Superposition applied to a continuous beam.

From this relationship, the value of R_C can be computed. The remaining reactions R_A and R_E can then be found in the conventional manner, allowing the creation of the shearing force and bending moment diagrams.

9–7 SUCCESSIVE INTEGRATION METHOD

In this section we show the mathematical relationships among the moment, slope, and deflection curves from which you can solve for the actual equations for a given beam with a given loading and support condition.

Figure 9–26 shows a small segment of a beam in its initial straight shape and its deflected shape. The sides of the segment remain straight as the beam deflects, but they rotate with respect to a point at the neutral axis. This results in compression in the top portion of the segment and tension in the bottom portion, a fact used in the development of the flexure formula in Chapter 7.

The rotated sides of the segment intersect at the center of curvature and form the small angle $d\theta$. Note also the radius of curvature, R, measured from the center of curvature to the neutral axis. From the geometry shown in the figure,

$$\Delta s = R(d\theta) \tag{9–3}$$

and

$$\delta = c(d\theta) \tag{9–4}$$

where Δs is the length of the segment at the neutral axis
δ is the elongation of the bottom line of the segment that occurs as the beam deflects
c has the same meaning as in the flexure formula, the distance from the neutral axis to the outer fiber of the section

FIGURE 9–26 Relation between radius R and deformation δ.

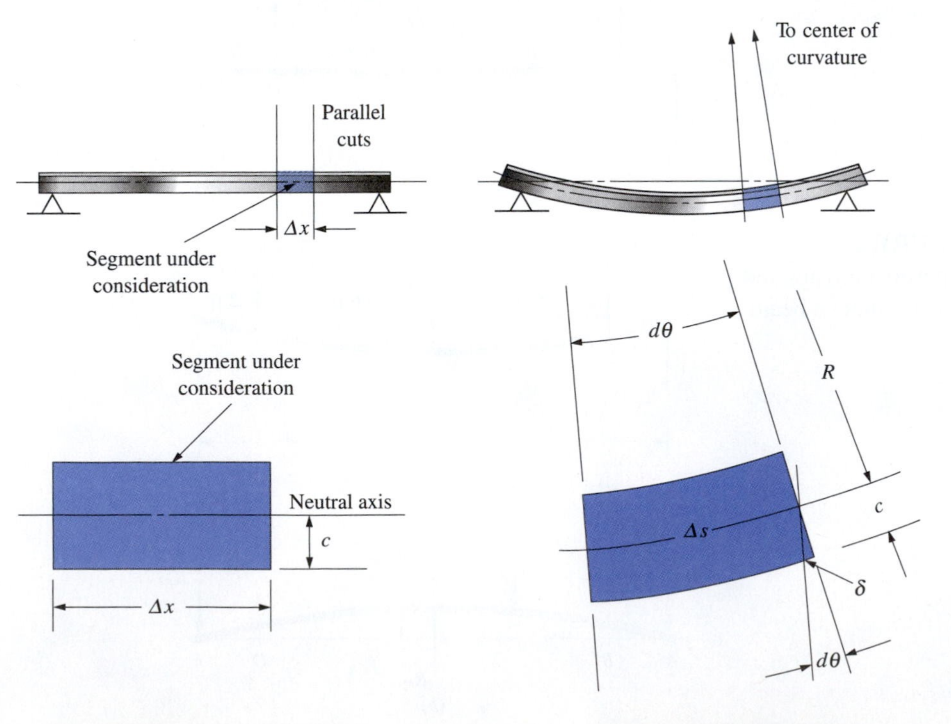

Recall that the definition of the neutral axis states that no strain occurs there. Then the length Δs in the deflected beam segment equals the length Δx in the undeflected segment, and Equation (9–3) can be written

$$\Delta x = R(d\theta) \tag{9–5}$$

Now both Equations (9–4) and (9–5) can be solved for $d\theta$.

$$d\theta = \frac{\delta}{c}$$

$$d\theta = \frac{\Delta x}{R}$$

Equating these values of $d\theta$ to each other gives

$$\frac{\delta}{c} = \frac{\Delta x}{R}$$

Another form of this equation is

$$\frac{c}{R} = \frac{\delta}{\Delta x}$$

The right side of this equation conforms to the definition of strain, ϵ. Then

$$\epsilon = \frac{c}{R} \tag{9–6}$$

Earlier it was shown that

$$\epsilon = \frac{\sigma}{E}$$

where σ, the stress due to bending, can be computed from the flexure formula,

$$\sigma = \frac{Mc}{I}$$

Then

$$\epsilon = \frac{\sigma}{E} = \frac{Mc}{EI}$$

Combining this with Equation (9–6) gives

$$\frac{c}{R} = \frac{Mc}{EI}$$

Dividing both sides by c gives

$$\frac{1}{R} = \frac{M}{EI} \tag{9–7}$$

Equation (9–7) is useful in developing the moment-area method for finding beam deflections. See Section 9–8.

In analytic geometry, the reciprocal of the radius of curvature, $1/R$, is defined as the *curvature* and denoted by κ, the lowercase Greek letter kappa. Then

$$\kappa = \frac{M}{EI} \tag{9–8}$$

Equation (9–8) indicates that the curvature gets greater as the bending moment increases, which stands to reason. Similarly, the curvature decreases as the beam stiffness, EI, increases.

Another principle of analytic geometry states that if the equation of a curve is expressed as $y = f(x)$, that is, y is a function of x, then the curvature is

$$\kappa = \frac{d^2y}{dx^2} \tag{9–9}$$

Combining Equations (9–8) and (9–9) gives

$$\frac{M}{EI} = \frac{d^2y}{dx^2} \tag{9–10}$$

or

$$M = EI\frac{d^2y}{dx^2} \tag{9–11}$$

Equations (9–10) and (9–11) are useful in developing the successive integration method for finding beam deflections, described next. Remember that this relationship applies to beams that are initially straight and where the curvature after loading is very small.

Successive Integration Method—General Approach. This section presents a general approach that allows the determination of deflection at any point on the beam. The advantages of this approach are the following:

1. The result is a set of equations for deflection at all parts of the beam. Deflection at any point can then be found by substitution of the beam stiffness properties of E and I, and the position on the beam.
2. Data are easily obtained from which a plot of the shape of the deflection curve may be made.
3. The equations for the *slope* of the beam at any point are generated, as are deflections. This is important in some machinery applications such as shafts at bearings and shafts carrying gears. An excessive slope of the shaft would result in poor performance and reduced life of the bearings or gears.
4. The fundamental relationships among loads, manner of support, beam stiffness properties, slope, and deflections are emphasized in the solution procedure. The designer who understands these relationships can make more efficient designs.

5. The method requires the application of only simple mathematical concepts.
6. The point of maximum deflection can be found directly from the resulting equations.

The basis for the successive integration method has been developed in Sections 9–3 and 9–7. The five beam diagrams will be prepared, in a manner similar to that shown in Figure 9–5, to relate the loads, shearing forces, bending moments, slopes, and deflections over the entire length of the beam.

The load, shearing force, and bending moment diagrams can be drawn using the principles from Chapter 5. Then equations for the bending moment are derived for all segments of the bending moment diagram.

Equation (9–11) is then used to develop the slope and deflection equations from the moment equations by integrating twice with respect to the position on the beam, x, as follows.

$$M = EI\frac{d^2y}{dx^2} \tag{9–11}$$

Now, integrating once with respect to x gives

$$\int M\,dx = EI\int \frac{d^2y}{dx^2}dx = EI\frac{dy}{dx} \tag{9–12}$$

Earlier, in Section 9–3, Equation (9–2), we showed that $dy/dx = \theta$, the slope of the deflection curve. Then,

$$\int M\,dx = EI\theta = \theta EI \tag{9–13}$$

Equation (9–12) can be integrated again, giving

$$\int EI\theta\,dx = EI\int \frac{dy}{dx}dx = EIy = yEI \tag{9–14}$$

After the final values for $EI\theta$ and EIy have been determined, they will be divided by the beam stiffness, EI, to obtain the values for slope, θ, and deflection, y.

The steps indicated by Equations (9–12) through (9–14) are to be completed for each segment of the beam for which the moment diagram is continuous. Also, because our objective is to obtain discrete equations for slope and deflection for particular beam-loading patterns, we will need to evaluate a constant of integration for each integration performed.

The development of the equations for the bending moment versus position is often accomplished by integrating the equations for the shearing force versus x, as shown in Chapter 5. This follows from the rule that the change in bending moment between two points on a beam is equal to the area under the shearing force curve between the same two points.

The step-by-step method used to find the deflection of beams using the general approach is as follows.

Steps in the Successive Integration Method for Beam Deflections

1. Determine the reactions at the supports for the beam.
2. Draw the shearing force and bending moment diagrams using the same procedures presented in Chapter 5, and identify the magnitudes at critical points.
3. Divide the beam into segments in which the shearing force diagram is continuous by naming points at the places where abrupt changes occur with the letters A, B, C, D, etc.
4. Write equations for the shearing force curve in each segment. In most cases, these will be equations of straight lines, that is, equations involving x to the first power. Sometimes, as for beams carrying concentrated loads, the equation will be simply of the form

$$V = \text{constant}$$

5. For each segment, perform the process,

$$M = \int V\,dx + C$$

To evaluate the constant of integration that ties the moment equation to the particular values already known for the moment diagram, insert known boundary conditions and solve for C.

6. For each segment, perform the process,

$$\theta EI = \int M\,dx + C$$

The constant of integration generated here cannot be evaluated directly right away. So each constant should be identified separately by a subscript such as C_1, C_2, C_3, etc. Then when they are evaluated (in step 9), they can be put in their proper places.

7. For each segment, perform the process,

$$yEI = \int \theta EI\,dx + C$$

Here again, the constants should be labeled with subscripts.

8. Establish *boundary conditions* for the slope and deflection diagrams. The same number of boundary conditions must be identified as there are unknown constants from steps 6 and 7. Boundary conditions express mathematically the special values of slope and deflection at certain points and the fact that both the slope curve and the deflection curve are continuous. Typical boundary conditions are:
   a. The deflection of the beam at each support is zero.
   b. The deflection of the beam at the end of one segment is equal to the deflection of the beam at the beginning of the next segment. This follows from the fact that the deflection curve is continuous; that is, it has no abrupt changes.
   c. The slope of the beam at the end of one segment is equal to the slope at the beginning of the next segment. The slope has no abrupt changes.
   d. For the special case of a cantilever, the slope of the beam at the support is also zero.
9. Combine all the boundary conditions to evaluate all the constants of integration. This typically involves the solution of a set of simultaneous equations in which the number of equations is equal to the number of unknown constants of integration. Equation-solving software or graphing calculators are very helpful for this step.
10. Substitute the constants of integration back into the slope and deflection equations, thus completing them. The value of the slope or deflection at any point can then be evaluated by simply placing the proper value of the position on the beam in the equation. Points of maximum deflection in any segment can also be found.

Example Problem 9–10 illustrates this method.

Example Problem 9–10 Figure 9–27 shows a beam used as a part of a special structure for a machine. The 20 K (20 000 lb) load at A and the 30 K (30 000 lb) load at C represent places where heavy equipment is supported. Between the two supports at B and D, the uniformly distributed load of 2 K/ft (2000 lb/ft) is due to stored bulk materials, which are in a bin supported by the beam. All loads are static. To maintain accuracy of the product produced by the machine, the maximum allowable deflection of the beam is 0.05 in. Specify an acceptable wide-flange steel beam, and also check the stress in the beam.

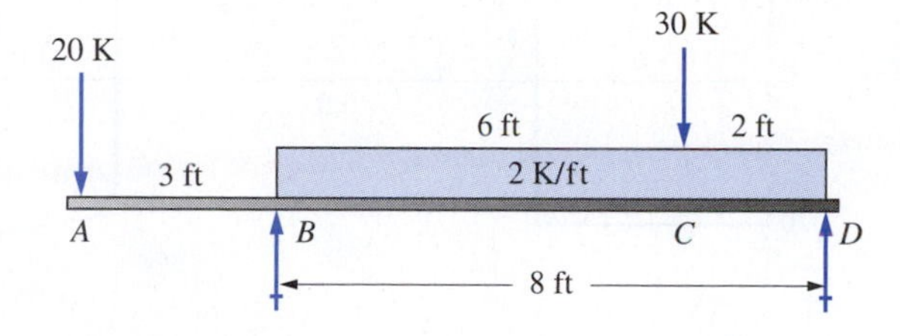

FIGURE 9–27 Beam for Example Problem 9–10.

Solution

Objective Specify a steel wide-flange beam shape to limit the deflection to 0.05 in. Check the stress in the selected beam to ensure safety.

Given Beam loading in Figure 9–27.

Analysis The beam will be analyzed to determine where the maximum deflection will occur. Then the required moment of inertia will be determined to limit the deflection to 0.05 in. A wide-flange beam that has the required moment of inertia will then be selected. The ten-step procedure discussed earlier will be used. The solution is shown in a programmed format. You should work through the problem yourself before looking at the given solution.

Results ***Steps 1 and 2*** call for drawing the shearing force and bending moment diagrams. Do this now before checking the following result. Figure 9–28 shows the results. Now do step 3.

Step 3. Three segments are required: AB, BC, and CD. These are the segments over which the shearing force diagram is continuous. Now do step 4 to get the shearing force curve equations.

Step 4. The results are:

$$V_{AB} = -20 \tag{a}$$

$$V_{BC} = -2x + 29 \tag{b}$$

$$V_{CD} = -2x - 1 \tag{c}$$

In the segments BC and CD, the shearing force curve is a straight line with a slope of -2 kip/ft, the same as the load. Any method of writing the equation of a straight line can be used to derive these equations.

Now do step 5 of the procedure.

Step 5. You should have the following for the moment equations. First,

$$M_{AB} = \int V_{AB}\,dx + C = \int -20\,dx + C = -20x + C$$

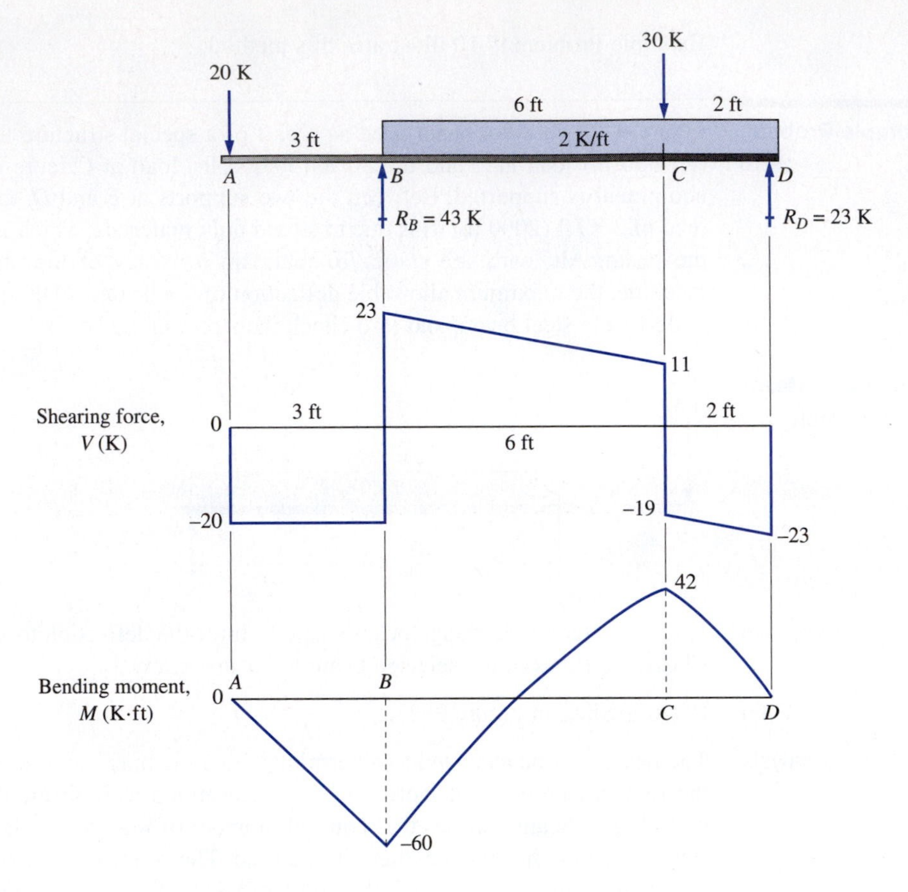

FIGURE 9–28 Load, shearing force, and bending moment diagrams for Example Problem 9–10.

At $x = 0$, $M_{AB} = 0$. Therefore, $C = 0$ and

$$M_{AB} = -20x \qquad \textbf{(d)}$$

Next.

$$M_{BC} = \int V_{BC}\,dx + C = \int (-2x + 29)\,dx + C = -x^2 + 29x + C$$

At $x = 3$, $M_{BC} = -60$. Therefore, $C = -138$ and

$$M_{BC} = -x^2 + 29x - 138 \qquad \textbf{(e)}$$

Finally,

$$M_{CD} = \int V_{CD}\,dx + C = \int (-2x - 1)\,dx + C = -x^2 - x + C$$

At $x = 9$, $M_{CD} = 42$. Therefore, $C = 132$ and

$$M_{CD} = -x^2 - x + 132 \qquad \textbf{(f)}$$

Now do step 6 to get equations for θEI.

Step 6. By integrating the moment equations,

$$\theta_{AB} EI = \int M_{AB}\, dx + C = \int (-20x)\, dx + C$$

$$\theta_{AB} EI = -10x^2 + C_1 \quad \textbf{(g)}$$

$$\theta_{BC} EI = \int M_{BC}\, dx + C = \int (-x^2 + 29x - 138)\, dx + C$$

$$\theta_{BC} EI = -x^3/3 + 14.5x^2 - 138x + C_2 \quad \textbf{(h)}$$

$$\theta_{CD} EI = \int M_{CD}\, dx + C = \int (-x^2 - x + 132)\, dx + C$$

$$\theta_{CD} EI = -x^3/3 - x^2/2 + 132x + C_3 \quad \textbf{(i)}$$

Now in step 7, integrate Equations (g), (h), and (i) to get the yEI equations.

Step 7. You should have

$$y_{AB} EI = \int \theta_{AB} EI\, dx + C$$

$$y_{AB} EI = -10x^3/3 + C_1x + C_4 \quad \textbf{(j)}$$

$$y_{BC} EI = \int \theta_{BC} EI\, dx + C$$

$$y_{BC} EI = -x^4/12 + 14.5x^3/3 - 69x^2 + C_2x + C_5 \quad \textbf{(k)}$$

$$y_{CD} EI = \int \theta_{CD} EI\, dx + C$$

$$y_{CD} EI = -x^4/12 - x^3/6 + 66x^2 + C_3x + C_6 \quad \textbf{(l)}$$

Step 8 calls for identifying boundary conditions. Six are required since there are six unknown constants of integration in Equations (g) through (l). Write them now.

Step 8. Considering zero deflection points and the continuity of the slope and deflection curves, we can say

1. At $x = 3$, $y_{AB} EI = 0$
2. At $x = 3$, $y_{BC} EI = 0$
3. At $x = 11$, $y_{CD} EI = 0$

} (zero deflection at supports)

4. At $x = 9$, $y_{BC} EI = y_{CD} EI$ (continuous deflection curve at C)
5. At $x = 3$, $\theta_{AB} EI = \theta_{BC} EI$
6. At $x = 9$, $\theta_{BC} EI = \theta_{CD} EI$

} (continuous slope curve at B and C)

We can now substitute these values of x into the proper equations and solve for C_1 through C_6. First make the substitutions and reduce the resulting equations to the form involving the constants.

For the previous six conditions listed, the following equations result:

1. $3C_1 + C_4 \qquad = 90$
2. $3C_2 + C_5 \qquad = 497.25$

$$
\begin{aligned}
&3.\ 11C_3 + C_6 &&= -6544.08\overline{33} = -78\,529/12\\
&4.\ 9C_2 - 9C_3 + C_5 - C_6 &&= 7290\\
&5.\ C_1 - C_2 &&= -202.5\\
&6.\ C_2 - C_3 &&= 1215
\end{aligned}
$$

The value of the quantity on the right side of Equation 3 is expressed in excessively high precision. This is not often necessary but is being done for this example to eliminate the accumulation of round-off errors as the problem solution proceeds. There are many steps to the final solution, and inaccuracies at this stage can result in significant variation in the results that might frustrate you as you follow the solution. Note that writing the constant in Equation 3 as $-6544.08\overline{33}$ indicates that the 3's repeat to infinity. Thus, this is an inherently inaccurate representation of the number. Entering the number as the exact fraction ($-78\,529/12$) into an equation solver would eliminate the error. The use of a computer-based equation solver such as MATHCAD, TK Solver, MATLAB, or MAPLE facilitates the laborious calculations involved in the balance of the procedure. Many high-level calculators having graphing capability also contain simultaneous equation solvers.

Now solve the six equations simultaneously for the values of C_1 through C_6.

Step 9. The results are:

$$
\begin{aligned}
C_1 &= 132.\overline{333} = 397/3 & C_2 &= 334.8\overline{33} = 4018/12\\
C_3 &= -880.1\overline{66} = -5281/6 & C_4 &= -307\ \text{(exact)}\\
C_5 &= -507.25\ \text{(exact)} & C_6 &= 3137.75\ \text{(exact)}
\end{aligned}
$$

Step 10. The final equations for θ and y can now be written by substituting the constants into Equations (g) through (l). The results are shown next.

$$
\begin{aligned}
\theta_{AB}EI &= -10x^2 + 132.\overline{333}\\
\theta_{BC}EI &= -x^3/3 + 14.5x^2 - 138x + 334.8\overline{33}\\
\theta_{CD}EI &= -x^3/3 - x^2/2 + 132x - 880.1\overline{66}\\
y_{AB}EI &= -10x^3/3 + 132.\overline{333}x - 307\\
y_{BC}EI &= -x^4/12 + 14.5x^3/3 - 69x^2 + 334.8\overline{33}x - 507.25\\
y_{CD}EI &= -x^4/12 - x^3/6 + 66x^2 - 880.1\overline{66}x + 3137.75
\end{aligned}
$$

Having the completed equations, we can now determine the point of maximum deflection, which is the primary objective of the analysis. Based on the loading, the probable shape of the deflected beam would be like that shown in Figure 9–29. Therefore, the maximum deflection could occur at point A at the end of the overhang, at a point to the right of B (upward), or at a point near the load at C (downward). It is probable that there are two points of zero slope at the points E and F, as shown in Figure 9–29. We would need to know where the slope equation $\theta_{BC}EI$ equals zero to determine where the maximum deflections occur.

Notice that this is a third-degree equation. The use of a graphing calculator and an equation solver facilitates finding the points where $\theta_{BC}EI = 0$. Figure 9–30 shows an expanded plot of the BC segment of the beam showing that the zero points occur at $x = 3.836$ ft and at $x = 8.366$ ft.

We can now determine the values for yEI at points A, E, and F to find out which is larger.

FIGURE 9–29 Slope and deflection curves for Example Problem 9–10.

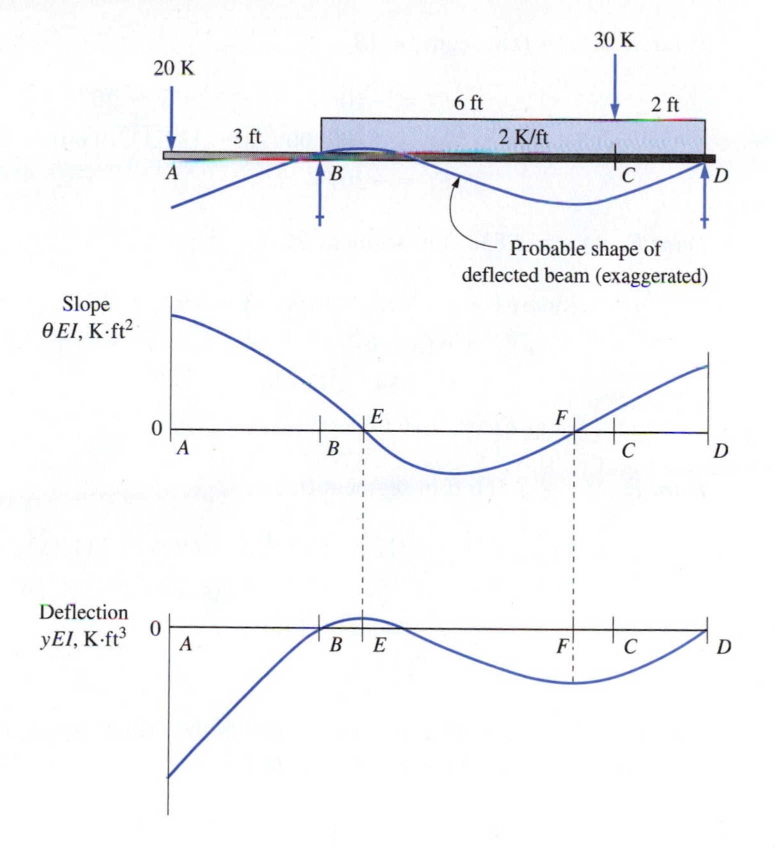

FIGURE 9–30 Graph showing points of zero slope.

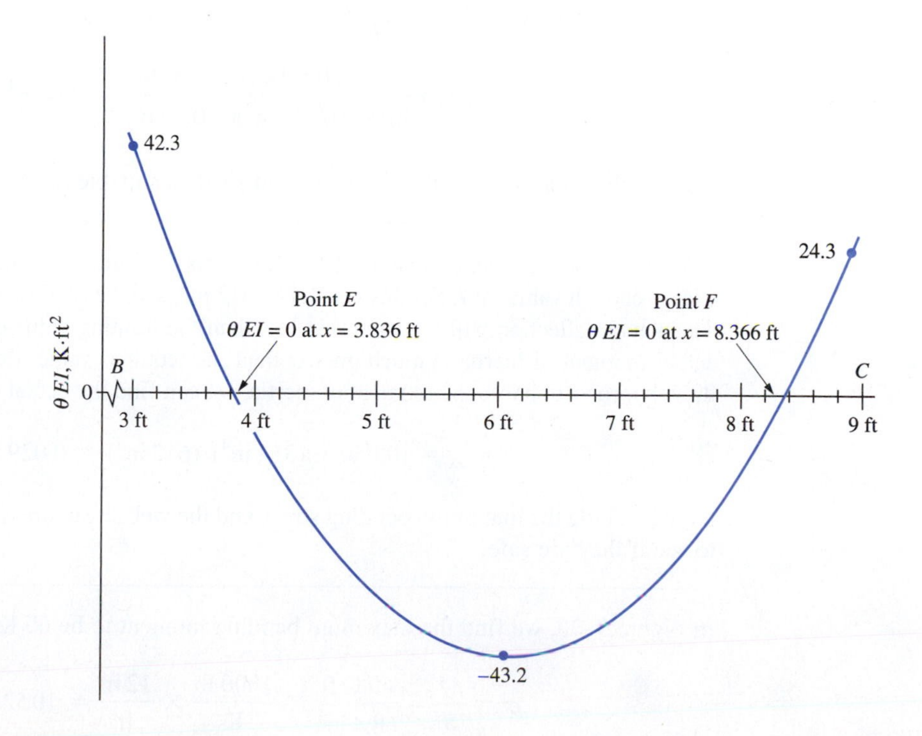

Point A. At $x = 0$ in segment AB,

$$y_{AB}EI = -10x^3/3 + 132.\overline{333}x - 307$$
$$y_A EI = -10(0.00)^3/3 + 132.\overline{333}(0.00) - 307$$
$$y_A EI = -307\ \text{K}\cdot\text{ft}^3$$

Point E. At $x = 3.836$ ft in segment BC,

$$y_{BC}EI = -x^4/12 + 14.5x^3/3 - 69x^2 + 334.8\overline{33}x - 507.25$$
$$y_E EI = -(3.836)^4/12 + 14.5(3.836)^3/3 - 69(3.836)^2 + 334.8\overline{33}(3.836) - 507.25$$
$$y_E EI = +16.62\ \text{K}\cdot\text{ft}^3$$

Point F. At $x = 8.366$ ft in segment BC,

$$y_{BC}EI = -x^4/12 + 14.5x^3/3 - 69x^2 + 334.8\overline{33}x - 507.25$$
$$y_F EI = -(8.366)^4/12 + 14.5(8.366)^3/3 - 69(8.366)^2 + 334.8\overline{33}(8.366) - 507.25$$
$$y_F EI = -113.5\ \text{K}\cdot\text{ft}^3$$

The largest value occurs at point A, so that is the critical point. We must choose a beam that limits the deflection at A to 0.05 in or less.

$$y_A EI = -307\ \text{K}\cdot\text{ft}^3$$

Let $y_A = -0.05$ in. Then the required I is

$$I = \frac{-307\ \text{K}\cdot\text{ft}^3}{Ey_A} \times \frac{1000\ \text{lb}}{\text{K}} \times \frac{(12\ \text{in})^3}{\text{ft}^3}$$
$$I = \frac{(-307)(1000)(1728)\ \text{lb}\cdot\text{in}^3}{(30 \times 10^6\ \text{lb/in}^2)(-0.05\ \text{in})} = 354\ \text{in}^4$$

Consult the table of wide-flange beams and select a suitable beam.

A W18×40 beam is the best choice from Appendix A–7 since it is the lightest beam that has a large enough value for I. For this beam, $I = 612\ \text{in}^4$, and the section modulus is $S = 68.4\ \text{in}^3$. The actual deflection will be somewhat less than the limiting value of 0.050 in because the actual moment of inertia is much greater than the required value. Because the deflection is directly proportional to the moment of inertia, we can find the actual deflection from

$$y_{max} = (0.050\ \text{in})(354\ \text{in}^4)/(612\ \text{in}^4) = 0.029\ \text{in}$$

Now compute the maximum bending stress and the web shear stress in the beam and check to see if they are safe.

In Figure 9–28, we find the maximum bending moment to be 60 K·ft. Then

$$\sigma = \frac{M}{S} = \frac{60\ \text{K}\cdot\text{ft}}{68.4\ \text{in}^3} \times \frac{1000\ \text{lb}}{\text{K}} \times \frac{12\ \text{in}}{\text{ft}} = 10\,526\ \text{psi}$$

We can specify that the beam be made from ASTM A992 structural steel having a yield strength of 50 000 psi. Using the AISC specifications for static loads, the allowable stress is

$$\sigma_d = 0.66\,s_y = (0.66)(50\,000\text{ psi}) = 33\,000\text{ psi}$$

This is much greater than the computed maximum stress in the beam. Therefore, the beam is safe for bending.

We can also check the shearing stress in the web. The maximum shearing force is 23 K or 23 000 lb at either points B or D. We use the web shear formula, $\tau = V/th$. From Appendix A–7 we find $t = 0.315$ in and $h = 17.90$ in. Then

$$\tau = V/th = (23\,000\text{ lb})/(0.315\text{ in})(17.90\text{ in}) = 4079\text{ psi}$$

The allowable web shearing stress is

$$\tau_d = 0.40\,s_y = (0.40)(50\,000\text{ psi}) = 20\,000\text{ psi}$$

Therefore the beam is safe for web shear. We will also have to check the AISC specifications to determine the need for lateral bracing.

Comment We have designed a beam to carry the load shown in Figure 9–27 that is safe for bending and shear and that will have a deflection no greater than 0.050 in. The beam is a W18×40 steel-rolled shape made from ASTM A992 structural steel. The maximum deflection occurs at point A at the left end of the overhang and has a value of 0.029 in. We also created complete equations for the shearing force, bending moment, slope, and deflection at all points on the beam.

9–8 MOMENT-AREA METHOD

The semigraphical procedure for finding beam deflections, called the *moment-area method,* is useful for problems in which a fairly complex loading pattern occurs or when the beam has a varying cross section along its length. Such cases are difficult to handle with the other methods presented in this chapter.

Shafts for mechanical drives are examples where the cross section varies throughout the length of the beam. Figure 9–31 shows a shaft designed to carry two gears where the changes in diameter provide shoulders against which to seat the gears and bearings to provide axial location. Note, also, that the bending moment decreases toward the ends of the shaft, allowing smaller sections to be safe with regard to bending stress.

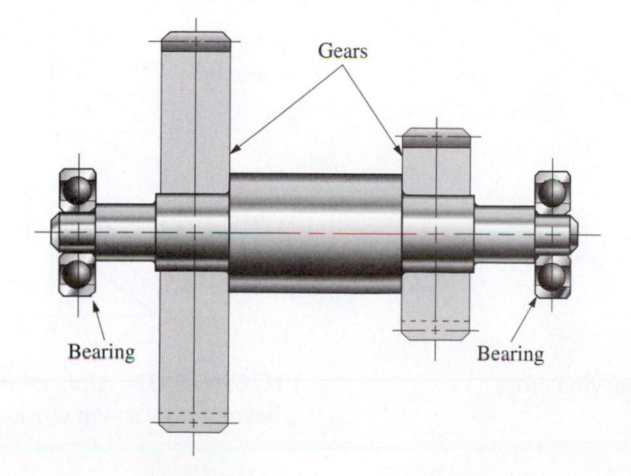

FIGURE 9–31 Shaft with varying cross sections.

In structural applications of beams, varying cross sections are often used to make more economical members. Larger sections having higher moments of inertia are used at sections where the bending moment is high, while decreased section sizes are used in places where the bending moment is lower. Figure 9–32 shows an example.

The moment-area method uses the quantity M/EI, the bending moment divided by the stiffness of the beam, to determine the deflection of the beam at selected points. Then it is convenient to prepare such a diagram as a part of the beam analysis procedure. If the beam has the same cross section over its entire length, the M/EI diagram looks similar to the familiar bending moment diagram except that its values have been divided by the quantity EI. However, if the moment of inertia of the cross section varies along the length of the beam, the shape of the M/EI diagram will be different. This is shown in Figure 9–33.

Recall Equation (9–10) from Section 9–7,

$$\frac{M}{EI} = \frac{d^2y}{dx^2} \tag{9–10}$$

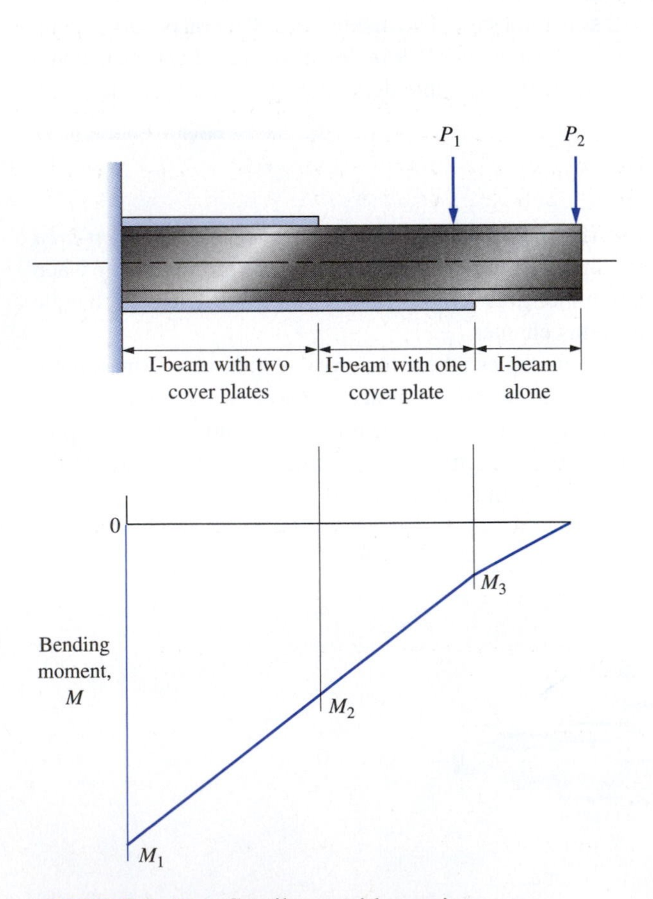

FIGURE 9–32 Cantilever with varying cross sections.

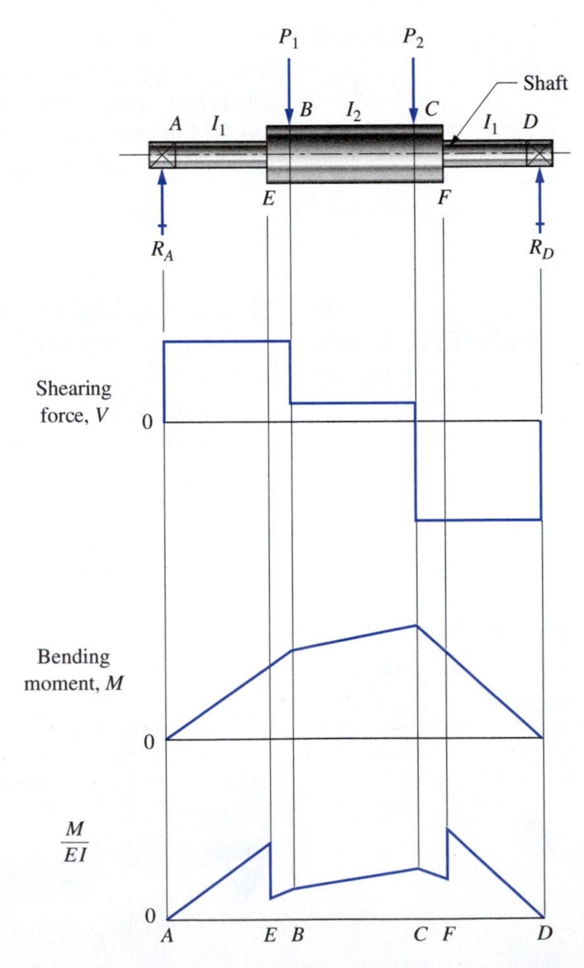

FIGURE 9–33 Illustration of M/EI diagrams for a beam with varying cross sections.

This formula relates the deflection of the beam, y, as a function of position, x, the bending moment, M, and the beam stiffness, EI. The right side of Equation (9–10) can be rewritten as

$$\frac{d^2y}{dx^2} = \frac{d}{dx}\left(\frac{dy}{dx}\right)$$

But note that dy/dx is defined as the slope of the deflection curve, θ; that is, $dy/dx = \theta$. Then

$$\frac{d^2y}{dx^2} = \frac{d\theta}{dx}$$

Equation (9–10) can then be written

$$\frac{M}{EI} = \frac{d\theta}{dx}$$

Solving for $d\theta$ gives

$$d\theta = \frac{M}{EI}dx \quad \textbf{(9–15)}$$

The interpretation of Equation (9–15) can be seen in Figure 9–34 in which the right side, $(M/EI)\,dx$, is the area under the M/EI diagram over a small length dx. Then $d\theta$ is the change in the angle of the slope over the same distance dx. If tangent lines are drawn to the deflection curve of the beam at the two points marking the beginning and the end of the segment dx, the angle between them is $d\theta$.

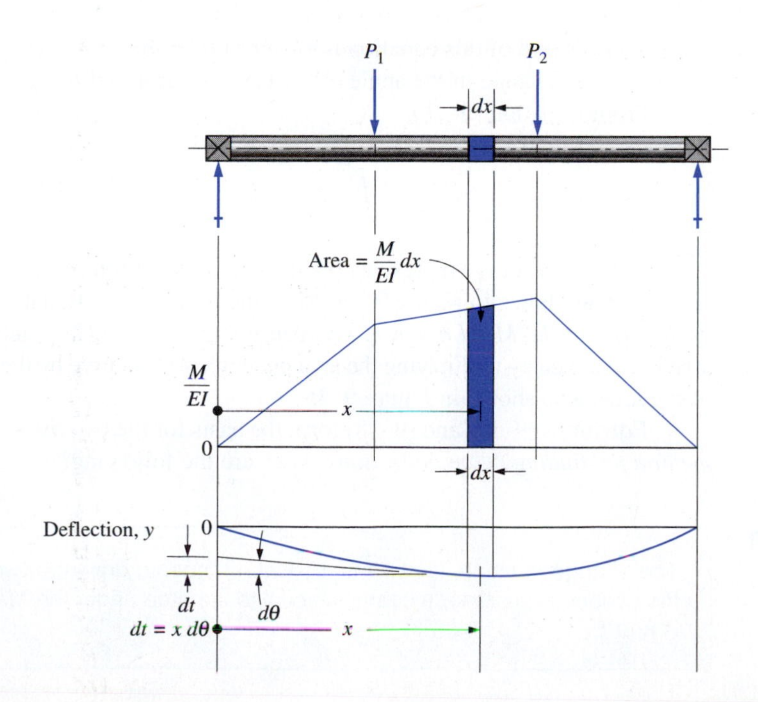

FIGURE 9–34 Principles of the moment-area method for beam deflections.

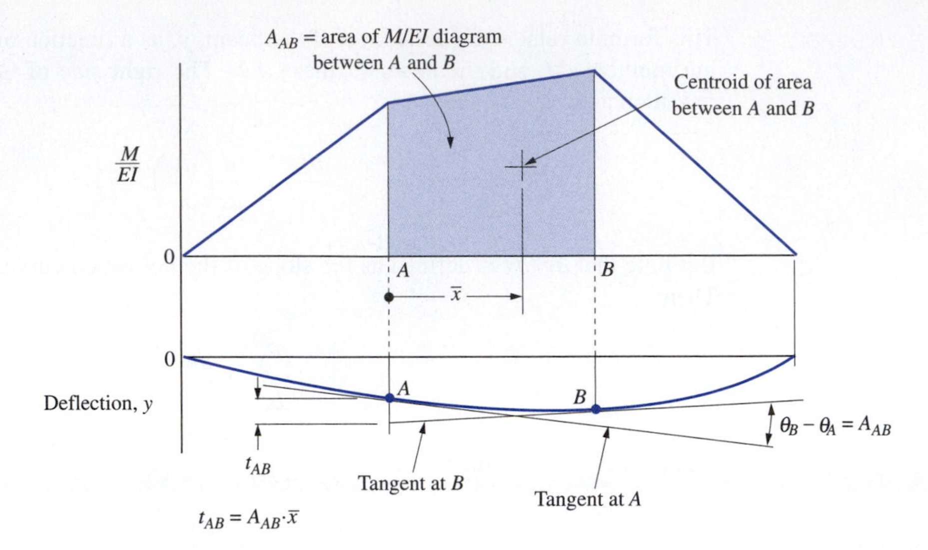

FIGURE 9–35 Illustrations of the two theorems of the moment-area method for beam deflections.

The change in angle $d\theta$ causes a change in the vertical position of a point at some distance x from the small element dx, as shown in Figure 9–34. Calling the change in vertical position dt, it can be found from

$$dt = x\,d\theta = \frac{M}{EI}x\,dx \qquad \textbf{(9–16)}$$

To determine the effect of the change in angle over a larger segment of the beam, Equations (9–15) and (9–16) must be integrated over the length of the segment. For example, over the segment A–B shown in Figure 9–35 from Equation (9–15),

$$\int_A^B d\theta = \theta_B - \theta_A = \int_A^B \frac{M}{EI}dx \qquad \textbf{(9–17)}$$

The last part of this equation is the area under the M/EI curve between A and B. This is equal to the change in the angle of the tangents at A and B, $\theta_B - \theta_A$.

From Equation (9–16),

$$\int_A^B dt = t_{AB} = \int_A^B \frac{M}{EI}x\,dx \qquad \textbf{(9–18)}$$

Here the term t_{AB} represents the tangential deviation of the point A from the tangent to point B, as shown in Figure 9–35. Also, the right side of Equation (9–18) is the *moment of the area of the M/EI diagram between points A and B*. In practice, the moment of the area is computed by multiplying the area under the M/EI curve by the distance to the centroid of the area, also shown in Figure 9–35.

Equations (9–17) and (9–18) form the basis for the two *theorems of the moment-area method for finding beam deflections*. They are the following:

Theorem 1

The change in angle, in radians, between tangents drawn at two points A and B on the deflection curve for a beam, is equal to the area under the M/EI diagram between A and B.

Theorem 2

The vertical deviation of point A on the deflection curve for a beam from the tangent through another point B on the curve is equal to the moment of the area under the M/EI curve with respect to point A.

Applications of the Moment-Area Method. In this section we show several examples of the use of the moment-area method for finding the deflection of beams. Procedures are developed for each class of beam according to the manner of loading and support. Considered are the following:

1. Cantilevers with a variety of loads
2. Symmetrically loaded simply supported beams
3. Beams with varying cross sections
4. Unsymmetrically loaded simply supported beams

Cantilevers. The definition of a cantilever includes the requirement that it is rigidly fixed to a support structure so that no rotation of the beam can occur at the support. Therefore, the tangent to the deflection curve at the support is always in line with the original position of the neutral axis of the beam in the unloaded state. If the beam is horizontal, as we usually picture it, this tangent is also horizontal.

The following procedure for finding the deflection of any point on a cantilever, uses the two theorems developed in Section 9–10 along with the observation that the tangent to the deflection curve at the support is horizontal.

Procedure for Finding the Deflection of a Cantilever—Moment-Area Method

1. Draw the load, shearing force, and bending moment diagrams.
2. Divide the bending moment values by the beam stiffness, EI, and draw the M/EI diagram. The dimension for the quantity M/EI is $(\text{length})^{-1}$; for example, m^{-1},ft^{-1}, or in^{-1}.
3. Compute the area of the M/EI diagram and locate its centroid. If the diagram is not a simple shape, break it into parts and find the area and centroid for each part separately. If the deflection at the end of the cantilever is desired, the area of the entire M/EI diagram is used. If the deflection of some other point is desired, only the area between the support and the point of interest is used.
4. Use Theorem 2 to compute the vertical deviation of the point of interest from the tangent to the neutral axis of the beam at the support. Because the tangent is coincident with the original position of the neutral axis, the deviation thus found is the actual deflection of the beam at the point of interest. If all loads are in the same direction, the maximum deflection occurs at the end of the cantilever.

Example Problem 9–11 Use the moment-area method to determine the deflection at the end of the steel cantilever shown in Figure 9–36.

Solution

Objective: Compute the deflection at the end of the cantilever.

Given: Beam and loading in Figure 9–36.

Analysis: Use the *Procedure for finding the deflection of a cantilever—moment-area method.*

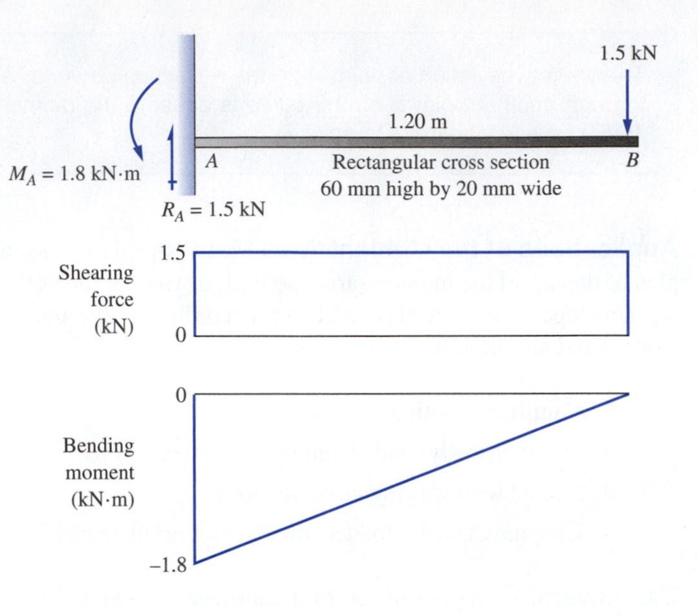

FIGURE 9–36 Load, shearing force and bending moment diagrams for Example Problems 9–11 and 9–12.

Results

Step 1. Figure 9–36 shows the load, shearing force, and bending moment diagrams.

Step 2. The stiffness is computed here:

$$E = 207\text{ GPa} = 207 \times 10^9\text{ N/m}^2$$

$$I = \frac{th^3}{12} = \frac{(0.02\text{ m})(0.06\text{ m})^3}{12} = 3.60 \times 10^{-7}\text{ m}^4$$

$$EI = (207 \times 10^9\text{ N/m}^2)(3.60 \times 10^{-7}\text{ m}^4) = 7.45 \times 10^4\text{ N}\cdot\text{m}^2$$

The M/EI diagram is drawn in Figure 9–37. Note that the only changes from the bending moment diagram are the units and the values because the beam has a constant stiffness along its length.

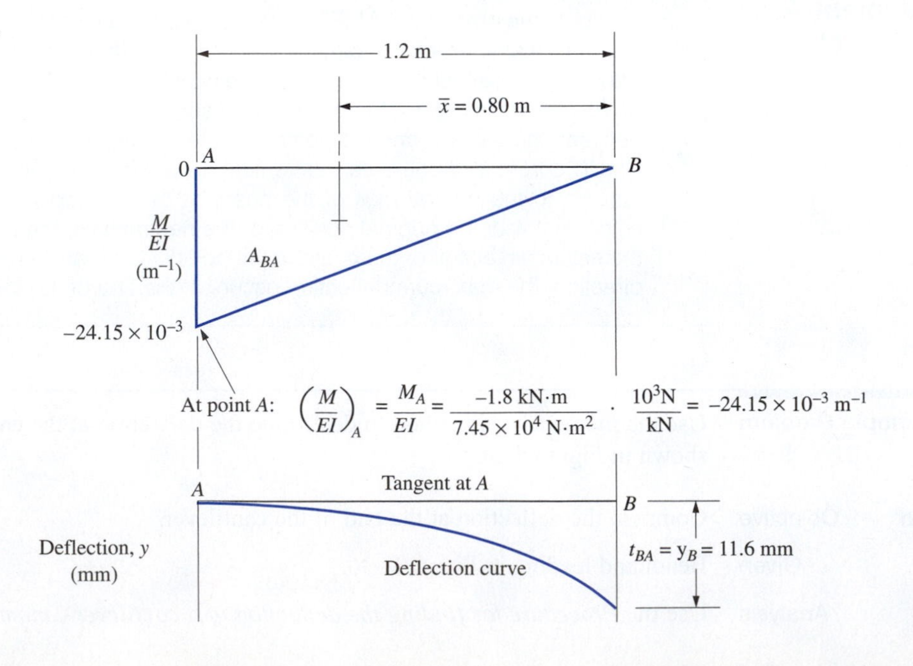

FIGURE 9–37 M/EI curve and deflection curve for Example Problem 9–11.

Step 3. The desired area is that of the entire triangular shape of the M/EI diagram, called A_{BA} here to indicate that it is used to compute the deflection of point B relative to A.

$$A_{BA} = (0.5)(24.15 \times 10^{-3}\,\text{m}^{-1})(1.20\,\text{m}) = 14.5 \times 10^{-3}\,\text{rad}$$

The centroid of this area is two-thirds of the distance from B to A, 0.80 m.

Step 4. To implement Theorem 2, we need to compute the moment of the area found in step 3. This is equal to t_{BA}, the vertical deviation of point B from the tangent drawn to the deflection curve at point A.

$$t_{BA} = A_{BA} \times \bar{x} = (14.5 \times 10^{-3}\,\text{rad})(0.80\,\text{m})$$
$$t_{BA} = y_B = 11.6 \times 10^{-3}\,\text{m} = 11.6\,\text{mm}$$

Because the tangent to point A is horizontal, t_{BA} is equal to the deflection of the beam at its end, point B.

Comment This result is identical to that which would be found by applying the formula for Case *a* in Appendix A–24. The value of the moment-area method is much more evident when multiple loads are involved or if the cantilever has a varying cross section along its length.

Example Problem 9–12 For the same beam used in Example Problem 9–11 and shown in Figure 9–36, compute the deflection at a point 1.0 m from the support.

Solution

Objective Compute the deflection 1.0 m from the left end of the cantilever.

Given Beam and loading in Figure 9–36.

Analysis Use the *Procedure for finding the deflection of a cantilever—moment-area method.* Steps 1 and 2 from Example Problem 9–11 are identical, resulting in the load, shearing force, bending moment, and M/EI diagrams shown in Figures 9–36 and 9–37. The solution procedure is continued at step 3.

Results ***Step 3.*** This step changes because only the area of the M/EI diagram between the support and the point 1.0 m out on the beam is used, as shown in Figure 9–38. Calling the point of interest, point C, we need to compute t_{CA}; that is, the vertical deviation of point C relative to the tangent drawn to point A at the support. The required area is a trapezoid, and it is convenient to break it into a triangle and a rectangle and treat them as simple shapes. The calculation then takes the form

$$t_{CA} = A_{CA} \times \bar{x} = A_{CA1} \times x_1 + A_{CA2} \times x_2$$

where the areas and x-distances are shown in Figure 9–38. Note that the distances are measured *from the point C to the centroid of the component area.* Then

$$A_{CA1} \times x_1 = (0.004\,025\,\text{m}^{-1})(1.0\,\text{m})(0.50\,\text{m}) = 2.01 \times 10^{-3}\,\text{m}$$
$$A_{CA2} \times x_2 = (0.5)(0.020\,125\,\text{m}^{-1})(1.0\,\text{m})(0.667\,\text{m}) = 6.71 \times 10^{-3}\,\text{m}$$
$$t_{CA} = y_C = (2.01 + 6.71)(10^{-3})\,\text{m} = 8.72\,\text{mm}$$

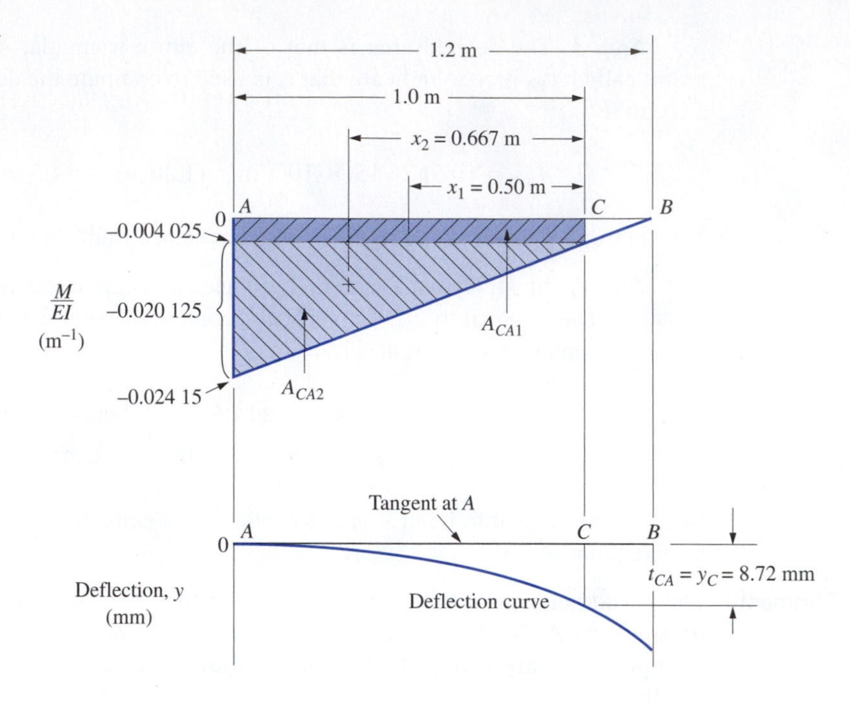

FIGURE 9–38 *M/EI* curve and deflection curve for Example Problem 9–12.

As before, because the tangent to point A is horizontal, the vertical deviation, t_{CA}, is the true deflection of point C.

Symmetrically Loaded Simply Supported Beams. This class of problems has the advantage that it is known that the maximum deflection occurs at the middle of the span of the beam. An example is shown in Figure 9–39, in which the beam carries two identical loads equally spaced from the supports. Of course, any loading for which the point of maximum deflection can be predicted can be solved by the procedure illustrated here.

Procedure for Finding the Deflection of a Symmetrically Loaded Simply Supported Beam—Moment-Area Method

1. Draw the load, shearing force, and bending moment diagrams.
2. Divide the bending moment values by the beam stiffness, EI, and draw the M/EI diagram.
3. If the maximum deflection, at the middle of the span, is desired, use that part of the M/EI diagram between the middle and one of the supports; that is, half of the diagram.
4. Use Theorem 2 to compute the vertical deviation of the point at one of the supports from the tangent to the neutral axis of the beam at its middle. Because the tangent is horizontal and because the deflection at the support is actually zero, the deviation thus found is the actual deflection of the beam at its middle.
5. To determine the deflection at some other point on the same beam, use the area of the M/EI diagram between the middle and the point of interest. Use Theorem 2 to compute the vertical deviation of the point of interest from the point of maximum deflection at the middle of the beam. Then subtract this deviation from the maximum deflection found in step 4.

Example Problem 9–13 Determine the maximum deflection of the beam shown in Figure 9–39. The beam is an aluminum channel, C6×2.834, made from 6061-T6 aluminum, positioned with the legs downward.

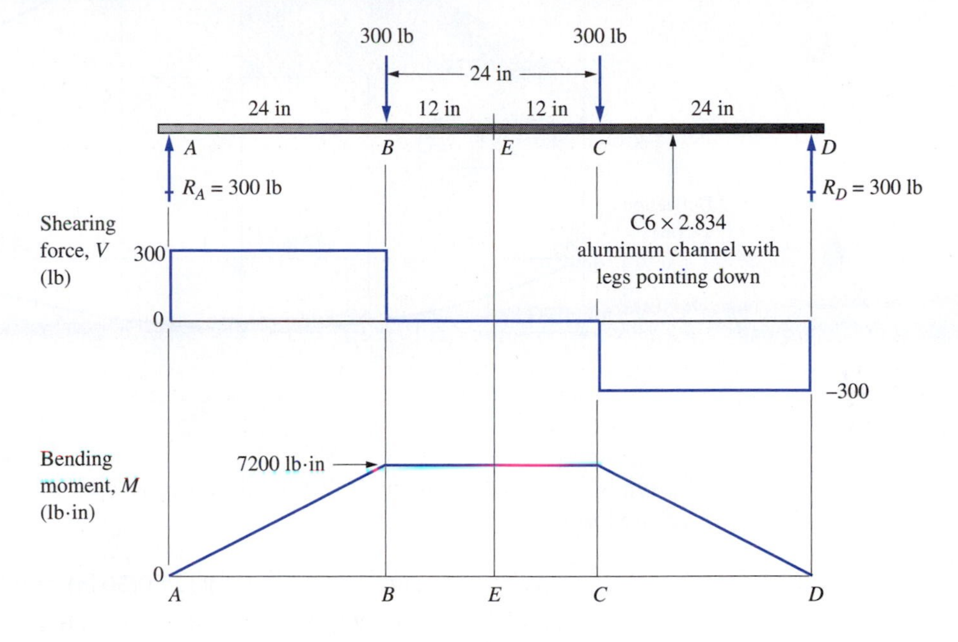

FIGURE 9–39 Load, shearing force, and bending moment diagrams for Example Problems 9–13 and 9–14.

Solution

Objective: Compute the maximum deflection of the beam.

Given: Beam and loading in Figure 9–39.

Analysis: Use the *Procedure for finding the deflection of a symmetrically loaded simply supported beam—moment-area method,* steps 1–4. Because the loading pattern is symmetrical, the maximum deflection will occur at the middle of the beam.

Results: ***Step 1.*** The load, shearing force, and bending moment diagrams are shown in Figure 9–39, prepared in the traditional manner. The maximum bending moment is 7200 lb·in between *B* and *C*.

Step 2. The stiffness of the beam, *EI*, is found using data from the Appendixes. From Appendix A–18, *E* for 6061-T6 aluminum is 10×10^6 psi. From Appendix A–10, the moment of inertia of the channel, with respect to the *Y-Y* axis, is 1.53 in^4. Then

$$EI = (10 \times 10^6\ \mathrm{lb/in^2})(1.53\ \mathrm{in^4}) = 1.53 \times 10^7\ \mathrm{lb \cdot in^2}$$

Because the stiffness of the beam is uniform over its entire length, the M/EI diagram has the same shape as the bending moment diagram, but the values are different, as shown in Figure 9–40. The maximum value of M/EI is $4.71 \times 10^{-4}\ \mathrm{in^{-1}}$.

Step 3. To determine the deflection at the middle of the beam, one-half of the M/EI diagram is used. For convenience, this is separated into a rectangle and a triangle, and the centroid of each part is shown.

Step 4. We need to find t_{AE}, the vertical deviation of point *A* from the tangent drawn to the deflection curve at point *E*, the middle of the beam. By Theorem 2,

$$t_{AE} = A_{AE1} \times x_{A1} + A_{AE2} \times x_{A2}$$

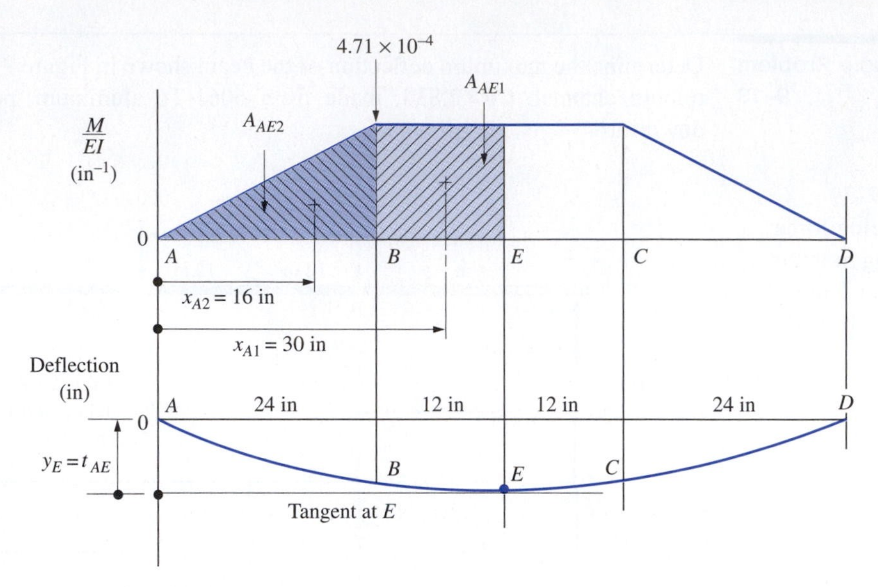

FIGURE 9–40 M/EI diagram and deflection curve for Example Problem 9–13.

The symbols x_{A1} and x_{A2} indicate that the distances to the centroids of the areas must be measured *from point A*.

$$A_{AE1} \times x_{A1} = (4.71 \times 10^{-4}\,\text{in}^{-1})(12\,\text{in})(30\,\text{in}) = 0.170\,\text{in}$$
$$A_{AE2} \times x_{A2} = (0.5)(4.71 \times 10^{-4}\,\text{in}^{-1})(24\,\text{in})(16\,\text{in}) = 0.090\,\text{in}$$
$$t_{AE} = y_E = 0.170 + 0.090 = 0.260\,\text{in}$$

This is the vertical deviation of point A from the tangent to point E. Because the tangent is horizontal and because the actual deflection of point A is zero, this represents the true deflection of point E relative to the original position of the neutral axis of the beam.

Comment This result is identical to that which would be found by applying the formula for Case c in Appendix A–23.

Example Problem 9–14 For the same beam used for Example Problem 9–13, shown in Figure 9–39, determine the deflection at point B under one of the loads.

Solution

Objective Compute the deflection at point B under one of the loads.

Given Beam and loading in Figure 9–39.

Analysis Use the *Procedure for finding the deflection of a symmetrically loaded simply supported beam—moment-area method,* steps 1–5. Steps 1–4 from Example Problem 9–13 are identical, resulting in the load, shearing force, bending moment, M/EI, and deflection diagrams shown in Figures 9–39 and 9–40. The solution procedure is continued at step 5.

Results ***Step 5.*** We can use the moment-area method to determine the vertical deviation, t_{BE}, of point B from the tangent to point E at the middle of the beam. Then, subtracting that from the value of t_{AE} found in Example Problem 9–13 gives the true deflection of point B. Shown in Figure 9–41 are the data needed to compute t_{BE}.

$$t_{BE} = A_{BE1} \times x_{B1} = (4.71 \times 10^{-4}\,\text{in}^{-1})(12\,\text{in})(6\,\text{in}) = 0.034\,\text{in}$$

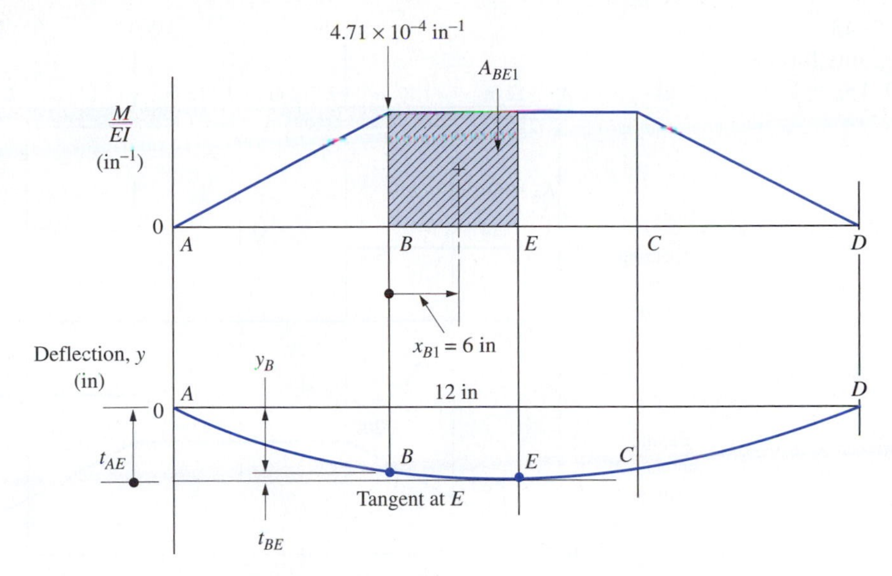

FIGURE 9–41 M/EI diagram and deflection curve for Example Problem 9–14.

Note that the distance x_{B1} must be measured from point B. Then the deflection of point B is

$$y_B = t_{AE} - t_{BE} = 0.260 - 0.034 = 0.226 \text{ in}$$

Beams with Varying Cross Sections. One of the major uses of the moment-area method is to find the deflection of a beam having a varying cross section along its length. Only one additional step is required as compared with beams having a uniform cross section, like those considered thus far.

An example of such a beam is illustrated in Figure 9–42. Note that it is a modification of the beam used in Example Problems 9–13 and 9–14 and shown in Figure 9–39.

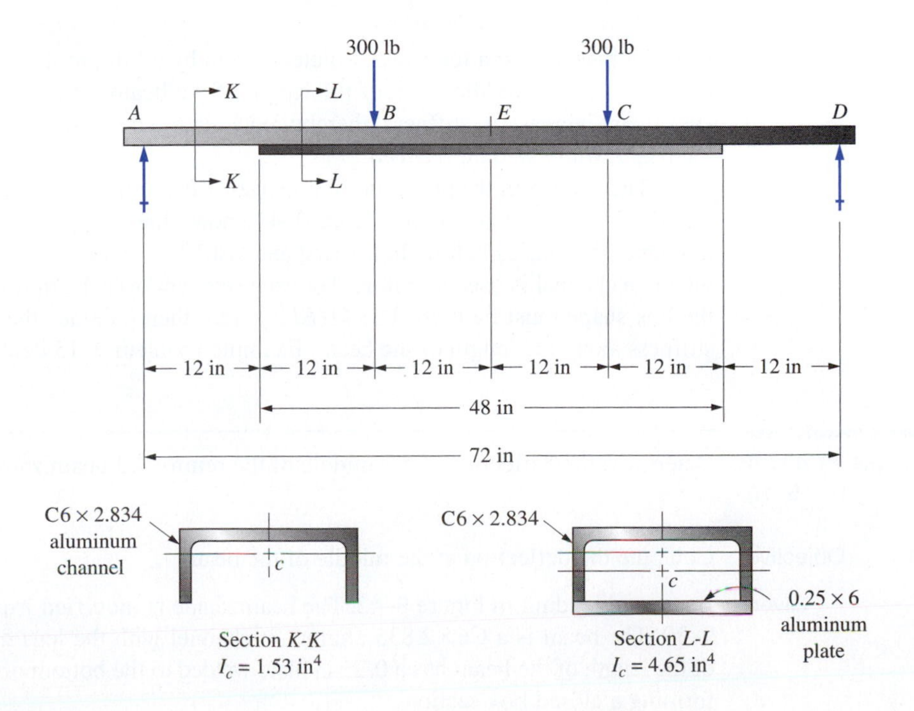

FIGURE 9–42 Beam for Example Problem 9–15.

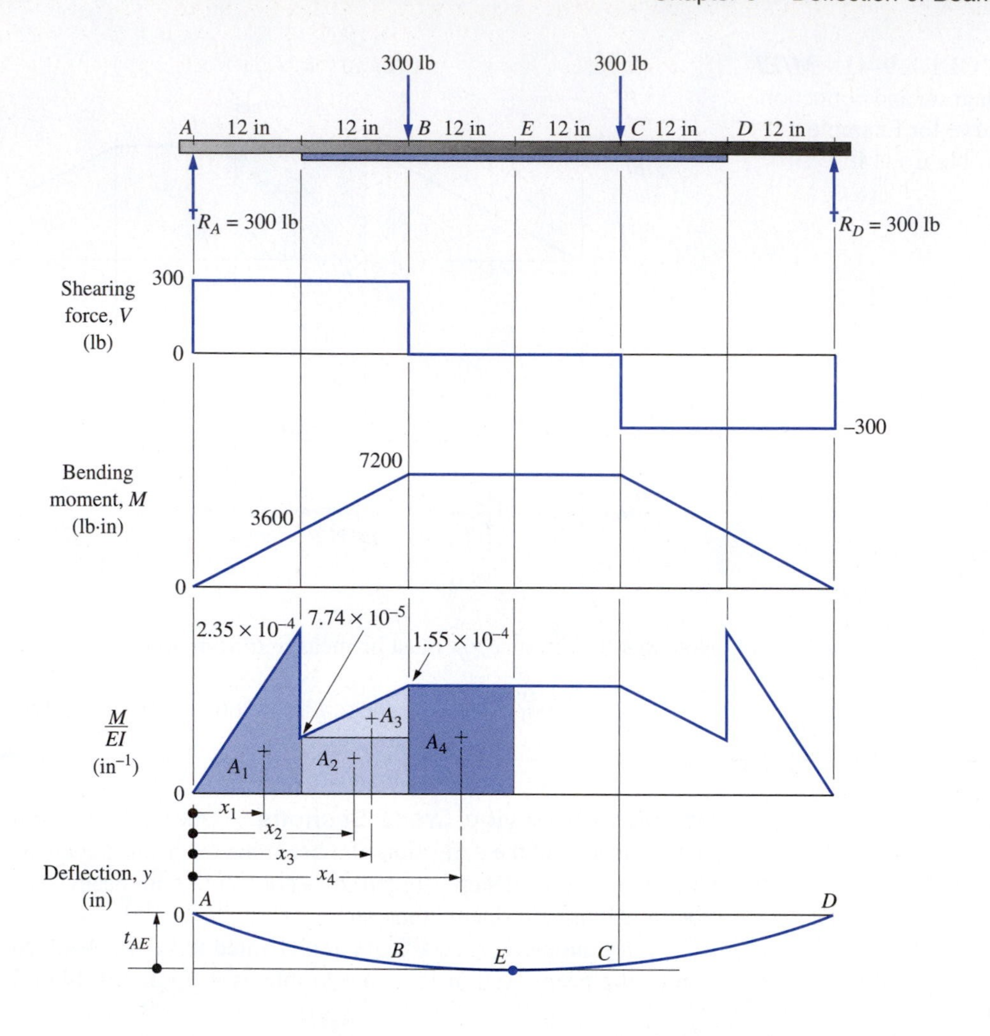

FIGURE 9–43 Beam diagrams for Problem 9–15.

Here we have added a rectangular plate, 0.25 in by 6.0 in, to the underside of the original channel over the middle 48 in of the length of the beam. The box shape would provide a significant increase in stiffness, thereby reducing the deflection of the beam. The stress in the beam would also be reduced.

The change in the procedure for analyzing the deflection of the beam is in the preparation of the M/EI diagram. Figure 9–43 shows the load, shearing force, and bending moment diagrams as before. In the first and last 12 in of the M/EI diagram, the stiffness of the plain channel is used as before. For each segment over the middle 48 in, the stiffness of the box shape must be used. The M/EI diagram then includes the effect of the change in stiffness along the length of the beam. Example Problem 9–15 demonstrates this process.

Example Problem 9–15 Determine the deflection at the middle of the reinforced beam shown in Figure 9–42.

Solution Objective Compute the deflection at the middle of the beam.

Given Beam and loading in Figure 9–42. The beam shape is modified from that shown in Figure 9–39. The beam is a C6×2.835 aluminum channel with the legs down. The middle 48 in of the length of the beam has a 0.25-in plate welded to the bottom of the legs of the channel, forming a closed box section.

Analysis Parts of the solution from Example Problem 9–13 are used. The M/EI diagram is adjusted to include the effect of the modified cross section of the beam.

Results ***Step 1.*** The load, shearing force, and bending moment diagrams are prepared in the traditional manner, as shown in Figure 9–43.

Step 2. To prepare the M/EI diagram, two values for the stiffness, EI, are needed. The plain channel has the same value used in previous problems, 1.53×10^7 lb·in^2. For the box shape,

$$EI = (10 \times 10^6 \text{ lb/in}^2)(4.65 \text{ in}^4) = 4.65 \times 10^7 \text{ lb}\cdot\text{in}^2$$

Then, at the point on the beam just before 12 in from A, where the bending moment is 3600 lb·in

$$\frac{M}{EI} = \frac{3600 \text{ lb}\cdot\text{in}}{1.53 \times 10^7 \text{ lb}\cdot\text{in}^2} = 2.35 \times 10^{-4} \text{ in}^{-1}$$

Just beyond 12 in from A,

$$\frac{M}{EI} = \frac{3600 \text{ lb}\cdot\text{in}}{4.65 \times 10^7 \text{ lb}\cdot\text{in}^2} = 7.74 \times 10^{-5} \text{ in}^{-1}$$

At point B, where $M = 7200$ lb·in and $EI = 4.65 \times 10^7$ lb·in^2,

$$\frac{M}{EI} = \frac{7200 \text{ lb}\cdot\text{in}}{4.65 \times 10^7 \text{ lb}\cdot\text{in}^2} = 1.55 \times 10^{-4} \text{ in}^{-1}$$

These values establish the critical points on the M/EI diagram.

Step 3. The moment area for the left half of the M/EI diagram will be used to determine the value of t_{AE}, as done in Example Problem 9–13. For convenience, the total area is divided into four parts, as shown in Figure 9–43, with the locations of the centroids indicated relative to point A. The distances are

$$x_1 = (\tfrac{2}{3})(12 \text{ in}) = 8 \text{ in}$$

$$x_2 = (\tfrac{1}{2})(12 \text{ in}) + 12 \text{ in} = 18 \text{ in}$$

$$x_3 = (\tfrac{2}{3})(12 \text{ in}) + 12 \text{ in} = 20 \text{ in}$$

$$x_4 = (\tfrac{1}{2})(12 \text{ in}) + 24 \text{ in} = 30 \text{ in}$$

Step 4. We can now use Theorem 2 to compute the value of t_{AE}, the deviation of point A from the tangent to point E, by computing the moment of each of the four areas shown crosshatched in the M/EI diagram of Figure 9–43.

$$t_{AE} = A_1x_1 + A_2x_2 + A_3x_3 + A_4x_4$$

$$A_1x_1 = (0.5)(2.35 \times 10^{-4} \text{ in}^{-1})(12 \text{ in})(8 \text{ in}) = 1.128 \times 10^{-2} \text{ in}$$

$$A_2x_2 = (7.74 \times 10^{-5} \text{ in}^{-1})(12 \text{ in})(18 \text{ in}) = 1.672 \times 10^{-2} \text{ in}$$

$$A_3x_3 = (0.5)(7.74 \times 10^{-5}\,\text{in}^{-1})(12\,\text{in})(20\,\text{in}) = 9.293 \times 10^{-3}\,\text{in}$$
$$A_4x_4 = (1.55 \times 10^{-4}\,\text{in}^{-1})(12\,\text{in})(30\,\text{in}) = 5.580 \times 10^{-2}\,\text{in}$$

Then

$$t_{AE} = y_E = \sum(A_ix_i) = 9.309 \times 10^{-2}\,\text{in} = 0.093\,\text{in}$$

Comment As before, this value is equal to the deflection of point E at the middle of the beam. Comparing it to the deflection of 0.260 in found in Example Problem 9–14, the addition of the cover plate reduced the maximum deflection by approximately 64%.

Unsymmetrically Loaded Simply Supported Beams. The major difference between this type of beam and the ones considered earlier is that the point of maximum deflection is unknown. Special care is needed to describe the geometry of the M/EI diagram and the deflection curve for the beam.

The general procedure for finding the deflection at any point on the deflection curve for an unsymmetrically loaded simply supported beam is outlined in the following steps. Because of the myriad of different loading patterns, the specifics of applying this procedure may have to be adjusted for any given problem. You are advised to check the fundamental principles of the moment-area method as the solution of a problem is completed. The method will be demonstrated in Example Problem 9–16.

Procedure for Finding the Deflection of an Unsymmetrically Loaded Simply Supported Beam—Moment-Area Method

1. Draw the load, shearing force, and bending moment diagrams.
2. Construct the M/EI diagram by dividing the bending moment at any point by the value of the beam stiffness, EI, at that point.
3. Sketch the probable shape of the deflection curve. Then draw the tangent to the deflection curve at one of the supports. Using Theorem 2, compute the vertical deviation of the other support from this tangent line. The moment of the entire M/EI diagram with respect to the second support is required.
4. Using proportions, compute the distance from the zero-axis to the tangent line from step 3 at the point for which the deflection is desired.
5. Using Theorem 2, compute the vertical deviation of the point of interest from the tangent line from step 3. The moment for that portion of the M/EI diagram between the first support and the point of interest will be used.
6. Subtract the deviation computed in step 5 from that found in step 4. The result is the deflection of the beam at the desired point.

Example Problem 9–16 Determine the deflection of the beam shown in Figure 9–44 at its middle, 1.0 m from the supports. The beam is an American Standard steel beam, S3×5.7.

Solution Objective Compute the deflection at the middle of the beam.

Given Beam and loading shown in Figure 9–44. Beam is steel.
Beam shape is an American Standard S3×5.7.

Anaysis Use the *Procedure for finding the deflection of an unsymmetrically loaded simply supported beam—moment-area method.*

Results ***Step 1.*** The load, shearing force, and bending moment diagrams are shown in Figure 9–44.

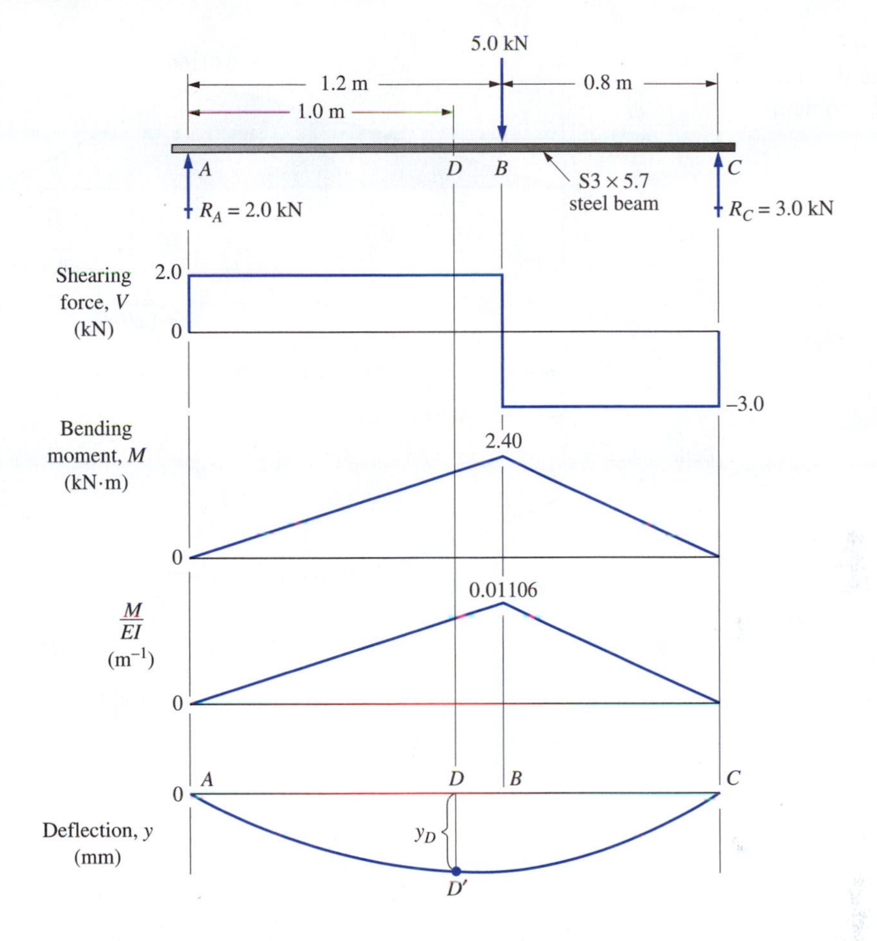

FIGURE 9–44 Beam for Example Problem 9–16.

Step 2. The beam has a uniform stiffness along its entire length resulting in the shape of the M/EI diagram being the same as the bending moment diagram. The value of M/EI at point B can be computed by dividing the bending moment there (2.40 kN·m or 2400 N·m) by EI. We will use $E = 207$ GPa for steel. From the Appendix we find $I = 2.50\ \text{in}^4$, and this must be converted to metric units.

$$I = \frac{(2.50\ \text{in}^4)(0.0254\ \text{m})^4}{1.0\ \text{in}^4} = 1.041 \times 10^{-6}\ \text{m}^4$$

Then the beam stiffness is

$$EI = (207 \times 10^9\ \text{N/m}^2)(1.041 \times 10^{-6}\ \text{m}^4) = 2.15 \times 10^5\ \text{N}\cdot\text{m}^2$$

The value of M/EI at point B on the beam can now be computed.

$$\left(\frac{M}{EI}\right)_B = \frac{2400\ \text{N}\cdot\text{m}}{2.15 \times 10^5\ \text{N}\cdot\text{m}^2} = 0.01114\ \text{m}^{-1}$$

The M/EI diagram is drawn in Figure 9–44. It is desired to compute the deflection of the beam at its middle, labeled point D.

Step 3. Figure 9–44 shows an exaggerated sketch of the deflection curve for the beam. It is likely that the maximum deflection will occur very close to the middle of the beam where

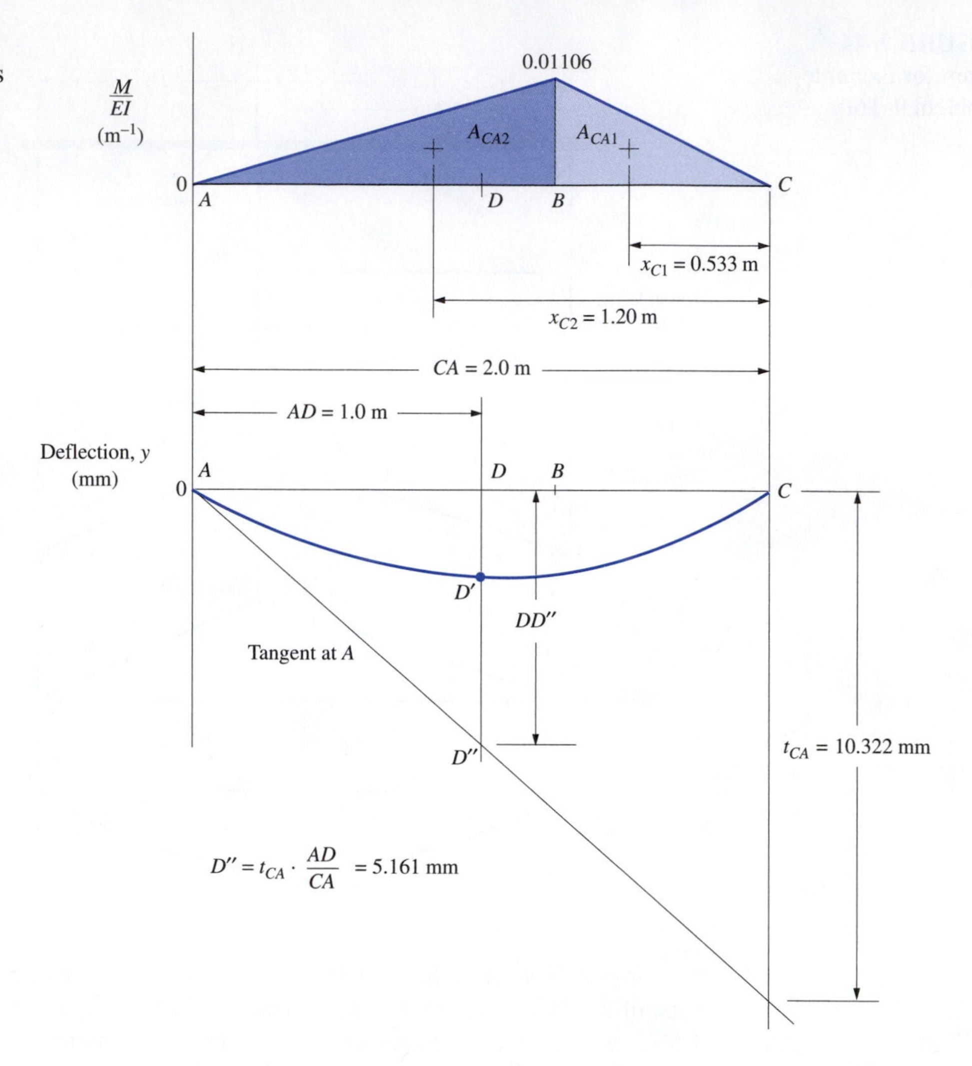

FIGURE 9–45 Moment-area diagrams for Example Problem 9–16.

we are to determine the deflection, point D. Figure 9–45 shows the tangent to the deflection curve at point A at the left support and the vertical deviation of point C from this line. Note that point C is a known point on the deflection curve because the deflection there is zero. Now we can use Theorem 2 to compute t_{CA}. The entire M/EI diagram is used, broken into two triangles.

$$t_{CA} = A_{CA1}x_{C1} + A_{CA2}x_{C2}$$
$$A_{CA1}x_{C1} = (0.5)(0.01106\ \text{m}^{-1})(0.8\ \text{m})(0.533\ \text{m}) = 0.002359\ \text{m}$$
$$A_{CA2}x_{C2} = (0.5)(0.01106\ \text{m}^{-1})(1.2\ \text{m})(1.2\ \text{m}) = 0.007963\ \text{m}$$

Then

$$t_{CA} = 0.002359 + 0.007963 = 0.010322\ \text{m} = 10.322\ \text{mm}$$

Step 4. Use the principle of proportion to determine the distance DD'' from D to the tangent line.

$$\frac{t_{CA}}{CA} = \frac{DD''}{AD}$$

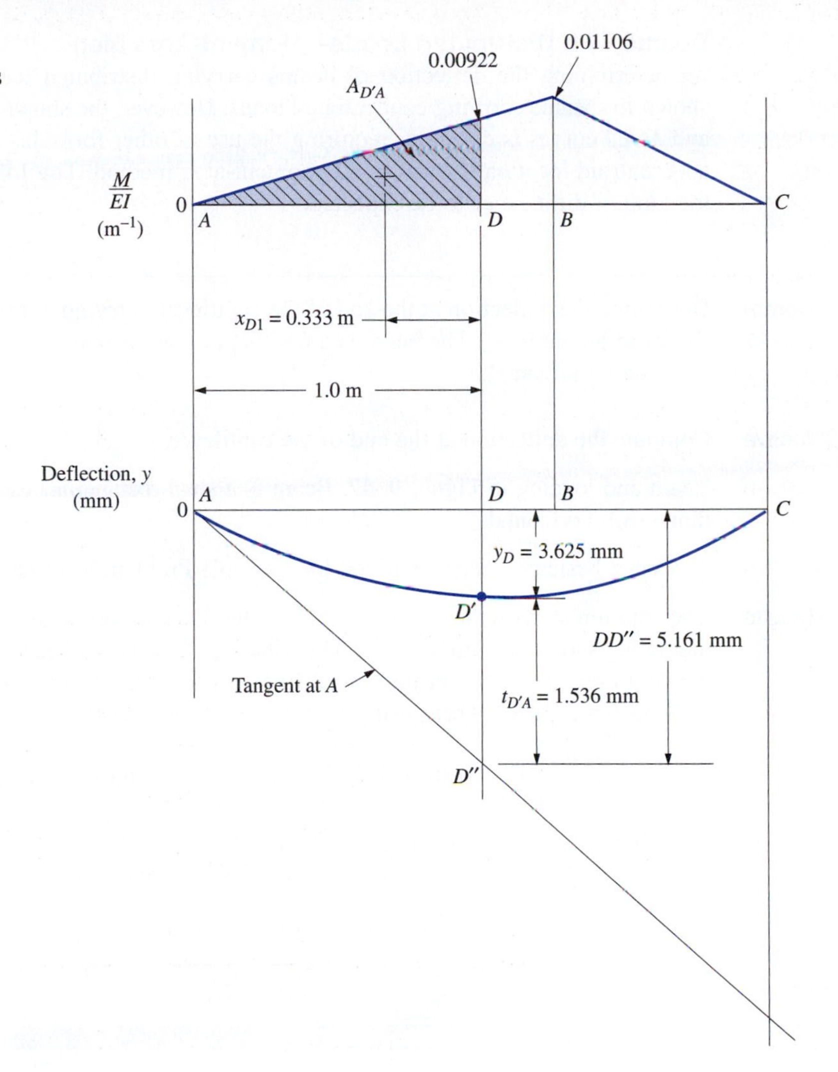

FIGURE 9–46 Moment-area diagrams for Example Problem 9–16.

or

$$DD'' = t_{CA} \times \frac{AD}{CA} = (10.322\text{ mm}) \times \frac{1.0\text{ m}}{2.0\text{ m}} = 5.161\text{ mm}$$

Step 5. Compute the deviation, $t_{D'A}$, of point D' from the tangent line drawn to point A using Theorem 2. The part of the M/EI diagram between D and A is used as shown in Figure 9–46.

$$t_{D'A} = A_{D'A}x_{D1} = (0.5)(0.00922\text{ m}^{-1})(1.0\text{ m})(0.333\text{ m}) = 0.001536\text{ m}$$
$$t_{D'A} = 1.536\text{ mm}$$

Step 6. From the geometry of the deflection diagram shown in Figure 9–46 the deflection at point D, y_D, is

$$y_D = DD' = DD'' - t_{D'A} = 5.161 - 1.536 = 3.625\text{ mm}$$

Beams with Distributed Loads—Moment-Area Method. The general procedure for determining the deflection of beams carrying distributed loads is the same as that shown for beams carrying concentrated loads. However, the shape of the bending moment and M/EI curves is different, requiring the use of other formulas for computing the area and centroid location for use in the moment-area method. The following example shows the kind of differences to be expected.

Example Problem 9–17 Determine the deflection at the end of the cantilever carrying a uniformly distributed load shown in Figure 9–47. The beam is a $6\times2\times\frac{1}{4}$ hollow rectangular steel tube with the 6.0-in dimension horizontal.

Solution

Objective: Compute the deflection at the end of the cantilever.

Given: Beam and loading in Figure 9–47. Beam is a steel rectangular tube, $6\times2\times\frac{1}{4}$, with 6.0-in dimension horizontal.

Analysis: The same basic procedure outlined for Example Problem 9–11 can be used.

Results: The solution starts with the preparation of the load, shearing force, and bending moment diagrams, shown in Figure 9–47. Then the M/EI curve will have the same shape as the bending moment curve because the stiffness of the beam is uniform. From Appendix A–9 we find $I = 2.21 \text{ in}^4$. Then, using $E = 30 \times 10^6$ psi for steel,

$$EI = (30 \times 10^6 \text{ lb/in}^2)(2.21 \text{ in}^4) = 6.63 \times 10^7 \text{ lb}\cdot\text{in}^2$$

Now Figure 9–48 shows the M/EI curve along with the deflection curve for the beam. The horizontal line on the deflection diagram is the tangent drawn to the beam shape at point A,

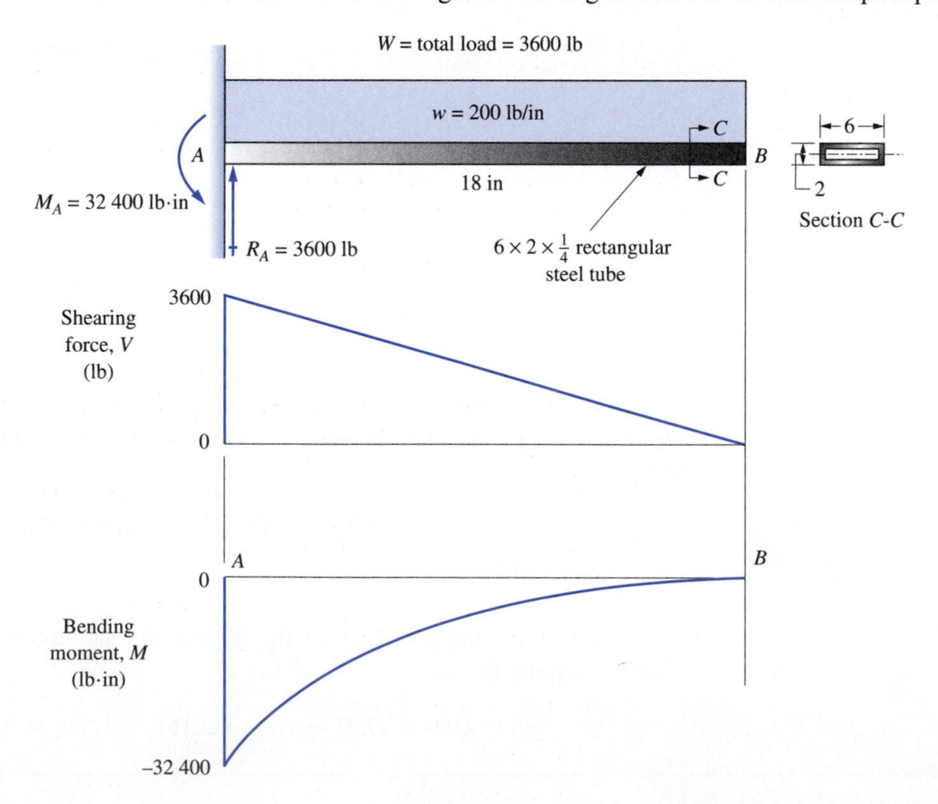

FIGURE 9–47 Load, shearing force, and bending moment diagrams for Example Problem 9–17.

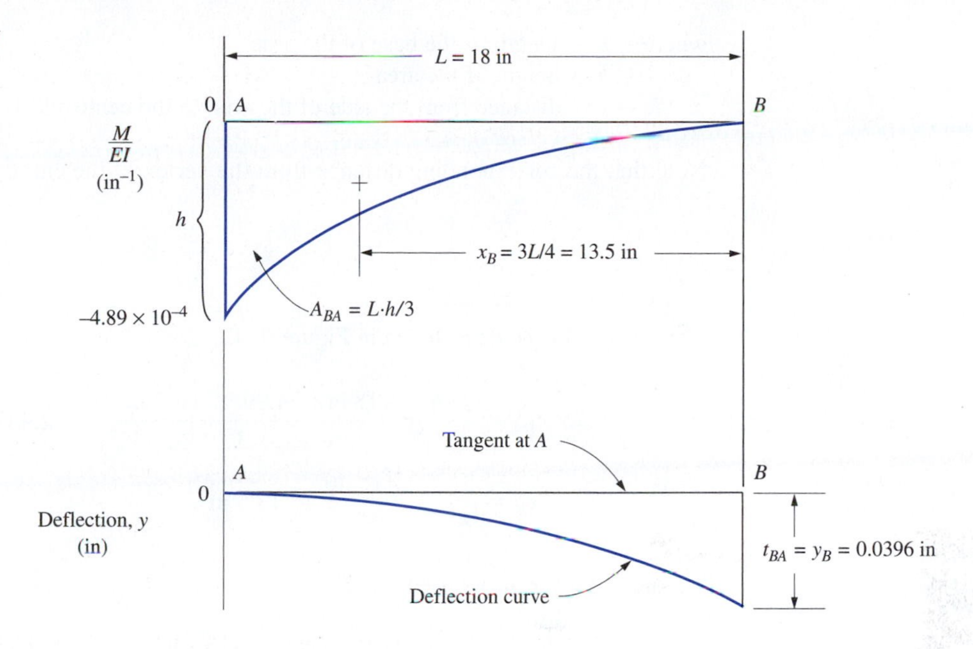

FIGURE 9–48 *M/EI* and deflection curves for Example Problem 9–17.

where the beam is fixed to the support. Then, at the right end of the beam, the deviation of the beam deflection curve from this tangent, t_{BA}, is equal to the deflection of the beam itself.

Using Theorem 2, the deviation t_{BA} is equal to the product of the area of the M/EI curve between B and A times the distance from point B to the centroid of the area. That is,

$$t_{BA} = A_{BA} \cdot x_B$$

Recalling that the load, shearing force, and bending moment diagrams are related to each other in such a way that the higher curve is the derivative of the curve below it, the following conclusions can be drawn:

1. The shearing force curve is a first-degree curve (straight line with constant slope). Its equation is of the form

$$V = m \cdot x + b$$

   where m is the slope of the line and b is its intercept with the vertical axis. The variable x is the position on the beam.
2. The bending moment curve is a second-degree curve, a parabola. The general equation of the curve is of the form

$$M = a \cdot x^2 + b$$

Appendix A–1 shows the relationships for computing the area and the location of the centroid for areas bounded by second-degree curves. For an area having the shape of the bending moment or M/EI curves,

$$\text{area} = \frac{L \cdot h}{3}$$

$$x = \frac{L}{4}$$

where L = length of the base of the area
h = height of the area
x = distance from the side of the area to the centroid

Note that the corresponding distance from the vertex of the curve to the centroid is

$$x' = \frac{3L}{4}$$

Now, for the data shown in Figure 9–42,

$$A_{BA} = \frac{L \cdot h}{3} = \frac{(18\text{ in})(-4.89 \times 10^{-4}\text{ in}^{-1})}{3} = 2.932 \times 10^{-3}$$

$$x_B = \frac{3L}{4} = \frac{3(18\text{ in})}{4} = 13.5\text{ in}$$

Theorem 2 can now be used:

$$t_{BA} = A_{BA}x_B = (2.932 \times 10^{-3})(13.5\text{ in}) = 0.0396\text{ in}$$

This is equal to the deflection of the beam at its end, y_B.

REFERENCES

1. Aluminum Association, *Aluminium Design Manual,* Washington, DC, 2005.
2. Blodgett, Omer W., *Design of Weldments,* James F. Lincoln Arc Welding Foundation, Cleveland, OH, 1963.
3. Mott, Robert L., *Machine Elements in Mechanical Design,* 4th ed., Prentice Hall, Upper Saddle River, NJ, 2004.
4. Popov, E. P., *Engineering Mechanics of Solids,* 2nd ed. Prentice Hall, Upper Saddle River, NJ, 1999.
5. Young, W. C. and R. G. Budynas, *Roark's Formulas for Stress and Strain,* 7th ed., McGraw-Hill, New York, 2002.

INTERNET SITES

1. MDSolids www.mdsolids.com Educational software for students taking mechanics of materials, strength of materials, or mechanics of deformable solids; comprised of several modules, one of which is beam analysis.
2. Orand Systems www.orandsystems.com Producer of the Beam2D software for beam analysis.
3. Creative Engineering www.suverkrop.com/winbuild.html Structural analysis software with the facility to select the least weight beam for a given loading.
4. MITCalc www.mitcalc.com Calculation software for mechanical, industrial, and technical analysis, including beam analysis and many other technical topics.
5. Engineers Edge www.engineersedge.com/CalculatorsOnline.shtml A variety of online calculators, including *Beam Deflections and Stress, Section Properties of Selected Shapes, Pressure Vessels,* and other mechanical design phenomena.
6. Free Structural Software www.structural-engineering.fsnet.co.uk/free.htm Listings of many software packages, many of which are free and others that have a free trial period.

PROBLEMS

Formula Method—Statically Determinate Beams

9–1.M A round shaft having a diameter of 32 mm is 700 mm long and carries a 3.0-kN load at its center. If the shaft is steel and simply supported at its ends, compute the deflection at the center.

9–2.M For the shaft in Problem 9–1, compute the deflection if the shaft is 6061-T6 aluminum instead of steel.

9–3.M For the shaft in Problem 9–1, compute the deflection if the ends are fixed against rotation instead of simply supported.

9–4.M For the shaft in Problem 9–1, compute the deflection if the shaft is 350 mm long rather than 700 mm.

9–5.M For the shaft in Problem 9–1, compute the deflection if the diameter is 25 mm instead of 32 mm.

9–6.M For the shaft in Problem 9–1, compute the deflection if the load is placed 175 mm from the left support rather than in the center. Compute the deflection both at the load and at the center of the shaft.

9–7.E A wide-flange steel beam, W12×16, carries the load shown in Figure P9–7. Compute the deflection at the loads and at the center of the beam.

FIGURE P9–7

9–8.E A standard steel $1\frac{1}{2}$-in schedule 40 pipe carries a 650-lb load at the center of a 28-in span, simply supported. Compute the deflection of the pipe at the load.

9–9.E An Aluminum Association Standard I-beam, I8×6.181, carries a uniformly distributed load of 1125 lb/ft over a 10-ft span. Compute the deflection at the center of the span.

9–10.E For the beam in Problem 9–9, compute the deflection at a point 3.5 ft from the left support of the beam.

9–11.E A wide-flange steel beam, W12×30, carries the load shown in Figure P9–11. Compute the deflection at the load.

9–12.E For the beam in Problem 9–11, compute the deflection at the load if the left support is moved 2.0 ft toward the load.

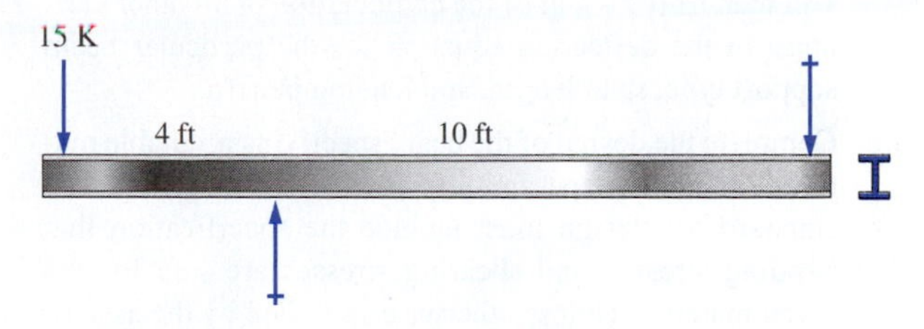

FIGURE P9–11

9–13.E For the beam in Problem 9–11, compute the maximum upward deflection and determine its location.

9–14.E A 1-in schedule 40 steel pipe is used as a cantilever beam 8 in long to support a load of 120 lb at its end. Compute the deflection of the pipe at the end.

9–15.M A round steel bar is to be used to carry a single concentrated load of 3.0 kN at the center of a 700-mm-long span on simple supports. Determine the required diameter of the bar if its deflection must not exceed 0.12 mm.

9–16.M For the bar designed in Problem 9–15, compute the stress in the bar and specify a suitable steel to provide a design factor of 8 based on ultimate strength.

9–17.E A flat strip of steel 0.100 in wide and 1.200 in long is clamped at one end and loaded at the other like a cantilever beam (as in Case *a* in Appendix A–24). What should be the thickness of the strip if it is to deflect 0.15 in under a load of 0.52 lb?

9–18.E A wood joist in a commercial building is 14 ft 6 in long and carries a uniformly distributed load of 50 lb/ft. It is 1.50 in wide and 9.25 in high. If it is made of southern pine, compute the maximum deflection of the joist. Also compute the stress in the joist due to bending and horizontal shear, and compare them with the allowable stresses for No. 2 grade southern pine.

Formula Method—Statically Indeterminate Beams

For Problems 9–19 through 9–45, use the formulas in Appendix A–25 to complete any of the following according to the instructions given for a particular assignment.

(a) Determine the reactions and draw the complete shearing force and bending moment diagrams. Report the maximum shearing force and the maximum bending moment and indicate where they occur.

(b) Where deflection formulas are available, also compute the maximum deflection of the beam expressed in the form,

$$y = C_d/EI$$

where EI is the beam stiffness, the product of the modulus of elasticity for the material of the beam and the moment of inertia of the cross section of the beam. The term C_d will then be the result of the computation of all other variables in the deflection equation for the particular beam support type, span length, and loading pattern.

(c) Complete the design of the beam specifying a suitable material, and the shape and size for the cross section. The standard for design must include the specification that bending stresses and shearing stresses are safe for the given material. Unless otherwise specified by the assignment, consider all loads to be static.

(d) Complete the design of the beam to limit the maximum deflection to some specified value as given by the assignment. In the absence of a specified limit, use $L/360$ as the maximum allowable deflection where L is the span between supports or the overall length of the beam. The design must specify a suitable material, and the shape and size for the cross section. This assignment may be linked to part b where the deflection was calculated in terms of the beam stiffness, EI. Then, for example, you may specify the material and its value for E, compute the limiting deflection, and solve for the required moment of inertia, I. The shape and size of the cross section can then be determined. Note that any design must also be shown to be safe with regard to bending stresses and shearing stresses as in part c.

9–19.M Use A–25(a) with $P = 35$ kN, $L = 4.0$ m.

9–20.M Use A–25(b) with $P = 35$ kN, $L = 4.0$ m, $a = 1.50$ m.

9–21.M Use A–25(b) with $P = 35$ kN, $L = 4.0$ m, $a = 2.50$ m.

9–22.E Use A–25(c) with $w = 400$ lb/ft, $L = 14.0$ ft.

9–23.E Use A–25(c) with $w = 50$ lb/in, $L = 16.0$ in.

9–24.E Use A–25(d) with $P = 350$ lb, $L = 10.8$ in, $a = 2.50$ in.

9–25.M Use A–25(e) with $P = 35$ kN, $L = 4.0$ m.

9–26.M Use A–25(f) with $P = 35$ kN, $L = 4.0$ m, $a = 1.50$ m.

9–27.M Use A–25(f) with $P = 35$ kN, $L = 4.0$ m, $a = 2.50$ m.

9–28.E Use A–25(g) with $w = 400$ lb/ft, $L = 14.0$ ft.

9–29.E Use A–25(g) with $w = 50$ lb/in, $L = 16.0$ in.

9–30.E Use A–25(h) with $w = 400$ lb/ft, $L = 7.0$ ft.

9–31.E Use A–25(h) with $w = 50$ lb/in, $L = 8.0$ in.

9–32.E Use A–25(i) with $w = 400$ lb/ft, $L = 56$ in.

9–33.E Use A–25(i) with $w = 50$ lb/in, $L = 5.333$ in.

9–34.E Use A–25(j) with $w = 400$ lb/ft, $L = 3.5$ ft.

9–35.E Use A–25(j) with $w = 50$ lb/in, $L = 4.0$ in.

9–36.E Use Figure P9–36.

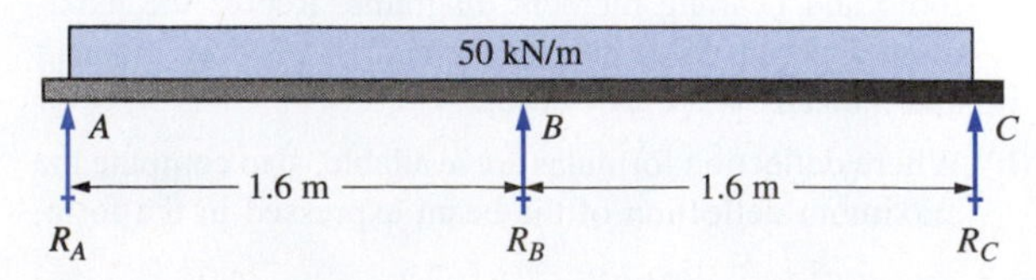

FIGURE P9–36

9–37.M Use Figure P9–37.

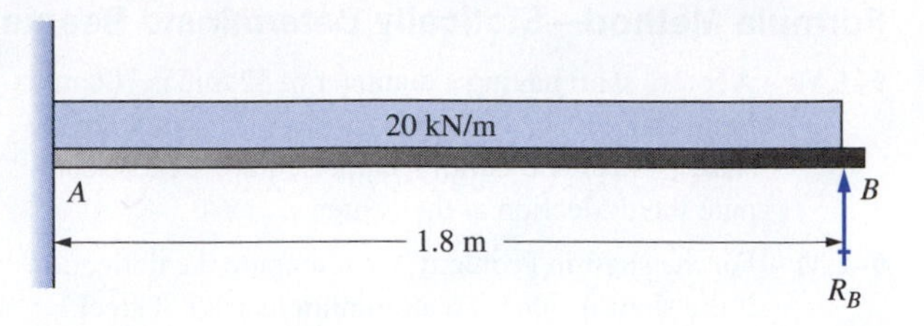

FIGURE P9–37

9–38.E Use Figure P9–38.

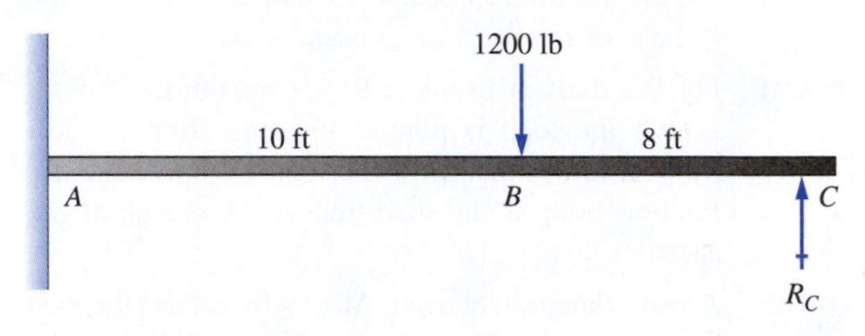

FIGURE P9–38

9–39.M Use A–25(d) with $P = 18$ kN, $L = 2.75$ m, $a = 1.40$ m.

9–40.E Use A–25(f) with $P = 8500$ lb, $L = 109$ in, $a = 75$ in.

9–41.E Use A–25(h) with $w = 4200$ lb/ft, $L = 16.0$ ft.

9–42.M Use A–25(i) with $w = 50$ kN/m, $L = 3.60$ m.

9–43.E Use A–25(j) with $w = 15$ lb/in, $L = 36$ in.

9–44.E Use A–25(e) with $P = 140$ lb, $L = 54$ in.

9–45.M Use A–25(b) with $P = 250$ N, $L = 55$ mm, $a = 15$ mm.

Comparisons of Beam Behavior

9–46.M Compare the behavior of the four beams shown in Figure P9–46 with regard to shearing force, bending moment, and maximum deflection. In each case, the beam is designed to support a uniformly distributed load across a given span. Complete the analysis in a manner similar to that used in Example Problems 9–3 through 9–7.

9–47.M Compare Problems 9–22, 9–28, 9–30, 9–32, and 9–34 with regard to the maximum values of shearing force, bending moment, and deflection.

9–48.M Compare Problems 9–23, 9–29, 9–31, 9–33, and 9–35 with regard to the maximum values of shearing force, bending moment, and deflection.

9–49.E Specify a suitable design for a wooden beam, simply supported at its ends, to carry a uniformly distributed load of 120 lb/ft for a span of 24 ft. The beam must be

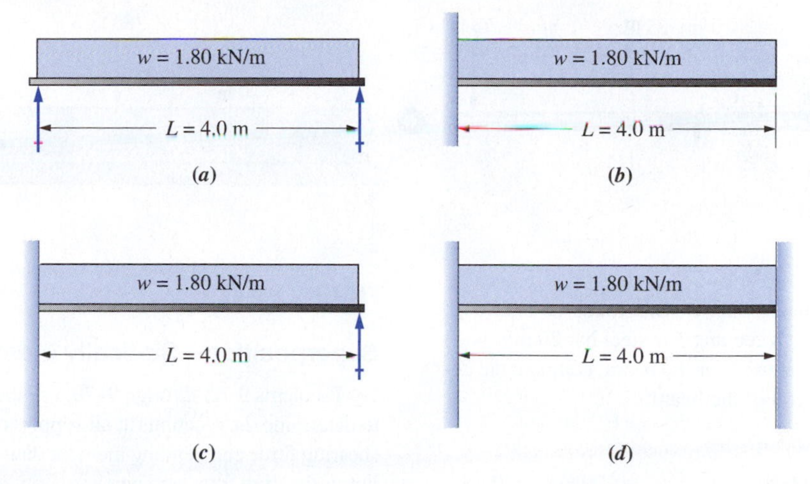

FIGURE P9–46 Beams for Problem 9–46. (a) Simply supported beam. (b) Cantilever. (c) Supported cantilever. (d) Fixed end beam.

safe for both bending and shear stresses when made from No. 2 grade southern pine. Then compute the maximum deflection for the beam you have designed.

9–50.E Repeat Problem 9–49 except place an additional support at the middle of the beam, 12 ft from either end.

9–51.E Repeat Problem 9–49 except use four supports 8 ft apart.

9–52.E Compare the behavior of the three beams designed in Problems 9–49, 9–50, and 9–51 with regard to shearing force, bending moment, and maximum deflection. Complete the analysis in the manner used in Example Problems 9–3 through 9–7.

Superposition—Statically Determinate Beams

9–53.M The load shown in Figure P9–53a is being carried by an extruded aluminum (6061-T6) beam having the shape shown in Figure 9–53b. Compute the deflection of the beam at each load.

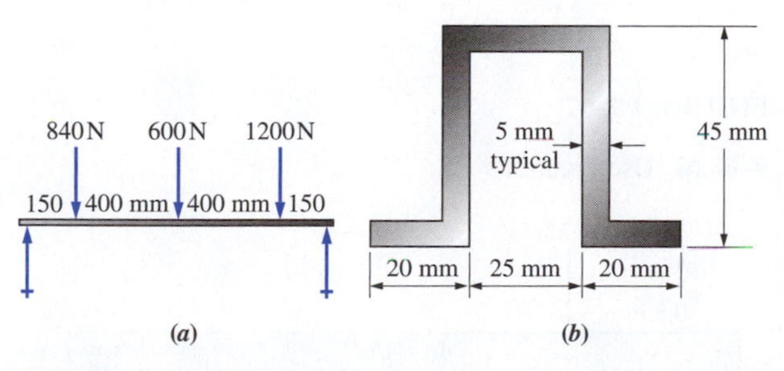

FIGURE P9–53

9–54.M The loads shown in Figure P9–54a represent the feet of a motor on a machine frame. The frame member has the cross section shown in Figure P9–54b, which has a moment of inertia of 16 956 mm^4. Compute the deflection at each load. The aluminum alloy 2014-T4 is used for the frame.

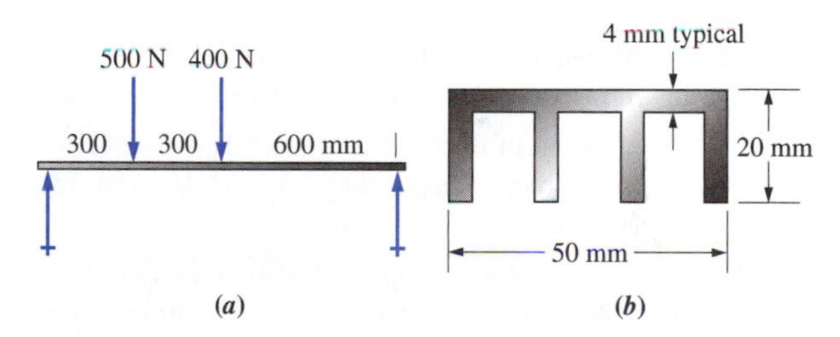

FIGURE P9–54

9–55.C Compute the deflection at the middle of a steel W460×82 (W18×55) beam when it carries the load shown in Figure P9–55.

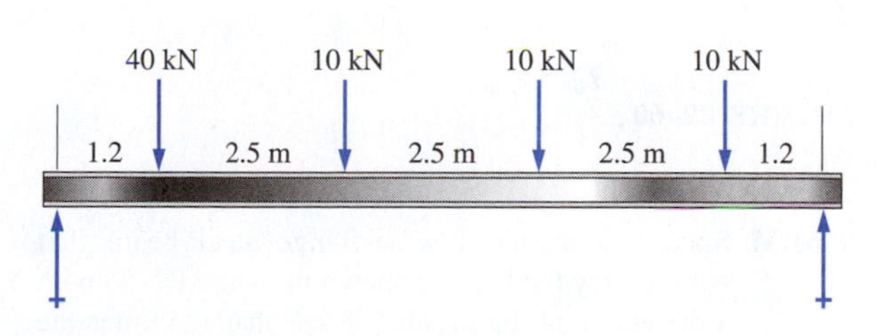

FIGURE P9–55

9–56.E A 1-in schedule 40 steel pipe carries the two loads shown in Figure P9–56. Compute the deflection of the pipe at each load.

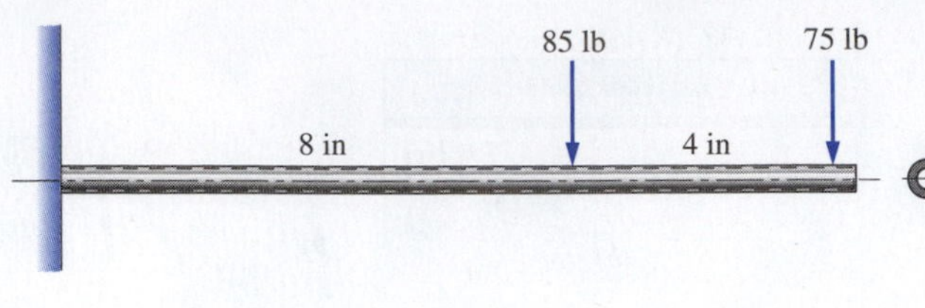

FIGURE P9–56

9–57.M A cantilever beam carries two loads as shown in Figure P9–57. If a rectangular steel bar 20 mm wide by 80 mm high is used for the beam, compute the deflection at the end of the beam.

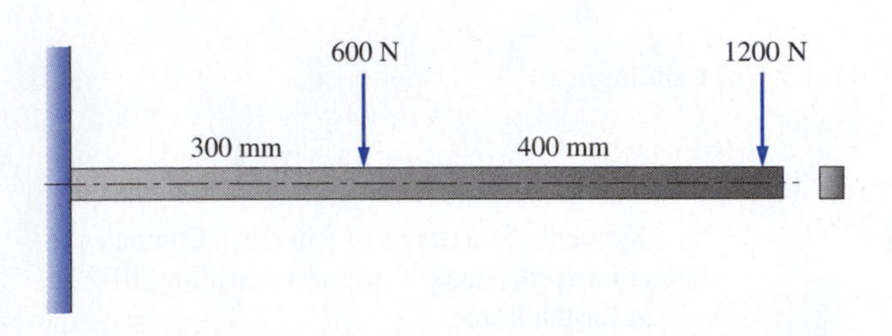

FIGURE P9–57

9–58.M For the beam in Problem 9–57, compute the deflection if the bar is aluminum 2014-T4 rather than steel.

9–59.M For the beam in Problem 9–57, compute the deflection if the bar is magnesium, ASTM AZ 63A-T6, instead of steel.

9–60.E The load shown in Figure P9–60 is carried by a round steel bar having a diameter of 0.800 in. Compute the deflection of the bar at the right end.

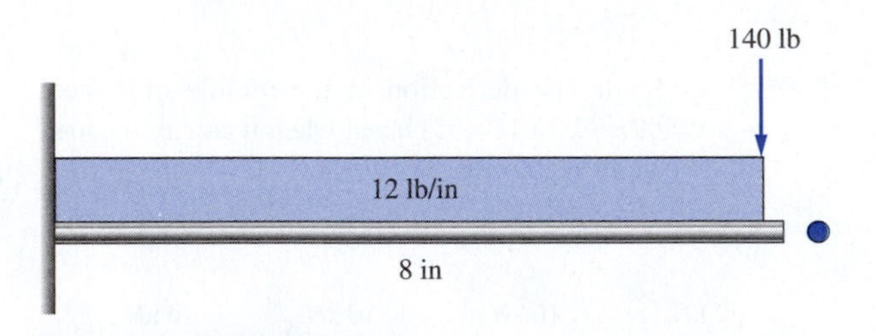

FIGURE P9–60

9–61.M Specify a standard wide-flange steel beam that would carry the loading shown in Figure P9–55 with a deflection at the middle of less than $\frac{1}{360}$ times the length of the beam.

9–62.E It is planned to use an Aluminum Association channel with its legs down to carry the loads shown in Figure P9–62 so that the flat face can contact the load. The maximum allowable deflection is 0.080 in. Specify a suitable channel.

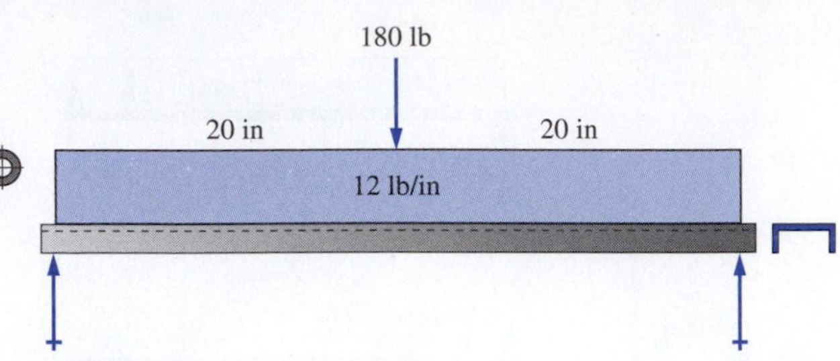

FIGURE P9–62

Superposition—Statically Indeterminate Beams

For Problems 9–63 through 9–70, use the superposition method to determine the reactions at all supports and draw the complete shearing force and bending moment diagrams. Indicate the maximum shearing force and bending moment for each beam.

9–63.M Use Figure P9–36.

9–64.M Use Figure P9–37.

9–65.E Use Figure P9–38.

9–66.E Use Figure P9–66.

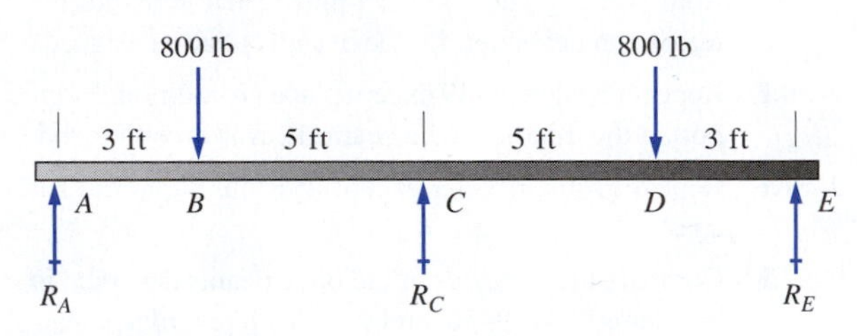

FIGURE P9–66

9–67.E Use Figure P9–67.

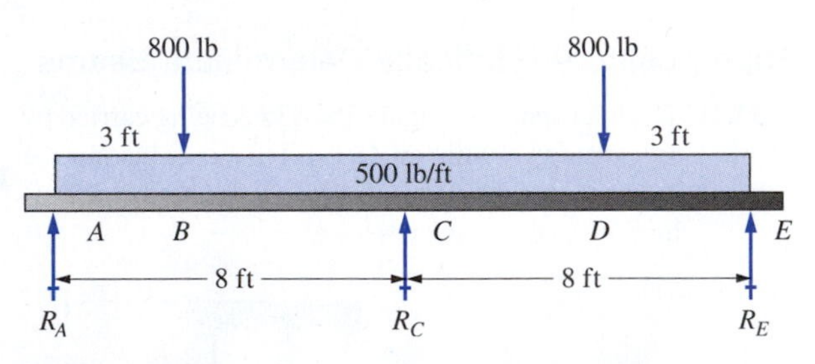

FIGURE P9–67

9–68.M Use Figure P9–68.

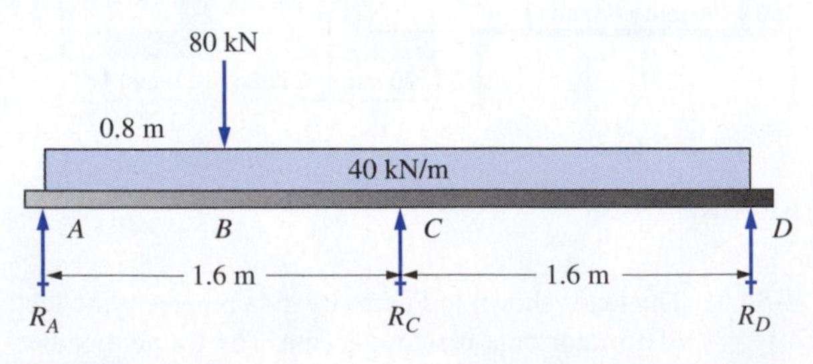

FIGURE P9–68

9–69.M Use Figure P9–69.

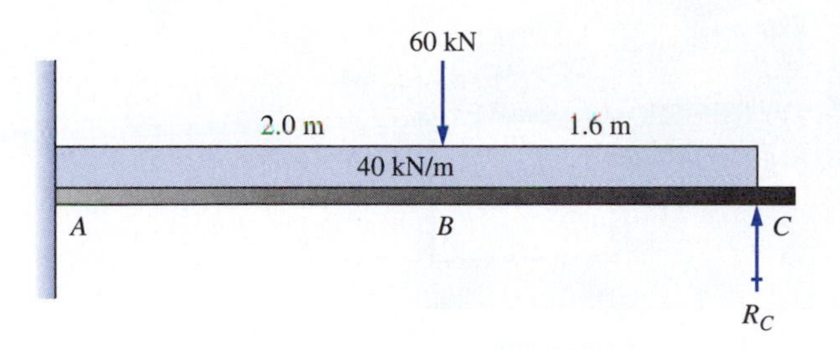

FIGURE P9–69

9–70.E Use Figure P9–70.

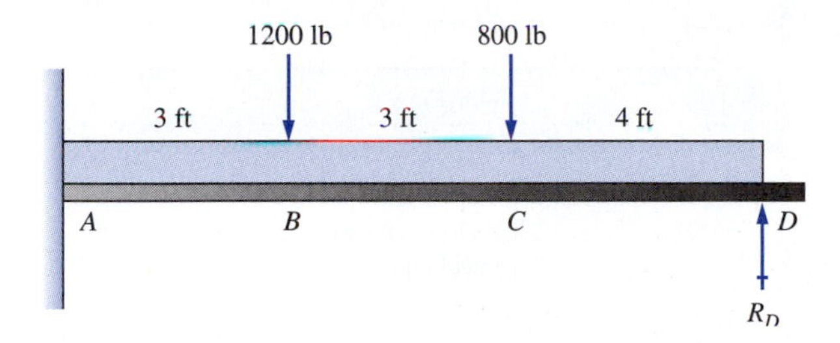

FIGURE P9–70

Design Problems—Stress and Deflection Limits

9–71.E A plank from a wooden deck carries a load like that shown in Case *j* in Appendix A–25 with $w = 100$ lb/ft and $L = 24$ in. The plank is a standard 2 × 6 having a thickness of 1.50 in and a width of 5.50 in with the long dimension horizontal. Would the plank be safe if it is made from No. 2 grade southern pine?

9–72.M Two designs for a diving board are proposed, as shown in Figure P9–72. Compare the designs with regard to shearing force, bending moment, and deflection. Express the deflection in terms of the beam stiffness, *EI*. Note that cases from Appendixes A–23 and A–25 can be used.

9–73.M For each of the proposed diving board designs described in Problem 9–72 and shown in Figure P9–72, complete the design, specifying the material, the cross section, and the final dimensions. The board is to be 600 mm wide.

9–74.M Figure P9–74 shows a roof beam over a loading dock of a factory building. Compute the reactions at the supports and draw the complete shearing force and bending moment diagrams.

9–75.M For the roof beam described in Problem 9–74, complete its design, specifying the material, the cross section, and the final dimensions.

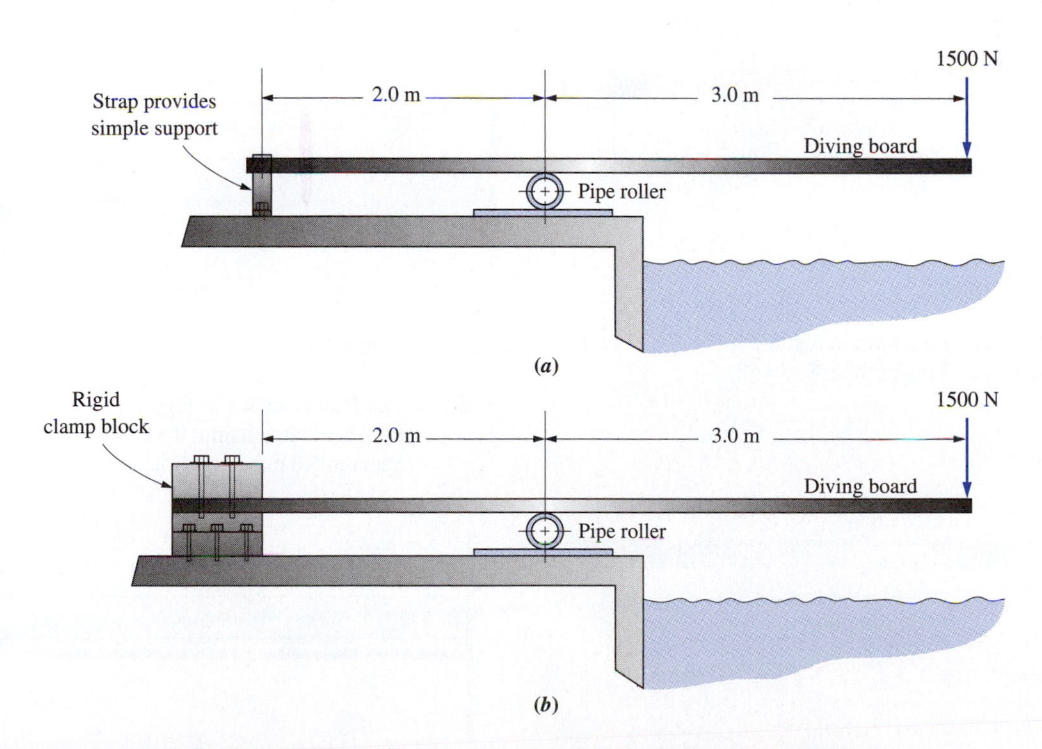

FIGURE P9–72 Proposals for diving boards for Problems 9–72 and 9–73. (a) Simple supports with overhang. (b) Supported cantilever with overhang.

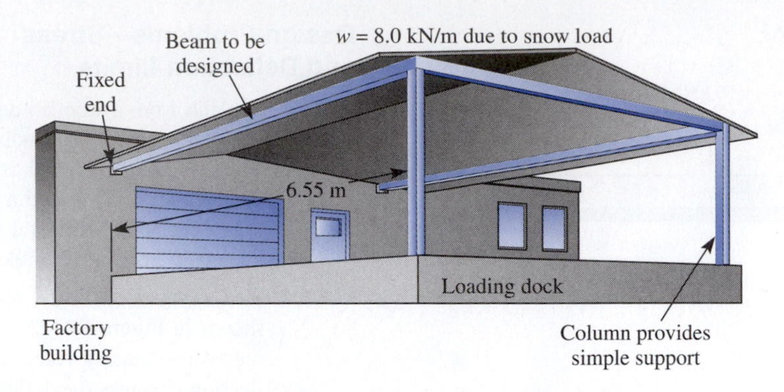

FIGURE P9–74 Roof beam for a loading dock for Problems 9–74 and 9–75.

Successive Integration Method

For the following problems, 9–76 through 9–84, use the general approach outlined in Section 9–7 to determine the equations for the shape of the deflection curves of the beams. Unless otherwise noted, compute the maximum deflection of the beam from the equations and tell where it occurs.

9–76.E The loading is shown in Figure P9–76. The beam is a rectangular steel bar, 1.0 in wide and 2.0 in high.

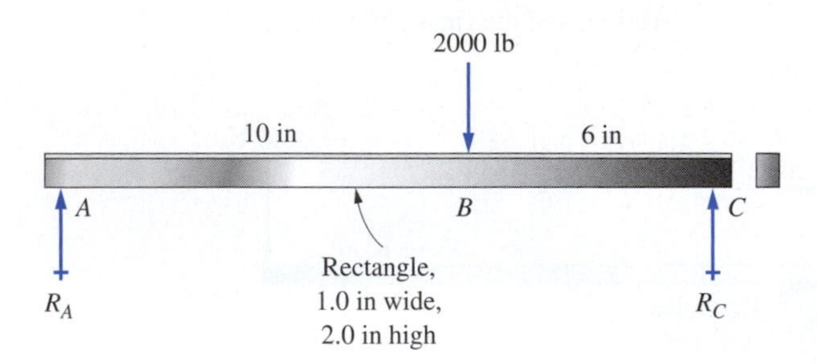

FIGURE P9–76

9–77.E The loading is shown in Figure P9–77. The beam is a steel wide-flange shape W18×55.

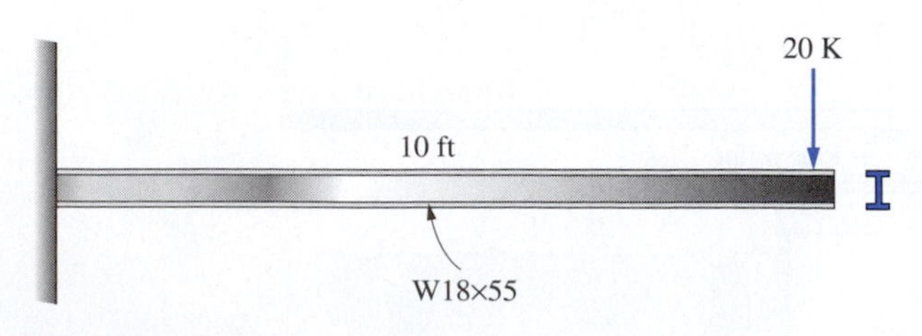

FIGURE P9–77

9–78.C The loading is shown in Figure P9–78. The beam is a $2\frac{1}{2}$-in schedule 40 steel pipe.

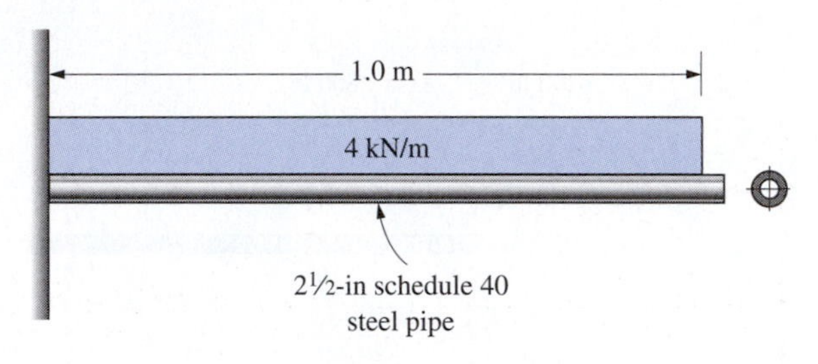

FIGURE P9–78

9–79.E The loading is shown in Figure P9–79. The beam is a steel wide-flange shape W24×76.

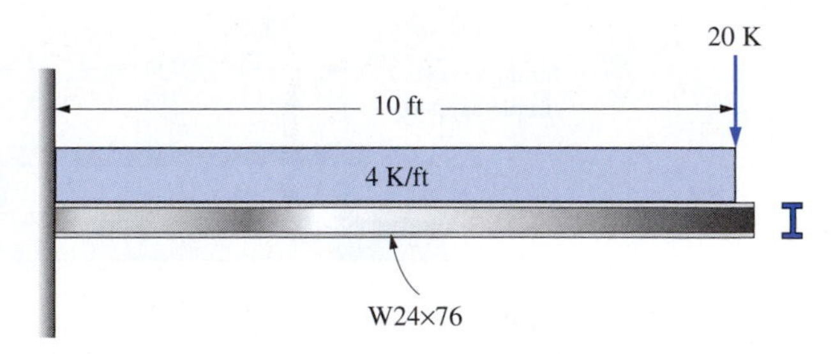

FIGURE P9–79

9–80.M The load is shown in Figure P9–80. Design a round steel bar that will limit the deflection at the end of the beam to 5.0 mm.

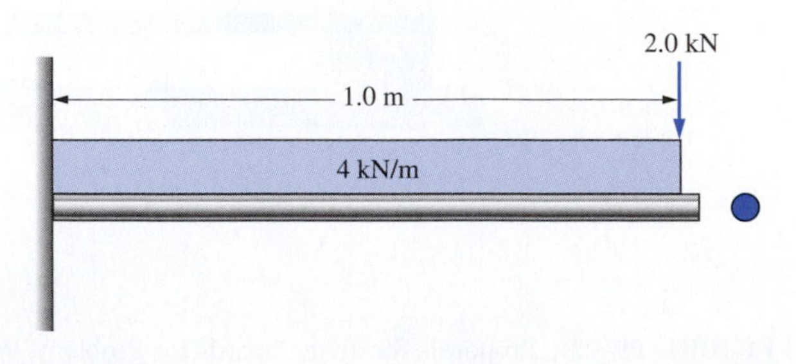

FIGURE P9–80

9–81.C The load is shown in Figure P9–81. Design a steel beam that will limit the maximum deflection to 1.0 mm. Use any shape, including those listed in the Appendix.

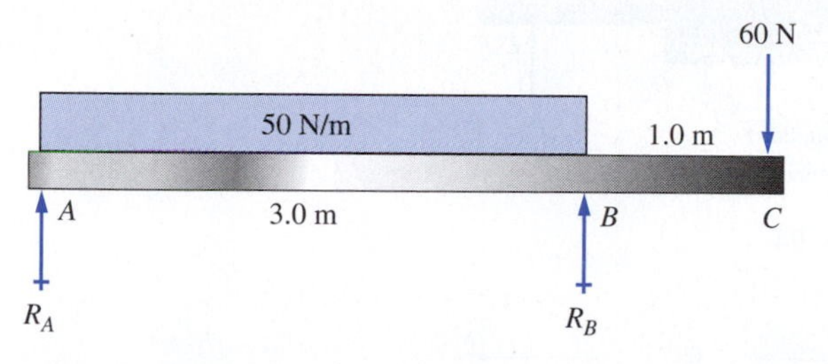

FIGURE P9–81

9–82.C The load is shown in Figure P9–82. Select an aluminum I-beam that will limit the stress to 120 MPa; then compute the maximum deflection in the beam.

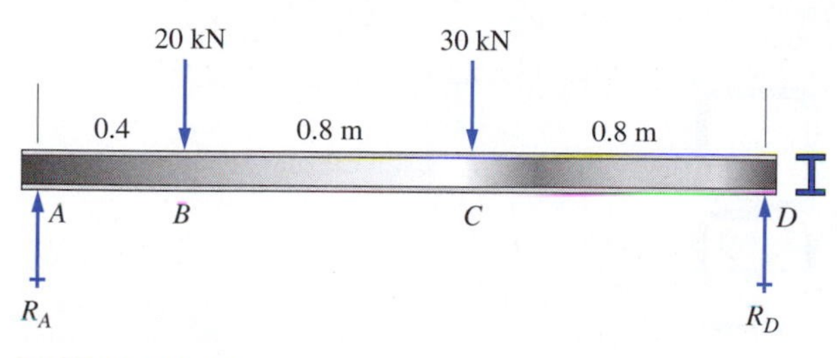

FIGURE P9–82

9–83.C A steel wide-flange beam W360×39 carries the loading shown in Figure P9–83. Compute the maximum deflection between the supports and at each end.

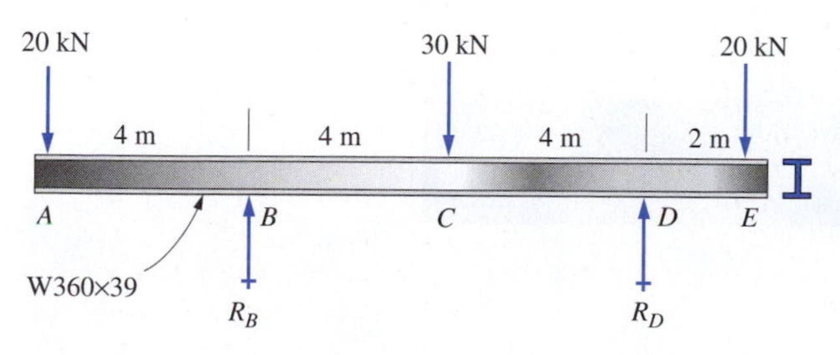

FIGURE P9–83

9–84.M The loading shown in Figure P9–84 represents a steel shaft for a large machine. The loads are due to gears mounted on the shaft. Assuming that the entire shaft will be the same diameter, determine the required diameter to limit the deflection at any gear to 0.13 mm.

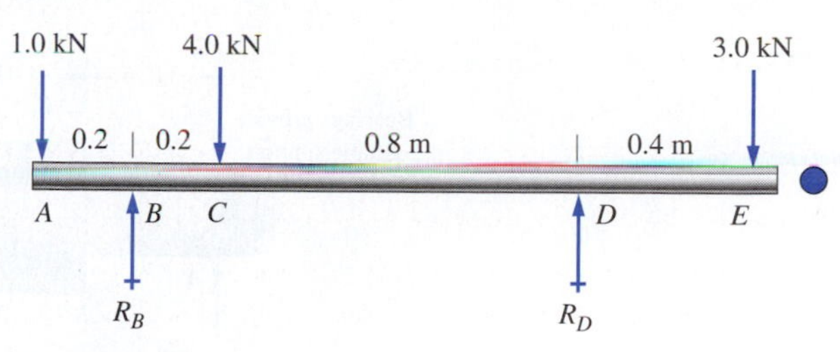

FIGURE P9–84

Moment-Area Method

Use the moment-area method to solve the following problems.

9–85.E For the beam shown in Figure P9–76, compute the deflection at the load. The beam is a rectangular steel bar 1.0 in wide and 2.0 in high.

9–86.E For the beam shown in Figure P9–76, compute the deflection at the middle, 8.0 in from either support. The beam is a rectangular steel bar 1.0 in wide and 2.0 in high.

9–87.E For the beam shown in Figure P9–77, compute the deflection at the end. The beam is a steel wide-flange shape W18×55.

9–88.C For the beam shown in Figure P9–78, compute the deflection at the end. The beam is a $2\frac{1}{2}$-in schedule 40 steel pipe (PIPE64STD).

9–89.E For the beam shown in Figure P9–79, compute the deflection at the end. The beam is a steel wide-flange shape W24×76.

9–90.M For the beam shown in Figure P9–80, compute the deflection at the end. The beam is a round aluminum bar 6061-T6 with a diameter of 100 mm.

9–91.C For the beam shown in Figure P9–81, compute the deflection at the right end, point *C*. The beam is a square steel structural tube $2\times2\times\frac{1}{4}$ (HSS51×51×6.4).

9–92.C For the beam shown in Figure P9–82, compute the deflection at point *C*. The beam is an aluminum I-beam I7×5.800 made from 6061-T6.

9–93.C For the beam shown in Figure P9–83, compute the deflection at point *A*. The beam is a steel wide-flange shape W14×26.

9–94.E Figure P9–94 shows a stepped round steel shaft carrying a single concentrated load at its middle. Compute the deflection under the load.

9–95.E Figure P9–95 shows a composite beam made from steel structural tubing. Compute the deflection at the load.

9–96.E Figure P9–96 shows a cantilever beam made from a steel wide-flange shape W18×55 with cover plates welded to the top and bottom of the beam over the first 6.0 ft. Compute the deflection at the end of the beam.

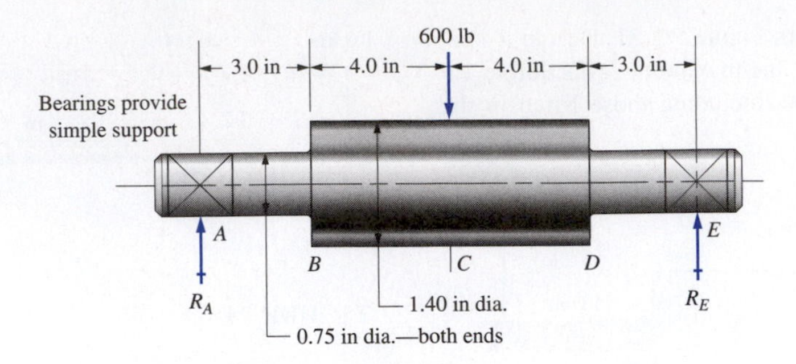

FIGURE P9–94

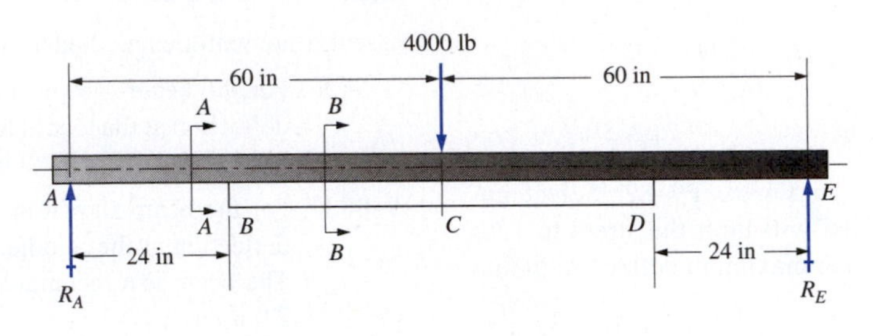

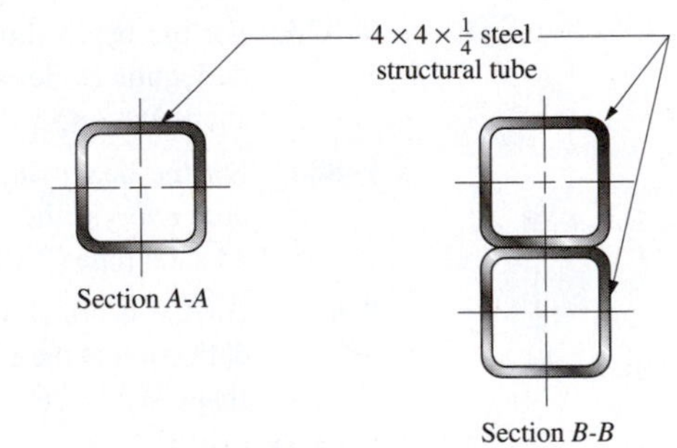

FIGURE P9–95

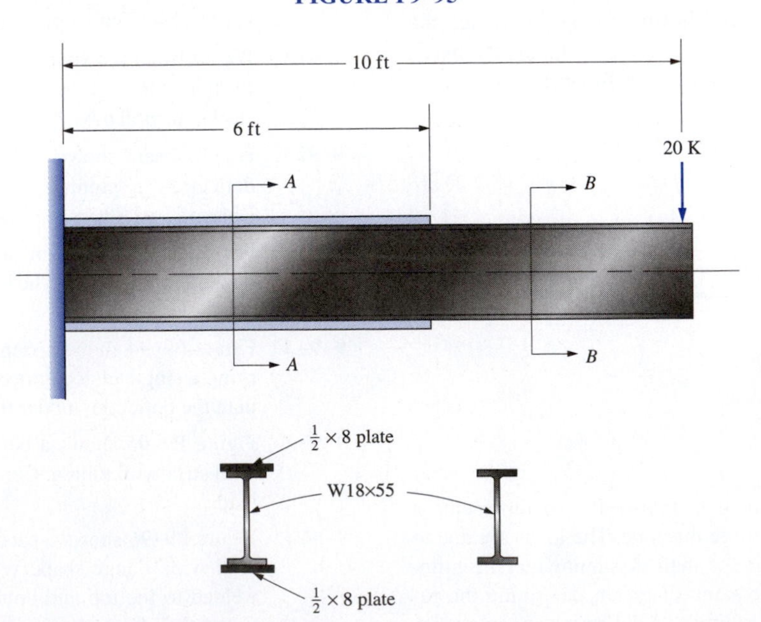

FIGURE P9–96

9–97.E The top member for a gantry crane is made as shown in Figure P9–97. The two end pieces are $3 \times 3 \times \frac{1}{4}$ steel structural tubing. The middle piece is $4 \times 4 \times \frac{1}{2}$. They are combined for a distance of 10.0 in with the 3-in tube fitted tightly inside the 4-in tube and brazed together. Compute the deflection at the middle of the beam.

9–98.M A crude diving board is made by nailing two wood planks together as shown in Figure P9–98. Compute the deflection at the end if the diver exerts a force of 1.80 kN at the end. The wood is No. 2 southern pine.

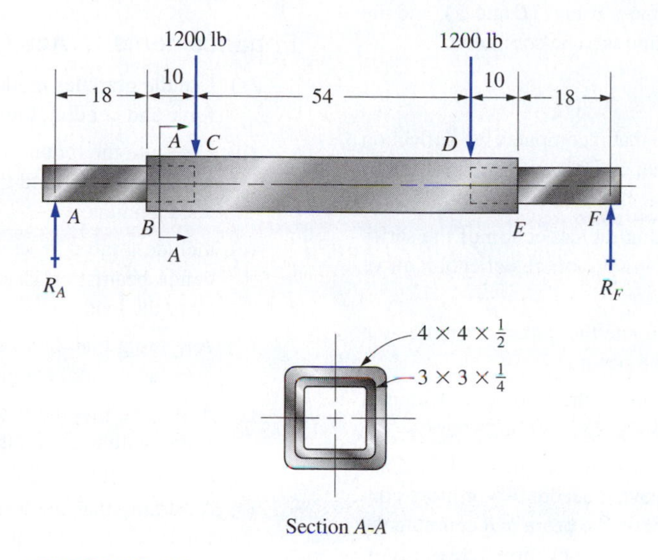

FIGURE P9–97 Gantry crane beam for Problem 9–97.

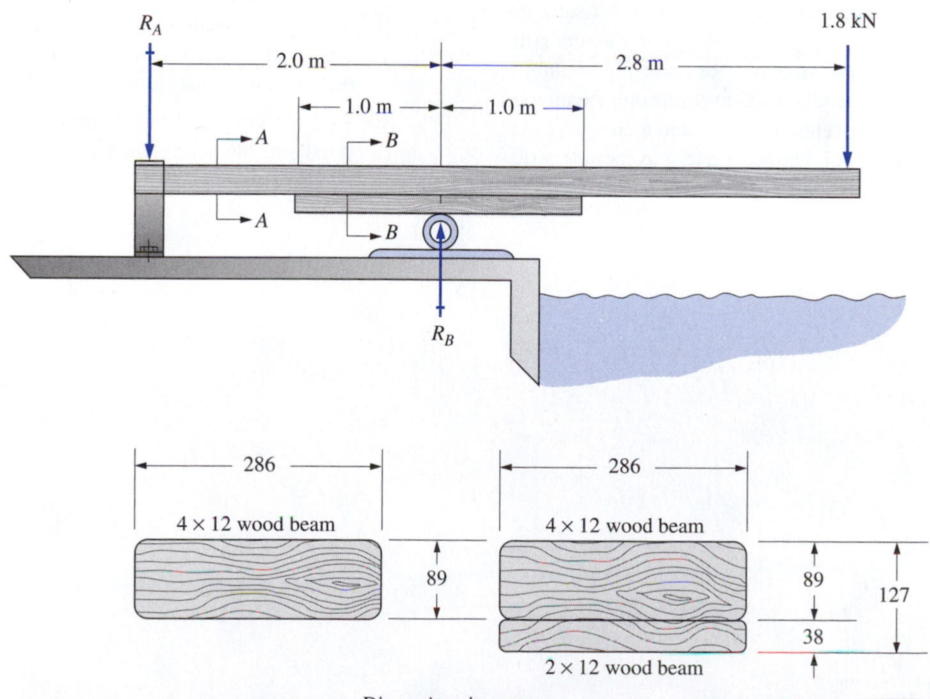

FIGURE P9–98 Diving board for Problem 9–98.

COMPUTER ASSIGNMENTS

1. Write a program or spreadsheet to evaluate the deflection of a simply supported beam carrying one concentrated load between the supports using the formulas given in Case *b* of Appendix A–23. The program should accept input for the length, position of the load, length of the overhang, if any, the beam stiffness values (E and I), and the point for which the deflection is to be computed.

Enhancements

(a) Design the program so that it computes the deflection at a series of points that could permit the plotting of the complete deflection curve.

(b) In addition to computing the deflection of the series of points, have the program plot the deflection curve on a plotter or printer.

(c) Have the program compute the maximum deflection and the point at which it occurs.

2. Repeat Assignment 1 for any of the loading and support patterns shown in Appendix A–23.
3. Write a program similar to that of Assignment 1 for Case *b* of Appendix A–23, but have it accept two or more concentrated loads at any point on the beam and compute the deflection at specified points using the principle of superposition.
4. Combine two or more programs that solve for the deflection of beams for a given loading pattern so that superposition can be used to compute the deflection at any point due to the combined loading. For example, combine Cases *b* and *d* of Appendix A–23 to handle any beam with a combination of concentrated loads and a complete uniformly distributed load. Or, add Case *g* to include a distributed load over only part of the length of the beam.
5. Repeat Assignments 1 through 4 for the cantilevers from Appendix A–24.
6. Write a program or spreadsheet to solve for the critical shearing forces and bending moments for any of the statically indeterminate beam types in Appendix A–25.

Enhancements to Assignment 6

(a) Use the graphics mode to plot the complete shearing force and bending moment diagrams for the beams.

(b) Compute the required section modulus for the beam cross section to limit the stress due to bending a specified amount.

(c) Include a table of section properties for steel wide-flange beams and search for suitable beam sizes to carry the load.

(d) Assuming that the beam cross section will be a solid circular section, compute the required diameter.

(e) Assuming that the beam cross section will be a rectangle with a given ratio of height to thickness, compute the required dimensions.

(f) Assuming that the beam cross section will be a rectangle with a given height *or* thickness, compute the other required dimension.

(g) Assuming that the beam is to be made from wood, in a rectangular shape, compute the required area of the beam cross section to limit the shearing stress to a specified value. Use the special shear formula for rectangular beams from Chapter 8.

(h) Add the computation of the deflection at specified points on the beam using the formulas of Appendix A–25.

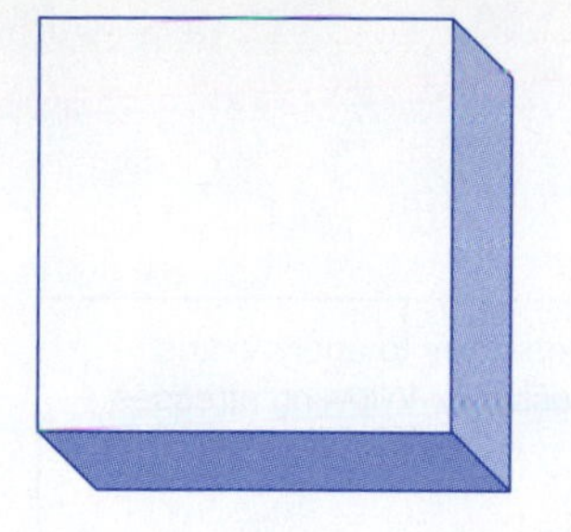

10

Combined Stresses

The Big Picture

Combined Stresses

Discussion Map

- In the previous chapters the objective was to help you acquire the competencies to analyze and design load-carrying members that were subjected to a single type of stress. The following stresses were included:
  - Direct tension stress
  - Direct compression stress
  - Direct shear stress
  - Bearing stress
  - Torsional shear stress
  - Bending stress

  The associated strains and deformations were also considered for the individual cases.
- In Chapter 10 you will consider cases in which more than one type of stress exists in the member at the same time. First we discuss special combinations of stresses that occur frequently in machines, structures, and vehicles. We will consider the following:
  - Combined normal stresses produced by bending, direct tension, and compression
  - Combined normal and shearing stresses that may be produced by bending, axial loads, and torsion.
- The successful analysis of combined stresses requires that you be able to visualize the stress distribution in the material from which the load-carrying member is made. The basic approach we use is to first visualize the stress distribution due to the *components* of the total load that can be analyzed with the same procedures that were developed in earlier chapters. We then combine these components at specific points within the cross section of the member where the maximum combined stress is likely to occur.
- There are many other cases in which you will not know the direction of the maximum stresses. You will have to make whatever analysis you can from the given data and then use additional techniques to compute the maximum stress and determine the direction in which it acts.
- Other cases may require that you compute the stress in a particular direction that may differ from that for which you can conveniently compute the stress distribution.

Activity Look around you to find examples of load-carrying members that are subjected to combined stresses.

- Try to find beams in bending that also carry axial tensile or compressive stresses.
- Power transmission shafts carrying gears, belt sheaves, or chain sprockets typically transmit torque and also experience bending forces. (Figure 1-3)
- Construction and agricultural machinery often contain members subjected to bending, axial, and torsional stresses simultaneously. [Figure 1-1(d)]
- Crane booms that extend upward from a truck bed are subjected to combined bending and axial compression.
- Bicycle and motorcycle frames, cranks, seat supports, and other components carry complex arrays of forces, moments, and torques. [Figure 1-1(b)]
- Automotive frames, suspension members, steering linkages, and drive train components experience multiple types of loading during different operating conditions

such as smooth straight-line driving, turning, braking, acceleration, and driving over rough roads or demanding off-road terrain.

- Interior and body components of automobiles such as seat adjusters, window lifts, windshield wiper systems, hinges, and latches for doors, hoods, or trunks experience combined stresses.
- Aircraft and aerospace vehicles employ stressed skin construction with specially shaped members that distribute loads throughout the framework, producing lightweight, highly optimized structures. [Figure 1-1(e)]
- Structural beams in buildings and bridges which carry loads that are offset from their neutral axes experience both bending and torsional stresses. (Figure 10-3)
- Components for furniture such as chairs, tables, and cabinets for audiovisual equipment typically must withstand combined axial compression and bending.
- Highway signs extending over roadways that must not only support their own weight, but also must resist high wind forces, subject their support posts to combined bending, axial compression, and torsional shear stresses.

FIGURE 10–1
(Source: Pearson Education/PH College.)

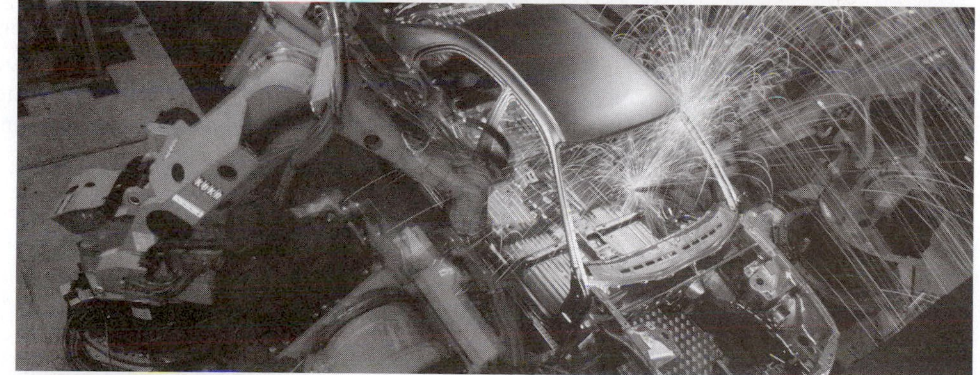

FIGURE 10–2
(Source: Getty Images, Inc.—Taxi.)

FIGURE 10–3
(Source: Omni-Photo Communications, Inc.)

- Frames and housings for industrial machinery, robots, engines, pumps, machine tools, measuring devices, and consumer products typically are subjected to complex stress fields that cannot be analyzed by the application of conventional stress analysis formulas. (Figure 10-2)

In this chapter you will learn methods of analyzing these types of load-carrying members. First you will consider several commonly encountered special cases where combinations, adaptations, and extensions of stress analysis techniques already learned from this book can be applied. That is, combinations of direct axial tensile or compressive normal stresses, bending stresses, direct shearing stress, vertical shearing stresses in beams, and torsional shear stress will be considered. Then, a more general method of combining stresses is developed and applied that is useful for more complex loading and component geometries.

In the general method, you will start with the known stress system at a particular point where any combination of normal stresses (tension or compression) and shearing stresses exist. The known stress element may be found from direct analysis, experimental stress analysis, or computer-aided finite element analysis. From the known stress element, you will learn how to compute the maximum and minimum principal (normal) stresses, the maximum shear stress, and the orientation of the stress element on which those stresses occur. For some stress analyses, it is useful to compute an additional combined stress, called the *von Mises* stress, from the principal stresses. The definition of the von Mises stress will be presented later.

These stresses can be compared with the inherent strength of the material from which the member is made to ensure that the member is safe and that it will perform to its expected requirements. The appropriate material strengths may include ultimate tensile strength, ultimate compressive strength, ultimate shear strength, tensile yield strength, compressive yield strength, yield strength in shear, or endurance (fatigue) strength.

Following courses of study, such as design of machine elements, finite element analysis, experimental stress analysis, aerospace structural design, structural steel design, design of concrete structures, or design of wood structures, extend these methods of stress analysis to even more complex cases. Furthermore, advanced courses in elasticity, plasticity, composite materials, material science, plates and shells, and fracture mechanics will extend your mastery of the wide range of applications you will encounter in your career.

10–1 OBJECTIVES OF THIS CHAPTER

After completing this chapter, you should be able to:

1. Recognize cases for which combined stresses occur.
2. Represent the stress condition on a stress element.
3. Recognize the importance of visualizing the stress distribution over the cross section of a load-carrying member and considering the stress condition at a point.
4. Recognize the importance of free-body diagrams of components of structures and mechanisms in the analysis of combined stresses.
5. Compute the combined normal stress resulting from the application of bending stress with either direct tensile or compressive stresses using the principle of superposition.
6. Evaluate the design factor for combined normal stress, including the properties of either isotropic or nonisotropic materials.

7. Optimize the shape and dimensions of a load-carrying member relative to the variation of stress in the member and its strength properties.
8. Analyze members subjected only to combined bending and torsion by computing the resulting maximum shear stress.
9. Use the maximum shear stress theory of failure properly.
10. Apply the equivalent torque technique to analyze members subjected to combined bending and torsion.
11. Consider the stress concentration factors when using the equivalent torque technique.
12. Understand the development of the equations for combined stresses, from which you can compute the following:
    a. The maximum and minimum principal stresses
    b. The orientation of the principal stress element
    c. The maximum shear stress on an element
    d. The orientation of the maximum shear stress element
    e. The normal stress that acts along with the maximum shear stress
    f. The normal and shear stress that occurs on the element oriented in any direction
13. Construct Mohr's circle for biaxial stress.
14. Interpret the information available from Mohr's circle for the stress condition at a point in any orientation.
15. Use the data from Mohr's circle to draw the principal stress element and the maximum shear stress element.

10–2 THE STRESS ELEMENT

In general, combined stress refers to cases in which two or more types of stress are exerted on a given point at the same time. The component stresses can be either *normal* (i.e., tension or compression) or *shear* stresses.

When a load-carrying member is subjected to two or more different kinds of stresses, the first task is to compute the stress due to each component. Then a decision is made about which point within the member has the highest *combination* of stresses, and the combined stress analysis is completed for that point. For some special cases, it is desired to know the stress condition at a given point regardless of whether or not it is the point of maximum stress. Examples would be near welds in a fabricated structure, along the grain of a wood member, or near a point of attachment of one member to another.

After the point of interest is identified, the stress condition at that point is determined from the classical stress analysis relationships presented in this book, if possible. At times, because of the complexity of the geometry of the member or the loading pattern, it is not possible to complete a reliable stress analysis entirely by computations. In such cases experimental stress analysis can be employed in which strain gages, photoelastic models, or strain-sensitive coatings give data experimentally. Also, with the aid of finite element stress analysis techniques, the stress condition can be determined using computer-based analysis.

Then, after using one of these methods, you should know the information required to construct the *stress element,* as shown in Figure 10–4. The element is assumed to be infinitesimally small and aligned with known directions on the member being analyzed. The complete element, as shown, could have a normal stress (either tensile or compressive) acting on each pair of faces in mutually perpendicular directions, usually named the x and y axes. As

FIGURE 10–4
Complete stress element.

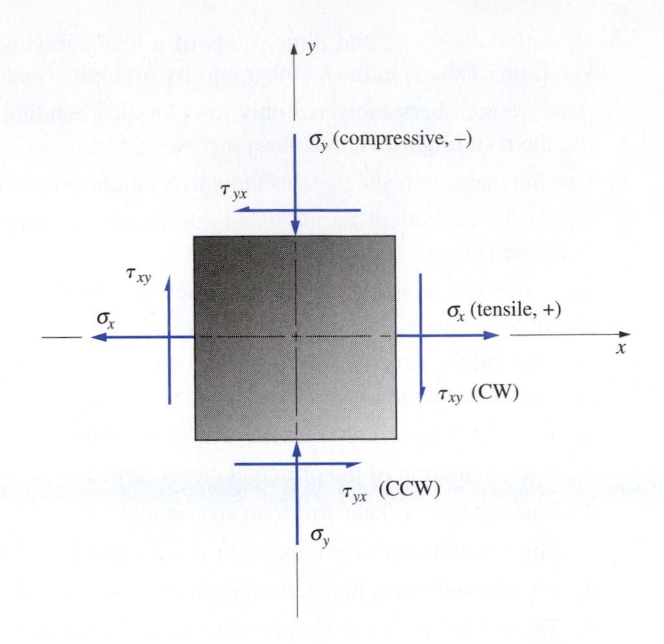

the name *normal stress* implies, these stresses act normal (perpendicular) to the faces. As shown, σ_x is aligned with the x-axis and is a tensile stress tending to pull the element apart. Recall that tensile stresses are considered positive. Then σ_y is shown to be compressive, tending to crush the element. Compressive stresses are considered negative.

In addition, there may be shear stresses acting along the sides of the element as if each side were being cut away from the adjacent material. Recall from the earlier discussion of shearing stresses that a set of four shears, all equal in magnitude, exist on any element in equilibrium. On any two opposite faces the shear stresses will act in opposite directions, thus creating a *couple* that tends to rotate the element. Then there must exist a pair of shear stresses on the adjacent faces producing an oppositely directed couple for the element to be in equilibrium. We will refer to each *pair* of shears using a double subscript notation. For example, τ_{xy} refers to the shear stress acting perpendicular to the x-axis and parallel to the y-axis. Conversely, τ_{yx} acts perpendicular to the y-axis and parallel to the x-axis. Rather than establish a convention for signs of the shear stresses, we will refer to them as *clockwise* (CW) or *counterclockwise* (CCW), according to how they tend to rotate the stress element.

10–3 STRESS DISTRIBUTION CREATED BY BASIC STRESSES

Figures 10–5, 10–6, and 10–7 show direct tension, direct compression, and bending stresses in which normal stresses were developed. It is important to know the distribution of stress within the member as shown in the figures. Also shown are stress elements subjected to these kinds of stresses. The following list contains the principal formulas for computing the value of these stresses.

Direct tension: (see Figure 10–5)	$\sigma = \dfrac{F}{A}$	Uniform over the area (Chapters 1, 3)
Direct compression: (see Figure 10–6)	$\sigma = \dfrac{-F}{A}$	Uniform over the area (Chapters 1, 3)

FIGURE 10–5 Distribution of normal stress for direct tension.

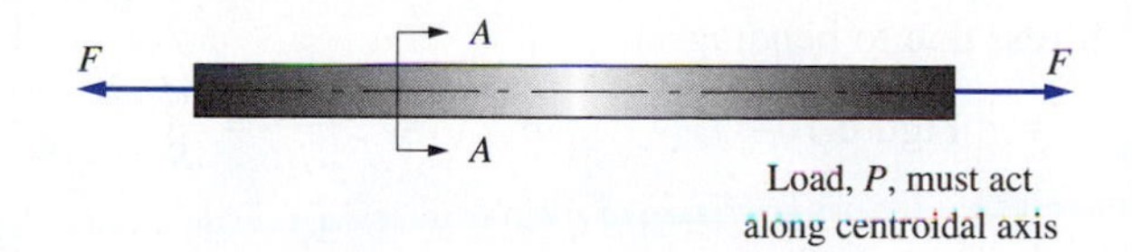

(*a*) Load condition—Direct tension

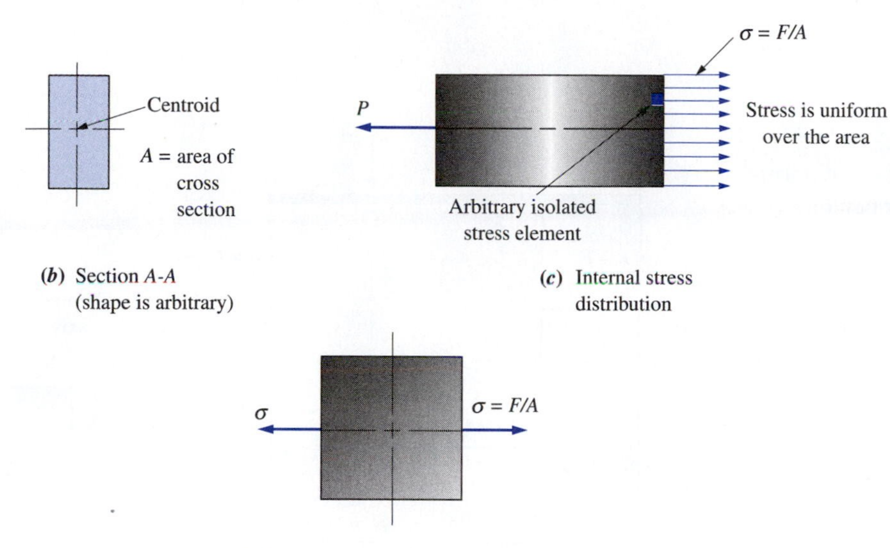

(*b*) Section *A-A* (shape is arbitrary)

(*c*) Internal stress distribution

(*d*) Stress element—Normal tensile stress

FIGURE 10–6 Distribution of normal stress for direct compression.

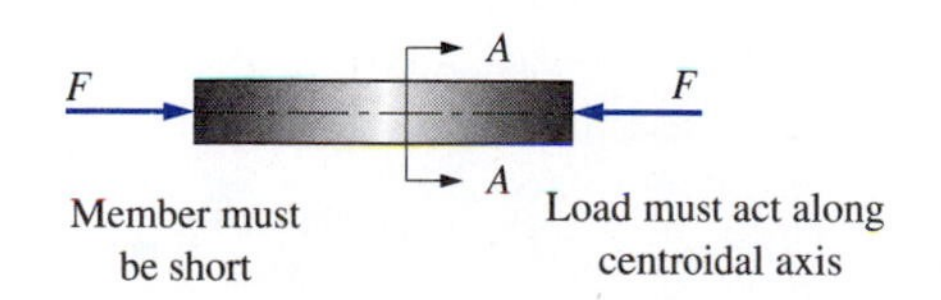

(*a*) Load condition—Direct compressive stress

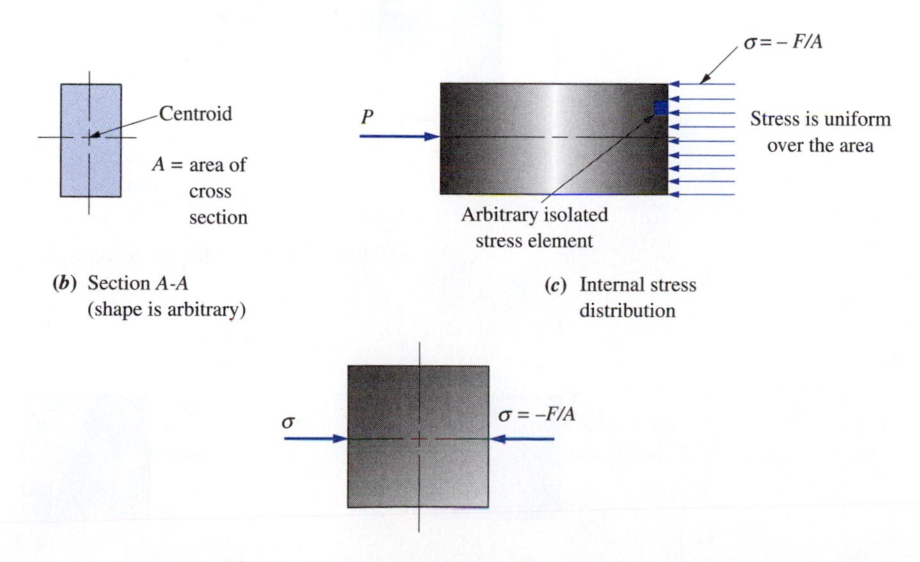

(*b*) Section *A-A* (shape is arbitrary)

(*c*) Internal stress distribution

(*d*) Stress element—Normal compressive stress

Stress due to bending: Maximum stress at outer surfaces

(see Figure 10–7) $$\sigma_{max} = \pm\frac{Mc}{I} = \frac{M}{S}$$ (Chapter 7)

$$\sigma = \pm\frac{My}{I}$$ Bending stress at any point y (Chapter 7)

FIGURE 10–7 Distribution of normal stress for bending.

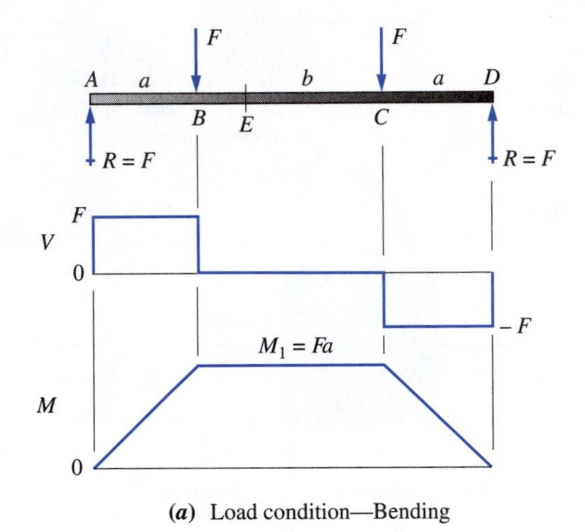

(*a*) Load condition—Bending

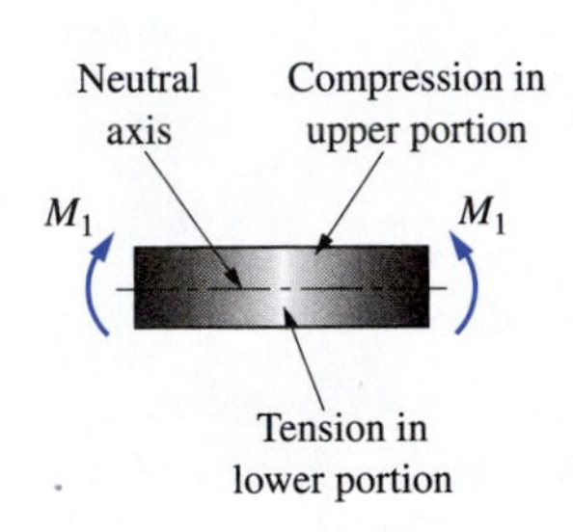

(*b*) Beam segment in positive bending

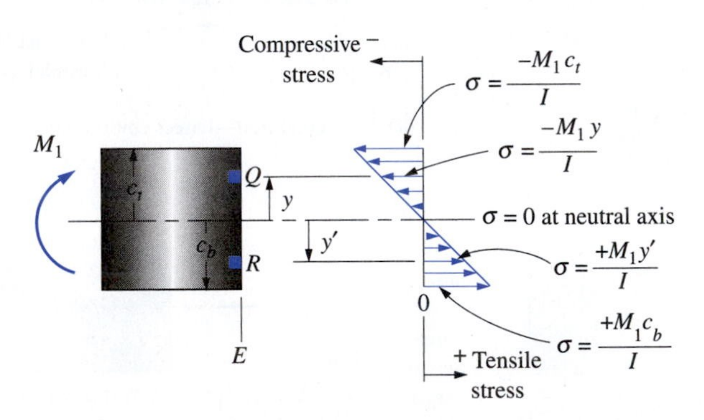

(*c*) Internal stress distribution at section *E*

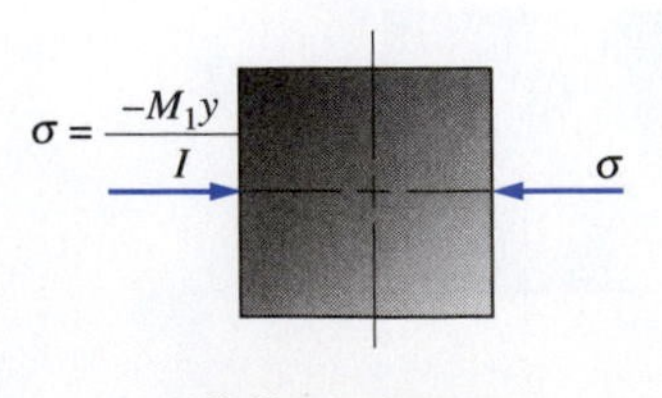

(*d*) Stress element *Q*

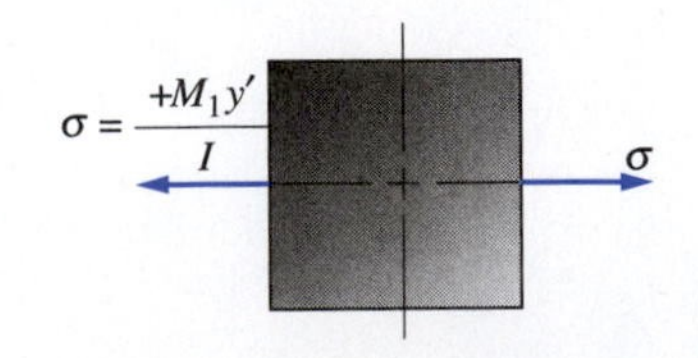

(*e*) Stress element *R*

Figures 10–8, 10–9, and 10–10 show three cases in which shearing stresses are produced along with the stress distributions and the stress elements subjected to these types of stress. The following list contains the principal formulas used to compute shearing stresses.

Direct shear: (see Figure 10–8) $\tau = \dfrac{F}{A_s}$ Uniform over the area (Chapter 3)

Torsional shear: (see Figure 10–9) $\tau_{max} = \dfrac{Tc}{J}$ Maximum at outer surface (Chapter 4)

$\tau = \dfrac{Tr}{J}$ Torsional shear at any radius (Chapter 4)

Shearing stresses in beams: (see Figure 10–10) $\tau = \dfrac{VQ}{It}$ (Chapter 8)

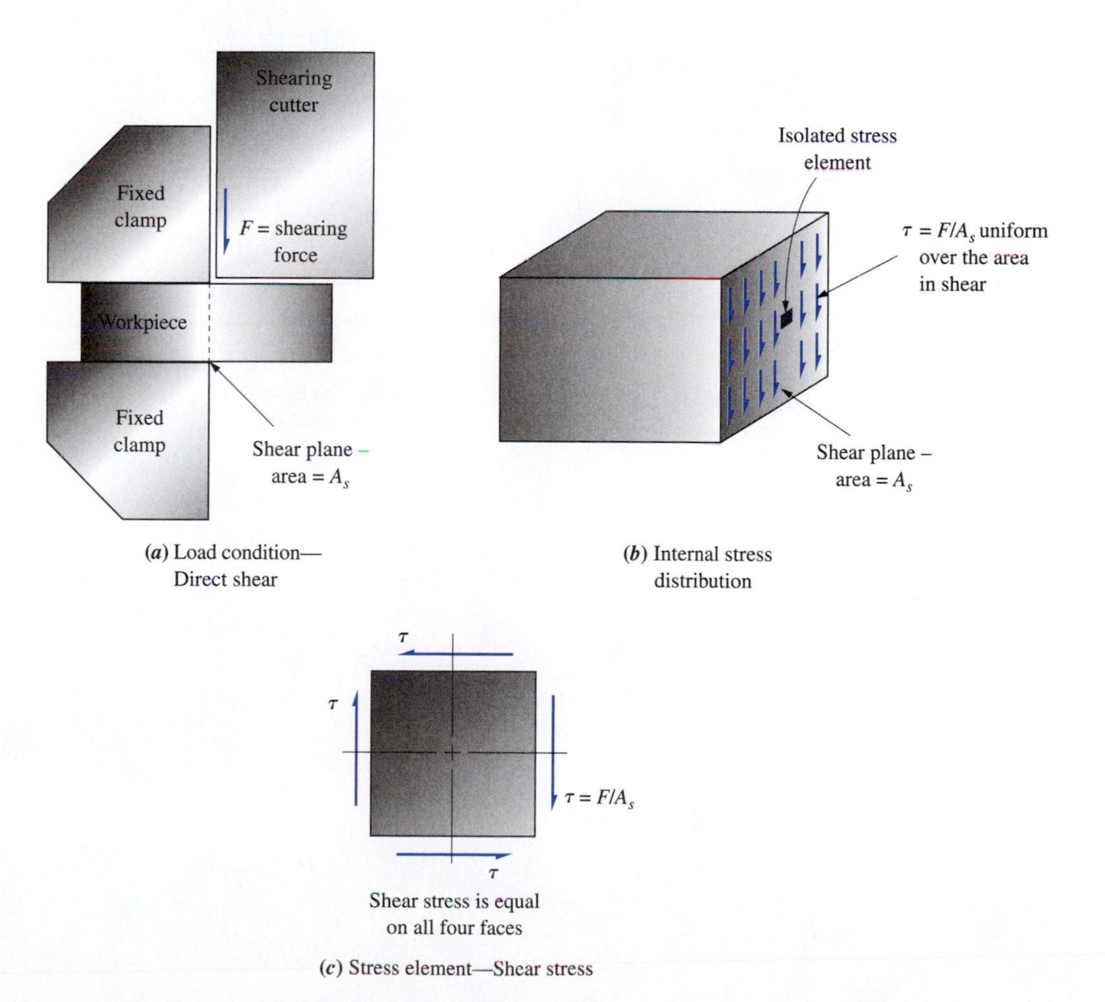

(*a*) Load condition—Direct shear

(*b*) Internal stress distribution

(*c*) Stress element—Shear stress

FIGURE 10–8 Distribution of shearing stress for direct shear.

FIGURE 10–9
Distribution of torsional shear stress in a solid round shaft.

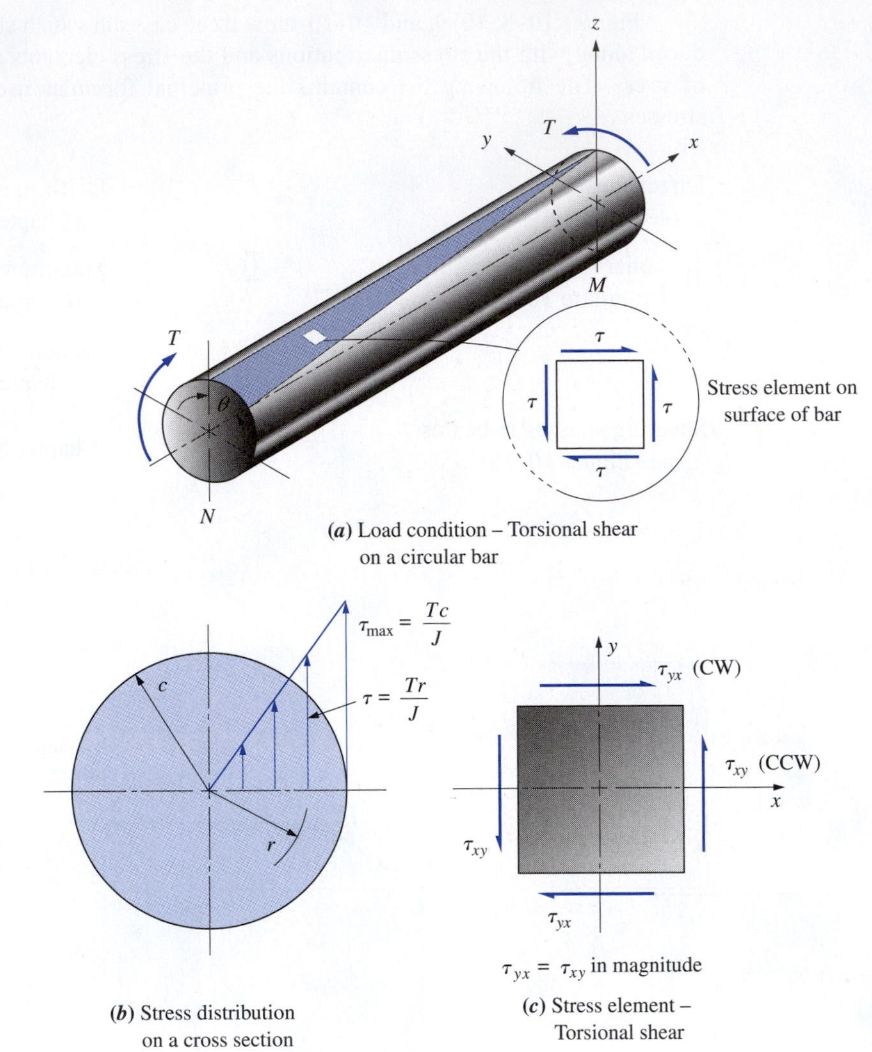

FIGURE 10–10 Distribution of shearing stress in a beam.

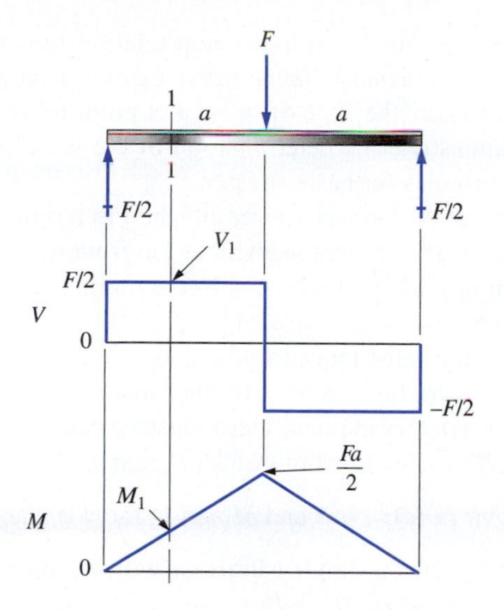

(*a*) Load condition — Beam in bending

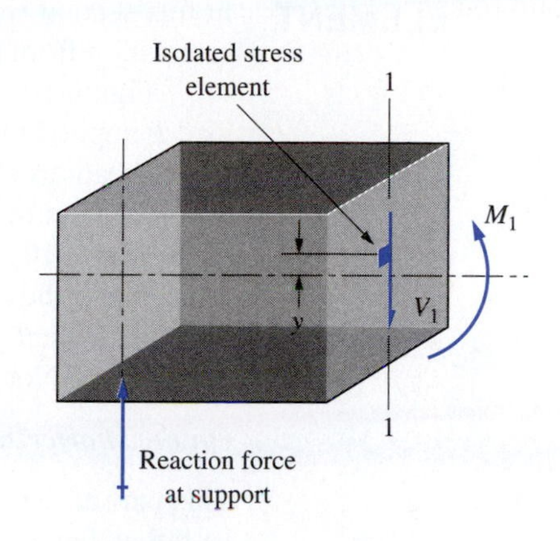

(*b*) Free-body diagram of segment to left of Section 1-1

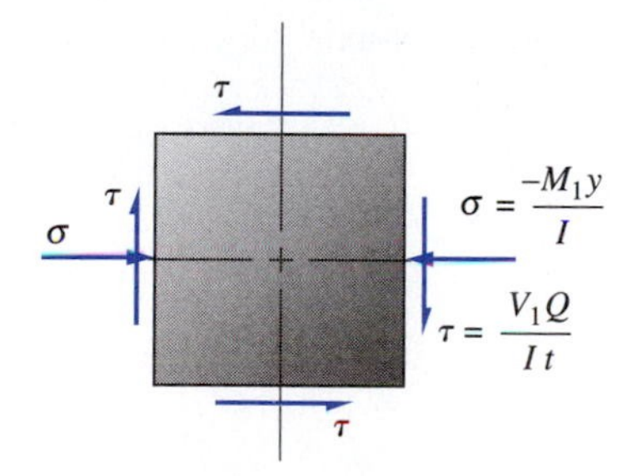

(*c*) Stress element

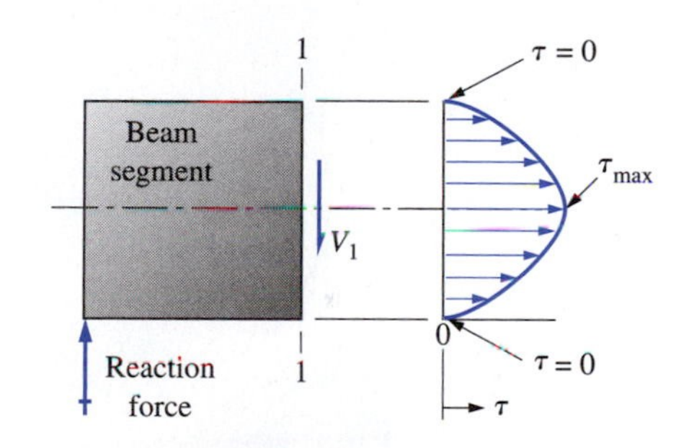

(*d*) Distribution of shearing stress on Section 1-1

10–4 CREATING THE INITIAL STRESS ELEMENT

One major objective of this chapter is to develop relationships from which the *maximum principal stresses* and the *maximum shear stress* can be determined. Before this can be done, it is necessary to know the state of stress at a point of interest in some orientation. In this section we demonstrate the determination of the initial stress condition by direct calculation from the formulas for basic stresses.

Figure 10–11 shows an L-shaped lever attached to a rigid surface with a downward load P applied to its end. The shorter segment at the front of the lever is loaded as a cantilever beam as shown in Figure 10–12(a), with the moment at its left end resisted by the other segment of the lever.

Figure 10–12(b) shows the longer segment as a free-body diagram. At the front, the force P and the torque T are the reactions to the force and moment at the left end of the shorter segment in (a). Then at the rear, there must be reactions M, P, and T to maintain equilibrium. Note carefully the directions of M, P, and T.

On the shorter segment in (a): P and M_1 act in the y–z plane

This part of the lever acts like a simple cantilever with the maximum bending moment at its left end and having a value of $M_1 = P \cdot a$.

On the longer segment in (b): P and M_2 act in the x–z plane
T acts in the y–z plane producing torsion about the x-axis

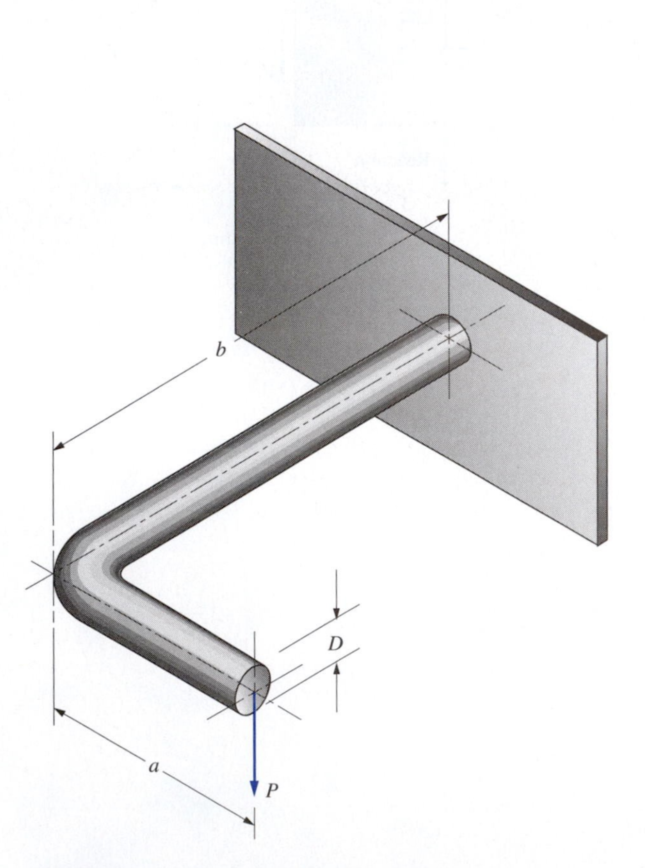

FIGURE 10–11 L-shaped lever illustrating combined stresses.

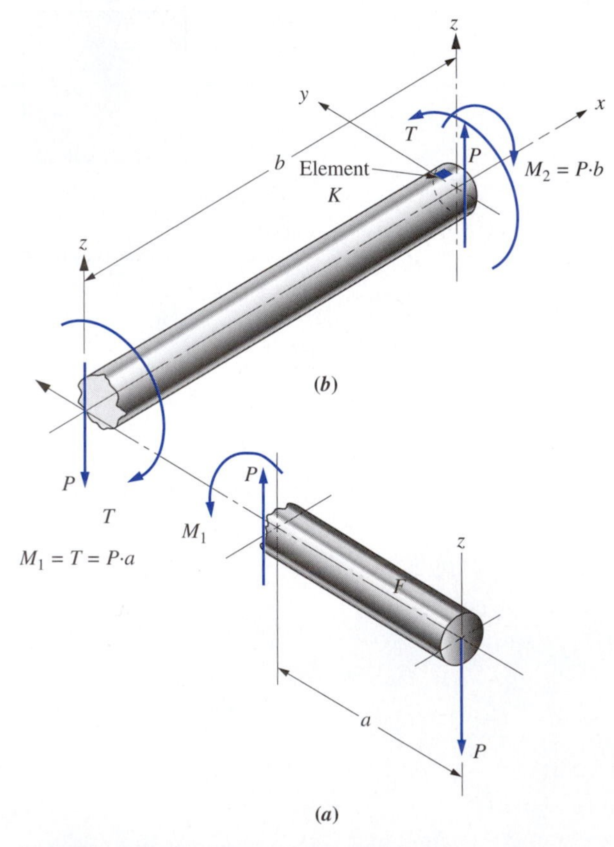

FIGURE 10–12 Free-body diagrams of segments of an L-shaped lever. (a) Shorter segment of lever. (b) Longer segment of lever.

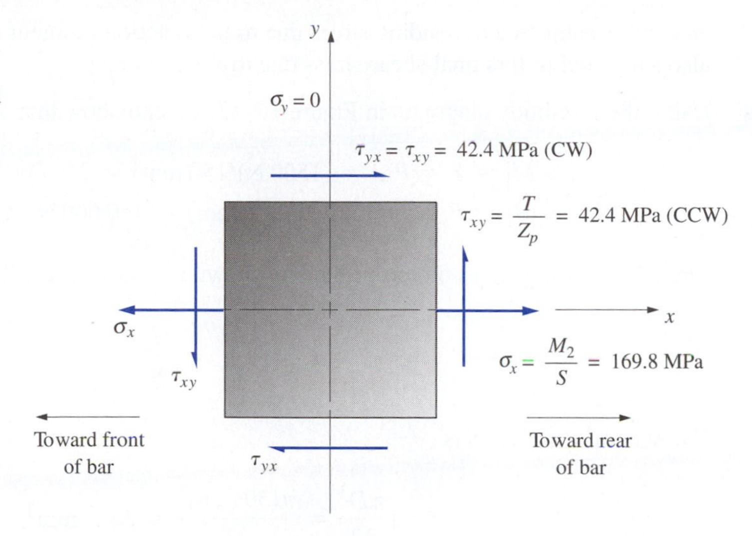

FIGURE 10–13 Stress on element *K* from Figure 10–12 with data from Example Problem 10–1.

This part of the lever acts like a cantilever with added torsion. The magnitude of the torque, *T*, is equal to $M_1 = P \cdot a$, and is uniform along the entire length of the longer segment of the bar. The magnitude of M_2 has the value: $M_2 = P \cdot b$. It is largest at the far end where the bar is attached rigidly to the support plate.

The bar is subjected to a combined stress condition with the following kinds of stresses:

Bending stress due to the bending moment

Torsional shear stress due to the torque

Shearing stress due to the vertical shearing force

We can conclude that one of the points where the stress is likely to be the highest is on the top of the long segment near the support. Called element *K* in Figure 10–12(b), it would see the maximum tensile bending stress and the maximum torsional shear stress. But the vertical shearing stress would be zero because it is at the outer surface away from the neutral axis.

Element *K* is then subjected to a combined stress, as shown in Figure 10–13. The tensile normal stress, σ_x, is directed parallel to the *x*-axis along the top surface of the bar. The applied torque tends to produce a shear stress τ_{xy} in the negative *y*-direction on the face toward the front of the bar and in the positive *y*-direction on the face toward the rear. Together they create a counterclockwise couple on the element. The element is completed by showing the shear stresses τ_{yx} producing a clockwise couple on the other faces.

Example Problem 10–1 includes illustrative calculations for the values of the stresses that are shown in Figure 10–13.

Example Problem 10–1 Figure 10–11 shows an L-shaped lever carrying a downward force at its end. Compute the stress condition that exists on a point on top of the lever near the support. Let $P = 1500$ N, $a = 150$ mm, $b = 300$ mm, and $D = 30$ mm. Show the stress condition on a stress element.

Solution

Objective: Compute the stress condition and draw the stress element.

Given: Geometry and loading in Figure 10–11. $P = 1500$ N
Dimensions: $a = 150$ mm, $b = 300$ mm, $D = 30$ mm.

Analysis: Figure 10–12 shows the lever separated into two free-body diagrams. The point of interest is labeled "Element *K*" at the right end of the lever where it joins the fixed support. The element is subjected to bending stress due to the reaction moment at the support, M_2. It is also subjected to torsional shear stress due to the torque, T.

Results Using the free-body diagrams in Figure 10–12, we can show that

$$M_1 = T = P \cdot a = (1500\ \text{N})(150\ \text{mm}) = 225\,000\ \text{N}\cdot\text{mm}$$
$$M_2 = P \cdot b = (1500\ \text{N})(300\ \text{mm}) = 450\,000\ \text{N}\cdot\text{mm}$$

Then the bending stress on top of the bar, shown as element K in Figure 10–12(b), is

$$\sigma_x = \frac{M_2 c}{I} = \frac{M_2}{S}$$

The section modulus S is

$$S = \frac{\pi D^3}{32} = \frac{\pi(30\ \text{mm})^3}{32} = 2651\ \text{mm}^3$$

Then

$$\sigma_x = \frac{450\,000\ \text{N}\cdot\text{mm}}{2651\ \text{mm}^3} = 169.8\ \text{N/mm}^2 = 169.8\ \text{MPa}$$

The torsional shear stress is at its maximum value all around the outer surface of the bar, with the value of

$$\tau = \frac{T}{Z_p}$$

The polar section modulus Z_p is

$$Z_p = \frac{\pi D^3}{16} = \frac{\pi(30\ \text{mm})^3}{16} = 5301\ \text{mm}^3$$

Then

$$\tau = \frac{225\,000\ \text{N}\cdot\text{mm}}{5301\ \text{mm}^3} = 42.4\ \text{N/mm}^2 = 42.4\ \text{MPa}$$

Comment The bending and shear stresses are shown on the stress element K in Figure 10–13. This is likely to be the point of maximum combined stress, which will be discussed later in this chapter. At a point on the side of the bar on the y-axis, a larger shearing stress would exist because the maximum torsional shear stress combines with the maximum vertical shear stress due to bending. But the bending stress there is zero. That element should also be analyzed.

10–5 COMBINED NORMAL STRESSES

This section examines one type of combined stress where the point of interest is subjected only to normal stress, that is, tension or compression. The following section addresses another special case of combined stress where combined normal and shear stresses act on the point of interest. Then the balance of the chapter develops and applies the more general case of combined stress where equations for the maximum normal stress, minimum normal stress, and maximum shear stress are computed for any combination of known stresses on a particular stress element. A semigraphical method called *Mohr's circle* is then used to facilitate the computations and to help you visualize the full range of possible stress conditions that can exist at a particular point in a load-carrying member.

The first combination to be considered is bending with direct tension or compression. In any combined stress problem, it is helpful to visualize the stress distribution caused by the various components of the total stress pattern. You should review Section 10–3 for summaries of the stress distribution for bending and direct tension and compression. Notice that bending results in tensile and compressive stresses, as do both direct tension and direct compression. Since the same kind of stresses are produced, a simple algebraic sum of the stresses produced at any point is all that is required to compute the resultant stress at that point. This process is called *superposition*.

Guidelines for Solving Problems with Combined Normal Stresses

These guidelines pertain to situations in which two or more loads or components of loads act in a manner that produces normal stresses (tension and/or compression) on the load-carrying member. In general, loads producing bending stress or direct tension or compression are to be included. The objective is to compute the maximum combined stress in the member. Let tensile stresses be positive (+) and compressive stresses be negative (−).

1. Draw the free-body diagram for the load-carrying member and compute the magnitude of all applied forces.
2. For any force that acts at an angle to the neutral axis of the member, resolve the force into components perpendicular and parallel to the neutral axis.
3. Forces or components acting in a direction coincident with the neutral axis will produce either direct tension or direct compression with uniform stress distribution across the cross section. Compute these stresses.
4. Forces or components acting perpendicular to the neutral axis cause bending stresses. Using the methods of Chapter 5, determine the bending moments caused by these forces, either individually or in combination. Then, for the section subjected to the largest bending moment, compute the bending stress from the flexure formula, $\sigma = M/S$. The maximum stress will occur at the outermost fibers of the cross section. Note at which points the stress is tensile and which is compressive.
5. Forces or components acting parallel to the neutral axis but whose line of action is away from that axis also cause bending. The bending moment is the product of the force times the perpendicular distance from the neutral axis to the line of action of the force. Compute the bending stress due to such moments at any section where the combined stress may be the maximum.
6. Considering all normal stresses computed from steps 1–5, use superposition to combine them at any point within any cross section where the combined stress may be the maximum. Superposition is accomplished by algebraically summing all stresses acting at a point, taking care to note if each component stress is tensile (+) or compressive (−). It may be necessary to evaluate the stress condition at two or more points if it is not obvious where the maximum combined stress occurs. In general, the superposition process can be expressed as,

$$\sigma_{comb} = \pm\frac{F}{A} \pm \frac{M}{S} \qquad (10\text{–}1)$$

where the term, $\pm F/A$, includes all direct tensile and compressive stresses acting at the point of interest and the term, $\pm M/S$, includes all bending stresses acting at that point. The sign for each stress must be logically determined from the load causing the individual stress.

7. The maximum combined stress on the member can then be compared with the design stress for the material from which the member is to be made to compute the resulting design factor and to assess the safety of the member. For isotropic materials, either tensile or compressive stress could cause failure, whichever is maximum. For nonisotropic materials having different strengths in tension and compression, it is necessary to compute the resulting design factor for both the tensile and the compressive stress to determine which is critical. Also, in general, it will be necessary to consider the stability of those parts of members subjected to compressive stresses by analyzing the tendency for buckling or local crippling of the member. See Chapter 11 for buckling of column-like compression members. The analysis for crippling and buckling of parts of members will require reference to other sources. See the references at the end of Chapters 7, 8, and 10.

An example of a member in which both bending and direct tensile stresses are developed is shown in Figure 10–14. The two horizontal beams support a 10 000-lb load by means of the cable assembly. The beams are rigidly attached to columns, so that they act as cantilever beams. The load at the end of each beam is equal to the tension in the cable. Figure 10–15 shows that the vertical component of the tension in each cable must be 5000 lb. That is,

$$F \cos 60° = 5000 \text{ lb}$$

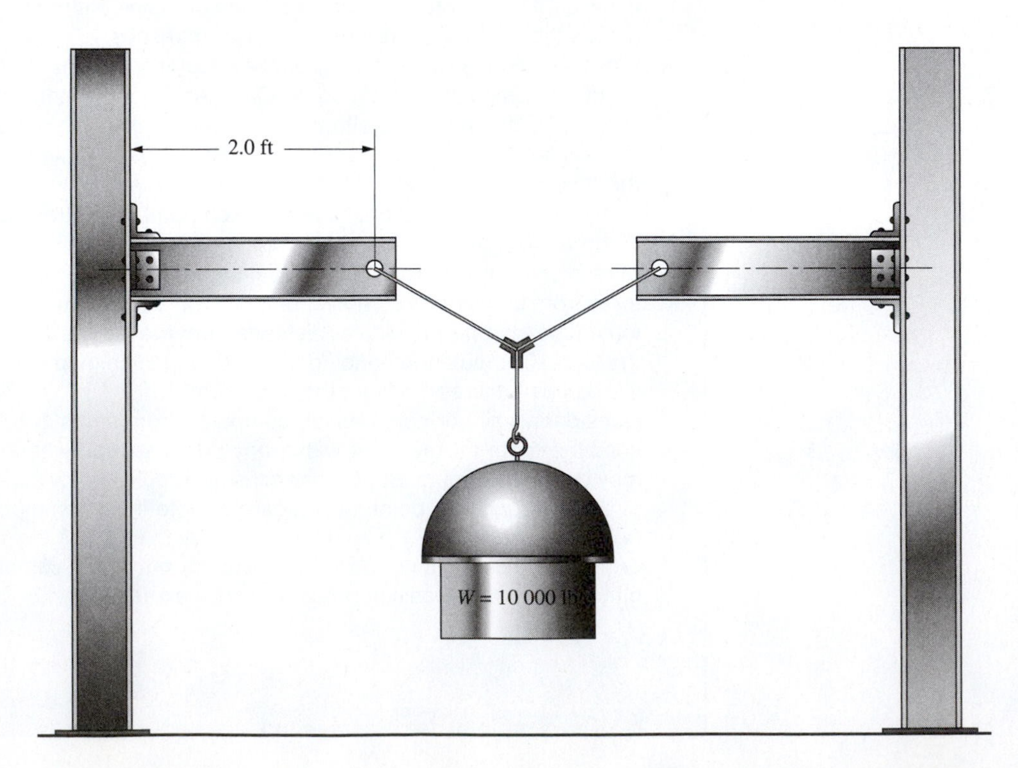

FIGURE 10–14 Two cantilever beams subjected to combined bending and axial tensile stresses.

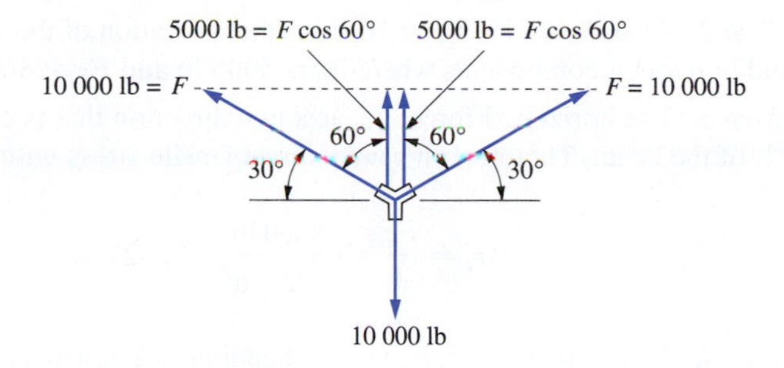

FIGURE 10–15 Force analysis on cable system.

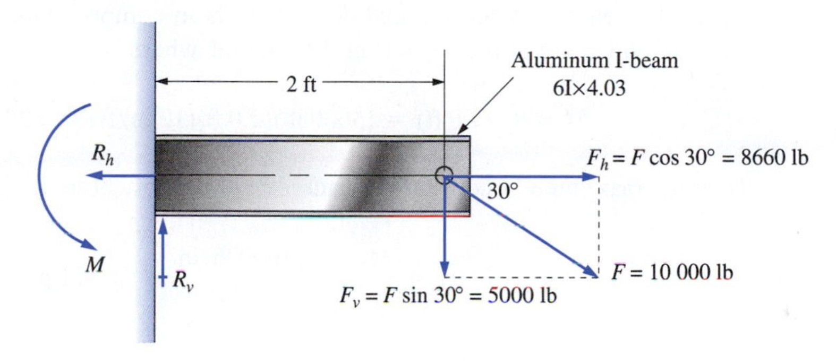

FIGURE 10–16 Forces applied to each beam.

and the total tension in the cable is

$$F = \frac{5000 \text{ lb}}{\cos 60°} = 10\,000 \text{ lb}$$

This is the force applied to each beam, as shown in Figure 10–16.

Example Problem 10–2 completes the analysis of the combined stress condition in each of the horizontal cantilever beams in Figure 10–14.

Example Problem 10–2 For the system shown in Figure 10–14, determine the maximum combined tensile or compressive stress in each horizontal cantilever beam when a static load of 10 000 lb is suspended from the cable system between them. The beams are standard Aluminum Association I-beams, 6I×4.03. Then, if the beams are to be made of 6061-T6 aluminum alloy, compute the resulting design factor.

Solution

Objective Compute the maximum combined stress and the resulting design factor for the horizontal cantilever beams in Figure 10–14.

Given Load weighs 10 000 lb. Force system detail shown in Figure 10–15. Tension in each cable is 10 000 lb. The 2.0-ft-long beams are aluminum 6I×4.03 made from 6061-T6 aluminum alloy. From Appendix A–11, the properties of the beam are $A = 3.427 \text{ in}^2$ and $S = 7.33 \text{ in}^3$.

Analysis The tension in the cable attached to the end of each cantilever beam will tend to cause direct tensile stress combined with bending stress in the beam. Use the *Guidelines for solving problems with combined normal stresses* outlined in this section.

Results ***Step 1.*** Figure 10–16 shows the free-body diagram of one beam with the 10 000 lb force applied by the cable to the end of the beam. The reaction at the left end where the beam is rigidly attached to the column consists of a vertical reaction force, a horizontal reaction force, and a counterclockwise moment.

Step 2. Also shown in Figure 10–16 is the resolution of the 10000 lb force into vertical and horizontal components where $F_v = 5000$ lb and $F_h = 8660$ lb.

Step 3. The horizontal force, F_h, acts in a direction that is coincident with the neutral axis of the beam. Therefore, it causes direct tensile stress with a magnitude of

$$\sigma_t = \frac{F_h}{A} = \frac{8660\ \text{lb}}{3.427\ \text{in}^2} = 2527\ \text{psi}$$

Step 4. The vertical force, F_v, causes bending in a downward direction such that the top of the beam is in tension and the bottom is in compression. The maximum bending moment will occur at the support at the left end, where

$$M = F_v(2.0\ \text{ft}) = (5000\ \text{lb})(2.0\ \text{ft})(12\ \text{in/ft}) = 120\,000\ \text{lb}\cdot\text{in}$$

Then the maximum bending stress caused by this moment is

$$\sigma_b = \frac{M}{S} = \frac{120\,000\ \text{lb}\cdot\text{in}}{7.33\ \text{in}^3} = 16\,371\ \text{psi}$$

A stress of this magnitude occurs as a tensile stress at the top surface and as a compressive stress at the bottom surface of the beam at the support.

Step 5. This step does not apply to this problem because there is no horizontal force offset from the netural axis.

Step 6. It can be reasoned that the maximum combined stress occurs at the top surface of the beam at the support because both the direct tensile stress computed in step 3 and the bending stress computed in step 4 are tensile at that point. Therefore, they will add together. Using superposition, as defined in Equation 10–1,

$$\sigma_{\text{top}} = 2527\ \text{psi} + 16\,371\ \text{psi} = 18\,898\ \text{psi tension}$$

For comparison, the combined stress at the bottom surface of the beam is

$$\sigma_{\text{bot}} = 2527\ \text{psi} - 16\,371\ \text{psi} = -13\,844\ \text{psi compression}$$

Figure 10–17 shows a set of diagrams that illustrate the process of superposition. Part (a) is the stress in the beam caused by bending. Part (b) shows the direct tensile stress due to F_h. Part (c) shows the combined stress distribution.

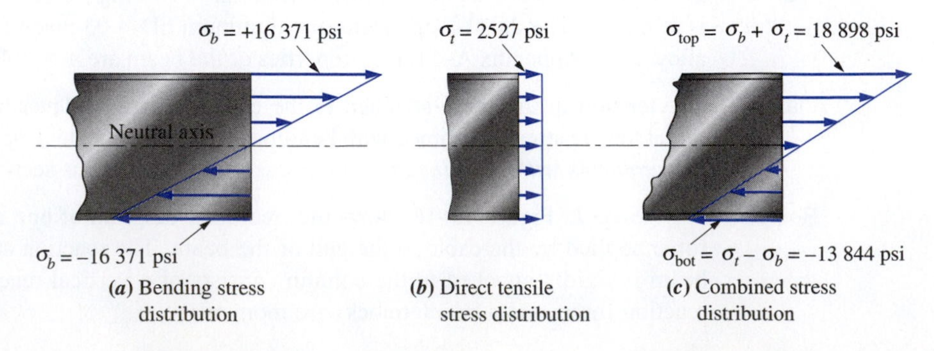

FIGURE 10–17 Diagram of the superposition principle applied to the beams of Figure 10–14.

Step 7. Because the load is static, we can compute the design factor based on the yield strength of the 6061-T6 aluminum alloy, where $S_y = 40\,000$ psi (Appendix A–18). Then,

$$N = s_y/\sigma_{top} = (40\,000 \text{ psi})/(18\,898 \text{ psi}) = 2.11$$

Comment This should be adequate for a purely static load. If there is uncertainty about the magnitude of the load or if there is a possibility of the load being applied with some shock or impact, a higher design factor would be preferred.

Example Problem 10–3 A picnic table in a park is made by supporting a circular top on a pipe that is rigidly held in concrete in the ground. Figure 10–18 shows the arrangement. Compute the maximum stress in the pipe if a person having a mass of 135 kg sits on the edge of the table. The pipe is made of an aluminum alloy and has an outside diameter of 170 mm and an inside diameter of 163 mm. If the aluminum is 6061-T4, compute the resulting design factor based on both yield strength and ultimate strength. Then comment on the suitability of the design.

FIGURE 10–18 Picnic table supported by a pipe.

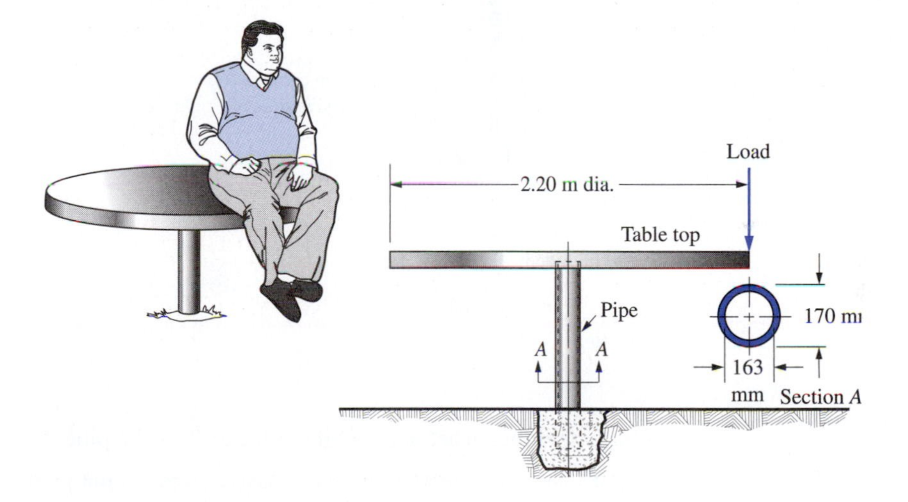

Solution Objective Compute the maximum stress in the pipe in Figure 10–18 and the design factor based on both yield strength and ultimate strength.

Given Loading and pipe dimensions shown in Figure 10–18. Load is the force due to a 135 kg mass on the edge of the table. Pipe is aluminum, 6061-T4; $s_y = 145$ MPa; $s_u = 241$ MPa.

Analysis The pipe is subjected to combined bending and direct compression as illustrated in Figure 10–19, the free-body diagram of the pipe. The effect of the load is to produce a downward force on the top of the pipe while exerting a moment in the clockwise direction. The moment is the product of the load times the radius of the table top. The reaction at the bottom of the pipe, supplied by the concrete, is an upward force combined with a counterclockwise moment. Use the *Guidelines for solving problems with combined normal stresses*.

Results ***Step 1.*** Figure 10–19 shows the free-body diagram. The force is the gravitational attraction of the 135-kg mass.

$$F = m \cdot g = (135 \text{ kg})(9.81 \text{ m/s}^2) = 1324 \text{ N}$$

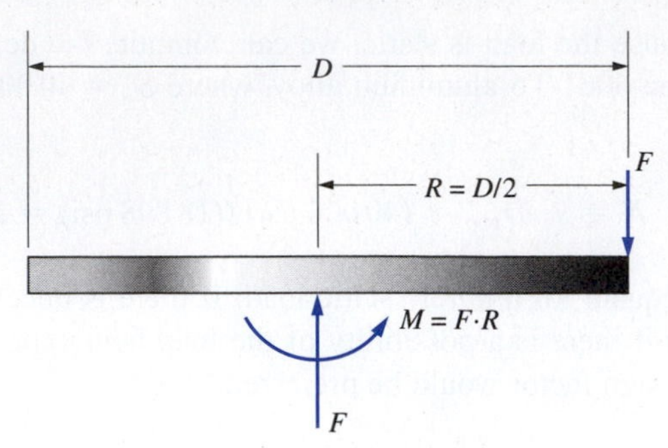

(*a*) Free-body diagram of table top

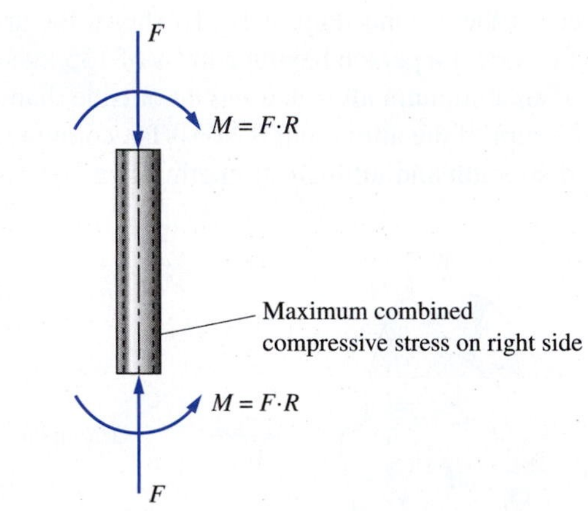

(*b*) Free-body diagram of pipe

FIGURE 10–19 Free-body diagrams for table top and pipe for Example Problem 10–3.

Step 2. No forces act at an angle to the axis of the pipe.

Step 3. Now the direct axial compressive stress in the pipe is

$$\sigma_a = -\frac{F}{A}$$

But

$$A = \frac{\pi(D_o^2 - D_i^2)}{4} = \frac{\pi(170^2 - 163^2)\,\text{mm}^2}{4} = 1831\ \text{mm}^2$$

Then

$$\sigma_a = -\frac{1324\ \text{N}}{1831\ \text{mm}^2} = -0.723\ \text{MPa}$$

This stress is a uniform, compressive stress across any cross section of the pipe.

Step 4. No forces act perpendicular to the axis of the pipe.

Step 5. Since the force acts at a distance of 1.1 m from the axis of the pipe, the moment is

$$M = FR = (1324\ \text{N})(1.1\ \text{m}) = 1456\ \text{N}\cdot\text{m}$$

The bending stress computation requires the application of the flexure formula,

$$\sigma_b = \frac{Mc}{I}$$

$$\text{where}\quad c = \frac{D_o}{2} = \frac{170\ \text{mm}}{2} = 85\ \text{mm}$$

$$I = \frac{\pi}{64}(D_o^4 - D_i^4) = \frac{\pi}{64}(170^4 - 163^4)\ \text{mm}^4$$

$$I = 6.347 \times 10^6\ \text{mm}^4$$

Then

$$\sigma_b = \frac{(1456\ \text{N}\cdot\text{m})(85\ \text{mm})}{6.347 \times 10^6\ \text{mm}^4} \times \frac{10^3\ \text{mm}}{\text{m}} = 19.5\ \text{MPa}$$

Step 6. The bending stress σ_b produces compressive stress on the right side of the pipe and tension on the left side. Since the direct compression stress adds to the bending stress on the right side, that is where the maximum stress would occur. The combined stress would then be, using Equation 10–1,

$$\sigma_c = -\sigma_a - \sigma_b = (-0.723 - 19.5)\ \text{MPa} = -20.22\ \text{MPa}$$

Step 7. The design factor based on yield strength is

$$N = \frac{s_y}{\sigma_c} = \frac{145\ \text{MPa}}{20.22\ \text{MPa}} = 7.17$$

The design factor based on ultimate strength is

$$N = \frac{s_u}{\sigma_c} = \frac{241\ \text{MPa}}{20.22\ \text{MPa}} = 11.9$$

Comment These values should be acceptable for this application based on the normal tensile and compressive stresses. Table 3–2 in Chapter 3 suggests $N = 2$ based on yield strength for static loading and $N = 12$ based on ultimate strength for impact. If the person simply sits on the edge of the table, the loading would be considered to be static; but a person jumping on the edge would apply an impact load. The design factor of 11.9 is sufficiently close to the recommended value of 12. However, additional analysis should be performed to evaluate the tendency for the pipe to buckle as a column, as discussed in Chapter 11. Also, Reference 1 defines the procedures for evaluating the tendency for local buckling of a hollow circular pipe subjected to compression.

LESSON 12

10–6 COMBINED NORMAL AND SHEAR STRESSES

Rotating shafts in machines transmitting power represent good examples of members loaded in such a way as to produce combined bending and torsion. Figure 10–20 shows a shaft carrying two chain sprockets. Power is delivered to the shaft through the sprocket at C and transmitted down the shaft to the sprocket at B, which in turn delivers it to another shaft. Because it is transmitting power, the shaft between B and C is subject to a torque and torsional shear stress, as you learned in Chapter 4. In order for the sprockets to transmit torque, they must be pulled by one side of the chain. At C, the back side of the chain must pull down with the force F_1 in order to drive the sprocket clockwise. Since the sprocket at B drives a mating sprocket, the front side of the chain would be in tension under the force F_2. The two forces, F_1 and F_2, acting downward cause bending of the shaft. Thus the shaft must be analyzed for both torsional shear stress and bending stress. Then, since both stresses act at the same place on the shaft, their combined effect must be determined. The method of analysis to be used is called the *maximum shear stress theory of failure,* which is described next. Then example problems will be shown.

Maximum Shear Stress Theory of Failure. When the tensile or compressive stress caused by bending occurs at the same place that a shearing stress occurs, the two kinds of stress combine to produce a larger shearing stress. The maximum stress can be computed from

$$\tau_{max} = \sqrt{\left(\frac{\sigma}{2}\right)^2 + \tau^2} \tag{10–2}$$

In Equation (10–2), σ refers to the magnitude of the tensile or compressive stress at the point, and τ is the shear stress at the same point. The result τ_{max} is the maximum shear stress at the point. The basis of Equation (10–2) is shown by using Mohr's circle later in this chapter.

The maximum shear stress theory of failure states that a member fails when the maximum shear stress exceeds the yield strength of the material in shear. This failure theory shows good correlation with test results for ductile metals such as most steels.

Equivalent Torque. Equation (10–2) can be expressed in a simplified form for the particular case of a circular shaft subjected to bending and torsion. Evaluating the bending stress separately, the maximum tensile or compressive stress would be

$$\sigma = \frac{M}{S}$$

where $S = \dfrac{\pi D^3}{32}$ = section modulus

D = diameter of the shaft

M = bending moment on the section

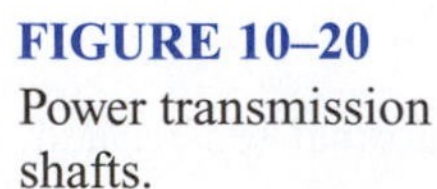
FIGURE 10–20 Power transmission shafts.

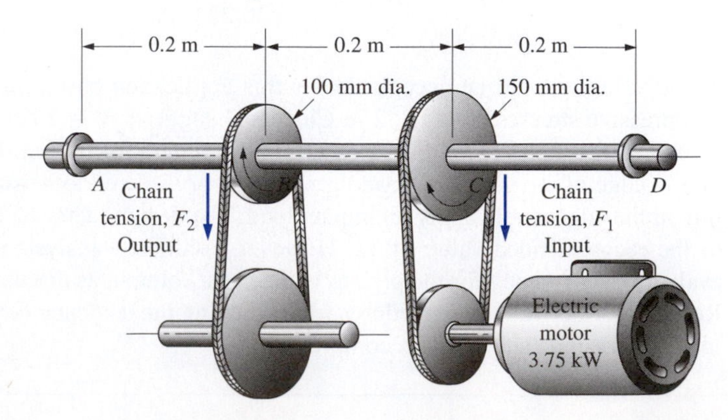

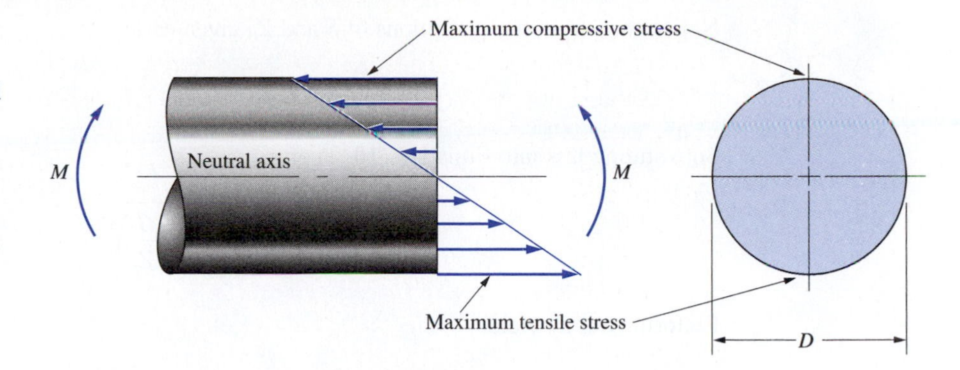

FIGURE 10–21 Bending stress distribution in a round shaft.

The maximum stress due to bending occurs at the outside surface of the shaft, as shown in Figure 10–21.

Now consider the torsional shear stress separately. In Chapter 4 the torsional shear stress equation was derived:

$$\tau = \frac{T}{Z_p}$$

where $Z_p = \dfrac{\pi D^3}{16}$ = polar section modulus

T = torque on the section

The maximum shear stress occurs at the outer surface of the shaft all the way around the circumference, as shown in Figure 10–22.

Thus, the maximum tensile stress and the maximum torsional shear stress both occur at the same point in the shaft. Now let's use Equation (10–2) to get an expression for the combined stress in terms of the bending moment M, the torque T, and the shaft diameter D.

$$\tau_{max} = \sqrt{\left(\frac{\sigma}{2}\right)^2 + \tau^2}$$

$$\tau_{max} = \sqrt{\left(\frac{M}{2S}\right)^2 + \left(\frac{T}{Z_p}\right)^2} \quad \textbf{(10–3)}$$

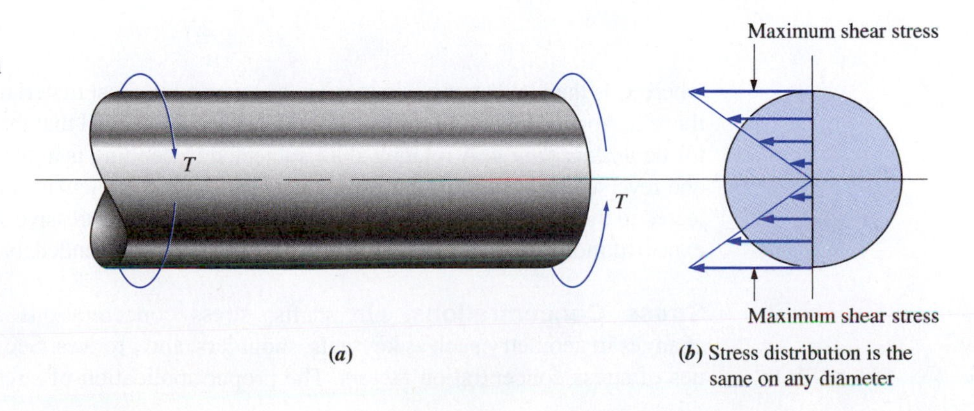

FIGURE 10–22 Shear stress distribution in a round shaft.

Notice that, from the definitions of S and Z_p given earlier,

$$2S = Z_p$$

Substituting this into Equation (10–3) gives

$$\tau_{max} = \sqrt{\left(\frac{M}{Z_p}\right)^2 + \left(\frac{T}{Z_p}\right)^2}$$

Factoring out Z_p yields

$$\tau_{max} = \frac{1}{Z_p}\sqrt{M^2 + T^2}$$

Sometimes the term $\sqrt{M^2 + T^2}$ is called the *equivalent torque* because it represents the amount of torque that would have to be applied to the shaft by itself to cause the equivalent magnitude of shear stress as the combination of bending and torsion. Calling the equivalent torque T_e,

$$T_e = \sqrt{M^2 + T^2} \quad \textbf{(10–4)}$$

and

$$\tau_{max} = \frac{T_e}{Z_p} \quad \textbf{(10–5)}$$

Equations (10–4) and (10–5) greatly simplify the calculation of the maximum shear stress in a circular shaft subjected to bending and torsion.

In the design of circular shafts subjected to bending and torsion, a design stress can be specified giving the maximum allowable shear stress. This was done in Chapter 4.

$$\tau_d = \frac{s_{ys}}{N}$$

where s_{ys} is the yield strength of the material in shear. Since s_{ys} is seldom known, the approximate value found from $s_{ys} = s_y/2$ can be used. Then

$$\tau_d = \frac{s_y}{2N} \quad \textbf{(10–6)}$$

where s_y is the yield strength in tension, as reported in most material property tables, such as those in Appendices A–14 through A–18. It is recommended that the value of the design factor be *no less than 4*. A rotating shaft subjected to bending is a good example of a repeated and reversed load. With each revolution of the shaft, a particular point on the surface is subjected to the maximum tensile and then the maximum compressive stress. Thus fatigue is the expected mode of failure, and $N = 4$ or greater is recommended, based on yield strength.

Stress Concentrations. In shafts, stress concentrations are created by abrupt changes in geometry such as keyseats, shoulders, and grooves. See Appendix A–22 for values of stress concentration factors. The proper application of stress concentration factors to the equivalent torque Equations (10–4) and (10–5) should be considered carefully. If the value of K_t at a section of interest is equal for both bending and torsion, then it can be applied directly to Equation (10–5). That is,

$$\tau_{max} = \frac{T_e K_t}{Z_p} \quad \textbf{(10–7)}$$

The form of Equation (10–7) can also be applied as a conservative calculation of τ_{max} by selecting K_t as the largest value for either torsion or bending.

To account for the proper K_t for both torsion and bending, Equation (10–4) can be modified as

$$T_e = \sqrt{(K_{tB}M)^2 + (K_{tT}T)^2} \quad \textbf{(10–8)}$$

Then Equation (10–5) can be used directly to compute the maximum shear stress.

Example Problem 10–4 Specify a suitable material for the shaft shown in Figure 10–20. The shaft has a uniform diameter of 55 mm and rotates at 120 rpm while transmitting 3.75 kW of power. The chain sprockets at B and C are keyed to the shaft with sled-runner keyseats. Sprocket C receives the power, and sprocket B delivers it to another shaft. The bearings at A and D provide simple supports for the shaft.

Solution

Objective: Specify a suitable material for the shaft.

Given: Shaft and loading in Figure 10–20.
Power $= P = 3.75$ kW. Rotational speed $= n = 120$ rpm.
Shaft diameter $= D = 55$ mm.
Sled-runner keyseats at B and C.
Simple supports at A and D.

Analysis: The several steps to be used to solve this problem are outlined here.

1. The torque in the shaft will be computed for the known power and rotational speed from $T = P/n$, as developed in Chapter 4.
2. The tensions in the chains for sprockets B and C will be computed. These are the forces that produce bending in the shaft.
3. Considering the shaft to be a beam, the shearing force and bending moment diagrams will be drawn for it.
4. At the section where the maximum bending moment occurs, the equivalent torque T_e will be computed from Equation (10–4).
5. The polar section modulus Z_p and the stress concentration factor K_t will be determined.
6. The maximum shear stress will be computed from Equation (10–7).
7. The required yield strength of the shaft material will be computed by letting $\tau_{max} = \tau_d$ in Equation (10–6) and solving for s_y. Remember, let $N = 4$ or more.
8. A steel that has a sufficient yield strength will be selected from Appendix A–14.

Results: ***Step 1.*** The desirable unit for torque is N·m. Then it is most convenient to observe that the power unit of kilowatts is equivalent to the units of kN·m/s. Also, rotational speed must be expressed in rad/s.

$$n = \frac{120\text{ rev}}{\text{min}} \times \frac{2\pi\text{ rad}}{\text{rev}} \times \frac{1\text{ min}}{60\text{ s}} = 12.57\text{ rad/s}$$

We can now compute torque.

$$T = \frac{P}{n} = \frac{3.75\ \text{kN}\cdot\text{m}}{\text{s}} \times \frac{1}{12.57\ \text{rad/s}} = 0.298\ \text{kN}\cdot\text{m}$$

Step 2. The tensions in the chains are indicated in Figure 10–20 by the forces F_1 and F_2. For the shaft to be in equilibrium, the torque on both sprockets must be the same in magnitude but opposite in direction. On either sprocket the torque is the product of the chain force times the *radius* of the pulley. That is,

$$T = F_1 R_1 = F_2 R_2$$

The forces can now be computed.

$$F_1 = \frac{T}{R_1} = \frac{0.298\ \text{kN}\cdot\text{m}}{75\ \text{mm}} \times \frac{10^3\ \text{mm}}{\text{m}} = 3.97\ \text{kN}$$

$$F_2 = \frac{T}{R_2} = \frac{0.298\ \text{kN}\cdot\text{m}}{50\ \text{mm}} \times \frac{10^3\ \text{mm}}{\text{m}} = 5.96\ \text{kN}$$

Step 3. Figure 10–23 shows the complete shearing force and bending moment diagrams found by the methods of Chapter 5. The maximum bending moment is 1.06 kN·m at section B, where one of the sprockets is located.

Step 4. At section B, the torque in the shaft is 0.298 kN·m and the bending moment is 1.06 kN·m. Then,

$$T_e = \sqrt{M^2 + T^2} = \sqrt{(1.06)^2 + (0.298)^2} = 1.10\ \text{kN}\cdot\text{m}$$

Step 5. $Z_p = \dfrac{\pi D^3}{16} = \dfrac{\pi(55\ \text{mm})^3}{16} = 32.67 \times 10^3\ \text{mm}^3$

For the keyseat at section B securing the sprocket to the shaft, we will use $K_t = 1.6$, as reported in Figure 4–14.

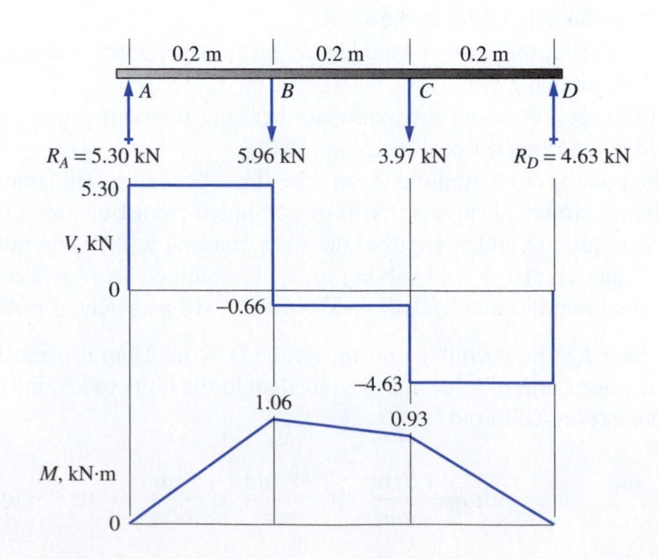

FIGURE 10–23 Shearing force and bending moment diagrams for Example Problem 10–4.

$$\textit{Step 6.}\ \tau_{max} = \frac{T_e K_t}{Z_p} = \frac{1.10 \times 10^3\ \text{N·m}(1.6)}{32.67 \times 10^3\ \text{mm}^3} \times \frac{10^3\ \text{mm}}{\text{m}} = 53.9\ \text{MPa}$$

$$\textit{Step 7.}\ \text{Let } \tau_{max} = \tau_d = \frac{s_y}{2N}$$

Then $s_y = 2N\tau_{max} = (2)(4)(53.9\ \text{MPa}) = 431\ \text{MPa}$

Step 8. Referring to Appendix A–14, we find that several alloys could be used. For example, AISI 1040, cold-drawn, has a yield strength of 490 MPa. Alloy AISI 1141 OQT 1300 has a yield strength of 469 MPa and also a very good ductility, as indicated by the 28% elongation. Either of these would be a reasonable choice.

10–7 EQUATIONS FOR STRESSES IN ANY DIRECTION

The *initial stress element* discussed in Section 10–4 was oriented in a convenient direction related to the member being analyzed. The methods of this section allow you to compute the stresses in any direction and to compute the maximum normal stresses and the maximum shear stress directly.

Figure 10–24 shows a stress element with orthogonal axes u and v superimposed on the initial element such that the u-axis is at an angle ϕ relative to the given x-axis. In general, there will be a normal stress σ_u and a shear stress τ_{uv} acting on the inclined surface AC. The following development will result in equations for those stresses.

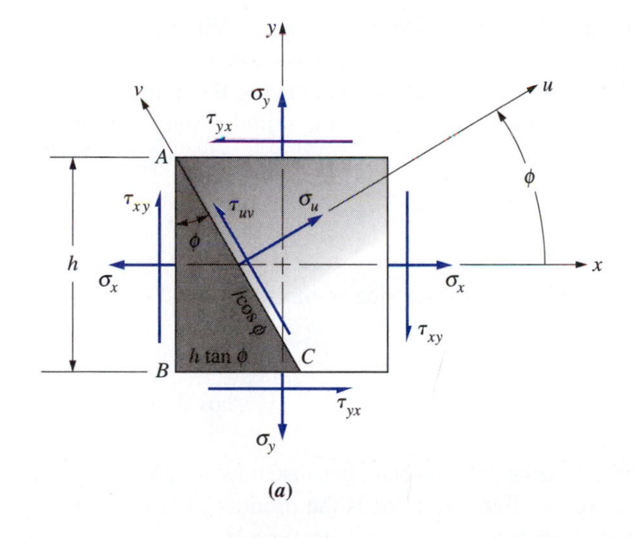

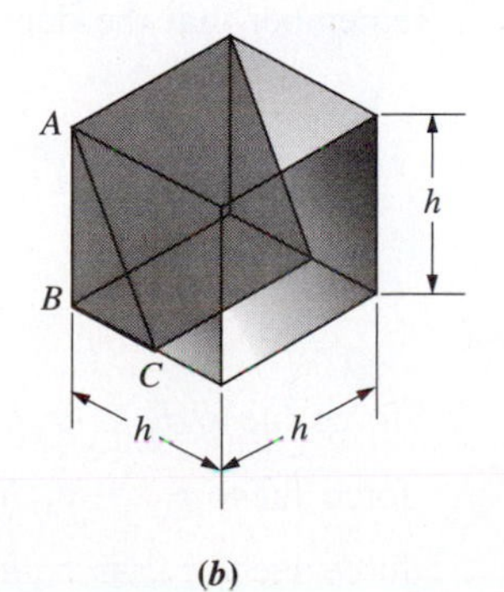

FIGURE 10–24 Initial stress element with u and v axes included. (a) Stress element with inclined surface. (b) 3-dimensional element showing wedge.

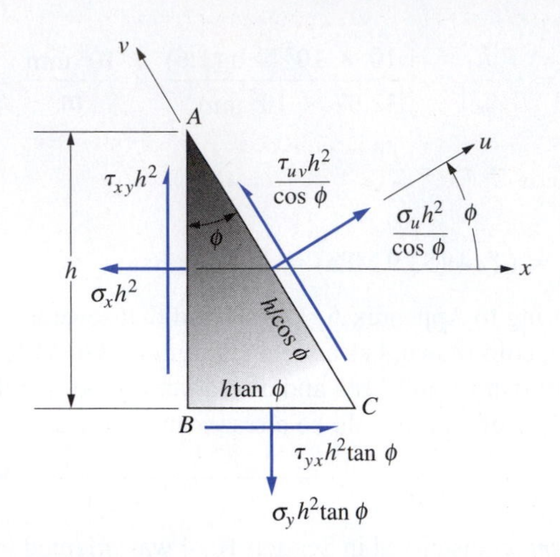

FIGURE 10–25 Free-body diagram of wedge showing forces acting on each face.

Before going on, note that Figure 10–24(a) shows only two dimensions of an element that is really a three-dimensional cube. Part (b) of the figure shows the entire cube with each side having the dimension h.

Normal Stress in the u-Direction, σ_u. Visualize a wedge-shaped part of the initial element as shown in Figure 10–25. Note how the angle ϕ is located. The side AB is the original left side of the initial element having the height h. The bottom of the wedge, side BC, is only a part of the bottom of the initial element where the length is determined by the angle ϕ itself.

$$BC = h \cdot \tan\phi$$

Also, the length of the sloped side of the wedge, AC, is

$$AC = \frac{h}{\cos\phi}$$

These lengths are important because now we are going to consider all the forces that act on the wedge. Because force is the product of stress times area, we need to know the area on which each stress acts. Starting with σ_x, it acts on the entire left face of the wedge having an area equal to h^2. Remember that the depth of the wedge perpendicular to the paper is also h. Then,

$$\text{force due to } \sigma_x = \sigma_x h^2$$

Using similar logic,

$$\text{force due to } \sigma_y = \sigma_y h^2 \tan\phi$$
$$\text{force due to } \tau_{xy} = \tau_{xy} h^2$$
$$\text{force due to } \tau_{yx} = \tau_{yx} h^2 \tan\phi$$

The stresses acting on the inclined face of the wedge must also be considered:

$$\text{force due to } \sigma_u = \frac{\sigma_u h^2}{\cos \phi}$$

$$\text{force due to } \tau_{uv} = \frac{\tau_{uv} h^2}{\cos \phi}$$

Now using the principle of equilibrium, we can sum forces in the u-direction. From the resulting equation, we can solve for σ_u. The process is facilitated by resolving all forces into components perpendicular and parallel to the inclined face of the wedge. Figure 10–26 shows this for each force except for those due to σ_u and τ_{uv} which are already aligned with the u and v axes. Then,

$$\sum F_u = 0 = \frac{\sigma_u h^2}{\cos \phi} - \sigma_x h^2 \cos \phi - \sigma_y h^2 \tan \phi \sin \phi + \tau_{xy} h^2 \sin \phi$$
$$+ \tau_{yx} h^2 \tan \phi \cos \phi$$

To begin solving for σ_u, all terms include h^2, which can be canceled. Also, we have noted that $\tau_{xy} = \tau_{yx}$ and we can let $\tan \phi = \sin \phi / \cos \phi$. The equilibrium equation then becomes

$$0 = \frac{\sigma_u}{\cos \phi} - \sigma_x \cos \phi - \frac{\sigma_y \sin \phi \sin \phi}{\cos \phi} + \tau_{xy} \sin \phi + \frac{\tau_{xy} \sin \phi \cos \phi}{\cos \phi}$$

Now multiply by $\cos \phi$ to obtain

$$0 = \sigma_u - \sigma_x \cos^2 \phi - \sigma_y \sin^2 \phi + \tau_{xy} \sin \phi \cos \phi + \tau_{xy} \sin \phi \cos \phi$$

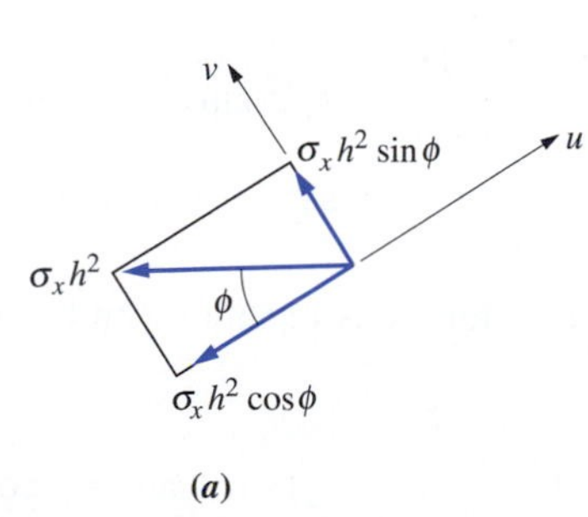

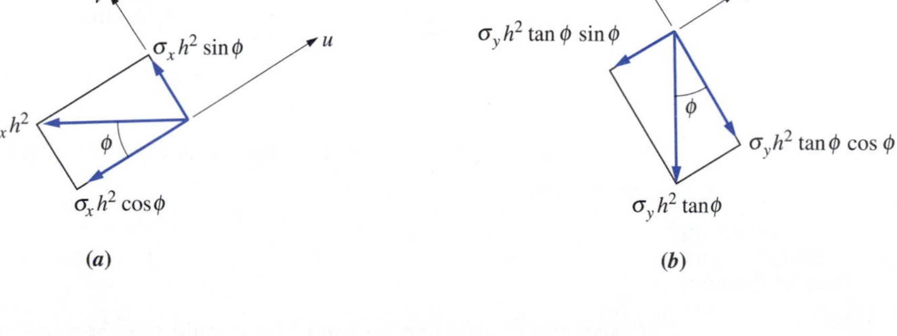

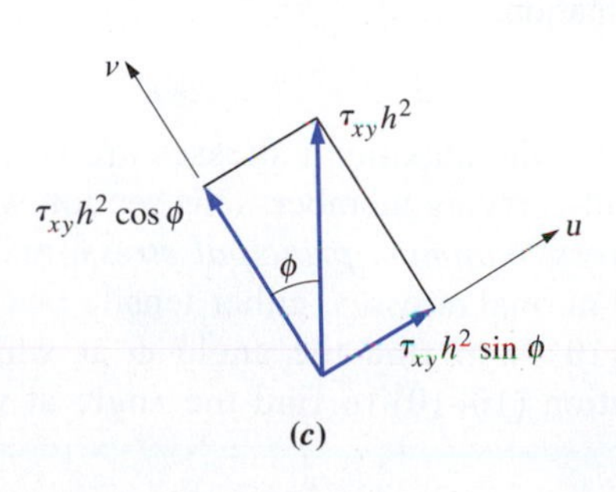

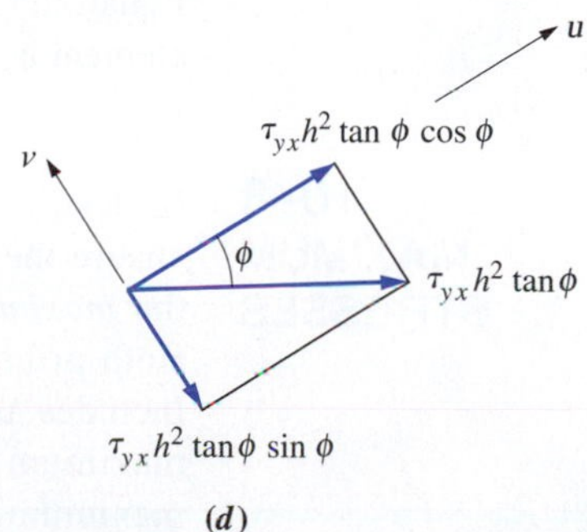

FIGURE 10–26 Resolution of forces along u and v directions. (a) Components of force due to σ_x. (b) components of force due to σ_y. (c) Components of force due to τ_{xy}. (d) Components of force due to τ_{yx}.

Combine the last two terms and solve for σ_u.

$$\sigma_u = \sigma_x \cos^2 \phi + \sigma_y \sin^2 \phi - 2\tau_{xy} \sin \phi \cos \phi$$

This is a usable formula for computing σ_u, but a more convenient form can be obtained by using the following trigonometric identities:

$$\cos^2 \phi = \tfrac{1}{2} + \tfrac{1}{2} \cos 2\phi$$

$$\sin^2 \phi = \tfrac{1}{2} - \tfrac{1}{2} \cos 2\phi$$

$$\sin \phi \cos \phi = \tfrac{1}{2} \sin 2\phi$$

Making the substitutions gives

$$\sigma_u = \tfrac{1}{2}\sigma_x + \tfrac{1}{2}\sigma_x \cos 2\phi + \tfrac{1}{2}\sigma_y - \tfrac{1}{2}\sigma_y \cos 2\phi - \tau_{xy} \sin 2\phi$$

Combining terms, we obtain

Normal Stress in u-Direction

$$\sigma_u = \tfrac{1}{2}(\sigma_x + \sigma_y) + \tfrac{1}{2}(\sigma_x - \sigma_y) \cos 2\phi - \tau_{xy} \sin 2\phi \qquad \textbf{(10–9)}$$

Equation (10–9) can be used to compute the normal stress in any direction provided that the stress condition in some direction, indicated by the x and y axes, is known.

Shearing Stress, τ_{uv}, Acting Parallel to the Cut Plane. Now the equation for the shearing stress, τ_{uv}, acting parallel to the cut plane and perpendicular to σ_u will be developed. Again referring to Figures 10–25 and 10–26, we can sum the forces on the wedge-shaped element acting in the v-direction.

$$\sum F_v = 0 = \frac{\tau_{uv}h^2}{\cos \phi} + \sigma_x h^2 \sin \phi - \sigma_y h^2 \tan \phi \cos \phi$$
$$+ \tau_{xy}h^2 \cos \phi - \tau_{yx}h^2 \tan \phi \sin \phi$$

Using the same techniques as before, this equation can be simplified and solved for τ_{uv}, resulting in

Shearing Stress, τ_{uv} on Face of Element

$$\tau_{uv} = -\tfrac{1}{2}(\sigma_x - \sigma_y) \sin 2\phi - \tau_{xy} \cos 2\phi \qquad \textbf{(10–10)}$$

Equation (10–10) can be used to compute the shearing stress that acts on the face of the element at any angular orientation.

10–8 MAXIMUM STRESSES

In design and stress analysis, the maximum stresses are typically desired in order to ensure the safety of the load-carrying member. This section will develop equations for the *maximum principal stress, minimum principal stress,* and *maximum shear stress.* Both principal stresses are normal stresses, either tensile or compressive. The process includes using Equation (10–9) to find the angle ϕ at which the normal stress is maximum and using Equation (10–10) to find the angle at which the shear stress is maximum.

Principal Stresses. From the study of calculus, we know that the value of the angle ϕ at which the maximum or minimum normal stress occurs can be found by differentiating the function and setting the result equal to zero, then solving for ϕ. Differentiating Equation (10–9) gives

$$\frac{d\sigma_u}{d\phi} = 0 = 0 + \tfrac{1}{2}(\sigma_x - \sigma_y)(-\sin 2\phi)(2) - \tau_{xy}\cos 2\phi(2)$$

Dividing by cos 2ϕ and simplifying gives

$$0 = -(\sigma_x - \sigma_y)\tan 2\phi - 2\tau_{xy}$$

Solving for $\tan 2\phi$ gives

$$\tan 2\phi = \frac{-2\tau_{xy}}{\sigma_x - \sigma_y} = \frac{-\tau_{xy}}{\frac{1}{2}(\sigma_x - \sigma_y)} \qquad \textbf{(10–11)}$$

The angle ϕ is then

Angle Locating Maximum Principal stress, σ_1

$$\phi = \tfrac{1}{2}\tan^{-1}\left[\frac{-\tau_{xy}}{\frac{1}{2}(\sigma_x - \sigma_y)}\right] \qquad \textbf{(10–12)}$$

If we substitute the value of ϕ defined by Equations (10–11) and (10–12) into Equation (10–9), we can develop an equation for the maximum normal stress on the element. In addition, we can develop the equation for the minimum normal stress. These two stresses are called the *principal stresses* with σ_1 used to denote the *maximum principal stress* and σ_2 denoting the *minimum principal stress.*

Note from Equation (10–9) that we need values for sin 2ϕ and cos 2ϕ. Figure 10–27 is a graphical aid to obtaining expressions for these functions. The right triangle has the opposite and adjacent sides defined by the terms of the tangent function in Equation (10–11).

Substituting into Equation (10–9) and simplifying gives

Maximum Principal Stress, σ_1

$$\sigma_{max} = \sigma_1 = \tfrac{1}{2}(\sigma_x + \sigma_y) + \sqrt{\left[\tfrac{1}{2}(\sigma_x - \sigma_y)\right]^2 + \tau_{xy}^2} \qquad \textbf{(10–13)}$$

The triangle is defined by Equation (10–11):

$$\tan 2\phi = \frac{-\tau_{xy}}{\frac{1}{2}(\sigma_x - \sigma_y)}$$

$$\sin 2\phi = \frac{-\tau_{xy}}{\sqrt{\left[\frac{1}{2}(\sigma_x - \sigma_y)\right]^2 + \tau_{xy}^2}}$$

$$\cos 2\phi = \frac{1/2(\sigma_x - \sigma_y)}{\sqrt{\left[\frac{1}{2}(\sigma_x - \sigma_y)\right]^2 + \tau_{xy}^2}}$$

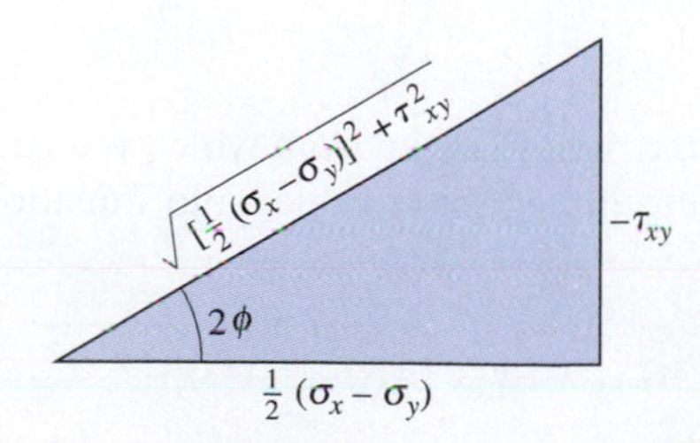

FIGURE 10–27
Development of sin 2ϕ and cos 2ϕ for principal stress formulas.

Because the square root has two possible values, + and −, we can also find the expression for the minimum principal stress, σ_2.

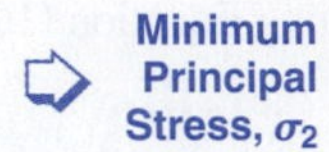

$$\sigma_{\min} = \sigma_2 = \tfrac{1}{2}(\sigma_x + \sigma_y) - \sqrt{[\tfrac{1}{2}(\sigma_x - \sigma_y)]^2 + \tau_{xy}^2} \qquad \textbf{(10–14)}$$

Conceivably, there would be a shear stress existing along with these normal stresses. It can be shown that by substituting the value of ϕ from Equation (10–12) into the shear stress Equation (10–10) that the result will be zero. In conclusion,

> On the element on which the principal stresses act, the shear stress is zero.

Maximum Shear Stress. The same technique can be used to find the maximum shear stress by working with Equation (10–10). Differentiating with respect to ϕ and setting the result equal to zero gives

$$\frac{d\tau_{uv}}{d\phi} = 0 = -\tfrac{1}{2}(\sigma_x - \sigma_y)(\cos 2\phi)(2) - \tau_{xy}(-\sin 2\phi)(2)$$

Dividing by cos 2ϕ and simplifying gives

$$\tan 2\phi = \frac{(\sigma_x - \sigma_y)}{2\tau_{xy}} = \frac{\tfrac{1}{2}(\sigma_x - \sigma_y)}{\tau_{xy}} \qquad \textbf{(10–15)}$$

Solving for ϕ yields

$$\phi = \tfrac{1}{2}\tan^{-1}\left[\frac{\tfrac{1}{2}(\sigma_x - \sigma_y)}{\tau_{xy}}\right] \qquad \textbf{(10–16)}$$

Obviously, the value for ϕ from Equation (10–16) is different from that of Equation (10–12). In fact, we will see that the two values are *always 45 deg apart.*

The triangle is defined by Equation (10–15):

$$\tan 2\phi = \frac{\tfrac{1}{2}(\sigma_x - \sigma_y)}{\tau_{xy}}$$

$$\sin 2\phi = \frac{\tfrac{1}{2}(\sigma_x - \sigma_y)}{\sqrt{[\tfrac{1}{2}(\sigma_x - \sigma_y)]^2 + \tau_{xy}^2}}$$

$$\cos 2\phi = \frac{\tau_{xy}}{\sqrt{[\tfrac{1}{2}(\sigma_x - \sigma_y)]^2 + \tau_{xy}^2}}$$

Figure 10–28 shows the right triangle, from which we can find sin 2ϕ and cos 2ϕ as we did in Figure 10–27. Substituting these values into Equation (10–10) gives the maximum shear stress.

$$\tau_{\max} = \pm\sqrt{[\tfrac{1}{2}(\sigma_x - \sigma_y)]^2 + \tau_{xy}^2} \qquad \textbf{(10–17)}$$

FIGURE 10–28
Development of sin 2ϕ and cos 2ϕ for maximum shear stress formula.

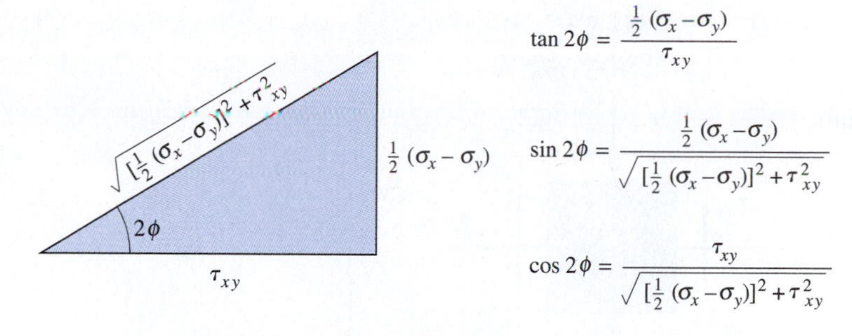

Here we should check to see if there is a normal stress existing on the element having the maximum shear stress. Substituting the value of ϕ from Equation (10–16) into the general normal stress Equation (10–9) gives

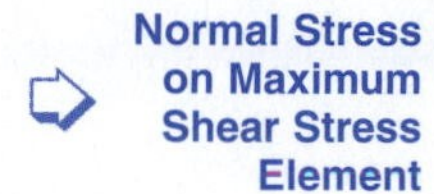
Normal Stress on Maximum Shear Stress Element

$$\sigma_{avg} = \tfrac{1}{2}(\sigma_x + \sigma_y) \tag{10–18}$$

This is the formula for the *average* of the initial normal stresses, σ_x and σ_y. Thus we can conclude:

> On the element on which the maximum shear stress occurs, there will also be a normal stress equal to the average of the initial normal stresses.

10–9 MOHR'S CIRCLE FOR STRESS

The use of Equations (10–9) through (10–18) often presents difficulties because of the many possible combinations of signs for the terms σ_x, σ_y, τ_{xy}, and ϕ. Also, two roots of the square root and the fact that the inverse tangent function can result in angles in any of the four quadrants presents difficulties. Fortunately, there is a graphical aid, called *Mohr's circle,* available that can help to overcome these problems. The use of Mohr's circle should give you a better understanding of the general case of stress at a point.

It can be shown that the two equations, (10–9) and (10–10), for the normal and shear stress at a point in any direction can be combined and arranged in the form of the equation for a circle. First presented by Otto Mohr in 1895, the circle allows a rapid and exact computation of:

1. The maximum and minimum principal stresses [Equations (10–13) and (10–14)]
2. The maximum shear stress [Equation (10–17)]
3. The angles of orientation of the principal stress element and the maximum shear stress element [Equations (10–12) and (10–16)]
4. The normal stress that exists along with the maximum shear stress on the maximum shear stress element [Equation (10–18)]
5. The stress condition at any angular orientation of the stress element [Equations (10–9) and (10–10)]

The procedure listed next can be used to draw Mohr's circle. An example is illustrated in Figures 10–29 to 10–31 as the steps are discussed. Steps 1–7 are shown in Figure 10–29 with the initial stress element required for step 1 included. For this example we are using a generalized initial stress element with σ_x tensile (positive), σ_y compressive (negative), and τ_{xy} acting counterclockwise (CCW). The circle will appear different for problems

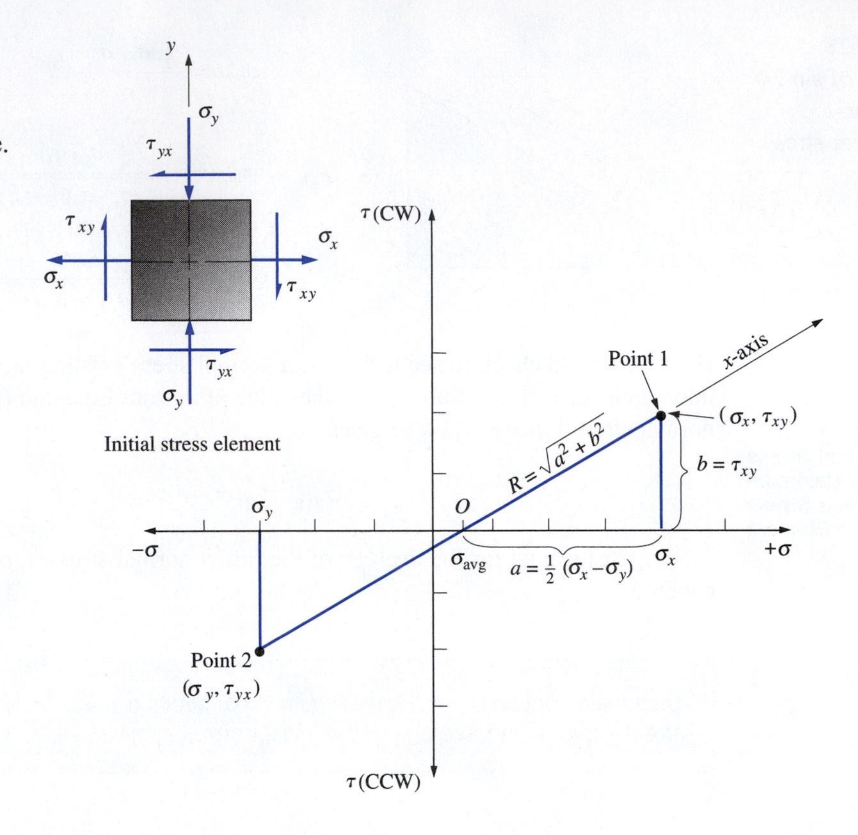

FIGURE 10–29 Steps 1 through 7 of Mohr's circle construction procedure.

having stresses acting in different directions. No numerical data are given in this example problem, and the results are shown in symbol form to demonstrate the nature of the quantities that make up the complete Mohr's circle. Several problems with actual numerical data are shown later in this chapter.

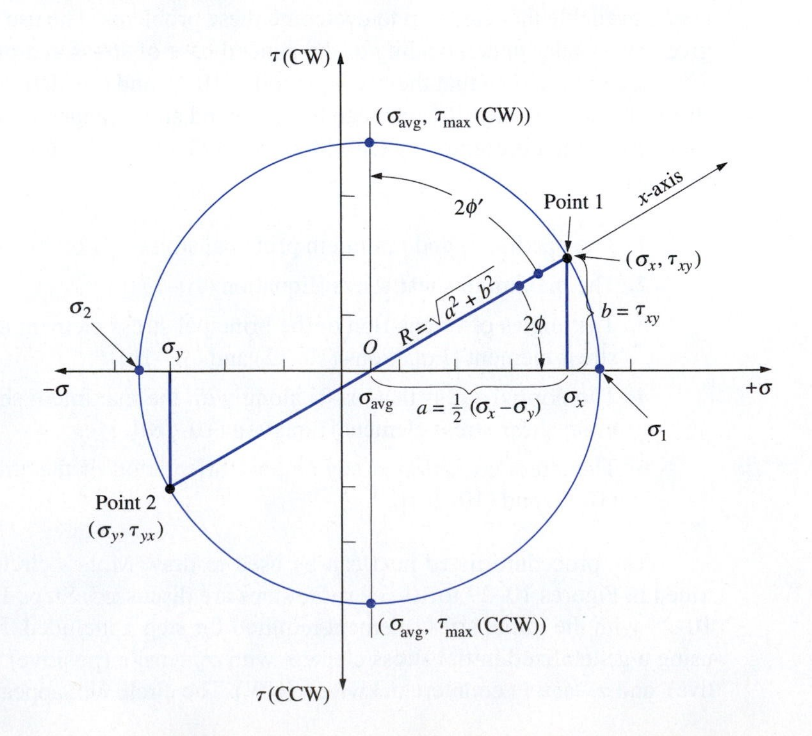

FIGURE 10–30 Completed Mohr's circle.

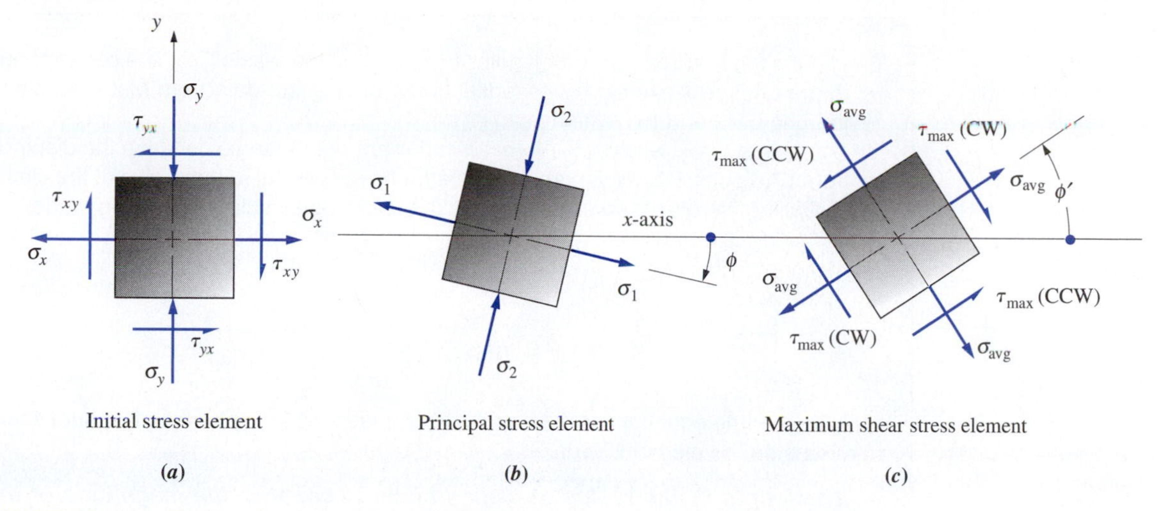

FIGURE 10–31 General form of final results from Mohr's circle analysis.

Mohr's circle is drawn on a set of perpendicular axes with shear stress, τ, plotted vertically and normal stresses, σ, horizontally, as shown in Figure 10–29. The following convention is used in this book.

Sign Conventions:

1. Positive normal stresses (tensile) are to the right.
2. Negative normal stresses (compressive) are to the left.
3. Shear stresses that tend to rotate the stress element clockwise (CW) are plotted upward on the τ-axis.
4. Shear stresses that tend to rotate the stress element counterclockwise (CCW) are plotted downward.

Procedure for Drawing Mohr's Circle

1. Identify the stress condition at the point of interest and represent it as the *initial stress element* in the manner shown in Figure 10–29.
2. The combination σ_x and τ_{xy} is plotted as *point 1* on the $\sigma - \tau$ plane.
3. The combination σ_y and τ_{yx} is then plotted as *point 2*. Note that τ_{xy} and τ_{yx} always act in opposite directions. Therefore, one point will be plotted above the σ-axis and one will be below.
4. Draw a straight line between the two points.
5. This line crosses the σ-axis at the center of Mohr's circle, which is also the value of the *average normal stress* applied to the initial stress element. The location of the center can be observed from the data used to plot the points or computed from Equation (10–18), repeated here:

$$\sigma_{avg} = \frac{1}{2}(\sigma_x + \sigma_y)$$

For convenience, label the center O.

6. Identify the line from *O* through point 1 (σ_x, τ_{xy}) as the *x*-axis. This line corresponds to the original *x*-axis and is essential to correlating the data from Mohr's circle to the original *x* and *y* directions.
7. The points *O*, σ_x, and point 1 form an important right triangle because the distance from *O* to point 1, the hypotenuse of the triangle, is equal to the radius of the circle, *R*. Calling the other two sides *a* and *b*, the following calculations can be made:

$$a = \tfrac{1}{2}(\sigma_x - \sigma_y)$$
$$b = \tau_{xy}$$
$$R = \sqrt{a^2 + b^2} = \sqrt{[\tfrac{1}{2}(\sigma_x - \sigma_y)]^2 + \tau_{xy}^2}$$

Note that the equation for *R* is identical to Equation (10–17) for the maximum shear stress on the element. Thus,

The length of the radius of Mohr's circle is equal to the magnitude of the maximum shear stress.

Step 8–11 are shown in Figure 10–30.

8. Draw the complete circle with the center at *O* and the radius *R*.
9. Draw the vertical diameter of the circle. The point at the top of the circle has the coordinates (σ_{avg}, τ_{max}), where the shear stress has the clockwise (CW) direction. The point at the bottom of the circle represents (σ_{avg}, τ_{max}) where the shear stress is counterclockwise (CCW).
10. Identify the points on the σ-axis at the ends of the horizontal diameter as σ_1 at the right (the maximum principal stress) and σ_2 at the left (the minimum principal stress). Note that the shear stress is zero at these points.
11. Determine the values for σ_1 and σ_2 from

$$\sigma_1 = \text{“}O\text{”} + R \qquad \textbf{(10–19)}$$
$$\sigma_2 = \text{“}O\text{”} - R \qquad \textbf{(10–20)}$$

where "*O*" represents the coordinate of the center of the circle, σ_{avg}, and *R* is the radius of the circle. Thus Equations (10–19) and (10–20) are identical to Equations (10–13) and (10–14) for the principal stresses.

The following steps determine the angles of orientation of the principal stress element and maximum shear stress element. An important concept to remember is that *angles obtained from Mohr's circle are double the true angles*. The reason for this is that the equations on which it is based; Equations (10–9) and (10–10), are functions of 2ϕ.

12. The orientation of the principal stress element is determined by finding the angle *from* the *x*-axis to the "σ_1"-axis, labeled 2ϕ in Figure 10–30. From the data on the circle you can see that

$$2\phi = \tan^{-1}\left(\frac{b}{a}\right)$$

The argument of this inverse tangent function is the same as the absolute value of the argument shown in Equation (10–12). Problems with signs for the resulting angle are avoided by noting the *direction from the x-axis to the σ_1-axis* on the circle, clockwise for the present example. Then the principal stress element is rotated *in the same direction* from the *x*-axis by an amount ϕ to locate the face on which the maximum principal stress σ_1 acts.

13. Draw the principal stress element in its proper orientation as determined from step 12 with the two principal stresses σ_1 and σ_2 shown [see Figure 10–31 (a) and (b)].
14. The orientation of the maximum shear stress element is determined by finding the angle *from* the *x*-axis *to* the τ_{max}-axis, labeled $2\phi'$ in Figure 10–30. In the present example,

$$2\phi' = 90° - 2\phi$$

From trigonometry it can be shown that this is equivalent to finding the inverse tangent of a/b, the reciprocal of the argument used to find 2ϕ. Thus it is a true evaluation of Equation (10–16), derived to find the angle of orientation of the element on which the maximum shear stress occurs.

Again problems with signs for the resulting angle are avoided by noting the *direction from the x-axis to the* τ_{max}*-axis* on the circle, counterclockwise for the present example. Then the maximum shear stress element is rotated *in the same direction* from the *x*-axis by an amount ϕ' to locate the face on which the clockwise maximum shear stress acts.
15. Draw the maximum shear stress element in its proper orientation as determined from step 14 with the shear stresses on all four faces and the average normal stress acting on each face [see Figure 10–31(c)].

As a whole, Figure 10–31 is the desired result from a Mohr's circle analysis. Shown are the initial stress element that establishes the *x* and *y* axes, the principal stress element drawn in proper rotation relative to the *x*-axis, and the maximum shear stress element also drawn in the proper rotation relative to the *x*-axis.

Example Problem 10–5 demonstrates this procedure with specific data for the stresses on the initial stress element.

Example Problem 10–5

It has been determined that a point in a load-carrying member is subjected to the following stress condition:

$$\sigma_x = 400\ \text{MPa} \qquad \sigma_y = -300\ \text{MPa} \qquad \tau_{xy} = 200\ \text{MPa (CW)}$$

Perform the following:

(a) Draw the initial stress element.

(b) Draw the complete Mohr's circle, labeling critical points.

(c) Draw the complete principal stress element.

(d) Draw the maximum shear stress element.

Solution

The 15-step *Procedure for drawing Mohr's circle* is used here to complete the problem. The numerical results from steps 1–12 are summarized here and shown in Figure 10–32.

Step 1. The initial stress element is shown at the upper left of Figure 10–32.

Step 2. Point 1 is plotted at $\sigma_x = 400$ MPa and $\tau_{xy} = 200$ MPa in quadrant 1.

Step 3. Point 2 is plotted at $\sigma_y = -300$ MPa and $\tau_{yx} = -200$ MPa in quadrant 3.

Step 4. The line from point 1 to point 2 has been drawn.

Step 5. The line from step 4 crosses the σ-axis at the average applied normal stress, called point O in Figure 10–32, computed from

$$\sigma_{avg} = \tfrac{1}{2}(\sigma_x + \sigma_y) = \tfrac{1}{2}[400 + (-300)] = 50 \text{ MPa}$$

Step 6. Point O is the center of the circle. The line from point O through point 1 is labeled as the x-axis to correspond with the x-axis on the initial stress element.

Step 7. The values of a, b, and R are found using the triangle formed by the lines from point O to point 1 to $\sigma_x = 400$ MPa and back to point O.
The lower side of the triangle,

$$a = \tfrac{1}{2}(\sigma_x - \sigma_y) = \tfrac{1}{2}[400 - (-300)] = 350 \text{ MPa}$$

The vertical side of the triangle, b, is completed from:

$$b = \tau_{xy} = 200 \text{ MPa}$$

The radius of the circle, R, is completed from:

$$R = \sqrt{a^2 + b^2} = \sqrt{(350)^2 + (200)^2} = 403 \text{ MPa}$$

FIGURE 10–32
Complete Mohr's circle for Example Problem 10–15.

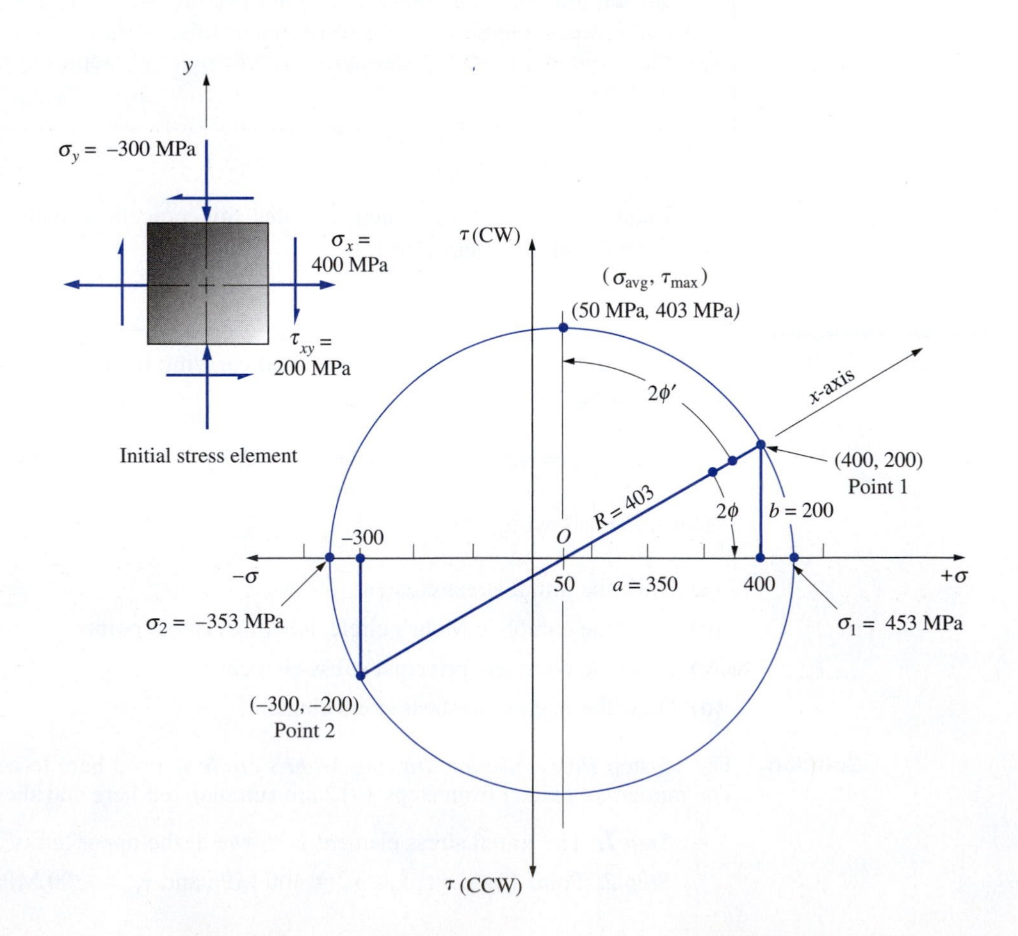

Step 8. This is the drawing of the circle with point O as the center at $\sigma_{avg} = 50$ MPa and a radius of $R = 403$ MPa.

Step 9. The vertical diameter of the circle has been drawn through point O. The intersection of this line with the circle at the top indicates the value of $\tau_{max} = 403$ MPa, the same as the value of R.

Step 10. The maximum principal stress, σ_1, is at the right end of the horizontal diameter of the circle and the minimum principal stress, σ_2, is at the left.

Step 11. The values for σ_1 and σ_2 are

$$\sigma_1 = O + R = 50 + 403 = 453 \text{ MPa}$$
$$\sigma_2 = O - R = 50 - 403 = -353 \text{ MPa}$$

Step 12. The angle 2ϕ is shown on the circle as the angle from the x-axis to the σ_1-axis, a clockwise rotation. The value is computed from

$$2\phi = \tan^{-1}\left(\frac{b}{a}\right) = \tan^{-1}\left(\frac{200}{350}\right) = 29.74°$$

Note that 2ϕ is CW from the x-axis to σ_1 on the circle.

$$\phi = \frac{29.74°}{2} = 14.87°$$

Step 13. Using the results from steps 11 and 12, the principal stress element is drawn as shown in Figure 10–33(b). The element is rotated 14.87° CW from the original x-axis to the face on which the tensile stress $\sigma_1 = 453$ MPa acts. The compressive stress $\sigma_2 = -353$ MPa acts on the faces perpendicular to the σ_1 faces.

Step 14. The angle $2\phi'$ is shown in Figure 10–32 drawn from the x-axis CCW to the vertical diameter that locates τ_{max} at the top of the circle. Its value can be found in either of two ways. First, using Equation (10–16), observe that the numerator is the same as the

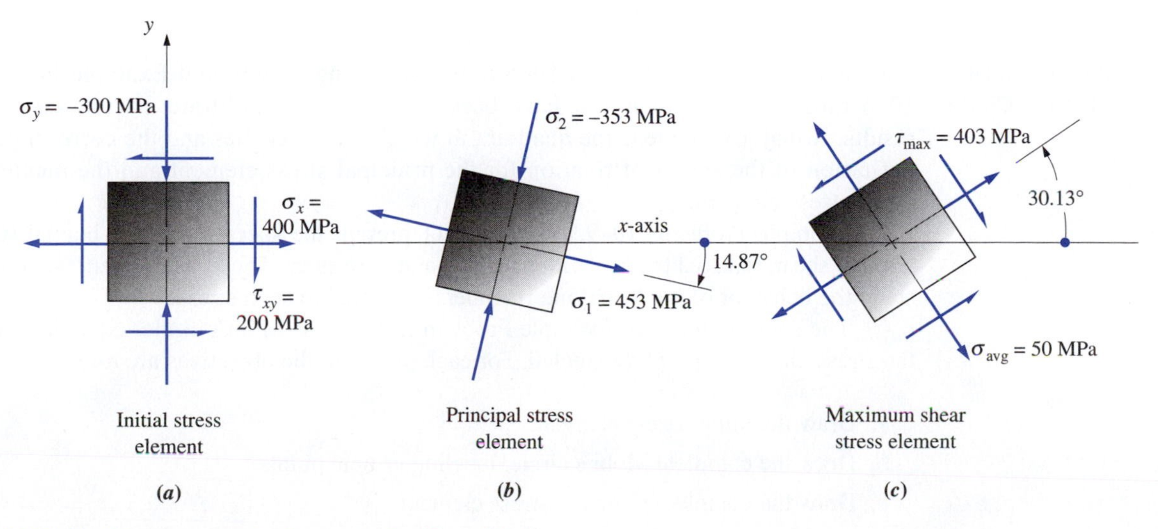

FIGURE 10–33 Results for Example Problem 10–5.

value of a and the denominator is the same as the value of b from the construction of the circle. Then

$$2\phi' = \tan^{-1}(a/b) = \tan^{-1}(350/200) = 60.26° \text{ CCW}$$

Or, using the geometry of the circle, we can compute

$$2\phi' = 90° - 2\phi = 90° - 29.74° = 60.26° \text{ CCW}$$

Then the angle ϕ' is one-half of $2\phi'$.

$$\phi' = \frac{60.26°}{2} = 30.13°$$

Step 15. The maximum shear stress element is drawn in Figure 10–33(c), rotated 30.13° CCW from the original x-axis to the face on which the positive τ_{max} acts. The maximum shear stress of 403 MPa is shown on all four faces with vectors that create the two pairs of opposing couples characteristic of shear stresses on a stress element. Also shown is the tensile stress $\sigma_{max} = 50$ MPa acting on all four faces of the element.

Summary of Results for Example Problem 10–5 Mohr's Circle

Given $\sigma_x = 400$ MPa $\quad \sigma_y = -300$ MPa $\quad \tau_{xy} = 200$ MPa CW

Results Figures 10–32 and 10–33.

$\sigma_1 = 453$ MPa $\quad \sigma_2 = -353$ MPa $\quad \phi = 14.87°$ CW from x-axis

$\tau_{max} = 403$ MPa $\quad \sigma_{avg} = 50$ MPa $\quad \phi' = 30.13°$ CCW from x-axis

Comment The x-axis is in the first quadrant.

Examples of the Use of Mohr's Circle

The data for Example Problem 10–5 in the preceding section and Example Problems 10–6 through 10–11 to follow have been selected to demonstrate a wide variety of results. A major variable is the quadrant in which the x-axis lies and the corresponding definition of the angles of rotation for the principal stress element and the maximum shear stress element.

Example Problems 10–9, 10–10, 10–11 present the special cases of biaxial stress with no shear, uniaxial tension with no shear, and pure shear. These should help you understand the behavior of load-carrying members subjected to such stresses.

The solution for each Example Problem to follow is Mohr's circle itself along with the stress elements, properly labeled. For each problem, the objectives are to:

a. Draw the initial stress element.
b. Draw the complete Mohr's circle, labeling critical points.
c. Draw the complete principal stress element.
d. Draw the complete shear stress element.

Example Problem 10–6 Mohr's Circle

Given $\sigma_x = 60 \text{ ksi}$ $\sigma_y = -40 \text{ ksi}$ $\tau_{xy} = 30 \text{ ksi CCW}$

Results Figure 10–34.

$\sigma_1 = 68.3 \text{ ksi}$ $\sigma_2 = -48.3 \text{ ksi}$ $\phi = 15.48° \text{ CCW}$

$\tau_{max} = 58.3 \text{ ksi}$ $\sigma_{avg} = 10.0 \text{ ksi}$ $\phi' = 60.48° \text{ CCW}$

Comment The *x*-axis is in the second quadrant.

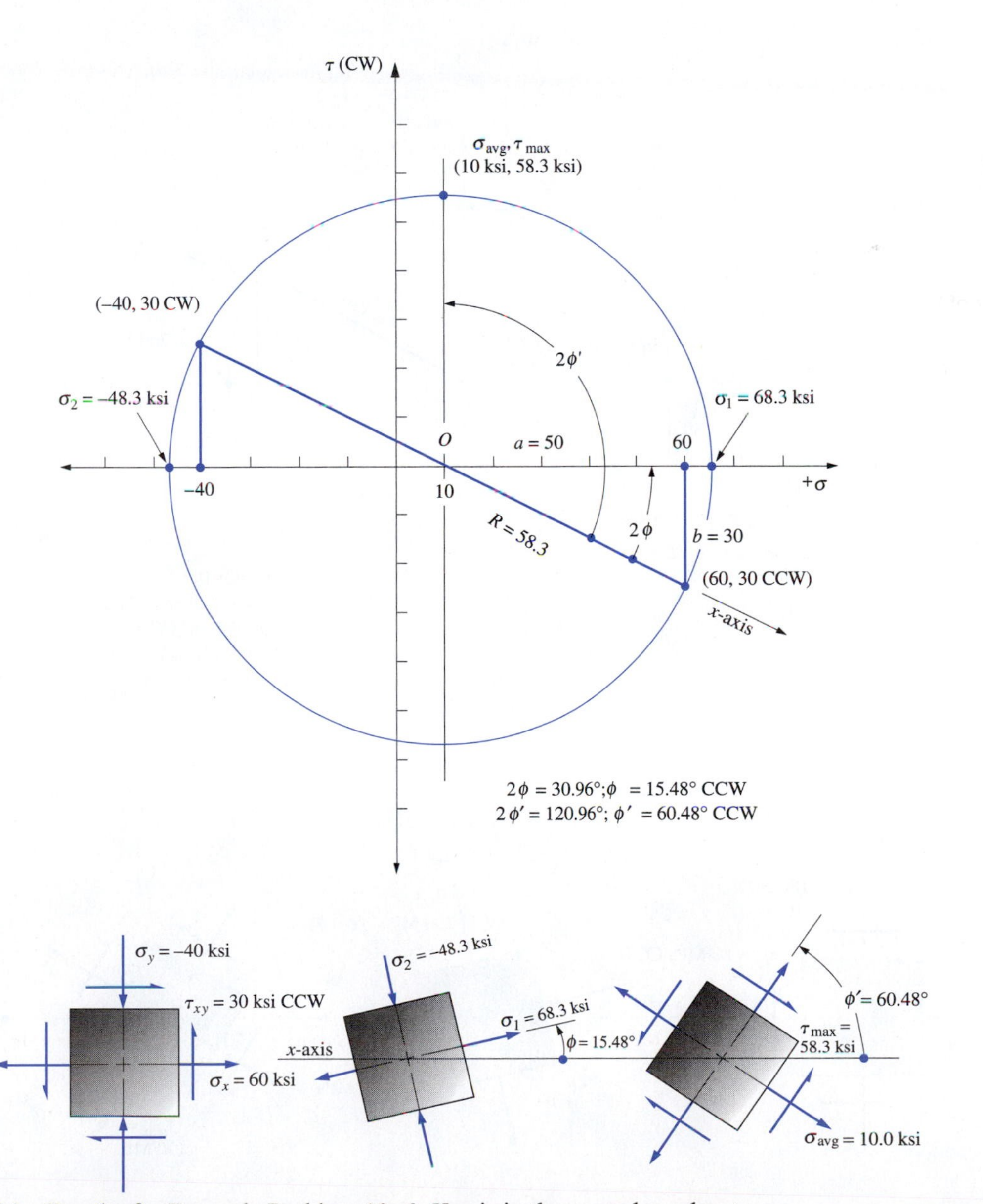

FIGURE 10–34 Results for Example Problem 10–6. *X*-axis in the second quadrant.

Example Problem 10–7 Mohr's Circle

Given $\sigma_x = -120$ MPa $\sigma_y = 180$ MPa $\tau_{xy} = 80$ MPa CCW

Results Figure 10–35.

$\sigma_1 = 200$ MPa $\sigma_2 = -140$ MPa $\phi = 75.96°$ CCW

$\tau_{max} = 170$ MPa $\sigma_{avg} = 30$ MPa $\phi' = 59.04°$ CW

Comment The x-axis is in the third quadrant.

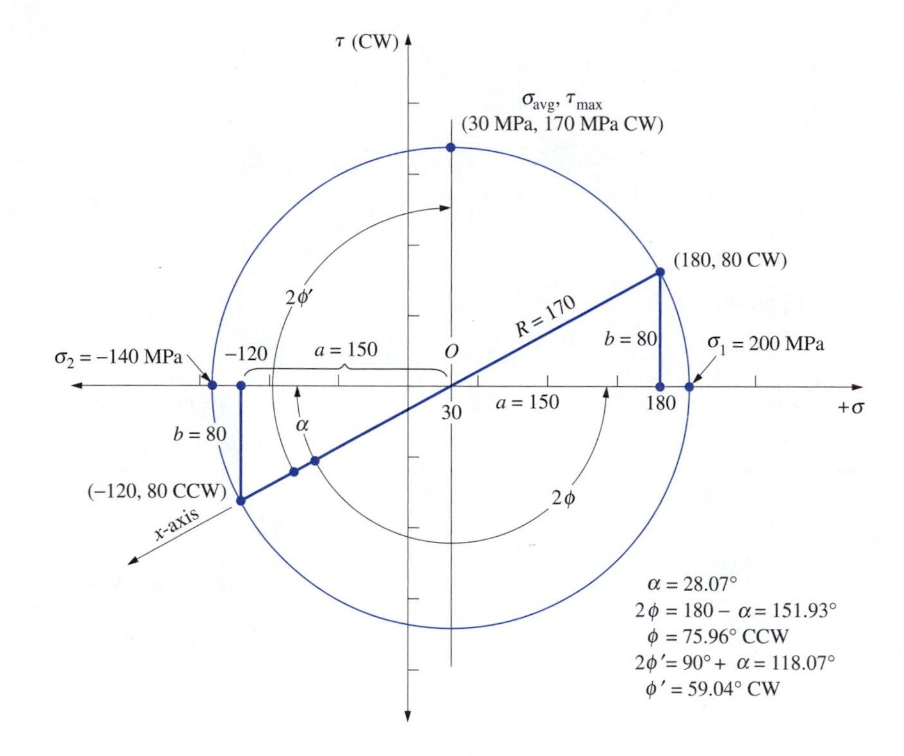

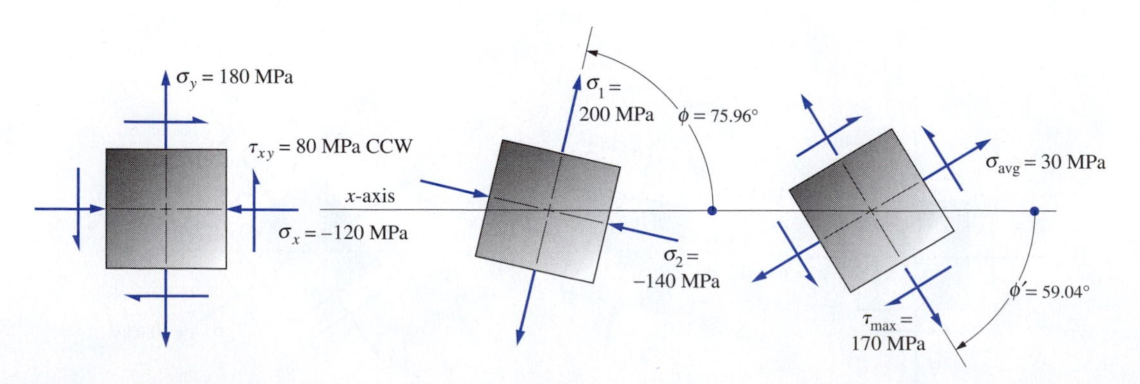

FIGURE 10–35 Results for Example Problem 10–7. X-axis in the third quadrant.

Example Problem 10–8 Mohr's Circle

Given $\sigma_x = -30$ ksi $\quad \sigma_y = 20$ ksi $\quad \tau_{xy} = 40$ ksi CW

Results Figure 10–36.

$\sigma_1 = 42.17$ ksi $\quad \sigma_2 = -52.17$ ksi $\quad \phi = 61.0°$ CW

$\tau_{max} = 47.17$ ksi $\quad \sigma_{avg} = -5.0$ ksi $\quad \phi' = 16.0°$ CW

Comment The x-axis is in the fourth quadrant.

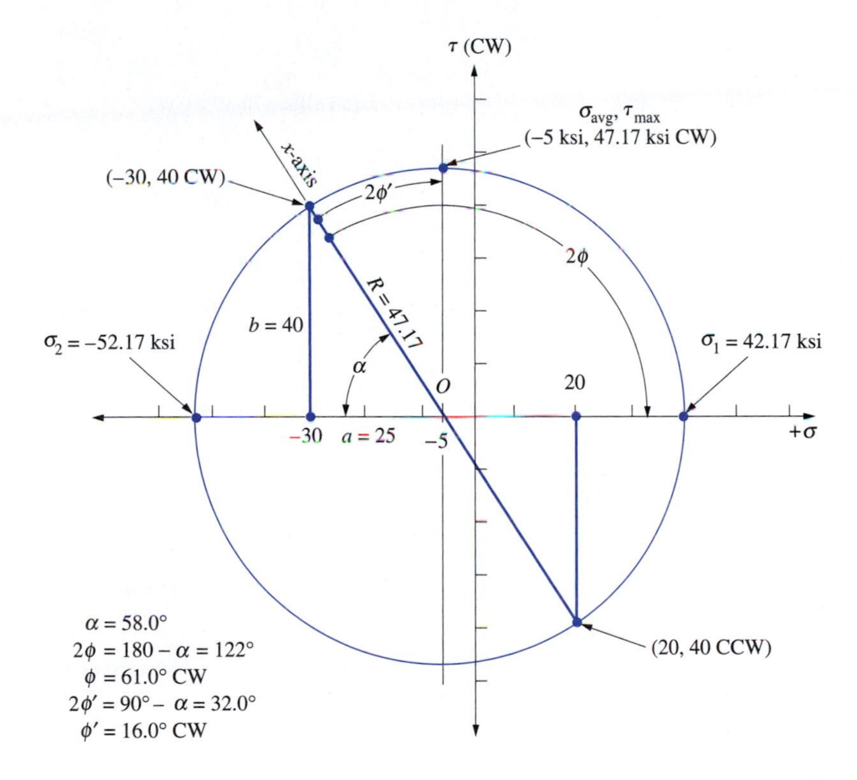

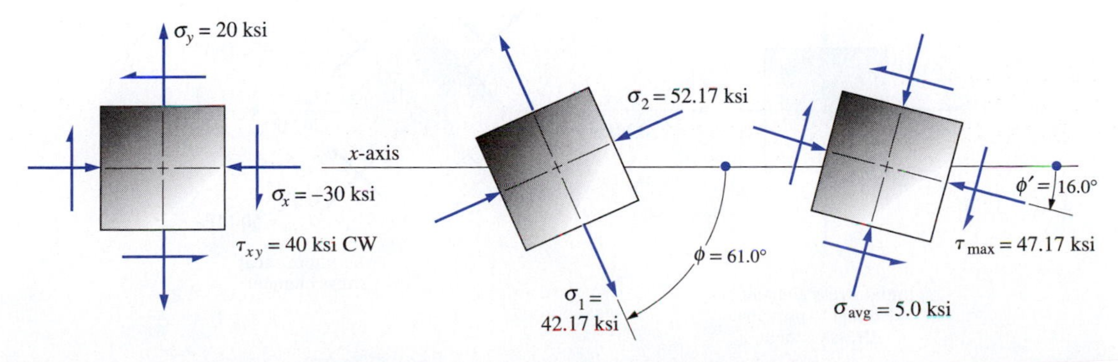

FIGURE 10–36 Results for Example Problem 10–8. X-axis in the fourth quadrant.

Example Problem 10–9 Mohr's Circle

Given $\sigma_x = 220\ \text{MPa}$ $\sigma_y = -120\ \text{MPa}$ $\tau_{xy} = 0\ \text{MPa}$

Results Figure 10–37.

$\sigma_1 = 220\ \text{MPa}$ $\sigma_2 = -120\ \text{MPa}$ $\phi = 0°$

$\tau_{max} = 170\ \text{MPa}$ $\sigma_{avg} = 50\ \text{MPa}$ $\phi' = 45.0°\ \text{CCW}$

Comment Special case of biaxial stress with no shear on the given element.

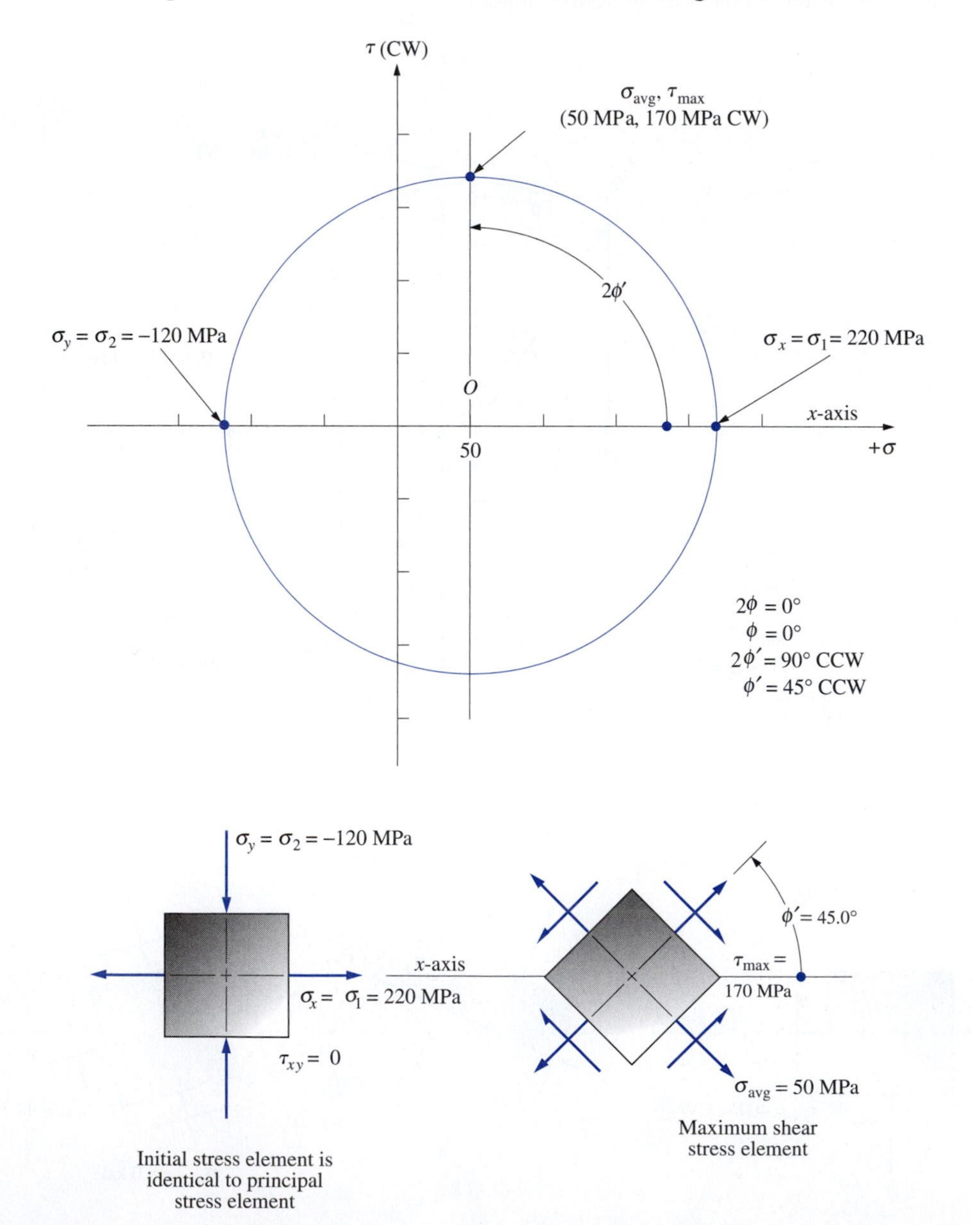

FIGURE 10–37 Results of Example Problem 10–9. Special case of biaxial stress with no shear.

Example Problem 10–10 Mohr's Circle

Given $\sigma_x = 40\text{ ksi}$ $\sigma_y = 0\text{ ksi}$ $\tau_{xy} = 0\text{ ksi}$

Results Figure 10–38.

$\sigma_1 = 40\text{ ksi}$ $\sigma_2 = 0\text{ ksi}$ $\phi = 0°$

$\tau_{max} = 20\text{ ksi}$ $\sigma_{avg} = 20\text{ ksi}$ $\phi' = 45.0°\text{ CCW}$

Comment Special case of uniaxial tension with no shear.

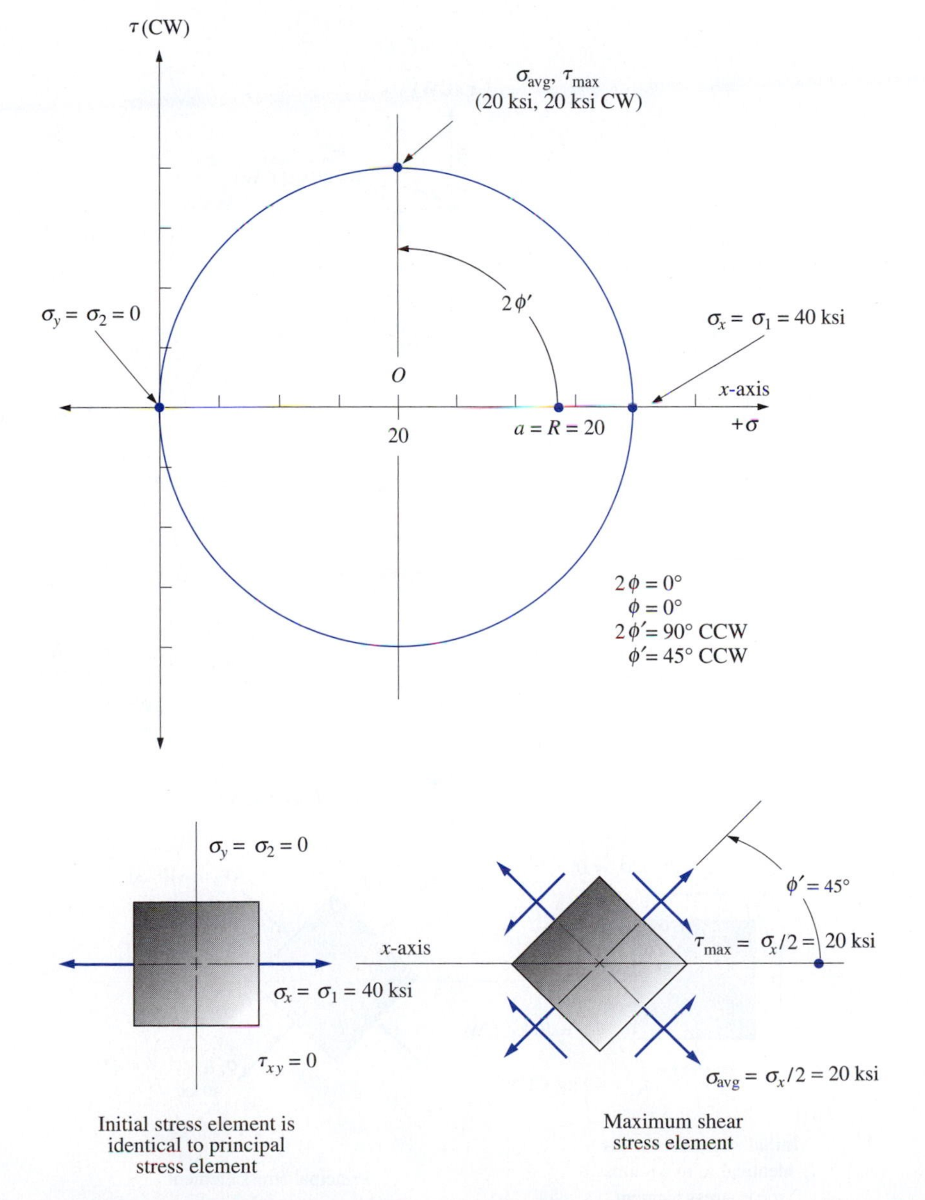

FIGURE 10–38 Results of Example Problem 10–10. Special case of uniaxial tension.

Example Problem 10–11 Mohr's Circle

Given $\sigma_x = 0\text{ ksi}$ $\sigma_y = 0\text{ ksi}$ $\tau_{xy} = 40\text{ ksi CW}$

Results Figure 10–39.

$\sigma_1 = 40\text{ ksi}$ $\sigma_2 = -40\text{ ksi}$ $\phi = 45°\text{ CW}$

$\tau_{max} = 40\text{ ksi}$ $\sigma_{avg} = 0\text{ ksi}$ $\phi' = 0°$

Comment Special case of pure shear.

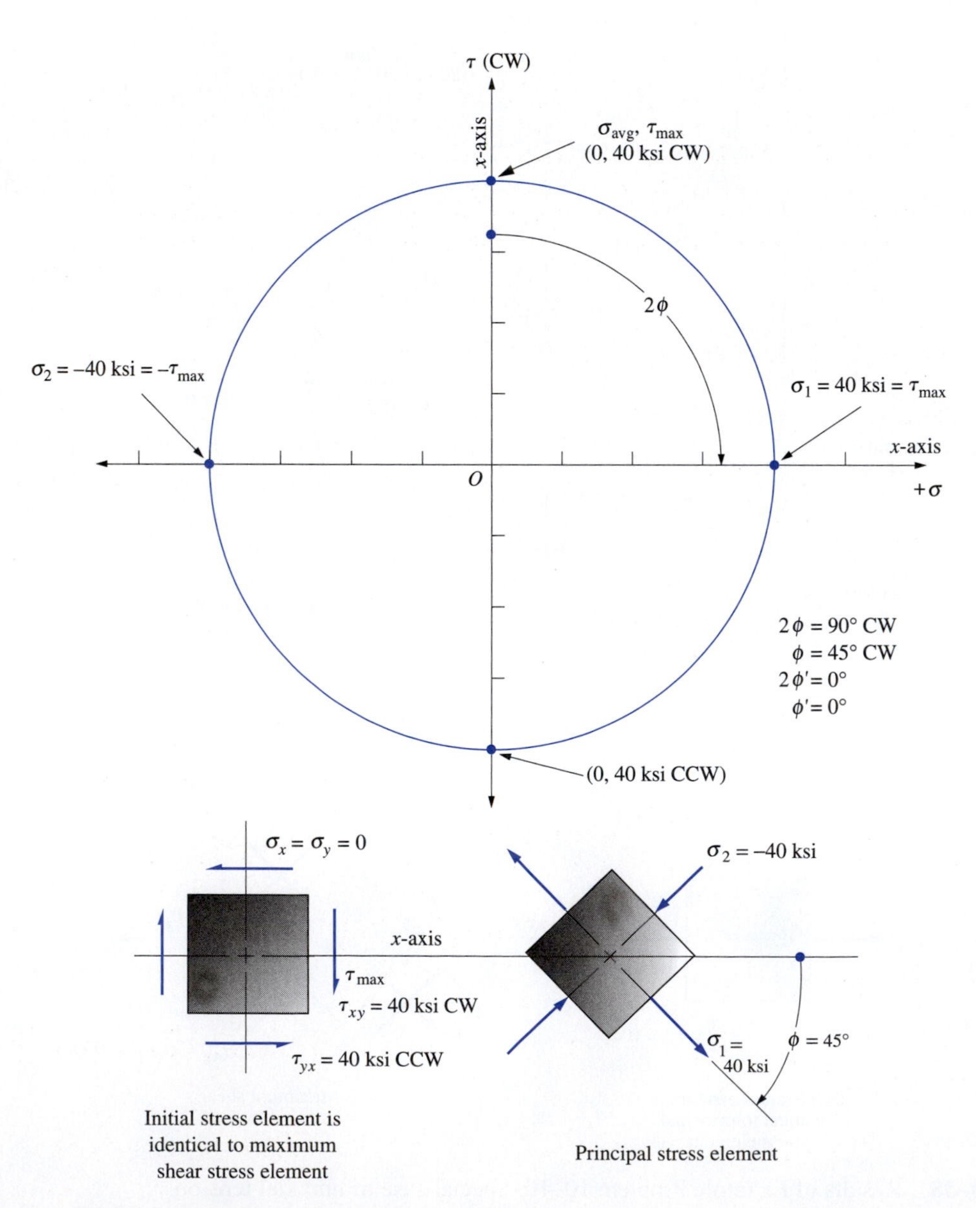

FIGURE 10–39 Results of Example Problem 10–11. Special case of pure shear.

Maximum Shear Stress Theory of Failure

One of the most widely used principles of design is the *maximum shear stress theory of failure,* which states:

> A ductile material can be expected to fail when the maximum shear stress to which the material is subjected exceeds the yield strength of the material in shear.

Of course, to apply this theory, it is necessary to be able to compute the magnitude of the maximum shear stress. If the member is subjected to pure shear, such as torsional shear stress, direct shear stress, or shearing stress in beams in bending, the maximum shear stress can be computed directly from formulas such as those developed in this book. But if a combined stress condition exists, the use of Equation (10–17) or Mohr's circle should be used to determine the maximum shear stress.

One special case of combined stress that occurs often is one in which a normal stress in only one direction is combined with a shear stress. For example, a round bar could be subjected to a direct axial tension while also being twisted. In many types of mechanical power transmissions, shafts are subjected to bending and torsion simultaneously. Certain fasteners may be subjected to tension combined with direct shear.

A simple formula can be developed for such cases using Mohr's circle or Equation (10–17). If only a normal stress in the x-direction, σ_x, combined with a shearing stress, τ_{xy} exists, the maximum shear stress is that given earlier as Equation (10–2).

$$\tau_{max} = \sqrt{(\sigma_x/2)^2 + \tau_{xy}^2} \qquad \textbf{(10–2)}$$

This formula can be developed from Equation (10–17) by letting $\sigma_y = 0$.

Example Problem 10–12

A solid circular bar is 45 mm in diameter and is subjected to an axial tensile force of 120 kN along with a torque of 1150 N·m. Compute the maximum shear stress in the bar.

Solution

Objective: Compute the maximum shear stress in the bar.

Given: Diameter = D = 45 mm.
Axial force = F = 120 kN = 120 000 N.
Torque = T = 1150 N·m = 1150 000 N·mm.

Analysis: Use Equation (10–2) to compute τ_{max}.

Results:

1. First, the applied normal stress can be found using the direct stress formula.

$$\sigma = F/A$$
$$A = \pi D^2/4 = \pi(45\text{ mm})^2/4 = 1590\text{ mm}^2$$
$$\sigma = (120\,000\text{ N})/(1590\text{ mm}^2) = 75.5\text{ N/mm}^2 = 75.5\text{ MPa}$$

2. Next, the applied shear stress can be found from the torsional shear stress formula.

$$\tau = T/Z_p$$
$$Z_p = \pi D^3/16 = \pi(45\text{ mm})^3/16 = 17\,892\text{ mm}^3$$
$$\tau = (1\,150\,000\text{ N}\cdot\text{mm})/(17\,892\text{ mm}^3) = 64.3\text{ N/mm}^2 = 64.3\text{ MPa}$$

3. Then using Equation (10–2) yields

$$\tau_{max} = \sqrt{\left(\frac{75.5\text{ MPa}}{2}\right)^2 + (64.3\text{ MPa})^2} = 74.6\text{ MPa}$$

Comment: This stress should be compared with the design shear stress.

10–10 STRESS CONDITION ON SELECTED PLANES

There are some cases in which it is desirable to know the stress condition on an element at some selected angle of orientation relative to the reference direction. Figures 10–40 and 10–41 show examples.

The wood block in Figure 10–40 shows that the grain of the wood is inclined at an angle of 30° CCW from the given *x*-axis. Because the wood is very weak in shear parallel to the grain, it is desirable to know the stresses in this direction.

Figure 10–41 shows a member fabricated by welding two components along a seam inclined at an angle relative to the given *x*-axis. The welding operation could produce a weaker material in the near vicinity of the weld, particularly if the component parts were made from steel that was heat-treated prior to welding. The same is true of many aluminum alloys. For such cases, the allowable stresses are somewhat lower along the weld line.

Additional examples include:

- A flat-sided tank made from welded steel plates for which some of the weld lines are inclined at some angle to the major axes along which loads are applied. The loads may be due to pressure in the tank, the weight of the structure, or accessory equipment mounted on the tank. In addition, there may be penetrations in the tank walls to install ports used to fill or drain the tank, for observation windows, or to install sensors. These elements may be welded in place at different angles, thus requiring the knowledge of stress levels along those angles. Many materials such as aluminum or heat-treated steel have significantly lower strength near welds in the heat-affected zone.
- Some large pipes are made from flat sheets of steel that are spiral wound into the desired tubular shape and then continuously welded along the seam. Knowing the stresses in line with the seam is important.
- Composite materials are inherently nonisotropic, being stronger in directions where the strong, stiff reinforcing fibers are laid and typically much weaker at other angles, depending on the manner of laying up the fibers. Analysis of stresses in several directions would be important. See Section 2–12 in Chapter 2.
- The environmental conditions to which a part is exposed during operation may also affect the material properties. For example, a furnace part may be subjected to

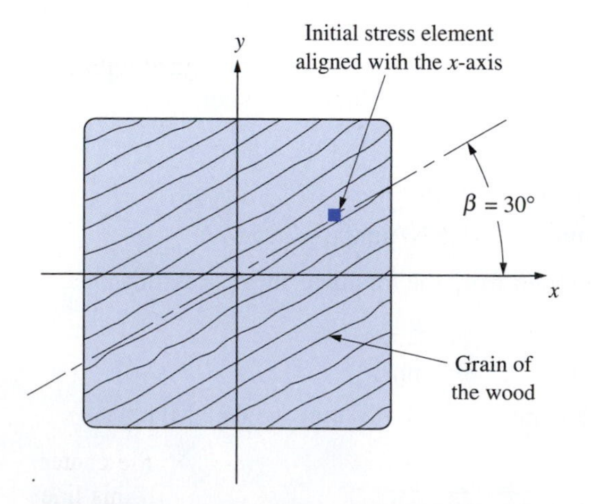

FIGURE 10–40 Cross section of a wood post with the grain running 30° with respect to the *x*-axis.

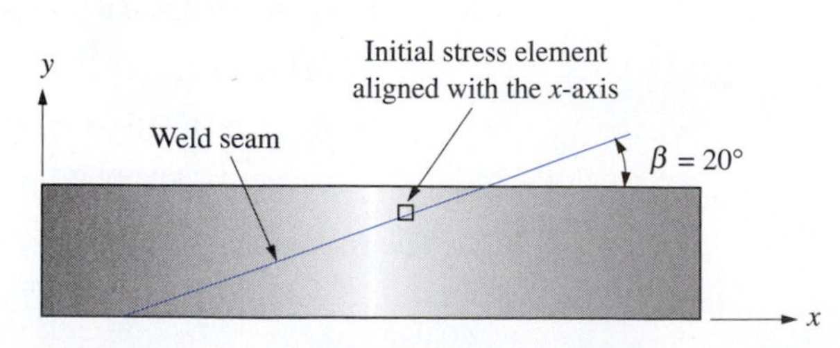

FIGURE 10–41 Flat bar welded along a 20° inclined seam.

local heating by radiant energy along a particular line. The strength of the heated material will be lower than that which remains cool, and thus it is desirable to know the stress condition along the angle of the heat-affected zone.

Mohr's circle can be used to determine the stress condition at specified angles of orientation of the stress element. The procedure is outlined here and demonstrated in Example Problem 10–13.

Procedure for Finding the Stress at a Specified Angle

Given: Stress condition on the given element aligned in the x and y directions.
Objective: Determine the normal and shear stresses on the element at a specified angle, β, relative to the given x direction.

Step 1: Draw the complete Mohr's circle for the element.

Step 2: Identify the line representing the x-axis on the circle.

Step 3: Measure the angle 2β from the x-axis and draw a line through the center of Mohr's circle extending to the two intersections with the circle. This line represents the axis aligned with the direction of interest.

Step 4: Using the geometry of the circle, determine the coordinates (σ and τ) of the first point of intersection nearest to the x-axis. The σ component is the normal stress acting on the element in the direction of β. The τ component is the shear stress acting on the faces of the element. The coordinates of the second point represent the normal and shear stresses acting on the faces of the element of interest that are parallel to the β-axis.

Step 5: Draw the element of interest showing the normal and shear stress acting on it.

Example Problem 10–13

For the flat bar in Figure 10–41 welded along a seam at an angle of 20° CCW from the x-axis, the stress element aligned parallel with the x- and y-axes is subjected to these stresses:

$$\sigma_x = 400 \text{ MPa} \qquad \sigma_y = -300 \text{ MPa} \qquad \tau_{xy} = 200 \text{ MPa CW}$$

Determine the stress condition on the element inclined at the 20° angle, aligned with the weld seam.

Solution

Objective: Draw the stress element aligned with the weld seam at 20° with the x-axis.

Given: Note that the given stress element is the same as that used in Example Problem 10–5. The basic Mohr's circle from that problem is shown in Figure 10–32 and it is reproduced in Figure 10–42 with additional constructor described next.

Analysis: Use the *Procedure for finding the stress at a specified angle.*

Results: ***Steps 1 and 2*** are shown on the original Mohr's circle.

Step 3. The desired axis is one inclined at 20° CCW from the x-axis. Recalling that angles in Mohr's circle are *double* the true angles, we can draw a line through the center of the circle at an angle of $2\beta = 40°$ CCW from the x-axis. The intersection of this line with the circle, labeled A in the figure, locates the point on the circle defining the stress condition of the desired element. The coordinates of this point, (σ_A, τ_A), give the normal and shear stresses acting on one set of faces of the desired stress element.

FIGURE 10–42 Complete Mohr's circle for Example Problem 10–13 showing stresses on an element inclined at 20° CCW from the x-axis.

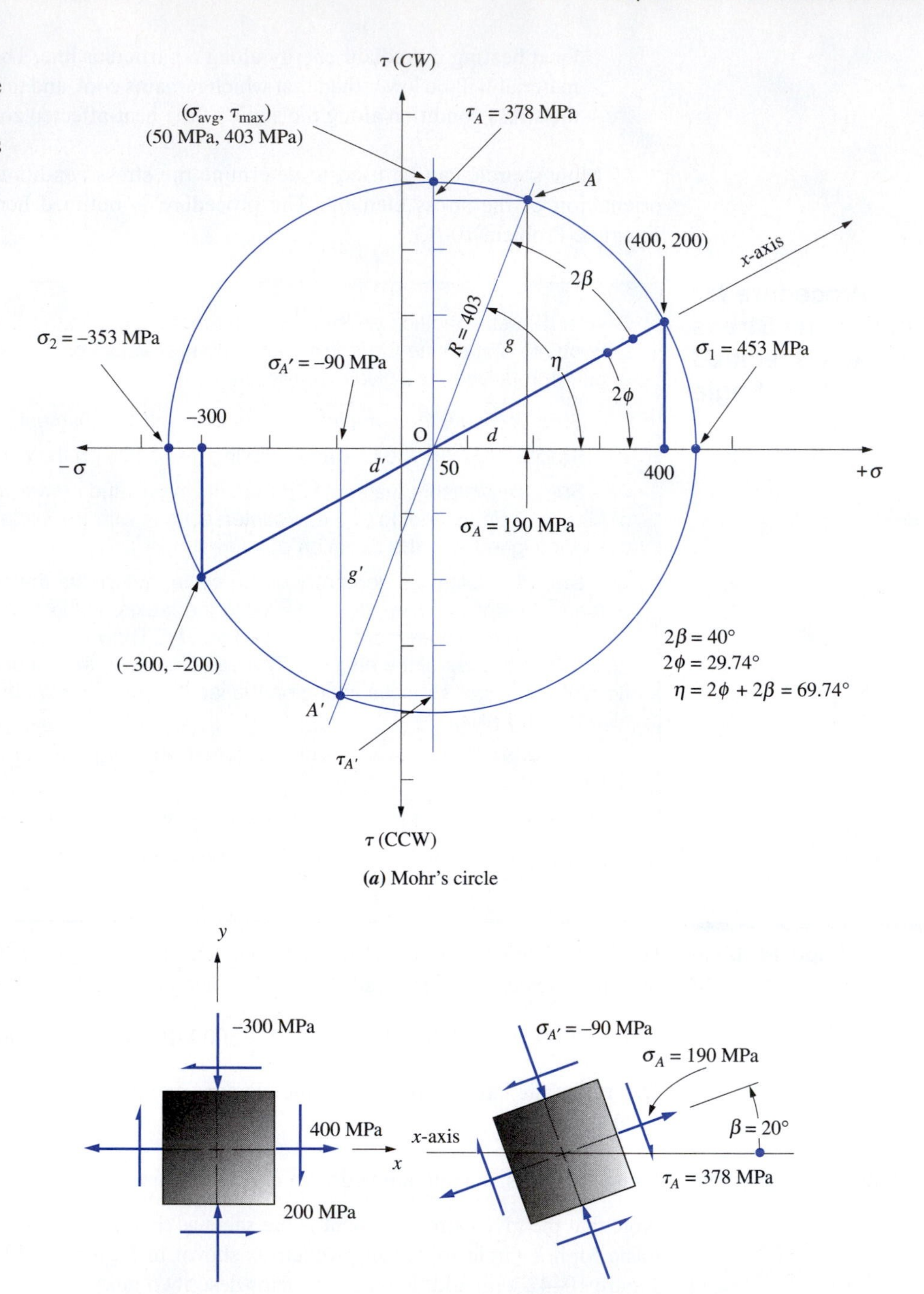

Step 4. Simple trigonometry and the basic geometry of the circle can be used to determine σ_A and τ_A by projecting lines vertically and horizontally from point A to the σ- and τ-axes, respectively. The total angle from the σ-axis to the axis through point A, called η (eta) in the figure, is the sum of 2ϕ and 2β. In Example Problem 10–5, we found $2\phi = 29.74°$. Then

$$\eta = 2\phi + 2\beta = 29.74° + 40° = 69.74°$$

A triangle has been identified in Figure 10–42 with sides labeled d, g, and R. From this triangle, we can compute

$$d = R\cos\eta = (403)\cos 69.74^\circ = 140$$
$$g = R\sin\eta = (403)\sin 69.74^\circ = 378$$

These values enable the computation of

$$\sigma_A = O + d = 50 + 140 = 190\text{ MPa}$$
$$\tau_A = g = 378\text{ MPa CW}$$

where O indicates the value of the normal stress at the center of Mohr's circle.

The stresses on the remaining set of faces of the desired stress element are the coordinates of the point A' located 180° from A on the circle and, therefore, 90° from the faces on which (σ_A, τ_A) act. Projecting vertically and horizontally from A' to the σ- and τ-axes locates σ_A and τ_A. By similar triangles we can say that $d' = d$ and $g' = g$. Then

$$\sigma_{A'} = O - d' = 50 - 140 = -90\text{ MPa}$$
$$\tau_{A'} = g' = 378\text{ MPa CCW}$$

Comment Figure 10–42(c) shows the final element inclined at 20° from the x-axis. This is the stress condition experienced by the material along the weld line.

10–11 SPECIAL CASE IN WHICH BOTH PRINCIPAL STRESSES HAVE THE SAME SIGN

In preceding sections involving Mohr's circle, we used the convention that σ_1 is the maximum principal stress and σ_2 is the minimum principal stress. This is true for cases of plane stress (stresses applied in a single plane) if σ_1 and σ_2 have opposite signs, that is, when one is tensile and one is compressive. Also, in such cases, the shear stress found at the top of the circle (equal to the radius, R) is the true maximum shear stress on the element.

But special care must be exercised when Mohr's circle indicates that σ_1 and σ_2 have the same sign. Even though we are dealing with plane stress, the true stress element is three dimensional and should be represented as a cube rather than a square, as shown in Figure 10–43. Faces 1, 2, 3, and 4 correspond to the sides of the square element and faces 5 and 6 are the "front" and "back." For plane stress the stresses on faces 5 and 6 are zero.

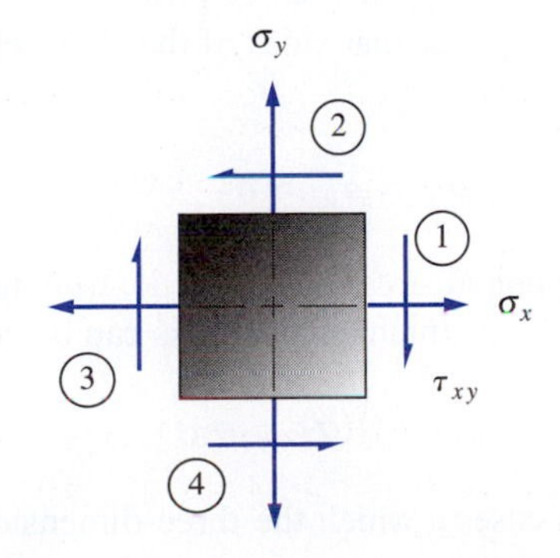

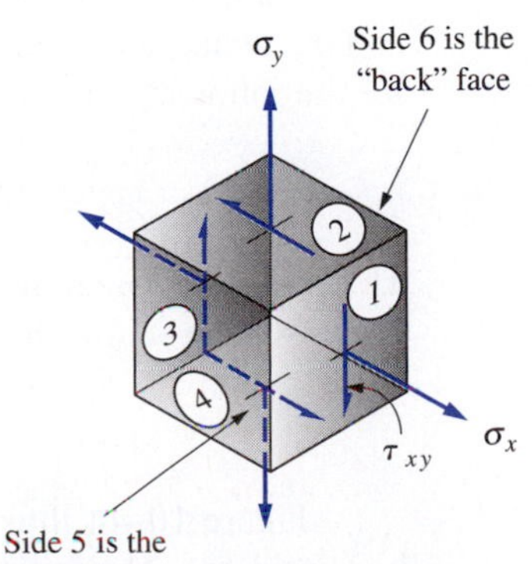

FIGURE 10–43 Plane stress shown as two-dimensional and three-dimensional stress elements. (a) Two-dimensional stress element. (b) Three-dimensional stress element.

FIGURE 10–44
Mohr's circle for which σ_1 and σ_2 are both positive.

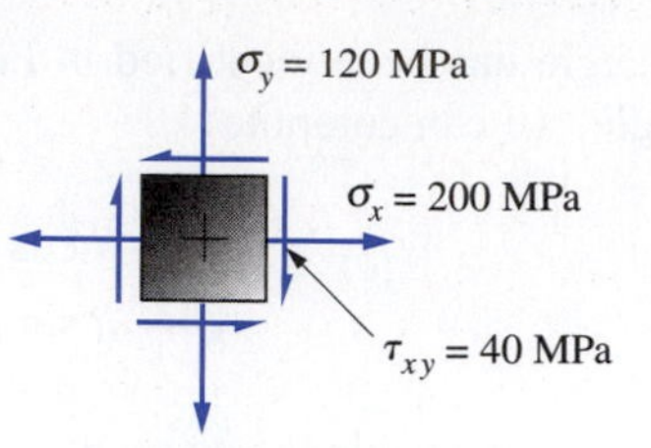

(*a*) Initial stress element

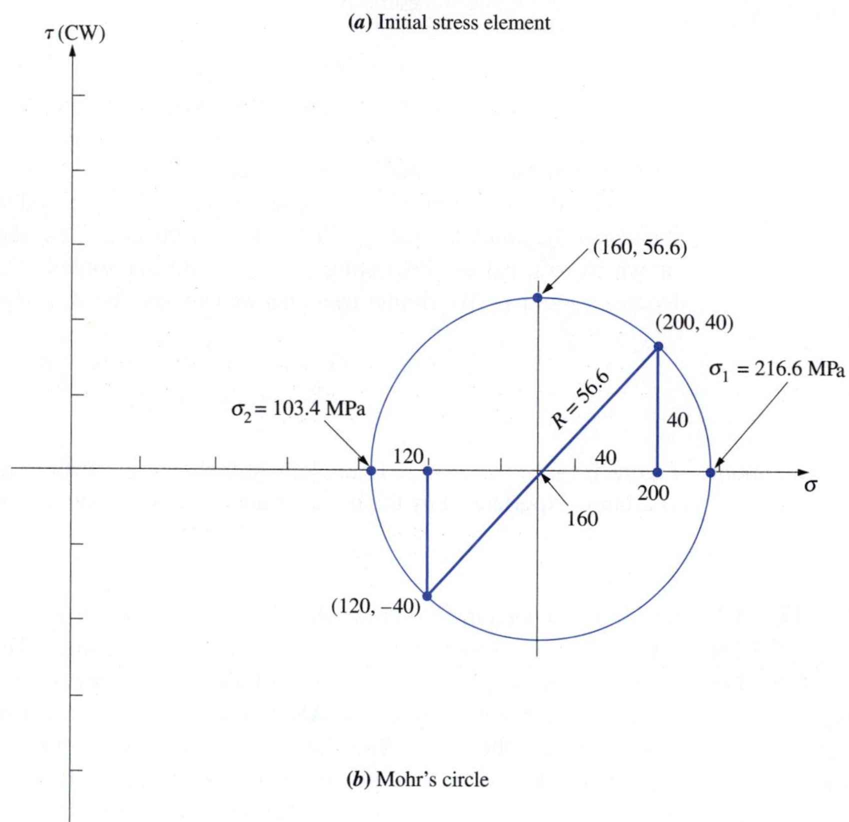

(*b*) Mohr's circle

For the three-dimensional stress element, three principal stresses exist, called σ_1, σ_2, and σ_3, acting on the mutually perpendicular sides of the stress element. Convention calls for the following order:

$$\sigma_1 > \sigma_2 > \sigma_3$$

Thus σ_3 is the true minimum principal stress and σ_1 is the true maximum principal stress. It can also be shown that the true maximum shear stress can be computed from

$$\tau_{max} = \tfrac{1}{2}(\sigma_1 - \sigma_3) \qquad \textbf{(10–21)}$$

Figure 10–44 illustrates a case in which the three-dimensional stress element must be considered. The initial stress element, shown in part (a), carries the following stresses:

$$\sigma_x = 200 \text{ MPa} \qquad \sigma_y = 120 \text{ MPa} \qquad \tau_{xy} = 40 \text{ MPa CW}$$

Part (b) of the figure shows the conventional Mohr's circle, drawn according to the procedure outlined in Section 10–9. Note that both σ_1 and σ_2 are positive or tensile. Then, considering that the stress on the "front" and "back" faces is zero, this is the true minimum principal stress. We can then say that

$$\sigma_1 = 216.6 \text{ MPa}$$
$$\sigma_2 = 103.4 \text{ MPa}$$
$$\sigma_3 = 0 \text{ MPa}$$

From Equation (10–21), the true maximum shear stress is

$$\tau_{max} = \tfrac{1}{2}(\sigma_1 - \sigma_3) = \tfrac{1}{2}(216.6 - 0) = 108.3 \text{ MPa}$$

These concepts can be visualized graphically by creating a set of three Mohr's circles rather than only one. Figure 10–45 shows the circle obtained from the initial stress element, a second circle encompassing σ_1 and σ_3, and a third encompassing σ_2 and σ_3. Thus each circle represents the plane on which two of the three principal stresses act. The point at the top of each circle indicates the largest shear stress that would occur in that plane. Then the largest circle, drawn for σ_1 and σ_3, produces the true maximum shear stess, and its value is consistent with Equation (10–21).

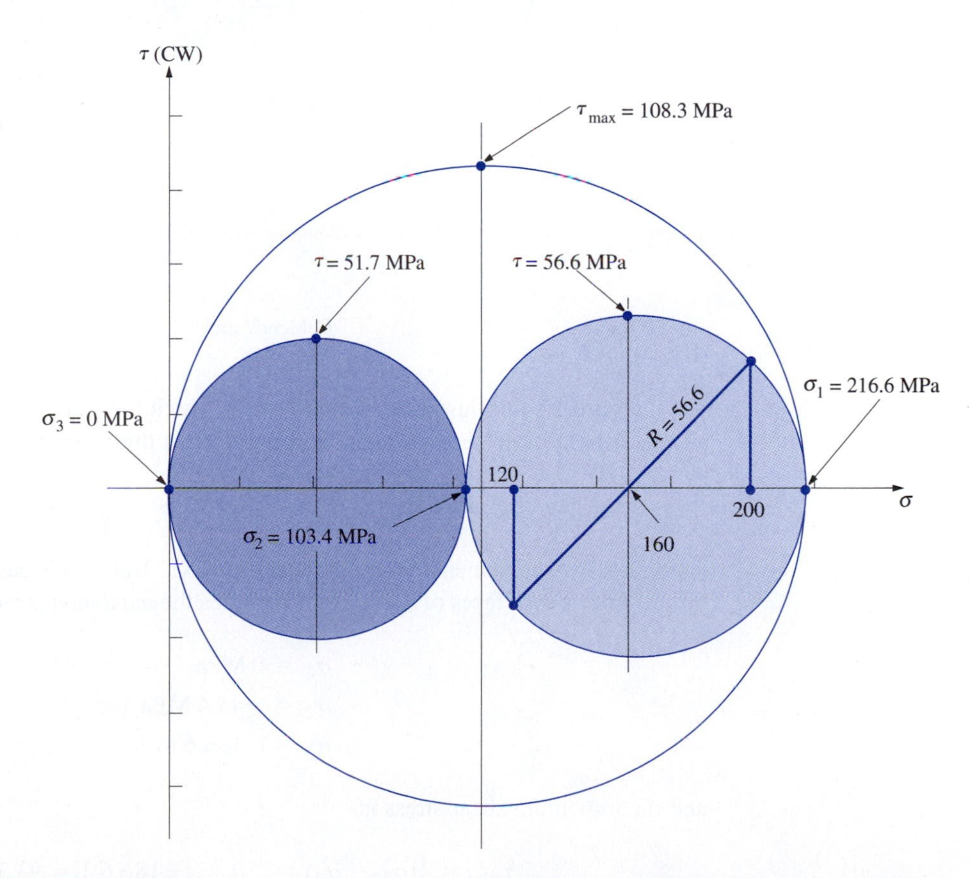

FIGURE 10–45 Three related Mohr's circles showing σ_1, σ_2, σ_3, and τ_{max}.

FIGURE 10–46 Three related Mohr's circles showing σ_1, σ_2, σ_3, and τ_{max}.

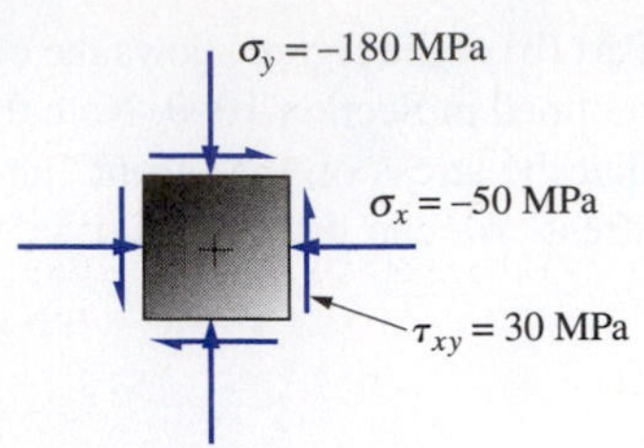

(*a*) Initial stress element

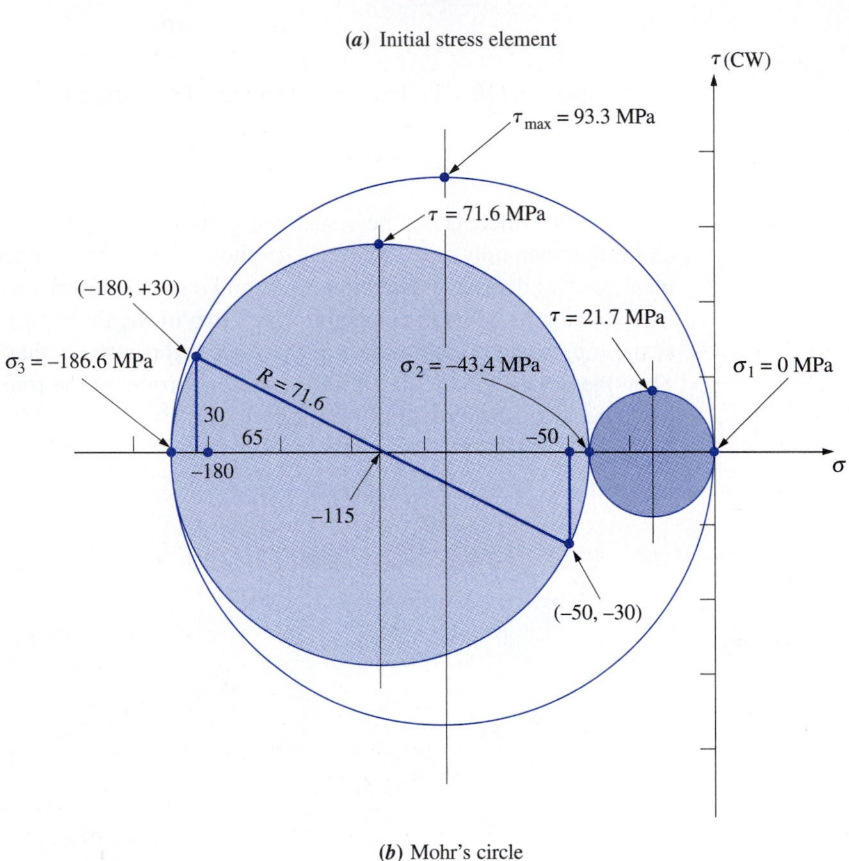

(*b*) Mohr's circle

Figure 10–46 illustrates another case in which the principal stresses from the initial stress element have the same sign, both negative in this case. The initial stresses are

$$\sigma_x = -50 \text{ MPa} \qquad \sigma_y = -180 \text{ MPa} \qquad \tau_{xy} = 30 \text{ MPa CCW}$$

Here, too, the supplementary circles must be drawn. But in this case, the zero stress on the "front" and "back" faces of the element becomes the *maximum* principal stress (σ_1). That is,

$$\sigma_1 = 0 \text{ MPa}$$
$$\sigma_2 = -43.4 \text{ MPa}$$
$$\sigma_3 = -186.6 \text{ MPa}$$

and the maximum shear stress is

$$\tau_{max} = \tfrac{1}{2}(\sigma_1 - \sigma_3) = \tfrac{1}{2}[0 - (-186.6)] = 93.3 \text{ MPa}$$

Summary—Procedures for Dealing with Cases in Which Principal Stresses Have the Same Sign

This summary pertains to the situation in which the Mohr's circle analysis of a stress element subjected to plane stress (stresses applied only in two dimensions) yields the result that both principal stresses (σ_1 and σ_2) are of the same sign; that is, both are tensile or both are compressive. In such cases, the following steps should be completed to provide a true picture of the stress condition on the three-dimensional element.

A. Draw the complete Mohr's circle for the plane stress condition, and identify the principal stresses σ_1 and σ_2.

B. If both principal stresses are tensile (positive) [see example in Figure 10–45]:
   1. Consider that the *zero stress* acting in the direction perpendicular to the initial stress element is the true minimum principal stress. Then it is necessary to define *three* principal stresses such that:

      σ_1 = Maximum principal stress from the first Mohr's circle.

      σ_2 = Minimum principal stress from the first Mohr's circle.

      σ_3 = Zero (true minimum principal stress).

   2. Draw a secondary Mohr's circle whose diameter extends from σ_1 to σ_3 on the σ-axis. The center of the circle will be at the average of σ_1 and σ_3, $(\sigma_1 + \sigma_3)/2$. But, because $\sigma_3 = 0$, the average is $\sigma_1/2$.
   3. The maximum shear stress is at the top of the secondary circle and its value is also $\sigma_1/2$.

C. If both principal stresses are compressive (negative) [see example in Figure 10–46]:
   1. Consider that the *zero stress* acting in the direction perpendicular to the initial stress element is the true maximum principal stress. Then it is necessary to define *three* principal stresses such that:

      σ_1 = Zero (true maximum principal stress).

      σ_2 = Maximum principal stress from the first Mohr's circle.

      σ_3 = Minimum principal stress from the first Mohr's circle

   2. Draw a secondary Mohr's circle whose diameter extends from σ_1 to σ_3 on the σ-axis. The center of the circle will be at the average of σ_1 and σ_3, $(\sigma_1 + \sigma_3)/2$. But, because $\sigma_1 = 0$, the average is $\sigma_3/2$.
   3. The maximum shear stress is at the top of the secondary circle, and its magnitude is also $\sigma_3/2$.

10–12 USE OF STRAIN-GAGE ROSETTES TO DETERMINE PRINCIPAL STRESSES

In the earlier sections of this chapter, data were given for the initial biaxial stress condition that existed on an element of a load-carrying member. Then the principal stresses and the maximum shear stress were calculated either with the equations developed in Sections 10–7 and 10–8 or by the use of Mohr's circle as discussed in Section 10–9. The stresses on the initial stress element may have been determined by direct calculation using the principles of stress analysis as demonstrated in Section 10–4.

An alternate method of determining the initial stress condition is to use experimental stress analysis techniques. This could be used to analyze, for example, the complex case for an automatic transmission for a car or the housing for an air compressor. Photoelastic techniques may first be used to determine where the regions of maximum stress are located. Then strain gages can be applied in those areas to measure more

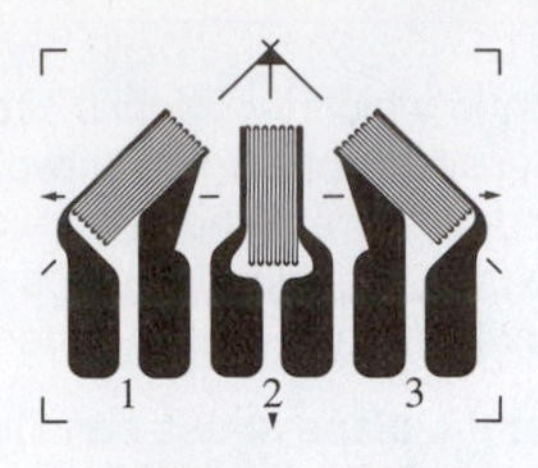

(*a*) 45-degree rosette

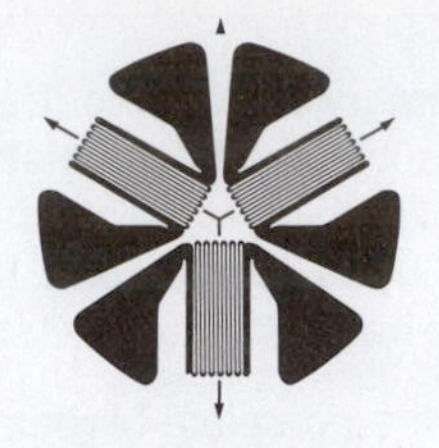
(*b*) 60-degree rosette

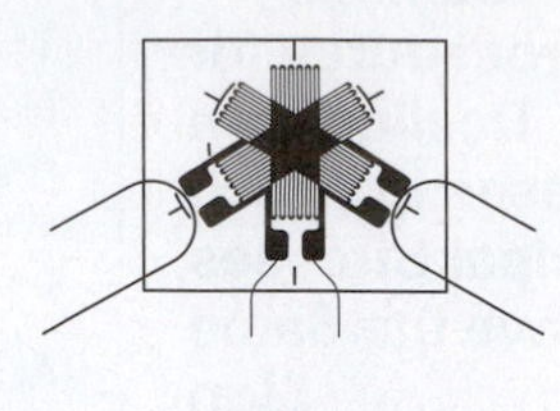
(*c*) Stacked rosette

FIGURE 10–47 Strain-gage rosettes (Source: Measurements Group, Inc., Raleigh, NC)

precisely the magnitudes of strains in particular directions. If known, the strain gages should be aligned as closely as possible with the directions of the principal stresses.

However, in cases of complex geometry and complex loading, the directions of principal stresses are not known. Then the use of a *strain-gage rosette* having an array of three gages with a precise geometric relationship among the gages is recommended. Figure 10–47 shows three different styles and others are available. Part (a) shows a 45-degree rosette, sometimes called a *rectangular rosette* or a 0°, 45°, 90° rosette to indicate the orientation of the three gages. Parts (b) and (c) show two styles of 60-degree rosettes, sometimes called *delta rosettes* or 0°, 60°, 120° rosettes. The stacked rosette in (c) is used where space is tight or where there are large gradients of stress. However, there are possible difficulties in mounting such gages because of their thickness and because the upper gages are displaced from the true surface where the strains are to be measured.

In general, the gage marked *number 1* is installed carefully aligned with some reference axis on the part to be tested. After mounting, the member is loaded and gage readings are taken from all three gages with their output typically called ϵ_1, ϵ_2, and ϵ_3. From these measurements it can be shown that the maximum and minimum strains can be computed from the following equations. Note that different equations are used for the 45-degree rosette and for the 60-degree rosette because the angle between the gages enters into the calculations. Also, it is necessary to determine the orientation of the principal strain axes relative to the direction of gage number one. The angle from gage number one to the nearest principal strain is called β and is computed from the equation that follows the strain equations.

45° Rosette:

$$\epsilon_{\max} = \frac{(\epsilon_1 + \epsilon_3)}{2} + \frac{\sqrt{(\epsilon_1 - \epsilon_2)^2 + (\epsilon_2 - \epsilon_3)^2}}{\sqrt{2}} \quad \textbf{(10–22)}$$

$$\epsilon_{\min} = \frac{(\epsilon_1 + \epsilon_3)}{2} - \frac{\sqrt{(\epsilon_1 - \epsilon_2)^2 + (\epsilon_2 - \epsilon_3)^2}}{\sqrt{2}} \quad \textbf{(10–23)}$$

$$\beta = \frac{1}{2}\tan^{-1}\left[\frac{(\epsilon_2 - \epsilon_3) - (\epsilon_1 - \epsilon_2)}{(\epsilon_1 - \epsilon_3)}\right] \quad \textbf{(10–24)}$$

60° Rosette:

$$\epsilon_{\max} = \frac{(\epsilon_1 + \epsilon_2 + \epsilon_3)}{3} + \frac{\sqrt{2}}{3}\sqrt{(\epsilon_1 - \epsilon_2)^2 + (\epsilon_2 - \epsilon_3)^2 + (\epsilon_1 - \epsilon_3)^2} \quad \textbf{(10–25)}$$

$$\epsilon_{\min} = \frac{(\epsilon_1 + \epsilon_2 + \epsilon_3)}{3} - \frac{\sqrt{2}}{3}\sqrt{(\epsilon_1 - \epsilon_2)^2 + (\epsilon_2 - \epsilon_3)^2 + (\epsilon_1 - \epsilon_3)^2} \quad \textbf{(10–26)}$$

$$\beta = \frac{1}{2}\tan^{-1}\left[\frac{\sqrt{3}(\epsilon_2 - \epsilon_3)}{(\epsilon_1 - \epsilon_2) + (\epsilon_1 - \epsilon_3)}\right] \quad \textbf{(10–27)}$$

The measured strains are typically very small. For example, with using U.S. Customary units, the order of magnitude is usually under 5000×10^{-6} in/in. For convenience, some experimental stress analysts write this as 5000 $\mu\varepsilon$, read "5000 microstrains." Calculations in equations such as those in Equations (10–22) through (10–27) are then completed using only the whole number while recognizing that the actual complete value must be considered.

Principal Stresses Obtained from Principal Strains. The ultimate goal is to determine the principal stresses from these measurements of strain using the strain-gage rosette. To accomplish that goal we need to review some concepts and relationships discussed in Chapters 1–3.

We discussed the relationship between stress and strain by defining the *modulus of elasticity, E.* You should recall that, for *uniaxial stress,* such as direct tension or compression,

$$\sigma = E\epsilon$$

That is, stress is the product of the strain times the modulus of elasticity of the material.

Here we are considering the more general case of *biaxial stress,* where stresses in two directions occur simultaneously. Now we must recall the fact that when a load-carrying member undergoes deformation in one direction, there also occurs deformation in the perpendicular directions. Figure 10–48 shows this concept as it was defined in Chapter 2. We use the term *Poisson's ratio,* ν, to represent the ratio of the normal strain in the direction of the applied stress and the lateral strain in the perpendicular directions. Poisson's ratio is a property of the material, and typical values are given in Table 2–1.

Now, if we know the strain in one direction, say ϵ_x, the strain in the y-direction, perpendicular to x, is

$$\epsilon_y = \nu\epsilon_x$$

FIGURE 10–48 Illustration of Poisson's ratio for an element in tension.

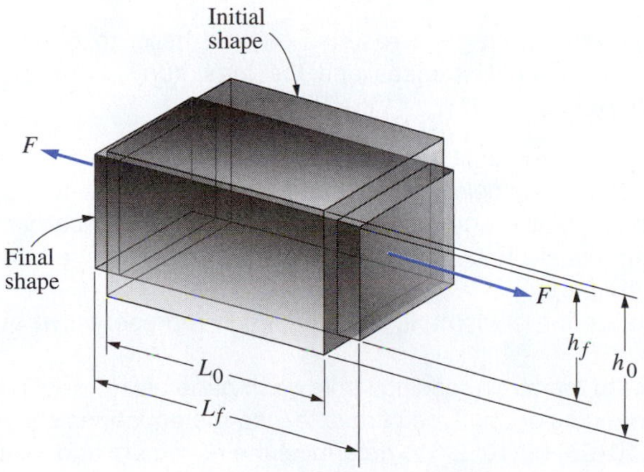

$$\text{Axial strain} = \frac{L_f - L_0}{L_0} = E_a$$

$$\text{Lateral strain} = \frac{h_f - h_0}{h_0} = E_L$$

$$\text{Poisson's ratio} = \frac{-E_L}{E_a} = \nu$$

Under the influence of biaxial strain, the strain in each direction will affect the strain in the other direction. Thus the calculation of the maximum stresses using the maximum strains is more complex than $\sigma = E\epsilon$. The development of the relationship is not shown here. The final result is

$$\sigma_{max} = \frac{E}{1 - \nu^2}(\epsilon_{max} + \nu\epsilon_{min}) \quad \textbf{(10–28)}$$

$$\sigma_{min} = \frac{E}{1 - \nu^2}(\epsilon_{min} + \nu\epsilon_{max}) \quad \textbf{(10–29)}$$

The maximum shearing strain in the plane of the original element, γ_{max}, can also be found from the absolute value of the difference between the maximum and minimum normal strains in the plane.

$$\gamma_{max} = |(\epsilon_{max} - \epsilon_{min})| \quad \textbf{(10–30)}$$

The units for γ are radians, considered to be dimensionless. Then the maximum shearing stress can be computed using the definition of G, the shear modulus of elasticity.

$$\tau_{max} = G\gamma_{max} \quad \textbf{(10–31)}$$

The units for τ will be the same as those for G. An alternate form for G was shown in Chapter 2. Then Equation 10–31 can also be written

$$\tau_{max} = \frac{E\gamma_{max}}{2(1 + \nu)} \quad \textbf{(10–32)}$$

The same precautions apply when the maximum and minimum principal stresses have the same sign as discussed in Section 10–11.

Procedure for Analyzing the Data from a Strain-Gage Rosette

In summary, strain-gage rosettes can be used to determine the maximum and minimum principal stresses and the maximum shearing stress by the following procedure:

1. Determine the area where maximum stresses are likely to occur. You may use judgment, photoelastic stress analysis, or finite element analysis.
2. Apply a strain-gage rosette to the area of the load-carrying member where measurements are to be taken, being careful to align gage number 1 with a known axis on the part.
3. Operate the equipment, applying loads representative of the largest expected in service.
4. Record the strain readings from each gage giving values for ϵ_1, ϵ_2, and ϵ_3.
5. Depending on the type of rosette, use the equations shown as (10–22) to (10–24) or (10–25) to (10–27) to determine the maximum and minimum strains at the point of interest and the angle of orientation of these strains relative to the alignment of gage number 1.
6. Use Equations (10–28) and (10–29) to compute the maximum and minimum principal stresses. The directions of the maximum principal stresses are the same as the directions of the principal strains.
7. Compute the maximum shearing strain from Equation (10–30).

8. Compute the maximum shearing stress from Equation (10–31) or (10–32).
9. Check to see if the maximum and minimum principal stresses have the same sign. If so, complete the additional analysis discussed in Section 10–11.

Example Problem 10–14

As part of the product development process for a new pump, the designer has instrumented a critical area of the housing with a rectangular strain-gage rosette like that shown in Figure 10–47(a). Gage 1 is aligned with the horizontal centerline of the pump inlet passage. During the test under high-capacity operating conditions, the following readings were taken for the strains from the three arms of the gage: $\epsilon_1 = 950 \times 10^{-6}$ in/in, $\epsilon_2 = -375 \times 10^{-6}$ in/in, $\epsilon_3 = 525 \times 10^{-6}$ in/in. The material for the pump housing is aluminum 2014-T6. Compute the maximum principal stress, the minimum principal stress, and the maximum shear stress at the location where the rosette is mounted.

Solution

Objective: Compute σ_{max}, σ_{min}, and τ_{max}.

Given: Strain readings from a 0°, 45°, 90° strain-gage rosette.

$$\epsilon_1 = 950 \times 10^{-6}\ \text{in/in},\ \epsilon_2 = -375 \times 10^{-6}\ \text{in/in},\ \epsilon_3 = 525 \times 10^{-6}\ \text{in/in}.$$

Aluminum 2014-T6: $E = 10.6 \times 10^6$ psi, $s_y = 60\,000$ psi (Appendix A–18) $\nu = 0.33$ (Table 2–1)

Analysis: Use the *Procedure for analyzing the data from a strain-gage rosette.*

Results: ***Steps 1–4*** have already been completed.

Step 5: In using Equations (10–22) to (10–24), we will show only the whole number part of the strain values. The unit is then microstrains, $\mu\epsilon$.

The maximum principal strain is

$$\epsilon_{max} = \frac{(\epsilon_1 + \epsilon_3)}{2} + \frac{\sqrt{(\epsilon_1 - \epsilon_2)^2 + (\epsilon_2 - \epsilon_3)^2}}{\sqrt{2}}$$

$$\epsilon_{max} = \frac{(950 + 525)}{2} + \frac{\sqrt{[(950 - (-375)]^2 + (-375 - 525)^2}}{\sqrt{2}} = 1870\ \mu\epsilon$$

The minimum principal strain is

$$\epsilon_{min} = \frac{(\epsilon_1 + \epsilon_3)}{2} - \frac{\sqrt{(\epsilon_1 - \epsilon_2)^2 + (\epsilon_2 - \epsilon_3)^2}}{\sqrt{2}}$$

$$\epsilon_{min} = \frac{(950 + 525)}{2} - \frac{\sqrt{[(950 - (-375)]^2 + (-375 - 525)^2}}{\sqrt{2}} = -395\ \mu\epsilon$$

The angle from gage number 1 to the nearest principal strain axis is

$$\beta = \frac{1}{2}\tan^{-1}\left[\frac{(\epsilon_2 - \epsilon_3) - (\epsilon_1 - \epsilon_2)}{(\epsilon_1 - \epsilon_3)}\right]$$

$$\beta = \frac{1}{2}\tan^{-1}\left[\frac{(-375 - 525) - [950 - (-375)]}{(950 - 525)}\right] = -39.6°$$

Step 6: The maximum principal stress is found from Equation (10–28).

$$\sigma_{max} = \frac{E}{1 - \nu^2}(\epsilon_{max} + \nu\epsilon_{min})$$

$$\sigma_{max} = \frac{10.6 \times 10^6 \text{ psi}}{1 - (0.33)^2}[1870 + 0.33(-395)](10^{-6}) = 20\,694 \text{ psi}$$

The minimum principal stress is found from Equation (10–29).

$$\sigma_{min} = \frac{E}{1 - \nu^2}(\epsilon_{min} + \nu\epsilon_{max})$$

$$\sigma_{min} = \frac{10.6 \times 10^6 \text{ psi}}{1 - (0.33)^2}[-395 + 0.33(1870)](10^{-6}) = 2642 \text{ psi}$$

Step 7: The maximum shearing strain is found from Equation (10–30).

$$\gamma_{max} = |(\epsilon_{max} - \epsilon_{min})| = |1870 - (-395)| = 2265 \times 10^{-6} \text{ rad}$$

Step 8: The maximum shearing stress in the plane of the initial element is found from Equation (10–32).

$$\tau_{max} = \frac{E\gamma_{max}}{2(1 + \nu)} = \frac{(10.6 \times 10^6 \text{ psi})(2265 \times 10^{-6})}{2(1 + 0.33)} = 9026 \text{ psi}$$

Step 9: We note that both the maximum principal stress and the minimum principal stress are positive or tensile. Therefore, we must draw a supplemental Mohr's circle to determine the true maximum shearing strain. See Figure 10–49. The circle is simply

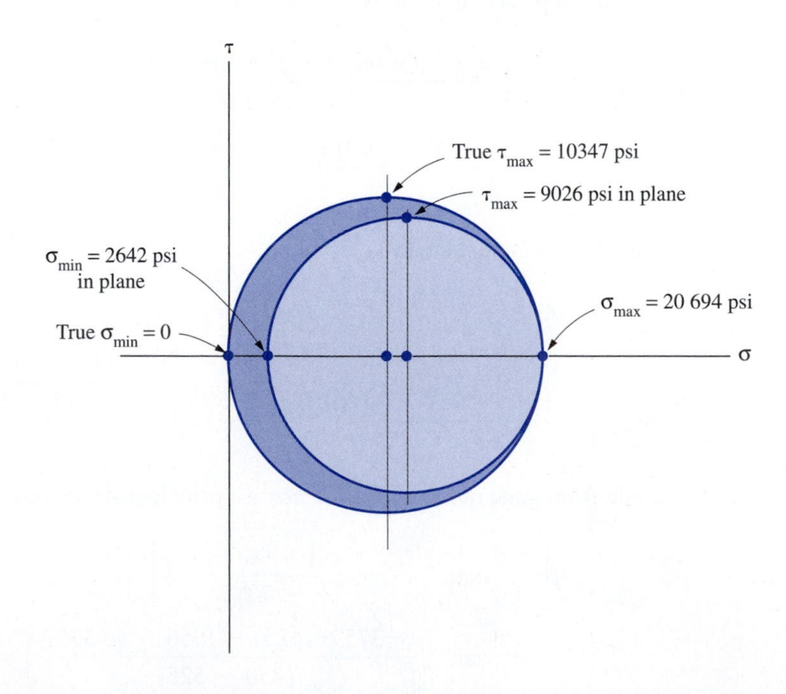

FIGURE 10–49 Mohr's circles for Example Problem 10–14.

drawn by plotting both principal stresses on the horizontal axis and drawing the circle to include both. The supplementary circle is drawn through σ_{max} and the origin of the axes because that represents a zero stress perpendicular to the plane of the initial stress element. The true maximum shearing stress is equal to the radius of this circle, found from

$$\tau_{max} = (\sigma_{max} - 0)/2 = (20\,694 \text{ psi})/2 = 10\,347 \text{ psi}$$

Summary The final results are:

$\sigma_{max} = 20\,694$ psi $\quad \sigma_{min} = 2642$ psi in the plane of the initial element

True $\sigma_{min} = 0$ perpendicular to the plane of the initial element

$\tau_{max} = 10\,347$ psi

Spreadsheet Solution for Calculations for Strain-Gage Rosettes. The calculations for stress and strain for rosettes is cumbersome, time-consuming, and prone to error. It is very desirable to use a spreadsheet, computer algebra routine, or programmable calculator to perform the calculations. Figure 10–50 shows a spreadsheet that accomplishes this objective. Here are some of its features.

1. The nine-step *Procedure for analyzing the data from a strain-gage rosette* is implemented.
2. It assumes that the stress condition is biaxial and that the stress perpendicular to the plane of the measured strains is zero.
3. It includes sections for both rectangular and delta-style rosettes.
4. It allows the use of either U.S. Customary units or SI metric units.
5. Input data required are highlighted by shading the appropriate cells.
6. Enter material property data for modulus of elasticity, E, and Poisson's ratio, ν. For E, enter only the significant digits as shown. It is assumed that E is in millions of psi or GPa (10^9 Pa).
7. Enter strain data as whole numbers of microstrains. The spreadsheet uses this value times 10^{-6} in/in or 10^{-6} m/m. Of course, the strain data are actually dimensionless.
8. The first set of data shown in the upper left of the spreadsheet is identical to the data from Example Problem 10–14.
9. Take care in interpreting the results for the maximum shear stress. Two values are calculated. One is the maximum shear stress in the plane of the initial stress element. This is the true maximum shear stress only if the maximum and minimum principal stresses have different signs. If they have the same signs, the alternate calculation labeled "True Max Shear Stress" is used. This recognizes that there is a larger shear stress in a plane other than the plane of the initial element. A supplementary Mohr's circle is drawn as discussed in Section 10–11 and as demonstrated in Example Problem 10–14 and Figure 10–49.

FIGURE 10–50
Spreadsheet with calculations for strain-gage rosettes.

SPREADSHEET FOR COMPUTING PRINCIPAL STRAINS AND STRESSES FROM STRAIN-GAGE ROSETTE OUTPUT DATA			
Refer to Section 10–12 for method. Input data in shaded elements			
Material Properties	*U.S. Customary Units*		*SI Metric Units*
Modulus of Elasticity	10.6×10^6 psi		73.1×10^9 Pa
Poisson's Ratio	0.33		0.33
Rectangular [0, 45, 90 degree] Rosette Data [Uses Equations 10–22 to 10–24]			
	U.S. Customary Units		*SI Metric Units*
Strain from Gage 1	950×10^{-6} in/in		950×10^{-6} m/m
Strain from Gage 2	-375×10^{-6} in/in		-375×10^{-6} m/m
Strain from Gage 3	525×10^{-6} in/in		525×10^{-6} m/m
	Results:		***Results:***
Max Principal Strain	1870×10^{-6} in/in		1870×10^{-6} m/m
Min Principal Strain	-395×10^{-6} in/in		-395×10^{-6} m/m
Angle, β, from the axis of gage 1 to the nearer principal axis			
Angle β	–39.6 degrees		–39.6 degrees
Max Principal Stress	20695 psi		143 MPa
Min Principal Stress	2641 psi		18.2 MPa
Max Shear Strain	2265×10^{-6} radians [Dimensionless]		2265×10^{-6} radians
Max Shear Stress	9027 psi [in plane of initial element]		62.3 MPa
*******Only in the case that Max and Min Principal Stresses have the same sign******			
True Max Shear Stress	10347 psi		71.4 MPa
[Assuming stress = 0 perpendicular to plane of initial element]			
Delta [0, 60, 120 degree] Rosette Data [Uses Equations 10–25 to 10–27]			
	U.S. Customary Units		*SI Metric Units*
Strain from Gage 1	1250×10^{-6} in/in		1250×10^{-6} m/m
Strain from Gage 2	-235×10^{-6} in/in		-235×10^{-6} m/m
Strain from Gage 3	645×10^{-6} in/in		645×10^{-6} m/m
	Results:		***Results:***
Max Principal Strain	1416×10^{-6} in/in		1416×10^{-6} m/m
Min Principal Strain	-309×10^{-6} in/in		-309×10^{-6} m/m
Angle, β, from the axis of gage 1 to the nearer principal axis			
Angle β	–18.1 degrees		–18.1 degrees
Max Principal Stress	15626 psi		108 MPa
Min Principal Stress	1882 psi		13.0 MPa
Max Shear Strain	1725×10^{-6} radians [Dimensionless]		1725×10^{-6} radians
Max Shear Stress	6872 psi [in plane of initial element]		47.4 MPa
*******Only in the case that Max and Min Principal Stresses have the same sign******			
True Max Shear Stress	7813 psi		53.9 MPa
[Assuming stress = 0 perpendicular to plane of initial element]			

REFERENCES

1. Aluminum Association, *Aluminum Design Manual,* Washington, DC, 2005.
2. American Institute of Steel Construction, *Steel Construction Manual,* 13th ed., Chicago, 2005.
3. Budynas, R. G. and J. K. Nisbett, *Shigley's Mechanical Engineering Design,* 8th ed., McGraw-Hill, New York, 2007.
4. Dally, James W. and W. F. Riley, *Experimental Stress Analysis,* 4th ed., College House Enterprises, Knoxville, TN, 2005.
5. Doyle, James F., *Modern Experimental Stress Analysis,* John Wiley & Sons, New York, 2004.
6. Doyle, James F. and J. W. Phillips, eds., *Manual on Experimental Stress Analysis,* 5th ed., Society for Experimental Stress Analysis, Westport, CT, 1989.
7. Mott, Robert L., *Machine Elements in Mechanical Design,* 4th ed., Prentice Hall, Upper Saddle River, NJ, 2004.
8. Young, W. C. and R. D. Cook, *Advanced Mechanics of Materials,* 2nd ed., Prentice Hall, Upper Saddle River, NJ, 1999.

INTERNET SITES

1. Vishay Micro-Measurements www.vishay.com Through its Measurement Group, Vishay produces strain gages and associated equipment, photoelastic stress measurement devices and materials, test and measurement systems, and accessories. From the complete listing of products, select *strain gages, PhotoStress® Plus,* or *Test and measurements.*
2. Vishay—Experimental Stress Analysis Notebook www.vishay.com/company/brands/measurements-group/schools/nbindex.htm A collection of 30 issues of the Notebook that covers details of strain-gage technology and many practical applications and student projects that use strain gages.
3. HBM www.hbm.com A member of the Spectris plc instrumentation and electronic controls group, HBM produces strain gages and a variety of sensors, transducers, and electronic instrumentation systems.
4. Society for Experimental Mechanics (SEM) http://sem.org SEM promotes research and application of engineering measurement using experimental methods for testing materials and the forces that affect them. Interest areas are materials, modeling and analysis, strain measurement, and structural testing.
5. Stress Photonics, Inc. www.stressphotonics.com Providers of full-field stress and strain measurement systems and real time nondestructive evaluation (NDE) and inspection systems, including the Grey-field Polariscope Photoelastic Strain Measurement System and the DeltaTherm Thermoelastic Stress Measurement System.
6. Omega.com www.omega.com/literature/transactions/volume3/strain.html An online introduction to strain-gage technology, sensor designs, measuring circuits, applications, and installation.
7. Geokon, Incorporated www.geokon.com Designers and manufacturers of geotechnical instrumentation for the construction industry, including embedment strain gages, weldable strain gages, rebar strain meters, joint meters, strand meters for cables, and many other devices. Select *Product Information.*

PROBLEMS

Combined Normal Stresses

10–1.E A $2\frac{1}{2}$-in schedule 40 steel pipe is used as a support for a basketball backboard, as shown in Figure P10–1. It is securely fixed into the ground. Compute the stress that would be developed in the pipe if a 230-lb player hung on the base of the rim of the basket.

10–2.M The bracket shown in Figure P10–2 has a rectangular cross section 18 mm wide by 75 mm high. It is securely attached to the wall. Compute the maximum stress in the bracket.

10–3.E The beam shown in Figure P10–3 carries a 6000-lb load attached to a bracket below the beam. Compute the stress at points M and N, where it is attached to the column.

10–4.E For the beam shown in Figure P10–3, compute the stress at points M and N if the 6000-lb load acts vertically downward instead of at an angle.

10–5.E For the beam shown in Figure P10–3, compute the stress at points M and N if the 6000-lb load acts back toward the column at an angle of 40 deg below the horizontal instead of as shown.

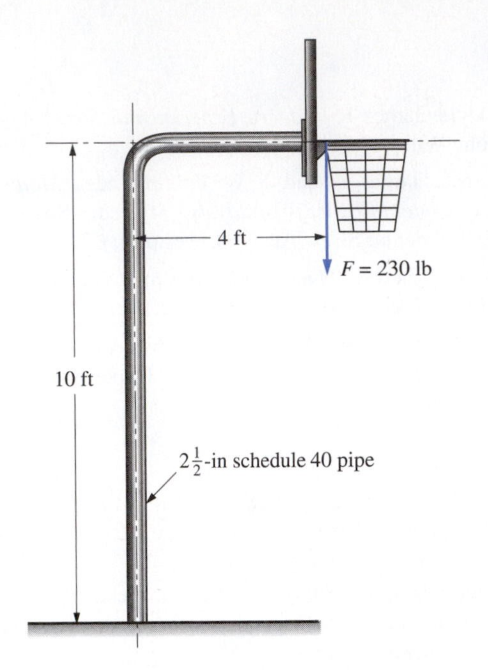

FIGURE P10–1 Basketball backboard for Problem 10–1.

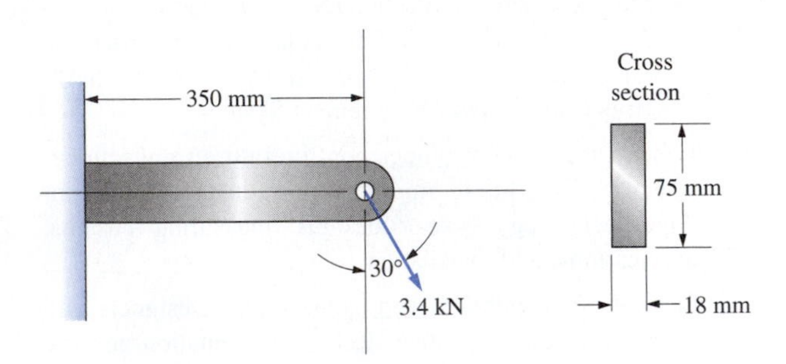

FIGURE P10–2 Bracket for Problem 10–2.

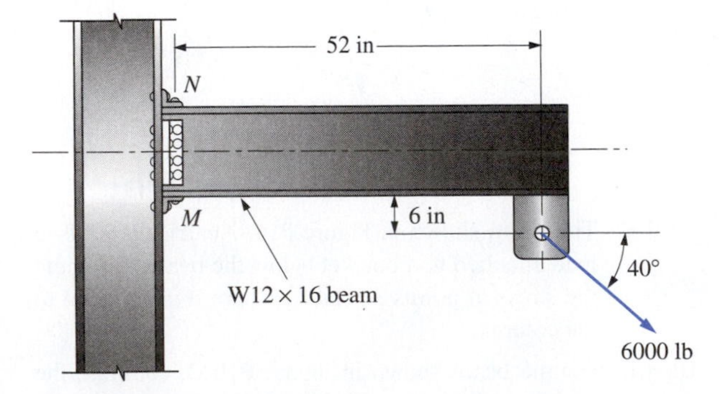

FIGURE P10–3 Beam for Problems 10–3, 10–4, and 10–5.

10–6.M Compute the maximum stress in the top portion of the coping saw frame shown in Figure P10–6 if the tension in the blade is 125 N.

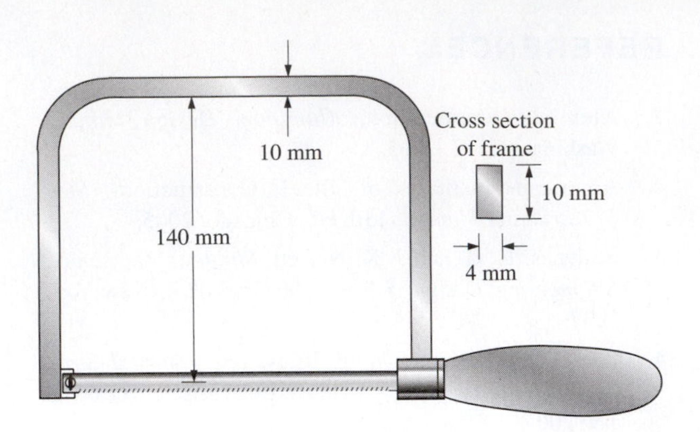

FIGURE P10–6 Coping saw frame for Problem 10–6.

10–7.M Compute the maximum stress in the crane beam shown in Figure P10–7 when a load of 12 kN is applied at the middle of the beam.

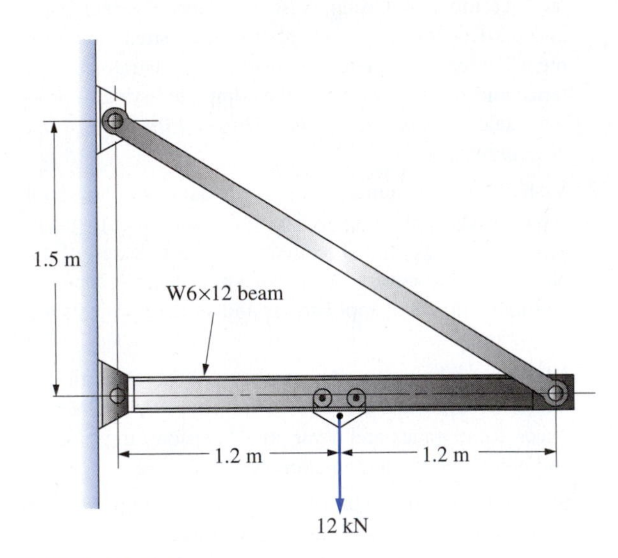

FIGURE P10–7 Crane beam for Problem 10–7.

10–8.M Figure P10–8 shows a metal-cutting hacksaw. Its frame is made of hollow tubing having an outside diameter of 12 mm and a wall thickness of 1.0 mm. The blade is pulled taut by the wing nut so that a tensile force of 160 N is applied to the blade. Compute the maximum stress in the top section of the tubular frame.

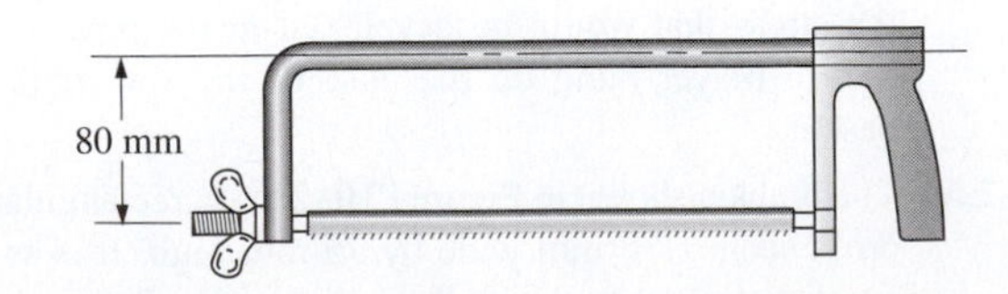

FIGURE P10–8 Hacksaw frame for Problem 10–8.

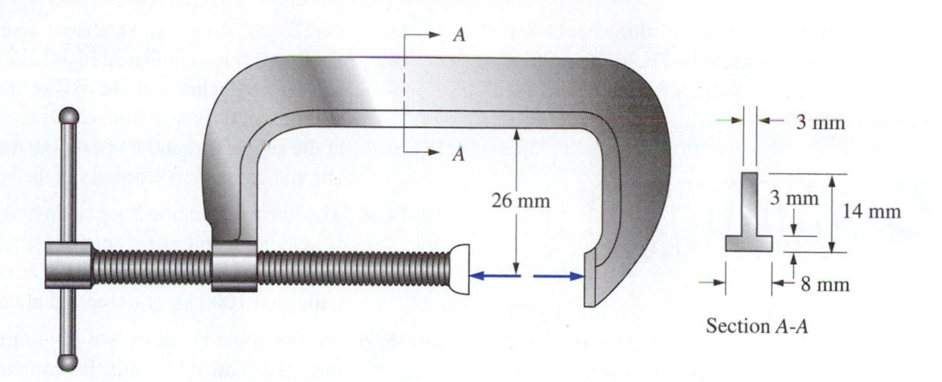

FIGURE P10–9 C-clamp for Problem 10–9.

10–9.M The C-clamp in Figure P10–9 is made of cast zinc, ZA12. Determine the allowable clamping force that the clamp can exert if it is desired to have a design factor of 4 based on ultimate strength in either tension or compression.

10–10.M The C-clamp shown in Figure P10–10 is made of cast malleable iron, ASTM A220 grade 45008. Determine the allowable clamping force that the clamp can exert if it is desired to have a design factor of 4 based on ultimate strength in either tension or compression.

10–11.M A tool used for compressing a coil spring to allow its installation in a car is shown in Figure P10–11. A force of 1200 N is applied near the ends of the extended lugs, as shown. Compute the maximum tensile stress in the threaded rod. Assume a stress

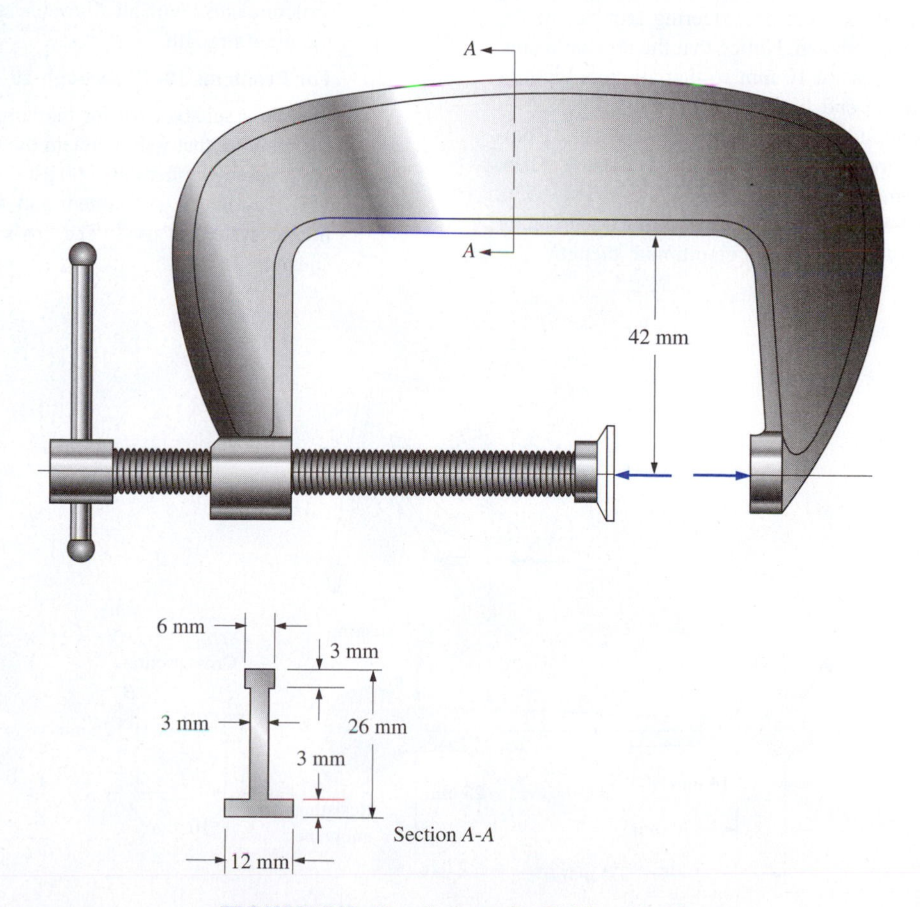

FIGURE P10–10 C-clamp for Problem 10–10.

concentration factor of 3.0 at the thread root for both tensile and bending stresses. Then, using a design factor of 2 based on yield strength, specify a suitable material for the rod.

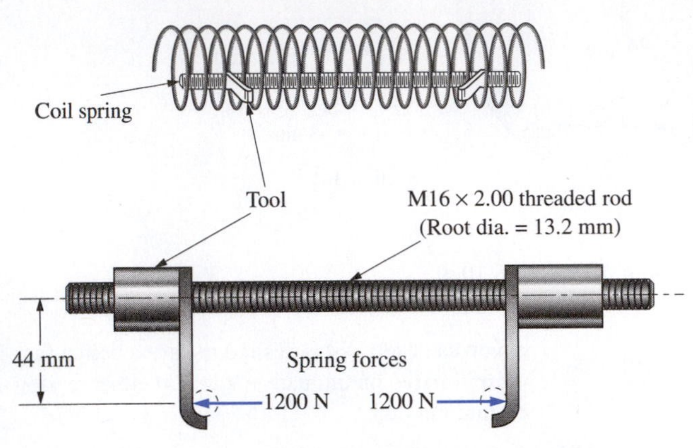

FIGURE P10–11 Tool for compressing springs for Problem 10–11.

10–12.M Figure P10–12 shows a portion of the steering mechanism for a car. The steering arm has the detailed design shown. Notice that the arm has a constant thickness of 10 mm so that all cross sections between the end lugs are rectangular. The 225-N force is applied to the arm at a 60-deg angle. Compute the stress in the arm at sections *A* and *B*. Then, if the arm is made of ductile iron, ASTM A536, grade 80-55-6, compute the minimum design factor at these two points based on ultimate strength.

10–13.E An S3×5.7 American Standard I-beam is subjected to the forces shown in Figure P10–13. The 4600-lb force acts directly in line with the axis of the beam. The 500-lb downward force at *A* produces the reactions shown at the supports *B* and *C*. Compute the maximum tensile and compressive stresses in the beam.

10–14.M The horizontal crane boom shown in Figure P10–14 is made of a hollow rectangular steel tube. Compute the stress in the boom just to the left of point *B* when a mass of 1000 kg is supported at the end.

10–15.M For the crane boom shown in Figure P10–14, compute the load that could be supported if a design factor of 3 based on yield strength is desired. The boom is made of AISI 1040 hot-rolled steel. Analyze only sections where the full box section carries the load, assuming that sections at connections are adequately reinforced.

10–16.E The member *EF* in the truss shown in Figure P10–16 carries an axial tensile load of 54 000 lb in addition to the two 1200-lb loads shown. It is planned to use two steel angles for the member, arranged back to back. Specify a suitable size for the angles if they are to be made of ASTM A36 structural steel with an allowable stress of 0.6 times the yield strength.

For Problems 10–17 through 10–20:

Specify a suitable size for the horizontal portion of the member that will maintain the combined tensile or compressive stress to 6000 psi if the data are in the U.S. Customary unit system and 42 MPa if the SI metric system is used. The cross section is to be square.

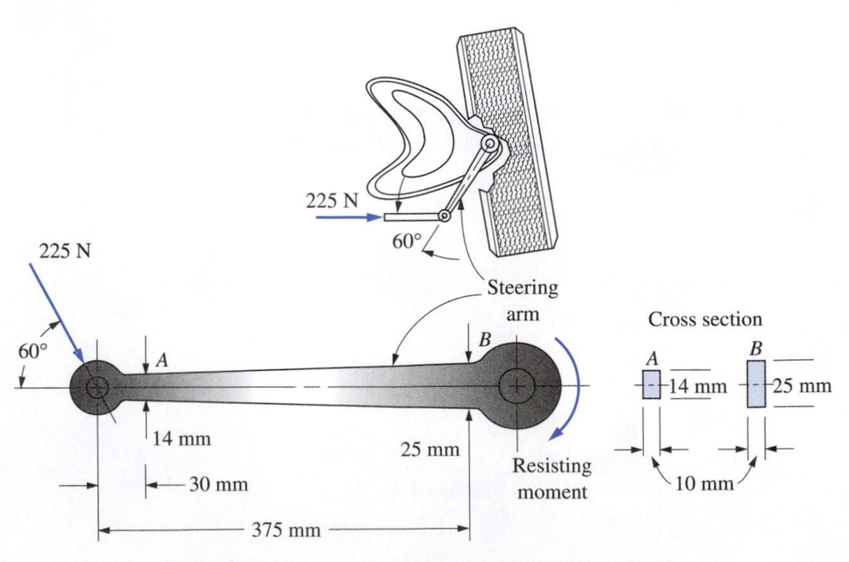

FIGURE P10–12 Steering arm for Problem 10–12.

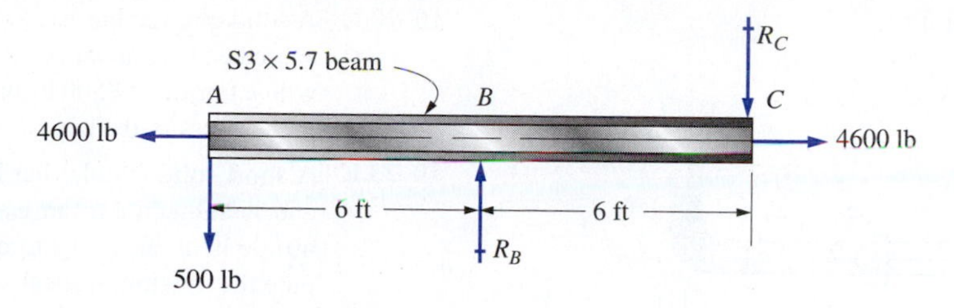

FIGURE P10–13 Beam for Problem 10–13.

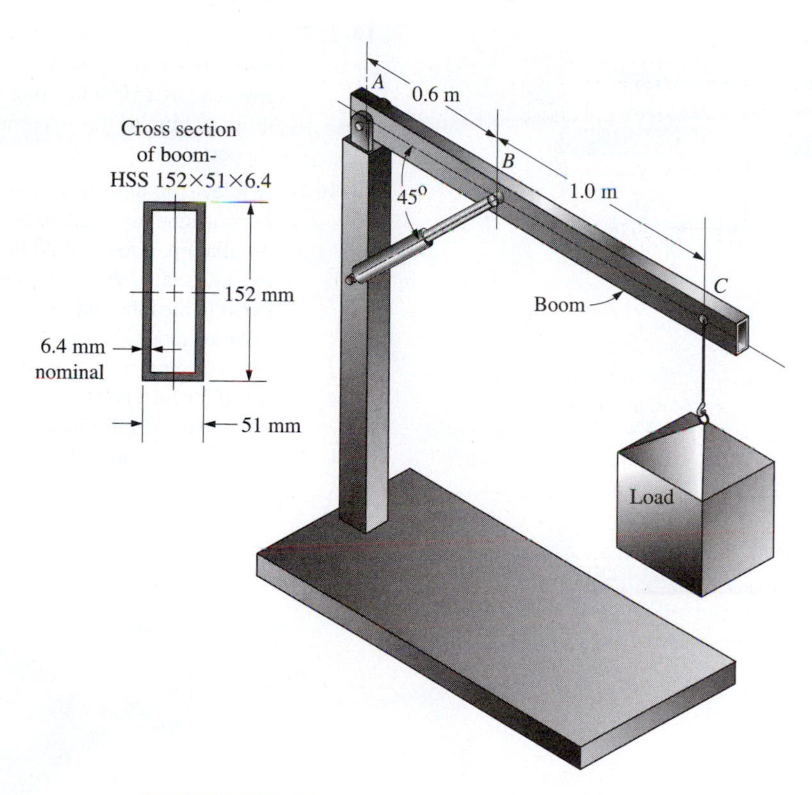

FIGURE P10–14 Crane boom for Problem 10–14.

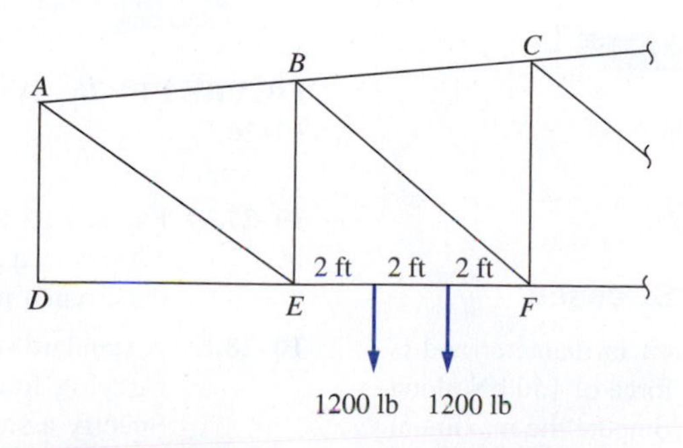

FIGURE P10–16 Truss for Problem 10–16.

10–17.M Use Figure P10–17.

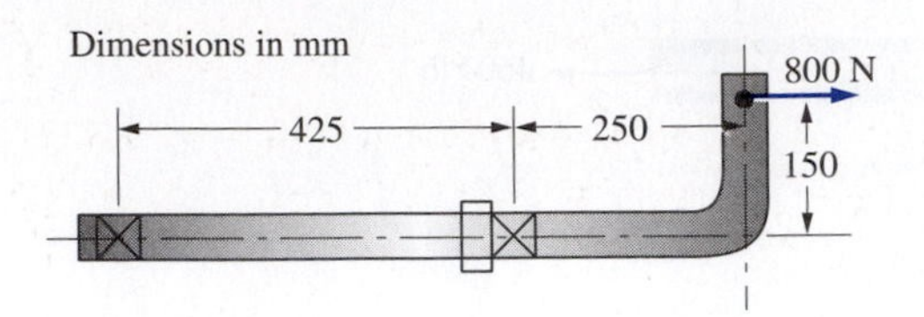

FIGURE P10–17

10–18.E Use Figure P10–18.

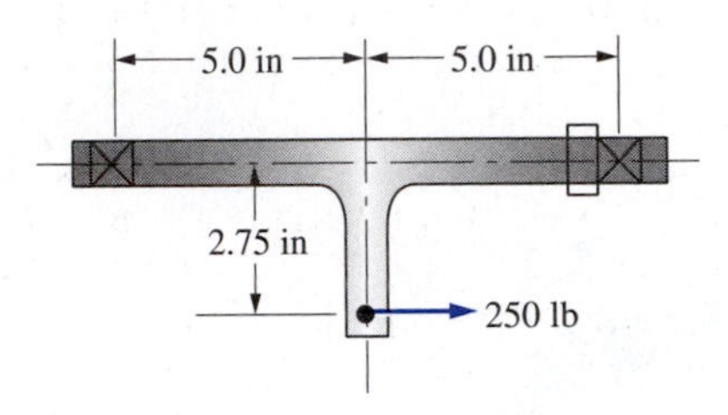

FIGURE P10–18

10–19.M Use Figure P10–19.

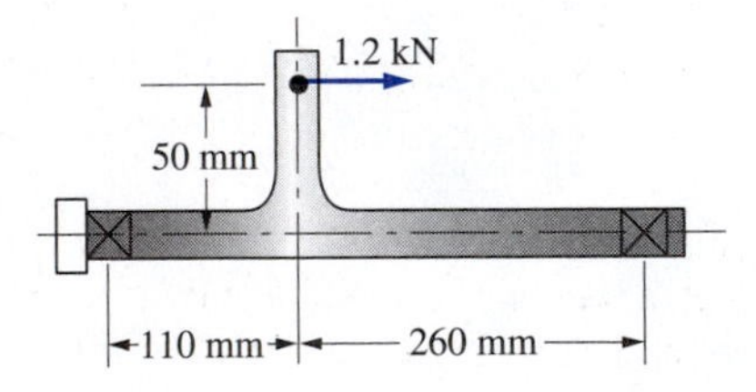

FIGURE P10–19

10–20.E Use Figure P10–20.

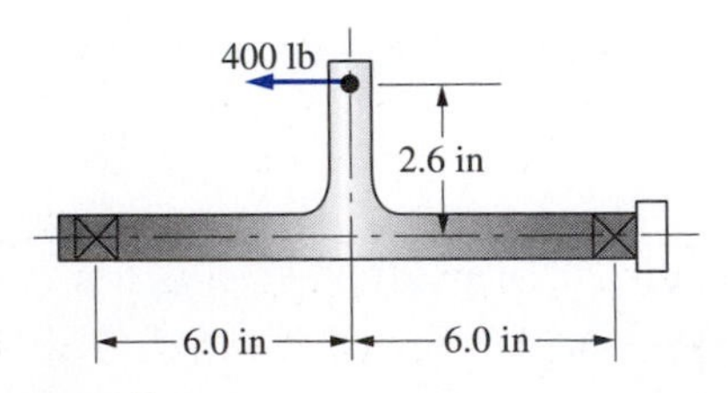

FIGURE P10–20

Combined Normal and Shear Stresses

10–21.M A solid circular bar is 40 mm in diameter and is subjected to an axial tensile force of 150 kN along with a torque of 500 N·m. Compute the maximum shear stress in the bar.

10–22.E A solid circular bar has a diameter of 2.25 in and is subjected to an axial tensile force of 47 000 lb along with a torque of 8500 lb·in. Compute the maximum shear stress in the bar.

10–23.E A short solid circular bar has a diameter of 4.00 in and is subjected to an axial compressive force of 40 000 lb along with a torque of 25 000 lb·in. Compute the maximum shear stress in the bar.

10–24.E A short hollow circular post is made from a 12-in schedule 40 steel pipe and it carries an axial compressive load of 250 000 lb along with a torque of 180 000 lb·in. Compute the maximum shear stress in the post.

10–25.E A short hollow circular support is made from a 3-in schedule 40 steel pipe and carries an axial compressive load of 25 000 lb along with a torque of 15 500 lb·in. Compute the maximum shear stress in the support.

10–26.E A television antenna is mounted on a hollow aluminum tube as sketched in Figure P10–26. During installation, a force of 20 lb is applied to the end of the antenna as shown. Calculate the torsional shear stress in the tube and the stress due to bending. Consider the tube to be simply supported against bending at the clamps, but assume that rotation is not permitted. If the tube is made of 6061-T6 aluminum, would it be safe under this load? The tube has an outside diameter of 1.50 in and a wall thickness of $\frac{1}{16}$ in.

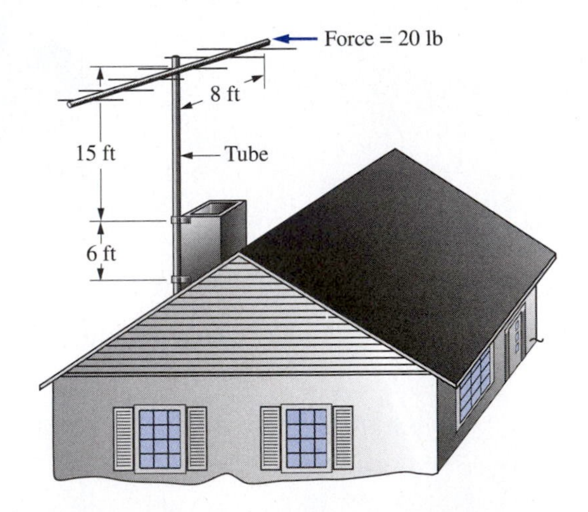

FIGURE P10–26 Antenna mounting tube for Problem 10–26.

10–27.M Figure P10–27 shows a crank to which a force F of 1200 N is applied. Compute the maximum stress in the circular portion of the crank.

10–28.E A standard steel pipe is to be used to support a bar carrying four loads, as shown in Figure P10–28. Specify a suitable pipe that would keep the maximum shear stress to 8000 psi.

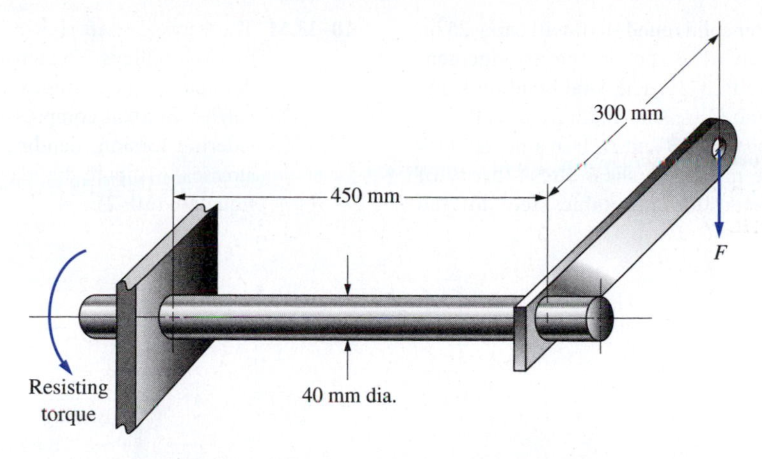

FIGURE P10–27 Crank for Problem 10–27.

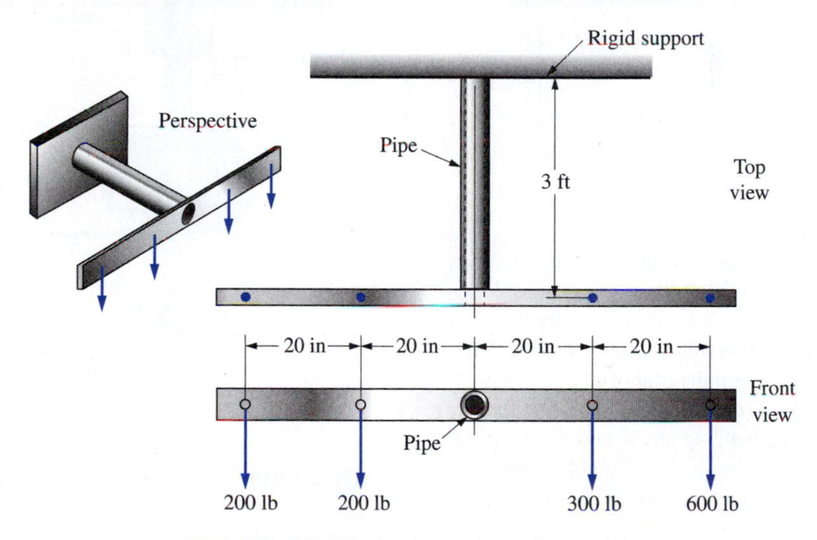

FIGURE P10–28 Bracket for Problem 10–28.

Rotating Shafts—Combined Torsional Shear and Bending Stresses

10–29.M A circular shaft carries the load shown in Figure P10–29. The shaft carries a torque of 1500 N·m between sections *B* and *C*. Compute the maximum shear stress in the shaft near section *B*.

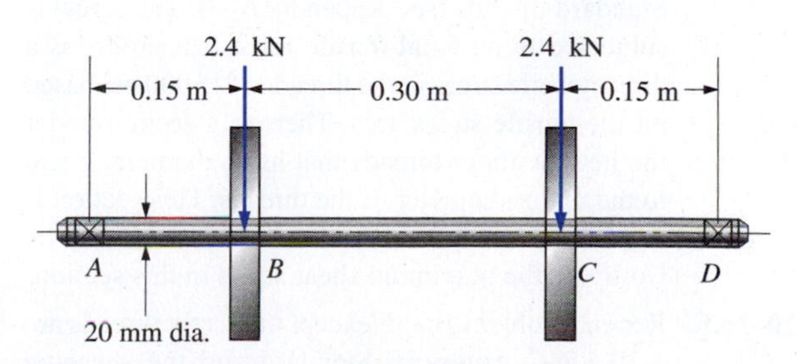

FIGURE P10–29 Shaft for Problem 10–29.

10–30.M A circular shaft carries the load shown in Figure P10–30. The shaft carries a torque of 4500 N·m between sections *B* and *C*. Compute the maximum shear stress in the shaft near section *B*.

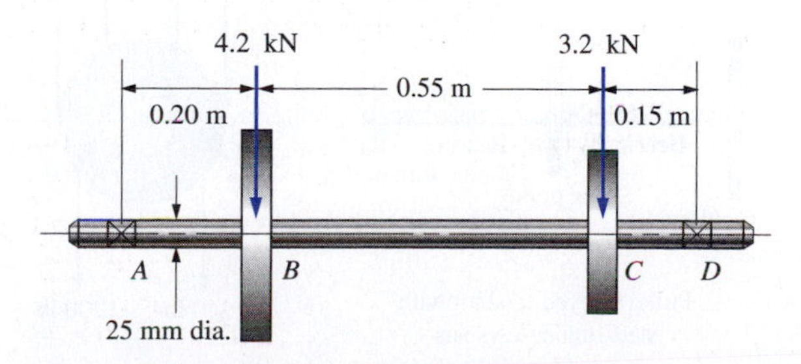

FIGURE P10–30 Shaft for Problem 10–30.

10–31.E A 1.0-in-diameter solid round shaft will carry 25 hp while rotating at 1150 rpm in the arrangement shown in Figure P10–31. The total bending loads at the gears A and C are shown, along with the reactions at the bearings B and D. Use a design factor of 6 for the maximum shear stress theory of failure, and determine a suitable steel for the shaft.

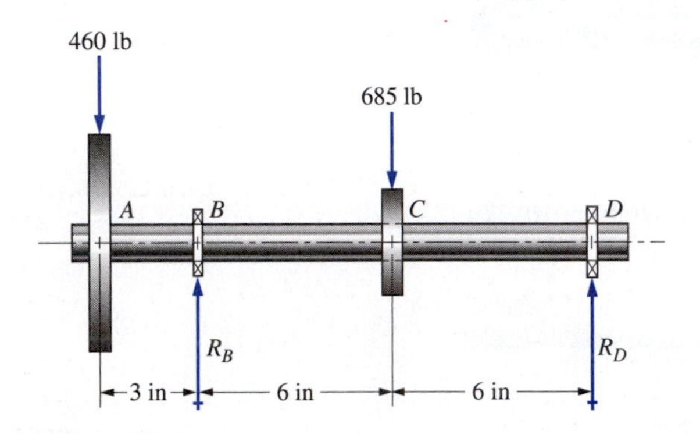

FIGURE P10–31 Shaft for Problem 10–31.

10–32.E Three pulleys are mounted on a rotating shaft, as shown in Figure P10–32. Belt tensions are as shown in the end view. All power is received by the shaft through pulley C from below. Pulleys A and E deliver power to mating pulleys above.

(a) Compute the torque at all points in the shaft.

(b) Compute the bending stress and the torsional shear stress in the shaft at the point where the maximum bending moment occurs if the shaft diameter is 1.75 in.

(c) Compute the maximum shear stress at the point used in step (b). Then specify a suitable material for the shaft.

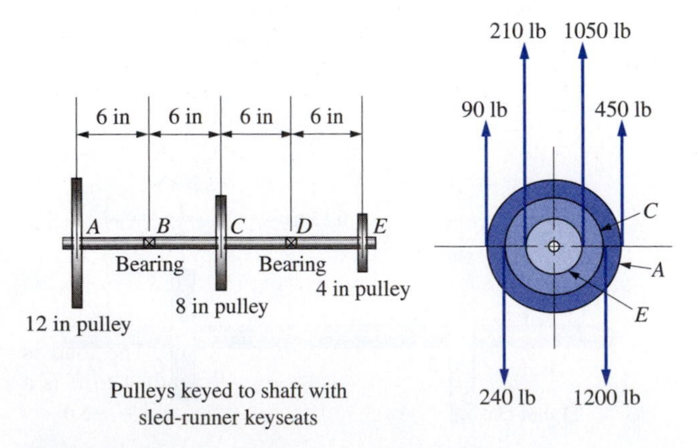

FIGURE P10–32 Shaft for Problem 10–32.

10–33.M The vertical shaft shown in Figure P10–33 carries two belt pulleys. The tensile forces in the belts under operating conditions are shown. Also, the shaft carries an axial compressive load of 6.2 kN. Considering torsion, bending, and axial compressive stresses, compute the maximum shear stress using Equation (10–2).

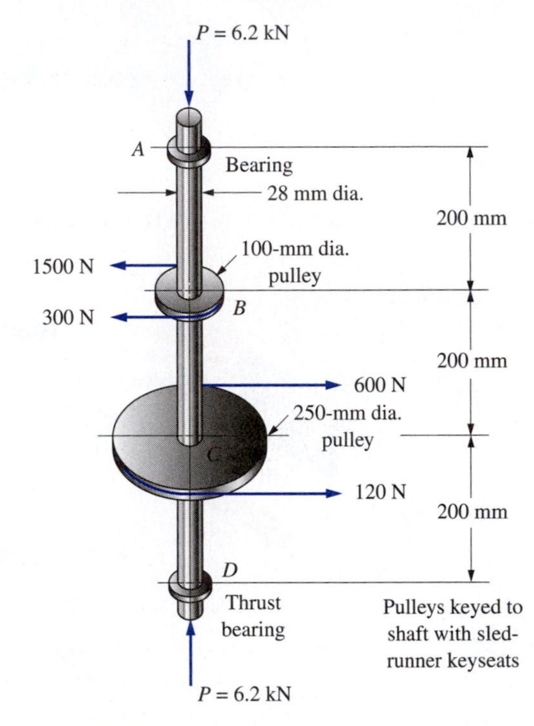

FIGURE P10–33 Shaft for Problem 10–33.

10–34.M For the shaft in Problem 10–33, specify a suitable steel that would provide a design factor of 4 based on yield strength in shear.

Combined Axial Tension and Direct Shear Stresses

10–35.E A machine screw has Number 8-32 UNC American Standard threads (see Appendix A–3). The screw is subjected to an axial tensile force that produces a direct tensile stress in the threads of 15 000 psi based on the tensile stress area. There is a section under the head without threads that has a diameter equal to the major diameter of the threads. This section is also subjected to a direct shearing force of 120 lb. Compute the maximum shear stress in this section.

10–36.E Repeat Problem 10–35 except the screw threads are $\frac{1}{4}$-20 UNC American Standard and the shearing force is 775 lb.

10–37.E Repeat Problem 10–35 except the screw threads are No. 4-48 UNF American Standard and the shearing force is 50 lb.

10–38.E Repeat Problem 10–35 except the screw threads are $1\frac{1}{4}$-12 UNF and the shearing force is 2500 lb.

10–39.M A machine screw has metric threads with a major diameter of 16 mm and a pitch of 2.0 mm (see Appendix A–3). The screw is subjected to an axial force that produces a direct tensile stress in the threads of 120 MPa based on the tensile stress area. There is a section under the head without threads that has a diameter equal to the major diameter of the threads. This section also is subjected to a direct shearing force of 8.0 kN. Compute the maximum shear stress in this section.

10–40.M A machine screw has metric threads with a major diameter of 48 mm and a pitch of 5.0 mm (see Appendix A–3). The screw is subjected to an axial tensile force that produces a direct tensile stress in the threads of 120 MPa based on the tensile stress area. There is a section under the head without threads that has a diameter equal to the major diameter of the threads. This section also is subjected to a direct shearing force of 80 kN. Compute the maximum shear stress in this section.

Combined Bending and Vertical Shear Stresses

10–41.E A rectangular bar is used as a beam carrying a concentrated load of 5500 lb at the middle of its 60-in length. The cross section is 2.00 in wide and 6.00 in high with the 6.00-in dimension oriented vertically. Compute the maximum shear stress that occurs in the bar near the load at the following points within the section:
- **(a)** At the bottom of the bar.
- **(b)** At the top of the bar.
- **(c)** At the neutral axis.
- **(d)** At a point 1.0 in above the bottom of the bar.
- **(e)** At a point 2.0 in above the bottom of the bar.

10–42.E Repeat Problem 10–41 except the beam is an aluminum I-beam, I6×4.692.

10–43.E Repeat Problem 10–41 except the load is a uniformly distributed load of 100 lb/in over the entire length. Consider cross sections near the middle of the beam, near the supports, and 15 in from the left support.

Noncircular Sections—Combined Normal and Torsional Shear Stresses

10–44.M A square bar is 25 mm on a side and carries an axial tensile load of 75 kN along with a torque of 245 N·m. Compute the maximum shear stress in the bar. (*Note:* Refer to Section 4–11 and Figure 4–27.)

10–45.M A rectangular bar is 30 mm by 50 mm in cross section and is subjected to an axial tensile force of 175 kN along with a torque of 525 N·m. Compute the maximum shear stress in the bar. (*Note:* Refer to Section 4–11 and Figure 4–27.)

10–46.M A bar has a cross section in the form of an equilateral triangle, 50 mm on a side. It carries an axial tensile force of 115 kN along with a torque of 775 N·m. Compute the maximum shear stress in the bar. (*Note:* Refer to Section 4–11 and Figure 4–27.)

10–47.E A link in a large mechanism is made from a square structural tube HSS3×3×$\frac{1}{4}$ (see Appendix A–9). It is originally designed to carry an axial tensile load that produces a design factor of 3 based on the yield strength of ASTM A500 cold-formed structural steel, shaped, grade C.
- **(a)** Determine this load and the maximum shear stress that results in the tube under this load.
- **(b)** In operation, a torque of 950 lb·ft is experienced by the tube in addition to the axial load. Compute the maximum shear stress under this combined loading and compute the resulting design factor based on the yield strength of the steel in shear. (See Section 4–11 and Figure 4–27.)

ADDITIONAL REVIEW AND PRACTICE PROBLEMS

10–48. Compute the maximum tensile and compressive stresses in the horizontal portion of the bracket shown in Figure P10–48.

10–49. Repeat Problem 10–48 with the load applied downward and to the left at an angle of 35° to the horizontal.

10–50. Repeat Problem 10–48 with the load applied to the right rather than to the left.

10–51. Repeat Problem 10–48 with the load applied upward and to the right at an angle of 65° to the horizontal.

10–52. The bracket shown in Figure P10–52 is made from structural steel tubing, HSS3×3×$\frac{1}{4}$. It is attached rigidly to the ceiling. Compute the maximum tensile and compressive stresses in the bracket and state where they occur.

10–53. Figure P10–53 shows a small crane. The load is 34.0 kN. The connection at each point A to F is a 30-mm-diameter pin in a clevis. Determine the maximum normal stresses in the horizontal and vertical members of the crane.

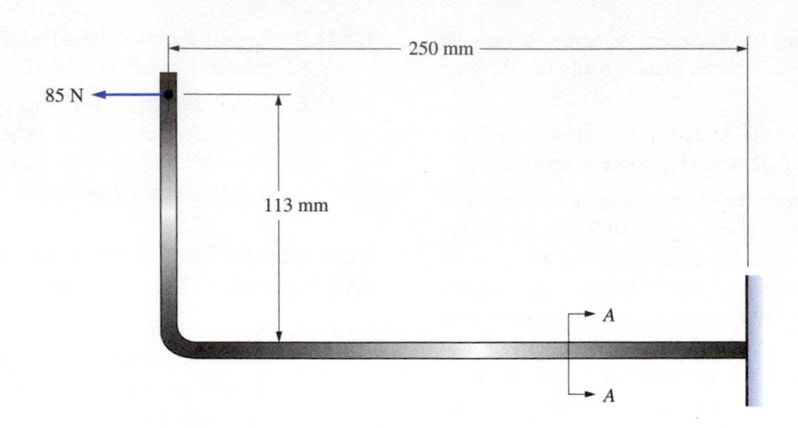

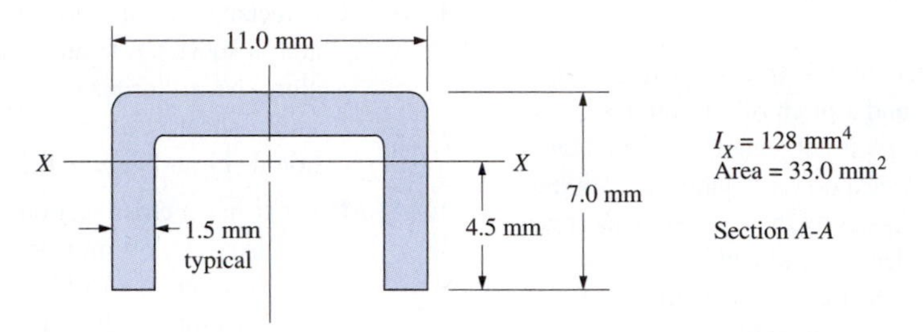

FIGURE P10–48

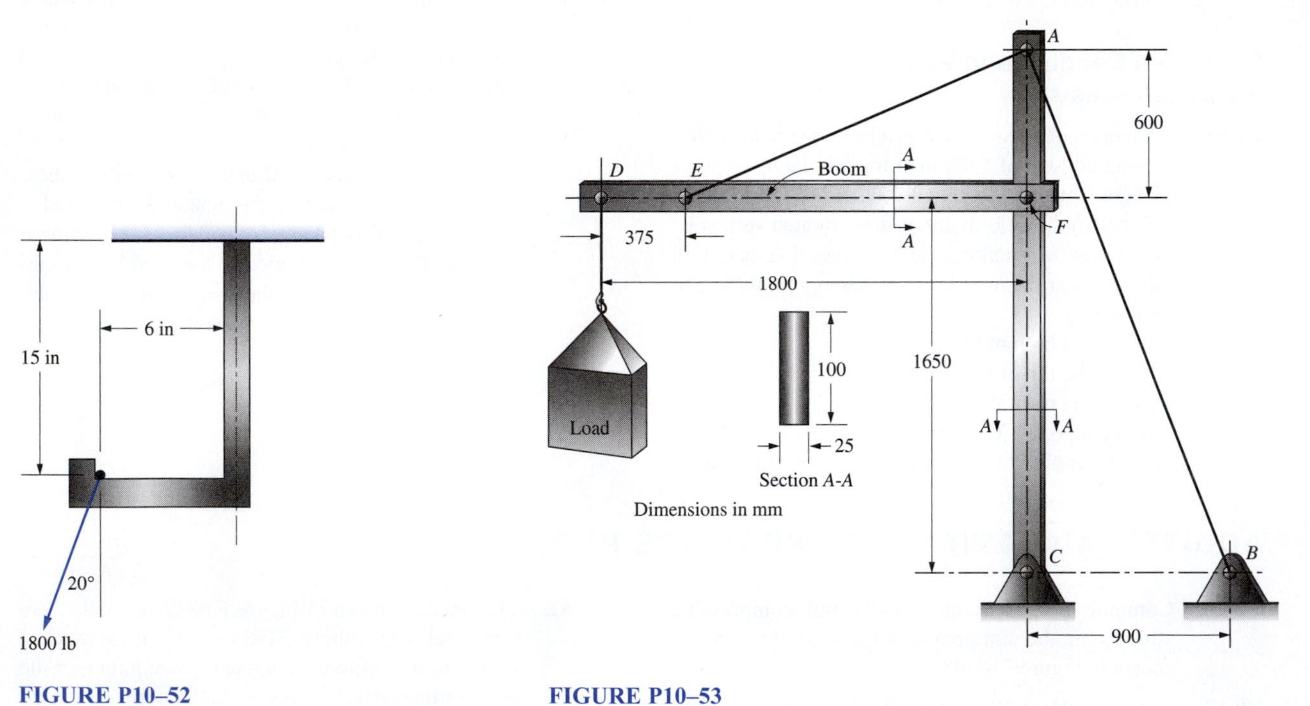

FIGURE P10–52

FIGURE P10–53

10–54. Figure P10–54 shows a rotating shaft carrying transverse loads. The distributed load between the bearings is provided by a large flat pulley. The load at the end of the overhang is applied by a chain drive. The shaft transmits 30.5 hp while rotating at 320 rpm. Compute the maximum shear stress in the shaft and specify a suitable material.

10–55. The cantilevered bar shown in Figure P10–55 carries an offset load as shown. Determine the stress condition that occurs on elements *M* and *N*.

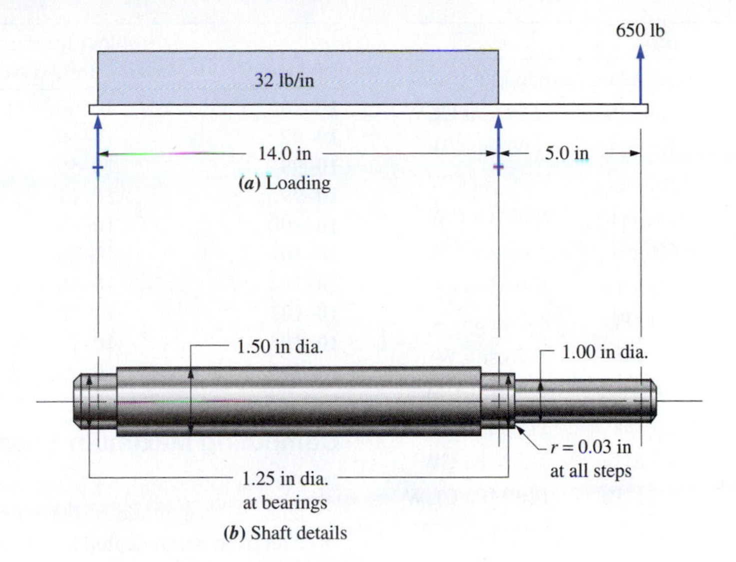

FIGURE P10–54

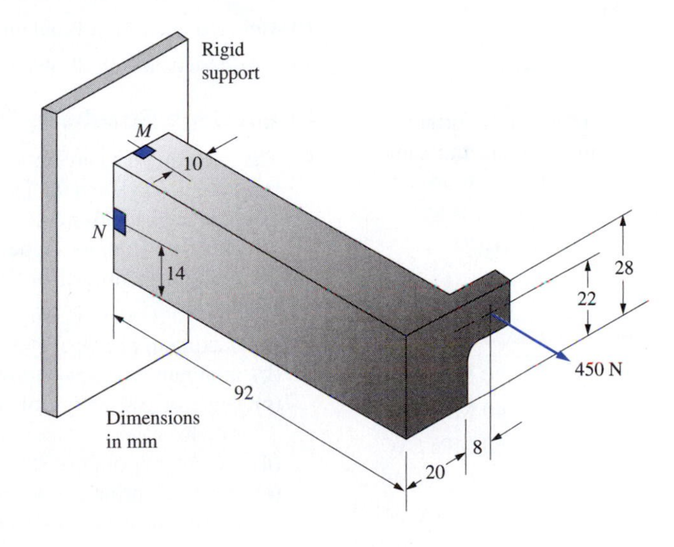

FIGURE P10–55

Mohr's Circle

A. For Problems 10–56 to 10–83, determine the principal stresses and the maximum shear stress using Mohr's circle. The following data sets give the stresses on the initial stress element. Perform the following operations.

(a) Draw the complete Mohr's circle, labeling critical points including σ_1, σ_2, τ_{max}, and σ_{avg}.

(b) On Mohr's circle, indicate which line represents the x-axis on the initial stress element.

(c) On Mohr's circle, indicate the angles from the line representing the x-axis to the σ_1-axis and the τ_{max}-axis.

(d) Draw the principal stress element and the maximum shear stress element in their proper orientation relative to the initial stress element.

Problem	σ_x	σ_y	τ_{xy}
10–56	300 MPa	−100 MPa	80 MPa CW
10–57	250 MPa	−50 MPa	40 MPa CW
10–58	80 MPa	−10 MPa	60 MPa CW
10–59	150 MPa	10 MPa	100 MPa CW
10–60	20 ksi	−5 ksi	10 ksi CCW
10–61	38 ksi	−25 ksi	18 ksi CCW
10–62	55 ksi	15 ksi	40 ksi CCW
10–63	32 ksi	−50 ksi	20 ksi CCW
10–64	−900 kPa	600 kPa	350 kPa CCW
10–65	−580 kPa	130 kPa	75 kPa CCW

Problem	σ_x	σ_y	τ_{xy}
10–66	−840 kPa	−35 kPa	650 kPa CCW
10–67	−325 kPa	50 kPa	110 kPa CCW
10–68	−1800 psi	300 psi	800 psi CW
10–69	−6500 psi	1500 psi	1200 psi CW
10–70	−4250 psi	3250 psi	2800 psi CW
10–71	−150 psi	8600 psi	80 psi CW
10–72	260 MPa	0 MPa	190 MPa CCW
10–73	1450 kPa	0 kPa	830 kPa CW
10–74	22 ksi	0 ksi	6.8 ksi CW
10–75	6750 psi	0 psi	3120 psi CCW
10–76	0 ksi	−28 ksi	12 ksi CW
10–77	0 MPa	440 MPa	215 MPa CW
10–78	0 MPa	260 MPa	140 MPa CCW
10–79	0 kPa	−1560 kPa	810 kPa CCW
10–80	225 MPa	−85 MPa	0 MPa
10–81	6250 psi	−875 psi	0 psi
10–82	775 kPa	−145 kPa	0 kPa
10–83	38.6 ksi	−13.4 ksi	0 ksi

B. For Problems 10–84 to 10–95 that result in the principal stresses from the initial Mohr's circle having the same sign, use the procedures from Section 10–11 to draw supplementary circles and find the following:

(a) The three principal stresses: σ_1, σ_2, and σ_3.
(b) The true maximum shear stress.

Problem	σ_x	σ_y	τ_{xy}
10–84	300 MPa	100 MPa	80 MPa CW
10–85	250 MPa	150 MPa	40 MPa CW
10–86	180 MPa	110 MPa	60 MPa CW
10–87	150 MPa	80 MPa	30 MPa CW
10–88	30 ksi	15 ksi	10 ksi CCW
10–89	38 ksi	25 ksi	8 ksi CCW
10–90	55 ksi	15 ksi	5 ksi CCW
10–91	32 ksi	50 ksi	20 ksi CCW
10–92	−840 kPa	−335 kPa	120 kPa CCW
10–93	−325 kPa	−50 kPa	60 kPa CCW
10–94	−1800 psi	−300 psi	80 psi CW
10–95	−6500 psi	−2500 psi	1200 psi CW

C. For Problems 10–96 to 10–105, use the data from the indicated problem for the initial stress element to draw Mohr's circle. Then determine the stress condition on the element at the specified angle of rotation from the given *x*-axis. Draw the rotated element in its proper relation to the initial stress element and indicate the normal and shear stresses acting on it.

Problem	Problem for the initial stress data	Angle of rotation from the *x*-axis
10–96	10–56	30 deg CCW
10–97	10–56	30 deg CW
10–98	10–59	70 deg CCW
10–99	10–61	20 deg CW
10–100	10–63	50 deg CCW
10–101	10–65	45 deg CW
10–102	10–68	10 deg CCW
10–103	10–70	25 deg CW
10–104	10–71	80 deg CW
10–105	10–73	65 deg CW

Computing Maximum Shear Stress

D. For the following problems, use Equation (10–2) to compute the magnitude of the maximum shear stress for the data from the indicated problem.

10–106. Use data from Problem 10–72.

10–107. Use data from Problem 10–73.

10–108. Use data from Problem 10–74.

10–109. Use data from Problem 10–75.

Strain Gage Rosettes—Rectangular

E. The following data are obtained from tests in which a rectangular [0°, 45°, and 90°] strain-gage rosette was applied to a product made from the given material. Locate material property data from the Appendix and from Table 2–1. Use the procedure outlined in Section 10–12 to compute the following:

(a) maximum principal strain
(b) minimum principal strain
(c) angle of orientation of the principal strains from the axis for gage 1
(d) maximum principal stress
(e) minimum principal stress
(f) maximum shearing strain
(g) maximum shearing stress

10–110. $\epsilon_1 = 1480\,\mu\epsilon$, $\epsilon_2 = 165\,\mu\epsilon$, $\epsilon_3 = 428\,\mu\epsilon$. Aluminum 6061–T6. SI metric units.

10–111. $\epsilon_1 = 853\,\mu\epsilon$, $\epsilon_2 = 406\,\mu\epsilon$, $\epsilon_3 = 641\,\mu\epsilon$. Aluminum 7075-T6. SI metric units.

10–112. $\epsilon_1 = 389\,\mu\epsilon$, $\epsilon_2 = 737\,\mu\epsilon$, $\epsilon_3 = -290\,\mu\epsilon$. AISI 1040 cold-drawn steel. SI metric units.

10–113. $\epsilon_1 = 925\,\mu\epsilon$, $\epsilon_2 = -631\,\mu\epsilon$, $\epsilon_3 = 552\,\mu\epsilon$. AISI 4140 OQT 900 steel. SI metric units.

10–114. $\epsilon_1 = 169\,\mu\epsilon$, $\epsilon_2 = -266\,\mu\epsilon$, $\epsilon_3 = 543\,\mu\epsilon$. Copper C14500 hard. U.S. Customary units.

10–115. $\epsilon_1 = 775\,\mu\epsilon$, $\epsilon_2 = 369\,\mu\epsilon$, $\epsilon_3 = -318\,\mu\epsilon$. Titanium Ti-6A1-4V, aged. U.S. Customary units.

10–116. $\epsilon_1 = 389\,\mu\epsilon$, $\epsilon_2 = 737\,\mu\epsilon$, $\epsilon_3 = -290\,\mu\epsilon$. Ductile iron, ASTM A536, 80-55-6. U.S. Customary units.

10–117. $\epsilon_1 = 1532\,\mu\epsilon$, $\epsilon_2 = -228\,\mu\epsilon$, $\epsilon_3 = 893\,\mu\epsilon$. AISI 501 OQT 1000 Stainless steel. U.S. Customary units.

Strain Gage Rosettes—Delta

Repeat the problems in section E, but the data are taken from a delta [0°, 60°, and 120°] strain-gage rosette.

10–118. Use data from Problem 10–110.

10–119. Use data from Problem 10–111.

10–120. Use data from Problem 10–112.

10–121. Use data from Problem 10–113.

10–122. Use data from Problem 10–114.

10–123. Use data from Problem 10–115.

10–124. Use data from Problem 10–116.

10–125. Use data from Problem 10–117.

COMPUTER ASSIGNMENTS

1. Write a program for a computer, spreadsheet, or a programmable calculator to aid in the construction of Mohr's circle. Input the initial stresses, σ_x, σ_y, and τ_{xy}. Have the program compute the radius of the circle, the maximum and minimum principal stresses, the maximum shear stress, and the average stress. Use the program in conjunction with freehand sketching of the circle for the data for Problems 10–56 through 10–79.
2. Enhance the program in Assignment 1 by computing the angle of orientation of the principal stress element and the angle of orientation of the maximum shear stress element.
3. Enhance the program in Assignment 1 by computing the normal and shear stresses on the element rotated at any specified angle relative to the original x-axis.
4. Enhance the program in Assignment 1 by causing it to detect if the principal stresses from the initial Mohr's circle are of the same sign; and in such cases, print out the three principal stresses in the proper order, σ_1, σ_2, σ_3. Also have the program compute the true maximum shear stress from Equation (10–21).
5. Produce a spreadsheet similar to that shown in Figure 10–50 to analyze strain-gage rosette data for either a rectangular or delta-style rosette. The results are to include the items identified in the problem set E.

11

Columns

The Big Picture

Columns

Discussion Map

A column is a relatively long, slender member loaded in compression.

The failure mode for a column is called *buckling,* a common term for the condition of *elastic instability,* when the load on an initially straight column causes it to bend significantly. If the load is increased a small amount from the buckling load, the column would collapse suddenly—a very dangerous situation.

- How do we determine when a compression member is long and slender?
- How do we determine the magnitude of the load at which buckling would occur?
- What kind of cross-sectional shapes are preferred for columns?
- What influence does the manner of holding the ends of a column have on the buckling load?
- What industry standards apply to columns?

These and other details about column analysis and design are presented in this chapter.

Activity

Let's explore this concept of a column. Find some examples of things around you that may fit that definition. Then describe each of them, giving its length, the shape and dimensions of its cross section, and the material from which it is made. Here are a few examples to get you started.

A meter stick: It is typically made from wood or aluminum. Obviously its length is 1.00 m. The cross section is typically about 30 mm wide and 4 mm thick. It even looks long and slender.

A steel 6-in (152 mm) rule: You might have used one of these in a class where you operated metal-cutting machine tools or wood lathes. Again its length is obviously 6.00 in (152 mm). It is made from a flat strip of steel, 0.75 in wide and 0.020 in thick (19 mm × 0.50 mm). Even though it is much shorter than the meter stick, it is thinner, that is, more slender. Both length and slenderness matter.

A wood dowel that you might get at a hardware store. Perhaps it is 36 in long and has a diameter of $\frac{3}{8}$ in (914 mm × 9.5 mm).

What examples have you found?

Now let's try to load one of these items with a direct axial compressive load. This means that the line of action of the load is in line with the long axis of the column. Just support it on a table or the floor and push down on it with your hand. Try to push straight down and not sideways at all, but don't hold it tight with your fingers. Be careful not to push too hard and break it!

What happened?

Here we describe the behavior of the wood meter stick. Loading it slowly, we find that it can support a very small load while remaining straight. But without much exertion, we can cause the stick to bend noticeably. This phenomenon is called *buckling*. Be careful! With just a modest increase in load after buckling occurs, the stick would break easily. Notice that the meter stick buckles about the thin dimension of its cross section. Figure 11–1 shows a sketch of what happened. Figure 11–2(a) shows a side view. You probably would have predicted that based on your own experience. Later, we will quantify why that happens.

FIGURE 11–1
Illustration of buckling of a meter stick.

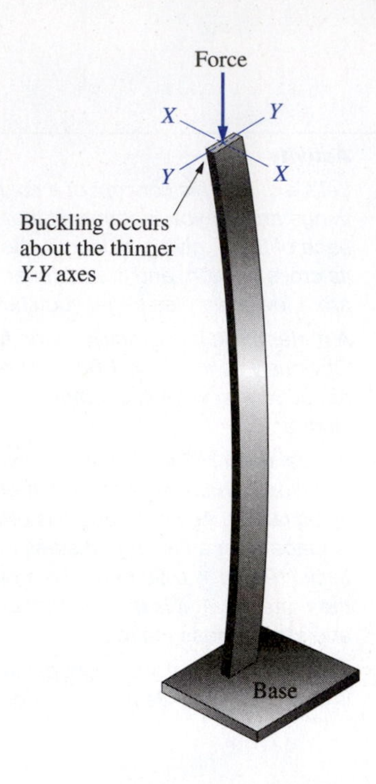

Now let's change the procedure a bit. It appears that the meter stick tends to bow out near the middle, say at the 0.50 m point. What if we provide some lateral support on the sides of the stick at that point? Try it yourself if you have a meter stick. Place your fingers lightly on either side to restrain the tendency for it to bow outward. Now push down on it again as you did before.

What happened?

Now we can apply a much higher load to the stick without it buckling. But there is a point where the load is high enough to see a quite different form of buckling. The lower half and the

FIGURE 11–2
Comparison of shapes for buckled columns.

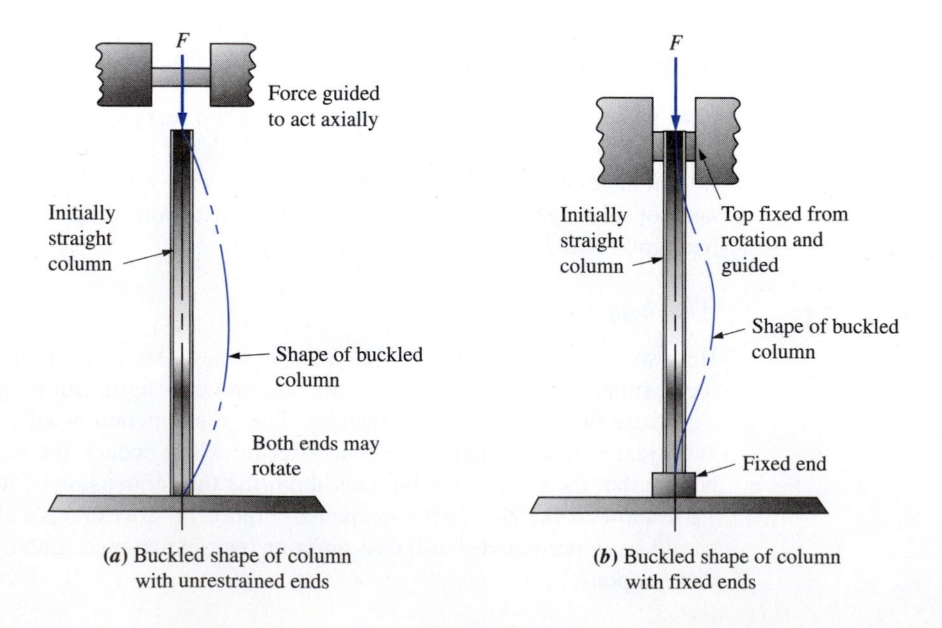

upper half of the stick buckle with one going one way and one going the other. In fact, it looks as if the stick takes the form of a complete sine wave. We will discuss this observation later.

Let's change the procedure again. Grab both ends of the meter stick with a firm grip and try as hard as you can to keep the stick from rotating while simultaneously applying an axial load that will cause buckling.

What happened?

First, you should notice that it takes a much higher force to cause buckling. Also, you should notice that the shape of the buckled stick is different from that when you did not restrain the ends. Two factors are working here. Grabbing the stick with your fists effectively shortens the column by about 90 mm (3.5 inches) on each end. Because the column is shorter, it takes a higher load to cause buckling. But also, your effort at keeping the ends from rotating caused the buckled shape to be similar to that shown in Figure 11–2(b). You essentially *fixed the ends*. We will discuss this phenomenon later also.

Are all of the columns you found perfectly straight?

Probably not. Of those tested here, most tended to buckle in a particular direction because they were initially crooked. Pushing down on them had the additional effect of bending the crooked section even more in the same direction. We will explore that phenomenon later also.

The examples discussed here are all items that were not meant to carry axial compressive loads. They do serve well as demonstrators for buckling.

What examples of columns can you find that are more substantial and that were designed to be sufficiently strong and stable to withstand sizable axial compressive loads?

You might not be able to carry all of these examples into a classroom, laboratory, or office, but here are a few.

- The vertical columns of a steel-framed building: The lower columns on a multistory building must be strong and stiff enough to hold up all the weight above them. Even in a one-story building, they must hold the roof structure and, possibly, a snow load on top of that.
- The steel posts holding up the beam across the length of the basement of a home: The beam supports the joists from the floor above and all the weight of the furniture and the people there. The posts carry that load to the basement floor or the foundation. The posts are likely made from steel pipe or round hollow structural sections (HSS). That is an efficient shape for a column as we will discuss later in this chapter.
- The cylinder rod of a hydraulic actuator: You may have seen these on a piece of construction equipment, agricultural machinery, or in an industrial automation system. Some of these cylinder rods push with great force and they must be designed not to buckle as they extend out from the cylinder.

What others have you found?

Now summarize the observations so far.

- We have demonstrated that a long, slender member tends to buckle when subjected to an axial compressive load. But when does a member become long and slender? We define the term *slenderness ratio* later to quantify that. It is a function of the length of the column, how its ends are held, and the shape and size of the cross section of the column.

- We demonstrated that a column can take some magnitude of axial load before buckling begins. Then the onset of buckling is quite sudden. At what load will it buckle? We show several methods of predicting that in this chapter.
- We demonstrated that the way the ends of the column are held affects the buckling load. We elaborate on that later using the term *end fixity*.
- The columns we found were possibly made from different materials, such as steel, aluminum, wood, or plastic. What effect does the material have on its tendency to buckle? We show in this chapter that the material's modulus of elasticity, E, has a major effect on the tendency for a long column to buckle. For shorter columns, the yield strength is also a factor.
- Some of the columns we found were initially crooked. It seemed that they buckled at a lower load than the straight ones and always in the direction of the initial crookedness. Can we quantify that? Yes, additional approaches to analyze crooked columns and those that have the load applied off the axis (called *eccentrically loaded columns*) will be presented.

11–1 OBJECTIVES OF THIS CHAPTER

After completing this chapter, you should be able to:

1. Define *column*.
2. Differentiate between a column and a short compression member.
3. Describe the phenomenon of *buckling,* also called *elastic instability*.
4. Define *radius of gyration* for the cross section of a column and be able to compute its magnitude.
5. Understand that a column is expected to buckle about the axis for which the radius of gyration is the minimum.
6. Define *end-fixity factor, K*, and specify the appropriate value depending on the manner of supporting the ends of a column.
7. Define *effective length,* L_e, *slenderness ratio,* and *transition slenderness ratio* (also called the *column constant,* C_c) and compute their values.
8. Use the values for the slenderness ratio and the column constant to determine when a column is *long* or *short*.
9. Use the *Euler formula* for computing the critical buckling load for long columns and the *J. B. Johnson formula* for short columns.
10. Apply a design factor to the critical buckling load to determine the *allowable load* on a column.
11. Recognize efficient shapes for column cross sections.
12. Design columns to safely carry given axial compression loads.
13. Apply the specifications of the American Institute of Steel Construction (AISC) and the Aluminum Association to the analysis of columns.
14. Analyze columns that are initially crooked to determine their critical buckling load.
15. Analyze columns for which the applied load acts eccentric to the axis of the column.

11–2 SLENDERNESS RATIO

A column has been described as a relatively long, slender member loaded in compression. This description is stated in relative terms and is not very useful for analysis.

The measure of the slenderness of a column must take into account the length, the cross-sectional shape and dimensions of the column, and the manner of attaching the ends of the column to the structures that supply loads and reactions to the column. The commonly used measure of slenderness is the *slenderness ratio,* defined as

$$SR = \frac{KL}{r} = \frac{L_e}{r} \tag{11–1}$$

where L = *actual length* of the column between points of support or lateral restraint
K = *end-fixity factor*
L_e = *effective length,* taking into account the manner of attaching the ends (note that $L_e = KL$)
r = *smallest radius of gyration* of the cross section of the column

Each of these terms is discussed here.

Actual Length, *L*. For a simple column having the load applied at one end and the reaction provided at the other, the actual length is, obviously, the length between its ends. But for components of structures loaded in compression where a means of restraining the member laterally to prevent buckling is provided, the actual length is taken between points of restraint. Each part is then considered to be a separate column.

End-Fixity Factor, *K*. The end-fixity factor is a measure of the degree to which each end of the column is restrained against rotation. Three classic types of end connections are typically considered: the pinned end, the fixed end, and the free end. Figure 11–3 shows these end types in several combinations with the corresponding values of K. Note that two values of K are given. One is the theoretical value and the other is the one typically used in practical situations, recognizing that it is difficult to achieve the truly fixed end, as discussed later. See also Section 11–6.

Figure 11–3(a) shows a commercially available demonstration device that illustrates the relative stiffness and resistance to buckling of four end-fixity conditions. The sizes of the stacks of weights on the upper end of the first three indicate the approximate load at which buckling is initiated. The very small load on the right end column (the fixed-free condition) does not represent the true buckling load because the column itself is unstable and bends to one side easily if there is any degree of misalignment of the load from the axis of the column.

Pinned ends of columns are essentially unrestrained against rotation. When a column with two pinned ends buckles, it assumes the shape of a smooth curve between its ends, as shown in Figure 11–3(b). This is the basic case of column buckling, and the value of $K = 1.0$ is applied to columns having two pinned ends. An ideal implementation of the pinned end is the frictionless spherical ball joint that would permit the column to rotate in any direction about any axis. For a cylindrical pin joint, free rotation is permitted about the centerline of the pin, but some restraint is provided in the plane perpendicular to the centerline. Care must be exercised in applying end-fixity factors to cylindrical pins for this reason. It is assumed that the pinned end is guided in some way so that the line of action of the axial load remains unchanged.

The combination of one fixed end and one pinned end is shown in Figure 11–3(c). Notice that the buckled shape approaches the fixed end with a zero slope while the pinned end rotates freely. The theoretical value of $K = 0.7$ applies to such an end-fixity while $K = 0.80$ is recommended for practical use.

Fixed ends provide theoretically perfect restraint against rotation of the column at its ends. As the column tends to buckle, the deflected shape of the axis of the column must approach the fixed end with a zero slope, as illustrated in Figure 11–3(d). The buckled

FIGURE 11–3
Values of K for effective length, $L_e = KL$, for four different end fixities.

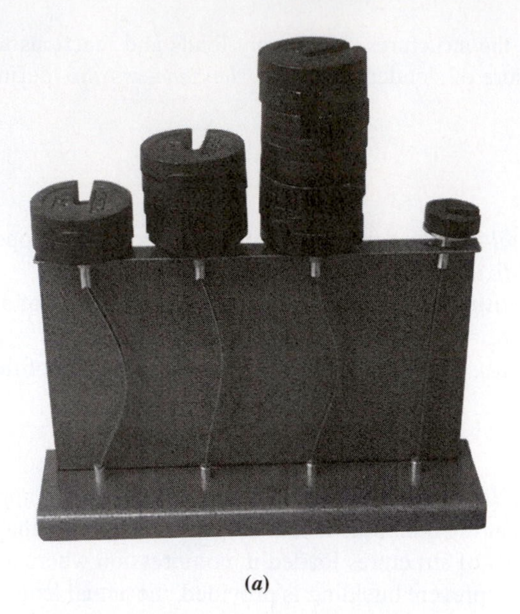

(*a*)

(**a**) Commercially available demonstrator for end fixity
(Source: P.A.Hilton Ltd. Hi-Tech, Hampshire, England)

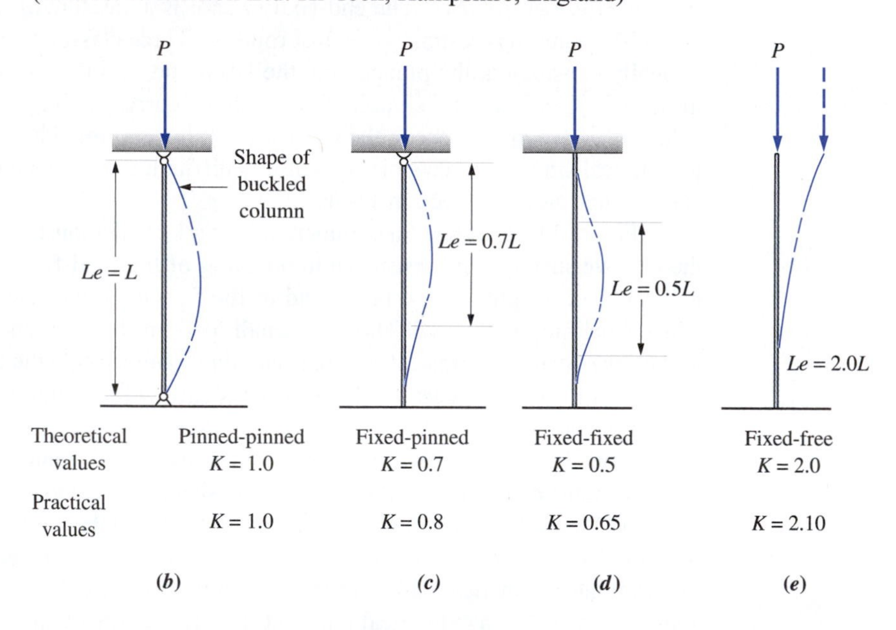

(*b*) (*c*) (*d*) (*e*)

shape bows out in the middle but exhibits two points of inflection to reverse the direction of curvature near the ends. The theoretical value of the end-fixity factor is $K = 0.5$, indicating that the column acts as if it were only one-half as long as it really is. Columns with fixed ends are much stiffer than pinned-end columns and can therefore take higher loads before buckling. It should be understood that it is very difficult to provide perfectly fixed ends for a column. It requires that the connection to the column is rigid and stiff and that the structure to which the loads are transferred is also rigid and stiff. For this reason, the higher value of $K = 0.65$ is recommended for practical use.

A free end for a column is unrestrained against rotation and also against translation. Because it can move in any direction, this is the worst case for column end-fixity. The only practical way of using a column with a free end is to have the opposite end fixed, as illustrated in Figure 11–3(e). Such a column is sometimes referred to as the *flagpole* case because the fixed end is similar to the flagpole inserted deeply into a tight-fitting socket while the other end is free to move in any direction. Called the fixed-free end condition, the theoretical value of K is 2.0. A practical value is $K = 2.10$.

Effective Length, L_e. Effective length combines the actual length with the end-fixity factor; $L_e = KL$. In the problems in this book we use the recommended practical values for the end-fixity factor, as shown in Figure 11–3. In summary, the following relationships will be used to compute the effective length:

Effective Length

1. Pinned-end columns: $L_e = KL = 1.0(L) = L$
2. Fixed-pinned columns: $L_e = KL = 0.80(L)$
3. Fixed-end columns: $L_e = KL = 0.65(L)$
4. Fixed-free columns: $L_e = KL = 2.10(L)$

Radius of Gyration, r. The measure of slenderness of the cross section of the column is its radius of gyration, r, defined as

$$r = \sqrt{\frac{I}{A}} \tag{11–2}$$

where I = moment of inertia of the cross section of the column with respect to one of the principal axes
A = area of the cross section.

Because both I and A are geometrical properties of the cross section, the radius of gyration, r, is also. Formulas for computing r for several common shapes are given in Appendix A–1. Also, r is listed with other properties for some of the standard shapes in the Appendix. For those for which r is not listed, the values of I and A are available and Equation (11–2) can be used to compute r very simply. Section 6–10 in Chapter 6 includes additional discussion of the radius of gyration with examples and practice problems.

Note that the value of the radius of gyration, r, is dependent on the axis about which it is to be computed. In most cases, it is required that you determine the axis for which the *radius of gyration is the smallest,* because that is the axis about which the column would likely buckle. Consider, for example, a column made from a rectangular section whose width is much greater than its thickness, as sketched in Figure 11–1. The meter stick demonstrates that when loaded in axial compression with little or no restraint at the ends, the column will always buckle with respect to the axis through the thinnest dimension.

Refer to Figure 11–4 to illustrate this point. Shown there are sketches of the thin rectangular cross section of the meter stick sketched in Figure 11–1. Part (a) shows the shape with respect to the centroidal axis Y-Y. The thickness of the rectangle is t and the width is h. Then the moment of inertia of the rectangle with respect to the Y-Y axis is

$$I_Y = ht^3/12$$

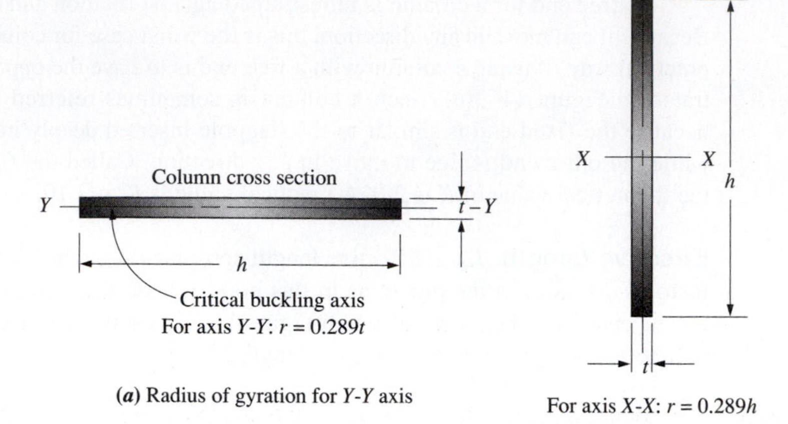

FIGURE 11–4 Radius of gyration of the cross section of a thin, rectangular column.

The area is simply

$$A = th$$

Now, using Equation (11–2) we can obtain a relationship for the radius of gyration, r_Y.

$$r_Y = \sqrt{\frac{I_Y}{A}} = \sqrt{\frac{ht^3/12}{th}} = \sqrt{\frac{t^2}{12}} = \frac{t}{\sqrt{12}} = 0.289t$$

Similarly, using Figure 11–4(b), we can obtain an equation for r_X.

$$I_X = th^3/12$$
$$A = th$$
$$r_X = \sqrt{\frac{I_X}{A}} = \sqrt{\frac{th^3/12}{th}} = \sqrt{\frac{h^2}{12}} = \frac{h}{\sqrt{12}} = 0.289h$$

Observe that because $h > t$, $r_X > r_Y$ and then r_Y is the smallest radius of gyration for the section.

For the wide-flange beams (Appendix A–7) and American Standard beams (Appendix A–8), the minimum value of r is that computed with respect to the Y-Y axis; that is,

$$r_{min} = r_Y = \sqrt{\frac{I_Y}{A}}$$

Similarly, for rectangular hollow structural sections (HSS) (Appendix A–9), the minimum radius of gyration is that with respect to the Y-Y axis. Values for r are listed in the table.

For structural steel angles, called L-shapes, neither the X-X nor the Y-Y axis provides the minimum radius of gyration. As illustrated in Appendix A–5, r_{min} is computed with respect to the Z-Z axis, with the values listed in the table.

For symmetrical sections, the value of r is the same with respect to any principal axis. Such shapes are the solid or hollow circular section and the solid or hollow square section.

Summary of the Method for Computing the Slenderness Ratio

1. Determine the actual length of the column, L, between endpoints or between points of lateral restraint.
2. Determine the end-fixity factor from the manner of support of the ends and by reference to Figure 11–3.
3. Compute the effective length, $L_e = KL$.
4. Compute the *smallest* radius of gyration for the cross section of the column.
5. Compute the slenderness ratio from

$$SR = \frac{L_e}{r_{min}}$$

11–3 TRANSITION SLENDERNESS RATIO

Transition Slenderness Ratio

When is a column considered long? The answer to this question requires the determination of the *transition slenderness ratio,* or column constant C_c.

$$C_c = \sqrt{\frac{2\pi^2 E}{s_y}} \quad \textbf{(11–3)}$$

The following rules are applied to determine if a given column is long or short.

If the actual effective slenderness ratio L_e/r is greater than C_c, then the column is long, and the Euler formula, defined in the next section, should be used to analyze the column.

If the actual ratio L_e/r is less than C_c, then the column is short. In these cases, either the J. B. Johnson formula, special codes, or the direct compressive stress formula should be used, as discussed in later sections.

Where a column is being analyzed to determine the load it will carry, the value of C_c and the actual ratio L_e/r should be computed first to determine which method of analysis should be used. Notice that C_c depends on the material properties of yield strength s_y and modulus of elasticity E. In working with steel, E is usually taken to be 207 GPa (30×10^6 psi). Using this value and assuming a range of values for yield strength, we obtain the values for C_c shown in Figure 11–5. Be aware that the value of E for structural steels is typically taken to be 200 GPa (29×10^6 psi), shifting the curve downward very slightly.

For aluminum, E is approximately 69 GPa (10×10^6 psi). The corresponding values for C_c are shown in Figure 11–6.

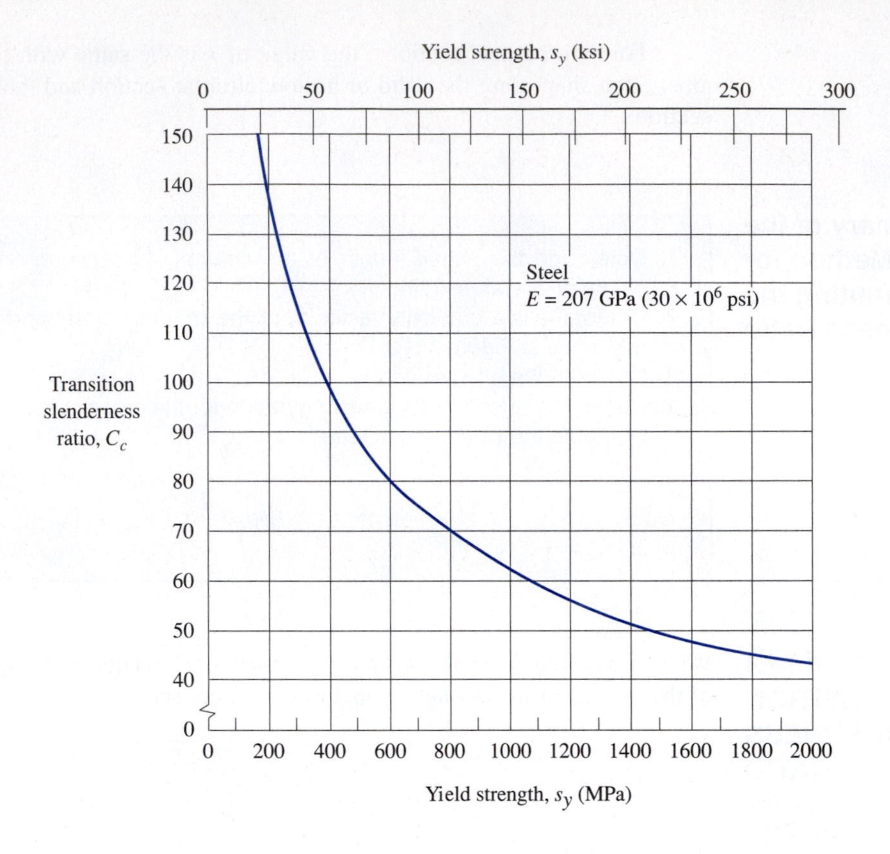

FIGURE 11–5 Transition slenderness ratio C_c versus yield strength for steel.

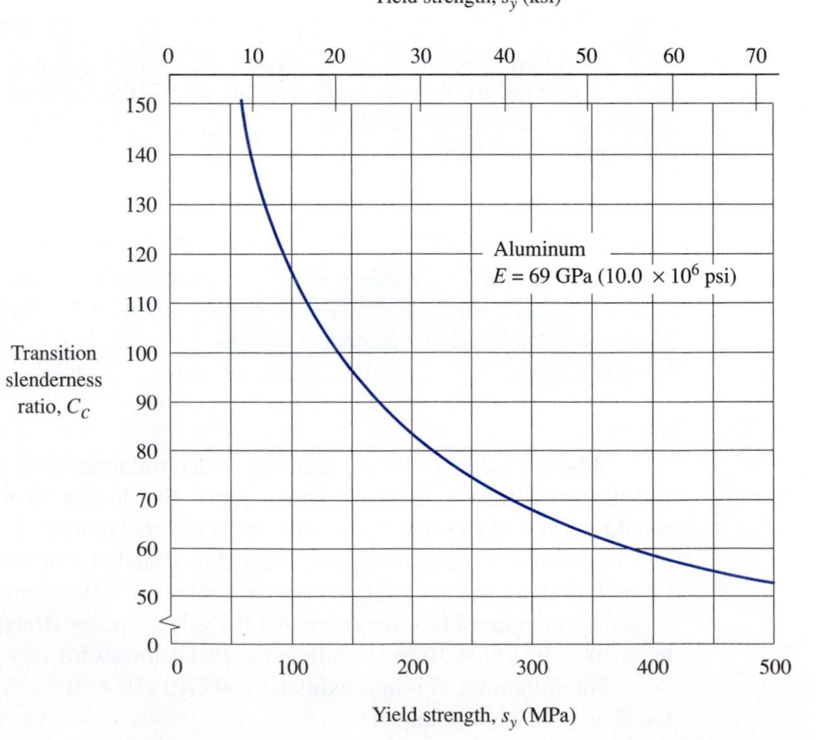

FIGURE 11–6 Transition slenderness ratio C_c versus yield strength for aluminum.

11–4 THE EULER FORMULA FOR LONG COLUMNS

For long columns having an effective slenderness ratio greater than the transition value C_c, the Euler formula can be used to predict the critical load at which the column would be expected to buckle. The formula is

$$P_{cr} = \frac{\pi^2 EA}{(L_e/r)^2} \quad \textbf{(11–4)}$$

where A is the cross-sectional area of the column. An alternative form can be expressed in terms of the moment of inertia by noting that $r^2 = I/A$. Then the formula becomes

Euler Formula for Long Columns

$$P_{cr} = \frac{\pi^2 EI}{L_e^2} \quad \textbf{(11–5)}$$

11–5 THE J. B. JOHNSON FORMULA FOR SHORT COLUMNS

If a column has an actual effective slenderness ratio L_e/r less than the transition value C_c, the Euler formula predicts an unreasonably high critical load. One formula recommended for machine design applications in the range of L_e/r less than C_c is the J. B. Johnson formula.

$$P_{cr} = As_y\left[1 - \frac{s_y (L_e/r)^2}{4\pi^2 E}\right] \quad \textbf{(11–6)}$$

This is one form of a set of equations called parabolic formulas, and it agrees well with the performance of steel columns in typical machinery.

J. B. Johnson Formula for Short Columns

The Johnson formula gives the same result as the Euler formula for the critical load at the transition slenderness ratio C_c. Then for very short columns, the critical load approaches that which would be predicted from the direct compressive stress equation, $\sigma = P/A$. Therefore, it could be said that the Johnson formula applies best to columns of intermediate length.

11–6 SUMMARY—BUCKLING FORMULAS

This section summarizes the basis for the Euler and Johnson formulas for column analysis and elaborates on the behavior of columns of varying materials and slenderness ratios.

The buckling phenomenon is not a failure of the material from which a column is made. Rather, it is a failure of the column as a whole to maintain its shape. This type of failure is called *elastic instability*.

Recall the exercises you did during the Big Picture section of this chapter. As you loaded a long, slender column, such as the meter stick, you observed that the column buckled at some moderate load. If the load was removed after observing the occurrence of buckling, you should notice that the column was not damaged. There was no yielding or fracture of the material.

To design a safe column, you must ensure that it remains elastically stable. The basis of both the Euler and Johnson formulas is developed from the field of stress analysis known as *elasticity*. Reference 6 is a useful source for this development.

The principle of elastic stability states that a column is stable if it maintains its straight shape as the load is increased. There is a level of load, though, at which the column is unable to maintain its shape. Then it buckles. We call the load at which buckling occurs the *critical buckling load, P_{cr}*. Obviously, as the designer of the column, you must ensure that the actual load applied to the column is well below P_{cr}.

When the axial load on a column is below the critical buckling load, it has sufficient stiffness to resist the tendency to deviate from the straight-line orientation of its neutral axis. Even when the load is slightly away from the axis, the column is able to maintain its shape. We can visualize this by referring to Figure 11–2. Part (a) shows a straight, pinned-end column carrying an axial compressive load. It also shows, in an exaggerated manner, that at the point of incipient buckling, the column takes the shape of one-half of a sine wave. Any deviation of the neutral axis from a straight line introduces bending in the column. When the applied force is less than P_{cr}, the stiffness of the column is sufficient to resist this deformation and to maintain the straightness of the column. It acts like a kind of spring, sometimes called a *leaf spring,* to return the column to its straight form whenever there is a tendency to deviate. The mathematical techniques of differential equations were used by the Swiss mathematician Leonhard Euler (1707–1783) in 1757 to analyze this condition and to produce what is now called the *Euler formula* in his honor [Equation (11–4)].

Recognizing that the deformed, buckled, pinned-end column takes the shape of one-half of a sine wave can also help us to visualize why different end-fixity conditions affect the magnitude of the critical buckling load. Refer to Figure 11–3 that shows the buckled shape of four columns with different end-fixity. The explanation of the theoretical value for K, the end-fixity factor, is given here.

a. Part (b) is the pinned-end design, and the entire length conforms to the sine wave form after buckling starts. This is the reason that $L_e = L$.

b. Part (d) shows the fixed-fixed design, and the sine wave form occurs only over the middle half of the column. Therefore, the theoretical effective length is $L_e = 0.5L$.

c. Part (e) shows the fixed-free design, and the deformed shape of the column is only one-fourth of a sine wave. To complete the one-half sine wave that characterizes the buckled shape, the curved line for the deformed neutral axis would have to be extended an equal distance *below* the bottom, fixed end of the column. Therefore the effective length is $L_e = 2.0\ L$.

d. Part (c) shows the fixed-pinned design. You should logically see that this design is somewhat midway between those shown in parts (b) and (d). Indeed, the sine wave form occurs in approximately the upper two-thirds of the column. Therefore, the theoretical effective length is $L_e = 0.7\ L$.

Again, because it is very difficult to produce a perfectly fixed end for a column, the design values for the end-fixity factor, K, are used in practical problems such as those in this book.

Comparison of the Euler and Johnson Formulas. We have noted that the Euler formula is only valid for columns having slenderness ratios greater than the transition slenderness ratio, C_c. When the actual L_e/r is less than C_c, the Johnson formula is recommended. Then, for very small values of L_e/r, the result from the Johnson formula approaches the load at which the material of the column would fail by yielding under direct axial compression.

Another way to examine these concepts is to divide both the Euler [Equation (11–4)] and Johnson [(Equation (11–6)] formulas by the cross sectional area, A. The left side of the formulas then have the form, P/A, and represent an average stress on the column cross section.

Figure 11–7 illustrates these concepts in graphical form. The vertical axis plots the average stress in the column when the axial load is equal to the critical buckling load. The horizontal axis is the slenderness ratio, L_e/r. The material properties of modulus of elasticity, E, and yield strength, s_y, affect the critical buckling load and so the graph presents a family of curves representative of different materials. For a given material, the right-most part is for large values of $L_e/r > C_c$. In this region, the behavior of columns made from any alloy of the material conforms to the Euler formula. At $L_e/r = C_c$, the Euler and Johnson formulas are tangent. Then, for $L_e/r < C_c$, the Johnson formula applies. Finally for very

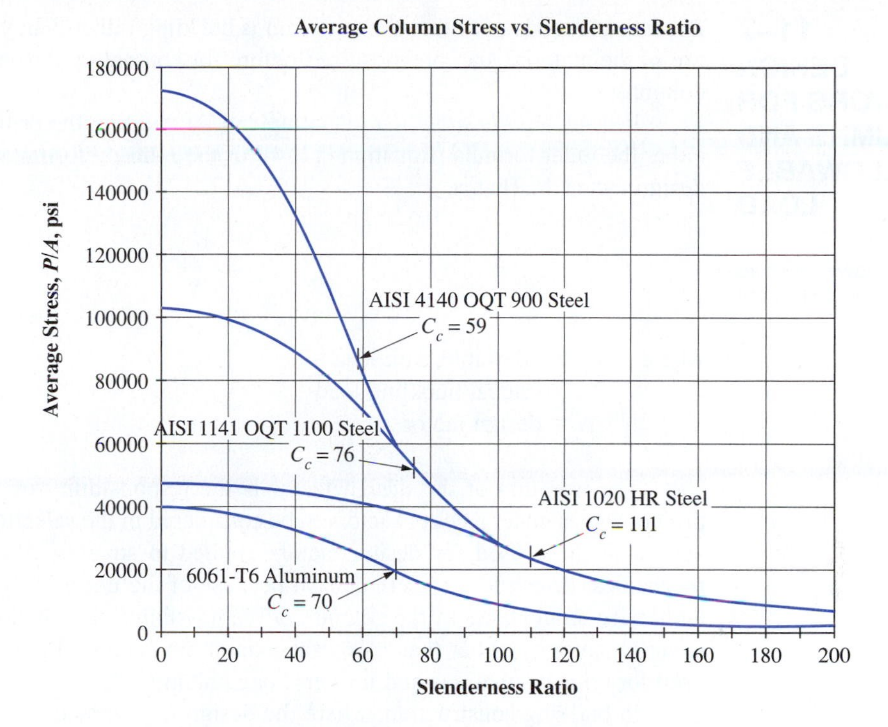

FIGURE 11–7 Average column stress vs. slenderness ratio.

short columns, the average stress approaches the yield strength of the material representing failure by direct compression without any buckling effect.

Laterally Braced Columns. Recall the exercise described in the Big Picture section. When the long, slender meter stick was held lightly at mid-length, the stick could withstand a much higher load before buckling. The lateral bracing effectively divides the column into two separate columns, one-half as long as the full column. The critical buckling load is then increased dramatically. When buckling does occur, each part of the column deforms in the shape of the one-half sine wave as shown in Figure 11–8. The whole column then takes the shape of a full sine wave. An example problem shown later illustrates this result.

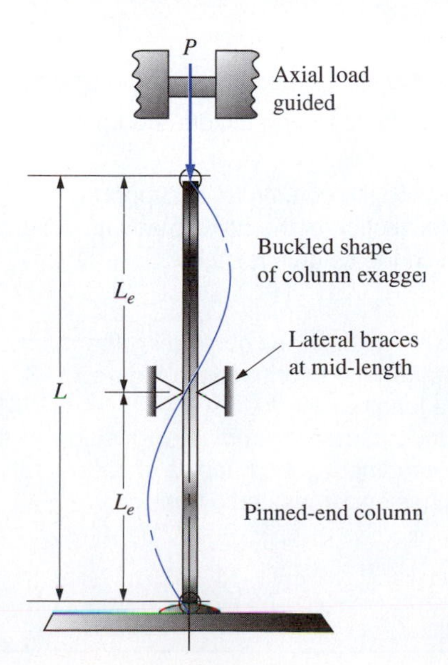

FIGURE 11–8 Laterally braced column.

11–7 DESIGN FACTORS FOR COLUMNS AND ALLOWABLE LOAD

Because the mode of failure for a column is buckling rather than yielding or ultimate failure of the material, the methods used before for computing design stress do not apply to columns.

Instead, an *allowable load* is computed by dividing the critical buckling load from either the Euler formula [Equation (11–4)] or the Johnson formula [Equation (11–6)] by a design factor, N. That is,

Allowable Load on a Column

$$P_a = \frac{P_{cr}}{N} \qquad \textbf{(11–7)}$$

where P_a = allowable, safe load
P_{cr} = critical buckling load
N = design factor

The selection of the design factor is the responsibility of the designer unless the project comes under a code. Factors to be considered in the selection of a design factor are similar to those used for design factors applied to stresses. A common factor used in mechanical design is $N = 3.0$, chosen because of the uncertainty of material properties, end-fixity, straightness of the column, or the possibility that the load will be applied with some eccentricity rather than along the axis of the column. Larger factors are sometimes used for critical situations and for very long columns.

In building construction, where the design is governed by the specifications of the American Institute of Steel Construction, AISC, a factor of 1.92 is recommended for long columns. The Aluminum Association calls for $N = 1.95$ for long columns. See Sections 11–12 and 11–13.

11–8 SUMMARY—METHOD OF ANALYZING COLUMNS

The purpose of this section is to summarize the concepts presented in Sections 11–3 through 11–7 into a procedure that can be used to analyze columns. It can be applied to a straight column having a uniform cross section throughout its length, and for which the compression load is applied in line with the centroidal axis of the column.

Method of Analyzing Columns

To start, it is assumed that the following factors are known:

1. The actual length, L
2. The manner of connecting the column to its supports
3. The shape of the cross section of the column and its dimensions
4. The material from which the column is made.

Then the procedure is:

1. Determine the end-fixity factor, K, by comparing the manner of connection of the column to its supports with the information in Figure 11–3.
2. Compute the effective length, $L_e = KL$.
3. Compute the minimum value of the radius of gyration of the cross section from $r_{min} = \sqrt{I_{min}/A}$; or determine r_{min} from tables of data.
4. Compute the maximum slenderness ratio from

$$SR_{max} = \frac{L_e}{r_{min}}$$

5. Using the modulus of elasticity, E, and the yield strength, s_y, for the material, compute the column constant,

$$C_c = \sqrt{\frac{2\pi^2 E}{s_y}}$$

6. Compare the value of SR with C_c.
   a. If $SR > C_c$, the column is long. Use the Euler formula to compute the critical buckling load,

$$P_{cr} = \frac{\pi^2 EA}{(SR)^2} \quad \textbf{(11–4)}$$

   b. If $SR < C_c$, the column is short. Use the Johnson formula to compute the critical buckling load,

$$P_{cr} = As_y\left[1 - \frac{s_y(SR)^2}{4\pi^2 E}\right] \quad \textbf{(11–6)}$$

7. Specify the design factor, N.
8. Compute the allowable load, P_a,

$$P_a = \frac{P_{cr}}{N}$$

Example Problem 11–1 A round compression member with both ends pinned and made of AISI 1020 cold-drawn steel is to be used in a machine. Its diameter is 25 mm, and its length is 950 mm. What maximum load can the member take before buckling would be expected? Also compute the allowable load on the column for a design factor of $N = 3$.

Solution

Objective: Compute the critical buckling load for the column and the allowable load for a design factor of $N = 3$.

Given: $L = 950$ mm. Cross section is circular; $D = 25$ mm. Pinned ends.
Column is steel; AISI 1020 cold-drawn.
From Appendix A–14: $s_y = 441$ MPa; $E = 207$ GPa $= 207 \times 10^9$ N/m^2

Analysis: Use the *Method of Analyzing Columns*.

Results: ***Step 1.*** Determine the end-fixity factor. For the pinned-end column, $K = 1.0$.

Step 2. Compute the effective length.

$$L_e = KL = 1.0(L) = 950 \text{ mm}$$

Step 3. Compute the smallest value of the radius of gyration. From Appendix A–1, for any axis of a circular cross section, $r = D/4$. Then,

$$r = \frac{D}{4} = \frac{25 \text{ mm}}{4} = 6.25 \text{ mm}$$

Step 4. Compute the slenderness ratio, $SR = L_e/r$.

$$SR = \frac{L_e}{r} = \frac{950 \text{ mm}}{6.25 \text{ mm}} = 152$$

Step 5. Compute the column constant, C_c.

$$C_c = \sqrt{\frac{2\pi^2 E}{s_y}} = \sqrt{\frac{2\pi^2(207 \times 10^9 \text{ N/m}^2)}{441 \times 10^6 \text{ N/m}^2}} = 96.3$$

Step 6. Compare C_c with SR and decide if column is long or short. Then use the appropriate column formula to compute the critical buckling load. Since SR is greater than C_c, Euler's formula applies.

$$P_{cr} = \frac{\pi^2 EA}{(SR)^2}$$

The area is

$$A = \frac{\pi D^2}{4} = \frac{\pi(25 \text{ mm})2}{4} = 491 \text{ mm}^2$$

Then

$$P_{cr} = \frac{\pi^2(207 \times 10^9 \text{ N/m}^2)(491 \text{ mm}^2)}{(152)^2} \times \frac{1 \text{ m}^2}{(10^3 \text{ mm})^2} = 43.4 \text{ kN}$$

Step 7. A design factor of $N = 3$ is specified.

Step 8. The allowable load, P_a, is

$$P_a = \frac{P_{cr}}{N} = \frac{43.4 \text{ kN}}{3} = 14.5 \text{ kN}$$

Example Problem 11–2 Determine the critical load on a steel column having a square cross section 12 mm on a side with a length of 300 mm. The column is to be made of AISI 1040, hot-rolled. It will be rigidly welded to a firm support at one end and connected by a pin joint at the other. Also compute the allowable load on the column for a design factor of $N = 3$.

Solution

Objective: Compute the critical buckling load for the column and the allowable load for a design factor of $N = 3$.

Given: $L = 300$ mm. Cross section is square; each side is $b = 12$ mm.
One pinned end; one fixed end. Column is steel; AISI 1040 hot-rolled.
From Appendix A–14: $s_y = 414$ MPa; $E = 207$ GPa $= 207 \times 10^9$ N/m^2

Analysis: Use the *Method of Analyzing Columns*.

Results: ***Step 1.*** Determine the end-fixity factor. For the fixed-pinned-end column, $K = 0.80$ is a practical value (Figure 11–3).

Step 2. Compute the effective length.

$$L_e = KL = 0.80(L) = 0.80(300\text{ mm}) = 240\text{ mm}$$

Step 3. Compute the smallest value of the radius of gyration. From Appendix A–1, for a square cross section, $r = b/\sqrt{12}$. Then,

$$r = \frac{b}{\sqrt{12}} = \frac{12\text{ mm}}{\sqrt{12}} = 3.46\text{ mm}$$

Step 4. Compute the slenderness ratio, $SR = L_e/r$.

$$SR = \frac{L_e}{r} = \frac{KL}{r} = \frac{(0.8)\,(300\text{ mm})}{3.46\text{ mm}} = 69.4$$

Step 5. Normally, we would compute the value of the column constant, C_c. But, in this case, let's use Figure 11–5. For a steel having a yield strength of 414 MPa, $C_c = 96$, approximately.

Step 6. Compare C_c with SR and decide if the column is long or short. Then use the appropriate column formula to compute the critical buckling load. Since SR is less than C_c, the Johnson formula, Equation (11–6), should be used.

$$P_{cr} = As_y\left[1 - \frac{s_y(SR)^2}{4\pi^2 E}\right]$$

The area of the square cross section is

$$A = b^2 = (12\text{ mm})^2 = 144\text{ mm}^2$$

Then,

$$P_{cr} = (144\text{ mm}^2)\left(\frac{414\text{ N}}{\text{mm}^2}\right)\left[1 - \frac{(414 \times 10^6\text{ N/m}^2)\,(69.4)^2}{4\pi^2(207 \times 10^9\text{ N/m}^2)}\right]$$

$$P_{cr} = 45.1\text{ kN}$$

Step 7. A design factor of $N = 3$ is specified.

Step 8. The allowable load, P_a, is

$$P_a = \frac{P_{cr}}{N} = \frac{45.1\text{ kN}}{3} = 15.0\text{ kN}$$

Example Problem 11–3 Refer to the results of Example Problem 11–1 in which a 25-mm-diameter steel column with a length of 950 mm was analyzed to determine the critical buckling load. The column was made from AISI 1020 cold-drawn steel. It was found that $P_{cr} = 43.4$ kN. Now consider a redesign of the structure for which this column is a part. It is decided to provide lateral bracing in all directions for the column at its mid-length. Determine the critical buckling load for the redesigned column.

Solution Objective Compute the critical buckling load for the braced column.

Given Data from Example Problem 11–1. Circular cross section, $D = 25$ mm. $L = 950$ mm. Pinned ends. AISI 1020 CD steel; $E = 207$ GPa; $s_y = 441$ MPa. Column braced at 450 mm from each end.

Analysis Use the *Method of Analyzing Columns.*

Results ***Step 1.*** End-fixity factor, $K = 1.0$ for pinned ends.

Step 2. Effective length: The unbraced length is now 450 mm. Then $L_e = KL = 1.0(450\text{ mm}) = 450$ mm

Step 3. Radius of gyration $= r = 6.25$ mm [from Example Problem 11–1]

Step 4. Slenderness ratio $= SR = L_e/r = (450\text{ mm})/(6.25\text{ mm}) = 72$

Step 5. Column Constant $= C_c = 96.2$ [from Example Problem 11–1]

Step 6. Because $SR < C_c$, use the Johnson formula. $A = 491\text{ mm}^2$

$$P_{cr} = As_y\left[1 - \frac{s_y(SR)^2}{4\pi^2 E}\right]$$

$$P_{cr} = (491\text{ mm}^2)\left(\frac{441\text{ N}}{\text{mm}^2}\right)\left[1 - \frac{(441 \times 10^6\text{ N/m}^2)(72)^2}{4\pi^2(207 \times 10^9\text{ N/m}^2)}\right] = 156\text{ kN}$$

Comment The critical buckling load has been increased from 43.4 kN to 156 kN, over $3\frac{1}{2}$ times. That is a significant improvement. The column behaves as if it were only one-half as long.

11–9 COLUMN ANALYSIS SPREADSHEET

Completing the process described in Section 11–8 using a calculator, pencil, and paper is tedious. A spreadsheet automates the calculations after you have entered the pertinent data for the particular column to be analyzed. Figure 11–9 shows the output of a spreadsheet used to solve Example Problem 11–1. The layout of the spreadsheet could be done in many ways, and you are encouraged to develop your own style. The following comments describe the features of the given spreadsheet.

1. At the top of the sheet, instructions to the user are given for entering data and for units. This sheet is for SI metric units only. A different sheet would be used if U.S. Customary data were to be used.
2. On the left side of the sheet are listed the various data that must be provided by the user to run the calculations. On the right are listed the output values. Formulas for computing L_e, C_c, KL/r, and allowable load are written directly into the cell where the computed values show. The output data for the message "Column is: *long*" and the critical buckling load are produced by *functions* set up within *macros* written in Visual Basic and placed on a separate sheet of the spreadsheet. Figure 11–10 shows the two macros used. The first (*LorS*) carries out the decision process to test whether the column is long or short as indicated by comparison of its slenderness ratio with the column constant. The second (*Pcr*) computes the critical buckling load using either the Euler formula or the J. B. Johnson formula, depending on the result of the *LorS* macro. These functions are called by statements in the cells where "long" and the computed value of the critical buckling load (43.42 kN) are located.

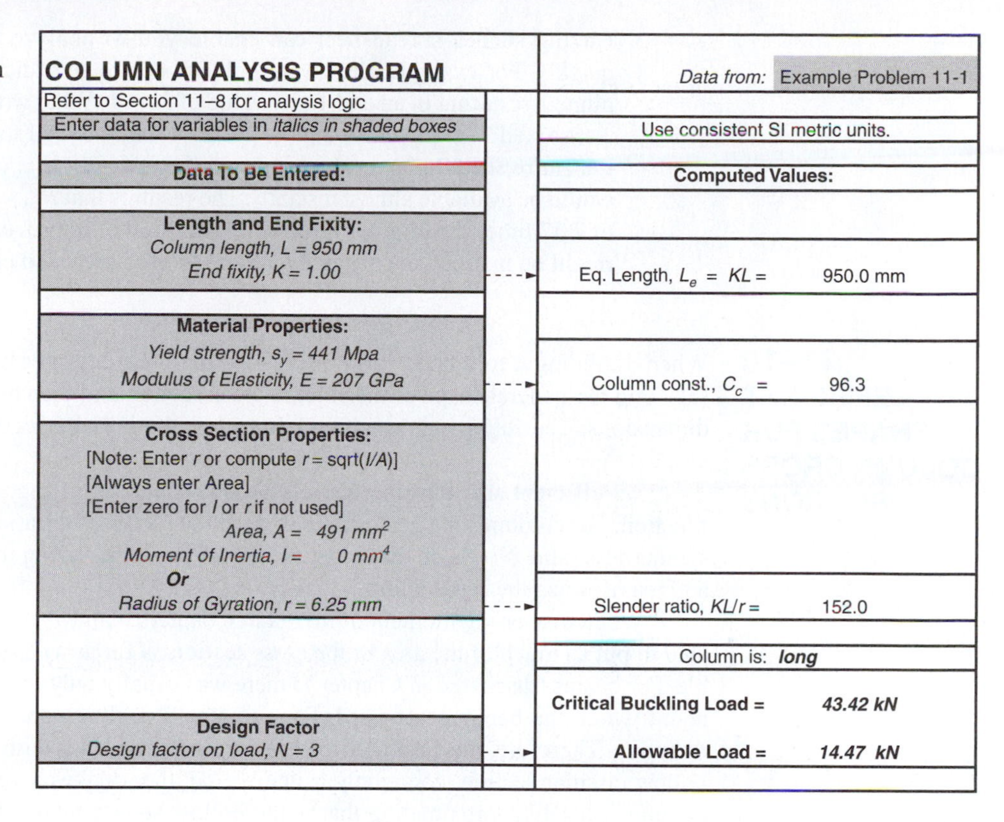

FIGURE 11–9 Spreadsheet for column analysis with data from Example Problem 11–1 [SI data].

FIGURE 11–10 Macros used in the column analysis spreadsheet.

```
'LorS Macro
'Determines if column is long or short.
Function LorS(SR,CC)
   If SR > CC Then
      LorS = "long"
   Else
      LorS = "short"
   End If
End Function

'Critical Load Macro
'Uses Euler formula for long columns
'Uses Johnson formula for short columns
Function Pcr(LorS, SR, E, A, Sy)
Const Pi = 3.1415926
   If LorS = "long" Then
      Pcr = Pi ^ 2 * E * A / SR ^ 2
       'Euler Equation; Eq. (11-4)
   Else
      Pcr = A * Sy(1-(Sy * SR ^ 2 / (4 * Pi ^ 2 * E))
       'Johnson Equation; Eq. (11-7)
   End If
End Function
```

3. Having such a spreadsheet can enable you to analyze several design options quickly. For example, the given problem statement indicated that the ends were pinned, resulting in an end-fixity value of $K = 1$. What would happen if both ends were fixed? Simply changing the value of that one cell to $K = 0.65$ would cause the entire sheet to be recalculated, and the revised value of critical buckling load would be available almost instantly. The result is that $P_{cr} = 102.76$ kN, an increase of 2.37 times the original value. With that kind of improvement, you, the designer, might be inclined to change the design to produce fixed ends.

11–10 EFFICIENT SHAPES FOR COLUMN CROSS SECTIONS

When designing a column to carry a specified load, the designer has the responsibility for selecting the general shape of the column cross section and then to determine the required dimensions. The following principles may aid in the initial selection of the cross-section shape.

An efficient shape is one that uses a small amount of material to perform a given function. For columns, the goal is to maximize the radius of gyration in order to reduce the slenderness ratio. Note also that because $r = \sqrt{I/A}$, maximizing the moment of inertia for a given area has the same effect.

When discussing moment of inertia in Chapters 6 and 7, it was noted that it is desirable to put as much of the area of the cross section as far away from the centroid as possible. For beams (discussed in Chapter 7) there was usually only one important axis, the axis about which the bending occurred. For columns, buckling can, in general, occur in any direction. Therefore, it is desirable to have uniform properties with respect to any axis. The hollow circular section, commonly called a pipe, then makes a very efficient shape for a column. Closely approximating that is the hollow square tube. Fabricated sections made from standard structural sections can also be used, as shown in Figure 11–11.

Building columns are often made from special wide-flange shapes called *column sections*. They have relatively wide, thick flanges as compared with the shapes typically selected for beams. This makes the moment of inertia with respect to the *Y-Y* axis more

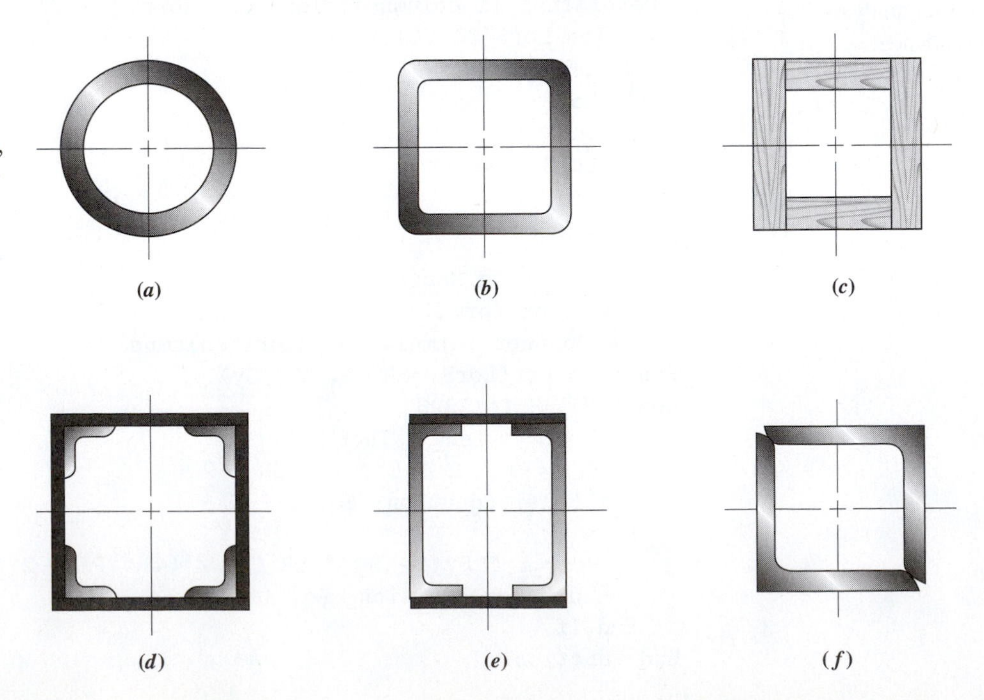

FIGURE 11–11 Examples of efficient column shapes. (a) Hollow circular section, pipe. (b) Hollow square tube. (c) Box section made from wood beams. (d) Equal-leg angles with plates. (e) Aluminum channels with plates. (f) Two equal-leg angles.

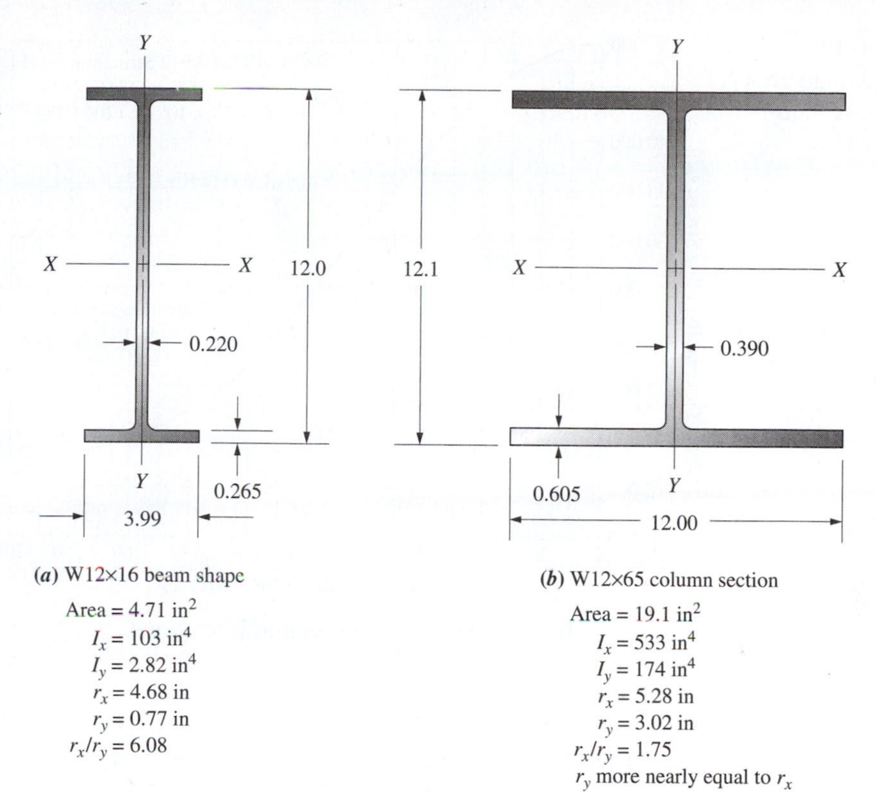

FIGURE 11–12 Comparison of a wide-flange beam shape with a column section.

nearly equal to that for the *X-X* axis. The result is that the radii of gyration for the two axes are more nearly equal also. Figure 11–12 shows a comparison of two 12-in wide-flange shapes: one a column section and one a typical beam shape. Note that the smaller radius of gyration should be used in computing the slenderness ratio.

11–11 SPECIFICATIONS OF THE AISC

Columns are essential elements of many structures. The design and analysis of columns in construction applications are governed by the specifications of the AISC, the American Institute of Steel Construction (Reference 2), briefly summarized here for column sections that are loaded through their centroidal axes and that do not exhibit local buckling of thin extended flanges. The method involves the following variables. Note that symbols used here are similar to those used in earlier sections of this chapter and are not necessarily the same as those in the AISC manual.

$$\textit{Transition slenderness ratio} = SR_t = 4.71\sqrt{E/s_y} \tag{11–8}$$

(SR_t is approximately 6% larger than C_c)

$$\textit{Elastic critical buckling stress} = s_e = \pi^2 E/(SR)^2 \tag{11–9}$$

(sometimes called *Euler stress*)

Flexural buckling stress $= s_{cr}$ whose value is dependent on the actual SR

If $SR \leq SR_t$, then the column is short and,

$$s_{cr} = [0.658^d]\, s_y \qquad \text{and the exponent } d = s_y/s_e \tag{11–10}$$

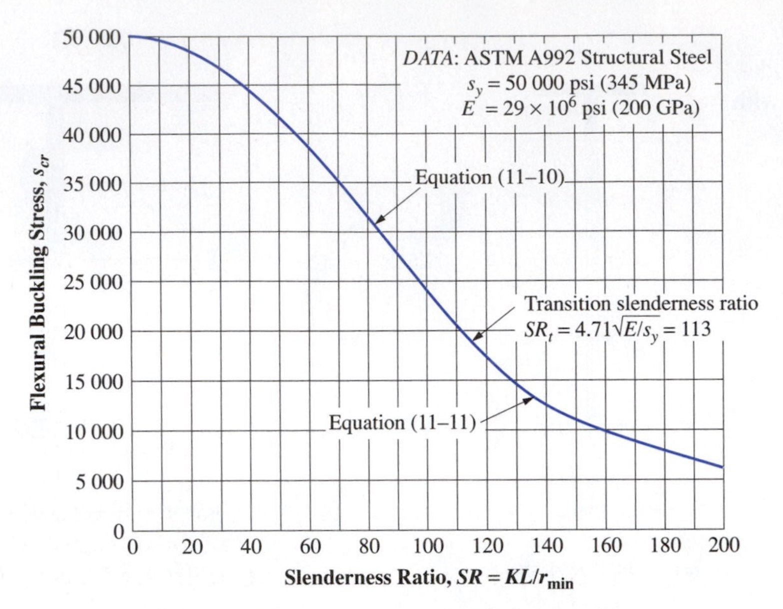

FIGURE 11–13 Flexure buckling stress vs. slenderness ratio—AISC method.

If $SR > SR_t$, then the column is long and,

$$s_{cr} = 0.877\, s_e \tag{11–11}$$

Then the *nominal buckling strength* $= P_n = s_{cr} A_g$ **(11–12)**

(A_g is the column's gross area)

Finally, the *allowable compressive strength* $= P_a = P_n/1.67.$ **(11–13)**

It should be noted that these formulas are simpler and somewhat different from those reported in previous editions of the AISC manual. However, the resulting values of P_a are within approximately 2% of the previous results.

Figure 11–13 shows a graph of the *flexural buckling stress,* s_{cr}, versus the actual slenderness ratio of a column. Data are for ASTM A992 structural steel, the most common steel for W-beams and column sections. Note how Equations (11–10) and (11–11) are tangent at the transition slenderness ratio value and how the buckling stress approaches the yield strength for very short columns. The AISC recommends that the largest usable slenderness ratio be 200.

Example Problem 11–4 Compute the allowable compressive strength, P_a, for a column made from steel rectangular structural tubing, HSS102×51×6.4. The material is ASTM A500, grade B, structural steel. The column length is 3050 mm and its ends are pinned.

Solution We will use Equations (11–8) to (11–13) with $E = 200$ GPa $= 200\,000$ MPa and $s_y = 290$ MPa. The tube is expected to buckle about the *Y-Y* axis so $r_{min} = r_y = 19.8$ mm [Appendix A–9(SI)] and $A_g = 1570\text{ mm}^2$.

$$\textit{Actual slenderness ratio} = SR = KL/r = 1.00(3050\text{ mm}/19.8\text{ mm}) = 154$$

$$\textit{Transition slenderness ratio} = SR_t = 4.71\sqrt{E/S_y} = 4.71\sqrt{(200\,000/290)} = 123.7$$

$$\textit{Elastic critical buckling stress} = s_e = \pi^2 E/(SR)^2 = \pi^2\,(200\,000\text{ MPa})/(154)^2$$

$$s_e = 83.23\text{ MPa}$$

Because $SR > SR_t$, the column is long and we use Equation (11–11):

$$s_{cr} = 0.877\, s_e = 0.877(83.23\ \text{MPa}) = 73.0\ \text{MPa}$$

Now, using Equation (11–12),
The *nominal buckling strength* $= P_n = s_{cr} A_g = (73.0\ \text{N/mm}^2)(1570\ \text{mm}^2) = 114.6\ \text{kN}$
Using Equation (11–13),
The *allowable compressive strength* $= P_a = P_n/1.67 = 114.6\ \text{kN}/1.67 = 68.6\ \text{kN}$

Example Problem 11–5 Compute the allowable compressive strength, P_a, for the column described in Example Problem 11–4 except it will be installed with both ends fixed instead of pinned.

Solution *Actual slenderness ratio* $= SR = KL/r = (0.65)(3050\ \text{mm}/19.8\ \text{mm}) = 100$
From Example Problem 11–4, *transition slenderness ratio* $= SR_t = 123.7$
Elastic critical buckling stress $= s_e = \pi^2 E/(SR)^2 = \pi^2\ (200\ 000\ \text{MPa})/(100)^2$

$$s_e = 197.4\ \text{MPa}$$

Then the column is short and Equation (11–10) is applied next.
The exponent $d = s_y/s_e = 290\ \text{MPa}/197.4\ \text{MPa} = 1.469$ and

$$s_{cr} = [0.658^{d}]\, s_y = [0.658^{1.469}](290\ \text{MPa}) = 156.8\ \text{MPa}$$

The *nominal buckling strength* $= P_n = s_{cr} A_g = (156.8\ \text{N/mm}^2)(1570\ \text{mm}^2) = 246.2\ \text{kN}$
Using Equation (11–13),
The *allowable compressive strength* $= P_a = P_n/1.67 = 246.2\ \text{kN}/1.67 = 147.4\ \text{kN}$

Comment The resulting allowable compressive strength for the fixed-end column is 2.15 times greater than the column design having pinned ends.

11–12 SPECIFICATIONS OF THE ALUMINUM ASSOCIATION

The Aluminum Association publication, *Aluminum Design Manual* (see Reference 1), defines allowable stresses for columns for each of several aluminum alloys and their heat treatments. Three different equations are given for short, intermediate, and long columns defined in relation to slenderness limits. The equations are of the form

$$\frac{P_a}{A} = \frac{s_y}{FS} \quad \text{(short columns)} \qquad \textbf{(11–14)}$$

$$\frac{P_a}{A} = \frac{B_c - D_c(L_e/r)}{FS} \quad \text{(intermediate columns)} \qquad \textbf{(11–15)}$$

$$\frac{P_a}{A} = \frac{\pi^2 E}{FS(L_e/r)^2} \quad \text{(long columns)} \qquad \textbf{(11–16)}$$

In all three cases, it is recommended that $FS = 1.95$ for buildings and similar structures. The short column analysis assumes that buckling will not occur and that safety is dependent on the yield strength of the material. Equation (11–16) for long columns is the Euler formula with a factor of safety applied. The intermediate column formula

[Equation (11–15)] depends on buckling constants B_c and D_c, which are functions of the yield strength of the aluminum alloy and the modulus of elasticity. The division between intermediate and long columns is similar to the C_c used previously in this chapter.

Following are the specific equations for the alloy 6061-T6 used in building structures in the forms of sheet, plate, extrusions, structural shapes, rod, bar, tube, and pipe.

Short columns and intermediate columns: $0 < L_e/r < 66$

$$\frac{P_a}{A} = \left(20.2 - 0.126\frac{L_e}{r}\right)\text{ksi} \tag{11–17a}$$

$$\frac{P_a}{A} = \left(139 - 0.869\frac{L_e}{r}\right)\text{MPa} \tag{11–17b}$$

Long columns: $L_e/r > 66$

$$\frac{P_a}{A} = \frac{51\,000}{(L_e/r)^2}\text{ksi} \tag{11–18a}$$

$$\frac{P_a}{A} = \frac{352\,000}{(L_e/r)^2}\text{MPa} \tag{11–18b}$$

See Reference 1 for column design stresses for other aluminum alloys.

11–13 NON-CENTRALLY LOADED COLUMNS

All of the analysis methods discussed in this chapter so far have been limited to loadings in which the compressive loads on the columns act in-line with the centroidal axis of the column cross section. Also, it is assumed that the column axis is perfectly straight prior to the application of the loads. We have used the term *straight centrally loaded column* to describe such a case.

Many real columns violate these assumptions to some degree. Figure 11–14 shows two such conditions. If a column is initially *crooked,* the applied compressive force on the column

FIGURE 11–14 Illustration of crooked and eccentric columns.

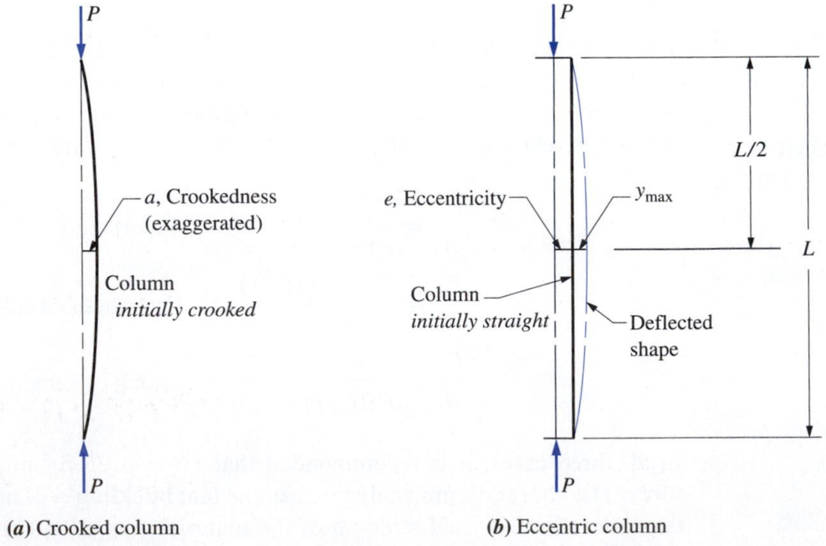

tends to cause bending in the column in addition to buckling, and failure would occur at a lower load than that predicted from the equations used so far in this chapter. An *eccentrically loaded column* is one in which there is some purposeful offsetting of the line of action of the compressive load from the centroidal axis of the column. Here, again, there is some bending stress produced in addition to the axial compressive stress that tends to cause buckling.

Crooked Columns. The crooked column formula allows an initial crookedness, a, to be considered (see References 4, 5, and 6):

Crooked Column Formula

$$P_a^2 - \frac{1}{N}\left[s_y A + \left(1 + \frac{ac}{r^2}\right)P_{cr}\right]P_a + \frac{s_y A P_{cr}}{N^2} = 0 \quad (11\text{–}19)$$

where c = distance from the neutral axis of the cross section about which bending occurs to its outer edge

P_{cr} is defined to be the critical load found from the *Euler formula*.

Although this formula may become increasingly inaccurate for shorter columns, it is not appropriate to switch to the Johnson formula as it is for straight columns.

The crooked column formula is a quadratic with respect to the allowable load P_a. Evaluating all constant terms in Equation (11–19) produces an equation of the form

$$P_a^2 + C_1 P_a + C_2 = 0$$

Then, from the solution for a quadratic equation,

$$P_a = 0.5\left[-C_1 - \sqrt{C_1^2 - 4C_2}\right]$$

The smaller of the two possible solutions is selected.

Example Problem 11–6 A column has both ends pinned and has a length of 32 in. It has a circular cross section with a diameter of 0.75 in and an initial crookedness of 0.125 in. The material is AISI 1040 hot-rolled steel. Compute the allowable load for a design factor of 3.

Solution

Objective: Specify the allowable load for the column.

Given: Solid circular cross section: $D = 0.75$ in; $L = 32$ in; use $N = 3$.
Both ends are pinned. Initial crookedness $= a = 0.125$ in.
Material: AISI 1040 hot-rolled steel.

Analysis: Use Equation (11–19). First evaluate C_1 and C_2. Then solve the quadratic equation for P_a.

Results:

$$s_y = 60\,000 \text{ psi}$$

$$A = \pi D^2/4 = (\pi)(0.75)^2/4 = 0.442 \text{ in}^2$$

$$r = D/4 = 0.75/4 = 0.188 \text{ in}$$

$$c = D/2 = 0.75/2 = 0.375 \text{ in}$$

$$KL/r = [(1.0)(32)]/0.188 = 171$$

$$P_{cr} = \frac{\pi^2 EA}{(KL/r)^2} = \frac{\pi^2(30\,000\,000)(0.442)}{(171)^2} = 4476 \text{ lb}$$

$$C_1 = \frac{-1}{N}\left[s_y A + \left(1 + \frac{ac}{r^2}\right)P_{cr}\right] = -12\,311$$

$$C_2 = \frac{s_y A P_{cr}}{N^2} = 1.319 \times 10^7$$

The quadratic is therefore

$$P_a^2 - 12\,311 P_a + 1.319 \times 10^7 = 0$$

Comment From this, $P_a = 1186$ lb is the allowable load.

Figure 11–15 shows the solution of Example Problem 11–6 using a spreadsheet. Whereas its appearance is similar to that of the earlier column analysis spreadsheet, the details follow the calculations needed to solve Equation (11–19). On the lower left, two special data values are needed: (1) the crookedness a and (2) the distance c from the neutral axis for buckling to the outer surface of the cross section. In the middle of the right part are listed some intermediate values used in Equation (11–19): C_1 and C_2 as defined in the solution to Example Problem 11–6. The result, the allowable load, P_a, is at the lower right of the spreadsheet. Above that, for comparison, the computed value of the critical buckling load is given for a straight column of the same design. Note that this solution procedure is most accurate for long columns. If the analysis indicates that the column is *short* rather than *long,* the designer should take note of how short it is by comparing the slen-

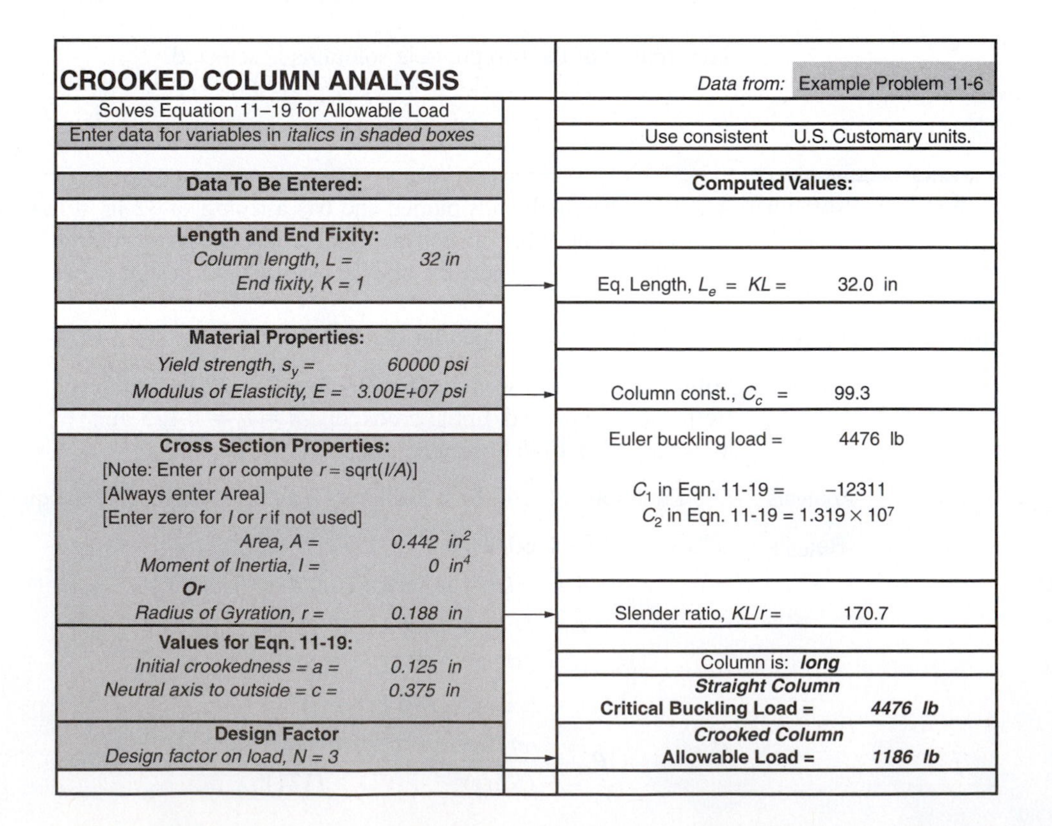

FIGURE 11–15 Spreadsheet for analysis of crooked columns—U.S. Customary units.

derness ratio, KL/r, with the column constant, C_c. If the column is quite short, the designer should not rely on the accuracy of the result from Equation (11–19).

Eccentrically Loaded Columns. An *eccentric load* is one that is applied away from the centroidal axis of the cross section of the column, as shown in Figure 11–14(b). Such a load exerts bending in addition to the column action that results in the deflected shape shown in the figure. The maximum stress in the deflected column occurs in the outermost fibers of the cross section at the midlength of the column where the maximum deflection, y_{max}, occurs. Let's denote the stress at this point as $\sigma_{L/2}$. Then, for any applied load, P,

Secant Formula for Eccentrically Loaded Columns

$$\sigma_{L/2} = \frac{P}{A}\left[1 + \frac{ec}{r^2}\sec\left(\frac{KL}{2r}\sqrt{\frac{P}{AE}}\right)\right] \quad \textbf{(11–20)}$$

(See Reference 6.) Note that this stress is *not* directly proportional to the load. When evaluating the secant in this formula, note that its argument in the parentheses is in *radians*. Also, because most calculators do not have the secant function, recall that the secant is equal to 1/cosine.

For design purposes, we would like to specify a design factor, N, that can be applied to the *failure load* similar to that defined for straight, centrally loaded columns. However, in this case, failure is predicted when the maximum stress in the column exceeds the yield strength of the material. Let's now define a new term, P_y, to be the load applied to the eccentrically loaded column when the maximum stress is equal to the yield strength. Equation (11–20) then becomes

$$s_y = \frac{P_y}{A}\left[1 + \frac{ec}{r^2}\sec\left(\frac{KL}{2r}\sqrt{\frac{P_y}{AE}}\right)\right]$$

Now, if we define the *allowable load* to be

$$P_a = P_y/N$$

or

$$P_y = NP_a$$

Design Equation for Eccentrically Loaded Columns

this equation becomes

$$\text{Required} \quad s_y = \frac{NP_a}{A}\left[1 + \frac{ec}{r^2}\sec\left(\frac{KL}{2r}\sqrt{\frac{NP_a}{AE}}\right)\right] \quad \textbf{(11–21)}$$

This equation cannot be solved for either N or P_a. Therefore, an iterative solution is required, as will be demonstrated in Example Problem 11–7.

Maximum Deflection in an Eccentrically Loaded Column

Another critical factor may be the amount of deflection of the axis of the column due to the eccentric load:

$$y_{max} = e\left[\sec\left(\frac{KL}{R}\sqrt{\frac{P}{AE}}\right) - 1\right] \quad \textbf{(11–22)}$$

Note that the argument of the secant is the same as that used in Equation (11–20).

Example Problem 11–7

For the column of Example Problem 11–6, compute the maximum stress and deflection if a load of 1075 lb is applied with an eccentricity of 0.75 in.

Solution Objective Compute the stress and the deflection for the eccentrically loaded column.

Given Data from Example Problem 11–6, but eccentricity = e = 0.75 in.
Solid circular cross section: $D = 0.75$ in; $L = 32$ in.
Both ends are pinned; $KL = 32$ in; $r = 0.188$ in; $c = D/2 = 0.375$ in.
Material: AISI 1040 hot-rolled steel; $E = 30 \times 10^6$ psi, $s_y = 60\ 000$ psi

Analysis Use Equation (11–20) to compute maximum stress. Then use Equation (11–22) to compute maximum deflection.

Results All terms have been evaluated before. Then the maximum stress is found from Equation (11–20):

$$\sigma_{L/2} = \frac{1075}{0.422}\left[1 + \frac{(0.75)(0.375)}{(0.188)^2}\sec\left(\frac{32}{2(0.188)}\sqrt{\frac{1075}{(0.422)(30 \times 10^6)}}\right)\right]$$

$$\sigma_{L/2} = 29\,300 \text{ psi}$$

The maximum deflection is found from Equation (11–22):

$$y_{max} = 0.75\left[\sec\left(\frac{32}{2(0.188)}\sqrt{\frac{1075}{(0.442)(30 \times 10^6)}}\right) - 1\right] = 0.293 \text{ in}$$

Comments The maximum stress is 29 300 psi at the midlength of the column. The deflection there is 0.293 in.

Example Problem 11–8 The stress in the column found in Example Problem 11–7 seems high for the AISI 1040 hot-rolled steel. Redesign the column to achieve a design factor of at least 3. Use only the preferred sizes listed in Appendix A–2

Solution Objective Redesign the eccentrically loaded column of Example Problem 11–7 to reduce the stress and achieve a design factor of at least 3.

Given Data from Example Problems 11–6 and 11–7.

Analysis Use a larger diameter. Use Equation (11–21) to compute the required strength. Then compare that with the strength of AISI 1040 hot-rolled steel. Iterate until the stress is satisfactory.

Results Appendix 3 gives the value for the yield strength of AISI 1040 HR to be 60 000 psi. If we choose to retain the same material, the cross-sectional dimensions of the column must be increased to decrease the stress. Equation (11–21) can be used to evaluate a design alternative.

The objective is to find suitable values for A, c, and r for the cross section such that $P_a = 1075$ lb; $N = 3$; $L_e = 32$ in; $e = 0.75$ in; and the value of the entire right side of the equation is less than 60 000 psi. The original design had a circular cross section with a diameter of 0.75 in. Let's try increasing the diameter to $D = 1.00$ in. Then

$$A = \pi D^2/4 = \pi(1.00 \text{ in})^2/4 = 0.785 \text{ in}^2$$
$$r = D/4 = (1.00 \text{ in})/4 = 0.250 \text{ in}$$
$$r^2 = (0.250 \text{ in})^2 = 0.0625 \text{ in}^2$$
$$c = D/2 = (1.00 \text{ in})/2 = 0.50 \text{ in}$$

Now let's call the right side of Equation (11–21) s'_y. Then

$$s'_y = \frac{3(1075)}{0.785}\left[1 + \frac{(0.75)(0.50)}{(0.0625)}\sec\left(\frac{32}{2(0.250)}\sqrt{\frac{(3)(1075)}{(0.785)(30 \times 10^6)}}\right)\right]$$

$$s'_y = 37\,740 \text{ psi} = \text{required value of } s_y$$

This is well below the value of $s_y = 60\,000$ psi for the given steel, and it gives the desired design factor of 3.0 or greater. Trying the only smaller preferred size from Appendix A–2 (7/8 in = 0.875 in), the required s_y is 65 825 psi, and that is too high. Therefore, specify $D = 1.00$ in.

Now we can evaluate the expected maximum deflection with the new design using Equation (11–22):

$$y_{max} = 0.75\left[\sec\left(\frac{32}{2(0.250)}\sqrt{\frac{1075}{(0.785(30 \times 10^6)}}\right) - 1\right]$$

$$y_{max} = 0.076 \text{ in}$$

Comments The diameter of 1.00 in is satisfactory. The maximum deflection for the column is 0.076 in.

Figure 11–16 shows the solution of the eccentric column problem of Example Problem 11–8 using a spreadsheet to evaluate Equations (11–21) and (11–22). It is a design aid that facilitates the iteration required to determine an acceptable geometry for a column to carry a specified load with a desired design factor. Note that the data are in U.S. Customary units. At the lower left of the spreadsheet, data required for Equations (11–21)

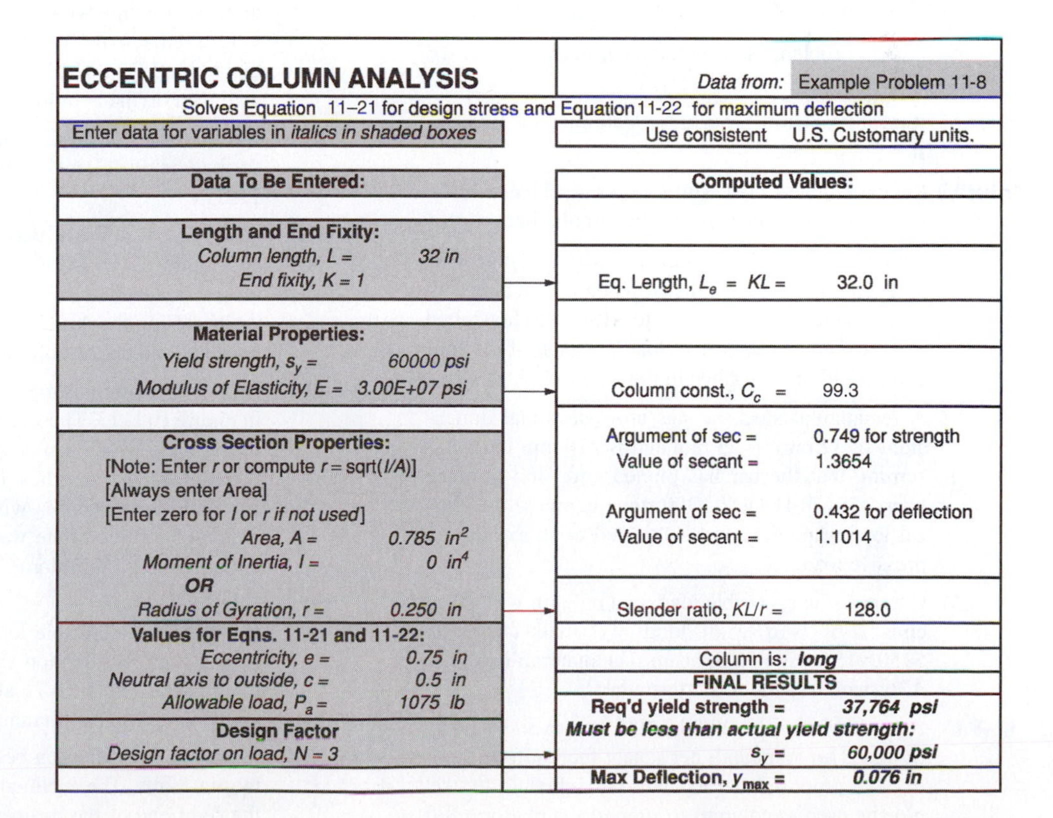

FIGURE 11–16 Spreadsheet for analysis of eccentric columns—U.S. customary units.

and (11–22) are entered by the designer, along with the other data discussed for earlier column analysis spreadsheets. The "**FINAL RESULTS**" at the lower right show the computed value of the required yield strength of the material for the column and compare it with the given value entered by the designer near the upper left. The designer must ensure that the actual value is greater than the computed value by trying different values for the diameter. The last part of the right side of the spreadsheet gives the computed maximum deflection of the column that occurs at its midlength.

REFERENCES

1. Aluminum Association, *Aluminum Design Manual,* Washington, DC, 2005.
2. American Institute of Steel Construction, *Steel Construction Manual,* 9th ed., Chicago, IL 2005.
3. Mott, Robert L., *Machine Elements in Mechanical Design,* 4th ed., Prentice Hall, Upper Saddle River, NJ, 2004.
4. Spotts, M. F., T. E. Shoup, and L. E. Hornberger. *Design of Machine Elements,* 8th ed. Prentice Hall, Upper Saddle River, NJ, 2004.
5. Timoshenko, S. *Strength of Materials,* Vol. 2, 2nd ed., Van Nostrand Reinhold, New York, 1941.
6. Timoshenko, S. and J. M. Gere, *Theory of Elastic Stability,* 2nd ed., McGraw-Hill, New York, 1961.

PROBLEMS

11–1.M Determine the critical load for a pinned-end column made of a circular bar of AISI 1020 hot-rolled steel. The diameter of the bar is 20 mm, and its length is 800 mm.

11–2.M Repeat Problem 11–1 with the length of 350 mm.

11–3.M Repeat Problem 11–1 with the bar made of 6061-T6 aluminum instead of steel.

11–4.M Repeat Problem 11–1 with the column ends fixed instead of pinned.

11–5.M Repeat Problem 11–1 with a square steel bar with the same cross-sectional area as the circular bar.

11–6.M For a PIPE 25STD (1-in schedule 40) steel pipe used as a column, determine the critical load if it is 2.05 m long. The material is similar to AISI 1020 hot-rolled steel. Compute the critical load for each of the four end conditions described in Figure 11–3.

11–7.M A rectangular steel bar has cross-sectional dimensions of 12 mm by 25 mm and is 210 mm long. Assuming that the bar has pinned ends and is made from AISI 1141 OQT 1300 steel, compute the critical load when the bar is subjected to an axial compressive load.

11–8.M Compute the allowable load on a column with fixed ends if it is 5.45 m long and made from an S150×18.6 (S6×12.5) beam. The material is ASTM A36 steel. Use the AISC formula.

11–9.E A raised platform is 20 ft by 40 ft in area and is being designed for 75 pounds per square foot uniform loading. It is proposed that standard 3-in schedule 40 steel pipe be used as columns to support the platform 8 ft above the ground with the base fixed and the top free. How many columns would be required if a design factor of 3.0 is desired? Use $s_y = 30\,000$ psi.

11–10.M An aluminum I-beam, I254×12.87 (I10×8.646), is used as a column with two pinned ends. It is 2.80 m long and made of 6061-T6 aluminum. Using Equations (11–17b) and (11–18b), compute the allowable load on the column.

11–11.M Compute the allowable load for the column described in Problem 11–10 if the length is only 1.40 m.

11–12.E A column is a W8×10 steel beam, 12.50 ft long, and made of ASTM A992 steel. Its ends are attached in such a way that L_e is approximately $0.80L$. Using the AISC formulas, determine the allowable load on the column.

11–13.E A built-up column is made of four angles, as shown in Figure P11–13. The angles are held together with lacing bars, which can be neglected in the analysis of geometrical properties. Using the standard Johnson or Euler equations with $L_e = L$ and a design factor of 3.0, compute the allowable load on the column if it is 18.4 ft long. The angles are of ASTM A36 steel.

11–14.E Compute the allowable load on a built-up column having the cross section shown in Figure P11–14. Use $L_e = L$ and 6061-T6 aluminum. The column is 10.5 ft long. Use the Aluminum Association formulas.

11–15.C Figure P11–15 shows a beam supported at its ends by pin-joints. The inclined bar at the top supports the right end of the beam, but also places an axial

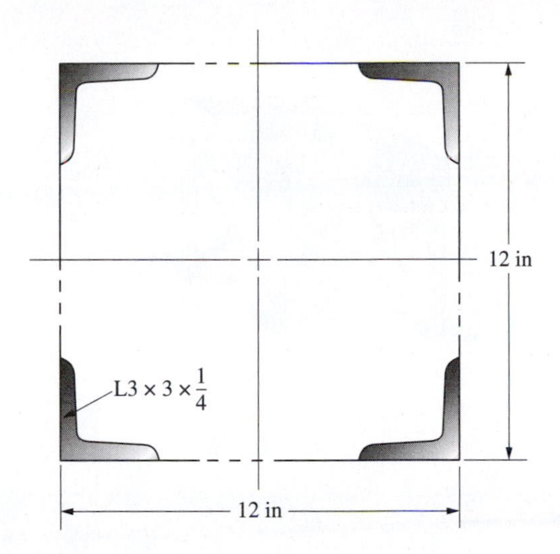

FIGURE P11–13 Built-up column cross section for Problem 11–13.

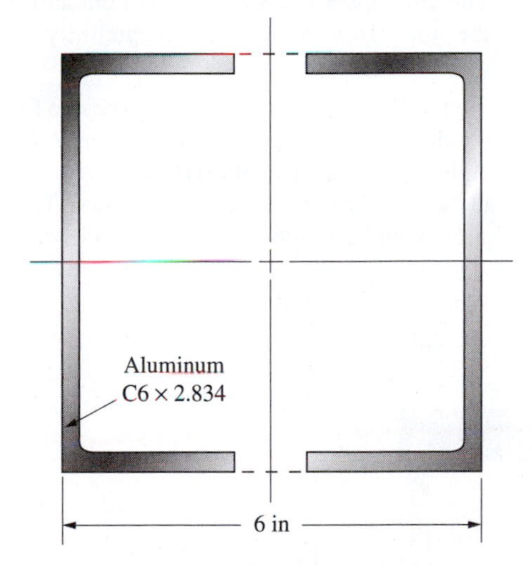

FIGURE P11–14 Built-up column cross section for Problem 11–14.

compressive force in the beam. Would a standard S150×18.6 (S6×12.5) beam be satisfactory in this application if it carries 1320 kg at its end? The beam is made from ASTM A36 steel.

11–16.E A link in a mechanism which is 8.40 in long, has a rectangular cross section $\frac{1}{4}$ in × $\frac{1}{8}$ in, and is subjected to a compressive load of 50 lb. If the link has pinned ends, is it safe from buckling? Cold-drawn AISI 1040 steel is used in the link.

11–17.M A piston rod on a shock absorber is 12 mm in diameter and has a maximum length of 190 mm outside the shock absorber body. The rod is made of AISI 1141 OQT 1300 steel. Consider one end to be pinned and the other to be fixed. What axial compressive load on the rod would be one-third of the critical buckling load?

11–18.E A stabilizing rod in an automobile suspension system is a round bar loaded in compression. It is subjected to 1375 lb of axial load and supported at its ends by pin-type connections, 28.5 in apart. Would a 0.800-in diameter bar of AISI 1020 hot-rolled steel be satisfactory for this application?

11–19.E A structure is being designed to support a large hopper over a plastic extruding machine, as sketched in Figure P11–19. The hopper is to be carried by four columns which share the load equally. The structure is cross-braced by rods. It is proposed that the columns be made from standard 2-in schedule 40 pipe. They will be fixed at the floor. Because of the cross-bracing, the top of each column is guided so that it behaves as if it were rounded or pinned. The material for the pipe is AISI 1020 steel, hot rolled. The hopper is designed to hold 20 000 lb of plastic powder. Are the proposed columns adequate for this load?

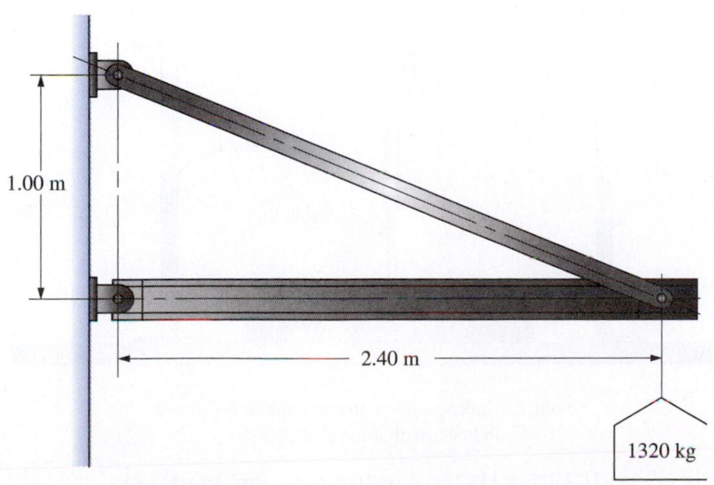

FIGURE P11–15 Structure for Problem 11–15.

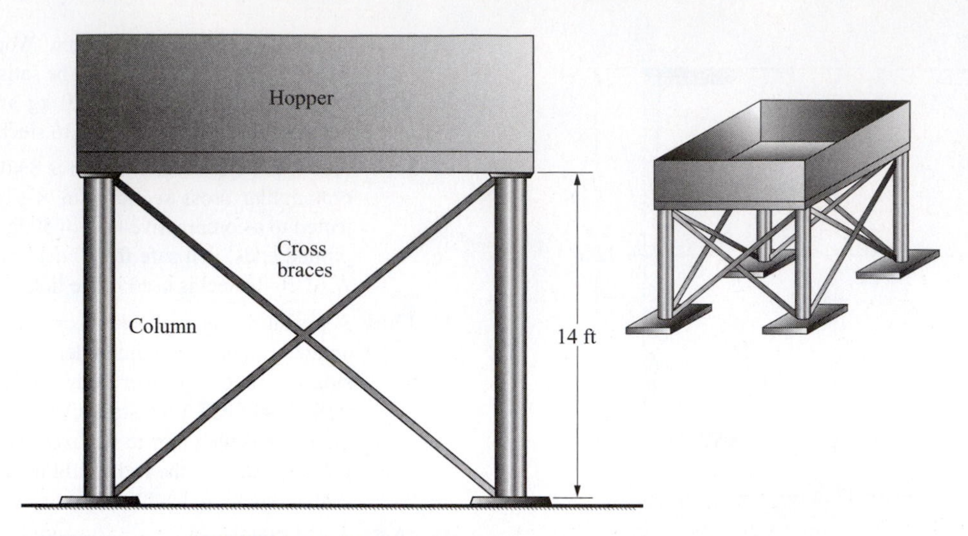

FIGURE P11–19 Hopper for Problems 11–19 and 11–20.

11–20.E Discuss how the column design in Problem 11–19 would be affected if a careless forklift driver runs into the cross braces and breaks them.

11–21.E The assembly shown in Figure P11–21 is used to test parts by pulling on them repeatedly with the hydraulic cylinder. A maximum force of 3000 lb can be exerted by the cylinder. The parts of the assembly of concern here are the columns. It is proposed that the two columns be made of $1\frac{1}{4}$-in square bars of aluminum alloy 6061-T6. The columns are fixed at the bottom and free at the top. Determine the acceptability of the proposal.

11–22.E Figure P11–22 shows the proposed design for a hydraulic press used to compact solid waste. A piston at the right is capable of exerting a force of 12 500 lb through the connecting rod to the ram. The rod is straight and centrally loaded. It is made from AISI 1040 WQT 1100 steel. Compute the resulting design factor for this design.

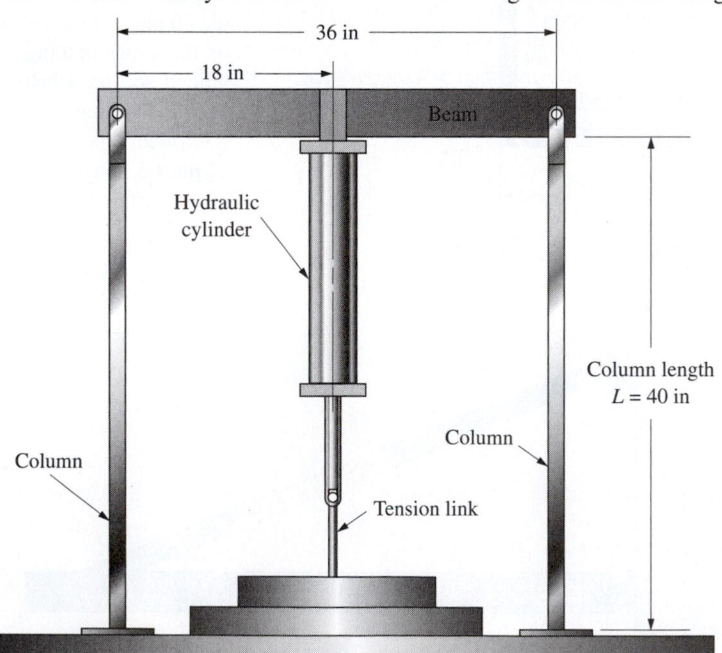

Note: Cylinder pulls up on tension link and down on beam with a force of 3000 lb.

FIGURE P11–21 Test fixture for Problem 11–21.

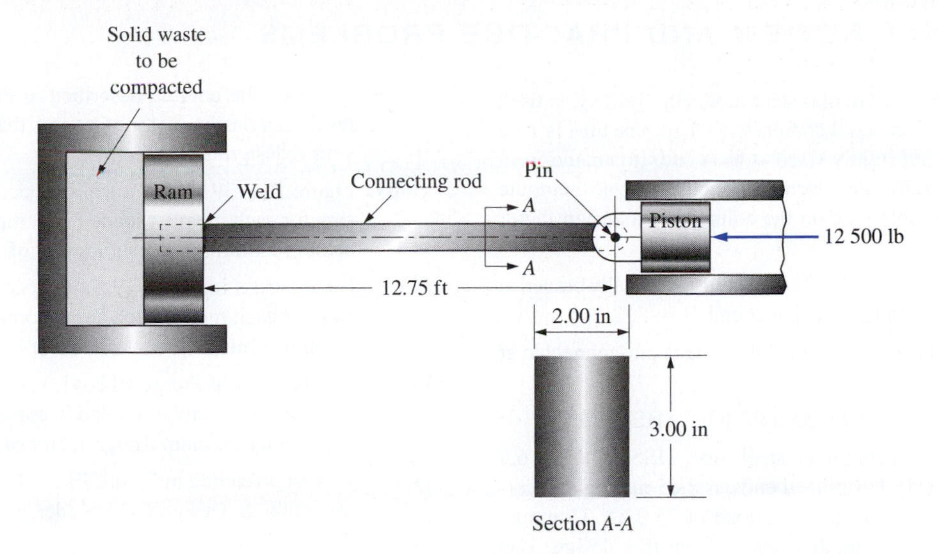

FIGURE P11–22 Solid waste compactor for Problems 11–22, 11–23, 11–24, and 11–25.

11–23.E For the conditions described in Problem 11–22, specify the required diameter of the connecting rod if it is made as a solid circular cross section. Use a design factor of 4.0.

11–24.E For the conditions described in Problem 11–22, specify a suitable standard steel pipe for use as the connecting rod. Use a design factor of 4.0. The pipe is to be made from ASTM A501 structural steel.

11–25.E For the conditions described in Problem 11–22, specify a suitable standard I-beam for use as the connecting rod. Use a design factor of 4.0. The I-beam is to be made from aluminum alloy 6061-T6. The connection at the piston is as shown in Figure P11–25.

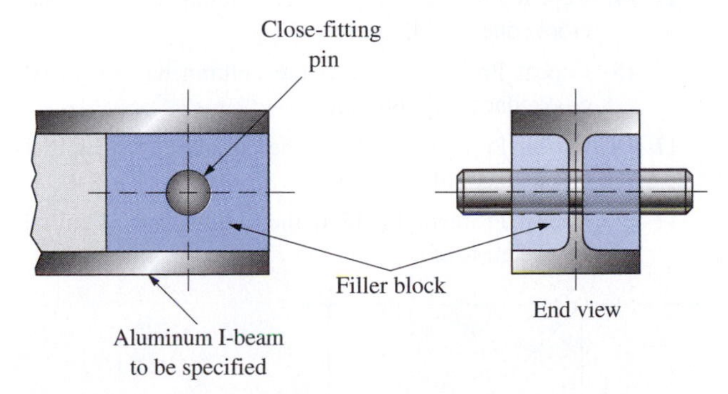

FIGURE P11–25 End connection for I-beam for Problem 11–25.

11–26.E A hollow square tube, HSS3×3×$\frac{1}{4}$, made from ASTM A500 steel, grade B, is used as a building column having a length of 16.5 ft. Using $L_e = 0.80L$, compute the allowable load on the column for a design factor of 3.0.

11–27.E A hollow rectangular tube, HSS4×2×$\frac{1}{4}$, made from ASTM A500 steel, grade B, is used as a building column having a length of 16.5 ft. Using $L_e = 0.80L$, compute the allowable load on the column for a design factor of 3.0.

11–28.E A column is made by welding two standard steel angles, 3×3×$\frac{1}{4}$, into the form shown in Figure 11–11(f). The angles are made from ASTM A36 structural steel. If the column has a length of 16.5 ft and $L_e = 0.8L$, compute the allowable load on the column for a design factor of 3.0.

11–29.M A rectangular steel bar, made of AISI 1020 hot-rolled steel, is used as a safety brace to hold the ram of a large punch press while dies are installed in the press. The bar has cross-sectional dimensions of 60 mm by 40 mm. Its length is 750 mm and its ends are welded to heavy flat plates which rest on the flat bed of the press and the flat underside of the ram. Specify a safe load that could be applied to the brace.

11–30.M It is planned to use an aluminum channel, C102×2.586(C4×1.738), as a column having a length of 4.25 m. The ends can be considered to be pinned. The aluminum is alloy 6061-T4. Compute the allowable load on the column for a design factor of 4.0.

11–31.M In an attempt to improve the load-carrying capacity of the column described in Problem 11–30, alloy 6061-T6 is proposed in place of the 6061-T4 to take advantage of its higher strength. Evaluate the effect of this proposed change on the allowable load.

11–32.E Compute the allowable load on the steel W12×65 column section shown in Figure 11–12(b) if it is 22.5 ft long, made from ASTM A992 steel, and installed such that $L_e = 0.8L$, Use the AISC code.

ADDITIONAL REVIEW AND PRACTICE PROBLEMS

11–33. A hollow rectangular steel tube, HSS3×2×$\frac{1}{4}$, is used to support an axial compressive load. The tube is 13.6 ft long and rigidly fixed at each end. It is made from ASTM A501, hot-formed structural steel. Compute the allowable load on the column for a design factor of 3.0.

11–34. Repeat Problem 11–33 if the tube is laterally braced at a point 80 in from its lower end.

11–35. Repeat Problem 11–33 if the tube is pin-connected at its top.

11–36. Repeat Problem 11–33 if the tube is HSS3×3×$\frac{1}{4}$.

11–37. A hollow rectangular steel tube, HSS102×51×6.4 (4×2×1/4), has pinned ends, is 2.65 m long and carries an axial compressive load of 75.0 kN. Compute the design factor that results from this design. Use ASTM A501 steel.

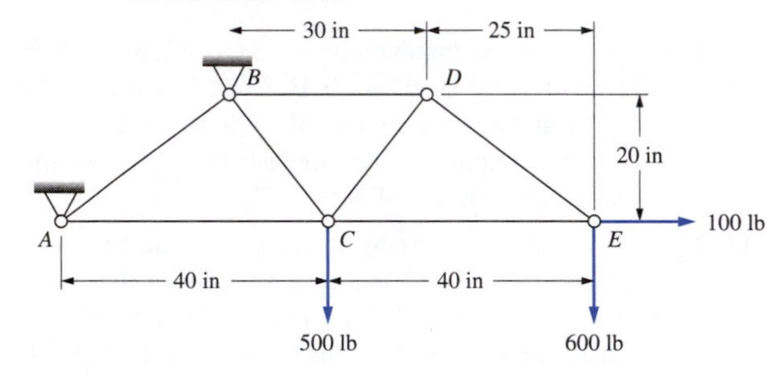

FIGURE P11–39 Truss for Problem 11–39.

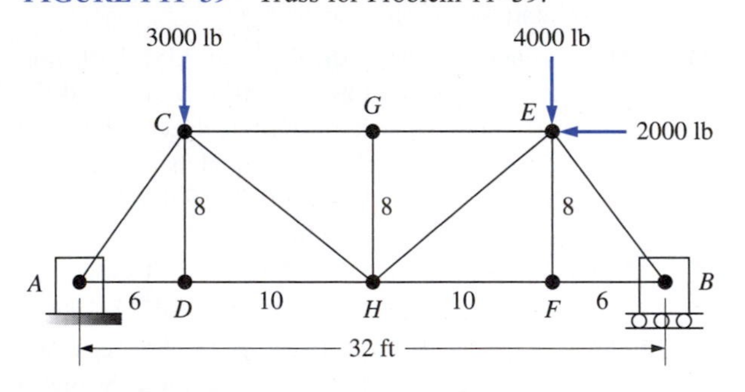

FIGURE P11–40

11–38. Redesign the column described in Problem 11–37 to result in a design factor of no less that 3.0. Verify that your redesign is satisfactory.

11–39. Figure P11–39 shows a truss. Specify a suitable design for each member loaded in compression that will achieve a minimum design factor of 3.0.

11–40. For the truss in Figure P11–40, specify a suitable design for each member loaded in compression that will result in a minimum design factor of 2.5.

11–41. For the truss in Figure P11–41, specify a suitable design for each member loaded in compression that will result in a minimum design factor of 2.5.

11–42. A sling, sketched in Figure P11–42, is to carry a load of 18 000 lb. Design the spreader.

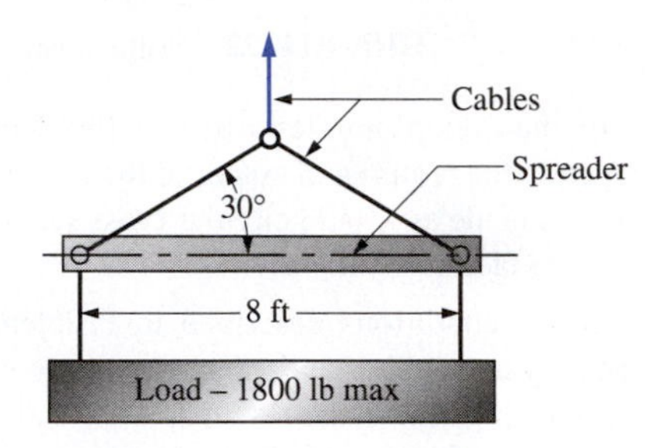

FIGURE P11–42

11–43. Repeat Problem 11–42 if the angle shown is changed from 30° to 15°.

Crooked Columns: Find P_a for $N = 3$.

11–44. Repeat Problem 11–1 if the column has an initial crookedness of 4.0 mm.

11–45. Repeat Problem 11–7 if the column has an initial crookedness of 1.60 mm.

11–46. Repeat Problem 11–10 if the column has an initial crookedness of 14.0 mm.

11–47. Repeat Problem 11–12 if the column has an initial crookedness of 0.75 in.

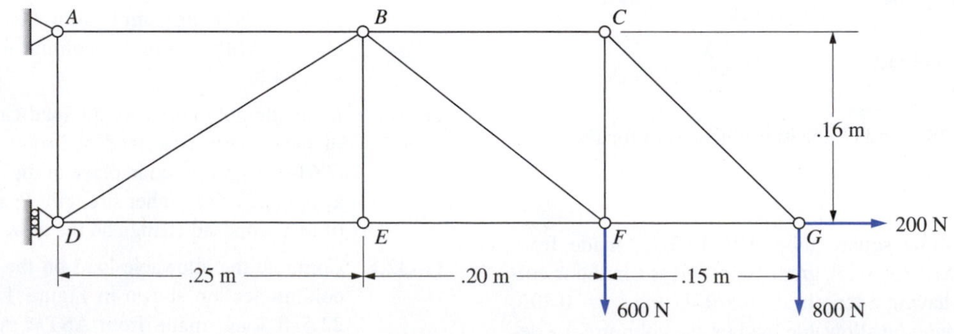

FIGURE P11–41

11–48. Repeat Problem 11–33 if the column has an initial crookedness of 1.25 in.

11–49. Repeat Problem 11–37 if the column has an initial crookedness of 32 mm.

Eccentrically Loaded Columns

11–50. An aluminum (6061-T4) column is 42 in long and has a square cross section, 1.25 in on a side. If it carries a compressive load of 1250 lb, applied with an eccentricity of 0.60 in, compute the maximum stress in the column and the maximum deflection.

11–51. A steel (AISI 1020 hot-rolled) column is 3.2 m long and is made from a standard PIPE 75STD (3-in schedule 40) steel pipe (see Appendix A–12). If a compressive load of 30.5 kN is applied with an eccentricity of 150 mm, compute the maximum stress in the column and the maximum deflection.

11–52. A connecting link in a mechanism is 14.75 in long and has a square cross section 0.250 in on a side. It is made from annealed AISI 301 stainless steel. Use $E =$ 28 000 000 psi. If it carries a compressive load of 45 lb with an eccentricity of 0.30 in, compute the maximum stress and the maximum deflection.

11–53. A hollow square steel tube, 40 in long, is proposed for use as a prop to hold up the ram of a punch press during installation of new dies. The ram weighs 75 000 lb. The prop is made from HSS4×4×$\frac{1}{4}$ structural tubing. It is made from steel similar to structural steel, ASTM A500, grade C. If the load applied by the ram could have an eccentricity of 0.50 in. would the prop be safe?

11–54. Compute the maximum stress and deflection that can be expected in the steel machine member carrying an eccentric load as shown in Figure P11–54. The load P is 1000 lb. If a design factor of 3 is desired, specify a suitable steel.

11–55. An axial load of 4000 lb is applied to a C5×9 steel channel that is 112 in long and made from ASTM A36 structural steel. The line of action of the load acts at the mid-depth of the web and the mid-width of the flanges. The ends are pinned. Would the channel be suitable if a design factor of 3.0 is desirable?

11–56. Figure P11–56 shows a HSS4×4×$\frac{1}{2}$ steel column made from ASTM A500, grade B, structural steel. To accommodate a special mounting restriction, the load is applied eccentrically as shown. Determine the amount of load that the column can safely support. The top end of the column is supported laterally by the structure.

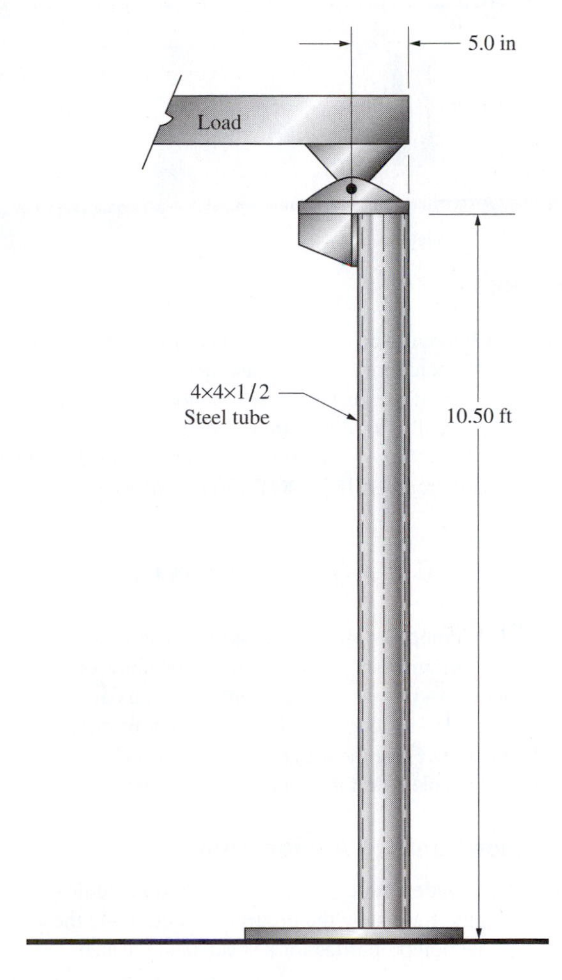

FIGURE P11–56

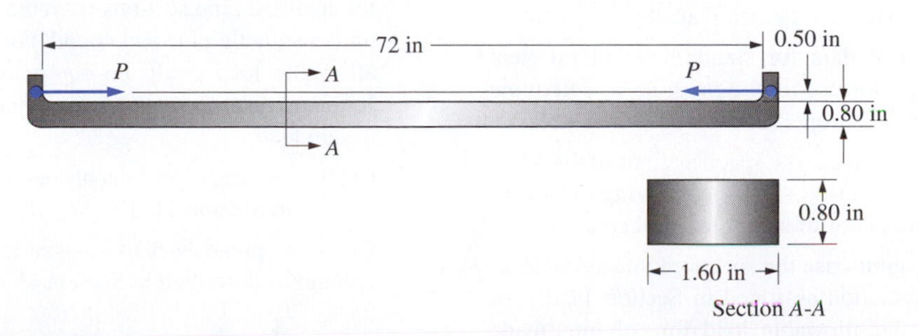

FIGURE P11–54

11–57. The device shown in Figure P11–57 is subjected to opposing forces F. Determine the maximum allowable load to achieve a design factor of 3. The device is made from aluminum 6061-T6.

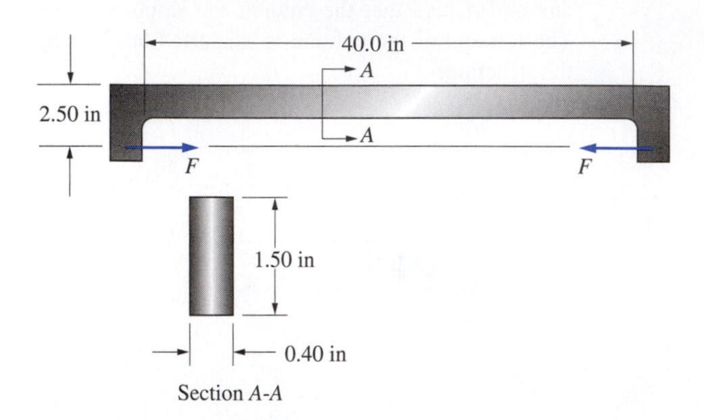

FIGURE P11–57

11–58. A hydraulic cylinder is capable of exerting a force of 5200 N to move a heavy casting along a conveyor. The design of the pusher causes the load to be applied eccentrically to the piston rod as shown in Figure P11–58. Is the piston rod safe under this loading if it is made from AISI501 OQT 1000 stainless steel?

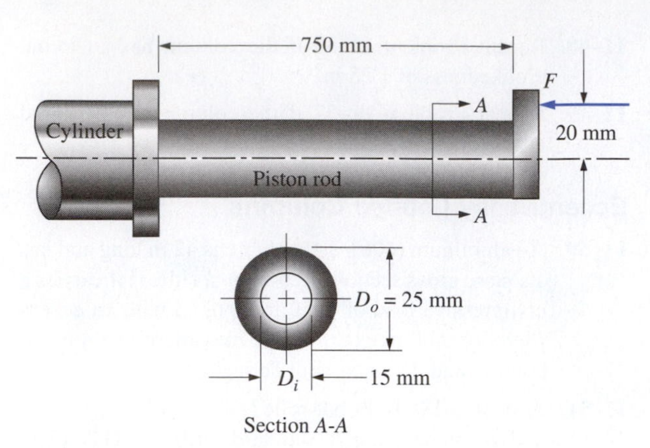

FIGURE P11–58

11–59. A standard 2-in schedule 40 steel pipe is proposed to be used to support the roof of a porch during renovation. Its length is 13.0 ft. The pipe is made from ASTM A501 structural steel.

(a) Determine the safe load on the pipe to achieve a design factor of 3 if the pipe is straight.

(b) Determine the safe load if the pipe has an initial crookedness of 1.25 in.

COMPUTER ASSIGNMENTS

1. Write a computer program or spreadsheet to analyze proposed column designs using the procedure outlined in Section 11–8. Have all the essential design data for material, end-fixity, length, and cross-section properties input by the user. Have the program output the critical load and the allowable load for a given design factor.

Enhancements to Assignment 1

(a) Include a table of data for standard schedule 40 steel pipe for use by the program to determine the cross-section properties for a specified pipe size.

(b) Design the program to handle columns made from solid circular cross sections and compute the cross-section properties for a given diameter.

(c) Add a table of data for standard structural steel square tubing for use by the program to determine the cross-section properties for a specified size.

(d) Have the program use the specifications of the AISC as stated in Section 11–11 for computing the allowable load and factor of safety for steel columns.

(e) Have the program use the specifications of the Aluminum Association as stated in Section 11–12 for computing the allowable load for columns made from 6061-T6.

2. Write a program to design a column with a solid circular cross section to carry a given load with a given design factor. Note that the program will have to check to see that the correct method of analysis is used, either the Euler formula for long columns or the Johnson formula for short columns, after an initial assumption is made.
3. Write a program to design a column with a solid square cross section to carry a given load with a given design factor.
4. Write a program to select a suitable schedule 40 steel pipe to carry a given load with a given design factor. The program could be designed to search through a table of data for standard pipe sections from the smallest to the largest until a suitable pipe was found. For each trial section, the allowable load could be computed using the Euler or Johnson formula, as required, and compared with the design load.
5. Create a spreadsheet to analyze a crooked column as described in Section 11–13.
6. Create a spreadsheet to analyze an eccentrically loaded column as described in Section 11–13.

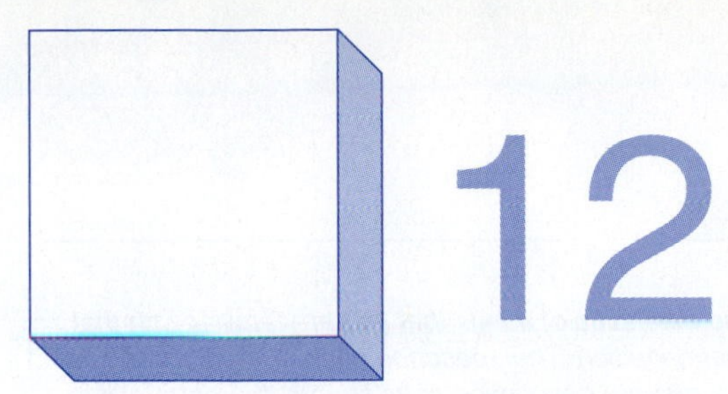

12

Pressure Vessels

The Big Picture

Pressure Vessels

Discussion Map

A pressure vessel is a container designed to hold liquids or gases under internal pressure.

- Typical pressure vessels are either spherical or cylindrical in shape.
- It is important for you to understand how a pressure vessel can fail so you can design it to be safe under a specific applied pressure.
- Failure occurs when the internal pressure causes an excessively high tensile stress in the walls of the vessel.
- The size of the vessel and the thickness of its wall are the primary variables that affect the level of stress.
- Two different types of analysis are discussed in this chapter, one for thin-walled vessels and one for thick-walled vessels. We quantify the difference between these two classifications of vessels later.
- Now let's explore further this concept of a pressure vessel.

Activity

Find some examples of things that you are familiar with that may fit that definition. Then describe each of them, giving its function, a general description of its shape, the material it is made from, its length, and the overall dimensions. Here are a few examples to get you started. See Figure 12–1.

- A tank to hold the pressurized propane fuel for a household gas grill: *It is typically a cylinder with domed ends, made from steel, with a diameter of about 300 mm (12 in). The length of the straight, cylindrical part is about 225 mm (9 in) and the overall length is about 350 mm (14 in). The cylindrical or spherical shape gives the tank a sturdy, stiff feel, but when struck lightly by something hard, it seems that the thickness of the steel making up its walls is relatively thin. There is a penetration in the top of the domed end for a valve to enable you to draw out some propane to cook the perfect steak, chicken, hamburger, or fish.*
- A compressed air tank: *You may have seen examples of these in an automobile service station, on a construction site, or in an industrial plant. They are connected to a compressor that delivers air at approximately 400 to 700 kPa (60 to 100 psi) to the tank where a relatively large volume is stored until needed to add air to a tire, to blow debris from the floor, or to raise a car on the lift. While they vary greatly in size and capacity, they are typically of a similar shape as the propane tank, being mostly cylindrical with domed ends. The tank incorporates a pressure switch that senses tank pressure and cuts the compressor off when the tank reaches the desired pressure setting. This is important because the tank is designed to hold safely a specified maximum pressure. The rupture of a compressed air tank is very dangerous. In series with the discharge line from the tank valve, there is usually a pressure regulator to limit the pressure at the working end of the line.*
- A spherical high-pressure storage tank: *Have you ever seen one of these in a photograph or video of a spacecraft? In this chapter we will see that the spherical shape is an optimum shape for a pressure vessel. They are used when there is a need to store gases or liquids at high pressure to be able to use the thinnest, lightest material. Light, strong materials such as titanium or advanced composites are often used in such applications.*

Continue your own search for examples of pressure vessels. Some you may be able to find in person and, if not, the Internet can help you find quite a variety. Check out the sites listed at the end of this chapter by going first to the group labeled *Pressure Vessel Manufacturers.* These are companies that specialize in the design and fabrication of pressure vessels for such applications as:

- Medical sterilization systems
- Chemical processing plants

- Food production facilities
- Compressed air tanks
- Petrochemical and pharmaceutical manufacturing and many others

Another interesting Internet search you could try is to investigate the several ways in which hydrogen can be stored for use in fuel cells for vehicle power or in electrical generation systems. When using hydrogen in gaseous form, it is important to store it at high pressure so that only a moderate volume is required to hold sufficient fuel to operate a vehicle for approximately 300 miles (480 km) or an electrical power system for several hours. Practical designs for cylindrical tanks have been proposed that are maintained at 3000 to 10 000 psi (20.7 MPa to 69.0 MPa) so that a vehicle hydrogen fuel tank would take approximately the same space as a typical gasoline fuel tank. Check out the designs by Quantum Technologies, Inc., at **www.qtww.com**. Other methods of storing hydrogen are being studied that allow operation at lower pressures.

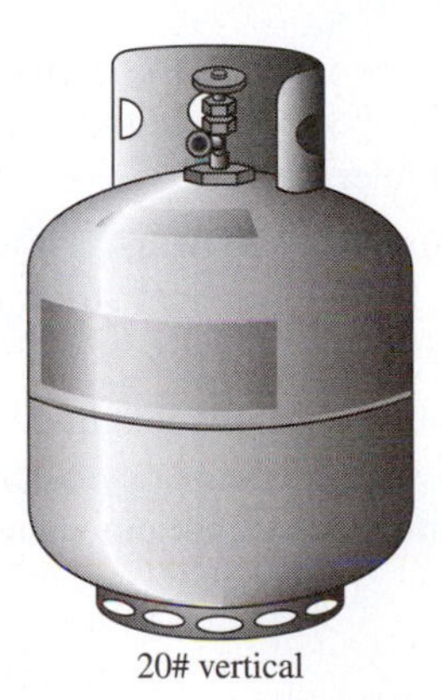

(*a*) Typical household propane cylinder

(*b*) General purpose air compressor with pressure tank

(*c*) High pressure spherical tank

FIGURE 12–1 Examples of pressure vessels. (Source: (*c*) Photo Researchers, Inc.)

What examples have you found?

In this chapter you will learn how the internal pressure causes stresses in the walls of the pressure vessel. This will enable you to design safe vessels yourself and to recognize the importance of protecting and maintaining existing tanks to keep them safe. Most of the pressure vessels you will encounter in your career are likely to be of the *thin-wall* type. But some, particularly those carrying high-pressure fluids, may be in the range of *thick-wall* designs. In this chapter we quantify the definition of when a vessel is considered to have a thin and thick wall. This is an important distinction because the method of analysis is very different for these two classes of vessels.

12–1 OBJECTIVES OF THIS CHAPTER

After completing this chapter, you should be able to:

1. Determine whether a pressure vessel should be classified as *thin-walled* or *thick-walled.*
2. Describe *hoop stress* as it is applied to spheres carrying an internal pressure and apply the hoop stress formula to compute the maximum stress in the wall of a thin-walled sphere.
3. Describe *hoop stress* as it is applied to cylinders carrying an internal pressure and apply the hoop stress formula to compute the maximum stress in the wall of a thin-walled cylinder.
4. Describe *longitudinal stress* as it is applied to cylinders carrying internal pressure and apply the longitudinal stress formula to compute the stress in the wall of a thin-walled cylinder that acts in the direction parallel to the axis of the cylinder.
5. Identify the hoop stress, longitudinal stress, and radial stress developed in the wall of a thick-walled sphere or cylinder due to internal pressure and apply the formulas for computing the maximum values of the stresses.

12–2 DISTINCTION BETWEEN THIN-WALLED AND THICK-WALLED PRESSURE VESSELS

In general, the magnitude of the stress in the wall of a pressure vessel varies as a function of position within the wall. A precise analysis enables the computation of the stress at any point. The formulas for making such a computation will be shown in a later section.

However, when the wall thickness of the pressure vessel is small, the assumption that the stress is uniform throughout the wall results in very little error. Also, this assumption permits the development of relatively simple formulas for stress.

First you need to understand the basic geometry of a pressure vessel and to define some terms. Figure 12–2 shows the definition of key diameters, radii, and the wall thickness for cylinders and spheres. Most are obvious from the figure. The mean radius, R_m, is defined as the average of the outside radius and the inside radius. That is,

Mean Radius

$$R_m = \frac{R_o + R_i}{2} \tag{12–1}$$

We can also define the *mean diameter* as

Mean Diameter

$$D_m = \frac{D_o + D_i}{2} \tag{12–2}$$

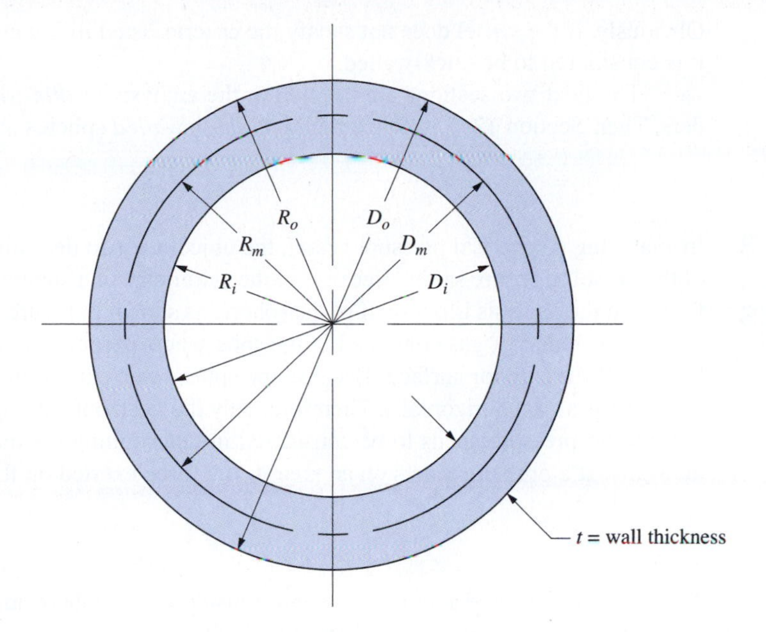

FIGURE 12–2 Definition of key diameters, radii, and wall thickness for cylinders and spheres.

Other useful formulas are

$$R_i = R_o - t \qquad D_i = D_o - 2t$$

$$R_m = R_o - \frac{t}{2} \qquad D_m = D_o - t$$

$$R_m = R_i + \frac{t}{2} \qquad D_m = D_i + t$$

The criterion for determining when a pressure vessel can be considered thin-walled is as follows:

> If the ratio of the mean radius of the vessel to its wall thickness is 10 or greater, the stress is very nearly uniform and it can be assumed that all the material of the wall shares equally to resist the applied forces. Such pressure vessels are called thin-walled vessels.

Then a pressure vessel is considered to be thin if

$$\frac{R_m}{t} \geq 10 \tag{12–3}$$

where t is the wall thickness of the vessel.

Because the diameter is twice the radius, the criterion for a vessel to be considered thin-walled is also written as

$$\frac{D_m}{t} \geq 20 \tag{12–4}$$

Obviously, if the vessel does not satisfy the criteria listed in Equations (12–3) and (12–4), it is considered to be thick-walled.

The next two sections are devoted to the analysis of *thin-walled* spheres and cylinders. Then Section 12–5 is concerned with *thick-walled* spheres and cylinders.

12–3 THIN-WALLED SPHERES

In analyzing a spherical pressure vessel, the objective is to determine the stress in the wall of the vessel to ensure safety. Because of the symmetry of a sphere, a convenient free body for use in the analysis is one-half of the sphere, as shown in Figure 12–3. The internal pressure of the liquid or gas contained in the sphere acts perpendicular to the walls, uniformly over all the interior surface. Because the sphere was cut through a diameter, the forces in the walls all act horizontally. Therefore, only the horizontal component of the forces due to the fluid pressure needs to be considered in determining the magnitude of the force in the walls. If a pressure p acts on an area A, the force exerted on the area is

$$F = pA \tag{12–5}$$

Taking all of the force acting on the entire inside of the sphere and finding the horizontal component, we find the resultant force in the horizontal direction to be

$$F_R = pA_p \tag{12–6}$$

where A_p is the *projected area* of the sphere on the plane through the diameter. Therefore,

$$A_p = \frac{\pi D_m^2}{4} \tag{12–7}$$

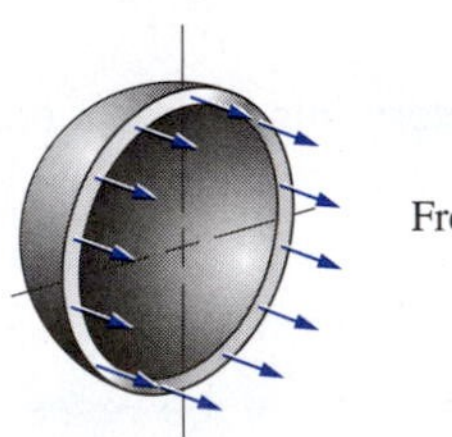

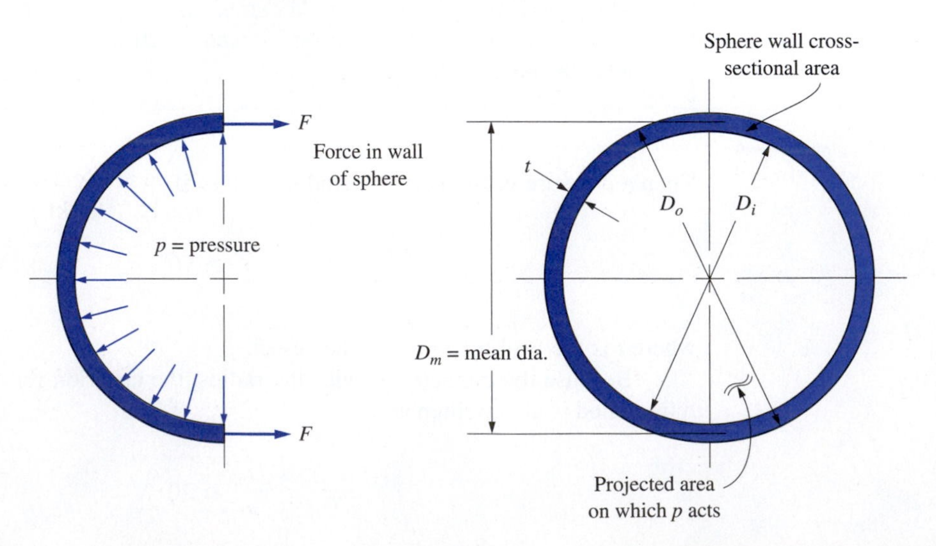

FIGURE 12–3 Free-body diagram for a sphere carrying internal pressure.

Because of the equilibrium of the horizontal forces on the free body, the forces in the walls must also equal F_R, as computed in Equation (12–6). These tensile forces acting on the cross-sectional area of the walls of the sphere cause tensile stresses to be developed. That is,

$$\sigma = \frac{F_R}{A_w} \tag{12–8}$$

where A_w is the area of the annular ring cut to create the free body, as shown in Figure 12–2. The actual area is

$$A_w = \frac{\pi}{4}(D_o^2 - D_i^2) \tag{12–9}$$

However, for thin-walled spheres with a wall thickness t, less than about $\frac{1}{10}$ of the radius of the sphere, the wall area can be closely approximated as

$$A_w = \pi D_m t \tag{12–10}$$

This is the area of a rectangular strip having a thickness t and a length equal to the mean circumference of the sphere, πD_m.

Equations (12–6) and (12–8) can be combined to yield an equation for stress,

$$\sigma = \frac{F_R}{A_w} = \frac{pA_p}{A_w} \tag{12–11}$$

Expressing A_p and A_w in terms of D_m and t from Equations (12–7) and (12–10) gives

Stress in a Thin-Walled Sphere

$$\sigma = \frac{p(\pi D_m^2/4)}{\pi D_m t} = \frac{pD_m}{4t} \tag{12–12}$$

This is the expression for the stress in the wall of a thin-walled sphere subjected to internal pressure. Very little error (less than 5%) will result from using either the outside or inside diameter in place of the mean diameter.

Example Problem 12–1 Compute the stress in the wall of a sphere having an inside diameter of 300 mm and a wall thickness of 1.50 mm when carrying nitrogen gas at 3500 kPa internal pressure.

Solution

Objective: Compute the stress in the wall of the sphere.

Given: $p = 3500$ kPa; $D_i = 300$ mm; $t = 1.50$ mm.

Analysis: We must first determine if the sphere can be considered to be thin-walled by computing the ratio of the mean diameter to the wall thickness.

$$D_m = D_i + t = 300\text{ mm} + 1.50\text{ mm} = 301.5\text{ mm}$$
$$D_m/t = 301.5\text{ mm}/1.50\text{ mm} = 201$$

Because this is far greater than the lower limit of 20, the sphere is thin. Then Equation (12–12) should be used to compute the stress.

Results $$\sigma = \frac{pD_m}{4t} = \frac{(3500 \times 10^3\ \text{Pa})(301.5\ \text{mm})}{4(1.50\ \text{mm})}$$

$$\sigma = 175.9 \times 10^6\ \text{Pa} = 175.9\ \text{MPa}$$

12–4 THIN-WALLED CYLINDERS

Cylinders are frequently used for pressure vessels, for example, as storage tanks, hydraulic and pneumatic actuators, and for piping of fluids under pressure. The stresses in the walls of cylinders are similar to those found for spheres, although the maximum value is greater.

Two separate analyses are shown here. In one case, the tendency for the internal pressure to pull the cylinder apart in a direction parallel to its axis is found. This is called *longitudinal stress*. Next, a ring around the cylinder is analyzed to determine the stress tending to pull the ring apart. This is called *hoop stress,* or *tangential stress*.

Longitudinal Stress. Figure 12–4 shows a part of a cylinder, which is subjected to an internal pressure, cut perpendicular to its axis to create a free body. Assuming that the end of the cylinder is closed, the pressure acting on the circular area of the end would produce a resultant force of

$$F_R = pA = p\left(\frac{\pi D_m^2}{4}\right) \quad \textbf{(12–13)}$$

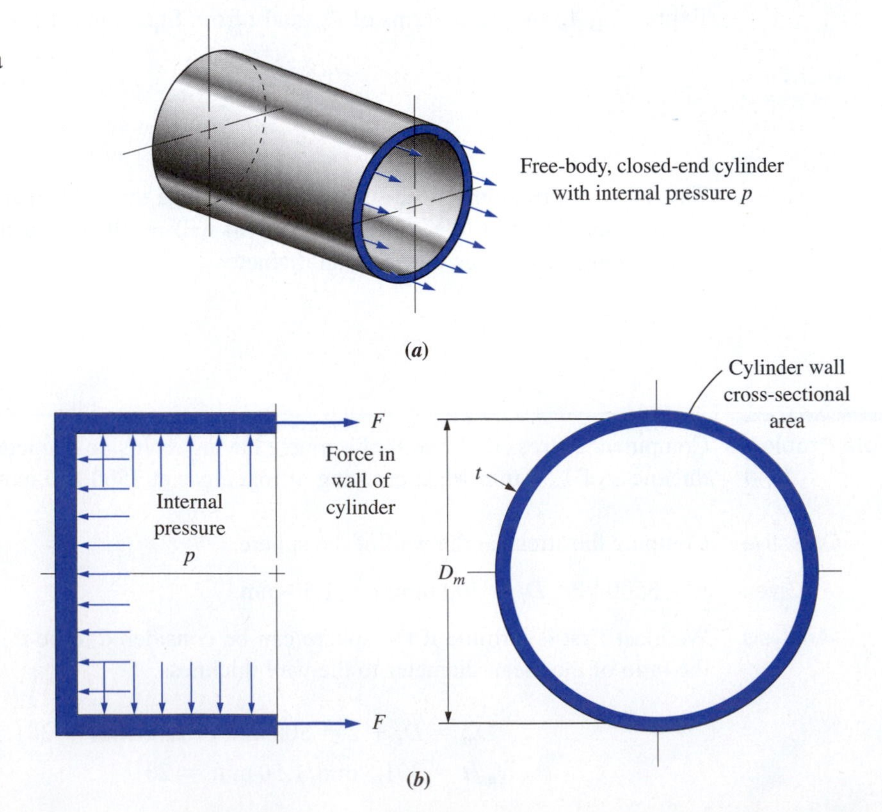

FIGURE 12–4 Free-body diagram of a cylinder carrying internal pressure showing longitudinal stress.

This force must be resisted by the force in the walls of the cylinder, which, in turn, creates a tensile stress in the walls. The stress is

$$\sigma = \frac{F_R}{A_w} \tag{12–14}$$

Assuming that the walls are thin, as we did for spheres,

$$A_w = \pi D_m t \tag{12–15}$$

where t is the wall thickness.

Now combining Equations (12–13), (12–14), and (12–15),

$$\sigma = \frac{F_R}{A_w} = \frac{p(\pi D_m^2/4)}{\pi D_m t} = \frac{pD_m}{4t} \tag{12–16}$$

This is the stress in the wall of the cylinder in a direction parallel to the axis, called the longitudinal stress. Notice that it is of the same magnitude as that found for the wall of a sphere. But this is not the maximum stress, as shown next.

Hoop Stress. The presence of the tangential or hoop stress can be visualized by isolating a ring from the cylinder, as shown in Figure 12–5. The internal pressure pushes outward evenly all around the ring. The ring must develop a tensile stress in a direction tangential to the circumference of the ring to resist the tendency of the pressure to burst the ring.

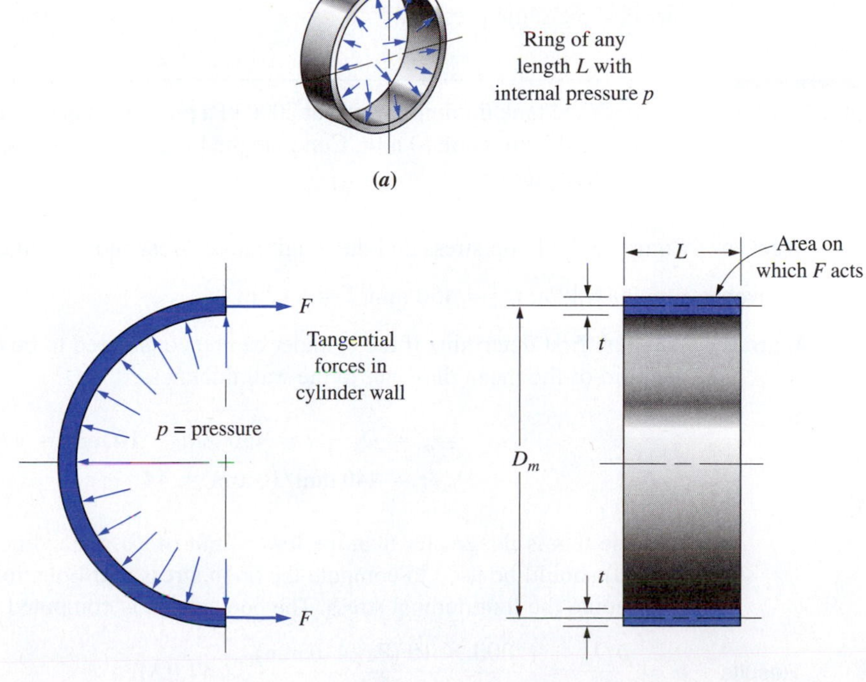

FIGURE 12–5 Free-body diagram of a cylinder carrying internal pressure showing hoop stress.

The magnitude of the stress can be determined by using half of the ring as a free body, as shown in Figure 12–5(b).

The resultant of the forces due to the internal pressure must be determined in the horizontal direction and balanced with the forces in the walls of the ring. Using the same reasoning as we did for the analysis of the sphere, we find that the resultant force is the product of the pressure and the *projected area* of the ring. For a ring with a mean diameter D_m and a length L,

$$F_R = pA_p = p(D_m L) \tag{12–17}$$

The tensile stress in the wall of the cylinder is equal to the resisting force divided by the cross-sectional area of the wall. Again assuming that the wall is thin, the wall area is

$$A_w = 2tL \tag{12–18}$$

Then the stress is

$$\sigma = \frac{F_R}{A_w} = \frac{F_R}{2tL} \tag{12–19}$$

Combining Equations (12–17) and (12–19) gives

Hoop Stress in a Thin-Walled Cylinder

$$\sigma = \frac{F_R}{A_w} = \frac{pD_m L}{2tL} = \frac{pD_m}{2t} \tag{12–20}$$

This is the equation for the hoop stress in a thin cylinder subjected to internal pressure. Notice that the magnitude of the hoop stress is *twice* that of the longitudinal stress. Also, the hoop stress is twice that of the stress in a spherical container of the same diameter carrying the same pressure.

Example Problem 12–2 A cylindrical tank holding oxygen at 2000 kPa pressure has an outside diameter of 450 mm and a wall thickness of 10 mm. Compute the hoop stress and the longitudinal stress in the wall of the cylinder.

Solution

Objective: Compute the hoop stress and the longitudinal stress in the wall of the cylinder.

Given: $p = 2000$ kPa; $D_o = 450$ mm; $t = 10$ mm.

Analysis: We must first determine if the cylinder can be considered to be thin-walled by computing the ratio of the mean diameter to the wall thickness.

$$D_m = D_o - t = 450\text{ mm} - 10\text{ mm} = 440\text{ mm}$$
$$D_m/t = 440\text{ mm}/10\text{ mm} = 44$$

Because this is far greater than the lower limit of 20, the cylinder is thin. Then Equation (12–20) should be used to compute the hoop stress and Equation (12–16) should be used to compute the longitudinal stress. The hoop stress is computed first.

Results: $$\sigma = \frac{pD_m}{2t} = \frac{(2000 \times 10^3\text{ Pa})(440\text{ mm})}{2(10\text{ mm})} = 44.0\text{ MPa}$$

The longitudinal stress, from Equation (12–12), is

$$\sigma = \frac{pD_m}{4t} = 22.0 \text{ MPa}$$

Example Problem 12–3 Determine the pressure required to burst a standard 8-in schedule 40 steel pipe if the ultimate tensile strength of the steel is 40000 psi.

Solution Objective Compute the pressure required to burst the steel pipe.

Given Ultimate tensile strength of steel $= s_u = 40000$ psi
Pipe is a standard 8-in schedule 40 steel pipe.

The dimensions of the pipe are found in Appendix A–12 to be

$$\text{outside diameter} = 8.625 \text{ in} = D_o$$
$$\text{inside diameter} = 7.981 \text{ in} = D_i$$
$$\text{wall thickness} = 0.322 \text{ in} = t$$

Analysis We should first check to determine if the pipe should be called a thin-walled cylinder by computing the ratio of the mean diameter to the wall thickness.

$$D_m = \text{mean diameter} = \frac{D_o + D_i}{2} = 8.303 \text{ in}$$

$$\frac{D_m}{t} = \frac{8.303 \text{ in}}{0.322 \text{ in}} = 25.8$$

Since this ratio is greater than 20, the thin-wall equations can be used. The hoop stress is the maximum stress and should be used to compute the bursting pressure.

Results Use Equation (12–20).

$$\sigma = \frac{pD_m}{2t} \quad \textbf{(12–20)}$$

Letting $\sigma = 40000$ psi and using the mean diameter gives the bursting pressure to be

$$p = \frac{2t\sigma}{D_m} = \frac{(2)(0.322 \text{ in})(40000 \text{ lb/in}^2)}{8.303 \text{ in}} = 3102 \text{ psi}$$

Comment A design factor of 6 or greater is usually applied to the bursting pressure to get an allowable operating pressure. This cylinder would be limited to approximately 500 psi of internal pressure.

12–5 THICK-WALLED CYLINDERS AND SPHERES

The formulas in the preceding sections for thin-walled cylinders and spheres were derived under the assumption that the stress is uniform throughout the wall of the container. As stated, if the ratio of the diameter of the container to the wall thickness is greater than 20, this assumption is reasonably correct. Conversely, if the ratio is less than 20, the walls are considered to be thick, and a different analysis technique is required.

The detailed derivation of the thick-wall formulas will not be given here because of their complexity. See Reference 12. But the application of the formulas will be shown.

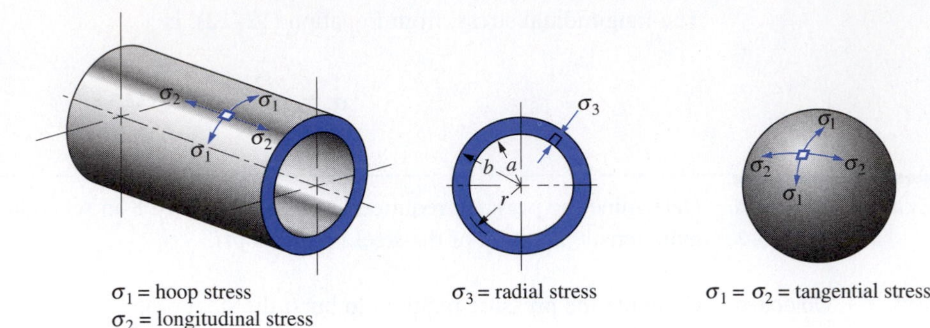

FIGURE 12–6 Notation for stresses in thick-walled cylinders and spheres.

For a thick-walled cylinder, Figure 12–6 shows the notation to be used. The geometry is characterized by the inner radius a, the outer radius b, and any radial position between a and b, called r. The *hoop stress* is called σ_1; the *longitudinal stress* is σ_2. These have the same meaning as they did for thin-walled vessels, except now they will have varying magnitudes at different positions in the wall. In addition to hoop and longitudinal stresses, a *radial stress* σ_3 is created in a thick-walled vessel. As the name implies, the radial stress acts along a radius of the cylinder or sphere. It is a compressive stress and varies from a magnitude of zero at the outer surface to a maximum at the inner surface, where it is equal to the internal pressure.

Table 12–1 shows a summary of the formulas needed to compute the three stresses in the walls of thick-walled cylinders and spheres subjected to internal pressure. The terms *longitudinal stress* and *hoop stress* do not apply to spheres. Instead, we refer to the *tangential stress,* which is the same in all directions around the sphere. Then

$$\text{tangential stress} = \sigma_1 = \sigma_2$$

TABLE 12–1 Stresses in thick-walled cylinders and spheres*

	Stress at position r	Maximum stress
Thick-walled cylinder		
Hoop (tangential)	$\sigma_1 = \dfrac{pa^2(b^2 + r^2)}{r^2(b^2 - a^2)}$	$\sigma_1 = \dfrac{p(b^2 + a^2)}{b^2 - a^2}$ (at inner surface)
Longitudinal	$\sigma_2 = \dfrac{pa^2}{b^2 - a^2}$	$\sigma_2 = \dfrac{pa^2}{b^2 - a^2}$ (uniform throughout wall)
Radial	$\sigma_3 = \dfrac{-pa^2(b^2 - r^2)}{r^2(b^2 - a^2)}$	$\sigma_3 = -p$ (at inner surface)
Thick-walled sphere		
Tangential	$\sigma_1 = \sigma_2 = \dfrac{pa^3(b^3 + 2r^3)}{2r^3(b^3 - a^3)}$	$\sigma_1 = \sigma_2 = \dfrac{p(b^3 + 2a^3)}{2(b^3 - a^3)}$ (at inner surface)
Radial	$\sigma_3 = \dfrac{-pa^3(b^3 - r^3)}{r^3(b^3 - a^3)}$	$\sigma_3 = -p$ (at inner surface)

*Symbols used here are as follows: a = inner radius; b = outer radius; r = any radius between a and b; p = internal pressure, uniform in all directions. Stresses are tensile when positive, compressive when negative.

12–6 ANALYSIS AND DESIGN PROCEDURES FOR PRESSURE VESSELS

Presented here is a summary of the principles discussed in this chapter related to the stress analysis of both thin-walled and thick-walled spheres and cylinders. The summary is given in the form of general procedures for analyzing and designing pressure vessels.

For design stresses, it is advised that Sections 3–3 to 3–6 be reviewed. It will be assumed here that the failure of a pressure vessel subjected to internal pressure is due to the tensile stresses occurring tangentially in the walls of the vessel. Design stresses should take into account the material from which the vessel is made, the operating environment, and whether the pressure is constant or varying in a cyclic fashion.

See also Section 12–7 for a discussion of other modes of failure in vessels having penetrations, structural supports, reinforcing rings, and other features that differ from the simple cylindrical or spherical shape.

Design Stresses. *For steady pressure,* the design stress can be based on the yield strength of the material

$$\sigma_d = s_y/N$$

The choice of the design factor, N, is often dictated by code because of the danger created when a pressure vessel fails. This is particularly true for vessels containing gases or steam under pressure because failures produce violent expulsion of the gas as the high level of stored energy is released. In the absence of a code, we will use $N = 4$ as a minimum and larger values should be used for critical applications or where uncertainty exists in the operating conditions or material properties. Another suggested guideline is to limit the pressure in a vessel to no more than $\frac{1}{6}$ of the pressure that would be predicted to burst it. This effectively calls for a design stress related to the ultimate tensile strength of the material of

$$\sigma_d = s_u/N = s_u/6$$

For cycling pressure, base the design stress on the ultimate strength,

$$\sigma_d = s_u/N$$

Use $N = 8$ as a minimum to produce a design stress related to the fatigue strength of the material.

A. Procedure for Analyzing Pressure Vessels

Given Internal pressure in the vessel, p.
Material from which the vessel is made. Ductile metal assumed.
Outside diameter, D_o; inside diameter, D_i; and wall thickness, t, for the vessel.

Objective Determine the maximum stress in the vessel and check the safety of that stress level with regard to the design stress in the material from which the vessel is made.

1. Compute the mean diameter, D_m, for the vessel from Equation (12–2), $D_m = (D_o + D_i)/2$.
2. Compute the ratio of the mean diameter to the wall thickness for the vessel, D_m/t.

3. If $D_m/t \geq 20$, the vessel can be considered to be thin-walled. Use Equation (12–12) for spheres or Equation (12–20) for cylinders to compute the maximum tangential stress in the walls of the vessel.

$$\sigma = pD_m/4t \quad \text{for spheres} \tag{12–12}$$

$$\sigma = pD_m/2t \quad \text{for cylinders} \tag{12–20}$$

4. If $D_m/t < 20$, the vessel must be considered to be thick-walled. Use equations from Table 12–1 to compute the maximum tangential or hoop stress in the walls of the vessel.

$$\sigma = \frac{p(b^3 + 2a^3)}{2(b^3 - a^3)} \quad \text{for spheres}$$

$$\sigma = \frac{p(b^2 + a^2)}{b^2 - a^2} \quad \text{for cylinders}$$

5. Compute the design stress for the material from which the vessel is made.
6. The actual maximum stress must be less than the design stress for safety.

B. Procedure for Designing Pressure Vessels for a Given Material

Given Internal pressure in the vessel, p.
Material from which the vessel is made. Ductile metal assumed.
Nominal internal diameter of the vessel based on the desired volumetric capacity.

Objective Specify the outside diameter, D_o; inside diameter, D_i; and wall thickness, t, for the vessel to ensure the safety of the vessel with regard to the design stress in the material from which the vessel is made.

1. Use the given diameter as an estimate of the mean diameter, D_m, for the vessel.
2. Assume at first that the vessel will be thin-walled and that the maximum stress can be computed from Equation (12–12) for a sphere or Equation (12–20) for a cylinder. This assumption will be checked later.
3. Compute the design stress for the material from which the vessel is to be made.
4. In the appropriate stress equation, substitute the design stress for the maximum stress and solve for the minimum required wall thickness, t.
5. Specify convenient values for t, D_i, and D_o, based on available material thicknesses. Appendix A–2 may also be used to specify preferred basic sizes.
6. Compute the actual mean diameter for the vessel using the specified dimensions.
7. Compute the ratio of the mean diameter to the wall thickness for the vessel, D_m/t.
8. If $D_m/t \geq 20$, the vessel is thin-walled as assumed and the design is finished.
9. If $D_m/t < 20$, the vessel must be considered to be thick-walled. Use equations from Table 12–1 to compute the maximum tangential or hoop stress in the walls of the vessel and compare that stress with the design stress. If the actual stress is less than the design stress, the design is satisfactory. If the actual maximum stress is greater than the design stress, increase the wall thickness and recompute the resulting stress. Continue this process until a satisfactory stress level and convenient dimensions for the vessel are obtained. Equation-solving computer software, spreadsheets, or a graphing calculator with equation-solving capability may facilitate this process.

C. Procedure for Specifying a Ductile Metal for a Pressure Vessel of a Given Size

Given Internal pressure in the vessel, p.
Outside diameter, D_o; inside diameter, D_i; and wall thickness, t, for the vessel.

Objective Specify a suitable ductile metal from which the vessel is to be made.

1. Compute the mean diameter, D_m, for the vessel from equation (12–2): $D_m = (D_o + D_i)/2$.
2. Compute the ratio of the mean diameter to the wall thickness for the vessel, D_m/t.
3. If $D_m/t \geq 20$, the vessel can be considered to be thin-walled. Use Equation (12–12) for spheres or Equation (12–20) for cylinders to compute the maximum tangential stress in the walls of the vessel.

$$\sigma = pD_m/4t \quad \text{for spheres} \qquad \textbf{(12–12)}$$

$$\sigma = pD_m/2t \quad \text{for cylinders} \qquad \textbf{(12–20)}$$

4. If $D_m/t < 20$, the vessel must be considered to be thick-walled. Use equations from Table 12–1 to compute the maximum tangential or hoop stress in the walls of the vessel.

$$\sigma = \frac{p(b^3 + 2a^3)}{2(b^3 - a^3)} \quad \text{for spheres}$$

$$\sigma = \frac{p(b^2 + a^2)}{b^2 - a^2} \quad \text{for cylinders}$$

5. Specify a suitable equation for the design stress from the discussion earlier in this section.
6. Let the design stress equal the computed maximum stress from step 3 or 4. Then solve for the appropriate material strength, either s_y or s_u from the design stress equation.
7. Specify a suitable material that has a strength greater than the minimum required value.

Example Problem 12–4 Compute the magnitude of the maximum longitudinal, hoop, and radial stresses in a cylinder carrying helium at a steady pressure of 10 000 psi. The outside diameter is 8.00 in and the inside diameter is 6.40 in. Specify a suitable material for the cylinder.

Solution

Objective Compute the maximum stresses and specify a material.

Given Pressure $= p = 10\ 000$ psi. $D_o = 8.00$ in. $D_i = 6.40$ in.

Analysis Use Procedure C from this section.

Results

Step 1. $D_m = (D_o + D_i)/2 = (8.00 + 6.40)/2 = 7.20$ in

Step 2. $t = (D_o - D_i)/2 = (8.00 - 6.40)/2 = 0.80$ in
$D_m/t = 7.20/0.80 = 9.00$

Step 3. This step does not apply. Cylinder is thick.

Step 4. Use equations from Table 12–1.

$$a = D_i/2 = 6.40/2 = 3.20 \text{ in}$$
$$b = D_o/2 = 8.00/2 = 4.00 \text{ in}$$

$$\sigma_1 = \frac{p(b^2 + a^2)}{b^2 - a^2} = \frac{(10\,000\ \text{psi})(4.00^2 + 3.20^2)\ \text{in}^2}{(4.00^2 - 3.20^2)\ \text{in}^2} = 45\,560\ \text{psi hoop}$$

$$\sigma_2 = \frac{pa^2}{b^2 - a^2} = \frac{(10\,000\ \text{psi})(3.20\ \text{in})^2}{(4.00^2 - 3.20^2)\ \text{in}^2} = 17\,780\ \text{psi longitudinal}$$

$$\sigma_3 = -p = -10\,000\ \text{psi radial}$$

All three stresses are a maximum at the inner surface of the cylinder.

Step 5. Let the design stress $= \sigma_d = s_y/4$.

Step 6. The maximum stress is the hoop stress, $\sigma_1 = 45\,560$ psi. Then the required yield strength for the material is

$$s_y = N(\sigma_2) = 4(45\,560\ \text{psi}) = 182\,200\ \text{psi} = 182\ \text{ksi}$$

Step 7. From Appendix A–14, we can specify AISI 4140 OQT 700 steel that has a yield strength of 212 ksi.

Example Problem 12–5

Compute the magnitude of the maximum tangential and radial stresses in a sphere carrying helium at a steady pressure of 10000 psi. The outside diameter is 8.00 in and the inside diameter is 6.40 in. Specify a suitable material for the cylinder.

Solution

Objective: Compute the maximum stresses and specify a material.

Given: Pressure $= p = 10\ 000$ psi. $D_o = 8.00$ in. $D_i = 6.40$ in.

Analysis: Use Procedure C from this section. These data are the same as those used in Example Problem 12–4. Some values will be carried forward.

Results: ***Steps 1, 2, 3.*** Sphere is thick-walled.

Step 4. Use equations from Table 12–1. $a = 3.20$ in. $b = 4.00$ in.

$$\sigma_1 = \sigma_2 = \frac{p(b^3 + 2a^3)}{2(b^3 - a^3)} = \frac{(10\,000\ \text{psi})[4.00^3 + 2(3.20)^3]\ \text{in}^3}{2(4.00^3 - 3.20^3)\ \text{in}^3}$$

$$\sigma_1 = \sigma_2 = 20\,740\ \text{psi tangential}$$

$$\sigma_3 = -p = -10\,000\ \text{psi radial}$$

Each of these stresses is a maximum at the inner surface.

Steps 5, 6. For a maximum stress of 20 740 psi, the required yield strength for the material is

$$s_y = N(\sigma_2) = 4(20\,740\ \text{psi}) = 82\,960\ \text{psi} = 83\ \text{ksi}$$

Step 7. From Appendix A–14, we can specify AISI 4140 OQT 1300 steel that has a yield strength of 101 ksi. Others could be used.

Comment: The maximum stress in the sphere is less than half that in the cylinder of the same size, allowing a material with a much lower strength to be used. Alternatively, it would be possible to design the sphere with the same material but with a smaller wall thickness.

Example Problem 12–6 A cylindrical vessel has an outside diameter of 400 mm and an inside diameter of 300 mm. For an internal pressure of 20.1 MPa, compute the hoop stress σ_1 at the inner and outer surfaces and at points within the wall at intervals of 10 mm. Plot a graph of σ_1 versus the radial position in the wall.

Solution

Objective: Compute the hoop stress at specified positions in the wall of the cylinder.

Given: Pressure $= p = 20.1\text{MPa}$. $D_o = 400$ mm. $D_i = 300$ mm.
Use 10 mm increments for radius within the wall from the outside surface to the inside surface.

Analysis: Use Steps 1–4 from Procedure A from this section.

Results:

Step 1. $D_m = (D_o + D_i)/2 = (400 + 300)/2 = 350$ mm

Step 2. $t = (D_o - D_i)/2 = (400 - 300)/2 = 50$ mm
$D_m/t = 350/50 = 7.00 < 20$; thick-walled cylinder

Step 3. This step does not apply.

Step 4. Use the equation for tangential stress from Table 12–1.

$$\sigma_1 = \frac{pa^2(b^2 + r^2)}{r^2(b^2 - a^2)}$$

$$a = D_i/2 = 300/2 = 150 \text{ mm}$$

$$b = D_o/2 = 400/2 = 200 \text{ mm}$$

The results are shown in tabular form below.

r (mm)	σ_2 (MPa)	
200	51.7	(Minimum at outer surface)
190	54.5	
180	57.7	
170	61.6	
160	66.2	
150	71.8	(Maximum at inner surface)

Comment: Figure 12–7 shows the graph of tangential stress versus position in the wall. The graph illustrates clearly that the assumption of uniform stress in the wall of a thick-walled cylinder would *not* be valid.

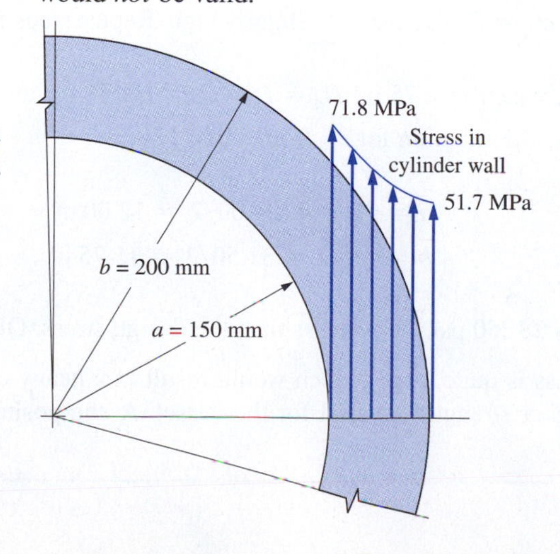

FIGURE 12–7 Variation of tangential stress in the wall of the thick-walled cylinder in Example Problem 12–6.

Example Problem 12–7 Design a cylinder to be made from aged titanium Ti-6Al-4V to carry compressed natural gas at 7500 psi. The internal diameter must be 24.00 in to provide the necessary volume. The design stress is to be $\frac{1}{6}$ of the ultimate strength of the titanium.

Solution

Objective: Design the cylinder.

Given: Pressure $= p = 7500$ psi. $D_i = 24.0$ in.
Titanium Ti-6Al-4V; $s_u = 170$ ksi (Appendix A–15)

Analysis: Use Procedure B from this section.

Results:

Step 1. Let $D_m = 24.00$ in.

Step 2. Assume thin-walled cylinder.

Step 3. Design stress,

$$\sigma_d = s_u/6 = (170\,000 \text{ psi})/6 = 28\,333 \text{ psi}$$

Step 4. Use the Equation (12–20) to compute nominal value for t.

$$t = \frac{pD_m}{2\sigma_d} = \frac{(7500 \text{ psi})(24.0 \text{ in})}{2(28\,333 \text{ psi})} = 3.18 \text{ in}$$

Step 5. Trial #1: $D_i = 24.00$; $t = 3.50$ in; $D_o = D_i + 2t = 31.00$ in

Step 6. $D_m = D_i + t = 24.00 + 3.50 = 27.50$ in

Step 7. $D_m/t = 27.50/3.50 = 7.86 < 20$; thick-walled.

Step 8. This step does not apply.

Step 9. Use the equation for σ_1 from Table 12–1.

$$a = D_i/2 = 24.00/2 = 12.00 \text{ in}$$

$$b = D_o/2 = 31.00/2 = 15.50 \text{ in}$$

$$\sigma_1 = \frac{p(b^2 + a^2)}{(b^2 - a^2)} = \frac{(7500 \text{ psi})(15.50^2 + 12.00^2)}{(15.50^2 - 12.00^2)}$$

$\sigma_1 = 29\,940$ psi Slightly high. Repeat steps 5 and 9.

Step 5. Increase $t = 3.75$ in; $D_o = D_i + 2t = 31.50$ in; thick-walled.

Step 9. Use the equation for σ_1 from Table 12–1.

$$a = D_i/2 = 24.00/2 = 12.00 \text{ in}$$

$$b = D_o/2 = 31.50/2 = 15.75 \text{ in}$$

Then $\sigma_1 = 28\,250$ psi. This is less than the design stress. OK.

Comment: The wall thickness is quite large, which would result in a heavy cylinder. Consider using a sphere and higher strength material for the vessel. A composite sphere may produce a lighter design.

12–7 SPREADSHEET AID FOR ANALYZING THICK-WALLED SPHERES AND CYLINDERS

The calculations called for in analyzing thick-walled spheres and cylinders can be tedious. The equations shown in Table 12–1 require numerous calculations. The use of spreadsheets, computer programs, programmable calculators, or computer algebra systems can simplify your task significantly.

Figure 12–8 shows an example spreadsheet that analyzes thick-walled spheres and cylinders for the maximum tangential, longitudinal, and radial stresses. Only the few data values in the shaded areas need to be entered by the user.

If the vessel is a sphere, the results are shown near the middle of the sheet. If the vessel is a cylinder, the maximum stresses are shown just below the middle of the sheet.

Also included in the lower part of the spreadsheet is the calculation of the tangential stress in a cylinder as a function of the radial position within the wall of the cylinder. This requires that the user specify the values of the radii for which the calculations are to be done.

The sample data included in the spreadsheet are the same as used in Example Problem 12–6 for a thick-walled cylinder and a graph of the distribution of tangential stress in the wall of the cylinder is shown in Figure 12–7.

Enhancements of this spreadsheet could be made to enable it to be used for designing pressure vessels using Procedures A, B, and C outlined earlier in this chapter. You are encouraged to produce such design aids.

FIGURE 12–8 Spreadsheet for analyzing stresses in thick-walled spheres and cylinders.

STRESSES IN THICK-WALLED CYLINDERS AND SPHERES

DATA REQUIRED:			
Pressure = p =	20100 kPa	Wall thickness = t =	50 mm
R_i = a =	150 mm	Mean Dia = D_m =	350 mm
R_O = b =	200 mm	Ratio: D_m/t =	7.0
		If Ratio < 20, Vessel is thick	

Analysis of a Sphere

Max Tangential Stress	32.05 MPa	At inner surface
Max Radial Stress	–20.10 MPa	At inner surface

Analysis of a Cylinder

Max Tangential Stress	71.79 MPa	At inner surface
Max Longitudinal Stress	25.84 MPa	Uniform through wall
Max Radial Stress	–20.10 MPa	At inner surface

Stress vs. Radius - Tangential Stress Only

Radius	***Stress***	
150	71.79 MPa	At inner surface
160	66.22 MPa	
170	61.61 MPa	
180	57.75 MPa	
190	54.48 MPa	
200	51.69 MPa	At outer surface

12–8 SHEARING STRESS IN CYLINDERS AND SPHERES

We have seen that internal pressure in a pressure vessel causes normal tensile stresses to be created in the walls of the cylinder or sphere. If these are the only externally generated stresses on the vessel, they will be the principal stresses on any planar surface element. No externally applied shearing stresses are created in the plane of the surface. This assumes that there is no torsion applied to the vessel and there are no external supports on or penetrations into the wall of the vessel in the area of interest.

Figure 12–9 shows the stress element at the surface of a cylinder along with the Mohr's circle for that element. It shows that there will be shearing stresses at angles other than 0° or 90° with respect to the *x*-axis where the longitudinal stress acts. The maximum shearing stress in the plane of the element is $\sigma_2/2$ and it occurs on an angle at 45° to the maximum principal stress axis.

Because both of the principal stresses are tensile, this is a classic case of both principal stresses having the same sign. It was observed in Section 10–11, during the discussion of Mohr's circle, that in such cases there is a need to consider the combination of the maximum principal stress with the stress perpendicular to the surface stress element to determine the true maximum shear stress on the element. The stress perpendicular to the plane of the cylinder, σ_3, is zero at the surface. Then, as we show in Figure 12–9, a supplementary Mohr's circle is drawn with σ_1 and σ_3 on the horizontal diameter. For this combination, the true maximum shearing stress is $\sigma_1/2$, acting in the *x-z* plane, at 45° to the maximum principal stress axis.

Pressure vessels are typically made from flat plates rolled into a cylindrical shape and welded along the seam. The design of the welds and the strength of the material in the vicinity of the welds are critical design parameters. The seam is often longitudinal for convenience of the rolling operation. The stress on the seam would be the tangential stress, which is the maximum principal stress.

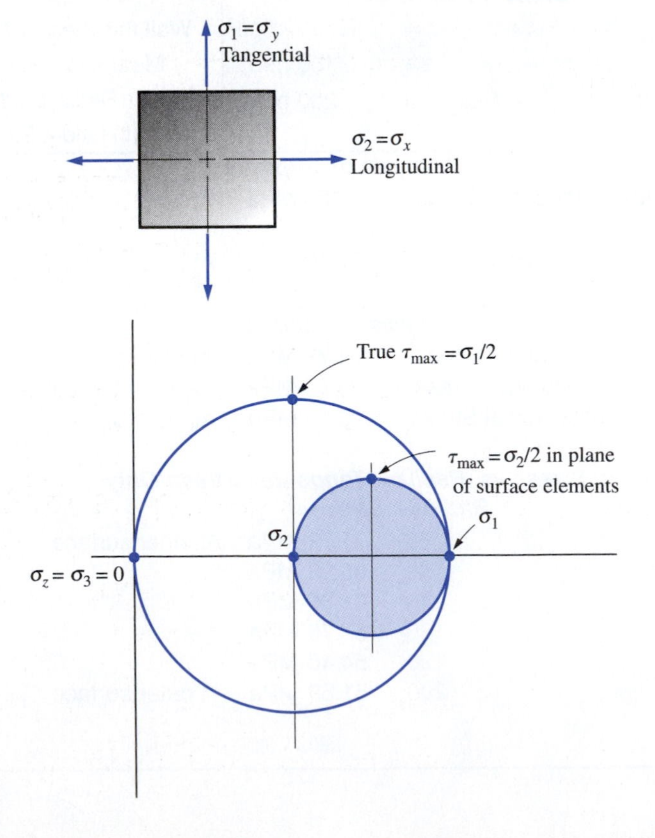

FIGURE 12–9 Principal stresses and shearing stresses in a pressurized cylinder.

Another method of making the cylindrical form is to roll the flat sheet along a spiral path as sketched in Figure 12–10. Then the entire spiral seam is welded. This is particularly attractive when making relatively long, large diameter tanks or pipes. In this case, you will need to know the stress condition on an element in line with the spiral seam. Mohr's circle can also be used for that task as demonstrated in Example Problem 12–8.

Example Problem 12–8

Figure 12–10 shows a tank that is to be made by rolling flat sheets of AISI 1040 cold-drawn steel into the spiral shape shown, where the spiral makes an angle of 65° with the horizontal axis of the tank. The design internal pressure is 1750 kPa. The inside diameter has been specified to be 900 mm to create the desired capacity in the tank. (a) Specify a suitable thickness for the steel sheet to provide a design factor of 4 based on yield strength or to produce a design factor of 6 based on ultimate strength. (b) For the final design of the tank, determine the stress condition on an element aligned with the weld.

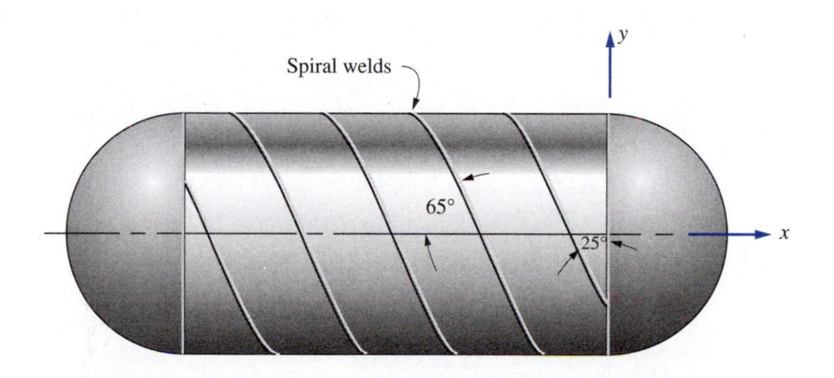

FIGURE 12–10 Tank for Example Problem 12–8.

Solution

Objective: Specify the tank thickness and determine the stress on the weld.

Given: Tank design in Figure 12–10. $D_i = 900$ mm. $p = 1750$ kPa.
AISI 1040 CD steel, $s_y = 565$ MPa, $s_u = 669$ MPa
$\sigma_d = s_y/4$ or $\sigma_d = s_u/6$, whichever is smaller.

Analysis: Use Procedure B from Section 12–6 to design the tank.
Then use Mohr's circle to determine the stress on the weld.

Results:

Steps 1 and 2. First assume $D_m = 900$ mm. Use Equation (12–20).

Step 3. Compute two design stresses and specify the smaller value.

$$\sigma_d = s_y/4 = (565\text{ MPa})/4 = 141.3\text{ MPa}$$
$$\sigma_d = s_u/6 = (669\text{ MPa})/6 = 111.5\text{ MPa}$$

Step 4. Solve for minimum required wall thickness, t. Let $\sigma_{max} = \sigma_d$.

$$\sigma_{max} = (pD_m)/(2t) \quad \text{Equation (12–20)}$$
$$t = \frac{pD_m}{2\sigma_{max}} = \frac{(1750 \times 10^3\text{ Pa})(900\text{ mm})}{2(111.5 \times 10^6\text{ Pa})} = 7.06\text{ mm}$$

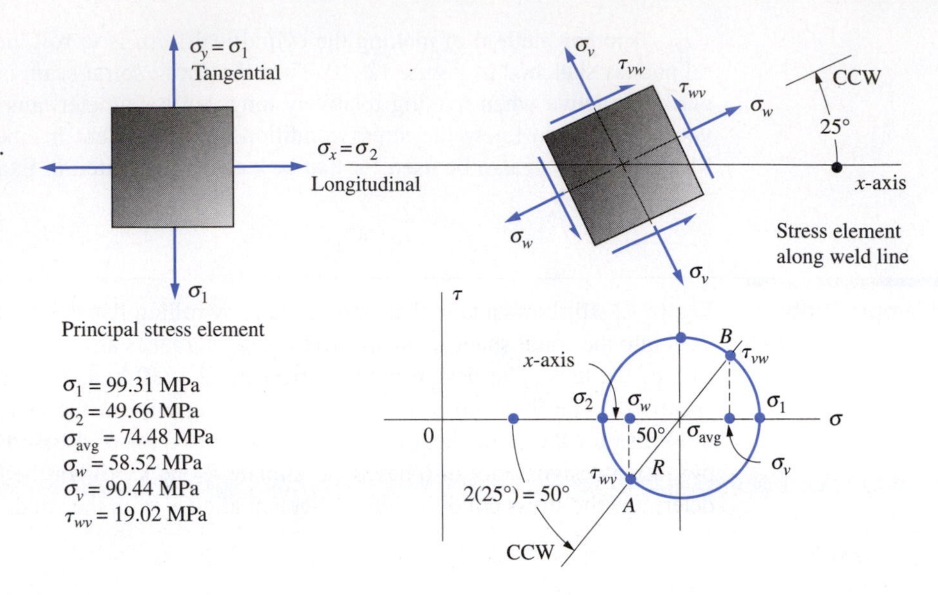

FIGURE 12–11 Stress elements and Mohr's circle for Example Problem 12–8.

Step 5. Specify $t = 8.0$ mm. Then,

$$D_i = 900\text{ mm}.$$
$$D_o = D_i + 2t = 900 + 2(8.0) = 916\text{ mm}$$

Step 6. $D_m = D_i + t = 900 + 8.0 = 908$ mm

Step 7. $D_m/t = (908\text{ mm})/(8.0\text{mm}) = 113.5$

Step 8. $D_m/t > 20$. The cylinder is very thin.

Mohr's Circle Analysis. See Figure 12–11.
The actual maximum principal stress, σ_1, is

$$\sigma_1 = (pD_m)/(2\,t)$$
$$\sigma_1 = \frac{(1750 \times 10^3\text{ Pa})(908\text{ mm})}{2(8.0\text{ mm})} = 99.31 \times 10^6\text{ Pa} = 99.31\text{ MPa}$$

The actual minimum principal stress, σ_2, is for this special case:

$$\sigma_2 = (pD_m)/(4t) = \sigma_1/2 = 49.66\text{ MPa}$$

The stress element in the plane of the tank is shown in Figure 12–11 with $\sigma_2 = \sigma_x$ and $\sigma_1 = \sigma_y$. No shear stress exists on the element in this orientation.

Mohr's circle is also shown in Figure 12–11. We need to find the stress on an element that is rotated 25° CCW from the x-axis.

In the Mohr's circle, the x-axis is the line from the center of the circle to σ_2. We then rotate CCW by an angle of $2(25°) = 50°$ to find the point on the circle that represents the stress condition along the weld. That point is labeled A on the circle, and it has the coordinates of σ_w, τ_{wv}. Here we are using the w subscript to indicate the direction perpendicular to the weld line and the v subscript to indicate the direction along the weld line. Then τ_{wv} is the shearing stress perpendicular to the w-direction and parallel to the v-direction.

Using the geometry of the circle we find

$$\sigma_w = \sigma_{avg} - R\cos(50°).$$

where R is the radius of the circle. We find R from,

$$R = (\sigma_1 - \sigma_2)/2 = (99.31 - 49.66)/2 = 24.83 \text{ MPa}$$

Then

$$\sigma_w = \sigma_{avg} - R(\cos 50°) = 74.48 - (24.83)(\cos 50°) = 58.52 \text{ MPa}$$

Also,

$$\tau_{wv} = R\sin(50°) = (24.83)(\sin 50°) = 19.02 \text{ MPa}$$

Finally, we find the normal stress parallel to the direction of the weld. That is found 180° from point A, called point B on the circle. Using similar logic,

$$\sigma_v = \sigma_{avg} + R(\cos 50°) = 74.48 + (24.83)(\cos 50°) = 90.44 \text{ MPa}$$

The final stress element oriented on the weld line is shown in Figure 12–11.

Summary The final design can be summarized as follows.

The tank is made from AISI 1040 CD steel.

$D_i = 900$ mm. $D_o = 916$ mm. $t = 8.0$ mm.

Normal stress perpendicular to the weld line is $\sigma_w = 58.52$ MPa.

Normal stress parallel to the weld line is $\sigma_v = 90.44$ MPa.

Shear stress parallel to the weld line is $\tau_{wv} = 19.02$ MPa.

12–9 OTHER DESIGN CONSIDERATIONS FOR PRESSURE VESSELS

Pressure vessel design and analysis techniques presented thus far have related only to the basic stress analysis of ideal cylinders and spheres without consideration of penetrations or other changes in geometry. Of course, most practical pressure vessels incorporate several types of features that cause the vessel to differ from the ideal shape. Figure 12–12 shows a computer-generated drawing of such a vessel. It has two cylindrical sections with different diameters, a conical transition section between them, spherical domed ends of two diameters, and three penetrations intersecting the shell with different geometries.

External loads are often applied that create stresses that combine with the stress due to internal pressure. For example,

- A spherical or cylindrical pressure vessel would typically have one or more ports used to fill or empty the vessel. The ports would often be welded into the vessel, causing a discontinuity in the geometry as well as modifying the material properties in the vicinity of the weld.
- Some pressure vessels used for chemical reactions or other material processing applications contain view ports for observation of the process. The view ports may contain flanges to hold the transparent window.
- Cylindrical vessels are often made with domed or hemispherical ends to provide a more optimum design to resist the internal pressure. But, because the tangential

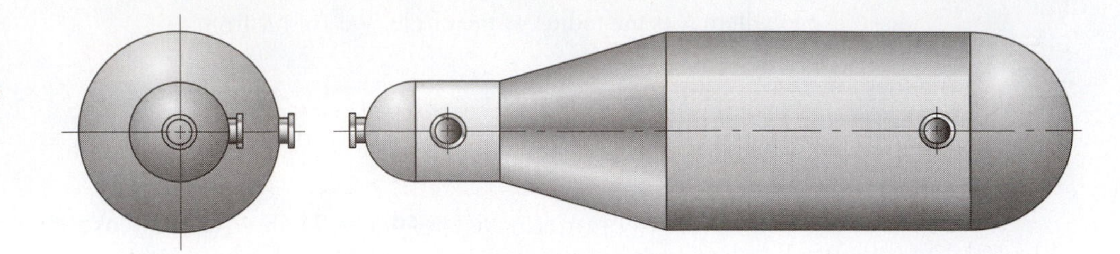

FIGURE 12–12 Computer-generated drawing of a pressure vessel having several geometry changes.

stress in the spherical end is less than that in the cylinder, special attention should be paid to the design at the intersection of the ends with the straight cylindrical portion.

- Large cylinders may have reinforcing bands or ribs applied to the inside or outside to stiffen the vessel structurally.
- Large cylinders and spheres may experience large stresses due to the actual weight of the vessel and its contents that combine with the stresses produced by the internal pressure. For example: a relatively long cylindrical tank laid horizontally and supported near its ends is subjected to bending stresses; a cylindrical tank positioned with its axis vertical is subjected to axial compressive stress.
- Large cylinders and spheres must be fitted with supports that transmit the weight of the vessel and its contents to a floor or the earth. Special stress conditions exist in the vicinity of such supports.
- Pressure vessels used in ground transportation equipment often experience dynamic loads caused by stopping, starting, product movement within the vessel, and vibrations caused by uneven roadways.
- Pressure vessels on aircraft and spacecraft are subjected to high acceleration forces during landings, takeoffs, launches, and rapid maneuvers.
- Joints between sections of pressure vessels made from two or more pieces often contain geometric discontinuities requiring special analysis techniques and careful fabrication.

Sources of Additional Information on Pressure Vessels. Analysis techniques for the conditions listed above are not covered in this book. The lists of references and Internet sites at the end of this chapter offer a wide variety of standards, guidelines, and computational aids to ensure safe and economical design of pressure vessels. The discussions that follow offer annotations about the references and Internet sites listed at the end of this chapter.

1. The primary reference for pressure vessel design standards in the U.S. is Reference 2, the *ASME Boiler and Pressure Vessel Code.* The most current revision should be used. Its major sections are:
   - a. I Power Boilers
   - b. II Materials
   - c. III Rules for Construction of Nuclear Facility Components
   - d. IV Heating Boilers
   - e. V Nondestructive Examination
   - f. VI Recommended Rules for the Care and Operation of Heating Boilers
   - g. VII Recommended Guidelines for the Care of Power Boilers
   - h. VIII Pressure Vessels
   - i. IX Welding and Brazing Qualifications

   j. X Fiber-Reinforced Plastic Pressure Vessels
   k. XI Rules for Inservice Inspection of Nuclear Power Plant Components
   l. XII Rules for Construction and Continued Service of Transport Tanks

   Numerous seminars, short courses, and specialized training programs based on the BPV Code are available, and many firms provide certified design, testing, and construction services in applying the code. Section VIII, *Pressure Vessels,* is the most pertinent to this book. See Internet site 1 for more information. Strict adherence to these standards is essential for ensuring protection of life and property.

2. Piping systems themselves perform as pressure vessels in conjunction with tanks, boilers, heat exchangers, and other special devices under pressure. Specialized requirements for materials, components (such as flanges, fittings, valves, and so forth), design, fabrication, assembly, erection, examination, inspection, and testing of piping are included in the following standards.
   a. Reference 3, ASME standard B31.1, provides standards for piping systems for electrical power generation stations, industrial plants, and central heating plants that typically operate at high temperatures and moderate to high pressures.
   b. Reference 4, ASME standard B31.3, covers piping typically found in petroleum refineries; chemical, pharmaceutical, textile, paper, semiconductor, and related processing plants; and terminals where shipping of the liquid or gaseous products occurs.
3. References 1, 5–8, 10, 11, and 14 are manuals and guidebooks that expand on the ASME standards and provide examples of the applications of provisions from the standards.
4. Reference 9 is part of a large, comprehensive manual for engineers, contractors, and vendors that design or provide equipment for use in the Lawrence Livermore National Laboratories, one of the major U.S. government research laboratories. The document on *Pressure Vessel and System Design* is the most pertinent part of the manual for this book.
   a. Section 4.0 discusses design controls for pressure vessels including materials selection, materials specifications (carbon and alloy steels, stainless steels, and titanium), design considerations, maximum allowable working pressure (MAWP), maximum operating pressure (MOP), test pressures, and calculation guide for thin- and thick-walled vessels and end-closures.
   b. Section 5.3 describes requirements for gas-pressure containment vessels that will contain toxic, radioactive, corrosive, or flammable materials.
   c. Section 6.3 provides guidelines and standards for pipe and tube support.
   d. Section 6.6 describes pressure relief devices that limit the pressure level in a pressure vessel or piping system.

   The manual can be viewed and downloaded from Internet site 3.
5. References 12 and 13 provide more background, derivation, and extension of the analytical methods and formulas used in this book for stress analysis and design of pressure vessels and piping systems and their components.
6. While the ASME code is most often used in the U.S., Internet site 2 describes a corresponding standard used in Great Britain and some other parts of Europe. Standards for other parts of the world exist and should be consulted when providing equipment for those areas.

7. Internet sites 4–9 connect to several providers of computer software packages for pressure vessel design that perform the complex computations required to design and analyze pressure vessels and their attachments. Most provide three-dimensional CAD capability, the ability to export the design drawings into other CAD systems, and complete analyses of the shell, stiffening rings, various styles of heads, nozzles, ports, flanges, and joints between elements. All stress analysis methods are correlated with the ASME codes and standards and consider material properties, design temperatures for vessel operation, welds, testing requirements, and documentation forms that must be completed and filed with the product designs.
8. Internet sites 10–14 connect to a small sample of geographically dispersed companies in the U.S. that offer a diversified array of products and services in the design and fabrication of pressure vessels, tanks, and piping systems. These sites may be useful to students and other users of this book who have limited experience with the details of such systems to help them visualize their complexity.

12–10 COMPOSITE PRESSURE VESSELS

The applications and examples presented in this chapter have emphasized the use of metals for the structural walls of pressure vessels. Other materials, particularly composite materials and reinforced plastics, are often used as well. The special characteristics of these materials must be understood when applying them to pressure vessels.

High-strength composite materials are well suited for use in fabricating pressure vessels. The fact that the primary stresses are either tangential (hoop) or longitudinal leads the designer of composite vessels to call for alignment of the composite fibers in the direction of the maximum stresses. Circumferential wrapping of a prepreg tape around a liner made from metal or plastic offers significant weight savings as compared with a design using only metal or plastic. To resist longitudinal stresses due to internal pressure along with other external forces, some tanks are wrapped in a helical fashion in addition to the circumferential wrapping. The thickness and direction of plies can be tailored to the specific set of loads expected in a particular application.

Materials selected for composite pressure vessels include E-glass/epoxy, structural glass/epoxy, and carbon/epoxy. Cost is a major factor in material specification.

Care must be exercised to ensure that the composite material bonds well and fits the geometry of any liner used in the vessel. Particular attention is needed in the domed ends of pressure cylinders and in the locations of ports. Ports are typically placed at the top or bottom at the poles of the domed ends in such a manner that the composite fibers are continuous. Placing ports in the sides of a tank would interrupt the integrity of the filament windings. Also, the geometry of the shape of the tank is often tailored to produce gradually varying stresses at joints between the cylindrical part and the domed ends. The thickness of composite plies is also varied to match the expected stresses.

The primary applications for composite pressure vessels include those in which light weight is a major design goal. The air supply tank for self-contained breathing apparatus (SCBA) used by firefighters is a good example because the lighter tank allows more mobility and less fatigue. Weight reductions in space and aeronautical applications allow greater payloads or higher performance of aerospace vehicles.

The development of ground vehicles using compressed natural gas (CNG) or hydrogen calls for the production of lightweight, compact cylinders to carry the fuel. Demonstration units incorporating fuel storage tanks made from advanced composites are in current use in buses, commercial fleet vehicles, automobiles, utility vehicles, and even power units for portable computers, remotely operated sensing systems, aerospace applications, and manufacturing equipment. Weight savings can be significant.

REFERENCES

1. Aluminum Association, *Aluminum Design Manual, Specifications for Aluminum Structures,* Aluminum Association, Washington, D.C., 2005.
2. American Society of Mechanical Engineers (ASME), *ASME Boiler and Pressure Vessel Code,* ASME, Fairfield, NJ, 2007.
3. American Society of Mechanical Engineers (ASME), *ASME Standard B31.1—2007 Power Piping,* ASME, Fairfield, NJ, 2007.
4. American Society of Mechanical Engineers (ASME), *ASME Standard B31.3—2006 Process Piping,* ASME, Fairfield, NJ, 2006.
5. Ball, B. E. and W. J. Carter, *CASTI Guidebook to ASME Section VIII Div. 1 Pressure Vessels,* 4th ed., CASTI Publishing, Edmonton, Alberta, Canada, 2005.
6. Chattopadhyay, S., *Pressure Vessels: Design and Practice,* CRC Press, Boca Raton, FL, 2004.
7. Ellenberger, P., R. Chuse, and B. E. Carson, *Pressure Vessels,* McGraw-Hill, New York, 2004.
8. Farr, J. R. and M. H. Jawad, *Guidebook for the Design of ASME Section VIII Pressure Vessels,* 3rd ed., American Society of Mechanical Engineers, Fairfield, NJ, 2006.
9. Lawrence Livermore National Laboratory (LLNL), *Environment, Safety and Health Manual,* Volume II, Part 18: Pressure/Noise/Hazardous Atmospheres, Document 18.2 *Pressure Vessel and System Design,* LLNL, Livermore, CA, 2005. (See Internet site 3.)
10. Moss, D. R., *Pressure Vessel Design Manual,* 3rd ed., Elsevier, New York, 2003.
11. Rao, K. R. (ed.), *Companion Guide to ASME Boiler & Pressure Vessel Code,* Volume 1 (2001), Volume 2 (2006), Volume 3 (2006), American Society of Mechanical Engineers, New York.
12. Young, W. C. and R. D. Cook, *Advanced Mechanics of Materials,* 2nd ed., Prentice Hall, Upper Saddle River, NJ, 1999.
13. Young, W. C. and R. G. Budynas, *Roark's Formulas for Stress and Strain,* 6th ed., McGraw-Hill, New York, 2002.
14. Zeman, F. L., F. Rsuscher, and S. Schindler, *Pressure Vessel Design: The Direct Route,* Elsevier, New York, 2006.

INTERNET SITES

General Information on Pressure Vessels

1. American Society of Mechanical Engineers www.asme.org/Codes/Publications/BPVC/ A listing of the ASME resources available related to the *ASME Boiler & Pressure Vessel Code.*
2. British Standards Institution www.bsi-global.com/PSS/About/index.xalter This page is the overall description of the BSI with links to sources of information about standards issued by BSI. Most relevant to this chapter is Standard BS EN 13445, *Unfired Pressure Vessels,* a widely used standard in Europe.
3. Lawrence Livermore National Laboratory (LLNL) www.llnl.gov/es_and_h/hsm/doc_18.02/doc18-02.html Part of LLNL's Environment, Safety and Health Manual, Volume II, Part 18: Pressure/Noise/Hazardous Atmospheres, Document 18.2 *Pressure Vessel and System Design,* 2005. This document is a good overview of a rigorous design process for pressure vessels, even though it is focused on LLNL's mission. Sections 4, 5, and 6 are most relevant to design considerations.

Computer Software for Pressure Vessel Design

4. Computer Engineering, Inc. www.computereng.com/products/advanced_pressure_vessel/ This software performs calculations in accordance with the ASME Section VIII of the ASME Boiler & Pressure Vessel Code, including the basic vessel, nozzles, heads, and accessories.
5. Algor—Center for Mechanical Design Technology www.algor.com/products/PVDesi1475/default.asp Producer of the PV/Designer software that facilitates the efficient design of pressure vessels in compliance with the ASME Boiler & Pressure Vessel Code.
6. COADE Engineering Software http://coade.com Producer of the PVElite software for pressure vessel design and the design of tall process towers in compliance with the ASME Boiler & Pressure Vessel Code. The companion CODECALC software assists in the design of heat exchangers, flanges, nozzles, and piping according to the ANSI B31.3 code.
7. Heat Transfer Consultants, Inc. www.htcsoftware.com Producer of the PVX-2007 Pressur Vessel & Heat Exchanger software.

8. Codeware, Inc. www.codeware.com Producer of the COMPRESS software for the design of pressure vessels in compliance with the ASME Boiler & Pressure Vessel Code.
9. Chempute Software www.chempute.com Supplier of software for the chemical, mechanical, and other engineering disciplines for chemical processing, power generation, and oil refining industries, including several packages dealing with pressure vessels and piping systems.

Pressure Vessel Manufacturers

The following list is a sampling of companies that offer a wide variety of services in the design and fabrication of pressure vessels, tanks, and piping systems.

10. PMF Industries, Inc. www.pmfind.com Producer of stainless steel pressure vessels for medical sterilizing and other sanitary applications; located in Williamsport, PA.
11. West Metal Works, Inc. www.westmetalworks.com Producer of pressure vessels according to the ASME code along with other containers and special fabrications; located in Buffalo, NY.
12. Bay Tank & Manufacturing Co., Inc. www.bavtankfab.com Producer of ASME code pressure vessels, reactors, columns, storage tanks, stacks, rotary kilns, scrubbers, and other products for petrochemical, pharmaceutical, power generation, and other industries; located in Panama City, FL.
13. Enerfab www.enerfab.com Producer of ASME pressure vessels, columns, reactors, fermenters, piping systems, and complete processing systems using a wide variety of materials; based in Cincinnati, OH, with operations in many other locations.
14. Roy E. Hanson, Jr. Manufacturing www.hansontank.com Hanson Tank manufacturers ASME pressure vessels, water storage tanks, air receivers, propane tanks, and a wide variety of other products; located in Los Angeles, CA.

PROBLEMS

12–1.M Compute the stress in a sphere having an outside diameter of 200 mm and an inside diameter of 184 mm if an internal pressure of 19.2 MPa is applied.

12–2.M A large, spherical storage tank for a compressed gas in a chemical plant is 10.5 m in diameter and is made of AISI 1040 hot-rolled steel plate, 12 mm thick. What internal pressure could the tank withstand if a design factor of 4.0 based on yield strength is desired?

12–3.M Titanium 6A1-4V is to be used to make a spherical tank having an outside diameter of 1200 mm. The working pressure in the tank is to be 4.20 MPa. Determine the required thickness of the tank wall if a design factor of 4.0 based on yield strength is desired.

12–4.M If the tank of Problem 12–3 was made of aluminum 2014-T6 sheet instead of titanium, compute the required wall thickness. Which design would weigh less?

12–5.E Compute the hoop stress in the walls of a 10-in schedule 40 steel pipe if it carries water at 150 psi.

12–6.M A pneumatic cylinder has a bore of 80 mm and a wall thickness of 3.5 mm. Compute the hoop stress in the cylinder wall if an internal pressure of 2.85 MPa is applied.

12–7.M A cylinder for carrying acetylene has a diameter of 300 mm and will hold the acetylene at 1.7 MPa. If a design factor of 4 is desired based on yield strength, compute the required wall thickness for the tank. Use AISI 1040 cold-drawn steel.

12–8.M The companion oxygen cylinder for the acetylene discussed in Problem 12–7 carries oxygen at 15.2 MPa. Its diameter is 250 mm. Compute the required wall thickness using the same design criteria.

12–9.M A propane tank for a recreation vehicle is made of AISI 1040 hot-rolled steel, 2.20 mm thick. The tank diameter is 450 mm. Determine what design factor would result based on yield strength if propane at 750 kPa is put into the tank.

12–10.M The supply tank for propane at the distributor is a cylinder having a diameter of 1800 mm. If it is desired to have a design factor of 4 based on yield strength using AISI 1040 hot-rolled steel, compute the required thickness of the tank walls when the internal pressure is 750 kPa.

12–11.M Oxygen on a spacecraft is carried at a pressure of 70.0 MPa in order to minimize the volume required. The spherical vessel has an outside diameter of 250 mm and a wall thickness of 18 mm. Compute the maximum tangential and radial stresses in the sphere.

12–12.M Compute the maximum longitudinal, hoop, and radial stresses in the wall of a standard $\frac{1}{2}$-in schedule 40 steel pipe when carrying an internal pressure of 1.72 MPa (250 psi).

12–13.M The barrel of a large field-artillery piece has a bore of 220 mm and an outside diameter of 300 mm. Compute the magnitude of the hoop stress in the barrel at points 10 mm apart from the inside to the outside surfaces. The internal pressure is 50 MPa.

12–14.M A $1\frac{1}{2}$-in schedule 40 steel pipe has a mean radius less than 10 times the wall thickness and thus should be classified as a thick-walled cylinder. Compute what maximum stresses would result from both the thin-wall and the thick-wall formulas due to an internal pressure of 10.0 MPa.

12–15.M A cylinder has an outside diameter of 50 mm and an inside diameter of 30 mm. Compute the maximum tangential stress in the wall of the cylinder due to an internal pressure of 7.0 MPa.

12–16.M For the cylinder of Problem 12–15, compute the tangential stress in the wall at increments of 2.0 mm from the inside to the outside. Then plot the results for stress versus radius.

12–17.M For the cylinder of Problem 12–15, compute the radial stress in the wall at increments of 2.0 mm from the inside to the outside. Then plot the results for stress versus radius.

12–18.M For the cylinder of Problem 12–15, compute the tangential stress that would have been predicted if thin-walled theory were used instead of thick-walled theory. Compare the result with the stress found in Problem 12–15.

12–19.M A sphere is made from stainless steel, AISI 501 OQT 1000. Its outside diameter is 500 mm and the wall thickness is 40 mm. Compute the maximum pressure that could be placed in the sphere if the maximum stress is to be one-fourth of the yield strength of the steel.

12–20.M A sphere has an outside diameter of 500 mm and an inside diameter of 420 mm. Compute the tangential stress in the wall at increments of 5.0 mm from the inside to the outside. Then plot the results. Use a pressure of 100 MPa.

12–21.M A sphere has an outside diameter of 500 mm and an inside diameter of 420 mm. Compute the radial stress in the wall at increments of 5.0 mm from the inside to the outside. Then plot the results. Use a pressure of 100 MPa.

12–22.M To visualize the importance of using the thick-walled formulas for computing stresses in the walls of a cylinder, compute the maximum predicted tangential stress in the wall of a cylinder from both the thin-walled and thick-walled formulas for the following conditions. The outside diameter for all designs is to be 400 mm. The wall thickness is to vary from 5.0 mm to 85.0 mm in 10.0 mm increments. Use a pressure of 10.0 MPa. Then compute the ratio of D_m/t and plot percent difference between the stress from the thick-walled and thin-walled theory versus that ratio. Note the increase in the percent difference as the value of D_m/t decreases, that is, as t increases.

12–23.M A sphere has an outside diameter of 400 mm and an inside diameter of 325 mm. Compute the variation of the tangential stress from the inside to the outside in increments of 7.5 mm. Use a pressure of 10.0 MPa.

12–24.M A sphere has an outside diameter of 400 mm and an inside diameter of 325 mm. Compute the variation of the radial stress from the inside to the outside in increments of 7.5 mm. Use a pressure of 10.0 MPa.

12–25.E Appendix A–12 lists the dimensions of American National Standard schedule 40 steel pipe. Which of the pipe sizes should be classified as thick-walled and which can be considered thin-walled?

12–26.E Design a cylindrical pressure vessel to carry compressed air for a self-contained breathing apparatus for use by firefighters when operating in smoke-filled buildings. The minimum inside diameter is to be 6.00 in and the length of the cylindrical portion of the tank is to be 15.0 in. It must withstand a service pressure of 450 psi. Use a design stress of $s_u/8$ to account for a large number of pressurization cycles. Also, check the final design for its ability to withstand a maximum pressure of 900 psi by computing the design factor based on yield strength. The tank is to be made from aluminum alloy 6061-T6. Compute the weight of just the cylindrical portion.

12–27.E Repeat Problem 12–26 but use titanium Ti-6A1-4V.

12–28.E Repeat Problem 12–26 but use stainless steel 17-4PH H900.

12–29.E For any of the designs for the SCBA air cylinder for Problems 12–26, 12–27, or 12–28, make a sketch of the complete tank by placing hemispherical heads on each end. Show a port at one end for attaching the pressure regulator and discharge device. Assuming that the wall thickness of the heads is the same as the wall thickness of the cylindrical portion, compute the approximate weight of the complete tank.

12–30.E Repeat Problem 12–26 but use the graphite/epoxy composite material listed in Table 2–13 in Chapter 2 that has a tensile strength of 278 ksi. Check the final design by computing the design factor with respect to tensile strength against the maximum pressure of 900 psi. The tank will be lined with a thin polymeric shell and wrapped fully with the unidirectional composite in a circumferential pattern to resist the hoop stress in the cylinder. Neglect the contribution of the liner in the stress analysis and in the weight calculation. (Note that the cylindrical shell would also likely require some plies of the composite to be placed in a helical fashion to resist the longitudinal stress and to permit the formation of the dome-shaped ends. Therefore, the final weight will be somewhat higher than computed for just the circumferentially wrapped part.)

12–31.E Design a spherical tank to carry oxygen at a pressure of 3000 psi with an internal diameter of 18.0 in. Use stainless steel AISI 501 OQT 1000 and a design factor of 6 based on the ultimate strength. Compute the weight of the tank.

12–32.E Repeat Problem 12–31 but use aluminum alloy 7075-T6.

12–33.E Repeat Problem 12–31 but use titanium alloy Ti-6A1-4V.

12–34.M Design a cylindrical tank for compressed natural gas at a pressure of 4.20 MPa. The minimum inside diameter is to be 450 mm. Use aluminum alloy 6061-T6 and a design factor of 8 based on the ultimate strength.

12–35.E Design a cylindrical tank for compressed air that will be used to provide remote service for truck tire repair. The air pressure will be 300 psi. The minimum internal diameter for the tank is to be 24 in. Use AISI 1040 cold-drawn steel and a design factor of 8 based on the ultimate strength. Check the final design for a maximum pressure of 600 psi by computing the design factor based on yield strength.

COMPUTER ASSIGNMENTS

1. Write a program or spreadsheet to compute the tangential stress in the wall of a thin-walled sphere. Include the computation of the mean diameter and the ratio of mean diameter to thickness to verify that it is thin-walled.
2. Write a program or spreadsheet to compute the tangential stress in the wall of a thin-walled cylinder. Include the computation of the mean diameter and the ratio of mean diameter to thickness to verify that it is thin-walled.
3. Write a program or spreadsheet to compute the longitudinal stress in the wall of a thin-walled cylinder. Include the computation of the mean diameter and the ratio of mean diameter to thickness to verify that it is thin-walled.
4. Combine the programs or spreadsheets of Assignments 2 and 3.
5. Combine the programs or spreadsheets of Assignments 1, 2, and 3, and let the user specify whether the vessel is a cylinder or a sphere.
6. Rewrite the programs or spreadsheets of Assignments 1, 2, and 5 so that the objective is to compute the required wall thickness for the pressure vessel to produce a given maximum stress for a given internal pressure.
7. Write a program or spreadsheet to compute the maximum longitudinal, hoop, and radial stress in the wall of a thick-walled cylinder using the formulas from Table 12–1.
8. Write a program or spreadsheet to compute the tangential stress at any radius within the wall of a thick-walled cylinder using the formulas from Table 12–1.
9. Write a program or spreadsheet to compute the radial stress at any radius within the wall of a thick-walled cylinder using the formulas from Table 12–1.
10. Write a program or spreadsheet to compute the tangential stress at any radius within the wall of a thick-walled sphere using the formulas from Table 12–1.
11. Write a program or spreadsheet to compute the radial stress at any radius within the wall of a thick-walled sphere using the formulas from Table 12–1.
12. Combine the programs or spreadsheets of Assignments 8 through 11.
13. Write a program or spreadsheet to compute the tangential stress distribution within the wall of a thick-walled cylinder using the formulas from Table 12–1. Start at the inside radius and specify a number of increments between the inside and the outside.
14. Write a program or spreadsheet to compute the radial stress distribution within the wall of a thick-walled cylinder using the formulas from Table 12–1. Start at the inside radius and specify a number of increments between the inside and the outside.
15. Write a program or spreadsheet to compute the tangential stress distribution within the wall of a thick-walled sphere using the formulas from Table 12–1. Start at the inside radius and specify a number of increments between the inside and the outside.
16. Write a program or spreadsheet to compute the radial stress distribution within the wall of a thick-walled sphere using the formulas from Table 12–1. Start at the inside radius and specify a number of increments between the inside and the outside.
17. Write a program or spreadsheet to perform the computations of the type called for in Problem 12–22.
18. Write a program or spreadsheet to perform the computations of the type called for in Problem 12–22, except do it for a sphere.
19. Write a program or spreadsheet to compute the maximum tangential stress in any standard schedule 40 pipe for a given internal pressure. Include a table of data for the dimensions of the pipe sizes listed in Appendix A–12. Include a check to see if the pipe is thick-walled or thin-walled.

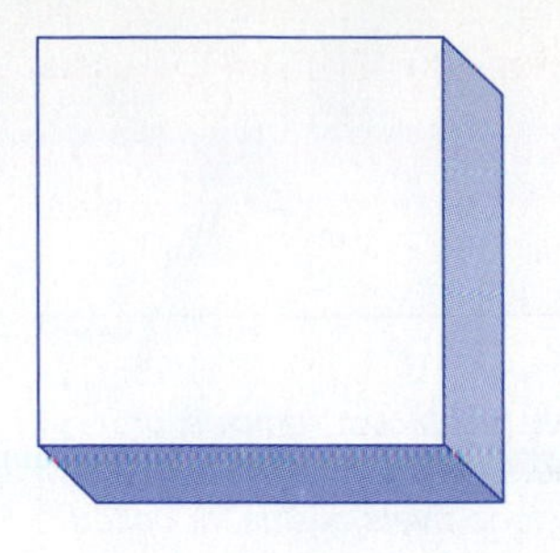

13

Connections

The Big Picture

Connections

Discussion Map

- Load-carrying members that make up structures and machines must act *together* to perform their desired functions. After completing the design or analysis of the primary members, it is necessary to specify suitable connections between them. As their name implies, connections provide the linkage between members.
- Structures and mechanical devices rely on the connections between load-carrying elements to maintain the integrity of the assemblies. The connections provide the path by which loads are transferred from one element to another.
- Three common types of connections are riveting, welding, and bolting. Figure 13–1 shows a bulk storage hopper supported by rectangular straps from a tee beam. During fabrication of the hopper, the support tabs were welded to the outside of the side walls. The tabs contain a pattern of holes, allowing the straps to be bolted on at the assembly site. Prior to installation of the tee beam, the straps were riveted to the web.
- The load due to the weight of the hopper and the materials in it must be transferred from the hopper walls into the tabs through the welds. Then the bolts transfer the load to the straps, which act as tension members. Finally, the rivets transfer the load into the tee.

Activity

Find some examples in structures, machines, vehicles, or consumer products where fasteners play a major role. Try to find at least ten examples that include rivets, welds, or bolted joints. In addition to the hopper shown in Figure 13–1, here are a few examples to get you started.

- The bolts that hold the alternator, air conditioning compressor, and other accessories to the engine of your car. *Each of these fasteners is critical to maintaining the location of the device with respect to the engine block and to the mating parts. The engine components themselves are fastened together with a variety of bolts that are very highly stressed and that must be tightened with a calibrated torque wrench. The bolts provide primarily a clamping function to ensure that the mating parts do not separate or move during loading. They are usually loaded in some combination of tension and shear.*
- The rivets that hold the steps of a stepladder to the side rails. *These critical fasteners ensure that the force transferred from your body weight to the steps is transferred safely into the side rails and thus to the ground. The other functional parts of the ladder are also typically fastened with tubular rivets that are inserted through holes in both parts and then upset with a tool that spreads the head against a solid surface. Rivets usually carry loads in shear.*
- The welds that hold the various parts of a bicycle frame together. *The frame members are usually tubular steel or some lighter metal that must be welded at the joints to obtain an integral, stiff, strong structure. The welds are subjected to combinations of bending, torsion, and direct shear as the cyclist executes maneuvers. Notice the detail differences in the frame among the touring, racing, and mountain bike styles.*
- Structural fasteners in building construction. *Find a construction site where a steel-framed building is being erected and look specifically at how the columns, beams, and other structural members are connected using bolting or welding. Examine the type of fasteners used, the arrangement of bolts in multiple-bolt joints, weld line patterns, and the design of the connected members. Figure 13–2 is a photograph of a sculpture that illustrates many of the types of structural joints used for instruction of students in civil engineering and technology, construction, and architecture. Notice the variety of angle clips, added plates, and modifications to the main structural members that facilitate fastening.*

Now, what other examples are on your list? Discuss them with your colleagues and the instructor and find out what different examples they have found.

Search the Internet

How much do you currently know about connections, fasteners, welding, and joining? This chapter will provide useful information on the functions of connections, their modes of failure, riveted connections, bolted connections, and welded joints. Before getting into the

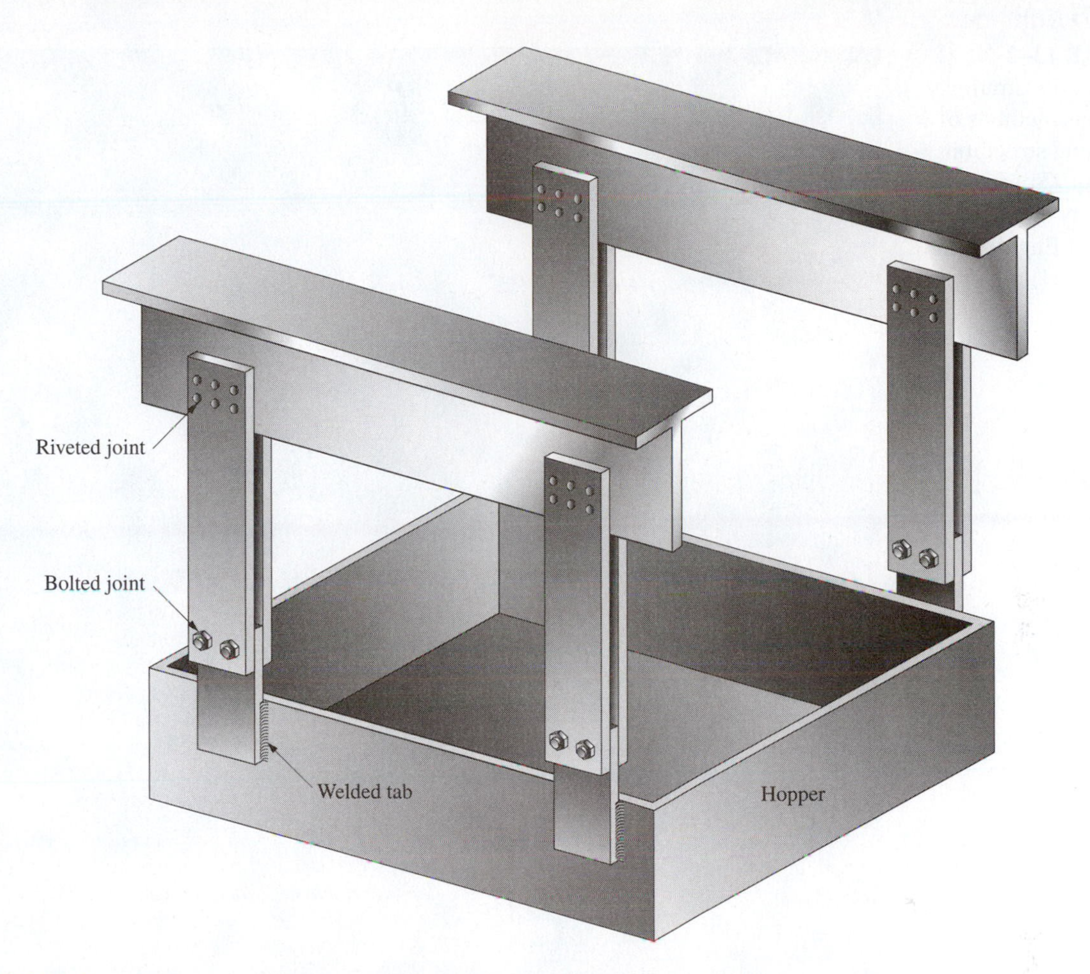

FIGURE 13–1 Illustration of three types of joints—riveted, bolted, and welded.

technical details, check out some of the Internet sites listed at the end of this chapter so you can get a good overview of:

- The technology of connection design and fabrication
- The standards that guide connection design
- Software that provides computer-assisted design and analysis of bolted joints
- A sampling of manufacturers of many types of fasteners used in building construction, automotive, aerospace, consumer products, agricultural, industrial, furniture, and other applications
- Welding technology and the design of welded joints.

13–1 OBJECTIVES OF THIS CHAPTER

The primary objective of this chapter is to provide data and methods of analysis for the safe design of riveted joints, bolted joints, and welded joints.

After completing this chapter, you should be able to:

1. Describe the typical geometry of riveted and bolted joints.
2. Identify the probable modes of failure for a joint.

FIGURE 13–2 Sculpture illustrating numerous methods of connecting structural members. (Source: University of Dayton School of Engineering, Dayton, OH)

3. Recognize typical styles of rivets.
4. Identify when a fastener is in single shear or double shear.
5. Analyze a riveted or bolted joint for shearing force capacity.
6. Analyze a riveted or bolted joint for tensile force capacity.
7. Analyze a riveted or bolted joint for bearing capacity.
8. Use the allowable stresses for steel structural connections as published by the American Institute of Steel Construction (AISC).
9. Describe the difference between a friction-type connection and a bearing-type connection and complete the appropriate analysis.
10. Use the allowable stresses for aluminum structural connections as published by the Aluminum Association.
11. Analyze both symmetrically loaded joints and eccentrically loaded joints.
12. Analyze welded joints with concentric loads.

13–2 MODES OF FAILURE FOR BOLTED JOINTS

Figure 13–3 shows simple lap joints in which two plates are connected with two bolts and nuts. The function of the joint is to clamp the plates together and to transfer a load applied to one plate through the bolt to the other plate. Follow the load path and visualize the types of stress created:

1. From plate 1, the load passes to the side surfaces of the bolts.
2. The bolts bear on the holes, tending to crush the plate material [Figure 13–3(a)].
3. The load passes through the bolt and into plate 2, producing bearing stress on the holes.
4. The opposing forces acting on plates 1 and 2 tend to cut (shear) the bolt at the interface between the two plates [Figure 13–3(b)].
5. The tensile forces on plates 1 and 2 tend to tear the material across the section with minimum area to resist the tensile force. That occurs through the section where the bolt holes are located [Figure 13–3(c)].
6. As the bolts bear on the side surfaces of the holes in the plates, there is a tendency to tear out the material from the bolt to the edge of the material in either plate [Figure 13–3(d)].

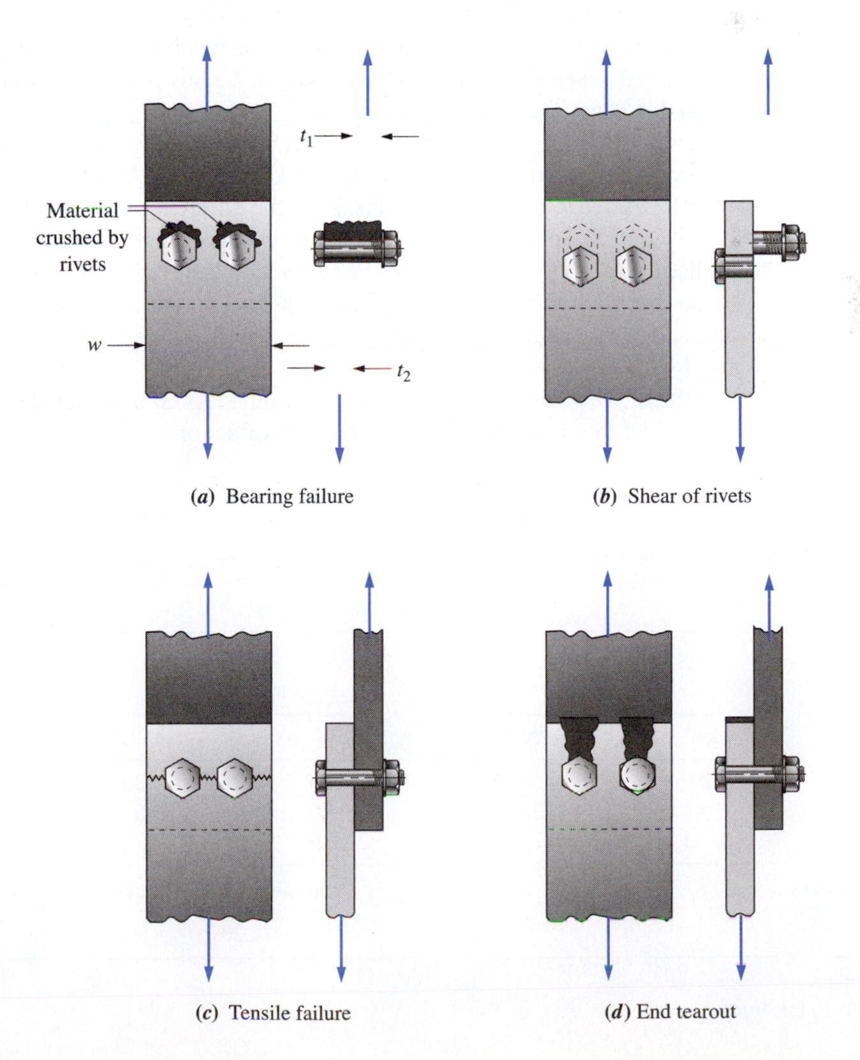

FIGURE 13–3 Types of failure of bolted connections.

Properly designed connections should have an edge distance from the centerline of the bolt to the edge of the plate being joined of at least three times the diameter of the bolt. The edge distance is measured in the direction of the bearing pressure. If this recommendation is heeded, then end tearout should not occur. Although this should be checked with analysis, it will be assumed to apply to example problems in this chapter. Thus, shear, bearing, and tensile failure modes only will be considered in evaluating joint strength.

13–3 DESIGN OF BOLTED CONNECTIONS

Two types of bolted connections are used in building construction:

- ***Bearing-type connections*** It is assumed that the joined plates are *not* clamped tightly enough to cause frictional forces between the plates to transmit loads. Therefore, the bolts do bear on the holes, and bearing failure must be investigated. Both shear failure and tensile failure could also occur.
- ***Slip-critical connections*** High clamping forces are produced in this type of joint to prevent slipping, and careful fabrication techniques are required to ensure that friction forces do share in the transmission of connection forces.

The design of slip-critical connections involves many variations and steps. Users are advised to see References 2 and 3 for details. Only bearing-type connections are discussed in this chapter.

Table 13–1 shows sample data for allowable stresses on bolted connections for structural steel, listing three types of bolts: ASTM A307, ASTM A325, and ASTM A490, with progressively increased strengths. Note the difference in allowable shearing stress, depending on whether threads are included in the shear plane or not. Figure 13–4 shows a standard hex-head bolt with a portion of its length threaded. It is preferred to design the joint so that the full-diameter part of the bolt is in the shear plane.

The methods of analyzing shear, bearing, and tensile failure modes are described next.

Shear Failure.

The bolt body is assumed to be in direct shear when a tensile load is applied to a joint, provided that the line of action of the load passes through the centroid

TABLE 13–1 Allowable stresses for steel structural connections for bearing-type connections.*

	Allowable shear stress					
	No threads in shear plane		Threads in shear plane		Allowable tensile stress	
Bolts	ksi	MPa	ksi	MPa	ksi	MPa
ASTM A307	12.0	82.5	12.0	82.5	22.5	155
ASTM A325	30.0	207	24.0	165	45.0	310
ASTM A490	37.5	260	30.0	207	56.5	390
Connected members			Allowable bearing stress		Allowable tensile stress‡	
All alloys			$1.20s_u$		$0.6s_y$	

*AISC specifications.

‡See Appendix A–16 for structural steels.

FIGURE 13–4
Bolted joint.

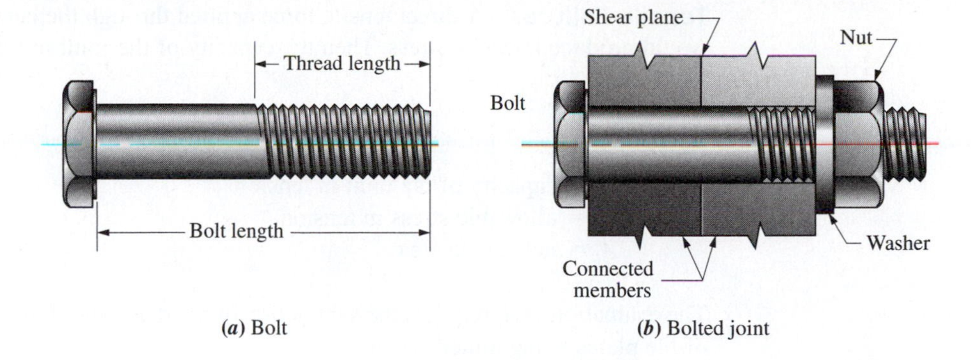

of the pattern of bolts. It is also assumed that the total applied load is shared equally among all the bolts. The capacity of a joint with regard to shear of the bolts is

$$F_s = \tau_a A_s \tag{13–1}$$

where F_s = capacity of the joint in shear
τ_a = allowable shear stress in bolts
A_s = area in shear

The area in shear is dependent on the number of cross sections of bolts available to resist shear. Calling this number N_s,

$$A_s = \frac{N_s \pi D^2}{4} \tag{13–2}$$

where D is the bolt diameter.

To determine N_s, it must be observed whether *single shear* or *double shear* exists in the joint. Figure 13–3 shows an example of single shear. Only one cross section of each bolt resists the applied load. Then N_s is equal to the number of bolts in the joint. The straps used to support the hopper in Figure 13–1 place the rivets and bolts in double shear. Two cross sections of each bolt resist the applied load. Then N_s is twice the number of bolts in the joint.

Bearing Failure. When a cylindrical bolt bears against the wall of a hole in the plate, a nonuniform pressure exists between them. As a simplification of the actual stress distribution, it is assumed that the area in bearing, A_b, is the rectangular area found by multiplying the plate thickness t by the diameter of the bolt D. This can be considered to be the *projected area* of the bolt hole. Then the bearing capacity of a joint is

$$F_b = \sigma_{ba} A_b \tag{13–3}$$

where F_b = capacity of the joint in bearing
σ_{ba} = allowable bearing stress
A_b = bearing area = $N_b D t$ **(13–4)**
N_b = number of bearing surfaces
t = plate thickness

The allowable bearing stress is typically based on the yield strength of the connected material because the fastener usually has higher strength. This should be checked.

Tensile Failure. A direct tensile force applied through the centroid of the bolt pattern would produce a tensile stress. Then the capacity of the joint in tension would be

$$F_t = \sigma_{ta} A_t \qquad \textbf{(13–5)}$$

where F_t = capacity of the joint in tension
σ_{ta} = allowable stress in tension
A_t = net tensile area

The evaluation of A_t requires the subtraction of the diameter of all the holes from the width of the plates being joined. Then

$$A_t = (w - ND_H)t \qquad \textbf{(13–6)}$$

where w = width of the plate
D_H = hole diameter (in structures use $D_H = D + \frac{1}{16}$ in or $D + 1.6$ mm)
N = number of holes at the section of interest
t = thickness of the plate

Example Problem 13–1 For the single lap bearing-type joint shown in Figure 13–3, determine the allowable load on the joint if the two plates are $\frac{1}{4}$ in thick by 2 in wide and joined by two steel bolts, $\frac{3}{8}$ in diameter, ASTM A490. The plates are ASTM A36 structural steel.

Solution

Objective: Compute the allowable load on the joint.

Given: Plate thickness = t = 0.25 in; plate width = w = 2.00 in
Plates are ASTM A36 structural steel. s_u = 58 ksi, s_y = 36 ksi
Bolts: Diameter = D = 0.375 in; ASTM A490
Bearing type connection; no threads in the shear plane.

Analysis: Possible failure in shear, bearing, and tension will be investigated. The lowest of the three values is the limiting load on the joint.

Results: ***Shear Failure***

$$F_s = \tau_a A_s \qquad \textbf{(13–1)}$$

$$\tau_a = 40\,000 \text{ psi} \quad \text{(Table 16–1)}$$

$$A_s = \frac{N_s \pi D^2}{4} = \frac{2\pi(0.375 \text{ in})^2}{4} = 0.221 \text{ in}^2 \qquad \textbf{(13–2)}$$

Then

$$F_s = (37\,500 \text{ lb/in}^2)(0.221 \text{ in}^2) = 8288 \text{ lb}$$

Bearing Failure

$$F_b = \sigma_{ba} A_b \qquad \textbf{(13–3)}$$

$$\sigma_{ba} = 1.20(58\,000 \text{ psi}) = 69\,600 \text{ psi}$$

$$A_b = N_b D t = (2)(0.375 \text{ in})(0.25 \text{ in}) = 0.188 \text{ in}^2 \qquad \textbf{(13–4)}$$

Then

$$F_b = (69\,600\ \text{lb/in}^2)(0.188\ \text{in}^2) = 13\,050\ \text{lb}$$

Tensile Failure

$$F_t = \sigma_{ta} A_t \quad \textbf{(13–5)}$$

$$\sigma_{ta} = 0.6(36\,000\ \text{psi}) = 21\,600\ \text{psi}$$

$$A_t = (W - ND_H)t = [2.0\ \text{in} - 2(0.375 + 0.063)\ \text{in}](0.25\ \text{in}) = 0.281\ \text{in}^2 \quad \textbf{(13–6)}$$

Then

$$F_t = (21\,600\ \text{lb/in}^2)(0.281\ \text{in}^2) = 6070\ \text{lb}$$

Comment Now the capacity in tension is the lowest, so the capacity of the joint is 6070 lb.

13–4 RIVETED JOINTS

Riveted joints are similar to those shown in Figure 13–3 except that the fastener takes one of the forms shown in Figure 13–5. The cylindrical body of the rivet is inserted into the holes in the members to be connected. With the preformed head held firmly from one side of the joint, the opposite end is pressed or hammered to form a head on the opposite side, clamping the members together. The holes for the rivets are typically quite close in size to the body diameter and the upsetting action during installation causes the body to swell, filling the hole and preventing relative motion between the connected members and the rivets. The blind rivet is unique in that the body part is inserted into the holes from one side and a special tool pulls on the mandrel, causing the head on the opposite side to spread out and grip the connected members. After forming, the mandrel snaps off and is discarded. The advantages of this technique are that it requires access to only one side of the connected members and the complete installation is quite rapid.

The basic analysis methods for riveted joints are similar to those described for bolted joints. Shear, bearing, and tensile failure modes must be analyzed. Equations (13–1) to (13–6) can be applied as illustrated in Example Problem 13–1. However, for tensile failure of the connected members, the hole diameter is taken to be equal to the body diameter of the rivet.

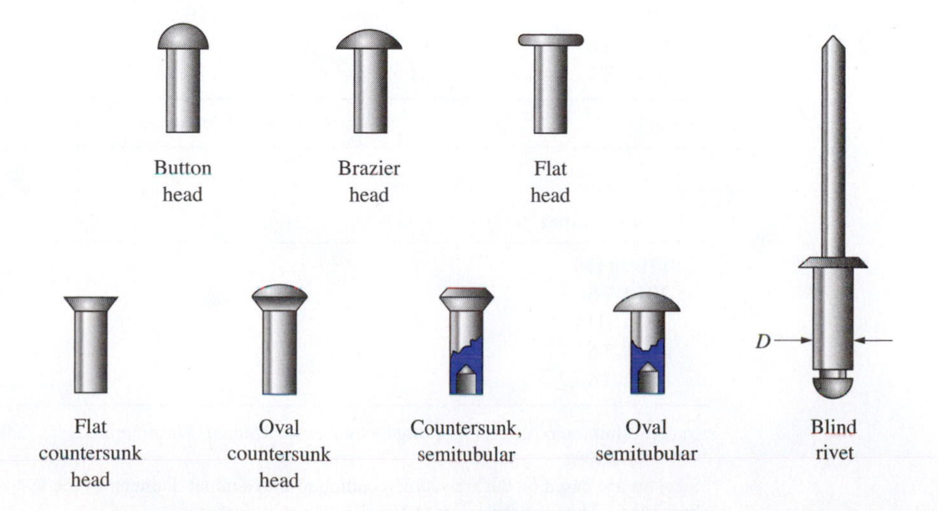

FIGURE 13–5 Examples of rivet styles.

TABLE 13–2 Allowable shear force capacity of typical rivets [Force in lb and (N)].

Body diameter [in (mm)]	Material			
	Aluminum	Carbon steel	Stainless steel	MONEL®
3/32 (2.38)	70 (311)	130 (578)	230 (1023)	200 (890)
1/8 (3.18)	120 (534)	260 (1156)	420 (1868)	350 (1557)
5/32 (3.99)	190 (845)	370 (1646)	650 (2891)	550 (2446)
3/16 (4.76)	260 (1156)	540 (2402)	950 (4226)	800 (3558)
1/4 (6.35)	460 (2046)	700 (3114)	1700 (7562)	1400 (6227)

Table 13–2 shows representative data for the allowable shear force capacity of rivets made from aluminum, carbon steel, stainless steel, and MONEL® (a nickel alloy with excellent corrosion resistance often used in marine and chemical plant applications; MONEL is a registered trademark of Special Metals Corporation). These data will be used in this book. However, strength data from specific suppliers should be sought for critical designs. See Internet site 13. Table 13–3 shows additional data for aluminum rivets, bolts, and connected members.

TABLE 13–3 Allowable stresses for aluminum structural connections for building-type structures.

Rivets

Alloy and temper before driving*	Allowable shear stress	
	ksi	MPa
1100-H14	4.1	28
2017-T4	14	96
6053-T61	8.5	58
6061-T6	10.5	72

Bolts

Alloy and temper	Allowable shears stress†		Allowable tensile stress†	
	ksi	MPa	ksi	MPa
2024-T4	16	110	26	179
6061-T6	11	76	18	124
7075-T73	18	124	29	200

Connected Members

Alloy and temper	Allowable bearing stress	
	ksi	MPa
1100-H14	16	110
2014-T6	62	427
3003-H14	21	145
6061-T6	39	269
6063-T6	31	214

Source: Aluminum Association, *Aluminum Design Manual.* Washington, D.C., 2000.

*All cold-driven.

†Stresses are based on the area corresponding to the nominal diameter of the bolt unless the threads are in the shear plane. Then the shear area is based on the root diameter.

13–5 ECCENTRICALLY LOADED RIVETED AND BOLTED JOINTS

Previously considered joints were restricted to cases in which the line of action of the load on the joint passed through the centroid of the pattern of rivets or bolts. In such cases, the applied load is divided equally among all the fasteners. When the load does not pass through the centroid of the fastener pattern, it is called an *eccentrically loaded joint,* and a nonuniform distribution of forces occurs in the fasteners.

In eccentrically loaded joints, the effect of the moment or couple on the fastener must be considered. Figure 13–6 shows a bracket attached to the side of a column and used to support an electric motor. The net downward force exerted by the weight of the motor and the belt tension act at a distance a from the center of the column flange. Then the total force system acting on the bolts of the bracket consists of the direct shearing force P plus the forces produced by the moment $P \times a$. Each of these components can be considered separately and then added together by using the principle of superposition.

Figure 13–7(a) shows that for the direct shearing force P, each bolt is assumed to carry an equal share of the load, just as in concentrically loaded joints. But in part (b) of the figure, because of the moment, each bolt is subjected to a force acting perpendicular to a radial line from the centroid to the bolt pattern. It is assumed that the magnitude of the force in a bolt due to the moment load is proportional to its distance r from the centroid. This magnitude is

$$R_i = \frac{Mr_i}{\sum r^2} \tag{13–7}$$

where R_i = shearing force in bolt i due to the moment M
r_i = radial distance to bolt i from the centroid to the bolt pattern
$\sum r^2$ = sum of the radial distances to *all* bolts in the pattern squared

If it is more convenient to work with horizontal and vertical components of forces, they can be found from

$$R_{ix} = \frac{My_i}{\sum r^2} = \frac{My_i}{\sum(x^2 + y^2)} \tag{13–8}$$

$$R_{iy} = \frac{Mx_i}{\sum r^2} = \frac{Mx_i}{\sum(x^2 + y^2)} \tag{13–9}$$

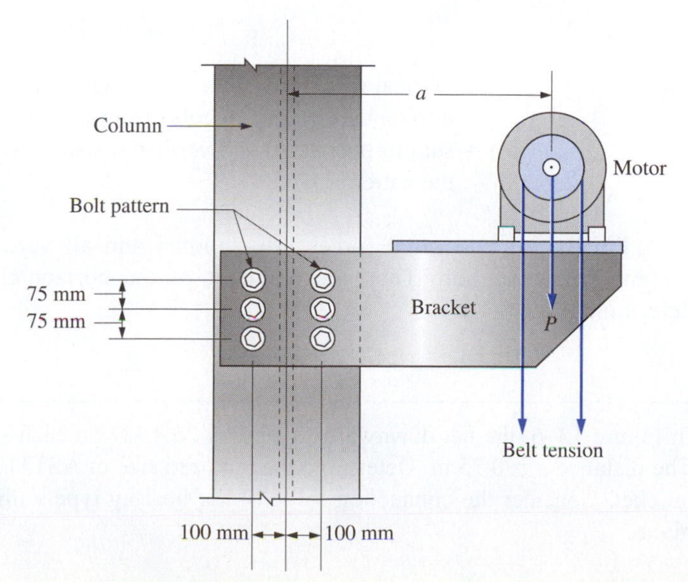

FIGURE 13–6 Eccentric load on a bolted joint.

FIGURE 13–7 Loads on bolts that are eccentrically loaded.

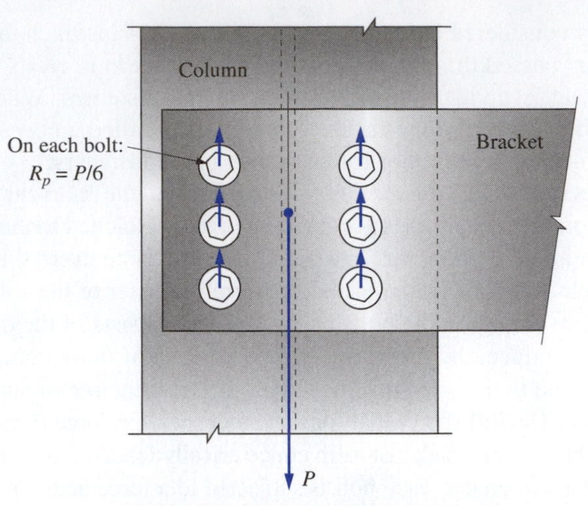

(*a*) Forces resisting *P*, the shearing force

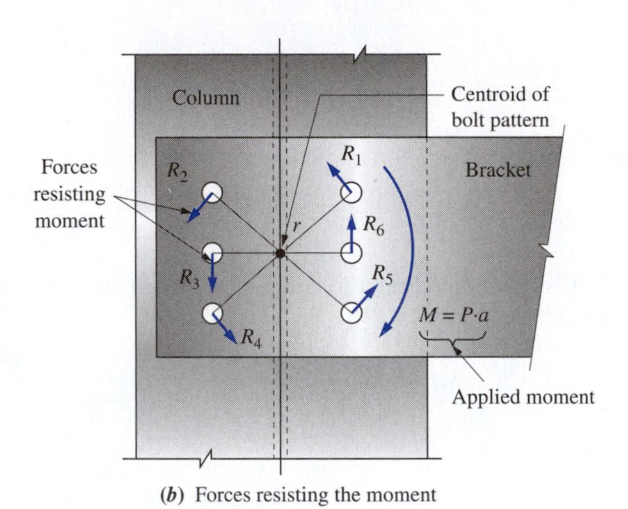

(*b*) Forces resisting the moment

where

y_i = vertical distance to bolt *i* from centroid

x_i = horizontal distance to bolt *i* from centroid

$\Sigma(x^2 + y^2)$ = sum of horizontal and vertical distances squared for all bolts in the pattern

Finally, all horizontal forces are summed and all vertical forces are summed for any particular bolt. Then the resultant of the horizontal and vertical forces is determined.

Example Problem 13–2 In Figure 13–6, the net downward force *P* is 26.4 kN on each side plate of the bracket. The distance *a* is 0.75 m. Determine the required size of ASTM A325 bolts to secure the bracket. Consider the connection to be of the bearing type with no threads in the shear plane.

Solution Objective Specify the size of the bolts in the joint.

Given Load = P = 26.4 kN downward. Moment arm = a = 0.75 m.
Bolt pattern in Figure 13–6. Bolts: ASTM A325
Bearing-type connection.

Analysis To determine the shearing force in each bolt to carry the direct vertical shearing force of $P = 26.4$ kN, each of the six bolts will be assumed to carry an equal share of the load. Then Equations (13–8) and (13–9) will be used to compute the forces in the most highly stressed bolt to resist the moment load, where

$$M = P \times a$$

The resulting forces will be combined vectorially to determine the resultant load on the most highly stressed bolt. Then the required size of that bolt will be computed based on the allowable shearing stress for the ASTM A325 bolts.

Results ***Direct Shearing Force***
The total downward shearing force is divided among six bolts. Then the load on each bolt, called R_p, is

$$R_p = \frac{P}{6} = \frac{26.4\text{ kN}}{6} = 4.4\text{ kN}$$

Figure 13–7(a) shows this to be an upward reaction force on each bolt.

Forces Resisting the Moment. In Equations (13–8) and (13–9), the following term is required:

$$\sum(x^2 + y^2) = 6(100\text{ mm})^2 + 4(75\text{ mm})^2 = 82\,500\text{ mm}^2$$

The moment on the joint is

$$M = P \times a = 26.4\text{ kN}\,(0.75\text{ m}) = 19.8\text{ kN}\cdot\text{m}$$

Starting first with bolt 1 at the upper right (see Figure 13–8),

$$R_{1x} = \frac{My_1}{\sum(x^2 + y^2)} = \frac{19.8\text{ kN}\cdot\text{m}(75\text{ mm})}{82\,500\text{ mm}^2} \times \frac{10^3\text{ mm}}{\text{m}}$$

$$R_{1x} = 18.0\text{ kN} \leftarrow \text{(acts toward the left)}$$

$$R_{1y} = \frac{Mx_1}{\sum(x^2 + y^2)} = \frac{(19.8\text{ kN}\cdot\text{m})(100\text{ mm})}{82\,500\text{ mm}^2} \times \frac{10^3\text{ mm}}{\text{m}}$$

$$R_{1y} = 24.0\text{ kN} \uparrow \text{(acts upward)}$$

Now the resultant of these forces can be found. In the vertical direction, R_p and R_{1y} both act upward.

$$R_p + R_{1y} = 4.4\text{ kN} + 24.0\text{ kN} = 28.4\text{ kN}$$

Only R_{1x} acts in the horizontal direction. Calling the total resultant force on bolt 1, R_{t1}.

$$R_{t1} = \sqrt{28.4^2 + 18.0^2} = 33.6\text{ kN}$$

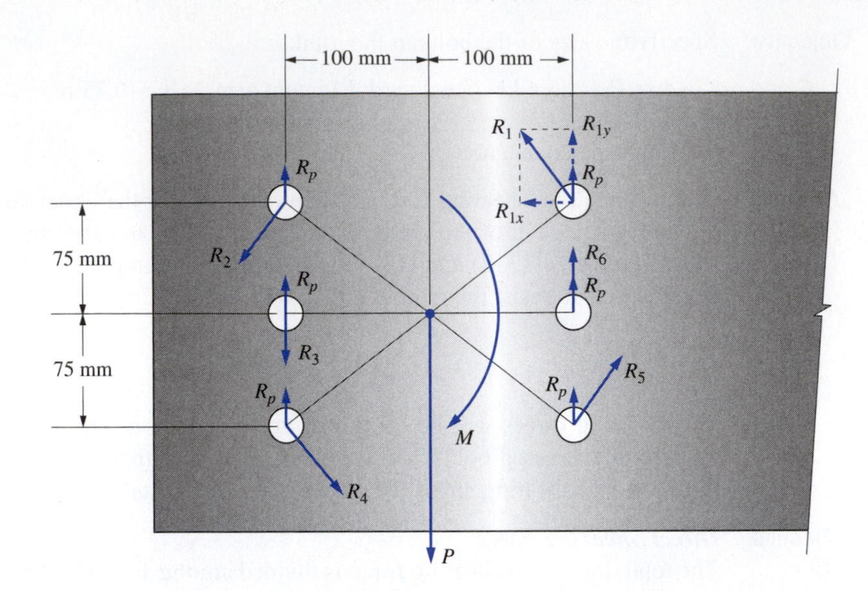

FIGURE 13–8
Forces on each bolt.

Investigating the other five bolts in a similar manner would show that bolt 1 is the most highly stressed. Then its diameter will be determined to limit the shear stress to 207 MPa (30.0 ksi) for the ASTM A325 bolts, as listed in Table 13–1.

$$\tau = \frac{R_{t1}}{A}$$

$$A = \frac{R_{t1}}{\tau_a} = \frac{33.6\text{ kN}}{207\text{ N/mm}^2} = 162\text{ mm}^2 = \frac{\pi D^2}{4}$$

$$D = \sqrt{\frac{4A}{\pi}} = \sqrt{\frac{4(162)\text{ mm}^2}{\pi}} = 14.4\text{ mm}$$

The nearest preferred metric size is 16 mm. If conventional inch units are to be specified,

$$D = 14.4\text{ mm} \times \frac{1\text{ in}}{25.4\text{ mm}} = 0.567\text{ in}$$

The nearest standard size is $\frac{5}{8}$ in (0.625 in).

Comment Specify either $D = 16$ mm or $D = \frac{5}{8}$ in bolts.

13–6 WELDED JOINTS WITH CONCENTRIC LOADS

Welding is a joining process in which heat is applied to cause two pieces of metal to become metallurgically bonded. The heat may be applied by a gas flame, an electric arc, a laser beam or by a combination of electric resistance heating and pressure.

Types of welds include butt, groove, fillet, and spot welds (as shown in Figure 13–9), and others. Groove and fillet welds are most frequently used in structural connections since they are readily adaptable to the shapes and plates that make up the structures. Spot welds are used for joining relatively light gauge steel sheets and cold-formed shapes.

The variables involved in designing welded joints are the shape and size of the weld, the choice of filler metal, the length of weld, and the position of the weld relative to the applied load.

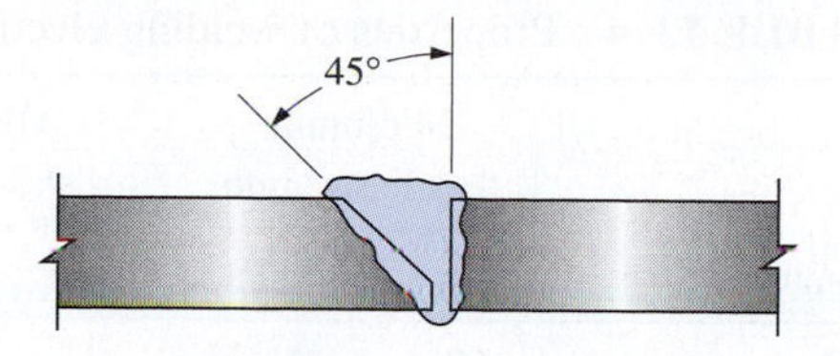

(*a*) Butt weld with single bevel groove

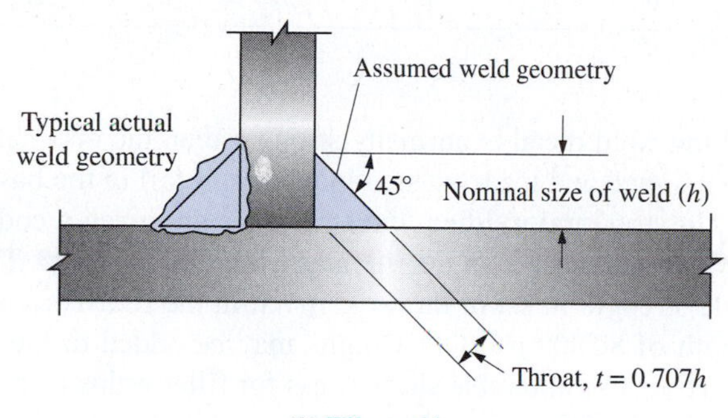

(*b*) Fillet weld

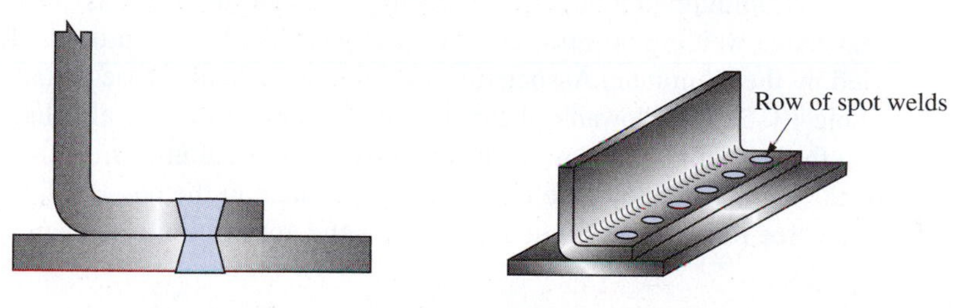

(*c*) Spot weld

FIGURE 13–9 Types of welds.

Fillet welds are assumed to have a slope of 45 degrees between the two surfaces joined, as shown in Figure 13–9(b). The size of the weld is denoted as the height of one side of the triangular-shaped fillet. Typical sizes range from $\frac{1}{8}$ in to $\frac{1}{2}$ in in steps of $\frac{1}{16}$ in. The stress developed in fillet welds is assumed to be *shear stress* regardless of the direction of application of the load. The maximum shear stress would occur at the throat of the fillet (see Figure 13–9), where the thickness is 0.707 times the nominal size of the weld. Then the shearing stress in the weld due to a load P is

$$\tau = \frac{P}{Lt} \quad \textbf{(13–10)}$$

where L = the length of weld
t = the thickness at the throat

Equation (13–10) is used *only* for concentrically loaded members. This requires that the line of action of the force on the welds passes through the centroid of the weld pattern. Eccentricity of the load produces a moment, in addition to the direct shearing force, which must be resisted by the weld metal. References 2, 3, 6, 11, and 15 at the end of this chapter contain pertinent information with regard to eccentrically loaded welded joints.

In electric arc welding, used mostly for structural connections, a filler rod is normally used to add metal to the welded zone. As the two parts to be joined are heated to a molten state, filler metal is added, which combines with the base metal. On cooling, the

TABLE 13–4 Properties of welding electrodes for steel.

Electrode type	Minimum tensile strength		Allowable shear stress		Typical metals joined
	ksi	MPa	ksi	MPa	
E60	60	414	18	124	A36, A500
E70	70	483	21	145	A572 Gr. 50 A913 Gr. 50, A992
E80	80	552	24	165	A913 Gr. 65

resulting weld metal is normally stronger than the original base metal. Therefore, a properly designed and made welded joint should fail in the base metal rather than in the weld.

In structural welding, the electrodes are given a code beginning with an E and followed by two or three digits, such as E60, E80, or E100. The number denotes the ultimate tensile strength in ksi of the weld metal in the rod. Thus an E80 rod would have a tensile strength of 80 000 psi. Other digits may be added to the code number to denote special properties. The allowable shear stress for fillet welds using electrodes is 0.3 times the tensile strength of the electrode according to AISC. Table 13–4 lists some common electrodes and their allowable stresses.

Aluminum products are welded using either the inert gas-shielded arc process or the resistance welding process. For the inert gas-shielded arc process, filler alloys are specified by the Aluminum Association for joining particular base metal alloys, as indicated in Table 13–5. The allowable shear stresses for such welds are also listed. It should be noted that the heat of welding lowers the properties of most aluminum alloys within 1.0 in (25 mm) of the weld, and allowance for this must be made in the design of welded assemblies.

See Reference 4 for additional data and application considerations for welded joints.

Example Problem 13–3 A lap joint is made by placing two $\frac{3}{8}$-in fillet welds across the full width of two $\frac{1}{2}$-in ASTM A36 steel plates, as shown in Figure 13–10. The shielded metal-arc method is used, using an E60 electrode. Compute the allowable load, P, that can be applied to the joint.

Solution **Objective** Compute the allowable load, P, on the joint.

Given Joint design in Figure 13–10. Plates are ASTM A36 steel.
E60 electrode used in shielded metal-arc method.

TABLE 13–5 Allowable shear stresses in fillet welds in aluminum building-type structures.

Base metal	Filler alloy 1100		4043		5356		5556	
	ksi	MPa	ksi	MPa	ksi	MPa	ksi	MPa
1100	3.2	22	4.8	33	—	—	—	—
3003	3.2	22	5.0	34	—	—	—	—
6061	—	—	5.0	34	7.0	48	8.5	59
6063	—	—	5.0	34	6.5	45	6.5	45

FIGURE 13–10
Welded lap joint.

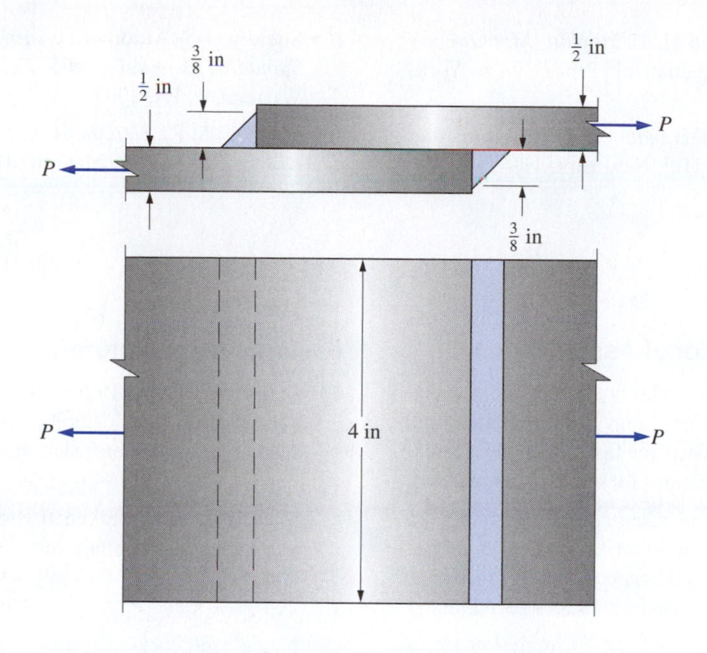

Analysis The load is assumed to be equally distributed on all parts of the weld, so Equation (13–10) can be used with $L = 8.0$ in.

$$\tau = \frac{P}{Lt}$$

Let τ equal the allowable stress of 18 ksi, listed in Table 13–4. The thickness t is

$$t = 0.707(\tfrac{3}{8}\text{ in}) = 0.265\text{ in}$$

Now we can solve for P.

$$P = \tau_a L t = (18\,000\text{ lb/in}^2)(8.0\text{ in})(0.265\text{ in}) = 38\,200\text{ lb}$$

REFERENCES

1. Aluminum Association, *Aluminum Design Manual,* Washington, DC, 2005.
2. American Institute of Steel Construction, *Specification for Structural Steel Buildings,* AISC, Chicago, IL, 2005.
3. American Institute of Steel Construction, *Allowable Stress Design Specification for Structural Joints Using ASTM A325 or A490 Bolts,* American Institute of Steel Construction, New York, 2001.
4. American Welding Society, *Standard AWS D1.1/D1.1M Structural Welding Code Steel,* American Welding Society, Miami, FL, 2006.
5. Bickford, J. H. *An Introduction to the Design and Behavior of Bolted Joints,* 3rd ed., Marcel Dekker, New York, 1995.
6. Blodgett, O. W., *Design of Weldments,* James F. Lincoln Arc Welding Foundation, Cleveland, OH, 1963.
7. Brockenbrough, R. L. and F. S. Merritt, *Structural Steel Designer's Handbook,* McGraw-Hill, New York, 2005.
8. Industrial Fasteners Institute, *Fastener Standards,* 7th ed., Industrial Fasteners Institute, Cleveland, OH, 2003.
9. Kissell, J. R. and R. Ferry, *Aluminum Structures: A Guide to Their Specifications and Design,* John Wiley & Sons, New York, 2002.
10. McCormac, J. C. and J. Nelson, *Structural Steel Design—LRED Method,* 4th ed., Prentice Hall, Upper Saddle River, NJ, 2007.
11. Mott, R. L., *Machine Elements in Mechanical Design,* 4th ed., Prentice Hall, Upper Saddle River, NJ, 2004.

12. Oberg E., F. D. Jones, and H. L. Horton, *Machinery's Handbook,* 27th ed., Industrial Press, New York, 2004.
13. Parmley, R. O., *Standard Handbook of Fastening and Joining,* 3rd ed., McGraw-Hill, New York, 1997.
14. Society of Automotive Engineers, *SAE Fastener Standards Manual—2005 Edition,* SAE International, Warrendale, PA, 2005.
15. Tamboli, A. R., *Handbook of Structural Steel Connection Design and Details,* McGraw-Hill, New York, 2000.

INTERNET SITES

Standards and Professional Associations

1. Aluminum Association www.aluminum.org The association of the aluminum industry. Source of information and documents on aluminum used for fasteners and information on the design of connections for aluminum structures.
2. American Institute of Steel Construction www.aisc.org Publisher of the *Specifications for Structural Steel Buildings* that contains extensive data on connections and procedures for designing and fabricating connections for steel buildings.
3. ASTM International www.astm.org Formerly known as the American Society for Testing and Materials. Developer of numerous standards for fasteners used in the building construction industry.
4. Industrial Fasteners Institute www.industrial-fasteners.org An association of manufacturers and suppliers of bolts, nuts, screws, rivets, and special formed parts and the materials and equipment to make them. The site includes standards, reference publications, and educational opportunities.
5. Research Council on Structural Connections (RCSC) www.boltcouncil.org An organization that stimulates and supports research on structural connections and prepares and publishes standards.
6. SAE International www.sae.org The Society of Automotive Engineers, the engineering society for advancing mobility on land or sea, in air or space. Publisher of the *SAE Handbook* and the *SAE Fastener Manual,* each containing useful information about the use of fasteners and their properties.
7. NASA Ames Research Center (ARC) www.windtunnels.arc.nasa.gov/strucjo.html Part of a larger site for ARC that lists standards for structural joints for any structure or equipment supplied for use in their wind tunnels. The list is a good overview of the design and fabrication considerations important to successful structural joints.

Software for Joint Design

8. Bolt Science www.boltscience.com Producer of software for the analysis of bolted joints. The site includes an informative *Tutorial on the Basics of Bolted Joints.*
9. Sensor Products, Inc. www.sensorproducts.com Developer of the BoltFAST software for bolted joints that includes joint analysis, thread analysis, and tightening torque.

Fastener Manufacturers

10. St. Louis Screw & Bolt Company www.stlouisscrewbolt.com Manufacturer of bolts, nuts, and washers to ASTM standards for the construction industry.
11. Nucor Fastener Division www.nucor-fastener.com Manufacturer of hex head cap screws in SAE, ASTM, and metric grades; hex nuts; and structural bolts, nuts, and washers.
12. Nylok Corporation www.nylok.com Manufacturer of Nylok® self-locking fasteners for automotive, aerospace, consumer products, agricultural, industrial, furniture, and many other applications.
13. The Fastener Group www.fastenergroup.com A supplier of bolts, cap screws, nuts, rivets, and numerous other types of fasteners for general industry uses.
14. SPS Technologies, Inc. www.spstech.com/unbrako Manufacturer of engineered fasteners under the Unbrako®, Flexloc®, and Durlok® brands, including socket head cap screws, locknuts, and vibration resistant nuts and bolts for industrial machinery, automotive, and aerospace applications. The site includes catalogs and engineering data.

Welding Technology and Standards

15. American Welding Society www.aws.org Publisher of the Standard AWS D1.1/D1.1M *Structural Welding Code Steel* and many other publications related to the design of welded structures, the process of welding, and the welding industry.
16. James F. Lincoln Foundation www.jflf.org An organization that promotes education and training in welding technology. The site includes much information about welding processes, joint design, and guides to welded steel construction.
17. Miller Electric Company www.millerwelds.com A manufacturer of welding equipment and accessories. The site includes a Training/Education section with information about welding processes.
18. Hobart Institute of Welding Technology www.welding.org An educational organization that provides instruction in the performance of welding techniques. The site includes welding tips and a glossary of welding terms.

PROBLEMS

13–1. Determine the allowable loads on the joints shown in Figure P13–1. The fasteners are all carbon steel rivets having the shearing force capacity shown in Table 13–2. The plates are ASTM A36 structural steel.

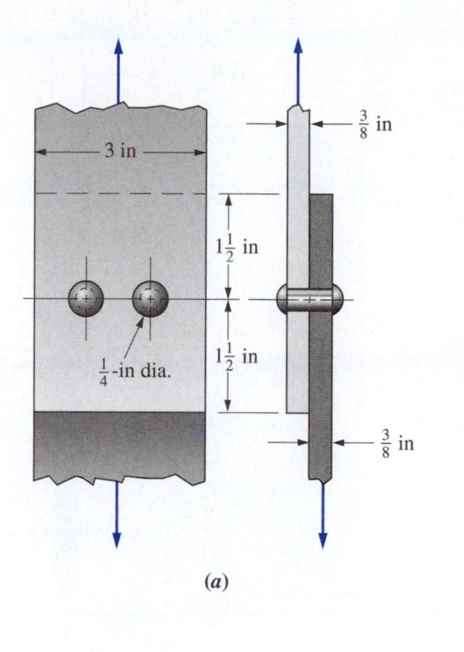

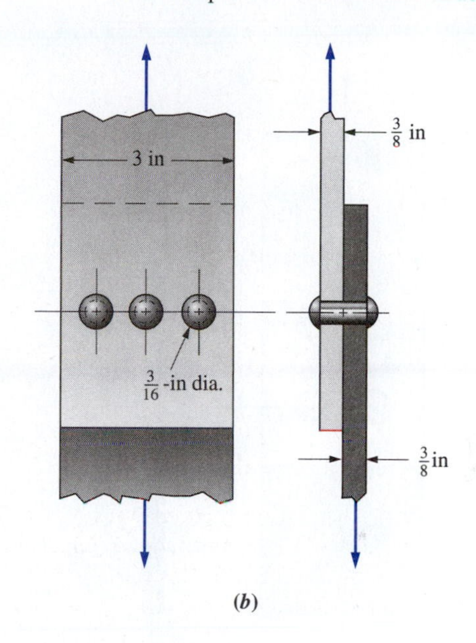

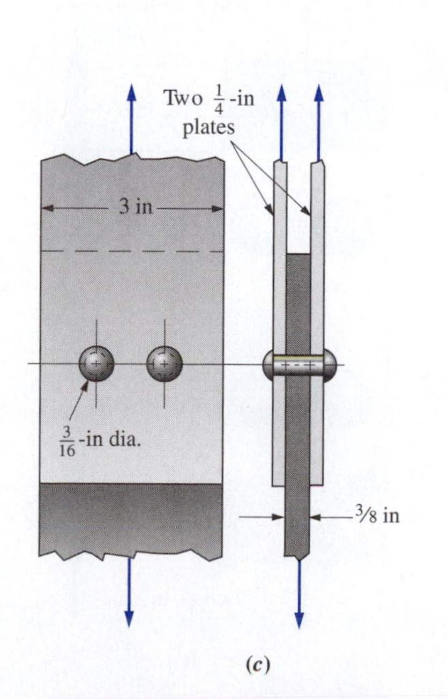

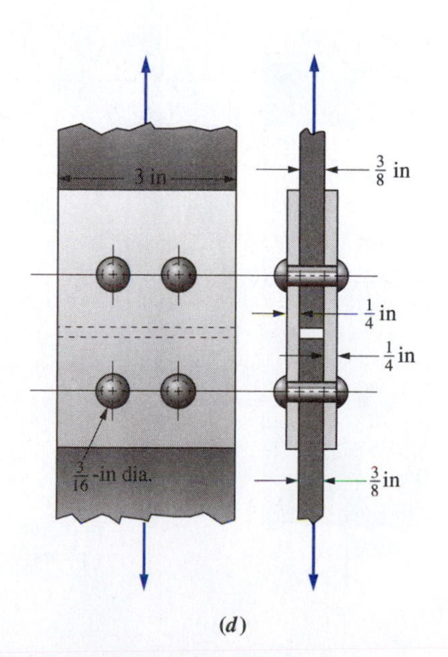

FIGURE P13–1 Joints for Problems 13–1 and 13–3.

13–2. Determine the allowable loads on the joints shown in Figure P13–2. The fasteners are all stainless steel rivets having the shearing force capacity shown in Table 13–2. The plates are AISI 430 stainless steel in the full hard condition.

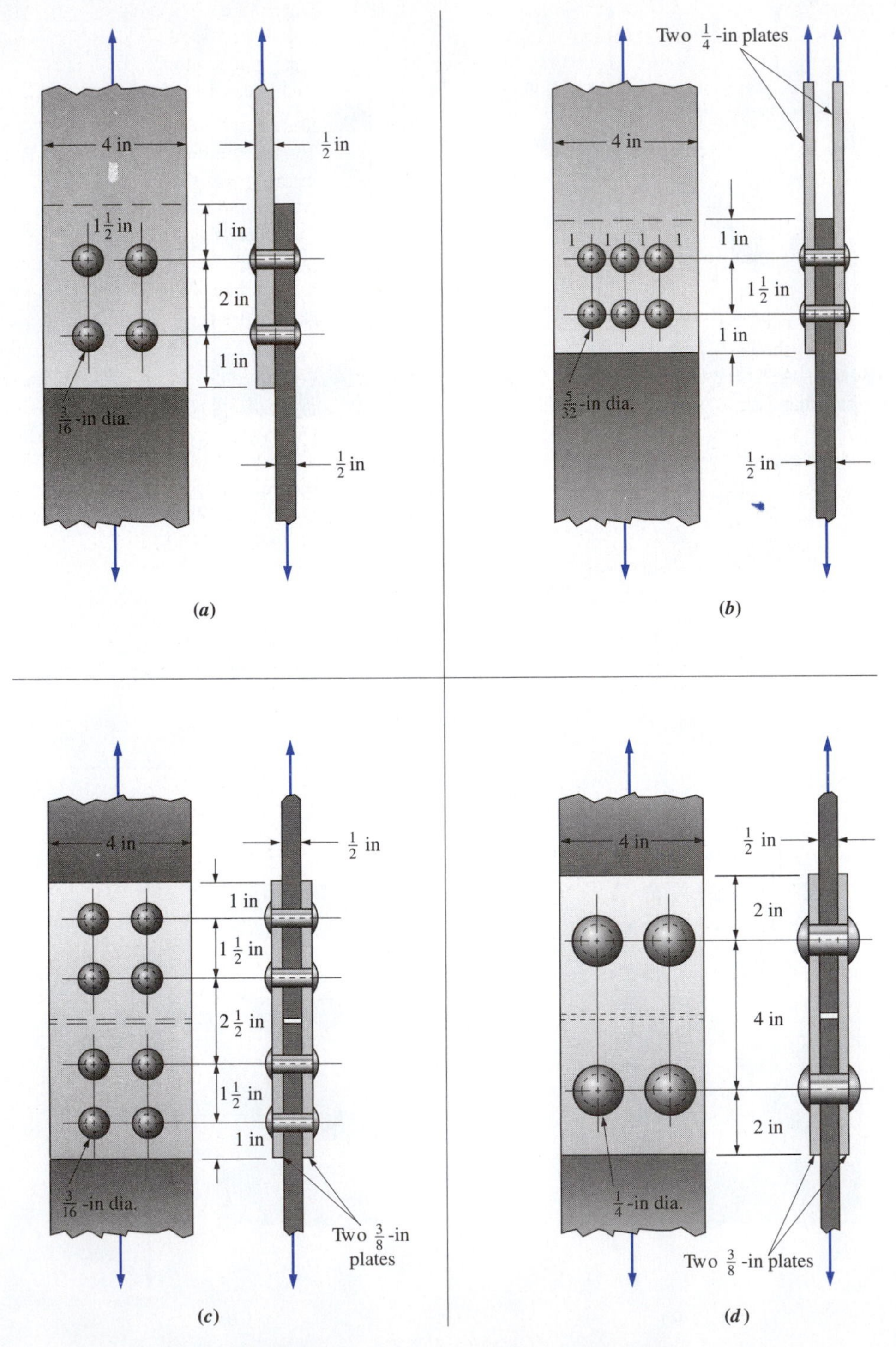

FIGURE P13–2 Joints for Problems 13–2 and 13–4.

13–3. Determine the allowable loads on the joints shown in Figure P13–1 if the fasteners are all ASTM A307 steel bolts providing a bearing-type connection. The plates are ASTM A242 HSLA corrosion resistant structural steel.

13–4. Determine the allowable loads on the joints shown in Figure P13–2 if the fasteners are all ASTM A325 high strength steel bolts providing a bearing-type connection. The plates are ASTM A514 quenched and tempered alloy structural steel.

13–5. Determine the required diameter of the bolts used to attach the cantilever beam to the column as shown in Figure P13–5. Use ASTM A325 bolts. The steel plate is made from ASTM A36 structural steel and the column is made from ASTM A992 structural steel.

13–6. Design the connection of the channel to the column for the bracket shown in Figure P13–6. The channel is made from ASTM A36 structural steel and the column is made from ASTM A992 structural steel. Specify the bolt material, the number of bolts, the bolt pattern (location and spacing), and the size of the bolts. Use the data from Table 13–1.

13–7.E For the connection shown in Figure P13–1(a), assume that, instead of the two rivets, the two plates were welded across the ends of the 3-in-wide plates using $\frac{5}{16}$-in welds. The plates are ASTM A36 steel and the electric arc welding technique is used with E60 electrodes. Determine the allowable load on the connection.

13–8.E Determine the allowable load on the joint shown in Figure P13–2(c) if $\frac{1}{4}$-in welds using E70 electrodes were placed along both ends of both cover plates. The plates are ASTM A572 grade 50 steel.

13–9.M Design the joint at the top of the straps in Figure 13–1 if the total load in the hopper is 15.0 megagrams (Mg). The beam is a WT12×34 made of ASTM A36 steel and has a web thickness of 10.6 mm. The clear vertical height of the web is about 275 mm. Use steel rivets and specify the pattern, number of rivets, rivet diameter, rivet material, strap material, and strap dimensions. Use data from Table 13–2.

13–10.M Design the joint at the bottom of the straps in Figure 13–1 if the total load in the hopper is 15.0 Mg. Use steel bolts and a bearing-type connection. Specify the pattern, number of bolts, bolt diameter, bolt material, strap material, and strap dimensions. You may want to coordinate the strap design with the results of Problem 13–9. The design of the tab in

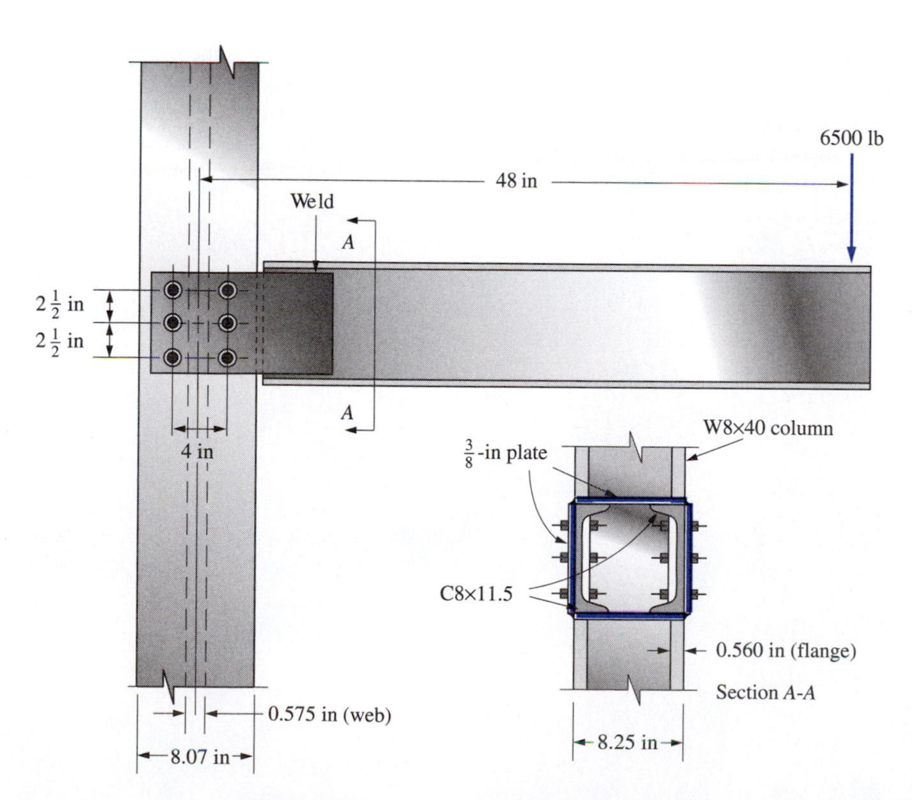

FIGURE P13–5 Connection for Problem 13–5.

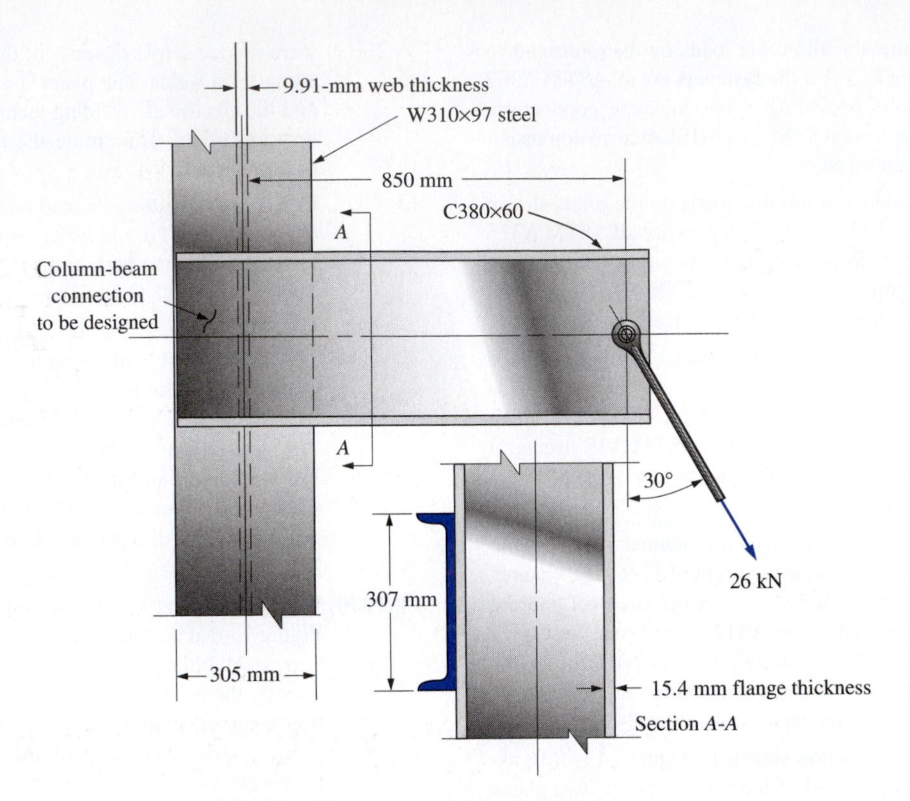

FIGURE P13–6 Connection for Problem 13–6.

Problem 13–11 is also affected by the design of the bolted joint.

13–11.M Design the tab to be welded to the hopper for connection to the support straps, as shown in Figure 13–1. The hopper load is 15.0 Mg. The material from which the hopper is made is ASTM A36 steel. Specify the width and thickness of the tab and the design of the welded joint. You may want to coordinate the tab design with the bolted connection called for in Problem 13–10.

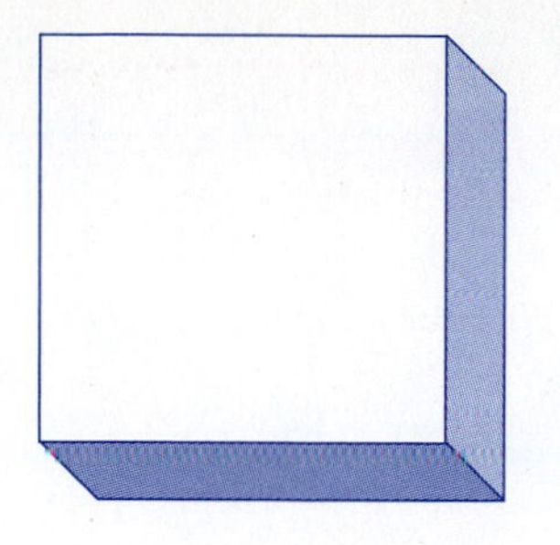

Appendix

List of Appendixes

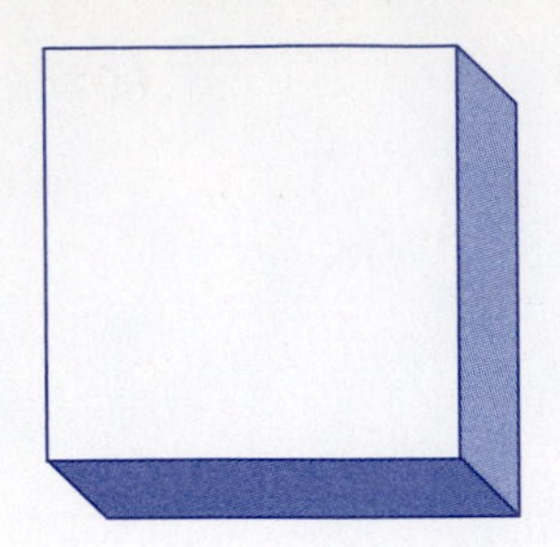

Appendix

A–1 Properties of areas.*

*Symbols used are:

A = area r = radius of gyration = $\sqrt{I/A}$
I = moment of inertia J = polar moment of inertia
S = section modulus Z_p = polar section modulus

Circle

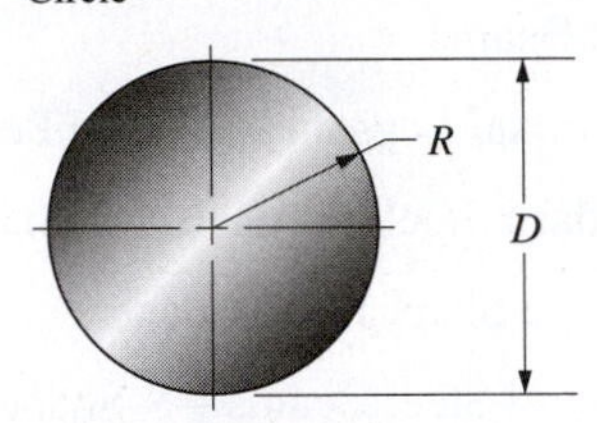

$$A = \frac{\pi D^2}{4} = \pi R^2 \qquad r = \frac{D}{4} = \frac{R}{2}$$

$$I = \frac{\pi D^4}{64} \qquad J = \frac{\pi D^4}{32}$$

$$S = \frac{\pi D^3}{32} \qquad Z_p = \frac{\pi D^3}{16}$$

Circumference = $\pi D = 2\pi R$

Hollow circle (tube)

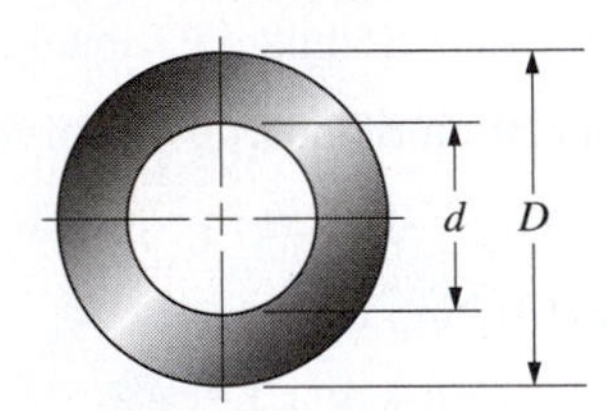

$$A = \frac{\pi(D^2 - d^2)}{4} \qquad r = \frac{\sqrt{D^2 + d^2}}{4}$$

$$I = \frac{\pi(D^4 - d^4)}{64} \qquad J = \frac{\pi(D^4 - d^4)}{32}$$

$$S = \frac{\pi(D^4 - d^4)}{32D} \qquad Z_p = \frac{\pi(D^4 - d^4)}{16D}$$

Square

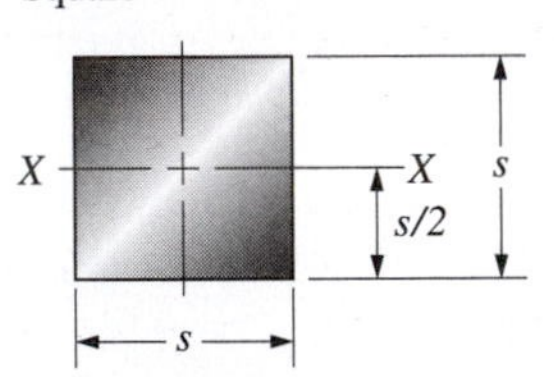

$$A = s^2$$

$$I_x = \frac{s^4}{12} \qquad r_x = \frac{s}{\sqrt{12}}$$

$$S_x = \frac{s^3}{6}$$

Rectangle

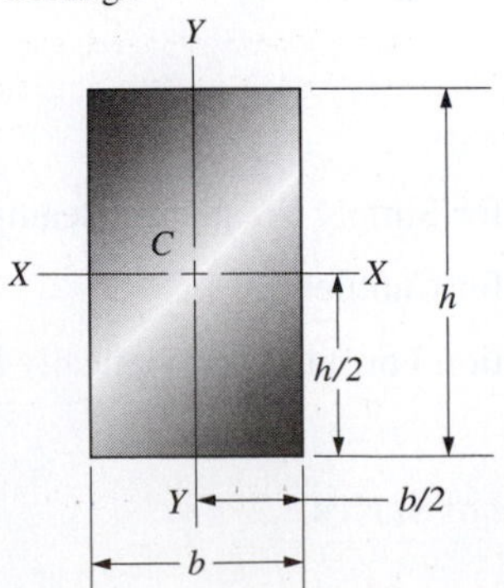

$$A = bh$$

$$I_x = \frac{bh^3}{12} \qquad S_x = \frac{bh^2}{6} \qquad r_x = \frac{h}{\sqrt{12}}$$

$$I_y = \frac{hb^3}{12} \qquad S_y = \frac{hb^2}{6} \qquad r_y = \frac{b}{\sqrt{12}}$$

Triangle

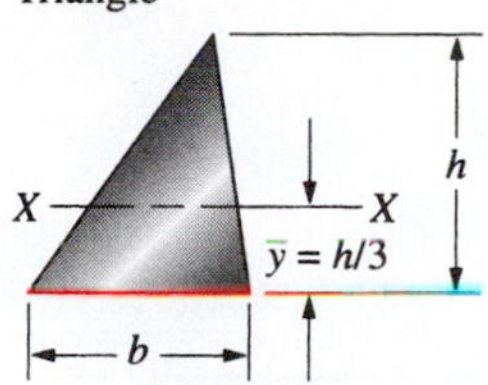

$$A = \frac{bh}{2}$$

$$I_x = \frac{bh^3}{36} \qquad r_x = \frac{h}{\sqrt{18}}$$

$$S_x = \frac{bh^2}{24}$$

Semicircle

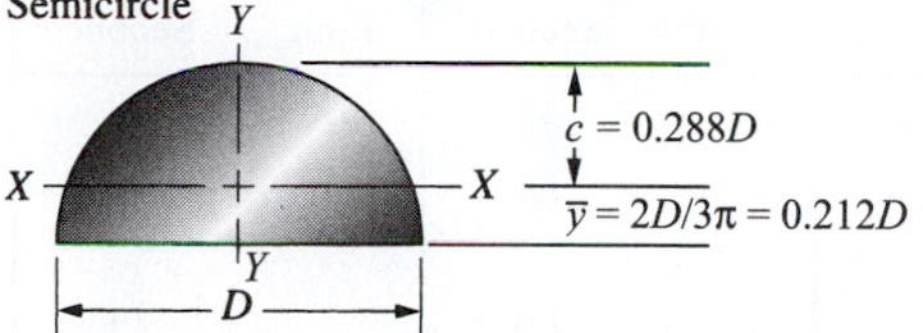

$$A = \frac{\pi D^2}{8}$$

$I_x = 0.00686D^4$ $\quad S_x = 0.0238D^3$ $\quad r_x = 0.132D$

$I_y = 0.0245D^4$ $\quad S_y = 0.0491D^3$ $\quad r_y = D/4 = R/2$

Regular hexagon

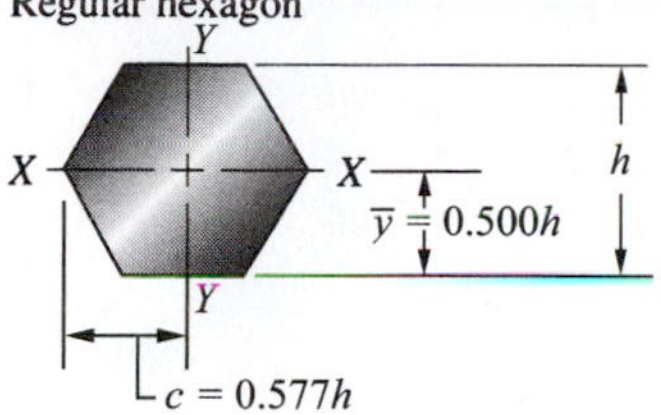

$A = 0.866h^2$

$I_x = 0.0601h^4$ $\quad S_x = 0.120h^3$ $\quad r_x = 0.264h$

$I_y = 0.0601h^4$ $\quad S_y = 0.104h^3$ $\quad r_y = 0.264h$

Area under a second-degree curve

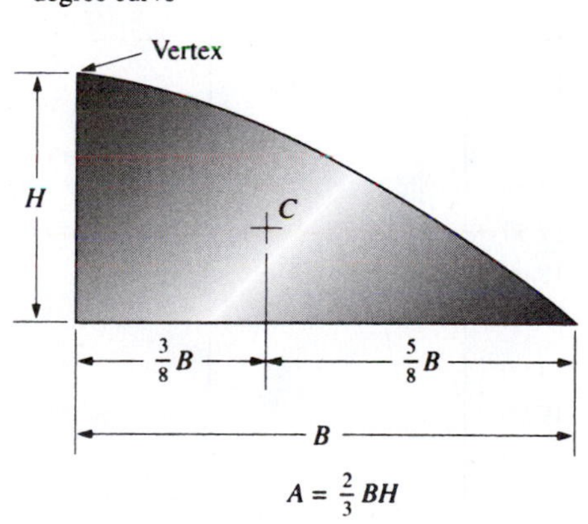

$A = \frac{2}{3}BH$

Area over a second-degree curve

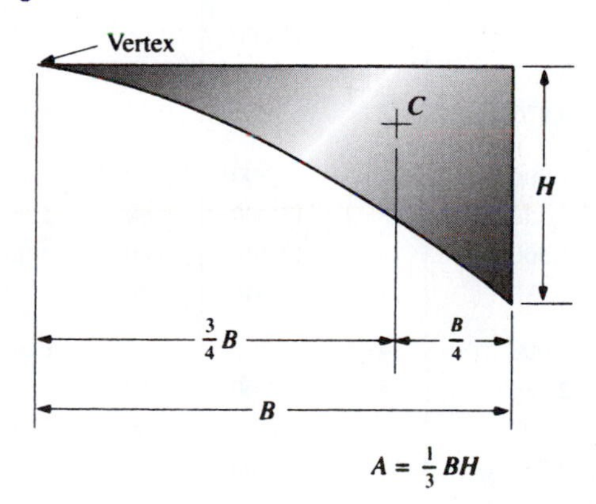

$A = \frac{1}{3}BH$

Area under a third-degree curve

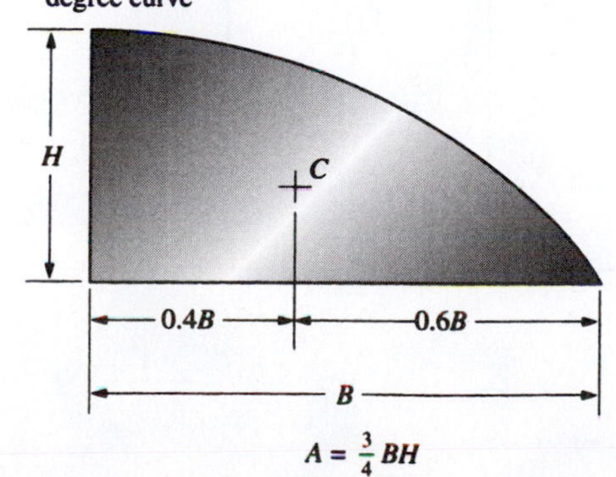

$A = \frac{3}{4}BH$

Area over a third-degree curve

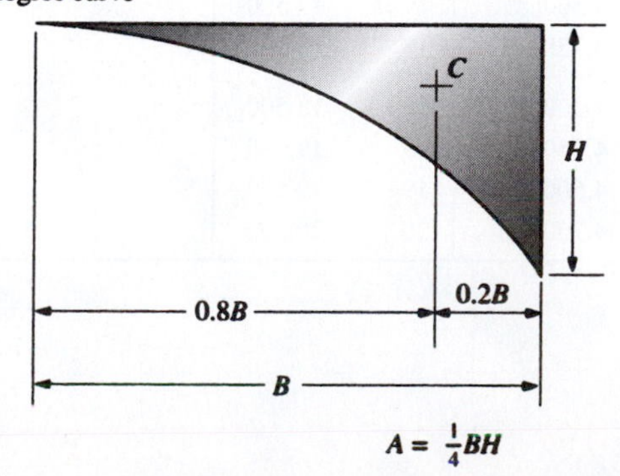

$A = \frac{1}{4}BH$

A–2 Preferred basic sizes.

Fractional (in)				Decimal (in)			Metric (mm)					
							First	Second	First	Second	First	Second
$\frac{1}{64}$	0.015 625	5	5.000	0.010	2.00	8.50	1		10		100	
$\frac{1}{32}$	0.031 25	$5\frac{1}{4}$	5.250	0.012	2.20	9.00		1.1		11		110
$\frac{1}{16}$	0.062 5	$5\frac{1}{2}$	5.500	0.016	2.40	9.50	1.2		12		120	
$\frac{3}{32}$	0.093 75	$5\frac{3}{4}$	5.750	0.020	2.60	10.00		1.4		14		140
$\frac{1}{8}$	0.125 0	6	6.000	0.025	2.80	10.50	1.6		16		160	
$\frac{5}{32}$	0.156 25	$6\frac{1}{2}$	6.500	0.032	3.00	11.00		1.8		18		180
$\frac{3}{16}$	0.187 5	7	7.000	0.040	3.20	11.50	2		20		200	
$\frac{1}{4}$	0.250 0	$7\frac{1}{2}$	7.500	0.05	3.40	12.00		2.2		22		220
$\frac{5}{16}$	0.312 5	8	8.000	0.06	3.60	12.50	2.5		25		250	
$\frac{3}{8}$	0.375 0	$8\frac{1}{2}$	8.500	0.08	3.80	13.00		2.8		28		280
$\frac{7}{16}$	0.437 5	9	9.000	0.10	4.00	13.50	3		30		300	
$\frac{1}{2}$	0.500 0	$9\frac{1}{2}$	9.500	0.12	4.20	14.00		3.5		35		350
$\frac{9}{16}$	0.562 5	10	10.000	0.16	4.40	14.50	4		40		400	
$\frac{5}{8}$	0.625 0	$10\frac{1}{2}$	10.500	0.20	4.60	15.00		4.5		45		450
$\frac{11}{16}$	0.687 5	11	11.000	0.24	4.80	15.50	5		50		500	
$\frac{3}{4}$	0.750 0	$11\frac{1}{2}$	11.500	0.30	5.00	16.00		5.5		55		550
$\frac{7}{8}$	0.875 0	12	12.000	0.40	5.20	16.50	6		60		600	
1	1.000	$12\frac{1}{2}$	12.500	0.50	5.40	17.00		7		70		700
$1\frac{1}{4}$	1.250	13	13.000	0.60	5.60	17.50	8		80		800	
$1\frac{1}{2}$	1.500	$13\frac{1}{2}$	13.500	0.80	5.80	18.00		9		90		900
$1\frac{3}{4}$	1.750	14	14.000	1.00	6.00	18.50					1000	
2	2.000	$14\frac{1}{2}$	14.500	1.20	6.50	19.00						
$2\frac{1}{4}$	2.250	15	15.000	1.40	7.00	19.50						
$2\frac{1}{2}$	2.500	$15\frac{1}{2}$	15.500	1.60	7.50	20.00						
$2\frac{3}{4}$	2.750	16	16.000	1.80	8.00							
3	3.000	$16\frac{1}{2}$	16.500									
$3\frac{1}{4}$	3.250	17	17.000									
$3\frac{1}{2}$	3.500	$17\frac{1}{2}$	17.500									
$3\frac{3}{4}$	3.750	18	18.000									
4	4.000	$18\frac{1}{2}$	18.500									
$4\frac{1}{4}$	4.250	19	19.000									
$4\frac{1}{2}$	4.500	$19\frac{1}{2}$	19.500									
$4\frac{3}{4}$	4.750	20	20.000									

A–3 Screw threads.

(a) American Standard thread dimensions, numbered sizes

		Coarse threads: UNC		Fine threads: UNF	
Size	Basic major diameter, D (in)	Threads per inch, n	Tensile stress area (in^2)	Threads per inch, n	Tensile stress area (in^2)
0	0.0600	—	—	80	0.00180
1	0.0730	64	0.00263	72	0.00278
2	0.0860	56	0.00370	64	0.00394
3	0.0990	48	0.00487	56	0.00523
4	0.1120	40	0.00604	48	0.00661
5	0.1250	40	0.00796	44	0.00830
6	0.1380	32	0.00909	40	0.01015
8	0.1640	32	0.0140	36	0.01474
10	0.1900	24	0.0175	32	0.0200
12	0.2160	24	0.0242	28	0.0258

(b) American Standard thread dimensions, fractional sizes

		Coarse threads: UNC		Fine threads: UNF	
Size	Basic major diameter, D (in)	Threads per inch, n	Tensile stress area (in^2)	Threads per inch, n	Tensile stress area (in^2)
$\frac{1}{4}$	0.2500	20	0.0318	28	0.0364
$\frac{5}{16}$	0.3125	18	0.0524	24	0.0580
$\frac{3}{8}$	0.3750	16	0.0775	24	0.0878
$\frac{7}{16}$	0.4375	14	0.1063	20	0.1187
$\frac{1}{2}$	0.5000	13	0.1419	20	0.1599
$\frac{9}{16}$	0.5625	12	0.182	18	0.203
$\frac{5}{8}$	0.6250	11	0.226	18	0.256
$\frac{3}{4}$	0.7500	10	0.334	16	0.373
$\frac{7}{8}$	0.8750	9	0.462	14	0.509
1	1.000	8	0.606	12	0.663
$1\frac{1}{8}$	1.125	7	0.763	12	0.856
$1\frac{1}{4}$	1.250	7	0.969	12	1.073
$1\frac{3}{8}$	1.375	6	1.155	12	1.315
$1\frac{1}{2}$	1.500	6	1.405	12	1.581
$1\frac{3}{4}$	1.750	5	1.90	—	—
2	2.000	$4\frac{1}{2}$	2.50	—	—

(c) Metric thread dimensions

Basic major diameter, D (mm)	Coarse threads Pitch (mm)	Coarse threads Tensile stress area (mm^2)	Fine threads Pitch (mm)	Fine threads Tensile stress area (mm^2)
1	0.25	0.460	—	—
1.6	0.35	1.27	0.20	1.57
2	0.4	2.07	0.25	2.45
2.5	0.45	3.39	0.35	3.70
3	0.5	5.03	0.35	5.61
4	0.7	8.78	0.5	9.79
5	0.8	14.2	0.5	16.1
6	1	20.1	0.75	22.0
8	1.25	36.6	1	39.2
10	1.5	58.0	1.25	61.2
12	1.75	84.3	1.25	92.1
16	2	157	1.5	167
20	2.5	245	1.5	272
24	3	353	2	384
30	3.5	561	2	621
36	4	817	3	865
42	4.5	1 121	—	—
48	5	1 473	—	—

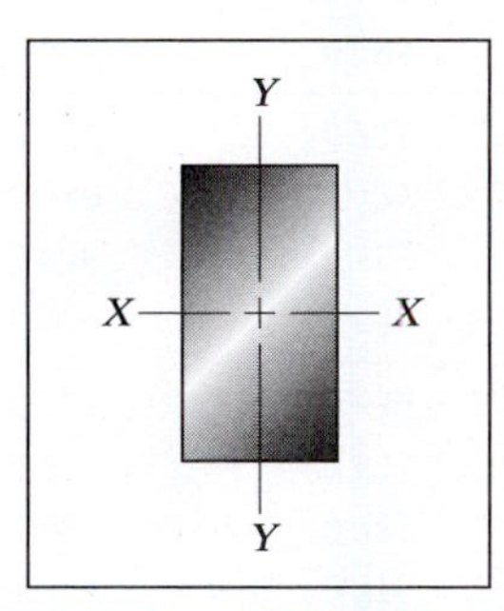

A–4 Properties of standard wood beams.

Nominal size	Actual size		Area of section		Moment of inertia, I_x		Section modulus, S_x	
	in	mm	in^2	mm^2	in^4	mm^4	in^3	mm^3
2 × 4	1.5 × 3.5	38 × 89	5.25	3.39×10^3	5.36	2.23×10^6	3.06	50.1×10^3
2 × 6	1.5 × 5.5	38 × 140	8.25	5.32×10^3	20.8	8.66×10^6	7.56	124×10^3
2 × 8	1.5 × 7.25	38 × 184	10.87	7.01×10^3	47.6	19.8×10^6	13.14	215×10^3
2 × 10	1.5 × 9.25	38 × 235	13.87	8.95×10^3	98.9	41.2×10^6	21.4	351×10^3
2 × 12	1.5 × 11.25	38 × 286	16.87	10.88×10^3	178	74.1×10^6	31.6	518×10^3
4 × 4	3.5 × 3.5	89 × 89	12.25	7.90×10^3	12.51	5.21×10^6	7.15	117×10^3
4 × 6	3.5 × 5.5	89 × 140	19.25	12.42×10^3	48.5	20.2×10^6	17.65	289×10^3
4 × 8	3.5 × 7.25	89 × 184	25.4	16.39×10^3	111.1	46.2×10^6	30.7	503×10^3
4 × 10	3.5 × 9.25	89 × 235	32.4	20.90×10^3	231	96.1×10^6	49.9	818×10^3
4 × 12	3.5 × 11.25	89 × 286	39.4	25.42×10^3	415	172×10^6	73.9	1211×10^3
6 × 6	5.5 × 5.5	140 × 140	30.3	19.55×10^3	76.3	31.8×10^6	27.7	454×10^3
6 × 8	5.5 × 7.5	140 × 191	41.3	26.65×10^3	193	80.3×10^6	51.6	846×10^3
6 × 10	5.5 × 9.5	140 × 241	52.3	33.74×10^3	393	164×10^6	82.7	1355×10^3
6 × 12	5.5 × 11.5	140 × 292	63.3	40.84×10^3	697	290×10^6	121	1983×10^3
8 × 8	7.5 × 7.5	191 × 191	56.3	36.32×10^3	264	110×10^6	70.3	1152×10^3
8 × 10	7.5 × 9.5	191 × 241	71.3	46.00×10^3	536	223×10^6	113	1852×10^3
8 × 12	7.5 × 11.5	191 × 292	86.3	55.68×10^3	951	396×10^6	165	2704×10^3
10 × 10	9.5 × 9.5	241 × 241	90.3	58.26×10^3	679	283×10^6	143	2343×10^3
10 × 12	9.5 × 11.5	241 × 292	109.3	70.52×10^3	1204	501×10^6	209	3425×10^3
12 × 12	11.5 × 11.5	292 × 292	132.3	85.35×10^3	1458	607×10^6	253	4146×10^3

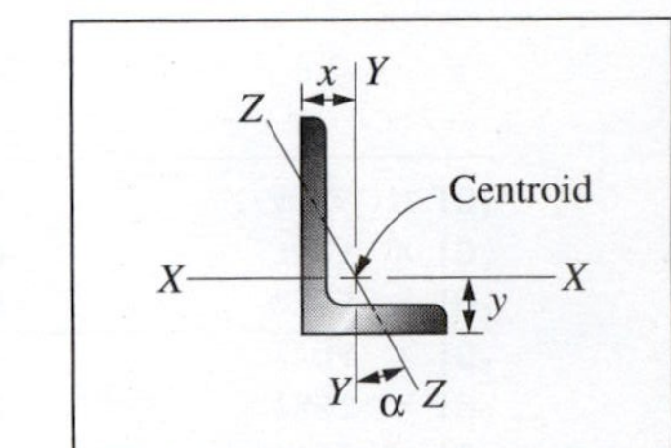

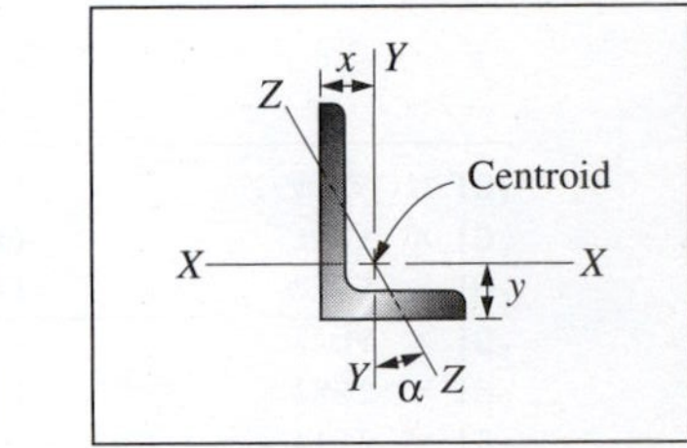

A–5 Properties of steel angles (L-shapes) U.S. Customary units.

Ref.	Shape (in) (in) (in)	Weight per foot (lb/ft)	Area, A (in^2)	Axis X-X: I_x (in^4)	Axis X-X: S_x (in^3)	Axis X-X: y (in)	Axis Y-Y: I_y (in^4)	Axis Y-Y: S_y (in^3)	Axis Y-Y: x (in)	Axis Z-Z: r (in)	Axis Z-Z: α (deg.)
a	L 8 × 8 × 1	51.3	15.1	89.1	15.8	2.36	89.1	15.8	2.36	1.56	45.0
b	L 8 × 8 × $\frac{1}{2}$	26.7	7.84	48.8	8.36	2.17	48.8	8.36	2.17	1.59	45.0
c	L 8 × 4 × 1	37.6	11.10	69.7	14.00	3.03	11.6	3.94	1.04	0.844	13.9
d	L 8 × 4 × $\frac{1}{2}$	19.7	5.80	38.6	7.48	2.84	6.75	2.15	0.854	0.863	14.9
e	L 6 × 6 × $\frac{3}{4}$	28.8	8.46	28.1	6.64	1.77	28.1	6.64	1.77	1.17	45.0
f	L 6 × 6 × $\frac{3}{8}$	14.9	4.38	15.4	3.51	1.62	15.4	3.51	1.62	1.19	45.0
g	L 6 × 4 × $\frac{3}{4}$	23.5	6.90	24.4	6.23	2.08	8.63	2.95	1.08	0.857	23.2
h	L 6 × 4 × $\frac{3}{8}$	12.2	3.58	13.4	3.30	1.94	4.84	1.58	0.940	0.871	24.1
i	L 4 × 4 × $\frac{1}{2}$	12.7	3.75	5.52	1.96	1.18	5.52	1.96	1.18	0.776	45.0
j	L 4 × 4 × $\frac{1}{4}$	6.58	1.93	3.00	1.03	1.08	3.00	1.03	1.08	0.783	45.0
k	L 4 × 3 × $\frac{1}{2}$	11.1	3.25	5.02	1.87	1.32	2.40	1.10	0.822	0.633	28.5
l	L 4 × 3 × $\frac{1}{4}$	5.75	1.69	2.75	0.988	1.22	1.33	0.585	0.725	0.639	29.2
m	L 3 × 3 × $\frac{1}{2}$	9.35	2.75	2.20	1.06	0.929	2.20	1.06	0.929	0.580	45.0
n	L 3 × 3 × $\frac{1}{4}$	4.89	1.44	1.23	0.569	0.836	1.23	0.569	0.836	0.585	45.0
o	L 3 × 2 × $\frac{1}{2}$	7.70	2.26	1.92	1.00	1.08	0.667	0.470	0.580	0.425	22.4
p	L 3 × 2 × $\frac{1}{4}$	4.09	1.20	1.09	0.541	0.980	0.390	0.258	0.487	0.431	23.6
q	L 2 × 2 × $\frac{3}{8}$	4.65	1.37	0.476	0.348	0.632	0.476	0.348	0.632	0.386	45.0
r	L 2 × 2 × $\frac{1}{4}$	3.21	0.944	0.346	0.244	0.586	0.346	0.244	0.586	0.387	45.0
s	L 2 × 2 × $\frac{1}{8}$	1.67	0.491	0.189	0.129	0.534	0.189	0.129	0.534	0.391	45.0

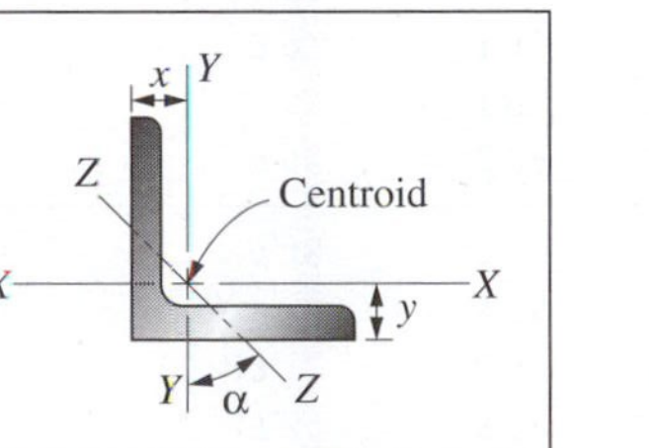

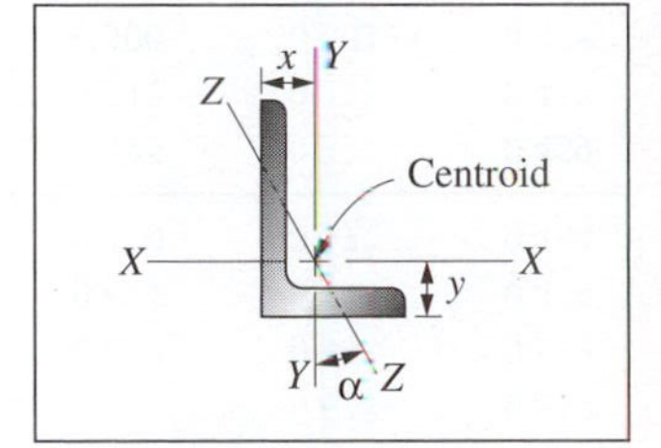

A–5(SI) Properties of steel angles (L-shapes) SI units.

		Mass	Weight		Section properties							
					Axis X-X			Axis Y-Y			Axis Z-Z	
Ref.	Shape (mm) (mm) (mm)	per m (kg/m)	per m (N/m)	Area, A (mm^2)	I_x (mm^4)	S_x (mm^3)	y (mm)	I_y (mm^4)	S_y (mm^3)	x (mm)	r (mm)	α (deg.)
a	L 203 × 203 × 25.4	76.3	749	9740	3.71E+07	2.59E+05	59.9	3.71E+07	2.59E+05	59.9	39.6	45.0
b	L 203 × 203 × 12.7	39.7	390	5060	2.03E+07	1.37E+05	55.1	2.03E+07	1.37E+05	55.1	40.4	45.0
c	L 203 × 102 × 25.4	55.9	549	7160	2.90E+07	2.29E+05	77.0	4.83E+06	6.46E+04	26.4	21.4	13.9
d	L 203 × 102 × 12.7	29.3	288	3740	1.61E+07	1.23E+05	72.1	2.81E+06	3.52E+04	21.7	21.9	14.9
e	L 152 × 152 × 19	42.9	420	5460	1.17E+07	1.09E+05	45.0	1.17E+07	1.09E+05	45.0	29.7	45.0
f	L 152 × 152 × 9.5	22.2	217	2830	6.41E+06	5.75E+04	41.1	6.41E+06	5.75E+04	41.1	30.2	45.0
g	L 152 × 102 × 19	35.0	343	4450	1.02E+07	1.02E+05	52.8	3.59E+06	4.84E+04	27.4	21.8	23.2
h	L 152 × 102 × 9.5	18.2	178	2310	5.58E+06	5.41E+04	49.3	2.01E+06	2.59E+04	23.9	22.1	24.1
i	L 102 × 102 × 12.7	18.9	185	2420	2.30E+06	3.21E+04	30.0	2.30E+06	3.21E+04	30.0	19.7	45.0
j	L 102 × 102 × 6.4	9.79	96.0	1250	1.25E+06	1.69E+04	27.4	1.25E+06	1.69E+04	27.4	19.9	45.0
k	L 102 × 76 × 12.7	16.5	162	2100	2.09E+06	3.06E+04	33.5	9.99E+05	1.80E+04	20.9	16.1	28.5
l	L 102 × 76 × 6.4	8.56	83.9	1090	1.14E+06	1.62E+04	31.0	5.54E+05	9.59E+03	18.4	16.2	29.2
m	L 76 × 76 × 12.7	13.9	136	1770	9.16E+05	1.74E+04	23.6	9.16E+05	1.74E+04	23.6	14.7	45.0
n	L 76 × 76 × 6.4	7.28	71.4	929	5.12E+05	9.33E+03	21.2	5.12E+05	9.33E+03	21.2	14.9	45.0
o	L 76 × 51 × 12.7	11.5	112	1460	7.99E+05	1.64E+04	27.4	2.78E+05	7.70E+03	14.7	10.8	22.4
p	L 76 × 51 × 6.4	6.09	59.7	774	4.54E+05	8.87E+03	24.9	1.62E+05	4.23E+03	12.4	10.9	23.6
q	L 51 × 51 × 9.5	6.92	67.9	884	1.98E+05	5.70E+03	16.1	1.98E+05	5.70E+03	16.1	9.80	45.0
r	L 51 × 51 × 6.4	4.78	46.9	609	1.44E+05	4.00E+03	14.9	1.44E+05	4.00E+03	14.9	9.83	45.0
s	L 51 × 51 × 3.2	2.48	24.4	317	7.87E+04	2.11E+03	13.6	7.87E+04	2.11E+03	13.6	9.93	45.0

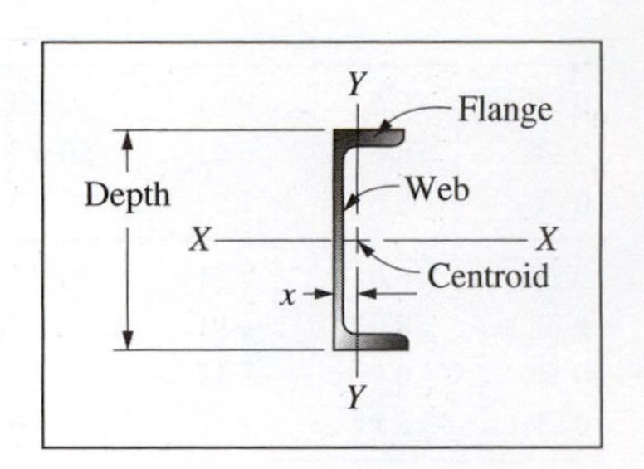

A–6 Properties of American Standard steel channels (C-shapes) U.S. Customary units.

					Flange		Section properties				
							Axis X-X		Axis Y-Y		
Ref.	Shape (in) (lb/ft)	Area, A (in^2)	Depth, d (in)	Web Thickness, t_w (in)	Width, b_f (in)	Thickness, t_f Average (in)	I_x (in^4)	S_x (in^3)	I_y (in^4)	S_y (in^3)	x (in)
a	C 15 × 50	14.7	15.0	0.716	3.72	0.650	404	53.8	11.0	3.77	0.799
b	C 15 × 40	11.8	15.0	0.520	3.52	0.650	348	46.5	9.17	3.34	0.778
c	C 12 × 30	8.82	12.0	0.510	3.17	0.501	162	27.0	5.12	2.05	0.674
d	C 12 × 25	7.35	12.0	0.387	3.05	0.501	144	24.0	4.45	1.87	0.674
e	C 10 × 30	8.82	10.0	0.673	3.03	0.436	103	20.7	3.93	1.65	0.649
f	C 10 × 20	5.88	10.0	0.379	2.74	0.436	78.9	15.8	2.80	1.31	0.606
g	C 9 × 20	5.88	9.0	0.448	2.65	0.413	60.9	13.5	2.41	1.17	0.583
h	C 9 × 15	4.41	9.0	0.285	2.49	0.413	51.0	11.3	1.91	1.01	0.586
i	C 8 × 18.75	5.51	8.0	0.487	2.53	0.390	43.9	11.0	1.97	1.01	0.565
j	C 8 × 11.5	3.38	8.0	0.220	2.26	0.390	32.5	8.14	1.31	0.775	0.572
k	C 7 × 14.75	4.33	7.0	0.419	2.30	0.366	27.2	7.78	1.37	0.772	0.532
l	C 7 × 9.8	2.87	7.0	0.210	2.09	0.366	21.2	6.07	0.957	0.617	0.541
m	C 6 × 13	3.83	6.0	0.437	2.16	0.343	17.3	5.78	1.05	0.638	0.514
n	C 6 × 8.2	2.40	6.0	0.200	1.92	0.343	13.1	4.35	0.687	0.488	0.512
o	C 5 × 9	2.64	5.0	0.325	1.89	0.320	8.89	3.56	0.624	0.444	0.478
p	C 5 × 6.7	1.97	5.0	0.190	1.75	0.320	7.48	2.99	0.470	0.372	0.484
q	C 4 × 7.25	2.13	4.0	0.321	1.72	0.296	4.58	2.29	0.425	0.337	0.459
r	C 4 × 5.4	1.59	4.0	0.184	1.58	0.296	3.85	1.92	0.312	0.277	0.457
s	C 3 × 6	1.76	3.0	0.356	1.60	0.273	2.07	1.38	0.300	0.263	0.455
t	C 3 × 4.1	1.21	3.0	0.170	1.41	0.273	1.65	1.10	0.191	0.196	0.437

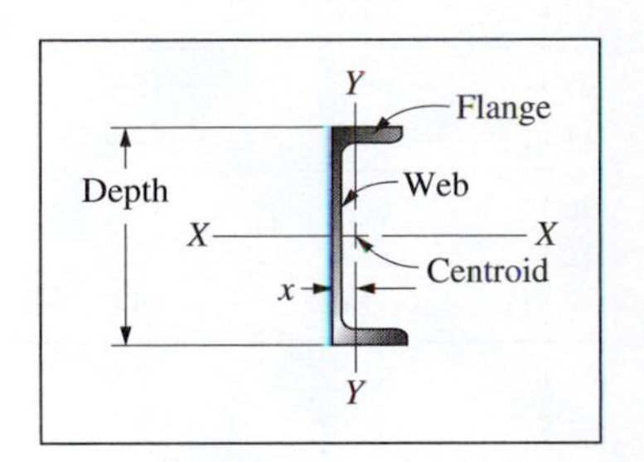

A–6 (SI) Properties of American Standard steel channels (C-shapes) SI units.

Ref.	Shape (mm) (kg/m)	Wt/m (KN/m)	Area, A (mm^2)	Depth, d (mm)	Web Thickness, t_w (mm)	Flange Width, b_f (mm)	Flange Thickness, t_f (mm)	Axis X-X I_x (mm^4)	Axis X-X S_x (mm^3)	Axis Y-Y I_y (mm^4)	Axis Y-Y S_y (mm^3)	x (mm)
a	C 380 × 74	0.730	9480	381	18.2	94.4	16.5	1.68E+08	8.82E+05	4.58E+06	6.18E+04	20.3
b	C 380 × 60	0.584	7610	381	13.2	89.4	16.5	1.45E+08	7.62E+05	3.82E+06	5.47E+04	19.8
c	C 300 × 45	0.438	5690	305	13.0	80.5	12.7	6.74E+07	4.43E+05	2.13E+06	3.36E+04	17.1
d	C 300 × 37	0.365	4740	305	9.83	77.4	12.7	5.99E+07	3.93E+05	1.85E+06	3.06E+04	17.1
e	C 250 × 45	0.438	5690	254	17.1	77.0	11.1	4.29E+07	3.39E+05	1.64E+06	2.70E+04	16.5
f	C 250 × 30	0.292	3790	254	9.63	69.6	11.1	3.28E+07	2.59E+05	1.17E+06	2.15E+04	15.4
g	C 230 × 30	0.292	3790	229	11.4	67.3	10.5	2.53E+07	2.21E+05	1.00E+06	1.92E+04	14.8
h	C 230 × 22	0.219	2850	229	7.24	63.1	10.5	2.12E+07	1.85E+05	7.95E+05	1.66E+04	14.9
i	C 200 × 27.9	0.274	3560	203	12.4	64.2	9.91	1.83E+07	1.80E+05	8.20E+05	1.66E+04	14.4
j	C 200 × 17.1	0.168	2180	203	5.59	57.4	9.91	1.35E+07	1.33E+05	5.45E+05	1.27E+04	14.5
k	C 180 × 22	0.215	2790	178	10.6	58.4	9.30	1.13E+07	1.28E+05	5.70E+05	1.27E+04	13.5
l	C 180 × 14.6	0.143	1850	178	5.33	53.1	9.30	8.82E+06	9.95E+04	3.98E+05	1.01E+04	13.7
m	C 150 × 19.3	0.190	2470	152	11.1	54.8	8.71	7.20E+06	9.47E+04	4.37E+05	1.05E+04	13.1
n	C 150 × 12.2	0.120	1550	152	5.08	48.8	8.71	5.45E+06	7.13E+04	2.86E+05	8.00E+03	13.0
o	C 130 × 13	0.128	1700	127	8.26	47.9	8.13	3.70E+06	5.83E+04	2.60E+05	7.28E+03	12.1
p	C 130 × 10.4	0.102	1270	127	4.83	44.5	8.13	3.11E+06	4.90E+04	1.96E+05	6.10E+03	12.3
q	C 100 × 10.8	0.106	1370	102	8.15	43.7	7.52	1.91E+06	3.75E+04	1.77E+05	5.52E+03	11.7
r	C 100 × 8	0.0788	1020	102	4.67	40.2	7.52	1.60E+06	3.15E+04	1.30E+05	4.54E+03	11.6
s	C 80 × 8.9	0.0876	1140	76.2	9.04	40.5	6.93	8.62E+05	2.26E+04	1.25E+05	4.31E+03	11.6
t	C 80 × 6.1	0.0598	777	76.2	4.32	35.8	6.93	6.87E+05	1.80E+04	7.95E+04	3.21E+03	11.1

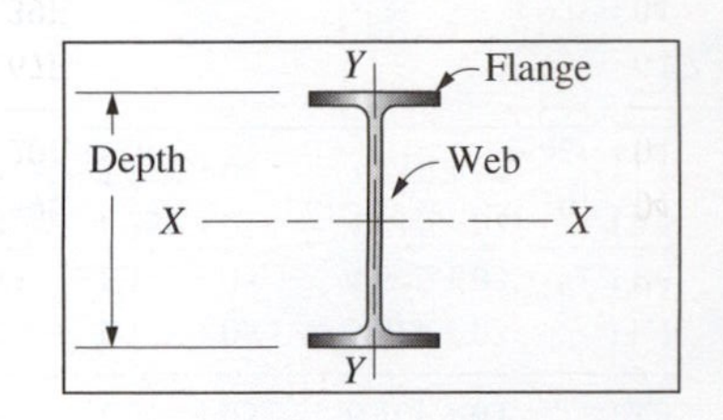

A–7 Properties of steel wide-flange shapes (W-shapes) U.S. Customary units.

Ref.	Shape (in) (lb/ft)	Area, A (in^2)	Depth, d (in)	Web Thickness, t_w (in)	Flange Width, b_f (in)	Flange Thickness, t_f (in)	Section properties Axis X-X I_x (in^4)	Axis X-X S_x (in^3)	Axis Y-Y I_y (in^4)	Axis Y-Y S_y (in^3)
a	W 30 × 173	51.0	30.4	0.655	15.0	1.070	8230	541	598	79.8
b	W 30 × 108	31.7	29.8	0.545	10.5	0.760	4470	299	146	27.9
c	W 27 × 146	43.1	27.4	0.605	14.0	0.975	5660	414	443	63.5
d	W 27 × 102	30.0	27.1	0.515	10.0	0.830	3620	267	139	27.8
e	W 24 × 76	22.4	23.9	0.440	8.99	0.680	2100	176	82.5	18.4
f	W 24 × 68	20.1	23.7	0.415	8.97	0.585	1830	154	70.4	15.7
g	W 21 × 73	21.5	21.2	0.455	8.30	0.740	1600	151	70.6	17.0
h	W 21 × 57	16.7	21.1	0.405	6.56	0.650	1170	111	30.6	9.35
i	W 18 × 55	16.2	18.1	0.390	7.53	0.630	890	98.3	44.9	11.9
j	W 18 × 40	11.8	17.9	0.315	6.02	0.525	612	68.4	19.1	6.35
k	W 14 × 43	12.6	13.7	0.305	8.00	0.530	428	62.7	45.2	11.3
l	W 14 × 26	7.69	13.9	0.255	5.03	0.420	245	35.3	8.91	3.54
m	W 12 × 65	19.1	12.1	0.390	12.0	0.605	533	87.9	174	29.1
n	W 12 × 30	8.79	12.3	0.260	6.52	0.440	238	38.6	20.3	6.24
o	W 12 × 16	4.71	12.0	0.220	3.99	0.265	103	17.1	2.82	1.41
p	W 10 × 60	17.60	10.2	0.420	10.1	0.680	341	66.7	116	23.0
q	W 10 × 30	8.84	10.5	0.300	5.81	0.510	170	32.4	16.7	5.75
r	W 10 × 12	3.54	9.87	0.190	3.96	0.210	53.8	10.9	2.18	1.10
s	W 8 × 40	11.70	8.25	0.360	8.07	0.560	146	35.5	49.1	12.2
t	W 8 × 21	6.16	8.28	0.250	5.27	0.400	75.3	18.2	9.77	3.71
u	W 8 × 10	2.96	7.89	0.170	3.94	0.205	30.8	7.81	2.09	1.06
v	W 6 × 15	4.43	5.99	0.230	5.99	0.260	29.1	9.72	9.32	3.11
w	W 6 × 12	3.55	6.03	0.230	4.00	0.280	22.1	7.31	2.99	1.50
x	W 5 × 19	5.54	5.15	0.270	5.03	0.430	26.2	10.2	9.13	3.63
y	W 5 × 16	4.68	5.01	0.240	5.00	0.360	21.3	8.51	7.51	3.00
z	W 4 × 13	3.83	4.16	0.280	4.06	0.345	11.3	5.46	3.86	1.90

A–7(SI) Properties of steel wide-flange shapes (W-shapes) SI units.

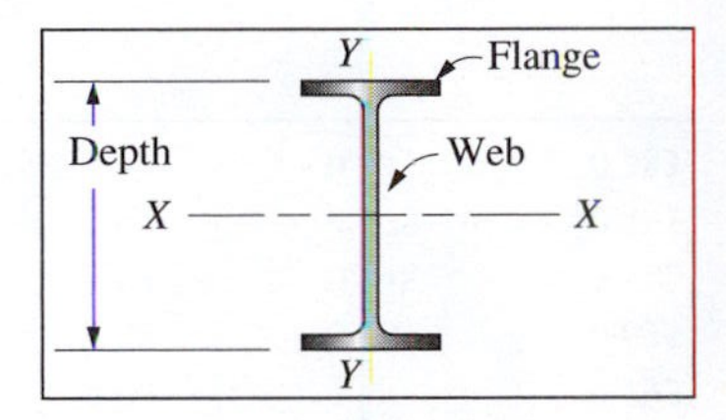

								Section properties			
						Flange		Axis X-X		Axis Y-Y	
Ref.	Shape (mm) (kg/m)	Wt/m (KN/M)	Area, A (mm^2)	Depth, d (mm)	Web Thickness, t_w (mm)	Width, b_f (mm)	Thickness, t_f (mm)	I_x (mm^4)	S_x (mm^3)	I_y (mm^4)	S_y (mm^3)
a	W 760 × 257	2.525	32 900	772	16.6	381	27.2	3.43E+09	8.87E+06	2.49E+08	1.31E+06
b	W 760 × 161	1.576	20 500	757	13.8	267	19.3	1.86E+09	4.90E+06	6.08E+07	4.57E+05
c	W 690 × 217	2.131	27 800	696	15.4	356	24.8	2.36E+09	6.79E+06	1.84E+08	1.04E+06
d	W 690 × 152	1.489	19 400	688	13.1	254	21.1	1.51E+09	4.38E+06	5.79E+07	4.56E+05
e	W 610 × 113	1.109	14 500	607	11.2	228	17.3	8.74E+08	2.88E+06	3.43E+07	3.02E+05
f	W 610 × 101	0.993	13 000	602	10.5	228	14.9	7.62E+08	2.52E+06	2.93E+07	2.57E+05
g	W 530 × 109	1.066	13 900	538	11.6	211	18.8	6.66E+08	2.47E+06	2.94E+07	2.79E+05
h	W 530 × 85	0.832	10 800	536	10.3	167	16.5	4.87E+08	1.82E+06	1.27E+07	1.53E+05
i	W 460 × 82	0.803	10 500	460	9.91	191	16.0	3.70E+08	1.61E+06	1.87E+07	1.95E+05
j	W 460 × 60	0.584	7610	455	8.00	153	13.3	2.55E+08	1.12E+06	7.95E+06	1.04E+05
k	W 360 × 64	0.628	8130	348	7.75	203	13.5	1.78E+08	1.03E+06	1.88E+07	1.85E+05
l	W 360 × 39	0.380	4960	353	6.48	128	10.7	1.02E+08	5.79E+05	3.71E+06	5.80E+04
m	W 310 × 97	0.949	12 300	307	9.91	305	15.4	2.22E+08	1.44E+06	7.24E+07	4.77E+05
n	W 310 × 44.5	0.437	5670	312	6.60	166	11.2	9.91E+07	6.33E+05	8.45E+06	1.02E+05
o	W 310 × 23.8	0.233	3040	305	5.59	101	6.73	4.29E+07	2.80E+05	1.17E+06	2.31E+04
p	W 250 × 89	0.876	11 400	259	10.7	257	17.3	1.42E+08	1.09E+06	4.83E+07	3.77E+05
q	W 250 × 44.8	0.439	5700	267	7.62	148	13	7.08E+07	5.31E+05	6.95E+06	9.42E+04
r	W 250 × 17.9	0.176	2280	251	4.83	101	5.33	2.24E+07	1.79E+05	9.07E+05	1.80E+04
s	W 200 × 59	0.579	7550	210	9.14	205	14.2	6.08E+07	5.82E+05	2.04E+07	2.00E+05
t	W 200 × 31.3	0.307	3970	210	6.35	134	10.2	3.13E+07	2.98E+05	4.07E+06	6.08E+04
u	W 200 × 15	0.146	1910	200	4.32	100	5.21	1.28E+07	1.28E+05	8.70E+05	1.74E+04
v	W 150 × 22.5	0.221	2860	152	5.84	152	6.60	1.21E+07	1.59E+05	3.88E+06	5.10E+04
w	W 150 × 18	0.175	2290	153	5.84	102	7.11	9.20E+06	1.20E+05	1.24E+06	2.46E+04
x	W 130 × 28.1	0.276	3570	131	6.86	128	10.9	1.09E+07	1.67E+05	3.80E+06	5.95E+04
y	W 130 × 23.8	0.233	3020	127	6.10	127	9.14	8.87E+06	1.39E+05	3.13E+06	4.92E+04
z	W 100 × 19.3	0.189	2470	106	7.11	103	8.76	4.70E+06	8.95E+04	1.61E+06	3.11E+04

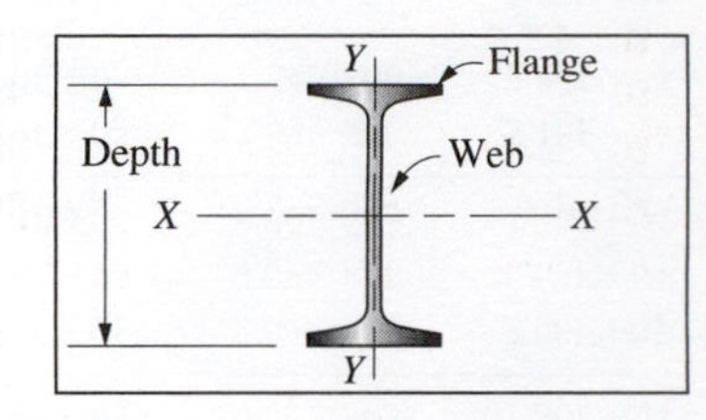

A–8 Properties of American Standard steel beams (S-shapes) U.S. Customary units.

Ref.	Shape (in) (lb/ft)	Area, A (in^2)	Depth, d (in)	Web Thickness, t_w (in)	Flange Width, b_f (in)	Flange Thickness, t_f (in)	Axis X-X I_x (in^4)	Axis X-X S_x (in^3)	Axis Y-Y I_y (in^4)	Axis Y-Y S_y (in^3)
a	S 24 × 121	35.5	24.5	0.800	8.05	1.090	3160	258	83.0	20.6
b	S 24 × 90	26.5	24.0	0.625	7.13	0.870	2250	187	44.7	12.5
c	S 20 × 96	28.2	20.3	0.800	7.20	0.920	1670	165	49.9	13.9
d	S 20 × 75	22.0	20.0	0.635	6.39	0.795	1280	128	29.5	9.25
e	S 20 × 66	19.4	20.0	0.505	6.26	0.795	1190	119	27.5	8.78
f	S 18 × 70	20.5	18.0	0.711	6.25	0.691	923	103	24.0	7.69
g	S 18 × 54.7	16.0	18.0	0.461	6.00	0.691	801	89.0	20.7	6.91
h	S 15 × 50	14.7	15.0	0.550	5.64	0.622	485	64.7	15.6	5.53
i	S 15 × 42.9	12.6	15.0	0.411	5.50	0.622	446	59.4	14.3	5.19
j	S 12 × 50	14.6	12.0	0.687	5.48	0.659	303	50.6	15.6	5.69
k	S 12 × 35	10.2	12.0	0.426	5.08	0.544	228	38.1	9.84	3.88
l	S 10 × 35	10.3	10.0	0.594	4.94	0.491	147	29.4	8.30	3.36
m	S 10 × 25.4	7.45	10.0	0.311	4.66	0.491	123	24.6	6.73	2.89
n	S 8 × 23	6.76	8.00	0.441	4.17	0.425	64.7	16.2	4.27	2.05
o	S 8 × 18.4	5.40	8.00	0.271	4.00	0.425	57.5	14.4	3.69	1.84
p	S 6 × 17.25	5.06	6.00	0.465	3.57	0.359	26.2	8.74	2.29	1.28
q	S 6 × 12.5	3.66	6.00	0.232	3.33	0.359	22.0	7.34	1.80	1.08
r	S 5 × 10	2.93	5.00	0.214	3.00	0.326	12.3	4.90	1.19	0.795
s	S 4 × 9.5	2.79	4.00	0.326	2.80	0.293	6.8	3.38	0.887	0.635
t	S 4 × 7.7	2.26	4.00	0.193	2.66	0.293	6.05	3.03	0.748	0.562
u	S 3 × 7.5	2.20	3.00	0.349	2.51	0.260	2.91	1.94	0.578	0.461
v	S 3 × 5.7	1.66	3.00	0.170	2.33	0.260	2.50	1.67	0.447	0.383

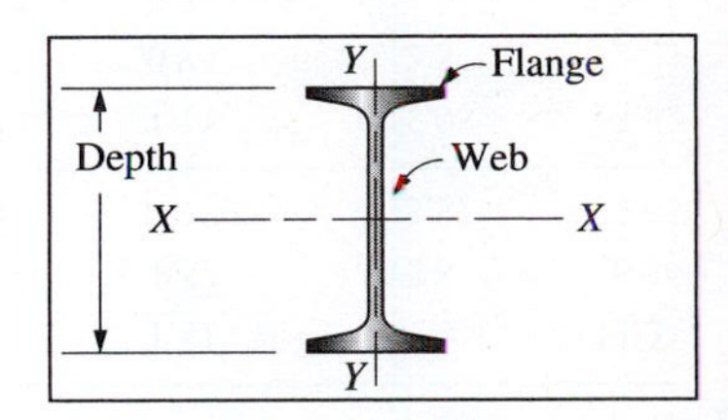

A–8(SI) Properties of American Standard steel beams (S-shapes) SI units.

Ref.	Shape (mm) (kg/m)	Wt/m (KN/m)	Area, A (mm^2)	Depth, d (mm)	Web Thickness, t_w (mm)	Flange Width, b_f (mm)	Flange Thickness, t_f (mm)	Axis X-X I_x (mm^4)	Axis X-X S_x (mm^3)	Axis Y-Y I_y (mm^4)	Axis Y-Y S_y (mm^3)
a	S 610 × 180	1.766	22 900	622	20.3	204	27.7	1.32E+09	4.23E+06	3.45E+07	3.38E+05
b	S 610 × 134	1.314	17 100	610	15.9	181	22.1	9.36E+08	3.06E+06	1.86E+07	2.05E+05
c	S 510 × 143	1.401	18 200	516	20.3	183	23.4	6.95E+08	2.70E+06	2.08E+07	2.28E+05
d	S 510 × 112	1.095	14 200	508	16.1	162	20.2	5.33E+08	2.10E+06	1.23E+07	1.52E+05
e	S 510 × 98.2	0.963	12 500	508	12.8	159	20.2	4.95E+08	1.95E+06	1.14E+07	1.44E+05
f	S 460 × 104	1.022	13 200	457	18.1	159	17.6	3.84E+08	1.69E+06	9.99E+06	1.26E+05
g	S 460 × 81.4	0.799	10 300	457	11.7	152	17.6	3.33E+08	1.46E+06	8.62E+06	1.13E+05
h	S 380 × 74	0.730	9480	381	14.0	143	15.8	2.02E+08	1.06E+06	6.49E+06	9.06E+04
i	S 380 × 64	0.626	8130	381	10.4	140	15.8	1.86E+08	9.74E+05	5.95E+06	8.51E+04
j	S 300 × 74	0.730	9420	305	17.4	139	16.7	1.26E+08	8.29E+05	6.49E+06	9.33E+04
k	S 300 × 52	0.511	6580	305	10.8	129	13.8	9.49E+07	6.24E+05	4.10E+06	6.36E+04
l	S 250 × 52	0.511	6650	254	15.1	125	12.5	6.12E+07	4.82E+05	3.45E+06	5.51E+04
m	S 250 × 37.8	0.371	4810	254	7.9	118	12.5	5.12E+07	4.03E+05	2.80E+06	4.74E+04
n	S 200 × 34	0.336	4360	203	11.2	106	10.8	2.69E+07	2.66E+05	1.78E+06	3.36E+04
o	S 200 × 27.4	0.269	3480	203	6.9	102	10.8	2.39E+07	2.36E+05	1.54E+06	3.02E+04
p	S 150 × 25.7	0.252	3260	152	11.8	90.7	9.1	1.09E+07	1.43E+05	9.53E+05	2.10E+04
q	S 150 × 18.6	0.182	2360	152	5.9	84.6	9.1	9.16E+06	1.20E+05	7.49E+05	1.77E+04
r	S 130 × 15	0.146	1890	127	5.4	76.2	8.3	5.12E+06	8.03E+04	4.95E+05	1.30E+04
s	S 100 × 14.1	0.138	1800	102	8.3	71.1	7.4	2.81E+06	5.54E+04	3.69E+05	1.04E+04
t	S 100 × 11.5	0.113	1460	102	4.9	67.6	7.4	2.52E+06	4.97E+04	3.11E+05	9.21E+03
u	S 80 × 11.2	0.110	1420	76.2	8.9	63.8	6.6	1.21E+06	3.18E+04	2.41E+05	7.56E+03
v	S 80 × 8.5	0.083	1070	76.2	4.3	59.2	6.6	1.04E+06	2.74E+04	1.86E+05	6.28E+03

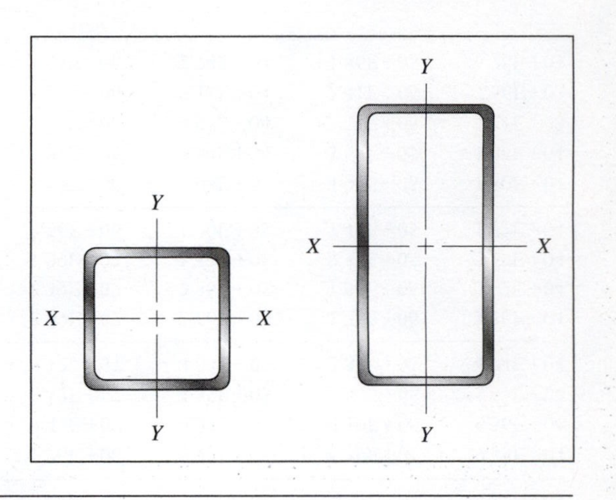

A–9 Properties of steel structural tubing (HSS-shapes) U.S. Customary units.

					Section properties							
					Axis X-X			Axis Y-Y			Torsional constants	
Ref.	Shape (in) (in) (in)	Design wall thickness t_w (in)	Weight per foot (lb/ft)	Area, A (in^2)	I_x (in^4)	S_x (in^3)	r_x (in)	I_y (in^4)	S_y (in^3)	r_y (in)	J (in^4)	C (in^3)
a	HSS 8 × 8 × 1/2	0.465	48.7	13.5	125	31.2	3.04	125	31.2	3.04	204	52.4
b	HSS 8 × 8 × 1/4	0.233	25.8	7.10	70.7	17.7	3.15	70.7	17.7	3.15	111	28.1
c	HSS 8 × 4 × 1/2	0.465	35.1	9.74	71.8	17.9	2.71	23.6	11.8	1.56	61.1	24.4
d	HSS 8 × 4 × 1/4	0.233	19.0	5.24	42.5	10.6	2.85	14.4	7.21	1.66	35.3	13.6
e	HSS 8 × 2 × 1/4	0.233	15.6	4.30	28.5	7.12	2.57	2.94	2.94	0.827	9.36	6.35
f	HSS 6 × 6 × 1/2	0.465	35.1	9.74	48.3	16.1	2.23	48.3	16.1	2.23	81.1	28.1
g	HSS 6 × 6 × 1/4	0.233	19.0	5.24	28.6	9.54	2.34	28.6	9.54	2.34	45.6	15.4
h	HSS 6 × 4 × 1/4	0.233	15.6	4.30	20.9	6.96	2.20	11.1	5.56	1.61	23.6	10.1
i	HSS 6 × 2 × 1/4	0.233	12.2	3.37	13.1	4.37	1.97	2.21	2.21	0.810	6.55	4.70
j	HSS 4 × 4 × 1/2	0.465	21.5	6.02	11.9	5.97	1.41	11.9	5.97	1.41	21.0	11.2
k	HSS 4 × 4 × 1/4	0.233	12.2	3.37	7.80	3.90	1.52	7.80	3.90	1.52	12.8	6.56
l	HSS 4 × 2 × 1/4	0.233	8.78	2.44	4.49	2.25	1.36	1.48	1.48	0.779	3.82	3.05
m	HSS 3 × 3 × 1/4	0.233	8.78	2.44	3.02	2.01	1.11	3.02	2.01	1.11	5.08	3.52
n	HSS 3 × 2 × 1/4	0.233	7.08	1.97	2.13	1.42	1.04	1.11	1.11	0.751	2.52	2.23
o	HSS 2 × 2 × 1/4	0.233	5.38	1.51	0.747	0.747	0.704	0.747	0.747	0.704	1.31	1.41

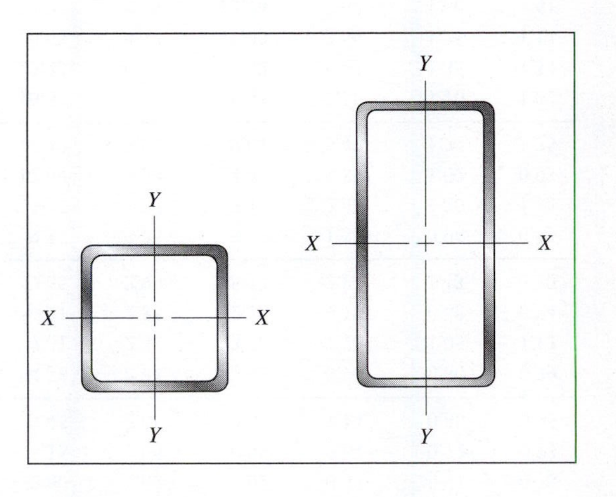

A–9(SI) Properties of steel structural tubing (HSS-shapes) SI units.

Ref.	Shape (mm) × (mm) × (mm)	Design wall thickness t_w (mm)	Mass per m (kg/m)	Weight per m (N/m)	Area A (mm²)	Axis X-X: I_x (mm⁴)	Axis X-X: S_x (mm³)	Axis X-X: r_x (mm)	Axis Y-Y: I_y (mm⁴)	Axis Y-Y: S_y (mm³)	Axis Y-Y: r_y (mm)	Torsional constants: J (mm⁴)	Torsional constants: C (mm³)
a	HSS 203 × 203 × 12.7	11.8	72.5	711	8710	5.20E+07	5.11E+05	77.2	5.20E+07	5.11E+05	77.2	8.49E+07	8.59E+05
b	HSS 203 × 203 × 6.4	5.92	38.4	377	4580	2.94E+07	2.90E+05	80.0	2.94E+07	2.90E+05	80.0	4.62E+07	4.61E+05
c	HSS 203 × 102 × 12.7	11.8	52.2	512	6280	2.99E+07	2.93E+05	68.8	9.82E+06	1.93E+05	39.6	2.54E+07	4.00E+05
d	HSS 203 × 102 × 6.4	5.92	28.3	277	3380	1.77E+07	1.74E+05	72.4	5.99E+06	1.18E+05	42.2	1.47E+07	2.23E+05
e	HSS 203 × 51 × 6.4	5.92	23.2	228	2770	1.19E+07	1.17E+05	65.3	1.22E+06	4.82E+04	21.0	3.90E+06	1.04E+05
f	HSS 152 × 152 × 12.7	11.8	52.2	512	6280	2.01E+07	2.64E+05	56.6	2.01E+07	2.64E+05	56.6	3.38E+07	4.61E+05
g	HSS 152 × 152 × 6.4	5.92	28.3	277	3380	1.19E+07	1.56E+05	59.4	1.19E+07	1.56E+05	59.4	1.90E+07	2.52E+05
h	HSS 152 × 102 × 6.4	5.92	23.2	228	2770	8.70E+06	1.14E+05	55.9	4.62E+06	9.11E+04	40.9	9.82E+06	1.66E+05
i	HSS 152 × 51 × 6.4	5.92	18.2	178	2170	5.45E+06	7.16E+04	50.0	9.20E+05	3.62E+04	20.6	2.73E+06	7.70E+04
j	HSS 102 × 102 × 12.7	11.8	32.0	314	3880	4.95E+06	9.78E+04	35.8	4.95E+06	9.78E+04	35.8	8.74E+06	1.84E+05
k	HSS 102 × 102 × 6.4	5.92	18.2	178	2170	3.25E+06	6.39E+04	38.6	3.25E+06	6.39E+04	38.6	5.33E+06	1.08E+05
l	HSS 102 × 51 × 6.4	5.92	13.1	128	1570	1.87E+06	3.69E+04	34.5	6.16E+05	2.43E+04	19.8	1.59E+06	5.00E+04
m	HSS 76 × 76 × 6.4	5.92	13.1	128	1570	1.26E+06	3.29E+04	28.2	1.26E+06	3.29E+04	28.2	2.11E+06	5.77E+04
n	HSS 76 × 51 × 6.4	5.92	10.5	103	1271	8.87E+05	2.33E+04	26.4	4.62E+05	1.82E+04	19.1	1.05E+06	3.65E+04
o	HSS 51 × 51 × 6.4	5.92	8.01	78.5	974	3.11E+05	1.22E+04	17.9	3.11E+05	1.22E+04	17.9	5.45E+05	2.31E+04

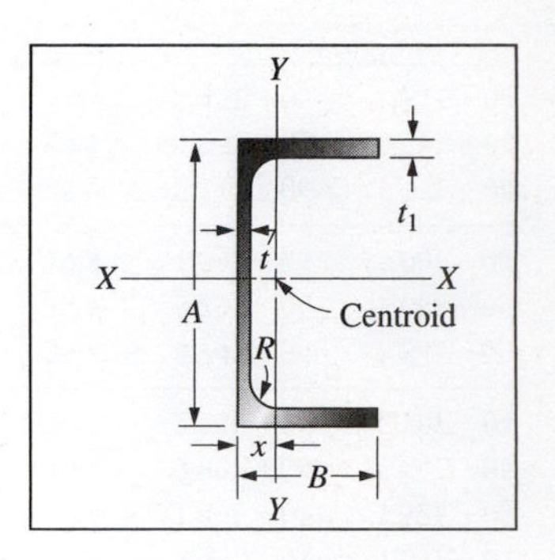

A–10 Properties of Aluminum Association standard channels U.S. Customary units.

Ref.	Shape (in) × (lb/ft)	Depth, A (in)	Width, B (in)	Area (in^2)	Flange Thickness, t_1 (in)	Web Thickness, t (in)	Axis X-X: I_x (in^4)	Axis X-X: S_x (in^3)	Axis X-X: r_x (in)	Axis Y-Y: I_y (in^4)	Axis Y-Y: S_y (in^3)	Axis Y-Y: r_y (in)	Axis Y-Y: x (in)
a	C 2 × 0.577	2.00	1.00	0.491	0.13	0.13	0.288	0.288	0.766	0.045	0.064	0.303	0.298
b	C 2 × 1.071	2.00	1.25	0.911	0.26	0.17	0.546	0.546	0.774	0.139	0.178	0.391	0.471
c	C 3 × 1.135	3.00	1.50	0.965	0.20	0.13	1.41	0.94	1.21	0.22	0.22	0.47	0.49
d	C 3 × 1.597	3.00	1.75	1.358	0.26	0.17	1.97	1.31	1.20	0.42	0.37	0.55	0.62
e	C 4 × 1.738	4.00	2.00	1.478	0.23	0.15	3.91	1.95	1.63	0.60	0.45	0.64	0.65
f	C 4 × 2.331	4.00	2.25	1.982	0.29	0.19	5.21	2.60	1.62	1.02	0.69	0.72	0.78
g	C 5 × 2.212	5.00	2.25	1.881	0.26	0.15	7.88	3.15	2.05	0.98	0.64	0.72	0.73
h	C 5 × 3.089	5.00	2.75	2.627	0.32	0.19	11.14	4.45	2.06	2.05	1.14	0.88	0.95
i	C 6 × 2.834	6.00	2.50	2.410	0.29	0.17	14.35	4.78	2.44	1.53	0.90	0.80	0.79
j	C 6 × 4.030	6.00	3.25	3.427	0.35	0.21	21.04	7.01	2.48	3.76	1.76	1.05	1.12
k	C 7 × 3.205	7.00	2.75	2.725	0.29	0.17	22.09	6.31	2.85	2.10	1.10	0.88	0.84
l	C 7 × 4.715	7.00	3.50	4.009	0.38	0.21	33.79	9.65	2.90	5.13	2.23	1.13	1.20
m	C 8 × 4.147	8.00	3.00	3.526	0.35	0.19	37.40	9.35	3.26	3.25	1.57	0.96	0.93
n	C 8 × 5.789	8.00	3.75	4.923	0.41	0.25	52.69	13.17	3.27	7.13	2.82	1.20	1.22
o	C 9 × 4.983	9.00	3.25	4.237	0.35	0.23	54.41	12.09	3.58	4.40	1.89	1.02	0.93
p	C 9 × 6.970	9.00	4.00	5.927	0.44	0.29	78.31	17.40	3.63	9.61	3.49	1.27	1.25
q	C 10 × 6.136	10.00	3.50	5.218	0.41	0.25	83.22	16.64	3.99	6.33	2.56	1.10	1.02
r	C 10 × 8.360	10.00	4.25	7.109	0.50	0.31	116.15	23.23	4.04	13.02	4.47	1.35	1.34
s	C 12 × 8.274	12.00	4.00	7.036	0.47	0.29	159.76	26.63	4.77	11.03	3.86	1.25	1.14
t	C 12 × 11.822	12.00	5.00	10.053	0.62	0.35	239.69	39.95	4.88	25.74	7.60	1.60	1.61

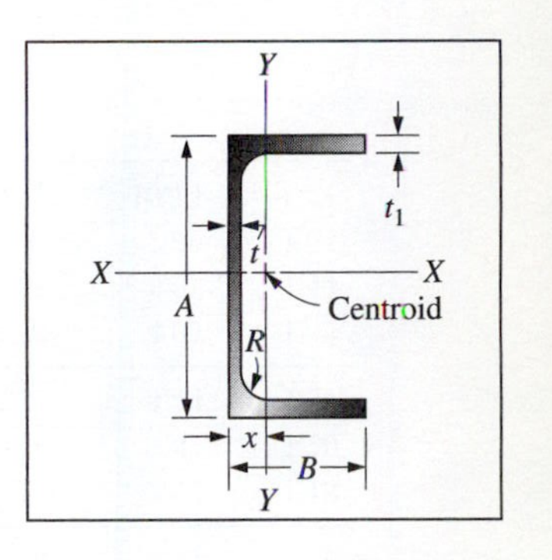

A–10(SI) Properties of Aluminum Association standard channels SI units.

Ref.	Shape (mm) (kg/m)	Wt/m (N/m)	Depth, A (mm)	Width, B (mm)	Area (mm²)	Flange Thickness, t_1 (mm)	Web Thickness, t (mm)	Axis X-X: I_x (mm⁴)	Axis X-X: S_x (mm³)	Axis X-X: r_x (mm)	Axis Y-Y: I_y (mm⁴)	Axis Y-Y: S_y (mm³)	Axis Y-Y: r_y (mm)	x (mm)
a	C 51 × 0.859	8.42	51	25	317	3.3	3.3	1.20E+05	4.72E+03	19.46	1.87E+04	1.05E+03	7.70	7.57
b	C 51 × 1.594	15.63	51	32	588	6.6	4.3	2.27E+05	8.95E+03	19.66	5.79E+04	2.92E+03	9.93	12.0
c	C 76 × 1.689	16.57	76	38	623	5.1	3.3	5.87E+05	1.54E+04	30.73	9.16E+04	3.61E+03	11.94	12.4
d	C 76 × 2.376	23.31	76	44	876	6.6	4.3	8.20E+05	2.15E+04	30.48	1.75E+05	6.06E+03	13.97	15.7
e	C 102 × 2.586	25.37	102	51	954	5.8	3.8	1.63E+06	3.20E+04	41.40	2.50E+05	7.38E+03	16.26	16.5
f	C 102 × 3.468	34.03	102	57	1279	7.4	4.8	2.17E+06	4.26E+04	41.15	4.25E+05	1.13E+04	18.29	19.8
g	C 127 × 3.291	32.29	127	57	1214	6.6	3.8	3.28E+06	5.16E+04	52.07	4.08E+05	1.05E+04	18.29	18.5
h	C 127 × 4.596	45.09	127	70	1695	8.1	4.8	4.64E+06	7.29E+04	52.32	8.53E+05	1.87E+04	22.35	24.1
i	C 152 × 4.217	41.37	152	64	1555	7.4	4.3	5.97E+06	7.83E+04	61.98	6.37E+05	1.48E+04	20.32	20.1
j	C 152 × 6.00	58.8	152	83	2211	8.9	5.3	8.76E+06	1.15E+05	62.99	1.56E+06	2.88E+04	26.67	28.4
k	C 178 × 4.77	46.8	178	70	1758	7.4	4.3	9.19E+06	1.03E+05	72.39	8.74E+05	1.80E+04	22.35	21.3
l	C 178 × 7.02	68.8	178	89	2587	9.7	5.3	1.41E+07	1.58E+05	73.7	2.14E+06	3.65E+04	28.70	30.5
m	C 203 × 6.17	60.5	203	76	2275	8.9	4.8	1.56E+07	1.53E+05	82.8	1.35E+06	2.57E+04	24.38	23.6
n	C 203 × 8.61	84.5	203	95	3176	10.4	6.4	2.19E+07	2.16E+05	83.1	2.97E+06	4.62E+04	30.48	31.0
o	C 229 × 7.41	72.7	229	83	2734	8.9	5.8	2.26E+07	1.98E+05	90.9	1.83E+06	3.10E+04	25.91	23.6
p	C 229 × 10.37	101.7	229	102	3824	11.2	7.4	3.26E+07	2.85E+05	92.2	4.00E+06	5.72E+04	32.26	31.8
q	C 254 × 9.13	89.6	254	89	3367	10.4	6.4	3.46E+07	2.73E+05	101.3	2.63E+06	4.20E+04	27.94	25.9
r	C 254 × 12.44	122.0	254	108	4587	12.7	7.9	4.83E+07	3.81E+05	102.6	5.42E+06	7.33E+04	34.29	34.0
s	C 305 × 12.31	120.8	305	102	4540	11.9	7.4	6.65E+07	4.36E+05	121.2	4.59E+06	6.33E+04	31.75	29.0
t	C 305 × 17.59	172.6	305	127	6486	15.7	8.9	9.98E+07	6.55E+05	124.0	1.07E+07	1.25E+05	40.64	40.9

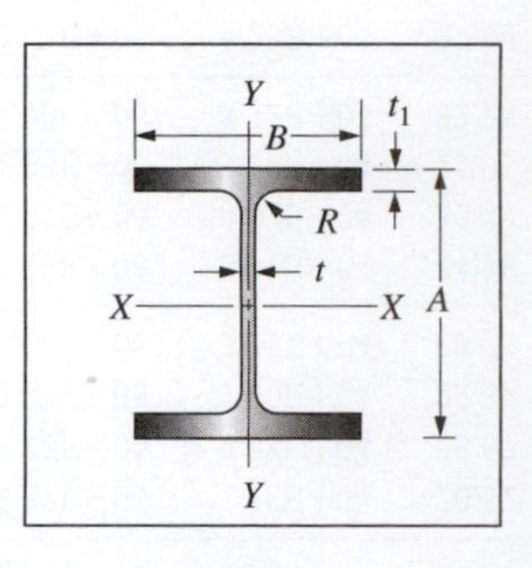

A–11 Properties of Aluminum Association standard I-beam shapes U.S. Customary units.

							Section properties					
							Axis X-X			Axis Y-Y		
Ref.	Shape (in) (lb/ft)	Depth, A (in)	Width, B (in)	Area (in^2)	Flange Thickness, t_1 (in)	Web Thickness, t (in)	I_x (in^4)	S_x (in^3)	r_x (in)	I_y (in^4)	S_y (in^3)	r_y (in)
a	I 3 × 1.637	3.00	2.50	1.392	0.20	0.13	2.24	1.49	1.27	0.52	0.42	0.61
b	I 3 × 2.030	3.00	2.50	1.726	0.26	0.15	2.71	1.81	1.25	0.68	0.54	0.63
c	I 4 × 2.311	4.00	3.00	1.965	0.23	0.15	5.62	2.81	1.69	1.04	0.69	0.73
d	I 4 × 2.793	4.00	3.00	2.375	0.29	0.17	6.71	3.36	1.68	1.31	0.87	0.74
e	I 5 × 3.700	5.00	3.50	3.146	0.32	0.19	13.94	5.58	2.11	2.29	1.31	0.85
f	I 6 × 4.030	6.00	4.00	3.427	0.29	0.19	21.99	7.33	2.53	3.10	1.55	0.95
g	I 6 × 4.692	6.00	4.00	3.990	0.35	0.21	25.5	8.50	2.53	3.74	1.87	0.97
h	I 7 × 5.800	7.00	4.50	4.932	0.38	0.23	42.89	12.25	2.95	5.78	2.57	1.08
i	I 8 × 6.181	8.00	5.00	5.256	0.35	0.23	59.69	14.92	3.37	7.30	2.92	1.18
j	I 8 × 7.023	8.00	5.00	5.972	0.41	0.25	67.78	16.94	3.37	8.55	3.42	1.20
k	I 9 × 8.361	9.00	5.50	7.110	0.44	0.27	102.02	22.67	3.79	12.22	4.44	1.31
l	I 10 × 8.646	10.00	6.00	7.352	0.41	0.25	132.09	26.42	4.24	14.78	4.93	1.42
m	I 10 × 10.286	10.00	6.00	8.747	0.50	0.29	155.79	31.16	4.22	18.03	6.01	1.44
n	I 12 × 11.672	12.00	7.00	9.925	0.47	0.29	255.57	42.60	5.07	26.90	7.69	1.65
o	I 12 × 14.292	12.00	7.00	12.153	0.62	0.31	317.33	52.89	5.11	35.48	10.14	1.71

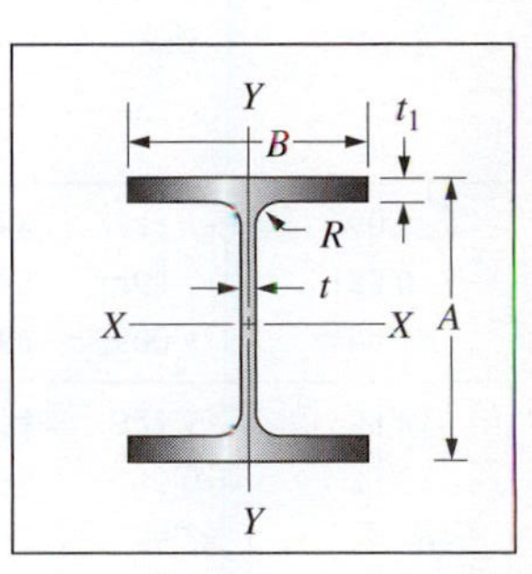

A–11(SI) Properties of Aluminum Association standard I-beam shapes SI units.

								Section properties					
								Axis X-X			Axis Y-Y		
Ref.	Shape (mm) (kg/m)	Wt/m (N/m)	Depth, A (mm)	Width, B (mm)	Area (mm^2)	Flange Thickness, t_1 (mm)	Web Thickness, t (mm)	I_x (mm^4)	S_x (mm^3)	r_x (mm)	I_y (mm^4)	S_y (mm^3)	r_y (mm)
a	I 76 × 2.436	23.90	76	64	898	5.1	3.3	9.32E+05	2.44E+04	32.26	2.16E+05	6.88E+03	15.49
b	I 76 × 3.021	29.63	76	64	1114	6.6	3.8	1.13E+06	2.97E+04	31.75	2.83E+05	8.85E+03	16.00
c	I 102 × 3.439	33.73	102	76	1268	5.8	3.8	2.34E+06	4.61E+04	42.93	4.33E+05	1.13E+04	18.54
d	I 102 × 4.156	40.77	102	76	1532	7.4	4.3	2.79E+06	5.51E+04	42.67	5.45E+05	1.43E+04	18.80
e	I 127 × 5.506	54.01	127	89	2030	8.1	4.8	5.80E+06	9.15E+04	53.59	9.53E+05	2.15E+04	21.59
f	I 152 × 5.997	58.83	152	102	2211	7.4	4.8	9.15E+06	1.20E+05	64.26	1.29E+06	2.54E+04	24.13
g	I 152 × 6.982	68.49	152	102	2574	8.9	5.3	1.06E+07	1.39E+05	64.26	1.56E+06	3.06E+04	24.64
h	I 178 × 8.630	84.66	178	114	3182	9.7	5.8	1.79E+07	2.01E+05	74.93	2.41E+06	4.21E+04	27.43
i	I 203 × 9.197	90.22	203	127	3391	8.9	5.8	2.48E+07	2.45E+05	85.60	3.04E+06	4.79E+04	29.97
j	I 203 × 10.45	102.5	203	127	3853	10.4	6.4	2.82E+07	2.78E+05	85.60	3.56E+06	5.61E+04	30.48
k	I 229 × 12.44	122.0	229	140	4587	11.2	6.9	4.25E+07	3.72E+05	96.27	5.09E+06	7.28E+04	33.27
l	I 254 × 12.87	126.2	254	152	4744	10.4	6.4	5.50E+07	4.33E+05	107.7	6.15E+06	8.08E+04	36.07
m	I 254 × 15.31	150.1	254	152	5644	12.7	7.4	6.48E+07	5.11E+05	107.2	7.50E+06	9.85E+04	36.58
n	I 305 × 44.50	436.5	305	178	6404	11.9	7.4	1.06E+08	6.98E+05	128.8	1.12E+07	1.26E+05	41.91
o	I 305 × 23.80	233.5	305	178	7841	15.7	7.9	1.32E+08	8.67E+05	129.8	1.48E+07	1.66E+05	43.43

A–12 Properties of steel pipe U.S. Customary units.

Ref.	Nominal size (in)	Outside Diameter (in)	Inside Diameter (in)	Wall thickness, t_w (in)	Area, A (in^2)	Section properties			Torsional constants	
						I (in^4)	S (in^3)	r (in)	J (in^4)	Z_p (in^3)
	Schedule 40 pipe									
a	1/8 in	0.405	0.269	0.068	0.072	1.06E-03	5.25E-03	0.122	2.13E-03	1.05E-02
b	1/4 in	0.540	0.364	0.088	0.125	3.31E-03	1.23E-02	0.163	6.62E-03	2.45E-02
c	3/8 in	0.675	0.493	0.091	0.167	7.29E-03	2.16E-02	0.209	1.46E-02	4.32E-02
d	PIPE 1/2 STD	0.840	0.622	0.109	0.250	1.71E-02	4.07E-02	0.261	3.42E-02	8.14E-02
e	PIPE 3/4 STD	1.050	0.824	0.113	0.333	3.70E-02	7.05E-02	0.334	7.41E-02	0.1411
f	PIPE 1 STD	1.315	1.049	0.133	0.494	8.73E-02	0.1328	0.421	0.1747	0.2657
g	PIPE 1-1/4 STD	1.660	1.380	0.140	0.669	0.1947	0.2346	0.540	0.3894	0.4692
h	PIPE 1-1/2 STD	1.900	1.610	0.145	0.799	0.3099	0.3262	0.623	0.6198	0.6524
i	PIPE 2 STD	2.375	2.067	0.154	1.075	0.6657	0.5606	0.787	1.331	1.121
j	PIPE 2-1/2 STD	2.875	2.469	0.203	1.704	1.530	1.064	0.947	3.059	2.128
k	PIPE 3 STD	3.500	3.068	0.216	2.228	3.017	1.724	1.164	6.034	3.448
l	PIPE 3-1/2 STD	4.000	3.548	0.226	2.680	4.788	2.394	1.337	9.575	4.788
m	PIPE 4 STD	4.500	4.026	0.237	3.174	7.233	3.214	1.510	14.47	6.429
n	PIPE 5 STD	5.563	5.047	0.258	4.300	15.16	5.451	1.878	30.32	10.90
o	PIPE 6 STD	6.625	6.065	0.280	5.581	28.14	8.496	2.245	56.28	16.99
p	PIPE 8 STD	8.625	7.981	0.322	8.399	72.49	16.81	2.938	145.0	33.62
q	PIPE 10 STD	10.750	10.020	0.365	11.908	160.7	29.90	3.674	321.5	59.81
r	12 in	12.750	11.938	0.406	15.745	300.2	47.09	4.367	600.4	94.18
s	16 in	16.000	15.000	0.500	24.347	731.9	91.49	5.483	1464	183.0
t	18 in	18.000	16.876	0.562	30.788	1171	130.2	6.168	2343	260.3

NOTE: All values shown are for standard schedule 40 steel pipe.
Rows d–q conform to AISC standards for dimensions of standard weight pipe. Rows a–c and r–t do not.
Many other sizes of round hollow structural sections (HSS) are available. See AISC Manual.

A–12(SI) Properties of steel pipe SI units.

						Section properties				
									Torsional constants	
Ref.	Nominal size (mm)	Outside Diameter (mm)	Inside Diameter (mm)	Wall Thickness, t_w (mm)	Area, A (mm^2)	I (mm^4)	S (mm^3)	r (mm)	J (mm^4)	Z_p (mm^3)
	Schedule 40 pipe									
a	3.2 mm	10.29	6.83	1.73	46.45	442.7	86.07	3.087	885.4	172.1
b	6.4 mm	13.72	9.25	2.24	80.62	1379	201.0	4.135	2757	402.1
c	9.5 mm	17.15	12.52	2.31	107.7	3035	354.0	5.308	6069	708.0
d	PIPE 13 STD	21.34	15.80	2.77	161.5	7114	666.9	6.637	14 228	1334
e	PIPE 19 STD	26.67	20.93	2.87	214.6	15416	1156	8.475	30 831	2312
f	PIPE 25 STD	33.40	26.64	3.38	318.6	36355	2177	10.68	72 710	4354
g	PIPE 32 STD	42.16	35.05	3.56	431.3	81044	3844	13.71	1.62E+05	7688
h	PIPE 38 STD	48.26	40.89	3.68	515.8	1.29E+05	5346	15.81	2.58E+05	10 691
i	PIPE 51 STD	60.33	52.50	3.91	693.2	2.77E+05	9187	19.99	5.54E+05	18 374
j	PIPE 64 STD	73.03	62.71	5.16	1099	6.37E+05	17 436	24.06	1.27E+06	34 873
k	PIPE 75 STD	88.90	77.93	5.49	1438	1.26E+06	28 253	29.55	2.51E+06	56 506
l	PIPE 89 STD	101.6	90.12	5.74	1729	1.99E+06	39 228	33.95	3.99E+06	78 457
m	PIPE 102 STD	114.3	102.3	6.02	2048	3.01E+06	52 676	38.34	6.02E+06	1.05E+05
n	PIPE 127 STD	141.3	128.2	6.55	2774	6.31E+06	89327	47.70	1.26E+07	1.79E+05
o	PIPE 152 STD	168.3	154.1	7.11	3601	1.17E+07	1.39E+05	57.04	2.34E+07	2.78E+05
p	PIPE 203 STD	219.1	202.7	8.18	5419	3.02E+07	2.75E+05	74.62	6.03E+07	5.51E+05
q	PIPE 254 STD	273.1	254.5	9.27	7683	6.69E+07	4.90E+05	93.32	1.34E+08	9.80E+05
r	305 mm	323.9	303.2	10.31	10 158	1.25E+08	7.72E+05	110.9	2.50E+08	1.54E+06
s	406 mm	406.4	381.0	12.70	15 708	3.05E+08	1.50E+06	139.3	6.09E+08	3.00E+06
t	457 mm	457.2	428.7	14.27	19 863	4.88E+08	2.13E+06	156.7	9.75E+08	4.27E+06

NOTE: All values shown are for standard schedule 40 steel pipe, converted to SI units.
Rows d–q conform to AISC standards for dimensions of standard weight pipe. Rows a–c and r–t do not.
Many other sizes of round hollow structural sections (HSS) are available. See AISC Manual.

A–13 Properties of steel mechanical tubing U.S. Customary units.

	Nominal size						Section properties			Torsional constants	
Ref.	OD (in)	Wall gauge	Outside Diameter (in)	Inside Diameter (in)	Wall thickness, t_w (in)	Area, A (in^2)	I (in^4)	S (in^3)	r (in)	J (in^4)	Z_p (in^3)
a	$\frac{1}{2}$	17	0.500	0.384	0.058	0.081	0.00200	0.00800	0.158	0.00400	0.0160
b	$\frac{1}{2}$	14	0.500	0.334	0.083	0.109	0.00246	0.00983	0.150	0.00491	0.0197
c	1	16	1.000	0.870	0.065	0.191	0.0210	0.0419	0.331	0.0419	0.0839
d	1	10	1.000	0.732	0.134	0.365	0.0350	0.0700	0.310	0.0700	0.140
e	$1\frac{1}{2}$	16	1.500	1.370	0.065	0.293	0.0756	0.101	0.508	0.151	0.202
f	$1\frac{1}{2}$	10	1.500	1.232	0.134	0.575	0.135	0.181	0.485	0.271	0.361
g	2	16	2.000	1.870	0.065	0.395	0.185	0.185	0.685	0.370	0.370
h	2	10	2.000	1.732	0.134	0.786	0.344	0.344	0.661	0.687	0.687
i	$2\frac{1}{2}$	10	2.500	2.232	0.134	0.996	0.699	0.559	0.838	1.398	1.119
j	$2\frac{1}{2}$	5	2.500	2.060	0.220	1.576	1.034	0.827	0.810	2.067	1.654
k	3	10	3.000	2.732	0.134	1.207	1.241	0.828	1.014	2.483	1.655
l	3	5	3.000	2.560	0.220	1.921	1.868	1.245	0.986	3.736	2.490
m	$3\frac{1}{2}$	10	3.500	3.232	0.134	1.417	2.010	1.149	1.191	4.020	2.297
n	$3\frac{1}{2}$	5	3.500	3.060	0.220	2.267	3.062	1.750	1.162	6.125	3.500
o	4	5	4.000	3.560	0.220	2.613	4.682	2.341	1.339	9.364	4.682
p	4	1	4.000	3.400	0.300	3.487	6.007	3.003	1.312	12.013	6.007
q	$4\frac{1}{2}$	5	4.500	4.060	0.220	2.958	6.791	3.018	1.515	13.583	6.037
r	$4\frac{1}{2}$	1	4.500	3.900	0.300	3.958	8.773	3.899	1.489	17.546	7.798
s	5	5	5.000	4.560	0.220	3.304	9.456	3.782	1.692	18.911	7.564
t	5	1	5.000	4.400	0.300	4.430	12.281	4.912	1.665	24.562	9.825

A–13(SI) Properties of steel mechanical tubing SI units.

Ref.	Nominal size		Outside Diameter (mm)	Inside Diameter (mm)	Wall thickness, t_w (mm)	Area, A (mm²)	Section properties			Torsional constants	
	OD (mm)	Wall gauge					I (mm⁴)	S (mm³)	r (mm)	J (mm⁴)	Z_p (mm³)
a	12.7	17	12.70	9.754	1.473	51.96	833	131.1	4.00	1665	262.3
b	12.7	14	12.70	8.484	2.108	70.15	1023	161.1	3.82	2045	322.1
c	25.4	16	25.40	22.098	1.651	123.2	8726	687.1	8.42	1.75E+04	1374
d	25.4	10	25.40	18.593	3.404	235.2	1.46E+04	1147	7.87	2.91E+04	2294
e	38.1	16	38.10	34.798	1.651	189.1	3.15E+04	1651	12.9	6.29E+04	3303
f	38.1	10	38.10	31.293	3.404	371.0	5.64E+04	2959	12.3	1.13E+05	5918
g	50.8	16	50.80	47.498	1.651	254.9	7.71E+04	3034	17.4	1.54E+05	6068
h	50.8	10	50.80	43.993	3.404	506.8	1.43E+05	5632	16.8	2.86E+05	1.13E+04
i	63.5	10	63.50	56.693	3.404	642.6	2.91E+05	9166	21.3	5.82E+05	1.83E+04
j	63.5	5	63.50	52.324	5.588	1017	4.30E+05	1.35E+04	20.6	8.60E+05	2.71E+04
k	76.2	10	76.20	69.393	3.404	778.4	5.17E+05	1.36E+04	25.8	1.03E+06	2.71E+04
l	76.2	5	76.20	65.024	5.588	1240	7.77E+05	2.04E+04	25.0	1.55E+06	4.08E+04
m	88.9	10	88.90	82.093	3.404	914.2	8.37E+05	1.88E+04	30.3	1.67E+06	3.76E+04
n	88.9	5	88.90	77.724	5.588	1463	1.27E+06	2.87E+04	29.5	2.55E+06	5.74E+04
o	101.6	5	101.60	90.424	5.588	1686	1.95E+06	3.84E+04	34.0	3.90E+06	7.67E+04
p	101.6	1	101.60	86.360	7.620	2250	2.50E+06	4.92E+04	33.3	5.00E+06	9.84E+04
q	114.3	5	114.30	103.124	5.588	1908	2.83E+06	4.95E+04	38.5	5.65E+06	9.89E+04
r	114.3	1	114.30	99.060	7.620	2554	3.65E+06	6.39E+04	37.8	7.30E+06	1.28E+05
s	127.0	5	127.00	115.824	5.588	2131	3.94E+06	6.20E+04	43.0	7.87E+06	1.24E+05
t	127.0	1	127.00	111.760	7.620	2858	5.11E+06	8.05E+04	42.3	1.02E+07	1.61E+05

A–14 Typical properties of carbon and alloy steels.*

Material AISI no.	Condition[†]	Ultimate strength, s_u ksi	Ultimate strength, s_u MPa	Yield strength, s_y ksi	Yield strength, s_y MPa	Percent elongation
1020	Annealed	57	393	43	296	36
1020	Hot-rolled	65	448	48	331	36
1020	Cold-drawn	75	517	64	441	20
1040	Annealed	75	517	51	352	30
1040	Hot-rolled	90	621	60	414	25
1040	Cold-drawn	97	669	82	565	16
1040	WQT 700	127	876	93	641	19
1040	WQT 900	118	814	90	621	22
1040	WQT 1100	107	738	80	552	24
1040	WQT 1300	87	600	63	434	32
1080	Annealed	89	614	54	372	25
1080	OQT 700	189	1303	141	972	12
1080	OQT 900	179	1234	129	889	13
1080	OQT 1100	145	1000	103	710	17
1080	OQT 1300	117	807	70	483	23
1141	Annealed	87	600	51	352	26
1141	Cold-drawn	112	772	95	655	14
1141	OQT 700	193	1331	172	1186	9
1141	OQT 900	146	1007	129	889	15
1141	OQT 1100	116	800	97	669	20
1141	OQT 1300	94	648	68	469	28
4140	Annealed	95	655	60	414	26
4140	OQT 700	231	1593	212	1462	12
4140	OQT 900	187	1289	173	1193	15
4140	OQT 1100	147	1014	131	903	18
4140	OQT 1300	118	814	101	696	23
5160	Annealed	105	724	40	276	17
5160	OQT 700	263	1813	238	1641	9
5160	OQT 900	196	1351	179	1234	12
5160	OQT 1100	149	1027	132	910	17
5160	OQT 1300	115	793	103	710	23

*Other properties approximately the same for all carbon and alloy steels:
Modulus of elasticity in tension = 30 000 000 psi (207 GPa)
Modulus of elasticity in shear = 11 500 000 psi (80 GPa)
Density = 0.283 lb_m/in^3 (7680 kg/m^3)

[†]OQT means oil-quenched and tempered. WQT means water-quenched and tempered.

A–15 Typical properties of stainless steels and nonferrous metals.

Material and condition		Ultimate strength, s_u		Yield strength, s_y		Percent elongation	Density		Modulus of elasticity, E	
		ksi	MPa	ksi	MPa		lb/in^{3†}	kg/m^3	psi	GPa
Stainless steels										
AISI 301 annealed		110	758	40	276	60	0.290	8030	28×10^6	193
AISI 301 full hard		185	1280	140	965	8	0.290	8030	28×10^6	193
AISI 430 annealed		75	517	40	276	30	0.280	7750	29×10^6	200
AISI 430 full hard		90	621	80	552	15	0.280	7750	29×10^6	200
AISI 501 annealed		70	483	30	207	28	0.280	7750	29×10^6	200
AISI 501OQT 1000		175	1210	135	931	15	0.280	7750	29×10^6	200
17-4PH H900		210	1450	185	1280	14	0.281	7780	28.5×10^6	197
PH 13-8 Mo H1000		215	1480	205	1410	13	0.279	7720	29.4×10^6	203
Copper and its alloys										
C14500 copper,	soft	32	221	10	69	50	0.323	8940	17×10^6	117
	hard	48	331	44	303	20				
C17200 Beryllium copper,	soft	72	496	20	138	20	0.298	8250	19×10^6	131
	hard	195	1344	145	1000	4				
C36000 brass,	soft	44	305	18	124	20	0.308	8530	16×10^6	110
	hard	70	480	35	240	4				
C54400 bronze,	hard	68	469	57	393	20	0.318	8800	17×10^6	117
Magnesium—cast										
ASTM AZ 63A-T6		40	276	19	131	5	0.066	1830	6.5×10^6	45
Zinc—cast-ZA 12		58	400	47	324	5	0.218	6030	12×10^6	83

A–15 *(continued)*

Titanium and its alloys									
Pure alpha Ti-65A									
Wrought	65	448	55	379	18	0.163	4515	15×10^6	103
Alpha alloy Ti-0.2Pd									
Wrought	50	345	40	276	20	0.163	4515	14.9×10^6	103
Beta alloy Ti-3Al-13V-11Cr									
Aged	185	1280	175	1210	6	0.176	4875	16.0×10^6	110
Alpha-beta alloy Ti-6A1-4V									
Aged	170	1170	155	1070	8	0.160	4432	16.5×10^6	114
Nickel-based alloys									
N06600—annealed									
70°F (21°C)	93	640	37	255	45	0.304	8420	30×10^6	207
800°F (427°C)	89	614	30	207	49				
1200°F (649°C)	65	448	27	186	39				
N06110—40% cold worked									
70°F (21°C)	175	1205	150	1034	18	0.302	8330	30×10^6	207
500°F (260°C)			130	896	18				
800°F (427°C)			120	827	18				
N04400—annealed [At 70°F (21°C]									
Annealed	80	550	30	207	50	0.318	8800	26×10^6	181
Cold drawn	100	690	75	517	30				

†This can be used as specific weight or mass density in lb_m/in^3.

A–16 Properties of structural steels.

Material ASTM no. and products	Ultimate strength, s_u*		Yield strength, s_y*		Percent elongation in 2 in
	ksi	MPa	ksi	MPa	
A36—Carbon steel: shapes,					
plates, and bars	58	400	36	248	21
A 53—grade B-pipe	60	414	35	240	—
A242—HSLA corrosion resistants:					
shapes, plates, and bars					
$\leq\frac{3}{4}$ in thick	70	483	50	345	21
$\frac{3}{4}$ to $1\frac{1}{2}$ in thick	67	462	46	317	21
$1\frac{1}{2}$ to 4 in thick	63	434	42	290	21
A500—Cold-formed structural tubing					
Round, grade B	58	400	42	290	23
Round grade C	62	427	46	317	21
Shaped, grade B	58	400	46	317	23
Shaped, grade C	62	427	50	345	21
A501—Hot-formed structural tubing,					
round or shaped	58	400	36	248	23
A514—Quenched and tempered					
alloy steel; plate					
$\leq 2\frac{1}{2}$ in thick	110	758	100	690	18
$2\frac{1}{2}$ to 6 in thick	100	690	90	620	16
A572—HSLA columbium-vanadium					
steel: shapes, plates and bars					
Grade 42	60	414	42	290	24
Grade 50	65	448	50	345	21
Grade 60	75	517	60	414	18
Grade 65	80	552	65	448	17
A913—HSLA, grade 65: shapes	80	552	65	448	17
A992—HSLA: W-Shapes only	65	448	50	345	21

*Minimum values; may range higher.

HSLA-High strength low-alloy

The American Institute of Steel Construction specifies $E = 29 \times 10^6$ psi (200 GPa) for structural steel.

A–17 Typical properties of cast iron.*

Material type and grade	Ultimate strength s_u† ksi	s_u† MPa	s_{uc}‡ ksi	s_{uc}‡ MPa	s_{us} ksi	s_{us} MPa	Yield strength s_{yt}‡ ksi	s_{yt}‡ MPa	Modulus of elasticity, E‡ psi	E‡ GPa	Percent elongation
Gray iron ASTM A48											
Grade 20	20	138	80	552	32	221	—	—	12.2×10^6	84	<1
Grade 40	40	276	140	965	57	393	—	—	19.4×10^6	134	<0.8
Grade 60	55	379	170	1170	72	496	—	—	21.5×10^6	148	<0.5
Ductile iron ASTM A536											
60-40-18	60	414	—	—	57	393	40	276	24×10^6	165	18
80-55- 6	80	552	—	—	73	503	55	379	24×10^6	165	6
100-70- 3	100	690	—	—	—	—	70	483	24×10^6	165	3
120-90- 2	120	827	180	1240	—	—	90	621	23×10^6	159	2
Austempered ductile iron (ADI)											
Grade 1	125	862	—	—	—	—	80	552	24×10^6	165	10
Grade 2	150	1034	—	—	—	—	100	690	24×10^6	165	7
Grade 3	175	1207	—	—	—	—	125	862	24×10^6	165	4
Grade 4	200	1379	—	—	—	—	155	1069	24×10^6	165	1
Malleable iron ASTM A220											
45008	65	448	240	1650	49	338	45	310	26×10^6	170	8
60004	80	552	240	1650	65	448	60	414	27×10^6	186	4
80002	95	655	240	1650	75	517	80	552	27×10^6	186	2

*The density of cast iron ranges from 0.25 to 0.27 lb_m/in^3 (6920 to 7480 kg/m^3).

†Minimum values; may range higher.

‡Approximate values; may range higher or lower by about 15%.

A–18 Typical properties of aluminum alloys.*

Alloy and temper	Ultimate strength, s_u ksi	s_u MPa	Yield strength, s_y ksi	s_y MPa	Percent elongation	Shear strength, s_{us} ksi	s_{us} MPa
1100-H12	16	110	15	103	25	10	69
1100-H18	24	165	22	152	15	13	90
2014-0	27	186	14	97	18	18	124
2014-T4	62	427	42	290	20	38	262
2014-T6	70	483	60	414	13	42	290
3003-0	16	110	6	41	40	11	76
3003-H12	19	131	18	124	20	12	83
3003-H18	29	200	27	186	10	16	110
5154-0	35	241	17	117	27	22	152
5154-H32	39	269	30	207	15	22	152
5154-H38	48	331	39	269	10	28	193
6061-0	18	124	8	55	30	12	83
6061-T4	35	241	21	145	25	24	165
6061-T6	45	310	40	276	17	30	207
7075-0	33	228	15	103	16	22	152
7075-T6	83	572	73	503	11	48	331
Casting alloys (permanent mold castings)							
204.0-T4	48	331	29	200	8	—	—
206.0-T6	65	445	59	405	6	—	—
356.0-T6	41	283	30	207	10	—	—

*Modulus of elasticity E for most aluminum alloys, including 1100, 3003, 6061, and 6063, is 10×10^6 psi (69.0 GPa). For 2014, $E = 10.6 \times 10^6$ psi (73.1 GPa). For 5154, $E = 10.2 \times 10^6$ psi (70.3 GPa). For 7075, $E = 10.4 \times 10^6$ psi (71.7 GPa). Density of most aluminum alloys is approximately 0.10 lb_m/in^3 (2770 kg/m^3).

A–19 Typical properties of wood.

Type and grade	Allowable stress										Modulus of elasticity	
	Bending		Tension parallel to grain		Horizontal shear		Compression: Perpendicular to grain		Compression: Parallel to grain			
	psi	MPa	psi	MPa	psi	MPa	psi	MPa	psi	MPa	ksi	GPa
Douglas fir—2 to 4 in thick, 6 in and wider												
No. 1	1750	12.1	1050	7.2	95	0.66	385	2.65	1250	8.62	1800	12.4
No. 2	1450	10.0	850	5.9	95	0.66	385	2.65	1000	6.90	1700	11.7
No. 3	800	5.5	475	3.3	95	0.66	385	2.65	600	4.14	1500	10.3
Hemlock—2 to 4 in thick, 6 in and wider												
No. 1	1400	9.6	825	5.7	75	0.52	245	1.69	1000	6.90	1500	10.3
No. 2	1150	7.9	675	4.7	75	0.52	245	1.69	800	5.52	1400	9.7
No. 3	625	4.3	375	2.6	75	0.52	245	1.69	500	3.45	1200	8.3
Southern pine—$2\frac{1}{2}$ to 4 in thick, 6 in and wider												
No. 1	1400	9.6	825	5.7	80	0.55	270	1.86	850	5.86	1600	11.0
No. 2	1000	6.9	575	4.0	70	0.48	230	1.59	550	3.79	1300	9.0
No. 3	650	4.5	375	2.6	70	0.48	230	1.59	400	2.76	1300	9.0

A–20 Typical properties of selected plastics.

Material	Type	Tensile strength		Tensile modulus		Flexural strength		Flexural modulus		Impact strength IZOD (ft·lb/in of notch)
		(ksi)	(MPa)	(ksi)	(MPa)	(ksi)	(MPa)	(ksi)	(MPa)	
Nylon 66	Dry	21.0	146	1200	8700	32.0	221	1100	7900	
30% Glass	50% R.H.	15.0	102	800	5500					
ABS	Medium-impact	6.0	41	360	2480	11.5	79	310	2140	4.0
	High-impact	5.0	34	250	1720	8.0	55	260	1790	7.0
Polycarbonate	General-purpose	9.0	62	340	2340	11.0	76	300	2070	12.0
Acrylic	Standard	10.5	72	430	2960	16.0	110	460	3170	0.4
	High-impact	5.4	37	220	1520	7.0	48	230	1590	1.2
PVC	Rigid	6.0	41	350	2410			300	2070	0.4–20.0 (varies widely)
Polyimide	25% graphite powder filler	5.7	39			12.8	88	900	6210	0.25
	Glass-fiber filler	27.0	186			50.0	345	3250	22 400	17.0
	Laminate	50.0	345			70.0	483	4000	27 580	13.0
Acetal	Copolymer	8.0	55	410	2830	13.0	90	375	2590	1.3
Polyurethane	Elastomer	5.0	34	100	690	0.6	4			No break
Phenolic	General	6.5	45	1100	7580	9.0	62	1100	7580	0.3
Polyester with glass-fiber mat reinforcement (approx. 30% glass by weight)										
	Lay-up, contact mold	9.0	62			16.0	110	800	5520	
	Cold press molded	12.0	83			22.0	152	1300	8960	
	Compression molded	25.0	172			10.0	69	1300	8960	

A–21 Design stress guidelines.

Direct Normal Stresses—General machine and structural design

Manner of loading	Ductile Materials (% Elongation > 5%)	Brittle Materials (% Elongation < 5%)
Static loads	$\sigma_d = s_y/2$	$\sigma_d = s_u/6$
Repeated loads	$\sigma_d = s_u/8$	$\sigma_d = s_u/10$
Impact or shock	$\sigma_d = s_u/12$	$\sigma_d = s_u/15$

Direct Normal Stresses—Static loads on steel members of building-like structures

AISC Code: $\sigma_d = s_y/1.67 = 0.60\ s_y$ or $\sigma_d = s_u/2.00 = 0.50\ s_u$
Whichever is lower.

Direct Normal Stresses—Static loads on aluminum members of building-like structures

Aluminum Association: $\sigma_d = s_y/1.65 = 0.61\ s_y$ or $\sigma_d = s_u/1.95 = 0.51\ s_u$
Whichever is lower.

Design Shear Stresses—For direct shear and for torsional shear stresses

Based on maximum shear stress theory of failure:

$$\tau_d = s_{ys}/N = 0.5\ s_y/N = s_y/2N$$

Manner of loading	Design factor	Design shear stress
Static loads	Use $N = 2$	$\tau_d = s_y/4$
Repeated loads	Use $N = 4$	$\tau_d = s_y/8$
Shock or impact	Use $N = 6$	$\tau_d = s_y/12$

Estimates for the Ultimate Strength in Shear

Formula	Material
$s_{us} = 0.65\ s_u$	Aluminum alloys
$s_{us} = 0.82\ s_u$	Steel—Plain carbon and alloy
$s_{us} = 0.90\ s_u$	Malleable iron and copper alloys
$s_{us} = 1.30\ s_u$	Gray cast iron

Allowable Bearing Stress

Steel—Flat surfaces or the projected area of pins in reamed, drilled, or bored holes:

$$\sigma_{bd} = 0.90\ s_y$$

Allowable Bearing Load, R_a,—Steel roller on flat steel plate

U.S. Customary Units	SI Metric Units
$R_a = (s_y - 13)(0.03\ dL)$	$R_a = (s_y - 90)(3.0 \times 10^{-5} dL)$

Where: R_a = Allowable bearing load in kips or kN
s_y = Yield strength of steel in ksi or MPa
d = Roller diameter in inches or mm
L = Length of roller in inches or mm

(continued)

A–21 *(continued)*

Allowable bearing stresses on masonry and soils for use in this book.

Material	Allowable bearing stress, σ_{bd} psi	MPa
Sandstone and limestone	400	2.76
Brick in cement mortar	250	1.72
Solid hard rock	350	2.41
Shale or medium rock	140	0.96
Soft rock	70	0.48
Hard clay or compact gravel	55	0.38
Soft clay or loose sand	15	0.10

Concrete: $\sigma_{bd} = Kf'_c = (0.34\sqrt{A_2/A_1})f'_c$ (But maximum $\sigma_{bd} = 0.68f'_c$)

Where: f'_c = Rated strength of concrete
A_1 = Bearing area
A_2 = Full area of the support

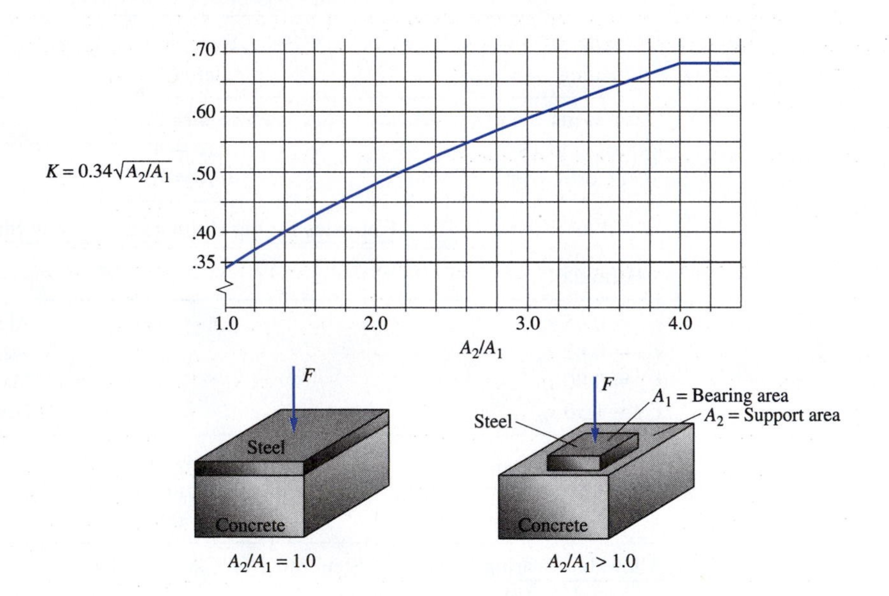

A–21 *(continued)*

Allowable Bearing Stresses on Masonry and Soils

Material	Allowable bearing stress, σ_{bd} psi	MPa
Sandstone and limestone	400	2.76
Brick in cement mortar	250	1.72
Solid hard rock	350	2.41
Shale or medium rock	140	0.96
Soft rock	70	0.42
Hard clay or compact gravel	55	0.38
Soft clay or loose sand	15	0.10

Concrete: $\sigma_{bd} = Kf'_c = (0.34\sqrt{A_2/A_1})f'_c$ (But maximum $\sigma_{bd} = 0.68f'_c$)
Where: f'_c = Rated strength of concrete
A_1 = Bearing area
A_2 = Full area of the support

Design Bending Stresses—General machine and structural design

Manner of loading	Ductile Materials (% Elongation >5%)	Brittle Materials (% Elongation <5%)
Static loads	$\sigma_d = s_y/2$	$\sigma_d = s_u/6$
Repeated loads	$\sigma_d = su/8$	$\sigma_d = s_u/10$
Impact or shock	$\sigma_d = su/12$	$\sigma_d = s_u/15$

Design Bending Stresses—AISC specifications for structural steel carrying static loads on building-like structures

$$\sigma_d = s_y/1.5 = 0.66\ s_y$$

Design Bending Stresses—Aluminum Association specifications for aluminum carrying static loads on building-like structures

$$\sigma_d = s_y/1.65 = 0.61\ s_y \quad \text{or} \quad \sigma_d = s_u/1.95 = 0.51\ s_u$$

Whichever is lower.

Design Shear Stresses for Beams in Bending
Rolled structural steel beam shapes—allowable web shear stress (AISC)

$$\tau_d = 0.40\ s_y$$

General ductile materials carrying static loads: Based on yield strength of the material in shear with design factor, $N = 2$:

$$\tau_d = s_{ys}/N = 0.5\ s_y/N = s_y/2N = s_y/2(2) = s_y/4 = 0.25\ s_y$$

A–22 Stress Concentration Factors

A–22–1 Axially loaded grooved round bar in tension.

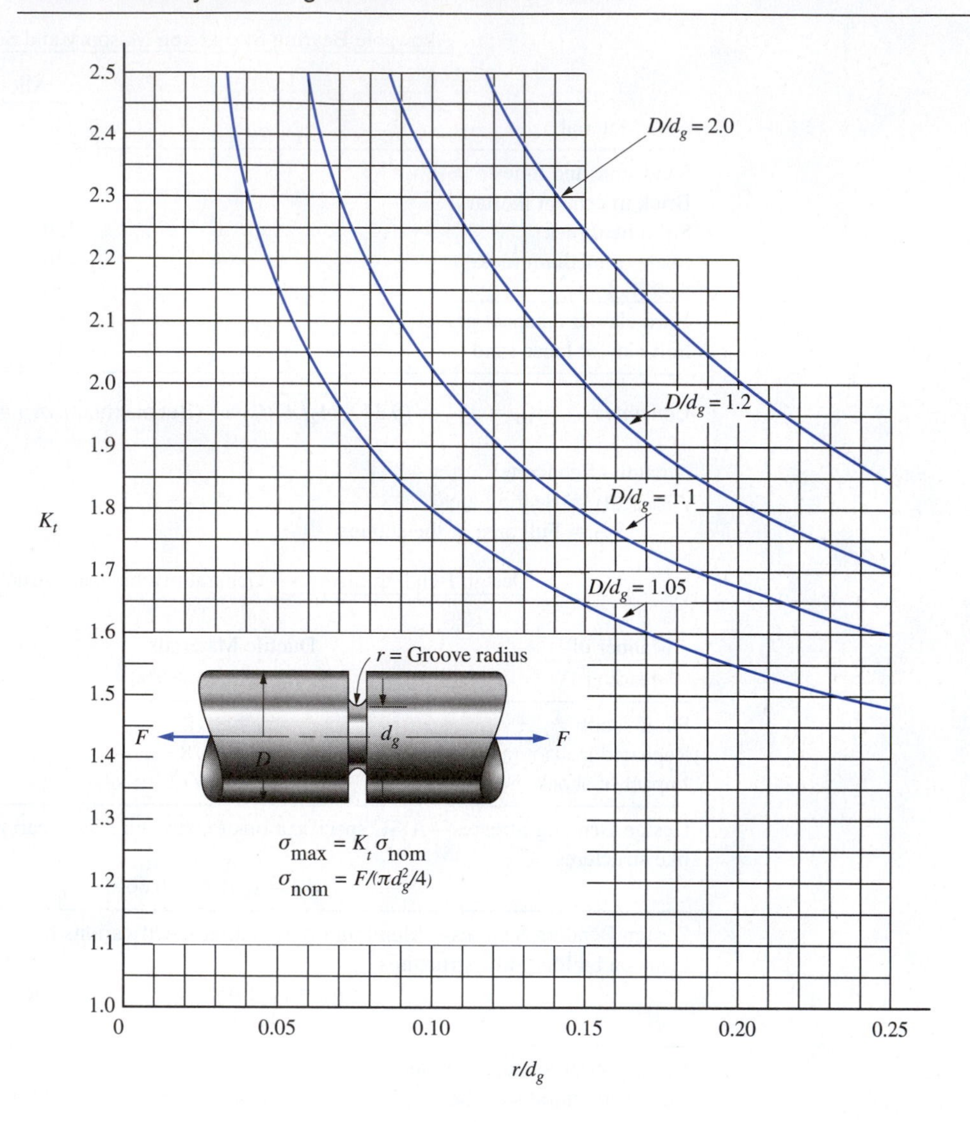

A–22–2 Axially loaded stepped round bar in tension.

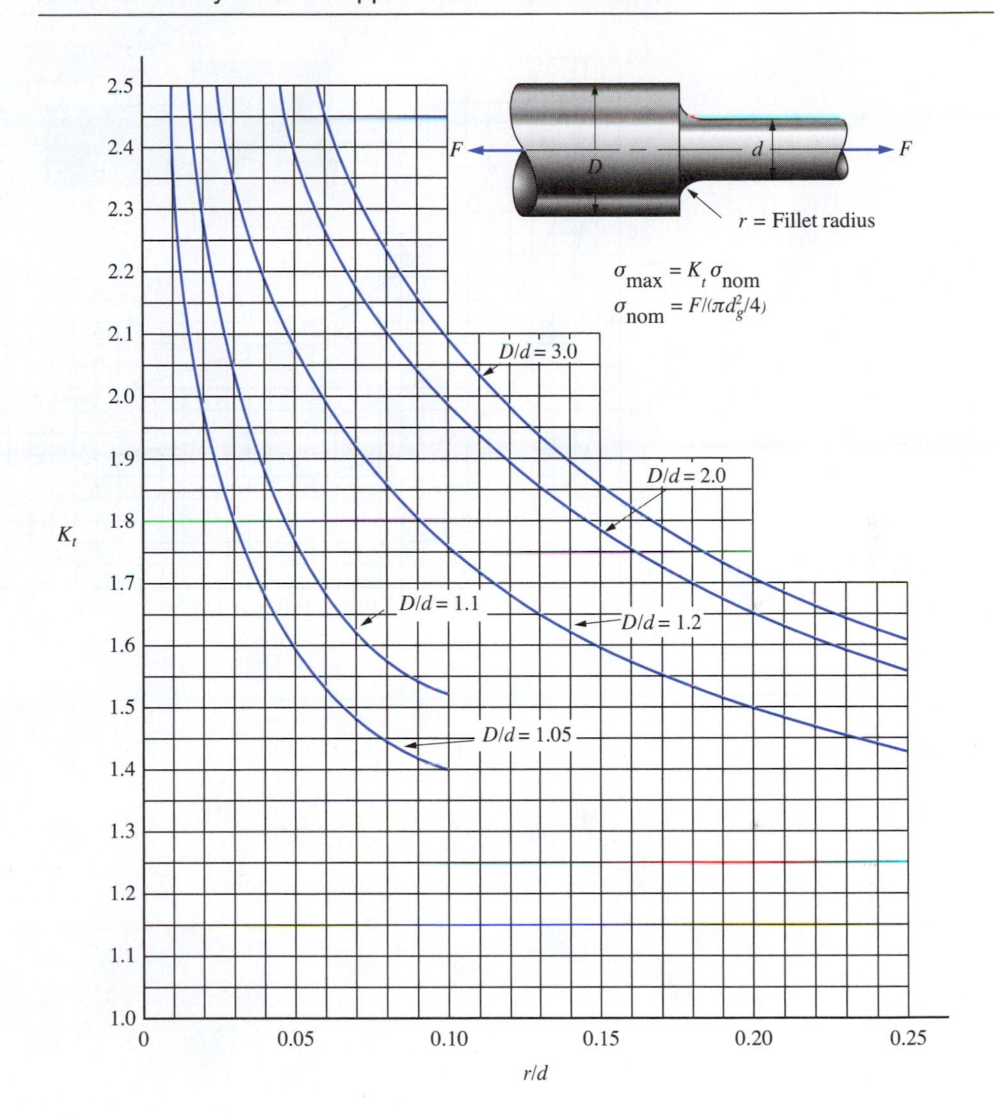

A–22–3 Axially loaded stepped flat plate in tension.

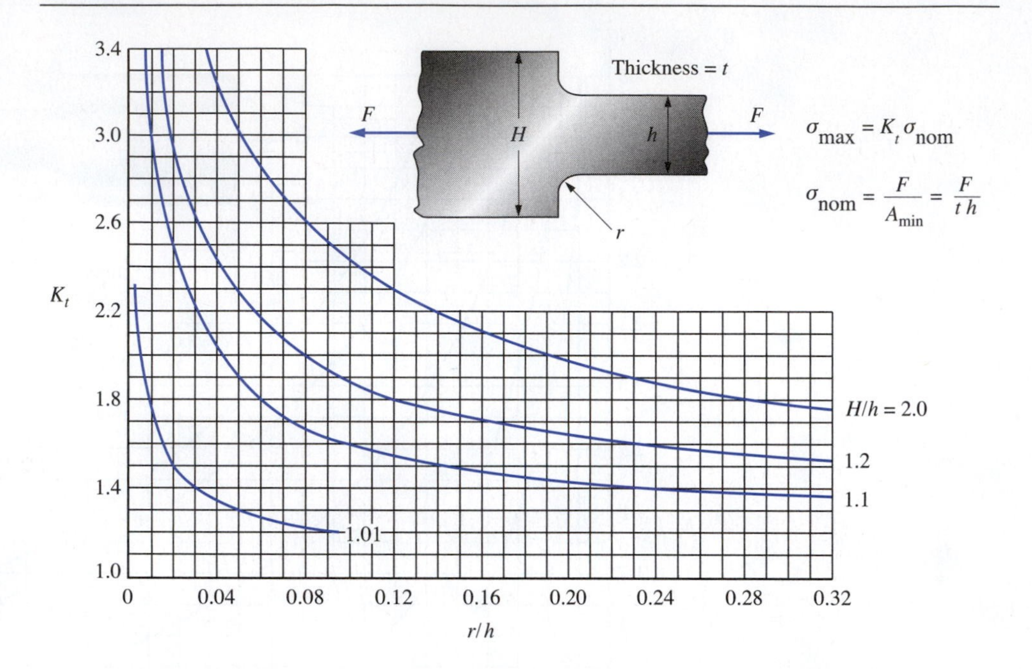

A–22–4 Flat plate with central hole in tension and bending

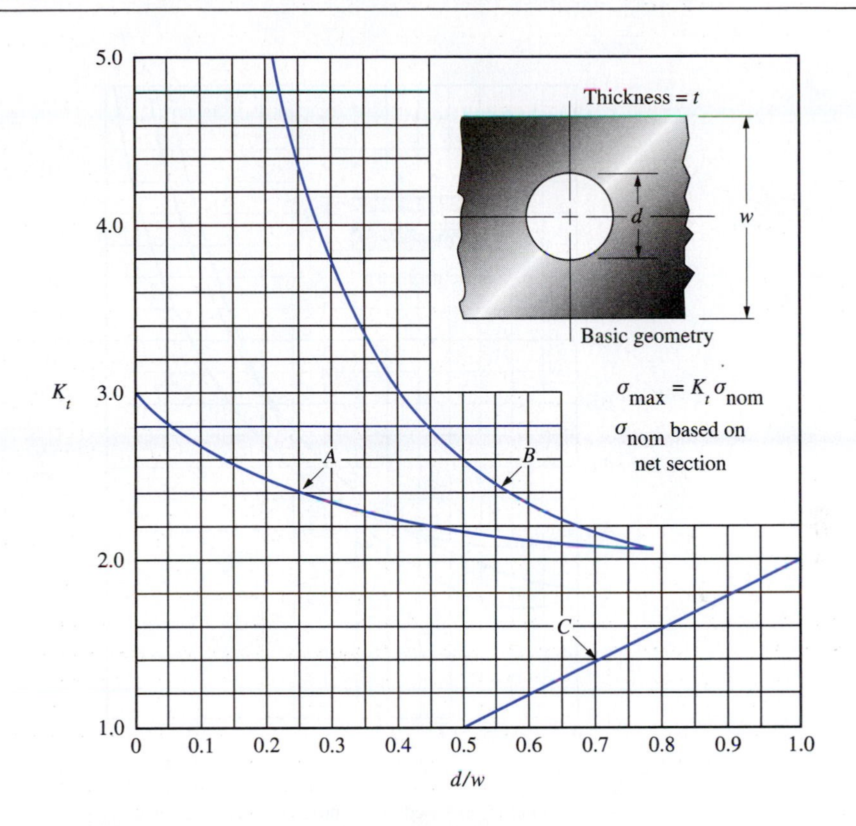

Curve A

Direct tension on plate

$$\sigma_{nom} = \frac{F}{A_{net}} = \frac{F}{(w-d)t}$$

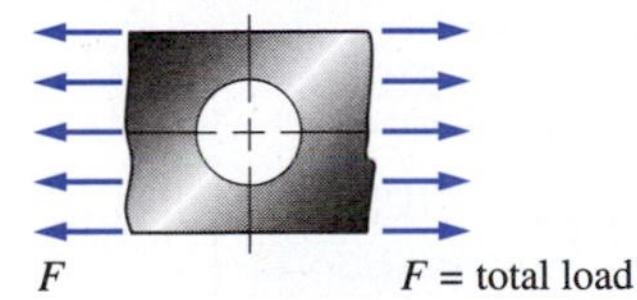

Curve B

Tension load applied through a pin in the hole

$$\sigma_{nom} = \frac{F}{A_{net}} = \frac{F}{(w-d)t}$$

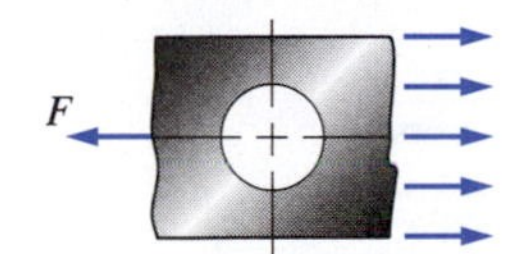

Curve C

Bending in the plane of the plate

$$\sigma_{nom} = \frac{Mc}{I_{net}} = \frac{6Mw}{(w^3-d^3)t}$$

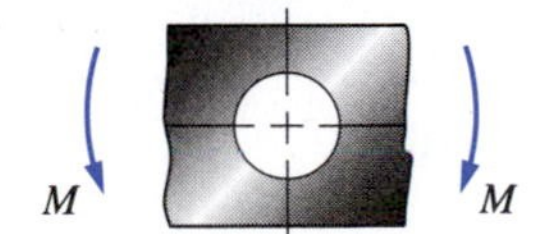

Note: $K_t = 1.0$ for $d/w < 0.5$

A–22–5 Round bar with transverse hole in tension, bending, and torsion.

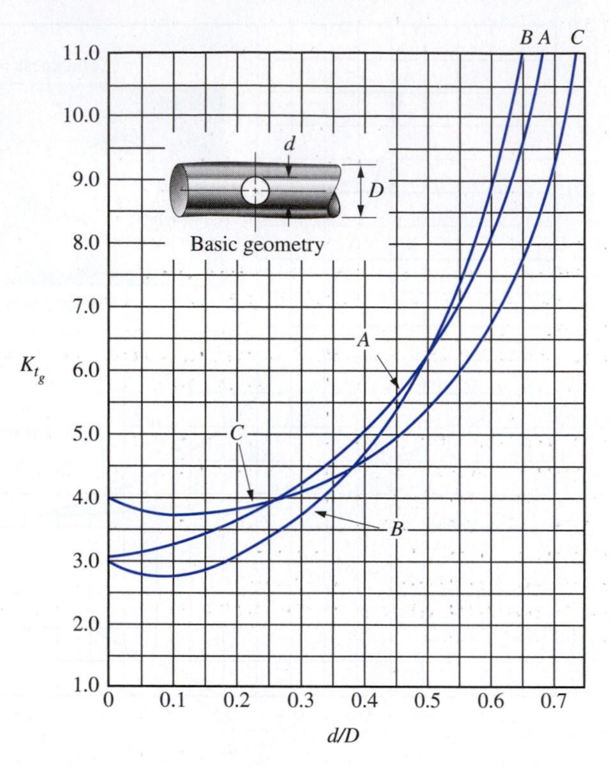

Note: K_{t_g} is based on the nominal stress in a round bar without a hole (gross section).

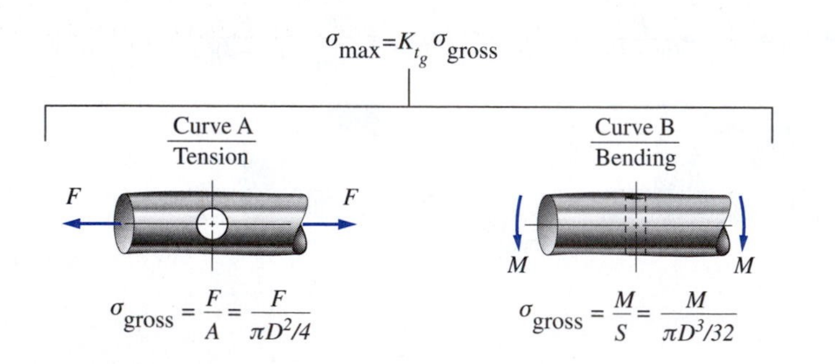

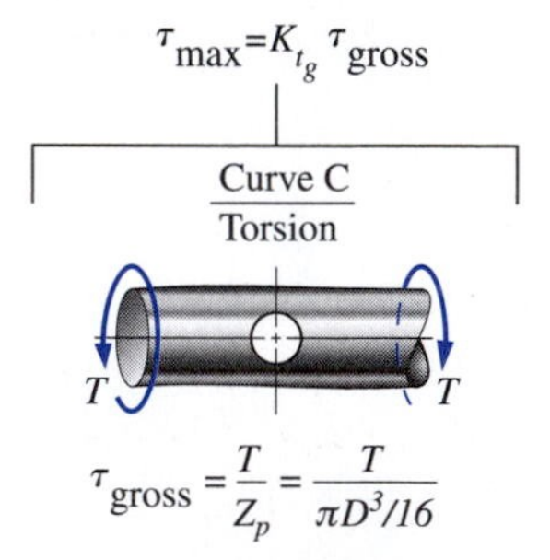

A–22–6 Grooved round bar in torsion.

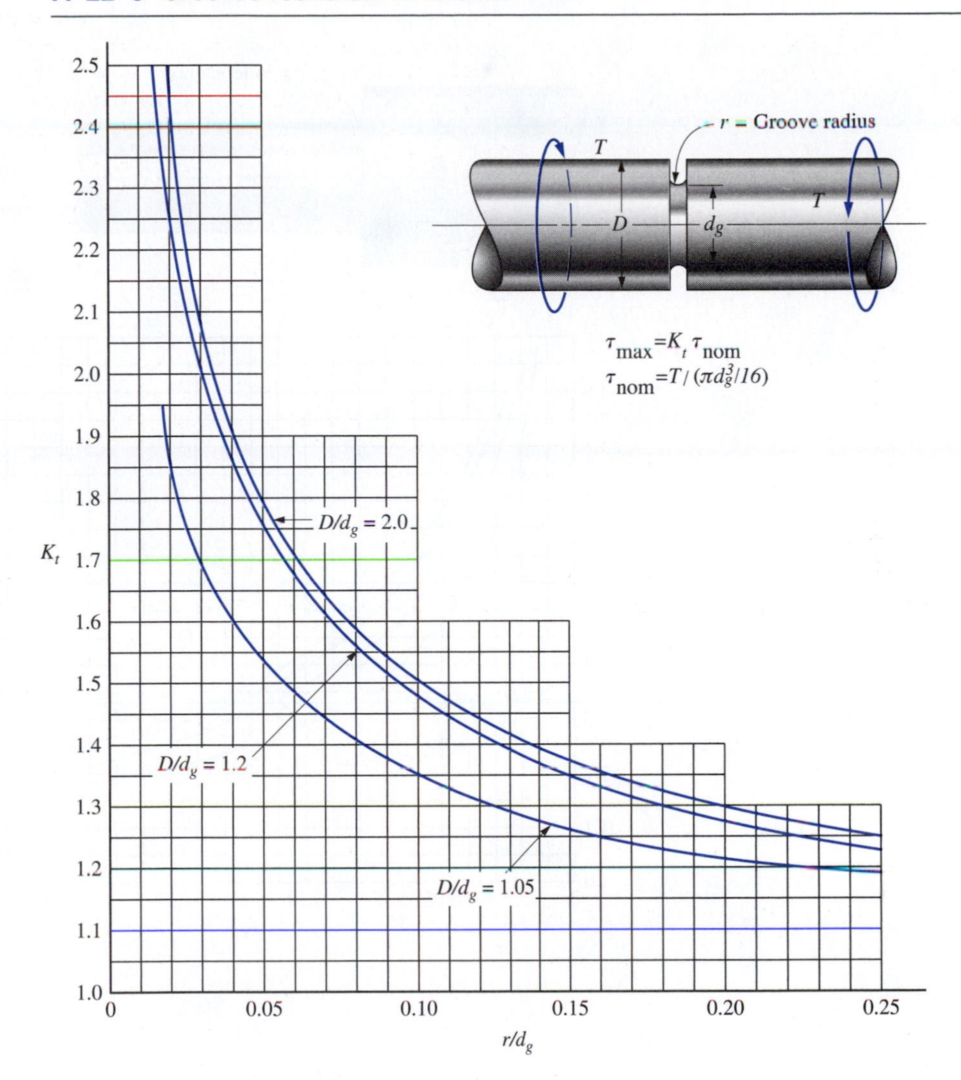

A–22–7 Stepped round bar in torsion.

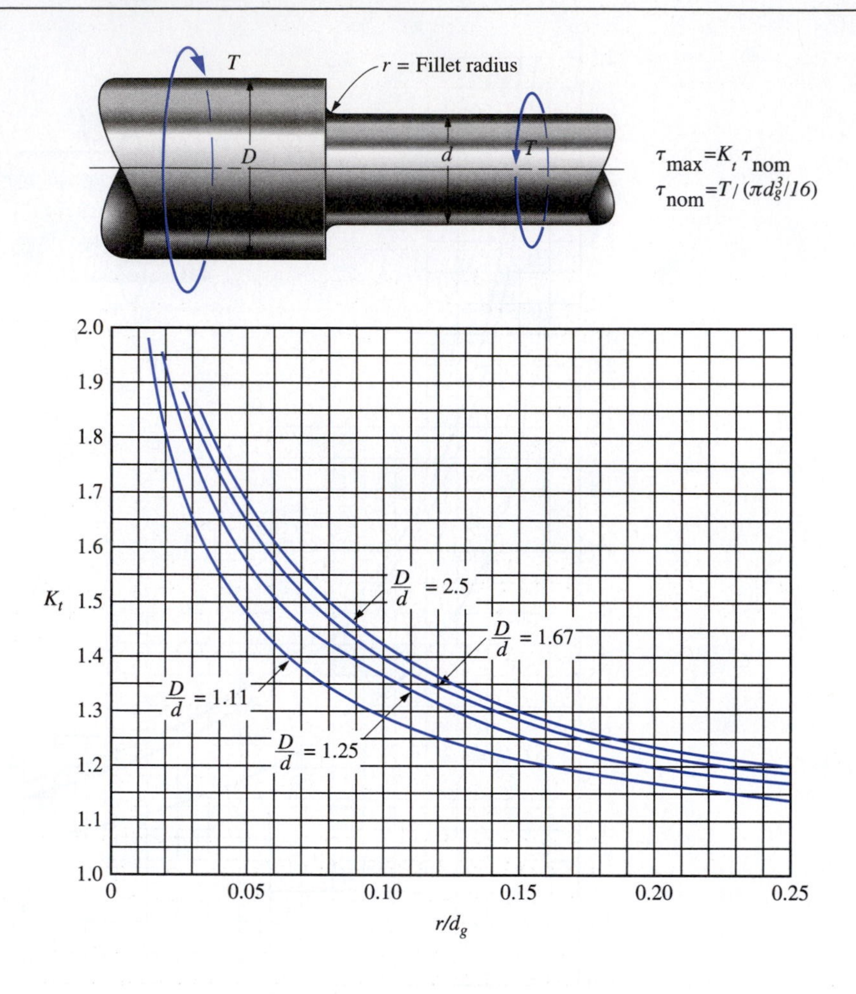

A–22–8 Grooved round bar in bending.

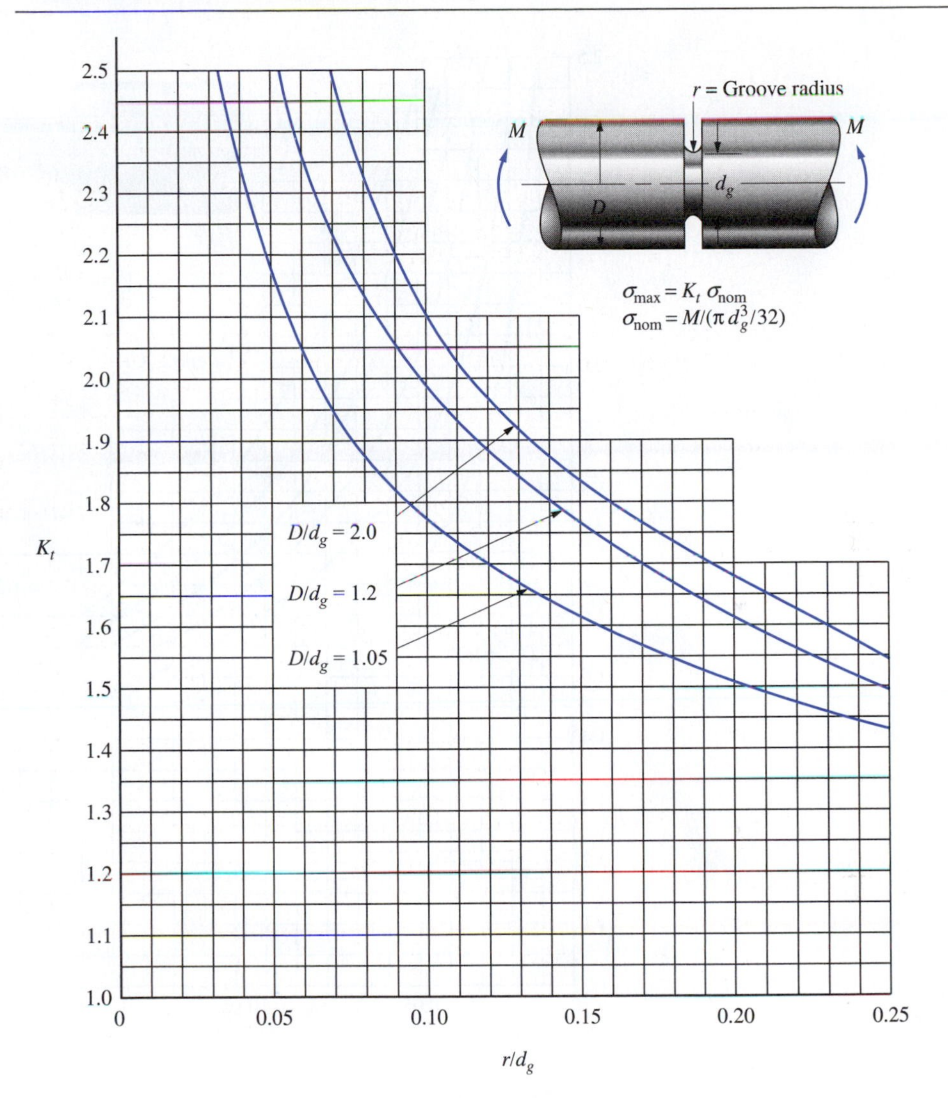

A–22–9 Stepped round bar in bending.

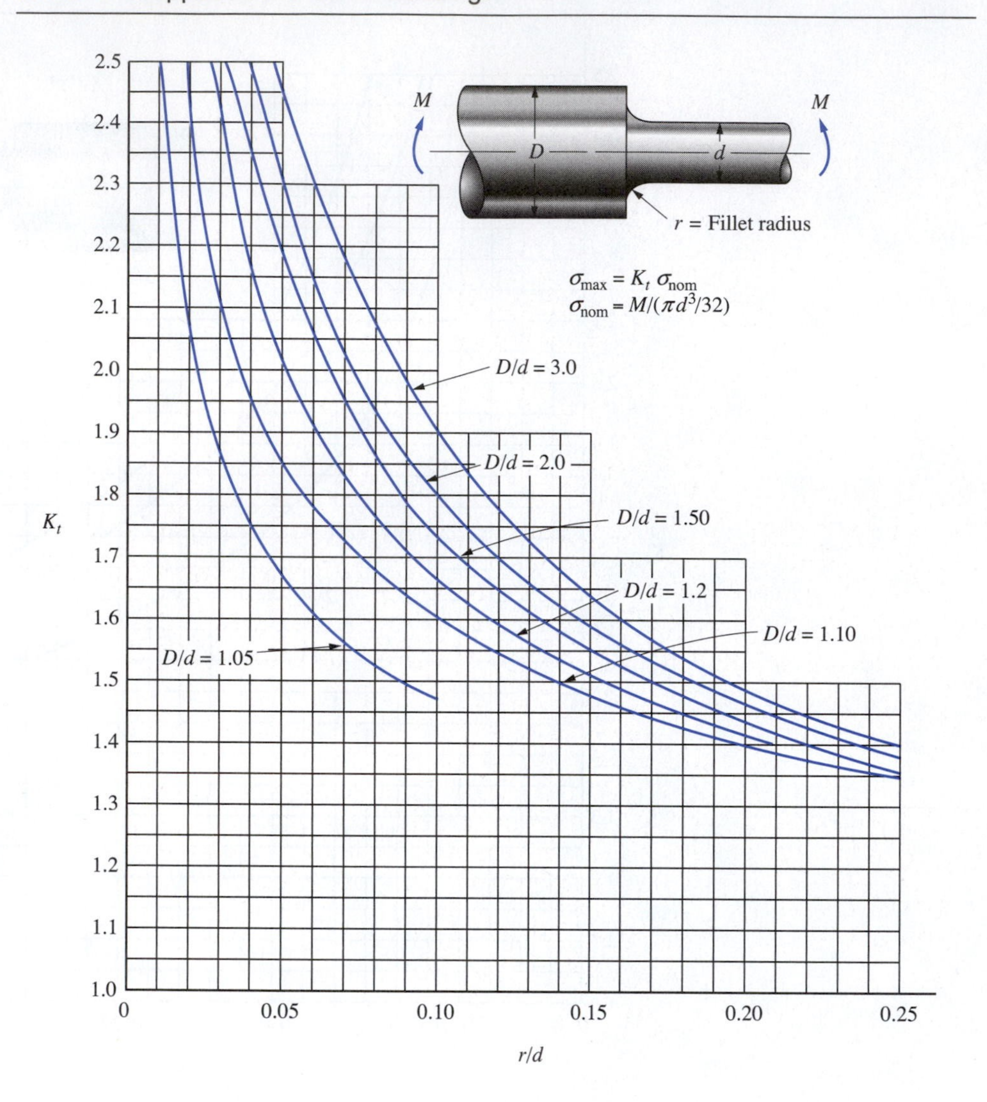

A–22–10 Stepped flat plate in bending.

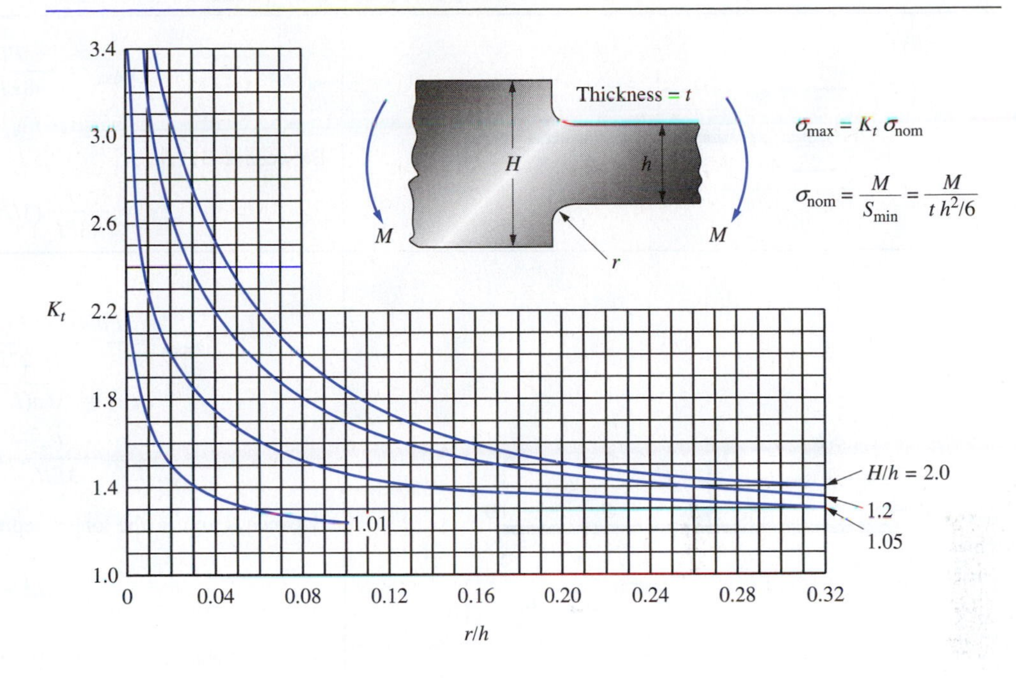

A–22–11 Shafts with keyseats—bending and torsion.

Type of keyseat	K_t^*
Sled-runner	1.6
Profile	2.0

*K_t is to be applied to the stress computed for the full nominal diameter of the shaft where the keyseat is located.

A–23 Beam deflection formulas for simply supported beams.

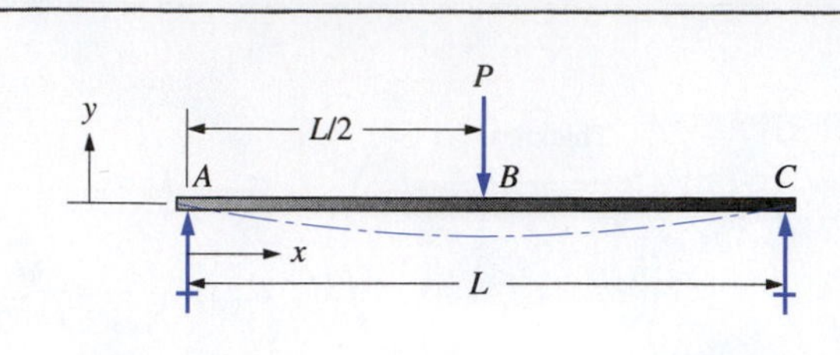

(a)

$$y_B = y_{max} = \frac{-PL^3}{48EI} \quad \text{at center}$$

Between A and B:

$$y = \frac{-Px}{48EI}(3L^2 - 4x^2)$$

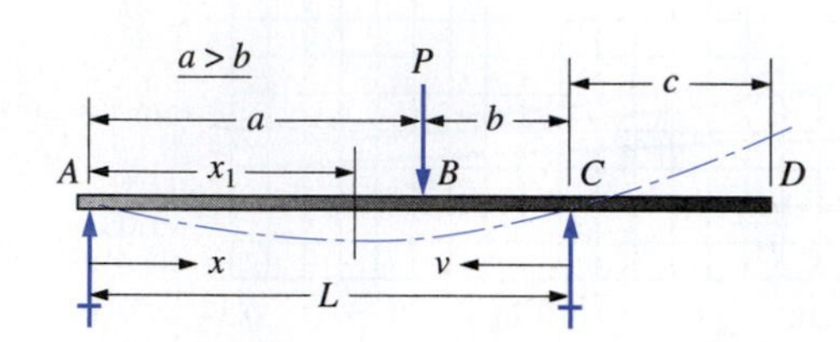

(b)

$$y_{max} = \frac{Pab(L + b)\sqrt{3a(L + b)}}{27EIL}$$

$$\text{at } x_1 = \sqrt{a(L + b)/3}$$

$$y_B = \frac{-Pa^2b^2}{3EIL} \quad \text{at load}$$

Between A and B (the longer segment):

$$y = \frac{-Pbx}{6EIL}(L^2 - b^2 - x^2)$$

Between B and C (the shorther segment):

$$y = \frac{-Pav}{6EIL}(L^2 - v^2 - a^2)$$

At end of overhang at D:

$$y_D = \frac{Pabc}{6EIL}(L + a)$$

(c)

$$y_E = y_{max} = \frac{-Pa}{24EI}(3L^2 - 4a^2) \quad \text{at center}$$

$$y_B = y_C = \frac{-Pa^2}{6EI}(3L - 4a) \quad \text{at loads}$$

Between A and B:

$$y = \frac{-Px}{6EI}(3aL - 3a^2 - x^2)$$

Between B and C:

$$y = \frac{-Pa}{6EI}(3Lx - 3x^2 - a^2)$$

A–23 *(continued)*

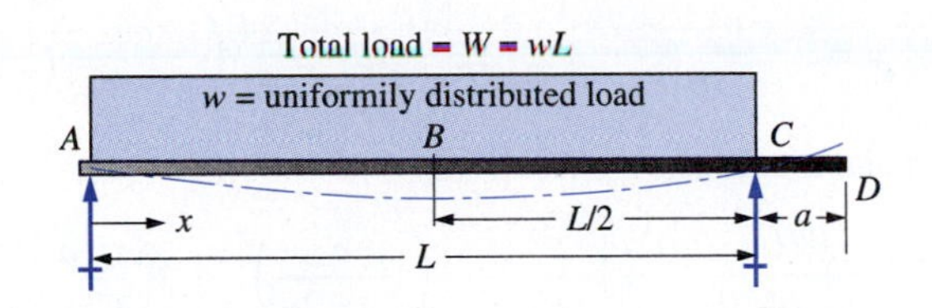

(*d*)

$$y_B = y_{\max} = \frac{-5wL^4}{384EI} = \frac{-5WL^3}{384EI} \text{ at center}$$

Between A and B:

$$y = \frac{-wx}{24EI}(L^3 - 2Lx^2 + x^3)$$

At D at end:

$$y_D = \frac{wL^3a}{24EI}$$

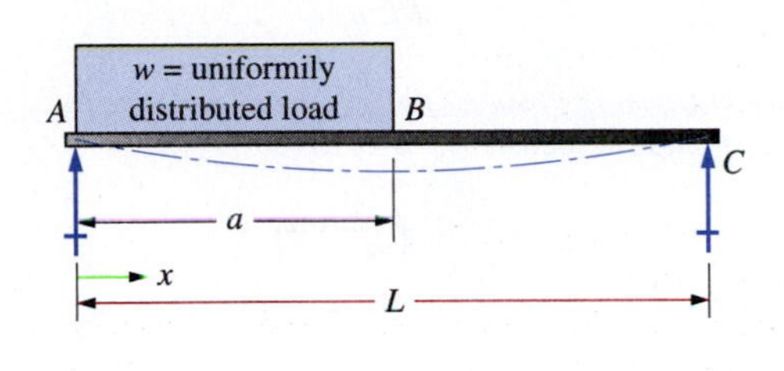

(*e*)

Between A and B:

$$y = \frac{-wx}{24EIL}[a^2(2L-a)^2 - 2ax^2(2L-a) + Lx^3]$$

Between B and C:

$$y = \frac{-wa^2(L-x)}{24EIL}(4Lx - 2x^2 - a^2)$$

(*f*)

M_B = concentrated moment at B

Between A and B:

$$y = \frac{-M_B}{6EI}\left[\left(6a - \frac{3a^2}{L} - 2L\right)x - \frac{x^3}{L}\right]$$

Between B and C:

$$y = \frac{M_B}{6EI}\left[3a^2 + 3x^2 - \frac{x^3}{L} - \left(2L + \frac{3a^2}{L}\right)x\right]$$

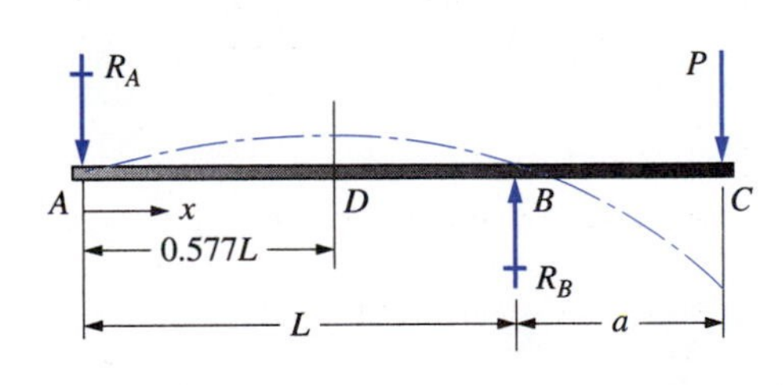

(*g*)

At C at end of overhang:

$$y_C = \frac{-Pa^2}{3EI}(L + a)$$

At D, maximum upward deflection:

$$y_D = 0.06415\frac{PaL^2}{EI}$$

A–23 *(continued)*

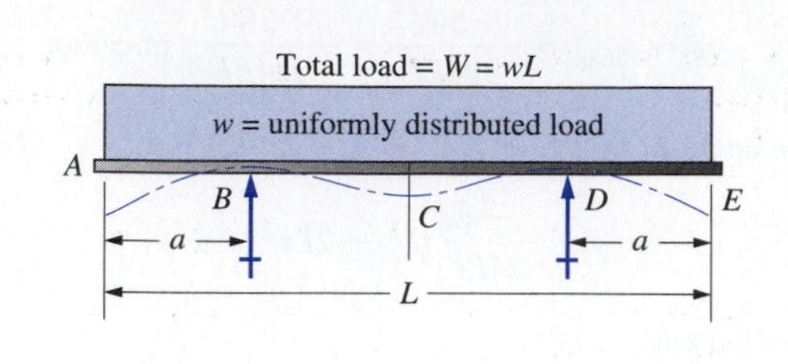

(*h*)

At C at center:

$$y = \frac{-W(L - 2a)^3}{384EI}\left[\frac{5}{L}(L - 2a) - \frac{24}{L}\left(\frac{a^2}{L - 2a}\right)\right]$$

At A and E at ends:

$$y = \frac{-W(L - 2a)^3 a}{24EIL}\left[-1 + 6\left(\frac{a}{L - 2a}\right)^2 + 3\left(\frac{a}{L - 2a}\right)^3\right]$$

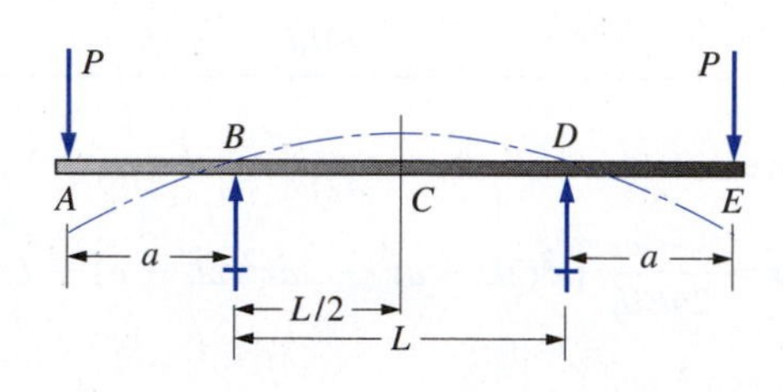

(*i*)

At C at center:

$$y = \frac{PL^2 a}{8EI}$$

At A and E at ends at loads:

$$y = \frac{-Pa^2}{3EI}\left(a + \frac{3}{2}L\right)$$

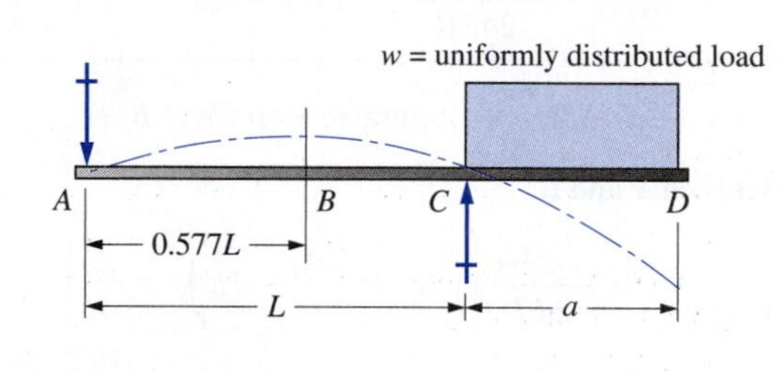

(*j*)

At B:

$$y = 0.03208\,\frac{wa^2L^2}{EI}$$

At D at end:

$$y = \frac{-wa^3}{24EI}(4L + 3a)$$

A–24 Beam deflection formulas for cantilevers.

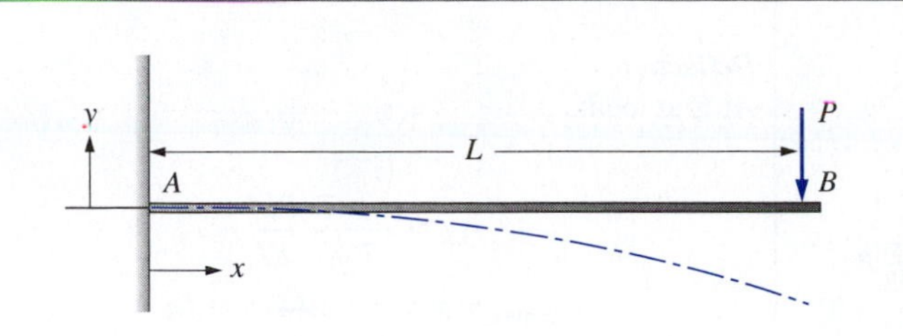

(a)

At B at end:

$$y_B = y_{\max} = \frac{-PL^3}{3EI}$$

Between A and B:

$$y = \frac{-Px^2}{6EI}(3L - x)$$

L
P
a
b
A
B
C
x

(b)

At B at load:

$$y_B = \frac{-Pa^3}{3EI}$$

At C at end:

$$y_C = y_{\max} = \frac{-Pa^2}{6EI}(3L - a)$$

Between A and B:

$$y = \frac{-Px^2}{6EI}(3a - x)$$

Between B and C:

$$y = \frac{-Pa^2}{6EI}(3x - a)$$

L
w = uniformly distributed load
A
B
x

(c)

W = total load = wL

At B at end:

$$y_B = y_{\max} = \frac{-WL^3}{8EI}$$

Between A and B:

$$y = \frac{-Wx^2}{24EIL}[2L^2 + (2L - x)^2]$$

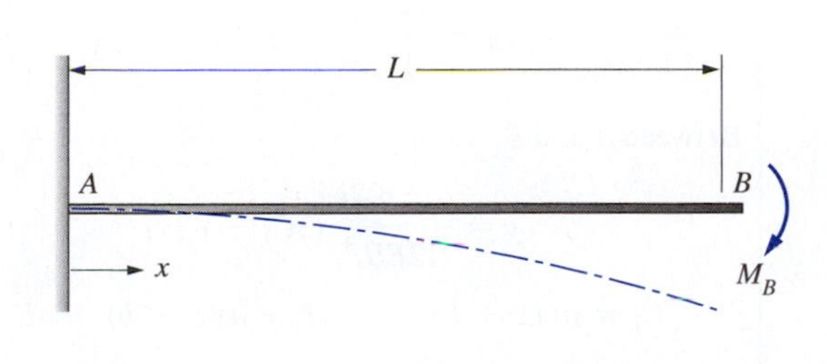

(d)

M_B = concentrated moment at end

At B at end:

$$y_B = y_{\max} = \frac{-M_B L^2}{2EI}$$

Between A and B:

$$y = \frac{-M_B x^2}{2EI}$$

A–25 Beam diagrams and deflection formulas for statically indeterminate beams.

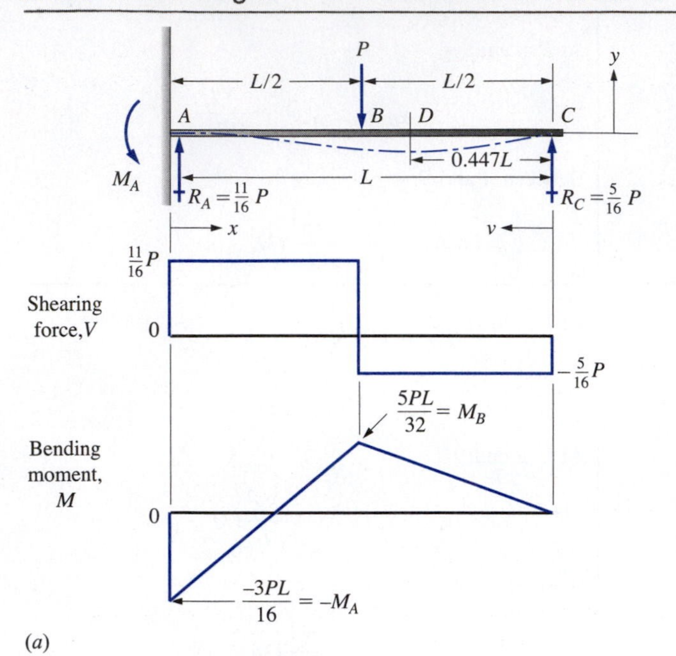

(a)

Deflections
At B at load:

$$y_B = \frac{-7}{768}\frac{PL^3}{EI}$$

y_{max} is at $v = 0.447L$ at D:

$$y_D = y_{max} = \frac{-PL^3}{107EI}$$

Between A and B:

$$y = \frac{-Px^2}{96EI}(9L - 11x)$$

Between B and C:

$$y = \frac{-Pv}{96EI}(3L^2 - 5v^2)$$

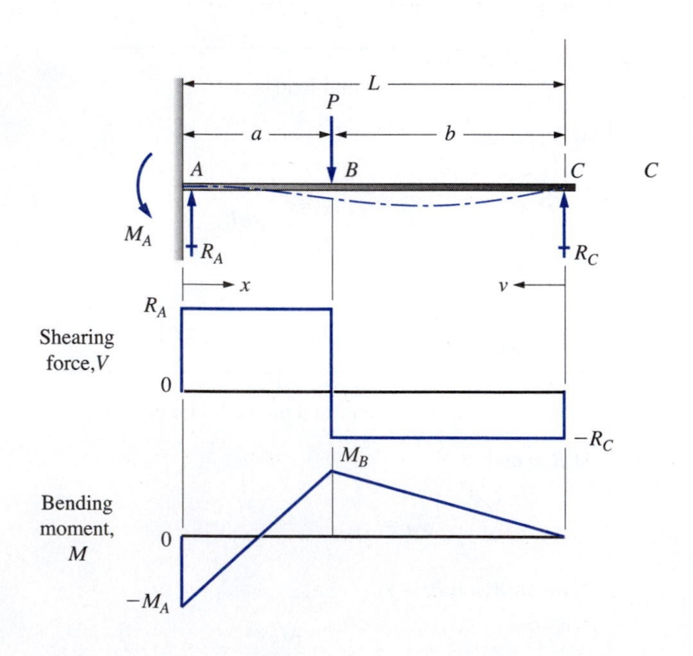

(b)

Reactions

$$R_A = \frac{Pb}{2L^3}(3L^2 - b^2)$$

$$R_C = \frac{Pa^2}{2L^3}(b + 2L)$$

Moments

$$M_A = \frac{-Pab}{2L^2}(b + L)$$

$$M_B = \frac{Pa^2b}{2L^3}(b + 2L)$$

Deflections
At B at load:

$$y_B = \frac{-Pa^3b^2}{12EIL^3}(3L + b)$$

Between A and B:

$$y = \frac{-Px^2b}{12EIL^3}(3C_1 - C_2x)$$

$$C_1 = aL(L + b);\ C_2 = (L + a)(L + b) + aL$$

Between B and C:

$$y = \frac{-Pa^2v}{12EIL^3}[3L^2b - v^2(3L - a)]$$

A–25 *(continued)*

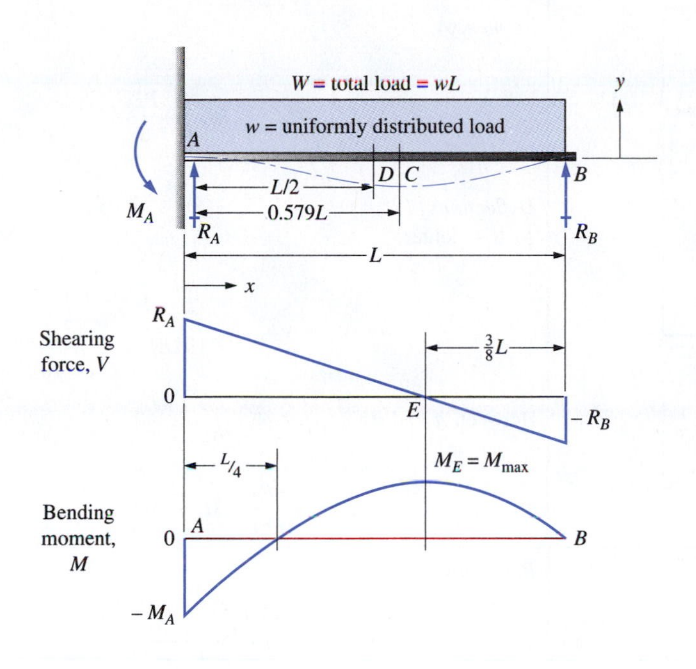

(*c*)

Ractions

$$R_A = \frac{5}{8}W$$

$$R_B = \frac{3}{8}W$$

Moments

$$M_A = -0.125WL$$

$$M_E = 0.0703WL$$

Deflections

At C at $x = 0.579L$:

$$y_C = y_{max} = \frac{-WL^3}{185EI}$$

At D at center:

$$y_D = \frac{-WL^3}{192EI}$$

Between A and B:

$$y = \frac{-Wx^2(L-x)}{48EIL}(3L-2x)$$

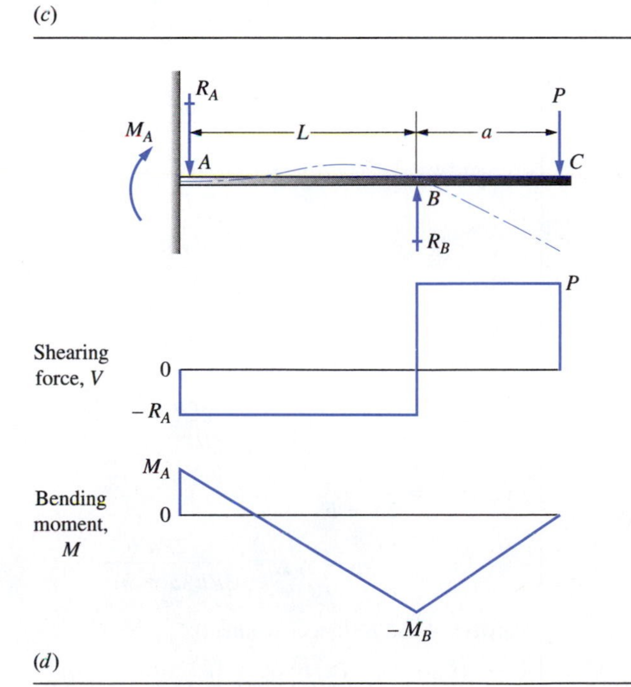

(*d*)

Reactions

$$R_A = \frac{-3Pa}{2L}$$

$$R_B = P\left(1 + \frac{3a}{2L}\right)$$

Moments

$$M_A = \frac{Pa}{2}$$

$$M_B = -Pa$$

Deflection

At C at end:

$$y_C = \frac{-PL^3}{EI}\left(\frac{a^2}{4L^2} + \frac{a^3}{3L^3}\right)$$

A–25 *(continued)*

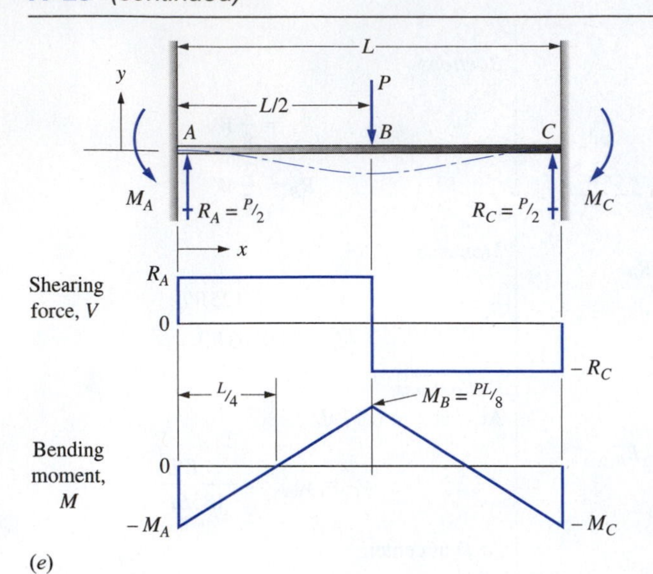

(e)

Moments

$$M_A = M_B = M_C = \frac{PL}{8}$$

Deflections

At B at center:

$$y_B = y_{\max} = \frac{-PL^3}{192EI}$$

Between A and B:

$$y = \frac{-Px^2}{48EI}(3L - 4x)$$

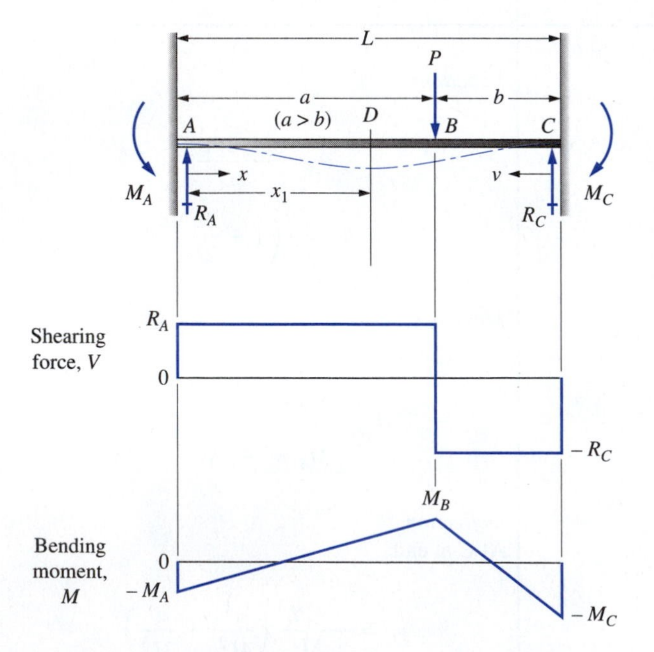

(f)

Reactions

$$R_A = \frac{Pb^2}{L^3}(3a + b)$$

$$R_C = \frac{Pa^2}{L^3}(3b + a)$$

Moments

$$M_A = \frac{-Pab^2}{L^2}$$

$$M_B = \frac{2Pa^2b^2}{L^3}$$

$$M_C = \frac{-Pa^2b}{L^2}$$

Deflections

At B at load:

$$y_B = \frac{-Pa^3b^3}{3EIL^3}$$

At D at $x_1 = \dfrac{2aL}{3a + b}$

$$y_D = y_{\max} = \frac{-2Pa^3b^2}{3EI(3a + b)^2}$$

Between A and B (longer segment):

$$y = \frac{-Px^2b^2}{6EIL^3}[2a(L - x) + L(a - x)]$$

Between B and C (shorter segment):

$$y = \frac{-Pv^2a^2}{6EIL^3}[2b(L - v) + L(b - v)]$$

A–25 *(continued)*

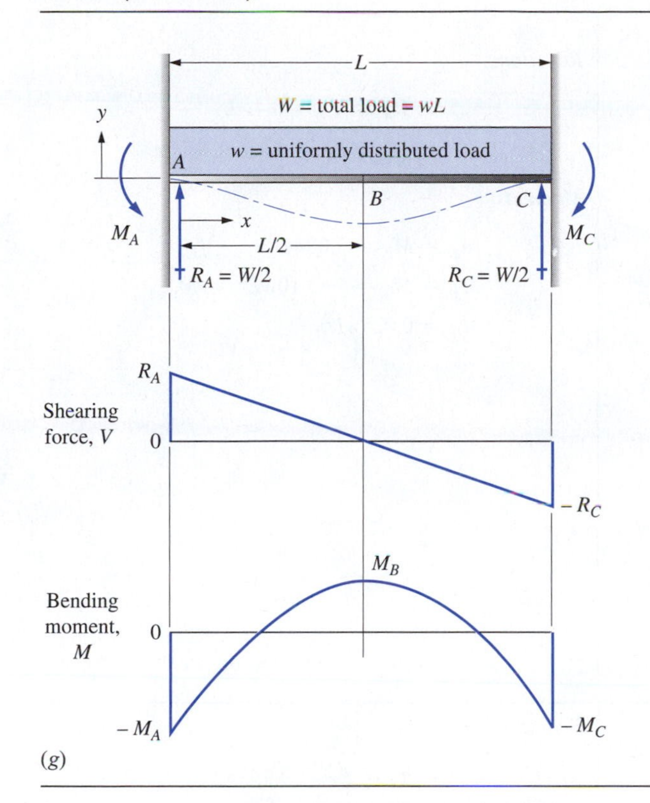

(g)

Moments

$$M_A = M_C = \frac{-WL}{12}$$

$$M_B = \frac{WL}{24}$$

Deflections

At B at center:

$$y_B = y_{max} = \frac{-WL^3}{384EI}$$

Between A and C:

$$y = \frac{-wx^2}{24EI}(L - x)^2$$

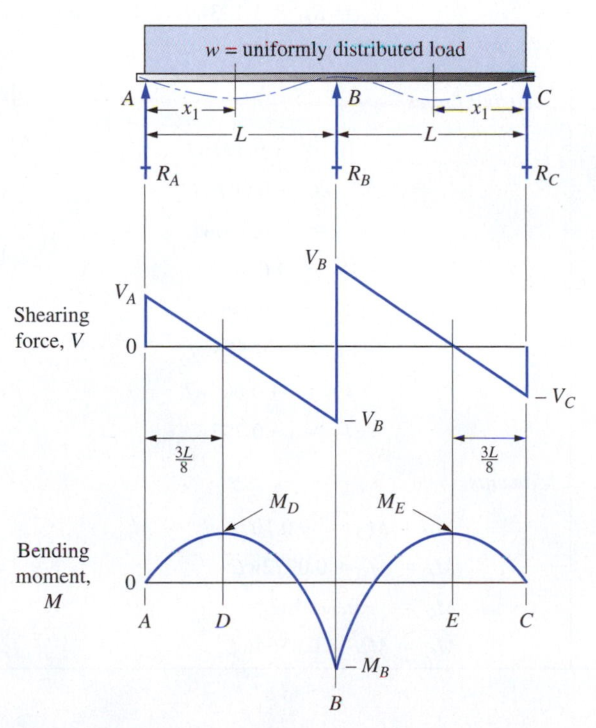

(h)

Reactions

$$R_A = R_C = \frac{3wL}{8}$$

$$R_B = 1.25wL$$

Shearing forces

$$V_A = V_C = R_A = R_C = \frac{3wL}{8}$$

$$V_B = \frac{5wL}{8}$$

Moments

$$M_D = M_E = 0.0703wL^2$$

$$M_B = -0.125wL^2$$

Deflections

At $x_1 = 0.4215L$ from A or C:

$$y_{max} = \frac{-wL^4}{185EI}$$

Between A and B:

$$y = \frac{-w}{48EI}(L^3x - 3Lx^3 + 2x^4)$$

A–25 (continued)

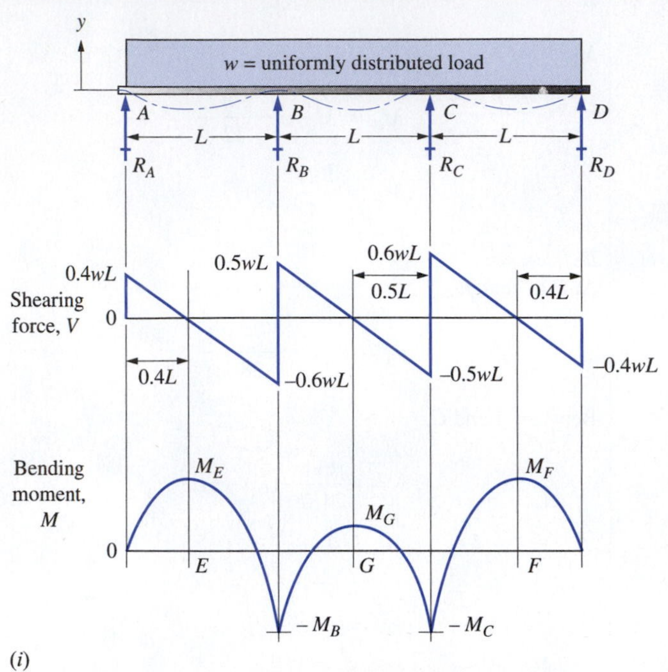

(i)

Reactions

$$R_A = R_D = 0.4wL$$
$$R_B = R_C = 1.10wL$$

Moments

$$M_E = M_F = 0.08wL^2$$
$$M_B = M_C = -0.10wL^2 = M_{max}$$
$$M_G = 0.025wL^2$$

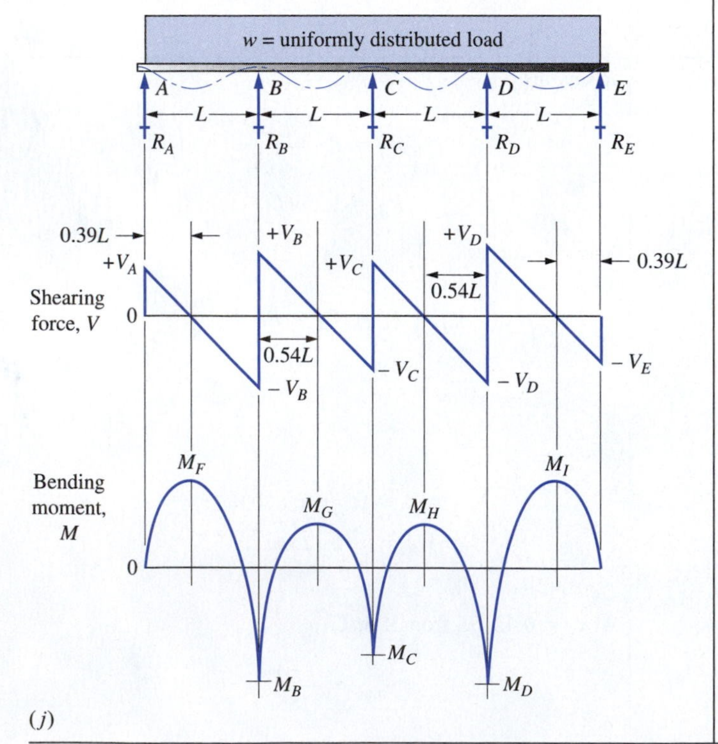

(j)

Reactions

$$R_A = R_E = 0.393wL$$
$$R_B = R_D = 1.143wL$$
$$R_C = 0.928wL$$

Shearing forces

$$V_A = +0.393wL$$
$$-V_B = -0.607wL$$
$$+V_B = +0.536wL$$
$$-V_C = +0.464wL$$
$$+V_C = +0.464wL$$
$$-V_D = -0.536wL$$
$$+V_D = +0.607wL$$
$$-V_E = -0.393wL$$

Moments

$$M_B = M_D = -0.1071wL^2 = M_{max}$$
$$M_F = M_I = 0.0772wL^2$$
$$M_C = -0.0714wL^2$$
$$M_G = M_H = 0.0364wL^2$$

A–26 Conversion factors.

Mass Standard SI unit: Kilogram (kg). Equivalent unit: $N \cdot s^2/m$.

$\frac{14.59 \text{ kg}}{\text{slug}}$ $\frac{32.174 \text{ lb}_m}{\text{slug}}$ $\frac{2.205 \text{ lb}_m}{\text{kg}}$ $\frac{453.6 \text{ grams}}{\text{lb}_m}$ $\frac{2000 \text{ lb}_m}{\text{ton}_m}$ $\frac{1000 \text{ kg}}{\text{metric ton}_m}$

Force Standard SI unit: Newton (N). Equivalent unit: $kg \cdot m/s^2$.

$\frac{4.448 \text{ N}}{\text{lb}_f}$ $\frac{10^5 \text{ dynes}}{\text{N}}$ $\frac{4.448 \times 10^5 \text{ dynes}}{\text{lb}_f}$ $\frac{224.8 \text{ lb}_f}{\text{kN}}$ $\frac{1000 \text{ lb}}{K}$

Length

$\frac{3.281 \text{ ft}}{\text{m}}$ $\frac{39.37 \text{ in}}{\text{m}}$ $\frac{12 \text{ in}}{\text{ft}}$ $\frac{25.4 \text{ mm}}{\text{in}}$ $\frac{1.609 \text{ km}}{\text{mi}}$ $\frac{5280 \text{ ft}}{\text{mi}}$

Area

$\frac{144 \text{ in}^2}{\text{ft}^2}$ $\frac{10.76 \text{ ft}^2}{\text{m}^2}$ $\frac{645.2 \text{ mm}^2}{\text{in}^2}$ $\frac{10^6 \text{ mm}^2}{\text{m}^2}$ $\frac{43{,}560 \text{ ft}^2}{\text{acre}}$ $\frac{10^4 \text{ m}^2}{\text{hectare}}$

Volume

$\frac{1728 \text{ in}^3}{\text{ft}^3}$ $\frac{231 \text{ in}^3}{\text{gal}}$ $\frac{7.48 \text{ gal}}{\text{ft}^3}$ $\frac{264 \text{ gal}}{\text{m}^3}$ $\frac{3.785 \text{ L}}{\text{gal}}$ $\frac{35.3 \text{ ft}^3}{\text{m}^3}$

Section Modulus

$\frac{1.639 \times 10^4 \text{ mm}^3}{\text{in}^3}$ $\frac{10^9 \text{ mm}^3}{\text{m}^3}$

Moment of Inertia or Second Moment of an Area

$\frac{4.162 \times 10^5 \text{ mm}^4}{\text{in}^4}$ $\frac{10^{12} \text{ mm}^4}{\text{m}^4}$

Density (Mass/Unit Volume)

$\frac{515.4 \text{ kg/m}^3}{\text{slug/ft}^3}$ $\frac{1000 \text{ kg/m}^3}{\text{gram/cm}^3}$ $\frac{32.17 \text{ lb}_m\text{/ft}^3}{\text{slug/ft}^3}$ $\frac{16.018 \text{ kg/m}^3}{\text{lb}_m\text{/ft}^3}$

Specific Weight (Weight/Unit Volume)

$\frac{157.1 \text{ N/m}^3}{\text{lb}_f\text{/ft}^3}$ $\frac{1728 \text{ lb/ft}^3}{\text{lb/in}^3}$

Bending Moment or Torque

$\frac{8.851 \text{ lb} \cdot \text{in}}{\text{N} \cdot \text{m}}$ $\frac{1.356 \text{ N} \cdot \text{m}}{\text{lb} \cdot \text{ft}}$

Pressure, Stress, or Loading Standard SI unit: Pascal (Pa). Equivalent units: N/m^2 or $kg/m \cdot s^2$.

$\frac{144 \text{ lb/ft}^2}{\text{lb/in}^2}$ $\frac{47.88 \text{ Pa}}{\text{lb/ft}^2}$ $\frac{6895 \text{ Pa}}{\text{lb/in}^2}$ $\frac{1 \text{ Pa}}{\text{N/m}^2}$ $\frac{6.895 \text{ MPa}}{\text{ksi}}$

Energy Standard SI unit: Joule (J). Equivalent units: $N \cdot m$ or $kg \cdot m^2/s^2$.

$\frac{1.356 \text{ J}}{\text{lb} \cdot \text{ft}}$ $\frac{1.0 \text{ J}}{\text{N} \cdot \text{m}}$ $\frac{8.85 \text{ lb} \cdot \text{in}}{\text{J}}$ $\frac{1.055 \text{ kJ}}{\text{Btu}}$ $\frac{3.600 \text{ kJ}}{\text{W} \cdot \text{hr}}$ $\frac{778 \text{ ft} \cdot \text{lb}}{\text{Btu}}$

Power Standard SI unit: Watt (W). Equivalent unit: J/s or $N \cdot m/s$.

$\frac{745.7 \text{ W}}{\text{hp}}$ $\frac{1.0 \text{ W}}{\text{N} \cdot \text{m/s}}$ $\frac{550 \text{ lb} \cdot \text{ft/s}}{\text{hp}}$ $\frac{1.356 \text{ W}}{\text{lb} \cdot \text{ft/s}}$ $\frac{3.412 \text{ Btu/hr}}{\text{W}}$ $\frac{1.341 \text{ hp}}{\text{kW}}$

General approach to application of conversion factors: Arrange the conversion factor from the table in such a manner that, when multiplied by the given quantity, the original units cancel out, leaving the desired units. See examples on next page.

Example 1. Convert a stress of 36 ksi to MPa.

$$\sigma = 36\,\text{ksi} \times \frac{6.895\,\text{MPa}}{\text{ksi}} = 248\,\text{MPa}$$

Example 2. Convert a stress of 1272 MPa to ksi.

$$\sigma = 1272\,\text{MPa} \times \frac{1.0\,\text{ksi}}{6.895\,\text{MPa}} = 184\,\text{ksi}$$

A–27 Review of the fundamentals of statics.

Introduction

The study of strength of materials depends on accurate knowledge of the forces acting on the load-carrying member being analyzed or designed.

It is expected that readers of this book have completed the study of a course in *statics* in which the principles of physics mechanics are used to determine the forces and moments acting on members of a structure or a machine.

Presented here is a brief review of the principles of statics to help readers recall fundamental principles and problem-solving techniques.

Forces

A force is a push or pull effort applied to a structure or a member of the structure. If the force tends to pull a member apart, it is called a *tensile force*. If the force tends to crush the member, it is called a *compressive force*. See Figure A–27–1 for examples of these kinds of forces applied in line with the axis of the members. These are called *axial forces*.

Forces on members in static equilibrium are always balanced in such a way that the member will not move. Thus in the two cases in Figure A–27–1, the two axial forces, F, are equal in magnitude but they act in opposite directions so they are balanced. You should also note that every part of these members experiences an *internal force* equal to the externally applied force, F. Figure A–27–2 shows this principle by illustrating a part of the tensile member cut anywhere between its ends. The force on the left is the externally applied force, F. The force on the right is the total internal force acting on the material of the member across its cross section.

Moments

A moment is the tendency for a force to cause *rotation* of a member about some point or axis. Figure A–27–3 shows two examples. Each of the forces shown would tend to rotate the member on which they act about the point identified as A.

The magnitude of the moment of a force is the product of the force times the perpendicular distance from the line of action of the force to the point about which the moment is being computed. That is,

$$M = \text{Force times distance} = F \times d$$

The direction is simply observed from the figure to be clockwise or counterclockwise.

FIGURE A–27–1
Types of axial forces.

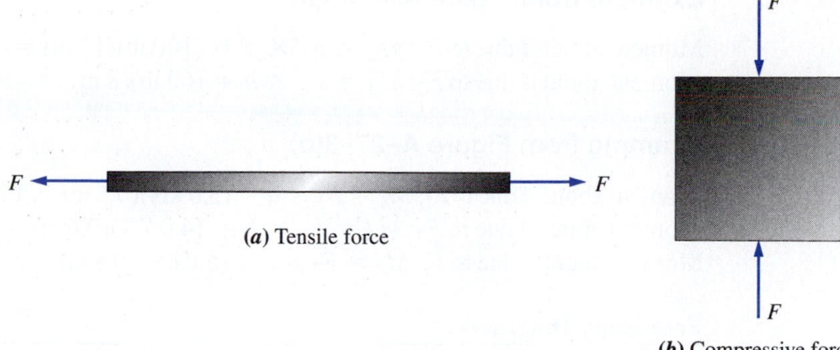

FIGURE A–27–2
Internal force.

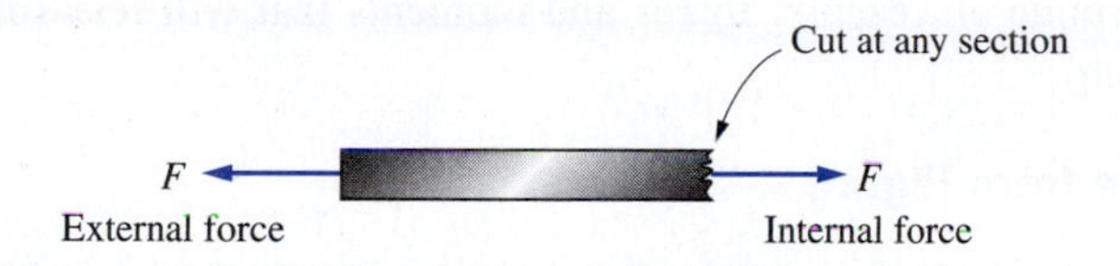

FIGURE A–27–3
Illustrations of moments.

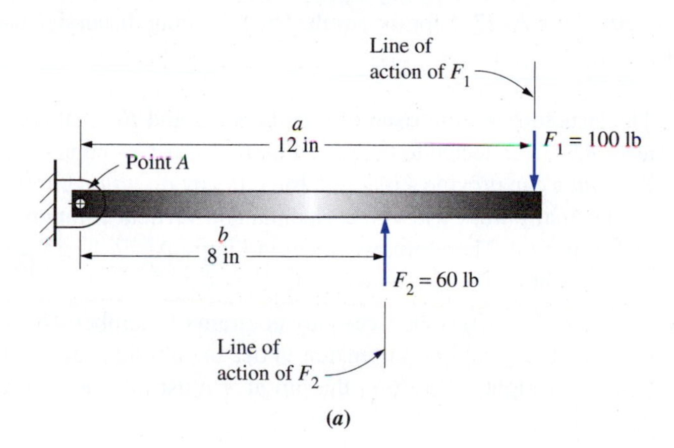

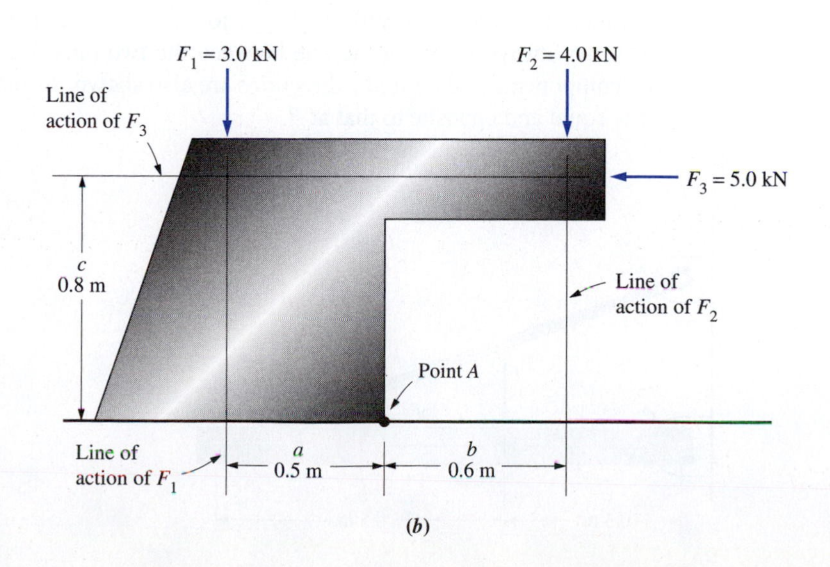

Example from Figure A–27–3(a)

Moment about A due to F_1: $M_A = F_1 \times a = (100\ \text{lb})(12\ \text{in}) = 1200\ \text{lb}\cdot\text{in}$ Clockwise
Moment about A due to F_2: $M_A = F_2 \times b = (60\ \text{lb})(8\ \text{in}) = 480\ \text{lb}\cdot\text{in}$ Counterclockwise

Example from Figure A–27–3(b)

Moment about A due to F_1: $M_A = F_1 \times a = (3.0\ \text{kN})(0.5\ \text{m}) = 1.5\ \text{kN}\cdot\text{m}$ Counterclockwise
Moment about A due to F_2: $M_A = F_2 \times b = (4.0\ \text{kN})(0.6\ \text{m}) = 2.4\ \text{kN}\cdot\text{m}$ Clockwise
Moment about A due to F_3: $M_A = F_3 \times c = (5.0\ \text{kN})(0.8\ \text{m}) = 4.0\ \text{kN}\cdot\text{m}$ Counterclockwise

Free-Body Diagrams

The ability to draw a complete free-body diagram of a structure and its members is an essential element of static analysis. You must show all externally applied forces and moments and determine all reaction forces and moments that will result in the structure being in equilibrium.

Example from Figure A–27–4

Show the free-body diagram for the complete structure and for each of the two members. The applied force is F_1 acting perpendicular to member BC.

See Figure A–27–5 for the result. The following discussion summarizes the important points.

a. The structure is comprised of members AB and BC that are connected by a pin joint at B. AB is connected to the pin support at A. BC is connected to the pin support at C. Pin joints can provide a reaction force in any direction but they cannot resist rotation. We normally work with the horizontal and vertical components of the reaction forces on a pin joint. Therefore, we show in Figure A–27–5(a) the two components, A_x and A_y, at A. Similarly, we show C_x and C_y at C.

b. Figure A–27–5(b) is the free-body diagram of member AB. You should recognize that the member would be in tension under the applied forces. The pin at B pulls down and to the right. Therefore, the pin at A must pull up and to the left to keep AB in equilibrium.

c. You should further recall that member AB is an example of a *two-force member* because it is loaded only through pin joints. The resultant forces on a two-force member always act along the line between the two pins. We label that force as AB. Its components in the x and y directions are also shown. Note that the force system at A is equal and opposite to that at B.

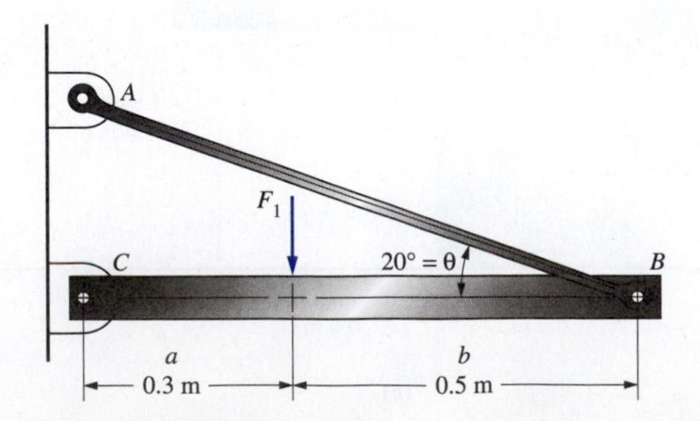

FIGURE A–27–4 Support structure.

FIGURE A–27–5
Free-body diagrams of structure and its components.

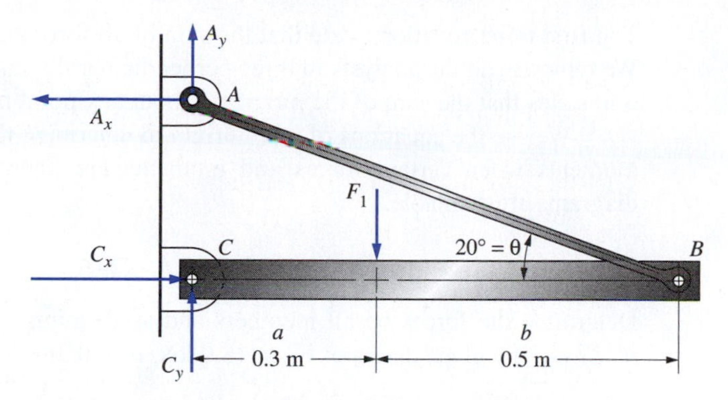

(*a*) Free-body diagram of entire structure

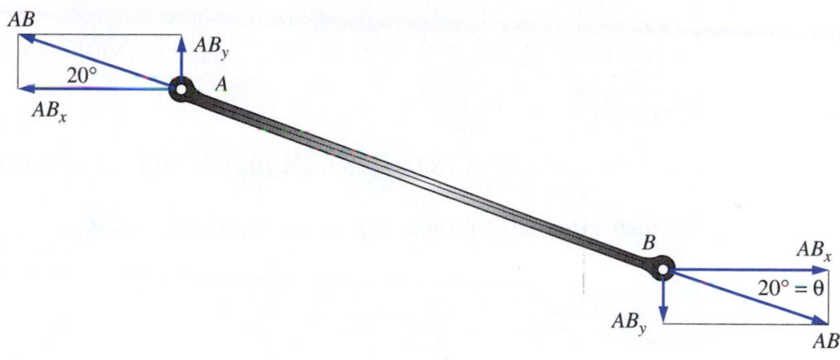

(*b*) Free-body diagram of member *AB*

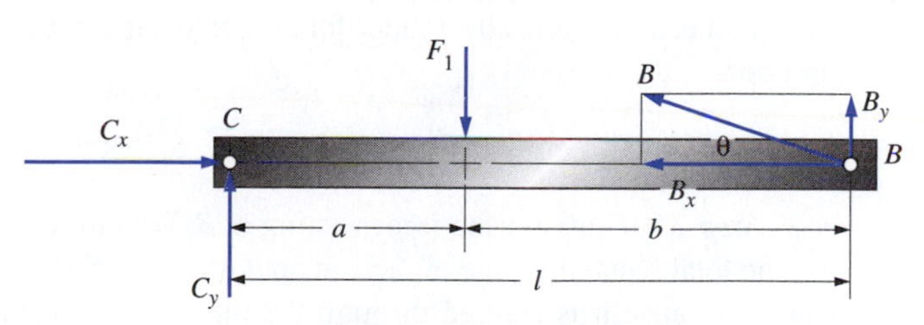

(*c*) Free-body diagram of member *BC*

d. Figure A–27–5(c) is the free-body diagram of member *BC*. This member is called a *beam* because it carries a load, F_1, acting perpendicular to its long axis. There must be an upward reaction force at both *B* and *C* to resist the downward force F_1. We call those forces B_y and C_y. Member *AB* exerts the supporting force on member *BC* at *B*. That force acts upward and to the left. We call the horizontal component of that force B_x. The total force at *B* is equal to the force *AB* described in (c). Finally, to balance the horizontal forces, there must be a force C_x acting toward the right at C.

Static Equilibrium

When a structure or a member is in static equilibrium, all forces and moments are balanced in such a way that there is no movement. The equations that describe static equilibrium are:

$$\sum F_x = 0 \qquad \sum F_y = 0 \qquad \sum F_z = 0$$

$$\sum M = 0 \text{ About any point}$$

The first three equations state that the sum of all forces in any direction must add to zero. We typically do the analysis in three perpendicular directions, x, y, and z. The fourth equation states that the sum of the moments about any point must be zero.

We use the equations of equilibrium to determine the values of unknown forces and moments when certain forces and moments are known and when suitable free-body diagrams are available.

Example from Figures A–27–4 and A–27–5

Determine the forces on all members and at all joints for the structure shown in Figure A–27–4. The given data are: $F_1 = 18.0$ kN, $a = 0.3$m, $b = 0.5$m, $\theta = 20°$.

Solution. We use the free-body diagrams shown in Figure A–27–5.

Step 1. Use part (c) first. We sum moments about point C to find the force B_y.

$$\sum M_C = 0 = F_1 a - B_y l = (18.0\,\text{kN})(0.3\,\text{m}) - B_y(0.8\,\text{m})$$

Then,

$$B_y = (18.0\,\text{kN})(0.3\,\text{m})/(0.8\,\text{m}) = 6.75\,\text{kN}$$

We then sum moments about point B to find the force C_y.

$$\sum M_B = 0 = F_1 b - C_y l = (18.0\,\text{kN})(0.5\,\text{m}) - C_y(0.8\,\text{m})$$

Then,

$$C_y = (18.0\,\text{kN})(0.5\,\text{m})/(0.8\,\text{m}) = 11.25\,\text{kN}$$

We can check to see if all vertical forces are balanced by summing forces in the vertical direction.

$$\sum F_y = C_y + B_y - F_1 = 11.25\,\text{kN} + 6.75\,\text{kN} - 18.0\,\text{kN} = 0 \qquad \text{(Check)}$$

Step 2. Consider the forces acting at B. We know that $B_y = 6.75$ kN. We also know that the total resultant force, B, acts at an angle of 20° above the horizontal toward the left. This is because it is applied through the pin by the force AB. The free-body diagram in (b) indicates that AB acts along the direction of the member AB because it is a two-force member. We can then say,

$$B_y = B \sin 20°$$

$$B_x = B \cos 20°$$

Then,

$$B = B_y/(\sin 20°) = (6.75\,\text{kN})/(\sin 20°) = 19.74\,\text{kN}$$

$$B_x = B \cos 20° = (19.74\,\text{kN})(\cos 20°) = 18.55\,\text{kN}$$

Step 3. The forces acting at pin B on both member AB and member BC must be equal and opposite because of the *principle of action-reaction.* Therefore, the axial force on member AB is: $AB = 19.74$ kN. The force AB acts at both A and B along the line between the two pins in a manner that places the member AB in tension. The components of AB at pin A are equal to the components at pin B, but they act in opposite directions.

Step 4. The only unknown now is C_x, the horizontal force acting at C on member BC. We can use the free-body diagram of member BC again from part (c) of the figure.

$$\sum F_x = 0 = C_x - B_x$$

Then,

$$C_x = B_x = 18.55 \text{ kN}$$

Equilibrium of Concurrent Force Systems

When the line of action of all forces acting on a member pass through the same point, the system is called a *concurrent force system*. For static equilibrium to exist in such a system, the vector sum of all forces must add to zero. Two methods can be used to analyze a concurrent force system to determine unknown forces.

The Component Method

This method calls for each force to be resolved into perpendicular components, usually horizontal and vertical. Then the classic equations of equilibrium are applied.

Example Using Figure A–27–6

Determine the force in each cable when the mass of the load is 1500 kg and the angle $\theta = 25°$

FIGURE A–27–6 Load carried by three cables showing force analysis.

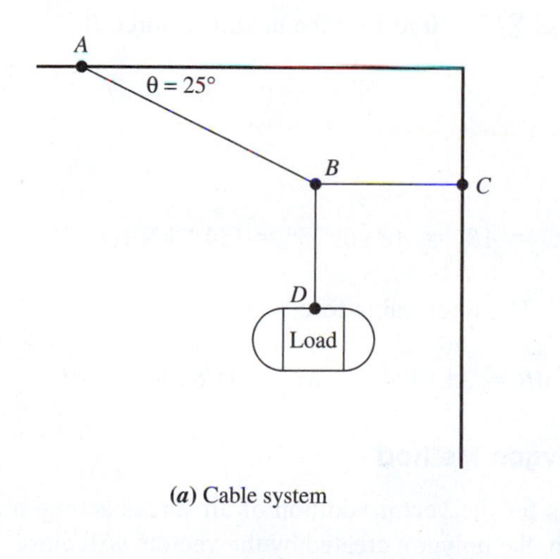

(*a*) Cable system

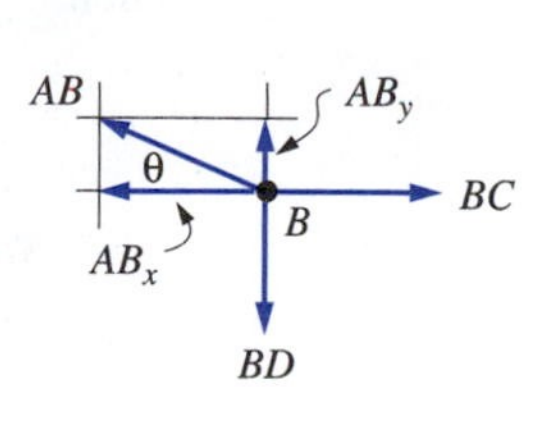

(*b*) Free-body diagram of B with components of AB

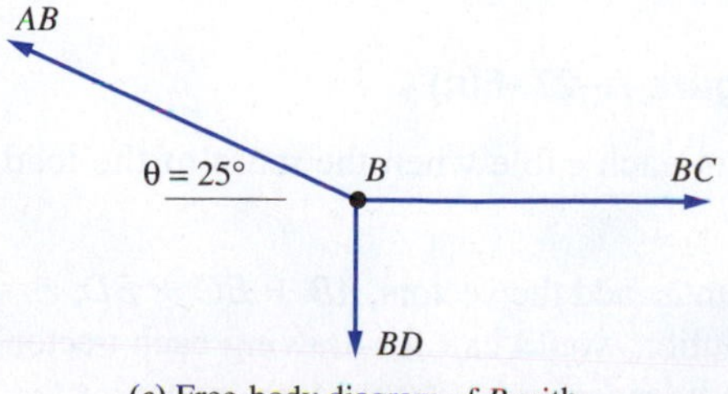

(*c*) Free-body diagram of B with vectors drawn to scale

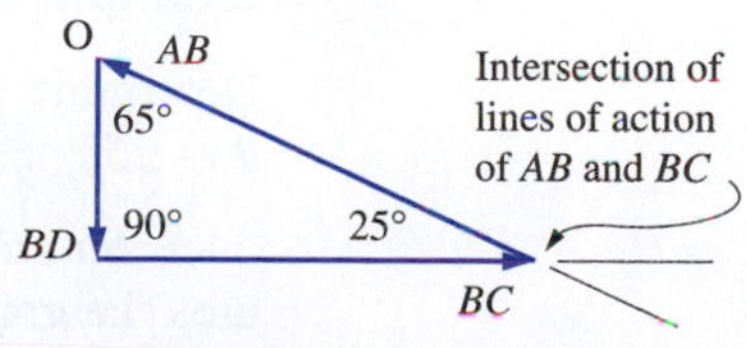

(*d*) Vector triangle showing vector sum $BD + BC + AB$

Solution. There are three cables that we will call AB, BC, and BD. All three are two-force members and pass through point B as shown in part (b). Therefore, they are concurrent.

Step 1. Determine the weight of the load. (See Section 1–5.)

$$w = mg = (1500\,\text{kg})(9.81\,\text{m/s}^2) = 14\,715\,\text{N} = 14.7\,\text{kN}$$

This is also the force in the cable BD.

Step 2. Draw the free-body diagram of point B and resolve each force into its x and y components. This is done in part (b) of the figure.

Step 3. Use $\sum F_y = 0$ to solve for the unknown force AB.

$$\sum F_y = 0 = AB_y - BD$$

Then,

$$AB_y = BD = 14.7\,\text{kN}$$

But AB_y is the vertical component of the cable force AB. Then,

$$AB_y = AB\sin\theta = AB\sin 25^\circ$$
$$AB = AB_y/(\sin 25^\circ) = (14.7\,\text{kN})/(\sin 25^\circ) = 34.8\,\text{kN}$$

Step 4. Use $\sum F_x = 0$ to find the unknown force BC.

$$\sum F_x = 0 = BC - AB_x$$

Then,

$$BC = AB_x = AB\cos 25^\circ = (34.8\,\text{kN})(\cos 25^\circ) = 31.6\,\text{kN}$$

Summary: The three cable forces are:

$$AB = 34.8\,\text{kN} \qquad BC = 31.6\,\text{kN} \qquad BD = 14.7\,\text{kN}$$

The Vector Polygon Method

This method calls for the vector addition of all forces acting at a point. When the forces are in equilibrium, the polygon created by the vectors will close, indicating that the vector sum is equal to zero.

Example Using Figure A–27–6(c)

Determine the force in each cable when the mass of the load is 1500 kg and the angle $\theta = 25^\circ$

Solution. We must add the vectors, $AB + BC + BD$, as shown in part (d) of the figure. The graphical solution would call for drawing each vector in its proper direction and with a length scaled to its magnitude. The vectors are connected "tip to tail." We will sketch a graphical vector diagram but solve for the required forces analytically.

Step 1. The sum can be done in any order and we could start at any point, say point O. We will actually sum the forces in the order $BD + BC + AB$. Let's first draw the known vector BD vertically downward to scale. The value is $BD = 14.7$ kN as found in the component method.

Step 2. Add vector BC from the tip of BD and acting horizontally to the right. Its length is unknown at this time, but its line of action is known. Draw the line of indefinite extent for now.

Step 3. Then, adding vector AB to the end of vector BC should cause the vector polygon to close by having the tip of AB fall right on point O. We can pass a line through point O in the direction of AB. Where this line crosses the line of action of vector BC establishes where the tip of BC is. Similarly, the tail of AB is also at that point.

Step 4. In the vector triangle thus formed, we know all three angles and the length of one side, BD. We can use the law of sines to find the lengths of the other two sides.

$$\frac{BD}{\sin 25^\circ} = \frac{AB}{\sin 90^\circ}$$

Then,

$$AB = (BD)(\sin 90^\circ)/(\sin 25^\circ) = (14.7\text{ kN})(\sin 90^\circ)/(\sin 25^\circ) = 34.8\text{ kN}$$

Also,

$$\frac{BD}{\sin 25^\circ} = \frac{BC}{\sin 65^\circ}$$

Then,

$$BC = (BD)(\sin 65^\circ)/(\sin 25^\circ) = (14.7\text{ kN})(\sin 65^\circ)/(\sin 25^\circ) = 31.6\text{ kN}$$

Summary. The results for the cable forces are identical to those found from the component method.

$$AB = 34.8\text{ kN} \qquad BC = 31.6\text{ kN} \qquad BD = 14.7\text{ kN}$$

Law of Cosines Applied to Force Analysis

In some vector triangle solutions, you will know the magnitudes of two forces and the angle between them. You can solve for the third force using the *law of cosines*.

Say for example, in Figure A–27–6(d), you know the magnitudes of $AB = 34.8$ kN, $BC = 31.6$ kN, and that the angle between them is 25°. You could solve for the magnitude of the force BD from:

$$(BD)^2 = (AB)^2 + (BC)^2 - 2(AB)(BC)\cos 25^\circ$$
$$= (34.8)^2 + (31.6)^2 - 2(34.8)(31.6)\cos 25^\circ = 216$$

Then,

$$BD = \sqrt{216} = 14.7\text{ kN}$$

You must carefully model the sides and angle to match the form of this equation.

Trusses

A truss is a structure comprised of only straight members connected by pin joints with loads applied only at joints. The result is that all members are two-force members carrying either tensile or compressive loads. Figure A–27–7 shows an example. Next we describe the *method of joints* for analyzing the forces in all members of a truss.

Example Using Figure A–27–7

Find the forces in all members of the truss shown in Figure A–27–7. Determine both the magnitude and direction (tension or compression) for each force.

FIGURE A–27–7
Forces on a truss and its joints.

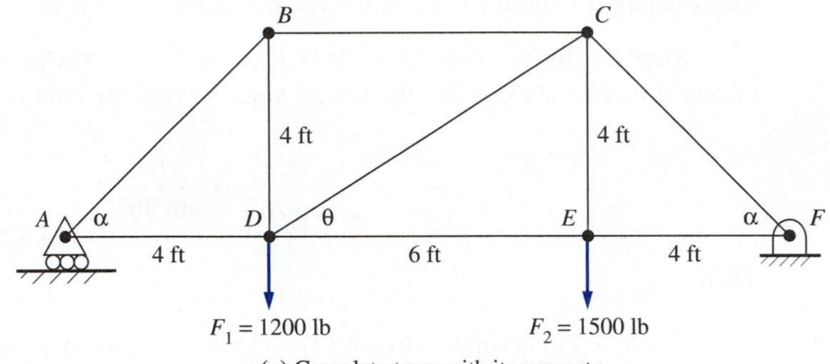

(*a*) Complete truss with its supports

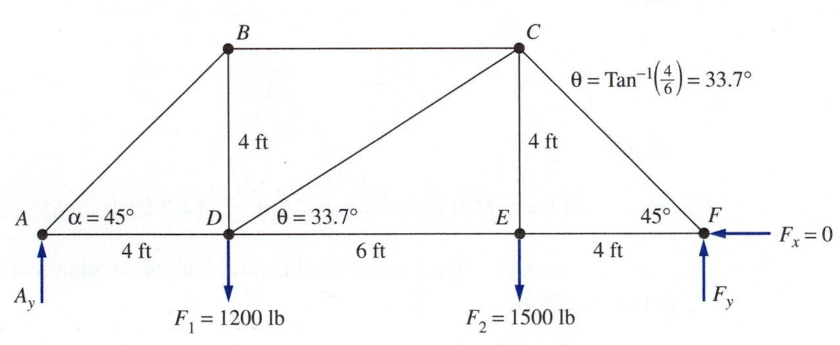

(*b*) Free-body diagram of complete truss

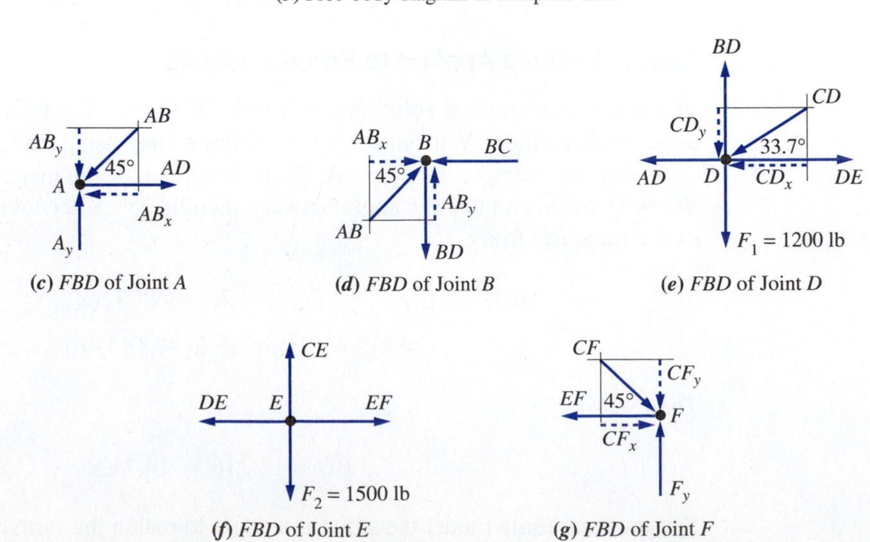

(*c*) *FBD* of Joint *A*

(*d*) *FBD* of Joint *B*

(*e*) *FBD* of Joint *D*

(*f*) *FBD* of Joint *E*

(*g*) *FBD* of Joint *F*

Analysis by Method of Joints

There are nine members in the truss. The general method for finding the forces in each member is described next.

a. Solve for the reactions at the supports for the entire truss.

b. Isolate one joint as a free body and show all forces acting on it. The selected joint must have at least one known force acting on it. It is recommended that there be no more than two unknown forces.

c. When a member is in tension, the force pulls out on the joint at either end. Conversely, a compression member pushes into a joint. Try to draw unknown forces in the proper direction that will ensure that the joint is in equilibrium.

d. Use the equations of static equilibrium for forces in the horizontal and vertical directions to determine the unknown forces at the selected joint.

e. Forces found at the first joint become known forces for analyzing other joints. Move to nearby joints and repeat steps b, c, and d repeatedly until the forces in all joints have been found.

Completion of the Analysis of the Truss in Figure A–27–7

Step 1. Using the entire truss as a free body, solve for the support reactions at joints A and F. See part (b) of the figure.
Sum moments about support A to find support force F_y at point F.

$$\sum M_A = F_1(4\text{ ft}) + F_2(10\text{ ft}) - F_y(14\text{ ft}) = (1200\text{ lb})(4\text{ ft}) + (1500\text{ lb})(10\text{ ft}) - F_y(14\text{ ft})$$

$$F_y = [(4800 + 15\,000)\text{ lb ft}]/14\text{ ft} = 1414\text{ lb} \qquad \text{Upward}$$

Sum moments about support F to find support force A_y at point A.

$$\sum M_F = F_1(10\text{ ft}) + F_2(4\text{ ft}) - A_y(14\text{ ft}) = (1200\text{ lb})(10\text{ ft}) + (1500\text{ lb})(4\text{ ft}) - A_y(14\text{ ft})$$

$$A_y = [(12\,000 + 6000)\text{ lb ft}]/14\text{ ft} = 1286\text{ lb} \qquad \text{Upward}$$

Step 2. Isolate joint A as a free body. See part (c) of the figure. Work with components of the force AB. $AB_x = AB\cos 45°$. $AB_y = AB\sin 45°$.

$$\sum F_y = 0 = A_y - AB_y$$

$$AB_y = A_y = 1286\text{ lb}$$

Then,

$$AB = AB_y/(\sin 45°) = (1286\text{ lb})/(\sin 45°) = 1818\text{ lb} \qquad \text{Compression}$$

$$\sum F_x = 0 = AD - AB_x$$

$$AD = AB_x = AB\cos 45° = (1818\text{ lb})(\cos 45°) = 1286\text{ lb} \qquad \text{Tension}$$

Step 3. Isolate joint B as a free body. See part (d) of the figure.

$$BD = 1286\text{ lb} \qquad \text{Tension}$$

$$\sum F_x = 0 = AB_x - BC = 1286\text{ lb} - BC$$

$$BC = 1286\text{ lb} \qquad \text{Compression}$$

Step 4. Isolate joint D as a free body. See part (e) of the figure.

$$\sum F_y = 0 = 1200\text{ lb} - BD + CD_y = 1200\text{ lb} - 1286\text{ lb} + CD_y = CD_y - 86\text{ lb}$$
$$CD_y = 86\text{ lb}$$

Then,

$$CD = CD_y/(\sin 33.7°) = (86\text{ lb})/(\sin 33.7°) = 155\text{ lb} \qquad \text{Compression}$$
$$\sum F_x = 0 = DE - AD - CD_x = DE - 1286\text{ lb} - (155\text{ lb})(\cos 33.7°)$$
$$DE = 1286\text{ lb} + 130\text{ lb} = 1416\text{ lb} \qquad \text{Tension}$$

Step 5. Isolate joint E as a free body. See part (f) of the figure.

$$\sum F_y = 0 = 1500\text{ lb} - CE$$
$$CE = 1500\text{ lb} \qquad \text{Tension}$$
$$\sum F_x = 0 = EF - DF = EF - 1416\text{ lb}$$
$$EF = 1416\text{ lb} \qquad \text{Tension}$$

Step 6. Isolate joint F as a free body. See part (g) of the figure.

$$\sum F_y = 0 = F_y - CF_y = 1414\text{ lb} - CF_y = 1414\text{ lb} - CF\sin 45°$$
$$CF = 1414\text{ lb}/(\sin 45°) = 2000\text{ lb} \qquad \text{Compression}$$

Summary of Forces in Members of the Truss

$AB = 1818$ lb (C)	$AD = 1286$ lb (T)	$BD = 1286$ lb (T)
$BC = 1286$ lb (C)	$CE = 1500$ lb (T)	$CD = 155$ lb (C)
$DE = 1416$ lb (T)	$EF = 1416$ lb (T)	$CF = 2000$ lb (C)

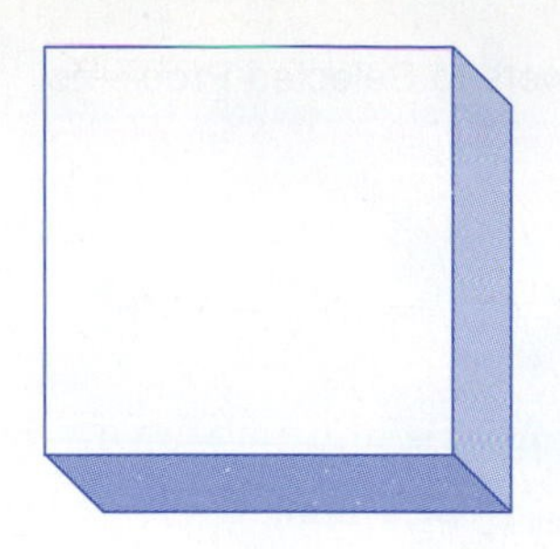

Answers to Selected Problems

Chapter 1

1–17. 7.85 kN front
11.77 kN rear

1–19. 54.5 mm

1–23. 1765 lb front
2646 lb rear

1–25. 55.1 lb
25.7 lb/in
2.14 in

1–27. 398 slugs

1–29. 8274 kPa

1–31. 96.5 to 524 MPa

1–33. 9097 mm^2

1–35. Area = 324 in^2
Area = 2.09×10^5 mm^2
Vol. = 3888 in^3
Vol. = 6.37×10^7 mm^3
Vol. = 6.37×10^{-2} m^3

1–37. 40.7 MPa

1–39. 5375 psi

1–41. 79.8 MPa

1–43. 803 psi

1–45. σ_{AB} = 107.4 MPa
σ_{BC} = 75.2 MPa
σ_{BD} = 131.1 MPa

1–47. σ_{AB} = 167 MPa tension
σ_{BC} = 77.8 MPa tension
σ_{CD} = 122 MPa tension

1–49. σ_{AB} = 20471 psi tension
σ_{BC} = 3129 psi tension

1–51. Forces: AD = CD = 10.5 kN
AB = BC = 9.09 kN
Stresses: $\sigma_{AB} = \sigma_{BC}$ = 25.3 MPa tension
σ_{BD} = 17.5 MPa tension
$\sigma_{AD} = \sigma_{CD}$ = 21.0 MPa compression

1–53. 50.0 MPa

1–55. 11 791 psi

1–57. 146 MPa

1–59. 151 MPa

1–61. 81.1 MPa

1–63. 24.7 MPa

1–65. Pin: τ = 50 930 psi
Collar: τ = 38 800 psi

1–67. 183 MPa

1–69. 73.9 MPa

1–71. 22.6 MPa

Chapter 2

2–15. 1020 HR

2–19. 16.4 lb

2–21. Magnesium

2–29. s_{ut} = 40 ksi; s_{uc} = 140 Ksi

2–31. Bending: σ_d = 1450 psi
Tension: σ_d = 850 psi
Compression: σ_d = 1000 psi parallel to grain
Compression: σ_d = 385 psi perpendicular to grain
Shear: τ_d = 95 psi

Problems 2–67 to 2–77: Data approximated from Figure P2–66

2–67. **(a)** s_y = 173 ksi—Yield point
(b) s_u = 187 ksi
(c) s_p = 162 ksi
(d) s_{el} = 168 ksi
(e) E = 29.0×10^6 psi
(f) 15% Elongation
(g) Ductile
(h) Steel
(i) AISI 4140 OQT 900

2–69. **(a)** s_y = 49 ksi—Yield point
(b) s_u = 65 ksi

(c) $s_p = 46$ ksi
(d) $s_{el} = 48$ ksi
(e) $E = 26.5 \times 10^6$ psi
(f) 36% Elongation
(g) Ductile
(h) Steel
(i) AISI 1020 CD

2–71. (a) $s_y = 53$ ksi—Offset
(b) $s_u = 59$ ksi
(c) $s_p = 31$ ksi
(d) $s_{el} = 42$ ksi
(e) $E = 12.0 \times 10^6$ psi
(f) 5.0% Elongation
(g) Borderline Brittle/Ductile
(h) Zinc
(i) Cast ZA-12

2–73. (a) $s_y = 19$ ksi—Offset
(b) $s_u = 40$ ksi
(c) $s_p = 14$ ksi
(d) $s_{el} = 17$ ksi
(e) $E = 6 \times 10^6$ psi
(f) 5% Elongation
(g) Borderline Brittle/Ductile
(h) Magnesium
(i) ASTM AZ 63A-T6

2–75. (a) $s_y = 40$ ksi—Offset
(b) $s_u = 45$ ksi
(c) $s_p = 30$ ksi
(d) $s_{el} = 35$ ksi
(e) $E = 10.0 \times 10^6$ psi
(f) 17% Elongation
(g) Ductile
(h) Aluminum
(i) 6061-T6

2–77. (a) $s_y = 80$ ksi—Offset
(b) $s_u = 95$ ksi
(c) $s_p = 55$ ksi
(d) $s_{el} = 68$ ksi
(e) $E = 26 \times 10^6$ psi
(f) 2.0% Elongation
(g) Brittle, but does yield
(h) Malleable Iron
(i) ASTM A220 Grade 80002

Chapter 3

3–1. Required $s_y = 216$ MPa

3–3. Required $s_u = 86\,000$ psi

3–5. No. Tensile stress too high.

3–7. $d_{min} = 0.824$ in

3–9. $d_{min} = 12.4$ mm

3–11. Required $\sigma_d > 803$ psi

3–13. 16.7 kN

3–15. $d_{min} = 0.412$ in

3–17. For sides B and H: $B_{min} = 22.2$ mm; $H_{min} = 44.4$ mm

3–19. Required $s_y = 360$ MPa

3–21. Required $s_u = 400$ MPa

Elastic Deformation

3–23. 0.041 in

3–25. Force = 2357 lb
$\sigma = 3655$ psi

3–27. (a) and (b) $d_{min} = 10.63$ mm; mass = 0.430 kg
(c) $d_{min} = 18.4$ mm; mass = 0.465 kg

3–29. (a) 0.857 mm
(b) 0.488 mm

3–31. Elongation = 0.0040 in
Compression = 0.00045 in

3–33. Force = 3214 lb; unsafe

3–35. $\delta = 0.016$ mm
$\sigma = 27.9$ MPa

3–37. 0.804 mm

3–39. 4.42 mm shorter

3–41. (a) $\delta = 0.276$ in; $\sigma = 37\,300$ psi (close to s_y)
(b) $\sigma = 62\,200$ psi—greater than s_u. Wire will break.

3–43. Force = 6737 lb
$\delta = 0.055$ in

3–45. Mass = 132 kg
$\sigma = 183$ MPa

3–47. 0.806 in

3–49. 180 MPa

3–51. (a) 0.459 mm (b) 213 MPa

3–53. 693 psi

3–55. 234.8°C

3–57. Brass: $\delta = 6.46$ mm
Stainless steel: $\delta = 3.51$ mm

3–59. 38.7 MPa

3–61. $\sigma = 37\,500$ psi compression. Bar should fail in compression and/or buckle.

3–63. 0.157 in

3–65. 154 MPa

3–67. $\sigma_c = 17.1$ MPa; $\sigma_s = 109$ MPa

3–69. 13.8 in

3–71. $d_{min} = 6.20$ mm

3–73. $\sigma_s = 427$ MPa; $\sigma_a = 49.4$ MPa

ADDITIONAL PRACTICE PROBLEMS

3–75. $F_{max} = 12.56$ kN; $\delta = 0.751$ mm

3–77. **(a)** $t = 110.6°C$
(b) $\sigma = 50.4$ MPa
(c) Safe for compression. Check buckling.

3–79. $\delta = 0.341$ mm. But stress is near yield strength.

3–81. For *AB*: $\delta = 15.2$ mm
For *BC*: $\delta = 15.2$ mm

Problems 3–83 to 3–89

	Length (in)	Strain (in/in)	Force (lb)
3–83.	2.0143	0.00714	42 840
3–85.	2.0137	0.00687	26 100
3–87.	2.0071	0.00353	8460
3–89.	2.0116	0.00581	27 900

Problems 3–91 to 3–99:

	Length (mm)	Strain (mm/mm)	Force (kN)
3–91.	50.716	0.01431	8.30
3–93.	50.486	0.00972	5.50
3–95.	50.148	0.00297	4.50
3–97.	50.455	0.00909	137.9
3–99.	50.083	0.00167	110.3

Stress Concentrations—Direct Axial Stresses

3–101. 20 140 psi

3–103. 224 MPa

3–105. 239 MPa

3–107. 52 800 psi

3–109. 33 127 psi

3–111. 63.7 MPa

3–113. 833 MPa

3–115. 34 240 psi

3–117. 34 020 psi

3–119. 531 MPa

3–121. 62.1 MPa

3–123. 15.9 MPa

3–125. 1905 N

Bearing Stress

3–127. A. 5869 psi, B. 181 psi
C. 80.2 psi, D. 20.1 psi

3–129. **(a)** 1610 psi, **(b)** 311 psi

3–131. **(a)** pin/tube interface, $\sigma_b = 106\ 700$ psi—Very high
(b) collar/tube interface, $\sigma_b = 31\ 950$ psi

3–133. **(a)** on middle part: $\sigma_b = 28.3$ MPa
(b) on outer parts: $\sigma_b = 21.25$ MPa

3–135. **(a)** steel plate: $\sigma_b = 2.25$ MPa—OK
(b) top of concrete: $\sigma_b = 0.563$ MPa—OK
(c) soil: $\sigma_b = 22.5$ kPa—OK

3–137. $R_a = 33.1$ kips

3–139. **(a)** $R_a = 27.6$ kips
(b) $R_a = 39.6$ kips

3–141. $R_a = 284$ kN

3–143. Required $A_b = 3.49$ in^2

3–145. 8633 lb

Direct Shear Stress

3–147. $a_{min} = 0.438$ in

3–149. $\tau = 151$ MPa; Required $s_y = 1208$ MPa

3–151. 185 kN

3–153. 256 kN

3–155. 18 300 lb

3–157. 62 650 lb

3–159. 119 500 lb

3–161. 119 700 lb

3–163. Pin A: $D_{A_{min}} = 13.8$ mm
Pins B and C: $D_{min} = 17.7$ mm

3–165. Lifting force = 2184 lb; $\tau_{axle} = 6275$ psi (double shear)

3–167. $D_{min} = 0.206$ in

3–169. 35 020 lb

3–171. 58.06 kN

3–173. 110.9 kN

3–175. 137.4 kN

Problems with More Than One Kind of Direct Stress and Design Problems

3–177. $D_{min} = 0.808$ in based on shear

3–179. **(a)** $F_{allow} = 6200$ lb Shear
(b) $F_{allow} = 35\ 340$ lb Bearing
(c) $F_{allow} = 2832$ lb Tension—Governs design

3–181. **(a)** F_{allow} = 3148 lb Shear—Governs design
(b) F_{allow} = 9375 lb Tension
(c) F_{allow} = 10 969 lb Bearing

3–183. **(a)** F_{allow} = 6626 lb Tension—Governs design
(b) F_{allow} = 10 713 lb Shear
(c) F_{allow} = 81 843 lb Bearing

Chapter 4

4–1. 178 MPa

4–3. 4042 psi

4–5. 83.8 MPa

4–7. τ = 6716 psi; safe

4–9. τ = 5190 psi; required s_y = 62 300 psi

4–11. τ = 65.5 MPa; θ = 0.0378 rad; required s_y = 262 MPa

4–13. D_i = 49.0 mm; D_o = 61.3 mm

4–15. d_{min} = 0.512 in

4–17. Power = 0.0686 hp; τ = 8488 psi; required s_y = 67 900 psi

4–19. D_i = 12.09 in; D_o = 15.11 in

4–21. 1.96 N·m

4–23. 0.1509 rad

4–25. 0.267 rad

4–27. 0.0756 rad

4–29. 0.278 rad

4–31. θ_{AB} = 0.0636 rad; θ_{AC} = 0.0976 rad

4–33. τ = 9.06 MPa; θ = 0.0046 rad

4–35. 49.0 MPa

4–37. 1370 lb·in

4–39. 2902 lb·in

4–41. 0.083 rad

4–43. 0.112 rad

4–45. 0.0667 rad

4–47. 1.82 MPa

4–49. 0.00363 rad

4–51. 0.0042 rad

4–53. 78 150 lb·in

4–55. 144 800 lb·in

4–57. τ_{tube}/τ_{pipe} = 1.100
$\theta_{tube}/\theta_{pipe}$ = 1.266

ADDITIONAL PRACTICE PROBLEMS

4–58. τ_{max} = 123 MPa

4–59. Required s_y = 493 MPa

4–61. Required s_y = 1480 MPa

4–63. D_{min} = 0.419 in; Specify D = 0.50 in

4–65. θ = 1.85 degrees

4–67. τ = 211 MPa at A at keyseat = τ_{max}
τ = 171 MPa at right of A at shoulder
τ = 55.7 MPa at right of bearing seat
τ = 39.6 MPa at retaining ring groove to left of B
τ = 26.4 MPa at B at keyseat
τ = 14.7 MPa at right of B at shoulder
τ = 10.7 MPa at step from 50 mm to 30 mm
τ = 10.3 MPa at left of bearing seat
τ = 22.5 MPa at left of C at shoulder
τ = 30.1 MPa at C at keyseat

4–69. L_{min} = 1.088 m

4–70. D_o = 21.79 mm; D_i = 14.53 mm

4–71. τ = 408.8 MPa

Chapter 5

NOTE: The following answers refer to Figures P5–1 through P5–84 and P5–93 through P5–110. For reactions, R_1 is the left; R_2 is the right. V and M refer to the maximum absolute values for shearing force and bending moment, respectively. The complete solutions require the construction of the complete shearing force and bending moment diagrams.

P5–1. R_1 = R_2 = 325 lb
V = 325 lb
M = 4550 lb·in

P5–3. R_1 = 11.43 K; R_2 = 4.57 K
V = 11.43 K
M = 45.7 K·ft

P5–5. R_1 = 575 N; R_2 = 325 N
V = 575 N
M = 195 N·m

P5–7. R_1 = 46.36 kN; R_2 = 23.64 kN
V = 46.36 kN
M = 71.54 kN·m

P5–9. R_1 = 1557 lb; R_2 = 1743 lb
V = 1557 lb
M = 6228 lb·in

P5–11. R_1 = 7.5 K; R_2 = 37.5 K
V = 20 K
M = 60 K·ft

P5–13. R_1 = R_2 = 250 N
V = 850 N
M = 362.5 N·m

P5–15. R_1 = 37.4 kN (down); R_2 = 38.3 kN (up)
V = 24.9 kN
M = 50 kN·m

P5–17. $R = 120$ lb
$V = 120$ lb
$M = 960$ lb·in

P5–19. $R = 24$ K
$V = 24$ K
$M = 168$ K·ft

P5–21. $R = 1800$ N
$V = 1800$ N
$M = 1020$ N·m

P5–23. $R = 120$ kN
$V = 120$ kN
$M = 240$ kN·m

P5–25. $R_1 = R_2 = 180$ lb
$V = 180$ lb
$M = 810$ lb·in

P5–27. $R_1 = 240$ lb; $R_2 = 120$ lb
$V = 240$ lb
$M = 640$ lb·in

P5–29. $R_1 = 99.2$ N; $R_2 = 65.8$ N
$V = 99.2$ N
$M = 9.9$ N·m

P5–31. $R_1 = 42$ kN; $R_2 = 50$ kN
$V = 50$ kN
$M = 152.2$ kN·m

P5–33. $R_1 = R_2 = 440$ lb
$V = 240$ lb
$M = 360$ lb·in

P5–35. $R_1 = 1456$ N; $R_2 = 644$ N
$V = 956$ N
$M = 125$ N·m

P5–37. $R_1 = 35.3$ N; $R_2 = 92.3$ N
$V = 52.2$ N
$M = 4.0$ N·m

P5–39. $R = 360$ lb
$V = 360$ lb
$M = 1620$ lb·in

P5–41. $R = 600$ N
$V = 600$ N
$M = 200$ N·m

P5–43. $R_1 = R_2 = 330$ lb
$V = 330$ lb
$M = 4200$ lb·in

P5–45. $R_1 = 36.6$ K; $R_2 = 30.4$ K
$V = 36.6$ K
$M = 183.2$ K·ft

P5–47. $R_1 = R_2 = 450$ N
$V = 450$ N
$M = 172.5$ N·m

P5–49. $R_1 = 180$ kN; $R_2 = 190$ kN
$V = 190$ kN
$M = 630$ kN·m

P5–51. $R_1 = 636$ lb; $R_2 = 1344$ lb
$V = 804$ lb
$M = 2528$ lb·in

P5–53. $R_1 = 4950$ N; $R_2 = 3100$ N
$V = 2950$ N
$M = 3350$ N·m

P5–55. $R = 236$ lb
$V = 236$ lb
$M = 1504$ lb·in

P5–57. $R = 1130$ N
$V = 1130$ N
$M = 709$ N·m

P5–59. $R = 230$ kN
$V = 230$ kN
$M = 430$ kN·m

P5–61. $R = 1400$ lb
$V = 1500$ lb
$M = 99\ 000$ lb·in

P5–63. $R = 1250$ N
$V = 1250$ N
$M = 1450$ N·m

P5–65. $R_1 = 1333$ lb; $R_2 = 2667$ lb
$V = 2667$ lb
$M = 5132$ lb·ft

P5–67. $R_1 = R_2 = 75$ N
$V = 75$ N
$M = 15$ N·m

P5–69. $R_1 = 8.60$ kN; $R_2 = 12.2$ kN
$V = 12.2$ kN
$M = 9.30$ kN·m

P5–71. $R_1 = R_2 = 5400$ lb
$V = 5400$ lb
$M = 19\ 800$ lb·ft

P5–73. $R = 10.08$ kN
$V = 10.08$ kN
$M = 8.064$ kN·m

P5–75. $R = 7875$ lb
$V = 7875$ lb
$M = 21\ 063$ lb·ft

For Problems P5–77 through P5–83, results are shown for the main horizontal section only.

P5–77. $R_1 = R_2 = 282$ N
$V = 282$ N
$M = 120$ N·m

P5–79. $R_1 = R_2 = 162$ N
$V = 162$ N
$M = 42.2$ N·m

P5–81. $R_1 = 165.4$ N; $R_2 = 18.4$ N
$V = 165.4$ N
$M = 16.54$ N·m

P5–83. $R_1 = 4.35$ N; $R_2 = 131.35$ N
$V = 127$ N
$M = 6.35$ N·m

ADDITIONAL PRACTICE PROBLEMS

Shearing Force and Bending Moment Using Free-Body Diagram Method

5–85. $V = -7.36$ kN; $M = 29.96$ kN·m

5–87. $V = 1000$ N; $M = -170$ N·m

5–89. $V = -1250$ N; $M = 2400$ N·m

5–91. $V = -4.20$ kN; $M = 8.20$ kN·m

For the format of answers for Problems 5–93 to 5–109, see note before P5–1.

5–93. $R_1 = 1200$ lb; $R_2 = 1200$ lb
$V = 1200$ lb
$M = 54\,000$ lb·in

5–95. $R_1 = 11\,300$ lb; $R_2 = 1250$ lb
$V = 6750$ lb
$M = 202\,500$ lb·in

5–97. $R_1 = 66.25$ kN; $R_2 = 128.75$ kN
$V = 78.75$ kN
$M = 132.5$ kN·m

5–99. $R_1 = 3100$ lb; $R_2 = 3100$ lb
$V = 3100$ lb
$M = 141\,000$ lb·in

5–101. $R_1 = 495$ lb; $R_2 = 1405$ lb
$V = 805$ lb
$M = 2450.25$ lb·in

5–103. $R_1 = 27.0$ kN; $R_2 = 9.0$ kN
V = 15.0 kN
$M = 18.0$ kN·m

5–105. $R_1 = 6200$ N; $R_2 = 36\,800$ N
$V = 18\,800$ N
$M = 3150$ N·m

5–107. $R_1 = 920$ lb; $R_2 = 520$ lb
$V = 920$ lb
$M = 6720$ lb·in

5–109. $R = 350$ N
$V = 350$ N
$M = 122.5$ N·m

Problems with Given Shearing Force Diagrams

Load and bending moment diagrams to be found.

5–111. Simple beam with three concentrated loads at B, C, and D; Reactions at A and E:
$R_A = 35$ kN, $R_E = 45$ kN; $F_B = 26$ kN, $F_C = 30$ kN, $F_D = 24$ kN
$M_A = 0$ kN·m, $M_B = 52.5$ kN·m, $M_C = 66.0$ kN·m (max), $M_D = 45.0$ kN·m, $M_E = 0$ kN·m

5–113. Cantilever, supported at A, with two concentrated loads at B and C:
$R_A = 80$ lb, $M_A = 500$ lb·ft CCW;
$F_B = 60$ lb, $F_C = 20$ lb
$M_A = -500$ lb·ft, $M_B = -100$ lb·ft

5–115. Simple beam with partial uniformly distributed load from B to C; Reactions at A and C:
$R_A = 4050$ lb, $R_C = 6750$ lb; Distributed load from B to C; $w = 1200$ lb/ft
$M_A = 0$ lb·ft, $M_B = 12\,150$ lb·ft,
$M_C = 0$ lb·ft,
$M_D = 18\,984.4$ lb·ft (max at $x = 6.375$ ft from A)

Problems with Given Bending Moment Diagrams

Load and shearing force diagrams to be found.

5–117. Simple beam with two concentrated loads at B and C; Reactions at A and D:
$R_A = 100$ lb, $R_D = 150$ lb; $F_B = 75$ lb, $F_C = 175$ lb
$V_{A\text{-}B} = 100$ lb, $V_{B\text{-}C} = 25$ lb, $V_{C\text{-}D} = -150$ lb

5–119. Simple beam with two concentrated loads at A and C; reactions at B and D:
$R_B = 19.17$ kN, $R_D = 10.0$ kN,
$F_A = 6.67$ kN, $F_C = 22.5$ kN
$V_{A\text{-}B} = -6.67$ kN, $V_{B\text{-}C} = 19.17$ kN, $V_{C\text{-}D} = -10.0$ kN

5–121. Simple beam with uniformly distributed load over entire length; reactions at A and B:
Uniformly distributed load = w = 900 lb/ft;
$R_A = 5400$ lb, $R_B = 5400$ lb
$V_A = 5400$ lb, $V_B = -5400$ lb,
$V_C = 0$ lb at 6.0 ft from A is point where maximum bending moment occurs

Continuous Beams—Theorem of Three Moments

5–124. $R_A = R_E = 371$ lb at ends of beam
$R_C = 858$ lb at middle support
$V_{max} = 429$ lb from B to D between two 800 lb loads
$M_{max} = 1113$ lb·ft under each load

5–126. $R_A = 56.5$ kN, $R_C = 135$ kN at middle support
$R_D = 16.5$ kN at right end
$V_{max} = -87.5$ kN at C
$M_{max} = 32.4$ kN·m at 80 kN load

5–128. $R_A = 8.90$ kN, $R_C = 34.1$ kN at middle support
$R_D = 6.00$ kN at right end
$V_{max} = 18.0$ kN at C
$M_{max} = 8.9$ kN·m at B at 25 kN load

5–130. $R_A = 21.1$ kN, $R_C = 101.8$ kN at middle support
$R_E = 37.1$ kN
$V_{max} = 62.9$ kN at C
$M_{max} = 31.7$ kN·m at B under 60 kN load

Chapter 6

NOTE: The following answers refer to Figures P6–1 through P6–39. The first number is the distance to the centroid from the bottom of the section unless noted otherwise. The second number is the moment of inertia with respect to the horizontal centroidal axis.

P6–1. 0.663 in; 0.3156 in^4
P6–3. 4.00 in; 184 in^4
P6–5. 35.0 mm; 2.66×10^5 mm^4
P6–7. 20.0 mm; 7.29×10^4 mm^4
P6–9. 20.0 mm; 1.35×10^5 mm^4
P6–11. 21.81 mm; 1.86×10^5 mm^4
P6–13. 23.33 mm; 1.41×10^5 mm^4
P6–15. 1.068 in; 0.3572 in^4
P6–17. 125 mm; 6.73×10^7 mm^4
P6–19. 0.9305 in; 1.2506 in^4
P6–21. 4.25 in; 151.4 in^4
P6–23. 2.25 in; 107.2 in^4
P6–25. 7.35 in; 831.5 in^4
P6–27. 7.40 in; 423.5 in^4
P6–29. 3.50 in; 89.26 in^4
P6–31. 3.00 in from center of either pipe; 17.87 in^4
P6–33. 2.717 in; 46.76 in^4
P6–35. 2.609 in; 58.25 in^4
P6–37. 3.50 in; 16.95 in^4
P6–39. 3.361 in; 44.34 in^4
P6–41. 8.172 in; 357.5 in^4
P6–43. 8.20 in; 376.1 in^4
P6–45. 5.023 in; 100.2 in^4
P6–47. 14.641 mm; 41 647 mm^4

Radius of Gyration—Horizontal Axis

6–49. 3.04 in
6–51. 70.35 mm
6–53. 8.78 mm
6–55. 15.67 mm
6–57. 6.11 mm
6–59. 0.72 in
6–61. 68.78 mm
6–63. 2.33 in
6–65. 4.02 in

Radius of Gyration—Vertical Axis

6–67. 1.424 in
6–69. 34.91 mm
6–71. 0.212 in
6–73. 0.809 in
6–75. 8.483 in
6–77. 2.065 in
6–79. 1.783 in
6–81. 0.732 in

Metric Versions of Previous Problems

*Figure numbers listed with **M** suffix.*

First number is distance to centroid from the bottom of section; second number is moment of inertia with respect to the centroidal axis.

P6–21M. 108 mm; 6.31×10^7 mm^4
P6–23M. 57.20 mm; 4.47×10^7 mm^4
P6–25M. 186 mm; 3.44×10^8 mm^4
P6–27M. 186.7 mm; 1.75×10^8 mm^4
P6–29M. 87 mm; 3.38×10^7 mm^4
P6–31M. 75 mm from center of either pipe; 7.11×10^6 mm^4
P6–33M. 68.07 mm; 1.84×10^7 mm^4

P6–35M. 85.39 mm; 2.36×10^7 mm^4

P6–37M. 889.25 mm; 7.08×10^6 mm^4

P6–39M. 85.16 mm; 1.83×10^7 mm^4

P6–41M. 207.3 mm; 1.49×10^8 mm^4

P6–43M. 207.9 mm; 1.56×10^8 mm^4

P6–45M. 127.5 mm; 4.18×10^7 mm^4

Chapter 7

7–1. 94.4 MPa

7–3. **(a)** 20 620 psi
(b) 41 240 psi

7–5. 21 050 psi

7–7. $\sigma_t = 6882$ psi
$\sigma_c = 12970$ psi

7–9. 5794 psi (39.9 MPa)

7–11. 13 963 psi

7–13. Required $S = 2636$ mm^3; for $h/b = 3.0$;
$b = 12.1$ mm; $h = 36.3$ mm
Convenient dimensions from Appendix A–2:
$b = 12$ mm; $h = 40$ mm; $S = 3200$ mm^3;
$A = 480$ mm^2; $h/b = 3.33$
$b = 14$ mm; $h = 35$ mm; $S = 2858$ mm^3;
$A = 490$ mm^2; $h/b = 2.5$

7–15. Required $s_u = 290$ MPa; possible material—6061-T6

7–17. Required $S = 88.2$ in^3; W20×66 steel beam

7–19. Required $s_u = 10.6$ ksi; OK for 6061-T4

7–21. Required $s_y = 284$ MPa; OK for 2014-T4

7–23. Required $S = 1.35$ in^3; 3-in schedule 40 steel pipe

7–24. Required $S = 5.84$ in^3; $6 \times 4 \times \frac{1}{4}$ or $8 \times 2 \times \frac{1}{4}$ steel tube

7–25. Required $S = 6.72$ in^3; 6I×4.030 aluminum beam

7–26. Required $S = 5.38$ in^3; W8×10 steel beam

7–27. Required $S = 7.47$ in^3; no suitable channel

7–28. Required $S = 7.47$ in^3; 6-in schedule 40 steel pipe

7–31. Required $S = 9.76$ in^3; W10×12 steel beam or required $S = 1.60 \times 10^5$ mm^3; W250×17.9

7–33. Required $S = 16.6$ in^3; W12×16 steel beam

7–35. Required $S = 1.89 \times 10^5$ mm^3; W310×223.8 steel beam

7–37. Required $S = 7.47 \times 10^5$ mm^3; W460×60 steel beam

7–39. Required $S = 14.5$ in^3; W12×16 steel beam

7–41. Required $S = 6.37 \times 10^3$ mm^3; W200×15 steel beam

7–43. Required $S = 16.6$ in^3; S10×25.4 steel beam

7–45. Required $S = 1.89 \times 10^5$ mm^3; S200×27.4 steel beam

7–47. Required $S = 7.47 \times 10^5$ mm^3; S380×64 steel beam

7–49. Required $S = 14.5$ in^3; S8×23 steel beam

7–51. Required $S = 6.37 \times 10^3$ mm^3; S80×8.5 steel beam

7–53. Required $S = 13.85$ in^3; W12×16 steel beam

7–55. Required $S = 1.58 \times 10^5$ mm^3; W250×17.9 steel beam

7–57. Required $S = 6.23 \times 10^5$ mm^3; W310×44.5 steel beam

7–59. Required $S = 12.12$ in^3; W12×16 steel beam

7–61. Required $S = 5.31 \times 10^3$ mm^3; W200×15 steel beam

7–63. Required $S = 18.8$ in^3; 2×10 wood beam

7–65. 2×8 wood beam

7–68. Required $S = 11.1$ in^3; 2×8 wood beam

7–69. Required $S = 25.0$ in^3; 2×12 wood beam

7–70. Required $S = 2.79$ in^3; 2×4 wood beam

7–71. Required $S = 12.5$ in^3; 2×8 wood beam

7–75. 10×12 wood beam; No. 2 southern pine
W8×10 steel beam; ASTM A992 steel

7–77. $\sigma_d = 4.3$ MPa
At A: $\sigma = 3.81$ MPa; OK
At B: $\sigma = 5.16$ MPa; unsafe
At C: $\sigma = 4.62$ MPa; unsafe

7–79. 4.86 N/mm

7–81. 1094 N

7–83. 676 lb

7–85. 102 lb

7–87. 6.77 lb/in

7–89. 3.80 ft from wall to joint
4-in pipe is safe at wall

7–91. 398 MPa at C

7–93. At fulcrum, $\sigma = 8000$ psi
At bottom hole, $\sigma = 5067$ psi

At next hole, $\sigma = 3800$ psi
At next hole, $\sigma = 2534$ psi
At next hole, $\sigma = 1267$ psi

7–94. At fulcrum, $\sigma = 8000$ psi
At bottom hole, $\sigma = 10\,000$ psi
At next hole, $\sigma = 7506$ psi
At next hole, $\sigma = 5004$ psi
At next hole, $\sigma = 2500$ psi

7–95. **(a)** With pivot in end hole as shown:
At fulcrum, $\sigma = 8000$ psi
At bottom hole, $\sigma = 8064$ psi
For pivot in any other hole, maximum stress is at fulcrum. For pivot in:
Hole 2: $\sigma = 6800$ psi
Hole 3: $\sigma = 5600$ psi
Hole 4: $\sigma = 4400$ psi
Hole 5: $\sigma = 3200$ psi

7–97. 109 MPa

7–98. 149 MPa

7–99. Required $s_u = 1195$ MPa
AISI 4140 OQT 900 (others possible)

7–100. 2513 N

7–101. 1622 N

7–103. Impossible

7–105. Yes. $d_{max} = 37.2$ mm

7–106. 118 MPa at first step (L_3)

7–107. Required $s_u = 946$ MPa
AISI 1141 OQT 900 (others possible)

7–109. $L_{1max} = 206$ mm
$L_{2max} = 83.4$ mm
$L_{3max} = 24.7$ mm

7–111.

x (mm)	σ (MPa)
0	0
40	52.1
80	76.5
120	87.9
160	92.6
200	93.8
240	112.5

7–113. $h_1 = 22$ mm; $h_2 = 22$ mm

7–115. Required $S = 16.36$ in^3
W12×16

7–117. For composite beam, $w = 4.18$ K/ft
For S-beam alone; $w = 3.58$ K/ft

7–118. 11.5 mm

7–120. 0.805 in

7–122. 25.5 mm

7–124. 46 mm from center

7–126. $e = 12.9$ mm

7–127. 4.94 lb/in

7–129. 822 N

7–131. 625 N

7–133. 21.0 kN

7–134. 6.94 kN/m

7–135. 9.69 kN/m

7–136. 48.0 kN

7–137. 126 kN

7–139. $\sigma_d = 46.0$ MPa; $\sigma_{max} = 68.4$ MPa; Unsafe

7–141. $\sigma_{max} = 47.0$ MPa at step 50 mm from R_1

7–143. Required $S = 66.7$ in^3; W18×40

7–145. Required $S = 3.25$ in^3; $4 \times 4 \times \frac{1}{4}$ or $6 \times 2 \times \frac{1}{4}$
Or: Required $S = 5.33 \times 10^4$ mm^3; HSS102×102×6.4 or HSS152×51×6.4

7–147. Required $S = 7.45$ in^3; $8 \times 2 \times \frac{1}{4}$

7–148. Allowable load = 3963 lb (Tension)

7–150. Required $S = 6.14$ in^3; W8×10;
Total weight = 83 lb

Chapter 8

8–1. 1.125 MPa

8–3. 1724 psi

8–5. 3.05 MPa

8–7. 3180 psi

8–9. 1661 psi

8–11. 69.3 psi

8–13. 7.46 MPa

8–15. 2.79 MPa

8–17. 10.3 MPa

8–19. 2342 psi

8–21. 245 lb

8–23. 788 lb

8–25. 5098 lb

8–27. 661 lb

8–29. 787 lb

8–31. Let y = distance from bottom of the I-shape:

y (in)	τ (psi)
0.0	0.0
0.5	5.65
1.0(−)	10.54
1.0(+)	63.25
1.5	67.39
2.0	70.78
2.5	73.42
3.0	75.30
3.5	76.43
4.0	76.81

8–33. 8733 psi

8–35. From web shear formula:
$\tau = 8017$ psi
Approximately 8% low compared with τ_{max} from 8–33.

8–37. Let y = distance from bottom of the I-shape:

y (in)	τ (psi)
0.0	0
0.175	155
0.35(−)	303
0.35(+)	6582
1.0	7073
2.0	7638
3.0	7976
4.0	8089

8–39. $\tau = 3788$ psi;
$\tau_d = 20000$ psi; OK
$\sigma = 21053$ psi;
$\sigma_d = 33000$ psi; Safe

8–41. W18×40; $\tau = 7892$ psi;
$\tau_d = 20000$ psi; OK

8–43. $1\frac{1}{4}$ in schedule 40 pipe;
$\tau = 2404$ psi;
$\tau_d = 8000$ psi; OK

8–45. 4 × 10 beam

8–47. 1256 lb

8–49. 116.3 MPa

8–51. **(a)** $\tau = 1125$ psi
(b) $\sigma = 6750$ psi
(c) $s_y = 20\,250$ psi; any steel

8–53. 35.4 mm; $\tau = 1.59$ MPa;
$N = 86.7$ for shear

8–55. 24.5 mm; $\tau = 1.28$ MPa;
$\sigma_d = 120$ MPa

8–57. Required $S = 0.45$ in^3; 2-in schedule 40 pipe

8–59. $q = 736$ lb/in; $\tau = 526$ psi

8–61. 433 lb/ft based on strength of rivets

8–63. 1722 lb based on bending

8–65. Maximum spacing = 4.36 in

8–67. Maximum spacing = 3.94 in

Web Shear

8–69. $\tau = 5185$ psi; $\tau_d = 20000$ psi; Safe

8–71. $\tau = 2821$ psi; $\tau_d = 20000$ psi; Safe

8–73. $\tau = 3599$ psi; $\tau_d = 20000$ psi; Safe

General Shear and Special Shear Formulas

8–75. $\tau = 66.2$ psi; $\tau_d = 70$ psi; Safe

8–77. $\tau = 4623$ psi; $\tau_d = 11\,500$ psi; Safe

8–79. $\tau = 209$ psi; $\tau_d = 10000$ psi; Safe

Shear Flow

8–81. $q = 112$ lb/in; Spacing = 2.41 in maximum

Chapter 9
Formula Method—Statically Determinate Beams

9–1. −2.01 mm

9–3. −0.503 mm

9–5. −5.40 mm

9–7. At loads: −0.251 in
At center: −0.385 in

9–9. −0.424 in

9–11. −0.271 in

9–13. +0.093-in deflection at $x = 69.2$ in

9–15. $D = 64.8$ mm

9–17. $t = 0.020$ in

Formula Method—Statically Indeterminate Beams

For Problems 9–19 through 9–45, the following values are reported:

Reactions at all supports
Shearing forces at critical points
Bending moments at critical points
Maximum deflection or deflection at selected points in the form, $y = C_d/EI$

When you specify the beam material, shape, and dimensions, you can compute the beam stiffness, *EI*, and use that to calculate the deflection. Deflections will be in the given unit of length when E and I are in the same units of force and length given in the answers. For example, in Problem 9 19, length is in m, force is in N. Then deflection is in m when E is in N/m^2 (Pa) and I is in m^4.

9–19. $R_A = V_A = 24063$ N, $R_C = -V_C = 10938$ N
$M_A = -26250$ N·m, $M_B = 21875$ N·m,
$M_C = 0$ N·m
$y_{max} = (-20934)/EI$ at 1.79 m from C

9–21. $R_A = V_A = 18760$ N, $R_C = -V_C = 16235$ N
$M_A = -22560$ N·m, $M_B = 24353$ N·m,
$M_C = 0$ N·m
$Y_B = (-21629)/EI$ at load

9–23. $R_A = V_A = 500$ lb, $R_B = -V_B = 300$ lb
$M_A = -1600$ lb·in, $M_E(\text{max}) = 900$ lb·in at $x = 10.0$ in, $M_B = 0$ lb·in
$y_{max} = (-17712)/EI$ at $x = 9.264$ in

9–25. $R_A = V_A = 17500$ N, $R_C = -V_C = 17500$ N
$M_A = -17500$ N·m, $M_B = 17500$ N·m,
$M_C = -17500$ N·m
$y_{max} = (-11667)/EI$ at B at center

9–27. $R_A = V_A = 11074$ N, $R_C = -V_C = 23926$ N
$M_A = -12305$ N·m, $M_B = 15381$ N·m,
$M_C = -20508$ N·m
$y_{max} = (-10127)E/EI$ at B at center

9–29. $R_A = V_A = 400$ lb, $R_B = -V_B = 400$ lb
$M_A = -1067$ lb·in, $M_B = 533$ lb·in,
$M_C = -1067$ lb·in
$y_{max} = (-8533)/EI$ at center

9–31. $R_A = V_A = 150$ lb, $R_B = 500$ lb,
$R_C = -V_C = 150$ lb
$M_A = 0$ lb·in, $M_B = -400$ lb·in, $M_C = 0$ lb·in
$M_D = M_E = 225$ lb·in at $x = 3.00$ in from A and C
$y_{max} = (-1107)/EI$ at $x = 3.372$ in from A or C

9–33. $R_A = R_D = 106.7$ lb, $R_B = R_C = 293.3$ lb
$V_A = -V_D = 106.7$ lb, $-V_B = V_C = 160$ lb
$M_A = M_D = 0$ lb·in, $M_B = M_C = -142.2$ lb·in,
$M_G = 35.6$ lb·in
$M_E = M_F = 113.8$ lb·in at $x = 2.13$ in from A and D

9–35. $R_A = R_E = 78.6$ lb, $R_B = R_D = 228.6$ lb,
$R_C = 185.6$ lb
$V_A = -V_E = 78.6$ lb, $V_B = V_D = 121.4$ lb,
$V_C = 92.8$ lb
$M_A = M_E = 0$ lb·in, $M_B = M_D = -85.7$ lb·in,
$M_C = -57.1$ lb·in
$M_F = M_I = 61.8$ lb·in at $x = 1.56$ in from A and E
$M_G = M_H = 29.1$ lb·in at $x = 2.16$ in from B and D

9–37. $R_A = V_A = 22500$ N, $R_B = V_B = 13500$ N
$M_A = -8100$ N·m, $M_E = 4556$ N·m at 0.675 in from B
$y_{max} = -1135/EI$ at 1.042 in from A

9–39. $R_A = V_A = -13750$ N, $R_B = 31750$ N
$V_B = -18000$ N
$M_A = 12600$ N·m, $M_B = -25200$ N·m, $M_C = 0$
$Y_C = -40719/EI$ at overhang at right end

9–41. $R_A = R_C = V_A = -V_C = 25200$ lb,
$R_B = 8400$ lb
$V_B = 42000$ lb
$M_A = M_C = 0$, $M_B = -134400$ lb·ft
$M_D = M_E = 75587$ lb·ft at $x = 6.0$ ft from A or C
$y_{max} = (-1.488 \times 10^6 \text{ lb·ft}^3)/EI$ at 6.74 ft from A or C

9–43. $R_A = R_E = V_E = -V_E = 212$ lb,
$R_B = R_D = 617$ lb, $R_C = 501$ lb
$V_B = V_D = 328$ lb, $V_C = 251$ lb
$M_A = M_E = 0$, $M_C = -1388$ lb·in
$M_B = M_D = -2082$ lb·in
$M_F = M_I = 1501$ lb·in at 14.04 in from A or E
$M_G = M_H = 708$ lb·in at 19.44 in from B or D

9–45. $R_A = V_A = 224.6$ N, $R_C = -V_C = 25.4$ N
$M_A = -2.355$ N·m, $M_B = 1.014$ N·m, $M_C = 0$
$y_B = (-1.386 \times 10^{-4})/EI$ at load

Comparisons of Beam Behavior

9–46. Comparison of four beam designs to carry a uniformly distributed load:

	V_{max}	V/V_1	M_{max}	M/M_1	y_{max}	y/y_1
(a)	3600 N	1.0	3600 N·m	1.0	$-6000/EI$	1.0
(b)	7200 N	2.0	−14 400 N·m	4.0	$-57600/EI$	9.6
(c)	4500 N	1.25	−3600 N·m	1.0	$-2490/EI$	0.415
(d)	3600 N	1.0	−2400 N·m	0.67	$-1200/EI$	0.20

9–48. Comparison of results of five problems:

	V_{max}	V/V_1	M_{max}	M/M_1	y_{max}	y/y_1
1. 9–23	500 lb	1.0	−1600 lb·in	1.0	$-17\,712/EI$	1.0
2. 9–29	400 lb	0.80	−1067 lb·in	0.667	$-8533/EI$	0.482
3. 9–31	250 lb	0.50	−400 lb·in	0.250	$-1107/EI$	0.0625
4. 9–33	160 lb	0.320	−142 lb·in	0.089	—	—
5. 9–35	121 lb	0.243	−85.7 lb·in	0.054	—	—

9–52. Comparison of results of three problems:

	V_{max}	V/V_1	M_{max}	M/M_1	y_{max}	y/y_1	A	A/A_1
1. 9–49	1440 lb	1.0	103 680 lb·in	1.0	$-896 \times 10^6/EI$	1.0	63.3 in^2	1.0
2. 9–50	900 lb	0.625	25 920 lb·in	0.25	$-372 \times 10^6/EI$	0.415	25.4 in^2	0.401
3. 9–51	576 lb	0.40	9 216 lb·in	0.089	—	—	13.87 in^2	0.172

Superposition—Statically Determinate Beams

9–53. $y_B = -1.291$ mm at 840 N load
$y_C = -3.055$ mm at 600 N load
$y_D = -1.353$ mm at 1200 N load

9–56. $y_B = -0.0140$ in at 85 lb load
$y_C = -0.0262$ in at 75 lb load at end

9–57. −0.869 mm

9–59. −3.997 mm

9–60. −0.0498 in

9–62. C5×2.212 Aluminum channel

Superposition—Statically Indeterminate Beams

9–64. $R_A = V_A = 22\,500$ N, $R_B = V_B = 13\,500$ N
$M_A = -8100$ N·m, $M_E = 4556$ N·m at 0.675 in from B
$y_{max} = -1135/EI$ at 1.042 in from A

9–66. $R_A = R_E = 371$ lb at ends of beam
$R_C = 858$ lb at middle support
$V_{max} = 429$ lb from B to D between two 800 lb loads
$M_{max} = 1113$ lb·ft under each load

9–68. $R_A = 56.5$ kN, $R_C = 135$ kN at middle support
$R_D = 16.5$ kN at right end
$V_{max} = -87.5$ kN at C
$M_{max} = 32.4$ kN·m at 80 kN load

9–70. $R_A = 4009$ lb at fixed end
$R_D = 1991$ lb at right support
$V_{max} = 4009$ lb at A
$M_{max} = -8490$ lb·ft at A

Design Problems—Stress and Deflection Limits

9–72. Comparison of two designs in Figure P9–72:
(a) $V = 2250$ N, $M = -4500$ N·m, $y = -22\,500/EI$
(b) $V = 3375$ N, $M = -4500$ N·m, $y = -20\,250/EI = 0.90\,y_a$
Little difference in the performance of the two designs.

9–74. $R_A = V_A = 32.75$ kN, $R_B = -V_B = 19.65$ kN
$M_A = -42.9$ kN·m, $M_E = 24.1$ kN·m

Successive Integration Method

9–76. $y = -0.0078$ in at $x = 8.56$ in

9–78. $y = -3.79$ mm

9–80. $D = 69.2$ mm

9–82. I178×8.630 (I7 × 5.800) aluminum beam
$y = -5.37$ mm at $x = 1.01$ m

9–84. $D = 109$ mm

Moment-Area Method

9–86. −0.0078 in

9–88. −3.79 mm

9–90. −3.445 mm

9–92. −5.13 mm

9–94. −0.01138 in

9–96. −0.257 in

9–98. −71.7 mm

Chapter 10

10–1. −10 510 psi

10–3. σ_N = 9480 psi;
σ_M = −7530 psi

10–5. σ_N = 13 980 psi;
σ_M = −15 931 psi

10–7. −64.3 MPa

10–9. 415 N

10–11. σ = 724 MPa; required s_y = 1448 MPa; AISI 4140 OQT 700

10–13. At *B*: σ = 24 328 psi tension on top of beam
σ = −18 785 psi compression on bottom of beam

10–15. Load = 9081 N;
Mass = 926 kg

10–17. 26 mm

10–19. 18.7 mm

10–21. 71.6 MPa

10–23. 2548 psi

10–25. 7189 psi

10–27. 51.6 MPa

10–29. 982 MPa

10–31. τ = 9923 psi, s_y = 119 ksi

10–33. 67.1 MPa

10–35. 7548 psi

10–37. 7149 psi

10–39. 61.5 MPa

10–41. (a) 3438 psi
(b) 3438 psi
(c) 344 psi
(d) 2300 psi
(e) 1186 psi

10–43.

	Middle of beam	Near supports	15 in from *A*
(a)	1875 psi	0 psi	1406 psi
(b)	1875 psi	0 psi	1406 psi
(c)	0 psi	375 psi	188 psi
(d)	1250 psi	208 psi	943 psi
(e)	625 psi	333 psi	498 psi

10–45. 75.3 MPa

10–47. (a) P = 40 667 lb;
τ_{max} = 8333 psi
(b) τ_{max} = 8925 psi; N = 2.80

ADDITIONAL PRACTICE PROBLEMS

10–49. σ_{max} = 397 MPa tensile; σ_{max} = 709 MPa compressive

10–51. σ_{max} = 821 MPa tensile; σ_{max} = 458 MPa compressive

10–55. At *M*: σ_{max} = 2.18 MPa tensile
At *N*: σ_{max} = 3.54 MPa compressive

NOTE: The complete solutions for Problems 10–56 through 10–82 require the construction of the complete Mohr's circle and the drawing of the principal stress element and the maximum shear stress element. Listed below are the significant numerical results.

Prob. No.	σ_1	σ_2	ϕ (deg)	τ_{max}	σ_{avg}	ϕ' (deg)
10–56	315.4 MPa	−115.4 MPa	10.9 CW	215.4 MPa	100.0 MPa	34.1 CCW
10–58	110.0 MPa	−40.0 MPa	26.6 CW	75.0 MPa	35.0 MPa	18.4 CCW
10–60	23.5 ksi	−8.5 ksi	19.3 CCW	16.0 ksi	7.5 ksi	64.3 CCW
10–62	79.7 ksi	−9.7 ksi	31.7 CCW	44.7 ksi	35.0 ksi	76.7 CCW
10–64	677.6 kPa	−977.6 kPa	77.5 CCW	827.6 kPa	−150.0 kPa	57.5 CW
10–66	327.0 kPa	−1202.0 kPa	60.9 CCW	764.5 kPa	−437.5 kPa	74.1 CW
10–68	570.0 psi	−2070.0 psi	71.3 CW	1320.0 psi	−750.0 psi	26.3 CW
10–70	4180.0 psi	−5180.0 psi	71.6 CW	4680.0 psi	−500.0 psi	26.6 CW
10–72	360.2 MPa	−100.2 MPa	27.8 CCW	230.2 MPa	130.0 MPa	72.8 CCW
10–74	23.9 ksi	−1.9 ksi	15.9 CW	12.9 ksi	11.0 ksi	29.1 CCW
10–76	4.4 ksi	−32.4 ksi	20.3 CW	18.4 ksi	−14.0 ksi	24.7 CCW
10–78	321.0 MPa	−61.0 MPa	64.4 CCW	191.0 MPa	130.0 MPa	68.6 CW
10–80	225.0 MPa	−85.0 MPa	0.0	155.0 MPa	70.0 MPa	45.0 CCW
10–82	775.0 kPa	−145.0 kPa	0.0	460.0 kPa	315.0 kPa	45.0 CCW

For Problems 10–84 through 10–94, Mohr's circle from the given data results in both principal stresses having the same sign. For this class of problems, the supplementary circle is drawn using the procedures discussed in Section 10–11 of the text. The results include three principal stresses where $\sigma_1 > \sigma_2 > \sigma_3$. Also, the maximum shear stress is found from the radius of the circle containing σ_1 and σ_3, and is equal to $\frac{1}{2}\sigma_1$ or $\frac{1}{2}\sigma_3$, whichever has the greatest magnitude. Angles of rotation of the resulting elements are not requested.

Prob. No.	σ_1	σ_2	σ_3	τ_{max}
10–84	328.1 MPa	71.9 MPa	0.0 MPa	164.0 MPa
10–86	214.5 MPa	75.5 MPa	0.0 MPa	107.2 MPa
10–88	35.0 ksi	10.0 ksi	0.0 ksi	17.5 ksi
10–90	55.6 ksi	14.4 ksi	0.0 ksi	27.8 ksi
10–92	0.0 kPa	−307.9 kPa	−867.1 kPa	433.5 kPa
10–94	0.0 psi	−295.7 psi	−1804.3 psi	902.1 psi

For Problems 10–96 through 10–104, Mohr's circles from earlier problems are used to find the stress condition on the element at some specified angle of rotation. The listed results include the two normal stresses and the shear stress on the specified element.

Prob. No.	σ_A	$\sigma_{A'}$	τ_A
10–96	130.7 MPa	69.3 MPa	213.2 MPa CW
10–98	−37.9 MPa	197.9 MPa	31.6 MPa CCW
10–100	3.6 ksi	−21.6 ksi	43.9 ksi CW
10–102	−2010.3 psi	510.3 psi	392.6 psi CW
10–104	8363.5 psi	86.5 psi	1421.2 psi CW

10–106. $\tau_{max} = 230.2$ MPa

10–108. $\tau_{max} = 12.9$ ksi

Strain Gage Rosettes

The following table lists answers for parts (a) through (g) for Problems 10–110 through 10–124

	(a)	(b)	(c)	(d)	(e)	(f)	(g)
10–110.	1902×10^{-6}	6.0×10^{-6}	−28.2°	147 MPa	49.1 MPa	1897 rad	49.2 MPa
10–112.	816×10^{-6}	-717×10^{-6}	31.9°	138 MPa	−109 MPa	1534 rad	123 MPa
10–114.	1006×10^{-6}	-294×10^{-6}	36.6°	17.3 ksi	0.731 ksi	1299 rad	8.30 ksi
10–116.	816×10^{-6}	-717×10^{-6}	31.9°	16.1 ksi	−12.9 ksi	1534 rad	14.5 ksi
10–118.	1494×10^{-6}	-112×10^{-6}	−5.4°	113 MPa	29.5 MPa	1607 rad	41.7 MPa
10–120.	882×10^{-6}	-324×10^{-6}	39.7°	178 MPa	−15.5 MPa	1206 rad	96.8 MPa
10–122.	616×10^{-6}	-319×10^{-6}	−43.8°	9.75 ksi	2.20 ksi	935 rad	5.98 ksi
10–124.	882×10^{-6}	-324×10^{-6}	39.7°	20.6 ksi	−2.24 ksi	1206 rad	11.4 ksi

Chapter 11

11–1. 25.1 kN

11–3. 8.35 kN

11–5. 26.2 kN

11–7. 111 kN

11–9. P_a = 7318 lb/column; use 9 columns

11–11. 499 kN

11–13. 65 300 lb

11–15. Axial force = 31.1 kN; critical load = 256.7 kN; N = 8.25 OK

11–17. 15.1 kN

11–19. Critical load = 10 914 lb; actual load = 5000 lb; N = 2.18; low

11–21. Critical load = 2849 lb; actual load = 1500 lb; N = 1.90; low

11–23. 2.68 in

11–25. Required I = 5.02 in^4; I7×5.800

11–27. 5649 lb

11–29. 245 kN

11–31. No improvement

ADDITIONAL PRACTICE PROBLEMS

11–33. 9420 lb

11–35. 6219 lb

11–37. N = 2.27

11–39. Design problem. Example solution
Compression members: ASTM A36 Structural steel
Round bar; Nearest 1/16 in
AC = 1925 lb; 40 in length; 15/16 in round bar; P_{all} = 2253 lb
CD = 750 lb: 25 in length; 9/16 in round bar; P_{all} = 750 lb
DE = 650 lb; 40 in length; 7/16 in round bar; P_{all} = 654 lb

11–41. Design problem. Example solution.
Compression members: Aluminum 6061–T6
Square bar; Nearest preferred mm size (App. 2)
DE = 2300 N; 250 mm length; 9.0 mm side; P_{all} = 2383 N
BD = 2597 N; 297 mm length; 11.0 mm side; P_{all} = 3767 N
EF = 2300 N; 200 mm length; 8.0 mm side; P_{all} = 2324 N
CF = 800 N; 160 mm length; 6.0 mm side; P_{all} = 1149 N
FG = 550 N; 150 mm length; 5.0 mm side; P_{all} = 630 N

11–43. Design problem. Example solution. Lightest steel tube.
Compressive force = 33 588 lb
HSS4×4×$\frac{1}{4}$ Steel tube; ASTM A501 Structural steel;
P_{all} = 42 441 lb.

11–45. P_a = 20.64 kN for N = 3

11–47. P_a = 8143 lb for N = 3

11–49. N = 1.068—Low

11–51. σ_{max} = 211 MPa (Unsafe), y_{max} = 25.8 mm (High)

11–53. Not safe. Required s_y = 103 ksi for N = 3.

11–55. Not safe. Required s_y = 104 ksi for N = 3.

11–57. F_a = 675 lb for eccentric loading in the plane of the drawing
F_a = 164 lb for buckling about the narrow thickness

11–59. **(a)** P_a = 2610 lb for straight pipe
(b) P_a = 1676 lb for crooked pipe

Chapter 12

12–1. 115 MPa

12–3. 4.70 mm minimum

12–5. 2134 psi

12–7. 1.80 mm

12–9. N = 3.80

12–11. σ_2 = 212 MPa; σ_3 = −70.0 MPa

12–13.

Radius (mm)	σ_2 (MPa)
110	166 inner surface
120	149
130	136
140	125
150	116 outer surface

12–15. 14.88 MPa

12–17.

Radius (mm)	σ_3 (MPa)
15	−7.00 inner surface
17	−4.58
19	−2.88
21	−1.64
23	−0.71
25	0.00 outer surface

12–19. 86.8 MPa

12–21.

Radius (mm)	σ_3 (MPa)
210	−100.0 inner surface
215	−83.3
220	−68.0
225	−54.1
230	−41.4
235	−29.7
240	−19.0
245	−9.1
250	0.00 outer surface

12–23.

Radius (mm)	σ_2 (MPa)
162.5	22.35 inner surface
170	20.99
177.5	19.84
185	18.88
192.5	18.06
200	17.35 outer surface

12–25. Sizes $\frac{1}{8}$ through 4 are thick-walled. Size 5 through 18 are thin-walled.

Chapter 13

13–1. **(a)** F_s = 1400 lb (Limit); F_b = 13 050 lb; F_t = 20 250 lb

(c) F_s = 2160 lb (Limit); F_b = 9788 lb; F_t = 21 263 lb

13–2. **(a)** F_s = 3800 lb (Limit); F_b = 40 500 lb; F_t = 87 000 lb

(c) F_s = 7600 lb (Limit); F_b = 40 500 lb; F_t = 87 000 lb

13–3. **(a)** F_s = 1178 lb (Limit); F_b = 15 750 lb; F_t = 26 708 lb

(c) F_s = 1325 lb (Limit); F_b = 11 813 lb; F_t = 28 114 lb

13–4. **(a)** F_s = 3313 lb (Limit); F_b = 49 500 lb; F_t = 112 485 lb

(c) F_s = 6627 lb (Limit); F_b = 49 500 lb; F_t = 104 970 lb

13–5. $\frac{3}{4}$-in bolts

13–7. 23 860 lb on welds

NOTE: Problems 13–9 to 13–11 are design problems for which there are no unique answers.

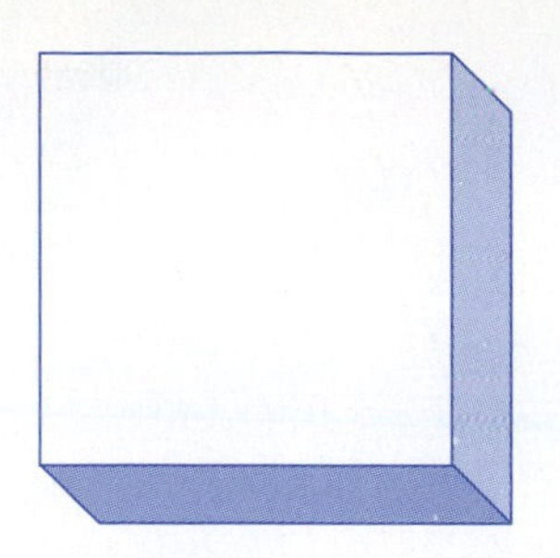

Index

D

N

O

P

Q

R

S